Jan Christian Habel • Thorsten Assmann
Editors

Relict Species

Phylogeography and Conservation Biology

Springer

Fonds National de la
Recherche Luxembourg

Editors
Dr. Jan Christian Habel
Musée National d'Histoire Naturelle Luxembourg
25, rue Münster
2160 Luxembourg
Janchristianhabel@gmx.de

Prof. Dr. Thorsten Assmann
Institute of Ecology and Environmental Chemistry
Leuphana University Lüneburg
Scharnhorststr. 1
21335 Lüneburg
Germany
assmann@uni.leuphana.de

ISBN 978-3-540-92159-2 e-ISBN 978-3-540-92160-8
DOI 10.1007/978-3-540-92160-8
Springer Heidelberg Dordrecht London New York

Library of Congress Control Number: 2009932420

Cover design: WMXDesign GmbH

Printed on acid-free paper

Springer is part of Springer Science+Business Media (www.springer.com)

Foreword

Mankind has evolved both genetically and culturally to become a most successful and dominant species. But we are now so numerous and our technology is so powerful that we are having major effects on the planet, its environment, and the biosphere. For some years prophets have warned of the possible detrimental consequences of our activities, such as pollution, deforestation, and overfishing, and recently it has become clear that we are even changing the atmosphere (e.g. ozone, carbon dioxide). This is worrying since the planet's life systems are involved and dependent on its functioning. Current climate change – global warming – is one recognised consequence of this larger problem.

To face this major challenge, we will need the research and advice of many disciplines – Physics, Chemistry, Earth Sciences, Biology, and Sociology – and particularly the commitment of wise politicians such as US Senator Al Gore.

An important aspect of this global problem that has been researched for several decades is the loss of species and the impoverishment of our ecosystems, and hence their ability to sustain themselves, and more particularly us! Through evolutionary time new species have been generated and some have gone extinct. Such extinction and regeneration are moulded by changes in the earth's crust, atmosphere, and resultant climate. Some extinctions have been massive, particularly those associated with catastrophic meteoric impacts like the end of the Cretaceous Period 65Mya. Each time, new species and ecosystems have evolved subsequently over millions of years. The rate of species loss today due to Man's activities is great, and has been classed as the next major extinction.

One way to study this problem is to focus on Relict Species. As species become rare through natural processes, they are reduced to few populations in few locations and are considered as threatened or endangered. Numerous species are now placed in this dangerous situation because of Man's expansion. This leads to great public concern and efforts for their conservation. To better inform these efforts, it is most valuable to study the origins, history, and status of such Relict Species.

For this, one clearly needs research on the species habitat requirements and ecological limits, particularly with the growing concern for current climate change. There is pertinent research on the last 2My, the Quaternary Period, with its cycles of increasingly severe ice ages that caused great changes in species abundance and distribution as seen in the fossil record. Our understanding of the origins and history of species have been greatly advanced by new DNA methods that allow the genetic relations and diversity across a species range to be revealed and analysed. This Phylogeographic approach is informing and sometimes challenging the views in Biogeography and Conservation, and permeating most areas of Evolutionary Biology.

A symposium on Relict Species seemed a good way forward. This would bring together the people studying the various relevant aspects, including topics in Ecology, Phylogeography, Climate Change, and Conservation. And so one was organised in Luxembourg in 2007. Europe was the theatre concerning many participants, where there is considerable interest in ice age refugia and postglacial colonisation to produce our present biota. Studies from Africa, North and South America, and Asia were also included, and the range of organisms covered fungi, plants, many insects, reptiles, and mammals. The relevance of phylogeography and climate change to the conservation of species was addressed in many contributions.

Museums were often considered as dusty shrines. But today the major ones are concerned in modern research in the evolution and management of biodiversity and in educating the next generation in such important matters. They also serve as rich archives and sources of material for such studies. The organisation of this symposium by the Luxembourg Museum is a signal example of such farsighted and crucial involvement.

April 2009 Godfrey Hewitt

Contents

Contributors

Thorsten Assmann Institute of Ecology and Environmental Chemistry, Leuphana University Lüneburg, Scharnhorststraße 1, D-21335 Lüneburg, Germany, assmann@uni.leuphana.de

Bettina Augenstein Biogeographie, Universität Trier, Am Wissenschaftspark 25-27, D-54296 Trier, Germany, augenstein@uni-trier.de

John C. Avise University of California, Irvine, CA 92697, USA, javise@uci.edu

Mónica Azpilicueta Amorín Department of Animal Biology and Ecology, University of Vigo, EUET Forestal, Campus Universitario, E-36006 Pontevedra, Spain, mazpilicueta@sek.es

Jiri Benes Faculty of Sciences, University Southern Bohemia & Institute of Entomology, Czech Academy of Sciences, Branisovska 31, 370 05 Ceske Budejovice, Czech Republic, benesjir@seznam.cz

Achille Casale Dipartimento di Zoologia e Antropologia Biologica, Università di Sassari, Via Muroni 25, I-07100 Sassari, Italy, casale@uniss.it

Anna Cassel-Lundhagen Department of Ecology, Swedish University of Agricultural Sciences, Ulls väg 16, SE-75007 Uppsala, Sweden, anna.lundhagen@ekol.slu.se

Adolfo Cordero Rivera Department of Animal Biology and Ecology, University of Vigo, EUET Forestal, Campus Universitario, E-36006 Pontevedra, Spain, adolfo.cordero@uvigo.es

Johannes Dambach Section of Lower Arthropods, Zoologisches Forschungsmuseum Alexander Koenig, D-53113 Bonn, Germany, j.dambach.zfmk@uni-bonn.de

Claudia Drees Institute of Ecology and Environmental Chemistry, Leuphana University Lüneburg, Scharnhorststraße 1, D-21335 Lüneburg, cdrees@uni-lueneburg.de

Götz Ellwanger Department I 2.2 "Habitats Directive/Natura 2000", German Federal Agency for Nature Conservation, Bonn, Germany, goetz.ellwanger@bfn.de

Aline Finger ETH Zürich, ITES-Ecosystem Management, Universitätstr. 16, CH-8092 Zürich, Switzerland, aline.finger@env.ethz.ch

Uwe Fritz Museum of Zoology (Museum für Tierkunde), Senckenberg Dresden, A. B. Meyer Building, D-01109 Dresden, Germany, uwe.fritz@senckenberg.de

József Geml Institute of Arctic Biology, University of Alaska Fairbanks, 311 Irving I Building, 902 N. Koyukuk Drive, P.O. Box 757000, Fairbanks, AK 99775-7000, USA, jozsef.geml@gmail.com

Philippe Goffart Observatoire de la Faune, de la Flore et des Habitats (OFFH), Département de l'Etude du Milieu naturel et agricole (DEMna), Service Public Wallon (SPW), DGARNE, Direction de la Nature et de l'Eau, 23, Avenue Maréchal Juin, B-5030 Gembloux, Belgium, Philippe.GOFFART@spw.wallonie.be

Markus Göker DSMZ – German Collection of Microorganisms and Cell Cultures, Inhoffenstraße 7 B, D-38124 Braunschweig, Germany, contact@goeker.org

Eckhard Gottschalk Centre for Nature Conservation, University of Göttingen, von-Siebold-Str. 2, D-37075 Göttingen, Germany, egottsc1@gwdg.de

Eva M. Griebeler Department of Ecology, Zoological Institute, University of Mainz, P.O. Box 3980, D-55099 Mainz, Germany, em.griebeler@uni-mainz.de

Guido W. Grimm Department of Palaeobotany, Swedish Museum of Natural History, P.O. Box 50007, S-10405 Stockholm, Sweden, guido.grimm@nrm.se

Daniela Guicking FB18, Systematik und Morphologie der Pflanzen, Universität Kassel, Heinrich-Plett Str. 40, D-34132 Kassel, Germany, guicking@uni-kassel.de

Jan C. Habel Section Zoologie des Invertébrés, Musée national d´histoire naturelle, Section Zoologie des Invertébrés, L-2160 Luxembourg, Luxembourg; Biogeographie, Universität Trier, Am Wissenschaftspark 25-27, D-54296 Trier, Germany, Janchristianhabel@gmx.de

Ghaiet M. Hajji Unité de Biologie Animale et Systématique Évolutive, Département de Biologie, Faculté des Sciences de Tunis, Campus Universitaire, 2092 El Manar II, Tunisia, gayahajji@hotmail.com

Thomas J. Hayden School of Biology and Environmental Science, University College Dublin, Belfield, Dublin 4, Ireland, tom.hayden@ucd.ie

Vera Hemleben Center for Plant Molecular Biology (ZMBP) – General Genetics, University of Tübingen, Auf der Morgenstelle 28, D-72076 Tübingen, Germany, vera.hemleben@uni-tuebingen.de

Godfrey Hewitt School of Biological Sciences, University of East Anglia, Norwich, Norfolk, NR4 7TJ, UK, g.hewitt@uea.ac.uk

San S. Hmwe No. 27 Swe' Taw Street, Koe' Lone Kwin Kyaung Quarter, Hmaw Bi Township, Yangon Division, Myanmar, sansan.hmwe@gmail.com

Axel Hochkirch Biogeographie, Universität Trier, Am Wissenschaftspark 25-27, D-54296 Trier, Germany, hochkirch@uni-trier.de

Ulrich Joger Staatliches Naturhistorisches Museum, Pockelsstr. 10, D-38106 Braunschweig, Germany, ulrich.joger@snhm.niedersachsen.de

Svetlana Kalyabina-Hauf Institut für Pharmazie und molekulare Biotechnologie, Universität Heidelberg, Im Neuenheimer Feld 364, D-69120 Heidelberg, Germany

Charlotte Klank ETH Zürich, ITES-Ecosystem Management, Universitätstr. 16, CH-8092 Zürich, Switzerland, charlotte.klank@env.ethz.ch

Martin Konvička Faculty of Sciences, University Southern Bohemia & Institute of Entomology, Czech Academy of Sciences, Branisovska 31, D-37005 Ceske Budejovice, Czech Republic, konva@entu.cas.cz

Gary A. Laursen Institute of Arctic Biology, University of Alaska Fairbanks, 311 Irving I Building, 902 N. Koyukuk Drive, P.O. Box 757000, Fairbanks, AK 99775-7000, USA, ffgal@uaf.edu

Stefan Lötters Department of Herpetology, Zoologisehes Forschungsmuseum Alexander Koenig, Adenaverallee 160, D-53113 Bonn, Germany; Department of Biogeography, Trier University, Am Wissenschaftspark 25-27, D-54286 Trier, Germany, loetters@uni-trier.de

Andrea Matern Institute of Ecology and Environmental Chemistry, Leuphana University Lüneburg, Scharnhorststraße 1, D-21335 Lüneburg, matern@leuphana.de

Marc Meyer Musée National d'Histoire Naturelle Luxembourg, Section Zoologie des Invertébrés, 25 rue Münster, L-2160 Luxembourg, Luxembourg

Allan D. McDevitt School of Biology and Environmental Science, University College Dublin, Belfield, Dublin 4, Ireland, amcdev@gmail.com

Christoph Muster Molecular Evolution and Animal Systematics, Institute of Zoology II, University of Leipzig, Talstraße 33, D-04103 Leipzig, Germany, muster@rz.uni-leipzig.de

Zoltan T. Nagy Royal Belgian Institute of Natural Sciences, JEMU, rue Vautier 29, B-1000 Brussels, Belgium, nagy@anl.gov

Gabriel Nève Institut Méditerranéen d'Ecologie et de Paléoécologie, UMR CNRS 6116, Case 36, Université de Provence, 3 Place Victor Hugo, F-13331 Marseille Cedex 3, France, gabriel.neve@univ-provence.fr

Markus Quante GKSS Research Center, Institute for Coastal Research, Geesthacht, Germany, Markus.quante@gkss.de

Ramugondo V. Rambau Department of Biology (Area 2), University of York, P.O. Box 373, York YO10 5YW, UK; Evolutionary Genomics Group, Department of Botany and Zoology, Stellenbosch University, Matieland 7602, Stellenbosch, South Africa; Department of Botany and Zoology, DST-NRF Centre of Excellence for Invasion Biology, Stellenbosch University, Matieland 7602, Stellenbosch, South Africa, rvr2@sun.ac.za

Thomas Reimann Institute of General Zoology and Genetics, University of Münster, Schlossplatz 5, D-48149 Münster, Germany; Present address: Projektträger Jülich Gmbh, Zimmerstr. 26-27, D-10969 Berlin, Germany, t.reimann@fz-juelich.de

Dennis Rödder Department of Herpetology, Zoologisches Forschungsmuseum Alexander Koenig, D-53113, Bonn, Germany; Department of Biogeography, University of Trier, D-54286, Trier, Germany, roedder@uni-trier.de

Nina A. Sazanova Laboratory of Botany, Institute of Biological Problems of the North, Far East Branch of Russian Academy of Sciences, Portovaya Str. 18, Magadan 685000, Russia, nsazanova@ibpn.ru

Matthias Schlee Center for Plant Molecular Biology (ZMBP) – General Genetics, University of Tübingen, Auf der Morgenstelle 28, D-72076 Tübingen, Germany, matthias.schlee@gmx.de

Andreas Schuldt Institute of Ecology and Environmental Chemistry, Leuphana University Lüneburg, Scharnhorststrabe 1, D-21335, Lüneburg, Germany

Andreas Schlüter Zoology Department, Staatliches Museum für Naturkunde Stuttgart, Rosenstein 1, D-70191 Stuttgart, Germany, schlueter.smns@naturkundemuseum-bw.de

Thomas Schmitt Biogeograpie, Universität Trier, Am Wissenschaftspark 25-27, D-54296 Trier, Germany, thsh@uni-trier.de

Peter Schönswetter Department of Biogeography, Faculty Centre Biodiversity, University of Vienna, Rennweg 14, A-1030 Wien, Austria, peter.schoenswetter@univie.ac.at

Eckhard Schröder Department I 2.2 "Habitats Directive/Natura 2000", German Federal Agency for Nature Conservation, Bonn, Germany, eckhard.schroeder@bfn.de

Nicolas Schtickzelle Biodiversity Research Centre, Unité d'écologie et de biogéographie, Université catholique de Louvain, 4-5 (Carnoy), Croix du Sud, B-1348 Louvain-la-Neuve, Belgium, nicolas.schtickzelle@uclouvain.be

Jeremy B. Searle Department of Biology (Area 2), University of York, P.O. Box 373, York YO10 5YW, UK, jbs3@york.ac.uk

Axel Ssymank Department I 2.2 "Habitats Directive/Natura 2000", German Federal Agency for Nature Conservation, Bonn, Germany, axel.ssymank@bfn.de

Lee Taylor Institute of Arctic Biology, University of Alaska Fairbanks, 311 Irving I Building, 902 N. Koyukuk Drive, P.O. Box 757000, Fairbanks, AK 99775-7000, USA, lee.taylor@iab.alaska.edu

Rodham E. Tulloss P.O. Box 57, Roosevelt, NJ 08555-0057, USA, Res. Assoc. (hons.), New York Botanical Garden, Bronx, NY, USA, ret@eticomm.net

Camille Turlure Biodiversity Research Centre, Unité d'écologie et de biogéographie, Université catholique de Louvain, 4-5 (Carnoy), Croix du Sud, B-1348 Louvain-la-Neuve, Belgium, camille.turlure@uclouvain.be

Zoltán Varga Dept. Evolutionary Zoology, Faculty of Science and Technology, University of Debrecen, Egyetem-tér 1, H-4010 Debrecen, Hungary, zvarga@tigris.unideb.hu

Régine Verlaque Institut Méditerranéen d'Ecologie et Paléoécologie, UMR CNRS 6116, Case 4, Université de Provence, 3 Place Victor Hugo, F-13331 Marseille Cedex 3, France, regine.verlaque@univ-provence.fr

Marta Vila Department of Molecular and Cell Biology, Faculty of Sciences, University of La Coruña Campus Zapateira, E-15008 La Coruña, Spain, mvila@udc.es

M. Vischer-Leopold Department I 2.2 "Habitats Directive/Natura 2000", German Federal Agency for Nature Conservation, Bonn, Germany mareike. vischer-leopold@bfn.de

Alfried P. Vogler Department of Entomology, The Natural History Museum, Cromwell Road, London, SW7 5BD, UK; Division of Biology, Imperial College London, Silwood Park Campus, Ascot, Berkshire SL5 7PY, UK, a.vogler@imperial.ac.uk

Goddert von Oheimb Institute of Ecology and Environmental Chemistry, Leuphana University of Lüneburg, Scharnhorststraße 1, D-21335 Lüneburg, vonoheimb@uni-lueneburg.de

Michael Wink Institut für Pharmazie und molekulare Biotechnologie, Universität Heidelberg, Im Neuenheimer Feld 364, D-69120 Heidelberg, Germany, wink@uni-hd.de

Glenn Yannic Department of Ecology and Evolution, University of Lausanne, CH-1015 Lausanne, Switzerland, glenn.yannic@gmail.com

Frank E. Zachos Zoological Institute, Christian-Albrechts-University, Olshausenstrasse 40, D-24118 Kiel, Germany, fzachos@zoologie.uni-kiel.de

Marco Zimmermann Institute of Forest Botany and Forest Zoology, TU Dresden, Tharandt, Germany, marcozimmermann53129@googlemail.com

Relict Species: From Past to Future

Jan C. Habel, Thorsten Assmann, Thomas Schmitt, and John C. Avise

1 What Are Relicts?

Dictionaries define a "relict" as something that has survived, usually as a trace, from the past. In biology, relicts are distinctive populations or species that typically are small in size or severely restricted in geographic range. Biologists distinguish between taxonomic and biogeographic relicts. Taxonomic relicts are a few or sole survivors of a once diverse taxonomic assemblage, whereas biogeographic relicts are descendants of once widespread taxa (or populations) that now have a narrow geographic distribution (Lomolino et al. 2006). Both categories sometimes coincide, as for example in the case of "living fossils" (such as ginko, lungfishes, crossopterygians, or marsupials) that closely resemble their ancient ancestors in overall phenotype (Futuyma 2005; Lomolino et al. 2006; Beierkuhnlein 2007). In the following, we focus on biogeographic relicts.

Climatic and other large-scale environmental changes can have fundamental impact on the distributional patterns of species and alter the composition of communities and ecosystems (e.g. Hewitt 1999; Schweiger et al. 2009). Species react by range shifts, or by local or regional extinctions when their *adaptive capacity* is exceeded (Agarwal 2001). The origins and distributions of modern-day relicts can often be related to environmental changes of the past. Glacial periods and warm interglacial periods of the Quaternary, in particular, often help us to understand how relict species and

J.C. Habel (✉)
Musée national d'histoire naturelle, Section Zoologie des Invertébrés, L-2160 Luxembourg, Luxembourg
e-mail: Janchristianhabel@gmx.de

J.C. Habel and T. Schmitt
Department of Biogeography, Trier University, D-54286 Trier, Germany

T. Assmann
Institute of Ecology and Environmental Chemistry, Leuphana University Lüneburg, D-21335 Lüneburg, Germany

J.C. Avise
University of California, CA 92697 Irvine, USA

J.C. Habel and T. Assmann (eds.), *Relict Species: Phylogeography and Conservation Biology*,
DOI 10.1007/978-3-540-92160-8_1, © Springer-Verlag Berlin Heidelberg 2010

populations arise and sometimes have served as important sources for re-colonizing much larger areas (de Lattin 1967). For example, many thermophilic species survived the last ice age in relatively small refugia on Europe's Southern peninsulas. During interglacial periods, these species often colonized large areas. Thus, many temperate species that are widespread today existed as small relict populations just a few thousand years ago (cf. Schmitt 2007). Conversely, many cold-adapted species show evidence of formerly wide distributions, followed by severe range restrictions during postglacial warming. Prominent examples of the latter are the butterflies *Proclossiana eunomia* pictured on this book (front side) and *Lycaena helle* (small picture on the back side), which currently resides only in high-elevation mountain enclaves and at more Northern latitudes that have been shown to coincide with distinctive genetic clusters in this species (Habel et al. 2009). Although such species today (and during the other warm periods between glacial periods) are restricted to small areas, they may have the potential to recolonize larger areas if and when the climate cools again. In general, the assumption that relict species inevitably are on their way to extinction may be erroneous; at least some relict populations undoubtedly retain the potential to adapt to a broad spectrum of environmental conditions (Hampe and Petit 2005).

Range restrictions and expansions are also known to result from human-mediated landscape changes. For example, woodlands covered large areas of Central Europe prior to anthropogenic deforestation. Especially during the Middle Ages and early modern times, these extensive woodlands were degraded to a few small remnants. After the establishment of modern forestry and changes in land use, woodland coverage again increased enormously. The former woodland remnants preserved forest-inhabiting plants and animals, some of which still show relict-like distributions but others of which were able to re-colonize large areas (e.g. Assmann 1999; Drees et al. 2008). Comparable changes are known for numerous habitat types and regions. These alternating scenarios of range expansions and restrictions remind us that species with relict-like versus wide distributions are *not* necessarily contradictory indicators of past environmental changes.

2 Conservation of Relict Species and Populations

Why focus special conservation attention on relict as opposed to non-relict taxa? One explanation may reside in the connotation for relict that appears from the word "trace" in the dictionary definition. Typically, little remains of any relict, implying in the current biological context that extant populations of a relict species are few in number or small in size. Thus, populations of a relict taxon might occupy disjunct mountain peaks or perhaps a few isolated lakes, for example. Rarity and restricted distributions obviously place relict species in special demographic jeopardy of extinction, especially in the face of ongoing climatic changes and other ecological perturbations.

Another more subtle connotation in the dictionary definition is that relicts are unchanged or "left behind" relative to their non-relict counterparts. It seems doubtful, however, that evolutionary genetic change has ceased or even slowed substantially in relict species. Indeed, if genetic drift is a potent evolutionary force (as expected for small populations), then the genetic differences of extant relicts from their ancestors or from their modern sister taxa might even be greater than the genetic differences of non-relict species from their respective ancestors or extant relatives. One of the most robust discoveries from the field of molecular evolution is that genetic change proceeds inexorably through time, in essentially all species, and sometimes at surprisingly steady mean rates. An interesting scientific question is whether relict species (as defined by small population size or restricted geographic distribution) tend to evolve at the genotypic or phenotypic levels at rates that differ consistently from those of their non-relict counterparts (García-Ramos and Kirkpatrick 1997).

Another set of scientifically interesting questions stems from the fact that various relict species and populations can have strikingly different ages, thus making them excellent objects for comparative evolutionary and ecological studies (Lesica and Allendorf 1995). In particular, phylogeographic surveys can give great insights into historical, ecological and evolutionary processes related to past shifts (restrictions and expansions) in the distributions of species (Avise 2000; Hewitt 2000, 2004; Schmitt 2007). Moreover, by helping to identify evolutionarily significant units (ESUs) and management units (MUs), phylogeographic and population genetic analyses can help to identify relict populations of special importance for conservation and management (cf. Moritz 1994; Avise and Hamrick 1996; Pérez-Tris et al. 2004; De Guia and Saitoh 2006). Finally, there is a huge scientific interest in stochastic processes as well as inbreeding in relict species that consist either of single populations or structured metapopulations (cf. Melbourne and Hastings 2008). These processes can reduce genetic variability (Petit et al. 2003; Chang et al. 2004) and diminish genetic fitness (cf. Berger 1990; Reed and Frankham 2003) within relict populations of both plants (Oostermeijer et al. 1996) and animals (Wynhoff et al. 1996; Madsen et al. 1999).

Many relict species are already endangered and referenced in *Red Lists*. Conservation must not ignore relicts, as they form an essential component of overall biodiversity. Relict populations and species are, almost by definition, "survivors". Thus, it is both sad and ironic that conservation biologists must now be concerned about the survival into the future of relict species that, by definition, are proven survivors from the evolutionary past. Nonetheless, such conservation concern is merited, because human actions are precipitating global climatic and other environmental changes that may fall outside the adaptive scope of many relict (and other) living species.

3 The Scope of This Book

This book focuses on relict species and describes their history, current status, and future trends. It presents a compilation of case studies representing different methodological approaches and different temporal and spatial scales, all meant to illustrate evolutionary processes and ecological traits of isolated remnant populations. The book is structured into five sections. Section 1 deals with the basics of climate change and the responses of species and ecosystems. Sections 2 and 3 deal with the effects of pre-glacial and glacial phenomena, and postglacial range expansions, on relict species. Section 4 includes conservation approaches to protect and preserve relict species. Finally, in Section 5, future trends of relict species and their projected distributional patterns are discussed. The book also includes four mini-reviews that highlight molecular techniques, the biogeography of Europe, cave species, and ecological niche modelling.

References

Agarwal AA (2001) Phenotypic plasticity in the interactions and evolution of species. Science 294:321–326

Assmann T (1999) The ground beetle fauna of ancient and recent woodlands in the lowlands of North-West Germany (Coleoptera, Carabidae). Biodivers Conserv 8:1499–1517

Avise JC (2000) Phylogeography. The history and formation of species. Harvard University Press, Cambridge

Avise JC, Hamrick JL (eds) (1996) Conservation genetics: case histories from nature. Chapman & Hall, New York

Beierkuhnlein C (2007) Biogeographie - Die räumliche Organisation des Lebens in einer sich verändernden Welt. UTB Ulmer, 397

Berger J (1990) Persistence of different-sized populations: an empirical assessment of rapid extinctions in bighorn sheep. Conserv Biol 4:91–98

Chang CS, Kim H, Park TY, Maunder M (2004) Low levels of genetic variation among Southern peripheral populations of the threatened herb, *Leontice microrhyncha* (Berberidaceae) in Korea. Biol Conserv 119:387–396

De Guia APO, Saitoh T (2006) The gap between the concept and definitions in the Evolutionarily Significant Unit: the need to integrate neutral genetic variation and adaptive variation. Ecol Res 22:1440–1703

Drees C, Matern A, Rasplus J-Y, Terlutter H, Assmann T, Weber F (2008) Microsatellites and allozymes as the genetic memory of habitat fragmentation and defragmentation in populations of the ground beetle *Carabus auronitens* (Col., Carabidae). J Biogeogr 35:1937–1949

Futuyma DJ (2005) Eolution. Sinauer Associates, Sunderland, MA

García-Ramos G, Kirkpatrick M (1997) Genetic models of adaptation and geneflow in peripheral populations. Evolution 51:21–28

Habel JC, Augenstein B, Nève G, Rödder D, Assmann T (2009) Population genetics and ecological niche modelling reveal high fragmentation and potential future extinction of the endangered relict butterfly *Lycaena helle* in. In: Habel JC, Assmann T (eds) Relict species, phylogeography and conservation biology. Springer, Heidelberg

Hampe A, Petit RJ (2005) Conserving biodiversity under climate change: the rear edge matters. Ecol Lett 8:461–467

Hewitt GM (2000) The genetic legacy of the ice ages. Nature 405:907–913

Hewitt GM (2004) Genetic consequences of climatic oscillations in the Quaternary. Philos Trans R Soc Lond B 359:183–195

Hewitt GM (1999) Postglacial re-colonization of European biota. Biol J Linn Soc 68:87–112

de Lattin G (1967) Grundriß der Zoogeographie. Verlag Gustav Fischer, Jena

Lesica P, Allendorf FW (1995) When peripheral populations are valuable for conservation. Cons Biol 9:753–760

Lomolino MV, Riddle BR, Brown JH (2006) Biogeography. Sinauer, Sunderland, MA
Madsen T, Shine R, Olsson M, Wittzell H (1999) Restoration of an inbred adder population. Nature 402:34–35
Melbourne BA, Hastings A (2008) Extinction risk depends strongly on factors contributing to stochasticity. Nature 454:100–103
Moritz C (1994) Defining evolutionarily significant units for conservation. Trends Ecol Evol 9:373–375
Oostermeijer JGB, Brugman ML, De Boer ER, Den Nijs HCM (1996) Temporal and spatial variation in the demography of *Gentiana pneumonanthe*, a rare perennial herb. J Ecol 84:166–174
Pérez-Tris J, Bensch S, Carbonell R, Helbig AJ, Tellería JL (2004) Historical diversification of migration patterns in a passerine bird. Evolution 58:1819–1832
Petit RJ, Aguinagalde I, de Beaulieu JL, Bittkau C, Brewe S, Cheddadi R (2003a) Glacial refugia: hotspots but not melting pots of genetic diversity Science 300:1563–1655
Petit RJ, Aguinagalde I, de Beaulieu JL, Bittkau C, Brewe S, Cheddadi R (2003b) Glacial refugia: hotspots but not melting pots of genetic diversity. Science 300:1563–1655
Reed DH, Frankham R (2003) Correlation between fitness and genetic diversity. Conserv Biol 17:230–237
Schmitt T (2007) Molecular biogeography of Europe: pleistocene cycles and postglacial trends. Front Zool 4:11
Schweiger O, Stettele J, Kudrna O, Klotz S, Kühn I (2009) Climate change can cause spatial mismatch of trophically interacting species. Ecology 89:3472–3479
Wynhoff I, Oostermeijer JGB, Scheper M, Van der Made JG (1996) Effects of habitat fragmentation on the butterfly *Maculinea alcon* in the Netherlands. In: Settele J, Margules C, Poschlod P, Henle K (eds) Species survival in fragmented landscapes. Geo J Libr 35:15–23

Part I
Climate and Ecosystems

The Changing Climate: Past, Present, Future

Markus Quante

Abstract Over the 4.6 billion years of its existence, the Earth has seen a large variety of climate states. During the evolution of our planet, its climate was characterized by periods of enhanced climate variability or even swings and some more or less stable – almost quiet – periods. Natural climate variability was the rule rather than an exception and the evolution of life on Earth was closely linked to climate and its change.

For about 250 years, mankind has interfered stronger with the climate system via the release of radiative gases and particles in substantial amounts into the atmosphere. A global mean near surface temperature increase – global warming – can be deduced from instrumental observations, which started in about 1860. The pace and amount of this temperature increase is unprecedented at least the past 1600 years, as reconstructions from proxy data indicate. The observed warming can be attributed to a large extent to human activities as the most recent report of Intergovernmental Panel on Climate Change states. The atmospheric temperature increase is accompanied by an increase in sea surface temperature and a rise of the sea level. Evidence is building that human-induced climate change has also a direct influence on changes in precipitation and the hydrological cycle.

Climate projection driven by socio-economic scenarios indicate that the global temperature and sea level rise will continue throughout the twenty-first century and beyond, the amount of which is strongly dependent on the underlying emission assumptions.

There are a few climate elements that may be sensitive to sudden, abrupt changes, when a set of conditioning parameters is overstepped or certain thresholds are passed; a prominent example is a possible collapse of the thermohaline circulation in the North Atlantic. Here, further research in necessary to quantify thresholds, effects and time horizons.

Overall it can be said that a certain amount of future climate change is unavoidable regardless which route of emission reductions mankind will follow and that emissions from the twenty-first century will noticeably affect climate over the entire millennium.

M. Quante
GKSS Research Center, Institute for Coastal Research, Geesthacht, Germany
e-mail: Markus.quante@gkss.de

J.C. Habel and T. Assmann (eds.), *Relict Species: Phylogeography and Conservation Biology*,
DOI 10.1007/978-3-540-92160-8_2,

1 Introduction

About 4.6 billion years ago the solar system, including planet Earth, was formed. During the evolution of our planet its climate was characterized by periods of enhanced variability or even swings and some more or less stable – almost quiet phases. The history of the Earth is a story of the co-development of the oceans, the atmosphere, crustal rocks, climate and life in a cosmic perspective (Cloud 1978). From the viewpoint of us human beings, one of the most striking questions in this context is: what is needed and in what quality and quantity to enable planets to support life? Whether planets are suitable for life depends to a large extent on their volatile abundances, especially water, and, as a precondition for their climate which is strictly determined by the availability of energy, on the distance from their central star.

The Earth's climate has remained conducive to life for at least the past 3.5 billion years, despite a much lower solar luminosity at the beginning of this time span (about 30% less, Gough 1981), which is known as faint young sun (Kasting et al. 1988; Kasting and Catling 2003). The fact that the global temperature varied only between values which allowed for liquid water during the early phase (faint young sun paradox) is most probably due to elevated concentrations of the greenhouse gases carbon dioxide (CO_2) and/or methane (CH_4), which are involved in long-term feedback loops, which stabilize climate in such a way as to allow a water cycle to be maintained (Broecker 1985; Pavlov et al. 2000; Kasting 2005).

The main external and internal natural drivers for climate variability which had more or less an influence over the entire time span of the Earth's history are related to changes in the sun's output on different time scales (e.g. Haigh 2007), orbital changes with different periodicities (e.g. Berger 1988), plate tectonics (Raymo and Ruddiman 1992), and volcanism (Lamb 1970; Robock 2000). After a purely physicochemical determined phase at the very beginning of the evolution of life, the climate started to be strongly influenced by the biosphere. With the emergence of photosynthesis in the oceans and on land, the biosphere contributed essentially to the high oxygen concentration in the atmosphere and thereby gained an important influence on the chemistry of trace gases. The biosphere started to become involved in many of the cycles of matter, and with the water cycle and the carbon cycle established the strongest signature in the climate system. The development and expansion of vegetation on land substantially changed the albedo of the Earth's surface, thereby impacting on the energy budget of the planet.

The stability of climate over several recent millennia has allowed land to be cleared, developed and activated to allow for the production of food needed by a growing population. With the development of agriculture some 1,000 years ago and the related release of greenhouse gases, mainly CO_2 and CH_4, and changes in land cover affecting the Earth's albedo and biogeophysical cycles, humans started to influence the global climate (Ruddiman 2003). With the advent of coal-burning factories and power plants in modern times, industrial societies began to release CO_2 on a large scale into the atmosphere. The development of the mobile society with an ever increasing number of motorised vehicles (locomotives, ships and cars) added to the emission portfolio. Crutzen and Stoermer (2000) called the industrial era with humans massively altering the greenhouse gas concentration in the atmosphere as the "Anthropocene"; the year 1800 has been assigned as the beginning of this epoch (see also Steffen et al. 2007). All relevant greenhouse gas concentrations have increased considerably over the last 200 years and the global mean temperature is rising and projected to rise throughout the next centuries (IPCC 2007a). Today, mankind finds itself in a position where urgent decisions need to be made and address the issues - in what form and amount emissions related to its activity have to be reduced - in order to *avoid* dangerous interferences with the climate system, as is postulated by the United Nations Framework Convention on Climate Change (UN 1992).

Stabilizing the carbon dioxide component of climate change is strongly related to changing *how* we produce our energy. Stabilization will require carbon emission-free primary energy sources in addition to a reduced end-use energy demand and improvements in energy efficiency. In a widely recognized paper, Pacala and Socolow (2004) build up a scenario with a target value of 500 ppm CO_2 over a 50-year-time scale ,which relies entirely on current technologies. In contrast, Hoffert et al. (2002) conclude from their assessment that only an advanced technology path can lead to a stabilized climate and that intensive research and development on this sector is urgently needed. Besides mitigation, adaptation measures (IPCC 2007b) are also being discussed, given the prospect that a certain change of the global climate is almost unavoidable (IPCC 2007a). In recent years, an essentially different and older idea to address global warming has been reanimated, it concentrates on an active, deliberate *interference* with the climate system in order to reduce the risks associated with climate change. The proposed measures for an attempt to transform the Earth climate on planetary scale are generally summarised by the expression geoengineering (Keith 2000; Schneider 2001; Weart 2008). Geoengineering is currently being broadly but controversially discussed, since many consequences of the proposed concepts are highly uncertain, and might stay uncertain for a long time, and the ethical, political and legal aspects are only barley touched upon (Fleming 2007; Robock 2008; Schneider 2008).

Most probably, climate change *is* and *will be* affecting the physiology, phenology and the distribution of plant and animal species all over the world. Observational and modelling studies are supporting this statement, with details being still under discussion. Plants and animals are responding in an overall unsurprising way, as they did at the end of the last ice age. Mainly in an effort to keep the temperature zone they are adapted to (but other parameters might also be important), they are shifting their range towards the higher latitudes and, if possible, towards higher altitudes such as mountainous regions. Such shifts have been observed in a wide range of species. European butterflies can serve as a well-documented example, and Parmesan et al. (1999) found that more than 2/3 of the species they followed had shifted their range Northward during the twentieth century. In the Alps and the rocky mountains, trees and grasses have been migrating upslope. Details on this topic can not be presented in the frame of this overview; however, some relevant information regarding species, biodiversity and climate change can be extracted from the following publications and the many cited references therein: Hughes 2000; Sala et al. 2000; Walther et al., 2002; Parmesan and Yohe 2003; Walther 2003; Parmesan 2006; Botkin et al. 2007; Fischlin et al. 2007; Hoegh-Guldberg et al. 2007, 2008; Kerr et al. 2007; Midgley et al. 2007; Thuiller 2007; Bonan 2008; Lee and Jetz 2008; Rosenzweig et al. 2008; Thuiller et al. 2008.

The science of climate change is touching upon a huge number of topics with many facets, which can not be fully accounted for here, and the selected aspects can not be elaborated in depth. Hence, an attempt has been made in the following sections to provide an extended selection of relevant references for further study, many of them in easily attainable journals. In the subsequent sections and after a short summary of the principal scientific fundamentals, the story of the Earth's climate over the past is recapitulated with enhanced emphasis on the Holocene. The section on the present-day climate builds on observations based on direct measurements of climate parameters, mainly temperature and precipitation. Finally, the outlook into the future reports on modelling studies of global and regional climate change assessments based on scenario-driven simulations. The limitations of model predictions, especially with respect to regionalisation, will also be touched upon. Regarding present and future climate, the results of the latest assessment report (AR4) published by the Intergovernmental Panel on Climate Change (IPCC 2007a) are used as a baseline. As the IPCC AR4 considers only published scientific papers up to 2005, for some aspects more recent work has been considered, in addition.

2 Scientific Basis

2.1 The Climate System and Relevant Processes

The usual definition of climate is that it encompasses the slowly varying aspects of the atmosphere–hydrosphere–land surface system. In some sense, climate is the average condition of the weather over several years to tens of years (averaging times need to be carefully chosen), as exemplified by the parameters viz., temperature, wind velocity, relative humidity, cloudiness and the amount of precipitation. Modern climate definitions include higher order statistics beyond mean values, such as the magnitudes of day-to-day or year-to-year variations, standard deviations or measures of shapes of parameter distributions.

Climate depends not only on atmospheric processes and composition, but also physical, chemical, and biological processes involving other components of the Earth system play a crucial role. In order to understand what the factors are which control the evolution of climate, the interactions among the different components of the Earth system need to be assessed (e.g. Brasseur et al. 1999). This adds to the complexity of the topic since the evolution and feedbacks involved run on a variety of different time scales. The atmosphere, the hydrosphere, the biosphere, the cryosphere and the lithosphere are the five different Earth system regimes with widely varying impacts and time scales which make up the climate system.

The *abundance of water* in its three states of aggregation and a functioning global water cycle is of utmost importance for the climate system (e.g. Pagano and Sorooshian 2006; Quante and Matthias 2006). The phase of water depends on the temperature and pressure it is exposed to. At the normal range of atmospheric pressures and temperatures on Earth water can exist in all three of its basic states, as is evident from its phase diagram in Fig. 1, which shows the phase transition curves as a function of temperature and partial pressure (see Webster 1994). The Earth's trajectory in Fig. 1, driven by an increasing water vapour greenhouse effect, intercepts the phase curves in the vicinity of the triple point of water (273.16 K), allowing the formation of a complex hydrological cycle. In contrast, because Venus is a star with a considerably warmer primitive surface temperature, the curve for Venus does not intercept any of the water phase transition lines at all. Water stays in the gaseous phase on this planet. The state curve for Mars starts at a relatively low temperature (~240 K; not shown here) and rapidly intercepts the vapour-ice phase transition.

All subcomponents of the climate system are involved in or maintain processes which can have a huge impact on climate. The interplay between radiation and convection in the atmosphere regulates the temperature at the Earth´s surface. The oceans, which cover about 72% of the surface area of our planet, influence climate by their large thermal inertia and their important role in taking up carbon dioxide from the atmosphere. If present, the cryosphere with extensive snow and

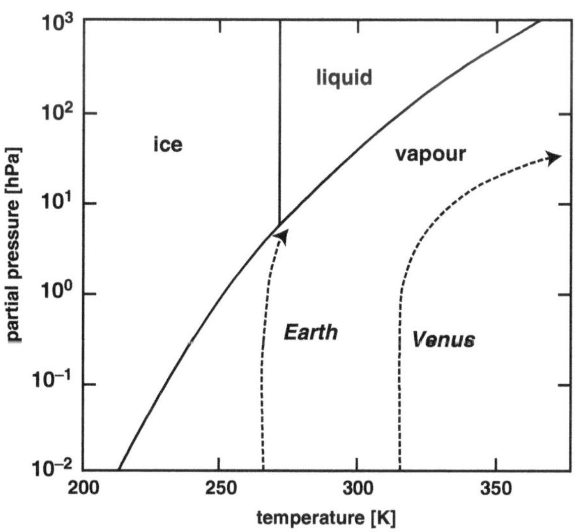

Fig. 1 Phase diagram of water illustrating the possible occurrence of the three states of water for the range of temperatures observed at the Earth's surface and in the lower atmosphere; ~190 K–325 K and 0–50 hPa partial pressure range at the surface. Figure adapted from Quante and Matthias (2006)

ice covered areas has a strong influence on the planetary albedo (Parkinson 2006). Besides their large influence on the Earth´s albedo with a corresponding net cooling effect, the clouds are also contributing to a warming of the surface by absorbing infrared radiation and emitting it partly back towards the ground (e.g. Quante 2004). Clouds cover at any given time between 60 and 70% of the globe. Living organisms on land and in the oceans are involved in liberating oxygen and sequestering carbon in the Earth's crust and thereby reducing the CO_2 concentration of the atmosphere. The evolution of vegetation is strongly coupled with that of soil and climate, and there is a myriad of interactions involved (e.g. Berry et al. 2005; Barth et al. 2005). Plate tectonics exerts an influence on climate on time scales of more than millions of years through continental drift, creation of mountains (Turcotte and Schubert 2002) and volcanism (Robock 2000). Of all the biogeochemical cycles, the hydrological (e.g. Quante and Matthias 2006; Oki and Kanae 2006) and the carbon cycle (e.g. Houghton 2007; Doney and Schimel 2007) are the most relevant for climate and its evolution.

The major external forcing of the climate system comes from the sun. Everything on Earth relies on a steady energy flow provided by our central star. The amount of radiation produced by the sun is not constant, especially in the short, ultraviolet wavelengths. Due to changes in the magnetic structure of the gaseous sun, the solar activity shows variations, which are manifested in an 11-year cycle. Although attempts have been made, a firm theoretical coupling of this short-term solar activity fluctuations with climate changes could not be found. A slightly enhanced energy deposition in the stratosphere is among the most recognized effects. Some evidence for the influence of solar activity variations on the lower atmosphere and climate is critically assessed by Bard and Frank (2006), Foukal et al. (2006) and Haigh (2007). These variations play some role in the discussion on modern global warming, since if climate changes due to the sun were significantly large, it would be more difficult to extract the anthropogenic signal from the climate record (see Sect. 4.4). The story is different, however, when dealing with the long-term evolution of the sun; in its infancy, the sun's intensity was about 30% less than what is observed today (Gough 1981), the relatively moderate climate under these conditions is generally referred to as "faint young sun paradox" (e.g. Sagan and Chyba 1997). A further possible external forcing of Earth's climate might come via galactic cosmic rays and their influence on clouds (Marsh and Svensmark 2000; Carslaw et al. 2002; Kristjánsson et al. 2004; Kirkby 2008). The related science and the potential magnitude of postulated effects is currently being debated and planned experiments at the Conseil Europëen pour la Recherche Nucléaire (CERN) should, at least, provide some insight into underlying cloud microphysical processes.

In summary, Fig. 2 sketches the Earth system and its interactions, encompassing the physical climate system, biogeochemical cycles, external forcing, and the effects of human activities. For a more rigorous treatment of the different aspects concerning the climate system and underlying processes, the books by Peixoto and Oort 1992; Graedel and Crutzen 1993; Hartmann 1994; Brasseur et al. 1999; Seinfeld and Pandis 2006; Ruddiman 2008; and Pierrehumbert 2009 are recommended.

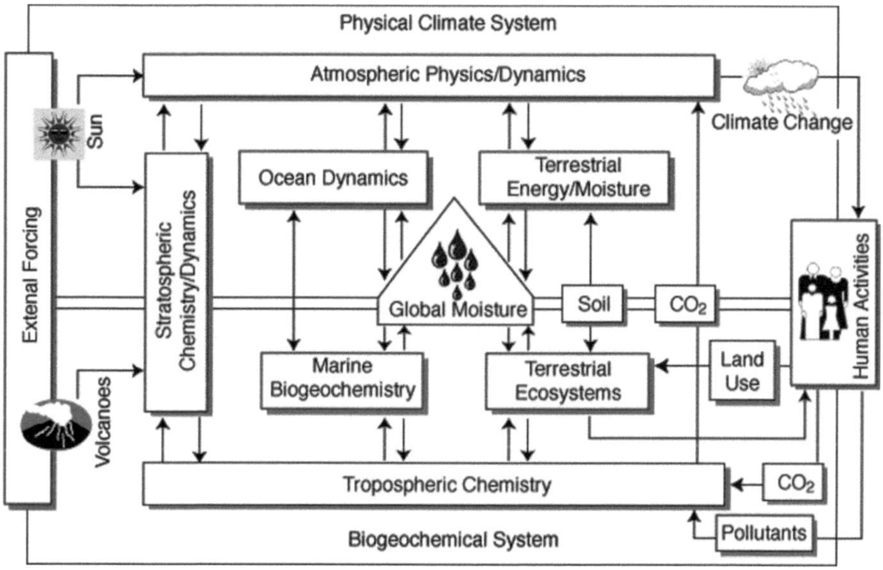

Fig. 2 Schematic diagram of the climate system and its interactions among different components of the Earth system (from Earth System Science: Overview, NASA)

2.2 Climate Variability

Climate varies on all time scales longer than its definition limit and on a wide range of spatial scales, from regional to global. The variability of climate can be expressed in terms of two basic modes: the forced mode with variations which are the response of the climate system to changes in external forcing, which themselves are not influenced by the climatic variables themselves, and the free mode variations due to internal instabilities and feedbacks leading to non-linear interactions among the various compounds of the climate system (Peixoto and Oort 1992; Ghil 2002a). Also, stochastic forcings could produce sudden impulses to the climate variables (Hasselmann 1976). It is due to this complex nature of the variability that a reliable identification of an anthropogenic influence on climate and especially its quantification turns out to be so difficult.

The temporal variability of the climate system over the last 10 million years can be visualized by reproducing a power spectrum for temperature near the surface (Fig. 3), which is a composite of several climatic time series. The sharp lines in the spectrum correspond to periodically forced variations (daily and yearly cycles), broader peaks arise from internal modes of variability, and the continuous segments of the spectrum reflect stochastically forced variations, as well as deterministic chaos resulting from the interplay of non-linear feedbacks (Ghil 2002b).

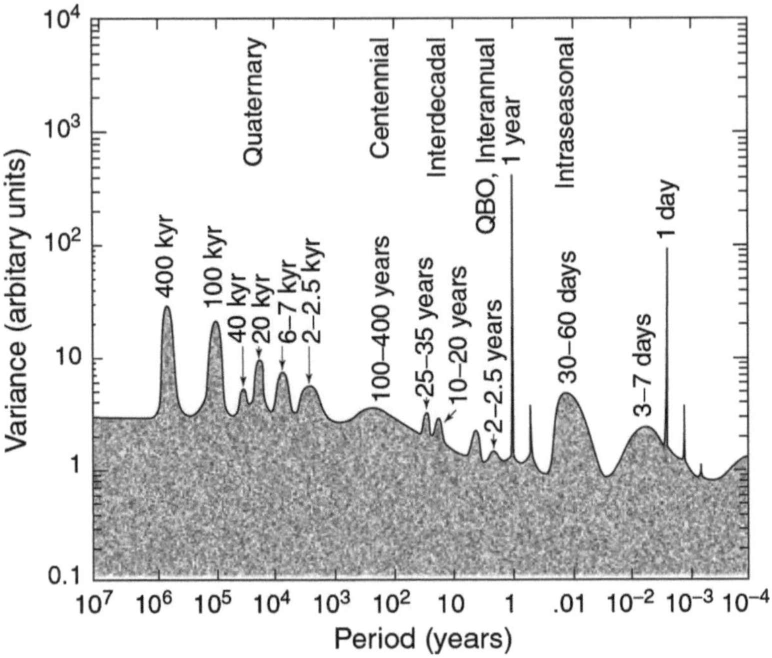

Fig. 3 Composite power spectrum of climate variability over the last 10 Myr (after Ghil 2002b)

The signature immediately to the left of the peak at the 1-year-period reflects the influence of an interannual variability, which is known as El Niño-Southern Oscillation (ENSO) (Neelin et al. 1998) and involves internal modes of the tropical Pacific ocean-atmosphere variability. Periods of several decades as they occur in the spectrum are the subject of ongoing research. As the interdecadal variability is related to climate system internal processes this part of the spectrum might hold some clues for the extension of the predictability of the ocean-atmosphere system.

The spectrum part with periods between 1,000 years (1 kyr) to about 1 million years (1 Myr) in Fig. 3 represents the paleoclimatic variability. The temperature data underlying the spectrum far outside the instrumental record come from proxy indicators such as coral reef records, tree rings, marine sediment cores, and ice core records. The time span captured by the spectrum is called *ice age era* with its main glaciation cycles (Imbrie and Imbrie 1979), it occupies no more than about one-tenth of the documented Earth history. The past 2.5 Myr with the marked climatic swings are called Quaternary period.

The external forcings comprise variations caused by smaller fluctuations in the sunlight reaching the Earth, which occur over longer timescales. These involve changes in the shape of the Earth's orbit around the sun and changes in the Earth-sun distance at the time of the equinoxes. Early in the twentieth century, in Belgrade, Milutin Milankovic´ produced a mathematical theory that became the basis for a general explanation of the dynamical behaviour of the climate system, i.e. the major changes in past and future (Berger 1988). In particular, the theory which builds on deterministic astronomical variations in the orbital parameters of the Earth should explain the recent ice ages on Earth. The theory was not fully accepted by his contemporary scholars, and today we know that orbits are *not* everything (Broecker and Kunzig 2008), but basic orbit variability can be used to reconstruct climate records.

The prominent peaks around 20, 40 and 400 kyr in the spectrum in Fig. 3 are a result of quasi periodic variations in the Earth's orbit (Berger 1988), which affect the summer insolation at high latitudes. In intervals with relatively weak summer insolation snow deposited during winter does not melt completely, so over thousands of years thick ice sheets can be accumulated (enforced by the ice-albedo-feedback). The cycle of about 100,000 years is due to changes in the eccentricity of the orbit (elliticity), variations of the tilt of the Earth's axis of rotation relative to the orbit plane (the obliquity) leads to the 41,000-year-cycle and the 23,000-year-(19,000 year) cycle marks the changing axial precession. Details of the orbital theory of climate variation are still under investigation (Paillard 2006).

Overall, it can be said that climate variability results from complex interactions of forced and free variations because the climate system is a highly non-linear, dissipative system with many sources of instabilities (Peixoto and Oort 1992). Feedback mechanisms act as internal controls of the system, they originate from special couplings between the open subcomponents and are of particular importance for the Earth's climate system. Feedbacks may either amplify or dampen the original distortions of the system. Important feedbacks are the water vapour-, cloud-, snow/ice-albedo-feedbacks, and the carbon-silicate cycle/carbon dioxide feedback (Held and Soden 2000; Stephens 2005; Kasting and Catling 2003).

2.3 *Energy Balance and Greenhouse Effect*

The Earth receives the overwhelming amount of its energy from the sun in the form of visible and near-infrared radiation; it is mainly this latter energy which warms its surface. Our planet is cooled by the emission of thermal infrared radiation into space. Treating the Earth as a blackbody radiator with an effective temperature T_e, the balance of incoming radiation and outgoing radiation leads to the following relationship (e.g. Wallace and Hobbs 2006): $\sigma T_e^4 = S/4(1-A)$, where σ is the Stefan–Boltzmann constant, where S is the solar flux at the distance Sun–Earth, and A is the planetary albedo. Using the appropriate numerical constants and quantities, solving the equation yields an effective emission temperature of 255 K. Reality is, however, slightly more complex, in that the Earth together with its atmosphere is *not* a perfect blackbody. The atmosphere warms the surface by the so-called greenhouse effect; parts of the infrared radiation emitted from the surface is selectively absorbed and re-emitted by infrared-active gases within the atmosphere. With the present-day incoming solar flux (~1,368 Wm^{-2}), albedo (~0.3) and atmospheric composition, a global mean surface temperature of 288 K results. The difference between the effective emission temperature and this surface temperature of 33 K is the magnitude of the actual greenhouse effect. A more complete treatment of the involved energy fluxes with emphasis on the energy budget resolved for additional processes is given in Kiehl and Trenberth (1997). The most difficult factor in the quantification of the energy fluxes through the atmosphere is the planetary albedo, which is determined by ocean and land surface characteristics (soil type, soil moisture, vegetation) and the three-dimensional cloud distribution in the atmosphere, which is responsible for the largest fraction. The relevant cloud (and aerosol) properties are difficult to predict. Thus, calculations of past or future climates based on energy principles (radiative and turbulent fluxes) are subject to high uncertainties, i.e. since clouds are involved in several feedback loops (Stephens 2005).

In the present-day atmosphere the most important greenhouse gases are water vapour and carbon dioxide, of which the former contributes about two-thirds of the associated warming. Lesser contributions come from methane, nitrous oxide, ozone and various chlorofluorocarbons. It is important to distinguish between long-lived greenhouse gases, which are removed slowly from the atmosphere on a time scale of hundreds to thousands of years, and short-lived greenhouse gases, which are removed within weeks to a year by condensation or fast chemical reactions. The short-lived greenhouse gases act primarily as a feedback mechanism. In the eighteenth century, the atmospheric concentration of most of these gases (with the exception of water vapour) began to be significantly altered due to emissions from power plants, industry, agriculture and animal farming as well as the mobility sector. The ongoing greenhouse effect discussion is driven by the anthropogenic emissions of CO_2, CH_4, N_2O, O_3 and CFC which show strong increases, especially over the last decades. It has to be noted that all of these gases have a relatively long residence time in the atmosphere and therefore, a high greenhouse warming potential.

Water vapour as the major player in the Earth's energy budget is buffered by the huge oceans on a time scale of a few weeks. This gas adjusts its atmospheric concentration in response to climate changes, and it has a strong positive feedback in the climate system, thus amplifying global warming caused by other forcings (Held and Soden 2000). Water vapour, although an important greenhouse gas, is not a prime driver of modern climate change (it plays an essential role via its positive feedback). As there are no significant anthropogenic emissions, water vapour has not become the subject of political regulatory protocols.

Of course human beings started influencing the atmospheric composition, and greenhouse gas concentrations, well before the massive industrialisation began. About 11,000 years ago stone-age farmers may have already altered Earth's climate by clearing forests and irrigating fields to grow crops. Besides resulting in changes to the albedo, these activities may have led to considerable amounts of CO_2 and CH_4 being emitted. It should be mentioned that some scholars have even put forward the hypothesis that this early interference with the climate system could possibly have averted the start of a new ice age (Ruddiman 2003, 2005). This controversial hypothesis is discussed, for example, by Claussen et al. (2005), since the concentrations of CO_2 during the Holcene could also be attributed to natural processes (e.g. Broecker and Stocker 2006; Steffen et al. 2007).

2.4 CO_2: The Flagship of Long-Lived Greenhouse Gases

CO_2 is currently the most important greenhouse gas (IPCC 2007a,b). It is involved in a complex carbon cycle (which also includes CH_4), which in turn is closely connected via feedbacks to the total climate system (Houghton 2007). CO_2 is a trace gas on Earth and it is present in the atmosphere in a relatively low concentration. However, this concentration has varied strongly over geological time (Doney and Schimel 2007). CO_2 is implicated in virtually all of the great climate shifts in the history of our planet, including the coming and going of ice ages, warm ice-free states and the collapse of the Earth into a globally frozen state some 600 million years ago, the latter event being known as *snowball earth* (Hoffman and Schrag 2002; Pierrehumbert 2005). It was also involved in preserving conditions favourable to life on the young Earth, when the sun was much fainter (25–30%) than it is today (Kasting and Catling 2003). Of course, there is a lot to say about CO_2 in the biogeochemical carbon cycle and its many relations to climate. For further study, the overview publications compiled by Doney and Schimel (2007), Hansen et al. (2007a), and Houghton (2007) and the articles cited by those authors can be referred to.

Immediately before the beginning of the industrial era, the atmospheric CO_2 concentration amounted to about 280 ppm and varied between 260 and 280 ppm in the 1,000 of years previously (Houghton 2007). Because there is relatively little CO_2 in the atmosphere to begin with, modern human activity has the prospect of doubling its concentration within the twenty-first century. A large mount of the emitted CO_2 is taken up by the oceans and the vegetation, but a significant part remains within the atmosphere. State-of-the-art assessments based in part on the oxygen measurement methods developed by Ralph Keeling, the son of famous Dave Keeling, allocate about 35% of the emitted CO_2 to an ocean uptake and about 15% to land sinks, mainly the large forests distributed over the continents (Sabine et al. 2004; Broecker and Kunzig 2008). Due to its long atmospheric residence time, CO_2 is well mixed over the hemispheres and its concentration increase over time can be assessed to first order by determining the relevant values at one location. Today, there are approximately 100 stations worldwide where weekly air samples are collected and analysed for CO_2. Keeling started the first systematic monitoring of CO_2 concentrations in 1958 at the Mauna Loa observatory, Hawaii, in the middle of the Pacific and far away from any large industrial sources (Keeling et al. 1976). The resulting concentration curve has become a sort of an icon of global warming in modern times, and is also known as *Keeling-curve* (Broecker and Kunzig 2008). In Fig. 4, the Mauna Loa CO_2 concentration is plotted, beginning in 1958 and including the most recent observed data. Atmospheric CO_2 rose from about 315 ppm when the measurements started to about 385 ppm by 2007, about

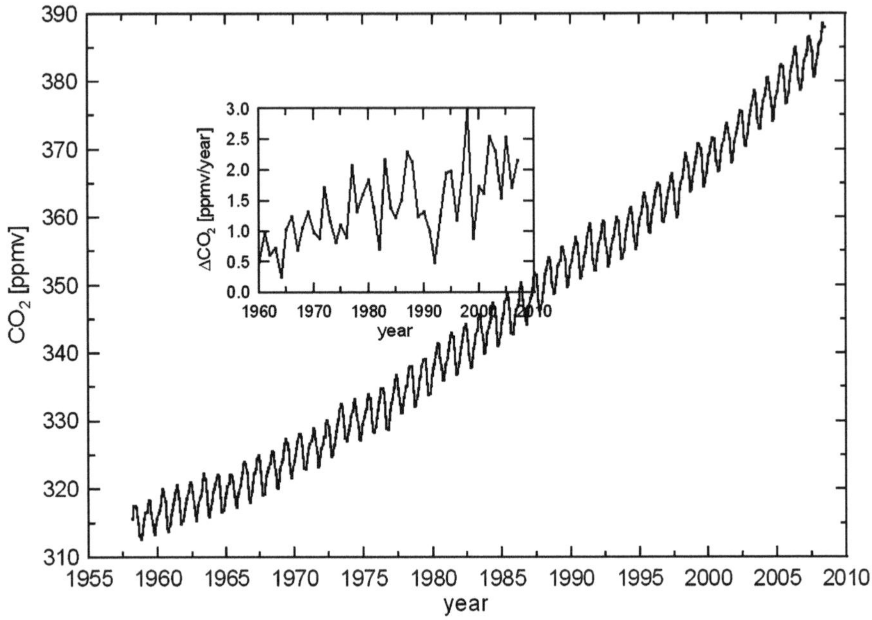

Fig. 4 Monthly mean carbon dioxide concentrations measured at Mauna Loa Observatory, Hawaii. The small inlay graph shows annual mean carbon dioxide growth rates for the Mauna Loa station. Data provided by Dr. Pieter Tans, NOAA/ESRL

38% above the pre-industrial value of 280 ppm. The present concentration is the highest, which has been recorded over the last 800,000 years (Lüthi et al. 2008) and is probably also the highest over the last 20 million years (Pearson and Palmer 2000). The annual fluctuations seen on the overall rising curve are due to seasonally changes in the uptake of CO_2 by vegetation mainly in the Northern hemisphere; the corresponding curve (not shown here) taken at the South Pole looks rather smooth in comparison. The inlay graph in Fig. 4 shows the CO_2 increase rate at Mauna Loa, with positive values throughout and a clear tendency for higher values in later years. The subtle variations in the annual increase rates indicate the response of terrestrial and marine processes to climate variability (Doney and Schimel 2007). The increase rate in the global average concentration of atmospheric CO_2 between 2000 and 2006 was 1.93 ppm year^{-1}. This rate is the highest since the beginning of continuous monitoring in 1958 and is a significant increase over the increase rates in earlier decades; the average growth rates for the 1980s and the 1990s were 1.58 and 1.49 ppm year^{-1}, respectively (Canadell et al. 2007). The authors also estimate that 35% of the increase in the atmospheric CO_2 growth rate between 1970–1999 and 2000–2006 was caused by a decrease in the efficiency of the land and ocean sinks in removing anthropogenic CO_2 (18%) and by an increase in carbon intensity of the global economy (17%). The remaining 65% was due to the increase in the global economy. Due to the still strongly increasing CO_2 emissions, the concentration of CO_2 will continue to rise in the coming decades, thus further enhancing the anthropogenic greenhouse effect. In relation to these prospects and for the treatment of other greenhouse gases, Section 4 and the most recent UN climate assessment report (IPCC 2007a,b) should be referred to.

3 The Past: As Seen Through Proxies

This view on the climate of the past, is in its kernel orientated along the condensed treatments offered by Graedel and Crutzen (1993), Wallace and Hobbs (2006), and Pierrehumbert (2009) and complemented by selected new findings and some addendums.

3.1 Deep Past: The Geological Record

During the first 700–800 million years of its history, the Earth was still under continual bombardement by planetesimals. Geologists call this early phase the *Hadean Epoch* (4.6–3.8 Gyr ago), for which the geological record is almost non-existent. Such impacts should have liberated water vapour and other volatile substances, thus resulting in the formation of a primordial atmosphere. Not surprisingly, the climate of the Hadean is very poorly understood, with only conceptual and model-based statements predominating (Kasting and Ono 2006).

A sparse geological record is available for the subsequent Archaean epoch (3.8–2.5 Gyr ago), and where Earth has become stable enough to allow for early life to develop in the oceans. The climate during the early Archaean was probably hot, as surface temperatures of $70\pm15°C$, reconstructed from isotopic ratios of mountain land in South Africa, suggest (Kasting and Ono 2006). From paleo-weathering analyses it can be concluded, that during the late Archaean moderate surface temperatures must have predominated (Holland 1984). Although there are large uncertainties here, the balance of evidence suggests a relatively cool early Earth with respect to potential ocean vapourizing events (Kasting and Ono 2006).

3.1.1 Faint Young Sun

It is known that the emission of radiation from stars increases gradually over their lifetime. The luminosity of the sun is believed to have increased more or less linearly by about 30% since the formation of the solar system (Gough 1981). This, in turn, means the solar radiation received by early Earth was considerably lower as compared to the present day value of 1,370 $W m^{-2}$ at the top of the atmosphere. Assessments of historical climate must account for this less energetic start and which is known as faint young sun problem (Kasting et al. 1988).

In spite of the faint young sun, geological evidence indicates that the oceans have been largely free of ice and that the Earth's surface was not perpetually frozen during its early phase of evolution. Furthermore, there is also no convincing evidence in the geological record, which suggests that widespread glaciations occurred before, about 2.3 Gyr ago. The atmosphere must have contained much higher concentrations of greenhouse gases compared to modern times. The response of the inorganic carbon cycle and the carbonate-silicate-cycle to changes in climate kept the CO_2 level in the atmosphere at a sufficiently elevated level, in this context volcanism and weathering were relevant processes within the negative feedback process on climate (Berner 2004).

3.1.2 Methane

In recent years, it has become obvious that CO_2 alone, could not have compensated for the remarkably reduced solar input. Due to a much lower oxygen concentration in the early atmosphere, methane could have had a considerably longer life span which, in addition to a possible early source for methane (methanogens), made CH_4 a candidate for the greenhouse gas responsible for stabilizing the Earth's early climate. The details of the role of methane in early climate are quite complicated and only partly settled. For a more in-depth discussion on methane the reader is referred to Kasting and Catling (2003), Kasting (2005), and Kasting and Ono (2006).

3.1.3 Snowball Earth

Acting in concert with orbit-steered periodic patterns of solar radiation, both the ebbing and rising oceans as well as the forming continents lead to changes in the Earth's reflectivity at the surface (albedo). Land areas can also influence the albedo considerably by the formation of ice caps and by vegetation growth. The world-wide distribution of types, conditions, and position of rocks permit a determination of the periods and extent of snow and ice appearance on the globe (Broecker and Kunzig 2008). This evidence leads to the conclusion that the Earth has passed through a number of glaciation cycles. The earliest verifiable extensive glacial epoch falls in the time interval 2.4–2.2 Gyr BP (Palaeoproterozoic glaciations). Major glaciations extending all the way into the tropics have been termed "snowball Earth" (Hoffman and Schrag 2002; Pierrehumbert 2002; Allen 2006). "Hard snowball Earth" events are characterized by global sea ice cover (Pierrehumbert 2005). The second extended glaciation period with at least two events occurred between 750 and 600 Myr ago and the third, the Permian glaciation about 280 Myr ago. The ice-albedo feedback mechanism is involved in building-up the snowball, whereas an increasing greenhouse gas effect, carbonate formation and windblown dust events are among the complex processes involved in reversing this phenomenon, with possible abrupt flips into an ice-free state (extended hothouse phase). The role of CO_2 during the Phanerozoic eon is explicitly treated by Crowley (2000b) and an excellent review on its paleo-proxies is given by Royer et al. (2001).

3.2 *The Past 100 Million Years*

During the Cretaceous epoch, which ended 65 Myr BP, surface air temperatures were higher than they are at present. This is especially true for higher latitudes. An extended interval of cooling followed the Cretacous epoch, which went along with declining CO_2 concentrations. The cooling culminated in the Pleistocene glaciation, which began about 2.5 Myr ago. Plate tectonics regulating the carbonate-silicate cycle along with the glaciation of the Arctic continent around 30–15 Myr ago and its drift toward higher latitudes are believed to have been responsible for this gradual trend in cooling. Since about 65 million years ago, Earth's climate has undergone a significant and complex evolution. Gradual shifts driven by tectonic processes and periodic cycles caused by orbital rhythms are responsible for rapid shifts and extreme transitions. An example is the Late Paleocene Thermal Maximum at about 55 Myr ago, which dominates the Cenozoic era (Zachos et al. 2001). This event could be the best ancient climate analogue for future increases in atmospheric CO_2, and although there has been much progress in recent years in revealing the "carbon mysteries" (Pagani et al. 2006) there are still many aspects which remain unknown. The sudden global warming around 55 Myr ago provides evidence for high climate sensitivity to atmospheric CO_2. The fast changes recorded in deep-sea sediment isotope cores have especially helped to improve our perspective on the mechanisms involved in rapid alterations in the climate system. A detailed treatment of the geological evidence and paleogeography can be found in Pierrehumbert (2009). The Cretaceous hothouse climate and the Pleistocene icehouse climate represent opposite extremes of the Earth's typical climate state over the past 500 million years.

3.3 *The Past Million Years*

In the past 2.5 million years, the so-called Quaternary period, the Earth's climate has been marked by temperature swings between extended glacial periods, which were characterized by thick ice sheets covering large parts of North America, Northern Europe and Siberia, and interglacial times characterised by an ice covering only in Antarctica and sometimes Greenland, as is the case today. A schematic illustrating these glacial and interglacial swings and putting the temperature trends of this period in an historical perspective is displayed in Fig. 5.

Drilling is a valuable way of gaining information about the Pleistocene period. Sediment cores reaching a few million years back in time and ice core records dating back to about 800,000 years have been extensively analysed. The ice at the base of the Antarctic is about one million years old. The composition of the past air is

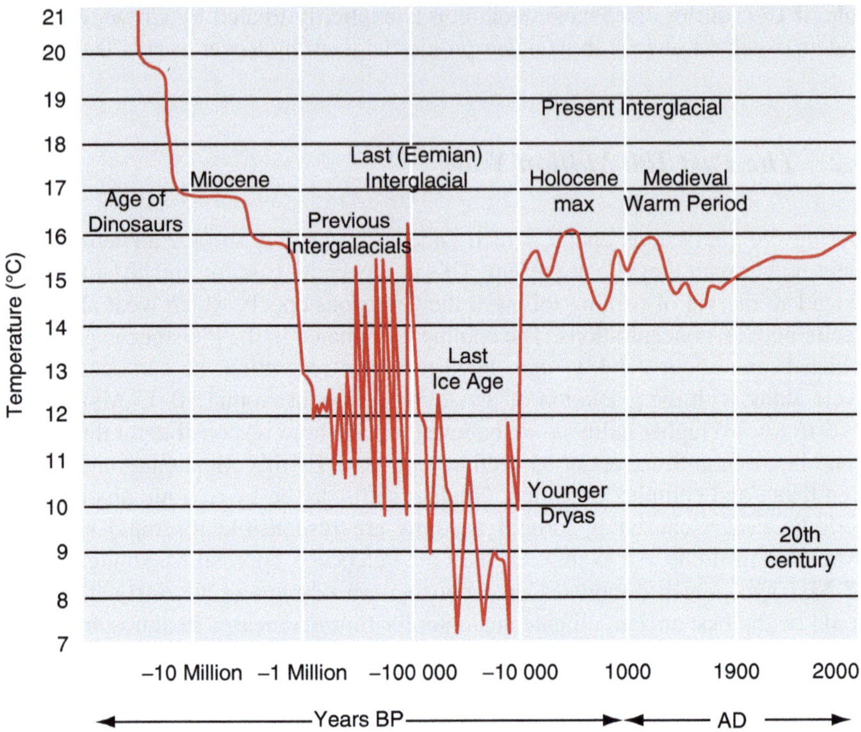

Fig. 5 Temperature trends for the last 10 Million years. Note the different scalings on the time axis: the twentieth century is shown in linear scale. Earlier periods are shown in terms of increasing powers of ten but are linear within each period (figure adapted from Bureau of Meteorology, Australia)

preserved in bubbles within the ice and stable isotopes, derived from the ice itself, are essentially recorders of the temperature at that time. Changes in the atmospheric CO_2 concentrations over the past 800 kyr, are illustrated as a composite of EPICA Dome C, Vostok and Taylor Dome ice cores together with the EPICA Dome C temperature anomaly record in Fig. 6; this figure is taken from Lüthi et al. (2008). Similar data for methane can be found in Loulergue et al. (2008). Based on the analysis of ice records like the ones shown in Fig. 6 and an alignment with many marine sediment cores, it is clear that on time scales of tens of thousands of years or longer, temperature and some other climate parameters vary coherently with one another and that the variations are global in extent. Atmospheric CO_2 and methane concentrations have risen and fallen synchronously with temperature. The correlation between CO_2 and temperature does not, however, fully determine the underlying causes and which as detailed time assessments reveal (Pierrehumbert 2009), for many parts of the record CO_2 is leading temperature by a few hundred years, thereby suggesting that there are different mechanisms acting as compared to those dominant in the present-day anthropogenic greenhouse warming. Over the last

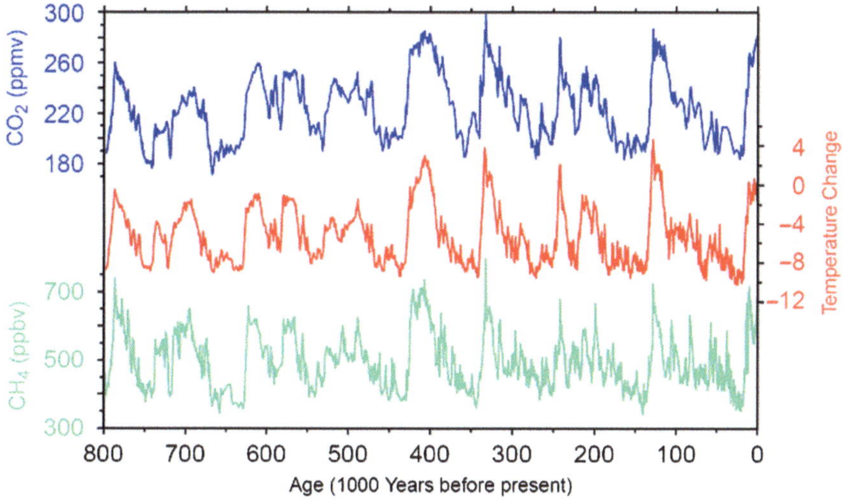

Fig. 6 Composite CO_2 (upper curve) and CH_4 (lower curve) concentrations from EPICA Dome C, Vostok and Taylor Dome ice cores cover the past 800 kyr. (from Lüthi et al. 2008 and Loulergue et al. 2008). The temperature anomaly (middle curve) is the deviation from the mean temperature of the last millennium (Jouzel et al. 2007)

800,000 years, the Milankowitch cycles were the dominant causes and pace-makers for climate variability. Global temperatures cooled at irregular rates during the extended glacial epochs and rose much faster at the beginning of the interglacials. These interglacials recurred during the recorded period at intervals of roughly 100,000 years and had durations typically of about 15,000–20,000 years. The last glacial maximum occurred around 20,000 years ago. The pronounced climatic swings during this Quarternary period are believed to be driven by subtle variations in the orbit parameters, which affect the summer insolation at high Northern latitudes. When the summer insolation was relatively weak for longer periods, the winter snow could partially survive, leading to thick ice sheets over a time span of thousands of years, with the ice-albedo effect amplifying the orbital forcing. Although its correlation with temperature is not totally resolved, because CO_2 is a greenhouse gas, it is fairly certain that a rise in CO_2 concentration warms the planet and reinforces the termination of ice ages, whereas a decrease in CO_2 concentration enhances cooling and reinforces the onset of glaciation.

The data shown in Fig. 6 are the longest and most carefully studied records of polar-ice climate histories published. The record tells us that the highest temperature occurred during this period around 130 kyr ago. The CO_2 concentration during the past 800 kyr never exceeded the present-day value of 385 ppm. For the many interesting details, such as short-term fluctuations as disclosed by the analysis of the sediment and ice core records, the articles by Jouzel et al. (2007), Lüthi et al. (2008), Loulergue et al. (2008) and the book by Pierrehumbert (2009) should be consulted. Interesting findings concerning the important Asian monsoon over the past quarter million years are reported by Wang et al. (2008).

The oceans, which cover more than two-thirds of our blue planet, definitely contribute as strongly as the large ice sheets to the timing of climate events. These waters move in a global circulation system, which is driven by subtle density differences and thereby transport huge amounts of heat energy. Ocean circulation is thus an active and highly nonlinear player in the global climate game. Increasingly clear evidence implicates ocean circulation in abrupt and dramatic climate shifts, such as sudden temperature changes in Greenland on the order of 5–10°C and massive surges of icebergs into the North Atlantic Ocean – events that have occurred repeatedly during the last glacial cycle (Rahmstorf 2002).

3.4 The End of the Last Ice Age and the Holocene

The records of climate proxies for the past 20,000 years, i.e. from about the end of the last glacial maximum, are so complete that not only the typical temperature and precipitation patterns are known but also features such as the geographical ranges of specific species of trees.

The transition from the last glacial to the current interglacial epoch was dramatic. The ice sheets started shrinking around 15,000 years ago. By 12,000 years ago, the Laurentide ice sheet was pouring large volumes of melt water into newly formed lakes and rivers setting the stage for flood events which shaped the landscape, as it is known today.

One of the most striking abrupt change events occurred as the Earth came out of its most recent ice age, the result was a sudden strong warming, called the Bolling warm period (Pierrehumbert 2009), which brought the climate back to a full interglacial warm period. However, this event did not last long and the climate reverted back abruptly to full glacial temperatures. This relapse into ice age conditions lasted about 800 years and ended about 11,700 years ago and is referred to as the Younger Dryas. The timescale of the Younger Dryas event is much shorter than that of the orbital cycles. One explanation for this sudden event, which is still under discussion but favoured by leading scientists (Broecker and Kunzig 2008), could be the massive draining of a huge glacial lake into the ocean, thus influencing the deep water formation causing a shutdown of the oceanic thermohaline circulation and therefore, reducing the poleward heat transport by the Gulf Stream. Whether the Younger Dryas event was truly global in extent, such as the climatic swings on the time scales of 10,000 years and longer have been, is still under debate. Signatures of this event, however, show up clearly in the Greenland ice cores and in European proxy climate records.

Compared to these strong changes, the Bolling warm period and the Younger Dryas in the first 1,000 years after the end of the ice age and the relatively smooth changes over the past 10,000 years, the years of our civilisation, and a tranquil warmth and cessation of high amplitude millennial variability appear, from the climate dynamics point of view, to be quite boring. However, this will most probably not remain like this since human beings started to interfere intensively with the system.

3.4.1 Holocene

The Holocene is the youngest part of the Earth's history. Its definition is given geologically and corresponds to the post-glacial warm period, beginning with the glacier retreat from the moraines in central Scandinavia. Its start is dated as 11,550 years ago. The Holocene has not witnessed the large temperature swings which characterized the previous glacial epochs. There were a few smaller scale temperature variations, and some of these appear to have had considerable impacts on some societies which settled and started agriculture within this epoch. The Little Ice Age, starting in the fourteenth century and lasting into the nineteenth century, is one popular example for this new type of climate impact. A Viking colony in Greenland failed, the population of Iceland declined substantially and farms were abandoned in some parts of Norway and the Alps (Wallace and Hobbs 2006). A comprehensive treatment of the dynamics of natural climate variability during the Holocene based on an integrative analysis combining theoretical assessments, observational data and climate modelling is provided by Fischer et al. (2004). Possible anthropogenic influence on the early Holocene climate is discussed by Ruddiman (2003, 2005) and Claussen et al. (2005).

Reconstructed palaeoclimate variabilities are based on ice cores, lake sediments, tree rings, corals and marine sediments. During the Holocene, the global distribution of rainfall has varied quite remarkably. Due to the changing orientation of the Earth's axis in orbit around the sun and a decreasing Sun–Earth distance, by the 9 kyr BP the average solar radiation over the Northern hemisphere was 8% higher in July and the same percentage lower in January as compared to today. An enhanced temperature contrast between ocean and land was the major effect, causing stronger oceanic evaporation and stronger precipitation over the continents. The gradual formation of a strong monsoon pattern probably also resulted, which in turn led to increased vegetation. In a Northern centric view, the years with the warmest summers started about 7,000 years ago and are sometimes called *Climatic Optimum*. However, this was certainly not optimal for the continental areas in the Southern hemisphere, which experienced a weak cooling.

As fossils from lake beds indicate, areas of the Sahara Desert were vegetated, the so-called "green Sahara", and some parts were even swampy around 6,000 years ago. Parts of the Middle East where the first agricultural settlements were formed have subsequently become extremely dry. In response to weakening insolation forcing, a reduced African monsoon and abrupt hydrological changes resulted in progressive drying of regional terrestrial ecosystems. There is an overall consensus that during the Holocene optimum, the Sahara was greener than what it is today (Claussen 2003).

Large scale dust mobilisation followed from around 4,300 years before the present. Today's deserts and regional meteorological systems were established at about 2,700 year BP (Kroepelin et al. 2008).

Variations on smaller timescales in the order of decades to a century were superimposed on longer term variability (Witt and Schumann 2005), and they also had their influence on humans relying on a suitable climate. The collapse of the Mayan and Anasazi civilizations has been attributed to the inability of these societies to adapt to trends towards drier climates.

The warmed continental interiors in Northern latitudes created a climate which was comfortable for humans and was also, without a doubt, comfortable for living organisms in general. These conditions definitely contributed to the expansion of human activity in these latitudes and were a major determining factor in the history of civilization (Cloud 1978; Lamb 1995).

A comprehensive study based on selected proxy-based reconstructions of different climate variables, together with state-of-the-art time series of natural forcings (orbital variations, solar activity variations, large tropical volcanic eruptions, land cover and greenhouse gases), can be found in Wanner et al. (2008). Here, the most recent results concerning all aspects ranging from desertification to monsoon shifts to Alpine glacier advances are presented and discussed. The work is underpinned by results from General Circulation Models (GCMs) and Earth System Models of Intermediate Complexity (EMICs) to establish a most complete explanatory framework. A meandering tour of Holocene paleoclimate research focussing on Europe and Northern Africa is provided by Battarbee and Binney (2008), these authors bring together an interdisciplinary group of paleoclimatologists.

3.4.2 The Last Millennium: The Late Holocene

Detailed records of climate change during the last millennium come from proxy indicators such as mountain glaciers, tree rings, and corals stored in data archives. Additionally, historical observations recorded by humans over several centuries for several regions are also available.

Up to the beginning of the last century climate changes were small and varied in pattern from region to region. A small cooling in high Northern latitudes surrounding the North Atlantic Ocean and know as the Little Ice Age is one of the climatic fluctuations of interest.

Several attempts have been made to synthesize high-resolution records from ice cores, glacier lengths, tree rings, corals and borehole temperatures into a single estimate of the Northern hemisphere or global temperature changes over the last millennium. These large-scale surface temperature reconstructions are not without problems. Proxy records are all linked in some way to temperature, and in addition, they are also affected in a complex way by other climatic, biological and ecological factors.

The first of these complex reconstructions were published in 1998 and 1999 by Mann, Bradley, and Hughes. The basic conclusion of Mann et al. (1998, 1999) was that the late twentieth century warmth in the Northern hemisphere was unprecedented during, at least, the last 1,000 years. This research received wide attention, in part because it was illustrated with a simple graphic, the so-called hockeystick curve, that many interpreted as definitive evidence of anthropogenic causes of recent climate change. The research was given prominence in the 2001 report of the Intergovernmental Panel on Climate Change. There is still an ongoing debate about the details of some of the published temperature records, the means of their reconstruction and the underlying statistical measures. Some basic points of concern were addressed by von Storch et al. (2004) and McIntyre and McKitrick (2005). More recent updates or evaluations of the temperature reconstructions, methods and statistical aspects can be found in Mann and Jones 2003; Jones and Mann 2004; D'Arrigo et al. 2006; NRC 2006; Rybski et al. 2006; and Mann 2007.

Despite the aforementioned scientific debate it can be stated that large-scale surface temperature reconstructions yield a generally consistent picture of temperature trends during the preceding millennium, including the relatively warm conditions centred around 1,000 B.P. (identified by some as the "Medieval Warm Period") and a relatively cold period (or "Little Ice Age") centred around 1,700. The existence of a Little Ice Age from roughly 1,500–1,850 is supported by a wide variety of evidence including ice cores, tree rings, borehole temperatures, glacier length records, and historical documents (NRC 2006). Figure 7 shows a compilation of large-scale surface temperature reconstructions from different research groups, each using its own methodology and selection of proxies, as well as the instrumental record (beginning in 1856) of global mean surface temperature. The reconstructed temperatures all show a more or less pronounced gradual decline for almost 900 years after the medieval warm period and a considerable warming

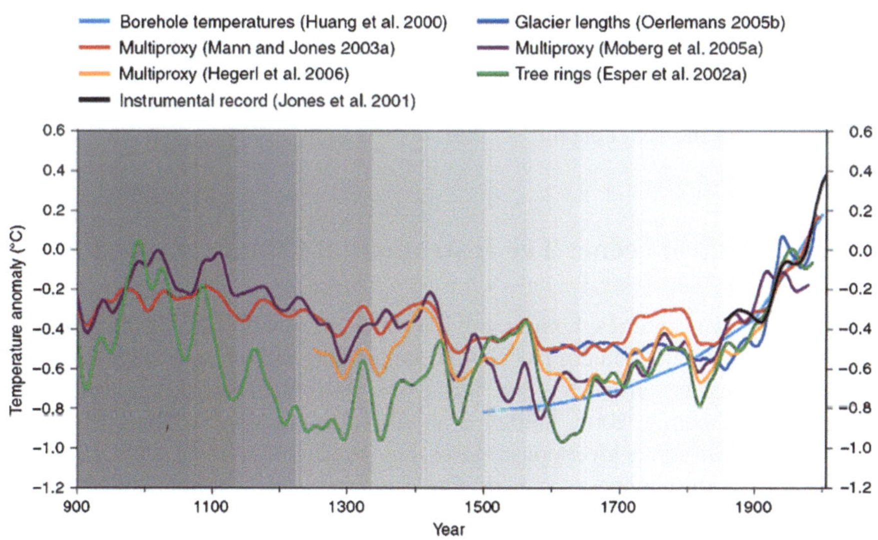

Fig. 7 Smoothed reconstructions of large-scale (Northern Hemisphere mean or global mean) surface temperature variations over the last 1,100 years from six different research teams are shown along with the instrumental record of global mean surface temperature. Each curve portrays a somewhat different history of temperature variations and is subject to a somewhat different set of uncertainties that generally increase going backward in time (as indicated by the *gray shading*). Figure adapted from NRC (2006)

which began in the middle of the nineteenth century. Comparing the results of the different methods reveals that there appears to be much more certainty in the reconstructed temperatures for the last 400 years.

The warming in the last century displayed in Fig. 7 stands out as *unusual* within the displayed temperature series.

The special assessment report of the National Research Council (NRC 2006) concludes that it can be said with a high level of confidence that global mean surface temperature was *higher* during the last few decades of the twentieth century than during any comparable period during the preceding four centuries. The NRC committee placed less confidence in large-scale surface temperature reconstructions for the period from A.D. 900 to 1600. Presently available proxy evidence indicates that temperatures at many, but not all, individual locations were higher during the past 25 years than during any period of comparable length since A.D. 900. Based on the analyses presented in the original papers by Mann et al. (1998, 1999) and newer supporting evidence, the NRC committee finds it plausible that the Northern hemisphere was warmer during the last few decades of the twentieth century than during any comparable period over the preceding millennium (NRC 2006). Reconstruction attempts for the past two millennia are presented in Mann and Jones (2003), Jones and Mann (2004), Mann (2007), and these proxy reconstructions suggest that the late twentieth century warmth is anomalous also in the context of the past 2,000 years.

Concluding a long-term persistence study on six of the reconstructed Northern hemisphere temperature records, Rybski et al. (2006) state: “Accordingly, the hypothesis that at least part of the recent warming cannot be solely related to natural factors, may be accepted with a very low risk, independently of the database used”. The unprecedented warming during the last 150 years will be discussed more closely in the next chapter.

4 The Anthropocene: The Instrumental Climate Record

Over the last two centuries, human activities have profoundly altered many aspects of the Earth’s system, including its climate and its biogeochemistry, and to the point where some argue that we are entering a new geological era, the Anthropocene (Crutzen and Stoermer 2000; Steffen et al. 2007). In other words, the atmospheric concentration of some greenhouse gases has been enhanced to such an extent (IPCC 2007a) during the course of the Anthropocene, that regional to global changes in surface temperature and other climate parameters are to be expected.

The instrumental era of meteorological observations began with the invention of the thermometer by Galileo in the late sixteenth century and the mercury barometer by Torricelli in the seventeenth century. Instrumental records are, by far, the most reliable of all climate data, as they are precisely dated and employ physical calibrations. The modern instrumental measurements of climate parameters include thermometer-based surface temperatures from land regions and from the oceans, sea level pressure, continental and oceanic precipitation, sea ice extent, wind, and humidity. Unfortunately, it has only been since the late 1800s that enough stations have been systemically recording these data to permit reasonable estimates of global means.

There is a huge body of literature available addressing many aspects of climate change over the instrumental observation period. The most comprehensive assessment is published in the fourth assessment report of the IPCC (IPCC 2007a, Chap. 3). Only a few *key points* can be highlighted by this review.

4.1 Temperature Changes

Reliable global means of surface temperature are available since 1850. Several compilations by different research groups provide gridded monthly mean temperature estimates back through the mid nineteenth century on a large scale (Jones and Mann 2004). The reconstructions by the different groups are very similar, although there are some disagreements mainly in the early 1900s when station coverage was still sparse.

Figure 8 shows a time series of estimated annual global mean surface temperatures for the time period from 1850 to 2007 based on thermometer measurements; the values displayed in the figure are deviations from the mean of the base period 1961–1990. This time series is being compiled jointly by the Climatic Research

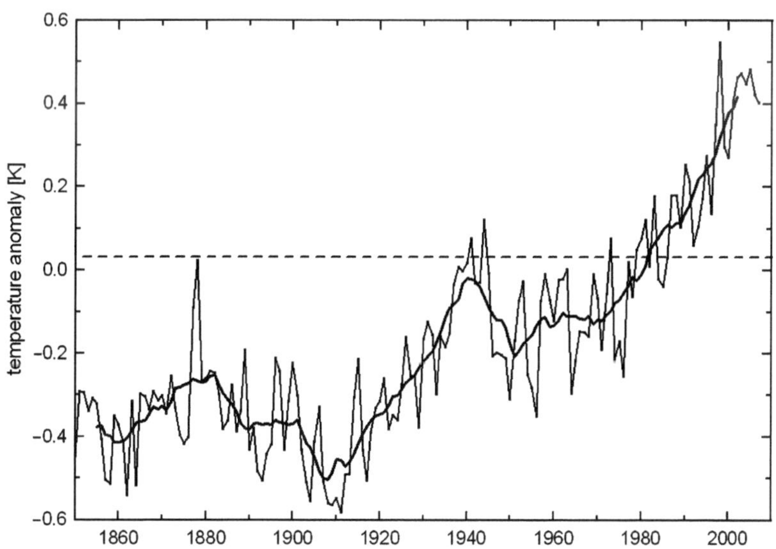

Fig. 8 Global annual mean temperature anomalies (from the mean of the base period 1961–1990) for the years 1850–2007, the smoother line shows an 11-year running average. Data provided by Climatic Research Unit, Norwich, UK, and Hadley Centre, UK Met Office

Unit and the UK Met. Office Hadley Centre; see Brohan et al. (2006) for details and error estimates. There is little temperature trend between 1850 and 1920, but between 1920 and 2007 the global mean temperature has risen by about 0.8°C. This has not been a steady rise though, temperature rose quickly during the 1920s to early 1940s, stabilized or fell slightly from the 1940s through to the late 1970s, and has again risen since 1980 to its present value. The 1990s were the warmest complete decade in the series. The warmest year of the entire series was 1998, with a temperature of 0.546 K above the mean of the current climatic period (1961–1990). Twelve of the thirteen warmest years in the series have now occurred in the past 13 years (1995–2007). The 10 warmest years on record are 1998, 2005, 2003, 2002, 2004, 2006, 2001, 2007, 2001, and 1997. Increased concentrations of greenhouse gases in the atmosphere (see e.g. Fig. 4 for CO_2) due to human activities are most likely the underlying cause of warming in the twentieth century (IPCC 2007a).

The warming of the Earth is not distributed evenly over the globe, there are quite a few regional and seasonal differences to report. Continents show more rapid temperature increases compared to the oceans. A distinct warming can be observed in winter and spring in higher Northern latitudes, i.e. in the Arctic. Average Arctic temperatures increased at almost twice the global average rate over the past 100 years.

Widespread changes in extreme temperatures have also been observed over the last 50 years. Cold days, cold nights and frost have become less frequent while hot days, hot nights, and heat waves have become more frequent (IPCC 2007a; Easterling et al. 2000a). Several studies have shown that anthropogenic warming trends in Europe imply an increased probability of very hot summers. For example, the summer of 2003 with its record-breaking central European summer temperatures was probably the hottest since 1,500 (Stott et al. 2004), and a shift towards a regime with an increased temperature variability in addition to increases in mean temperature may have been responsible for this remarkable heat wave (Schär et al. 2004).

In addition to the changes in surface temperatures with a general warming tendency, analyses of balloon-borne and satellite measurements show a cooling of the stratosphere, which is consistent within greenhouse gas-driven climatic change. The observed pattern of tropospheric warming and stratospheric cooling is *very likely* due to the combined influences of greenhouse gas increases and stratospheric ozone depletion.

4.2 Other Changes

4.2.1 Hydrological Parameters

Due to a high variability of the water cycle on all time scales, significant trends in hydrological quantities are difficult to substantiate. Nevertheless, Huntington (2006) concludes a review of available studies that although data are often incomplete in spatial and temporal sense, the weight of evidence indicates an ongoing intensification of the water cycle. One of the central questions is, however, whether global warming leads to more evaporation and hence to an increased water vapour content in the atmosphere. Trenberth et al. (2005) observed increases in precipitable water over the global oceans since the mid-1980s. But caution is advised by Trenberth et al. (2005) for extracting information on trends in tropospheric water vapour from global reanalysis products, since they suffer from spurious variability and trends related to changing data quality and coverage. Satellite data presented by Soden et al. (2005) supports column-integrated moistening trends for the years from the mid-1980s. A most recently published evaluation of in situ surface air and dew point temperature data found very significantly increasing trends in global and Northern hemispheric specific humidity (Dai 2006). More details on evaporation changes are given in Quante and Matthias (2006).

The increased atmospheric moisture content associated with a warming might be expected to lead to increased global mean precipitation. However, precipitation is also strongly influenced by changes in the tropospheric energy budget and the atmospheric circulation, so spatio-temporal patterns in precipitation changes are likely to be complex. Concerning precipitation, the most recent IPCC reports states (IPCC 2007a): Global terrestrial annual mean precipitation showed a small upward trend over the twentieth century of approximately 2.1 mm per decade (based on the Global Historic Climatology Network, GHCN, data). Long-term trends from 1900 to 2005 have been observed in the amount of precipitation over many large regions. Significantly increased precipitation has been observed in Eastern parts of North and South America, Northern Europe and Northern and central Asia. The frequency of heavy precipitation events has increased since 1950 over most land areas, consistent with warming and observed increases of atmospheric water vapour. Drying has been observed in the Sahel, the Mediterranean, Southern Africa and parts of Southern Asia. Precipitation is highly variable spatially and temporally, and data are limited in some regions. More intense and longer droughts have been observed over wider areas since the 1970s, particularly in the tropics and subtropics. Increased drying linked with higher temperatures and decreased precipitation have contributed to changes in drought (IPCC 2007a).

4.2.2 Cryogenic Components

The warming in the Arctic caused a decrease in snow cover since 1980, i.e. in spring time, as well as a decrease in sea ice of about 2.7% per decade, the latter is especially pronounced in late summer (September) showing a trend of 7.4% per decade. A record minimum has been observed in September 2007, which is not included in the reported trends. Also, in other parts of the globe changes in cryospheric components can be observed. Mountain glaciers and snow cover have declined on average in both hemispheres. Losses from the ice sheets of Greenland and Antarctica have very likely contributed to the observed sea level rise between 1993 and 2003 (IPCC 2007a). Flow speed has increased for some Greenland and

Antarctic outlet glaciers, which drain ice from the interior of the ice sheets. The corresponding increased ice sheet mass loss has often followed thinning, reduction or loss of ice shelves or loss of floating glacier tongues.

4.2.3 Sea Level Rise

Global average sea level rose at an average rate of 1.8 mm $year^{-1}$ (range 1.3–2.3) over the years 1961–2003. The rate was faster over the years 1993–2003, with about 3.1 mm $year^{-1}$ (range 2.4–3.8). Whether the faster rate for 1993–2003 reflects decadal variability or an increase in the longer-term trend is unclear. There is *high confidence* that the rate of observed sea level rise increased from the nineteenth to the twentieth century. The total twentieth century rise is estimated to be 0.17 m (range 0.12–0.22) (IPCC 2007a). An extended evaluation by Church, and White (2006) concluded a global mean sea-level rise from January 1870 to December 2004 of 0.195 m, a twentieth century rate of sea-level rise of 1.7 ± 0.3 mm $year^{-1}$ and a significant acceleration of sea-level rise of 0.013 ± 0.006 mm $year^{-2}$.

4.2.4 Wind

Mid-latitude westerly winds have strengthened in both hemispheres since the 1960s. There is observational evidence for an increase in intense tropical cyclone activity in the North Atlantic since about 1970, correlated with increases of tropical sea surface temperatures (IPCC 2007a). There are also suggestions of increased intense tropical cyclone activity in some other regions (Emanuel 2005; Webster et al. 2005) where concerns over data quality are greater. Multi-decadal variability and the quality of the tropical cyclone records prior to routine satellite observations in about 1970 complicate the detection of long-term trends in tropical cyclone activity. There is no clear trend in the annual numbers of tropical cyclones.

4.3 Attribution

A central topic within the climate debate of the recent years was whether, or to what an extent, the observed changes in climate can be attributed to anthropogenic emission of greenhouse gases or are due to natural variability. This extensively conducted discussion can not be fully summarized here; however, some aspects could be emphasized.

One central argument of the so-called "climate sceptics" is based on observed changes in the sun's activity and a rise in solar radiation intensity. The question arises: to what extent are changes in the sun responsible for the recent warming? Although the rarity of the current episode of high average sunspot numbers (Solanki et al. 2004) may indicate that the sun has contributed to the unusual climate change during the twentieth century, the work of Solanki and Krivova (2003) reveals that solar variability is unlikely to have been the dominant cause of the strong warming observed during the past three decades. A model-based exhaustive study by Allen et al. (2006) concludes that increasing *anthropogenic gas concentrations* produced a 0.3–0.5 K per century warming from 1906 to 1996 and are the dominant cause of global warming after 1976. In contrast, Scafetta and West (2006, 2007) report that empirical analyses suggest that *solar variability* accounts for as much as 69% of the twentieth century warming and 25 to 35% of recent warming, globally. The IPCC report (AR4) compares observed continental- and global-scale changes in surface temperature with results simulated by climate models using natural and anthropogenic forcings. One relevant conclusion is (IPCC 2007a): "The observed widespread warming of the atmosphere and ocean, together with ice mass loss, support the conclusion that it is *extremely unlikely* that global climate change of the past fifty years can be explained without external forcing, and *very likely* that it is not due to known natural causes alone".

A multivariate analysis using the most reliable estimates of the relevant parameters together with the observed surface temperature record from 1889 to 2006 enabled Lean and Rind (2008) to assess the global and regional responses to different forcings. Their conclusion with respect to the solar influence is that solar forcing contributed *negligibly* to long-term warming over the past 25 years and about 10% of the warming over the last century. This result is in strong contrast to the 69% estimate of Scafetta and West (2006).

5 The Future: The "Era" of Model Assessments

The only rational way to attain reasonable estimates of future climate change is by the use of complex physical models of the climate system. In order to produce climate projections for decades or centuries to come, reasonable anthropogenic greenhouse gas and aerosol emission data as well as land use scenarios are needed in conjunction with sophisticated computer codes as forcing boundary values. Climate models and emission scenarios are briefly introduced in the following sections before the results from such model assessments of future climate situations are presented.

5.1 Climate Models

Climate models are based on mathematical representations, which attempt to reproduce the behaviour of the Earth's climate system (Trenberth 1992). As outlined in Sect. 2.1 the climate system is composed of a number of subcompartments (e.g. the atmosphere, hydrosphere, cryosphere, and biosphere), each component is a nonlinear system in itself and is associated with a characteristic time scale.

The representations of climate models attempt to account for the most important external and internal forcings, they also try to include the most important processes which are involved in feedbacks within the system. The models are based on physical principles such as the conservation of energy, momentum, and mass. Although the fundamental principles appear to be robust, computational limits preclude their numerical solution on all scales which include the many processes that are important in the climate system. For these processes, modellers have to rely on parameterisations which try to capture the fundamental phenomenology of a small scale process. For more details on the state-of-the-art of climate modelling see, e.g. Washington and Parkinson (2005), Kiehl and Ramanathan (2006), or Donner and Large (2008).

Current climate models simulate the atmosphere, with a prescribed atmospheric composition and incoming solar radiation, and include explicit modelling of the ocean's general circulation and sea-ice dynamics, and relevant land processes. By way of interfaces the subcomponents can exchange energy, momentum, and mass. A typical horizontal grid resolution of today's global atmospheric models is about 180 km; regional climate models on the other hand, capture only part of the globe, and typically work with a 25–50 km horizontal grid resolution. In the vertical dimension, the resolution decreases with the distance from the Earth's surface and ranges from a few decametres to a few kilometres. Among the most important feedbacks climate models have to cope with is the water vapour and the cloud feedbacks as well as the snow, sea and land-ice albedo feedbacks, which also need to be mentioned. Cloud feedbacks can be substantial for the overall performance of a climate model, but the sign and the magnitude of the global mean cloud feedback depends on so many factors that it remains very uncertain (Stephens 2005; Bony et al. 2006).

Climate models are far from perfect, but they are unmatched in their ability to quantify otherwise qualitative hypotheses. More than a dozen facilities worldwide develop and maintain state-of-the-art climate models and comprehensive model comparisons have been conducted using the present-day climate as a primary test bed. These models have successfully captured fundamental aspects of air, ocean, and sea-ice circulations as well as their variability. In recent climate studies (i.e. IPCC 2007a), results of ensembles of a number of models have been reported (such as outlined in Sect. 5.3.). Interestingly however, comparison studies have revealed that the average across a number of models outperformed any given single model. Climate models are evolving more and more into Earth system models, which include chemical and biological processes (e.g. vegetation dynamics) as well as simple socio-economic sub-models.

5.2 Emission Scenarios

A scenario is a description of potential future conditions produced to inform and assist in what would otherwise be decision-making with an even greater uncertainty.

However, more than just a reliable model of the physical climate system is needed for climate projections of the future. Here, there is also the need to supply the models with estimates of the development of the forcing parameters. Of central interest are future anthropogenic modifications of atmospheric greenhouse gasses and aerosol concentrations as well as characteristics of the Earth's surface. The IPCC has produced three sets of scenarios of twenty-first century greenhouse-gas emissions, the most ambitious and important of which were produced for the *Special Report on Emissions Scenarios* (SRES) (IPCC 2000). SRES produced four qualitative storylines on which six "marker" scenarios were based – one model quantification of each storyline plus two technological variants of one storyline which stressed fossil-intensive and low-carbon energy supply technologies – and each produced by a different energy-economic model. Scenarios are descriptions of (different) possible futures, they are a series of alternative visions of the future (storylines), which are possible, plausible, internally consistent but *not* necessarily probable (von Storch 2008). The information which is really required is based on scenarios of emissions of radiactively active substances, and these scenarios themselves depend on a variety of future developments in different sectors and in particular on population growth, changes in technology and the energy system and its efficiency. Many of theses factors are not easy to predict and so a variety of assumptions have entered into these scenarios. It is not surprising, therefore, that critical discussions arose as a result (e.g. Tol 2008; Pielke et al. 2008). The SRES scenarios do not include additional climate initiatives, which means that no scenarios are included which explicitly assume the implementation of the United Nations Framework Convention on Climate Change or the emissions targets of the Kyoto Protocol. In other words, these scenarios do not anticipate any specific mitigation policies for avoiding climate change. The IPCC scenarios are grouped in the following way:

(A1) represents a world of rapid economic growth and a rapid introduction of new and more efficient technology. The A1 storyline and scenario family describes a future world of very rapid economic growth, a global population which peaks in mid-century and declines thereafter and the rapid introduction of new and more efficient technologies. Major underlying themes are convergence among regions, capacity building and increased cultural and social interactions, with a substantial reduction in regional differences in per capita income.

(A2) represents a very heterogeneous world with an emphasis on family values and local traditions (high-CO_2). The A2 storyline and scenario family describes a very heterogeneous world. The underlying theme is self-reliance and preservation of local identities. Fertility patterns across regions converge very slowly, which results in a continuously increasing population. Economic development is, primarily, regionally oriented and per capita economic growth and technological change are more fragmented and slower than in other storylines.

(B1) represents a world of "dematerialization" and the introduction of clean technologies (low-CO_2). The B1 storyline and scenario family describes a convergent world with the same global population, which peaks in mid-century and declines thereafter, as is the case in the A1 storyline, but with rapid change in economic structures toward a service and information economy, with reductions in material intensity and the introduction of clean and resource-efficient technologies. The emphasis is on global solutions to economic, social and environmental sustainability, including improved equity, but without additional climate initiatives.

(B2) represents a world with an emphasis on local solutions to economic and environmental sustainability. It is a world with a continuously increasing global population, at a rate lower than A2, intermediate levels of economic development, and less rapid and more diverse technological change compared to the B1 and A1 storylines. While the scenario is also oriented towards environmental protection and social equity, it focuses on local and regional levels.

An illustrative scenario was chosen for each of the six scenario groups A1B, A1FI, A1T, A2, B1 and B2, and all should be considered equally sound. The A1 scenario is broken down into some variants: in recent years, the A1B (balanced – not relying too heavily on one particular energy source and similar improvement rates for all energy supply and end-use technologies) was widely used together with the B2 scenario. A1F represents a world, which still relies heavily on the use of fossil fuels (worst case) and A1T is a scenario, which incorporates the use of advanced technologies. A selection of results presented in the next section will refer to these scenarios.

5.3 *Climate of the Twenty-First Century*

For about two decades now, coupled ocean–atmosphere climate models have been in use for making sophisticated climate change projections for the current century. Most of the published model studies are reviewed by the successive assessment reports of the IPCC starting in 1990. Over the years, the underlying simulations have become more advanced, either through the inclusion of additional dynamic sub-models, improved parameterisations or increased spatial resolution. The results of the newest simulations are the basis of the climate projections and impacts discussed in the fourth IPCC assessment report (IPCC 2007a,b). The results presented in this section will, to a large extent, build on these studies.

5.3.1 Temperature

One of the most communicated graphs of the IPCC AR4 is reproduced in Fig. 9. The graph shows the time evolution of the multi-model global-mean surface temperature in transient emission scenario driven experiments according to the most comprehensive model study. As can be seen in Fig. 9, the projected temperature changes depend decisively on the underlying scenario. Based on all six scenarios and taking the model ranges, expressed by the vertical bars on the right hand side of Fig. 9, within single scenarios into account, global warming between 1.8°C

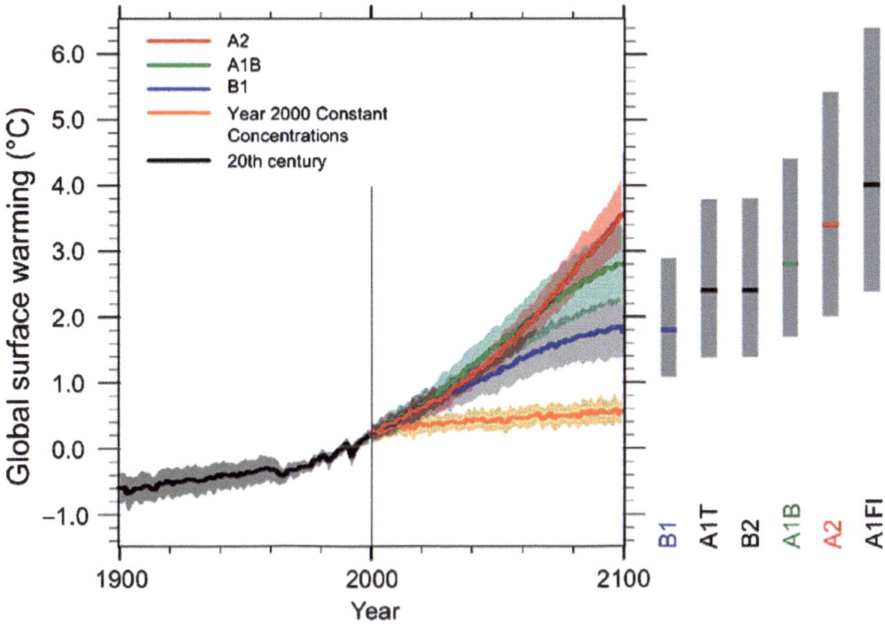

Fig. 9 Multi-model means of surface warming (relative to 1980–1999) for several scenarios A2, A1B and B1, shown as continuations of the 20th-century simulation. Lines show the multi-model means, shading denotes the ± 1 standard deviation range of individual model annual means. Note that not all scenarios are run by the same number of models, thus different standard deviations are not suited for interpretation across scenarios (IPCC 2007a)

(1.1–2.9°C) for B1 and 4.0°C (2.4–6.4°C) for A1FI has to be anticipated by the end of the twenty-first century (2090–2099 relative to 1980–1999). In the underlying climate simulations the observed temperature changes of the last century are reproduced (left part of Fig. 9) and amplified in the course of the twenty-first century. Uncertainty in the climate projections is due not only to the unknown path society will take with respect to socio-economic aspects and technological development, as expressed by the scenarios, but also to considerable differences between the participating models. These inter-model differences are, to a large extent,due to the representation of feedback processes in the atmospheric general circulation models, and here especially the cloud feedback (Bony et al. 2006). However, all the models show a considerable warming by the end of the century, and even climate simulations keeping the same greenhouse gas concentrations as the year 2000 show an increase in global mean temperature of another half of a degree, reflecting the thermal inertia in the climate system. Table 1 provides the numerical values for global warming at the end of the century. The best estimate as well as the likely range from the multi-model evaluation for the different scenarios are given.

The geographical distribution of climate change is of paramount interest for many climate impact studies. Considerable spatial variation exists in modelled

Table 1 Projected global average surface warming at the end of the twenty-first century ((in °C at 2090–2099 relative to 1980–1999)

Scenario	Best estimate	Likely range
B1	1.8	1.1–2.9
A1T	2.4	1.4–3.8
B2	2.4	1.4–3.8
A1B	2.8	1.7–4.4
A2	3.4	2.0–5.4
A1FI	4.0	2.4–6.4

Data from IPCC (2007a)

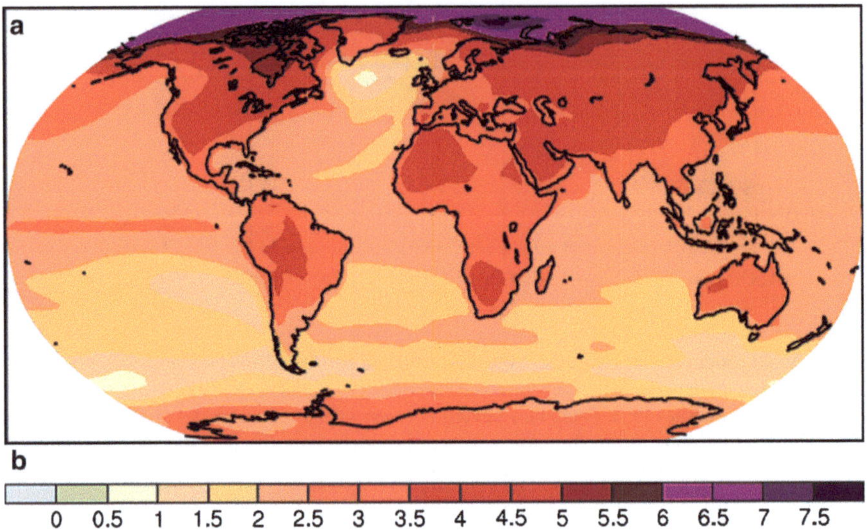

Fig. 10 Multi-model mean of annual mean surface warming (surface air temperature change relative to 1980–1999 in °C) for the scenario A1B (medium emission scenario) and the time period 2080–2099. Figure extracted from IPCC (2007a)

changes in climate. There is a certain pattern of climate change over the globe, which is consistently reproduced by all contributing models. As an example, the multi-model surface temperature distribution at the end of the century resulting from scenario A1B runs is shown in Fig. 10; all other scenario simulations yield the same principle pattern. In general, however, continents will warm faster than the oceans. The strongest warming is evident over the Northern land masses and the polar regions due to the positive ice-albedo effect. In these regions, the warming trend is about twice that of the global mean, and is similar therefore to that observed for the previous century. The lowest temperature change might be expected over the Southern oceans in the latitude band between 40 and 50°S and over North Atlantic ocean. Oceans respond with a delay, since the vertical mixing distributes the additional heat over a much larger volume compared to land masses. Vertical mixing is especially efficient in the North Atlantic and the Southern Ocean. It has to be mentioned, that there are remarkable differences between the single models when regional temperature distributions produced by the global models are more closely inspected (Fig. 10).

5.3.2 Precipitation

The direct implications of global warming are higher evaporation rates and therefore, higher precipitation rates. As a global and multi-model average, an increase in precipitation of about 4–5% can be expected towards the end of the twenty-first century (Meehl et al. 2005), with a net change over land accounting for 24% of the global mean increase in precipitation. It should also be pointed out that for precipitation over land especially, large inter-model deviations exist, and these occur not only in the regional mean but also in the global mean. Figure 11 shows the distribution of changes in the annual mean precipitation as a scenario A1B multi-model mean expressed in mm per day for the last two decades of the present twenty-first century. It can be seen that a large general increase in precipitation is not to be expected, but that there will be regions on the globe where more water will be available and others where drying might occur. More rain is simulated for the equatorial belt and higher geographical latitudes. Less precipitation (up to 20%) is projected especially for the subtropics (oceans, Mediterranean sea, South Africa, Australia). This projected decrease is considered especially robust, and *very likely* to occur (IPCC 2007a). From the results of inter-annual evaluations of the model a shift in the timing and strengths of precipitation events has been diagnosed, and this could be of huge importance for the future management of drinking water supply and the fate of ecosystems where water is critical.

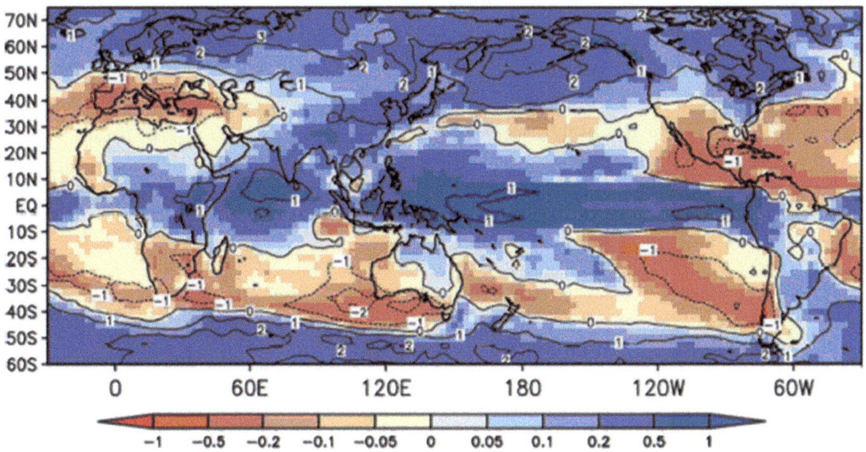

Fig. 11 Mean changes in precipitation (in mm day^{-1}). Changes are annual means for the A1B scenario for the period 2080–2099 relative to 1980–1999. Figure from Daisuke Nohara

5.3.3 Sea Level Rise

It can be considered as a robust result of the IPCC Model projection that the sea level will continue to rise in the twenty-first century because of thermal expansion and loss of land ice. Sea level rise was not geographically uniform in the past and will not be in the future. Projected warming due to emissions of greenhouse gases during the twenty-first century will continue to contribute to sea level rise for many centuries to come. Sea level rise due to thermal expansion and loss of mass from ice sheets will continue for centuries or millennia even if the radiative forcing were to be stabilised. Table 2 provides the numerical values for the sea level rise at the end of the century, and the likely range from the multi-model evaluation for the different scenarios is also provided. The range excludes future rapid dynamical changes in ice flow, which is not parameterized in the current ice sub-models.

In all scenarios, the average rate of sea level rise during the twenty-first century is *very likely* to exceed the 1961–2003 average rate (1.8 ± 0.5 mm year^{-1}). During 2090–2099 under A1B, the central estimate of the rate of rise is 3.8 mm year^{-1}. For an average model, the scenario spread in sea level rise is only 0.02 m by the middle of the century, and by the end of the century it is 0.15 m. Thermal expansion is the largest component, contributing 70–75% of the central estimate in these projections for all scenarios. Glaciers, ice caps and the Greenland ice sheet are also projected to contribute positively to sea level. The Antarctic Ice Sheet will receive increased snowfall without experiencing substantial surface melting, thus gaining mass and contributing negatively to sea level. Further accelerations in ice flow of the kind recently observed in some Greenland outlet glaciers and West Antarctic ice streams could substantially increase the contribution from the ice sheets (IPCC 2007a). 2005 was the deadline for scientific input into the IPCC AR4 report; however, there has been some further work carried out on sea level rise which was motivated by the observed acceleration of the Greenland ice melting. The models used for the AR4 sea level assessment have certain shortcomings concerning the small scale processes in melting physics. All newer estimates of the possible sea level rise towards the end of the twenty-first century come up with higher estimates compared to those reported in IPCC (2007a). For example, Rahmstorf (2007) uses a semi-empirical relation which connects global sea-level rise to global mean surface temperature in the twentieth century. The use of such a semi-empirical relationship is, however, not without its critics. When the established correlations are applied to

Table 2 Projected global sea level rise at the end of the twenty-first century (in m between 2090–2099 relative to 1980–1999; model-based range excluding future rapid dynamical changes in ice flow)

Scenario	Likely range
B1	0.18–0.38
A1T	0.20–0.45
B2	0.20–0.43
A1B	0.21–0.48
A2	0.23–0.51
A1FI	0.26–0.59

Data from IPCC (2007a)

the future warming scenarios of the IPCC, the relationship results in a projected sea-level rise in 2100 of 0.5–1.4 m above the 1990 level. A comparison of the past sea level rise from studies of the last interglacial period with presently projected conditions, led Rohling et al. (2008) to the statement "We find average rates of sea-level rise of 1.6 m per century. As global mean temperatures during MIS-5e were comparable to projections for future climate change under the influence of anthropogenic greenhouse-gas emissions, these observed rates of sea-level change inform the ongoing debate about high versus low rates of sea-level rise in the coming century". An extreme possible sea level rise of up to 5 m is brought into discussion by Hansen (2007c) who also refers to palaeoclimate records containing numerous examples of ice sheets disintegrations yielding sea level rises of several metres per century in situations when forcings were smaller than that of the business-as-usual scenario (close to SRES A2).

Overall, it should be stated that future sea level rise is still very uncertain, but the numbers given in the IPCC AR4 seem to mark the absolute lower boundary of possibilities, and a possible rise in sea level of a meter or slightly more by the end of the century seems to be more likely.

5.3.4 Extreme Events

Recent years have seen a number of weather events causing large losses of life (Easterling et al. 2000b). Climate model improvements have resulted in an enhanced ability to simulate many aspects of climate variability and extremes. However, they are still characterized by systematic errors and limitations in accurately simulating regional climate conditions. Nevertheless, an analysis of the frequency of occurrence of selected parameters have shown that besides a shift in the mean, a broadening of the probability density distribution is also to be expected, thereby implying a higher variability and more extreme values. Heat waves will become more frequent and longer lasting in a future warmer climate (IPCC 2007a).

Extremes of daily precipitation are *likely* to increase in many regions. The increase is considered as *very likely* in Northern Europe, South Asia, East Asia, Australia and New Zealand. Multi-model climate projections for the twenty-first century show both increases in precipitation intensity and the number of consecutive dry days in many regions (Tebaldi et al. 2006). Precipitation intensity increases almost everywhere, but particularly at mid- and high latitudes, where mean precipitation also increases. There are many regions in the sub-tropics and lower mid-latitudes with an increased number of consecutive dry days, the sign of this parameter is reversed at higher mid-latitudes and high latitudes, where mean precipitations are expected to increase.

Although somewhat counter-intuitive, a projected increase in the risk of intense precipitation and flooding is also associated with the risk of drying. Intense and heavy episodic rainfall events with high runoff are interspersed with longer relatively dry periods with increased evapotranspiration especially in the sub-tropics (Bates et al. 2008).

5.3.5 Regional Scale Information: Dynamical Downscaling

Downscaling is a method which derives local-to-regional-scale information from larger-scale models. The dynamical downscaling method uses the output from global climate models, which typically run on a coarse resolution, as a driver for regional climate models, which again can be setup in a nested manner down to the desired spatial resolution. The dynamical model on the regional scale allows for the added value of the development of meteorological phenomena, which evolve because of regional flow and surface flux features. Of course, the quality of the downscaled product depends to a very large extent on the quality of the driving model.

It is beyond the scope of this overview to discuss all the results from the published regional climate projection studies; however, some major references will be provided here. For example, Räisänen et al. (2004) and Giorgi et al. (2004), and Bhend and von Storch (2008) each highlight different aspects of climate change projections for entire Europe. In the following, a few selected regional results are summarized:

Alps

Raible et al. (2006) evaluated reconstructed data series from the past and performed climate model projections for the Atlantic-European and Alpine region. Time-slice simulations using a regional model were carried out. In the A2 scenario simulation (2071–2100) the winter and summer temperatures show an average increase of about 3 and 4°C, respectively. Comparing the control time-slice 1961–1990 with the A2 scenario, the interannual variability is unchanged in winter, whereas a strong increase on these time scales in summer is observed (Schär et al. 2004). Precipitation behaves differently; in winter, a small average increase is simulated in the A2 scenario. The simulated trend in winter precipitation for the control period 1961–1990 is consistent with the observed trend and the reconstruction; thus this behaviour may be explained by internal variability. In summer, the mean precipitation shows a small decrease compared with the mean of the control simulation. Comparing the A2 with the control simulation, the inter-annual variability of precipitation remains unchanged in winter, but increases in summer.

Baltic Region

The climate projections reported in BACC (2008) focussed on an *Assessment of Climate Change in the Baltic Sea Basin*. The main results are on the time horizon of 2100, and show that without mitigation measures the surface temperatures are expected to rise by about 4–6°C in the Northern Baltic region and by about 3–5°C in the Southern part of the Baltic including parts of Poland and Northern Germany. This temperature increase will lead to a reduction in the winter ice cover of the

Baltic Sea by 50–80%. Winter precipitation is projected to rise by 25–70% and in summer precipitation is expected to decrease up to 45% by the end of the century.

Extremes Over Europe

Beniston et al. (2007) evaluated regional climate model projections for future extreme events in the European climate. They explored how heat waves, heavy precipitation, drought, wind storms, and storm surges change between the present (1961–1990) and future (2071–2100).

The main results of the study can be summarized as follows: Heat waves – regional surface warming causes the frequency, intensity and duration of heat waves to increase over Europe; Precipitation – heavy winter precipitation increases in central and Northern Europe and decreases in the South; heavy summer precipitation increases in North-Eastern Europe and decreases in the South. Mediterranean droughts start earlier in the year and last longer; Winter storms – extreme wind speeds increase between 45° N and 55° N, except over and South of the Alps, and become more North-westerly than they are at present.

Schär et al. (2004) simulated possible future European climate using a regional climate model and a scenario with increased atmospheric greenhouse-gas concentrations. The assessment of European heat waves in relation to the extreme event in 2003 (strongest heat wave for more than 500 years) was at the centre of the study. The authors found that under a certain emission scenario (SRES A2) temperature variability over Europe increases by up to 100%, with maximum changes in central and Eastern Europe. The European summer climate may also experience a pronounced increase in year-to-year variability in response to greenhouse-gas forcing. The regional climate model simulations suggest that towards the end of the century – under the given scenario assumptions – about every second summer could be as warm or warmer (and as dry or dryer) than that of 2003. Thus, the conditions experienced during the serious heat wave of 2003 will be typical summer conditions by the end of the century.

Also, Diffenbaugh et al. (2007) found that elevated greenhouse gas concentrations dramatically increase heat stress risk in the Mediterranean region, with the occurrence of hot extremes increasing by 200–500% throughout the region.

The work of Räisänen et al. (2004), for example, and model evaluation work by the present author underpin the idea that the quality of the driving models predominantly determine the output of the regional climate models. This statement holds especially for water cycle quantities (Graham et al. 2007). In this case, the differences between the results obtained using two driving models coupled to one regional climate model with each simulating the same emission scenario is larger than those observed between one large scale model and one subsequent regional model simulating two extremely opposite scenarios. With the present-day suite of models it has to be stated that regional climate change results derived using only one driving model feeding one regional model need to be interpreted with caution.

5.3.6 Abrupt Changes

The Earth's climate system is highly non-linear: inputs and outputs are not proportional, change is often episodic and abrupt rather than slow and gradual, and multiple equilibria are the norm (Rial et al. 2004). Most of the studies and debates on potential climate change, along with its ecological and economic impacts, have focused on the ongoing build-up of industrial greenhouse gases in the atmosphere and a following *gradual* increase in global temperatures. This line of thinking, however, fails to consider another potentially disruptive climate scenario. It ignores recent and rapidly advancing evidence that the Earth's climate has repeatedly shifted *abruptly* and *dramatically* in the past, and is capable of doing so in the future. Fossil evidence clearly demonstrates that the Earth's climate can shift gears *within a decade*, establishing new and different patterns, which can persist from decades to centuries (NRC 2002).

The evidence also shows that the Earth's climate system has sensitive thresholds. Pushed past these thresholds and the system can jump quickly from one stable operating mode to a completely different one. These thresholds have been communicated in recent literature under the concept of "tipping points" (Lenton et al. 2008).

Many triggers for abrupt changes have been identified in the climate system and some of them are discussed in NRC (2002), Alley et al. (2003), and IPCC (2007a). One of the more prominent effects of abrupt changes is related to the deep water formation in the Northern Atlantic. The past 30 years have witnessed an intrusion of low-salinity deep waters which spread over the entire sub-polar North Atlantic Ocean and the seas between Greenland and Europe, and in just those regions which are critical for abrupt shifts in the thermohaline circulation, which has been implicated in many abrupt climate-change events in the past (Alley et al. 2003).

A salient feature of terrestrial climate change is its *asymmetry*. Warming is rapid, usually followed by a slower descent into colder climate. Given the symmetry of orbital forcings, the cause of rapid warming at glacial 'terminations' must lie in a climate feedback. Clearly, the asymmetric feedback is the albedo flip of ice and snow which occurs when they become warm enough to begin melting (Hansen et al. 2007a). A climate forcing which 'flips' the albedo of a sufficient portion of an ice sheet can initiate a cataclysm. This positive feedback can lead—and has led—to a relatively fast increase in surface temperatures.

5.4 Climate on the Millennium Time Scale

The present chapter will close with a short outlook on the projected climate change for the actual millennium. Most of the published climate projections address the twenty-first century and only a few studies exist, which look further into the future. Because of the considerable inertia of the climate system – caused by the long residence times of many greenhouse gases, the large heat capacity of the oceans and the long-term memory of other components of the system, such as ice sheets – in the absence of appropriate mitigation measures, human modifications of the climate system are expected to persist for many centuries to come. For example, it is likely that a substantial fraction of the projected CO_2 emissions will stay in the atmosphere for millennia and therefore, the related climate changes are to be expected in the longterm (Archer and Brovkin, 2008).

Hasselmann et al. (2003) report on simulations up to the year 3,000 using a nonlinear impulse response climate model coupled to an elementary economical model. Their results depend strongly on the assumed emission scenarios, for estimated peak CO_2 concentrations between about 1,200 and 4,000 ppm, and considering the use of all conventional and exotic resources, global temperature increases of 4–9°C are predicted for the second half of this millennium. The subsequent sea level rise will range from 3 to 9 m. The conclusion of Hasselmann et al. (2003) is that the global warming anticipated in the following centuries is a potentially far more serious problem compared to projected climate change in this century, i.e. if greenhouse gas emissions continue to increase unabated.

Based on simulations with eight Earth system models of intermediate complexity (climate-carbon cycle models) and different emission scenarios with assumed stabilized greenhouse gas concentrations after the year 2,100, Plattner et al. (2008) assess global temperature changes and sea level rise up to the year 3,000. The authors have identified a substantial climate change commitment for global temperature increase and sea level rise even after the stabilization of atmospheric greenhouse gas concentrations. For an initially high-CO_2 A2 plus stabilization scenario, the additional warming by the year 3,000 is calculated to range from 1.3 to 2.2 K and for a low-CO_2 B1 plus stabilization scenario from 0.6 to 1.6 K. The sea level continues to rise due to thermal expansion and the respective values for the A2 and B1 scenarios are 0.5–5.5 m and 0.3–1.1 m. The climate-carbon cycle models revealed that by the year 3,000 most of the anthropogenic carbon emissions will be eventually taken up by the ocean, but that a substantial fraction (15–28%) will still remain in the atmosphere. Even if the carbon emissions are hypothetically reduced to zero after the year 2,100, sea level will rise further due to ongoing thermal expansion. In this instance, a rise of 0.1–0.7 m by the year 3,000 has been calculated. Global surface temperatures remain high in the zero-post- 2,100 scenario, and compared to the year 2,100 a reduction of only a few tenths Kelvin can be expected by the year 3,000.

In summary, it can be said that emissions from the twenty-first century will *noticeably* affect the climate over the entire millennium.

6 Concluding Remarks

Over the 4.6 billion years of its evolution, the Earth has seen a large variety of climate states, which in their extremes have ranged from either the *snowball Earth* or *hot house*. Extended periods of an intermediate climate with only moderate changes have also been deduced from proxy data. However, climate variability is the *rule* and not the exception and the evolution of life on Earth was closely linked to climate and its change (Kasting and Catling 2003).

For about 250 years, mankind has interfered stronger with the climate system in the sense that, in addition to land use changes, radiative active gases and particles have been released in substantial amounts into the atmosphere. Over this period especially, the atmospheric concentrations of the greenhouse gases CO_2 and methane have risen significantly and account for an additional radiative forcing of more than 2 $W m^{-2}$ (IPCC 2007a). A global mean temperature increase can be deduced from instrumental observations for the same time span. The recent warming recorded in the instrumental observations over the last decade exceeds the temperatures (including the uncertainty ranges) of at least the past 1,600 years, as reconstructed for the Northern hemisphere from an expanded set of proxy data (Mann et al. 2008). Climate model projections driven by greenhouse gas and particle emissions according to a set of socio-economic scenarios emphasise a further increase in global temperature for the present century and beyond (IPCC 2007a). These projections encouraged some scientists to speak of a dangerous interference with the climate system, which could place the Earth perilously close to dramatic climate change that could run out of our control, resulting in great dangers for humans, other creatures and ecosystems (Hansen et al. 2007a,b). Based on the likelihood of strong El Niños in the future, Hansen et al. 2006 argue that a global warming of more than about 1°C, relative to the year 2000, would constitute a "dangerous" climate change for some regions, as judged from the likely effects on sea level and extermination of species. Although the term "*dangerous interference*" holds some subjective notion, there does appear to be a widely acknowledged awareness that, besides emission mitigation actions, adaptation measures have to be implemented.

Even if there have been hotter times in the deep history of our planet, there are two aspects which enhance the relevance of the current change in temperature and other climate parameters. Firstly, there are many *more people* than ever which live in regions of the world which are vulnerable to climate impacts. Secondly, the *pace* of the temperature increase in relation to the amplitude is almost unprecedented in the analysed climate records and will lead to a geographical shift in climate conditions, which may make it difficult for some ecosystems to adapt to or for some species to migrate in time (IPCC 2007b).

There are a few climate elements which may be sensitive to sudden, abrupt changes when the boundary of a set of conditioning parameters is overstepped or when certain thresholds are passed. Large, abrupt, and widespread climate changes with major impacts have occurred repeatedly in the past (Alley et al. 2003). An often cited example of an abrupt change is that of a possible collapse of the thermohaline circulation originating in the North Atlantic.

The discussion on abrupt changes has, generally, only considered the implications from an economic and ecological points of view; however, Hulme (2003) also discusses the implications from a more *social* and *behavioural* science perspective. Although abrupt changes might be of enhanced relevance for the global society, a detailed knowledge of the geophysical processes involved is *not* complete and a scientific consensus regarding the overall relevance and expected timing has *not* yet been fully reached. Therefore, extending our research into the area of abrupt climate change is urgently needed. Lenton et al. (2008) define tipping points in the climate system and report transition times of between 10 and 100 years. Among the discussed tipping elements are those of the thermohaline circulation, the Arctic sea-ice, the Greenland and West Antarctic ice sheets, the Indian summer monsoon, the Sahara/Sahel, West African monsoon and the Amazon rainforest (not the complete list). Lenton et al. (2008) also note that society may be lulled into a false sense of security by smooth projections of global change. According to their study, a synthesis of our present knowledge suggests that a variety of tipping elements could reach their critical point within this century. In their more general analysis, Steffen et al (2007) conclude with the statement: "Whatever unfolds, the next few decades will surely be a tipping point in the evolution of the Anthropocene".

Acknowledgement The author thanks Dr. Jan Habel and Professor Thorsten Assmann for having given the opportunity to contribute to the symposium on relict species in Luxembourg. Prof. Raymond Pierrehumbert, University of Chicago, USA, Prof. Leo Donner, Princeton University, USA, and Dr. Victor Brovkin, MPI for Meteorology, Hamburg, Germany, provided yet unpublished material, thank you all for the kindness. Valuable comments on the manuscript provided by Dr. Desmond Murphy are gratefully acknowledged. A part of the work is based on lectures on environmental meteorology and climate physics by the author at the University of Lüneburg, Germany.

References

Allen PA (2006) Snowball Earth on trial. Eos Trans Am Geophys Union 87(45):495

Allen MR, Gillett NP, Kettleborough JA, Hegerl G, Schnur R, Stott PA, Boer G, Covery G, Delworth TL, Jones GS, Mitchell JFB, Barnett TP (2006) Quantifying anthropogenic influence on recent near-surface temperature change. Surv Geophys 27(5):491–454

Alley RB, Marotzke J, Nordhaus WD, Overpeck JT, Peteet DM, Pielke RA Jr, Pierrehumbert RT, Rhines PB, Stocker TF, Talley LD, Wallace JM (2003) Abrupt climate change. Science 299:2005–2010

Archer D, Brovkin V (2008) Millennial atmospheric lifetime of anthropogenic CO_2. Clim Change 90:283–297

The BACC author team (2008) Assessment of climate change in the Baltic Sea basin. Springer, Berlin

Bard E, Frank M (2006) Climate change and solar variability: what's new under the Sun? Earth Planet Sci Lett 248:1–14

Barth M, McFadden J, Sun J, Wiedinmyer C, Chuang P, Collins D, Griffin R, Hannigan M, Karl T, Kim S, Lasher-Trapp S, Levis S, Litvak M, Mahowald N, Moore K, Nandi S, Nemitz E, Nenes A, Potosnak M, Raymond TM, Smith J, Stroud C, Still C (2005) The coupling between land ecosystems and the atmospheric hydrological cycle. Bull Am Meteorol Soc 86(12):1738–1742

Bates BC, Kundzewicz ZW, Wu S, Palutikof JP (eds) (2008) Climate change and water. Technical Paper of the Intergovernmental Panel on Climate Change, IPCC Secretariat, Geneva, pp. 210

Battarbee RW, Binney HA (eds) (2008) Natural climate variablility and global warming: A Holocene perspective. Wiley-Blackwell, p 288

Beniston M, Stephenson DB, Christensen OB, Ferro CAT, Frei C, Goyette S, Halsnaes K, Holt T, Jylhü K, Koffi B, Palutikoff J, Schöll R, Semmler T, Woth K (2007) Future extreme events in

European climate; an exploration of Regional Climate Model projections. Clim Change 81:71–95

Berger A (1988) Milankovitch theory and climate. Rev Geophys 26:624–657

Berner RA (2004) The Phanerozoic carbon cycle: CO_2 and O_2. Oxford University Press, New York, p 158

Berry SL, Farquhar GD, Roderick ML (2005) Co-evolution of climate, vegetation, soil and air. In: Blöschl G, Sivapalan M (eds) Encyclopedia of hydrological sciences, vol 1, Theory, organisation and scale. Wiley, Chichester, pp 177–192

Bhend J, von Storch H (2008) Consistency of observed winter precipitation trends in Northern Europe with regional climate change projections. Clim Dyn 31:17–28

Bonan GB (2008) Forests and climate change: Forcings, feedbacks, and the climate benefits of forests. Science 320:1444–1449

Bony S, Colman R, Kattsov VM, Allan RP, Bretherton CS, Dufresne J-L, Hall A, Hallegatte S, Holland MM, Ingram W, Randall DA, Soden DJ, Tselioudis G, Webb MJ (2006) How well do we understand and evaluate climate change feedback processes? J Clim 19:3445–3482

Botkin D, Saxe H, Araújo MB, Betts R, Bradshaw R, Cedhagen T, Chesson P, Davis MB, Dawson T, Etterson J, Faith DP, Guisan A, Ferrier S, Hansen AS, Hilbert D, Kareiva P, Margules C, New M, Skov F, Sobel MJ, Stockwell D (2007) Forecasting effects of global warming on biodiversity. Bioscience 57:227–236

Brasseur GP, Orlando JJ, Tyndall GS (eds) (1999) Atmospheric chemistry and global change. Oxford University Press, New York, p 654

Broecker WS (1985) How to built a habitable planet?. Eldigio Press, New York

Broecker WC, Stocker TF (2006) The Holocene CO_2 rise: anthropogenic or natural? Eos 87(3):27–29

Broecker WS, Kunzig R (2008) Fixing climate: what past climate changes reveal about the current threat – and how to counter it. Simon & Schuster, New York, p 272

Brohan P, Kennedy JJ, Harris I, Tett SFB, Jones PD (2006) Uncertainty estimates in regional and global observed temperature changes: a new dataset from 1850. J Geophys Res 111:D12106. doi:10.1029/2005JD006548

Canadell JG, Le Quéré C, Raupach M, Field C, Buitehuis E, Ciais P, Conway T, Gillett N, Houghton R, Marland G (2007) Contributions to accelerating atmospheric CO_2 growth from economic activity, carbon intensity, and efficiency of natural sinks. Proc Natl Acad Sci USA 104(47):18866–18870

Carslaw KS, Harrison RG, Kirkby J (2002) Cosmic rays, clouds and climate. Science 298:1732–1737

Church JA, White NJ (2006) A 20th century acceleration in global sea-level rise. Geophys Res Lett 33(1):L01602

Claussen M (2003) Simulation of Holocene climate change using climate-system models. In: Mackay A, Battarbee R, Birks J, Oldfield F (eds) Global change in the Holocene. Arnold, London, pp 422–434

Claussen M, Brovkin V, Calov R, Ganopolski A, Kubatzki C (2005) Did humankind prevent an early glaciation? Comment on Ruddiman's hypothesis of a pre-historic anthropocene. Clim Change 69:409–417

Cloud P (1978) Cosmos, Earth, and man: a short history of the universe. Yale University Press, New Haven, CT, p 372

Crowley TJ (2000a) Causes of climate change over the past 1000 years. Science 289:270–277

Crowley TJ (2000b) Carbon dioxide and Phanerozoic climate. In: Huber BT, MacLeod KG, Wing SL (eds) Warm climates in Earth history. Cambridge University Press, Cambridge, pp 425–444

Crutzen PJ, Stoermer EF (2000) The "Anthropocene". Global Change Newslett 41:12–13

Dai A (2006) Recent climatology, variability and trends in global surface humidity. J Clim 19:3589–3606

D'Arrigo R, Wilson R, Jacoby G (2006) On the long-term context for late twentieth century warming. J Geophys Res 111:D03103. doi:10.1029/2005JD006352

Diffenbaugh NS, Pal JS, Giorgi F, Gao X (2007) Heat stress intensification in the Mediterranean climate change hotspot. Geophys Res Lett 34:L11706. doi:10.1029/2007GL030000

Donner LJ, Large WG (2008) Climate modeling. Annu Rev Environ Resour 33:2.1–2.17
Doney SC, Schimel DS (2007) Carbon and climate system coupling on timescales from. The Precambrian to the Anthropocene. Annu Rev Environ Resourc 32:31–66
Easterling DR, Evans JL, PYa G, Karl TR, Kunkel KE, Ambenje P (2000a) Observed variability and trends in extreme climate events: a brief review. Bull Am Meteorol Soc 81(3):417–425
Easterling DR, Meehl GA, Parmesan C, Changnon SA, Karl TR, Mearns LO (2000b) Climate extremes: observations, modeling, and impacts. Science 289:2068–2074
Emanuel KA (2005) Increasing destructiveness of tropical cyclones over the past 30 years. Nature 436:686–688
Fischlin A, Midgley GF, Price JT, Leemans R, Gopal B, Turley C, Rounsevell MDA, Dube OP, Tarazona J, Velichko AA (2007) Ecosystems, their properties, goods, and services. In: Parry ML, Canziani OF, Palutikof JP, van der Linden PJ, Hanson CE (eds) Climate Change 2007: Impacts, Adaptation and Vulnerability. Contribution of Working Group II to the Fourth Assessment Report of the Intergovernmental Panel on Climate Change. Cambridge University Press, Cambridge, pp 211–272
Fischer H, Kumke T, Lohmann G, Flöser G, Miller H, von Storch H, Negendank JFW (eds) (2004) The climate in historical times. Towards a synthesis of Holocene proxy data and climate models. Springer, Berlin, p 487
Fleming JR (2007) The climate engineers. Wilson Q 31:46–60
Foukal P, Fröhlich C, Spruit H, Wigley TML (2006) Physical mechanisms of solar luminosity variation, and its effect on climate. Nature 443:161–166
Ghil M (2002a) Climate variability: nonlinear aspects. In: Holton JR, Pyle J, Curry JA (eds) Encyclopedia of atmospheric sciences. Academic, San Diego, CA, pp 432–438
Ghil M (2002b) Natural climate variability. In: Munn T (ed) Encyclopedia of global environmental change, vol 1. Wiley, Chichester, pp 544–49
Giorgi F, Bi X, Pal JS (2004) Mean, interannual variability and trends in a regional climate change experiment over Europe. II: climate change scenarios (2071–2100). Clim Dyn 23:839–858
Graham LP, Hagemann S, Jaun S, Beniston M (2007) On interpreting hydrological change from regional climate models. Clim Change 81:97–122
Gough DO (1981) Solar interior structure and luminosity variations. Sol Phys 74:21–34
Graedel TE, Crutzen PJ (1993) Atmospheric change: an earth system perspective. WH Freeman and Company, New York, p 446
Haigh JD (2007) The Sun and the Earth's climate", Living Rev Solar Phys. 4 http://www.livingreviews.org/lrsp-2007-2, assesed 16.07.2008
Hansen J, Mki S, Ruedy R, Lo K, Lea DW, Medina-Elizade M (2006) Global temperature change. Proc Natl Acad Sci USA 103:14288–14293
Hansen J, Sato M, Kharecha P, Russell G, Lea DW, Siddall M (2007a) Climate change and trace gases. Phil Trans R Soc A 365:1925–1954
Hansen J, Mki S, Ruedy R, Kharecha P, Lacis A, Miller RL, Nazarenko L, Lo K, Schmidt GA, Russell G, Aleinov I, Bauer S, Baum E, Cairns B, Canuto V, Chandler M, Cheng Y, Cohen A, Del Genio A, Faluvegi G, Fleming E, Friend A, Hall T, Jackman C, Jonas J, Kelley M, Kiang NY, Koch D, Labow G, Lerner J, Menon S, Novakov T, Oinas V, Ja P, Perlwitz Ju, Rind D, Romanou A, Schmunk R, Shindell D, Stone P, Sun S, Streets D, Tausnev N, Thresher D, Unger N, Yao M, Zhang S (2007b) Dangerous human-made interference with climate: a GISS modelE study. Atmos Chem Phys 7:2287–2312
Hansen J (2007) Climate catastrophe. New Sci 195:30–34
Hartmann DL (1994) Global physical climatology. Academic, San Diego, CA, p 411
Hasselmann K (1976) Stochastic climate models, Part I: Theory. Tellus 28:473–485
Hasselmann K, Latif M, Hooss G, Azar C, Edenhofer O, Jaeger CC, Johannessen OM, Kemfert C, Welp M, Wokaun A (2003) The challenge of long-term climate change. Science 302:1923–1925
Held IM, Soden BJ (2000) Water vapor feedback and global warming. Annu Rev Energy Environ 25:441–475
Hoegh-Guldberg O, Mumby PJ, Hooten AJ, Steneck RS, Greenfield P, Gomez E, Harvell CD, Sale PF, Edwards AJ, Caldeira K, Knowlton N, Eakin CM, Iglesias-Prieto R, Muthiga N,

Bradbury RH, Dubi A, Hatziolos ME (2007) Coral reefs under rapid climate change and ocean acidification. Science 318:1737–1742
Hoegh-Guldberg O, Hughes L, McIntyre S, Lindenmayer DB, Parmesan C, Possingham HP, Thomas CD (2008) Assisted colonization and rapid climate change. Science 321:345–346
Hoffert MI, Caldeira K, Benford G, Criswell DR, Green C, Herzog H, Katzenberger JW, Kheshgi HS, Lackner KS, Lewis JS, Manheimer W, Mankins JC, Marland G, Mauel ME, Perkins LJ, Schlesinger ME, Volk T, Wigley TML (2002) Advanced technology paths to global climate stability: energy for a greenhouse planet. Science 295:981–987
Hoffman PF, Schrag DP (2002) The snowball earth hypothesis: testing the limits of global change. Terra Nova 14:129–155
Holland HD (1984) The chemical evolution of the atmosphere and oceans. Princeton University Press, Princeton, NJ 598
Houghton RA (2007) Balancing the global carbon budget. Annu Rev Earth Planet Sci 35:313–347
Hughes L (2000) Biological consequences of global warming: is the signal already apparent? Trends Ecol Evol 15:56–61
Hulme M (2003) Abrupt climate change: can society cope? Phil Trans R Soc Lond A 361:2001–2019
Huntington TG (2006) Evidence for intensification of the global water cycle: review and synthesis. J Hydrol 319:83–95
Imbrie J, Imbrie KP (1979) Ice ages: solving the mystery. Harvard University Press, Cambridge, MA, p 224
IPCC (2000) Special report on emission scenarios. In: Nakicenovic N, Swart R (eds) Cambridge University Press, Cambridge, p 570
IPCC (2007a) Climate change 2007: the physical science basis. In: Solomon S, Qin D, Manning M, Chen Z, Marquis M, Averyt KB, Tignor M, Miller HL (eds) Contribution of working group I to the fourth assessment report of the intergovernmental panel on climate change. Cambridge University Press, Cambridge, p 996
IPCC (2007b) Climate change 2007: impacts, adaptation and vulnerability. In: Parry ML, Canziani OF, Palutikof JP, van der Linden PJ, Hanson CE (eds) Contribution of working group II to the fourth assessment report of the intergovernmental panel on climate change. Cambridge University Press, Cambridge, p 976
Jones PD, Mann ME (2004) Climate over past millennia. Rev Geophys 42:RG2002. doi:doi: 10.1029/2003RG000143
Jouzel J, Masson-Delmotte V, Cattani O, Dreyfus G, Falourd S, Hoffmann G, Minster B, Nouet J, Barnola JM, Chappellaz J, Fischer H, Gallet JC, Johnsen S, Leuenberger M, Loulergue L, Luethi D, Oerter H, Parrenin F, Raisbeck G, Raynaud D, Schilt A, Schwander J, Selmo E, Souchez R, Spahni R, Stauffer B, Steffensen JP, Stenni B, Stocker TF, Tison JL, Werner M, Wolff EW (2007) Orbital and millennial Antarctic climate variability over the past 800, 000 years. Science 317(5839):793–796
Kasting JF (2005) Methane and climate during the Precambrian era. Precambrian Res 137:119–129
Kasting JF, Toon OB, Pollack JB (1988) How climate evolved on the terrestrial planets. Sci Am 258(2):90–97
Kasting JF, Catling DC (2003) Evolution of a habitable planet. Annu Rev Astron Astrophys 41:429–463
Kasting JF, Ono S (2006) Palaeoclimates: the first two billion years. Phil Trans R Soc B 361:917–929
Keeling CD, Bacastow RB, Bainbridge AE, Ekdahl CA, Guenther PR, Waterman LS (1976) Atmospheric carbon dioxide variations at Mauna Loa Observatory, Hawaii. Tellus 28:538–551
Kerr JT, Kharouba H, Currie DJ (2007) The macroecological contribution to global change solutions. Science 316:1581–1584
Kiehl JT, Trenberth KE (1997) Earth's annual global mean energy budget. Bull Am Meteorol Soc 78:197–208
Kiehl JT, Ramanathan V (eds) (2006) Frontiers of climate modeling. Cambridge University Press, Cambridge, p 367

Keith DW (2000) Geoengineering the climate: history and prospect. Annu Rev Energy Environ 25:245–284
Kirkby J (2008) Cosmic rays and climate. Surv Geophys 28:333–375
Kristjánsson JE, Kristiansen J, Kaas E (2004) Solar activity, cosmic rays, clouds and climate – An update. Adv Space Sci 34:407–415
Kroepelin S, Verschuren D, Lezine AM, Eggermont H, Cocquyt C, Francus P, Cazet JP, Fagot M, Rumes B, Russell JM, Conley D, Schuster M, von Suchodoletz H, Engstrom DR (2008) Climate-driven ecosystem succession in the central Sahara: the last 6000 years. Science 320:765–768
Lamb HH (1970) Volcanic dust in the atmosphere, with a chronology and assessment of its meteorological significance. Philos Trans R Soc London, Ser A 266:425–533
Lamb HH (1995) Climate, history and the modern world, 2nd edn. Routledge, New York, p 464
Lean JL, Rind DH (2008) How natural and anthropogenic influences alter global and regional surface temperatures: 1889 to 2006. Geophys Res Lett 35:L18701. doi:10.1029/2008GL034864
Lee TM, Jetz W (2008) Future battlegrounds for conservation under global change. Proc R Soc B 275:1261–1270
Lenton TM, Held H, Kriegler E, Hall JW, Lucht W, Rahmstorf S, Schellnhuber HJ (2008) Tipping elements in the Earth system. Proc Natl Acad Sci 105:1786–1793
Loulergue L, Schilt A, Spahni R, Masson-Delmotte V, Blunier T, Lemieux B, Barnola JM, Raynaud D, Stocker TF, Chappellaz J (2008) Orbital and millennial-scale features of atmospheric CH4 over the last 800, 000 years. Nature 453:383–386
Lüthi D, Le Floch M, Bereiter B, Blunier T, Barnola JM, Siegenthaler U, Raynaud D, Jouzel J, Fischer H, Kawamura K, Stocker TF (2008) High-resolution carbon dioxide concentration record 650,000–800,000 years before present. Nature 453:379–382
Mann ME (2007) Climate over the past two millennia. Annu Rev Earth Planet Sci Lett 35:111–136
Mann ME, Bradley RS, Hughes MK (1998) Global-scale temperature patterns and climate forcing over the past six centuries. Nature 392:779–787
Mann ME, Bradley RS, Hughes MK (1999) Northern hemisphere temperatures during the past millennium: inferences, uncertainties, and limitations. Geophys Res Lett 26:759–762
Mann ME, Jones PD (2003) Global surface temperature over the past two millennia. Geophys Res Lett 30(15):1820. doi:doi: 10.1029/2003GL017814
Mann ME, Zhang Z, Hughes MK, Bradley RS, Miller SK, Rutherford S (2008) Proxy-based reconstructions of hemispheric and global surface temperature variations over the past two millennia. Proc Natl Acad Sci USA 105:13252–13257
Marsh N, Svensmark H (2000) Cosmic rays clouds and climate. Space Sci Rev 94:215–230
McIntyre S, McKitrick R (2005) Hockey sticks, principal components and spurious significance. Geophys Res Lett 32:L03710
Meehl GA, Arblaster JM, Tebaldi C (2005) Understanding future patterns of increased precipitation intensity in climate models. Geophys Res Lett 32:L18719
Midgley GF, Thuiller W, Higgins SI (2007) Plant species migration as a key uncertainty in predicting future impacts of climate change on ecosystems: progress and challenges. In: Canadell J, Pataki DE, Pitelka LF (eds) Terrestrial ecosystems in a changing World. Springer, Berlin, pp 129–137
NRC (2002) Abrupt climate change: inevitable surprises, US National Academy of Sciences, National Research Council Committee on Abrupt Climate Change, National Academy Press
NRC (2006) Surface temperature reconstructions for the last 2,000 Years. National Research Council, Committee on Surface Temperature Reconstructions for the Last 2,000 Years. National Academies Press, Washington, DC
Neelin JD, Battisti DS, Hirst AC, Jin FF, Wakata Y, Yamagata T, Zebiak SE (1998) ENSO theory. J Geophys Res 103:14261–14290
Oki T, Kanae S (2006) Global hydrological cycles and world water resources. Science 313:1068–1072
Pacala SW, Socolow RH (2004) Stabilization wedges: solving the climate problem for the next 50 years with current technologies. Science 305:968–972

Paillard D (2006) What drives the ice age cycle? Science 313:455–456
Pagani M, Caldeira K, Archer D, Zachos JC (2006) An ancient carbon mystery. Science 314:1556–1557
Pagano T, Sorooshian S (2006) Global water cycle (fundamentals, theory, mechanisms). In: Anderson MG (ed) Encyclopedia of hydrological sciences, vol 5. Wiley, Chichester, pp 2697–2711
Parkinson CL (2006) Earth's Cryosphere: current state and recent changes. Annu Rev Environ Resour 31:33–60
Parmesan C, Ryrholm N, Stefanescu C, Hill JK, Thomas CD, Descimon H, Huntley B, Kaila L, Kullberg J, Tammaru T, Tennent WJ, Thomas JA, Warren M (1999) Poleward shifts in geographical ranges of butterfly species associated with regional warming. Nature 399:579–583
Parmesan C, Yohe G (2003) A globally coherent fingerprint of climate change impacts across natural systems. Nature 421:37–42
Parmesan C (2006) Ecological and evolutionary responses to recent climate change. Annu Rev Ecol Evol Syst 37:637–669
Pavlov AA, Kasting JF, Brown LL, Rages KA, Freedman R (2000) Greenhouse warming by CH4 in the atmosphere of early Earth. J Geophys Res 105:11981–11990
Pearson PN, Palmer MR (2000) Atmospheric carbon dioxide concentrations over the past 60 million years. Nature 406:695–699
Peixoto JP, Oort AH (1992) The physics of climate. American Institute of Physics, New York, p 522
Pielke R Jr, Wigley T, Green C (2008) Dangerous assumptions. Nature 452:531–532
Pierrehumbert RT (2002) The hydrologic cycle in deep time climate problems. Nature 419:191–198
Pierrehumbert RT (2005) Climate dynamics of a hard snowball Earth. J Geophys Res 110:D01111. doi:10.1029/2004JD005162
Pierrehumbert RT (2009) Principles of planetary climates. Cambridge University Press, Cambridge (in press)
Plattner GK, Knutti R, Joos F, Stocker TF, von Bloh W, Brovkin V, Cameron D, Driesschaert E, Dutkiewiz S, Eby M, Edwards NR, Fichefet T, Hargreaves JC, Jones CD, Loutre MF, Matthews HD, Mouchet A, Müller SA, Nawrath S, Price A, Sokolov A, Strassmann KM, Weaver AJ (2008) Long-term climate commitments projected with climate - carbon cycle models. J Clim 21:2721–2751
Quante M (2004) The role of clouds in the climate system. J Phys 121:61–86
Quante M, Matthias V (2006) Water in the Earth´s atmosphere. J Phys 139:37–61
Rahmstorf S (2002) Ocean circulation and climate during the past 120,000 years. Nature 419:207–214
Rahmstorf S (2007) A semi-empirical approach to projecting future sea-level rise. Science 315:368–370
Raible CC, Casty C, Luterbacher J, Pauling A, Esper J, Frank DC, Büntgen U, Roesch AC, Tschuck P, Wild M, Vidale PL, Schär C, Wanner H (2006) Climate variability-observations, reconstructions, and model simulations for the Atlantic-European and Alpine region from 1500–2100 AD. Clim Change 79(1–2):9–29
Räisänen J, Hansson U, Ullerstig A, Döscher R, Graham LP, Jones C, Meier HEM, Samuelsson P, Willen U (2004) European climate in the late twenty-first century: regional simulations with two driving global models and two forcing scenarios. Clim Dyn 22:13–31
Raymo ME, Ruddiman WF (1992) Tectonic forcing of late cenozoic climate. Nature 359(6391):117–122
Rial JA, Pielke RA Sr, Beniston M, Claussen M, Canadell J, Cox P, Held H, de Noblet-Ducoudré N, Prinn R, Reynolds J, Salas JD (2004) Nonlinearities, feedbacks and critical thresholds within the Earth's climate system. Clim Change 65(1–2):11–38
Robock A (2000) Volcanic eruptions and climate. Rev Geophys 38:191–219
Robock A (2008) 20 reasons why geoengineering may be a bad idea. Bull At Sci 64(2):14–18. doi:doi: 10.2968/064002006 59

Rohling EJ, Grant K, Hemleben Ch, Siddall M, Hoogakker BAA, Bolshaw M, Kucera M (2008) High rates of sea-level rise during the last interglacial period. Nat Geosci 1:38–42
Rosenzweig C, Karoly D, Vicarelli M, Neofotis P, Wu Q, Casassa G, Menzel A, Root TL, Estrella N, Seguin B, Tryjanowski P, Liu C, Rawlins S, Imeson A (2008) Attributing physical and biological impacts to anthropogenic climate change. Nature 453:353–357
Royer DL, Bemer RA, Beerling DJ (2001) Phanerozoic atmospheric CO_2 change: evaluating geochemical and paleobiological approaches. Earth-Sci Rev 54:349–392
Ruddiman WF (2003) The anthropogenic green-house era began thousands of years ago. Clim Change 61:261–293
Ruddiman WF (2005) Plows, plagues and petroleum: how humans took control of climate. Princeton University Press, Princeton, p 272
Ruddiman WF (2008) Earth's climate: past and future, 2nd edn. W.H. Freeman & Company, New York, p 465
Rybski D, Bunde A, Havlin S, von Storch H (2006) Long-term persistence in climate and the detection problem. Geophys Res Lett 33:L06718. doi:10.1029/2005GL025591
Sabine C, Feely R, Gruber N, Key R, Lee K, Bullister J, Wanninkhof R, Wong C, Wallace D, Tilbrook B, Millero F, Peng T, Kozyr A, Ono T, Rios A (2004) The ocean sink for anthropogenic CO_2. Science 305:367–371
Sagan C, Chyba C (1997) The early faint sun paradox. Science 276:1217–1221
Sala OE et al (2000) Global biodiversity scenarios for the year 2100. Science 287:1770–1774
Scafetta N, West BJ (2006) Phenomenological solar contribution to the 1900–2000 global surface warming. Geophys Res Lett 33 doi:10.1029/2005GL025539
Scafetta N, West BJ (2007) Phenomenological reconstructions of the solar signature in the NH surface temperature records since 1600. J Geophys Res 112:D24S03. doi:doi:10.1029/2007JD008437
Schär C, Vidale PL, Lüthi D, Frei C, Haeberli C, Liniger MA, Appenzeller C (2004) The role of increasing temperature variability in European summer heatwaves. Nature 427:332–336
Schneider SH (2001) Earth systems: engineering and management. Nature 409:417–421
Schneider SH (2008) Geoengineering: could we or should we make it work? Phil Trans R Soc 366:3843–3862
Seinfeld JH, Pandis SN (2006) Atmospheric chemistry and physics: From air pollution to climate change, 2nd edn. Wiley, New York, p 1203
Soden BJ, Jackson DL, Ramaswamy V, Schwarzkopf MD, Huang X (2005) The radiative signature of upper tropospheric moistening. Science 310:841–844
Solanki SK, Krivova N (2003) Can solar variability explain global warming since 1970? J Geophys Res 108. doi:10.1029/2002JA009753
Solanki SK, Usoskin IG, Kromer B, Schussler M, Beer J (2004) Unusual activity of the Sun during recent decades compared to previous 11, 000 years. Nature 431:1084–1087
Steffen W, Crutzen PJ, McNeill JR (2007) The anthropocene: Are humans now overwhelming the great forces of nature? Ambio 36:614–621
Stephens GL (2005) Cloud feedbacks in the climate system: a critical review. J Clim 18:237–272
Stott PA, Stone AD, Allen MR (2004) Human contribution to the European heatwave of 2003. Nature 432:610–614
Tebaldi C, Hayhoe K, Arblaster JM, Meehl GA (2006) Going to the extremes- an intercomparison of model-simulated historical and future changes in extreme events. Clim Change 79:185–211
Thuiller W (2007) Biodiversity-climate change and the ecologist. Nature 448:550–552
Thuiller W, Albert C, Araújo MB, Berry PM, Cabeza M, Guisan G, Hickler T, Midgley GF, Paterson J, Schurr FM, Sykes MT, Zimmermann NE (2008) Predicting global change impacts on plant species distributions: future challenges. Perspect Plant Ecol Evol Syst 9:137–152
Tol RSJ (2008) Economic scenarios for global change. In: von Storch H, Tol RSJ, Flöser G (eds) Envioronmental crises – science and policy. Springer, Berlin, pp 17–35
Trenberth KE (1992) Climate system modeling. Cambridge University Press, Cambridge, p 788

Trenberth KE, Fasullo J, Smith L (2005) Trends and variability in column-integrated atmospheric water vapor. Clim Dyn 24:741–758
Turcotte DL, Schubert G (2002) Geodynamics, 2nd edn. Wiley, New York, p 456
UN (1992) United Nations Framework Convention on Climate Change. Report FCCC/INFORMAL/84, GE.05-62220 (E), United Nations http://unfccc.int/resource/docs/convkp/conveng.pdf
Wallace JM, Hobbs PV (2006) Atmospheric sciences: an introductory survey, 2nd edn. Academic, San Diego, CA 483
Walther GR, Post E, Convey P, Menzel A, Parmesan C, Beebee TJC, Fromentin JM, Hoegh-Guldberg O, Bairlein F (2002) Ecological responses to recent climate change. Nature 416:389–395
Walther GR (2003) Plants in a warmer world. Perspect Plant Ecol Evol Syst 6(3):169–185
Wang Y, Cheng H, Edwards RL, Kong X, Shao X, Chen S, Wu J, Jiang X, Wang X, An Z (2008) Millennial- and orbital-scale changes in the East Asian monsoon over the past 224, 000 years. Nature 451:1090–1093
Wanner H, Beer J, Bütikofer J, Crowley TJ, Cubasch U, Flückiger J, Goosse H, Grosjean M, Joos F, Kaplan JO, Küttel M, Müller SA, Prentice IC, Solomina O, Stocker TF, Tarasovk P, Wagner M, Widmann M (2008) Mid- to late Holocene climate change: an overview. Quat Sci Rev 27:1791–1828
Washington WM, Parkinson CL (2005) An introduction to three dimensional climate modeling. University Science Books, Sausalito, CA, p 354
Weart S (2008) The discovery of global warming. Revised and expanded edition. Harvard University Press, Harvard 190
Webster PJ (1994) The role of hydrological processes in ocean-atmosphere interactions. Rev Geophys 32:427–476
Webster PJ, Holland GJ, Curry JA, Chang HR (2005) Changes in tropical cyclone number, duration and intensity in a warming environment. Science 309(5742):1844–1846
Witt A, Schumann AY (2005) Holocene climate variability on millennial scales recorded in Greenland ice cores. Nonlin Processes Geophys 12:345–352
von Storch H (2008) Climate change scenarios – purpose and construction. In: von Storch H, Tol RSJ, Flöser G (eds) Envioronmental crises – science and policy. Springer, Berlin, pp 5–15
von Storch H, Zorita E, Jones J, Dimitriev Y, González-Rouco F, Tett S (2004) Reconstructing past climate from noisy data. Science 306:679–682
Zachos J, Pagani M, Sloan L, Thomas E, Billups K (2001) Trends, rhythms, and aberrations in global climate 65 Ma to Present. Science 292:686–693

Extra-Mediterranean Refugia, Post-Glacial Vegetation History and Area Dynamics in Eastern Central Europe

Zoltán Varga

Abstract Evidences from fossil records and genetic research suggest that the arboreal refugia were not restricted to Southern Europe and in particular to the Mediterranean peninsulas during the full-glacials. Fossil pollen data and macrofossil remains indicate that several tree species have survived also at the Southern edge of the cold-dry steppe-tundra area in Central and Eastern Europe. Recent results of surveys on the Late Pleistocene Mammalian fauna clearly contradict to the "*tree-less tundra*" models for Europe North of the transverse mountain ranges of the Pyrenees, Alps and Carpathians. It was pointed out that the carrying capacity to feed the herds of large herbivores demands a rather productive environment. The presence of Northern temperate refugia is also supported by the "non-analogue" assemblages of small mammals discovered from the Late Pleistocene of unglaciated areas of Eastern Central Europe. The assembly of species today typifying the tundra, steppe and semi-desert habitats seems to include also species from deciduous woodland. Extra-Mediterranean core areas were identified also in widely dispersed cold-tolerant frogs and reptiles. Some of their core areas had been at least near the Carpathians and/or marginal areas of the Carpathian Basin. The close faunal connections of the Carpathians suggest the existence of highly dynamic contacts and exchanges with mountains of the Balkan Peninsula during the climatic fluctuations of the Upper Pleistocene. The Eastern and Southern Carpathians, together with the mountains of Western Transylvania, can be considered as core areas of survival and autochtonous evolution in some invertebrate groups with limited mobility. The post-glacial re-population of the Carpathian Basin from different directions has been supported by Illyrian versus Dacian vicarious pairs of sister species/subspecies. In mobile insect groups, peripherically isolated sibling species/subspecies have only been evolved, which display manifold biogeographic connections, e.g. to the Balkan Peninsula, Asia Minor or Southern Russia. The organisation of community-complexes of the Pannonian forest-steppe

Z. Varga
Faculty of Science and Technology, Department of Evolutionary Zoology,
University of Debrecen, Egyetem-tér 1, H-4010 Debrecen, Hungary
e-mail: zvarga@tigris.unideb.hu

J.C. Habel and T. Assmann (eds.), *Relict Species: Phylogeography and Conservation Biology*,
DOI 10.1007/978-3-540-92160-8_3, © Springer-Verlag Berlin Heidelberg 2010

connected by habitat ecotones resulted in the overlap of several different faunal types, e.g. Mediterranean, Balkanic, Siberian, Ponto-Caspian, Ponto-Pannonian, Turano-Eremic and Xeromontane elements.

1 Introduction: Refugia and Core Areas of Dispersal

1.1 The Basic Components of the European Fauna

Traditional biogeography has been loaded with a seemingly endless discussion on the principles and methods of the biogeographical characterisation of species. In the geobotanical literature, the purely chronological (*geoelements*) and the historical approaches (*geno-*, *migro-* and *chrono-elements*) have regularly been clearly disentangled (e.g. Walter and Straka 1970). In the zoogeographical literature, the term "faunal element" or "faunal type" has repeatedly been used for both approaches, often with a mixture of points of view. Faunal types have often been differentiated on the basis of the shape and the extension of the ranges (e.g. the "Central European", "Palearctic" or "Holarctic", etc. faunal elements in many faunistical publications). Quantitative methods of biogeography based on large databases have only rarely been used. Dennis et al. (1991) differentiated some types of endemic versus "extent" (i.e. widespread) species of butterflies, and have outlined some "*faunal structures*", based on multivariate methods. As opposed to these methods, the European fauna has traditionally been subdivided into a "*holothermic*" refugial and a "*holopsychric*" invasion types (Rebel 1931). In addition, these faunal types have been characterised by some "*core areas*" (in German: *Arealkerne*, *chorologische Zentren*: Reinig 1950; de Lattin 1957, 1967) and interpreted as areas of survival (refugia) and, consequently, as "*centres of dispersal*" ("*Ausbreitungszentren*" in the German tradition).

The "*holothermic*" type was differentiated according to the secondary subdivision of the large Mediterranean refugial area (de Lattin 1949, 1957, 1967; Reinig 1950). Recently, this view was strongly confirmed and modulated by the growing molecular results and re-formulated as repetitive patterns of the generalised "paradigms" of core areas and tracks of post-glacial re-population. A general conclusion was that temperate species mainly derive from Mediterranean refugial populations that underwent range expansion in the late glacial and early post-glacial periods (Hewitt 1996, 1999, 2000, 2001, 2004; Taberlet et al. 1998; Schmitt and Hewitt 2004; Schmitt 2007). The other main group, the "*holopsychric*" type has been considered for a long time as a result of the "Siberian" invasion, suggested by some "classical", monoglacialistic biogeographical works (Hofmann 1873; Scharff 1899) and despite the evidences which have revealed the taxonomical differentiation of North-Eastern "boreal" and Southern European montane populations, especially in some butterfly species (Varga 1975, 1977; Nève 1996). Several authors (e.g. Schmitt and Seitz 2001; Steward and Lister 2001; Surget-Groba et al. 2001; Babik et al. 2004;

Ursenbacher et al. 2006; Saarma et al. 2007), however, have suggested an additional mode of colonisation of central and Northern Europe by *non-Mediterranean* populations, coming from one or more "continental" refugia: central Europe, Southern Ural, Caucasus and Western Asia. Other species exhibit mixed patterns, where different parts of Europe have been colonised both from Mediterranean and/or non-Mediterranean refugia (e.g. Fumagalli et al. 1996; Deffontaine et al. 2005; Kotlík et al. 2006).

1.2 The Geographical Projection of the Faunal Type Frequency Data

The faunal composition of formerly glaciated and therefore, nearly exclusively post-glacially re-populated Northern Europe is practically identical with North-Eastern Europe and Western Siberia which were considered as a main argument for being re-populated mostly from the East (Rebel 1931; de Lattin 1957, 1967). This major part of Europe clearly shows a high percentage of the "Siberian" faunal type in all mobile groups of animals such as birds (Stegmann 1932, 1938; Voous 1960, 1963) and butterflies (Reinig 1950; de Lattin 1964, 1967, see: Fig. 59, in Kostrowicki 1969; Fig. 1). This faunal type is also strongly represented in Central Europe, North of the Alps, being the prevailing faunal type mostly in the mountainous regions (e.g. Harz Mts., Bavarian and Bohemian forest, Sudetic Mts., etc.). In contrast, South of the large transversal chains of the Pyrenées, the Alps and the Northern Carpathians, the pre-dominance of the Mediterranean faunal type *sensu lato* has been demonstrated, with a decreasing gradient into Northern direction being extremely steep at the Pyrenees and partly North of the Alps and the Carpathians, but much more gently sloping from the Balkan peninsula to the Carpathian Basin. As a consequence of the overlapping of the different biogeographical influences, there is a transitional belt in Southern Central Europe including a large part of the Carpathian Basin where the proportions of the different faunal types are rather balanced (de Lattin 1967; Varga and Gyulai 1978; Varga 1995, 2003b, 2006).

Several consequences follow from this general biogeographical setting of Europe.

- In Europe, the highest number of endemic species is confined to some, mostly mountainous parts of the Mediterranean peninsulas (e.g. Williams et al. 1999; Finnie et al. 2007). These areas, more or less, regularly overlap in the different taxonomical groups; thus they can be considered as hotspots for endemism, and at the same time, they were the main areas of survival during the Quaternary glaciations. These areas often show also a high level of "multi-species genetic divergence" (Petit et al. 2003).
- On the other hand, there are ecologically transitional regions with high numbers of species, but without a high proportion of endemism (Williams et al. 1999). These are characterised by an overlap of the ranges of species of different

geographical origins caused by dispersal processes along gradients, e.g. the overlap of species belonging to different zonobiomes and azonal communities in the forest-steppe areas of the Carpathian Basin (Varga 1995). Recently, these areas have been identified as "melting pots" of genetic diversity (Petit et al. 2003) due to the secondary accumulation, re-distribution and re-combination of genotypes.

Another important observation was made by de Lattin (1957, 1964) who indirectly defined the "Siberian" species by their absence in the supposed Mediterranean refugial areas. These species show a peculiar "crowding" (de Lattins' "*Stauungslinie*") at the Northern boundary of the Mediterranean region. The re-interpretation of this biogeographical line subsequently followed in two steps. First, it was recognised that accumulated occurrence of marginal sub-species of continental species was shown around this line (Varga 1975, 1977), and later on, they were confirmed as multiple extra-Mediterranean refugia of continental species (Schmitt 2007; Schmitt et al. 2007) (Fig. 1).

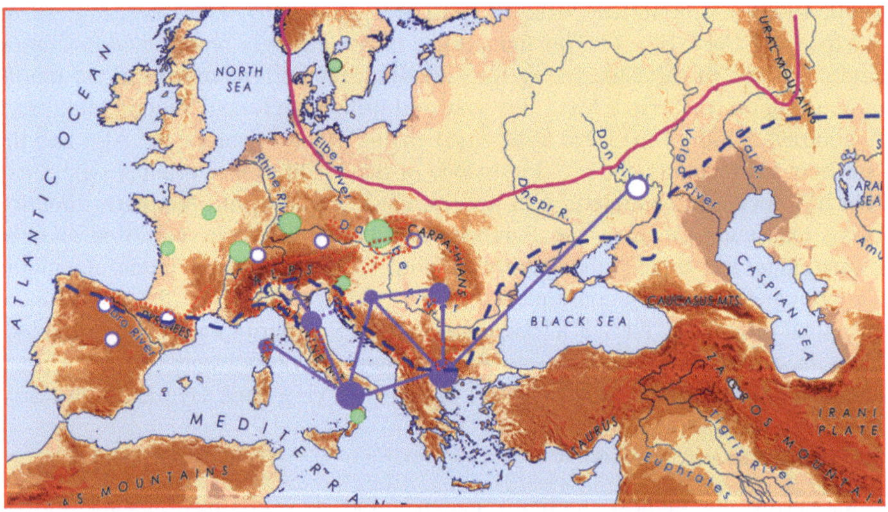

Fig. 1 Important characteristics of biodiversity in Europe, compiled from various sources. *Purple line*: Southern and Western border of the high (>40%) representation of the Boreo-Continental (= Siberian) faunal elements (de Lattin 1967; Kostrowicki 1969). *Blue line*: Southern border of the expansion ("Stauungslinie") of the Boreo-Continental (= Siberian) faunal elements (de Lattin 1967). *Red dots*: important mountain arboreal refugia (compiled from different sources). *Light blue circles*: core areas of the multispecies genetic divergence of the 25 European forests. *Full circles*: divergence higher than average, *empty circles*: lower than average. *Continuous lines* indicate high divergence, *dotted lines*, intermediate divergence (re-drawn and simplified from Petit et al. 2003). *Green circles*: mean number of haplotypes per forest, averaged across species. Diversity is highest at relatively high latitudes, North of the three European peninsulas (re-drawn and simplified from Petit et al. 2003)

1.3 The Geographical Projection of the Genetic Diversity Data

It is not only the number of the species, but also some characteristics of the genetic diversity that show the highest values in the "transitional" belt mentioned above. Based on the chloroplast DNA variation in 22 widespread European trees and shrubs, it was pointed out that despite their "individualistic" migration behaviour, diverse ecological requirements (from the Southern temperate to the Southern boreal zone) and different modes of seed dispersal, significant species of the woody flora of Northern Central Europe exhibited an essentially congruent pattern of genetic divergence (Petit et al. 2003; Fig. 1). Their genetically most unique populations are found in Southern and central Italy, Corsica, and the Balkan Peninsula extending into Northern Italy, the Northern Balkans and the Southern peripheries of the Carpathian Basin. However, this increased diversity is obtained mostly through the redistribution ("melting pot") of the genetic information already present in the populations in refugia (the actual "hot spots", i.e. areas where the diversity was evolved). Recent work on the genetics of oak populations has revealed that the existence of particular haplotypes in Romania resulted either from older indigenous relict populations or from crossing of populations originating from more than one refugium. There were several low or medium mountains North of the Balkan Peninsula that could have offered favourable climatic conditions during the Younger Dryas period to support viable populations of oak trees. At the beginning of the post-glacial times, the Carpathian Basin may have acted as a meeting point of several colonisation routes (Bordács et al. 2002; Petit et al. 2002). These genetic data are also supported by recent fossil findings of tree remains from the last full glacial period, North of the Mediterranean peninsulas. Fossil pollen data and macrofossil remains such as charcoals from the time of the last glacial maximum, i.e. 25,000–17,000 years ago, indicate that several tree species remained in small favourable pockets not only within the Mediterranean region, but also at the Southern edge of the cold and dry steppe-tundra area in Eastern, central, and South-Western Europe (Willis et al. 1995, 2000; Carcaillet and Vernet 2001).

Research on small-mammal mtDNA has also questioned the universality of Mediterranean refugia as the areas from which all temperate taxa colonised central and Northern Europe at the beginning of the present interglacial period. It was suggested that the Mediterranean "sanctuaria" (Widmer and Lexer 2001) in general were *not* also core areas of post-glacial expansion into deglaciated areas. Thus, a need for new paradigms for the phylogeography of cold- resistant arboreal species has been formulated, as well (Steward and Lister 2001). It was shown that Mediterranean populations of the shrews *Sorex minutus*, *Sorex araneus* and bank voles (*Chlethrionomys* spp.) did not contribute to the present-day gene pools in Central Europe (Fumagalli et al. 1996; Bilton et al. 1998). Instead, populations in Central Europe and Western Asia have participated suggesting the existence of glacial refugia in these regions. The central European evidences fit well with the botanical results from Hungary and the records of temperate mammals within cold-stage assemblages of central and Northern Europe (Fig. 2).

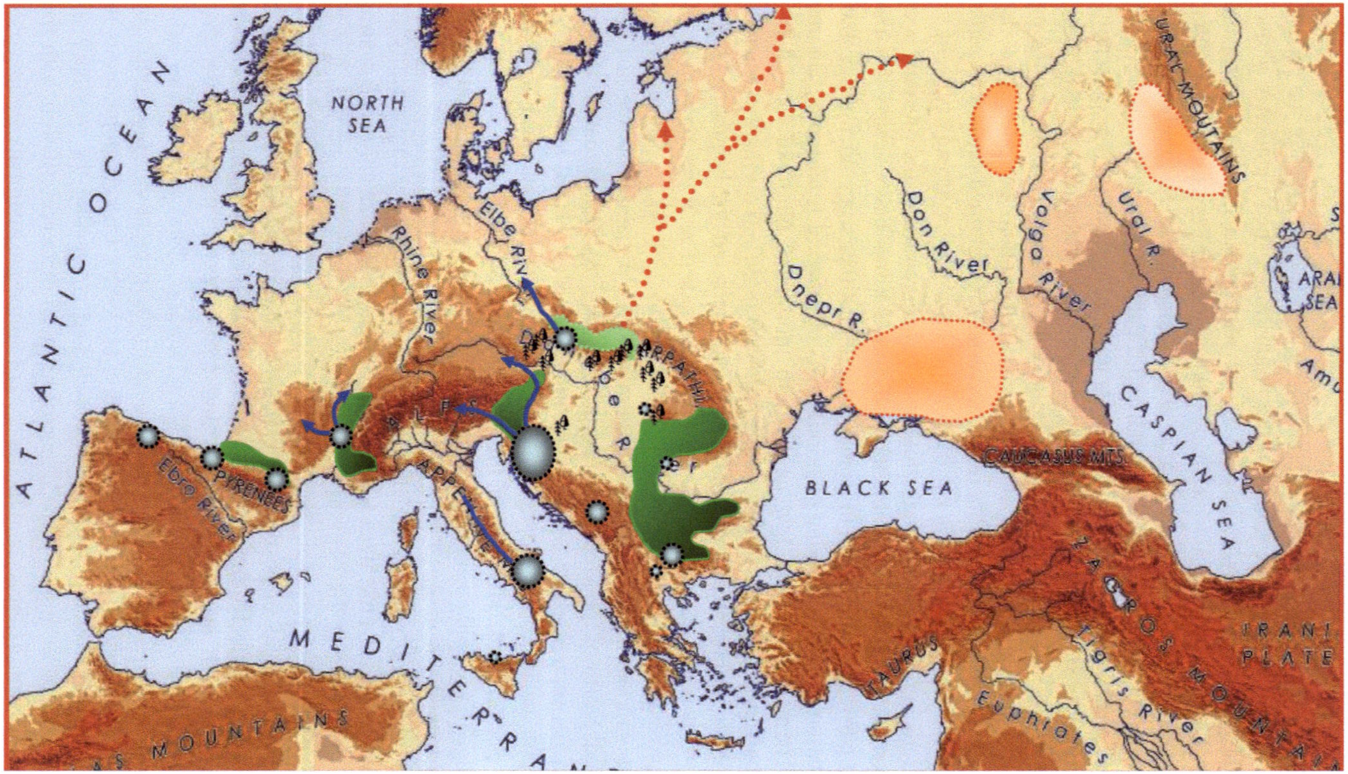

Fig. 2 Important extra-Mediterranean refugia, compiled from various sources. *Dark green fields*: European coniferous forest refugia as revealed by Schmitt and Haubrich (2008). *Light green field* with *brown arrows*: the Northern Carpathian arboreal refugium with expansion routes of boreal species (coniferous trees, brown bear, etc.), from various sources. *Grey circles* with *blue arrows*: refugia of the beech (Fagus sylvatica) with important post-glacial expansion routes (re-drawn from Magri et al 2006; Magri 2007). *Light brown patches*: Eastern European boreal forest refugia during the LGM (from various sources)

2 The Basic Types of Zonal Setting in Europe: Glacial Vs. Interglacial

2.1 The Notion of the Boreal Forest-Steppe as the Macro-Ecotone of the Mammoth Steppe and the Boreal Forest

The Pleistocene glacial-interglacial cycles have resulted in the "*antagonistic dynamics*" of biota belonging to contrasting macrohabitats (de Lattin 1967, p. 356). Thus, principally, two basic types of zonal setting can be distinguished. The inter-/post-glacial (but also interstadial!) type can be characterised by the expansion of the *Arboreal* and regression of the "*non-Arboreal*" macrohabitats. As a result, the "open" macrohabitats (tundra, steppe, semi-desert and high mountains) have been separated by wooded zonobiomes, such as boreal and nemoral forests. Significant belts of macro-ecotones have been developed at the edges of the forested belts, e.g. forest-steppe as a transitional habitat type between the nemoral and steppic zones, subdivided into a sub-mediterranean (Pannonian) and a continental sub-type (Zólyomi 1949; Sjörs 1963; Walter and Straka 1970; Varga 1995; Fekete and Varga 2006).

Thus, the glacial periods (stadials) have been characterised by a regression plus fragmentation of wooded habitats and consequently, by a broad contact of the tundra and the steppe zonobiomes with some forested "pockets" North of the refugial belt of the Mediterranean area. The transitional "macro-ecotones" developed at the forest-belt fringes: tundra-taiga and boreal forest-steppe (e.g. in Pannonian region). Recently, cold steppe vegetation with scattered boreal coniferous forests have been revealed during the cold stages in S Moravia, the SE margin of the Alps, E Hungary, parts of Romania (Carpathians), etc. (Rybnicková and Rybnicek 1991; Willis et al. 2000; Pokorny and Jankovská, 2000; Sümegi and Rudner 2001; Wohlfahrt et al. 2001; Björkman et al. 2003; Tantau et al. 2003; Willis and Andel 2004; Feurdean et al. 2007a,b). The presence of these Northern temperate refugia justifies the arguments for the so-called *non-analogue* assemblages of mammals described from the Late Pleistocene in unglaciated areas of the Northern hemisphere.

The Würm glaciation had at least six maxima, the last one taking place about 22.000–18.000 years BP. The Last Glacial Maximum (LGM) was extremely continental and ended in rapid climatic fluctuations. The late Würm *kryoxerotic* period was characterised by the expansion of cold-continental *heliophyta* and steppic species (Iversen 1958; Tarasov et al. 2000; Velichko et al. 2002). Pollen-based "treeless tundra" models for Europe North of the transverse mountain ranges of the Pyrenees, Alps and Carpathians (e.g. Frenzel 1992; Huntley and Birks 1983; Huntley and Allen 2003) have repeatedly been questioned by researchers of the late Pleistocene mammalian fauna (Kretzoi 1977; Guthrie 1990, 2000; Guthrie and van Kolfschoten 1999; Simakova 2001; Yurtsev 2001; Steward and Lister 2001), because the carrying capacity sufficient to feed numerous large herbivores such as the mammoth, the woolly rhinoceros, the reindeer, the giant deer, the bison or the auerochs demands a very productive environment ("steppe-tundra" or "mammoth steppe", Guthrie 1990), like the cold-continental meadow steppes in Southern Siberia and Northern Mongolia (Varga et al. 1989).

Although the East-West faunal movements between the Western and Eastern temperate refugia (see: nemoral disjunctions!) were hindered during the last glaciation, the tundro-steppic areas of E Europe and the mountain belts of Central Asia were connected to each other. Thus, *non-analogue communities* were composed by mixing tundral, steppie and eremic-oreal elements (e.g. *Lemmus* and *Dicrostonyx* spp. together with *Citellus* spp. and *Allactaga, Sicista, Lagurus, Marmota* and *Ochotona*). This boreal forest-steppe habitat type appears to have included also cold-tolerant species of temperate habitats, e.g. *Clethrionomys glareolus, Apodemus sylvaticus* and *S. araneus.* Extra-Mediterranean core areas have also been revealed in several, widely dispersed cold-tolerant frogs and reptiles, as *Rana arvalis, Zootoca vivipara, Vipera berus* (Surget-Groba et al. 2001; Babik et al. 2004; Ursenbacher et al. 2006; Kotlík et al. 2006; Saarma et al. 2007).

3 European Refugia of Boreal Species

3.1 Biogeographical Definition and Sub-Division of Boreal Species

The species of the boreal zone show a significant diversity of extension and taxonomical structure of ranges. Most boreal taiga-species are widely distributed from the Far East across Siberia to Scandinavia, often without significant geographical variations. They are partly inhabitants of forests but are often connected with peat bogs, e.g. due to the food-plant specialisation (e.g. the butterflies and moths *Colias palaeno, Boloria aquilonaris, Plebejus optilete, Anarta cordigera, Arichanna melanaria*). Vicarious taxa often occur in Southern Himalayan coniferous forests with dense scrub layer and undergrowth, rich in evergreen plant species (e.g. *Rhododendron, Vaccinium* spp.). Also, some boreal plant species complexes have a huge Holarctic or Eurasiatic range while their closest relatives are restricted to some Southern mountainous areas, as parts of the Himalaya, Tibet, etc., e.g. the boreal species complex consisting of the circum-boreal *Saxifraga hirculus* and the related spot-like Central Asiatic species: *S. diversifolia*, *S. przewalskii* and further species in Tibet. In other cases, the zonal boreal "taiga" distribution is combined with considerable disjunct ranges in the sub-alpine mixed coniferous forests of the Himalayas (e.g. several small passerine birds as *Dumeticola thoracica, Luscinia calliope, Muscicapa sibirica, Phylloscopus proregulus, Tarsiger cyanurus*) and/or with several larger or smaller spots in the European mountain coniferous belts, e.g. in many taiga-inhabiting birds, as *Tetrao urogallus, Tetrastes bonasia, Strix uralensis, Aegolius funereus, Picoides tridactylus*, etc. (Stegmann 1938; Voous 1963).

Such species are most richly represented in the coniferous zone of the South Siberian high mountains.

The number of (nearly) exclusively European boreal and boreo-montane species is relatively low. However, a molecular biogeographical analysis of such species can unravel the European coniferous forest refugia (Schmitt and Haubrich 2008). The existence of European coniferous forest refugia is also supported by the *Western-Eastern* sub-division of several boreal species with the Western populations obviously having European refugia during the last glaciations, e.g. *Picea abies* most probably in the mountains of the Balkans and in the Eastern and Southern Carpathians. The North-Eastern part of European Russia was populated from the East by the sister (sub-)species *Picea obovata*. In the catchment area of the Northern Dvina, a hybrid belt has been formed between them. The European *Pinus cembra* survived the LGM in Southern Alpine and Carpathian refugia (Willis et al. 2000; Wohlfahrt et al. 2001). Its sister species is the SiberianNorthern Mongolian *Pinus sibirica,* a dominant species of the light-penetrated mountain taiga, often mixed with *Larix sibirica* (Walter and Breckle 1986).

3.2 Carpathian–Balkanic Boreo-Montane Arboreal Refugia

It is known since several decades that the Southern part of the Carpathian Mts. was a refugial area for temperate and mountain forest taxa during the last glacial period (Huntley and Birks 1983). The Eastern and the Southern Carpathians have been repeatedly pointed out as important glacial refugia, from which trees started to expand at the beginning of the Holocene (e.g. Huntley and Birks 1983; Bennett et al. 1991; Willis 1994). In particular, Willis et al. (2000) indicated that temperate refugia in Europe during cold periods might not have been restricted to the three Southern peninsulas (Iberia, Italy and the Balkans) because trees were undoubtedly present in Central Europe in areas such as the Carpathian Basin during the last cold stage. Radiocarbon data indicate the continuous presence of coniferous woodland not only during the relatively mild period from 35 to 25 kyr BP (thousands of radiocarbon years before the present), but also into the more severe last glacial maximum 25–17 kyr BP (Sümegi and Rudner 2001; Willis and Andel 2004).

The existence of full-glacial forests in Eastern Europe during the LGM (Willis et al. 2000) has often been questioned, but it has also been repeatedly confirmed by several macrofossil and palynological studies from the Carpathians (e.g. Farcaş et al. 1999; Björkman et al. 2003; Tantau et al. 2003; Willis and Andel 2004; Feurdean et al. 2007a) demonstrating the survival of most coniferous and cold-resistant deciduous trees or even the beech (Magri et al. 2006, Magri 2007) in refugial "pockets" in the hilly areas of the Carpathians or North of them. Similarly, forest refugia were also found in Southern Moravia, in the Eastern Alps, in the Dinaric Mountains and in the Eastern Balkans, etc. (Willis 1994; Willis and Niklas 2004).

The climatic conditions in some favourable localities of these regions could not have been as severe as those in Northern and central Europe during the LGM and during the beginning of the last deglaciation.

In the NE Carpathians (Mt Gutaiului), *Pinus mugo, P. sylvestris* and later on also *Larix* became established from 14,500 year BP onwards (Wohlfahrt et al. 2001; Björkman et al. 2003). Between 14,150 and 13,950 years BP, *P. cembra* have replaced *P. mugo* and *P. sylvestris.* At 13,950 cal year BP, the tree cover increased and *Picea* appeared for the first time, together with *P. cembra, P. mugo* and *Larix decidua.* The analysis of the sediments of the peat bog "Mohos" in the Eastern Carpathians (Tantau et al. 2003) has shown that an open boreal forest was dominated by *Pinus* during the last phase of the LGM. The presence of *Picea* pollen refers to its existence in a local refugium. The pollen data at the end of the late Weichsel indicate that beside the dominant *Pinus* and *Betula* species, *Alnus, Ulmus* and *Picea* also occurred in the area. The increased representation of *Alnus* at about 14,150 year BP and that of *Picea* and *Ulmus* between 13,750 and 13,200 years BP is interpreted as a consequence of nearby refugia of these trees. Beech pollen was dated in the North- Eastern Carpathians (Semenic Mts.) to 9,500 year BP while in most other diagrams for the Romanian Carpathians, it was registered between 7,500 and 8,000 years BP. From the late expansion of *Quercus, Tilia, Fraxinus, Acer* and *Corylus* between 10,750 and 10,200 years BP, it was concluded that these trees had to immigrate into the area from refugia further away. However, these refugia cannot have been situated very far from the Carpathians. It seems most likely that these refugia were located further to the South in Romania, or in lower hilly areas of the Carpathian Basin (Feurdean et al. 2007b). Several surveys across Hungary (Rudner and Sümegi 2001; Willis et al. 2001) from about 30 sites clearly demonstrate that cyclically recurring forested habitat developed in the Carpathian Basin during the late Weichselian, although this forest type might have been of an open forest or steppe forest/forest-steppe type. These forests or forest-steppes were of the *boreal* type mostly with cold-tolerant trees as *Picea, Larix, Pinus sylvestris* and *P. cembra, Betula, Carpinus, Salix* and *Juniperus communis.*

In Bátorliget, North-Eastern Hungary, a light-penetrated taiga forest and a cold steppe with *Artemisia* spp. dominated during the last cold phase of the Weichselian glaciation. However, it has also been demonstrated that some deciduous trees such as *Quercus, Ulmus, Alnus* and *Tilia* survived in meso/microclimatically favourable "pockets" of this mosaic landscape (Willis et al. 1995). Later, *Tilia* dominated woodland and a mixture of deciduous trees including *Ulmus, Corylus, Tilia, Fraxinus* and coniferous species (*Pinus* and *Picea*) (Willis et al. 1995, 2000; Magyari et al. 1999) were present in the early Holocene. In a neighbouring area of Romania (Turbuta), pollen of *Quercus* was not found before 12,000 year BP implying that *Quercus* did not survive in the proximity of the study site during the late glacial. Local stands of *Quercus* pollen became established between 12,000 and 11,000 years BP, but its local expansion did not occur until about 11,000 cal year BP and peaked between 10,000 and 8,500 years BP. This is generally simultaneous with the time of *Quercus* expansion in North-Western, Western and South-Western Romania, though in the Eastern Carpathians its expansion was dated to ca.

500 years later (Farcaş et al. 1999, 2004; Björkman et al. 2003; Tantau et al. 2003, 2006; Feurdean and Bennike 2004; Feurdean et al. 2007a).

The fossil and genetic data unanimously indicate that the beech (*Fagus sylvatica*) also survived the last glacial period in multiple refuge areas. Recent publications (Magri et al. 2006; Magri 2007) clearly demonstrate that the central European refugia of the beech were separated from the Mediterranean ones. In addition, the Mediterranean core areas did not contribute to the colonisation of central and Northern Europe. The Illyrian-Slovenian population migrated only to a limited extent Northward to colonise the rest of Europe. This population proved to be genetically similar to the population in Southern Moravia–Southern Bohemia, which is considered to be a possible refuge area on the basis of fossil data. This core area might have been the source for the colonisation of the Carpathians. From a genetic point of view, the populations in the Apuseni Mts. have been characterised by rather high allelic richness (Gömöry et al. 2003). Thus, it is possible that a secondary refugium of the beech was located in this area, which, however, did not significantly contribute to the colonisation of the Carpathian arc.

3.3 Refugia of Cold-Tolerant Invertebrates and Exothermic Vertebrates in Eastern Central Europe

Vertebrate and Mollusc remains show that the lowest Weichselian loess layers in the Carpathian Basin were formed during the first cool/dry phase of the last glacial, between 50,000 and 70,000 years BP (Hertelendy et al. 1992). The dominance of woodland species decreased, but the tree cover was able to survive the unfavourable environmental changes because, as a result of the mosaic-like environment, some mild and humid micro-climatic areas developed in the foothill zone. On the Tokaj hill, for example, a coniferous forest-steppe was revealed with wildfires in a taiga-like environment during the LGM. Charcoal from *Picea* and podsolic soils were identified (Sümegi and Rudner 2001).

Between 22,000 and 20,000 years BP, a decline of thermophilous gastropod species and expansion of a kryo-xerophilous, *xeromontane* element were observed (Sümegi and Krolopp 2002; Füköh et al. 1995). The occurrence of *Vallonia tenuilabris* was shown together with the boreo–Alpine/montane *Columella columella*. The dominance of the kryophilous land snails reached values of 80% in the Northern parts of the Carpathian Basin while the occurrence of the same group was about 40% in the Southern parts. Data of mollusc analysis refer to a large-scale heterogeneity in both palaeogeographic and palaeobiogeographic conditions for the area examined, creating some sort of a meeting point of faunal elements adapted to different environmental conditions in the region. The "gastropod-thermometer" developed by Krolopp and Sümegi (1995) showed a mean July temperature about 11–12°C in the Northern and 13–14°C in the Southern part of the basin during the LGM. Several further land snails and slugs (*Arianta arbustorum, Trochoidea geyeri, Arion fuscus* and *Arion* spp.) were also able to survive harsh climatic conditions in small spots in the central European periglacial (Haase et al. 2003; Pinceel et al. 2005) or even within the Alpine permafrost area. DNA and allozyme data on *A. fuscus* show that its glacial survival was possible in or at the periphery of the Alps as well as in other European mountain ranges (e.g. the Tatra Mountains and Southern Balkans).

In the group of continental species, the woodland ringlet (*E. medusa*) represents a particularly interesting and well-studied case. Good evidences have been found based on allozyme data that this specie had multiple Würm ice-age differentiation centres around the glaciated Alps, in the Carpathian region and in the Balkan Peninsula (Schmitt and Seitz 2001; Schmitt et al. 2007). This glacial distribution pattern of *E. medusa* shows several parallel features with cold-tolerant vertebrates such as the adder, the bank vole, etc. (Deffontaine et al. 2005; Kotlík et al. 2006; Ursenbacher et al. 2006), most probably due to the particular climatic conditions of Eastern Europe during the last ice-ages. Supposedly, the decline of the temperature was, especially in the summer period, less dramatic as in the Atlantic part of Europe. This was combined with a decrease in the precipitation causing a transitional climate and mosaic-like vegetation similar to the recent conditions in the cold-continental forest-steppe belts of Southern Siberia and Northern Mongolia. Three of the four groups of *E. medusa* confined to the Carpathians most probably had their differentiation centres at the low elevations of the Southern-South-eastern Carpathians. The fourth lineage, composed of populations from the Western Carpathians and the North-Eastern Carpathians, might have had a core area of larger extent near to the Northern Carpathians or in the Carpathian Basin. The genetic diversity values of these populations suggest that the differentiation centre in the Carpathian Basin was as large and stable as the ones in the Southern Carpathians (Schmitt et al. 2007). Evidence for Central European refugia was also found for several other woodland species (Steward and Lister 2001). Recently, Schmitt and Haubrich (2008) indicated several Eastern European refugia of the large ringlet (*Erebia euryale*), a butterfly species strictly restricted to the European mountain coniferous forest zone. They concluded that the mountains of South-Eastern Europe were the most important reservoirs of the coniferous forests in Europe.

The Pleistocene glacial refugia of the European *Bombina* toads were located both in the "classical" refugial areas of the Appenins and the Balkans (core areas of *B. pachypus* and sub-species of *B. variegata*, Szymura et al. 2000; Canestrelli et al. 2006) as well as more to the North, in the Carpathians and the adjoining lowlands (Vörös et al. 2006; Hofman et al. 2007). Strong genetic evidences have been provided that *B. variegata* survived the LGM in the Carpathians. The mtDNA and allozyme data suggest two separate refugia. One clade probably had its refugium in the Southeastern edge of the Carpathians while the most likely refugium of the other clade was in the Southern Carpathians, where the haplotype diversity is the highest. However, the deep genetic divergence among European *Bombina* lineages suggests their pre-glacial origin.

The European populations of the moor frog are subdivided into three genetic lineages from which two are exclusively found in the Carpathian Basin. They, most probably, survived the LGM in the Carpathian Basin and have expanded to the

North to a rather limited extent only (Babik et al. 2004). The survival of *R. arvalis* in the Carpathian Basin was also demonstrated with fossil records (Venczel 1997). This view is supported by further recent data suggesting that this region might have been an important LGM refugial area in other amphibians and reptiles, as well. In the lizard *Z. vivipara*, a haplotype restricted to Northern Hungary and Austria was found in a species-wide survey of mtDNA variation (Surget-Groba et al. 2001). This ovoviviparous form shows peculiar karyotypic characters and demonstrates that the ovovivipary independently evolved in the Western European and Pannonian (*Z. vivipara pannonica*) populations (Kupriyanova et al. 2006; Surget-Groba et al. 2006). The suggested refugia of the adder (*V. berus*) would be in the Southern pre-Alpine lowland (Italian clade), in the "Illyrian" core area in the Western Balkan peninsula (Balkanic clade) and several core areas of the Northern clade: in the Carpathian region (Carpathian sub-clade), East of the Carpathians (Eastern subclades) and in a location between Great Britain and Poland (Central European subclade). The survival of two mitochondrial lineages in the Carpathian Basin throughout the LGM was postulated in this species, one East of the Alps, and the other in the North-Eastern part of the basin or even in the Eastern Carpathians (Carlsson 2003; Ursenbacher et al. 2006).

3.4 Refugia of Boreal Birds and Mammals in Eastern Central Europe

Comparison of phylogeographic structures in several Eurasiatic boreal species has shown that species associated with the taiga forest revealed essentially similar patterns. In the wood lemming (*Myopus schisticolor*), and also in most other boreal forest species, no substantial phylogeographic divisions across Northern Eurasia have been reported (Zink et al. 2002; Fedorov et al. 2008). The contraction of the range of these species to a single, probably Southern Siberian refugial area during the late Pleistocene followed by demographic expansion seems to be a general background for their shallow phylogeographic structure. The most important genetic discontinuity has usually been observed between the " Northern" Eurasiatic and the "Far East" clades. The limited distribution range of the South-eastern lineages suggests that their core areas ("Manchurian refugium" of de Lattin 1967) could not play an essential role in the post-glacial colonisation of Northern Eurasia by boreal forest species. A weak phylogeographic structure was also discovered in the flying squirrel (*Pteromys volans*), in which the divergence between the Far Eastern and Northern Eurasian groups may have been initiated during the Weichselian glaciation (Oshida et al. 2005). Mountain ranges in Southern Siberia and North-Eastern China may have isolated the Far Eastern group. Although the largest part of North-Eastern Europe and Siberia was not covered with ice during the last glaciation, most parts were overgrown by cold tundra-steppe ("mammoth steppe") under extremely cold and dry conditions (Svendsen et al. 1999, 2004; Simakova 2001; Schirrmeister et al. 2002). At that time, multiple isolated refugia of *P. volans* could be formed in

Eurasia. After the last glaciation, *P. volans* might have expanded throughout Northern Eurasia, along both sides of the Ural mountains.

Other boreal species show clear phylogeographical structures with well-differentiated populations in the coniferous forest zone of the South European mountains. They belong to rather different taxonomical groups. The brown bear complex can be mentioned as a classical example, (e.g. Taberlet and Bouvet 1994; Valdiosera et al. 2007). The discovery of 18,000-year-old charcoal of yew (*Taxus baccata*) and Scots pine (*P. sylvestris*) in Western Slovakia (Litynska-Zajac 1995) indicates that the climatic conditions at the Western Carpathians might have been suitable for the brown bear. The Eastern Carpathians served also as a refuge area for brown bears during the last glacial phase. Nevertheless, recolonization from the Eastern Carpathian refuge appears to have been less effective compared to the migration that began from the North-Western Carpathian refuge. This recolonization pattern can be explained by the more Northern position of the North-Western Carpathians (leading edge).

Southern European montane refugia have also been revealed in the capercaillie (*T. urogallus*). It was suggested that the Southern European *aquitanus* lineage had expanded throughout Europe before the LGM, and the Eastern, *urogallus* lineage expanded in Asia and North-Eastern Europe. During the LGM, the two lineages were restricted into separate refugia (*aquitanus*: Iberia and Balkans, *urogallus*: Southern Siberia). During the post-glacial re-forestation, the *urogallus* lineage replaced *aquitanus* in Europe and forced them to the South-west into their refugia in the Pyrenees and Cantabrian Mts. (Duriez et al. 2007).

Different phylogeographical structures have been revealed in the root voles (Brunhoff et al. 2003), lemmings (Fedorov et al. 1999), collared lemmings (Fedorov and Stenseth 2002), and common voles (Haynes et al. 2003; Fink et al. 2004). In these cases, the Ural Mountains separated the Northern European and Siberian lineages. Root voles in Europe form a Northern and a central mtDNA phylogroup. Fossil records from the last glaciation have demonstrated that collared lemmings (*Dicrostonyx*) and true lemmings (*Lemmus*) used to be the most common small mammals in the periglacial tundro-steppe of central Europe. These species were regularly accompanied by several vole species, including the root vole. These species assembly always remained North of the "classical" Southern European refugia. Thus, central European root vole populations can be considered the sources of Northward expansion during the deglaciation. This model is corroborated by the data demonstrating that the fragmented populations in the Netherlands, Slovakia and Hungary all belong to a single mtDNA phylogroup. Consequently, the threatened populations in Hungary, Austria, Slovakia and the Netherlands represent glacial and post-glacial relicts (van de Zande et al. 2000).

Bilton et al. (1998) show that the Mediterranean populations of *S. minutus*, *S. araneus* and *Clethrionomys* spp. did not contribute to the present-day gene pools of the Central and Northern European populations. This view has also been supported by the high rate of endemism in the Mediterranean bank vole phylogroups. In contrast with some forest rodents (*A. sylvaticus*, *A. flavicollis*), bank voles would not be restricted to the Southern forest refugia during the last glaciations. Therefore, bank vole populations might have survived the Quaternary glaciations in their Northern refugia. This may have resulted in their present complex phylogeographic pattern including multiple glacial refugia in central Europe, in the Mediterranean mountains and possibly also in several Eastern regions. Based on fossil records, the Carpathian region has been suggested as a glacial refugial area for this species. These refugia were most likely to be located near the Alps or in the Carpathians and possibly at the network of streams in the marginal areas of the Carpathian Basin. Phylogeographic analyses of the bank vole have already suggested that glacial refugia located in central and Eastern Europe made a major contribution to the modern population of this species in Europe (Deffontaine et al. 2005; Kotlík et al. 2006). The analysis of nucleotide diversity also demonstrates that these Eastern Central European regions acted as core areas of expansion for the Western lineage of the bank vole. Another refugial area was proposed in the regions of Eastern Romania, Western Azov Sea and the Crimea (Jaarola and Searle 2002).

4 Area Dynamics, Evolution and Diversity Patterns in the Carpathian Basin

4.1 Endemic Taxa and Autochtonous Evolution in the Carpathian Basin

The level of endemism generally correlates with the geological age of the refugia where relict-like taxa have been evolved and/or could survive. The Carpathian Basin belongs to the geologically young areas of Europe. Its relief developed under the influence of the Alpine orogenesis and by retreat of the Paratethys and the Pannonian inland sea. Moreover, the phylogeography of some freshwater invertebrates (e.g. Neritidae snails, see: Bunje 2007; Fehér et al. 2007) is clearly connected with the evolution of the Ponto-Pannonian water basin and of the Danube catchment area. In addition, there are several taxonomical groups with considerable proportion of endemic species, e.g. the land gastropods (Soós 1943) the earthworms (Lumbricidae: Csuzdi and Pop 2007) or some soil arthropods (e.g. Opiliones, Diplopoda: Korsós 1994; Collembola: Dányi and Traser 2007). Their core areas clearly coincide with the younger tertiary land masses within and near the Carpathian Basin.

Most endemic species are narrow specialists, inhabiting extreme habitats, such as thermal springs, karstic caves and karstic springs (Table 1). Several endemic troglobionta have been described in gastropods, pseudo-scorpions, harvestmen, spiders and springtails, often occurring within a single or a few caves of karstic mountains. Several species of earthworms, millipedes, centipedes and assels can be considered as holo-endemic species of the Western Transylvanian (Apuseni) mountains (Csuzdi and Pop 2007; Varga and Rakosy 2008).

Endemic terrestrial insects of the Carpathians are, as a rule, short-winged, flightless species such as the bush-crickets *Isophya, Poecilimon* spp.; some stenotopic relict grasshoppers (*Capraiuscola ebneri*, *Podismopsis transsylvanica*, *Uvarovitettix transsylvanica*, *Zubovskia banatica*; Kis 1965, 1980); numerous species of the

Table 1 Examples of stenotopic species of extreme habitats endemic to the Carpathians and the Carpathian basin

Taxonomic group	Species	Habitat type	Occurrence
Gastropoda	*Melanopsis parreysi*	Thermal springs	Baile Felix (Oradea)
Gastropoda	*Theodoxus prevostianus*	Thermal springs	Hungarian Middle Range
Gastropoda:	*Paladilhia hungarica*	Karstic water in caves	Mecsek Mts. (Abaliget)
Gastropoda	*Paladilhiopsis transsylvanica*	Karstic water in caves	Mţi. Apuseni
Gastropoda:	*Paladilhiopsis leruthi*	Karstic water in caves	Mţi. Apuseni
Amphipoda:	*Niphargus tatrensis*	Karstic water in caves	Calcareous mts. in the Northern Carpathians
Gastropoda	*Bythinella* (*Sadleriana*) *pannonica*	Karstic springs	Calcareous mts. in the Northern Carpathians
Palpigradi:	*Eukoenenia vagvoelgyii*	Karstic caves	Aggtelek karst, Slovakian karst
Araneae	*Centromerus chappuisi*	Karstic caves	Mţi. Apuseni
Araneae	*Troglohypantes racovitzai*	Karstic caves	Mţi. Apuseni
Araneae	*Nesticus biroi*	Karstic caves	Mţi. Apuseni
Diplopoda	*Haasea hungarica*	Karstic caves	Mecsek Mts. (Abaliget)
Diplopoda	*Hungarosoma bokori*	Karstic caves	Mecsek Mts. (Abaliget)
Isopoda	*Mesoniscus graniger*	Karstic caves	Aggtelek karst, Slovakian karst
Collembola	*Pumilinura dudichi*	Karstic caves	Aggtelek karst
Collembola	*Protaphorura kadici*	Karstic caves	Aggtelek karst
Collembola	*Arrhopalites dudichi*	Karstic caves	Aggtelek karst
Collembola	*Arrhopalites hungaricus*	Karstic caves	Aggtelek karst
Collembola	*Arrhopalites buekkensis*	Karstic caves	Bükk Mts. (N. Hungary)
Coleoptera, Carabidae	*Duvalius bokori*	Karstic caves	Slovakian karst
Coleoptera, Carabidae	*Duvalius gebhardti*	Karstic caves	Bükk Mts.
Coleoptera, Carabidae	*Duvalius microphthalmus*	Karstic caves	Calcareous mts. in the Northern Carpathians
Coleoptera, Carabidae	*Duvalius hungaricus*	Karstic caves	Aggtelek karst, Slovakian karst
Coleoptera, Carabidae	*Duvalius redtenbacheri*	Karstic caves	Mţi. Apuseni
Coleoptera, Carabidae	*Duvalius biroi*	Karstic caves	Mţi. Apuseni

Table 2 Examples of endemic species of forest soils in Transylvania

Lumbricida	*Allolobophora dacica*	Forest soil+litter	Mţi. Apuseni
Lumbricida	*Octodrilus* spp. (12 spp.!)	Forest soil+litter	Mţi. Apuseni
Pseudoscorpiones	*Neobisium brevipes*	Forest soil+litter	Mţi. Apuseni
Opiliones	*Ischyropsalis dacica*	Forest soil+litter	Mţi. Apuseni
Diplopoda	*Trachysphaera biharica*	Forest soil+litter	Mţi. Apuseni
Diplopoda	*Typhloiulus serbani*	Forest soil+litter	Mţi. Apuseni
Myriopoda	*Monotarsobius paucioculatus*	Forest soil+litter	Mţi. Apuseni
Myriopoda	*Monotarsobius spelaeus*	Forest soil+litter	Mţi. Apuseni
Myriopoda	*Monotarsobius burzenlandicus*	Forest soil+litter	Mţi. Apuseni
Myriopoda	*Lithobius dacicus*	Forest soil+litter	Mţi. Apuseni
Isopoda	*Cylisticus major*	Forest soil+litter	Mţi. Apuseni
Isopoda	*Trachelipus spinulatus*	Forest soil+litter	Mţi. Apuseni
Isopoda	*Biharoniscus racovitzai*	Forest soil+litter	Mţi. Apuseni

ground beetles (*Duvalius, Trechus, Patrobus, Morphocarabus* spp.) and weevils (e.g. *Otiorrhynchus* spp.). A bulk of these endemic taxa is confined to the Eastern and Southern Carpathians, to the Apuseni Mts.and to the mountains of Banat, which could preserve relict species (e.g. the tertiary relict gastropods *Chilostoma banaticum, Pomatias rivulare*) or some narrow endemic species of Isopoda and Diplopoda (Table 2) in refugia without permafrost phenomena during the last glaciations (Bennett et al. 1991; Krolopp and Sümegi 1995; Willis et al. 1995).

In the more mobile insect groups, the proportion of endemism lies rather low (e.g. in *Odonata* no endemic taxa occur in the Carpathian Basin). Most endemic Lepidoptera of the Carpathian Basin belong to Microlepidoptera, which have flightless females and are strictly specialised to some food plants living on halophyta in the saline grasslands of the Fertő-Neusiedlersee area (Kasy 1965) and those of the Great Hungarian plain (Kiskunság and Hortobágy). Endemic subspecies of *Geometridae* and *Noctuidae* evolved as peripheric isolates of turano-eremic species from the late-glacial, *kryoxerotic* periods, e.g. *Narraga tessularia kasyi, Saragossa porosa kenderesensis* (on food plants: *Artemisia santonicum, A. pontica*) and *Hadula dianthi hungarica* (on *Gypsophila muralis*). Some endemic taxa in the sandy areas of the Pannonian lowland are specialised predators or parasitoids, e.g. the spider *Dictyna szaboi* and the pompilid wasp *Cryptocheilus szabopatayi*. Further species described as endemics later proved widely-dispersed steppicolous species. In their majority, the endemics of the lower, hilly parts of the Carpathian Basin, however, represent thermophilous post-(inter?-) glacial relicts with connections to the Balkan Peninsula, Asia Minor or Southern Russia (e.g. *Apamea sicula tallosi* in warm-humid alluvial areas, *Dioszeghyana schmidtii schmidtii* and *Asteroscopus syriacus decipulae* in Pannonian xerothermic oak forests, *Polymixis rufocincta isolata* in the Villányi Mts.; *Chersotis fimbriola fimbriola, Euxoa vitta vitta, E. hastifera pomazensis* and *Cucullia mixta lorica* in the dolomitic areas of the Transdanubian Middle Range, *Chersotis fimbriola baloghi* in the Aggtelek Karst). Balkanic connections have also been observed in butterfly species, which are restricted to special, Pontic-Pannonian steppic food plants, e.g. *Plebeius sephirus* (feeding on *Astragalus exscapus, A. dasyanthus*), *Melitaea telona kovacsi* (on *Cirsium pannonicum*).

Several endemic Macrolepidoptera subspecies of the Carpathians belong to the genera *Erebia* and *Glacies* (Table 3). A few endemic taxa are only widespread in the Carpathians and in the neighbouring mountainous areas, e.g. *Aricia artaxerxes issekutzi, Photedes captiuncula delattini*, while others are confined to the Southern Carpathians, often with Balkanic connections: *Erebia cassioides neleus, Coenonympha rhodopensis schmidtii* (Varga 1975, 2003a). The subspecies of *Erebia melas* inhabit island-like, calcareous mountain stocks, *E. melas runcensis* in the Apuseni Mts., *E. melas melas* in the Cernei Mts. and *E. melas carpathicola* in the Eastern Carpathians). All these data clearly demonstrate that the Carpathians, especially the Eastern and Southern parts, together with the mountains of Western Transylvania (Apuseni Mts. and Banat) can be considered as core areas for the survival and autochtonous evolution in many invertebrate groups of limited mobility.

4.2 *The Carpathian–Balkanic Connections*

The close geological and faunal connections of the Carpathians suggest the existence of highly dynamic contacts with the mountains of the Balkan Peninsula during the Upper Pleistocene. These connections show a contrasting picture compared to the refugia of the Iberian and the Appenine peninsula which have been much more sheltered by the glaciated mountains of the Pyrenées and the Alps, respectively. At least two major arboreal refugia can be traced here: the Illyrian refugium related to the Dinarids and its foothills and the Carpatho-Dacian refugium related to the Carpathians and its foothills. Some areas attached to these refuges served as periodic habitats over climatically favourable periods. These are regarded as fluctuation zones (Varga 1995; Sümegi et al. 1998; Deli and Sümegi 1999).

Since the Carpathian Basin occupied a transitional position between the Balkanic refugia and the cold-continental tundro-steppe zone during the glacial periods, the post-glacial re-population of the Carpathian Basin proceeded (1) by long-distance dispersal from the more remote (atlanto- and ponto-) Mediterranean and Southern Continental refugia, and (2) also from some adjacent local survival areas, e.g. from North-Western Balkanic ("Illyrian") versus South Transylvanian ("Dacian") arboreal refugia. In such cases, the arrows of the Northwards dispersal of the South-Western and South-Eastern populations surround the arid central part of the basin. These components of the flora and fauna extend Northwards through the foothills of the Eastern Alps and Southwest-Pannonian hilly regions on the one hand, and through the hilly regions of the Banat area and the Western foothills of the Transylvanian "Island" mountains (Apuseni Mts.), on the other. In some cases, the populations of the South-Western and South-Eastern "strains" do not display any significant taxonomical differentiation, e.g. the silver lime (*Tilia tomentosa*) or

Table 3 Endemic subspecies of *Erebia* (Nymphalidae, Satyrinae) and *Glacies* (Geometridae, Ennominae) in the Carpathians (maps of *Erebia* spp.: Varga 1975; Varga and Schmitt 2008)

Name	Typical locality	Range
Erebia manto traianus (Hormuzaki, 1895)	Rareu (NE Carpathians)	NE-SE-S Carpathians
Erebia epiphron transsylvanica (Rebel, 1908)	Bucegi (SE Carpathians)	the whole Carpathian arc
Erebia pharte belaensis v. Goltz, 1937 (= *E. pharte carpatina* A. Popescu-Gorj & A. Szabó, 1986)	"Belaer Kalkalpen" (= Belanské Tatry) Rodna Mts. (NE Carpathians)	N+E Carpathians, S Carpathians
Erebia sudetica radnaënsis (Rebel, 1915)	Rodna Mts. (NE Carpathians)	NE Carpathians, S Carpathians
Erebia gorge rudkowskii (Bang-Haas, 1933)	Tatra Mts.	Tatra Mts., NE Carpathians
Erebia gorge fridericikoenigi (Varga 1998)	Piatra Craiului (SE Carpathians)	SE-S Carpathians
Erebia cassioides neleus (Freyer, 1833)	"Banater Gebirge"	S Carpathians
Erebia melas melas (Herbst, 1796)	"Transylvania" (Mţi Cernei)	Mţi Cernei (Domogled)
Erebia melas carpathicola (Alexinschi and Popescu-Gorj, 1959)	Haghmaşul Mare (E Carpathians)	E-SE Carpathians
Erebia melas runcensis (König, 1965)	Cheile Runcului (Mţi Apuseni)	Mţi Apuseni
Erebia pronoë regalis (Hormuzaki, 1937)	Bucegi (SE Carpathians)	SE Carpathians
Erebia pandrose roberti (Peschke, 1920).	Tatra Mts.	Tatra Mts, NE Carpathians
Erebia pandrose cibiniaca (Dannehl, 1927)	"Cibinsgebirge"	SE-S Carpathians
Glacies canaliculata schwingenschussi (Wehrli, 1921)	Tatra Mts.	higher mountains of the whole Carpathian arc
Glacies noricana carpathica (Schwingenschuss, 1915)	"Belaer Kalkalpen" (= Belanské Tatry), Mt. Bujacie	Belanské Tatry, Bucegi, Retezat
Glacies bentelii retyezatensis (Bartha, 1933) – subspecific status dubious!	Retezat Mts. (Lac. Zenoaga)	Retezat, Bucegi (ssp.?)
Glacies coracina dioszeghyi (Schmidt, 1930)	Retezat Mts. (Lac. Zenoaga)	Retezat, Bucegi (ssp.?)

some butterflies and moths (*Pyronia tithonus, Aplasta ononaria, Idaea nitidata, Zanclognatha tenuialis*). Much more evidence is provided by the re-population of the Carpathian Basin from different directions in the cases of vicarious pairs of closely related species or in subspecies of polycentric species. Such cases can mostly be mentioned in land gastropods, e.g. *Pomatias elegans – P. rivulare,*

Chilostoma illyricum – Ch. banaticum, or in flightless insects, e.g. short-winged Orthoptera: *Odontopodisma schmidti – O. rubripes, Isophya modestior – I. stysi* (Orci et al. 2005).

The Western Balkanic ("*Illyrean*") influences are most significant in the Southern and South-Western parts of Transdanubia. These areas are characterised by a humid sub-Mediterranean climate and do not have a significant rainfall deficit in the summer period. They belong to the belt of mesophilous zonal forests of *Fagion illyricum* and *Querco-Carpinion illyricum* and the Illyrean-Pannonian hardwood gallery forests (*Fraxino pannonicae-Ulmetum*) characterised by a richness in tertiary/inter-glacial relict, often geophytic plant species (Horvat et al. 1974).

The Transylvanian ("Dacian") influences are connected with the forested areas of the Eastern Carpathians and often transmitted by the Western Transylvanian mountains (Mahunka 1993, 2007; Varga 1989, 1995, 2003a,b). The occurrence of Dacian elements is typical of the Eastern part of the Hungarian Middle Range, especially in the higher parts of the volcanic Eperjes-Tokaj range and in the Karst areas of N Hungary and S Slovakia. Eastern Balkanic influences reach also the Hungarian Middle Range by relict-like occurrences of some Balkanic and Balkanic-Anatolian elements (e.g. *Noctuidae*: *A. syriacus* and *D. schmidtii*), especially in the warm foothill zone where the sub-Mediterranean influences are also significant. Relict occurrences of Dacian elements (bush-crickets: *Isophya stysi*, *Leptophyes discoidalis*, *Pholidoptera transsylvanica*; ground-beetles: *Carabus hampei ormayi*) have been recently discovered on the small, island-like volcanic hills of the Bereg lowland.

The influences of the Northern Carpathians are also significant in the NE part of the Hungarian Middle Range. There is a characteristic difference between the Eperjes-Tokaj volcanic chain on the one hand, and the limestone plateau of the Bükk Mts. and the N Hungarian karst on the other. The biotic contact of the Eperjes-Tokaj range with the Carpathians is young, obviously post-glacial, and can be characterised mostly by the presence of species, which are either typical of the montane forest belts of the Carpathians (e.g. numerous land snails: *Bielzia coerulans*, *Vestia gulo* and ground-beetles: *Carabus obsoletus*, *C. zawadszkyi*, *Abax schueppeli*) or widely dispersed in the Northern part of Central Europe, often having a Euro-Siberian distribution. The Bükk Mts., however, display an insular character. Its Carpathian and de-Alpine elements (e.g. land snails: *Spelaeodiscus triaria*, *Phenacolimax annularis*, the Geometrid moth *Entephria cyanata gerennae*) are isolated relicts. In the Aggtelek Karst area, the immediate contact with the higher limestone plateaus of Slovakia is combined with the occurrence of Carpathian (land snails: *B. coerulans*, *Cochlodina cerata, Trichia unidentata;* ground-beetles: *C. obsoletus*, *C. zawadszkyi*, *A. schueppeli*, *Trichotichnus laevicollis carpathicus*), boreal and xeromontane species at surprisingly low altitudes, influenced by the conspicuous meso-climatic and geomorphological features of this area. Some influences of the Northern and the Eastern Carpathians are to be observed at the NE marginal areas of the Pannonian lowland, i.e. along the upper course of the river Tisza and its tributaries (e.g. occurrence of land gastropods *Vitrea diaphana*, *B. coerulans*, *Balea stabilis*, *Perforatella dibothrion*, *P. vicina*).

4.3 The Forest: Steppe Dynamics in the Carpathian Basin

The Carpathian Basin belongs to the regions of Europe with the highest biodiversity (Williams et al. 1999). Due to its transitional position during the Quaternary climatic fluctuations, the overlap and accumulation of floristic and faunistic elements of contrasting habitats occurred here. This overlapping of different climatic provinces, enhanced by the varied relief, edaphic and hydrographic conditions, has resulted in suitable conditions for the survival of a large number of species belonging to different core areas and displaying various patterns of long-distance and short-distance re-populations.

It is typical for South-Eastern Central Europe that the large-scaled zonal settling of vegetation, characteristically developed in the East-European table-land, breaks down. In the Carpathian Basin, the concentric arrangement of vegetation belts is influenced by numerous climatic, orographic, hydrographic and edaphic factors (Varga 1995, 2003b). The forest-steppe, which is typical in the major, central lowland and hilly parts of the basin, is represented by a number of regional variants showing distinct geological, edaphic and meso-climatic characters. The forest, skirt and grassland compartments of each regional variant of the forest-steppe are highly intercorrelated. The Carpathians transmit (e.g. boreal) also filter certain different (e.g. steppic) biogeographical influences. Populations passing through the Carpathians will often be isolated and differentiated from the populations inhabiting other parts of their range of distribution.

The geographically transitional position of the Carpathian Basin resulted in a conspicuous mixture of faunal elements of diverse origins and geographical histories. The compartment structure of the vegetation complexes, typical for the Pannonian forest-steppe, has promoted the survival of very different faunal elements. Especially, the hilly areas of transitional climatic conditions surrounding the Pannonian lowland are populated by numerous, biogeographically important species and communities. The Southern, xerothermic slopes and foothills of the Hungarian Middle Range served both as refuges for thermo-xerophilous elements during several cold and cool-humid climatic phases of the Quaternary and as centres of their dispersal (Soó 1940, 1959; Wendelberger 1954, 1959; Zólyomi 1949, 1953, 1964). Thus, many thermophilous elements probably populated the Carpathian Basin not only by long-distance colonisation from remote, large glacial refuges, but also from numerous meso- or microclimatically favourable sites lying at the fluctuating borderlines of the Mediterranean refugial and periglacial belts. The varied and fine biostratigraphical structure of the Hungarian young Pleistocene, often characterised by a coexistence of forest and non-forest faunal elements (e.g. Jánossy 1979; Kordos 1977; Kretzoi 1969, 1977), provides evidences to support this view and demonstrates the transitional biogeographical character of this region during the whole time-span of the Quaternary period. New palynological data from the Eastern part of the Pannonian lowland (Bátorliget) also suggest the presence of forest refuges during the last glacial period (see also: 3.2.).

4.4 Relict Species with Long-Distance Disjunctions in the Carpathian Basin

There are several Mediterranean-Manchurian bicentric faunal elements with disjunct range occurring in the Carpathian Basin. The distribution of this species group is connected with the Ponto-Caspian waterway-system, and displays long-distance disjunctions from the vicarious Eastern Asiatic taxa, which often are only subspecifically differentiated (Lepidoptera: *Apatura metis metis* – *Apatura m. substituta*, *Chariaspilates formosarius hungaricus* – *Chariaspilates f. formosarius*, *Rhyparioides m. metelkanus* – *Rhyparioides metelkanus flavidus*, *Arytrura musculus* ssp. – *Arytrura m. musculus*). These and also some other species of this group (*Polypogon gryphalis*, *Herminia tenuialis*, *Diachrysia nadeja*) occur at the lower course of the Danube and the Drava as well as in swampy-boggy areas of the lowlands in Transdanubia, in the Banat and Eastern Hungary. The refugia of these faunal elements were probably along the lower courses of the Danube and its tributaries. Gallery forests of the Illyrian and the Pannonian type and alluvial wetlands accompanying the large rivers of the Pannonian lowland served as corridors for the Northwards expansion of these species.

Different types of long-distance disjunctions have been observed in the relict-like steppe and semi-desert species. The polytypic butterfly *Melanargia russiae*, which is widespread in West and Central Asia, South Siberia and in the mountains of Italy and the Balkan Peninsula, occurred locally – as *M. russiae clotho* – on tall-grass clearings of birch gallery forests of the sandy lowland in Kiskunság. Its extinction was partly due to the consequence of overcollecting, and mostly because of destroying the habitats (re-forestation with *Robinia pseudoacacia*). The habitats of *Chondrosoma fiduciarium* (Kasy 1965) are also tall-grass lowland and hilly steppes often mixed with slightly saline patches. Other species are confined to open dolomitic rocky swards (e.g. *Phyllometra culminaria, Lignyoptera fumidaria, C. mixta lorica*) or open sandy and rupicolous grasslands (*Oxytripia orbiculosa*, vanishing).

Eremic species are restricted to semi-desert-like habitats of the lowland with extreme edaphic conditions. This faunal type is represented by very few vertebrates: only the small rodent *Sicista subtilis* and the short-toed lark *Calandrella brachydactyla* belong to this group. Abundant examples can be found in strictly localised phytophagous insects, which are often connected with special halophytic plant communities. They are often represented by endemic Pannonian subspecies or allopatric sibling species of Turanian origin, e.g. the Noctuid moths *Saragossa porosa kenderesiensis* and *H. dianthi hungarica* or the Microlepidoptera: *Coleophora hungariae, C. klimeschiella, C. magyarica, C. peisoniella, Holcophora statices, Stenodes coenosana, Agriphila tersella hungarica*, etc. The dispersal of this species group could have taken place in the late glacial phases on the Pannonian lowland, with a subsequent isolation as a result of the post-glacial expansion of the forested belts.

Last but not the least, *xeromontane* elements are present also in the Carpathian Basin. Their two main groups are: the Mediterranean-xeromontane species, represented by a few vertebrates (e.g. *Monticola saxatilis* or the secondarily more expanded *Phoenicurus ochrurus*). A larger number of species, however, belongs to some insect groups, e.g. Noctuidae (*E. vitta, E. decora, E. birivia, Dichagyris candelisequa, Yigoga nigrescens, Chersotis margaritacea, Ch. fimbriola, Apamea platinea*, etc.) and Orthoptera (e.g. *Paracaloptenus caloptenoides*). The continental-xeromontane type is represented by some members of widely distributed Asiatic mountain steppe species as *Euxoa recussa, Dichagyris musiva* (*Noctuidae*) and by some relict-like inhabitants of the rocky dolomit grasslands as *P. culminaria, L. fumidaria* (*Geometridae*). It seems to be very probable that numerous genera, typical for the steppe biome, might have a xero-montane origin (especially Lycaenidae: e.g. the subgenus *Agrodiaetus* and other species groups of *Polyommatus* and *Plebeius;* Satyridae: *Chazara, Pseudochazara, Hyponephele;* Noctuidae: *Euxoa, Agrotis, Dichagyris, Yigoga, Rhyacia, Chersotis, Eugnorisma*, etc.). The same has to be supposed in the case of some endemic elements of the Pannonian flora (*Linum dolomiticum, Seseli leucospermum, Ferula sadleriana, Onosma tornense*, etc.).

5 Summary and Conclusions

The European fauna has traditionally been subdivided into a "*holothermic*" refugial and a "*holopsychric*" invasion type. The former type was differentiated according to the secondary subdivision of the Mediterranean refugial area. This view was confirmed and modulated by molecular results. A general conclusion was that temperate species mainly derive from Mediterranean refugial populations that underwent range expansion in the late glacial and early post-glacial periods. The other main group has been considered for a long time as a result of the "Siberian" invasion despite the evidences, which have revealed the taxonomical differentiation of North-Eastern "boreal" and Southern European montane populations. Several authors have suggested an additional mode of colonisation of Central and Northern Europe by *non-Mediterranean* populations, coming from one or more "continental" refugia. Fossil pollen data and macrofossil remains from the time of the Last Glacial Maximum indicate that several tree species remained in small favourable spots at the Southern edge of the steppe-tundra area. Research on small mammals has also questioned the universality of Mediterranean refugia. It was suggested that the Mediterranean "sanctuaria" in general were *not* also core areas of post-glacial expansion into deglaciated areas.

The Pleistocene glacial-interglacial cycles have resulted in the "*antagonistic dynamics*" of biota pertaining to contrasting macrohabitats. Principally, two basic types of zonal setting can be distinguished. The glacial periods have been characterised by a regressive fragmentation of wooded habitats and, consequently, by a broad contact of the tundra and the steppe zonobiomes with some forested "pockets" North of the refugial belt of the Mediterranean area. Transitional *zono-ecotones* developed at the forest-belt fringes: tundra-taiga and boreal forest-steppe. Pollen-based "tree-less tundra" models for Europe, North of the transverse mountain ranges, have repeatedly been questioned by researchers of the late Pleistocene mammalian fauna because the carrying capacity sufficient to feed numerous large herbivores demands a very productive environment ("mammoth steppe"). Thus, non-analogue communities were composed by mixing tundral, steppic and eremic-oreal elements. This boreal forest-steppe habitat type appears to have included also cold-tolerant species of temperate habitats.

The species of the boreal zone show a significant diversity of extension and of taxonomical structure of ranges. Comparison of phylogeographic structures in several Eurasiatic boreal species has shown that species associated with the taiga forest revealed essentially similar patterns. The number of exclusively European boreal and boreo-montane species is relatively low. However, a molecular biogeographical analysis of such species can unravel the European coniferous forest refugia. The existence of European coniferous forest refugia is also supported by the East–West subdivision of several boreal species. Temperate refugia in Europe during cold periods might not have been restricted to the three Southern peninsulas. These refugia were most likely to be located near the Alps or in the Carpathians and, possibly, at the network of streams in the marginal areas of the Carpathian Basin.

The level of endemism generally correlates with the geological age of the refugia where relict-like taxa have been evolved and/or could survive. The Carpathian Basin belongs to the geologically youngest areas of Europe. Its relief developed under the influence of the Alpine orogenesis and by retreat of the Paratethys and the Pannonian inland sea. There are, however, some taxonomical groups which show considerable proportion of endemic species (land gastropods, earthworms or some soil arthropods). Most endemic species are narrow specialists inhabiting extreme habitats, e.g. thermal springs, karstic caves and karstic springs. A bulk of these endemic taxa is confined to the Eastern and Southern Carpathians, to the Apuseni Mts. and to the mountains of Banat, which could preserve relict species or some narrow endemics in refugia without permafrost phenomena during the last glaciations. Since the Carpathian Basin occupied a transitional position between the Balkanic refugia and the cold-continental tundro-steppe zone during the glacial periods, the post-glacial re-population of the Carpathian Basin was preceeded (1) by long-distance dispersal from the more remote (atlanto- and ponto-) Mediterranean and Southern continental refugia, and (2) also from some adjacent local survival areas, e.g. from North-Western Balkanic ("Illyrian") versus South Transylvanian ("Dacian") arboreal refugia. In such cases, the arrows of the Northwards dispersal of the South-Western and South-Eastern populations surround the arid central part of the basin.

In the Carpathian Basin, the concentric arrangement of vegetation belts is influenced by numerous climatic, orographic, hydrographic and edaphic factors. The geographically transitional position of the Carpathian Basin resulted in a conspicuous mixture of faunal elements of diverse origins and geographical histories. The compartment structure of the vegetation complexes, typical of the Pannonian forest-steppe, has promoted the survival of very different faunal elements. Thus, many thermophilic elements probably populated the Carpathian Basin not only by long-distance colonisation from remote, large glacial refuges, but also from numerous meso- or microclimatically favourable sites lying at the fluctuating borderlines of the Mediterranean refugial and periglacial belts. Eremic species are restricted to semi-desert-like habitats of the lowland with extreme edaphic conditions. Abundant examples can be found in strictly localised phytophagous insects, which are often connected with special halophytic plant communities. They are often represented by endemic Pannonian subspecies or allopatric sibling species of Turanian origin. The dispersal of this species group could have taken place in the late glacial phases on the Pannonian lowland with a subsequent isolation as a result of the post-glacial expansion of the forested belts. Two main groups of *xeromontane* elements are present in the Carpathian Basin. A larger number of species of the Mediterranean-xeromontane species belongs to some insect groups, e.g. Noctuidae and Orthoptera. The continental-xeromontane type is represented by some members of widely distributed Asiatic mountain steppe species and by some relict-like inhabitants of the rocky dolomit grasslands. It seems to be very probable that numerous genera, typical of the steppe biome, might have a xeromontane origin.

Acknowledgements I am deeply indebted to the precursors of modern phylogeographic thoughts: to the late Gustaf de Lattin and Willy Reinig who inseminated the biogeography by genetic insights and shaped my ideas. The Alexander von Humboldt Foundation repeatedly supported my research fellowships in Germany. The survey of the faunal history of Hungary was partly supported by the grant NKFP-3 B/023/2004.

References

Babik W, Branicki W, Sandera M, Litvinchuk S, Borkin LJ, Irwin JT, Rafinski J (2004) Mitochondrial phylogeography of the moor frog, *Rana arvalis*. Mol Ecol 13:1469–1480

Bennett KD, Tzedakis PC, Willis KJ (1991) Quaternary refugia of North European trees. J Biogeogr 18:103–115

Bilton DT, Mirol PM, Mascheretti S, Fredga K, Zima J, Searle JB (1998) Mediterranean Europe as an area of endemism for small mammals rather than a source for Northwards post-glacial colonization. Proc R Soc Lond B Biol Sci 265:1219–1226

Björkman L, Feurdean A, Wohlfarth B (2003) Late-Glacial and Holocene forest dynamics at Steregoiu in the Gutaiului Mountains, Northwest Romania. Rev Palaeobot Palynol 124:79–111

Bordács S, Popescu F, Sladed D, Csaikl UM et al (2002) Chloroplast DNA variation of white oaks in Northern Balkans and in the Carpathian Basin. For Ecol Manage 156:197–209

Brunhoff C, Galbreath KE, Fedorov VB, Cook JA, Jaarola M (2003) Holarctic phylogeography of the root vole (*Microtus oeconomus*): implications for late Quaternary biogeography of high latitudes. Mol Ecol 12:957–968

Bunje PME (2007) Fluvial range expansion, allopatry and parallel evolution in a Danubian snail lineage (Neritidae: Theodoxus). Biol J Linn Soc 90:603–617

Canestrelli D, Cimmaruta R, Costantini V, Nascetti G (2006) Genetic diversity and phylogeography of the Apennine yellow-bellied toad *Bombina pachypus*, with implications for conservation. Mol Ecol 15:3741–3754

Carcaillet C, Vernet JL (2001) Comments on "the full-glacial forests of central and South-Eastern Europe" by Willis et al. Quat Res 55:385–387
Carlsson M (2003) Phylogeography of the adder (*Vipera berus*). Ph.D. Theses. Uppsala University
Csuzdi CS, Pop VV (2007) The earthworms of the Carpathian basin (in Hung). In: Forró L (ed) The genesis of the fauna of the Carpathian Basin (in Hung.). Natural History Museum, Budapest, pp 13–20
Dányi L, Traser G (2007) The springtails of Hungary (in Hung.). In: Forró L (ed) The genesis of the fauna of the Carpathian Basin (in Hung.). Natural History Museum, Budapest, pp 21–28
Deffontaine V, Libois R, Kotlík P, Sommer R, Nieberding C, Paradis E, Searle JB, Michaux JR (2005) Beyond the Mediterranean peninsulas: evidence of Central European glacial refugia for a temperate forest mammal species, the bank vole (*Clethrionomys glareolus*). Mol Ecol 14:1727–1739
Deli T, Sümegi P (1999) Biogeographical characterisation of Szatmár-Bereg plain based on the mollusc fauna. In: Hamar J, Sárkány-Kiss A (eds) The Upper Tisza Valley. Szeged, Tiscia Monograph Series, pp 471–477
Dennis RLH, Williams WR, Shreeve TG (1991) A multivariate approach to the determination of faunal structures among European butterfly species (Lep.: Rhopalocera). Zool J Linn Soc 101:1–49
Duriez O, Sachet JM, Menoni E, Pidancier N, Miquel Ch, Taberlet P (2007) Phylogeography of the capercaillie in Eurasia: what is the conservation status in the Pyrenees and Cantabrian Mounts? Cons Genet 8:513–526
Farcaş S, de Beaulieu JL, Reille M, Coldea G, Diaconeasa B, Goeury C, Goslar T, Jull T (1999) First 14C datings of Late Glacial and Holocene pollen sequences from Romanian Carpathes. C R Acad Sci Paris, Sciences de la vie 322:799–807
Farcaş S, Miclaus M, Tantau I (2004) Correlations between the actual hilly and plain vegetation from Transylvania and recent–sub-recent palynological spectra. Studii Cercet Biol Bistrita 9:99–111
Fedorov VB, Goropashnaya AV, Jarrell GH, Cook JA (1999) Phylogeographic structure and mitochondrial DNA variation in true lemmings (*Lemmus*) from the Eurasian Arctic. Biol J Linn Soc 66:357–371
Fedorov VB, Goropashnaya AV, Boeskorov GG, Fredga K (2008) Comparative phylogeography and demographic history of the wood lemming (*Myopus schisticolor*): implications for late Quaternary history of the taiga species in Eurasia. Mol Ecol 17:598–610
Fedorov VB, Stenseth NC (2002) Multiple glacial refugia in the North American Arctic: inference from phylogeography of the collared lemming (*Dicrostonyx groenlandicus*). Proc R Soc Lond B Biol Sci 269:2071–2077
Fehér Z, Varga A, Deli T, Domokos T, Szabó K, Bozsó M, Pénzes ZS (2007) Phylogenetic surveys on protected molluscs (in Hung.). In: Forró L (ed) The genesis of the fauna of the Carpathian Basin (in Hung.). Natural History Museum, Budapest, pp 183–200
Fekete G, Varga Z (2006) Pannonian vegetation (in Hung.). In: Fekete G, Varga Z (eds) Vegetation and wildlife of landscapes of Hungary (in Hung.). MTA Társadkut Közp, Budapest, pp 78–80
Feurdean A, Bennike O (2004) Late Quaternary palaeoecological and paleoclimatological reconstruction in the GutaiuluiMountains, NWRomania. J Quat Sci 19:809–827
Feurdean A, Wohlfarth B, Björkman L, Tantau I, Bennike O, Willis KJ, Farcas S, Robertsson AM (2007a) The influence of refugial population on Lateglacial and early Holocene vegetational changes in Romania. Rev Palaeobot Palynol 145:305–320
Feurdean A, Mosbrugger V, Onac BP, Polyak V, Veres D (2007b) Younger Dryas to mid-Holocene environmental history of the lowlands of NW Transylvania, Romania. Quat Res 68:364–378
Fink S, Excoffier L, Heckel G (2004) Mitochondrial gene diversity in the common vole *Microtus arvalis* shaped by historical divergence and local adaptations. Mol Ecol 13:3501–3514
Finnie TJR, Preston CD, Hill MO, Uotila P, Crawley MJ (2007) Floristic elements in European vascular plants: an analysis based on Atlas Florae Europaeae. J Biogeogr 34:1848–1872
Frenzel B (1992) European climate reconstructed from documentary data, methods and results. Gustav Fischer, Stuttgart

Fumagalli L, Hausser J, Taberlet P, Gielly L, Steward DT (1996) Phylogenetic structure of the Holarctic *Sorex araneus* group and its relationship with *S. samniticus*, as inferred from mtDNA sequences. Hereditas 125:191–199

Füköh L, Krolopp E, Sümegi P (1995) Quaternary malacostratigraphy in Hungary. Malacol Newsl 1(Suppl):1–219

Gömöry D, Paule L, Shvadchak IM, Popescu F, Sukowska M, Hynek V, Longauer R (2003) Spatial patterns of the genetic differentiation in European beech (*Fagus sylvatica* L.) at allozyme loci in the Carpathians and the adjacent regions. Silvae Genet 52:78–83

Guthrie RD (1990) Frozen fauna of the mammoth steppe: the story of blue babe. University of Chicago Press, London

Guthrie RD (2000) Origin and cause of the mammoth steppe: a story of cloud cover, woolly mammal tooth pits, buckles, and inside-out Beringia. Quat Sci Rev 20:549–574

Guthrie D, van Kolfschoten T (1999) Neither warm and moist nor cold and arid: the ecology of the Mid Upper Palaeolithic. In: Roebroeks W, Mussi W, Svoboda J, Fennema K (eds) Hunters of the golden age, 31. Analecta Prehistorica, Leiden, pp 13–20

Haase M, Misof B, Wirth T, Baminger H, Baur B (2003) Mitochondrial differentiation in a polymorphic land snail: evidence for Pleistocene survival within the boundaries of permafrost. J Evol Biol 16:415–428

Haynes S, Jaarola M, Searle JB (2003) Phylogeography of the common vole (*Microtus arvalis*) with particular emphasis on the colonization of the Orkney archipelago. Mol Ecol 12:951–956

Hertelendy E, Sümegi P, Szöör G (1992) Geochronological and paleoclimatic characterisation of Quaternary sediments in the Great Hungarian Plain. Radiocarbon 34:833–839

Hewitt GM (1996) Some genetic consequences of ice ages and their role in divergence and speciation. Biol J Linn Soc 58:247–276

Hewitt GM (1999) Post-glacial re-colonisation of European biota. Biol J Linn Soc 68:87–112

Hewitt GM (2000) The genetic legacy of the Quaternary ice ages. Nature 405:907–913

Hewitt GM (2001) Speciation, hybrid zones and phylogeography – or seeing genes in space and time. Mol Ecol 10:537–549

Hewitt GM (2004) Genetic consequences of climatic oscillation in the Quaternary. Phil Trans R Soc Lond B 359:183–195

Hofmann E (1873) Die Isoporien der europäischer Tagfalter. Dissertatio, Phil. Fac., Jena.

Hofman S, Spolsky Ch, Uzzel Th, Cogalniceanu D, Babik W, Szymura JM (2007) Phylogeography of the fire-bellied toads Bombina :independent Pleistocene histories inferred from mitochondrial genomes. Mol Ecol 16:2301–2316

Horvat I, Glavac V, Ellenberg H (1974) Die Vegetation Südosteuropas. Gustav Fischer, Stuttgart

Huntley B, Birks HJB (1983) An Atlas of past and present pollen maps for Europe: 0–13, 000 years ago. Cambridge University Press, Cambridge

Huntley B, Allen JRM (2003) Glacial environments III. Palaeovegetation patterns in late glacial Europe. In: van Andel TH, Davies SW (eds) Neanderthals and modern humans in the European landscape during the last glaciation. McDonald Institute for Archaeological Research, Cambridge, pp 79–102

Iversen J (1958) The bearing of glacial and interglacial epochs on the formation and extinction of plant taxa. In: Hedberg O (ed.): Systematics of to-day. Uppsala Univ. Arsskr. 6:210–215

Jaarola M, Searle JB (2002) Phylogeography of field voles (*Microtus agrestis*) in Eurasia inferred from mitochondrial DNA sequences. Mol Ecol 11:2613–2621

Jánossy D (1979) Subdivision of the Hungarian Pleistocene on the basis of vertebrate fauna (in Hung.). Akadémiai Kiadó, Budapest

Kasy F (1965) Zur Kenntnis der Schmetterlingsfauna des östlichen Neusiedlersee-Gebietes. Wissensch Arb Burgenland (Eisenstadt) 34:75–211

Kis B (1965) Zubovskia banatica, eine neue Orthoptera-Art aus Rumänien. Reichenbachia, Abhandl Mus Tierkunde, Dresden 5:5–8

Kis B (1980) Die endemischen Orthopteren in der Fauna von Rumänien. Muz Brukenthal, Stud Comunic 24:421–431

Kordos L (1977) A sketch of the biostratigraphy of the Hungarian Holocene (in Hung.). Földr Közl 25:144–160

Korsós Z (1994) Checklist, preliminary distribution maps and bibliography of Millipedes of Hungary (Diplopoda). Misc zool hung 9:29–82

Kostrowicki AS (1969) Geography of the Palearctic papilionoidea (Lepidoptera). Panstwowe Wydawnictwo Naukowe, Warszawa

Kotlík P, Deffontaine V, Mascheretti S, Zima J, Michaux JR, Searle JB (2006) A Northern glacial refugium for bank voles (Chlethrionomys glareolus). Proc Natl Acad Sci USA 103(40):14860–14864

Kretzoi M (1969) Sketch of the late Cenozoic terrestrial biostratigraphy of Hungary (in Hung.). Földr Közl 16(92):179–198

Kretzoi M (1977) Ecological Conditions of the "Loess" Period in Hungary as Revealed from the Vertebrate Fauna (in Hung.). Földr Közl 25:75–93

Krolopp E, Sümegi P (1995) Palaeoecological reconstruction of the late Pleistocene, based on loess malacofauna in Hungary. GeoJournal 36(2–3):213–222

Kupriyanova LA, Mayer W, Böhme W (2006) Karyotype diversity of the Eurasian lizard *Zootoca vivipara* (Jacquin, 1787) from Central Europe and the evolution of viviparity. In: Vences M, Köhler J, Ziegler T, Böhme W (eds): Herpetologia Bonnensis II. Proceedings of the 13th of the Congress Society European Herpetology pp. 67–72

Lattin G de (1949) Beiträge zur Zoogeographie des Mittelmeergebietes. Verh dtsch zool Ges Kiel Zool Anz Suppl 13 (1948): 143–151

Lattin G de (1957) Die Ausbreitungszentren der holarktischen Landtierwelt. Verh dtsch zool Ges Hamburg Zool Anz Suppl 21 (1956): 380–410

de Lattin G (1964) Die Verbreitung des sibirischen Faunenelements in der Westpaläarktis. Nat Mus 94:104–125

de Lattin G (1967) Grundriss der Zoogeographie. Gustav Fischer, Stuttgart

Litynska-Zajac M (1995) Anthracological analysis. In: Hromada J, Kozlowski J (eds) Complex of upper palaeolithic sites near. Jagellonian University Press, Moravany, Western Slovakia, pp 74–79

Magri D (2007) Patterns of post-glacial spread and the extent of glacial refugia of European beech (Fagus sylvatica). J. Biogeogr . doi:10.1111/j.1365-2699.2007.01803.x

Magri D, Vendramin GG, Comps B, Dupanloup I et al (2006) A new scenario for the Quaternary history of European beech populations: palaeobotanical evidence and genetic consequences. New Phytol 171:199–221

Mahunka S (1993) *Hungaromotrichus baloghi* gen. et sp. n. (Acari: Oribatida), and some suggestions to the faunagenesis of the Carpathian basin. Folia ent hung 54:75–83

Mahunka S (2007) The oribatid mites of the Carpathian basin (in Hung.). In: Forró L (ed) The genesis of the fauna of the Carpathian Basin. Natural History Museum, Budapest, pp 37–44

Magyari E, Jakab G, Rudner E, Sümegi P (1999) Palynologicaland plant macrofossil data on Late Pleistocene short term climatic oscillations in North-east Hungary. Acta Palaeobot, Suppl 2:491–502

Michaux JR, Magnanou E, Paradis E, Nieberding C, Libois RM (2003) Mitochondrial phylogeography of the woodmouse (Apodemus sylvaticus) in the Western Palearctic region. Mol Ecol 12:685–697

Nève G (1996) Dispersion chez un espèce à habitat fragmenté: Proclossiana eunomia (Lepidoptera, Nymphalidae). Dissertation, University of catholique de Louvain, Louvain-la-Neuve, pp. 1–128

Orci KM, Nagy B, Szövényi G, Rácz IA, Varga Z (2005) A comparative study on the song and morphology of *Isophya stysi* Cejchan, 1958 and *I. modestior* Brunner von Wattenwyl, 1882. Zool Anz 244:31–42

Oshida T, Abramov A, Yanagava H, Masuda R (2005) Phylogeography of the Russian flying squirrel (*Pteromys volans*): implication of refugia theory in arboreal small mammal of Eurasia. Mol Ecol 14:1191–1196

Petit RJ, Csaikl UM, Bordács S et al (2002) Chloroplast DNA variation in European white oaks. Phylogeography and patterns of diversity based on data from over 2600 populations. For Ecol Manage 156:5–26

Petit RJ, Aguinagalde I, de Beaulieu J-L, Bittkau C, Brewer S, Cheddadi R, Ennos R, Fineschi S, Grivet D, Lascoux M, Mohnty A, Müller-Starck G, Demesure-Musch B, Palmé A, Pedro Martin J, Rendell S, Vendramin GG (2003) Glacial refugia: hotspots but not melting pots of genetic diversity. Science 300:1563–1565
Pinceel J, Jordaens K, Pfenninger M, Backeljau T (2005) Rangewide phylogeography of a terrestrial slug in Europe: evidence for Alpine refugia and rapid colonization after the Pleistocene glaciations. Mol Ecol 14:1133–1150
Pokorny P, Jankovská V (2000) Long-term vegetation dynamics and the infilling process of a former lake (Svarcenberk, Czech Republic). Folia Geobot 35:433–457
Rácz GR, Gubányi A, Vozár Á (2005) Morphometric differences among root vole (Muridae: *Microtus oeconomus*) populations in Hungary. Acta Zool Hung 51:39–53
Rebel H (1931) Zur Frage der europäischen Faunenelemente. Ann Nathist Mus Wien 46:49–55
Reinig W (1950) Chorologische Voraussetzungen für die Analyse von Formenkreisen. Syllegomena biol. Festschr Kleinschmidt, Leipzig, pp 346–378
Rudner ZE, Sümegi P (2001) Recurring Taiga forest-steppe habitats in the Carpathian Basin in the Upper Weichselian. Quat Int 76(77):177–189
Rybnıcková E, Rybnıcek K (1991) In: Kovar-Eder J (ed.): Palaeovegetational development in Europe and regions relevant to its plaeofloristic evolution. Mus Nat Hist Vienna, pp. 73–79
Saarma U, Ho SYW, Pybus OG, Kaljuste M et al (2007) Mitogenetic structure of brown bears (*Ursus arctos* L.) in North-Eastern Europe and a new time frame for the formation of European brown bear lineages. Mol Ecol 16:401–413
Scharff RF (1899) The history of the European fauna. Walter Scott, London
Schirrmeister L, Siegert C, Kuznetsova T et al (2002) Paleoenvironmental and paleoclimatic records from permafrost deposits in the Arctic region of Northern Siberia. Quat Int 89:97–118
Schmitt T (2007) Molecular biogeography of Europe: Pleistocene cycles and postglacial trends. Front Zool 4:11. doi:10.1186/1742-9994-4-11
Schmitt T, Hewitt G (2004) The genetic pattern of population threat and loss: a case study of butterflies. Mol Ecol 13:21–31
Schmitt T, Haubrich K (2008) The genetic structure of the mountain forest butterfly *Erebia euryale* unravels the late Pleistocene and Postglacial history of the mountain forest biome in Europe. Mol Ecol 17:2194–2207
Schmitt T, Rákosy L, Abadjiev S, Müller P (2007) Multiple differentiation centres of a non-Mediterranean butterfly species in South-Eastern Europe. J Biogeogr 34:939–950
Schmitt T, Seitz A (2001) Intraspecific allozymatic differentiation reveals the glacial refugia and the postglacial expansions of European *Erebia medusa* (Lepidoptera, Nymphalidae). Biol J Linn Soc 74:429–458
Simakova AA (2001) The vegetation and mammoth distribution during the second half of the Late Pleistocene on the Russian Plain (33–17 ka). The World of Elephants, International Congress, Rome 2001, pp. 355–358
Sjörs H (1963) Amphi-Atlantic zonation, from Nemoral to Arctic. In: Löve A, Löve D (eds) North Atlantic biota and their history. Pergamon Press, Oxford, pp 109–126
Soó R (1940) Vergangenheit und Gegenwart der pannonischen Flora und Vegetation. Nova Acta Leopold 9:1–49
Soó R (1959) Streitfragen über die Entstehung der Vegetation des Alföld und ihre heutige Beurteilung. Földr Ért 8:1–26
Soós L (1943) The mollusca fauna of the carpathian basin (in Hung) természettud. Társulat, Budapest
Stegmann B (1932) Herkunft der paläarktischen Taiga-Vögel. Arch Naturgesch (NF) 1:392–397
Stegmann B (1938) Grundzüge der ornithogeographischen Gliederung des paläarktischen Gebietes. Fauna SSSR, Moscow Leningrad
Steward JR, Lister AM (2001) Cryptic Northern refugia and the origins of the modern biota. TREE 16:608–613
Sümegi P, Hertelendi E, Magyari E et al (1998) Evolution of the environment in the Carpathian Basin during the last 30,000 BP years and its effects on the ancient habits of the different

cultures. In: Költő L, Bartosiewicz L (eds.): the Archeological Inst. of the Hungarian Academy of Sciences (HAS), vol II, Budapest, pp. 183–197
Sümegi P, Krolopp E (2002) Quatermalocological analyses for modelling the Upper Weichselian palaeoenvironmental changes in the Carpathian basin. Quat Int 91:53–63
Sümegi P, Rudner ZE (2001) In situ charcoal fragments as remains of natural wild fires in the upper Würm of the Carpathian Basin. Quat Int 76(77):165–176
Surget-Groba Y, Heulin B, Guillaume C-P, Thorpe R et al (2001) Intraspecifi c phylogeography of *Lacerta vivipara* and the evolution of viviparity. Mol Phylogenet Evol 18:449–459
Surget-Groba Y, Heulin B, Guillaume C-P, Puky M, Semenov B, Orlova V, Kupriyanova L, Ghira I, Smajda B (2006) Multiple origins of viviparity, or reversal from viviparity to oviparity and the evolution of parity. Biol J Linn Soc 87:1–11
Svendsen JI, Astakhov VI, Bolshiyanov DY et al (1999) Maximum extent of the Eurasian ice sheets in the Barents and Kara Sea region during the Weichselian. Boreas 28:234–242
Svendsen JI, Alexanderson H, Astakhov VI et al (2004) Late Quaternary ice sheet history of Northern Eurasia. Quat Sci Rev 23:1229–1271
Szymura JM, Uzzell T, Spolsky C (2000) Mitochondrial DNA variation in the hybridising fire-bellied toads, *Bombina bombina* and *B. variegata*. Mol Ecol 9:891–899
Taberlet P, Bouvet J (1994) Mitochondrial DNA polymorphism, phylogeography and conservation genetics of the brown bear (Ursus arctos) in Europe. Proc R Soc Lond B 255:195–200
Taberlet P, Fumagalli L, Wust-Saucy A-G, Cosson J-F (1998) Comparative phylogeography and postglacial colonisation routes in Europe. Mol Ecol 7:453–464
Tantau I, Reille M, de Beaulieu J-L, Farcaş S, Goslar T, Paterne M (2003) Vegetation history in the Eastern Romanian Carpathians: pollen analysis of two sequences from the Mohos crater. Veget Hist Archaeobot 12:113–125
Tantau I, Reille M, de Beaulieu J-L, Farcaş S (2006) Late Glacial and Holocene vegetation history in the Southern part of Transylvania (Romania): pollen analysis of two sequences from Avrig. J Quat Sci 21:49–61
Tarasov PE, Volkova VS, Webb T (2000) Last glacial maximum biomes reconstructed from pollen and plant macrofossil data from Northern Eurasia. J Biogeogr 27:609–620
Ursenbacher S, Carlsson M, Helfer V, Tegelström H, Fumagalli L (2006) Phylogeography and Pleistocene refugia of the adder (*Vipera berus*) as inferred from mitochondrial DNA sequence data. Mol Ecol 15:3425–3437
Valdiosera CE, Garcia N, Anderlung C, Dalen L, Cregut-Bonnoure E, Kahlke R-D, Stiller M, Brandström M, Thomas MG, Arsuaga J-L, Götherström A, Barnes I (2007) Staying out in the cold: glacial refugia and mitochondrial DNA phylogeography in ancient European brown bears. Mol Ecol 16:5140–5148
Van de Zande L, Van Apeldoorn RC, Blijdenstein AF, De Jong D, Van Delden W, Bijlsma R (2000) Microsatellite analysis of population structure and genetic differentiation within and between populations of the root vole, *Microtus oeconomus* in the Netherlands. Mol Ecol 9:1651–1656
Varga Z (1975) Geographische Isolation und Subspeziation bei den Hochgebirgslepidopteren der Balkanhalbinsel. Acta Ent Jugosl 11:5–39
Varga Z (1977) Das Prinzip der areal-analytischen Methode in der Zoogeographie und die Faunenelemente-Einteilung der europäischen Tagschmetterlinge. Acta Biol Debr 14:223–285
Varga Z (1989) Die Waldsteppen des pannonischen Raumes aus biogeographischer Sicht. Düsseldorfer geobot Kolloq 6:35–50
Varga Z (1995) Geographical patterns of biodiversity in the Palearctic and in the carpathian basin. Acta Zool Hung 41:71–92
Varga Z (2002): Post-glacial dispersal strategies of Orthoptera and Lepidoptera in Europe and in the Carpathian basin. In: Proceedings of 13th International Colloquium of EIS, Leiden, pp. 93–105
Varga Z (2003a) The geographical distribution of high mountain macrolepidoptera in Europe. In: Nagy L, Grabherr G, Körner C, Thompson DBA (eds) Alpine biodiversity in Europe. Springer, Berlin, pp 239–257
Varga Z (2003b) Zoogeography of the Carpathian basin (in Hung.). In: Láng I, Bedő Z, Csete L (eds) Flora, Fauna, Habitats, vol III, Magyar Tudománytár. MTA Társadkut Közp, Budapest, pp 89–119

Varga Z (2006) Faunal history and biogeography of the Carpathian basin. In: Fekete G, Varga Z (eds) Vegetation and wildlife of landscapes of Hungary (in Hung.). MTA Társadkut Közp, Budapest, pp 44–75

Varga Z, Gyulai I (1978) Die Faunenelemente-Einteilung der Noctuiden Ungarns und die Verteilung der Faunenelemente in den Lokalfaunen. Acta Biol Debr 15:257–295

Varga Z, Peregovits L, Ronkay L (1989) Zoogeographical survey of the mongolian noctuidae fauna. Nota Lepid 12(Suppl. 1):63–64

Varga Z, Rákosy L (2009) Biodiversität der Karstgebiete im Karpatenbecken am Beispiel der Gross-Schmetterlingsfauna der Turzii-Schlucht bzw. des Aggteleker Karstgebietes. In: Proceedings of Congress SIEEC Cluj-Napoca (2007), (in press)

Velichko AA, Catto N, Drenova AN, Klimanov VA, Kremenetski KV, Nechaev VP (2002) Climate changes in East Europe and Siberia at the Late glacial–holocene transition. Quat Int 91:75–99

Venczel M (1997) Amphibians and reptiles from the lower Pleistocene of Osztramos (Hungary). Nymphaea 23–25:77–88

Voous KH (1960) Atlas of European birds. Nelson, London

Voous KH (1963) The concept of faunal elements or faunal types. Proceedings of the 13 international congress ornithological, pp. 1104–1108

Vörös J, Alcobendas M, Martínez-Solano I, García-París M (2006) Evolution of *Bombina bombina* and *Bombina variegata* (Anura: Discoglossidae) in the Carpathian Basin: a history of repeated mt-DNA introgression across species. Mol Phylogen Evol 38:705–718

Walter H, Breckle S-W (1986) Ökologie der Erde. Spezielle Ökologie der Gemäßigten und Arktischen Zonen Euro-Nordasiens. G, Fischer, Stuttgart

Walter H, Straka H (1970) Arealkunde. Floristisch-historische Geobotanik. Ulmer, Stuttgart

Wendelberger G (1954) Steppen, Trockenrasen und Wälder des pannonischen Raumes. Festschr Aichinger 1:574–634

Wendelberger G (1959) Die Waldsteppen des pannonischen Raumes. Veröff geobot Inst Rübel 35:77–113

Widmer A, Lexer CH (2001) Glacial refugia: sanctuaries for allelic richness, but not for gene diversity. TREE 16(No.6):267–269

Williams P, Humphries C, Araújo M (1999) Mapping Europe's biodiversity. In: Delbaere B (ed) Facts and figures on Europe's biodiversity, State and Trends 1998–1999. ECNC, Tilburg, pp 12–20

Willis KJ (1994) The vegetational history of the Balkans. Quat Sci Rev 13:769–788

Willis KJ, Sümegi P, Braun M, Tóth A (1995) The late Qarternary environmental history of Bátorliget, NE Hungary. Palaeogeogr Palaeoclimatol Palaeoecol 118:25–47

Willis KJ, Rudner ZE, Sümegi P (2000) Full glacial forests of Central and South Eastern Europe. Quat Res 53:203–213

Willis KJ, van Andel TH (2004) Trees or no trees? The environments of central and Eastern Europe during the Last Glaciation. Quat Sci Rev 23:2369–2387

Willis KJ, Niklas K (2004) The role of Quaternary environmental change in plant macroevolution: the exception or the rule? Phil Trans R Soc Lond B 359:159–173

Wohlfarth B, Hannon G, Feurdean A, Ghergari L, Onac BP, Possnert G (2001) Reconstruction of climatic and environmental changes in NW Romania during the early part of the last deglaciation (15,000–13,600 cal yr BP). Quat Sci Rev 20:1897–1914

Yurtsev BA (2001) The Pleistocene "Tundra-Steppe" and the productivity paradox: the landscape approach. In: Elias SA, Brigham-Grette, J (eds.): Beringean Paleoenvironments. Quat Sci Rev 20:165–174

Zink RM, Drovetski SV, Rohwer S (2002) Phylogeographic patterns in the great spotted woodpecker *Dendrocopos major* across Eurasia. J Avian Biol 33:175–178

Zólyomi B (1949) Die Mitteldonau-Florenscheide und das Dolomitphänomen. Bot Közl 39:209–224

Zólyomi B (1953) Die Entwicklungsgeschichte der Vegetation Ungarns seit dem letzten Interglazial. Acta Biol Hung 4:367–413

Zólyomi B (1964) Pannonische vegetationsprobleme. Verh Zool-Bot Ges Wien 103–104:144–151

Part II
Ancient Relicts

Review
The Dark Side of Relict Species Biology: Cave Animals as Ancient Lineages

Thorsten Assmann, Achille Casale, Claudia Drees, Jan C. Habel, Andrea Matern, and Andreas Schuldt

Abstract Due to their fascinating biology and phenomena belonging to the realm of scientific curiosity, cave animals have been objects of study for zoologists for numerous decades. This chapter not only focuses on the extremes (e.g., absence of eyes, specialization to extreme environments), but also serves as an introduction to understand the geographic distribution patterns and history of these highly diverse ecological groups with their relict characteristics. After an introduction to the subterranean environment in Sect. 1, we briefly review the biology and ecology of cave animals with their regressive and progressive evolutionary tendencies in order to understand the innate reasons for restricted distribution patterns (Sect. 2). In Sect. 3, we summarize the main aspects of our knowledge regarding the distribution of these species, especially in the Holarctic; and finally in Sect. 4, we highlight the relict characteristics of cave animal distribution and the ancient phylogenetic splits between cave and surface lineages.

T. Assmann (✉), C. Drees, A. Matern, and A. Schuldt
Institute of Ecology and Environmental Chemistry, Leuphana University Lüneburg, Scharnhorststraße 1, D-21335 Lüneburg, Germany
e-mail: assmann@uni.leuphana.de

A. Casale
Dipartimento di Zoologia e Genetica Evoluzionistica, Università di Sassari, Via Muroni 25, I-07100 Sassari, Italy

J.C. Habel
Musée national d'histoire naturelle, Section Zoologie des Invertébrés, 2160 Luxembourg, Luxembourg;
University Trier, Biogeographie, D-54296 Trier, Germany

J.C. Habel and T. Assmann (eds.), *Relict Species: Phylogeography and Conservation Biology*,
DOI 10.1007/978-3-540-92160-8_4, © Springer-Verlag Berlin Heidelberg 2010

1 The Subterranean Environment

The environmental conditions of caves and other cavities in massifs have frequently been investigated in conjunction with studies on cave animals. The first studies were carried out in the 19th century in a region known as Karst or Kars (in Slovenia), located between Ljubljana and Trieste. It was rapidly understood that the so-called karst processes tend to create specific geomorphology and landscape with crevices, caves, dolines, or karrens when the land is formed by soluble rocks. In the case of the Slovenian Karst, the land is formed by carbonated rocks (mainly, limestone and dolomite), the type of rock which is most important worldwide for the development of caves. Other soluble rocks, such as evaporate rocks (e.g., gypsum, salty rocks, some quarzites), can also undergo karst processes, sometimes, however, only slowly. Today, a landscape in which the main features of rock, soil, and hydrology are predominantly caused by dissolution of bedrock, is called "karst", from the Slovenian word for the above-mentioned region (Ginés and Ginés 1992).

Rock solubility is one of the main prerequisites for the formation of crevices and cavities. In the case of limestone, the main constituent is calcite which is not very soluble in pure water. However, under natural conditions water contains some acids, and this increases the solubility of calcite. The most important of these acids is carbonic acid, formed from water and carbon dioxide present in the atmosphere and soil (as a consequence of biological activity of roots, animals, protozoa, and bacteria). Other acids from the soil or from other rocks (e.g., humic acids) also have the potential to solubilize calcite. The dissolution of calcite and thus of limestone depends on several chemical dissociation and hydration reactions, both involving complex interplay of physico-chemical equilibria (Herman 2005) which, under specific conditions, cause secondary deposits in caves, e.g., stalagmites, flowstones, helictites, gypsum flowers, and columns.

The progressive dissolution of rock results in the development of fissures allowing infiltration of water into the inner parts of the karst massif, a process which tends to disarticulate the superficial hydrological network. This feeds the karst aquifer, and a subterranean hydrological network develops which promotes progressive dissolution and erosion of the inner parts of the karst massifs. Together with breakdown processes, rock solubility is the main force involved in the formation of characteristic karst phenomena such as caves, karrens, dolines, crevices, fissures, and karst springs (Culver and White 2005).

Caves can also be developed by other geological processes (e.g., pseudokarst, cryptokarst, or volcanism), but most caves are located on limestone in karstic areas. Since most work on cave animals has been conducted in limestone caves in temperate and Mediterranean climates, the following section also focuses on this type of habitat.

In general, the physical environment in caves varies much less than the surrounding and/or superficial habitats (Culver 1982; Culver et al. 1995; Wilkens et al. 2000; Culver and White 2005). The lack (or at least reduction) of diurnal and annual rhythms results in an environment without (or with strongly reduced) temporal coordination of animal activity. Especially, in the inner part of large caves or karst massifs the temperature varies less than 1°C in the course of the year and is, moreover, close to annual mean temperature (e.g., Juberthie 1969). Relative humidity and physico-chemical parameters of cave water are also relatively stable. Additionally, due to the decelerating effect of fissures and crevices, drainage basins in karst massifs have moderate flood peaks compared to other drainage basins (Culver 1982).

In the 1980s, the so-called superficial underground compartment (="milieu souterrain superficiel," MSS) was discovered in the Pyrenees and subsequently in many other regions with a temperate or Mediterranean climate. The MSS corresponds to both the underground compartment (caves) and the deeper horizons of the top soil. It consists of a system of heterogenous macroporal voids in rocky material underneath the soil (mesocavernous habitats; for a detailed description of the MSS, see Juberthie et al. 1980). Due to its intercrossing nature, between the underground and the surface, the MSS shows numerous intermediate conditions; however, the microclimate in some MSS seems to be more stable than in some large caves, due to the absence of air circulations which change in caves from winter to summer.

Due to the general absence of photoautotrophic organisms, cave communities and their food web are based on decomposers. Food enters caves only as organic matter in streams and vertical shafts (also via the MSS, see Gers 1998, as dissolved organic matter in water), and by feces of animals which regularly visit caves (e.g., bats). Only a few chemoautotrophically-based cave ecosystems have primary producers (e.g., Sarbu et al. 1996). In the underground and the superficial underground compartment, Collembola and Acarina function as "underground plankton" because of their basal position in many subterranean food webs. In general, decomposers (e.g., Diplopoda and Diplura) and first- and second-order predators (e.g., Coleoptera, Chilopoda, and Aranea) occur. For many cave animals, food is scarce and its availability is more stochastic and unpredictable in superficial habitats; its pattern of occurrence also seems to be more clustered (Gers 1998).

2 Biology and Ecology of Cave Animals

Cave animals are adapted to the above-mentioned special habitat conditions, which include reduced environmental variation and darkness. Loss of eyes, loss of wings in insects, and reduction of pigments are elements of regressive evolution; prolongation of extremities and increased sensory systems are elements of progressive evolution which occur in cave animals. The development of such features associated with cave life is known as troglomorphy (Christiansen 2005). In many cases, these features evolved convergently (e.g., the head flattening in cave vertebrates, a strong enlargement of elytra, but not of the abdomen – known as pseudophysogastry – in different lineages of beetles) and resulted in a special habitus of many cave-adapted animals (Figs. 1–3). These adaptations to the special environment are the reason why true cave animals are not able to survive outside of their underground habitats. The members of this ecologically characterized group which live exclusively in caves (or the MSS) during their whole life cycle are termed *troglobionts* or troglobites.

Fig. 1 European cave salamander or olm (*Proteus anguinus*), larva (drawn by Peter Schüle)

Troglophiles are capable of spending their entire life cycles either in underground or in surface habitats. *Trogloxenes* are vagrants or animals which can spend part of their life in caves, but are not able to develop in the underground. The latter terms are mainly used for terrestrial species. *Stygobionts* or stygobites live exclusively in subterranean waters (including cave and interstitial habitats), whereas stygophiles are able to develop in subterranean and in surface water bodies (cf. Culver and White 2005).

Regressive evolution, in particular, is not yet well understood. The simple neutral mutation hypothesis implies that regression is caused by random mutations in genes involved in the development of a given character. Under the relaxed selective pressure of the cave environment, these random mutations can accumulate and lead to the loss of the given character. Jeffery (2005, 2008) reviewed studies of regressive evolution in the Mexican Blind Cavefish (*Astyanax mexicanus*) and concluded that pigmentation regression is probably best explained by the neutral mutation hypothesis, whereas eye degeneration may be caused by adaptive evolution and pleiotropic effects of the genes also involved in the development of eyes. A review on the regressive evolution of cave fishes, in general, is given by Jeffery (2009).

As a consequence of eye reduction, the circadian rhythm of many cave animals is also reduced. Extremely evolved cave animals exhibit no or only weakly developed circadian rhythms of locomotory activity and resting. However, it was shown that light can induce clear activity rhythms in troglobite species (Lamprecht and Weber 1979, 1992).

Phenomena of progressive evolution characterize many cave animals. The prolongation of extremities in arthropods and the increased number of receptors for environmental information (especially, for olfactory stimuli) are features of most cave animals (Juberthie 1979; Culver 1982; Christiansen 2005). The spectrum of

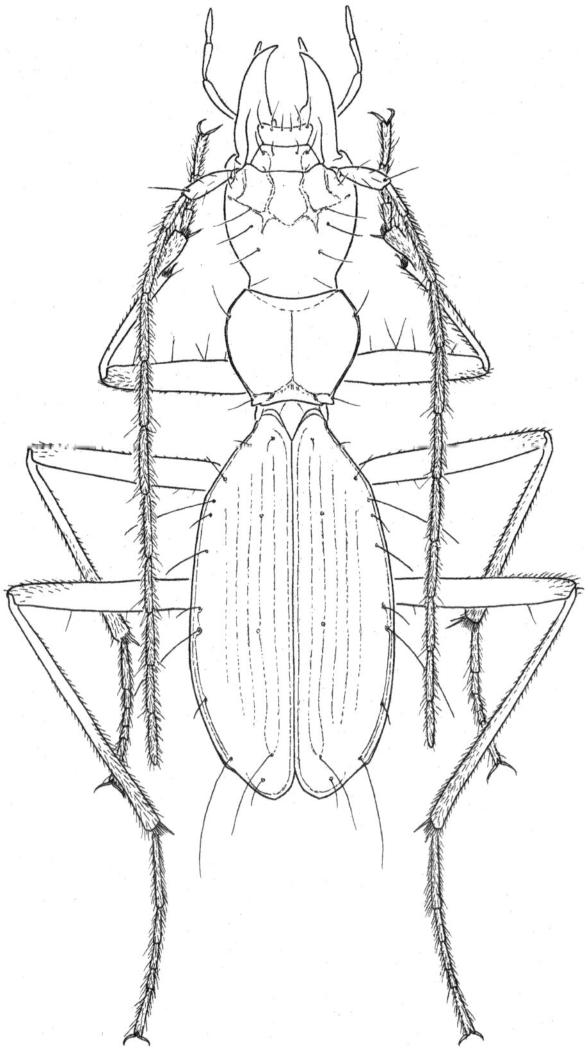

Fig. 2 Carabid cave beetle *Aphaenopidius kamnikensis*. Adapted from Assmann and Lompe (2006) and Drovenik et al. (2008), modified

consumed food ranges from feces, litter, and rotten wood to living prey. Some predators are specialized (e.g., *Neaphaenops tellkampfii* with its preference for eggs of the cave cricket *Hadenoecus subterraneus*, Kane and Poulson 1976), while some others are polyphagous (e.g., *Aphaenops crypticola*, Gers 1996). The reduced and stochastic availability of food seems to have resulted in selection pressure to a lower metabolic rate which allows both greater resistance to starvation and a comparably higher availability of energy for reproduction. Generally, but not in all

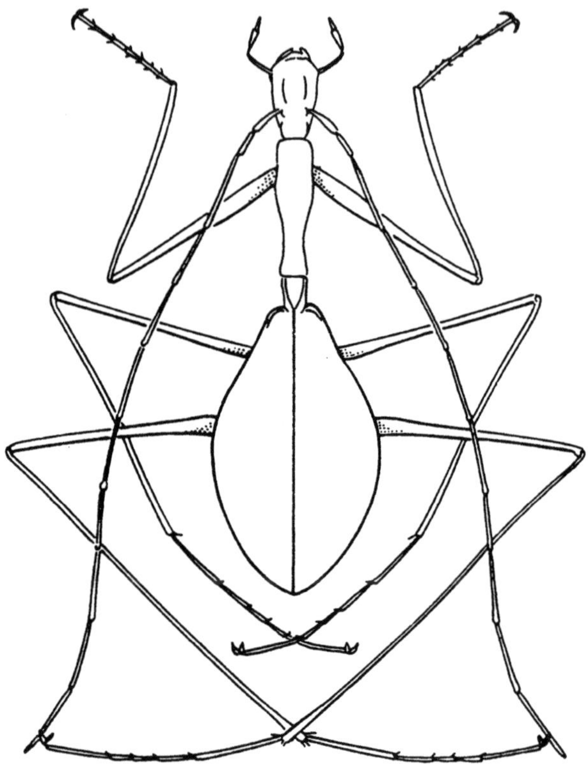

Fig. 3 Cholevid cave beetles *Antroherpon dombrowski*. Adapted from Jeannel (1924)

animal groups, cave species show a lower metabolic rate than their relatives that inhabit epigeic habitats (Vandel 1965; Culver 1982; Hüppop 2005).

The life span of many cave animals seems to be prolonged: The olm or European cave salamander (*Proteus anguinus*) can become older than a century (Durand 1998), and mark-recapture experiments with the microphthalmic cave ground beetle *Laemostenus schreibersi* in an Austrian cave estimated an age of more than 8 years for some individuals (Rusdea 1994, 2000). Moreover, cave animals have the tendency to produce fewer, but larger eggs and/or offsprings (Hüppop 2005). K-selection (cf. longevity, reproduction biology) seems to act on animals in the underground (see Stearns 1977 for a critique on the concept of r- and K-selection). Some members of the cholevid beetle genus *Speonomus* from the Pyrenees are an extreme example for the tendency to invest more yolk in fewer eggs. In the most extremely adapted species, only one egg ripens in a female at any given time. The hatched larva does not feed, grow, or molt, but pupates rapidly. Thus, the cycle of these species comprises only a single larval stage. This means that female investment in one egg enables the development of a whole beetle (Fig. 4). Transitions to the classical life cycle of cholevid beetles with three successive larval stages are also known, however (Glacon-Deleurance 1963).

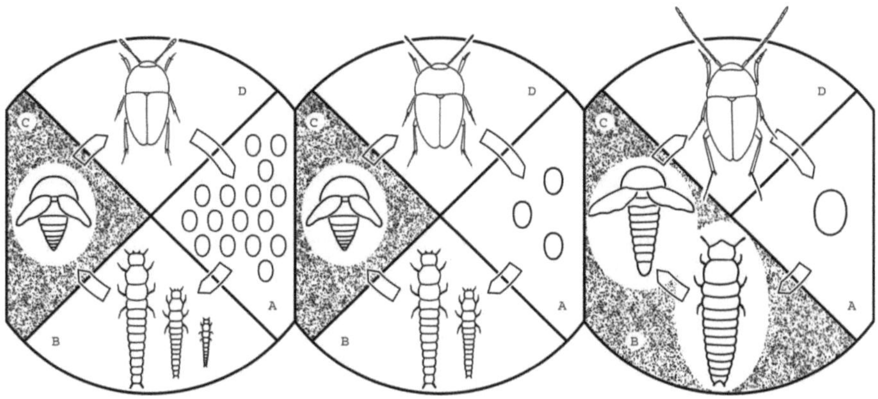

Fig. 4 Life cycles of different Cholevidae, Leptodirinae. Adapted from Glacon-Deleurance (1963) and Giachino et al. (1998), redrawn by Peter Schüle

3 Biogeography and Diversity Hotspots

The marked differences between the distribution patterns of aquatic and terrestrial troglobites were recognized by biologists several decades ago (Thienemann 1950; Vandel 1965; Lamoreux 2004). The Northern distribution limit of terrestrial troglobites does not reach as far North as that of aquatic troglobites, at least in Europe and North America. For example, members of *Niphargus*, an aquatic amphipod genus with blind species which inhabit groundwater in caves and other geological settings, are distributed Northwards as far as Ireland. Terrestrial forms (e.g., some genera and many species of blind cave Trechini beetles), however, reach their Northern limits in the so-called “massifs de refuge” at the Southern border of the Alps, in small and isolated refugia at the North-Western and North-Eastern border of the Alps, and in the Carpathians (Holdhaus 1954; Vandel 1965). The reasons for the different distribution patterns are still under discussion (cf. Lamoreux 2004).

The Southern edge of permafrost ground coincides notably with the Northern edge of the distribution range of terrestrial troglobites (Holdhaus line; see for a detailed discussion Drees et al. 2009). Due to the narrow adaptations to their habitats, terrestrial cave animals are not able to leave the massifs in which they live. Power of dispersal is, therefore, extremely restricted and it is highly probable that terrestrial cave animals survived the glacial periods within the massifs in which they are still distributed. Exceptions are known where the species have been able to colonize from nearby caves or massifs, probably using the MSS (cf. Daffner 1993).

Cave animals belong to the trophic groups of decomposers and carnivores, and their food chain depends on organic input from epigeic parts of the ecosystem. Survival of terrestrial cave animals during the glacial periods indicates a long-lasting persistence of ecosystems in the given regions.

The postglacial colonization of Eastern North America seems to be very complex, and many questions remain unanswered. However, it is known that glacial refuges were localized in areas Northwards up to close to the Laurentide Ice Sheet (Soltis et al. 2006), and terrestrial cave animals were restricted to non-glaciated areas, with a concentration in Alabama (Culver et al. 2000). Some regions of the world lack a terrestrial cave fauna (e.g., Northern Asia), while in others their existence has been discovered only in recent decades (e.g., in the tropics, Howarth 1983). A general overview of the worldwide distribution of terrestrial and aquatic cave animals is given by Juberthie and Decu (1998).

Bizarre disjunct distribution patterns in North America and in Europe can be found in higher amphibian taxa with troglobitic or at least troglophilic species: The family of cave salamanders (Proteidae) comprises the genus *Necturus* in surface waterbodies of North America and the European cave salamander or olm (*P. anguinus*) Southeast of the Alps in Europe; troglophilic lungless taxa inhabit disjunct distribution areas in Western North America (*Hydromantes*) and in the North-Western Mediterranean (*Speleomantes* and *Atylodes*; Durand 1998; Weisrock et al. 2005). Disjunct distributions are also known from groups of blind ground beetles which occur in North America and in the Palearctic (the so-called "série phylétique" of *Aphaenops*, Trechini; Casale and Laneyrie, 1982). Such disjunct small distribution areas on two continents are strong arguments for ancient splits in the phylogeny of the given groups. Without doubt, these taxa are ancient relicts of a former wider distribution.

One of the "hottest hotspots" of terrestrial cave fauna is located in the Western Palearctic realm, probably with a center in the Northern part of the Mediterranean region (Culver and Sket 2000; Gibert and Culver 2005; Culver et al. 2006). Ground beetles are a good example of "clustered" cave species richness: Casale et al. (1998) cite more than 50 genera of troglobitic ground beetles of the tribe Trechini from the Mediterranean realm (mainly from Southern Europe; and 13, 4, and 7 genera from North America, South America, and Australia, respectively). Another important hotspot seems to be Southeast Asia (more than 30 genera listed by Casale et al. 1998, and numerous newly described taxa, cf. Lorenz 2005). More than 600 of a total of more than 1,000 troglobitic trechine species from the Western Palearctic have been described.

The species number of some cave genera is low and there are even some monotypic genera (e.g., *Ameroduvalius*), whereas others are extremely rich in species, e.g., troglobitic ground beetles with >250 *Pseudanophthalmus* species in North America, and a similar number of *Duvalius* species in the Palearctic (Casale et al. 1998; Culver et al. 2003). Extraordinary species richness is also a feature of some aquatic cave animal groups, especially of the amphipod families Niphargidae and Crangonyctidae (both *Niphargus* and *Stygobromus* comprising about 200 species in Europe and North America, respectively, Peck 1998; Gibert and Culver 2005). In general, however, the species richness of aquatic fauna is lower than that of terrestrial species (Culver et al. 2003). The reasons for the different species richness patterns have not yet been conclusively ascertained (cf. Lamoreux 2004).

The invertebrates inhabiting the underground are still poorly studied, from a systematic point of view. Species richness of ground beetles inhabiting underground habitats in the Palearctic has not yet reached an asymptote (Fig. 5), whereas overall species richness of carabids in most European and North African countries reached a clear asymptote during the 20th century (Schuldt and Assmann 2009).

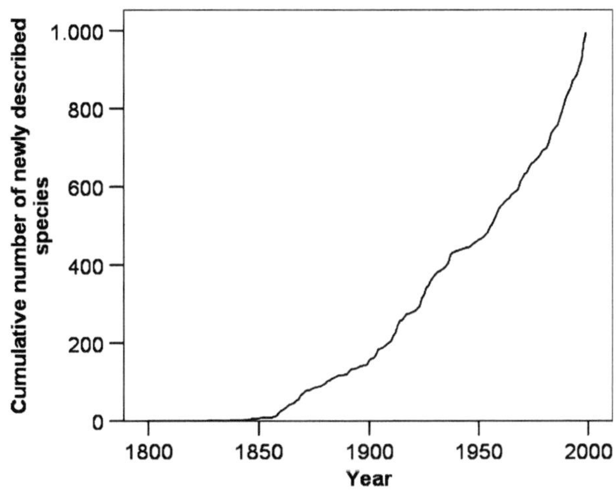

Fig. 5 Cumulative number of carabid taxa inhabiting underground habitats (troglobionts and troglophiles, including blind species from the superficial underground compartment) described from the Palearctic (1800–1999). Generated from the Catalog of Palearctic Coleoptera, Löbl and Smetana (2003)

4 Relict Status of Cave Animals

Biospeleologists agree that cave animals share a common ancestor with epigeic species. These ancestors lack features of progressive and regressive evolution which is related to the progress of troglobite evolution. According to the "Pleistocene-effect theory", ancestral forms are thought to have adapted ("taken refuge") to caves when "postglacial climates became warmer and drier, and thus inimical to ancestors preadapted to cool, wet microhabitats" (Vandel 1965; Culver 1982; Barr and Holsinger 1985 p. 331; Moldovan 2005).

This scenario does not seem to be true, however, from modern molecular studies on many cave animals. The DNA divergence of studied taxa indicates that biogeographic events which led to the separation of the gene pools date back some millions of years (Caccone and Sbordoni 2001; Porter 2007). Even surface and cave *Astyanax* populations seem to have diverged at least a few million years ago (Strecker et al. 2003; Strecker et al. 2004; 1.8–4.5 million years: Porter et al. 2007).

At least one population of the isopod *Asellus aquaticus* is an up to 3.9 million years old PrePleistocene "relict" of one lineage which invaded the underground. Other cave populations diverged during the glacial periods (Verovnik et al. 2004). However, despite the fact that we have molecular data only from a few related cave and surface populations/species we have to assume that the origin of many terrestrial and aquatic cave animals dates back to times (long) before the glacial periods.

Thanks to molecular studies, we are beginning to understand the evolution of speciation and colonization processes (e.g., Verovnik et al. 2005), but our knowledge of these processes in terrestrial species is still poor. Erwin (1985) developed the taxon pulse model to explain the divergence of the hyper-diverse group of ground beetles and Casale et al. (1998) adopted it to stress the origin of subterranean lineages. However, despite its sophisticated and outstanding character, a molecular test using arboreal and epigeic ground beetles did not support the taxon pulse model (Ober 2003).

Cave animals are treated as relict species for good reasons, because in many cases they are taxa with no ancestral stock in the immediate area (Barr 1985). Taking the long-lasting divergence since the PrePleistocene times between many cave and surface lineages into account, we can denominate many ancient relicts among cave animal species. This also seems to be true for taxa with striking disjunct distributions (see biogeography and biodiversity hotspots, Sect. 3): The split from the common ancestor of *Hydromantes* (North America) on the one hand and *Speleomantes* and *Atylodes* (Europe) on the other hand occurred during the Paleocene-Eocene and both lineages can be named relicts, not only because of their ancient origin but also because of their restricted distribution areas (Vieites et al. 2007).

Despite our understanding of the biogeographic history of cave animals, numerous questions are still unanswered. Some have already been mentioned above, such as the reasons for different distribution and species richness patterns between terrestrial and aquatic troglobionts or the reason for the existence of regions without cave fauna (South of permafrost grounds during glacial periods). In addition to these gaps in knowledge, the phylogeny of many vertebrate, and even more invertebrate, species remains to be studied. For example, phylogeny is still unresolved for the European cave salamander or olm (*P. anguinus*) (Weisrock et al. 2005) described as the first eyeless cave animal by Laurenti in 1768, even though this bizarre species was already studied by Cuvier, Lamarck, and even Darwin in his famous book on "The origin of species," already one and a half centuries ago.

References

Assmann T, Lompe A (2006) *Aphaenopidius* Müller, 1913. In: Müller-Motzfeld G (ed) Carabidae. Elsevier, Heidelberg, p 149

Barr TC (1985) Pattern and process in speciation of trechine beetles in Eastern North America (Coleoptera: Carabidae: Trechinae). In: Ball GE (ed) Taxonomy, phylogeny and zoogeography of beetles and ants. Dr. W. Junk Publishers, Dordrecht, pp 350–407

Barr TC, Holsinger JR (1985) Speciation in cave animals. Annu Rev Ecol Syst 16:313–337

Caccone A, Sbordoni V (2001) Molecular biogeography of cave life: a study using mitochondrial DNA from Bathysciine beetles. Evolution 55:122–130

Casale A, Laneyrie R (1982) Trechodinae et Trechinae du monde: tableau des sous-familles, tribus, séries phylétiques, genres, et catalogue général des espèces. Mémoires de Biospéologie 9:1–226

Casale A, Vigna-Taglianti A, Juberthie C (1998) Coleoptera Carabidae. In: Juberthie C, Decu V (eds) Encyclopedia biospeologica. Tome II. Société de Biospéologie, Moulis (France), pp 1047–1081

Christiansen K (2005) Morphological adaptations. In: Culver DC, White WB (eds) Encyclopedia of caves. Elsevier, Amsterdam, pp 386–397

Culver D (1982) Cave life: evolution and ecology. Harvard University Press, Cambridge, MA

Culver D, Sket B (2000) Hotspots of subterranean biodiversity in caves and wells. J Cave Karst Stud 62:11–17
Culver D, White WB (2005) Encyclopedia of caves. Elsevier, Burlington, MA
Culver DC, Kane TC, Fong DW (1995) Adaptation and natural selection in caves. Harvard University Press, Cambridge, MA
Culver DC, Master LL, Christman MC, Hobbs HH (2000) Obligate cave fauna of the 48 contiguous United States. Conserv Biol 14:386–401
Culver DC, Christman MC, Elliott WR, Hobbs HH, Reddell JR (2003) The North American obligate cave fauna: regional patterns. Biodivers Conserv 12:441–468
Culver DC, Deharveng L, Bedos A, Lewis JJ, Madden M, Reddell JR, Sket B, Trontelj P, White D (2006) The mid-latitude biodiversity ridge in terrestrial cave fauna. Ecography 29:120–128
Daffner H (1993) Die Arten der Gattung *Arctaphaenops* Meixner, 1925 (Coleoptera: Carabidae). Koleopterologische Rundschau/Coleopterological Revue 63:1–18
Drees C, Matern A, von Oheimb G, Reimann T, Assmann T (2009) Multiple glacial refuges of unwinged ground beetles in Europe – molecular data support classical phylogeographic models. In: Habel JC, Assmann T (eds) Relict species – phylogeography and conservation biology. Springer, Heidelberg. yet to update
Drovenik B, Weber F, Paill WG, Assmann T (2008) *Aphaenopidius kamnikensis* Drovenik, 1987 in Kärnten. Angewandte Carabidologie 8:73–76
Durand JP (1998) Amphibia. In: Juberthie C, Decu V (eds) Encyclopedia Biospeologica II. Société de Biospéologie, Moulis, pp 1215–1243
Erwin T (1985) The taxon pulse: a general pattern of lineage radiation and extinction among carabid beetles. In: Ball GE (ed) Taxonomy, phylogeny and zoogeography of beetles and ants. Dr. W. Junk Publishers, Dordrecht, pp 437–493
Gers C (1996) Stratégies alimentaires des Coléoptères troglobies du genre *Aphaenops* (Coleoptera, Trechinae). Mémoires de Biospéologie 17:35–45
Gers C (1998) Diversity of energy fluxes and interactions between arthropod communities: from soil to cave. Acta Oecol Int J Ecol 19:205–213
Giachino PM, Decu V, Juberthie C (1998) Coleoptera Cholevidae. In: Juberthie C, Decu V (eds) Encyclopaedia biospeologia. Société de Biospéologie, Moulis (France), pp 1083–1122
Gibert J, Culver D (2005) Diversity patterns in Europe. In: Culver D, White WB (eds) Encyclopedia of caves. Elsevier, Amsterdam, pp 196–201
Ginés A, Ginés J (1992) Karst phenomena and biospeleological environments. Monografias Museo Nacional de Ciencias Naturales 7:27–56
Glacon-Deleurance S (1963) Recherches sur les colépteres troglobies de la sous-famille des Bathyscininae. Ann Sci Nat 5:1–172
Herman JS (2005) Water chemistry in caves. In: Culver D, White WB (eds) Encyclopedia of caves. Elsevier, Burlington, MA, pp 609–614
Holdhaus K (1954) Die Spuren der Eiszeit in der Tierwelt Europas. Wagner, Innsbruck
Howarth FG (1983) Ecology of cave arthropods. Annu Rev Entomol 28:365–389
Hüppop K (2005) Adaptation to low food. In: Culver D, White WB (eds) Encyclopedia of caves. Elsevier, Amsterdam, pp 4–10
Jeannel R (1924) Monographie des Bathysciinae. Archives de zoologie expérimentale et générale 63:1–436
Jeffery WR (2005) Adaptive evolution of eye degeneration in the Mexican blind cavefish. J Hered 96:185–196
Jeffery WR (2008) Emerging model systems in evo-devo: cavefish and microevolution of development. Evol Dev 10:265–272
Jeffery WR (2009) Evolution of cave fish. Annual Review of Genetics 43:25–47
Juberthie C (1969) Relations entre le climat, le microclimat et les *Aphaenops cerberus* dans le grotte de Sainte-Catherine (Ariege). Annales de Spéléologie 24:75–104
Juberthie C (1979) L'évolution des coléopteres Trechinae souterrains (Coleoptera, Carabidae). In: Den Boer PJ, Thiele HU, Weber F (eds) On the evolution of behaviour in carabid beetles. Veenman and Zonen, Wageningen, pp 83–102

Juberthie C, Decu V (eds) (1998) Encyclopaedia Biologica, Tome II. Société Internationale de Biospéologie, Moulis and Bucarest

Juberthie C, Delay B, Bouillon M (1980) Extension du milieu souterrain en zone non calcaire: description d'un nouveau milieu et de son peuplement par les Coléoptères troglobies. Mémoires de Biospéologie 7:19–52

Kane TC, Poulson TL (1976) Foraging by cave beetles: spatial and temporal heterogeneity of prey. Ecology 57:793–800

Lamoreux J (2004) Stygobites are more wide-ranging than troglobites. J Cave Karst Stud 66:18–19

Lamprecht G, Weber F (1979) The regressive evolution of the circadian system controlling locomotion in cavernicolous animals. In: Den Boer PJ, Thiele HU, Weber F (eds) On the evolution of behaviour in carabid beetles. Veenman and Zonen, Wageningen, pp 69–82

Lamprecht G, Weber F (1992) Spontaneous locomation behaviour in cavernicolous animals: the regression of the endogenous circadian system. In: Camacho AI (ed) The natural history of biospeleology. Museo Nacional de Ciencias Naturales, Madrid, pp 225–262

Löbl I, Smetana A (eds) (2003) Catalogue of Palearctic Coleoptera Vol. 1: Archostemata, Myxophaga, Adephaga. Apollo Books, Stenstrup

Lorenz W (2005) A systematic list of extant ground beetles of the world (Coleoptera "Geadephaga": Trachypachydiae and Carabidae, incl. Paussinae, Cicindelinae, Rhysodinae). 2nd edition. Lorenz, Tutzing

Moldovan OT (2005) Beetles. In: Culver D, White WB (eds) Encyclopedia of caves. Elsevier, Amsterdam, pp 45–51

Ober KA (2003) Arboreality and morphological evolution in ground beetles (Carabidae: Harpalinae): testing the taxon pulse model. Evolution 57:1343–1358

Peck SB (1998) A summary of diversity and distribution of the obligate cave-inhabiting faunas of the United States and Canada. J Cave Karst Stud 60:18–26

Porter ML (2007) Subterranean biogeography: what have we learned from molecular techniques? J Cave Karst Stud 69:179–186

Porter ML, Dittmar K, Perez-Losada M (2007) How long does evolution of the troglomorphic form take? Estimating divergence times in *Astyanax mexicanus*. Time in Karst – 2007 (Slovenska Akademija Znanosti Umetnosti): 173–182

Rusdea E (1994) Population dynamics of *Laemostenus schreibersi* (carabidae) in a cave in Carinthia (Austria). In: Desender K, Dufrene M, Loreau M, Luff ML, Maelfait J-P (eds) Carabid beetles – ecology and evolution. Kluwer, Dordrecht, pp 207–212

Rusdea E (2000) Langlebigkeit der Adulten – ein Faktor, der die Populationsgröße stabilisieren kann. Erläutert am Beispiel einer Langzeituntersuchung am Höhlenlaufkäfer *Laemostenus schreibersi* (Coleoptera, Carabidae). Mitteilungen der Deutschen Gesellschaft für Allgemeine und Angewandten Entomologie 12:517–521

Sarbu SM, Kane TC, Kinkle BK (1996) A chemoautotrophically based cave ecosystem. Science 272:1953–1955

Schuldt A, Assmann T (2009) Environmental and historical effects on richness and endemism patterns of carabid beetles in the Western Palearctic. Ecography . doi:10.1111/j.1600-0587.2009.05763.x

Soltis DE, Morris AB, McLachlan JS, Manos PS, Soltis PS (2006) Comparative phylogeography of unglaciated Eastern North America. Mol Ecol 15:4261–4293

Stearns SC (1977) The evolution of life history traits: a critique of the theory and a review of the data. Ann Rev Ecol Syst 8:145–171

Strecker U, Bernatchez L, Wilkens H (2003) Genetic divergence between cave and surface populations of *Astyanax* in Mexico (Characidae, Teleostei). Mol Ecol 12:699–710

Strecker U, Faundez VH, Wilkens H (2004) Phylogeography of surface and cave *Astyanax* (Teleostei) from Central and North America based on cytochrome b sequence data. Mol Phylogen Evol 33:469–481

Thienemann A (1950) Die Binnengewässer XVIII: Verbreitungsgeschichte der Süsswassertierwelt Europas. Verlag Schweizerbart, Stuttgart

Vandel A (1965) Biospeleology: The biology of cavernicolous animals. Pergamon Press, Oxford

Verovnik R, Sket B, Trontelj P (2004) Phylogeography of subterranean and surface populations of water lice *Asellus aquaticus* (Crustacea: Isopoda). Mol Ecol 13:1519–1532
Verovnik R, Sket B, Trontelj P (2005) The colonization of Europe by the freshwater crustacean *Asellus aquaticus* (Crustacea: Isopoda) proceeded from ancient refugia and was directed by habitat connectivity. Mol Ecol 14:4355–4369
Vieites DR, Min MS, Wake DB (2007) Rapid diversification and dispersal during periods of global warming by plethodontid salamanders. Proc Natl Acad Sci USA 104:19903–19907
Weisrock DW, Harmon LJ, Larson A (2005) Resolving deep phylogenetic relationships in salamanders: analyses of mitochondrial and nuclear genomic data. Syst Biol 54:758–777
Wilkens H, Culver DC, Humphreys WF (2000) Subterranean ecosystems. Elsevier, Amsterdam

Relicts Within the Genus Complex *Astragalus/Oxytropis* (Fabaceae), and the Comparison of Diversity by Objective Means

Matthias Schlee, Markus Göker, GuidoW. Grimm, and Vera Hemleben

Abstract *Astragalus* and *Oxytropis* represent one of the largest angiosperm genus complexes. It is widely discussed why – and if at all – it contains so many species. Here, we discuss how the genetic divergence in a postglacial relict species, *Oxytropis pilosa*, can provide insights into the relative genetic variance of taxa. Methods are introduced to obtain sequence clusters of similar genetic diversity and to detect statistically significant differences. These approaches provide objective means to identify taxa that deserve the same protection efforts, independent of a fixed species concept. *O. pilosa* ITS sequences display an inter- and intra individual variance, comparable to clusters comprising numerous accepted species. Regarding genetic diversity, some *O. pilosa* populations represent relicts that deserve the same or a higher protection status than some species of *Astragalus*. We discuss the need of objectivity in identifying the evolutionary hot spots as well as the measures to define and protect them.

1 Introduction

Astragalus (~2.500 species; Podlech 1986) and *Oxytropis* (~300 species) are closely related genera, representing one of the largest angiosperm genus complexes. Taxonomic instability within the genus complex comprising both *Astragalus* and

M. Schlee (✉) and V. Hemleben
Center for Plant Molecular Biology (ZMBP) – General Genetics, University of Tübingen, Auf der Morgenstelle 28, D-72076 Tübingen, Germany
e-mail: matthias.schlee@gmx.de

M. Göker
DSMZ – German Collection of Microorganisms and Cell Cultures, Inhoffenstraße 7B, D-38124 Braunschweig, Germany

G.W. Grimm
Department of Palaeobotany, Swedish Museum of Natural History, Box 50007, S-10405 Stockholm, Sweden

J.C. Habel and T. Assmann (eds.), *Relict Species: Phylogeography and Conservation Biology*,
DOI 10.1007/978-3-540-92160-8_5, © Springer-Verlag Berlin Heidelberg 2010

Oxytropis results in a high number of similarities, which may differ in their assignment to the genus. In fact, the separation of both genera is based on a single morphological criterion only: *Astragalus* exhibits a mucronate keel along the adaxial side of the legume, whereas *Oxytropis* is characterized by an abaxial carinate legume (Candolle 1802; Barneby 1964; Chater 1968). Molecular studies encountered difficulties in providing sufficient resolution at the species level, independent of the sequence marker used (nuclear and/or chloroplast DNA spacer: Sanderson 1991; Sanderson and Doyle 1993; Wojciechowski et al. 1993, 1999; Kazempour Osaloo et al. 2003). Efforts in chemotaxonomy were promising (Williams and Barneby 1977), but have been only sporadically applied (Tin et al. 2007).

It has been widely discussed why so many species are recognized for *Astragalus* (Sanderson and Wojciechowski 1996), but it has seldom been addressed whether this large number of species, which is continuing to increase in local floras, optimally represent the natural relationships. *Astragalus* underwent a considerable radiation in all its main centers of biodiversity: Eurasia (mainly Southwest Asia), Africa, North America, and South America. The high number of morphologically defined species – and varieties – is caused by numerous endemics, often referred to as "relict" species because of their isolated mountainous stands and small populations. It is rare to find widespread *Astragalus* species. *Astragalus* covers various climates within its biogeographical areas, such as continental to arctic steppes, stony alpine slopes, and Mediterranean mountainous rocky places or even tropical African highland (Barneby 1964; Gillett 1964; Chater 1968). The preference for mountainous habitats, in conjunction with (post)glacial rapid environmental changes, may have facilitated differentiation processes leading to a split up into numerous species and varieties.

Oxytropis shows a quite similar distribution area in Eurasia and North America compared to *Astragalus*, but is absent from South America. This distribution is comparable to that of another large genus – *Pedicularis* (Sauer et al. 1990). *Oxytropis* exceeds more to the (arctic) North of the Northern Hemisphere than *Astragalus*, and some remarkable species such as *O. campestris* with its distinct sub species can be found on both continents. *Oxytropis pilosa* (Fig. 1) is a rare relict species exhibiting a highly fragmented distribution area. It is mainly found in the Pannonian Basin (Hungary to Austria) and in the Alps with prompt occurrences in Scandinavia (including the seashore of Gotland), the Baltic as well as in the Abruzzo (Hegi 1924; Meusel et al. 1965; M. Schlee, pers. observation). In a number of other accounts including *O. pilosa* (M. Schlee, G.W. Grimm, M. Göker, V. Hemleben, in prep.), the internal transcribed spacers (ITS1 and ITS2) of the nuclear ribosomal DNA (nrDNA) allowed reconstruction of detailed evolutionary pathways (Schlee et al. 2003; Denk and Grimm 2005; Grimm et al. 2007) as they are biparentally inherited, part of a multicopy gene region and undergo (to some degree incomplete) concerted evolution as well as frequent intragenomic recombination (Arnheim et al. 1980; Hemleben et al. 1988; Baldwin et al. 1995; Volkov et al. 2007). We detected significant intraspecific and intraindividual genetic variability in *O. pilosa* by cloning and sequencing a large number of nrDNA internal transcribed spacer sequences (ITS1 and ITS2; M. Schlee, G.W. Grimm, M. Göker,

Fig. 1 *Oxytropis pilosa*, xerothermic habitat on gypsum, Ascoli Piceno, Marche, Italy

V. Hemleben, in prep.). The variability found allows discussion of migratory and evolutionary patterns for this species, which will form the subject of an upcoming study.

Here, we make use of the large amount of sequence data obtained for *O. pilosa* to conduct a comparison of the genetic diversity of clades by objective means. This is done to find an alternative to classical methods, such as morphotaxonomy-based approaches for the identification of units that represent hot spots of biodiversity (here, genetic resources). We apply non-hierarchical single-linkage clustering to the combined *Astragalus* and *Oxytropis* ITS dataset after determining the distance threshold that optimally delimits *O. pilosa*. Thus, the resulting clusters represent estimates for species boundaries obtained by the use of an operational and fully-automated method treating *O. pilosa* as the standard. The clusters are expected to be uniform, given the distance threshold and the clustering algorithm (i.e., non-hierarchical single-linkage, NHSL). The obtained clusters are discussed with respect to the taxonomic affiliation (assigned species) of the comprised sequences. A character-based approach is used to determine whether the genetic diversity of selected clusters is significantly distinct. If one agrees that in conservation management major emphasis needs to be laid on equal protection efforts for taxa that are equally diverse (Watt 1947; Stock et al. 1999; Potthast 2000; Schlee 2004), it is crucial to compare diversity by appropriate means. Taking into account that taxonomic affiliations and taxonomic ranks often only poorly reflect relative genetic variance, we wonder whether they are sufficient to determine valuable genetic resources. Here, we thus address the question if biodiversity assessment based on substantial molecular data is superior to estimating diversity from data linked to the recognition of morphospecies and if the proposed techniques are valuable to estimate the relative need for conservation efforts as a means to discriminate protection units. Finally, we briefly discuss the conservation measures that are appropriate in the case of *O. pilosa*, once the most valuable genetic resources have been identified.

2 Material and Methods

In addition to our own dataset of *Oxytropis* ITS clones comprising the ITS1, 5.8S rDNA, and ITS2 regions, and flanking 18S and 25S rDNA (also including some European *Astragalus*; EMBL accession numbers AM401376 to AM401574; AM943374 to AM943384; FM205750 to FM205773), we added all ITS sequences stored in DDBJ, EMBL, and NCBI gene banks (sequences usually obtained by direct sequencing, not by cloning; downloaded on January 2008) for the genera *Astragalus* and *Oxytropis*. Solitary ITS1 and ITS2 sequences were combined if they belonged to the same source according to the original literature/description. The sequences were aligned using the POA program (Lee et al. 2002). The 5.8S rDNA was excluded from the analyses. Phylogenetic inference was performed under the maximum likelihood (ML) optimality criterion using RAxML 7.0 (Stamatakis 2006; Stamatakis et al. 2008). The program implements a new fast ML bootstrapping and subsequent search for the best topology. Duplicate sequences were eliminated prior to ML analyses. Tree inference and 100 bootstrap replicates were conducted under the CAT approximation (Stamatakis 2006), but final parameter optimization was done under a GTR+Γ model. A plain text (NEXUS) file containing the complete alignment, GenBank accession numbers, and information on the sets of identical sequences reduced by RAxML to a single one, respectively, is available at http://www.goeker.org/mg/clustering/.

To obtain a non-nested sequence classification based on the given distance (or similarity) threshold, we use NHSL clustering, as in the case of the popular tool blastclust (ftp://ftp.ncbi.nih.gov/blast/documents/blastclust.html; regarding hierarchical single-linkage clustering, see, e.g., Legendre and Legendre 1998, pp. 308–312). NHSL is based on the notion of a "link". A link is defined as any distance between two objects (here, sequences) that is smaller than or equal to the predefined threshold. NHSL starts by assigning the first object to the first cluster. For each of following objects in turn, it assigns that object to the same cluster than a previously clustered object if the distance between them is a link. If no such previously classified object is found, the current object is assigned to a new cluster. If several such previously classified objects are found that belong to distinct clusters, these clusters are joined.

Obviously, larger threshold values will result in larger but less numerous clusters, whereas small thresholds will lead to numerous small clusters. A biologically sensible threshold is usually not known a priori. However, a particularly well studied (and monophyletic) taxon (here, *O. pilosa*), can be used as a standard by determining the lowest possible threshold that results in all sequences obtained from that taxon being assigned to a single cluster. It is easy to implement an algorithm that calculates this value for a predefined group. Wirth et al. (1966, p. 61) who used similarities instead of distances define "...a similarity value, *c*, which is the largest fixed linking similarity value for which the cluster is still an interlinked aggregate of specimens". For each of the *n* objects within the group, the distance to the least distant object that belongs to the same group is determined; the largest of these *n* values represents the result, analogous to *c*. This is due to the fact that in NHSL, a *single* link is sufficient for an object to be assigned to a cluster. This algorithm and NHSL have been implemented in the program OPTSIL (Göker et al. 2009) which is available upon request. The determination of the standard threshold and NHSL relied on uncorrected distances (also called "Hamming" or "*p*" distances; e.g., Swofford et al. 1996, p. 455), which represent the relative number of deviations between two sequences. For downloaded sequences, we relied on the NCBI taxonomy provided in the same files to assign them to genera and species. The optimal threshold calculated for *O. pilosa* was then applied in NHSL.

A disadvantage of the NHSL approach is that clustering methods such as single-linkage clustering cannot be considered as valid methods of phylogenetic inference, mainly because lower pair wise distances (or higher similarities) do not necessarily indicate a closer phylogenetic relationship (e.g., Felsenstein 2004, pp. 165–167). This issue has led to the widespread avoidance of UPGMA clustering (Sokal and Michener 1958) in phylogenetic studies. However, the difficulties may be tempered or even disappear if non-nested clustering is applied and if species are to be distinguished, because within-species sequence dissimilarities are expected to be rather low. The same rationale applies to the usual arguments against uncorrected distances because unobserved, superimposed nucleotide substitutions only play a role if distances are large (e.g., Felsenstein 2004, p. 158).

Göker and Grimm (2008) used the well-known Shannon entropy formula (Shannon 1948) to calculate the character data of *hosts* (plant individuals) from the character data of their *associates* (cloned sequences obtained from the respective plant individuals). For all sequences belonging to the same individual, the entropy of each alignment column was calculated to represent the amount of genetic divergence within each individual using the program G2CEF designed and implemented by M. Göker (downloadable from http://www.goeker.org/mg/distance/). An alignment of length *n* will thus result in *n* corrected entropy values per group. Because the variance in nucleotide characters may depend on the number of sequences, each entropy value was corrected by division through the maximum possible entropy for the given number of associates, which is 0 in the case of a single associate. Accordingly, the corrected entropy values for individuals present with a single sequence only are undefined.

For each pair of groups, *n* differences between the *n* corresponding corrected entropy values can be determined. Subsequently, a non-parametric Wilcoxon signed-rank test or a parametric t-test (or any other appropriate statistical test) can be applied to assess whether the distribution of these differences significantly deviates from 0 and, thus, the two original entropy distributions are significantly different from each other. This procedure is similar to paired-site tests used to assess whether the scores of two phylogenetic trees are significantly different, given a sequence alignment of the same taxa (Felsenstein 2004, p. 364 ff.). Pair wise tests were conducted with R (R Development Core Team 2005) and restricted to those clusters obtained by NHSL that comprised more than five sequences.

3 Results

Our final data matrix comprised 437 ITS sequences of *Astragalus* and *Oxytropis* species comprising 464 characters from ITS1 and ITS2. Of these 437 sequences, 184 were exact sequential duplicates of others; 291 of the 464 characters were variable and 195 were parsimony-informative. After elimination of duplicates, the alignment analyzed under ML comprised 358 distinct alignment patterns, including a proportion of gaps and completely undetermined characters of 7.48%. The best ML tree obtained is shown in Fig. 2. The tree was rooted along an unresolved backbone polytomy giving respect to a bipartition that separated all *Astragalus* s.str. from the remaining *Astragalus* lineages and *Oxytropis* (in 98% of ML bootstrap replicates). The remaining backbone could be considered to be rather well-resolved, indicating several lineages with bootstrap support (BS) between 70 and 100. Towards the tip of the tree, a general decrease in BS was observed (details not shown). Particularly, the well-known *Neo-Astragalus* clade received little BS. Multiple accessions of the same *Astragalus* species (also listed in Table 1) were found within the same clade (with varying support), as were all ITS sequences of *O. pilosa* (BS=74, new cloned data plus one directly sequenced NCBI GenBank accession). In contrast, the two other extensively sampled species of *Oxytropis*, *O. arctica* and *O. campestris*, were intermixed with each other and with several other *Oxytropis* species.

Optimal distance thresholds calculated for the present species are shown in Table 1. We obtained an optimal value for *O. pilosa* of about 0.0069. Results of NHSL obtained with the optimal threshold are shown together with the ML tree in Fig. 2.

The largest clusters and their properties are shown in Table 2. *Oxytropis campestris* appeared inseparable from *O. arctica* as well as from a number of less intensively sampled *Oxytropis* species. The clusters 39, 42, 44 and 45 each contained a number of distinct *Astragalus* species, indicating that genetic divergence within many accepted *Astragalus* species is much lower than within either *O. pilosa* or the *O. arctica–O campestris* complex.

The results of the Wilcoxon signed-rank and t-tests, which are shown in Table 3, confirmed this conclusion. According to the Wilcoxon test, genetic divergence of *O. pilosa* was not significantly different from the *O. arctica–O. campestris* complex, but significantly different (lower) from *Astragalus* clusters 44 and 45 (species numbers given in Table 2), and significantly different (higher) from *Astragalus* cluster 42 and *Astragalus* cluster 39, respectively. According to the t-test, genetic divergence of *O. pilosa* was not significantly different from any of the other clusters, whereas the *O. arctica–O. campestris* complex was significantly more divergent than *Astragalus* clusters 42 and 39, respectively.

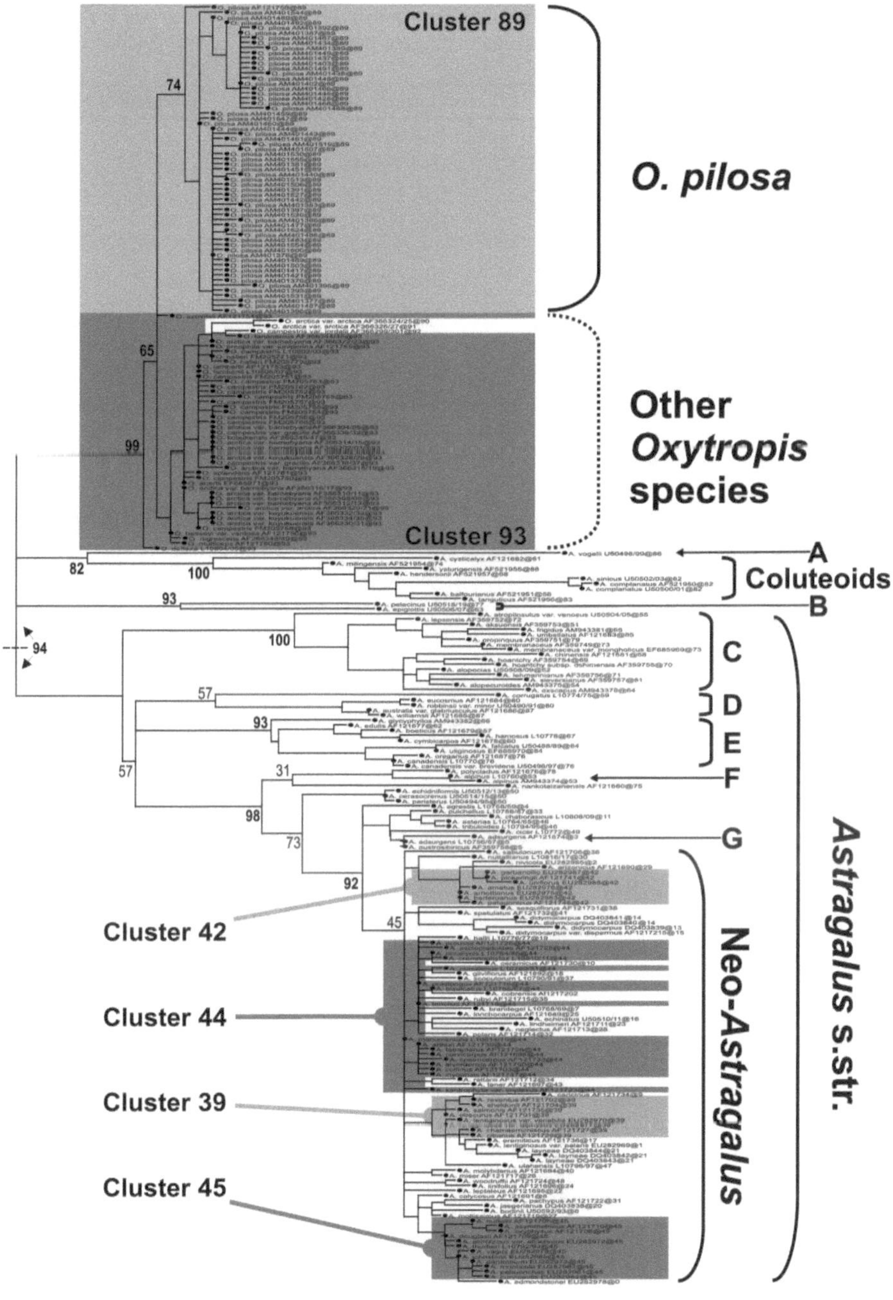

Fig. 2 ML phylogram based on ITS data of *Astragalus* and *Oxytropis*, with major clusters (see text) and clades annotated. Numbers at nodes indicate non-parametric bootstrap support based on 100 ML bootstrap replicates. Clusters (see text) are highlighted by distinct background shading, clade names are given at right. A, *A. vogelii;* B, *A. epiglottis* and *A. pelecinus;* C, clade of Eurasian *Astragalus;* D, *A. corrugatus* and allies*;* E, clade of Eurasian and North American *Astragalus*; F, *A. alpinus*; G, *A. adsurgens* (cf. Table 1)

Table 1 Optimal distance thresholds (i.e., smallest thresholds for which the group is retained in a non-hierarchical single-linkage clustering; NHSL) for the *Astragalus* and *Oxytropis* taxa present in the molecular dataset

Taxon	Optimal threshold	Number of sequences
A. hoantchy, clade C	0.01609195	3
A. didymocarpus, *Neo-Astragalus* clade	0.01395349	4
A. adsurgens, G	0.01173709	2
O. campestris	0.01162791	23
A. lentiginosus, *Neo-Astragalus* clade	0.00697674	6
O. arctica	0.00692841	18
O. pilosa	0.00688073	200
A. membranaceus, clade C	0.00458716	3
A. alpinus, F	0.00440529	2
A. layneae, *Neo-Astragalus* clade	0.00232558	3
O. halleri	0.00229358	7
A. complanatus, Coluteoid clade	0.00229358	2
Others	0	1

An optimal threshold of 0 is due to the presence of no more than a single sequence for the respective taxon or due to all sequences being identical. Single uppercase letters (C, F, G) after the taxon name refer to Fig. 2

Table 2 Clusters containing more than five sequences obtained by using non-hierarchical single-linkage clustering and the optimal distance threshold determined for *O. pilosa*

Cluster no.	Number of species	Number of sequences contained	Taxonomy	Average corrected entropy
89	1	200	*O. pilosa*	0.005865
93	16	59	*O. arctica*, *O. campestris*, *Oxytropis* spp.	0.009686
44	19	19	*A.* spp. (*Neo-Astragalus*)	0.008344
45	14	16	*A.* spp. (*Neo-Astragalus*)	0.008122
42	13	16	*A.* spp. (*Neo-Astragalus*)	0.002533
39	12	16	*A.* spp. (*Neo-Astragalus*)	0.002326

The reference numbers of each cluster, the numbers, and the taxonomic affiliations of the contained sequences, and the average corrected entropy values are shown

Table 3 Results of the Wilcoxon signed-rank tests (lower left half) and t-tests (upper right half) applied to the five largest clusters found by NHSL

Cluster no.	89	93	44	45	42	39
89	–	0.1466	0.4906	0.4288	0.08392	0.2013
93	0.4882	–	0.7527	0.6992	0.0005*	0.0279*
44	0.0099*	0.7671	–	0.9497	0.0291*	0.0694
45	0.0025*	0.4939	0.7987	–	0.0326*	0.0853
42	<0.0001*	<0.0001*	0.1052	0.0393*	–	0.9373
39	<0.0001*	<0.0001*	0.004*	0.0160*	0.3964	–

Differences significantly ($p \leq 0.05$) deviating from 0 are marked with an asterisk

4 Discussion and Conclusion

Genetic variability of populations within the *Astragalus-Oxytropis* complex has most often been addressed by techniques, such as allozyme polymorphisms (Karron et al. 1988; two endemic and two widespread species), RFLP (Kulshreshtha et al. 2004; one *Astragalus* species comprising six varieties, two *Oxytropis* species, pea as outgroup), AFLP, or RAPD (Travis et al. 1996; one variety). Only a few studies used cloned ITS sequences to compare populations (Dong et al. 2003, eight widespread species, one comprising two varieties, another comprising two sub species; Yip and Kwan 2006, one widespread species from 23 localities divided into at least two undefined varieties). Both studies found the marker highly conserved, and proposed the use of 5S rDNA spacers for the future. In contrast, our results indicate that if a sufficiently large number of cloned sequences are retrieved and if populations are sampled from a broad biogeographic range, valuable insights into the genetic divergence of populations can be obtained. Such substantially sampled taxa (here, *O. pilosa*) can be used as standards to optimize distance thresholds for non-hierarchical clustering algorithms. The adapted clustering techniques will result in a division of sequences into subsets of comparable genetic divergence as a basis to define conservation units. Statistical tests based on multiple sequence alignments allow us to compare the molecular diversity of different groups by objective means and to identify the significant differences regarding genetic variance. In contrast, traditional approaches such as AFLP, RAPD, RFLP, and microsatellite studies are often methodologically restricted to the species level and are statistically difficult to interpret (Felsenstein 2004, pp. 240f, 246f).

Results of our molecular comparison approach are in agreement with previous studies. For instance, the fact that *O. campestris* and *O. arctica* are hardly separable has been addressed by Jorgensen et al. (2003). Likewise, the considerably high genetic variance observed in our *O. pilosa* data confirms the interpretation of that species as a relict species. Particularly, the comparison of *O. pilosa* and the *Neo-Astragalus* species underlines how much divergence is preserved within the former. In fact, single populations of *O. pilosa* are as divergent as or even considerably more divergent than some *Astragalus* species. As a consequence, they deserve the same conservation efforts as the latter. This result is striking regarding the tendency in literature to suggest the protection of even single alleles within a variety of *A. cremnophylax* (Travis et al. 1996; for further examples, see the discussion in Soltis and Gitzendanner 1999), one of the numerous species included in our cluster 44 (Fig. 2 and Tables 1–3). Already the limited data available on *Astragalus* indicates that some species within the *Neo-Astragalus* clade (Table 1) are genetically more diverse than others, which should be taken into account in course of conservation.

In our view, not each unique individual and/or morphotype should per se be considered a relict or endemic. Biodiversity hotspots and genetic resources can be detected by objective means, similar to the approach presented here. Suitable nature conservation efforts should, therefore, also rely on molecular markers and algorithms that identify those relict stocks with the highest potential for ongoing speciation processes, and those stocks that are endangered to become extinct. A highly adaptive species complex will frequently evolve (and lose) new, unique, and locally restricted morphotypes, but will not necessarily significantly increase its genetic diversity (Table 3). As a consequence, such a species complex will show a high biodiversity as reflected by the number of accepted (endemic) "species", with only little genetic divergence between them. On the other hand, true relict species such as *O. pilosa* are morphologically and ecologically restricted or conservative (Hegi 1924), but can harbor a high genetic diversity (cf. Tables 1–3).

Focusing on the genetic diversity within the ITS regions may only be part of the story. Most likely, amplifying additional genes will result in additional insight into ecology and morphology. However, because of a lack of both time and funding, it is hardly realistic to do such multi gene sequencing for a broad sample size, and to detect and protect each individual that is a potential founder organism. Rather, one should focus on the identification and protection of at least some of the remaining true relicts in our rapidly changing world. Even though the cost of sequencing is continuously decreasing (because of recent advances in high-throughput techniques), there will always remain a trade-off between the number of loci amplified per individual and the number of individuals examined. As demonstrated above, it is easy to quantitatively determine units with similar genetic diversity, as long as a sufficient number of sequences is amplified for an appropriate reference taxon (here, *O. pilosa*). Accordingly, focusing on a single molecular locus, preferably one for which a large number of sequences are already deposited in public databases, is likely to remain the method of choice in many future studies. Once the most valuable genetic resources have been identified, using the methods described above (or any other appropriate means), protection measures for these resources can be laid out. Phytosociological studies can provide valuable insights into a species' habitat needs. Such studies are necessary, particularly if the genetic divergence within single individuals is so high that they deserve considerable conservation efforts, as confirmed by the present study for *O. pilosa*. In such a case, conservation should aim at preserving as much of the estimated genetic richness as is possible. The monitoring of relicts should be intensified, in general, using the combination of molecular and phytosociological methods to detect the evolutionary hotspots and endangered genetic resources as well as the means to protect them. Phytosociological surveys were conducted for *O. pilosa* using the Braun-Blanquet approach to quickly obtain detail-rich relevées (Schlee et al. 2003; M. Schlee, G.W. Grimm, M. Göker, V. Hemleben, in prep.). As a conclusion from these surveys, conservation efforts need to include more suitable land-use of the habitats. A management system should prefer a grazing over a cutting regime. The toxicity of *Astragalus* and *Oxytropis* species, in general, is here of lesser importance because food aversions can be conditioned if they do not already exist in grazing animals (Ralphs and Provenza 1999). Hairiness of the plants (e.g., *O. pilosa*) or spines (many *Astragalus* species) then remain the species' natural means of protection. If disturbances are too weak, networking concepts for habitats (e.g., NATURA 2000 for Europe) fail. However, if the natural linkage between habitats cannot be restored, networking is not necessarily desirable when speciation is in progress because it might be thwarted by artificial hybridization or homogenization among populations. The protection of ecological processes is, therefore, highly recommended instead of pursuing a networking concept at all cost. As a consequence, conservation efforts must hastily be improved to protect the disjunct populations of *O. pilosa* and to retain their genetic diversity as a valuable natural resource.

Acknowledgments Special thanks go to Angela Dressel and Karin Stögerer for technical assistance, and to Wilhelm Sauer for discussion. Reiner Schmahl, Eberhard Koch, Natalie Schmalz and Matthias Kropf, Ludmilla Barjaewa, Ludmilla Mantschenkowa, and Erich Kalmbach, Friedrich Schlee, Evgenia Zeidler-Schewzowa, and the University of Barnaul (Altai, Siberia) are kindly acknowledged for contributing plant material, GZU, LY, M, and MA for sending specimens for revision. Hospitality of the visited herbaria BC, BOLO, C, DR, E, FB, FI, FR, G, HBG, K, LE, LI, LJU, M, MPU (incl. SIGMA), P, RO, S, STU, TUB, UPS, W, WU, and Z was enjoyed with great pleasure. The project was supported by a grant of the "Landesgraduiertenförderung Baden-Württemberg"; grants for travelling were obtained from "Reinhold-und-Maria-Teufel-Stiftung" and the German Academic Exchange Service (DAAD).

References

Arnheim N, Krystal M, Schmickel R, Wilson G, Ryder O, Zimmer E (1980) Molecular evidence for genetic exchanges among ribosomal genes on nonhomologous chromosomes in man and apes. Proc Natl Acad Sci USA 77(12):7323–7327

Baldwin BG, Sanderson MJ, Porter JM, Wojciechowski MF, Campbell CS, Donoghue MJ (1995) The ITS region of nuclear ribosomal DNA: a valuable source of evidence of angiosperm phylogeny. Ann Mo Bot Gard 82(2):247–277

Barneby RC (1964) Atlas of North American *Astragalus*. Mem NY Bot Gard 13(1–2):1–1188

Candolle AP de (1802) [Astragalogia] Augustini-Pyrami Decandolle, Astragalogia nempe Astragali, Biserrulae et Oxytropidis, nec non Phacae, Colutae et Lessertiae, Historia Iconibus Illustrata. Joann. Bapt. Garnery, Parisiis

Chater AO (1968) *Astragalus*. In: Tutin TG, Heywood VH, Burges NA, Moore DM, Valentine DH, Walters SM, Webb DA (eds) Flora europaea, vol 2, 1st edn, Rosaceae to Umbelliferae. Cambridge University Press, Cambridge, pp 108–124

Denk T, Grimm GW (2005) Phylogeny and biogeography of *Zelkova* (Ulmaceae s.str.) as inferred from leaf morphology, ITS sequence data and the fossil record. Bot J Linn Soc 147(2):129–157

Dong TTX, Ma XQ, Clarke C, Song ZH, Ji ZN, Lo CK, Tsim KWK (2003) Phylogeny of *Astragalus* in China: molecular evidence from the DNA sequences of 5S rRNA spacer, ITS, and 18S rRNA. J Agric Food Chem 51(23):6709–6714

Felsenstein J (2004) Inferring phylogenies. Sinauer Associates, Sunderland, MA

Gillett JB (1964) *Astragalus* L. (Leguminosae) in the highlands of Tropical Africa. Kew Bull 17(3):413–423

Göker M, Grimm GW (2008) General functions to transform associate data to host data, and their use in phylogenetic inference from sequences with intra-individual variability. BMC Evol Biol 8:86

Grimm GW, Denk T, Hemleben V (2007) Evolutionary history and systematics of *Acer* section *Acer* – a case study of low-level phylogenetics. Plant Syst Evol 267(1–4):215–253

Hegi G (1924) Flora von Mitteleuropa. IV. Band, 3. Teil: Dicotyledones 2. Teil Leguminosae – Tropaeolaceae, 1st edn. Carl Hanser, München

Hemleben V, Ganal M, Gerstner J, Schiebel K, Torres RA (1988) Organization and length heterogeneity of plant ribosomal RNA genes. In: Kahl G (ed) Architecture of eukaryotic genes. VCH, Weinheim, pp 371–383

Jorgensen JL, Stehlik I, Brochmann C, Conti E (2003) Implications of ITS sequences and RAPD markers for the taxonomy and biogeography of the *Oxytropis campestris* and *O. arctica* (Fabaceae) complexes in Alaska. Am J Bot 90(10):1470–1480

Karron JD, Linhart YB, Chaulk CA, Robertson CA (1988) Genetic structure of populations of geographically restricted and widespread species of *Astragalus* (Fabaceae). Am J Bot 75(8):1114–1119

Kazempour Osaloo S, Maassoumi AA, Murakami N (2003) Molecular systematics of the genus *Astragalus* L. (Fabaceae): Phylogenetic analyses of nuclear ribosomal DNA internal transcribed spacers and chloroplast gene *ndhF* sequences. Plant Syst Evol 242(1–4):1–32

Kulshreshtha S, Creamer R, Sterling TM (2004) Phylogenetic relationships among New Mexico *Astragalus mollissimus* varieties and *Oxytropis* species by restriction fragment analysis. Weed Sci 52(6):984–988

Lee C, Grasso C, Sharlow M (2002) Multiple sequence alignment using partial order graphs. Bioinformatics 18(3):452–464

Legendre P, Legendre L (1998) Numerical ecology, 2nd English edn. Elsevier, Amsterdam

Meusel H, Jäger E, Weinert E (1965) Vergleichende Chorologie der zentraleuropäischen Flora. Band 1, Teil 1: Text; Band 1, Teil 2: Karten. Gustav Fischer, Jena

Podlech D (1986) Taxonomic and phytogeographical problems in *Astragalus* of the Old World and South-West Asia. Proc R Soc Edinb B 89:37–43

Potthast T (2000) Funktionssicherung und/oder Aufbruch ins Ungewisse? Anmerkungen zum Prozeßschutz. In: Jax K (ed) Funktionsbegriff und Unsicherheit in der Ökologie. Beiträge zu einer Tagung des Arbeitskreises “Theorie” in der Gesellschaft für Ökologie vom 10. bis 12. März 1999 im Heinrich-Fabri-Institut der Universität Tübingen in Blaubeuren (= Theorie in der Ökologie; 2). Peter Lang, Frankfurta a.M, pp 65–81

Ralphs MH, Provenza FD (1999) Conditioned food aversions: principles and practices, with special reference to social facilitation. Proc Nutr Soc 58(4):813–820

R Development Core Team (2005) R: A language and environment for statistical computing. R foundation for statistical computing, Vienna, Austria

Sanderson MJ (1991) Phylogenetic relationships within North American *Astragalus* L. (Fabaceae). Syst Bot 16(3):414–430

Sanderson MJ, Doyle JJ (1993) Phylogenetic relationships in North American *Astragalus* (Fabaceae) based on chloroplast DNA restriction site variation. Syst Bot 18(3):395–408

Sanderson MJ, Wojciechowski MF (1996) Diversification rates in a temperate legume clade: are there “so many species” of *Astragalus* (Fabaceae)? Am J Bot 83(11):1488–1502

Sauer W, Gebert S, Shu-Kun C, Li H (1990) Beitrag zur Kenntnis chinesischer *Pedicularis*-Arten (Scrophulariaceae). Razprave IV. razreda SAZU [Razprave / Slovenska Akademija Znanosti in Umetnosti, Razred za Naravoslovne Vede = Dissertationes / Academia Scientiarum et Artium Slovenica, Classis 4]. Ljubljana 31(19):291–316

Schlee M (2004) Probleme der Erhaltung biologischer Vielfalt in der Kulturlandschaft – Ökologische Schäden durch verfehlte Pflegekonzepte. In: Potthast T (ed) Ökologische Schäden – begriffliche, methodologische und ethische Aspekte (= Theorie in der Ökologie; 10). Peter Lang, Frankfurt a.M, pp 95–120

Schlee M, Sauer W, Hemleben V (2003) Molekulare und pflanzensoziologische Analyse von pontisch-pannonischen Reliktarten aus wärmebegünstigten Saum-Gesellschaften Süddeutschlands und benachbarter Gebiete. Nova Acta Leopold NF 87(328):379–387

Shannon C (1948) A mathematical theory of information. Bell System Tech J 27:379–423

Sokal RR, Michener CD (1958) A statistical method for evaluating systematic relationships. Univ Kansas Sci Bull 38(Pt 2, No 22):1409–1438

Soltis PS, Gitzendanner MA (1999) Molecular systematics and the conservation of rare species [Review]. Conserv Biol 13(3):471–483

Stamatakis A (2006) RAxML-VI-HPC: maximum likelihood-based phylogenetic analyses with thousands of taxa and mixed models. Bioinformatics 22(21):2688

Stamatakis A, Hoover P, Rougemont J (2008) A rapid bootstrap algorithm for the RAxML web-servers. Syst Biol 57(5):758–771

Stock M, Eskildsen K, Gätje C, Kellermann A (1999) Evaluation procedure for nature conservation in a national park – a proposal for the protection of ecological processes. Z Ökologie Naturschutz 8(1):81–95

Swofford DL, Olsen GJ, Waddell PJ, Hillis DL (1996) Phylogenetic inference. In: Hillis DM, Moritz C, Mable BK (eds) Molecular systematics. Sinauer Associates, Sunderland, MA, pp 407–514

Tin MMY, Cho C-H, Chan K, James AE, Ko JKS (2007) *Astragalus* saponins induce growth inhibition and apoptosis in human colon cancer cells and tumor xenograft. Carcinogenesis 28(6):1347–1355

Travis SE, Maschinski J, Keim P (1996) An analysis of genetic variation in *Astragalus cremnophylax* var. *cremnophylax*, a critically endangered plant, using AFLP markers. Mol Ecol 5(6):735–745

Volkov R, Komarova N, Hemleben V (2007) Ribosomal DNA in plant hybrids: inheritance, rearrangement, expression. Syst Biodiver 5(3):261–276

Watt AS (1947) Pattern and process in the plant community. J Ecol 35(1/2):1–22

Williams MC, Barneby RC (1977) The occurrence of nitro-toxins in North American *Astragalus* (Fabaceae). Brittonia 29(3):310–326

Wirth M, Estabrook GF, Rogers DJ (1966) A graph theory model for systematic biology, with an example for the Oncidiinae (Orchidaceae). Syst Zool 15(1):59–69

Wojciechowski MF, Sanderson MJ, Baldwin BG, Donoghue MJ (1993) Monophyly of aneuploid *Astragalus* (Fabaceae): Evidence from nuclear ribosomal DNA internal transcribed spacer sequences. Am J Bot 80(6):711–722

Wojciechowski MF, Sanderson MJ, Hu JM (1999) Evidence on the monophyly of *Astragalus* (Fabaceae) and its major subgroups based on nuclear ribosomal DNA ITS and chloroplast DNA *trn*L intron data. Syst Bot 24(3):409–437

Yip PY, Kwan HS (2006) Molecular identification of *Astragalus membranaceus* at the species and locality levels. J Ethnopharmacol 106(2):222–229

Relict Populations and Endemic Clades in Palearctic Reptiles: Evolutionary History and Implications for Conservation*

Ulrich Joger, Uwe Fritz, Daniela Guicking, Svetlana Kalyabina-Hauf, Zoltan T. Nagy, and Michael Wink

Abstract The phylogeographic history of eight species complexes of West Palearctic reptiles was reconstructed using mitochondrial and nuclear markers. Cryptic endemic taxa were detected in the Southern European peninsulas (Iberia, Southern Italy/Sicily, and Greece) as well as in North Africa, Anatolia, Iran, and the Caucasus. These endemics are mainly of Tertiary or early Pleistocene age. Only part of them can be categorized as relicts of a former, more widely distributed taxon, having survived in ice age refugia; others have probably always remained in a restricted area, close to their origin.

Typical Pleistocene relict populations are phylogenetically related to larger conspecific populations elsewhere. They may be restricted to mountain habitats like *Vipera berus* in the Alps and in Greek mountains, or to islands like *Lacerta bilineata* on Cres and *Natrix tessellata* in Crete.

Holocene relict populations stem from a postPleistocene range expansion (usually in Southeast–Northwest direction). They reached their largest distribution during the Holocene climatic optimum and were later restricted to small areas in the climatically most favorable regions. Examples are *N. tessellata*, *Zamenis longissimus*, *L. bilineata*,

U. Joger (✉)
Staatliches Naturhistorisches Museum, Pockelsstr. 10, D-38179 Braunschweig, Germany
e-mail: ulrich.joger@snhm.niedersachsen.de

U. Fritz
Museum für Tierkunde, Natural History State Collection, A.B. Meyer Building, D-01109 Dresden, Germany

D. Guicking
Universität Kassel, FB1, Systematik und Morphologie der Pflanzen, Heinrich-Plett Str. 40, D-34132 Kassel, Germany

S. Kalyabina-Hauf and M. Wink
Institut für Pharmazie und molekulare Biotechnologie, Universität Heidelberg, Im Neuenheimer Feld 364, D-69120 Heidelberg, Germany

Z.T. Nagy
Royal Belgian Institute of Natural Sciences, JEMU, rue Vautier 29, B-1000 Brussels, Belgium

* We dedicate this chapter to the memory of Peter Lenk (1964–2005)

J.C. Habel and T. Assmann (eds.), *Relict Species: Phylogeography and Conservation Biology*, DOI 10.1007/978-3-540-92160-8_6,

L. viridis, and *Emys orbicularis* in Germany. These populations are only poorly differentiated genetically; hence the risk of losing genetic diversity is not an appropriate argument for their protection. They may react to global warming by a range expansion.

Southern European, North African, and Middle Eastern relicts may, on the contrary, be threatened by global warming. These populations are in need of more studies, and should be brought to the attention of national and international conservation agencies.

1 Introduction

Reptiles are known as excellent indicators of the biogeographical history of a given region. This is due to their limited mobility (no long-distance dispersal), effectiveness of ecological barriers, and strong reaction to climate change due to ectothermic metabolism. The same three factors are also effective in isolating populations over a shorter or longer time; thus relict populations may be frequently isolated, and where favorable conditions prevail, they may survive up to the present. Dependent on the isolation time, separate populations develop genetic differences, but particular populations may also retain ancestral character states while other ancestral populations have gone extinct.

The importance of conserving genetic diversity is an internationally accepted goal since the 1992 Rio conference. With this in mind, molecular phylogeographic analyses serve two goals:

1. To reconstruct the evolutionary history of a group of organisms
2. To explore the spatial distribution of genetic diversity with special focus on unique genotypes restricted to particular regional populations; these populations deserve special attention in conservation programs.

A general requirement for a serious phylogeographic study must be that it should cover (ideally) the whole geographic range of a given species complex (Avise 1994).

The results summarized here have been obtained during a large-scale research project to elucidate the phylogeography of eight European reptile species complexes (overviews in Joger et al. 2006, 2007): The European pond turtles (*Emys orbicularis* complex), the water snakes of the genus *Natrix*, the European whip snake (*Hierophis viridiflavus*), the Aesculapian snakes (*Zamenis longissimus*/*Z. lineatus*), the adders (*Vipera berus* complex), the green lizards (*Lacerta viridis*/*L. bilineata*), and the sand lizard (*L. agilis*). These species complexes were selected so that similar phylogeographic patterns can be expected.

2 Materials and Methods

2.1 Laboratory Techniques

Blood samples were taken in the field (or from captive specimens of known collection locality) by caudal puncture (see Joger and Lenk 1997). The animals were released after photographic documentation. DNA was also extracted from ethanol preserved tissues. We used a standard method for obtaining total genomic DNA (Sambrook and Russell 2001).

As a gene tree is not necessarily identical to the species tree (see Avise 1994), we analyzed both mitochondrial and nuclear markers, wherever possible. Mitochondrial DNA, predominantly cytochrome b, was amplified by PCR and directly sequenced. (see the cited original publications for primer sequences). The applied nuclear markers were either allozymes and plasma proteins, or the inter-spaces between microsatellites – ISSRs (see Joger et al. 2006, 2007).

Both kinds of markers supplement each other and their combination thus enhances the capacity to resolve the phylogeographical history. Mitochondrial sequences are powerful tools to reconstructing subsequent multiple splitting events. Moreover, as they are inherited only maternally, without recombination, introgressions lead to the coexistence of several matrilines (haplotypes), thus preserving multiple origins of populations. On the other hand, hybridization (e.g., in secondary contact zones) cannot be detected by mitochondrial DNA. Bisexually inherited nuclear markers do not have this restriction.

2.2 *Phylogenetic Reconstruction*

For phylogenetic reconstruction from the cytochrome b sequences, three methods were employed – maximum parsimony (MP), maximum likelihood (ML), and Bayesian inference of phylogeny (BI). MP and ML analyzes were used as implemented in the program Paup 4.0b10 (Swofford 2002). Bayesian phylogenetic analysis was performed using the program MrBayes 3.0b4 (Huelsenbeck and Ronquist 2001). Because prior analyzes have shown that saturation effects are negligible in our data set, all characters were equally weighted in the parsimony analysis. ML trees were calculated using the models selected under the Akaike Information Criterion (AIC) in Modeltest 3.06 (Posada and Crandall 1998), using heuristic searches with tree-bisection-reconnection branch swapping and ten random addition sequence replicates. Bayesian analysis was performed in four chains for 1–2 millions of generations with the same parameter settings as for the ML calculation.

Support for clades from both data sets was assessed by bootstrapping (Felsenstein 1985). Bootstrap analyzes for neighbor joining and MP were performed with 1,000 replicates and under the same parameter settings as for the original trees. Under the ML criterion, bootstrap analyzes were performed with 100 replicates and without branch swapping to save computation time.

3 Results

3.1 *European Pond Turtle (Emys orbicularis Complex)*

Emys orbicularis (Linnaeus, 1758) is a polytypic species composed of a number of morphologically defined subspecies (Fritz 1996).

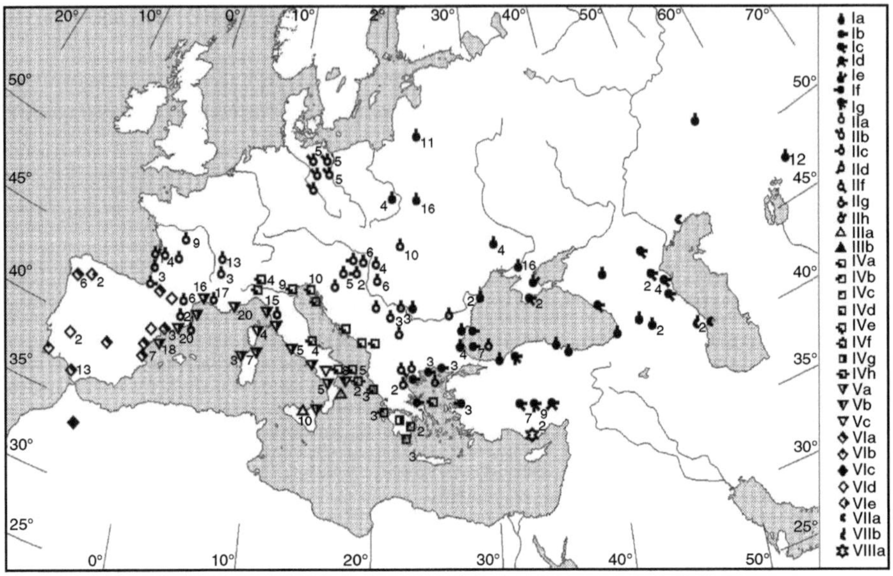

Fig. 1 Geographical distribution of *Emys haplotypes* (only natural populations). Some rare haplotypes are not shown. Note that only haplotypes of groups I and II occur naturally North of the Alps

We sequenced the mitochondrial cytochrome b gene of more than 1,100 individuals of pond turtles from about 200 localities, and identified 56 different haplotypes (see Fritz et al., 2007). These could be grouped into nine monophyletic haplotype groups (Figs. 1 and 2).

Group I is composed of ten haplotypes (Ia to Ij). They are distributed in Eastern Europe, in the Aegean region and in Anatolia. Lineage If from Izmir province seems to represent the most ancient surviving haplotype. Only haplotype Ia is found North of the Black Sea and the Crimea – all others occur further South. The considerable variation of group I in Turkey points to the possibility of several refugial areas in the Anatolian peninsula. From there, postglacial expansion most likely occurred both eastwards and westwards, around the Black Sea.

Group II is the sister group of group I. Both constitute the nominate subspecies *E. o. orbicularis*. Group II occupies an arc-shaped range from Northern Greece across Danube lowlands, parts of Central Europe, and France to North-Eastern Spain. The haplotype with the widest range- IIa- is the direct ancestor of all other haplotypes II, except the more ancestral haplotype IIf which is found in Macedonia. This indicates a possible Pleistocene refugial area in the Southern Balkan Peninsula. From there, haploype IIa spread along the Danube valley, westwards North of the Alps (see Fritz 1996). Haplotype IIb, found in Eastern Germany only, derived from IIa by one mutational step like other local haplotypes (see Fig. 2). The question whether small German populations of haplotype IIa are indigenous relicts or

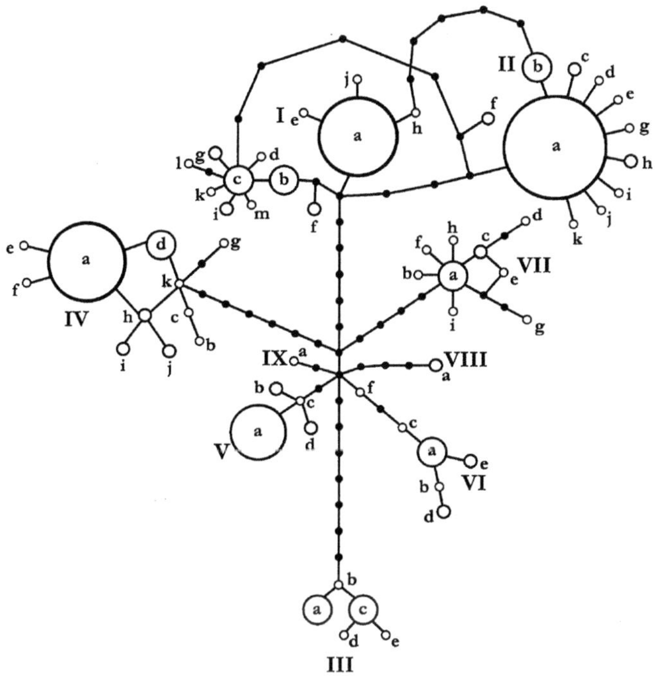

Fig. 2 TCS parsimony network of the *Emys* haplotypes. Haplotype group III=*Emys trinacris*. *Open circles* symbolize known haplotypes, *circle* size corresponds to the frequency of haplotypes as detected; *dots* on lines show single substitution steps between haplotypes (most parsimonious connection)

derived from introduced individuals, is debated among conservationists (see Fritz et al. 2004).

Group III inhabits Sicily and possibly the extreme South of the Italian mainland. As this group is highly differentiated from all others both genetically and morphologically, it has been described as a separate species, *E. trinacris* (Fritz et al. 2005). It must have remained in isolation for a long time.

Group IV is found around the Adriatic Sea, from Southern Greece to Italy, East of the Apennines. The Northernmost parts of the range are occupied by haplotype IVa only. On the island of Evvia, haplotypes IVa and Ib occur together.

Group V replaces group IV West of the Apennines. It reaches around the coast of the Tyrrhenian Sea to Catalonia and also inhabits Sardinia and Corsica. In Northeastern Spain it occurs in mixed populations with haplotypes IIa, VIa, and VId, in Apulia with IVb and IVd,and in Sicily with IIIa.

Group VI is confined to the Iberian Peninsula and Northwestern Africa. Its most ancestral haplotypes, VIc, and VIf, are found in Morocco. North Africa should therefore be considered as the probable origin of group VI.

Group VII is only found in the Caspian area and Iran, and group VIII is restricted to Southern Anatolia. The geographic origin of group IX is unknown.

3.2 *Green Lizards (Lacerta viridis/L. bilineata)*

Lacerta viridis has been subdivided into two biological species, Western European *L. bilineata* Daudin, 1802 and Eastern European *L. viridis* (Laurenti, 1768). The hybrids between them show reduced fertility (Rykena 1991, 1996, 2001)

Amann et al. (2001) identified a hybrid zone in Northeastern Italy and the adjacent part of Slovenia, in which a restricted and asymmetrical gene flow (predominantly from the East to the West) could be detected (Joger et al. 1998). This type of hybrid zone of "species in statu nascendi" will remain spatially restricted, but may be stable for a long time, if it is stabilized by equivalent invasion and evasion, hybridization, and selection (Barton and Hewitt 1988).

Our cytochrome b tree (Fig. 3) is concordant with a previously published allozyme tree (Amann et al. 1997, Joger et al. 2001) by grouping in two main clusters – the two species *L. bilineata* and *L. viridis*. Both trees group the populations from the hybrid zone (Udine, Trieste) to *viridis*, which contradicts a recent analysis by Böhme et al. (2006). These authors included those populations (and Slovenian populations) in a "Western Balkan group" extending South to Greece. As this group clustered with *L. bilineata* in their tree, they assigned it to *L. bilineata*, in order to prevent a paraphyletic *viridis*. Mayer and Beyerlein (2001), using 12s and 16s RNA genes, also found Western Greek *viridis* associated with *bilineata*, while Brückner et al. (2001) found that cytochrome b from the same animals clustered with *viridis*. These contradictory results were confirmed by Godinho et al. (2005). However none of the teams had included Turkish *L. viridis* in their analysis. In our tree (Fig. 7), with Turkish lizards included, there is no paraphyly of *L. viridis*. The lizards from Trieste and Udine cluster with *viridis*, as well as the Western Balkan group. The Croatian population from the island of Cres clusters with the Western *bilineata*. This was also found by Brückner et al. (2001) and by Godinho et al. (2005). Cres may harbor a relict population of a once more easterly distributed *bilineata*, recently restricted by a Western advance of *viridis*.

For both *L. viridis* and *L. bilineata*, a number of subspecies have been described (Nettmann 2001). Most of them are found in Southern Italy (*L. bilineata*), Greece, or Turkey (*L. viridis*) and point to multiple glacial refuges in those areas. However, the molecular data are not yet sufficient to decide upon this issue.

3.3 *Sand Lizards (Lacerta aglis Complex)*

Lacerta agilis Linnaeus, 1758, is a widespread species in Europe and Northwestern Asia. Nine subspecies have been described (Bischoff 1988). A phylogeographic analysis based on cytochrome b sequences (Kalyabina et al. 2001, Kalyabina-Hauf et al., 2004a) confirmed most of the subspecies, and revalidated a 10th one, *L.a.tauridica* Suchov, 1926, from Crimea. If the branching events (Fig. 4) are linked with the geographical distribution of the subspecies, the Caucasian/Black Sea region

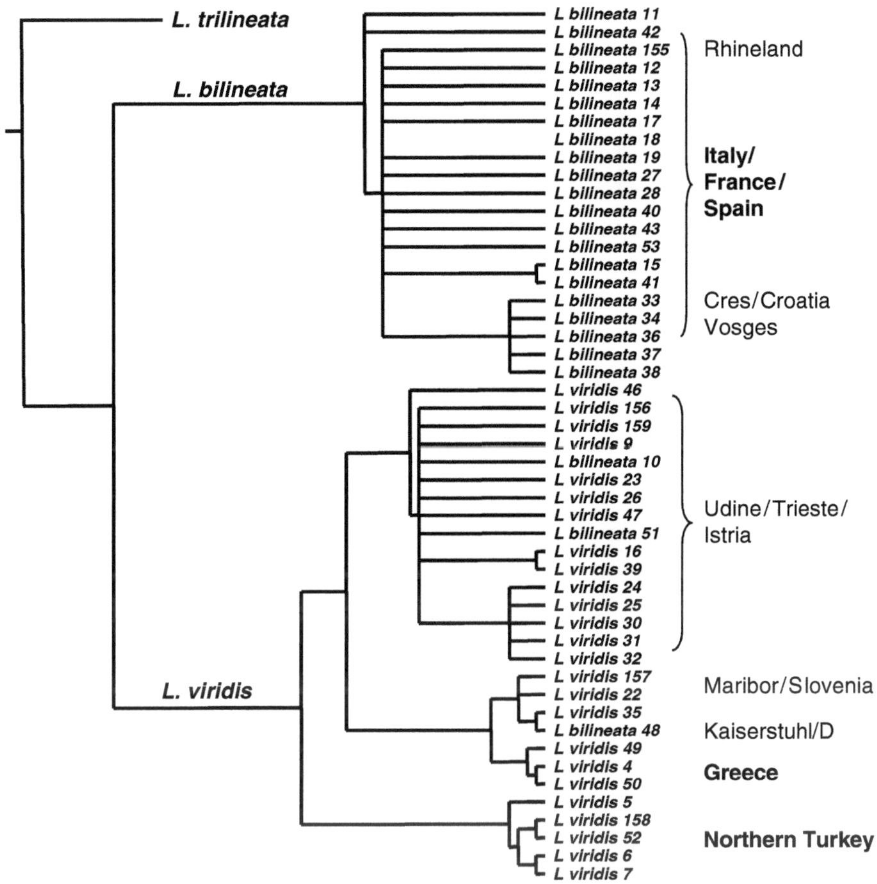

Fig. 3 Maximum parsimony tree for Green lizards, derived from mitochondrial cytochrome b sequences. Note that several individuals were initially determined as *L. bilineata* (from Trieste, Italy, and Kaiserstuhl, Germany), but have haplotypes of *L. viridis*, whereas lizards from the Croatian island of Cres are unambiguously *L. bilineata*

emerges as the group's most probable area of origin (Joger et al. 2007). The most ancestral surviving clade is *L. a. boemica* Suchov, 1929, from the Eastern Caucasus. It may deserve species status.

The Western clade (distributed from Western Russia to the Pyrenees), in which the earliest internal branches lead to the easternmost subspecies – *L. a. chersonensis* and the next branching to an unnamed Carpathian group – must have moved gradually westwards.

The Eastern clade (from the Caucasus West to Crimea and East to Lake Baikal), and a Balkan subspecies (*L. a. bosnica*) are the other two major clades. As *L. a. bosnica* is sister of the Eastern clade, while the Western clade branched off from the common stem earlier, the most probable phylogeographic hypothesis must take into account several waves of colonization from East to West. A molecular

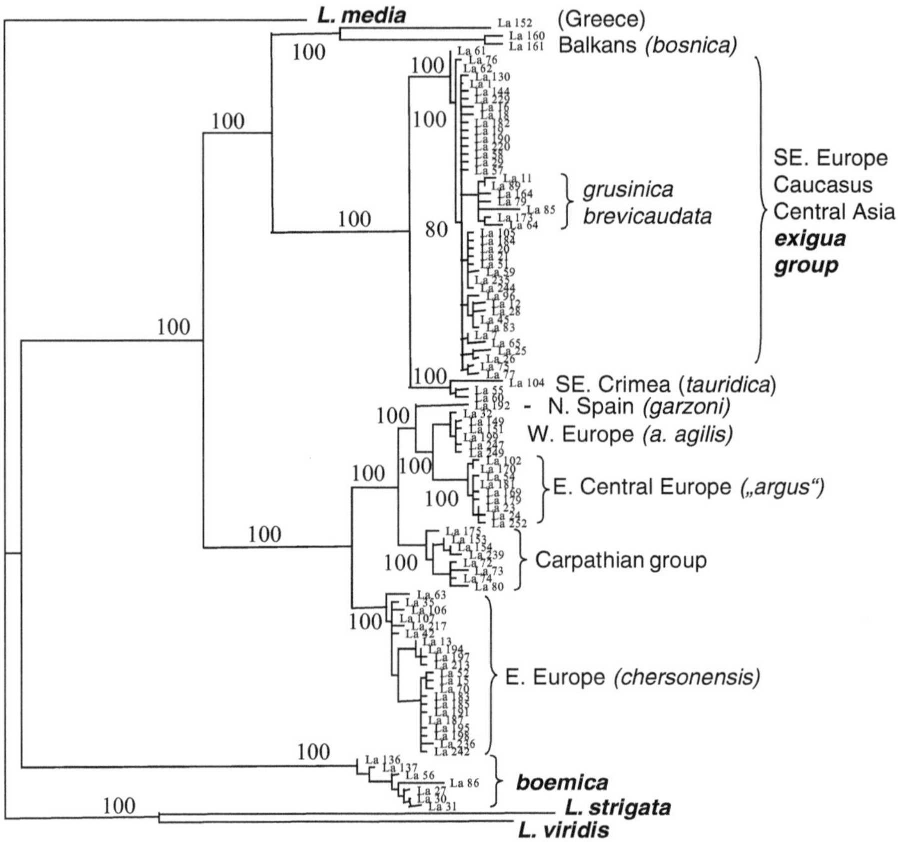

Fig. 4 Maximum Parsimony phylogram (consensus of 300 equally parsimonious trees) of the *Lacerta agilis* complex, derived from mitochondrial cytochrome b sequences. *L. media*, *L. strigata*, and *L. viridis* are used as outgroups. Bootstrap values indicated. The *L. exigua* group unites subspecies from Caucasus (*L. a. grusinica, L. a. brevicaudata*) as well as Southeastern European *exigua* s.str. and various central Asian *agilis*

clock places these waves into different periods, from upper Pliocene to middle Pleistocene. The colonization of Central Asia appears to be more recent (maybe even Holocene).

3.4 European Whip Snake (Hierophis viridiflavus)

Hierophis viridiflavus (Lacépède, 1789) is distributed over most of France, Italy, and Slovenia and reaches into South-Western Switzerland, Luxemburg, Belgium, Northernmost Spain, and Croatia. An Eastern subspecies, *H. v. carbonarius* (Bonaparte, 1833), was mainly based on coloration and refuted by Schätti and Vanni (1986). However, our molecular analysis based on both cytochrome b sequences and ISSR fingerprints, confirmed it (Nagy et al. 2002).

In general, a Western and an Eastern clade are discernable, the border between them being the Apennine chain (Fig. 5). A third clade, found in Sicily and Southern Calabria, is sister to the Eastern clade. Its subspecific name should be *H. v. xanthurus* (Rafinesque, 1810). ISSR fingerprints revealed distinctive loci for these three clades, but also identified several specimens of hybrid origin which share bands characteristic of Eastern and Western clade.

In this species, the phylogeographic history is also documented by fossil records from Austria, Southern Germany, Czech Republic, and Poland (Szyndlar and Böhme 1993, Ivanov 1997). These areas were reached in warmer periods of the Pleistocene, as well as in the Pliocene, but in glacial periods the distribution area

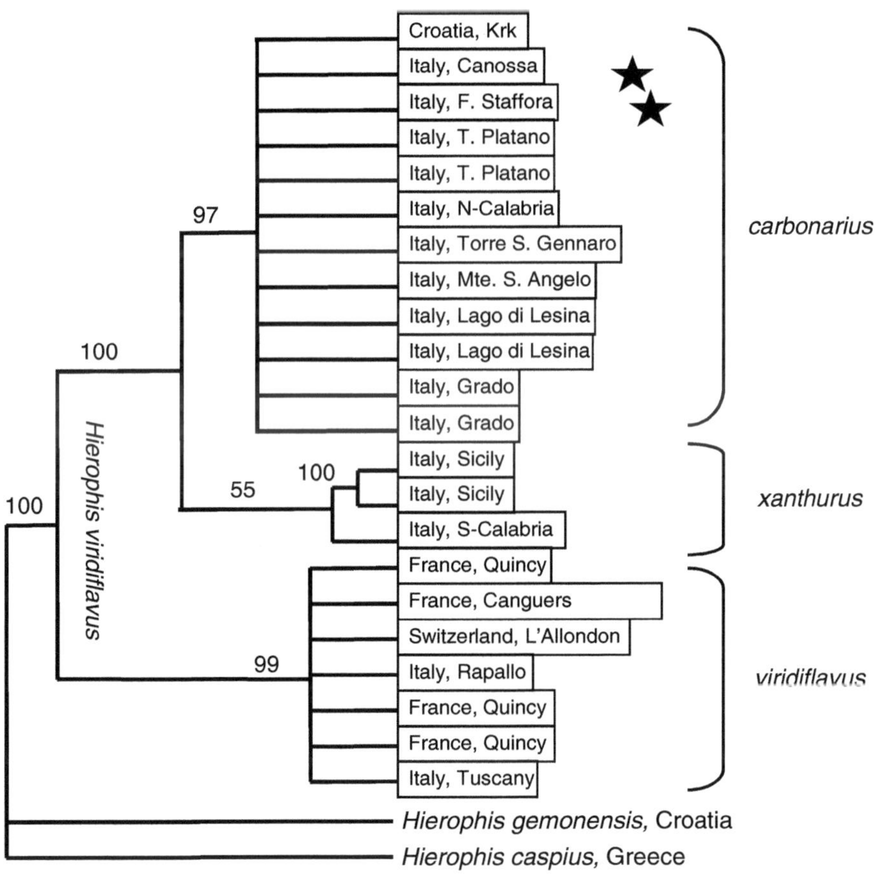

Fig. 5 Phylogeography of *Hierophis viridiflavus* derived from cytochrome b sequences. Maximum Parsimony tree with bootstraps values for the main haplotype groups. *=Populations East of the Apennines, in which alleles of the Western group were recorded by ISSR-PCR (putative hybrids)

shrunk to its Italian core area where the genetic differentiation occurred. No relicts remained outside the main distribution area.

3.5 Aesculapian Snake (*Zamenis longissimus* Complex)

Zamenis longissimus (Laurenti, 1768), formerly *E. longissima*, is found in France, Italy, Austria, the Balkans, and Turkey, with isolated relict populations in Southern Germany, Czech Republic, and Poland.

Protein electrophoretic comparisons (Lenk and Joger 1994) revealed only slight geographic structuring in *Z. longissimus* which, for example, showed an association of the Central European relicts to the Balkan populations and not to the Italian populations (thus refuting the hypothesis that the Romans had imported the Aesculapian snake to Germany). Moreover Peters (1977) had already reported fossil *Z. longissimus* from deposits of the early Holocene climatic optimum in North-Eastern Germany. However, a strong difference in blood protein electrophoretic patterns was found between Southern Italian Aesculapian snakes and all others. A morphological analysis (Lenk and Wüster 1991) concluded that the Southern Italian population must be regarded as a separate species, *Z. lineatus* (Camerano, 1891). A cytochrome b sequence analysis (Fig. 6) also shows the external position of *Z. lineatus* with regard to *Z. longissimus* (sequence difference of 7%). Moreover, it shows that in *Z. longissimus*, an Eastern haplotype group (found along the Aegean and Black Seas) can be distinguished from a Western haplotype group. Both groups nearly meet in Greece. They are separated by the PindosMountains and must have had different Pleistocene refugia. The Western haplotype group is composed of a Danubian haplotype (which includes the German and Czech populations) and a Mediterranean haplotype, from which a Western haplotype (hitherto found in

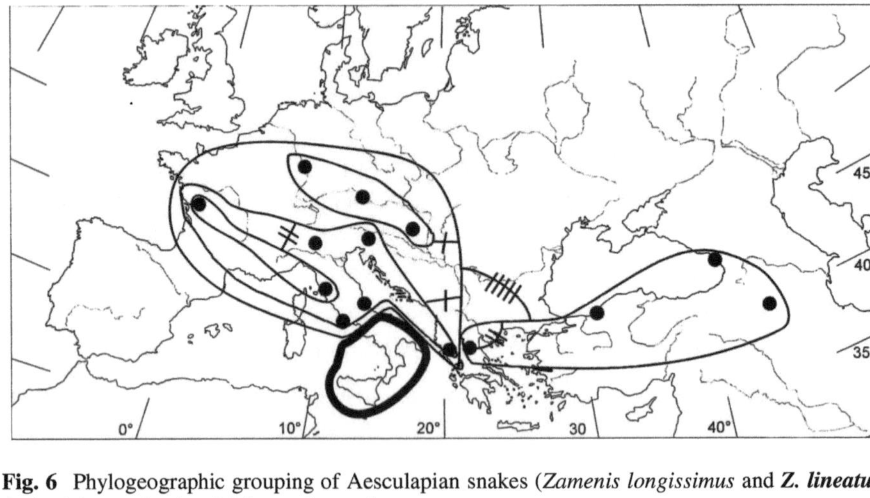

Fig. 6 Phylogeographic grouping of Aesculapian snakes (*Zamenis longissimus* and ***Z. lineatus***) derived from mitochondrial cytochrome b sequences

France and in Tuscany) derived. Three microrefuges may therefore be assumed for the Western haplotype group.

3.6 *Viperine Snake (Natrix maura)*

Natrix maura (Linnaeus, 1758) inhabits aquatic habitats in the Western part of the Mediterranean, from Italy to Morrocco and Tunisia, North to Central France, and Lake Geneva. Two phylogeographic analyzes were done on the basis of cytochrome b sequences (Barata et al. 2008, Guicking et al. 2008) and ISSR fingerprints (Guicking et al. 2008). Both data sets showed identical groupings (Fig. 7).

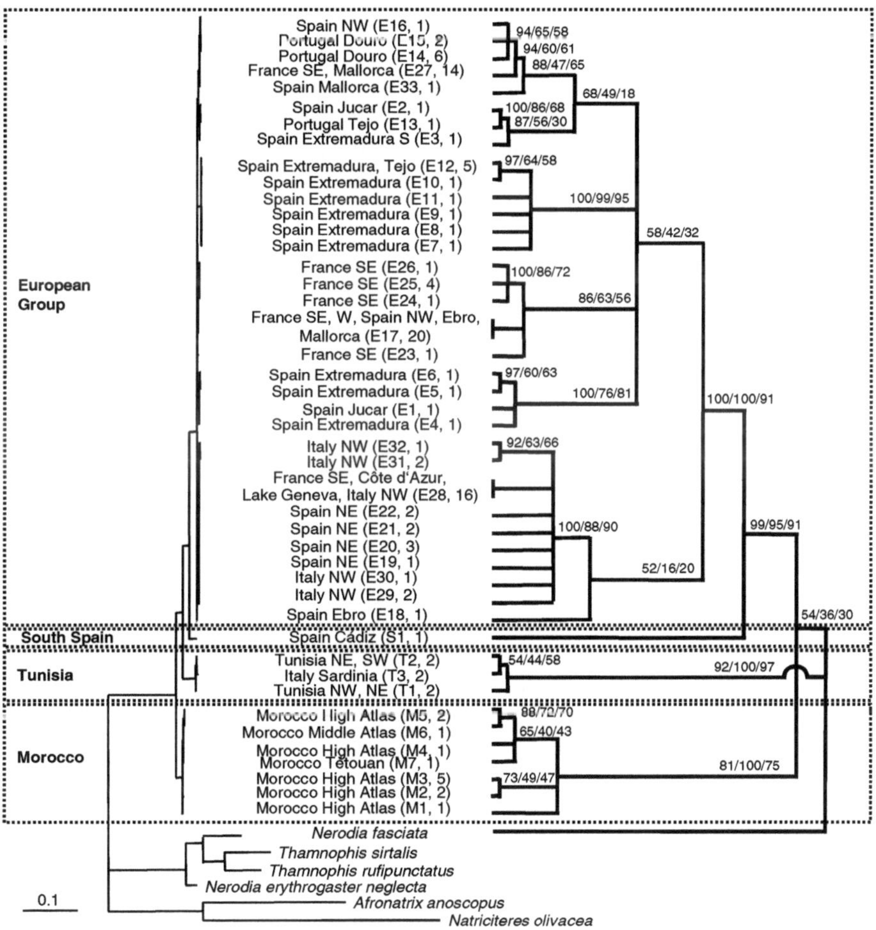

Fig. 7 Phylogenetic trees of *N. maura* reconstructed from cytochrome b haplotypes, *left*: maximum likelihood phylogram, *right*: Bayesian cladogram. Sequences from *N. natrix* and *N. tessellata* have been used for outgroup rooting. Support values are indicated in the order Bayesian/ML bootstrap values/MP bootstrap values

3 major haplotype groups were found – two North African (Tunisian, Moroccan) and one European. Eastern Moroccan and Algerian *Natrix* cluster with the Tunisian group (Barata et al. 2008). The origin of the species can be assumed in North Africa. The fact that one of the Tunisian haplotypes is also found in Sardinia may be explained by human introduction, probably in Roman times. The Viperine snakes of Mallorca were probably introduced, too, as they share a haplotype which also occurs on the French and Spanish mainland (Guicking et al. 2006b).

In the European group, a haplotype from extreme Southern Spain is sister to all other European haplotypes. It may therefore be regarded a relict of the first European settlement of *N. maura*. A standard molecular clock dates the colonization of the Iberian Peninsula as Pliocene (ca. 3 m.y.b.p.). A comparable phylogeographical pattern exists in the turtle *Mauremys leprosa* (Fritz et al. 2005).

In Europe, genetic diversity is lowest among Northeastern populations (Italy, France, Switzerland, NE Spain), while distinct genetic groups occur in Southern Spain and Portugal (Extremadura, Tejo). This is paralleled by other Iberian taxa, such as salamanders (Joger and Steinfartz 1994). Southern Iberian populations are more isolated and most differentiated. Northern Iberian populations are the sources of post-Pleistocene colonizations of France and Italy, but an additional micro-refuge in Southern France or Western Italy cannot be excluded.

3.7 Dice Snake (*Natrix tessellata*)

Natrix tessellata (Laurenti, 1768) is the sister species of *N. natrix* (Guicking et al. 2006a). The Dice snake is the ecological equivalent of the Viperine snake further East, occupying a vast distributional range from Germany, Switzerland, and Italy to Central Asia, Iran, and Arabia. The phylogeographic analysis of cytochrome b sequence data revealed nine distinct haplotype groups, of which two occur in Greece and only one in the rest of Europe. The Middle East is the most probable area of origin of this species. Early branching events led to highly differentiated haplotype groups found in Iran (up to 10% sequence difference), Greece, and Northern Arabia (Figs. 8 and 9). Four haplotype groups occuring around the Black Sea in Turkey, and Central Asia are related to each other; ISSR fingerprints show that populations of these groups hybridize. A distinct group is found in Crete, and a rather homogeneous group occupies all the European range except Greece.

The existence of two separate Pleistocene refugial areas in the Southern Balkans can be deduced from the data. Only one of them served as a source for the recolonization of the rest of the European area. The genetic diversity among European populations is low and shows little geographic structuring. The invasion of Italy could be dated back to the last interglacial maximally, and thus a separate microrefuge in the Apennine peninsula is possible for the last glacial (Guicking et al. 2009).

A number of microrefuges must have existed in Anatolia, in the Caucasus, and in West Central Asia.

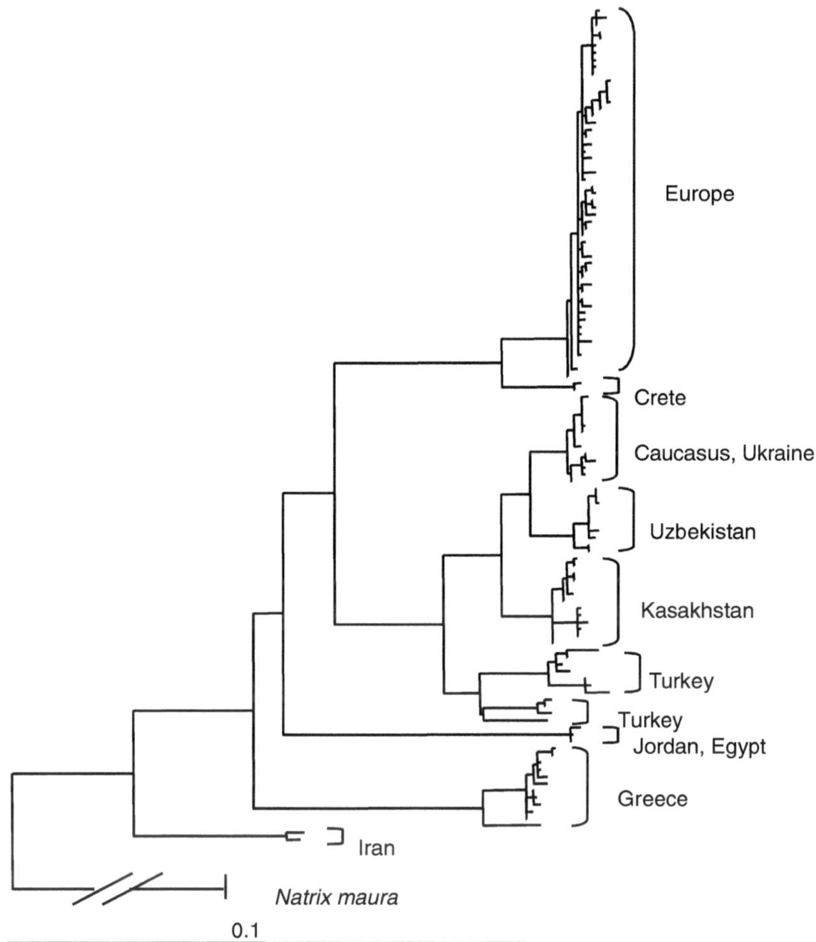

Fig. 8 Maximum Likelihood Phylogram for *Natrix tessellata,* based on complete cytochrome b sequences. 9–10 different haplotype groups can be distinguished. *Natrix maura* served as outgroup

3.8 Adders (*Vipera berus* Group)

In contrast to the species treated earlier, adders have a more Northern distribution, occupying a vast territory from Northern Spain to the Polar circle in Scandinavia, and to the Pacific Ocean in the East. Yet here too, genetic differentiation and speciation are concentrated in the Southern margin of the distribution area. The Iberian Adder, *V. seoanei* Lataste, 1879, has been regarded as a separate species for long. In the Black Sea area, *V. nikolskii* Vedmederya et al. 1986 (Ukraine, Southern Russia) and *V. barani* Joger and Böhme 1984 (Turkey) have been described, but their species status is doubtful. In the Balkans, a subspecies *V. berus bosniensis*

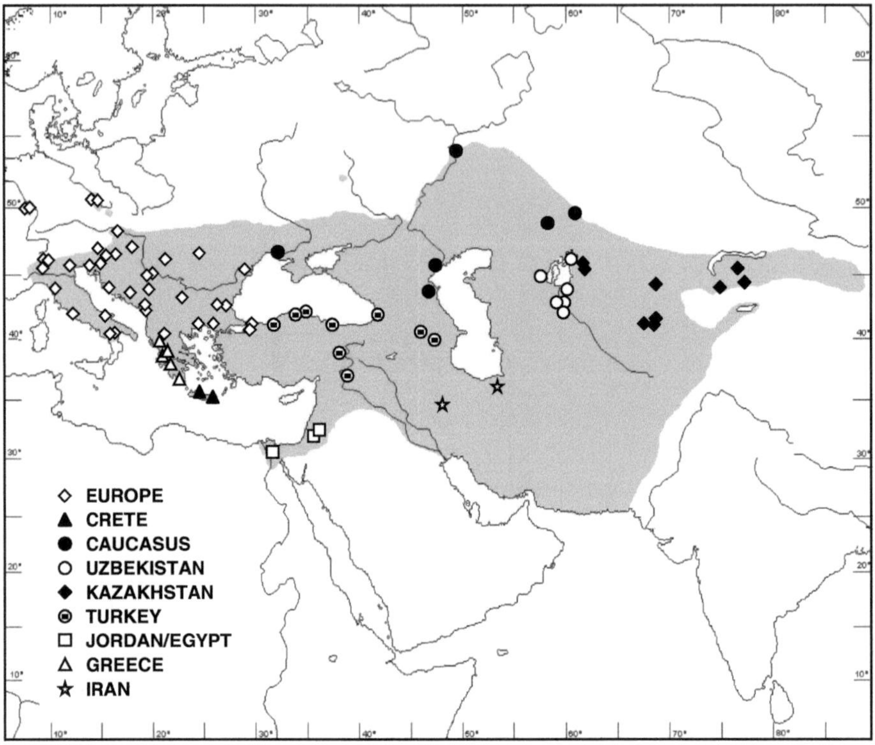

Fig. 9 Geographical distribution of haplotype groups in *Natrix tessellata,* modified from Guicking et al. (2009.)

Boettger, 1889, is known, and in the Far East, *V. berus sachalinensis* Tsarevski, 1916, has been distinguished. The relationships between these taxa have been studied by Kalyabina-Hauf et al. (2004b) and by Ursenbacher et al. (2006a).

In the cytochrome b tree (Fig. 10), *V. seoanei* appears as a sister to all other adders. *V. nikolskii* and *V. barani* cluster into one clade, though not with strong support. The genetic separation of Alpine *V. berus* from the rest of the species points to a separate glacial refuge South of the Alps (probably in Italy, where a population of lowland adders became extinct just several decades ago [see Scali and Gentilli, 1999]). It is also supported by ISSR fingerprint data (Kalyabina-Hauf et al. 2004b) and by mitochondrial control region sequences (Ursenbacher et al. 2006a). The refugia of Northern *berus* are difficult to determine, but the separate position of the Hungarian samples points to the Carpathian basin as one possible refuge. According to Ursenbacher et al. (2006a), additional refugia may have been located in France, Slovakia, or Southern Russia. Scandinavia was recolonized post-glacially from two sides, West and East, so several glacial refugia to this fairly cold tolerant snake are reasonable.

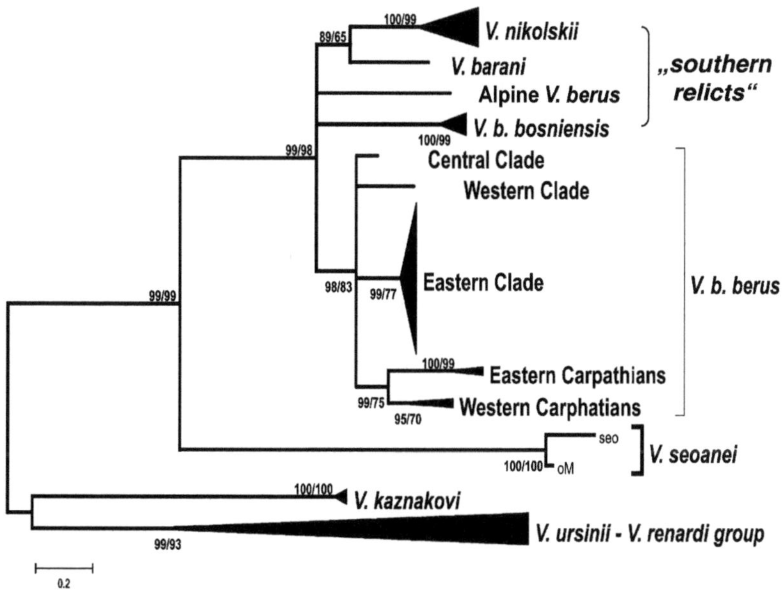

Fig. 10 Maximum Parsimony strict consensus tree of *Vipera berus* and relatives derived from cytochrome b haplotypes (1,003 bp). Bootstraps values indicated for Maximum Parsimony and Maximum Likelihood. *Vipera kaznakovi, V. renardi and V. ursinii* were used as outgroups

4 Discussion

4.1 Phylogeographic Patterns

The degree of genetic differentiation is fairly high in all studied species complexes, regardless of morphological variability. A species like *E. orbicularis*, which has a high number of geographically restricted morphological subspecies, shows a similar amount of genetic variation as *N. tessellata* and *N. maura* – species that have not been subdivided previously, due to morphological uniformity. These three species are more or less aquatic.

A general observation in all species complexes is the fact that genetic variation is geographically biased: It is generally lower in the North and much higher in the South. This reflects the effects of climatic change during the Pleistocene: Northern territories were invaded during interglacial periods (such as the present one), but the newly acquired territories had to be given up when the next glacial advanced (Taberlet et al. 1998). Thus in most taxa, today's occupants of the areas North of the Alps have a short history: The molecular trees show that their ancestor came from Southern refugia after the end of the last glaciation, i.e., about 10,000 years ago. This period was obviously too short for major genetic changes; only some private haplotypes, differing by a few mutations, have been found in the Northern territories (such as haplotype IIb in Central European *E. orbicularis*).

Moreover as a rule, only one or two of several existing haplotype groups of any given species managed to conquer the Northern part of the species' territory, and these haplotype groups also persist in their putative area of origin further South.

There are a few exceptions to this rule. In *V. berus*, the Northern part of the range is occupied by a typical haplotype group that does not occur South of the Alps. It has however, a wide distribution to the East, reaching the pacific coast of Asia. Therefore, although it is a boreal species, it probably did not survive the last glacial in Northern Europe, but rather in Eastern Europe or Asia, where a number of tree species had their refugia too (Gliemeroth 1995).

Another case is *L. agilis*, which is not a Mediterranean species. Its evolutionary center can be identified in the Caucasus. There are a few distinct Northern haplotype groups, two of them in Central Europe. *L.a.agilis*, the Western group, extends from South-central France to Western Germany, whereas *L.a.argus*, its Eastern sister, is found from Eastern Germany to Romania. This pattern parallels the distribution of *L. bilineata* (West) and *L. viridis* (East). Both taxon pairs may have had two glacial refugia: one in the Southeast and one in the Southwest, and invaded Central Europe from both sides simultaneously. The difference lies in the degree of speciation. The Central European *L. agilis* subspecies are quite young and may be interpreted as a product of the latest glacial fluctuations, whereas *L. viridis* and *L. bilineata* have had time to proceed much further in the speciation process. Their genetic differentiation must have started in the early Pleistocene (Joger et al. 2007).

4.2 The Time Scale of Evolution

The average evolutionary rate for Viperidae cytochome b has been estimated to 1.4% sequence difference per 1 million year (Ursenbacher et al. 2006a). For *Natrix*, we calculated a similar rate of 1–1.35% per 1 million year (Guicking et al. 2006a). For *L. agilis*, we estimated a higher rate of up to 2.5% per 1 million year (Kalyabina et al. 2001), but for *E. orbicularis*, a lower level of 0.3–0.4% per 1 million year (Lenk et al, 1999). These differences are interpreted as reflecting the different metabolic activity in reptiles, which is lowest in turtles and highest in lizards. Life expectancy is reciprocal to metabolic rates.

If these estimates are correct, sequence differences of 1% or more in turtles, 3% or more in snakes, 6% or more in lizards are due to prePleistocene branching events and hence, cannot be associated with Ice Age refugia. All species groups investigated here originated in the Tertiary and speciation events are of Pliocene age (Table 1). This is in accordance with the fossil record, as remains of today's snake species were found in Pliocene sediments as old as 3.4 Million years (Szyndlar and Böhme 1993). A mid-Miocene record of *Natrix* was considered a direct ancestor of the extant *N. natrix*, suggesting that the three clades in *Natrix* might have already been separated in the middle Miocene (Ivanov 2001). According to our estimate, the main clades of *Natrix* species are of late Miocene age, the Southern Spanish clade of *N. maura* is dated Pliocene, and the differentiation between the lineages of the main

Table 1 Speciation, sequence differences and approximate time estimates

Group	% sequence difference between sister species (cyt b)	Time estimate for species (m.y.)	Time estimate for major groups within species (m.y.)	Reference
Emys	1.4–1.7[a]	3–4	1.4–3	Lenk et al. (1999)
Lacerta agilis	6.5–7.3[b]	2.6–2.9	0.7–2.4	Kalyabina et al. (2001)
L. viridis complex	6.6–8.4[c]	2.6–3.4	0.5–2.5	Böhme et al. (2006)
Hierophis	10.5[d]	8–9	0.9–2.7	Nagy et al. (2002, 2004)
Zamenis (*Elaphe*)	7[e]	5–7	0.5–1	Lenk et al. (2001)
Natrix	18[f]	13–22	2–8	Guicking et al. (2006a)
Vipera berus group	5[g]	4	1.1–1.6	Kalyabina-Hauf et al. (2004b)

[a]Distance between *E. orbicularis* und *E. trinacris*
[b]Distance between *L. agilis* sspp. and *L.* (*a.*) *boemica* (assuming species status)
[c]Distance between *L. viridis* and *L. bilineata*
[d]Distance between *H. viridiflavus* and *H. gemonensis*
[e]Distance between *Z. longissimus* and *Z. lineatus*
[f]Distance between *N. natrix* and *N. tessellata*
[g]Distance between *V. berus* and *V. seoanei*

European clade is Pleistocene (Guicking et al. 2006a). On the other hand, Nagy et al. (2003) estimated evolutionary rates for colubrid snake mitochondrial genes as twice the rate as we assume here, hence, of the same magnitude as in *Lacerta*. If we accept this alternative, *Hierophis* and *Zamenis* would show similar time estimates for sister species as *Lacerta* (Pliocene); but *Natrix* speciation would still be Miocene.

Apart from *Natrix*, the main intraspecific clades in our selection are of late Pliocene or Pleistocene age. As pronounced climatic oscillations started already in late Pliocene, it seems appropriate to consider these intraspecific radiations as effects of climate change and associated range restrictions.

4.3 Pleistocene Refugia and Recolonization

Table 2 gives an overview of Pleistocene refugia inferred for the studied groups. The classical Mediterranean refugia (Iberia, Italy, and the Balkans) were used by all species groups, but in many cases, more than one microrefugium was located in each peninsula.

North African and Asian refugia did not play a role in recolonization after Pleistocene cold phases, but have acted as long time independent speciation centers. This observation may however, be biased by our selection of species, as we took into account only species which were able to colonize Central Europe (at least its climatically most favorable regions).

Table 2 Pleistocene refugia inferred from cytochrome b haplotype partitions

Group	Iberian	Apennine	Balkans	Pontic	Caspian	Other
Emys	several	several	several	several	1	North African
Lacerta agilis	1	–	2	1	several	Carpathian
L. viridis complex	–	2	several	1	–	
Hierophis viridiflavus	–	several	–	–	–	
Zamenis longissimus	–	2	2	?	?	
Natrix maura/ N. tessellata	several	?	2	several	several	North African West Asian
Vipera berus group	1(*V. Seoanei*)	1	1	2	–	Carpathian French

Among these species, the two most cold-tolerant, *V. berus* and *L. agilis*, were probably able to survive in non-Mediterranean refuges like central France and the Carpathian basin. A glacial refuge in France was also inferred for two species related to the aforementioned, *V. aspis* (Ursenbacher et al. 2006b) and *L. (Zootoca) vivipara* (Guillaume et al., 2000). A glacial refuge in the Carpathian basin has been assumed for two species of considerable cold tolerance, the Moor frog *Rana arvalis* (Babik et al. 2004) and the Field vole, *Microtus agrestis* (Jaarola and Searle 2002).

In genera which had refugia in all three European peninsulas (*Emys*, *Natrix*), four paradigms for post-glacial range expansion are possible (Hewitt 1999, 2000; Habel et al. 2005). Yet only one of them was used: The Eastern refugia in particular were the source areas for the colonization of Northern areas such as Germany. The reasons for this may be both geographical and ecophysiological. The lack of mountain barriers in Eastern Europe allowed a quick range extension once the climatic situation became favorable. On the other hand, Eastern refugia had a continental climate with more severe winters than Western refugia. Populations adapted to that climate could start their Northward expansion earlier and reach more Northern locations. In *E. orbicularis*, the Balkan-originated haplotype II crossed Central Europe North of the Alps and France to Northeastern Spain, while the Western Mediterranean haplotypes stayed where they were. This is paralleled by the Eastern shrew *Sorex araneus*, which reached Northeastern Spain post-glacially, but had to retreat into montane refuges there while the warmer climate favored the Western species *S. coronatus* (Lopez-Fuster and Ventura 1996).

4.4 What is a Relict Population?

Generally, a relict population can be defined as a part of an evolutionary lineage which has been isolated geographically.

Three types of relicts have to be considered:

1. *Geographical relicts* without significant genetic differentiation. This type of relict is usually found in "exclaves" beyond the Northern margin of a species' range. Typical representatives of this type are the minute German and Czech populations of *N. tessellata* and *Z. longissimus*. Both occupy geographically restricted, climatically favorable habitats along river valleys, preferably the South-facing (sunny) banks and adjacent hills of rivers flowing in latitudinal direction. The same areas are occupied by other "submediterranean" relict species such as *L. bilineata* and *Podarcis muralis*. Other such relicts (*L. viridis*, *E. orbicularis*) are found in subcontinental areas of Eastern Germany, which have more insolation in summer. All these species must have reached their presently isolated Northern territories in post-glacial times. In climatically favorable times, such as the climatic optimum 5,000–7,000 years ago, they must have extended their range to a much larger territory (e.g., *Z. longissimus* reached the Baltic Sea Coast (Peters 1977)). Yet the territories they occupy today are small and isolated. They are classified as *Holocene relict populations* (Table 3). Due to their very restricted occurrence and low population sizes today, these populations must be considered true relicts in Germany, and they are consequently classified in the highest national conservation categories. Yet, if the available genetic data are considered, their extinction would not mean significant loss to the species' total genetic diversity. With the exception of the above-mentioned East German haplotype of *E. orbicularis* and some single mutations in other species, all major haplotypes found in the isolated German populations also occur in the larger territories of the respective species further South. Of course, caution has to be taken not to rely solely on a limited amount of genetic data – usually of mitochondrial genes. Theoretically, a wide genomic screening could change the observed pattern.
2. *Genetic relicts*: This type of relicts is usually found at the Southern or Eastern edge of a species' range. Genetic relicts can be defined as possessing a unique, significantly differentiated haplotype or a haplotype group of their own. To qualify as a relict, a historical loss of territory is also necessary. Typically, these relicts have a long evolutionary history in one particular region and were unable to extend their range post-glacially. In some cases, such as the Alpine population of *V. berus*, the range probably shifted from lowland (Northern Italy) to highland (the Swiss and Austrian Alps), but the range extension to the North was accompanied by loss of territory in the South. This type of relict is older than Holocene. It is classified as *Pleistocene relict populations* (Table 3), if it is not considered taxonomically distinct. Pleistocene relict populations have often been isolated by ecological change (sometimes with human interference, as the aquatic *N. tessellata* in the desertificated Aral Sea region, and *E. orbicularis* in North Africa). Others were isolated from the main distribution area by a competing species, as the *L. bilineata* population of Cres Island, surrounded by *L. viridis*, which is thought to have expanded its territory. Another case is *V. berus* in Greece, restricted to high mountain habitats by both climatic warming and competing *V. ammodytes*.
3. *Endemic relict taxa*: If the genetic isolation has prevailed over long enough time, a taxonomic recognition of the isolated population as an endemic subspecies or

Table 3 Endemics and relicts among the investigated taxa

Meta-Taxon	Regional endemic subtaxa	Endemic relict subtaxa	Pleistocene relict populations	Holocene relict populations
Emys orbicularis complex	*E. trinacris* Sicily	*E.o.persica* Caspian Area *E.o.*ssp. S. Anatolia	*E. orbicularis* North Africa *E.o.orbicularis* Turkey	*E.o.orbicularis* E. Germany, Lithuania, Kazakhstan
Lacerta viridis complex	*L. bilineata* ssp. S. Italy, Sicily *L. viridis* ssp.Balkans	*L. viridis* ssp.Turkey	*L. bilineata* Balkans, Cres Isl.	*L. bilineata* W. Germany *L. viridis* E. Germany
Lacerta agilis complex	*L.a.tauridica* Crimea	*L.(a.)boemica* Caucasus *L.a.bosnica* Balkans *L.a.garzonii* Pyrenees	*L.a.chersonensis* Carpathians	*L. agilis*Britain, Sweden
Hierophis viridiflavus	*H.v.xanthurus* S. Italy, Sicily			
Zamenis longissimus complex	*Z. lineatus* S. Italy, Sicily			*Z. longissimus* Germany, Czech Rep.
Natrix maura	undescribed North Africa	undescribed Andalucia	*Natrix maura* Southern Spain	
N. tessellata	undescribed Middle East	undescribed Greece, Middle East	*N. tessellata* Crete, Aral Sea region	*N. tessellata* W. Germany, Czech Rep.
Vipera berus complex	*V.b.bosniensis* Balkans	*V. barani* Turkey *V. nikolskii* Black Sea area *V. seoanei* Pyrenees	*V. berus* Alpine Region *V. berus* Greece *V.b.berus* Carpathians	

species may be justified. Many of these taxa may also be termed Pleistocene relicts. Table 3 shows some examples. A distinction must be made between an endemic relict taxon and an endemic taxon which is not considered a relict. This decision must be based on the phylogeographic position of the taxa. For example, *E. trinacris* is an early offspring of the Tertiary *Emys* population in Europe. It was isolated in Sicily, where it differentiated genetically, but there is no evidence that it ever had a larger range, or that it had descendants which migrated elsewhere.So it is an endemic but not a relict. On the other hand, *E. orbicularis persica* (group VII in Figs 1 and 2) and the Southern Turkish subspecies (group VIII) must have had a larger range in the past. They were probably linked to Mediterranean groups before their range was restricted by Pleistocene expansion of *E.o.orbicularis*. A comparable fate may be assumed for *L. viridis* ssp. from Turkey and *L. (agilis) boemica* from Caucasus. Their phylogenetic position at the base of their respective radiation suggests that these taxa remained in the area of origin of those species, where they survived in a small area only, while more derived subspecies successively occupied new territories. On the other hand, several *L. viridis* ssp. from Balkans and *L. agilis tauridica* from Crimea are endemics, but not isolated, and are not likely to have experienced shrinking of their range. In *Natrix*, a number of distinct haplotype groups have been identified at the Southern margin of the ranges of *N. maura* (Andalucia, North Africa) and of *N. tessellata* (Greece, Iran, Arabia), but no taxonomic consequences have yet been drawn from that observation. We can expect both endemics and endemic relicts among those undescribed taxa, which occur in or near the probable areas of origin of those species (Guicking et al. 2006a). Finally in the *V. berus* complex, relicts occur at the Southern margin of the distribution area of this cold-adapted vipers. *V. barani* is completely isolated in Turkey, and *V. nikolskii* has apparently lost territory by intrograding *V.b.berus* from the North (Zinenko, unpublished data). *V. seoanei*, retreating in the Pyrenees, may be under pressure from two warm-adapted species which occur parapatrically: *V. aspis* and *V. latastei* (Stümpel et al. 2005).

How will the expected change of climate (global warming) affect relict reptile species? In short, there will be advantages for relicts of thermophilic species in the North, which may enlarge their territories, as is already observed with some plants and invertebrates.

Pleistocene relicts in the South, now confined to mountain areas like *V. seoanei* in the Pyrenees and *V. berus* in the Greek mountains, will have to retreat to higher altitudes, while lowland species invade into their former territory. Relicts of European species in North Africa (*E. orbicularis*) or Asia Minor (*L. viridis*, *V. barani*) may die out, while other species take their place.

Thus future climate change will favor some species to the disadvantage of others.

4.5 Conclusions for Conservation

Conservation priorities are set under national and international criteria. International goals were formulated at the Rio 1992 conference and include worldwide conservation of genetic heritage. This would mean that Holocene relict populations – genetically still very close to their relatives in the main distribution area – deserve less attention than endemics, regardless of whether they are relicts or not. Pleistocene relicts would occupy an intermediate position.

Thus in international conservation strategies, PostPleistocene relicts can be neglected. However on a national basis, there is usually much emphasis on rarities such as Mediterranean relicts in Central Europe. Although they are common and not very cared about in the Southern part of their range, they attract much attention in the North. Moreover, they are "flagships" of thermophilic biocenoses which contributea significant amount to the national species diversity in the depauperated Northern countries. *L. agilis*, a species which is very common in Southern Germany or Eastern Europe, was placed in the highest category of the European Fauna-Flora-Habitat directive because of its rarity in Britain and Sweden.

The main point to clarify is whether isolated populations are indigenous or introduced by humans. Genetic methods can provide useful data to solve this problem, but if the indigenous haplotype also occurs in the main range, they cannot provide a certain answer. For example in German *E. orbicularis*, the Eastern haplotype IIb is indigenous, whereas turtles from Western Germany -IIa – are found among many non-indigenous Southern European haplotypes (Fritz et al. 2004). Turtles with haplotype IIa could be introduced from France or Hungary, or could represent the last survivors of the indigenous population. The isolated West German Kaiserstuhl population of *L. viridis* (see Fig. 3) is now under suspicion of having been introduced, as *L. bilineata* would be the indigenous species expected there (Joger et al. 2001).

One main result of our genetic studies was the discovery of unique haplotypes at the Southern margin of most species ranges. Red lists and conservation plans in the Mediterranean and Caucasian countries must be directed to take special care of formerly ignored local endemics such as *E. trinacris*, *Z. lineatus,* and *H. viridiflavus xanthurus* in Sicily and Cantabria, *L.* (*agilis*) *boemica* in Armenia, Southern Spanish *N. maura* ssp. and Greek *N. tessellata* ssp.

Outside Europe, in Turkish Anatolia, Iran, and the Maghreb countries, increased research activities (particularly phylogeographic studies with molecular techniques) are needed to evaluate the high genetic diversity, identify relicts and cryptic species which are still unknown, and define conservation targets.

Acknowledgments We like to thank many colleagues – too numerous to mention individually – who provided tissues from different parts of Europe and non-European parts of distribution ranges. Robin Lawson helped with sequencing *Natrix* spp, Nik Stümpel and Alexander Zinenko sequenced additional *Vipera* samples and Toni Amann did initial work on the *Lacerta viridis* complex. Our work was financed with grants provided by the Deutsche Forschungsgemeinschaft (DFG Jo-134/7, Jo-134/9 and Wi-319/18).

References

Amann T, Razzetti E, Joger U (2001) La zona di contatto tra *Lacerta bilineata* (Daudin, 1802) e *Lacerta* viridis (Laurenti, 1768) in Italia. Atti della 3° Congresso Nazionale, Societas Herpetologica Italica. Pianura (science e storia dell'ambiente padano) 13:261–264

Amann T, Rykena S, Joger U, Nettmann HK, Veith M (1997) Zur artlichen Trennung von *Lacerta bilineata* Daudin, 1802, und *L. viridis* (Laurenti, 1768). Salamandra 33:255–268

Avise JC (1994) Molecular markers, natural history and evolution. Chapman & Hall, New York

Babik W, Branicki W, Sandera M et al (2004) Mitochondrial phylogeography of the moor frog, *Rana arvalis*. Mol Ecol 13:1469–1480

Barata M, Harris DJ, Castilho R (2008) Comparative phylogeography of Northwest African *Natrix maura* (Serpentes: Colubridae) inferred from mtDNA sequences. Afr Zool 43:1–7

Barton NH, Hewitt GM (1988) Analysis of hybrid zones. Annu Rev Ecol Syst 16:113–148

Bischoff W (1988) Zur Verbreitung und Systematik der Zauneidechse, *Lacerta agilis* Linnaeus 1758. Mertensiella 1:11–30

Böhme W, Joger U (1984) Eine neue Art des Vipera berus – Komplexes ausder Türkei. Amphibia-Reptilia 4:265–271

Böhme MU, Fritz U, Kotenko T, Dzukic G, Ljubisavljevic K, Tzankov N, Berendonk TU (2006) Phylogeography and cryptic variation within the *Lacerta viridis* complex (Lacertidae, Reptilia). Zool Scr 36:119–131

Brückner M, Klein B, Düring A, Mentel T, Rabus S, Soller JT (2001) Phylogeographic analysis of the *Lacerta viridis/bilineata* complex: molecular patterns and distribution. Mertensiella 13:45–51

Felsenstein J (1985) Confidence limits on phylogenies: an approach using the bootstrap. Evolution 39:783–791

Fritz U (1996) Zur innerartlichen Variabilität von *Emys orbicularis* (Linnaeus, 1758). 5. Innerartliche Hierarchie und Zoogeographie. Zool Abh Staatl Museum Tierkunde Dresden 49:31–71

Fritz U, Guicking D, Lenk P, Joger U, Wink M (2004) When turtle distribution tells European history: mtDNA haplotypes of *Emys orbicularis* reflect in Germany former division by the Iron Curtain. Biologia 59(Suppl.14):19–25

Fritz U, Fattizzo T, Guicking D, Tripepi S, Pennisi MG, Lenk P, Joger U, Wink M (2005) A new cryptic species of pond turtle from South Italy, the hottest spot in the range of the genus *Emys* (Reptilia: Testudines: Emydidae). Zool Scr 34:1–21

Fritz U, Guicking D, Kami H, Arakelyan M, Auer M, Ayaz D, Ayrez Fernandez C, Bakiev A, Celani A, Dzukic G, Fahd S, Havas P, Joger U, Khabibullin VN, Mazanaeva LF, Siroky P, Tripepi S, Velez AV, Velo Antón G, Wink M (2007) Mitochondrial phylogeography of European pond turtles (*Emys orbicularis*, *Emys trinacris*) – an update. Amphib Reptil 28:418–426

Gliemeroth AK (1995) Paläoökologische Untersuchungen über die letzten 22.000 Jahre in Europa. Paläoklimaforschung 18. Gustav Fischer, Stuttgart

Godinho R, Crespo EG, Ferrand N, Harris DJ (2005) Phylogeny and evolution of the green lizards, *Lacerta* spp. (Squamata: Lacertidae) based on mitochondrial and nuclear DNA sequences. Amphib Reptil 26:271–285

Guicking D, Joger U, Wink M (2009) Cryptic diversity in a Eurasian water snake (Natrix tessellata, Serpentes: Colubridae): Evidence from mitochondrial sequence data and nuclear ISSR-PCR finger printing. Org Divers Evol 9:201–214

Guicking D, Lawson R, Joger U, Wink M (2006a) Evolution and phylogeny of the genus *Natrix* (Serpentes: Colubridae). Biol J Linn Soc 87:127–143

Guicking D, Griffiths RA, Moore RD, Joger U, Wink M (2006b) Introduced alien or persecuted native? Resolving the origin of the viperine snake (*Natrix maura*) on the island of Mallorca. Biodivers Conserv 15:3045–3054

Guicking D, Joger U, Wink M (2008) Mitochondrial and nuclear phylogeography of the viperine snake (*Natrix maura*): evidence for strong intraspecific differentiation. Org Divers Evol 8:130–145

Guillaume CO, Heulin B, Attayago MJ, Bea A, Brana F (2000) Refuge areas and suture zones in the Pyrenean and Cantabrian regions: geographic variation of the female MPI sex-linked alleles among oviparous populations of the lizard *Lacerta* (*Zootoca*) *vivipara*. Ecography 23:3–10

Habel JC, Schmitt T, Müller P (2005) The fourth paradigma of postglacial range expansion of European terrestrial species: The phylogeography of the Marbled White butterfly (Satyrinae, Lepidoptera). J Biogeogr 32:1489–1497

Hewitt GM (1999) Post-glacial re-colonization of European biota. Biol J Linn Soc 58:247–276

Hewitt GM (2000) The genetic legacy of the Quaternary ice ages. Nature 405:907–913

Huelsenbeck JP, Ronquist F (2001) MrBayes: Bayesian inference of phylogenetic trees. Bioinformatics 17:754–755 Applications Note

Ivanov M (1997) Vývoj kenozické hadí fauny evropy. (The evolution of European Cenozoic snake fauna). In: Hladilová S (ed) Dynamics of interaction between marine and continental environments. Masaryk University, Brno, pp 59–91

Ivanov M (2001) Changes in the composition of the European snake fauna during the early miocene and at the Early/Middle Miocene transition. Paläontologische Z 74:563–573

Jaarola M, Searle JB (2002) Phylogeography of field voles (*Microtus agrestis*) in Eurasia inferred from mitochondrial DNA sequences. Mol Ecol 11:2613–2621

Joger U, Lenk P (1997) Entnahme und Behandlung von Blutproben bei Amphibien und Reptilien im Freiland. Mertensiella 7:329–340

Joger U, Steinfartz S (1994) Electrophoretic investigations in the evolutionary history of the West Mediterranean *Salamandra*. Mertensiella 4:241–254

Joger U, Amann T, Lenk P, Willand U (1998) Molekulare Merkmale und das phylogenetische Artkonzept. Zool Abh Staatl Museum Tierkunde Dresden 50(Suppl.) 100 Jahre Artkonzepte in der Zoologie:109–123

Joger U, Amann T, Veith M (2001) Phylogeographie und genetische Differenzierung im *Lacerta viridis-bilineata*-Komplex. Mertensiella 13:60–68

Joger U, Guicking D, Kalyabina-Hauf S, Lenk P, Nagy ZT, Wink M (2006) Phylogeographie, artbildung und postpleistozäne einwanderung mitteleuropäischer reptilien. Z Feldherpetol 10(Suppl.):29–59

Joger U, Fritz U, Guicking D, Kalyabina-Hauf S, Lenk P, Nagy ZT, Wink M (2007) Phylogeography of Western palaractic reptiles – Spatial and temporal speciation patterns. Zool Anz 246:293–313

Kalyabina SA, Milto KD, Ananjeva NB, Legal L, Joger U, Wink M (2001) Phylogeography and systematics of *Lacerta agilis* based on mitochondrial cytochrome b gene sequences: first results. Russ J Herpetol 8:149–159

Kalyabina-Hauf SA, Milto KD, Ananjeva NB, Joger U, Kotenko TI, Wink M (2004a) Reevaluation of the status of *Lacerta agilis tauridica* Suchov, 1926. Russ J Herpetol 11:65–72

Kalyabina-Hauf S, Schweiger S, Joger U, Mayer W, Wink M (2004b) Phylogeny and systematics of adders (*Vipera berus* complex). In: Joger U, Wollesen R (eds) Verbreitung, Ökologie und Schutz der Kreuzotter (*Vipera berus* [Linnaeus, 1758]). Mertensiella 15. Rheinbach, pp. 7–16

Lenk P, Joger U (1994) Genetic relationships between populations and intraspecific subdivision of *Elaphe longissima* (Laurenti, 1768). Amphib Reptil 15:363–373

Lenk P, Wüster W (1991) A multivariate approach to the systematics of Italian ratsnakes of the *Elaphe longissima* complex (Reptilia: Colubridae): Revalidation of Camerano's *Callopeltis longissimus* var. *lineata*. Herpetol J 9:153–162

Lenk P, Joger U, Fritz U, Heidrich P, Wink M (1999) Phylogeographic patterns in the mitochondrial cytochrome b gene of the European pond turtle, *Emys orbicularis* (Linnaeus). Mol Ecol 8:1911–1912

Lenk P, Joger U, Wink (2001) Phylogenetic relationstips among European ratsnakes of the genus *Elaphe* FITZINGER based on mitochondrial DNA sequence comparisons. Amphibia-Reptilia 22:329–339

Lopez-Fuster MJ, Ventura J (1996) A morphometrical review of the *Sorex araneus-arcticus* species group from the Iberian Peninsula (Insectivora, Soricidae). Bonn Zool Beitr 46:327–337

Mayer W, Beyerlein P (2001) Genetische Differenzierung von *Lacerta viridis/bilineata* und *L. trilineata* in Griechenland. Mertensiella 13:52–59

Nagy ZT, Joger U, Guicking D, Amann T, Wink M (2002) Phylogeography of the European whip snake, *Coluber (Hierophis) viridiflavus* Lacépède, 1789, inferred from nucleotide sequences of the mitochondrial cytochrome b gene and ISSR genomic fingerprinting. Biota (Race) 3:109–118

Nagy ZT, Joger U, Wink M, Glaw F, Vences M (2003) Multiple colonization of Madagascar and Socotra by colubrid snakes: evidence from nuclear and mitochondrial gene phylogenies. Proc R Soc Lond B 270:2613–2621

Nagy ZT, Lawson R, Joger U, Wink M (2004) Molecular systematics of racers, whipsnakes and relatives (Reptilia: Colubridae) using mitochondrial and nuclear markers. J Zool Syst Evol Res 42:223–233

Nettmann HK (2001) Die Smaragdeidechsen (Lacerta s. str.) – Eine Übersicht über Verwandtschaft und Formenvielfalt. Mertensiella 13:11–32

Peters G (1977) Die Reptilien aus den fossilen Tierbautensystemen von Pisede bei Malchim. Wiss Z Humboldt-Univ Berlin, Mathem-Naturwiss Reihe 26:307–327

Posada D, Crandall KA (1998) Modeltest: testing the model of DNA substitution. Bioinformatics 14:817–818

Rykena S (1991) Kreuzungsexperimente zur Prüfung der Artgrenzen im Genus *Lacerta* sensu stricto. Mitt Zool Mus Berlin 67:55–68

Rykena S (1996) Experimental interspecific hybridization in the genus *Lacerta*. Isr J Zool 42:171–184

Rykena S (2001) Experimental hybridization in Green Lizards (*Lacerta* s. str.), a tool to study species boundaries. Mertensiella 13:78–88

Sambrook J, Russell DW (2001) Molecular cloning. A laboratory manual. Cold Spring Harbor Laboratory Press, New York

Scali S, Gentilli A (1999) Morphometric analysis and sexual dimorphism of extinct adders (*Vipera berus*) of the Po plane (Northern Italy). In: Miaud C, Guyetant R (eds) Current studies in herpetology. Proceedings of the 9th Ordinary General Meeting of the SEH. Le Bourget du Lac, pp. 391–396

Schätti B, Vanni S (1986) Intraspecific variation in *Coluber viridiflavus* Lacepède, 1789, and validity of its subspecies (Reptilia, Serpentes, Colubridae). Rev Suisse Zool 93:219–232

Stümpel N, Brito JC, Saint Girons H (2005) *Vipera* (*Vipera*) *latastei* Boscá, 1878 – Stülpnasenotter. In: Joger U, Stümpel N (eds) Handbuch der Amphibien und Reptilien Europas, 3/IIB. Aula, Wiesbaden, pp 187–212

Swofford DL (2002) PAUP*. Phylogenetic Analysis Using Parsimony (*and Other Methods), Version 4.0b10. Sinauer Associates, Sunderland, MA

Szyndlar Z, Böhme W (1993) Die fossilen schlangen deutschlands: geschichte der faunen und ihrer erfoschung. Mertensiella 3:381–431

Taberlet P, Fumagalli L, Wust-Saucy A-G, Cosson JF (1998) Comparative phylogeography and postglacial colonization routes in Europe. Mol Ecol 7:453–464

Ursenbacher S, Carlsson M, Helfer V, Tegelström H, Fumagalli L (2006a) Phylogeography and pleistocene refugia of the adder (*Vipera berus*) as inferred from mitochondrial DNA sequence data. Mol Ecol 15:4325–3437

Ursenbacher S, Conelli A, Golay P, Monney J-C, Zuffi MAL, Thiery G, Durand T, Fumagalli L (2006b) Phylogeography of the asp viper (*Vipera aspis*) inferred from mitochondrial DNA sequence data: evidence for multiple Mediterranean refugial areas. Mol Phyl Evol 38:546–552

Vedmederja VI, Grubant VN, Rudaeva AV (1986) K voprosu o nazvanii chiornoy gadyuki lesostepi evropeiskoy chasti SSR. Vestnik Kharkovskogo Universiteta 288:83–85

Niche Conservatism among Allopatric Species of the Grasshopper Genus *Afrophlaeoba* Jago, 1983 in the Eastern Arc Mountains (Tanzania)

Axel Hochkirch

Abstract The persistence of species is strongly determined by their ecological niches. It is therefore, of major importance to obtain knowledge on the niche attributes of relict species in order to conserve these species and understand their current distribution patterns. Niche conservatism has been proposed to explain distribution patterns of many taxa, particularly in the tropics. Indeed, the high gamma diversity in the tropics seems to be influenced by allopatric assemblages of shared genera, whereas species overlap is often relatively low. Here, the niche overlap of allopatric *Afrophlaeoba* grasshopper species from the Eastern Arc Mountains (Tanzania) is studied and compared with niche overlaps among sympatric but distantly related species in the East Usambara Mountains. The results show that the genus *Afrophlaeoba* exhibits a strong degree of niche conservatism. All congeneric species are confined to grassy forest edges. Niche overlap within this genus was generally higher than expected by chance and higher than among sympatric species, whereas the variance of niche overlaps between species pairs was low. While the sympatric species could be distinguished ecologically using a few microhabitat variables, a comprehensive data set for the *Afrophlaeoba* species revealed only little resolution. The results support the hypothesis that diversification within tropical grasshoppers is not driven by niche divergence, but by allopatric isolation. Based on the niche quality of *Afrophlaeoba*, it is proposed that the persistence of the genus (and probably other forest taxa) might also be possible in riverine forests. There is a strong need for studies on the potential role of gallery forests as corridors for gene flow among mountain forest taxa.

A. Hochkirch
Department of Biogeography, Trier University
D-54296 Trier, Germany
e-mail: hochkirch@uni-trier.de

J.C. Habel and T. Assmann (eds.), *Relict Species: Phylogeography and Conservation Biology*, DOI 10.1007/978-3-540-92160-8_7, © Springer-Verlag Berlin Heidelberg 2010

1 Introduction

There is little doubt that the survival of populations is strongly influenced by their ecological niches (Thuiller et al. 2005). The occurrence of a relict species is, therefore, also determined by its niche quality, breadth and flexibility. Species or populations with a narrow and conserved niche are likely to be more prone to extinction than those with a broad niche or strong adaptability. Ecological niches of species are generally thought to be highly conserved (Peterson et al. 1999) and this seems to be particularly true for the tropics (Wiens and Graham 2005). Indeed, tropical mountain forests often maintain a species-rich fauna, and different mountain blocks have a low species overlap but a high number of shared genera (Hochkirch 1998). This pattern leads to high gamma-diversity (i.e. a high total biodiversity in larger tropical regions) with many relict species inhabiting single tropical mountains. The Eastern Arc (a mountain chain in East Africa) is an excellent area to study niche divergence of tropical relict species, as the mountains in this chain are covered by evergreen forests and separated by savannahs and dry woodlands (Lovett and Wasser 1993). The Eastern Arc belongs to the global hotspots of biodiversity (Myers et al. 2000) and is known for its great species richness with a high degree of endemism (Rodgers and Homewood 1982; Lovett and Wasser 1993; Burgess et al. 2007). This is particularly true for flightless insect species, such as most forest grasshoppers (Hochkirch 1998). One major reason for the high degree of endemism in the Eastern Arc fauna seems to be the comparatively stable climate during Pleistocene climatic fluctuations (Scharff 1993). Eastern Arc endemics are, therefore, thought to be relicts of formerly widespread forest taxa, which survived periods of drought at high elevations (Hochkirch 2001).

Studies on niche conservatism usually deal with climatic variables that might explain the geographic range of species (e.g. climate envelopes of congeneric species; Peterson et al. 1999). Other niche dimensions, such as the microhabitat preferences of tropical sibling species have rarely been considered. However, at a small spatial scale, sympatric insect species often partition niche space in terms of their microhabitat utilisation (Hochkirch 1995). Hence, it is necessary to compare the microhabitat requirements of allopatric species in order to address the question of niche conservatism. The flightless grasshopper genus *Afrophlaeoba* Jago 1983 consists of four allopatric species, each of which is endemic to a single mountain block of the Eastern Arc. Genetic and morphometric data suggest that the genus has been formerly more widespread in East Africa and that speciation took place after fragmentation of lowland and riverine forests during the early Pleistocene (Hochkirch 2001). The four species are morphologically rather similar but can be distinguished genetically and by multivariate morphometrics (Hochkirch 2001, 2005). Morphologically, the genus *Afrophlaeoba* represents the graminicolous type *sensu* Uvarov (1977). Most grasshopper species are specialised on a particular vegetation structure and many species exhibit behavioural adaptations to their microhabitat (Sänger 1977; Uvarov 1977). In many cases, there are also intersexual differences in microhabitat utilisation due to sex-specific differences in energy budgets (Hochkirch et al. 2007).

In the present paper, the microhabitat utilisation of all *Afrophlaeoba* species is studied to obtain information on the niche overlap among these allopatric species. Given that the hypothesis of niche conservatism is true, one would expect a high niche overlap among the four species. For comparison, the niche overlap of sympatric grasshopper species of the East Usambara Mountains is analysed. If niche overlap was greater among the sympatric (distantly related) species than among the allopatric (closely related) species, it would be likely that the specific vegetation structure of the mountain blocks has a stronger influence on niche overlap than the genetic relationship. The results might also help to reconstruct the ancestral niche of the species - information which is needed to draw conclusions on the former habitat connectivity.

2 Methods

2.1 Study Sites

The Eastern Arc Mountains are Pre-cambrian basement rocks, which have been faulted during the Miocene (Iversen 1991). The area consists of several mountain blocks with a high number of endemic species and genera (Burgess et al. 1998a). The more prominent rocks of the Arc are the Taita Hills (Kenya), North and South Pare Mts., East and West Usambara Mts., Nguu Mts., Nguru Mts., Ukaguru Mts., Rubeho Mts., Uluguru Mts., Malundwe Hill, Udzungwa Mts. and Mahenge Mts. (all in Tanzania). About 25% of the ca. 2,000 plant species and 66% of the 82 amphibian and reptile species occurring in the Eastern Arc are endemic (Lovett 1988; Howell 1993). The area is influenced by the Indian Ocean monsoon (Lovett 1993a) and the mountain blocks are separated by dry woodland and savannah. Marine drill-core data suggest that the coastal waters of East Africa were less influenced by Pleistocene climatic fluctuations (1–2°C) than other oceans (Prell et al. 1980). The stable temperatures of the Indian Ocean may have supported the high rainfall pattern of the Eastern Arc and the persistence of evergreen forests. The present climate of the Eastern Arc is mainly influenced by the seasonal movements of the intertropical convergence zone, which crosses the equator twice a year (Hamilton 1982). Subsequently, two rainy seasons occur in the East Usambara Mountains, while the rainy seasons merge to one long rainy season further South (e.g. in the Udzungwa Mts.). Secondly, the rainfall pattern is modified by oceanic currents and the positions of the mountains. The currents of the Indian Ocean (South Equatorial current) bring warm and humid waters to the East African coast, increasing the rainfall close to the coast and at the Eastern slopes of the mountains (Wasser and Lovett 1993). The Eastern Arc Mountains are known to receive the highest rainfall in East Africa (up to 2,000–3,000 mm per year). The temperatures are exceptionally low, with frosts occurring frequently above altitudes of ca. 2,000 m during July and August (Lovett 1998). The cool climate is mainly caused by cloudiness, which decreases the quantities of incoming radiation (Kenworthy 1966). Parts of the Usambaras and Ulugurus have an almost per-humid climate (Lovett 1993a).

The vegetation zones of the Eastern Arc mainly follow the altitudinal and moisture gradients (Lovett 1993b). The floristic variation with the altitude is continuous without any abrupt changes (Hamilton 1998). Forests naturally extend from the lowlands of the Eastern slopes up to 2,400 m above sea level. However, at the lower slopes many forests have been cleared. At low altitudes the forest vegetation is rather similar to the lowland coastal forests, which occur fragmented along the Eastern coast of Africa (Burgess et al. 1998b). The lowlands between the Southern (and inland) parts of the mountain chain are mainly covered by miombo woodland (composed of the tree genera *Brachystegia*, *Isoberlinia* and *Julbernardia*) and receive an annual rainfall of 750–1,000 mm with a long dry season in between (Hamilton 1982).

2.2 Study Objects

The genus *Afrophlaeoba* consists of four flightless grasshopper species, each of which is endemic to a single mountain block (Jago 1983; Hochkirch 2001). *Afrophlaeoba usambarica* occurs in the East Usambara Mountains, *Afrophlaeoba euthynota* in the Uluguru Mountains, *Afrophlaeoba nguru* in the Nguru Mountains and *Afrophlaeoba longicornis* in the Rubeho and Ukaguru Mountains. The species have a continuous life cycle and adults can be found throughout the year (Jago 1983; Hochkirch 1996). They mainly inhabit grassy forest edges and clearings and seem to be graminivorous (Hochkirch 1996). All other species studied are endemics to the East Usambara Mountains or occur from this mountain block to the coast. They include distantly related taxa as well as a sister genus of *Afrophlaeoba* – *Parodontomelus arachniformis* – which is another graminicolous species that belongs to the *Afrophlaeoba* genus group. It occurs in the coastal forests of Tanzania, the East Usambara Mountains and in Zanzibar (Jago 1983). *Rhainopomma usambaricum* is distributed from the East Usambaras to the Shimba Hills in South-Eastern Kenya (Jago 1981). It occurs at forest edges, paths and clearings in dense and high foliage, where it mainly feeds on forbs (Hochkirch 1995). *Parepistaurus pygmaeus* is known from the East Usambaras and the Nguru Mountains (Green 1998), where it inhabits warm and sunny forest edges with low and dense vegetation (Hochkirch 1995). *Ixalidium transiens* is endemic to the East Usambara Mountains. It can be found among litter at the forest floor and feeds mainly on plant debris (Hochkirch 1995).

2.3 Data Collection

The practical fieldwork was performed from 06 December 1997 to 13 February 1998 (*Afrophlaeoba* and *Parodontomelus*). The data was recorded during a short period of time to avoid seasonal influences. In order to collect data on the microhabitat preferences, the method described by Hochkirch (1995, 2001) was used. All data were obtained between 8:00 and 18:00 at the exact location of undisturbed individuals, which were haphazardly chosen. For each grasshopper the following information was recorded: date, time, study site, species, sex, behaviour, vegetation cover (divided into grasses, forbs, mosses, litter and bare ground, estimated in a circle of 25 cm radius around the individual), vegetation height (highest plant in a circle of 25 cm radius) and the substrate, on which the insect perched (including the categories grasses, forbs, bare ground, leaf litter, grass litter). Temperature and radiation was also recorded, but not included in the analyses as these values depend too much on weather conditions. In total, 300 records were available for the four *Afrophlaeoba* species and for *P. arachniformis*. For comparison, data of sympatric species of the East Usambara Mts. were included, which were recorded from 11 June 1994 to 13 November 1994 (*R. usambaricum*, *P. pygmaeus*, *I. transiens*). The data was obtained similarly to the above mentioned, but vegetation cover was not divided into different categories. In total, 732 records were obtained for *R. usambaricum*, 637 records for *P. pygmaeus* and 184 records for *I. transiens*.

2.4 Statistical Analysis

The Czechanowski index (Gotelli and Entsminger 2001) was applied to quantify niche overlap between the species for all microhabitat parameters. The Czechanowski index ranges from 0 (no shared resource states) to 1 (identical habit utilisation). To test whether the observed niche overlap differed from a random pattern we carried out null model analyses in EcoSim 7.0 (Gotelli and Entsminger 2001). EcoSim simulates patterns of niche overlap and compares these randomised results with the observed data matrix. We used the algorithm RA3 (Winemiller and Pianka 1990) to test for non-random niche overlap. In this procedure, the observed niche breadth of each species is retained, but the utilisation of any resource state is allowed, including categories that were available but not used by the species. For each data set 10,000 replicates were created in the simulation.

To test for correlations between vegetation parameters, multivariate ordination analyses were performed. Because of the long gradient length in the data set, a correspondence analysis (CCA) was performed as implemented in the community ecology package vegan 1.13-1 for R (Oksanen et al. 2005). To evaluate the most important variables for the species, the locations of the species were fitted onto the CCA plot, using the function "envfit" with 1,000 random permutations. This method also generates an R^2 measure and significance values based on the probability that random permutations would yield a higher degree of fit than the true data (Oksanen et al. 2005). The multivariate analyses were carried out in R 2.7.1 (R Development Core Team 2008).

3 Results

3.1 Niche Overlap

The Ecosim analyses revealed that for all variables examined, niche overlap among *Afrophlaeoba* species was significantly greater than expected by chance, whereas for several variables the variance of the niche overlaps was smaller than expected (Table 1, Fig. 1). Averaged across all variables, niche overlap varied between 80.7 and 85.0% between congeneric species pairs. Compared with sympatric species from the East Usambara Mts., niche overlap was generally higher among *Afrophlaeoba* (Fig. 2). In the East Usambaras, niche overlap was also more variable (variance was greater than expected by chance), ranging from 28.0 to 85.4% (mean: 59.1%).The greatest niche overlap among the sympatric species was found between *A. usambarica* and *P. pygmaeus* which occur syntopically at forest edges in the East Usambaras, but have a different diet (Hochkirch 2001).

3.2 Multivariate Analysis

The first function of the correspondence analysis (CCA) for the vegetation parameters of the *Afrophlaeoba* data set (including *P. arachniformis*) explained 64.7% of the variance and the second function 13.1% (Fig. 3). The grasshopper species were significantly correlated with the vegetation matrix ($P<0.001$), but the resolution was relatively low ($R^2=0.106$). The *Afrophlaeoba* species were plotted gradually along the first axis, while *P. arachniformis* was mainly found at patches with a great amount of leaf litter. For the complete data set (East Usambara grasshoppers combined with the *Afrophlaeoba* species), the CCA explained 68.6% of the variance and the second axis 31.4% (Fig. 4). Again, the species were significantly correlated with the vegetation data ($P<0.001$), but with a greater resolution ($R^2=0.276$). In the ordination plot (Fig. 4), the distance between the sympatric species was greater than between the *Afrophlaeoba* species.

4 Discussion

The results support the niche conservatism hypothesis. Compared to the marked niche differences among sympatric grasshopper species of the East Usambara Mts., niche overlap was high among the four allopatric, congeneric species. For all microhabitat variables analysed, niche overlap was significantly greater than expected by chance. A strong niche overlap was also reported for other traits such as phenology, diet and microclimatic factors (Hochkirch 2001). The results support equilibrium models of biodiversity, suggesting that niche specialisation plays a major role in the

Table 1 Czechanowski indices for each species pair of *Afrophlaeoba* and results of a null model analysis*

Czechanowski index	Vegetation cover	Gras cover	Forb cover	Leaf litter	Bare ground	Vegetation height	Perch height	Substrate
A. euthynota-A. longicornis	0.893	0.687	0.757	0.763	0.917	0.880	0.799	0.863
A. euthynota-A. nguru	0.807	0.860	0.817	0.617	0.907	0.863	0.800	0.774
A. euthynota-A. usambarica	0.840	0.913	0.913	0.760	0.977	0.817	0.888	0.888
A. longicornis-A. nguru	0.737	0.723	0.753	0.687	0.983	0.810	0.872	0.911
A. longicornis-A. usambarica	0.793	0.697	0.717	0.893	0.910	0.840	0.768	0.837
A. nguru-A. usambarica	0.900	0.870	0.850	0.783	0.900	0.733	0.779	0.764
Observed mean	0.828	0.792	0.801	0.751	0.932	0.824	0.818	0.840
Simulated mean	0.265	0.664	0.545	0.303	0.297	0.384	0.448	0.361
P (obs>exp)	<0.001	<0.001	<0.001	<0.001	<0.001	<0.001	<0.001	<0.001
observed variance	0.0003	0.0101	0.0053	0.0087	0.001	0.0027	0.0025	0.0036
simulated variance	0.0260	0.0064	0.0085	0.0263	0.1296	0.0107	0.0120	0.0466
P (obs<exp)	0.031	0.864	0.298	0.051	0.013	0.046	0.029	0.013

*The analyses were performed in Ecosim 7.0 Gotelli and Entsminger (2001). For each data set 10,000 replicates were created in the simulation

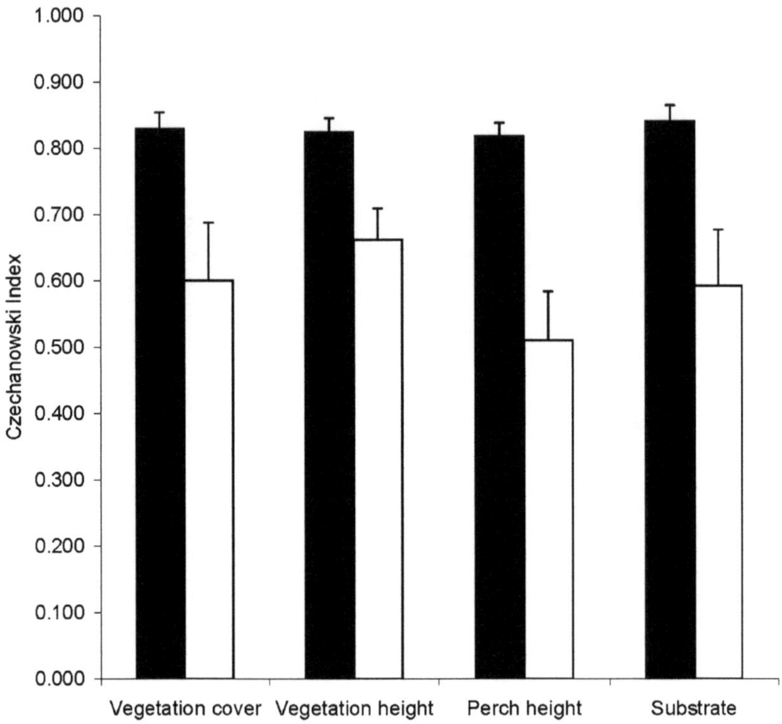

Fig. 1 Mean observed niche overlaps (*black*) and mean simulated niche overlaps (*white*) among allopatric *Afrophlaeoba* species. Error bars are variances. All observed niche overlaps were significantly higher than expected by chance

generation of the high alpha biodiversity in the evergreen mountain forests (Linsenmair 1990). Indeed, the five species from the East Usambara Mts. can be clearly distinguished ecologically: *I. transiens* occurs in the ground litter of the forest, where it feeds on litter; *R. usambaricum* occurs in the dense and high vegetation of small forest gaps, feeding on forbs; *P. arachniformis* occurs in the sparse and low grassy vegetation of forest gaps, feeding on grasses; *P. pygmaeus* and *A. usambarica* occur syntopic in the dense and low vegetation of the forest edge (which is also illustrated by a strong niche overlap), but *P. pygmaeus* is a forb feeder, whereas *A. usambarica* is graminivorous. Although the niche overlap between *P. arachniformis* and *A. usambarica* is also relatively high (78.8%), these two species never co-occur in the East Usambaras as *P. arachniformis* is confined to the forest interior, while *A. usambarica* is found at the forest edge (Hochkirch 1996, 2001).

The Eastern Arc is believed to have served as a major refuge area during periods of drought. This hypothesis is supported not only by a high number of endemic species, but also by phylogenetic studies (Hochkirch 2001). Some taxa have their closest relatives in West Africa (e.g. the flightless grasshopper genus *Loveridgacris* Rehn, 1954), some in Madagascar (e.g. the groundhopper genus *Cryptotettix*

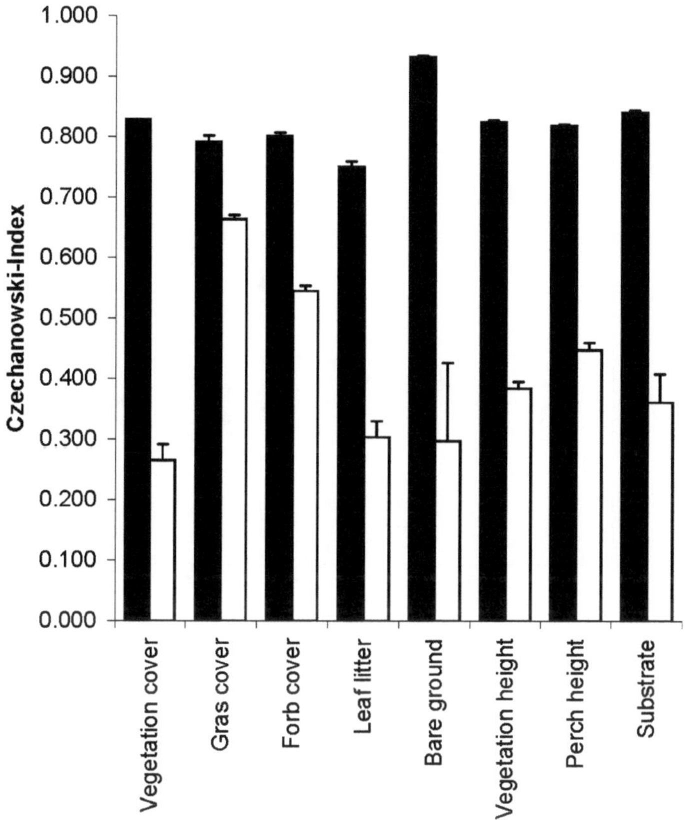

Fig. 2 Mean observed niche overlap (Czechanowski Index) among allopatric *Afrophlaeoba* species (*black*) and sympatric grasshopper species in the East Usambara Mts. (*white*). Error bars are standard errors. Missing error bars are too small to be displayed

Hancock, 1900) and some even in Southeast Asia (e.g. the Udzungwa forest partridge *Xenoperdix udzungwensis* Dinesen, 1994). Climatic stability is of major importance for the survival of such relicts (Lovett and Friis 1996), but interestingly the majority of neoendemics are located in the same areas as relict species (Fjeldså and Lovett 1997). The low number of endemic mammals in the Eastern Arc (Stanley et al. 1998) suggests that the forest refuges in East Africa were rather small.

The genus *Afrophlaeoba* represents a relatively young radiation in the Eastern Arc. Molecular phylogenetic analyses suggest that the oldest split within the genus originated during a period of drought during the early Pleistocene (Hochkirch 2001). Two vegetation types are thought to have provided corridors for gene flow among the formerly widespread populations of these flightless insects: coastal forests and riverine forests (Hochkirch 2005). A scenario of the evolutionary divergence of the genus (Hochkirch 2001) suggests that the ancestral distribution comprised at least the formerly widespread coastal forests, adjacent riverine forests

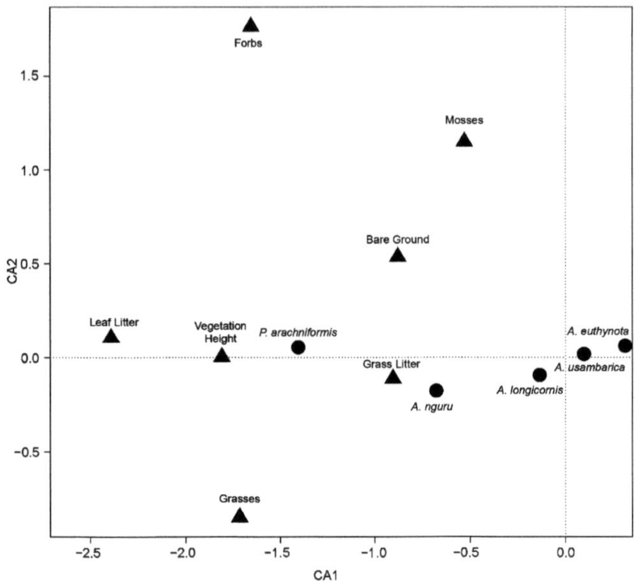

Fig. 3 Plot of the first two axes of a correspondence analysis of the microhabitat data of *Afrophlaeoba*. The first axis explains 64.7% of the variance; the second axis explains 13.1% of the variance

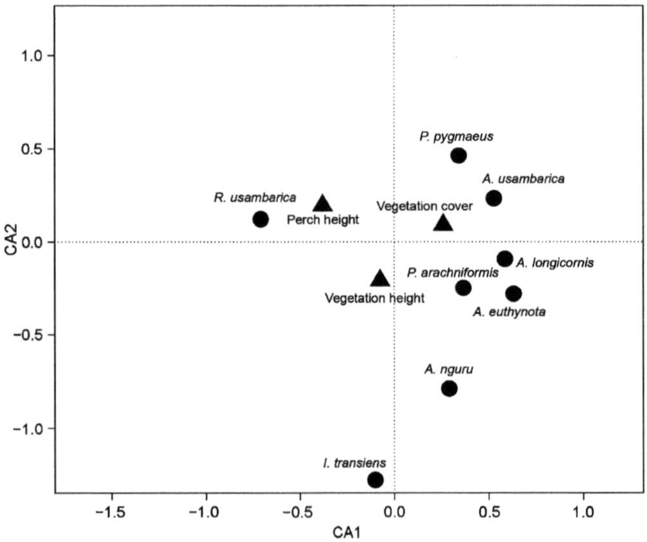

Fig. 4 Plot of the first two axes of a correspondence analysis of the microhabitat data of all grasshopper species examined. The first axis explains 68.6% of the variance and the second axis explains 31.4% of the variance. Sympatric species are represented by black dots; the three allopatric species (*A. nguru, A. longicornis, A. euthynota*) are shown in grey

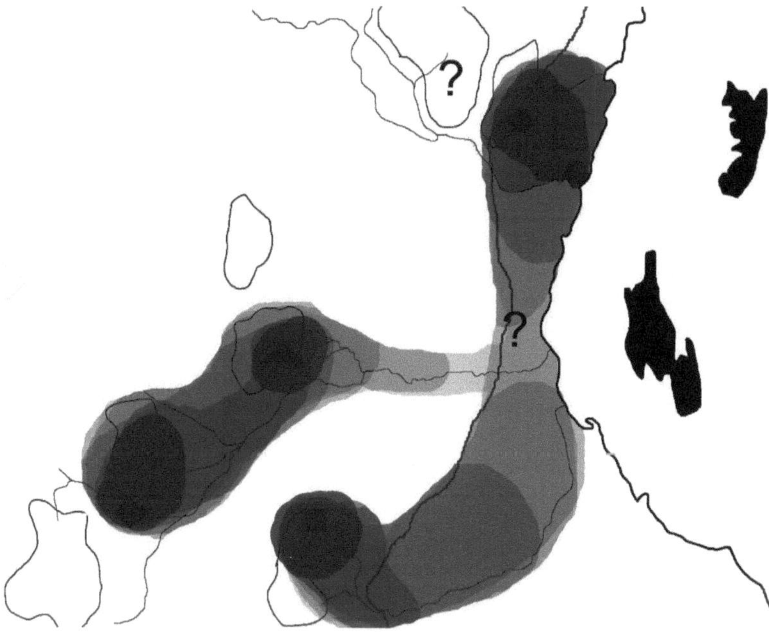

Fig. 5 Vicariance scenario for the fragmentation of populations of *Afrophlaeoba* (question marks represent unconfirmed recent records, lighter shadings represent higher ages; the off-shore islands of Zanzibar and Pemba are shown in black; modified from Hochkirch (2005)

and parts of the Eastern Arc (Figs. 5). The Wami River basin might have acted as corridor for the mountain and coastal populations. The phylogenetic relationships suggest an initial fragmentation of the "Wami River Connection" close to the coast, probably in the Zanzibar rain shadow which is the driest area at the coast (Hawthorne 1993). The breakdown of this habitat connection was followed by fragmentation of the coastal forests shortly after that. It is well known that coastal forests were once more widespread in areas that are now woodland (Clarke and Karoma 2000). Interestingly, the Zanzibar rain shadow represents an area where a number of other taxa show North-South disjunctions (Hawthorne 1993; Robbrecht 1996; Tattersfield 1998; Lindqvist and Albert 1999). The "Wami River Connection" between the Nguru Mountains and the Rubeho Mountains must have remained stable for a longer time and might even still exist.

Ecologically, the *Afrophlaeoba* species are not fully restricted to dense evergreen forests as suggested by Jago (1983); rather, they occur at forest edges and are able to persist in plantations of mango trees or in gallery forests, which might have served as corridors for gene flow between the mountain populations (Hochkirch 2005). It has also been assumed for other taxa with a disjunct distribution in the East African Mountains, that these might have once occurred continuously in lowland and riverine forests (Hamilton 1982; Brandl et al. 1996). Indeed, several taxa are known to be shared by the Coastal Forests, the Eastern Arc and several gallery forests, such as the Pongwe and Wami River valleys (Hoffman 1993; Kellman et al. 1994; Clarke 2000). These riverine forests might also have served as refuges for forest taxa in periods of drought, since they are less dependent on rainfall but more on groundwater (Medley 1998). Even small remnants of gallery forests might have been large enough to maintain populations of arthropods (Scharff 1992). There is a strong need for research on the possible interconnections between mountain forest populations, which might be provided by riverine forests.

Acknowledgements I thank my supervisor, Prof. Dr. Dietrich Mossakowski (Bremen, Germany), and my local contact, Prof. Kim Howell (Dar es Salaam, Tanzania), for their help and advice. This work was financially supported by the "Stifterverband für die Deutsche Wissenschaft" and by a "DAAD Doktorandenstipendium im Rahmen des gemeinsamen Hochschulsonderprogramms III von Bund und Ländern". The Commission for Science and Technology (COSTECH, Dar es Salaam, Tanzania) provided the research clearance for this study.

References

Brandl R, Bagine RNK, Kaib M (1996) The distribution of *Schedorhinotermes lamanianus* (Isoptera: Rhinotermitidae) and its termitophile *Paraclystis* (Lepidoptera: Tineidae) in Kenya: its importance for understanding East African biogeography. Glob Ecol Biogeogr 5:143–148

Burgess ND, Fjeldså J, Botterweg R (1998a) Faunal importance of the Eastern Arc mountains of Kenya and Tanzania. J East Afr Nat Hist 87:37–58

Burgess ND, Clarke GP, Rodgers WA (1998b) Coastal forests of Eastern Africa: status, endemism patterns and their potential causes. Biol J Linn Soc 64:337–367

Burgess ND, Butynski TM, Cordeiro NJ, Doggart NH, Fjeldså J, Howell KM, Kilahama FB, Loader SP, Lovett JC, Mbilinyi B, Menegon M, Moyer DC, Nashanda E, Perkin A, Rovero F, Stanley WT, Stuart SN (2007) The biological importance of the Eastern Arc mountains of Tanzania and Kenya. Biol Conserv 134:209–231

Clarke GP (2000) Defining the Eastern African Coastal Forests. In: Burgess ND, Clarke GP (eds) Coastal forests of Eastern Africa. IUCN, Gland, pp 9–26

Clarke GP, Karoma NJ (2000) History of anthropic disturbance. In: Burgess ND, Clarke GP (eds) Coastal forests of Eastern Africa. IUCN, Gland, pp 251–261

Fjeldså J, Lovett JC (1997) Geographical patterns of old and young species in African forest biota: the significance of specific montane areas as evolutionary centres. Biodivers Conserv 6:325–346

Green SV (1998)' Revision of the African grasshopper genus *Parepistaurus* Karsch 1896 (Orthoptera Acrididae Coptacridinae). Trop Zool 11:259–332

Gotelli NJ, Entsminger GL (2001) EcoSim: null models software for ecology. Version 7.0. Acquired Intelligence Inc. and Kesey-Bear. http://homepages.together.net/~gentsmin/ecosim.htm. Cited 11 July 2008

Hamilton AC (1982) Environmental history of East Africa: a study of the Quaternary. Academic, London

Hamilton AC (1998) Vegetation, climate and soil: altitudinal relationships on the East Usambara mountains, Tanzania. J East Afr Nat Hist 87:85–89

Hawthorne WD (1993) East African coastal forest botany. In: Lovett JC, Wasser SK (eds) Biogeography and ecology of the rain forests of Eastern Africa. Cambridge University Press, Cambridge, pp 57–99

Hochkirch A (1995) Habitatpräferenzen dreier Heuschreckenarten im submontanen Regenwald der Ost-Usambaraberge, NO-Tansania (Orthoptera; Acridoidea). Mitt Dtsch Ges Allg Angew Entomol 10:297–300

Hochkirch A (1996) Habitat preferences of grasshoppers (Orthoptera: Acridoidea, Eumastacoidea) in the East Usambara Mts., NE Tanzania, and their use for bioindication. Ecotropica 2:195–217

Hochkirch A (1998) A comparison of the grasshopper fauna (Orthoptera: Acridoidea) of the Uluguru Mts. and the East Usambara Mts. (Tanzania). J East Afr Nat Hist 87:221–232

Hochkirch A (2001) A phylogenetic analysis of the East African grasshopper genus *Afrophlaeoba* Jago, 1983 (Orthoptera: Acridoidea: Acrididae). PhD Thesis, University of Bremen

Hochkirch A (2005) Morphometric differentiation in the East African grasshopper genus *Afrophlaeoba* Jago, 1983 (Orthoptera: Acrididae). In: Huber BA, Sinclair BJ, Lampe K-H (eds) African biodiversity: molecules, organisms. Ecosystems. Springer, Berlin, pp 109–118

Hochkirch A, Gröning J, Krause S (2007) Intersexual niche segregation in Cepero's Groundhopper, *Tetrix ceperoi*. Evol Ecol 21:727–738

Hoffman RL (1993) Biogeography of East African montane forest millipedes. In: Lovett JC, Wasser SK (eds) Biogeography and ecology of the rain forests of Eastern Africa. Cambridge University Press, Cambridge, pp 103–114

Howell KM (1993) Herpetofauna of the Eastern African forests. In: Lovett JC, Wasser SK (eds) Biogeography and ecology of the rain forests of Eastern Africa. Cambridge University Press, Cambridge, pp 173–201

Iversen ST (1991) The Usambara mountains, N.E. Tanzania: phytogeography of the vascular plant flora. Symb Bot Ups 29:1–234

Jago ND (1981) A revision of the genus *Usambilla* Sjöstedt (Orthoptera: Acridoidea) and its allies. Bull Br Mus Nat Hist Entomol 43:1–38

Jago ND (1983) Flightless members of the *Phlaeoba* Genus group in Eastern and North-Eastern Africa and their evolutionary convergence with the Genus *Odontomelus* and its allies (Orthoptera, Acridoidea, Acrididae, Acridinae). T Am Entomol Soc 109:77–126

Kellman M, Tackaberry R, Brokaw N, Meave J (1994) Tropical gallery forests. Nat Geogr Res Explor 10:92–103

Kenworthy JM (1966) Temperature conditions in the tropical highland climates of East Africa. East Afr Geogr Rev 4:1–11

Lindqvist C, Albert VA (1999) Phylogeny and conservation of African violets (*Saintpaulia*: Gesneriaceae): new findings based on nuclear ribosomal 5S non-transcribed spacer sequences. Kew Bull 54:363–377

Linsenmair KE (1990) Tropische Biodiversität: Befunde und offene Probleme. Verh Dtsch Zool Ges 83:245–261

Lovett JC (1988) Endemism and affinities of the Tanzanian montane forest flora. Monogr Syst Bot Missouri Bot Gard 25:491–496

Lovett JC (1993a) Eastern Arc moist forest flora. In: Lovett JC, Wasser SK (eds) Biogeography and ecology of the rain forests of Eastern Africa. Cambridge University Press, Cambridge, pp 33–57

Lovett JC (1993b) Temperate and tropical floras in the mountains of Eastern Tanzania. Opera Bot 121:217–227

Lovett JC (1998) Importance of the Eastern Arc mountains for vascular plants. J East Afr Nat Hist 87:59–74

Lovett JC, Friis I (1996) Patterns of endemism in the woody flora of North-east and East Africa. In: Van der Maesen LJG et al (eds) The biodiversity of African plants. Kluwer Academic Press, Amsterdam, pp 582–601

Lovett JC, Wasser SK (1993) Biogeography and ecology of the rain forests of Eastern Africa. Cambridge University Press, Cambridge

Medley KE (1998) Landscape change and resource conservation along the Tana River, Kenya. In: Zimmerer KS, Young KR (eds) Nature's Geography – New Lessons for Conservation in Developing Countries. University of Wisconsin Press, Wisconsin, pp 39–55

Myers N, Mittermeier RA, Mittermeier CG, da Fonseca GAB, Kent J (2000) Biodiversity hotspots for conservation priorities. Nature 403:853–858

Oksanen J, Kindt R, O'Hara B (2005) The vegan package. http://cc.oulu.fi/~jarioksa/softhelp/vegan.html. Cited 11 July 2008

Peterson AT, Soberón J, Sánchez-Cordero V (1999) Conservatism of ecological niches in evolutionary time. Science 285:1265–1267

Prell WL, Hutson WH, Williams DF, Be AWH, Geitzenauer K, Molfino B (1980) Surface circulation of the Indian Ocean during the last glacial maximum, approximately 18,000 yr. B.P. Quat Res 14:309–336

R Development Core Team (2008) R: a language and environment for statistical computing. http://www.R-project.org. Cited 11 July 2008

Robbrecht E (1996) Generic distribution patterns in subsaharan African Rubiaceae (Angiospermae). J Biogeogr 23:311–328

Rodgers WA, Homewood KM (1982) Species richness and endemism in the Usambara mountain forests, Tanzania. Biol J Linn Soc 18:197–242

Sänger K (1977) Über die Beziehungen zwischen Heuschrecken und der Raumstruktur ihrer Habitate. Zool Jahrb Abt Syst 104:433–488

Scharff N (1992) The linyphiid fauna of Eastern Africa (Araneae: Linyphiidae) – distribution pattern, diversity and endemism. Biol J Linn Soc 45:117–154

Scharff N (1993) The Linyphiid spider fauna (Araneae: Linyphiidae) of mountain forests in the Eastern Arc mountains. In: Lovett JC, Wasser SK (eds) Biogeography and ecology of the rain forests of Eastern Africa. Cambridge University Press, Cambridge, pp 115–132

Stanley WT, Kihaule PM, Howell KM, Hutterer R (1998) Small mammals of the Eastern Arc mountains. J East Afr Nat Hist 87:91–100

Tattersfield P (1998) Patterns of diversity and endemism in East African land snails, and the implication for conservation. J Conchol 2:77–86 Special Publication

Thuiller W, Lavorel S, Araujo MB (2005) Niche properties and geographical extent as predictors of species sensitivity to climate change. Glob Ecol Biogeogr 14:347–357

Uvarov BP (1977) Grasshoppers and locusts – a handbook of general acridology. Volume 2. Cambridge University Press, Cambridge

Wasser SK, Lovett JC (1993) Introduction to the biogeography and ecology of the rain forests of Eastern Africa. In: Lovett JC, Wasser SK (eds) Biogeography and ecology of the rain forests of Eastern Africa. Cambridge University Press, Cambridge, pp 3–7

Wiens JJ, Graham CH (2005) Niche conservatism: Integrating evolution, ecology, and conservation biology. Annu Rev Ecol Evol Syst 36:519–539

Winemiller KO, Pianka ER (1990) Organization in natural assemblages of desert lizards and tropical fishes. Ecol Monogr 60:27–56

Conservation Genetics and Phylogeography of the Threatened Corsican and Barbary Red Deer *(Cervus elaphus corsicanus* and *C. e. barbarus)*

Frank E. Zachos, Ghaiet M. Hajji, and San S. Hmwe

Abstract The two red deer subspecies *C. e. corsicanus* (Corsican red deer, endemic to the Tyrrhenian islands of Sardinia and Corsica) and *C. e. barbarus* (Barbary red deer, presently confined to a small area along the Tunisian-Algerian border) are classified by the IUCN as "Endangered" and "Lower risk (near threatened)", respectively (until recently, the Barbary red deer's status was "Vulnerable"). Both subspecies underwent severe bottlenecks in the twentieth century. We review our results on genetic variability and differentiation in these two subspecies at polymorphic microsatellite loci and sequences of the mitochondrial control region and discuss their bearing on conservation and phylogeography. Our analyses yielded very low genetic variability for *C. e. corsicanus*. The Barbary red deer also showed low variability values in a European comparison, but diversity was not as low as might have been feared in the light of the population's history. Analyses of molecular variance, assignment tests and factorial correspondence analysis did not yield any signs of differentiation among the Tunisian subgroups studied, whereas there was clear evidence of a differentiation between the Sardinian and the Corsican red deer population. The phylogeographic analysis showed close relationships between *C. e. corsicanus* and autochthonous Italian deer at two unlinked sets of nuclear loci, suggesting as the most likely scenario an introduction of Italian mainland red deer to the Tyrrhenian islands, from where animals were probably translocated to North-Africa.

F.E. Zachos (✉)
Zoological Institute, Christian-Albrechts-University, Olshausenstrasse 40, D-24118 Kiel, Germany
e-mail: fzachos@zoologie.uni-kiel.de

G.M. Hajji
U. R. de Biologie Animale et Systématique Évolutive, Département de Biologie, El Manar, Tunisia
e-mail: gayahajji@hotmail.com

S.S. Hmwe
Three Quarter Kamayut Township Yangon University, (92) Khyne Shwe War Street (Baho Road), Post Office Yangon, Myanmar
e-mail: sansan.hmwe@gmail.com

J.C. Habel and T. Assmann (eds.), *Relict Species: Phylogeography and Conservation Biology*, 159
DOI 10.1007/978-3-540-92160-8_8,

1 Introduction

The red deer (*Cervus elaphus*) is not an endangered species as a whole. Two genetically and morphologically distinct Western-Palearctic populations, however, are listed by the World Conservation Union as "Endangered" (*C. e. corsicanus*, the Tyrrhenian or Corsican red deer) and "Lower Risk" (near threatened) (*C. e. barbarus*, the North-African Barbary red deer). In the present review, we shall give a summary on our work on these two subspecies with regard to (1) genetic variability and differentiation of subpopulations and (2) the phylogeographic origin of the Tyrrhenian red deer and its relationship to the Barbary red deer.

2 Conservation Genetics

2.1 Barbary Red Deer

Apart from recent introductions, the Barbary red deer (*C. e. barbarus* Bennett, 1833) is the only African deer taxon. The historical range of this deer covered large parts of Algeria, Tunisia and possibly Morocco, but at present the Barbary deer is restricted to a small area of cork oak and pine forest on the Tunisian-Algerian border (Fig. 1) – a concentration which makes it particularly susceptible to diseases and forest fires, which are not uncommon in the region (Kock and Schomber 1961; Cowan and Holloway 1973; Dolan 1988; Oumani 2006). The Barbary red deer is listed in Appendix III by CITES (a recent proposal to list it in Appendix I was rejected in June 2007), and its IUCN status has recently been changed from "Vulnerable" to "Lower Risk (near threatened)" (Wemmer 1998; IUCN Red List 2007). Barbary red deer are characterised by a simplified antler structure (frequent lack of the bez tine) and by white spots running in longitudinal lines (Dolan 1988; Geist 1998). They suffered a severe bottleneck with only seven specimens left in Tunisia in 1962, two of them in Ain Baccouch and five in Ain Seraouia (Van Dijk 1972), which, however, were augmented by animals from Algeria crossing the border (Dolan 1988; Wemmer 1998). Consequently, hunting was banned, and in 1966, a reserve of 417 hectares with 17 individuals was established at El Feidja, about 190 km Southwest of Tunis, from where the deer can emigrate freely (Direction Générale des Forêts – hereafter abbreviated as DGF – 1994). This reserve is located within the El Feidja National Park, which also includes an additional enclosure for the Barbary deer. Subsequently, an area of ca. 200 km along the Mediterranean in Northern Tunisia was colonised (Meyer 1972; Dolan 1988).

Apart from Algerian immigrants, the El Feidja population is the origin of all Barbary deer present in Tunisia. Between 1978 and 1996, several reintroductions of this species into different reserves, mainly Ain Baccouch (Tabarka) and Mhebès (Séjnane), were carried out (DGF 1994). Population size at El Feidja seems to have been about 100 individuals since 1972 (Van Dijk 1972; Dhouib 1998; Oumani 2006).

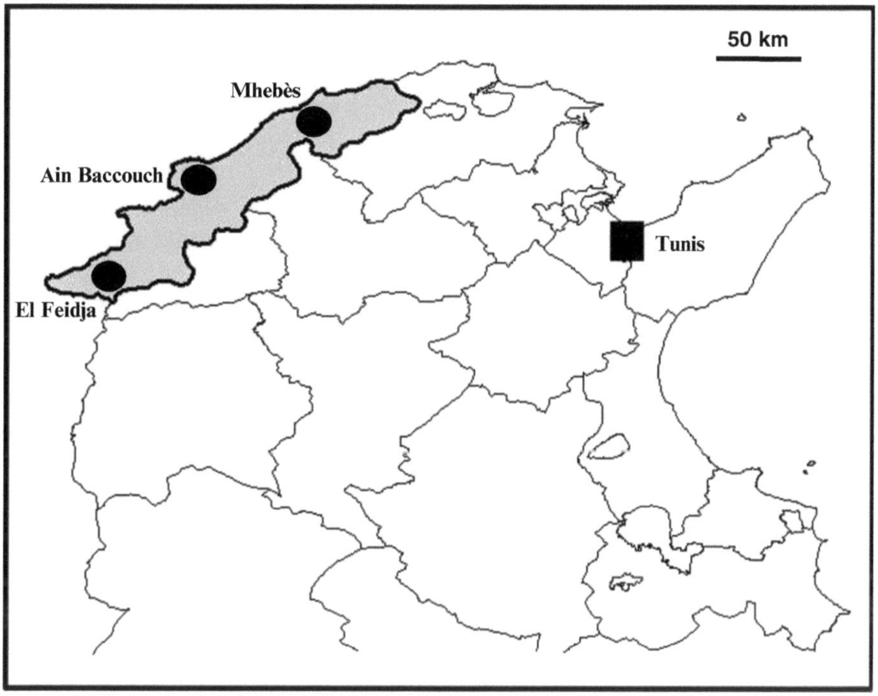

Fig. 1 Distribution of the Barbary red deer in Tunisia (*shaded area*) and location of sample sites. Figure taken from Hajji et al. (2007)

For Mhebès, Oumani recorded 80 individuals in 2002 and a similar number (70) in 2006 (Oumani 2006). The present number of deer at Mhebès is rather low compared with the number of founders (39 individuals) in 1979 (DGF 1994). The fact that neither of the two reserve populations (El Feidja and Mhebès) has increased is explained by Oumani (2006) with the dispersal of the deer from the reserve areas into adjacent habitats. Sadly, poaching is still also an important factor in limiting the population growth of the Barbary deer in Tunisia.

The total population size of *C. e. barbarus* in Tunisia was estimated by Oumani (pers. comm. 2006) to be about 800 individuals. This is in clear contrast to the census results obtained by the DGF according to which there are 2,000 deer in Tunisia (DGF 1994; see also Dolan 1988 who gives the same number). In addition to protection through government programmes, the WWF, in the context of the Regional Euro-Mediterranean Programme for the Environment (SMAP), is coordinating a project at El Feidja and Mhebès including studies on the population size and ecology of the deer in Northwestern Tunisia and the establishment of a deer enclosure at Mhebès. These conservation measures, however, are still sometimes thwarted by the government as was shown by the sale of 10 red deer from an enclosure in El Feidja in 2007 to a prince who wanted them for his private zoo in Morocco. In addition to these 10 animals, many more panicked during the capturing, jumped over the enclosure's fence and have not returned since.

Apart from the inclusion of some African samples in phylogenetic or phylogeographic studies (Polziehn and Strobeck 2002; Ludt et al. 2004; Hajji et al. 2008) there has only been one genetic study of Barbary red deer. Hajji et al. (2007) analysed 30 Tunisian Barbary red deer from El Feidja ($n=12$), Ain Baccouch ($n=8$) and Mhebès ($n=10$). Samples were collected in the field (faeces) or came from skin and hair of captured or accidentally killed or injured deer.

A stretch of 680 bp of the mitochondrial control region was successfully amplified and sequenced in 20 individuals. Amplification of 13 polymorphic microsatellites was successful in 19 red deer, a subset of 8 loci was amplified in all 30 individuals. Two distinct haplotypes were found differing by 16 transitions and being distributed evenly among the 20 specimens, occurring 11 and 9 times, respectively. This resulted in a low or intermediate haplotype diversity but in a rather high nucleotide diversity (Table 1), which is typical of populations having undergone a transient bottleneck where many haplotypes from the polymorphic original gene pool are lost through drift and those remaining are usually not closely related to one another (Nei et al. 1975).

The microsatellite loci analysed did not show any significant signs of linkage disequilibrium. Neither did tests for distortion of our results through stutter bands, large allele dropout or null alleles. Observed and expected heterozygosity values were 0.46 and 0.78, respectively, allelic diversity was 5.60 (Table 1).

A statistical test for a signature of the bottleneck in the microsatellite data (based on heterozygosity values relative to those expected under mutation-drift equilibrium and implemented in the Bottleneck software by Cornuet and Luikart 1996) was highly significant ($p=0.00018$).

We particularly tested our three sample sites El Feidja, Ain Baccouch and Mhebès for signs of (incipient) differentiation by calculating overall and pairwise F_{ST} and R_{ST} values (the latter being a microsatellite-specific analogue of F_{ST}) and by carrying out assignment tests in which individuals are assigned to the population of which their genotype is most typical. Additionally, a factorial correspondence analysis (FCA) was also conducted to see if this approach would yield any clusters of individuals related to their geographical origin. Positive signs of differentiation were not found in any of these calculations in that none of the F_{ST} or R_{ST} values were significantly different from zero; the assignment tests showed very high error rates (about 60%, which is not much lower than expected under random assignment to three populations); and there were no clusters of individuals in the FCA (not shown).

In Ain Baccouch, only one of the two haplotypes was found, but this is probably an artefact of the small sample size (control region amplification was successful in seven animals from Ain Baccouch) and may not be interpreted as evidence of differentiation. Consequently, the Tunisian red deer can still be considered a single, genetically homogeneous population.

The genetic data clearly reflect the population's bottleneck and show low genetic diversity for both mitochondrial control region sequences and nuclear microsatellite loci. However, compared to other populations known to be bottlenecked and/or inbred (e.g., Hasselbusch, Mesola or Corsica/Sardinia), variability values were not

Table 1 Genetic variability at microsatellite loci and the mitochondrial control region in the Tunisian and European red deer populations

Population	H_O	H_E	AD	HD	π [%]
Tunisia	0.46	0.78	5.6	0.529	1.266
Sardinia	0.48	0.66	–	0.872	0.473
Corsica	0.54	0.64	–	–	–
Mesola	0.51	0.62	–	0	0
Hasselbusch	0.58	0.66	4.9	0.493	0.337
Val di Susa	0.75	0.85	8.0	0.591	0.733
Tarvis	0.76	0.81	–	0.750	0.650
Southern Spain	0.65	0.80	–	0.652	1.318
Bulgaria	0.74	0.85	–	0.667	0.986
Romania	0.54	0.88	–	0.500	0.644
Spain	0.66	0.78	7.5	–	–
Scotland	0.44	0.81	–	0.478	0.536

H_O and H_E: observed and expected heterozygosity, AD allelic diversity (only given when sample size was comparable, i.e. between 16 and 22), HD haplotype diversity, π: nucleotide diversity. Mesola, Val di Susa and Tarvis are on the Italian mainland. The values for Corsica are from Hajji et al. 2008. Hasselbusch is a strongly inbred and isolated population in Northern Germany (see Zachos et al. 2007 for details). The data for Sardinia and Mesola are from Hmwe et al. (2006a), those for Val di Susa, Tarvis, Southern Spain and Bulgaria from Zachos et al. (2003), and the Spanish, Romanian and Scottish values are taken from Martinez et al. (2002), Feulner et al. (2004) and Hmwe et al. (2006b). For details of calculations see Hajji et al. (2007)

as low as might have been expected in the light of the population's demographic history (Table 1, Hmwe et al. 2006a). This may be due partly to immigration of Algerian animals. The Algerian population is estimated to have been about 300–400 head in the early 1950s before the Algerian war for independence and 400–600 in the mid-1970s (Salez 1959; Kock and Schomber 1961; Wemmer 1998; for an estimate on population density in Algeria cf. Burthey et al. 1992). The total Algerian population was estimated to be approximately 2,000 in the late 1980s (Dolan 1988) but since the official number for Tunisia is dubious, the same might also hold for Algeria.

Conservation measures should concentrate on the reduction of poaching, which still plays a major role in the mortality rates of Barbary red deer. At the moment, efforts are being made by the WWF to strengthen anti-poaching legislation. In addition, the distribution range of the Barbary red deer should be further extended to reduce its susceptibility to environmental stochasticity such as forest fires and diseases.

The analysis of Hajji et al. (2007) and ongoing field studies in Tunisia have yielded valuable data on *C. e. barbarus* but information about these deer outside Tunisia is much needed. There are still rumours of Barbary red deer having survived in Morocco. This sounds unlikely but should be investigated nonetheless. The most important future steps, however, must concentrate on Algeria. The population size in Algeria and the number of migrants between Algeria and Tunisia need to be known. An assessment of the genetic variability in Algerian red deer and whether they are differentiated from the Tunisian population or not are pre-requisites of an effective bilateral management which must be the ultimate goal in the conservation of these last African red deer. Fortunately, plans for such bilateral protection measures are under way at present.

2.2 *Tyrrhenian or Corsican Red Deer*

The Corsican red deer (*C. e. corsicanus* Erxleben, 1777) is the smallest of the existing red deer (Dolan 1988). It is confined to the Tyrrhenian islands Corsica and Sardinia and classified as "Endangered" by the IUCN (Wemmer 1998; IUCN Red List 2007). In the second half of the twentieth century, it underwent a severe bottleneck with only 100–200 animals remaining (Krumbiegel 1982; Dolan 1988; Kidjo et al. 2007 and references therein), but has recovered well since so that recently it has been suggested that the Tyrrhenian red deer be downgraded to "near threatened" in the IUCN Red List (Kidjo et al. 2007). Present numbers are given as at least 2,700 head by Lovari et al. (2007), but in line with Banwell (1998), who gives an estimate of only 1,500 specimens, another recent figure is "slightly >1,000" (Kidjo et al. 2007). After its extinction around 1970, the Corsican population was re-founded with altogether 14 Sardinian animals (of which one male died shortly after its introduction) between 1985 and 1994 (Dolan 1988; Gonzalez and Kidjo 2002; Kidjo et al. 2007). The introduced deer were at first kept in enclosures, but since 1998 there have been repeated releases into the wild so that in 2004, there were about 200 free-ranging animals on Corsica and another 50 in enclosures (see Kidjo et al. 2007 who give a detailed report on the extirpation and reintroduction history of the red deer on Corsica).

The first assessment of genetic variability based on five microsatellite loci and 332 bp of the mitochondrial control region yielded very low variability values for 17 Sardinian red deer in a European comparison (Zachos et al. 2003). A more recent study, based on a larger sample size of 29 Sardinian red deer and an extended panel of 12 microsatellite loci as well as a larger fraction of the mitochondrial control region (531 bp) confirmed these results (Hmwe et al. 2006a, see values in Table 1). Microsatellite variability was nearly as low as for Mesola, a relict population of truly autochthonous Italian mainland red deer which mark the lower end of genetic diversity in European-type red deer. The comparatively high number of eight haplotypes found (defined by seven polymorphic sites) may be a consequence of more than one colonisation of or introduction to the island. Only a few females would be enough to have caused this mitochondrial pattern without significantly increasing nuclear genetic variability. We have recently analysed a panel of eight further microsatellite loci, which show the same picture with respect to variability (Hajji et al. 2008), but also included 10 red deer from Corsica in our analysis. Genetic variability of the Corsican animals was still a little lower than that for Sardinia (observed and expected heterozygosity and allelic richness were 0.54, 0.64 and 3.50 for Corsica and 0.53, 0.71 and 4.40 for Sardinia), but the differences were

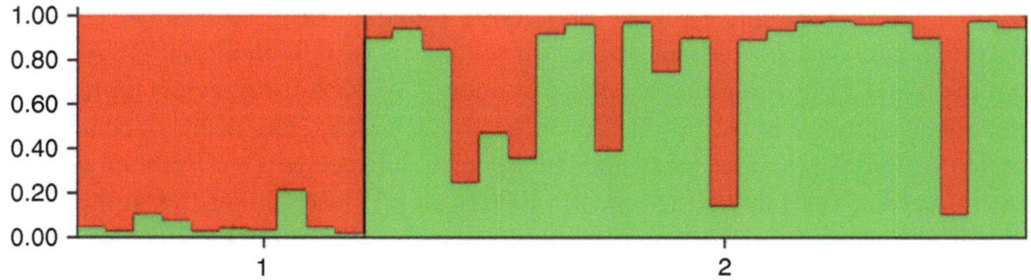

Fig. 2 Bayesian structure analysis of Corsican (1) and Sardinian (2) red deer with the STRUCTURE software (Pritchard et al. 2000). All Corsican animals were assigned to one cluster (*dark grey*) with high probabilities, whereas all but six Sardinian deer were assigned to a second cluster (*light grey*) with similarly high probabilities. When the data were tested for $K=3$ clusters (which yielded a higher overall probability but is not shown), only two of these six Sardinian specimens were grouped with the Corsican red deer, while the other 21 Sardinian red deer were assigned to two "non-Corsican" clusters

not significant. Interestingly, however, signs of considerable differentiation were found between the Corsican red deer and their population of origin on Sardinia. Pairwise F_{ST} for these two populations was about 15% and significantly different from zero, and the number of loci not in Hardy-Weinberg-Equilibrium due to an excess of homozygotes increased considerably (from two and zero in the single populations to five in the pooled data set), which is indicative of a Wahlund effect, i.e. a reduction in homozygosity relative to Hardy-Weinberg expectations due to substructuring of the total population. Further, assignment tests assigned more than 90% of the Corsican and Sardinian animals to their correct population of origin, and a Bayesian structure analysis without a priori assumptions of geographical origin also yielded distinct clusters roughly equivalent to the two islands (Fig. 2). Finally, a tree based on genetic distances among all individual red deer from Corsica and Sardinia also showed a non-random pattern in that one of the two main branches comprised exclusively Sardinian animals while the other comprised pre-dominantly Corsican deer (not shown).

3 Phylogeography of the Corsican Red Deer

The origin of the Corsican red deer has been a matter of considerable debate (reviewed in Vigne 1988 and Zachos and Hartl 2006). The two main questions are: (1) where did it come from and (2) how did it reach the Tyrrhenian islands – naturally during a glacial regression of the sea-level or through human introduction? The latter hypothesis was already put forward at the end of the nineteenth century (Lydekker 1898), and Groves (pers. comm. to Geist 1998) thought that the Tyrrhenian red deer were introduced from North-Africa by the Phoenicians. It has recently been shown that hares (*Lepus* sp.) were introduced to Sardinia from Africa (Scandura et al. 2007), so this might also hold for red deer. In line with this,

Ludt et al. (2004) and Skog et al. (2009) found close relationships between Sardinian and Barbary red deer based on sequences of mitochondrial DNA (cytochrome b and control region; see also Sommer et al. 2008 for a review on red deer phylogeography). Flerov (1952) regarded the Corsican red deer to be closely related to the Barbary red deer and the Spanish subspecies *C. e. hispanicus*, and Zachos et al. (2003) and Hmwe et al. (2006a) found close similarities at the mitochondrial control region between Sardinian and Spanish red deer. This was, however, not corroborated in more comprehensive studies (Ludt et al. 2004; Skog et al. (2009). An early genetic hypothesis based on mitochondrial RFLP markers linked *C. e. corsicanus* with Bulgaria and the Near East (Hartl et al. 1995) but none of the subsequent analyses yielded further evidence in favour of this assumption.

The only other genetically substantiated hypothesis as to the origin of the Tyrrhenian red deer is based on nuclear microsatellite loci. Zachos et al. (2003) and Hmwe et al. (2006a) found close relationships between Sardinian and European mainland red deer, in particular those from the Mesola wood in the Po delta (the relationship of mitochondrial control region sequences derived from Sardinian and Mesola red deer was ambiguous). The Mesola population is important from a phylogeographic point of view as it is the only surviving, truly autochthonous red deer population in the Italian peninsula, thus very probably at least partially representing the original Italian gene pool (Mattioli 1990). This hypothesis matches with morphological analyses which showed close similarities between Holocene red deer from Southern Italy and samples from the extinct original Corsican red deer population (Vigne 1988). According to this hypothesis red deer either naturally colonised the Tyrrhenian islands from the Italian mainland during a glacial period in the Pleistocene when the sea-level was considerably lower (for details see Hmwe et al. 2006a) or were introduced to Sardinia and Corsica by man. Putative Pleistocene red deer fossils are known from Corsica (*C. e. rossii*, Pereira 2001), but the extant Tyrrhenian red deer can hardly be considered a descendant of these early deer because, in spite of a very well-analysed fossil record, no red deer were found between the early Holocene (ca. 11500 cal. BP) and the mid-Holocene (5000 cal. BP or later) (Vigne et al. 1997; Vigne 1999; Costa et al. 2003). Sites from the Late Neolithic (4500 cal. BP) in Sardinia and from Classical Antiquity in Corsica, however, have frequently yielded red deer bones (Fonzo 1987; Vigne 1988). This is clearly indicative of human introduction after the indigenous deer populations had become extinct (Vigne 1992, 1999; Schüle 1993; Masseti 1998, 2002a). The Corsican population, possibly much younger than the Sardinian one, could have been founded with introduced animals from Sardinia or directly from the Italian mainland. Such introduction events were not uncommon at the time, and deer are known to have been introduced to many islands as early as the eleventh millennium BP (Vigne 1993; Masseti 1996, 2002a,b).

To shed further light on this issue, red deer from Corsica and Sardinia together with a selective sample of numerous European-type red deer were analysed for the first time, among them both Mesola and North-Africa (Hajji et al. 2008; see Fig. 3).

The study, based on eight nuclear microsatellite loci, yielded close relationships of *C. e. corsicanus* to red deer from both Mesola and North-Africa, as reflected, for

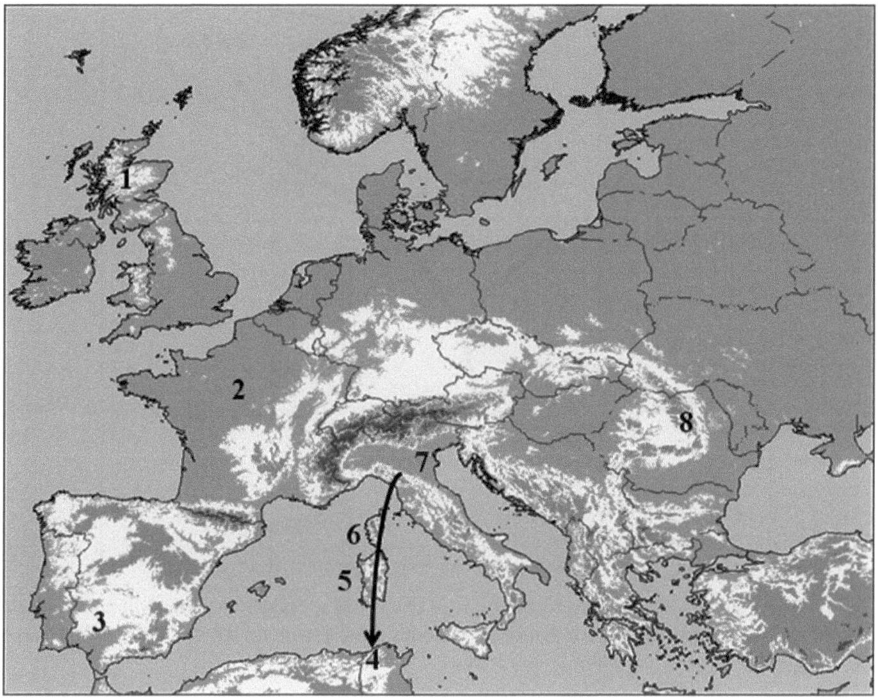

Fig. 3 Geographical location of sample sites studied in a phylogeographic analysis by Hajji et al. (2008). 1: Scotland (*C. e. scoticus*, $n=9$), 2: France (*C. e. hippelaphus*, $n=14$), 3: Southern Spain (*C. e. hispanicus*, $n=21$), 4: Tunisia (*C. e. barbarus*, $n=30$), 5: Sardinia (*C. e. corsicanus*, $n=23$), 6: Corsica (*C. e. corsicanus*, $n=10$), 7: Mesola (*C. e. hippelaphus*, $n=25$), 8: Romanian Carpathians (*C. e. montanus*, $n=17$). The arrow denotes the most probable direction of introductions

example, by the fact that more than 50% of all alleles found in Mesola and Tunisia, respectively, were alleles also recorded in the Tyrrhenian red deer. However, the analyses favoured Mesola as the closest relative of the Tyrrhenian red deer. This was evident from allele frequencies and distributions at single loci, from different parameters of genetic distance and from the resulting trees (see Fig. 4). Nei's standard distances (Nei 1972), chord distances (Cavalli-Sforza and Edwards 1967) and proportion-of-shared-alleles distances (Bowcock et al. 1994, calculated as 1-Dps) between *C. e. corsicanus* and Mesola were 0.745, 0.117 and 0.585, while the respective values between *C. e. corsicanus* and Tunisia were 1.015, 0.126 and 0.640. Assignment tests did not yield misassignments of *C. e. corsicanus* specimens except for the individuals incorrectly assigned between Sardinia and Corsica (see above).

Rather unexpectedly, the study also yielded a quite close genetic proximity between *C. e. corsicanus* and the French red deer population. This may be due to introductions from France to the Tyrrhenian islands (which cannot be totally ruled out, although there are no records confirming this) or, more probably, due to the limited number of loci (eight) used in the analyses and/or length homoplasy.

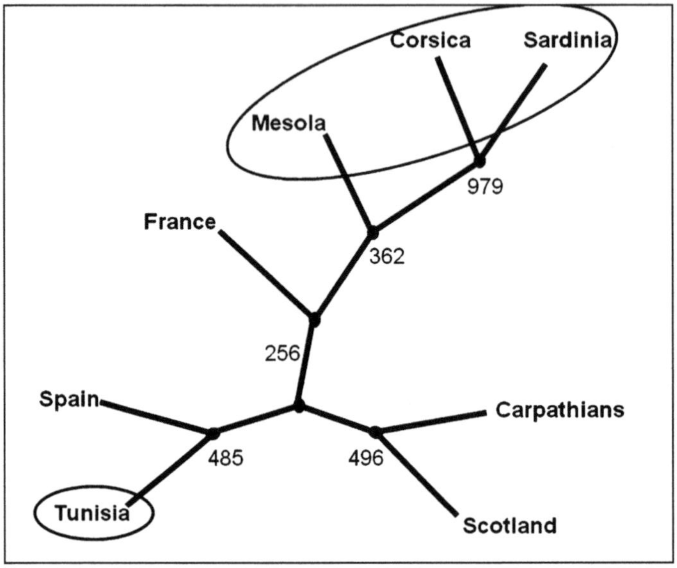

Fig. 4 Neighbour-joining tree based on pairwise chord distances derived from allele frequencies at eight nuclear microsatellite loci. Numbers at nodes refer to support from 1,000 bootstrap replicates. Figure modified from Hajji et al. (2008)

Perhaps the emerging picture would have been clearer had the animals been genotyped for more loci. This would also explain the rather low bootstrap support for nodes in our trees. Null alleles, stutter bands and allelic dropout, potential error sources known to be able to distort analyses based on microsatellites, were tested for with the MICRO-CHECKER software (van Oosterhout et al. 2004) and could be shown not to have influenced the data (see Hajji et al. 2008 for details).

In general, however, the results are in accordance with a priori expectations. They confirm those of Hmwe et al. (2006a) in that Mesola turned out to be the probably closest extant relative of the Tyrrhenian red deer. The recent study (Hajji et al. 2008) was based on eight completely different and unlinked nuclear loci, which further adds to the credibility of the main conclusion.

Nevertheless, although not as unequivocal as with mitochondrial sequence data, nuclear genetic analyses also confirmed a close relationship of Tyrrhenian and Barbary red deer. It is still unclear if there are true *Cervus* fossils in the Late Pleistocene of North-Africa (Cheylan 1991), and it is considered more probable that the modern Barbary red deer go back to introductions by humans in Holocene times (Kowalski and Rzebik-Kowalska 1991), although Hamilton (1978) thought that the Barbary red deer reached Africa via Sicily. Putting together both the palaeontological and the genetic results, it seems very likely that the Tyrrhenian red deer were introduced to Corsica and Sardinia from mainland Italy and that these deer were subsequently translocated to North-Africa (Fig. 3, see also Dobson 1998). This would neatly explain the seeming contradiction between nuclear and mitochondrial genetic data. The fact that the Mesola red deer do not show a mitochondrial haplotype unequivocally linked with those of the Tyrrhenian and the Barbary red deer is probably the result of drift in the small, long-time isolated and bottlenecked Mesola population. Effective population size is four times smaller for the mitochondrial than for the nuclear gene pool, which is why drift has a fourfold smaller effect on (nuclear) microsatellite allele frequencies than on control region sequences.

Apart from presenting a very probable solution to the debate about the origin of the Tyrrhenian red deer and its relative relationships with mainland Italian and African Barbary red deer, the study of Hajji et al. (2008) also highlights the need for taking into consideration all available sources – palaeontological, archaeozoological and genetic, and it is also a corroboration of previous studies, which warned against the exclusive use of mitochondrial DNA in phylogeographic analyses (e.g., Taberlet et al. 1998).

References

Banwell B (1998) Description of *Cervus elaphus corsicanus*. IUCN Deer Specialist Group Newsletter 14:9

Bowcock AM, Ruiz-Linares A, Tomfohrde J, Minch E, Kidd JR, Cavalli-Sforza LL (1994) High resolution human evolutionary trees with polymorphic microsatellites. Nature 368:455–457

Burthey F, Burthey A, Sennaoui F, Bensefia N (1992) The *Cervus elaphus barbarus* (Benett, 1833) in Algeria. Status and first elements of eco-ethology. Results and prospects after three years of investigation. In: Spitz F, Janeau G, Gonzalez G, Aulagnier S (eds) Ongulés/Ungulates 91. S. F. E. P. M. – I. R. G. M, Paris Toulouse, pp 271–276

Cavalli-Sforza LL, Edwards AWF (1967) Phylogenetic analysis: models and estimation procedures. Am J Hum Genet 19:233–257

Cheylan M (1991) Patterns of pleistocene turnover, current distribution and speciation among mediterranean mammals. In: Groves RH, Di Castri F (eds) Biogeography of mediterranean invasions. Cambridge University Press, Cambridge

Cornuet JM, Luikart G (1996) Description and power analysis of two tests for detecting recent population bottlenecks from allele frequency data. Genetics 144:2001–2014

Costa L, Vigne J-D, Bocherens H, Desse-Berset N, Heintz C, de Lanfranchi F, Magdeleine J, Ruas M-P, Thiébault S, Tozzo C (2003) Early settlement on Tyrrhenian islands (8th millennium cal. BC): mesolithic adaptation to local resources in Corsica and Northern Sardinia. In: Larsson L, Kindgre H, Knutsson K, Loeffler D, Åkerlund A (eds) Mesolithic on the mour. Oxbow Books, Oxford, pp 3–10

Cowan I McT, Holloway CW (1973) Threatened deer of the world: conservation status. Biol Conserv 5:243–250

Dhouib S (1998) Contribution à l'étude du cerf de Barbarie (*Cervus elaphus barbarus* Bennet, 1833), espèce préservée du parc national d'El Feidja: régime alimentaire et recensement de la population. Mémoire de DEA Fac Sci Bizerte

Direction Générale des Forêts (1994). Gestion de la Faune Sauvage et des Parcs Nationaux en Tunisie. Réintroduction, Gestion et Aménagement. Direction Générale des Forêts, Ministère de l'Agriculture, tunis

Dobson M (1998) Mammal distributions in the Western Mediterranean: the role of human intervention. Mammal Rev 28:77–88

Dolan JM (1988) A deer of many lands – a guide to the subspecies of the red deer *Cervus elaphus* L. Zoonooz LXII(10):4–34

Feulner PGD, Bielfeldt W, Zachos FE, Bradvarovic J, Eckert I, Hartl GB (2004) Mitochondrial DNA and microsatellite analyses of the genetic status of the presumed subspecies *Cervus elaphus montanus* (Carpathian red deer). Heredity 93:299–306

Flerov CC (1952) Musk deer and deer. In: Fauna of the USSR. Mammals, vol. 1,2. Academy of Sciences, Moscow

Fonzo O (1987) Reperti faunistici in Marmilla e Campidano nell'Età del Bronzo e nella prima Età del Ferro. La Sardegna nel Mediterraneo tra il secondo et il primo millennio a.C. Amministrazione Provinciale, Cagliari

Geist V (1998) Deer of the world. Their evolution, behavior, and ecology. Stackpole Books, Mechanicsburg, PA

Gonzalez G, Kidjo N (2002) A research project on conservation biology of Corsican red deer (*Cervus e. corsicanus*). IUCN Deer Specialist Group Newsletter 17:1–3

Hajji GM, Charfi-Cheikrouha F, Lorenzini R, Vigne J-D, Hartl GB, Zachos FE (2008) Phylogeography and founder effect of the endangered corsican red deer (*Cervus elaphus corsicanus*). Biodivers Conserv 17:659–673

Hajji GM, Zachos FE, Charfi-Cheikrouha F, Hartl GB (2007) Conservation genetics of the imperilled barbary red deer in Tunisia. Anim Conserv 10:229–235

Hamilton WR (1978) Cervidae and paleomerycidae. In: Maglio VJ, Cooke HBS (eds) Evolution of African mammals. Harvard University Press, Cambridge, MA

Hartl GB, Nadlinger K, Apollonio M, Markov G, Klein F, Lang G, Findo S, Markowski J (1995) Extensive mitochondrial-DNA differentiation among European red deer (*Cervus elaphus*) populations: implications for conservation and management. Z Saeugetierkd 60:41–52

Hmwe SS, Zachos FE, Eckert I, Lorenzini R, Fico R, Hartl GB (2006a) Conservation genetics of the endangered red deer from Sardinia and Mesola with further remarks on the phylogeography of *Cervus elaphus corsicanus*. Biol J Linn Soc 88:691–701

Hmwe SS, Zachos FE, Sale JB, Rose HR, Hartl GB (2006b) Genetic variability and differentiation in red deer (*Cervus elaphus*) from Scotland and England. J Zool 270:479–487

Kidjo N, Feracci G, Bideau E, Gonzalez G, Mattéi C, Marchand B, Aulagnier S (2007) Extirpation and reintroduction of the Corsican red deer *Cervus elaphus corsicanus* in Corsica. Oryx 41:488–494

Kock D, Schomber HW (1961) Beitrag zur Kenntnis der Verbreitung und des Bestandes des Atlashirsches (*Cervus elaphus barbarus*) sowie eine Bemerkung zu seiner Geweihausbildung. Saeugetierkundl Mit 9:51–54

Kowalski K, Rzebik-Kowalska B (1991) Mammals of Algeria. Polish Academy of Science, Wroklaw Warsaw Krakov

Krumbiegel I (1982) Der Korsika-Rothirsch (*Cervus elaphus corsicanus*, Erxleben 1777) und sein Biotop. Saeugetierkdl Mitt 30:281–286

Lovari S, Cuccus P, Murgia A, Murgia C, Soi F, Plantamura G (2007) Space use, habitat selection and browsing effects of red deer in Sardinia. Ital J Zool 74:179–189

Ludt CJ, Schroeder W, Rottmann O, Kuehn R (2004) Mitochondrial DNA phylogeography of red deer (*Cervus elaphus*). Mol Phylogenet Evol 31:1064–1083

Lydekker R (1898) The deer of all lands. A history of the family Cervidae living and extinct. Rowland Ward, London

Martinez JG, Carranza J, Fernandez-Garcia JL, Sanchez-Prieto CB (2002) Genetic variation of red deer populations under hunting exploitation in Southwestern Spain. J Wildlife Manage 66:1273–1282

Masseti M (1996) The postglacial diffusion of the genus Dama Frisch, 1775, in the Mediterranean region. Suppl Ric Biol Selvag 25:7–29

Masseti M (1998) Holocene and anthropochorous wild mammals of the Mediterranean islands. Anthropozoologica 28:3–20

Masseti M (2002a) Uomini e (non solo) topi. Firenze University Press, Florence

Masseti M (2002b) The red deer of Lampedusa (Pelagian islands, Italy): literary references and osteological evidence. Arch Nat Hist 29:51–66

Mattioli S (1990) Red deer in the Italian peninsula with particular reference to the Po delta population. Deer 8:95–98

Meyer P (1972) Zur Biologie und Ökologie des Atlashirsches *Cervus elaphus barbarus*, 1833. Z Saeugetierkunde 37:101–116
Nei M (1972) Genetic distance between populations. Am Nat 106:283–292
Nei M, Maruyama T, Chakraborty R (1975) The bottleneck effect and genetic variability in populations. Evolution 29:1–10
Oumani A (2006) Ecobiologie du cerf de Barbarie (*Cervus elaphus barbarus*) en Kroumirie-Mogods (Tunisie). PhD thesis, Université du 7 Novembre à Carthage
Pereira E (2001) *Cervus elaphus rossii* (Mammalia, Artiodactyla), a new endemic sub-species from the middle Pleistocene in Corsica. Palaeovertebrata 30:189–213
Polziehn RO, Strobeck C (2002) A phylogenetic comparison of red deer and wapiti using mitochondrial DNA. Mol Phylogenet Evol 22:342–356
Pritchard JK, Stephens M, Donnelly P (2000) Inference of population structure using multilocus genotype data. Genetics 155:945–959
Salez M (1959) Notes sur la distribution et la biologie du cerf de Barbarie (*Cervus elaphus barbarus*). Mammalia 23:132–138
Scandura M, Iacolina L, Ben Slimen H, Suchentrunk F, Apollonio M (2007) Mitochondrial CR-1 variation in Sardinian hares and its relationships with other old world hares (Genus *Lepus*). Biochem Genet 45:305–323
Schüle W (1993) Mammals, vegetation and the initial human settlement of the Mediterranean islands: a palaeoecological approach. J Biogeogr 20:399–412
Skog A, Zachos FE, Rueness EK, Feulner PGD, Mysterud A, Langvatn R, Lorenzini R, Hmwe SS, Lehoczky I, Hartl GB, Stenseth NC, Jakobsen KS (2009) Phylogeography of red deer (*Cervus elaphus*) in Europe. J Biogeogr 36:66–77
Sommer RS, Zachos FE, Street M, Jöris O, Skog A, Benecke N (2008) Late Quaternary distribution dynamics and phylogeography of the red deer (*Cervus elaphus*) in Europe. Quaternary Sci Rev 27:714–733
Taberlet P, Fumagalli L, Wust-Saucy A-G, Cosson J-F (1998) Comparative phylogeography and postglacial colonization routes in Europe. Mol Ecol 7:453–464
Van Dijk G (1972) Le cerf de Barbarie (*Cervus elaphus barbarus* Bennet, 1833) dans la forêt d'El Feidja en Tunisie. Rapport de Stage. Direction Générale des Forêts, Tunisie
van Oosterhout C, Hutchinson WF, Wills DPM, Shipley P (2004) MICRO-CHECKER (Version 2.2.3): software for identifying and correcting genotyping errors in microsatellite data. Mol Ecol Notes 4:535–538
Vigne J-D (1988) Les Mammifères post-glaciaires de Corse, étude Archéozoologique (XXVIe suppl. Gallia Préhistoire). CNRS, Paris
Vigne J-D (1992) Zooarchaeology and the biogeographical history of the mammals of Corsica and Sardinia since the last ice age. Mammal Rev 22:87–96
Vigne J-D (1993) Domestication ou appropriation pour la chasse: histoire d'un choix socio-culturel depuis le Néolithique. L'exemple des cerfs (*Cervus*). In: Desse J, Audoin-Rouzeau F (eds) Exploitation des animaux sauvages à travers le temps. APDCA éd, Antibes, France, pp 201–220
Vigne J-D (1999) The large "true" Mediterranean islands as a model for the Holocene human impact on the European vertebrate fauna? Recent data and new reflections. In: Benecke N (ed) The holocene history of the European vertebrate fauna. Modern aspects of research. Verlag Marie Leihdorf, Rahden, pp 295–321
Vigne J-D, Bailon S, Cuisin J (1997) Biostratigraphy of amphibians, reptiles, birds and mammals in Corsica and the role of man in the holocene turnover. Anthropozoologica 25–26:587–604
Wemmer C (1998) Deer. Status survey and conservation action plan. IUCN/SSC Deer Specialist Group, Gland, Switzerland
Zachos FE, Hartl GB (2006) Island populations, human introductions and the limitations of genetic analyses: the case of the Sardinian red deer (*Cervus elaphus corsicanus*). Hum Evol 21:177–183
Zachos FE, Althoff C, Steynitz VY, Eckert I, Hartl GB (2007) Genetic analysis of an isolated red deer (*Cervus elaphus*) population showing signs of inbreeding depression. Eur J Wildlife Res 53:61–67
Zachos F, Hartl GB, Apollonio M, Reutershan T (2003) On the phylogeographic origin of the corsican red deer (*Cervus elaphus corsicanus*): evidence from microsatellites and mitochondrial DNA. Mamm biol 68:284–298

Phylogeographic Analyses of a Boreal-Temperate Ectomycorrhizal Basidiomycete, *Amanita muscaria*, Suggest Forest Refugia in Alaska During the Last Glacial Maximum

József Geml, Rodham E. Tulloss, Gary A. Laursen, Nina A. Sazanova, and D. Lee Taylor

Abstract Beringia, including Alaska and North-Eastern Siberia, has long been a focal point for biogeographical research in a wide range of plant and animal taxa. Whether boreal forest refugia existed in Beringia during the Last Glacial Maximum (LGM) is a major but unanswered question in Quaternary science. We analysed DNA sequence data from *Amanita muscaria* using phylogenetic and coalescent methods to test whether this boreal-temperate ectomycorrhizal fungus was present in Alaska during the LGM. Our data indicates that there are at least two endemic groups showing no detectable migration from or to populations that survived the LGM in Southern refugia. One of these occurs in the boreal forests of Interior Alaska, while the other group is endemic to the maritime rainforests from Southeast Alaska to the Olympic peninsula. These results suggest genetic isolation preceding the LGM and support the hypothesis of glacial forest refugia in Interior Alaska and along the coast of Southeast Alaska and British Columbia.

J. Geml (✉)
Institute of Arctic Biology, University of Alaska Fairbanks, 311 Irving I Building, 902 N. Koyukuk Drive, P.O. Box 757000, Fairbanks, AK99775-7000, USA
e-mail: jgeml@iab.alaska.edu

R.E. Tulloss
Res. Assoc. (hons.), New York Botanical Garden, P.O. Box 57, Roosevelt, Bronx, NJ 08555-0057, USA

G.A. Laursen
Institute of Arctic Biology, University of Alaska Fairbanks, 311 Irving I Building, 902 N. Koyukuk Drive, P.O. Box 757000, Fairbanks, AK 99775-7000, USA

N.A. Sazanova
Laboratory of Botany, Institute of Biological Problems of the North, Far East Branch of Russian Academy of Sciences, Portovaya Str. 18, Magadan, 685000, Russia

D.L. Taylor
Institute of Arctic Biology, University of Alaska Fairbanks, 311 Irving I Building, 902 N. Koyukuk Drive, P.O. Box 757000, Fairbanks, AK 99775-7000, USA

J.C. Habel and T. Assmann (eds.), *Relict Species: Phylogeography and Conservation Biology*, 173
DOI 10.1007/978-3-540-92160-8_9,

1 Introduction

Climatic and geological changes in the Tertiary and Quaternary dramatically influenced the distribution of flora and fauna in North America. In the Tertiary, North America became separated from Eurasia, resulting in vicariant populations of many species of the ancient flora and fauna (Budantsev 1992; Graham 1999). During glacial maxima, plants, fungi and animals were forced to unglaciated refugia, from where they expanded to newly exposed areas in warmer interglacial periods. Reconstruction of these historical events is of paramount importance because they had major influences on past speciation events and are reflected in present day population structures.

Beringia, including Alaska and North-Eastern Siberia, has long been a focal point for biogeographical research in a wide range of plant and animal taxa. This high level of interest arises for two principal reasons. First, due to its diverse landscape and climate and the fact that much of the region remained ice-free during glacial maxima, Beringia served as a refugium for Arctic and subarctic flora and fauna (Adams and Faure 1997; Brubaker et al. 2005; Edwards et al. 2000; Hultén 1968). Second, during much of the Tertiary and the Quaternary periods, Beringia was the major land connection between Asia and North America and provided migration routes to a wide variety of organisms (for example, see Elias et al. 2000; Qian 1999; Swanson 2003). High genetic diversity and evidence for glacial refugia in Beringia have been reported in several, predominantly arctic-alpine animal and plant taxa, such as the Tundra Vole (*Microtus oeconomus*) (Brunhoff et al. 2003), Bighorn Sheep (*Ovis canadensis*) and Dall Sheep (*O. dalli*) (Loehr et al. 2005), Columbian Ground Squirrel (*Spermophilus columbianus*) (MacNeil and Strobeck 1987), Old World Swallowtail (*Papilio machaon*) (Sperling and Harrison 1994), *Amara alpina*, an arctic-alpine ground beetle (Reiss et al. 1999), *Paranoplocephala arctica*, a parasitic cestode of collared lemmings (Wickström et al. 2003), Entire-leaved Avens (*Dryas integrifolia*) (Tremblay and Schoen 1999), Purple Saxifrage (*Saxifraga oppositifolia*) (Abbott and Comes 2003), Arctic Bell-heather (*Cassiope tetragona*) (Eidesen et al. 2007) and the Bog Blueberry (*Vaccinium uliginosum*) (Alsos et al. 2005).

Whether fragments of boreal forest existed in Beringia during the Last Glacial Maximum (LGM) is a major, but, as yet, unanswered question in Quaternary science. Although earlier pollen-based biome reconstructions suggest that all of Beringia was covered by arctic tundra with no conifers (Edwards et al. 2000; Kaufman et al. 2004; Swanson 2003), phylogeographic analyses of DNA sequences and the most recent pollen data support the theory that *Picea glauca* and *P. mariana* existed in small refugia in Alaska (Anderson et al. 2006; Brubaker et al. 2005).

Ectomycorrhizal (ECM) fungi are obligate root-associated mutualistic symbionts of trees in boreal, temperate and tropical forests (Smith and Read 1997; Bas 1969; Härkönen et al. 1994; Simmons et al. 2002) and their past and present populations likely have been shaped by the distribution of their host trees. The likely importance of host trees in the distribution of ECM fungi has been repeatedly noted.

However, despite their ecological importance, very little is known about ECM phylogeography. Furthermore, learning about the phylogeographic history of ECM taxa not only provides information related to the organism in question, but may also help to resolve whether or not their hosts survived the LGM in Beringian refugia. To test the hypothesis of persistent LGM forest refugia in Alaska, we analysed molecular data from *Amanita muscaria*, an ECM fungus native to temperate and boreal forest regions of the Northern Hemisphere.

2 Materials and Methods

Multi-locus DNA sequence data for *A. muscaria* specimens were generated in previous studies (Geml et al. 2006, 2008, and Oda et al. 2004). Geml et al. (2006) reported that there are at least three phylogenetic species clades within the *A. muscaria* species complex that occur in Alaska (referred to as Clades I, II, and III). To our knowledge, Clade I is restricted to North America, while Clades II and III occur in Eurasia and in Alaska. For our tests in this study, 114 ITS rDNA sequences were chosen, which represent the three phylogenetic species. Because the phylogenetic species mentioned are non-interbreeding entities, population-level analyses were conducted separately for each species clade. Multiple sequence alignments were made using Clustal W (Thompson et al. 1997) and subsequently were corrected manually. Identical sequences were collapsed into haplotypes using SNAP Map (Aylor and Carbone 2003) after recoding insertion or deletions (indels) and excluding infinite-sites violations. The analyses presented here assume an infinite sites model, under which a polymorphic site is caused by exactly one mutation and there can be no more than two bases segregating. Base substitutions were categorised as phylogenetically uninformative or informative, and as transitions or transversions. Site compatibility matrices were generated from each haplotype dataset using SNAP Clade and SNAP Matrix (Markwordt et al. 2003) to examine compatibility/incompatibility among all variable sites, with four resultant incompatible sites removed from the data set. This was important as subsequent coalescent analyses assume that all variable sites are fully compatible. Two migration models were used. First, MDIV (Nielsen and Wakeley 2001) was used to determine whether there was any evidence of migration between Alaskan and non-Alaskan populations in each species clade, i.e. to test whether Alaskan populations could have survived the LGM in local forest refugia. For this purpose, specimens were assigned either to the 'Alaskan' or the 'non-Alaskan' group based on their localities. In Clades I and II, the 'non-Alaskan' groups consisted of specimens collected in the contiguous states of the US or in Eurasia, respectively. These represented populations that survived the LGM in Southern refugia. MDIV implements both likelihood and Bayesian methods using Markov chain Monte Carlo (MCMC) coalescent simulations to estimate the migration rate (*M*), population mean mutation rate (*Theta*), divergence time (*T*) and the time since the most recent common ancestor (TMRCA). This approach assumes that all populations descended from one panmictic population that may or may not have been followed by migration. For each dataset, the data was simulated assuming an infinite sites model with uniform prior. We used 2,000,000 steps in the chain for estimating the posterior probability distribution and an initial 500,000 steps to ensure that enough genealogies were simulated before approximating the posterior distribution. Second, if MDIV showed evidence of migration, MIGRATE was used to estimate migration rates assuming equilibrium migration rates (symmetrical or asymmetrical) in the history of the populations (Beerli and Felsenstein 2001). We applied the following specifications for the MIGRATE maximum-likelihood analyses: *M* and *Theta* generated from the FST calculation, migration model with variable *Theta* and constant mutation rate. Subsequently, we reconstructed the genealogy with the highest root probability, the ages of mutations and the TMRCA of the sample using coalescent simulations in Genetree v. 9.0 (Griffiths and Tavaré 1994). Ages were measured in coalescent units of 2*N*, where *N* is the population size.

3 Results

Two of the three species clades (Clades I and II) possessed sufficient variation in the ITS to test whether the Alaskan populations were genetically differentiated from non-Alaskan populations. The third, Clade III, was only represented by ten individuals forming three haplotypes with minimal variation. In this species, all Alaskan specimens carried the most frequent haplotype that was shared with Asian specimens as well. The datasets consisted of 653 and 640 characters for Clade I and Clade II, respectively, including gaps.

In Clade I, the 'North American' clade, 15 haplotypes were found, of which only one, haplotype D, was found in Alaska (Tables 1 and 2). MDIV showed no evidence of gene flow (M~0) and estimated a divergence time of T~0.9 between the Alaskan and non-Alaskan populations (Fig. 1a). Therefore, in simulations using Genetree, we assumed subdivided population structure and a very low level of migration (close to 0) for the starting backward migration matrix. The coalescent-based genealogy was informative with respect to the resolution of the splitting event between the Alaskan and non-Alaskan populations and inference of the mutational history with respect to variation between and within populations (Fig. 2a).

In Clade II, the 'Eurasian' clade, 14 haplotypes were found, six of which occur in Alaska (Tables 1-2). Of these six, five were found only in Alaska (haplotypes C, D, G, H, I), while one (haplotype A) represented the most abundant haplotype in the sample, spanning Eurasia and Alaska. It must be noted, that haplotype D (or subclade II/A in Geml et al. 2006), a lineage endemic to the maritime rainforests of the Pacific Northwest, was considered 'Alaskan' as opposed to 'Eurasian', despite not being strictly restricted by the political boundaries and occurring as far South as Northwestern Washington state. Apart from this, no other Clade II haplotypes have been found in North America, outside Alaska. Despite the very similar values for Theta (Fig. 1), the population structure in Clade II proved to be very different from

Table 1 Identity of *A. muscaria* haplotypes in our sample. Haplotype letters correspond to those in Table 2 and Fig. 2

Haplotypes	Specimen	Geographic origin
Clade I		
A	45785	Hampshire, Massachusetts, USA
B	45820	Bronx, New York, USA
C	CMP3143	Cochise Co., Arizona, USA
C	RET145-2	Tlaxcala, Mexico
C	CMP0648	Cochise Co., Arizona, USA
C	RET144-10	Tlaxcala, Mexico
D	GAL15330	Fairbanks, Alaska, USA
D	GAL16775	Fairbanks, Alaska, USA
D	GAL15776	Bonanza Creek LTER site, Alaska, USA
D	GAL16654	Fairbanks, Alaska, USA
D	GAL15336	Fairbanks, Alaska, USA
D	GAL17984	Fairbanks, Alaska, USA
D	GAL17982	Fairbanks, Alaska, USA
D	GAL15335	Fairbanks, Alaska, USA
E	RET145-1	Tlaxcala, Mexico
F	CMP1345	Cochise Co., Arizona, USA
G	49100	Cascade, Idaho, USA
H	45840	Lawrence, Massachusetts, USA
I	45843	Hampshire, Massachusetts, USA
J	45060	Amador, California, USA
J	LG382	Santa Cruz Island, California, USA
K	44761	Alpine, California, USA
L	45863	Mendocino, California, USA
L	RET320-1	Fremont Co., Idaho, USA
L	LG882	Santa Cruz Island, California, USA
M	RET271-2	Sussex Co., New Jersey, USA
N	45883	Piscataquis, Massachusetts, USA
N	RET271-3	Somerset Co., New Jersey, USA
O	RET383-3	Newfoundland, Canada
O	RET124-2	Suffolk Co., Massachusetts, USA
O	RET338-9	Clallam Co., Washington, USA
O	RET289-3	Cape May Co., New Jersey, USA
O	RET158-7	Burlington Co., New Jersey, USA
O	RET303-4	Monmouth Co., New Jersey, USA
Clade II		
A	30962	Kitakoma-gun, Yamanashi, Japan
A	30963	Kitakoma-gun, Yamanashi, Japan
A	30976	Kiso-gun, Nagano, Japan
A	30977	Ohno-gun, Gifu, Japan
A	30985	Ohno-gun, Gifu, Japan
A	30964	Gdynia, Poland
A	31445	Surrey, England, UK
A	80048	Surrey, England, UK

(*continued*)

Table 1 (continued)

Haplotypes	Specimen	Geographic origin
A	N10	Magadan district, Magadan region, Russia
A	GAL14284	Denali National Park, Alaska, USA
A	GAL16735	Fairbanks, Alaska, USA
A	GAL5895	Serpentine Hot Springs, N. of Nome, Alaska, USA
A	GAL15453	North Pole, Alaska, USA
A	GAL15461	North Pole, Alaska, USA
A	GAL5505	Denali National Park, Alaska, USA
A	RET036-3	Zürich canton, Switzerland
A	RET309-3	Sogn og Fjordane, Norway
A	RET143-5	Kamchatka, Russia
A	RET141-2	Highlands & Islands Reg., Scotland, UK
A	N3	Ola district, Magadan region, Russia
A	N12	Ola district, Magadan region, Russia
A	N5	Magadan district, Magadan region, Russia
A	N13	Omsukchan district, Magadan region, Russia
A	N6	Magadan district, Magadan region, Russia
A	MP23	Bird Creek Campground, Anchorage, Alaska, USA
A	N15	Tenka district, Magadan region, Russia
A	MP24	Bird Creek Campground, Anchorage, Alaska, USA
A	N16	Anadyr district, Chukot Autonomous Region, Russia
A	N1	Ola district, Magadan region, Russia
A	N11	Ola district, Magadan region, Russia
A	GAL8950	Denali National Park, Alaska, USA
A	GAL5900	Serpentine Hot Springs, N. of Nome, Alaska, USA
A	GAL5946	Quartz Creek, E. of Nome, Alaska, USA
A	GAL16735-2	Fairbanks, Alaska, USA
A	GAL18071	Fairbanks, Alaska, USA
A	GAL16735-3	Fairbanks, Alaska, USA
A	GAL18076	Fairbanks, Alaska, USA
A	GAL15454	North Pole, Alaska, USA
A	GAL18134	Bonanza Creek LTER site, Alaska, USA
A	GAL18136	Bonanza Creek LTER site, Alaska, USA
A	GAL17899	Fairbanks, Alaska, USA
A	GAL18012-6	Fairbanks, Alaska, USA
A	GAL18012-2	Fairbanks, Alaska, USA
A	GAL18012-4	Fairbanks, Alaska, USA
A	GAL18810-2	Homer, Alaska, USA
A	GAL18810-1	Homer, Alaska, USA
A	GAL17691	Fairbanks, Alaska, USA
B	RET107-6	Highlands & Islands Reg., Scotland, UK
C	GAL2005	Bonanza Creek LTER site, Alaska, USA
D	RET264-7	Skamania Co., Washington, USA
D	GAL4247	Glacier Hwy, mile 27, Juneau, Alaska, USA
D	GAL4302	Glacier Hwy, mile 27, Juneau, Alaska, USA
D	RET136-2	Whatcom Co., Washington, USA

(*continued*)

Table 1 (continued)

Haplotypes	Specimen	Geographic origin
D	GAL3643	Glacier Hwy, mile 30, Juneau, Alaska, USA
E	RET149-1	Southern Highlands Prov., Tanzania
E	RET149-2	Southern Highlands Prov., Tanzania
F	N8	Magadan district, Magadan region, Russia
G	GAL18122	Bonanza Creek LTER site, Alaska, USA
H	GAL2814	Dalton Hwy, mile 122, Alaska, USA
I	GAL4810	Denali National Park, Alaska, USA
J	31452	Hampshire, England, UK
J	30965	Gdansk, Poland
J	RET152-6	Baden-Württemberg, Germany
J	RET152-8	Bavaria, Germany
K	30961	Aomori-shi, Aomori, Japan
L	30987	Queenstown, New Zealand
M	N4	Ola district, Magadan region, Russia
N	N14	Tenka district, Magadan region, Russia

that of Clade I discussed above. MDIV showed evidence for intermediate levels of gene flow (M~2.5) between Alaskan and Eurasian populations and estimated no population divergence (T~0) (Fig. 1b). Our analyses using MIGRATE suggested that migration between Alaska and Eurasia has been asymmetrically bidirectional with Alaska receiving more migrants (M_{12}=9,482.98 and M_{21}=33.53; M_{12} referring to the migration from Eurasia to Alaska, and M_{21} *vice versa*). Therefore, in simulations using Genetree, we assumed panmictic population structure. As expected, the coalescent-based genealogy did not show any historical population division (Fig. 2b). However, based on the fact that the specimens from the maritime rainforest ecoregion of the Pacific Northwest of North America formed a well-supported sister group (II/A) to the rest of Clade II in our previous multi-locus phylogenetic analyses (Geml et al. 2006, 2008), we tested in a separate analysis whether or not there was any gene flow between this group inhabiting a special habitat with a restricted distribution and the rest of Clade II, this latter including boreal Alaskan and Eurasian specimens. In this case, MDIV showed no migration (M~0) and estimated a divergence time of T~1.5 between the two populations, i.e. subclade II/A vs. Clade II (Fig. 1c). Nonetheless, the coalescent-based genealogy estimated under an assumption of population subdivision did not differ significantly from the one generated assuming panmixia (Fig. 2c).

4 Discussion

The main goal of this project was to estimate migration between Alaskan and non-Alaskan populations within distinct phylogenetic species in the *A. muscaria* species complex in order to test whether or not these fungi could have survived the

Table 2 Polymorphic sites in haplotypes collapsed after recoding indels and excluding infinite-sites violations from the original dataset for coalescent analyses. Position refers to that in the original alignment, site number is the designation of the given mutation as shown on Fig. 1, site type refers to transition (t), transversion (v), or deletion (-) change with regard to the consensus sequence. Character type indicates either parsimony-informative (i) or non-informative (-) sites. Capital letters refer to haplotype designations on Fig. 2

Phylogenetic species	Clade I	Clade II
Position	1111122223344455 16660136600132712668 00134960218778400483	1111111233456 244667233455870506 1 366231200182485654280
Site number	11111111112 12345678901234567890	111111111122 123456789012345678901
Site Type	ttt-vt-ttttttttttttt	tttttttttvttvvvttvvtt
Character Type	-i-i-ii--------i----	----------i--i-------
Haplotypes:		
A	................C...	....
B	T..2................	G.......CT...........
C	.T...T..............	.G...................
D	.T.................C	..GC.CA............T.
E	.T...T.C............	....T..T....CC.......
F	.T...T...T....G..A..	........C............
G	..A...2........T....	.........T...........
H	...2................	..........A..........
I	....A...............	...........C.A.......
J	......2........T....	..............T......
K	......2.G..G.C......	................A....
L	......2.............	.................A...
M	.........T.G........	..................A..
N	..................T.	....................A
O	....	

LGM in Alaska. Because non-Alaskan populations most likely survived the LGM in refugia South of the major ice shields, the lack of migration between these and the Alaskan ones would suggest local survival of the latter, implying forest refugia in Alaska.

Overall, we found support for at least two endemic regional populations (haplotype D in Clade I, and haplotype D in Clade II) in different parts of Alaska, both of which exhibited genetic isolation and differentiation from other conspecific populations. Unfortunately, we could not estimate the times of divergence of the clades with certainty due to the lack of fossils and the great variance in nucleotide substitution rates in fungi. Nonetheless, it seems very likely that the major lineages (Clades I-III) separated well before the Pleistocene glacial cycles (Geml et al. 2006). On the other hand, the endemic regional populations likely represent groups that became isolated from the rest of their species in the Pleistocene and survived one or more glacial maxima in local refugia. Our findings support the existence of at least two independent such glacial forest refugia: (1) boreal forest in Interior Alaska; and

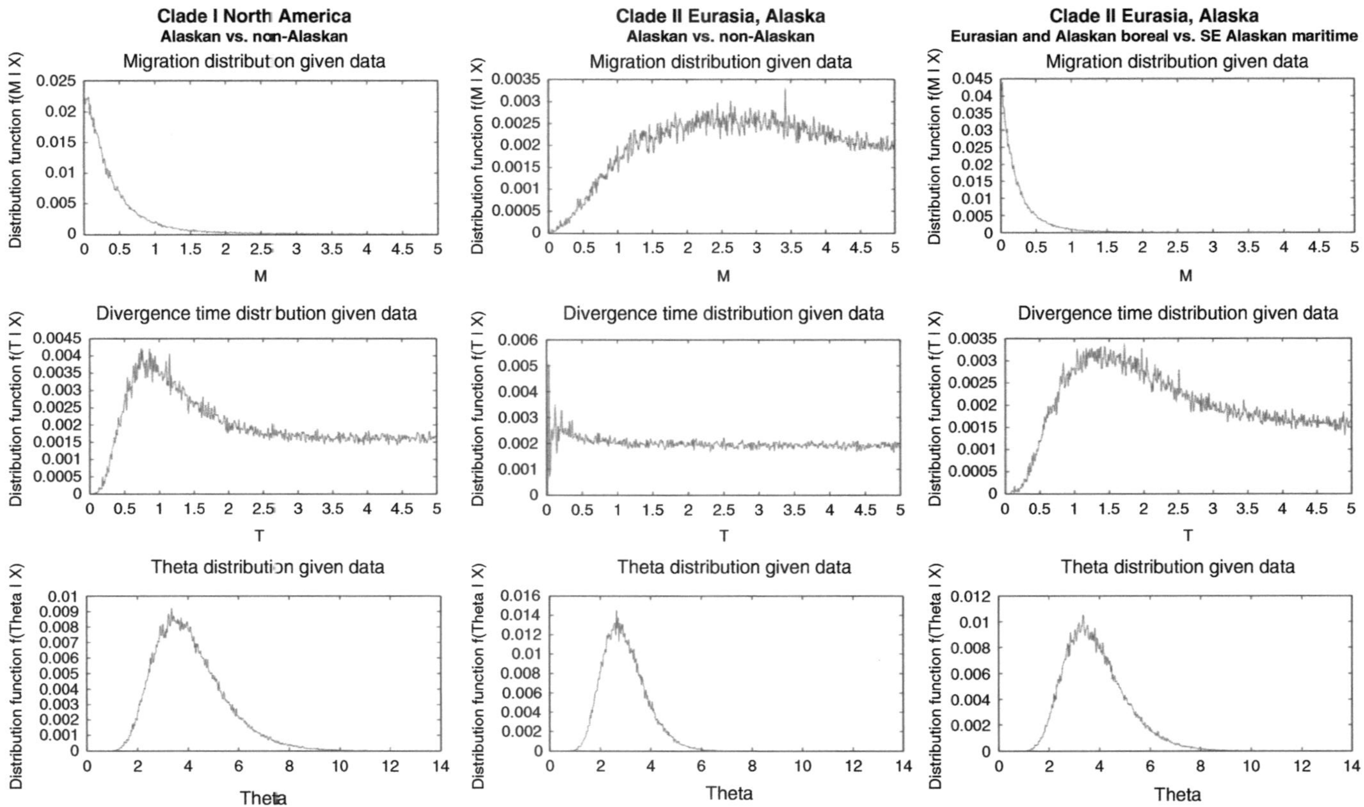

Fig. 1 Posterior probability distributions of migration (*M*), divergence time (*T*) and population mean mutation rate (*Theta*) between population pairs using MDIV. Estimates (axis X) with the highest distribution function (axis Y) values have the greatest probabilities. A and C show examples of zero migration and non-zero divergence time, while graphs in B indicate recent gene flow and no divergence between population pairs

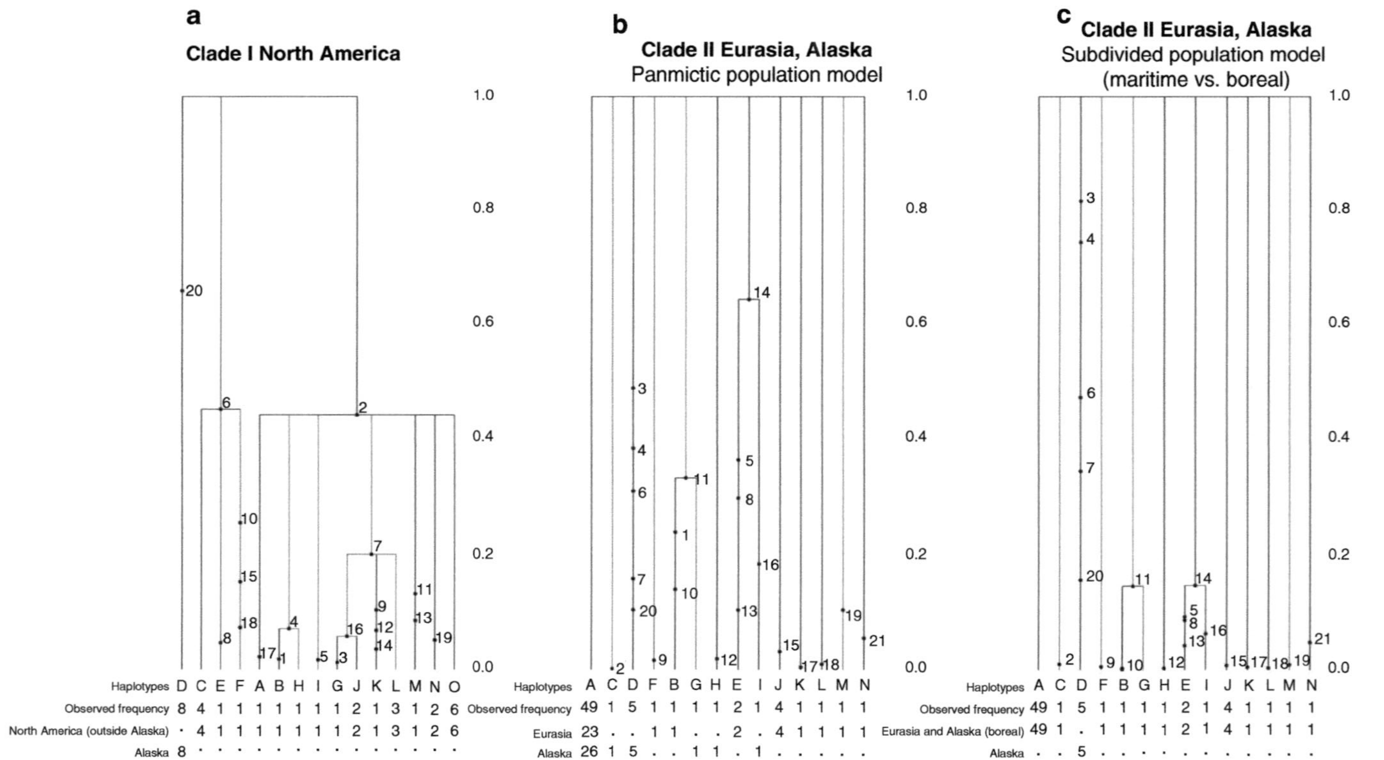

Fig. 2 Coalescent-based genealogies with the highest root probabilities showing the distribution of mutations. The inferred genealogies are based on 2 million simulations of the coalescent. The time scale is in coalescent units of 2*N*, where *N* is the population size. Mutations and bifurcations are time ordered from the top (past) to the bottom (present). Mutation designations correspond to the site numbers in Table 2. The numbers below the tree designate the distinct haplotypes, their observed frequencies in total and in the different geographic regions

(2) maritime rainforest in Southeast Alaska and the Pacific Northwest (Fig. 3). Apart from this, many Alaskan haplotypes in Clade II were shared with specimens from Eurasia. Our results suggesting high migration, most likely both pre- and post-glacial, across the Bering Strait and the similar observed numbers of haplotypes endemic to either Eurasia or Alaska leave open the question whether boreal members of Clade II survived the LGM in Alaska or colonised Alaska from Asia in multiple times in postglacial events.

The possible existence of forest refugia in Interior and Southeast Alaska is also supported by several other independent lines of evidence. For example, although previous palynological biome reconstructions suggest that all of

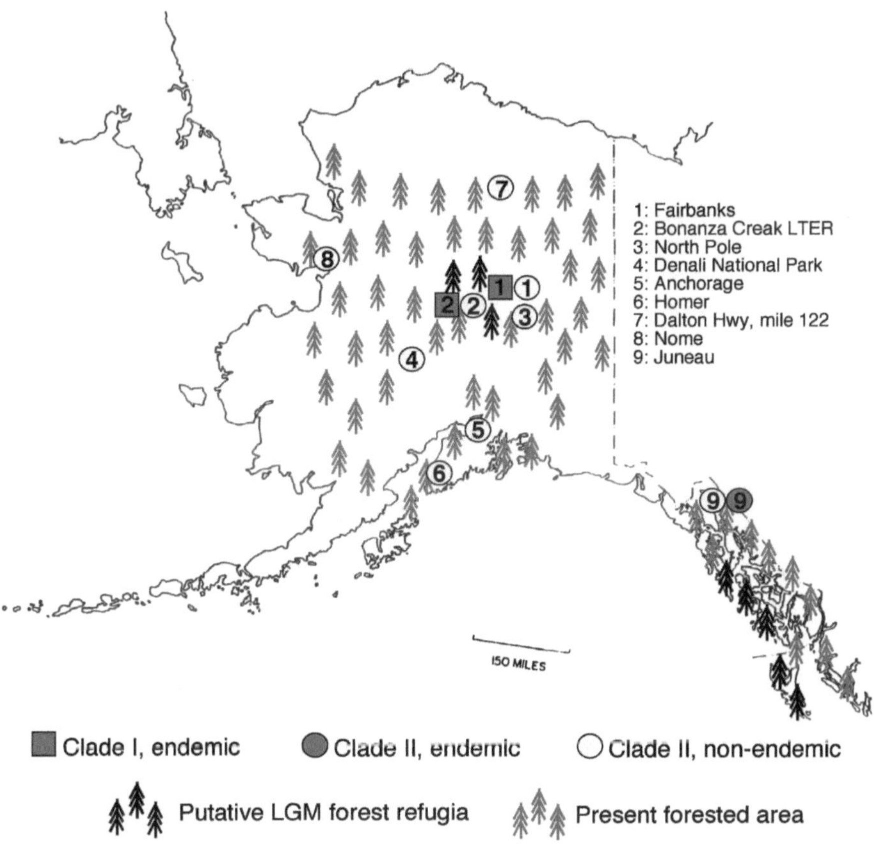

Fig. 3 Outline map of Alaska showing the collecting locations for the 43 Alaskan specimens included in this study. *Dark grey squares* and *circles* indicate sampling localities for endemic haplotypes, while *light grey circles* refer to haplotypes shared between Alaska and Eurasia. Black stylised spruce drawings mark putative forest refugia during the Last Glacial Maximum (LGM) suggested by this and other studies mentioned in the discussion. *Grey* tree figures indicate present forested area, without making a distinction between boreal and maritime forests of Interior and Southeast Alaska, respectively

Beringia was covered by arctic tundra (Edwards et al. 2000; Kaufman et al. 2004; Swanson 2003), recent pollen data by Brubaker et al. (2005) supports the theory that *Picea* forests were present in Eastern Beringia, at least in small fragments restricted to habitats with favourable microclimates. The Northernmost distribution of *Picea*, unlike that of *Betula* and *Populus*, is restricted, by definition, to the boreal region. Therefore, Brubaker's findings indicate the existence of the boreal forest biome in Alaska during the LGM. Similarly, Anderson et al. (2006), surveying forest stands across Northwestern North America, found several chloroplast DNA haplotypes of *Picea glauca* that were unique to Alaska, suggesting local survival. Maroja et al. (2007) analysed mitochondrial DNA sequence data from the spruce beetle (*Dendroctonus rufipennis*) and suggested that the postglacial range expansion of these beetles occurred from three refugia, one of which was in Interior Alaska.

Forest refugia in Southeast Alaska are also supported by genetic studies in several different taxa. For example, this is a region with high mammalian endemism due to its fragmented landscape and complex glacial history. Even animals, as mobile as wolves, have been shown to exhibit a strong signal of independent histories for the coastal and continental populations (Weckworth et al. 2005). Also, phylogenetic studies of Fleming and Cook (2002) on ermine (*Mustela erminea*) showed that there was at least one lineage strictly endemic to Southeast Alaska that likely survived locally during the Wisconsin glaciation. Locations for such glacial refugia in Southeast Alaska have been proposed by Carrarra et al. (2003) based on geological evidence.

In this paper, we demonstrated the use of ECM fungal taxa to indirectly infer glacial vegetation history. Therefore, our results are not only relevant to the mycological community, but to a broader biological community in general, particularly to those with a biogeographic interest.

References

Abbott RJ, Comes HP (2003) Evolution in the Arctic: a phylogeographic analysis of the circumarctic plant, *Saxifraga oppositifolia* (Purple saxifrage). New Phytol 161:211–224

Adams JM, Faure H (1997) Preliminary vegetation maps of the World since the last glacial maximum: An aid to archaeological understanding. J Archeol Sci 24:623–647

Alsos IG, Engelskjøn T, Gielly L, Taberlet P, Brochmann C (2005) Impact of ice ages on circumpolar molecular diversity: insights from an ecological key species. Mol Ecol 14:2739–2753

Anderson LL, Hu FS, Nelson DM, Petit RJ, Paige KN (2006) Ice-age endurance: DNA evidence for a white spruce refugium in Alaska. Proc Natl Acad Sci USA 103:12447–12450

Aylor D, Carbone I (2003) SNAP combine and map. Department of Plant Pathology, North Carolina State University, Raleigh, NC (http://www.cals.ncsu.edu/plantpath/faculty/carbone/home.html).

Bas C (1969) Morphology and subdivision of *Amanita* and a monograph of its section *Lepidella*. Persoonia 5:285–579

Beerli P, Felsenstein J (2001) Maximum-likelihood estimation of a migration matrix and effective population sizes in n subpopulations by using a coalescent approach. Proc Natl Acad Sci USA 98:4563–4568

Brubaker LB, Anderson PM, Edwards ME, Lozhkin AV (2005) Beringia as a glacial refugium for boreal trees and shrubs: new perspectives from mapped pollen data. J Biogeogr 32: 833–848

Brunhoff C, Galbreath KE, Fedorov VB, Cook JA, Aarola M (2003) Holarctic phylogeography of the root vole (*Microtus oeconomus*): implications for late Quaternary biogeography of high latitudes. Mol Ecol 12:957–968

Budantsev LY (1992) Early stages of formation and dispersal of the temperate flora in the boreal region. Bot Rev 58:1–48

Carrarra PE, Ager TA, Baichtal JF, VanSistine DP (2003) Map of glacial limits and possible refugia in the Southern Alexander Archipelago, Alaska, during the late Wisconsin Glaciation, Miscellaneous Field Studies Map, MF-2424. US Geological Service, Denver, CO

Edwards ME, Anderson PM, Brubaker LB et al (2000) Pollen-based biomes for Beringia 18, 000, 6000 and 0 14C yr BP. J Biogeogr 27:521–554

Eidesen PB, Carlsen T, Molau U, Brochmann C (2007) Repeatedly out of Beringia: *Cassiope tetragona* embraces the Arctic. J Biogeogr 34:1559–1574

Elias SA, Berman D, Alfimov A (2000) Late Pleistocene beetle faunas of Beringia: where East met West. J Biogeogr 27:1349–1363

Fleming MA, Cook JA (2002) Phylogeography of endemic ermine (*Mustela erminea*) in Southeast Alaska. Mol Ecol 11:795–807

Geml J, Laursen GA, O'Neill K, Nusbaum HC, Taylor DL (2006) Beringian origins and cryptic speciation events in the Fly Agaric (*Amanita muscaria*). Mol Ecol 15:225–239

Geml J, Tulloss RE, Laursen GA, Sazanova NA, Taylor DL (2008) Evidence for strong inter- and intracontinental phylogeographic structure in *Amanita muscaria*, a wind-dispersed ectomycorrhizal basidiomycete. Mol Phylogenet Evol 48:694–701. doi:doi:10.1016/j.ympev.2008.04.029

Graham A (1999) Late cretaceous and cenozoic History of North American Vegetation. Oxford University Press, Oxford

Griffiths RC, Tavaré S (1994) Ancestral inference in population genetics. Stat Sci 9:307–319

Härkönen M, Saarimäki T, Mwasumbi L (1994) Tanzanian mushrooms and their uses 4. Some reddish edible and poisonous Amanita species. Karstenia 34:47–60

Hey J, Wakeley J (1997) A coalescent estimator of the population recombination rate. Genetics 145:833–846

Hultén E (1968) Flora of Alaska and neighboring territories. Stanford University Press, Stanford

Kaufman DS, Ager TA, Anderson NJ et al (2004) Holocene thermal maximum in the Western Arctic (0–180°W). Quat Sci Rev 23:529–560

Loehr J, Worley K, Grapputo A, Carey J, Veitch A, Coltman DW (2005) Evidence for cryptic glacial refugia from North American mountain sheep mitochondrial DNA. J Evol Biol 19:419–430

MacNeil D, Strobeck C (1987) Evolutionary relationships among colonies of Columbian ground squirrels as shown by mitochondrial DNA. Evolution 41:873–881

Markwordt J, Doshi R, Carbone I (2003) SNAP clade and matrix. Department of Plant Pathology, North Carolina State University, Raleigh, NC (http://www.cals.ncsu.edu/plantpath/faculty/carbone/home.html).

Maroja LS, Bogdanowicz SM, Wallin KF, Raffa KF, Harrison RG (2007) Phylogeography of spruce beetles (*Dendroctonus rufipennis* Kirby) (Curculionidae: Scolytinae) in North America. Mol Ecol 16:2560–2573

Nei M, Li WH (1979) Mathematical model for studying genetic variation in terms of restriction endonucleases. Proc Natl Acad Sci USA 76:5269–5273

Nielsen R, Wakeley J (2001) Distinguishing migration from isolation: a Markov chain Monte Carlo approach. Genetics 158:885–896

Oda T, Tanaka C, Tsuda M (2004) Molecular phylogeny and biogeography of the widely distributed *Amanita* species A. muscaria and A. pantherina. Mycol Res 108:885–896

Qian H (1999) Floristic analysis of vascular plant genera of North America North of Mexico: characteristics of phytogeography. J Biogeogr 26:1307–1321

Reiss AR, Ashworth AC, Schwert DP (1999) Molecular genetic evidence for the post-Pleistocene divergence of populations of the arctic-alpine ground beetle *Amara alpina* (Paykull) (Coleoptera: Carabidae). J Biogeogr 26:785–794

Simmons C, Henkel T, Bas C (2002) The genus *Amanita* in the Pakaraima mountains of Guyana. Persoonia 17:563–582

Smith SE, Read DJ (1997) Mycorrhizal symbiosis, 2nd edn. Academic, Cambridge

Sperling FH, Harrison RG (1994) Mitochondrial DNA variation within and between species of the *Papilio machaon* group of swallowtail butterflies. Evolution 48:408–422

Swanson DK (2003) A comparison of taiga flora in North-Eastern Russia and Alaska/Yukon. J Biogeogr 30:1109–1121

Thompson JD, Gibson TJ, Plewniak F, Jeanmougin F, Higgins DG (1997) The CLUSTAL X windows interface: flexible strategies for multiple sequence alignment aided by quality analysis tools. Nucleic Acids Res 25:4876–4882

Tremblay NO, Schoen DJ (1999) Molecular phylogeography of *Dryas integrifolia*: glacial refugia and postglacial recolonization. Mol Ecol 8:1187–1198

Weckworth BV, Talbot S, Sage GK, Person DK, Cook J (2005) A signal for independent coastal and continental histories among North American wolves. Mol Ecol 14:917–931

Wickström LM, Haukisalmi V, Varis S, Hantula J, Fedorov VB, Henttonen H (2003) Phylogeography of the circumpolar *Paranoplocephala arctica* species complex (Cestoda: Anoplocephalidae) parasitizing collared lemmings (*Dicrostomyx spp.*). Mol Ecol 12:3359–3371

Part III
Glacial Relicts

Review
Refugial Areas and Postglacial Colonizations in the Western Palearctic

Jan C. Habel, Claudia Drees, Thomas Schmitt, and Thorsten Assmann

1 Introduction

Past climatic oscillations, such as the glacial-interglacial cycles of the Pleistocene (Williams et al. 1998) led to severe worldwide altitudinal and latitudinal range shifts of taxa and ecosystems (Hewitt 2004). For a reconstruction of the glacial and postglacial history of species, data obtained from methods ranging from evolutionary genetics to the analysis of distribution patterns and studies of pollen and subfossil remains of invertebrates and vertebrates are essential (de Lattin 1967; Coope 1970, 1978, 1994; Varga 1977; Huntley and Birks 1983; Taberlet et al. 1998; Willis et al. 1995; Hewitt 1999, 2000; Willis and van Andel 2004; Schmitt 2007; Varga and Schmitt 2008). The synthesis of all these analyses reveals the existence of different paradigm patterns in the Western Palearctic. In the following, we give a brief overview of these patterns.

2 Mediterranean Glacial Refugia

During the glacial periods, most thermophilic species became extinct over major parts of their former distribution range and only survived in "refugia" at lower latitudes which presented suitable climatic conditions. In Europe, these areas, located South of the transversal European high mountain systems, are the three well-studied refugial regions of the Mediterranean peninsulas (Iberia, Italy, and Balkans), all

J.C. Habel (✉)
Musée national d'histoire naturelle, Section Zoologie des Invertébrés, L-2160, Luxembourg, Luxembourg;
University of Trier, Department of Biogeography, D-54296 Trier, Germany
e-mail: Janchristianhabel@gmx.de

T. Schmitt
University of Trier, Biogeography, D-54296 Trier, Germany

C. Drees and T. Assmann
Institute of Ecology and Environmental Chemistry, Leuphana University Lüneburg, Scharnhorststrasse 1, D-21335 Lüneburg, Germany

J.C. Habel and T. Assmann (eds.), *Relict Species: Phylogeography and Conservation Biology*,
DOI 10.1007/978-3-540-92160-8_10,

more or less isolated from each other (Reinig 1937; de Lattin 1949; Huntley and Birks 1983; Bennett et al. 1991; Comes and Kadereit 1998; Hewitt 1996, 1999; Taberlet et al. 1998). The Maghreb and Asia Minor represent two further important areas in which thermophilic organisms survived the glacial periods (Busack 1986; Seddon et al. 2001, 2002; Harris et al. 2002; Seddon et al. 2002; Cosson et al. 2005; Fritz et al. 2006; Horn et al. 2006; Marmi et al. 2006; Weingartner et al. 2006; Habel et al. 2008). During these glacial periods, species differentiated into distinct genetic lineages, and repeated allopatric disjunctions reinforced these genetic differentiations, sometimes resulting in speciation events (Santucci et al. 1998; Habel et al. 2005). Subcenters have been postulated for these Mediterranean refugia (e.g., Reinig 1950); these have been tested by recent molecular analyses for the Balkans (Poulakakis et al. 2003, 2005; Parmakelis et al. 2006a, b Schmitt et al. 2006), Iberia (Harris et al. 2004; Martínez-Solano 2004; Carranza et al. 2006b), Southern Italy (Steinfartz et al. 2000; Seddon et al. 2001; Podnar et al. 2005; Canestrelli et al. 2006), and the Maghreb (Cosson et al. 2005; Carranza et al. 2006a,b; Fritz et al. 2006).

In the wake of climate warming during the postglacial period, the species expanded their distribution ranges from these refugia into higher altitudes and latitudes. These movements generally followed four paradigm patterns (Hewitt 2000; Habel et al. 2005): (1) postglacial expansion from all three Southern European differentiation centers, (2) expansion of the Western and Eastern lineages, with the Adriatic–Mediterranean lineage being trapped in Italy by the Alps, (3) expansion of the Adriatic– and Pontic–Mediterranean lineages, but trapping of the Atlantic-Mediterranean lineage by the Pyrenees, and (4) major expansion to Central Europe only from the Balkans, and trapping of the Atlantic– and Adriatic–Mediterranean lineages by the Pyrenees and Alps, respectively. These patterns are frequently repeated in many animal and plant species.

These recolonizations resulted in secondary contact and hybrid zones which can be traced by sensitive molecular markers (Hewitt 1988, 1993). Five regions of Europe are known for their high number of hybrid zones: (1) the Alps, (2) the Pyrenees, (3) Western Central Europe, (4) Eastern Central Europe, and (5) central Scandinavia (Taberlet et al. 1998; Hewitt 1999; Schmitt 2007). Further processes which occurred during colonization processes, such as the loss of genetic diversity due to stepping stone or leptokurtic range expansion, can be identified that help in the interpretation of the mode of past range change dynamics and the core areas of historical distribution areas (Hewitt 2001) (Fig. 1).

3 Extra-Mediterranean Refugia of Cold-Adapted and Temperate Species

During glacial periods, the regions between the Alps, Pyrenees, and Balkan mountains in the South and the Northern European ice shield in the North were largely covered by tundra and cold steppe on permafrost. Only cold-adapted species were

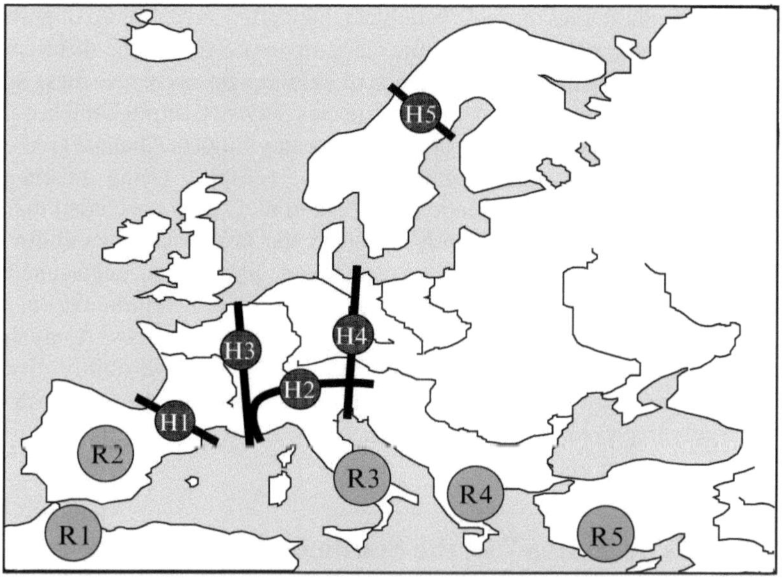

Fig. 1 The five major Mediterranean refugia and differentiation centers of the South European peninsulas (R1: Atlantic-Mediterranean in the Maghreb, R2: Atlantic-Mediterranean in Iberia, R3: Adriatic-Mediterranean, R4: Pontic-Mediterranean at the Balkans, R5: Pontic-Mediterranean in Asia Minor) and secondary contact- and hybridization zones during the postglacial range expansion (H1: Pyrenees, H2: Alps, H3 Western Central Europe, H4 Eastern Central Europe, H5: central Scandianvia). After Schmitt (2007), modified

able to survive under these extreme conditions. Many of these species are now restricted to the high mountain systems and/or the high latitudes of Eurasia, though they sometimes even reached North America (Holdhaus and Lindroth 1939; Holdhaus 1954; Varga and Schmitt 2008). Some, however, also have relict populations in cold habitats of lower mountain ranges. Stony debris formations with their exceptionally cold microclimatic conditions (at some sites with permafrost ground) host such species (Molenda 1996). Some of these species live exclusively North of the Alps (and not in the Alps!), thus making them Central European endemics (e.g., the ground beetle *Pterostichus neglectus*). However, many typical mountain species also showed discontinuous distribution patterns around the predominantly glaciated high mountain systems during the ice ages.

Furthermore, growing evidence supports the survival of numerous species which require woodlands or temperate climatic conditions North of the Mediterranean refugia scattered over large areas of Central Europe (Englbrecht et al. 2000; Jaarola and Searle 2002; Babik et al. 2004; Fink et al. 2004; Pinceel et al. 2005). This mainly coincides with the detection of woodland or at least tree species refugia in Central Europe during glacial periods (e.g., Willis and van Andel 2004). The Black Forest (and/or an area in its vicinity), for example, is discussed by Fink et al. (2004) and Drees et al. (2009) as a possible glacial refugium for woodland dwelling species.

Phylogeographic studies on some vertebrates supported the existence of previously unknown, or cryptic, refugia (e.g., in the Carpathians). The genetic differentiation of seaweeds, coupled with the concordance of similar patterns across these species, gives reason to conclude that some marine species persisted through the last ice age in flooded depressions on the floor of the present-day English Channel (see Provan and Bennett 2008 for a review of these cryptic refugia). Using an integrative approach with both fossil and genetic data, Petit et al. (2008) concluded that some temperate and boreal trees apparently survived the last glacial maximum under periglacial conditions, probably close to ice sheets and even in unglaciated areas North of ice sheets. All these recent results are beginning to revolutionize our understanding of glacial refugia for cold-adapted and temperate species. Analyzing the species distribution models with moderate to good predictive ability, Svenning et al. (2008) concluded that the view of Central European landscape North of the Alps as largely treeless needs to be revised.

4 "Massifs de Refuge" at the Southern Margin of the Alps

The former heavily glaciated inner regions of the Alps are highly differentiated from the Southern and Southeastern margins of these mountains, which were not strongly glaciated during the glacial periods. The latter regions are today characterized by an extraordinarily high biodiversity (cf. Ozenda 1982; Varga and Schmitt 2008). These glacial refugia at the Southern margin of the Alps have been described as "Massifs de Refuge" by botanists and zoologists lasting for over a century (e.g., Chodat and Pampanini 1902; Holdhaus 1904). Similar phenomena are also known for the Jura, the Northern Kalkalpen, and also large parts of the Pyrenees and Carpathians. Blind cave species were able to survive in these refugia (Assmann et al. 2009), and many other species, both plants and animals, with highly restricted distribution areas were scattered along the Southern and Southeastern margin of the Alps. Because some glaciers extended into the lowlands of Northern Italy, the Massifs de Refuge were interrupted and therefore lined up like "pearls on a necklace". Many of the species restricted to these areas are true relict species which had wider ranges in the past – as evidenced by their recent disjunct distribution pattern (e.g., *Pterostichus ziegleri*, Fig. 2).

On the other hand, some species which survived in these Massifs de Refuge also (re-)colonized the previously glaciated areas as e.g. known from weevils and moths. These species have two types of populations, diploid (with sexual reproduction) and polyploid (with parthenogenetic reproduction), with most species belonging to the genera *Otiorrhynchus* and *Solenobia* (Jahn 1941; Seiler 1943, reviewed in Suomalainen et al. 1976). The bisexual populations, from which the parthenogenetic populations evolved, represent the plesiomorphic group. In some weevil species, the diploid populations are mostly restricted to the regions of the Massifs de Refuge, and the polyploid populations exist in the parts of the Alps which were covered by ice sheets during the glacial periods, and/or in Central and Northern Europe. Thus, it is most likely that polyploidy evolved during postglacial range expansions.

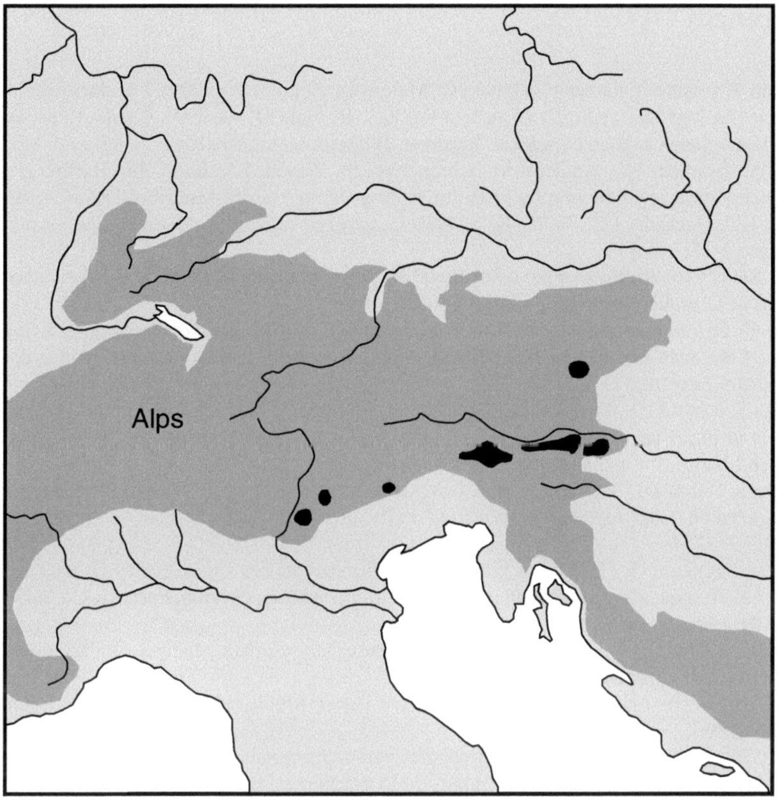

Fig. 2 Distribution pattern of the relict species *Pterostichus ziegleri* in the "Massifs de Refuge" at the Southern border of the Eastern Alps. After Holdhaus (1954) and Magistretti (1965), modified. *Grey*: Elevations above 500 m

5 Glacial Refugia in Northern Europe

Northern Europe with only few ice-free areas during the glacial periods appears to be an unlikely refugial area. However, some strongly differentiated species with several subspecies occur in Northern Europe, and at least some of these species were related to areas unglaciated during the ice ages. Consequently, Lindroth (1968, 1969, 1970) developed the hypothesis that some species survived the last glacial period in ice-free areas of Scandinavia. The majority of his contemporaries rejected this hypothesis. Evidence from recent genetic studies on rock ptarmigans (*Lagopus mutus*) and lemmings (*Lemmus lemmus*), however, supports the hypothesis of glacial refugia in Northern Europe (e.g., in Iceland and Northwestern Norway; Holder et al. 1999; Federov and Stenseth 2001). Thus, the existence of whole ecosystems with plant and animal communities seems probable, although more intensive investigations, especially in the field of strongly differentiated species are needed to confirm this.

References

Assmann T, Casale A, Drees C, Habel JC, Matern A, Schuldt A (2009) The dark side of relict species biology: cave animals as ancient lineages. In: Habel J, Assmann T (eds) Relict species – phylogeography and conservation. Springer, Heidelberg, pp. 91–104

Babik W, Branicki W, Sandera M, Litvinchuk S, Borkin LJ, Irwin JT, Rafinki J (2004) Mitochondrial phylogeography of the moor frog, *Rana arvalis*. Mol Ecol 13:1469–1480

Bennett KD, Tzekadis PC, Willis KJ (1991) Quaternary refugia of Northern European trees. J Biogeogr 18:103–115

Busack SD (1986) Biogeographic analysis of the herpetofauna separated by the formation of the Strait of Gibraltar. Natl Geogr Res 2:17–36

Canestrelli D, Cimmaruta R, Costantini V, Nascetti G (2006) Genetic diversity and phylogeography of the apennine yellow-bellied toad *Bombina pachypus*, with implications for conservation. Mol Ecol 15:3741–3754

Carranza S, Arnold EN, Pleguezuelos JM (2006a) Phylogeny, biogeography, and evolution of two Mediterranean snakes, *Malpolon monspessulanus* and *Hemorrhois hippocrepis* (Squamata, Colubridae), using mtDNA sequences. Mol Phyl Evol 40:532–546

Carranza S, Harris DJ, Arnold EN, Batista V, Gonzalez de la Vega JP (2006b) Phylogeography of the lacertid lizard, *Psammodromus algirus*, in Iberia and across the Strait of Gibraltar. J Biogeogr 33:1279–1288

Chodat R, Pampanini R (1902) Sur la distribution des plantes des Alpes Austro-Orientales, et plus particulierement d'un coix des plantes des Alpes cadoriques et venitiennes. Le Globe (Geneve) 41:109–121

Comes HP, Kadereit JW (1998) The effect of Quarternary climatic changes on plant distribution and evolution. Trends Plant Sci 3:432–438

Coope GR (1970) Interpretation of Quaternary insect fossils. Annual Revue of Entomology 15:97–120

Coope GR (1978) Constancy of insect species versus inconstancy of Quaternary environments. In: Mound LA, Waloff N (eds) Diversity of insect faunas. Blackwell, Oxord, pp 176–187

Coope GR (2004) The response of insect faunas to glacial-interglacial climatic fluctuations. Phil Trans R Soc Lond B 344:19–26

Cosson JF, Hutterer R, Libois R, Sarà M, Taberlet P, Vogel P (2005) Phylogeographical footprints of the Strait of Gibraltar and Quaternary climatic fluctuations in the Western Mediterranean: a case study with the greater white-toothed shrew, *Crocidura russula* (Mammalia: Soricidae). Mol Ecol 14:1151–1162

De Lattin G (1949) Beiträge zur Zoogeographie des Mittelmeergebietes. Zool Verh Kiel 143–151

De Lattin G (1967) Grundriß der Zoogeographie. Verlag Gustav Fischer, Jena

Drees C, Matern A, von Oheimb G, Reimann T, Assmann T (2009) Multiple glacial refuges of unwinged ground beetles in Europe – molecular data support classical phylogeographic models. In: Habel J, Assmann T (eds) Relict species – phylogeography and conservation biology. Springer, Heidelberg, pp. 199–216

Englbrecht CC, Freyhof J, Nolte A, Rassmann K, Schliewen U, Tautz D (2000) Phylogeography of the bullhead *Cottus gobio* (Pisces: Teleostei: Cottidae) suggests a pre-Pleistocene origin of the major Central European populations. Mol Ecol 9:709–722

Fedorov VB, Stenseth NC (2001) Glacial survival of the Norwegian lemming (*Lemmus lemmus*) in Scandinavia: inference from mitochondrial DNA variation. Proc Royal Soc London B 268:809–814

Fink S, Excoffier L, Heckel G (2004) Mitochondrial gene diversity in the common vole *Microtus arvalis* shaped by historical divergence and local adaptations. Mol Ecol 13:3501–3514

Fritz U, Barata M, Busack SD, Fritsch G, Castilho R (2006) Impact of mountain chains, sea straits and peripheral populations on genetic and taxonomic structure of a freshwater turtle, *Mauremys leprosa* (Reptilia, Testudines, Geoemydidae). Zool Scr 35:97–108

Habel JC, Meyer M, Mousadik AE, Schmitt T (2008) The complete phylogeography of the Marbled White butterfly species complex *Melanargia galathea/M. lachesis* (Lepidoptera: Satyridae), Organisms Divers Evol 8:121–129

Habel JC, Schmitt T, Müller P (2005) The fourth paradigm pattern of postglacial range expansion of European terrestrial species: the phylogeography of the Marbled White butterfly (Satyrinae, Lepidoptera). J Biogeogr 32:1489–1497

Harris DJ, Batista V, Carretero MA (2004) Assessment of genetic diversity within *Acanthodactylus erythrurus* (Reptilia: Lacertidae) in Morocco and the Iberian Peninsula using mitochondrial DNA sequence data. Amphibia-Reptil 25:227–232

Harris DJ, Carranza S, Arnold EN, Pinho C, Ferrand N (2002) Complex biogeographical distribution of genetic variation within *Podarcis* Wall lizards across the Strait of Gibraltar. J Biogeogr 29:1–6

Hewitt GM (1988) Hybrid zones – natural laboratories for evolutionary studies. Trends Ecol Evol 3:158–167

Hewitt GM (1993) Postglacial distribution and species substructure: lessons from pollen, insects and hybrid zones. Linnean Soc 14:97–123

Hewitt GM (1996) Some genetic consequences of ice ages, and their role in divergence and speciation. Biol J Linn Soc 58:247–276

Hewitt GM (1999) Post-glacial re-colonization of European biota. Biol J Linn Soc 68:87–112

Hewitt GM (2000) The genetic legacy of the Quaternary ice ages. Nature 405:907–913

Hewitt GM (2001) Speciation, hybrid zones and phylogeography – or seeing genes in space and time. Mol Ecol 10:537–549

Hewitt GM (2004) Genetic consequences of climatic oscillation in the Quaternary. Phil Trans R Soc Lond B 359:183–195

Holder K, Montgomerie R, Friesen VL (1999) A test of the glacial refugium hypothesis using patterns mitochondrial and nuclear DNA sequence variation in rock ptarmigan (*Lagopus mutus*). Evolution 53:1936–1950

Holdhaus K (1904) Über die Verbreitung der Coleopteren in den mitteleuropäischen Hochgebirgen. Verh zool-bot Ges Wien 56:629

Holdhaus K (1954) Die Spuren der Eiszeit in der Tierwelt Europas. Universitätsverlag Wagener, Innsbruck

Holdhaus K, Lindroth CH (1939) Die europäischen Koleopteren mit boreoalpiner Verbreitung. Ann Naturhist Mus Wien 50:123–309

Horn A, Roux-Morabito G, Lieutier F, Kerdelhue C (2006) Phylogeographic structure and past history of the circum-Mediterranean species *Tomicus destruens* Woll. (Coleoptera: Scolytinae). Mol Ecol 15:1603–1615

Huntley B, Birks HJB (1983) An atlas of past and present pollen maps of Europe: 0–130000 years ago. Cambridge University Press, Cambridge

Jaarola M, Searle JB (2002) Phylogeography of field voles (*Microtus agrestis*) in Eurasia inferred from mitochondrial DNA sequences. Mol Ecol 11:2613–2621

Jahn E (1941) Über Parthenogenese bei forstschädlichen *Otiorhynchus*-Arten in den während der Eiszeit vergletscherten Gebieten der Ostalpen. Zeitschr Ang Entomol 28:366–372

Lindroth CH (1968) The Icelandic form of *Carabus problematicus* Hbst (Col. Carabidae). Opuscula Entomol 33:157–182

Lindroth CH (1969) The theory of glacial refugia in Scandinavia – comments on present opinions. Not Entomol 39:178–192

Lindroth CH (1970) Survival of animals and plants on ice-free refugia during Pleistocene glaciations. Endeavour 29:129–134

Magistretti M (1965) Coleoptera, Cicindelidae, Carabidae – Catalogo topografico, Fauna d'Italia. Edizioni Calderini, Bologna

Marmi J, López-Giráldez F, Macdonald DW, Calafell F, Zholnerovskaya E, Domingo-Roura X (2006) Mitochondrial DNA reveals a strong phylogeographic structure in the badger across Eurasia. Mol Ecol 15:1007–1020

Martínez-Solano I (2004) Phylogeography of Iberian *Discoglossus* (Anura: Discoglossidae). J Zool Sys Evol Res 42:223–233

Molenda R (1996) Zoogeographische Bedeutung Kaltluft erzeugender Blockhalden im außeralpinen Mitteleuropa: Untersuchungen an Arthropoda, insbesondere Coleoptera. Abh Nan Ver Hamburg (NF) 35:5–93

Ozenda P (1982) Die Vegetation der Alpen im europäischen Gebirgsraum. Verlag Gustav Fischer, Stuttgart

Parmakelis A, Stathi I, Chatzaki M, Simaiakis S, Spanos L, Louis C, Mylonas M (2006a) Evolution of *Mesobuthus gibbosus* (Brullé, 1832) (Scorpiones: Buthidae) in the Northeastern Mediterranean region. Mol Ecol 15:2883–2894

Parmakelis A, Stathi I, Spanos L, Louis C, Mylonas M (2006b) Phylogeography of *Iurus dufoureius* (Brullé, 1832) (Scorpiones, Iuridae). J Biogeogr 33:251–260

Petit RJ, Shen Hu F, Dick CW (2008) Forests of the past: a window to future changes. Science 320:1450–1452

Pinceel J, Jordaens K, Pfenninger M, Backeljau T (2005) Rangewide phylogeography of a terrestrial slug in Europe: evidence for Alpine refugia and rapid colonization after the Pleistocene glaciations. Mol Ecol 14:1133–1150

Podnar M, Mayer W, Tvrtkovic N (2005) Phylogeography of the Italian wall lizard, *Podarcis sicula*, as revealed by mitochondrial DNA sequences. Mol Ecol 14:575–588

Poulakakis N, Lymberakis P, Antoniou A, Chalkia D, Zouros E, Mylonas M, Valakos E (2003) Molecular phylogeny and biogeography of the wall-lizard *Podarcis erhardii* (Squamata: Lacertidae). Mol Phyl Evol 28:38–46

Poulakakis N, Lymberakis P, Valakos E, Zouros E, Mylonas M (2005) Phylogeographic relationships and biogeography of *Podarcis* species from the Balkan Peninsula, by bayesian and maximum likelihood analyses of mitochondrial DNA sequences. Mol Phyl Evol 37:845–857

Proven J, Bennett KD (2008) Phylogeographic insights into cryptic glacial refugia. Trends Ecol Evol 23:564–571

Reinig W (1937) Die Holarktis. Verlag Gustav Fischer, Jena

Reinig WF (1950) Chorologische Voraussetzungen für die Analyse von Formenkreisen. Syllegomena Biologica, Festschrift für O. Kleinschmidt 346–378

Santucci F, Emerson B, Hewitt GM (1998) Mitochondrial DNA phylogeography of European hedgehogs. Mol Ecol 7:1163–1172

Schmitt T (2007) Molecular biogeography of Europe: pleistocene cycles and postglacial trends. Front Zool 4:11

Schmitt T, Habel JC, Zimmermann M, Müller P (2006) Genetic differentiation of the Marbled White butterfly, *Melanargia galathea*, accounts for glacial distribution patterns and postglacial range expansion in Southeastern Europe. Mol Ecol 15:1889–1901

Seddon JM, Santucci F, Reeve N, Hewitt GM (2002) Caucasus Mountains divide postulated postglacial colonization routes in the white-breasted hedgehog, *Erinaceus concolor*. J Evol Biol 15:463–467

Seddon JM, Santucci F, Reeve NJ, Hewitt GM (2001) DNA footprints of European hedgehogs, *Erinaceus europaeus* and *E. concolor*. Pleistocene refugia, postglacial expansion and colonization routes. Mol Ecol 10:2187–2198

Seiler J (1943) Über den Ursprung der Parthenogenese und Polyploidie bei Schmetterlingen. Archiv Julius Klaus-Stiftung 18:691–699

Steinfartz S, Veith M, Tautz D (2000) Mitochondrial sequence analysis of *Salamandra* taxa suggests old splits of major lineages and postglacial recolonization of Central Europe from distinct source populations of *S. salamandra*. Mol Ecol 9:397–410

Suomalainen E, Saura A, Lokki J (1976) Evolution of parthenogenetic insects. Evol Biol 9:209–257

Svenning J-C, Normand S, Kageyama M (2008) Glacial refugia of temperate trees in europe: insights from species distribution modelling. J Ecol 96:1117–1127

Taberlet P, Fumagalli L, Wust-Saucy A-G, Cosson J-F (1998) Comparative phylogeography and postglacial colonization routes in Europe. Mol Ecol 7:453–464

Varga Z (1977) Das Prinzip der areal-analytischen Methode in der Zoogeographie und die Faunenelement-Einteilung der europäischen Tagschmetterlinge (Lepidoptera: Diurna). Acta Biol Debr 14:223–285

Varga Z, Schmitt T (2008) Types of areal and oreotundral disjunctions in the Western Palearctic. Biol J Linn Soc 93:415–430

Weingartner E, Wahlberg N, Nylind S (2006) Speciation in *Pararge* (Satyrinae: Namphalidae) butterflies – North Africa is the source of ancestral populations of all *Pararge* species. Syst Entomol 31:621–632

Williams D, Dunkerley D, DeDecker P, Kershaw P, Chappell M (1998) Quaternary environments. Arnold, London

Willis KJ Sümegi P, Braun M, Toth A (1995) The late Quaternary environmental history of Bátorliget, NE Hungary. Palaegeogr Palaeoclimatol Palaeoecol 118:25–47

Willis KJ, van Andel TH (2004) Trees or no trees? The environments of central and Eastern Europe during the last glaciation. Quat Sci Rev 23:2369–2387

Multiple Glacial Refuges of Unwinged Ground Beetles in Europe: Molecular Data Support Classical Phylogeographic Models

Claudia Drees, Andrea Matern, Goddert von Oheimb, Thomas Reimann, and Thorsten Assmann

Abstract Since the 1930s, several European zoologists have developed scenarios for glacial refuges and postglacial expansions, mainly based on studies of the morphological differentiation of populations and distribution patterns of species. For example, Holdhaus described the distribution of blind euedaphic and troglobitic beetles restricted to an area South of a well-defined line crossing the Southern Europe from West to East. In these areas, where many endemic animal and plant species occur, other species that are currently more widely distributed in Europe were probably able to survive the glacial period(s). Molecular analyses of 77 populations of the silvicolous ground beetle *Carabus auronitens* support the existence of these postulated refuge areas. Genetic differentiation of *C. auronitens* provides good evidence for multiple refuges, which are, however, situated further North than previously assumed. Furthermore, genetic differentiation is more pronounced in the areas South of the "Holdhaus line" than in the areas North of it.

Dedication

We dedicate this work to Prof. em. Dr. Friedrich Weber, our former academic supervisor and colleague, without whose tenacity in the study of *C. auronitens* this paper would not have been possible.

C. Drees (✉), A. Matern, G. von Oheimb, and T. Assmann
Institute of Ecology and Environmental Chemistry, Leuphana University Lüneburg, Scharnhorststraße 1, D-21335 Lüneburg, Germany
e-mails: cdrees@uni-lueneburg.de; matern@uni-lueneburg.de; vonoheimb@uni-lueneburg.de; assmann@uni.leuphana.de

T. Reimann
Institute of General Zoology and Genetics, University of Münster, Schlossplatz 5, D-48149 Münster, Germany
Present address: Projekträger Jülich, Forschungszentrum Jülich GmbH, Zimmerstr. 26-27, D-10969 Berlin, Germany
e-mail: t.reimann@fz-juelich.de

J.C. Habel and T. Assmann (eds.), *Relict Species: Phylogeography and Conservation Biology*, DOI 10.1007/978-3-540-92160-8_11,

1 Introduction

The Evolutionary Synthesis of Dobzhansky, Mayr, Rensch, Wright, Simpson, and others communicated the idea of population genetics to other fields of biology. These authors argued convincingly that mutation, recombination, selection, and gene flow operating within species ("microevolution" in Dobzhansky's term) account for the origin of new species and for the long-term effects of evolution ("macroevolution" in Dobzhansky's term) (Dobzhansky 1937). The European exponent of Evolutionary Synthesis, Bernhard Rensch (Futuyma 2005), was very interested in the processes of genetic drift as can be studied easily in small populations, such as on islands and in glacial refuges (Rensch 1929, 1939, 1951). A lively discussion on the processes of differentiation, leading to racial formation and speciation, was held in these years (Rensch 1929, 1939, 1960; Krumbiegel 1932, 1936; De Lattin 1959). While Krumbiegel (1936) described and measured morphological differences as a result of these processes, the genetic consequences of the climate-induced range shifts, including postglacial recolonization, were hypothesized for the first time by Reinig (1938, 1939). Recently, this idea was confirmed by compilations of studies using molecular data (Hewitt 1996, 1999; recent review by Schmitt 2007).

Biogeography profited from the development of both evolution-orientated population biology and the classical models of biogeography by pioneers, such as Holdhaus (Holdhaus and Lindroth 1939) and De Lattin (1957, 1959, 1967). Holdhaus showed that the distribution of blind euedaphic and troglobitic (cave-inhabiting) beetles is restricted to a well-defined area in Europe (Holdhaus 1954). In the South of the "Holdhaus line", defined by the Northern distribution limit of these species, many endemic species occur. These areas, which coincide with the occurrence areas of those for endemic plant species of the Alps, have been known as "Massifs de Refuge" after Chodat and Pampanini (1902), and were discussed by Holdhaus (1906) as refuge areas also suitable for animals.

Surprisingly, Holdhaus' ideas on glacial refuges and postglacial colonization processes have never been rejected although contemporary Anglophone scientific work and biogeography primers have mostly ignored them. One reason for this might be that these basal works have mainly been published in German.

Over the last decades, our knowledge of processes of glacial and postglacial periods has increased enormously, mainly due to the progress in evolutionary genetics and the use of a variety of genetic markers (e.g., Stauffer et al. 1999; Seddon et al. 2001; Michaux et al. 2003; Berggren et al. 2005; Joger et al. 2007) and the combination of these results with other methods (e.g., analysis of pollen and fossil records) to reconstruct the past events (Terhurne-Berson et al. 2004; Cheddadi et al. 2006; Magri et al. 2006; Magri 2008). Interestingly, not only historical but also recent hypotheses and postglacial recolonization scenarios are frequently based on studies of insect species, especially on those with low dispersal power (e.g., Hewitt 1996).

The aims of our study are to test whether

1. the Holdhaus line is congruent with the current knowledge on the distribution of endemics in Europe. We use ground beetles as a taxonomic group for this reevaluation as they are well studied in Europe at most relevant levels (especially faunistics and taxonomy, Assmann et al. 2008). This beetle family comprises a vast number of species in Europe (over 3,600), many of which are blind. Thus, this taxon also seems appropriate to test the validity of Holdhaus' biogeographic description of the Northern limit of blind beetles.
2. the Holdhaus line can be used for analytical procedures in present-day phylogeography of European species which aim to distinguish between relict and (re-)colonizing species (and populations), and to delimit glacial refuges. We use the stenotopic flightless woodland ground beetle species *Carabus auronitens* as a model species.

2 The "Holdhaus Line": A Biogeographically Important Border in Europe

2.1 The Concept

Caves are very special habitats for animals due to their constant temperature and humidity conditions, their permanent darkness, and very low densities of prey. Consequently, troglobitic species often show striking morphological (long appendices, reduced eyes) and physiological (reduced metabolism, long life-cycles) adaptations to these habitats. Their dispersal ability is extremely low and limited. Most of the troglobitic species lack the ability to leave the massifs in which they occur and are, thus, among the species with the lowest power of dispersal (Culver and White 2005). In situations with such pronounced isolation between populations, numerous species were developed. For example, the world diversity hotspot of troglomorphic genera of Trechini (a species-rich tribe of ground beetles with many anopthalmic (eyeless) species) is located in the Western Palearctic realm (more than 50 genera) as compared with Asia (more than 30), the Nearctic (more than 15), and the Neotropics and Australia (about four and seven genera, respectively) (Casale et al. 1998; see also Lorenz 2005 for additional records on blind ground beetle genera in Southern Asia).

Holdhaus mapped the distribution ranges of two groups of beetles: the true cave-dwelling (troglobitic) beetles and the blind endogeic beetle species. He drew lines North of the distribution ranges of both groups, and stated that because of the very poor dispersal power of these species North of this border none of them could have survived glacial periods. We reevaluate the congruence between the Holdhaus line (as published in 1954) and our knowledge on Northern limits of cave and endogeic beetles.

Most of the cave-dwelling beetles inhabit not only caves but also systems of crevices and lacunae. In karstic regions, where most such species occur, they

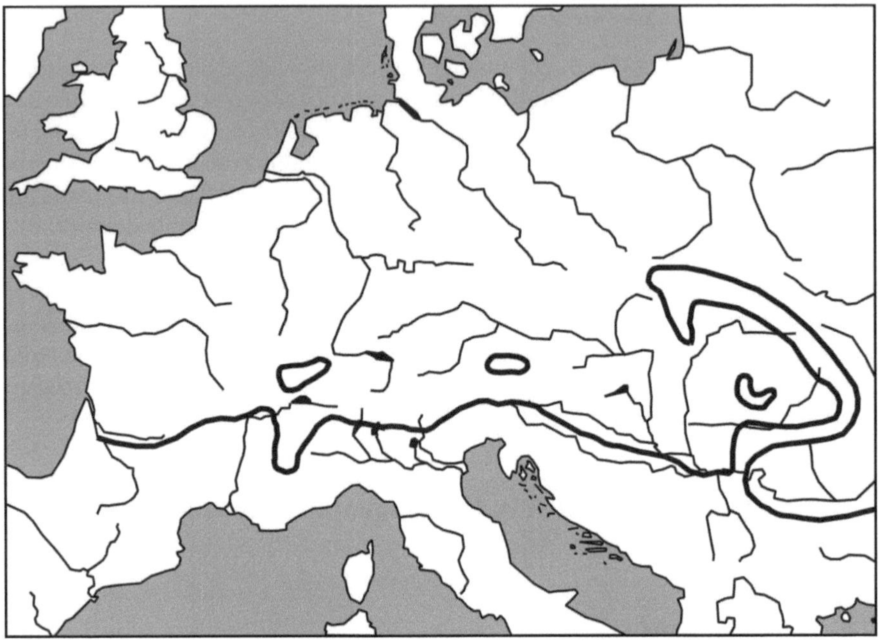

Fig. 1 Northern boundary of distribution ranges of anophthalmic cave beetle taxa in Europe: Holdhaus line. Modified after Holdhaus (1954)

frequently inhabit even whole massifs. Juberthie calls this special habitat "superficial underground compartment" (SUC) (Juberthie et al. 1980a,b), pointing out that there is no ecological difference between the so-called troglobitic and the so-called anophthalmic and endogeic species. Indeed, extraordinarily large blind trechine beetles which have been regarded as true cave-dwelling species were also recorded from the SUC (Drovenik et al. 2008).

For this reason, it seems appropriate to combine both the lines described by Holdhaus (which are mainly congruent) in the "Holdhaus line", named after its discoverer. The Holdhaus line runs from Bordeaux, via Lyon, the South of Alps, the Carpathians to the Black Sea (Fig. 1) (Holdhaus 1954) and is in good accordance with the Southern permafrost limit during the last glacial period (Lang 1994; Koenigswald 2002).

2.2 *The Holdhaus Line: Still up-to-Date?*

Since Holdhaus' basic work, numerous additional species and a vast number of additional records of previously described endogeic and cave-dwelling species have come to be known (e.g., Löbl and Smetana 2003). We analyzed the faunistic and taxonomic literature (e.g., Bonadona 1971; Vigna-Taglianti 1982; Kryzhanovskij et al. 1995; Avon 1997; Löbl and Smetana 2003; Casale and Vigna-Taglianti 2005) and verified the geographic information of the records regarding potential deviations from Holdhaus' concept.

In general, we found a remarkable congruence between the new findings and the "old" Holdhaus line. This is true for most faunistic records of anophthalmic or microphthalmic (with reduced eyes) ground beetles from a wide range of subfamilies and tribes (e.g., Trechini, Bembidiini, Anillina, Pterostichini, Platynini). We detected only three exceptions:

1. In the Northeastern Alps (Austria) and Northeast of the continuous line indicated by Holdhaus, there is an area with records of blind beetles. New records of the genus *Arctaphaenops* are outside the area marked by Holdhaus (cf. the records listed by Daffner 1993). The range of the three species which are regarded as valid species seems to be a little larger than the refuge area proposed by Holdhaus. This may be due to the postglacial expansions of the colonized area over distances of a few kilometers.
2. A similar situation exists in the Western Alps and its foothills (incl. the Jura) where the genera *Agostinia* and *Trichaphaenops* exhibit few populations (and species) (Bonadona 1971; Marggi 1992). We suggest uniting the isolated enclave North of the continuous Holdhaus line to an isolated range of an insignificant larger area (Fig. 1, Northwestern enclave).
3. Two records of blind carabid beetles are also known from sites far North of the Holdhaus line in Belgium (Desender 1986) and Germany (Malzacher 2000). Both records were taken in parks and gardens within cities. More than a century ago, trees (particularly, plane and chestnut) from Southern France were planted at both sites known to host blind endogeic Anillina beetles (*Anillus caecus*). Thus, it is very likely that the beetles were imported from the Mediterranean region along with earth and roots of young trees. Moreover, the survival of beetles proved the fact that they have the potential to survive far away from their usual distribution area and that their dispersal is limited.

These data support an extensive concordance between the distributions of blind species known today and those known in Holdhaus' time. Only very few corrections are needed to update the Holdhaus line.

A preliminary check of the distribution data of anophthalmic beetles from other families (e.g., Curculionidae, Cholevidae) supports the findings based on ground beetles, so that the Holdhaus line stands up to this initial perusal (cf. Casale et al. 1991; Giachino et al. 1998; Osella and Zuppa 1998).

3 The Holdhaus Line as a Tool for the Localization of Refuge Areas? The Case of *C. auronitens*

In this analysis, we investigate the hypothesis that the Holdhaus line helps to locate glacial refuges of species which are much more widely distributed today. As the Holdhaus line was derived from the distribution patterns of cave and endogeic beetles, this hypothesis may at least hold true for species of cold and humid habitats. We test this by compiling and reanalyzing the existing data from different population genetic analyses of *C. auronitens*.

3.1 *C. auronitens*

C. auronitens Fabricus, 1792 is a flightless ground beetle species of mostly cool and humid woodland habitats in both lowlands and montane to subalpine regions. Its distribution range covers wide areas of temperate Central Europe (Turin et al. 2003), with the Southernmost populations found in the Pyrenees, the Cevennes, Eastern parts of the Alps, and the Carpathians; it thus lives on either side of the Holdhaus line. In the Southwestern part of its range, the distribution is disjunct as populations exist in several areas in the Pyrenees (Forel and Leplat 1995) and the mountains further North (Montagne Noire and Cevennes, Puisségur 1964).

Both the biology and long-term population dynamics of the species (e.g., Weber and Heimbach 2001; compilation by Turin et al. 2003) and the genetic differentiation over its entire range (Assmann et al. 1994; Assmann and Weber 1997; Reimann et al. 2002) were investigated in detail.

3.2 *Testing the Hypothesis*

3.2.1 Datasets

In order to test our hypothesis, we reanalyzed the population genetic data by Assmann et al. (1994), Assmann and Weber (1997), and Reimann et al. (2002). Allozyme polymorphisms at five loci (Table 1) were obtained by electrophoretic separation using acrylamide gels. Standard methods (as described in Assmann and Weber 1997) and consistent naming of the alleles (Reimann et al. 2002) allowed data compilation and reanalysis. Only samples with a minimum sample size of 15 individuals were used. We used two different datasets (Table 1). Dataset A contains

Table 1 Population genetic datasets used for the reanalysis to localize the glacial refuge areas of *C. auronitens*

Dataset	Sample no. in this study	No. of samples	Loci (EC number)	No. alleles found	Sources (sample numbers in this study)
A	1–61	61	AAT (EC 2.6.1.1)	8	Assmann et al. (1994) (30–61); Assmann and Weber (1997) (1–29)
			EST (EC 3.1.1.1)	17	
			GPI (EC 5.3.1.9)	19	
			6PGDH (EC 1.1.1.44)	7	
			ME[1] (EC 1.1.1.40)	2[a]	
B	30–77	48	AAT	4	Assmann et al. (1994) (30–61); Reimann et al. (2002) (62–77)
			EST	8	
			GPI	15	
			6PGDH	3	

[a]Data for polymorphisms at the ME locus partly unpublished: Samples 1–29 are monomorphic for one allele, samples 30–61 monomorphic for another allele

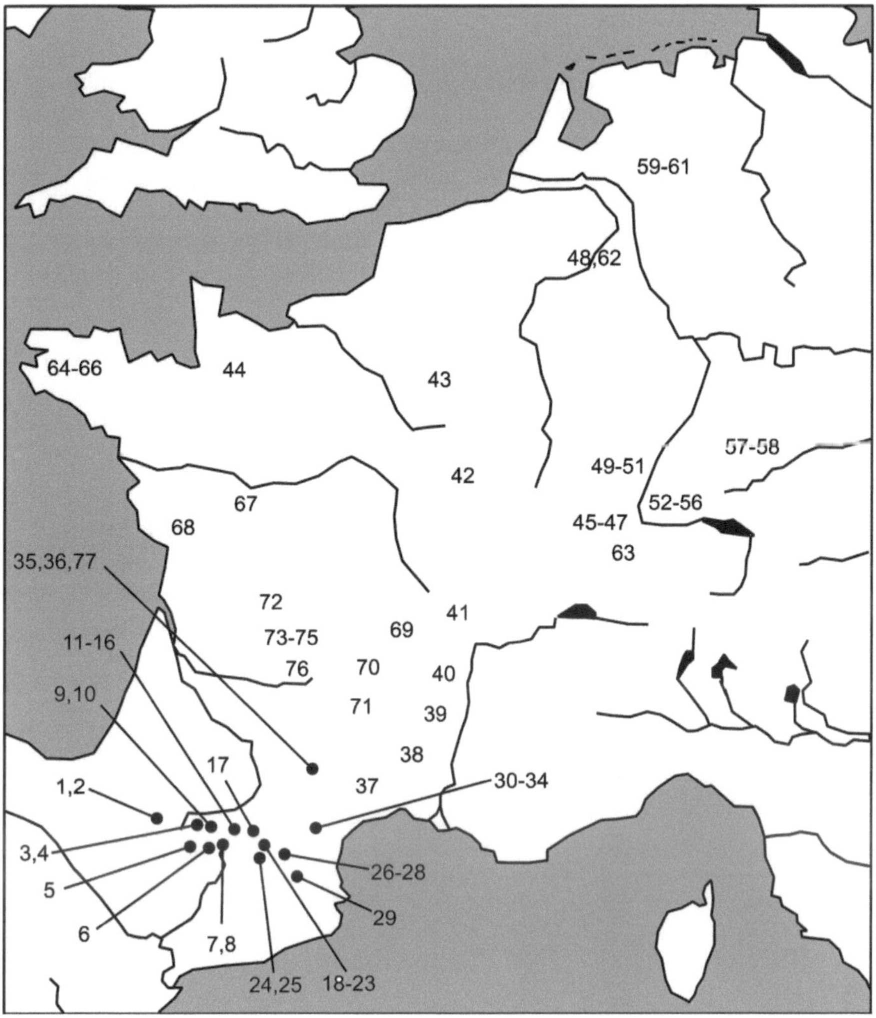

Fig. 2 Sample sites covering the Western part of the distribution area of *C. auronitens*. Renumbered after Assmann et al. (1994), Assmann and Weber (1997) and Reimann et al. (2002), see Table 1

information on a total of 1,771 individuals from 61 sample sites in the Western and Southwestern part of the distribution range of *C. auronitens* (including the Pyrenees) at five loci, whereas dataset B contains information from 48 samples with a total of 1,517 individuals at four loci. Dataset B comprises additional samples, mainly from Western France but lacks the samples from the Pyrenees used in dataset A. Sample sites were renumbered and are shown in Fig. 2.

3.2.2 Statistical Analyses

Nei's standard genetic distances (Nei 1978) were calculated from allele frequencies and phenograms (using neighbor-joining, Saitou and Nei 1987) were constructed (with PHYLIP, Felsenstein 2005) for both datasets. Principal component analyses (PCA) based on the arc-sin transformed allele frequencies were performed for both datasets using STATISTICA, version 7.1 (StatSoft Inc.). For the latter analyses rare alleles were omitted, thus only alleles with a minimum mean frequency of 10% were used. Private alleles (defined as alleles found in one or two neighboring sample sites only) were counted per sample.

3.3 Genetic Differentiation and Localization of Glacial Refuges

The analysis of the allele frequencies of 61 samples of dataset A by both cluster analysis and PCA reveals concordant patterns. Whereas the samples from the Pyrenees – a comparably small geographic range – exhibit a striking genetic structure, partly with large genetic distances, the remaining populations sampled in a large part of the distribution area of *C. auronitens* show comparably small genetic differentiation (Fig. 3).

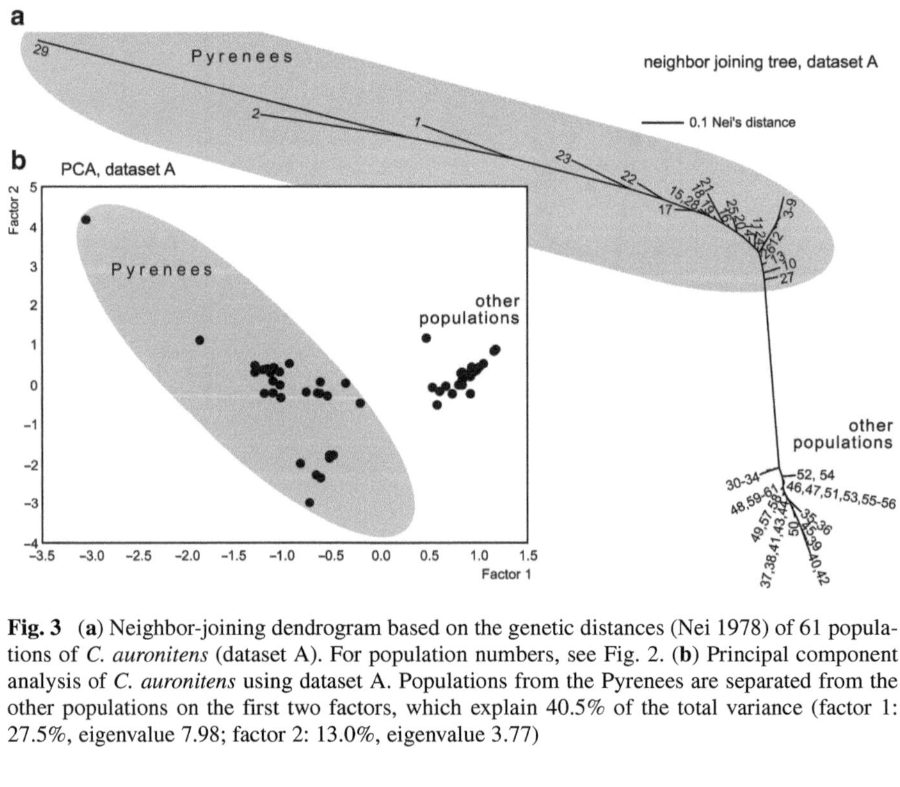

Fig. 3 (**a**) Neighbor-joining dendrogram based on the genetic distances (Nei 1978) of 61 populations of *C. auronitens* (dataset A). For population numbers, see Fig. 2. (**b**) Principal component analysis of *C. auronitens* using dataset A. Populations from the Pyrenees are separated from the other populations on the first two factors, which explain 40.5% of the total variance (factor 1: 27.5%, eigenvalue 7.98; factor 2: 13.0%, eigenvalue 3.77)

Moreover, the latter samples are clearly separated from those from the Pyrenees. A special case provides sample no. 29 from the easternmost part of the Pyrenees (Mt. Canigou). Whereas Assmann and Weber (1997) could not decide to which of the two groups (Pyrenees or non-Pyrenees) this sample clustered, we find evidence for a similarity to the Pyrenean *C. auronitens* populations but with a relatively strong genetic distance.

The analysis of dataset B (excluding the populations from the Pyrenees) offers more detailed information. Despite a generally lower genetic differentiation, three groups can be distinguished by means of PCA that can also be found in the tree: the samples originating from the disjunct areas of the Montagne Noire and the region around Rodez form their own groups, whereas the other populations cluster together (Fig. 4). Populations originating from sample sites that are localized South of the Holdhaus line are highlighted in Fig. 4a to give an impression of possible glacial refuge areas. In contrast to samples from Rodez and Montagne Noire, populations from the Cevennes (nos. 37–40) and populations from the Southern part of the Auvergne (nos. 70–71) do not cluster together but can be found – from a genetic point of view – in the midst of the samples from all over the Western part of the large distribution range of the study species.

The results support the hypothesis of multiple glacial refuge populations of *C. auronitens* (already given by Assmann et al. 1994), most of which behaved

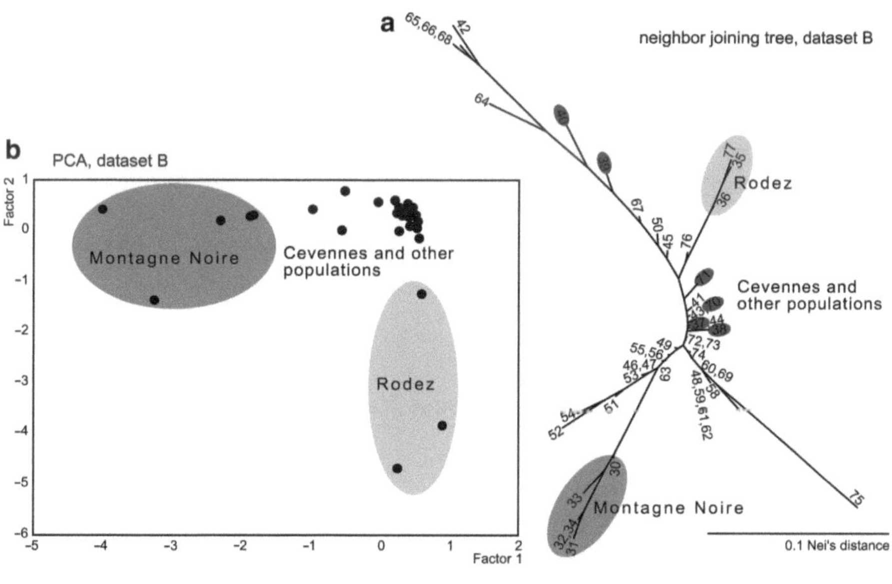

Fig. 4 (**a**) Neighbor-joining dendrogram based on the genetic distances (Nei 1978) of 48 populations of *C. auronitens* (dataset B). For population numbers, see Figure 2. (**b**) Principal component analysis of *C. auronitens* using dataset B. Populations from Montagne Noire and Rodez are separated from the other populations on the first two factors, which explain 42.0% of the total variance (factor 1: 23.0%, eigenvalue 2.98; factor 2: 19.0 %, eigenvalue 2.47)

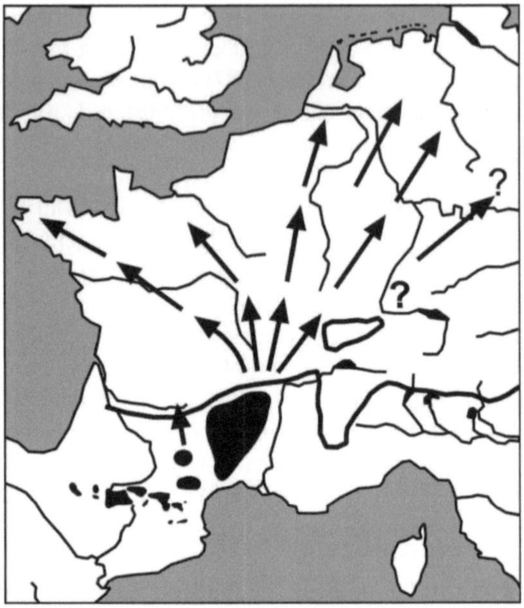

Fig. 5 Glacial refuge areas of *C. auronitens*. *Arrows* show postglacial recolonization from two of these refuges, "?" indicates a putative additional refuge area or unknown migration routes (see text). The black line indicates the Holdhaus line

endemically or showed only restricted dispersal (i.e., in the Pyrenees or the Montagne Noire), whereas the populations from the Cevennes and the Southern Auvergne recolonized large parts of Western and Central Europe (Fig. 5). The strong power of dispersal of the populations at the edge of the distribution range is demonstrated by a recent recolonization event of *C. auronitens* populations in Northwestern Germany after anthropogenic landscape changes (Drees et al. 2008).

3.4 Private Alleles: Evidence for a Refuge Area North of the Holdhaus Line

Private alleles are, from a population genetic point of view, of special interest. They developed either during the course of a long-lasting isolation of the respective population (i.e., the variant originates from mutation) or they originally occurred in many populations but were lost at all but one site, e.g., as a consequence of repeated extinction and recolonization (Hewitt 1996). These two sources of origin imply that private alleles occur predominantly in populations in or close to refuge areas.

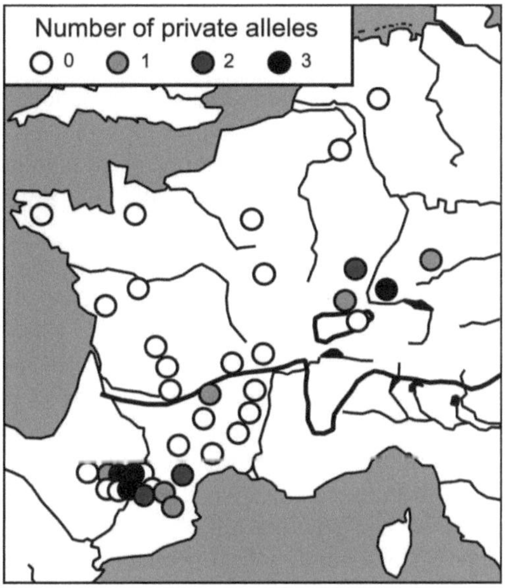

Fig. 6 Numbers of private alleles in *C. auronitens* samples investigated at four allozyme loci. Groups of samples are indicated by a single circle only (compare Fig. 2), private alleles for these groups are summed over the respective populations

Altogether 24 private alleles were found in the *C. auronitens* populations, 14 in the populations from the Pyrenees and another three in the remaining samples from Southern France (Fig. 6). Additionally, seven private alleles occur in the region around the border triangle of Switzerland, France, and Germany (including the Vosges, the Black Forest, and part of the Swiss and French Jura). These findings suggest the existence of at least one other glacial refuge area in this region North of the Holdhaus line.

3.5 Allele Numbers in Populations North and South of the Holdhaus Line

Altogether 50 different alleles were recorded in populations situated South of the Holdhaus line, whereas only 23 different alleles were found in populations North of it. This is also reflected by the allele numbers corrected for the number of investigated populations (rarefied to 34 populations, Krebs 1999): The populations situated South of the Holdhaus line show 41.5 alleles, whereas the populations North of it have only 22.0 different alleles.

With the exception of the seven private alleles in the Black Forest, the Vosges, and parts of the Swiss and French Jura (see above), all alleles that were found in populations North of the Holdhaus line are also found in populations located South of it. The Northern populations are, therefore, a section of the genetic richness of the Southern populations.

4 Discussion

Although the Holdhaus line was introduced more than a half century ago, it is still up-to-date. The brief overview of the most recent literature on European cave beetles (including the comprehensive faunistic investigations in Northern Italy and Southern France) shows that only a few amendments, but no basic corrections, had to be made to the Holdhaus line.

Of what importance is the Holdhaus line for the identification of refugial areas for other endemic taxa? Recent studies of endemic-rich taxa indicate that the Holdhaus line is not only a remarkable Northern border for the distribution of blind cave- or soil-dwelling beetles, but is also congruent with the Northern distribution limit of several endemics (e.g., from the genus *Leistus*, Assmann and Heine 1993).

If highly specialized endogeic species were able to survive the glacial period(s) in Southern France and Northern Italy, i.e., in regions close to the Holdhaus line, it is likely that other species which are currently more widely distributed in Europe would also have had a survival chance in these "Massifs de Refuge." Molecular analyses of populations of the ground beetle genus *Carabus* are a suitable method to localize former refuge areas, as shown for *Carabus solieri*, an endemic ground beetle species with a small postglacially recolonized area and a hybrid zone formed after a secondary contact of individuals from different glacial refuge areas (Garnier et al. 2004).

In our test species, *C. auronitens*, we found genetically rich but strongly diverging populations – rear edge populations according to Hampe and Petit (2005) and therefore of special importance for nature conservation. Moreover, these populations are obviously still situated close to their assumed glacial refuges South of the Holdhaus line (Fig. 5). Our analysis confirms the earlier ideas of multiple glacial refuges of *C. auronitens* in the Pyrenees, the Montagne Noire, close to Rodez, and in the Cevennes (Assmann et al. 1994; Assmann and Weber 1997; Reimann et al. 2002). Some of these refuge areas coincide with the putative refuges of the beech (*Fagus sylvatica*), as shown by a combination of palaeobotanical and genetic data (Magri et al. 2006). This example clearly illustrates the importance of the Holdhaus line in separating populations living in former glacial refuges from populations inhabiting postglacially recolonized areas. The Holdhaus line may also help to localize glacial refuge areas of other species, such as the fire salamander (*Salamandra salamandra*, Steinfartz et al. 2000), which prefers cold and humid conditions, and which recolonized parts of Europe postglacially.

The results of the genetic differentiation of *C. auronitens* do not mirror the subspecific taxonomy of *C. auronitens* as there are at least two subspecies described (*C. a. auronitens* and *C. a. festivus*, Turin et al. 2003) which are not separated genetically. The morphometric analysis by Terlutter (1991) does not support this taxonomic system either. For *C. auronitens*, there are numerous taxonomic works in which several subspecies are described (e.g., Breuning 1932; Deuve 1994) which are partly contradictory and seem to lack any objective criteria. Thus, we strongly recommend renouncing the use of taxonomic units below species level for *C. auronitens* (for a general discussion, see Assmann et al. 2008).

Our results show that populations that are situated North of the line are genetically poorer than those located South of it. These populations display – from a genetic point of view – only a section of the refuge populations (with the exception of the populations from the French-German-Swiss border triangle, see below), which is in concordance with Reinig's (1938) idea of stepwise allele elimination in the course of a set of postglacial expansion processes.

As an exception, we even found populations in the Southwestern part of Central Europe (Jura, Vosges, and Black Forest) which contain private alleles. It seems improbable that these alleles originate from one of the putative refuge populations in Southern France as the populations situated in between these areas are genetically impoverished and display only a section of the allelic richness of the populations from the Cevennes. Surprisingly, a set of endemic taxa was described for this Southwestern part of Central Europe just after Holdhaus had published his compendium. It is noteworthy that some of these species share the same habitat with *C. auronitens*: five diplopod species (Spelda 1991), one carabid species (Huber and Molenda 2004), and *Lumbricus badensis* (Kobel-Lamparski and Lamparski 2004). All these species prefer humid and cold habitats. Furthermore, the Black Forest region is explicitly named as a potential glacial refuge area for the vole *Microtus arvalis* (Jaarola and Searle 2002).

Normally, there is a striking connection between blind beetles and certain bedrock or soil types. Although limestone bedrocks are not colonized exclusively, there is a main occurrence of blind beetles in these soils. Therefore, the lack of endogeic or troglobitic beetles in den Black forest region point to a certain shortcoming of the Holdhaus line.

Refuge areas on non-limestone bedrocks can thus be overlooked due to the lack of this special group of indicator species. It seems, then, that the limits demarcated by the Holdhaus line are further South than the possible geographic position of some glacial refuges for species of cold and humid habitats. Holdhaus seems to have recognized part of this problem as he explicitly named the records of blind troglobitic beetles in the Jura which caused him to draw an isolated enclave North of the continuous Northern distribution limit of blind troglobitic beetles. The putative glacial refuge area in the Black Forest is situated close to this enclave.

The data for *C. auronitens* as well as for many other species of cold-humid habitats (review by Schmitt 2007) further support Holdhaus' idea (Holdhaus 1906, 1954) that glacial refuge areas are located in the extra-Mediterranean. These results are in line with the findings of Willis et al. (2000) and Willis and van Andel (2004) who provide evidence for the existence of trees in many parts of Central Europe throughout the cold stages of the full and late glacial.

References

Assmann T, Buse J, Drees C, Habel J, Härdtle W, Matern A, von Oheimb G, Schuldt A, Wrase DW (2008) From Latreille to DNA systematics – towards a modern synthesis for carabidology. In: Penev L, Erwin T, Assmann T (eds) Back to the roots or back to the future? Towards a new synthesis amongst taxonomic, ecological and biogeographical approaches in carabidology. Pensoft, Sofia, pp 41–76

Assmann T, Heine S (1993) Die *Leistus*-Arten der *Oreobius*-Gruppe: Systematik, Taxonomie und Verbreitung (Coleoptera, Carabidae: Nebriinae). Mitteilungen aus der Entomogischen Gesellschaft Basel 43:42–68

Assmann T, Nolte O, Reuter H (1994) Postglacial colonization of middle Europe by *Carabus auronitens* as revealed by population genetics (Coleoptera, Carabidae). In: Desender K (ed) Carabid beetles: ecology and evolution. Kluwer, Dordecht, pp 3–9

Assmann T, Weber F (1997) On the allozyme differentiation of *Carabus punctatoauratus* Germar (Coleoptera, Carabidae). J Zoolog Syst Evol Res 35:33–43

Avon C (1997) Les *Duvalius* et leur compexe evolutif: Espèces Francaise. Italiennes and Parasites. Laboratoire d'Entomologie Faune Hypogee et Endogee, Cagnes dur Mer (France)

Berggren KT, Ellegren H, Hewitt GM, Seddon JM (2005) Understanding the phylogeographic patterns of European hedgehogs, *Erinaceus concolor* and *E. europaeus* using the MHC. Heredity 95:84–90

Bonadona P (1971) Catalogue des Coléoptères Carabiques de France. Nouvelle Revue d'Entomologie Suppl 1:1–177

Breuning S (1932) Monographie der Gattung *Carabus* L. Reitter, Troppau

Culver DC, White WB (2005) Encyclopedia of caves. Elsevier, Burlington, VT

Casale A, Giachino PM, Vailati D (1991) Brevi considerazioni per una sistematica filogenetica dei Bythysciinae (Coleoptera: Cholevidae). Atti XII Congresso Nazionale Italiano Entomologia Bari, Società Entomologica Italiana, Bari 857–865

Casale A, Vigna-Taglianti A (2005) Coleotteri Caraboidei della Alpi e Prealpi centrali e orientali, e loro significato biogeografico. Biogeographia 26:129–201

Casale A, Vigna-Taglianti A, Juberthie C (1998) Coleoptera Carabidae. In: Juberthie C, Decu V (eds) Encyclopedia biospeologica. Tome II. Société de Biospéologie, Moulis, pp 1047–1081

Cheddadi R, Vendramin GG, Litt T, Francois L, Kageyama M, Lorentz S, Laurent JM, de Beaulieu JL, Sadori L, Jost A, Lunt D (2006) Imprints of glacial refugia in the modern genetic diversity of *Pinus sylvestris*. Glob Ecol Biogeogr 15:271–282

Chodat R, Pampanini R (1902) Sur la distribution des plantes des Alpes austro-orientales. Le globe: revue genevoise de géographie 41:63–132

Daffner H (1993) Die Arten der Gattung *Arctaphaenops* Meixner, 1925 (Coleoptera: Carabidae). Koleopterologische Rundschau 63:1–18

De Lattin G (1957) Die Ausbreitungszentren der holarktischen Landtierwelt. Zool Anz 20 (Supplement):380–410

De Lattin G (1959) Postglaziale Disjunktionen und Rassenbildung bei europäischen Lepidopteren. Zool Anz 22 (Supplement):392–403

De Lattin G (1967) Grundriß der Zoogeographie. Fischer, Jena

Desender K (1986) Distribution and ecology of carabid beetles in Belgium (Coleoptera, Carabidae). part 2. Studiedocumenten (Bruxelles) 27:1–24

Deuve T (1994) Une classification du genre *Carabus*. Sciences Nat, Venette

Dobzhansky T (1937) Genetics and the origin of species. Columbia University Press, New York

Drees C, Matern A, Rasplus JY, Assmann T, Weber F (2008) Microsatellites and allozymes as the genetic memory of habitat fragmentation and defragmentation in populations of the ground beetle *Carabus auronitens* (Col., Carabidae). J Biogeogr 35:1937–1949

Drovenik B, Weber F, Paill W, Assmann T (2008) *Aphaenopidius kamnikensis* Drovenik, 1987 in Kärnten. Angewandte Carabidologie 8:73–76

Felsenstein J (2005) PHYLIP (Phylogeny Inference Package) Version 3.6. Department of Genome Sciences, University of Washington, Seattle

Forel J, Leplat J (1995) Les Carabes de France. Science Nat, Venette

Futuyma DJ (2005) Evolution. Sinauer Associates, Sunderland, MA

Garnier S, Alibert P, Audiot P, Prieur B, Rasplus J-Y (2004) Isolation by distance and sharp discontinuities in gene frequencies: implications for the phylogeography of an alpine insect species, *Carabus solieri*. Mol Ecol 13:1883–1897

Giachino PM, Decu V, Juberthie C (1998) Coleoptera Cholevidae. In: Juberthie C, Decu V (eds) Encyclopaedia biospeologia. Société de Biospéologie, Moulis, pp 1083–1122
Hampe A, Petit RJ (2005) Conserving biodiversity under climate change: the rear edge matters. Ecol Lett 8:461–467
Hewitt GM (1996) Some genetic consequences of ice ages, and their role in divergence and speciation. Biol J Linn Soc 58:247–276
Hewitt GM (1999) Post-glacial re-colonization of European biota. Biol J Linn Soc 68:87–112
Holdhaus K (1906) Verbreitung der Koleopteren in den mitteleuropäischen Hochgebirgen. Verh Zool Bot Ges Wien 56:629–639
Holdhaus K, Lindroth CH (1939) Die europäischen Koleopteren mit boreoalpiner Verbreitung. Naturhistorisches Museum, Wien
Holdhaus K (1954) Die Spuren der Eiszeit in der Tierwelt Europas. Wagner, Innsbruck
Huber C, Molenda R (2004) *Nebria* (*Nebriola*) *praegensis* sp. nov., ein Periglazialrelikt im Südschwarzwald/Deutschland, mit Beschreibung der Larven (Insecta, Coleoptera, Carabidae). Beiträge des Naturhistorischen Museums Bern 2004:1–28
Jaarola M, Searle JB (2002) Phylogeography of field voles (*Microtus agrestis*) in Eurasia inferred from mitochondrial DNA sequences. Mol Ecol 11:2613–2621
Joger U, Fritz U, Guicking D, Kalyabina-Hauf S, Nagy ZT, Wink M (2007) Phylogeography of Western Palearctic reptiles – spatial and temporal speciation patterns. Zool Anz 246:293–313
Juberthie C, Delay B, Bouillon M (1980a) Sur l'existence d'un milieu souterrain superficiel en zone non calcaire. Comptes rendues de l'academie des sciences - Series III - Life sciences 290:49–52
Juberthie C, Delay B, Ruffat G (1980b) Extension du milieu souterrain en zone non-calcaire: description d'un nouveau milieu et de son peuplement par le Coléoptères troglobies. Mémoires de biospéologie 7:19–52
Kobel-Lamparski A, Lamparski F (2004) *Lumbricus badensis* - ein Franzose, der in Baden groß wurde. Mitteilungen des badischen Landesvereins für Naturkunde und Naturschutz N.F. 18:91–109
von Koenigswald W (2002) Lebendige Eiszeit – Klima und Tierwelt im Wandel. Theiss-Verlag, Stuttgart
Krebs CJ (1999) Ecological methodology, 2nd edn. Addison Wesley, Menlo Park, CA
Krumbiegel I (1932) Untersuchungen über physiologische Rassenbildung. Ein Beitrag zum Problem der Artbildung und der geographischen Variation. Zoologisches Jahrbuch für Systematik und Ökologie 63:183–280
Krumbiegel I (1936) Morphologische Untersuchungen über Rassenbildung, ein Beitrag zum Problem der Artbildung und der geographischen Variation. Zoologisches Jahrbuch für Systematik und Ökologie 68:105–178
Kryzhanovskij OL, Belousov IA, Kabak II, Kataev BM, Makarov KV, Shilenkov VG (1995) A checklist of the ground-beetles of Russia and adjancent lands (Insecta, Coleoptera, Carabidae). Pensoft, Sofia, Moscow
Lang G (1994) Quartäre Vegetationsgeschichte Mitteleuropas. Fischer, Stuttgart
Löbl I, Smetana A (eds) (2003) Catalogue of Palearctic Coleoptera Vol. 1: Archostemata, Myxophaga, Adephaga. Apollo Books, Stenstrup
Lorenz W (2005) A systematic list of extant ground beetles of the world (Coleoptera "Geadephaga": Trachypachydiae and Carabidae, incl. Paussinae, Cicindelinae, Rhysodinae). Lorenz, Tutzingen
Magri D (2008) Patterns of post-glacial spread and the extent of glacial refugia of European beech (*Fagus sylvatica*). J Biogeogr 35:450–463
Magri D, Vendramin GG, Comps B, Dupanloup I, Geburek T, Gomory D, Latalowa M, Litt T, Paule L, Roure JM, Tantau I, van der Knaap WO, Petit RJ, de Beaulieu J-L (2006) A new scenario for the Quaternary history of European beech populations: palaeobotanical evidence and genetic consequences. New Phytol 171:199–221
Malzacher P (2000) Erster Nachweis einer blinden Laufkäfer-Art in Deutschland (Bembidiinae, Anillini). Angewandte Carabidologie 2(3):71–72

Marggi W (1992) Faunistik der Sandlaufkäfer und Laufkäfer der Schweiz (Cicindelidae and Carabidae). Text und Verbreitungskarten. Documenta Faunistica Helvetiae 13/1 and 13/2: 1–477 and 471–243

Michaux JR, Magnanou E, Paradis E, Nieberding C, Libois R (2003) Mitochondrial phylogeography of the Woodmouse (*Apodemus sylvaticus*) in the Western Palearctic region. Mol Ecol 12:685–697

Nei M (1978) Estimation of average heterozygosity and genetic distance from a small number or individuals. Genetics 89:583–590

Osella G, Zuppa AM (1998) Coleoptera Curculionoidea. In: Juberthie C, Decu V (eds) Encyclopaedia Biospeologia. Tome II. Société de Biospéologie, Moulis, pp 1122–1130

Puisségur C (1964) Recherches sur la génétique des Carabes (*Chrysocarabus* et *Chrysotribax*). Vie et Milieu 18:1–288

Reimann T, Assmann T, Nolte O, Reuter H, Huber C, Weber F (2002) Palaeogeography and palaeoecology of *Carabus auronitens* (Coleoptera): characterization and localization of glacial refugia in Southern France and reconstruction of postglacial expansion routes by means of allozyme polymorphisms. Abhandlungen des Naturwissenschaftlichen Vereins in Hamburg (NF) 35:1–151

Reinig WF (1938) Elimination und Selektion: Eine Untersuchung über Merkmalsprogressionen bei Tieren und Pflanzen auf genetisch- und historisch-chorologischer Grundlage. Fischer, Jena

Reinig WF (1939) Die genetisch-chorologischen Grundlagen der gerichteten geographischen Variabilität. Zeitschrift für induktive Abstammungs- und Vererbungslehre 76:260–308

Rensch B (1929) Das Prinzip geographischer Rassenkreise und das Problem der Artbildung. Gebrüder Borntraeger, Berlin

Rensch B (1939) Typen der Artbildung. Biol Rev 14:180–222

Rensch B (1951) Die Verteilung der Tierwelt im Raum. In: von Bertalanfey (ed) Handbuch der Biologie. Akademische Verlagsgesellschaft Athenaion, Potsdam, pp. 125–172

Rensch B (1960) Evolution above species level. Columbia University Press, New York

Saitou N, Nei M (1987) The Neighbor-joining method: a new method for reconstructing phylogenetic trees. Mol Biol Evol 4:406–425

Schmitt T (2007) Molecular biogeography of Europe: Pleistocene cycles and postglacial trends. BMC Frontiers in Zoology 4:11

Seddon JM, Santucci F, Reeve NJ, Hewitt GM (2001) DNA footprints of European hedgehogs, *Erinaceus europaeus* and *E. concolor*. Pleistocene refugia, postglacial expansion and colonization routes. Mol Ecol 10:2187–2198

Spelda J (1991) Zur Faunistik und Systematik der Tausendfüßler (Myriapoda) Südwestdeutschlands. Jahreshefte der Gesellschaft für Naturkunde in Württemberg 146:211–232

Stauffer C, Lakatos F, Hewitt GM (1999) Phylogeography and postglacial colonization routes of *Ips typographus* L. (Coleoptera, Scolytidae). Mol Ecol 8:763–773

Steinfartz S, Veith M, Tautz D (2000) Mitochondrial sequence analysis of *Salamandra* taxa suggests old splits of major lineages and postglacial recolonizations of Central Europe from distinct source populations of *Salamandra salamandra*. Mol Ecol 9:397–410

Terhurne-Berson R, Litt T, Cheddadi R (2004) The spread of *Abies* throughout Europe since the last glacial period: combined macrofossil and pollen data. Vegetation History and Archaeobotany 13:257–268

Terlutter H (1991) Morphometrische und elektrophoretische Untersuchungen an westfälischen und südfranzösischen *Carabus auronitens*-Populationen (Col. Carabidae): Zum Problem der Eiszeitüberdauerung in Refugialgebieten und der nacheiszeitlichen Ausbreitung. Abhandlungen des Westfälischen Museums für Naturkunde 53:3–111

Turin H, Penev L, Casale A (2003) The Genus *Carabus* in Europe. Pensoft, Sofia

Vigna-Taglianti A (1982) Le attuali conoscenze sui Coleotteri Carabidi cavernicolo italiani. Lavori della Societa' Italiana di Biogeografia. N.S 7:339–430

Weber F, Heimbach U (2001) Behavioural, reproductive and developmental seasonality in *Carabus auronitens* and *Carabus nemoralis* (Col., Carabidae). A demographic comparison between two co-existing spring breeding populations and tests for intra- and interspecific competition and for synchronizing weather events. Mitteilungen aus der Biologischen Bundesanstalt für Land- und Forstwirtschaft 382:1–192

Willis KJ, Rudner E, Sumegi P (2000) The full-glacial forests of central and Southeastern Europe. Quatern Res 53:203–213

Willis KJ, van Andel TH (2004) Trees or no trees? The environments of central and Eastern Europe during the Last Glaciation. Quatern Sci Rev 23:2369–2387

Postglacial Recolonization of Continental Europe by the Pygmy Shrew (*Sorex minutus*) Inferred From Mitochondrial and Y Chromosomal DNA Sequences

Allan D. McDevitt, Glenn Yannic, Ramugondo V. Rambau, Thomas J. Hayden, and Jeremy B. Searle

Abstract The postglacial recolonization of the pygmy shrew (*Sorex minutus*, $n=68$) across continental Europe was examined using control region sequences (342 bp) and concatenated sequences (2,939 bp) of four paternally inherited introns, specific for the Y chromosome. Phylogenetic analysis (NJ, MP, ML, and Bayesian inference) of the control region retrieved five major lineages (Northern, Western, Italian, Balkan, and Spanish) in Europe, in agreement with the previous cytochrome *b* data. Our study identifies a new refugium for the pygmy shrew located in France during the Pleistocene glacial cycles and the most important refugium with regards to the recolonization of Eurasia was probably located in the Carpathians. In contrast, Y chromosome intron sequences retrieved four major lineages (Northern, Western, Italian, and Balkan). This may indicate contact between Western and Spanish mtDNA lineages after expansion from their respective refugia post-LGM. Our results also revealed differential introgression of mtDNA and Y chromosome markers at a local level when populations from different glacial refugia come into contact. These results highlight the importance of adopting a multiple marker approach when inferring phylogeographic structure and colonization history of species.

A.D. McDevitt (✉) and T.J. Hayden
School of Biology and Environmental Science, University College Dublin,
Belfield, Dublin 4, Ireland
e-mail: amcdev@gmail.com

G. Yannic
Department of Ecology and Evolution, University of Lausanne, CH-1015, Lausanne, Switzerland

R.V. Rambau and J.B. Searle
Department of Biology, University of York, PO Box 373, York YO10 5YW, UK

R.V. Rambau
Evolutionary Genomics Group, Department of Botany and Zoology, Stellenbosch University, Matieland 7602, Stellenbosch, South Africa and DST-NRF Centre of Excellence for Invasion Biology, Department of Botany and Zoology, Stellenbosch University, Matieland 7602, Stellenbosch, South Africa

J.C. Habel and T. Assmann (eds.), *Relict Species: Phylogeography and Conservation Biology*,
DOI 10.1007/978-3-540-92160-8_12,

1 Introduction

The Pleistocene glaciations and intervening warming periods have undoubtedly played a major role in the formation of distinct genetic lineages across the geographic distributions of species (Hewitt 2000). The Last Glacial Maximum (LGM) occurred approximately from 23,000 to 19,000 calendar years before present (y BP; Mix et al. 2001). During this period, extensive areas of Europe were covered by ice and permafrost. Many species could only survive in relatively sheltered areas and would, therefore, have retreated to refugia with more suitable environmental conditions. Phylogeographic studies of European mammals have revealed several recolonization patterns from these refugia during the Pleistocene climatic fluctuations (Taberlet et al. 1998; Hewitt 1999). Glacial refugia that have been identified from a wide variety of studies of mammalian fauna include the Iberian Peninsula, Italy, and the Balkans (see Seddon et al. 2001; Taberlet et al. 1998; Mascheretti et al. 2003; Michaux et al. 2003). However the Mediterranean regions would have been more important as hotspots for endemism rather than a source of Northward recolonization for the rest of Europe (Bilton et al. 1998). The importance of refugia other than the Mediterranean peninsulas has become apparent through studies of a variety of mammals (Bilton et al. 1998; Stewart and Lister 2001). These include the Ural (Jaarola and Searle 2002; Brunhoff et al. 2003; Deffontaine et al. 2005) and Carpathian (Pazonyi 2004; Kotlík et al. 2006; Sommer and Nadachowski 2006; Saarma et al. 2007) mountain ranges. Some species of small mammals (e.g., bank vole *Myodes glareolus*) tolerated the climate conditions present during the LGM (Kotlík et al. 2006) to survive and recolonize Europe from these more Northern latitudes. Glacial refugia may also have been present further North in Scandinavia. For instance, populations of root voles, *Microtus oeconomus*, may have persisted through the LGM as far North as Andøya, an island situated off Northern Norway (Brunhoff et al. 2006).

Since the conception of phylogeographical studies, the most popular molecular markers have been mitochondrial DNA (mtDNA) sequences (Avise 2000). There are many reasons for this; mtDNA has many favorable properties, such as haploidy, a lack of recombination, rapid evolution compared to nuclear DNA (Brown et al. 1979), and is predominantly maternally-inherited (Nishimura et al. 2006). But perhaps the main reason for its widespread application is its relative ease of use (Petit et al. 2002) since universal primers were developed to amplify across a wide range of mammalian families (Kocher et al. 1989; Irwin et al. 1991). Despite the many obvious benefits, the sole use of mtDNA data can potentially lead to the construction of an incorrect species history because only the maternally-inherited mitochondrial lineage is being investigated (Ballard and Whitlock 2004). This is particularly important when it is evident that males and females do not play 'symmetrical roles' in establishing and maintaining population structure (Greenwood 1980).

Recently, the human genome sequencing project provided Y-specific genetic markers for humans (Petit et al. 2002). This has allowed numerous studies on the colonization history of humans (Underhill et al. 2000; Hammer et al. 2003) based on male-inherited Y chromosome genes and the results have been consistent with the 'Out of Africa' hypothesis of modern humans based on mitochondrial, X-linked, and autosomal regions of DNA (Templeton 2002). However, there have been few phylogeographic studies in non-human mammalian species using Y chromosome markers. This may be due to the low levels of nucleotide diversity found in Y chromosome regions (Hellborg and Ellegren 2004). Recently, there have been studies on hybrid zones using Y chromosome structure (Jaarola et al. 1997) and Y-linked microsatellites (Balloux et al. 2000). Furthermore, combined datasets of mitochondrial, X and Y chromosomal genes have detected 'cryptic species' in the African elephant, *Loxodonta africana* (Roca et al. 2005), field vole, *Microtus agrestis* (Hellborg et al. 2005), and ground squirrels, *Spermophilus* (Gündüz et al. 2007). Brändli et al. (2005) used a similar set of markers to clarify the evolutionary history of the greater white-toothed shrew, *Crocidura russula* and its subspecies. Most of these studies have shown concordant tree topologies of mitochondrial, X and Y chromosomal DNA sequences when inferring evolutionary history, at the species- or subspecies-level. Therefore, the main finding of previous studies has been the reproductive isolation of genetic lineages. Little attention has been focused on sex-biased gene admixture when inferring the colonization history of a particular species.

The purpose of the present study is to investigate the female- and male-biased historical gene flow of the pygmy shrew, *Sorex minutus*. This study utilized a female-only inherited sequence (segment of the mitochondrial control region) and male-inherited sequences (four Y chromosome introns), and was centred on continental Europe and integrated in the context of the postglacial colonization history of the species. The pygmy shrew is the most widely distributed shrew species found in Europe (Mitchell-Jones et al. 1999) and it is also the oldest, with a fossil record dating back to the Pliocene, ca. 2 million years ago (Rzebik-Kowalska 1998). Its range extends from Western Iberia and Ireland to Siberia in the East, but is absent from Southern Iberia and most of the Mediterranean coasts and islands (Mitchell-Jones et al. 1999).

The pygmy shrew is an ideal choice for such a study because of its widespread distribution and fossil evidence of possible pygmy shrew survival in two refugia; with dates of approximately 17,000 y BP in Northern Spain (Pokines 2000) and 12–15,000 y BP in the Carpathian Basin (Pazonyi 2004). The pygmy shrew has also been the subject of previous postglacial colonization studies. Bilton et al. (1998) sequenced a 447 basepair (bp) segment of the cytochrome *b* gene from 14 specimens to reveal the importance of an unidentified glacial refugia located further North than the classic Mediterranean refugia. A similar result was found by Mascheretti et al. (2003) using a much larger dataset and a near-complete sequence (1,110 bp) of the cytochrome *b* gene when investigating the colonization history of Ireland. Glacial refugia identified thus far are the Iberian Peninsula, Italy, the Balkans, and a refugium located further North. Here we extend these findings using a combination of sex-linked and mitochondrial markers in an attempt to provide a more complete understanding of the pygmy shrew's colonization history.

2 Materials and Methods

2.1 Sampling and DNA Extraction

A total of 68 samples were analyzed in this study. Forty eight of these samples were obtained from collections in the Universities of York, United Kingdom, and Lausanne, Switzerland. Live pygmy shrews were also sampled in France (n = 5), including the island of Belle Île (n =10), using Trip Traps (Procter Bros. Ltd.). Details of sample locations are given in Table 1. DNA was extracted from ethanol-preserved tissue from tail tips, toe clippings or liver using the DNeasy Blood & Tissue Kit (Qiagen Ltd.) according to the manufacturer's protocol.

2.2 Mitochondrial DNA

The left domain of the control region was amplified using primers $tRNA^{Pro}$ (Stewart and Baker 1994a) and PS-CD (5′ - GGCGAGATTAAATGTTAGCTGG - 3′). PCR reactions consisted of 2 µl of DNA (40–200 ng/µl), 1.5 µl (0.375 µM; Sigma Aldrich) of forward and reverse primer and 45 µl of MegaMix~Blue (Microzone Ltd.). PCR amplification was carried out in a Peltier Thermal Cycler (MJ Research) using an initial denaturation of 95°C for 3 min followed by 30 cycles of 95°C for 45 s, annealing temperature of 60°C for 1 min, and an extension of 72°C for 90 s. This was followed by a final extension step of 10 min. Fragment sizes ranged from approximately 500–800 bp, depending on the number of tandem repeats as previously reported in other *Sorex* species (Stewart and Baker 1994a; Fumagalli et al. 1996).

2.3 Y Chromosome Introns

Five primer sets were chosen from Y chromosome introns which amplified in the common shrew, *Sorex araneus* (Hellborg and Ellegren 2003). The introns chosen were DBY1, DBY3, DBY7, UTY11, and ZFY4 generating products totaling approximately 3.2 kbp. Two other introns (DBY8, DBY14), which amplified in *S. araneus* (Hellborg and Ellegren 2003) were not chosen due to a lack of polymorphism in several races of the common shrew (G. Yannic, *unpublished data*). Amplification of the Y chromosome introns was based on the protocol used by Hellborg and Ellegren (2003). Briefly, PCR reactions consisted of 3 µl DNA (60–300 ng/µl), 5.0 µl 10×Buffer (Invitrogen), 2.5 µl of $MgCl_2$ (2.5 mM; Invitrogen), 1.0 µl dNTPs (0.2 mM; Invitrogen), 1.0 µl (0.25 µM; Sigma Aldrich) of forward and reverse primer for each locus and 0.3 µl of Platinum *Taq* Polymerase (5 U/µl; Invitrogen) to make up a 50 µl reaction.

PCR amplification was carried out using an initial denaturation at 95°C for 10 min followed by 20 cycles of 95°C for 30 s, a touchdown annealing temperature

Table 1 List of samples and haplotypes (both mtDNA and concatenated Y chromosome introns) present in each locality. The number of individuals per locality is represented by *n*. The coordinates of each sampling locality (numbered from 1 to 38) is plotted in Figs. 3 and 4

Location	Country	Coordinates	Map No.	*n*	mtDNA haplotype	Y haplotype
Rascafria	Spain	40°54′ N 03°53′ W	1	3	Spa1,2	Spa1,2
Encamp	Andorra	42°35′ N 01°35′ E	2	2	And1,2	Fra3
Le Mas-d'Artige	France	45°42′ N 02°08′ E	3	1	Fra1	–
Nexon	France	45°45′ N 01°15′ E	4	1	Fra2	Fra1
Belle Île	France	47°20′ N 03°11′ W	5	10	BI1,2,3,4,5	BI1
Fresseneville	France	50°05′ N 01°34′ E	6	1	Fra3	–
Broualan	France	48°28′ N 01°37′ W	7	2	Fra4,5	Fra2,3
Plancoet	France	48°30′ N 02°15′ W	8	1	Fra6	Fra4
Morlaix	France	48°35′ N 03°50′ W	9	2	Fra6	Fra3
Boxtel	Belgium	51°36′ N 05°20′ E	10	1	Bel1	–
Wageningen	Holland	51°58′ N 05°40′ E	11	1	Hol1	–
Hartz Mountains	Germany	51°45′ N 10°40′ E	12	2	Ger1	Ger1,2
Eberswalde	Germany	52°10′ N 13°45′ E	13	1	Ger2	Ger3
Kobbero	Denmark	56°43′ N 08°20′ E	14	1	Den1	North1
Askland	Norway	58°40′ N 08°35′ E	15	3	Nor1,2,3	Nor1
Klockkarvik	Norway	60°07′ N 05°00′ E	16	2	Nor4	–
Steinkjer	Norway	64°00′ N 13°53′ E	17	1	Nor5	North1
Tromso	Norway	68°50′ N 19°00′ E	18	1	–	North1
Karlskrona	Sweden	56°09′ N 15°15′ E	19	1	Swe1	Swe1
Vehkalahti	Finland	60°15′ N 27°00′ E	20	1	Fin1	–
Val d'Illiez	Switzerland	46°16′ N 06°43′ E	21	1	Swz1	Swz1
Bretolet	Switzerland	46°13′ N 06°40′ E	22	1	Swz2	Swz1
Bassin	Switzerland	46°32′ N 06°39′ E	23	1	Swz3	–
Trento	Italy	46°15′ N 11°50′ E	24	2	Ita1,2	–
Abruzzo	Italy	42°00′ N 14°00′ E	25	4	Ita3,4,5,6	Ita1
Calabre	Italy	37°07′ N 16°38′ E	26	1	Ita7	Ita2
Reggio	Italy	44°32′ N 11°21′ E	27	1	Ita8	Ita3
Cesky Jiretin	Czech Republic	50°41′ N 13°34′ E	28	2	Cze1,2	North1, Cze1
Donnerskirchen	Austria	47°56′ N 16°40′ E	29	1	Aut1	Aut1
Bratislava	Slovakia	48°07′ N 17°00′ E	30	5	Svk1,2,3,4	North1, Svk1
Blizocin	Poland	51°36′ N 22°16′ E	31	2	Pol1,Pol2	North1,Pol1
Vilnius	Lithuania	54°40′ N 25°19′ E	32	2	Lit1	North1
Kanev	Ukraine	49°41′ N 32°00′ E	33	1	Ukr1	North1
Brjansk	Russia	52°20′ N 34°00′ E	34	1	–	North1
Novosibirsk	Siberia	54°49′ N 83°06′ E	35	1	Sib1	North1
Lake Baikal	Siberia	53°40′ N 108°00′ E	36	1	–	Sib1
Pelister Mountain	Macedonia	41°00′ N 21°10′ E	37	1	Mac1	–
Strandzha Mountains	Turkey	41°45′ N 27°41′ E	38	2	Tur1,2	Tur1

from either 60–50°C (DBY1, DBY7) or 55–45°C (DBY3, UTY11) for 1 min, decreasing by 0.5°C per cycle and then an extension of 72°C for 90 s. This was followed by 20 cycles of 95°C for 30 s, 50°C (DBY1, DBY7) or 45°C (DBY3, UTY11) for 1 min, and 72°C for 90 s and a final extension step at 72°C for 10 min. All PCR reactions included a known female sample as a negative control. Amplification of a fragment in the female sample was used as evidence that the gene fragment was not specific to the Y chromosome. For this reason, intron ZFY4 was disregarded after amplification in several female samples.

2.4 Sequencing

PCR products were purified using Novagen® Spinprep™ PCR Clean-up Kit (Merck Biosciences) according to the manufacturer's protocol. Control region fragments were sequenced using the forward ($tRNA^{Pro}$) and reverse primers (PS-CD). Y chromosome introns, DBY3 and DBY7, were sequenced using forward primers only while introns, DBY1 and UTY11, were sequenced using forward and reverse primers. Sequencing was carried out by Macrogen Inc. (Seoul, Korea). Control region fragment and Y chromosomal intron sequences were deposited in GenBank (Accession nos. for control region fragments: EU564340–EU564393; DBY1 intron: EU564394–EU564407 and EF636561–EF636562; DBY3 intron: EU564408–EU564414 and EF636573; DBY7 intron: EU564415–EU564422 and EF636584; UTY11 intron: EU564423–EU564434 and EF636606. Concatenated Y chromosomal intron sequences are available upon request).

2.5 Data Analysis

Chromatogram contiguous sequences were assembled in Sequencer 4.5 (GeneCodes) and sequences were aligned manually using Se-Al 2.0 (Rambaut 1996). The 5′ end of the *Sorex* control region contains a tandem repeat region (Stewart and Baker 1994a,b; Fumagalli et al. 1996), with repeat motifs ranging from 78 to 80 bp in length. The number of tandem repeat copies ranged from 3 to 7 per individual and length heteroplasmy due to repeat numbers within and between individuals was present. All tandem repeat copies within an individual were screened for mutations between repeats (Rhymer et al. 2004). For analyses, in accordance with previous phylogenetic studies on *Sorex* species (Stewart and Baker 1994a,b, 1997; Rhymer et al. 2004) only the last tandem repeat and the unique flanking regions at the 5′ end containing the hypervariable region were used, in addition to the 3′ end of the repeat region (including part of the conserved Central Domain). This resulted in 342 bp (including indels) being used in the analyses. Chromatogram contigs for the four Y introns were assembled and analyzed individually before being combined in Se-Al 2.0 (Rambaut 1996) to form a concatenated dataset (2,939 bp including indels).

Control region and concatenated Y intron haplotypes were identified using DAMBE 4.5.50 (Xia and Xie 2001). Nucleotide diversity (π) was calculated using DnaSp 4.10.9 (Rozas et al. 2003). Net divergence (*Da*) was calculated in MEGA v. 3.1 (Kumar et al. 2004) using Kimura's 2-parameter distances (Kimura 1980). Standard errors (SE) were calculated by bootstrapping with 10,000 replicates.

Neighbor-joining (NJ), Maximum Parsimony (MP) and Maximum Likelihood (ML) phylogenetic trees derived from control region haplotypes were constructed in PAUP* 4.0b10 (Swofford 1998). ML and NJ trees were constructed using the HKY85+I+G (Hasegawa et al. 1985) model ($I=0.39$; $\alpha=0.7831$), selected by the Akaike Information Criterion (AIC) in ModelTest 3.06 (Posada and Crandall 1998). Indels were treated as missing data when constructing trees. Of the 342 bp that were sequenced, 72 characters were parsimony informative. Bootstrap support was calculated from 1,000 NJ, MP, and ML parametric replicates, respectively. Bayesian posterior probabilities were estimated in MrBayes 3.1.2 (Ronquist and Huelsenbeck 2003) from 1,000,000 generations sampled every 1,000th generation excluding a burn-in of 100,000 steps. *Sorex volnuchini* was used as the outgroup species, in accordance with previous phylogeographic studies on *S. minutus* (Bilton et al. 1998; Mascheretti et al. 2003).

The amount of genetic variation among and within phylogenetic lineages was determined by an analysis of molecular variance (AMOVA) implemented in Arlequin version 3.11 (Excoffier et al. 2005).

Recent population expansion from small founder populations for control region data was tested using mismatch distributions and Fu's (1997) F_S. Mismatch distributions of pairwise nucleotide differences of each lineage were calculated and compared with expected values for an expanding population (Rogers and Harpending 1992). Tests for goodness-of-fit statistics based on the sum of square deviations (SSD) for a model of sudden expansion were determined in Arlequin version 3.11 (Excoffier et al. 2005). Fu's F_S tests were performed to detect an excess of low frequency mutations in each lineage conforming to a model of sudden expansion. Tests for goodness-of-fit and Fu's F_S were generated in Arlequin v. 3.11 using parametric bootstrapping with 10,000 replicates.

Because sequence regions of the control region evolve at different rates (Stewart and Baker 1994a; Pesole et al. 1999), expansion times for lineages conforming to the sudden population expansion model were estimated using the unique sequence region at the 5′ end from the mismatch distribution (τ) of each lineage, using the formula $\tau = 2\ \mu t$ (where τ is the time since expansion in units of 1/2 μ generations), using a generalized least squares approach adapted from Rogers (1995), as described in Schneider and Excoffier (1999). Generation time was assumed to be 1 year (Churchfield and Searle 2008). An average evolutionary rate of 23% (15–31%) per million years (My^{-1}) was used based on the rate derived from other *Sorex* species for the unique sequence region at the 5′ end (Stewart and Baker 1994a). The divergence time (*T*) between lineages for control region data was estimated by $T=Da/2\mu$, where 2μ is the divergence rate of 23% My^{-1} for the unique sequence region at the 5′ end with 95% confidence intervals (CI) estimated according to Jaarola and Searle (2004).

A median-joining network of Y chromosomal introns was constructed using Network 4.2 (Bandelt et al. 1999) to illustrate relationships among haplotypes. Indels were included as informative characters (Hellborg et al. 2005). Bootstrap support for lineages was calculated separately in PAUP* 4.0b10 from 1,000 NJ and MP replicates, respectively. NJ analysis was performed using Kimura's 2-parameter distances (Kimura 1980) due to differences in the substitution patterns of the four introns (Hellborg et al. 2005).

3 Results

3.1 Mitochondrial DNA Data

We examined 63 pygmy shrews from continental Europe for 342 bp of the mitochondrial control region. Of the 342 nucleotides, 108 were polymorphic (including indels). Fifty one (47.2%) polymorphic sites were found in the 5′ unique flanking region. Thirty (27.8%) polymorphic sites were found in the last tandem repeat which was 78–80 bp in size and 27 (25%) polymorphic sites were found in the 3′ end and part of the Central Domain. Fifty three unique haplotypes were identified (see Table 1). Only one haplotype was shared between localities: Fra6 between two localities in Northern France. NJ, MP, ML, and Bayesian analysis of the 53 haplotypes revealed concordant topologies resulting in five major lineages (Fig. 1), with poor to relatively good statistical support. The Western lineage included individuals from Andorra, central France, and the French island Belle Île. The Northern lineage included individuals from Northern France and Norway in the West to Siberia in the East. The Spanish lineage was confined to Spain. The Balkan lineage was present as far North as Austria and Slovakia, and the Italian lineage was present in Italy and the Alps in Switzerland and France. One outlier haplotype, Ita7, was identified from this region (Fig. 1). The Northern lineage is separated from both Western and Italian lineages by two indels, 3 bp and 4 bp in size. The Spanish lineage is separated from the Northern lineage by an 8 bp indel, and the Spanish and Balkan lineages are separated by a 2 bp indel. This corroborates the lineages identified by Mascheretti et al. (2003) using cytochrome *b* but with certain lineages more widespread because of better sampling coverage. An AMOVA showed that the majority of control region variation (62.8%) was distributed among lineages with a little variation (10.13%) found among populations within lineages.

The nucleotide diversity (π) for lineages ranged from 0.01786 to 0.0331 (Table 2). Net divergence ($Da \pm$SE%) between lineages and estimated divergence in 1,000 s of calendar years before present (ky BP) are given in Table 3. Mismatch distributions for four out of the five lineages significantly conformed to the sudden population expansion model (Table 4). The Spanish lineage was excluded because of the small sample size ($n=3$). Fu's F_S values and estimates for time since expansion for the four lineages based on Rogers (1995) τ are given in Table 4.

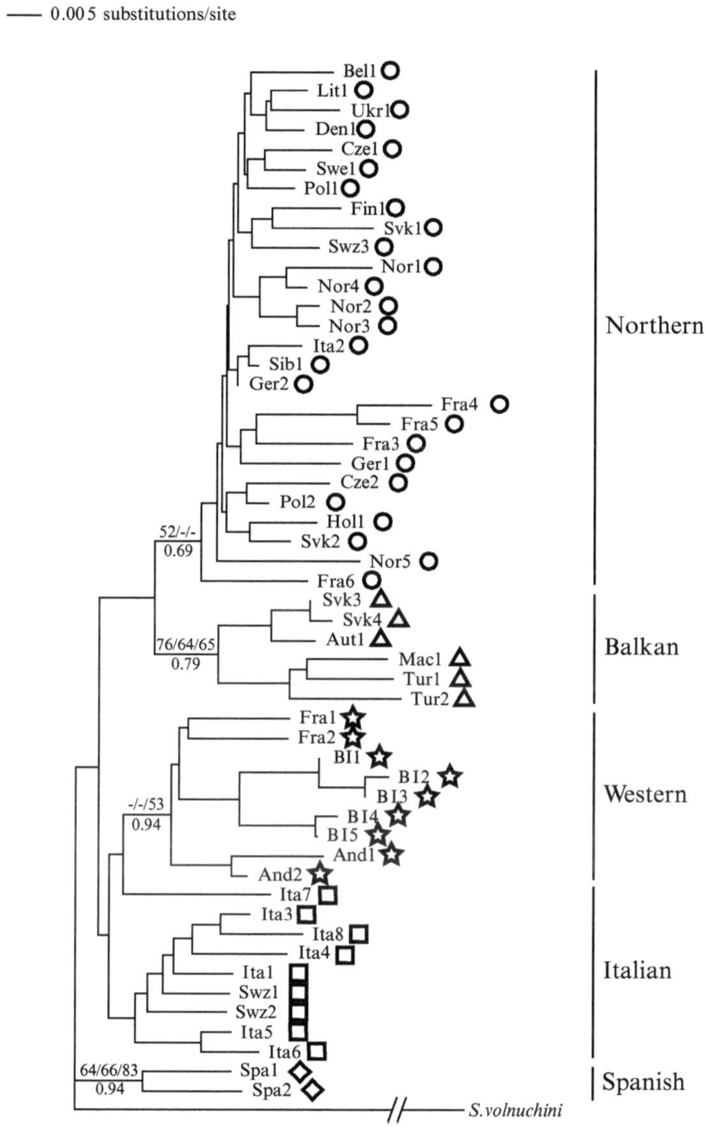

Fig. 1 Neighbour-joining tree using the HKY85+I+G model of substitution of 53 control region haplotypes (342 bp) showing the 5 lineages found in European *S. minutus*. See Table 1 for haplotype designations. Bootstrap values obtained by NJ, MP, ML (>50%) are shown above nodes and Bayesian probabilities (>0.50) are shown below nodes for each lineage

3.2 Y Chromosome Intron Data

We examined 39 pygmy shrews for 2,939 bp (including indels) of concatenated Y introns (DBY1: 1098 bp, DBY3: 596 bp, DBY7: 441 bp, UTY11: 804 bp), which resulted in 23 unique haplotypes (see Table 1 and Fig. 2). Sixty polymorphic sites

Table 2 Nucleotide diversity ($\pi \pm SD$) of and number of individuals (n) in each mtDNA and concatenated Y chromosomal intron lineage

		mtDNA	Y chromosome introns	
Lineage	n	$\pi \pm SD$	n	$\pi \pm SD$
Northern	30	0.02706±0.00176	23	0.00033±0.00009
Western	14	0.03193±0.00334	10	0.00084±0.00017
Italian	9	0.02746±0.00276	5	0.00136±0.00048
Balkan	7	0.02863±0.00516	1	–
Spanish	3	0.01786±0.00842	–	–
–	–	–	–	–
Total	63	0.04406±0.00181	39	0.00384±0.00034

Table 3 Estimates of percentage divergence (Da±95% CI) between control region lineages (upper diagonal) and estimates of divergence times [ky BP (±95% CI)] (lower diagonal) using an evolutionary rate of 23% My^{-1}

ky/Da±SE%	Northern	Western	Italian	Balkan	Spanish
Northern	–	6.06±2.13%	5.09±1.97%	5.53±2.21%	5.45±2.14%
Western	263 (171–356)	–	3.08±1.45%	8.22±2.73%	3.62±1.74%
Italian	221 (136–307)	134 (71–197)	–	6.93±2.58%	3.03±1.73%
Balkan	240 (144–337)	357 (239–476)	301 (189–413)	–	6.49±2.26%
Spanish	237 (144–330)	157 (82–233)	132 (57–207)	282 (184–380)	–

Table 4 Estimates of SSD, F_S, τ and time since expansion (TSE) in y BP for each lineage

mtDNA lineage	SSD	F_S	τ (95% CI)	TSE (95% CI)
Northern	0.00375	−15.945*	4.088 (2.801–5.260)	72,844 (49,911–93,728)
Western	0.07546	0.680	4.578 (0.420–8.410)	87,300 (8,009–160,374)
Italian	0.00367	−2.661	3.363 (1.408–6.920)	64,698 (27,087–133,128)
Balkan	0.03880	0.083	3.234 (0.025–6.068)	54,499 (421–102,258)

*Indicates significance ($p < 0.05$)

(including indels) were found in the concatenated Y chromosome dataset. Twenty-two (36.7%) polymorphic sites were found in DBY1; 8 (13.3%) in DBY3; 8 (13.3%) in DBY7, and 22 (36.7%) in UTY11. A 4 bp indel separated the Northern lineage from all other samples in DBY1 and the Northern lineage also possessed an extra repeat motif (CAAA) in the UTY11 intron.

Four major lineages were identified from the concatenated dataset by the median-joining network (Fig. 2). The Western Y lineage consisted of individuals from the Western and Spanish mtDNA lineages plus individuals from the Northern mtDNA lineage found in Northern France (Fig. 2). The Northern Y lineage included all individuals found in the Northern mtDNA lineage with the exception of those from Northern France and also included individuals from Austria and Slovakia, which belonged to the Balkan mtDNA lineage. The Italian Y lineage contained the same individuals as the Italian mtDNA lineage. Bootstrap support was high for

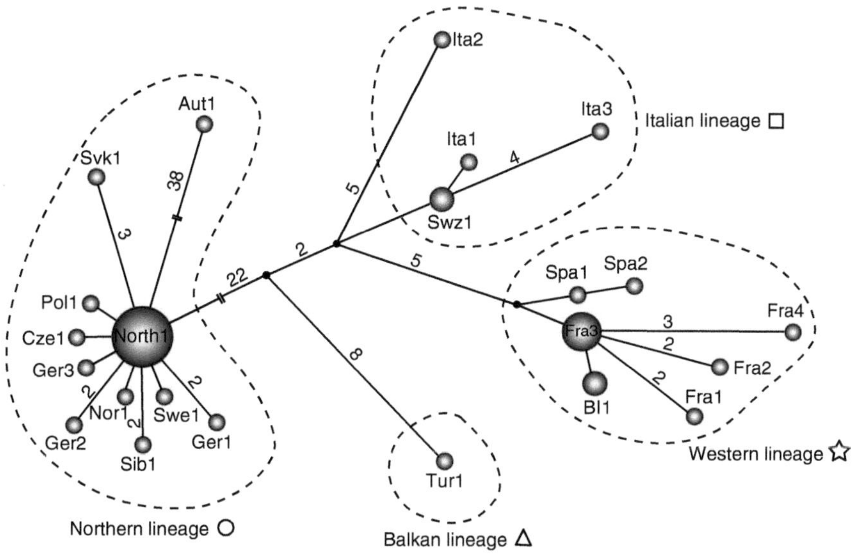

Fig. 2 Median-joining network of 23 concatenated Y chromosomal intron haplotypes (2,939 bp). Numbers on branches indicate more than one mutation. Inferred haplotypes are represented by black dots. Bootstrap support for the Northern Y lineage is 100% and support for the Western Y lineage is 96% using both NJ and MP methods, respectively. See Table 1 for haplotype designations

Western (NJ/MP: 96/96%) and Northern (NJ/MP: 100/100%) Y lineages, with weak support for the Italian Y lineage. Only one individual belonged to the Balkan Y lineage. These lineages were also identified from each Y intron individually, with the exception of DBY3 which failed to separate the Western and Italian lineages (data not shown). The nucleotide diversity (π) of each lineage is given in Table 2.

Eleven haplotypes were found in 23 individuals in the Northern Y lineage (see Table 1 and Fig. 2), with Y haplotype North1 found in 10 localities spread from Norway to Siberia. The Y haplotype Aut1 had several large indels. Seven haplotypes were found in the Western Y lineage and four in the Italian Y lineage. However, the Y haplotype Ita2 was quite different from other Italian individuals (Fig. 2). This was also the case for this individual in the control region analysis (haplotype Ita7; Fig. 1).

4 Discussion

4.1 Postglacial Colonization History Inferred From mtDNA

The five lineages identified from control region data agrees with previously published data for the cytochrome *b* gene (Mascheretti et al. 2003). However, statistical support for all but two (Balkan and Spanish lineages) of these control region lineages were

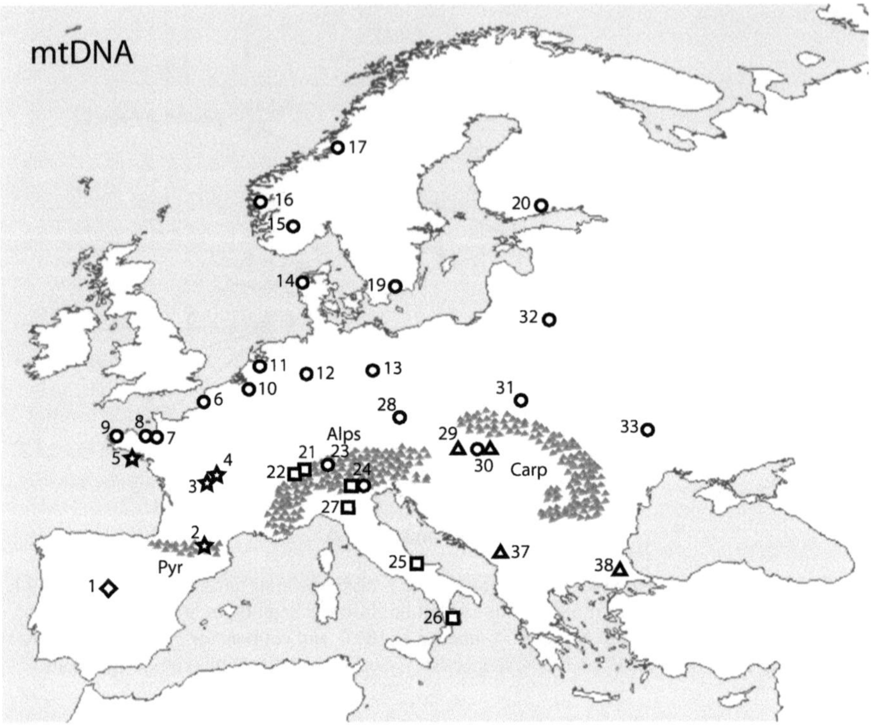

Fig. 3 Distribution of control region lineages of continental Europe (see Fig. 1 for symbol designation of lineages). Samples 34–36 belonging to the Northern mtDNA lineage are not shown. Clustered grey triangles represent the major mountain ranges; Pyr = Pyrenees and Carp = Carpathians

generally very low (<75%, Fig. 1), and this may be due to the large amount of variation and indels present in the *Sorex* control region. Although the Mediterranean peninsulas were important refugia for the pygmy shrew, they contributed relatively little to the spread of the pygmy shrew throughout Europe (Bilton et al. 1998).

From Fig. 3, the Spanish lineage appears to be confined to Spain which could indicate that the Pyrenees separated the glacial refugia for the Spanish lineage from that of the Western lineage in Southern France. The Andorran samples belong to the Western lineage as reported previously (Mascheretti et al. 2003) and could represent a Southern colonization after the LGM. Likewise, the Alps could have acted as a barrier between the Italian and other lineages at the LGM, with a subsequent spread of the Italian lineage into the mountains from an Italian refugium after the LGM.

With regards to the possible glacial refugium for the pygmy shrew Western lineage in Southern France, this is a new refugial location not previously suggested for the pygmy shrew. The common vole, *Microtus arvalis*, is proposed to have persisted North of the Pyrenees (Heckel et al. 2005) and fossil finds of the common shrew (*S. araneus*) or Millet's shrew (*S. coronatus*) indicate that it, along with other temperate species, survived in the Dordogne in South-Western France during the LGM (Sommer and Nadachowski 2006). From such a refugium, the Western lineage could have easily spread further North to the small French island, Belle Île, which only became separated from the French mainland approximately 8,000 years ago (A. Tresset, pers. comm.). This accounts for the relatively high nucleotide diversity (data not shown) present on the island compared to large islands, such as Ireland (McDevitt et al. 2009) which was probably colonized by a small number of individuals introduced by humans (Mascheretti et al. 2003; McDevitt et al. 2009). The Balkan lineage is present as far North as Slovakia and Austria to which it presumably spread from a Balkan refugium. Five individuals from the same locality in Slovakia include individuals representing two separate lineages, the Northern and Balkan. This mtDNA contact zone is located near the Carpathians which is also likely to have been a glacial refugium during the LGM for the pygmy shrew (Pazonyi 2004) as well as other mammalian species (Kotlík et al. 2006; Saarma et al. 2007). We propose that it was the Northern lineage that occupied this refugium. However, further sampling with mtDNA is needed to provide more support for this and other refugial locations.

The most widespread mtDNA lineage of the pygmy shrew found in Europe is the Northern lineage which occurs from Northern France in the West to Siberia in the East to the Scandinavian countries in the North. Mismatch distribution and a negative and significant Fu's F_S value indicate that this lineage originated from a single refugium and spread rapidly. The extensive distribution of this lineage may be due to a selective advantage of this mtDNA type in North temperate and arctic environments. Fontanillas et al. (2005) showed that certain mtDNA haplotypes in *C. russula* may be better adapted to colder environments than others, and this was particularly evident in females. A similar selective advantage has also been shown in humans (Mishmar et al. 2003) and been advocated in hares (Melo-Ferreira et al. 2005). Pygmy shrews inhabiting regions, such as the Carpathians during glacial cycles may have been better adapted to the temperature regimes found in North temperate and arctic regions than those present in refugia in the Mediterranean peninsulas and Southern France.

The approximate divergence times between the mtDNA lineages of the pygmy shrew indicate that separation may have occurred several glacial cycles ago (Renssen and Vandenberghe 2003). This is also the case with other small mammals in Europe (Jaarola and Searle 2004; Deffontaine et al. 2005). Larger, more mobile mammals show a lack of phylogeographical structuring pre-LGM (Hofreiter et al. 2004) but small mammals are more limited in their dispersal abilities and would, therefore, have remained isolated between glacial cycles, promoting divergence in mtDNA sequences. On the assumption of the molecular clock used, the Balkan and Northern lineages diverged from all other lineages approximately 240–357,000 and 221–263,000 y BP, respectively. The Western, Italian, and Spanish lineages diverged more recently (approximately 132–157,000 y BP).

Estimates of time since population expansion vary among lineages. The Northern and Italian lineages expanded 72,844 and 64,498 y BP, respectively, with lower 95% CI values (49,911 and 27,087 y BP, respectively) before the LGM

(19–23,000 y BP). Estimates for the other two lineages were somewhat similar. The estimate for the Western lineage is 87,300 y BP with a lower 95% CI value (8,009 y BP) which stretches into the Holocene. This was affected by the large sample size from Belle Île as the estimate was lower when excluding individuals from the island (57,952 y BP; 95% CI: 11,575–108,009 y BP). The Balkan lineage has an estimate of 54,499 y BP, but with a large 95% CI range (421 years BP–102,258 y BP). These large 95% CI ranges (approximately 50–100,000 y) are obviously affected by the low sample size used for all lineages apart from the Northern lineage. It is also worth noting that only the Northern lineage had a negative and significant F_S value. The data here support the view that lineages might have expanded during or after the Early Pleniglacial (74–59,000 y BP; Renssen and Vandenberghe 2003) but for all except the Northern and Italian lineages a later date for the expansion, after the LGM, cannot be excluded.

However, the results presented here should be treated with caution as the evolutionary rate varies greatly among different regions of the *Sorex* control region (Stewart and Baker 1994a). A divergence rate of 15–20% My^{-1} has been used previously for *S. araneus* (Andersson et al. 2005) based on an error in the abstract of the original paper (D. Stewart, *pers. comm.*). The true divergence rate quoted in the discussion section of Stewart and Baker (1994a) is 8.3–14.3% My^{-1} for the unique sequence and repeat regions combined. The value used here (23% My^{-1}) for the 5′ end of the unique flanking sequence may in fact underestimate the divergence between lineages and their expansion times. Evolutionary rates as high as 60.4% My^{-1} for the control region have been reported in *M. arvalis* (Heckel et al. 2005). The use of this value places the expansion of lineages as follows; Northern: 27,738 y BP; Western: 33,243 y BP (22,067 y BP without individuals from Belle Île); Italian: 24,636 y BP, and Balkan: 20,753 y BP with lower 95% CI values occurring at the end or after the LGM (160 y BP–19,006 y BP). This interpretation fits much better with the expectation of small refugial populations at the LGM and extensive expansion thereafter.

4.2 *Postglacial Colonization History Inferred From Y Chromosomal Introns*

Four major Y chromosome lineages were identified throughout Europe as opposed to five identified using mtDNA (Fig. 2). One of the major differences between female-mediated (mtDNA) and male-mediated (Y introns) lineages was the strong grouping of all French and Spanish shrews to form a Western Y lineage (Fig. 4). Pygmy shrews from Northern France belong to the Western Y lineage as opposed to the Northern lineage for mtDNA. The well supported Western Y lineage is possibly caused by the slower mutation rate in Y chromosome genes compared to faster evolving mitochondrial genes (Hellborg and Ellegren 2004). According to the mtDNA data, the Western, Spanish, and Italian mtDNA lineages diverged more recently from each other than the Northern and Balkan mtDNA lineages (Table 3),

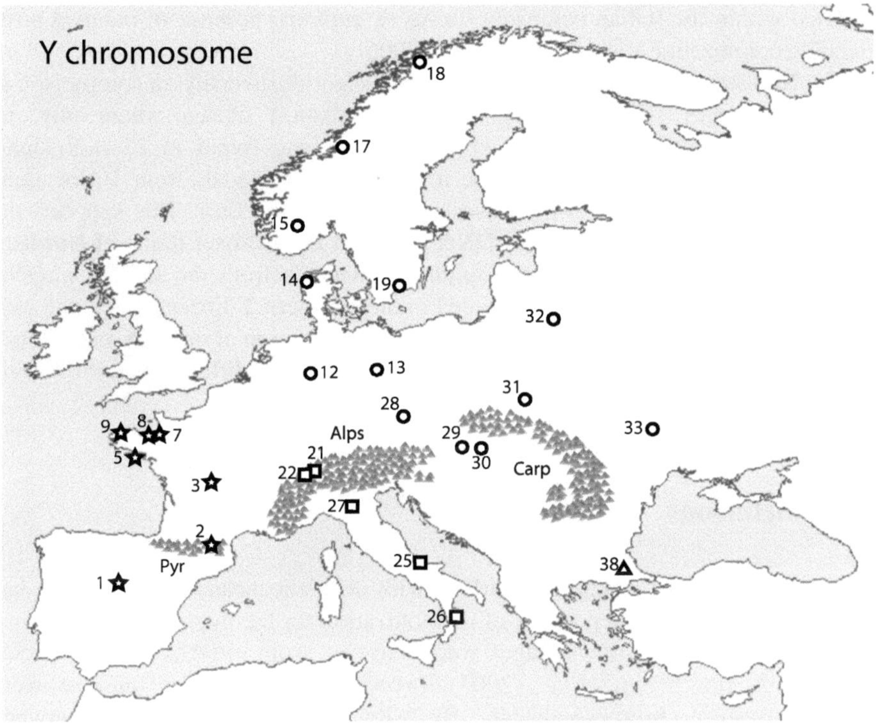

Fig. 4 Distribution of concatenated Y chromosomal intron lineages of continental Europe (see Fig. 2 for symbol designation of lineages). Samples 34–36 belonging to the Northern Y lineage are not shown. Clustered grey triangles represent the major mountain ranges; Pyr = Pyrenees and Carp = Carpathians

and the grouping of French and Spanish Y haplotypes in the same Y chromosome lineage may reflect a more ancient association. However, it could be expected that French, Spanish, and Italian individuals would group together in a Y lineage if this was the case, and although the Italian Y lineage is poorly supported, it is clearly distinct from the Western Y lineage (Fig. 2). Therefore, it is possible that there was gene flow between Spanish and French refugia more recently than the mtDNA data suggests, perhaps after they came into contact after the range expansion following the LGM.

The Y chromosome data supports the existence of a separate glacial refugium in Italy for the pygmy shrew, as already suggested by the mtDNA data. Interestingly, the individual from Southern Italy (Y haplotype Ita2, mtDNA haplotype Ita7; Figs. 3 and 4) did not group with the other Italian haplotypes for both Y chromosome introns and mtDNA. There may have been multiple refugia within Italy during the Pleistocene glaciations, causing isolation of certain populations from others. Further samples are needed to fully resolve this. Genetic differentiation has been reported within the Italian peninsula for *Sorex antinorii* because of putative post-glacial recolonization routes (Basset et al. 2006).

The Northern Y lineage is typified by low nucleotide diversity, in comparison to the other lineages (with the exception of the Balkan Y lineage where only one individual was sequenced). Only 11 haplotypes were found in 23 individuals (Fig. 3), with the North1 Y haplotype found in 13 individuals from 10 locations ranging from Norway in the Northwest to Siberia in the East. This supports the contention that expansion of the mtDNA Northern lineage over much of Northern Eurasia occurred from a single refugium. Individuals from the Balkan mtDNA lineage in Slovakia and Austria belonged to the Northern Y lineage. As is the case for Northern France, this indicates differential introgression of mtDNA and Y chromosome markers at a local level when populations from different glacial refugia come into contact.

5 Conclusions

Analyses of multiple molecular markers with different means of inheritance has demonstrated contrasting patterns of recolonization for the pygmy shrew in continental Europe. Five major lineages were retrieved from mtDNA data, in accordance with Mascheretti et al. (2003). In contrast, four major lineages were retrieved from Y chromosome data. By using mtDNA alone, contact between French and Spanish refugia would not have been identified and this study has highlighted the importance of adopting a multiple marker approach in phylogeographic studies.

The pygmy shrew is the oldest shrew found in Europe (based on fossil data; Rzebik-Kowalska 1998) and is relatively rare on the European mainland (often restricted to high altitudes) in comparison to other shrew species because of competition with species which occupy similar niches and as a result of habitat fragmentation. The genetic lineages described in this study diverged several glacial cycles ago and therefore each has an ancient evolutionary history (Jaarola and Searle 2004). Some of these lineages are relatively rare on the European continent in comparison to others. For example, the Northern lineage is spread throughout Northern and Eastern parts of Europe while the other lineages are relatively restricted in their distribution, often confined to the Southern refugial regions. Fully identifying refugial locations and protecting reservoirs of genetic diversity is vital for the long-term persistence and conservation of pygmy shrew populations (Bhagwat and Willis 2008).

Acknowledgments We thank Damien and Anne McDevitt, Jacqueline Fox, and Marine Pascal for field assistance in France. Thanks to Vitaly Volobouev and Michel Pascal for providing the appropriate licences for trapping and Alain Butet, Anne Tresset, and Carlos Nores for trapping advice. We particularly thank John O'Brien, James Carolan and everyone in Research Lab 4 (UCD) for technical assistance in the laboratory. We thank Don Stewart, John O'Brien, Rodrigo Vega, and Jerry Herman for their help and comments on the earlier drafts of the manuscript and Olaya Astudillo for help with figures. A.D. McDevitt was funded by the Irish Research Council for Science, Engineering and Technology (IRCSET) and a travel grant to France awarded by Enterprise Ireland. G. Yannic was supported by a grant from the Fondation Agassiz (Switzerland). J.B. Searle received field work funding from the Arts and Humanities Research Council (UK).

References

Andersson A-C, Alstrom-Rapaport C, Fredga K (2005) Lack of mitochondrial DNA divergence between chromosome races of the common shrew, *Sorex araneus*, in Sweden. Implications for interpreting chromosomal evolution and colonization history. Mol Ecol 14:2703–2716

Avise JC (2000) Phylogeography: the history and formation of species. Havard University Press, Cambridge, MA

Ballard JWO, Whitlock MC (2004) The incomplete natural history of mitochondria. Mol Ecol 13:729–744

Balloux F, Brünner H, Lugon-Moulin N, Hausser J, Goudet J (2000) Microsatellites can be misleading: an empirical and simulation study. Evolution 54:1414–1422

Bandelt HJ, Forster P, Röhl A (1999) Median-joining networks for inferring intraspecific phylogenies. Mol Biol Evol 16:37–48

Basset P, Yannic G, Hausser J (2006) Genetic and karyotypic structure in the shrews of the *Sorex araneus* group: are they independent? Mol Ecol 15:1577–1587

Bhagwat SA, Willis KJ (2008) Species persistence in Northerly glacial refugia of Europe: a matter of chance or biogeographical traits? J Biogeogr 35:464–482

Bilton DT, Mirol PM, Mascheretti S, Fredga K, Zima J, Searle JB (1998) Mediterranean Europe as an area of endemism for small mammals rather than a source of northwards postglacial colonization. Proc R Soc Lond Ser B 265:1219–1226

Brändli L, Handley LJ, Vogel P, Perrin N (2005) Evolutionary history of the greater white-toothed shrew (*Crocidura russula*) inferred from analysis of mtDNA, Y and X chromosome markers. Mol Phylogenet Evol 37:832–844

Brown WM, George M, Wilson AC (1979) Rapid evolution of animal mitochondrial DNA. Proc Natl Acad Sci USA 76:1967–1971

Brunhoff C, Galbreath KE, Federov VB, Cook JA, Jaarola M (2003) Holarctic phylogeography of the root vole (*Microtus oeconomus*): implications for late Quaternary biogeography of high altitudes. Mol Ecol 12:957–968

Brunhoff C, Yoccoz NG, Ims RA, Jaarola M (2006) Glacial survival or late glacial colonization? Phylogeography of the root vole (*Microtus oeconomus*) in North-West Norway. J Biogeogr 33:2136–2144

Churchfield S, Searle JB (2008) Pygmy shrew. In: Harris S, Yalden DW (eds) Mammals of the British Isles Handbook. 4th edition. The Mammal Society, London. pp. 267–271

Deffontaine V, Libois R, Kotlík P, Sommer R, Nieberding C, Paradis E, Searle JB, Michaux JR (2005) Beyond the Mediterranean peninsulas: evidence of central European glacial refugia for a temperate forest mammal species, the bank vole (*Clethrionomys glareolus*). Mol Ecol 14:1727–1739

Excoffier L, Laval G, Schneider S (2005) Arlequin ver. 3.0: an integrated software package for population genetic data analysis. Evol Bioinform Online 1:2611–2620

Fontanillas P, Dépraz A, Giorgi MS, Perrin N (2005) Nonshivering thermogenesis capacity associated to mitochondrial haplotypes and gender in the greater white-toothed shrew, *Crocidura russula*. Mol Ecol 14:661–670

Fu Y-X (1997) Statistical tests of neutrality of mutations against population growth, hitchhinking and background selection. Genetics 147:915–925

Fumagalli L, Taberlet P, Favre L, Hausser J (1996) Origin and evolution of homologous repeated sequences in the mitochondrial DNA control region of shrews. Mol Biol Evol 13:31–46

Greenwood PJ (1980) Mating systems, philopatry and dispersal in birds and mammals. Anim Behav 28:1140–1162
Gündüz I, Jaarola M, Tez C, Yeniyurt C, Polly PD, Searle JB (2007) Multigenic and morphometric differentiation of ground squirrels (*Spermophilus*, Sciuridae, Rodentia) in Turkey, with a description of a new species. Mol Phylogenet Evol 43:916–935
Hammer MF, Blackmer F, Garrigan D, Nachmann MW, Wilder JA (2003) Human population structure and its effects on sampling Y chromosome sequence variation. Genetics 164:1495–1509
Hasegawa M, Kishino K, Yano T (1985) Dating the human-ape splitting by a molecular clock of mitochondrial DNA. J Mol Evol 22:160–174
Heckel G, Burri R, Fink S, Desmet J-F, Excoffier L (2005) Genetic structure and colonization processes in European populations of the common vole, *Microtus arvalis*. Evolution 59:2231–2242
Hellborg L, Ellegren H (2003) Y chromosome conserved anchored tagged sequences (YCATS) for the analysis of mammalian male-specific DNA. Mol Ecol 12:283–291
Hellborg L, Ellegren H (2004) Low levels of nucleotide diversity in mammalian Y chromosomes. Mol Biol Evol 21:158–163
Hellborg L, Gündüz I, Jaarola M (2005) Analysis of sex-linked sequences supports a new mammal species in Europe. Mol Ecol 14:2025–2031
Hewitt GM (1999) Post-glacial re-colonization of European biota. Biol J Linn Soc 68:87–112
Hewitt GM (2000) The genetic legacy of the Quaternary ice ages. Nature 405:907–913
Hofreiter M, Serre D, Rohland N, Rabeder G, Nagel D, Conard N, Munzel S, Pääbo S (2004) Lack of phylogeography in European mammals before the last glaciation. Proc Natl Acad Sci USA 101:12963–12968
Irwin DM, Kocher TD, Wilson AC (1991) Evolution of the cytochrome *b* gene of mammals. J Mol Evol 32:128–144
Jaarola M, Tegelström H, Fredga K (1997) A contact zone with noncoincident clines for sex-specific markers in the field vole (*Microtus agrestis*). Evolution 51:241–249
Jaarola M, Searle JB (2002) Phylogeography of field voles (*Microtus agrestis*) in Eurasia inferred from mitochondrial DNA sequences. Mol Ecol 11:2613–2621
Jaarola M, Searle JB (2004) A highly divergent mitochondrial DNA lineage of *Microtus agrestis* in Southern Europe. Heredity 92:228–234
Kimura M (1980) A simple method for estimating evolutionary rates of base substitutions through comparative studies of nucleotide sequences. J Mol Evol 16:111–120
Kocher TD, Thomas WK, Meyer A, Edwards SV, Pääbo S, Villablanca FX, Wilson AC (1989) Dynamics of mitochondrial DNA evolution in animals: amplification and sequencing with conserved primers. Proc Natl Acad Sci USA 86:6196–6200
Kotlík P, Deffontaine V, Mascheretti S, Zima J, Michaux JR, Searle JB (2006) A Northern glacial refugium for bank voles (*Clethrionomys glareolus*). Proc Natl Acad Sci USA 103:14860–14864
Kumar S, Tamura K, Nei M (2004) MEGA3: integrated software for molecular evolutionary genetics analysis and sequence alignment. Brief Bioinform 5:150–163
Mascheretti S, Rogatcheva MB, Gündüz I, Fredga K, Searle JB (2003) How did pygmy shrews colonize Ireland? Clues from a phylogenetic analysis of mitochondrial cytochrome *b* sequences. Proc R Soc Lond Ser B 270:1593–1599
McDevitt AD, Rambau RV, O'Brien J, McDevitt CD, Hayden TJ, Searle JB (2009) Genetic variation in Irish pygmy shrews *Sorex minutus* (Soricomorpha: Soricidae): implications for colonization history. Biol J Linn Soc 97:918–927
Melo-Ferreira J, Boursot P, Suchentrunk F, Ferrand N, Alves PC (2005) Invasion from the cold past: extensive introgression of mountain hare (*Lepus timidus*) mitochondrial DNA into three other hare species in Northern Iberia. Mol Ecol 14:2459–2464
Michaux JR, Magnanou E, Paradis E, Nieberding C, Libois R (2003) Mitochondrial phylogeography of the woodmouse (*Apodemus sylvaticus*) in the Western Palearctic region. Mol Ecol 12:685–697
Mishmar D, Ruiz-Pensini E, Golik P et al (2003) Natural selection shaped regional mtDNA variation in humans. Proc Natl Acad Sci USA 100:171–176

Mitchell-Jones AJ, Amori G, Bogdanowicz W et al (1999) The atlas of European mammals. Poyser, London

Mix AC, Bard E, Schneider R (2001) Environmental processes of the ice age: land, oceans, glaciers (EPILOG). Quaternary Sci Rev 20:627–657

Nishimura Y, Yoshinari T, Naruse K, Yamada T, Sumi K, Mitani H, Higashiyama T, Kuroiwa T (2006) Active digestion of sperm mitochondrial DNA in single living sperm revealed by optical tweezers. Proc Natl Acad Sci USA 103:1382–1387

Pazonyi P (2004) Mammalian ecosystem dynamics in the carpathian basin during the last 27, 000 years. Palaeogeogr Palaeoclimatol Palaeoecol 212:295–314

Pesole G, Gissi C, De Chirico A, Saccone C (1999) Nucleotide substitution rate of mammalian mitochondrial genomes. J Mol Evol 48:427–434

Petit E, Balloux F, Excoffier L (2002) Mammalian population genetics: why not Y? Trends Ecol Evol 17:28–33

Pokines JT (2000) Microfaunal research design in the cantabrian Spanish paleolithic. J Anthropol Res 56:95–111

Posada D, Crandall KA (1998) MODELTEST: testing the model of DNA substitution. Bioinformatics 14:817–818

Rambaut A (1996) Se-Al, sequence alignment editor (version 2.0). Department of Zoology, University of Oxford. [http://evolve.zoo.ox.ac.uk]

Renssen H, Vandenberghe J (2003) Investigation of the relationship between permafrost distribution in NW Europe and extensive winter sea-ice cover in the North Atlantic Ocean during the cold phases of the Last Glaciation. Quat Sci Rev 22:209–223

Rhymer JD, Barbay JM, Givens HL (2004) Taxonomic relationship between *Sorex dispar* and *S. gaspensis*: inferences from mitochondrial DNA sequences. J Mammal 85:331–337

Roca AL, Georgiadis N, O'Brien SJ (2005) Cytonuclear genomic dissociation in African elephant species. Nat Genet 37:96–100

Rogers AR (1995) Genetic evidence for a pleistocene population explosion. Evolution 49:608–615

Rogers AR, Harpending H (1992) Population-growth makes waves in the distribution of pairwise genetic distances. Mol Biol Evol 9:552–569

Ronquist F, Huelsenbeck JP (2003) MrBayes 3: Bayesian phylogenetic inference under mixed models. Bioinformatics 19:1572–1574

Rozas J, Sánchez-DelBarrio JC, Messeguer X, Rozas R (2003) DnaSP, DNA polymorphism analysis by coalescent and other methods. Bioinformatics 19:2496–2497

Rzebik-Kowalska B (1998) Fossil history of shrews in Europe. In: Wójcik JM, Wolsan M (eds) Evolution of shrews. Mammal Research Institute, Polish Academy of Sciences, Bialowieza, pp 23–92

Saarma U, Ho SYW, Pybus OG et al (2007) Mitogenetic structure of brown bears (*Ursus arctos* L.) in Northeastern Europe and a new time frame for the formation of European brown bear lineages. Mol Ecol 16:401–413

Schneider S, Excoffier L (1999) Estimation of demographic parameters from the distribution of pairwise differences when the mutation rates vary among sites: application to human mitochondrial DNA. Genetics 152:1079–1089

Seddon JM, Santucci F, Reeve NJ, Hewitt GM (2001) DNA footprints of European hedgehogs, *Erinaceus europaeus* and *E. concolor*: pleistocene refugia, potglacial expansion and colonization routes. Mol Ecol 10:2187–2198

Sommer RS, Nadachowski A (2006) Glacial refugia of mammals in Europe: evidence from fossil records. Mammal Rev 36:251–265

Stewart DT, Baker AJ (1994a) Patterns of sequence variation in the mitochondrial D-loop region of shrews. Mol Biol Evol 11:9–21

Stewart DT, Baker AJ (1994b) Evolution of mtDNA D-loop sequences and their use in phylogenetic studies of shrews in the subgenus *Otisorex* (Sorex: Soricidae: Insectivora). Mol Phylogenet Evol 3:38–46

Stewart DT, Baker AJ (1997) A phylogeny of some taxa of masked shrews (*Sorex cinereus*) based on mitochondrial-DNA, D-loop sequences. J Mammal 78:361–376

Stewart JR, Lister AM (2001) Cryptic Northern refugia and the origins of modern biota. Trends Ecol Evol 16:608–613

Swofford DL (1998) PAUP*. Phylogenetic Analysis Using Parsimony (*and other methods). Version 4.0b10. Sinauer Associates, Sunderland, MA

Taberlet P, Fumagalli L, Wust-Saucy A-G, Cosson J-F (1998) Comparative phylogeography and post-glacial colonization routes in Europe. Mol Ecol 7:453–464

Templeton AR (2002) Out of Africa again and again. Nature 416:45–51

Underhill PA, Shen P, Lin AA et al (2000) Y chromosome sequence variation and the history of human populations. Nat Genet 26:358–361

Xia X, Xie Z (2001) DAMBE: data analysis in molecular biology and evolution. J Heredity 92:371–373

Part IV
Relicts Since Postglacial Times

Are Disjunct Alpine and Arctic-Alpine Animal and Plant Species in the Western Palearctic Really "Relics of a Cold Past"?

Thomas Schmitt, Christoph Muster, and Peter Schönswetter

Abstract The climatic fluctuations during the Pleistocene as well as the Holocene warming caused numerous disjunctions of cold-adapted, arctic-alpine, and alpine biota. However, the depths of the genetic splits among the disjunct parts of the species distributions vary considerably. The arctic ranges are usually weakly differentiated, and great similarity with at least some areas in more Southern regions is frequently found. Likewise, major mountain ranges in geographic proximity often share genetically similar populations. However, the genetic constitution of populations from more remote (predominantly Southern) mountain systems is strongly different from all other populations. This suggests recent vicariance events in the two former groups, but long-lasting isolation in the latter group, which apparently is mostly composed of relics of a more distant cold past.

1 Introduction

The world's climate has undergone dramatic changes during the history of the planet. The most recent period of climatic instability, the Pleistocene, was characterized by particularly strong and more or less regular oscillations between warm and predominantly moist, and cold and more often dry periods (Williams et al. 1998). During the cold stages, thermophilic species (e.g., Mediterranean or temperate taxa)

T. Schmitt (✉)
Department of Biogeography, Trier University, D-54286 Trier, Germany
e-mail: thsh@uni-trier.de

C. Muster
Molecular Evolution and Animal Systematics, Institute of Zoology II, University of Leipzig, Talstraße 33, D-04103 Leipzig, Germany
e-mail: muster@rz.uni-leipzig.de

P. Schönswetter
Department of Plant Biogeography, Faculty Centre Biodiversity, University of Vienna, A-1030 Wien, Austria
e-mail: peter.schoenswetter@univie.ac.at

J.C. Habel and T. Assmann (eds.), *Relict Species: Phylogeography and Conservation Biology*, DOI 10.1007/978-3-540-92160-8_13, © Springer-Verlag Berlin Heidelberg 2010

were restricted to Southern refugia in the Southern European peninsulas where they have survived in refugia and expanded their ranges during warm periods such as the interglacials or the postglacial (de Lattin 1967; Müller 1980; Hewitt 1996, 2000, 2004a,b; Taberlet et al. 1998; Schmitt 2007). For temperate species, this view has, however, been challenged by the recent combination of pollen data and genetic analyses (Steward and Lister 2001), suggesting that refugia of, e.g., the Common Beech (*Fagus sylvatica*), were situated much further North than previously suggested (Magri et al. 2006).

The cycles of area regression and expansion of species currently restricted to high mountain habitats and/or high latitudes were likely diametrical: these species were potentially widely distributed during cold stages and experienced range restrictions during warm periods (de Lattin 1967). Consequently, current distribution patterns of species adapted to high altitudes or high latitudes are often strongly disjunct and many of the isolated area fragments are small and widely separated from the core of their distribution area (Varga and Schmitt 2008).

The application of molecular techniques to reconstruct genealogical relationships above, at or even below the species level, has created a powerful tool to unravel the biogeographic history of taxa in space and time (Hewitt 1999, 2001, 2004a; Taberlet et al. 1998; Avise 2000; Schmitt 2007). We here review phylogeographic data on high mountain and high latitude species with disjunct distribution areas hitherto available for the Western Palearctic. In the following, we distinguish between two distributional types of species: (1) alpine, montane, and dinodal species of the European mountain system (EMS) of temperate regions sensu Ozenda (1988) and (2) arctic-alpine species. For each of these two above categories we summarize published studies and some unpublished results, and we search for congruencies as well as differences among the observed phylogeographic patterns. Based on the degree of genetic divergence between isolated "relic" taxa or populations and those from the main distribution areas, we aim to unravel if these entities represent relics of a cold past or are rather descendants of fairly young range expansions.

2 Species of the European Mountain System (EMS) of Temperate Regions

The distributions of species of the EMS as defined here include the Cordillera Cantábrica and range from there over the Pyrenees, Alps, and Apennines to the Carpathians and the mountains of the Northern and central Balkan Peninsula. They normally do not occur South of the Pyrenees or of the Italian Abruzzo, and are also lacking from the mountain ranges of the Southern Balkan Peninsula South of the Adamović line (Ozenda 1988). Few species of this category occur throughout the entire outlined area. While some of these mountain species are restricted to a part of the EMS where they are fairly frequent, others exhibit much wider but scattered distribution areas.

2.1 Alpine Species

During the last few years, an ample body of literature has accumulated exploring the phylogeography of alpine plants of the EMS. A review of hitherto available studies on Alpine plants was presented by Schönswetter et al. (2005). Therefore, we here consider only studies on species that show disjunctions among mountain ranges. For this group of alpine species, it could be shown by various examples in plants and animals that (1) the Alps often own more than one genetic lineage, (2) mountain areas in geographic proximity to the Alps may share genetic lineages with the adjacent regions of the Alps, and (3) more remote mountain areas mostly show strongly differentiated genetic lineages underpinning long geographic separation of these remote mountain areas and old relic status of their populations.

In the Alpine-Pyrenean disjunct *Phyteuma globulariifolium*, the populations from the Pyrenees were found to be genetically indistinguishable from Western Alpine plants, whereas a deep split was found in the Alps separating an Eastern and a Western group along River Etsch/Adige (Schönswetter et al. 2002). In the highly disjunct *Androsace halleri*, the most internal cpDNA haplotype is found in the Cordillera Cantábrica and in the Eastern Pyrenees, whereas the haplotypes from Massif Central and Alsace are derived by one and two mutational steps, respectively, indicating eastward range expansion (Dixon et al. 2007). A similar pattern was suggested in an amplified fragment length polymorphism (AFLP) study of *Pritzelago alpina* (Kropf et al. 2003), where populations from the Cordillera Cantábrica formed the sister group to the remaining populations. Some lineage diversification was observed in the Pyrenees, whereas one lineage was shared among Northeastern Alps, Tatras, and Southern Carpathians. A further clear example of colonization of the Carpathians from the Alps is provided by *Senecio carniolicus* (Suda et al. 2007), where diploid, tetraploid, and hexaploid cytotypes were found in the Eastern Alps, whereas the Carpathian populations (Tatras, Southern Carpathians) were uniformly hexaploid. In the Alpine-Carpathian *Hypochaeris uniflora*, Mráz et al. (2007) detected three strongly differentiated lineages in the Alps and the Western and the Eastern Carpathians, indicating long-term isolation of these three regions from each other. Most of the Alpine range was apparently colonized post-glacially from an Eastern Alpine refugium. By applying AFLP fingerprinting and chloroplast DNA data to *Campanula alpina* distributed in the Alps and Carpathians (mountains of the Balkan Peninsula were not sampled), Ronikier et al. (2008a) detected a deep split between the Alps and the Carpathians. Further splits separated Southwestern, Southeastern, Northeastern, and Northwestern parts of the Carpathian arch, indicating that the phylogeographical pattern of *C. alpina* has been mainly shaped by vicariance events. Virtually no AFLP differentiation was found among Pyrenean, alpine, and Tatra populations of *Oxytropis campestris* (Schönswetter et al. 2004a). In *Bupleurum stellatum*, mainly distributed in the Western Alps and the Southern part of the Central Alps, a deep split was found within the continuous distribution area roughly halfway between Lago di Garda and Lago di Como (Schönswetter and Tribsch 2005). The isolated population(s) from

Corsica clustered with high statistical support within the Eastern main group instead of the geographically closer Western group. The presence of several private AFLP fragments in the Corsican population, however, indicated that more time has elapsed since the dispersal event than in cases like *P. globulariifolium* and *O. campestris*. A strongly idiosyncratic genetic structure was revealed by a cpDNA study of *Pulsatilla vernalis* (Ronikier et al. 2008b), distributed over the EMS, parts of the Northerly adjacent lowlands, and Southern Fennoscandia. Two haplotype families separated by several mutational steps were found, both of which occurred throughout the species' distribution area except for the northernmost parts. Interestingly, the centralmost haplotype is essentially found over the entire distribution area, but was lacking from the Alps and Southern Finland. Specific microsatellite variation indicates the occurrence of periglacial lowland refugia North of the Alps, acting as source for post-glacial colonization of Scandinavia. The presence of different fixed haplotypes in Eastern and Western Scandinavia, however, suggests independent post-glacial colonization of these two areas, with possible founder effects.

In comparison with plants, much fewer animal species with disjunct alpine distributions have been analysed so far. A good example, resembling patterns already found in several plant species, was provided by allozyme polymorphisms in the butterfly *Erebia epiphron* (Schmitt et al. 2006). The lineage of *E. epiphron* from the Northern Alps is shared with the Jeseník population in Northern Czech Republic, thus strongly supporting a glacial link between the Northern Alps and the more Eastern Sudety Mountains. However, being currently separated by hundreds of kilometres from the Alps in a geographically small mountainous area, the *E. epiphron* populations in the Jeseník apparently represent a relic of a formerly wider distribution of the species. A link between the Western Alps and the Pyrenees is suggested by a genetic lineage shared by both regions. The Eastern Pyrenees, however, host an endemic lineage of this species underlining the independent evolution of one genetic entity most probably at the Southeastern slopes of the Pyrenees. Interestingly, the same area hosts a strongly differentiated lineage of the plant *A. halleri* (Dixon et al. 2007). Two further genetic groups in *E. epiphron* are endemic to the Southern and Eastern Alps (Fig. 1).

In summary, whereas some phylogeographic studies on species of the EMS provided evidence for highly dynamic scenarios involving unintuitive links among geographically widely separated mountain ranges (such as the link between the Eastern Alps and Corsica in *B. stellatum*, Schönswetter et al. 2005), others suggested vicariance events caused by disruptions during warm stages of the Pleistocene, resulting in splits of varying depths (shallow: e.g., *O. campestris*, Schönswetter et al. 2004a; deep: e.g., *H. uniflora*; Mráz et al. 2007).

2.2 Montane Species

Montane species show strong ecologic affinities to the forest belt habitats of the EMS and therefore occur at lower altitudes than the alpine species. However, this biogeographical group is much less studied than the one of alpine disjunct species.

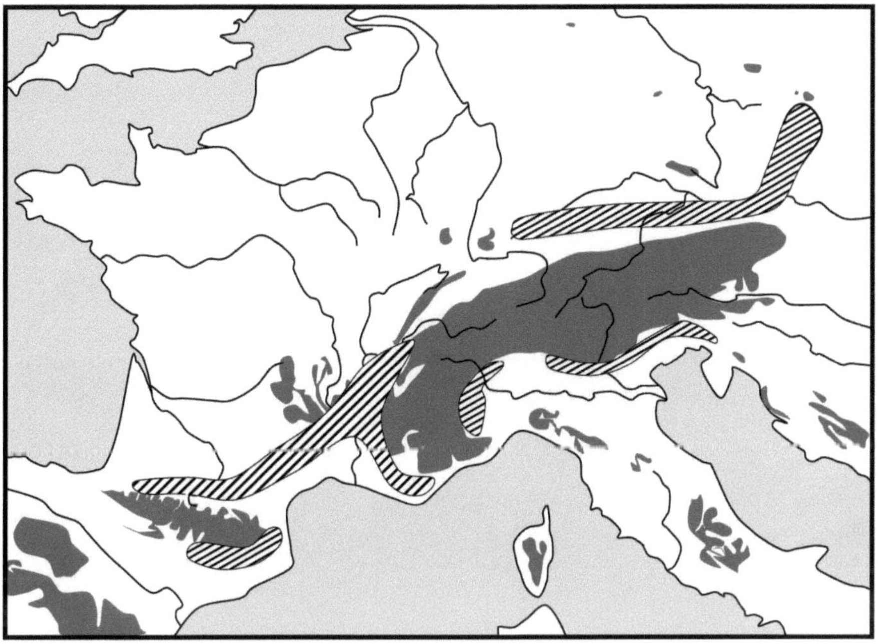

Fig. 1 Hypothetical distribution patterns of the alpine disjunct butterfly *Erebia epiphron* in Western Europe during the last glaciation. Redrawn from Schmitt et al. (2006)

A good example for a widespread montane species is represented by the butterfly *Erebia euryale*, which is a character species of the mountain forest biome of Southern Europe (Sonderegger 2005). Schmitt and Haubrich (2008) analysed allozyme patterns of 11 populations from four EMS (Pyrenees, Alps, Southern Carpathians, Rila) and found four well-supported groups (i.e., Pyrenees, Western Alps, Eastern Alps and Southeastern Europe, represented by Southern Carpathians and Rila). The genetic diversity of the populations was the highest in the Southeastern European group and decreased step-wise towards the West. Interestingly, the populations from Bulgaria and Romania were almost identical; therefore, the authors assume that they were not separated during the last ice age. In contrast, the differentiation among the three Western Alpine populations was considerable. Consequently, (1) the most important refugial area for this species has been located in Southeastern Europe including at least parts of the Carpathians and the Bulgarian mountains, (2) important refugial areas existed at the Southeastern or Eastern edge of the Alps, (3) scattered populations survived along the Southwestern Alps, and (4) relatively small populations have persisted at the foothills of the Pyrenees (Fig. 2). More studies in species of this distribution type are urgently needed to clarify whether the phylogeographic structure found in *E. euryale*, a character species of the European mountain forest biome, is representative for this biome as a whole or not.

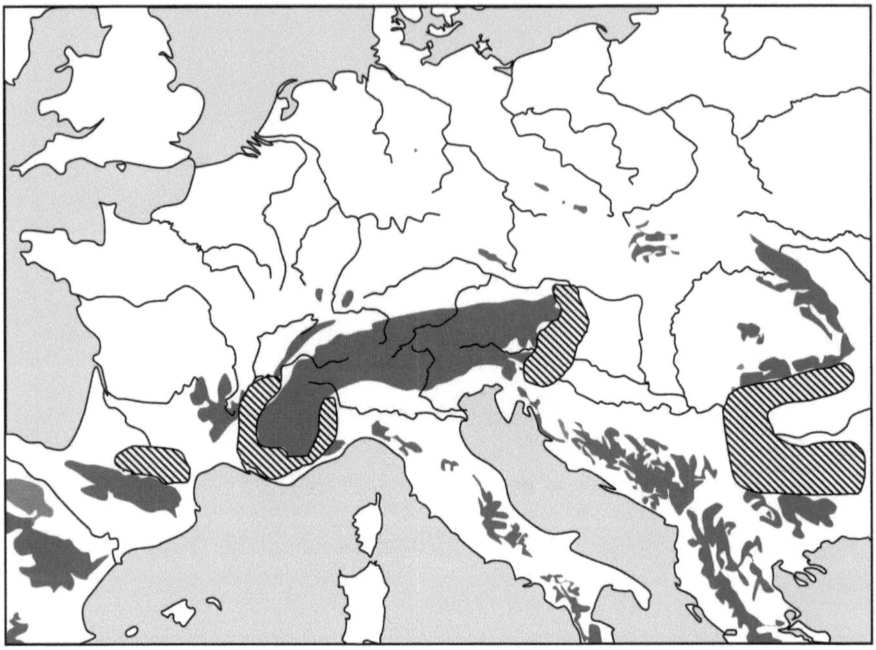

Fig. 2 Hypothetical distribution pattern of the montane disjunct butterfly *Erebia euryale* in Europe during the last glaciation. Note that these centres only refer to the ones detected in this survey; thus, further centres of survival are possible and reliable, e.g., in peninsular Italy. Redrawn from Schmitt and Haubrich (2008)

2.3 Dinodal Mountain Species

Species of headwaters of mountain areas are often classified as belonging into an own biogeographic unit, the dinodal (Malicky 1983, 2000). Due to their water dwelling larvae, many Trichoptera species are considered part of this group. The most extensive genetic study of mountain species in this unit was performed on the mitochondrial COI gene of the caddisfly *Drusus discolor* (Pauls et al. 2006), which occurs in the mountain areas of Southern and central Europe.

No close relationships among mountain systems were detected (i.e., strongly differentiated endemic genetic lineages were detected in the Cantabrian Mts., Pyrenees, Massif Central, Eastern Sudety Mts., Tatras, Eastern and Southern Carpathians, as well as the high mountain systems of the Balkan Peninsula). Thus, long-lasting isolation and survival of populations in all these mountain systems over several glacial and interglacial cycles is the most likely scenario, and the strong liaison of the *Drusus* larvae with cold headwater mountain streams might have successfully impeded the exchange of individuals between different high mountain systems throughout time. Genetic differentiation and structuring in *D. discolor* had some similarities with the one observed in the mountain forest species

E. euryale (Schmitt and Haubrich 2008), i.e., a Western and an Eastern Alpine lineage, an endemic Pyrenean group, and endemic lineages in Southeastern Europe.

Interestingly, in *D. discolor* all relic populations from the low mountain ranges of central Germany (i.e., Rothaargebirge, Rhön, Harz, Thüringer Wald, Fichtelgebirge, Erzgebirge) share the same genetic lineage endemic to this region. This underlines the inference of an independent centre of survival and differentiation in this region, followed by post-glacial retraction to these mountain areas from lower areas. The same phenomenon was observed for the Jura, Vosges, and Black Forest populations. Finally, the populations from the Bohemian Massif harbour the Eastern Alps lineage, thus giving evidence for a Würm glacial link between these two adjacent regions.

The phylogeographic patterns observed in *D. discolor* show two interesting characteristics: (1) Different high mountain systems do not share closely related lineages, thus rejecting the hypothesis of continuous lowland areas between such widely separated mountains during glaciations; (2) Lower mountain ranges being in some geographic proximity to each other share haplotypes of the same genetic lineage, thus supporting the idea that the lower areas between them allowed high levels of gene flow at least during the last glaciation. Maybe the intermediate lowlands even are the refugia for such genetic lineages.

Another representative of headwaters of mountain areas is the freshwater flatworm *Crenobia alpina*. Allozyme electrophoresis and mtDNA sequencing of this species in the Eastern Alps and Germany unravelled a considerable number of strongly differentiated genetic lineages with their divergence maybe dating back as far as the late Tertiary. Most of these lineages most probably survived the glaciations of the Alps in small peripheral refugia to which some of them might still be restricted. Several of these refugia had their most likely location at the Northern margin of the Alps (Brändle et al. 2007).

3 Arctic-Alpine Species

Arctic-alpine distribution ranges comprise the arctic as well as the alpine belt of more Southern mountain ranges. Typically, arctic-alpine distributions are large and often encompass the entire Northern Hemisphere (circumpolar distribution); more restricted ranges are, e.g., amphi-Atlantic or amphi-Pacific (Abbott and Brochmann 2003). The last few years have seen increasing interest in phylogeographic studies of arctic-alpine higher plants, which are by now the doubtlessly best studied group. In the following, we focus on the relationships between the disjunct distribution areas of arctic-alpine taxa in the EMS and their main distribution area in the Arctic.

A close relationship between the central and Northern European populations of a species appears intuitively self-evident and has often been suggested by classical biogeographers (e.g., Vierhapper 1918; Holdhaus 1954; de Lattin 1967; Ozenda 1988). Good examples for such a connection are *Ranunculus glacialis* (Schönswetter et al. 2003), *Arabis alpina* (Ehrich et al. 2007), and *Veronica alpina* (Albach et al. 2006), which all colonized the North Atlantic region from source populations in the Alps. *Dryas octopetala* likely colonized Northern Europe from source populations located between the Scandinavian ice shield and the Alps (Skrede et al. 2006). In contrast, the Scandinavian populations formed monophyletic groups in *Minuartia biflora* (Schönswetter et al. 2006b) and *Carex atrofusca* (Schönswetter et al. 2006a), indicating that Northern European and Alpine populations have been separated for a certain period of time. In *Carex bigelowii* (Schönswetter et al. 2008), the single Alpine (meta-)population has strong links to the phylogeographical group dominant in the Hercynic mountains, the Tatras, Scandinavia, Iceland, and Scotland. The Hercynic populations were genetically variable, while only a few clones were detected in the Alps, indicating a strong founder effect. Whereas the lack of phylogeographic structure within the four detected phylogeographic groups of *Juncus biglumis* did not allow tracing the closest relatives for Alpine and Scandinavian populations (Schönswetter et al. 2007), *Ranunculus pygmaeus* populations from the two latter areas have their closest relatives on the Taymyr Peninsula in Northern Siberia and in the Urals, respectively, (Schönswetter et al. 2006b). In *Comastoma tenellum* (Schönswetter et al. 2004b), two unrelated lineages were found in the Alps. The by far most common one was endemic to the Alps, while the second, encountered only in two Eastern Alpine populations, exhibited unresolved relationships to Carpathian and Scandinavian populations. Thus, this species has apparently colonized the Alps at least twice.

Similar arctic-alpine distribution patterns like in plants are also observed in many animal species. Holdhaus (1954) listed several dozens of invertebrate species with this distribution pattern in Europe in his seminal monograph on the legacy of the Pleistocene glaciations preserved in contemporary animal distributions. However, while most molecular studies are from plants, mitochondrial sequence variation has been studied in not more than three arthropod taxa of this disjunction type until now.

An illustrative example is provided by wolf spiders of the *Pardosa saltuaria* group (Muster and Berendonk 2006). This is a complex of six closely related species (Wunderlich 1984; Marusik et al. 1996) whose taxonomic status needs revision in the light of the recent genetic findings. The species group is patchily distributed from the Western Palearctic over Siberia to the Altai (Fig. 3a). Focussing on the European range, Muster and Berendonk (2006) found three clades of deep mitochondrial divergence, i.e., a Pyrenean clade, a Balkan clade, and a "Northern clade." The latter included Scandinavia, the Alps, the Carpathians, the Giant Mountains, and the Bohemian Forest. Recent results from a nuclear rDNA marker (spanning ITS1, 5.8S rDNA, ITS2) generally confirmed this pattern of genetic divergence (Muster et al. unpublished). A detailed population genetic analysis using the information of both loci (mitochondrial and nuclear) in a coalescent-based Bayesian Markov chain Monte Carlo (MCMC) framework (Hey and Nielsen 2007) revealed a pattern of extensive but differential late glacial gene flow among Northern clade areas (Muster et al. 2009). The inferred gene flow was mainly unidirectional with a prevailing direction from the North to the South.

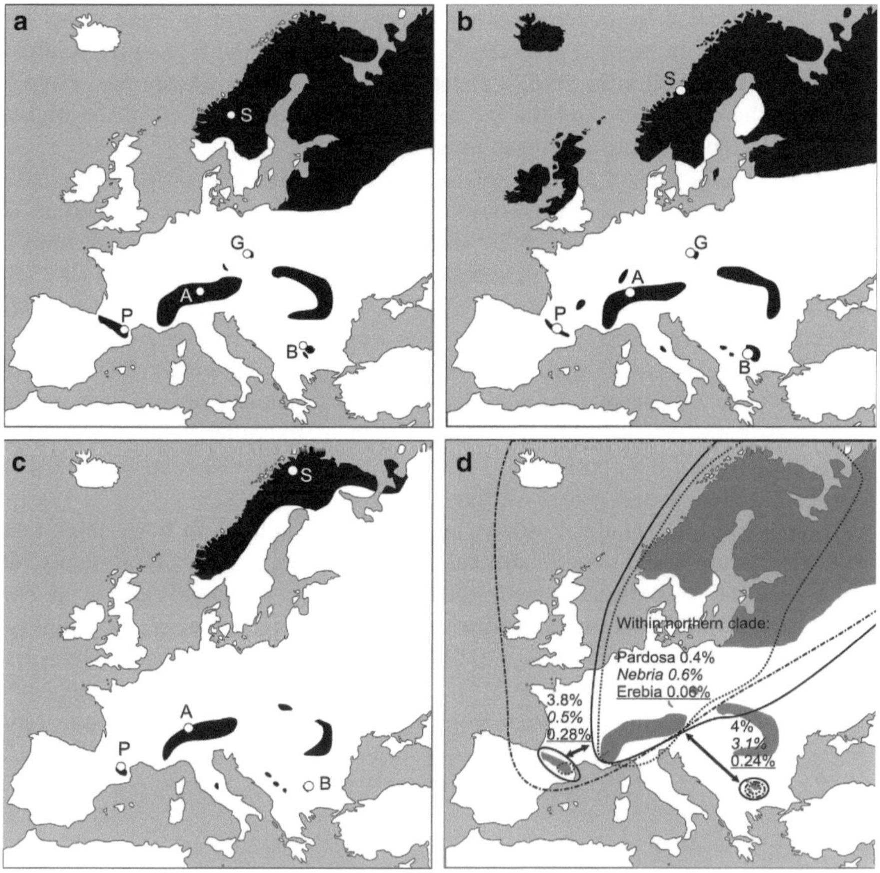

Fig. 3 European distribution areas and sampling localities of three arctic-alpine invertebrates: the wolf spiders of the *Pardosa saltuaria* group (**a**) redrawn from Muster and Berendonk (2006), the ground beetle *Nebria rufescens* (**b**) modified from Holdhaus and Lindroth (1939), and the butterfly *Erebia pandrose* (**c**) modified from Varga (1971). The white dots indicate the sampling locations. Abbreviations: *A* Alps; *B* Balkans; *G* Giant Mountains; *P* Pyrenees; *S* Scandinavia. In Fig. 3d the geographic extension of major mitochondrial clades is shown, together with the Kimura 2 parameter distance (in %) between clades and within the Northern clade. *Solid line Pardosa saltuaria* group, *chain line N. rufescens*, *dotted line Erebia pandrose*

In order to test the phylogeographic hypothesis derived from the wolf spider data, preliminary results from two further arctic-alpine arthropod taxa are presented here. The ground beetle *Nebria rufescens* (Fig. 3b) occupies an even larger arctic area than the *Pardosa* spiders and includes North America (Holdhaus 1954). The area of the butterfly *Erebia pandrose*, on the other hand, is mainly restricted to Europe (Fig. 3c), with a few populations in the mountains of Central Asia (Cupedo 2007). Within Europe, the distributions of these species are also different (Fig. 3a–c), with *N. rufescens* resembling more the boreomontane distribution type of de Lattin

(1967), as indicated by occurrences on the British Isles and in some rather low mountain ranges. In marked contrast, *E. pandrose* shows the typical arctic-alpine disjunction sensu de Lattin (1967), and even in Scandinavia the species' range is limited to the mountains and the far North, while it is restricted to the highest elevations in the Southern European mountains.

In a pilot study, Muster et al. (2009) sequenced the complete mitochondrial ND1 gene from five individuals of each of four (in *E. pandrose*) and five populations (in *N. rufescens*) as shown in Fig. 3. To allow the direct comparison of the level of genetic variation, the data set of the wolf spiders was reduced to the first five sequences of each of five similar areas. Both striking similarities and idiosyncratic differences in the resulting genetic structures were found. First, the overall level of sequence divergence was highly different (K2P-distance: *Pardosa* 2.4%, *Nebria* 1.3%, *Erebia* 0.2%) despite the fact that identical genes in co-distributed species were investigated. Analysis of molecular variance (Excoffier et al. 1992, 2005) clearly showed that the populations in the Southern mountains contributed disproportionally to the overall genetic differentiation of these taxa.

Despite the different life histories and migration abilities in these three taxa, great similarities in the genetic structures existed (Fig. 4). Most importantly, the haplotypes from the Balkans formed a distinct and highly supported clade in all three study systems. Another common feature was the low divergence among Scandinavia, the Alps, and several rather low Central European mountain ranges (Hercynian Mountains), which always clustered in a "Northern clade" with little geographic structure (except for *N. rufescens*, where each area has its own unique mutations).

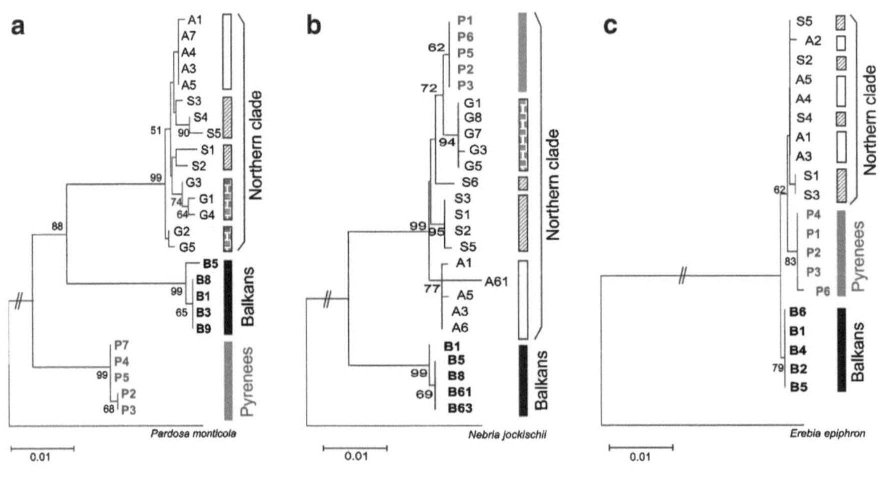

Fig. 4 Neighbour-joining trees showing the phylogenetic relationships of 25 individuals of the *Pardosa saltuaria* group (**a**), 25 individuals of *Nebria rufescens* (**b**), and 20 individuals of *Erebia pandrose* (**c**). All three trees are shown on the same scale, highlighting the different degree of overall sequence divergence. Bootstrap values>50% from 1,000 replicates are shown at the nodes. Letter abbreviations are the same as in Fig. 3

The position of the Pyrenees differed among the investigated taxa. While the Pyrenean haplotypes formed a clade as distinct from the Northern populations as the Balkan clade in *Pardosa* and *Erebia*, the Pyrenees clearly fell within the Northern clade in *Nebria*.

It is most likely that the populations from Scandinavia, the Alps, and the Hercynian Mountains are descendants of large populations that have dwelled in the periglacial tundra belt during the last glaciations, or at least that substantial levels of late glacial gene flow among mountain ranges occurred. Similarly, allozyme data from the diving beetle *Hydroporus glabriusculus* suggest a common late glacial origin of the British and the Scandinavian populations from source populations South of the Northern European ice sheet (Bilton 1994). In the *Nebria* beetles and in the arctic-alpine burnet moth *Zygaena exulans* (Schmitt and Hewitt 2004), such late glacial connections obviously existed also between the Alps and the Pyrenees. Thus, the traditional concept of enormous continuous distributions in the periglacial tundra belt during the glaciations (Holdhaus 1954; de Lattin 1967) could be partly corroborated. On the other hand, the Balkans and in some taxa also the Pyrenees have been isolated, and populations there consequently evolved independently for fairly long periods of time, probably since several glacial cycles. They thus represent old relic populations of high conservation importance. Altogether, the populations from disjunct parts of the distribution areas of arctic-alpine disjunct arthropods can be regarded as relics, which originated at very different time horizons.

4 Relics or not Relics: That's the Question

The genetic analyses of different species with alpine and arctic-alpine disjunct distribution patterns unravel a great variety of different biogeographical patterns and considerable differences in the time frames of the vicariance and/or dispersal events. Most often, separations in arctic-alpine species between the Northern populations and the high mountain systems in central and Southern Europe (often the Alps) are recent (i.e., post-glacial) phenomena. In alpine disjunct and in arctic-alpine species, adjoining high mountain systems often own the same genetic lineage (as, e.g., shown by multiple genetic links between the Western Alps and the Pyrenees, the Eastern Alps and the Carpathians, and the Southeastern Alps and the Dinaric mountain systems), thus underlining a recent (in most cases post-glacial) vicariance or dispersal event in the respective genetic lineage. A similar situation is often found in high and low mountain systems in geographic proximity (e.g., Northeastern Alps and Bayerischer Wald) or in neighbouring low mountains, e.g., in Germany. These areas thus harbour young relic populations that may result from large zonal distributions of such taxa in the tundra belt during the last ice age.

The most basal and thus oldest splits in arctic-alpine and alpine disjunct species are commonly found in the Southernmost European high mountain systems with typical alpine zonation. The Cordillera Cantábrica in this respect represents one important mountain area for the survival of old genetic entities, but also the high mountain systems of the Balkan Peninsula and to some degree the Pyrenees show importance as areas for the survival of old lineages, which can be interpreted as relics of cold stages of a more distant past.

References

Abbott RJ, Brochmann C (2003) History and evolution of the arctic flora: in the footsteps of Eric Hultén. Mol Ecol 12:299–313

Albach DC, Schönswetter P, Tribsch A (2006) Comparative phylogeography of closely related species of the *Veronica alpina* complex in Europe and North America. Mol Ecol 15:3269–3286

Avise JC (2000) Phylogeography. The history and formation of species. Harvard University Press, Cambridge

Bilton DT (1994) Phylogeography and recent historical biogeography of *Hydroporus glabriusculus* Aubé (Coleoptera: Dytiscidae) in the British Isles and Scandinavia. Biol J Linn Soc 51:293–307

Brändle M, Heuser R, Marten A, Brandl R (2007) Population structure of the freshwater flatworm *Crenobia alpina* (Dana): old lineages and low gene flow. J Biogeogr 34:1183–1192

Cupedo F (2007) Geographical variation and Pleistocene history of the *Erebia pandrose – sthennyo* complex (Nymphalidae; Satyrinae). Nota lepid 30:329–353

de Lattin G (1967) Grundriss der Zoogeographie. Fischer, Stuttgart

Dixon C, Schönswetter P, Schneeweiss GM (2007) Traces of ancient range shifts in a mountain plant group (*Androsace halleri* complex, Primulaceae). Mol Ecol 16:3890–3901

Ehrich D, Gaudeul M, Assefa A, Koch M, Mummenhoff K, Nemomissa S, IntraBioDiv Consortium, Brochmann C (2007) Genetic consequences of pleistocene range shifts: contrast between the Arctic, the Alps and the East African mountains. Mol Ecol 6:2542–2559

Excoffier L, Laval G, Schneider S (2005) Arlequin ver. 3.0: An integrated software package for population genetics data analysis. Evol Bioinform Online 1:47–50

Excoffier LP, Smouse E, Quattro JM (1992) Analysis of molecular variance inferred from metric distance among DNA haplotypes: application to human mitochondrial DNA restriction data. Genetics 131:479–491

Hewitt GM (1996) Some genetic consequences of ice ages, and their role in divergence and speciation. Biol J Linn Soc 58:247–276

Hewitt GM (1999) Post-glacial re-colonization of European biota. Biol J Linn Soc 68:87–112

Hewitt GM (2000) The genetic legacy of the Quaternary ice ages. Nature 405:907–913

Hewitt GM (2001) Speciation, hybrid zones and phylogeography – or seeing genes in space and time. Mol Ecol 10:537–549

Hewitt GM (2004a) Genetic consequences of climatic oscillation in the Quaternary. Phil Trans R Soc Lond B Biol Sci 359:183–195

Hewitt GM (2004b) The structure of biodiversity – insights from molecular phylogeography. Front Zool 1:4

Hey J, Nielsen R (2007) Integration within the Felsenstein equation for improved Markov chain Monte Carlo methods in population genetics. Proc Natl Acad Sci USA 104:2785–2790

Holdhaus K (1954) Die Spuren der Eiszeit in der Tierwelt Europas. Abh zool-bot Ges Wien 18:1–493

Holdhaus K, Lindroth CH (1939) Die europäischen Koleopteren mit boreoalpiner Verbreitung. Ann Nat Mus Wien 50:123–293

Kropf M, Kadereit JW, Comes HP (2003) Differential cycles of range contraction and expansion in European high mountain plants during the Late Quaternary: insights from *Pritzelago alpina* (L.) O. Kuntze (Brassicaceae). Mol Ecol 12:931–949

Magri D, Vendramin GG, Comps B, Dupanloup I, Geburek T, Gömöry D, Latałowa M, Litt T, Paule L, Roure JM, Tantau I, van der Knaap WO, Petit RJ, de Beaulieu J-L (2006) A new

scenario for the Quaternary history of European beech populations: palaeobotanical evidence and genetic consequences. New Phytol 171:199–221
Malicky H (1983) Chorological patterns and biome types of European Trichoptera and other freshwater insects. Arch Hydrobiol 96:223–244
Malicky H (2000) Arealdynamik und Biomgrundtypen am Beispiel der Köcherfliegen (Trichoptera). Entomol Basil 22:235–259
Marusik YM, Hippa H, Koponen S (1996) Spiders (Araneae) from the Altai area, Southern Siberia. Acta Zool Fenn 201:11–45
Mráz P, Gaudeul M, Rioux D, Gielly L, Choler P, Taberlet P (2007) Genetic structure of *Hypochaeris uniflora* (*Asteraceae*) suggests vicariance in the Carpathians and rapid post-glacial colonization of the Alps from an Eastern Alpine refugium. J Biogeogr 34:2100–2114
Müller P (1980) Biogeographie. UTB, Eugen Ulmer, Stuttgart
Muster C, Berendonk TU (2006) Divergence and diversity: lessons from an arctic-alpine distribution (*Pardosa saltuaria* group, Lycosidae). Mol Ecol 15:2921–2933
Muster C, Maddison WP, Uhlmann S, Berendonk TU, Vogler AP (2009) Arctic-Alpine distributions-metapopulations on a continental scale? Am Nat 173:313–326
Ozenda P (1988) Die Vegetation der Alpen im europäischen Gebirgsraum. Fischer, Stuttgart
Pauls SU, Lumbsch HT, Haase P (2006) Phylogeography of the montane cadddisfly *Drusus discolor*: evidence for multiple refugia and periglacial survival. Mol Ecol 15:2153–2169
Ronikier M, Cieślak E, Korbecka G (2008a) High genetic differentiation in the alpine plant *Campanula alpina* Jacq. (Campanulaceae): evidence for glacial survival in several Carpathian regions and long-term isolation between the Carpathians and the Eastern Alps. Mol Ecol 17:1763–1775
Ronikier M, Costa A, Fuertes Aguilar J, Nieto Feliner G, Küpfer P, Mirek Z (2008b) Phylogeography of *Pulsatilla vernalis* (L.) Mill. (Ranunculaceae): chloroplast DNA reveals two evolutionary lineages across central Europe and Scandinavia. J Biogeogr 35:1650–1664
Schmitt T (2007) Molecular biogeography of Europe: pleistocene cycles and postglacial trends. Front Zool 4:11
Schmitt T, Haubrich K (2008) The genetic structure of the mountain forest butterfly *Erebia euryale* unravels the late Pleistocene and Postglacial history of the mountain forest biome in Europe. Mol Ecol 17:2194–2207
Schmitt T, Hewitt GM (2004) Molecular biogeography of the arctic-alpine disjunct burnet moth species *Zygaena exulans* (Zygaenidae, Lepidoptera) in the Pyrenees and Alps. J Biogeogr 31:885–893
Schmitt T, Hewitt GM, Müller P (2006) Disjunct distributions during glacial and interglacial periods in mountain butterflies: *Erebia epiphron* as an example. J Evol Biol 19:108–113
Schönswetter P, Tribsch A (2005) Vicariance and dispersal in the alpine perennial, *Bupleurum stellatum* L. (Apiaceae). Taxon 54:725–732
Schönswetter P, Paun O, Tribsch A, Niklfeld H (2003) Out of the Alps: colonisation of the Arctic by East Alpine populations of *Ranunculus glacialis* (Ranunculaceae). Mol Ecol 12:3371–3381
Schönswetter P, Popp M, Brochmann C (2006a) Central Asian origin of and strong genetic differentiation among populations of the rare and disjunct *Carex atrofusca* (Cyperaceae) in the Alps. J Biogeogr 33:948–956
Schönswetter P, Popp M, Brochmann C (2006b) Rare arctic-alpine plants of the European Alps have different immigration histories: the snowbed species *Minuartia biflora* and *Ranunculus pygmaeus*. Mol Ecol 15:709–720
Schönswetter P, Stehlik I, Holderegger R, Tribsch A (2005) Molecular evidence for glacial refugia of mountain plants in the European Alps. Mol Ecol 14:3547–3555
Schönswetter P, Suda J, Popp M, Weiss-Schneeweiss H, Brochmann C (2007) Circumpolar phylogeography of *Juncus biglumis* (Juncaceae) inferred from AFLP fingerprints, cpDNA sequences, nuclear DNA content and chromosome numbers. Mol Phyl Evol 42:92–103
Schönswetter P, Tribsch A, Niklfeld H (2004a) Amplified Fragment Length Polymorphism (AFLP) reveals no genetic divergence of the Eastern Alpine endemic *Oxytropis campestris* subsp. *tiroliensis* (Fabaceae) from widespread subsp. *campestris*. Plant Syst Evol 244:245–255
Schönswetter P, Tribsch A, Niklfeld H (2004b) Amplified fragment length polymorphism (AFLP) suggests old *and* recent immigration into the Alps by the arctic-alpine annual *Comastoma tenellum* (Gentianaceae). J Biogeogr 31:1673–1681

Schönswetter P, Tribsch A, Barfuss M, Niklfeld H (2002) Several Pleistocene refugia detected in the high alpine plant *Phyteuma globulariifolium* in the European Alps. Mol Ecol 11:2637–2647

Schönswetter P, Elven R, Brochmann C (2008) Trans-Atlantic dispersal and large-scale lack of genetic structure in the circumpolar, arctic-alpine sedge *Carex bigelowii* s. lat. (Cyperaceae). Am J Bot 95:1006–1014

Skrede I, Eidesen PB, Portela RP, Brochmann C (2006) Refugia, differentiation and postglacial migration in arctic-alpine Eurasia, exemplified by the mountain avens (*Dryas octopetala* L.). Mol Ecol 15:827–1840

Sonderegger P (2005) Die Erebien der Schweiz. Selbstverlag, Biel/Bienne

Steward JR, Lister AM (2001) Cryptic Northern refugia and the origins of the modern biota. Trends Ecol Evol 16:608–613

Suda J, Weiss-Schneeweiss H, Tribsch A, Schneeweiss G, TrávníČek P, Schönswetter P (2007) Complex distribution patterns of di-, tetra- and hexaploid cytotypes in the European high mountain plant *Senecio carniolicus* Willd. (Asteraceae). Am J Bot 94:1391–1401

Taberlet P, Fumagalli L, Wust-Saucy A-G, Cosson J-F (1998) Comparative phylogeography and postglacial colonization routes in Europe. Mol Ecol 7:453–464

Varga ZS (1971) Die Erebien (Lep.: Satyridae) der Balkanhalbinsel und der Karpaten III. Die Verbreitung und subspezifische Gliederung von *Erebia pandrose* (Bkh., 1788) nebst Beschreibung einer neuen Unterart: *E. pandrose ambicolorata* ssp. nova. Acta Biol Debrecina 9:227–235

Varga ZS, Schmitt T (2008) Types of oreal and oreotundral disjunctions in the Western Palearctic. Biol J Linn Soc 93:415–430

Vierhapper F (1918) *Juncus biglumis* L. in den Alpen. Österr Bot Zeitschr 67:49–51

Williams D, Dunkerley D, DeDecker P, Kershaw P, Chappell M (1998) Quaternary environments. Arnold, London

Wunderlich J (1984) Seltene und bisher unbekannte Wolfspinnen aus Mitteleuropa und revision der *Pardosa saltuaria* – Gruppe (Arachnida: Araneae: Lycosidae). Verh nat-wiss Ver Hamburg (NF) 27:417–442

Linking Genetics and Ecology: Reconstructing the History of Relict Populations of an Endangered Semi-Aquatic Beetle

Andrea Matern, Claudia Drees, Alfried P. Vogler, and Thorsten Assmann

Abstract Population ecology and genetic studies are complementary approaches to address central questions of conservation biology and can provide information for the protection of biodiversity and the improvement of conservation measures which may otherwise be unattainable. This contribution highlights the benefit of combining diverse approaches for obtaining knowledge on a relict species and for implementing suitable conservation measures. We use the example of *Carabus variolosus*, a flightless European beetle that is listed as a Species of Community Interest in the EU Habitats and Species Directive. Ongoing investigations established the species' habitat choice, population ecology and population genetics, and yield first results from mitochondrial DNA analysis (COI). In small isolated habitat patches we found small sized populations that are genetically strongly differentiated even if adjacent. Evidence for a number of glacial refugia was found. However, all *C. v. nodulosus* populations studied North of the Alps share the same haplotype. We will discuss our findings in the light of the history of the species, of its taxonomic classification and of their implications for conservation.

1 Introduction

The flightless ground beetle *Carabus variolosus* Fabricius, 1787 (Coleoptera: Carabidae) is a rare and threatened habitat specialist of headwater areas and swamps in deciduous woodlands. Both larvae and adults live close to the water edge where

A. Matern (✉), C. Drees, and T. Assmann
Institute of Ecology and Environmental Chemistry, Leuphana University Lüneburg, Scharnhorststraße 1, D-21335 Lüneburg, Germany
e-mail: matern@uni-lueneburg.de

A.P. Vogler
Department of Entomology, The Natural History Museum, Cromwell Road, London, SW7 5BD, UK and Division of Biology, Imperial College London, Silwood Park Campus, Ascot, Berkshire, SL5 7PY, UK

J.C. Habel and T. Assmann (eds.), *Relict Species: Phylogeography and Conservation Biology*, DOI 10.1007/978-3-540-92160-8_14,

they forage (Sturani 1963). Populations persist in localised areas under pristine habitat conditions of which *C. variolosus* is considered to be an indicator (Turin et al. 2003). Remaining populations are spread in a disjunct range throughout Central and South Eastern Europe. While already confined to a restricted range, the distribution of *C. variolosus* is further contracting presumably due to habitat destruction and pollution (Pavicevic and Mesaros 1997; Turin et al. 2003; Matern and Assmann 2004), resulting in its status as a "relict species" (sensu Fryxell 1962).

Taxonomically, *C. variolosus* has been subdivided into two allopatric subspecies, which can be distinguished reliably only by the different tip shape of the male genitalia (Breuning 1926). The Western *C. v. nodulosus* Creutzer, 1799 extends to Germany and France and along the Dinaric mountains Southwards to Macedonia (Breuning 1926; Turin et al. 2003). The Eastern *C. variolosus* s. str. ranges from the Czech Republic and South Poland along the Carpathian mountains to the Western edge of the Balkan mountains in Bulgaria (Breuning 1926; Turin et al. 2003). Over the past 20 years some authors considered both taxa as sister species (e.g. Casale et al. 1982; Deuve 1994), triggering a debate about their taxonomic rank. In 2004, *C. variolosus* was listed as a "species of Community interest" under the EU Habitat and Species Directive, which requires the European member states, to secure its long-term survival. This may be achieved by strict measures for protection, including the designation of reserves, the application of suitable management plans, and the establishment and maintenance of a "favorable conservation status" for the species (The Council of the European Communities 2004). However, expanding basic knowledge about this rarely studied beetle is necessary to assess and monitor its conservation status and devise effective management measures.

Population ecology and population genetics are complementary approaches to address questions about species conservation management. Combined, they offer valuable insights into viable population size and structure, population dynamics and isolation, dispersal and colonization ability, while addressing different spatial and temporal scales (e.g. Ranius 2006; Schmeller and Merilä 2007; Schwartz et al. 2007). This is of great importance for the protection of biodiversity and the improvement of conservation measures, providing information on, for example, the habitat requirements of species (e.g. Gröning et al. 2007), the levels of genetic variability in populations (e.g. Böhme et al. 2007), and the recognition of distinct gene pools and evolutionary lineages (e.g. Palsbøll et al. 2007).

Here, we review our ongoing investigations on the population ecology and population genetics of *C. variolosus*, including recent results from mtDNA data. We discuss the findings with a specific focus on the (1) postglacial history of the Western subspecies *C. v. nodulosus* and (2) the implications of the results for conservation issues showing the benefit of combining field-based and genetic approaches.

2 Habitat Choice and Demography

Demography and habitat choice of *C. variolosus* were investigated in a 2 year mark-recapture study at two representative relict populations of *C. v. nodulosus* in North-Western Germany. Parameters determining the habitat preferences of adult beetles were estimated from habitat suitability models (e.g. Matern et al. 2007). These studies have highlighted and further refined which specific requirements determine habitat associations (Matern et al. 2007). Key habitat variables governing the occurrence of the adult beetles were found to be: short distance to water, high soil moisture, open woodland vegetation cover, a near-neutral pH of the soil and the lack of ground cover. Accordingly, the extent of suitable habitat was rather restricted (Fig. 1). Continuous observations of individual beetles confirmed their confinement to regions of wet soil and close to, or covered by, water (Drees et al. 2008).

Mark-recapture data showed *C. variolosus* to be a spring breeder in the sense of Larsson (1939), i.e. reproduction took place after the emergence of overwintering adults between April and June. Estimates of population sizes at both sites were very low (Matern et al. 2008). During the spring activity period in 2004 the total number of different adults captured amounted to 63 (Site 1) and 161 (Site 2) individuals, corresponding to a maximum of 150 (Site 1) and 215 (Site 2) individuals when corrected using the Jolly–Seber estimate (Jolly 1965; Seber 1965). Population densities, however, estimated at 1.75 (Site 1) and 0.85 (Site 2) individuals per 10m², were normal to fairly high in comparison to more common species of *Carabus*

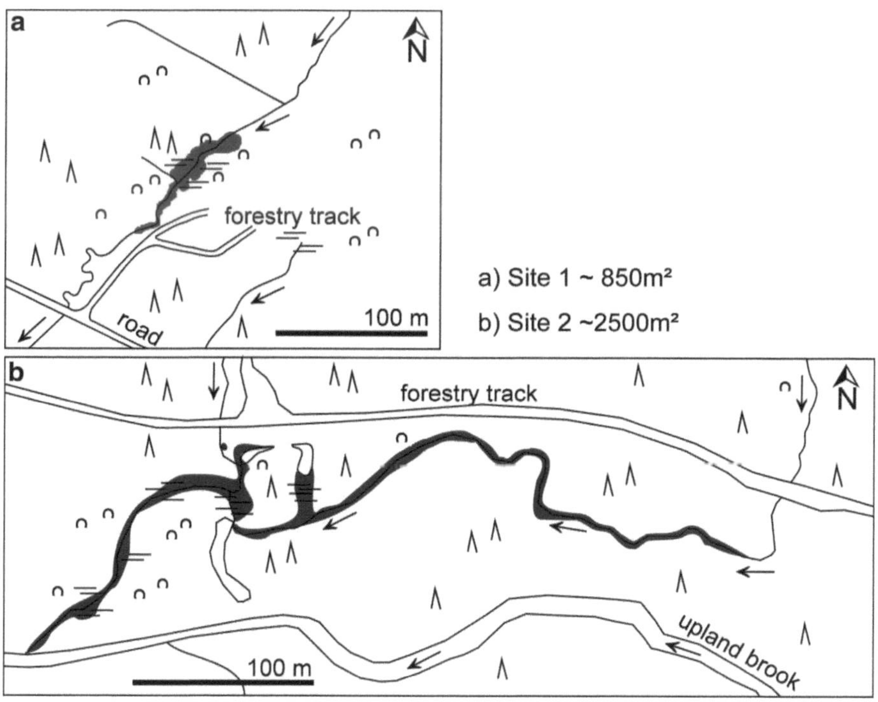

Fig. 1 Habitat preference of two *C. variolosus nodulosus* populations in the Forest of Arnsberg, Northwestern Germany. Occupied habitat is represented in grey. Figure 2 shows the location of these sites. (**a**) Population 1, (**b**) Population 2

(Matern et al. 2008). Hence, the total population sizes appear to be limited by the extent of suitable habitat, rather than low density. This illustrates the imminent threat to population survival from mechanisms operating in small habitat patches (extinction vortex, Gilpin and Soulé 1986). However, adults were found to be long-lived and almost 30% of the reproductively active individuals were at least in their second breeding period (Matern et al. 2008). This reduces risks of reproductive failure from climatic variation or other stochastic factors in any given year.

3 Genetic Population Structure According to Nuclear Genes

Genetic analysis using cellulose acetate electrophoresis was performed on 12 populations of *C. v. nodulosus* throughout most of its distributional range, and including those used for mark-recapture studies above. A total of 308 specimens were included in the analysis for a mean sample size of 25 individuals and 16 allozyme loci (Matern et al. 2009; Fig. 2). Despite small population sizes (inferred from small areas of suitable habitats at most sampling sites), we found no evidence for inbreeding, as populations did not deviate significantly from Hardy–Weinberg equilibrium (GENEPOP 3.4, Raymond and Rousset 1995). However, allele numbers and the amount of polymorphism per population were low, with only eight of the investigated loci found to be polymorphic and a maximum of seven polymorphic loci per population, indicating genetic impoverishment.

Population differentiation and correlation between genetic and geographical distances between samples were quantified with the Arlequin package (Excoffier et al. 2005). Despite the low number of private alleles, populations were highly differentiated (overall $F_{ST}>0.45$) owing to locally differentiated allele frequencies among populations, as exemplified by Fig. 2. This finding supports the possibility that populations were affected by genetic drift due to small population sizes. Neither clinal variation nor a directional loss of alleles that could hint at any postglacial recolonization scenario according to the allele elimination hypothesis (Reinig 1938; Hewitt 1999) could be detected. Generally, geographic distance only had a marginal influence (< 10%) on the isolation of populations (Mantel test). A neighbor-joining tree based on genetic distances among all *C. v. nodulosus* populations confirms this result, showing poor agreement between genetic and geographical distance across the entire range (not shown). Even neighboring populations at distances of only 2–3 km in the same forest and drainage system were found to be virtually independent of each other as shown, for example, by significant pairwise F_{ST} values of 0.27 and 0.14 between populations 1 and 2 and populations 5 and 6, respectively.

Strongly separated were two populations from Southern Slovenia (sites 11 and 12) which showed deviating allele frequencies and some private alleles compared to the other investigated populations. This is confirmed by Bayesian structure analysis (STRUCTURE, Pritchard et al. 2000) which generally assigns the individuals of these two populations to one distinct genetic cluster if no a priori assumptions of geographical origin are made (Matern et al. 2009). This pattern

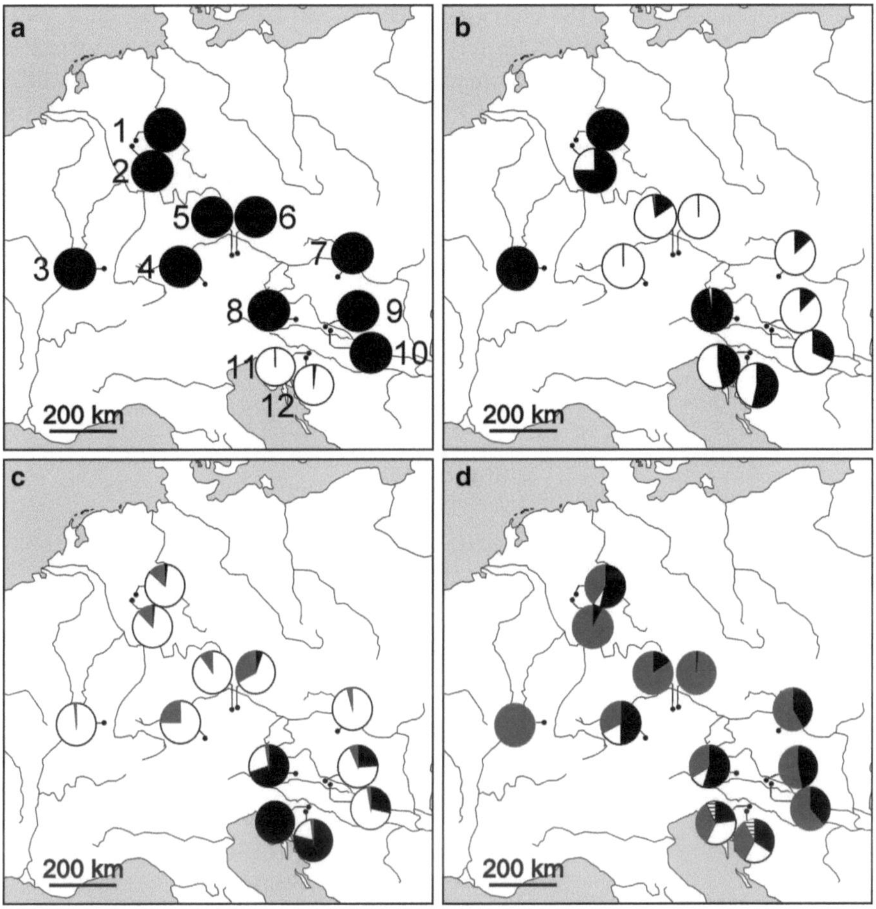

Fig. 2 Allele frequencies of (**a**) alcohol dehydrogenase, (**b**) glucose-6-phosphate isomerase, (**c**) aspartate aminotransferase, and (**d**) mannose-6-phosphate isomerase in the investigated populations of *C. variolosus nodulosus*. Different colours indicate various alleles at the respective locus

may be caused by genetic drift and an efficient present isolation between the Southern Slovenian samples and the remaining Central European populations that might have existed since the survival of these populations in different glacial refugia (cf. Chaps. 4 and 5.1, this contribution).

4 Mitochondrial DNA Data

Mitochondrial DNA sequences largely confirmed the results from allozyme studies, also indicating minimal within-population variation and the separation of the South Slovenian samples. We amplified a fragment of the cytochrome oxidase I (COI) gene using primers C1-J-2183 (Jerry) and TL2-N-3014 (Pat) (Simon et al. 1994) and successfully sequenced 608 bp for 114 specimens of *C. v. nodulosus* and six individuals from three Romanian populations of *C. v. variolosus*. Only nine different haplotypes were identified (Fig. 3). There was hardly any haplotype variation within populations of *C. v. nodulosus*. With the exception of two individuals deviating by one and three nucleotide changes, respectively, all remaining 92 individuals

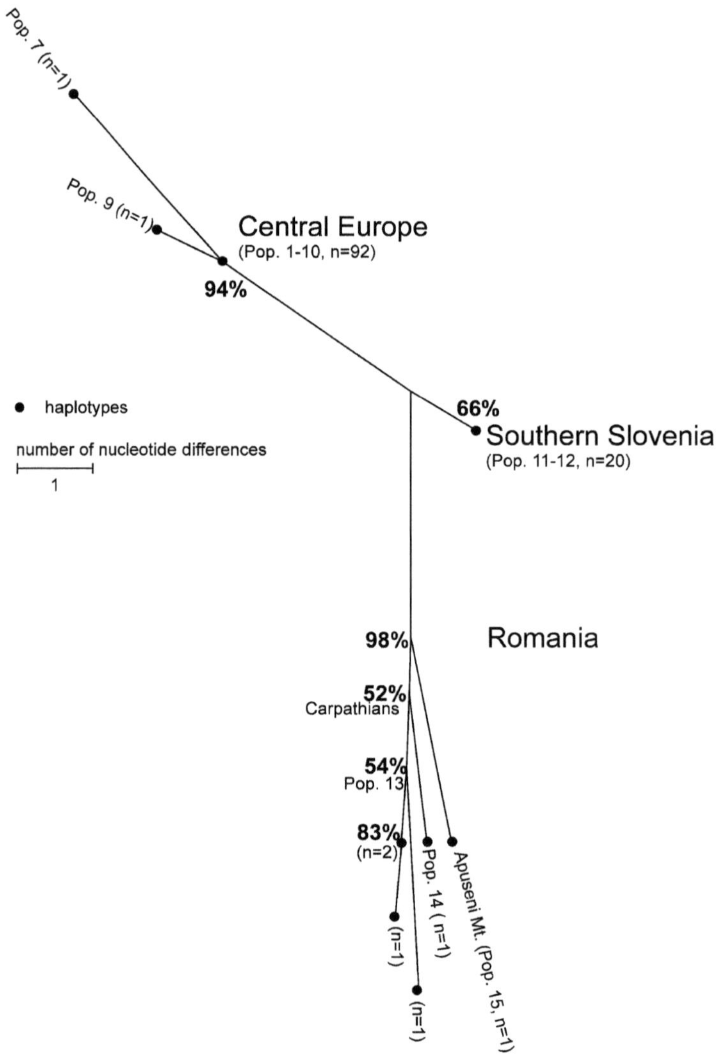

Fig. 3 Neighbor-joining tree (Saitou and Nei 1987) based on the number of nucleotide changes between 120 sequences of COI fragment. Numbers at nodes refer to support from 1,000 bootstrap replicates. Central European and Southern Slovenian specimens: *C. v. nodulosus*, Romanian specimens: *C. v. variolosus*. Phylogenetic analyses were conducted in MEGA4 (Tamura et al. 2007)

in the Central European populations (1–10) shared the same haplotype. The 20 individuals from Southern Slovenian (populations 11 and 12) shared a single haplotype that differed from the former by four nucleotide changes. All of these were separated from *C. v. variolosus* haplotypes by at least four nucleotides (Fig. 3). This finding corroborates that the Southern Slovenian populations of *C. v. nodulosus* survived in a different glacial refugium than the remaining European samples (cf. Chap. 5.1, this contribution).

5 Synthesis

5.1 *C. variolosus: A Relict Species*

In contrast to other weak dispersers that survived the ice ages in Southern refugia and remained largely confined to these areas ever since, *C. variolosus* expanded its range subsequent to postglacial climate changes despite being flightless. As a stenotopic woodland species specialised on water margins, *C. variolosus* must have suffered considerably from the climate changes of the glacial periods. Given the species' current habitat preference (e.g. Sturani 1962), its confinement to lowland or lower montane altitudes (e.g. Breuning 1926; Barloy et al. 2004) and its strict stenotopy established by the habitat models (Matern et al. 2007), it is likely that populations survived the ice ages in sheltered, wet deciduous woodlands. These habitats may have persisted during the Last Glacial Maximum in Iberia, Italy and the Balkans (Zagwijn 1992), and in the Carpathians (Kotlík et al. 2006). Deffontaine et al (2005) furthermore suggest the existence of mixed coniferous-deciduous forests in the river systems present near the Alps and in the Hungarian plain and Slovakia. Paleobotanical and genetic evidence also corroborates the survival of beech populations as far North as Southern France, the Eastern Alps, Slovenia, Istria and possibly Southern Moravia-Bohemia (Magri et al. 2006), but so far no refuges of forest-forming tree species are known from further North. Recent faunistic records of *C. variolosus*, however, exist from numerous regions North of the Alps such as the presence of *C. v. variolosus* at the foothills of the Sudetic Mountains, Upper Silesia and the Lublin Upland in Southern Poland (Pawlaczyk et al. 2004) and as far as the Hamburg area, where *C. v. nodulosus* occurred until about 100 years ago (Breuning 1926).

The homogeneous mtDNA in the Central European populations of *C. v. nodulosus* is consistent with a recent colonization of the region North of the Alps. Under the standard molecular clock for insect mtDNA of approximately 2.3% divergence per million years (Brower 1994), all Central European populations of *C. v. nodulosus* and the Southern Slovenian lineage separated no more than 0.3 Myrs ago (0.7% pairwise average sequence divergence). Likewise, the Central European populations and haplotypes of the subspecies *C. v. variolosus* separated approximately 0.5 Myrs ago (1.2% divergences). The existence of these lineages which likely predate the most recent glacial cycles may suggest the existence of more than one glacial refuge or source area within or near to the investigated area.

Post glacial colonization processes are not reflected in allozyme variability. We did not observe a directional loss of alleles according to the allele elimination hypothesis (Reinig 1938; Hewitt 1999) nor any regionally specific sets of allozymes and their combination in intermediate populations, which could suggest potential contact zones. The largely random geographic structure of allozyme variants, the lack of private alleles, and the lack of correlation in allele frequencies with geographic distance (Matern et al. 2009) suggests that the variation in Central European populations was more similar at first, but that populations lost variability subsequently due to stochastic processes. Hence, the case of *C. variolosus* could represent an example where postglacial colonization, in combination with subsequent geographic isolation, produced genetic differentiation of a similar magnitude as that resulting from survival in different glacial refuges. For example, the postglacial *C. variolosus* populations from Germany and the Vosges display an F_{ST} of 0.45 (7 polymorphic enzyme loci). This is similar to values obtained for 29 populations of *Carabus punctatoauratus* in the Pyrenees that survived in different glacial refugia (mean F_{ST} of 0.23, enzyme loci, Assmann and Weber 1997) or to values found in *C. solieri*, another flightless silvicolous ground beetle, that remains largely confined to its glacial refuges in the Southwestern Alps close to the border between France and Italy (F_{ST} of 0.34, microsatellite markers, Garnier et al. 2004). Although F_{ST} values obtained with different makers are not directly comparable, the latter example also supports the depth of population separation within the Northwestern lineage of *C. variolosus*.

It is likely that the present interpopulation differentiation of *C. v. nodulosus* is only partly due to low dispersal power of these flightless beetles, but was exacerbated by small population sizes. Ecological knowledge helps to understand better what affects the recent population history of *C. v. nodulosus*: its high habitat specificity not only increases the barrier function of the landscape structure surrounding suitable habitat patches but may also result in very small populations, as demonstrated for the above two populations in Northwestern Germany (Matern et al. 2008). Descriptions of patch sizes in additional populations of *C. v. nodulosus* suggest that these may exist or have existed under similar spatial conditions (e.g. Gersdorf and Kuntze 1957; Perraudin 1960; Morati and Huet 1995). While the investigated populations do not appear to be inbred, they are genetically impoverished in comparison with other carabid beetles (Matern et al. 2009). This renders them highly susceptible to genetic stochasticity that is associated with extinction vortices, in particular in the face of expected climatic changes.

While *C. v. nodulosus* was distributed much more widely until about a century ago, it has since suffered from a marked decline and nowadays only persists in very localised relict populations (Breuning 1926; Turin et al. 2003; Matern and Assmann 2004). Intensive human land use with its massive impact on the natural hydrological budget is thought to be responsible for the decline (e.g. Perraudin 1960; Weber and Weber 1966; Pavicevic and Mesaros 1997; Turin et al. 2003). The resulting destruction and degeneration of suitable habitat has aggravated the geographical separation of headwater areas and pristine forest swamps that already affected

C. variolosus populations under natural conditions but is now increasing the effect of habitat fragmentation. Currently, *C. v. nodulosus* is considered endangered throughout its entire range (Turin et al. 2003), and its highly restricted distribution is further contracting. Therefore, it is unquestionably a relict in the sense of Fryxell (1962) meeting also further specified criteria such as a reduced variability and narrow ecological tolerance, which potentially indicate a reduction of adaptability, and may ultimately be a candidate for extinction. While the beginning of the decline of *C. v. nodulosus* cannot be clearly identified, the relict status of *C. v. nodulosus* in Central Europe, however, is relatively recent, as the beetle successfully colonized this area post glacially. While the situation of *C. v. variolosus* apparently is less alarming (Turin et al. 2003; Pawlaczyk et al. 2004), this subspecies is also characterized by a narrow ecological tolerance and a restricted range that is further declining in some regions (Turin et al. 2003).

5.2 Implications for Conservation and Future Work

The example of a species that survived past climate change in refuges and benefited from reverse climatic changes by range expansion despite low dispersal power may give hope for conservation. However, *C. v. nodulosus*, like many other postglacial populations, shows reduced genetic variability (Hewitt 2001), which implies certain risks for long-term survival. Hence, the maintenance of genetic diversity within species is one important goal of biodiversity conservation.

As the Southern Slovenian populations are a distinct genetic unit holding genetic diversity not found in Central European populations, they should consequently be granted special protection alongside the Central European samples. To conserve the remaining genetic diversity, as many of the relict populations as possible should be targeted, since more than 45% of the total allozyme variance of *C. v. nodulosus* is found among populations (Matern et al. 2009). Where possible, a cautious enlargement of the habitat patches may also be attempted.

If further populations within or close to the core of the subspecies range could be found that display high genetic diversity (owing to their location in glacial refuges, for example), these of course would have an especially high conservation priority. Locations of all current relict populations of *C. variolosus* (and of other flightless, sensitive species) may also be an indicator of unique environmental conditions not affected by disturbance and demonstrate the general continuity (stability) of habitats. These occurrences can therefore be used to identify valuable habitats for the conservation of entire species communities.

This study provides an example of increased knowledge and sensible conservation strategies that were derived from a combination of diverse methodological approaches offering different perspectives. The contribution of ecological knowledge is essential where knowledge on a species is scant. Details on the habitat preference and habitat use of *C. variolosus* are crucial for an assessment of factors influencing the incidence and distribution pattern of the species. In the case of

C. variolosus this information especially emphasises the urgency to assess population sizes and the necessity to enlarge populations. The analysis of COI provides valuable information on long-term population history that was not available from allozyme analysis. Moreover, it strongly supports the allozyme results which already suggested that the Southern Slovenian populations are a differentiated entity, with relevance for conservation (Moritz 1994). The identification of differentiated population groups is considered crucial for the conservation of the genetic diversity of species (e.g. Ryder 1986; Moritz 1995) and provides important data in rescue efforts of species by means of translocations (e.g. Avise 1996). While mtDNA sequence data contribute additional information on population differentiation and genetic diversity, they should not be used exclusively in the discussion of conservation priorities. It is the population genetic approach with nuclear markers that offers an advanced insight into the isolation of populations and the level and overall distribution of genetic diversity within and across populations. The allozyme results underline the need for recognition of distinct Management Units even at adjacent sites within the same drainage system.

In the context of the EU Habitats Directive and Natura 2000 Network it is now important to extend analyses to the Eastern subspecies *C. v. variolosus*. Further allozyme and DNA analyses to unravel the phylogeny and discover subspecific population structure will show whether the conservation measures established for *C. v. nodulosus* are also valid for *C. v. variolosus*. One important clue, however, is already given by the current COI analysis: sequence divergence between the two taxa is surprisingly low. As the amount of sequence divergence is extremely variable in different lineages, it is impossible to apply a general threshold value to justify taxonomic distinction at the species level (Vogler et al. 1993). Some molecular taxonomists have (controversially) used a 3 or 2% COI divergence for species delimitation across a wide range of organisms including insects (cf. Duckett and Kjer 2003; Hebert et al. 2003, 2004; Moritz and Cicero 2004). Comparisons of mtDNA studies find sequence divergence of less than 2% primarily on an intraspecific level (Vogler and DeSalle 1994; Avise 2000; Woodcock et al. 2007). This is also the case among populations of another ground beetle species, *Nebria rufescens*, which are not formally distinguished at the species or subspecies level and display a mean mtDNA-sequence divergence (ND1) of 1.3% (Schmitt et al. this book). Although the two taxa of *C. variolosus* are "diagnosable" under a phylogenetic species concept (Vogler and DeSalle 1994), their divergence is low and only slightly greater than the striking separation of the Central European and Southern Slovenian populations. This finding may support the older classification as subspecies (cf. Breuning 1926), settle the past discussion on the potential species rank of *C. v. nodulosus* and counteract taxonomic inflation (cf. Isaac et al. 2004).

Future genetic assessment of *C. variolosus* will be necessary to study the success of conservation recommendations such as population increase or population connectivity and thus augment ecological monitoring. Moreover, it may also enhance our knowledge on the relationship between population sizes and the maintenance of genetic diversity and probably their influence on the persistence of these populations.

Acknowledgements We are grateful to Božidar Drovenik, Daniela Eisenacher, Jens Günther, Michael Franzen, Istvan Mathé, Wolfgang Paill, Laszlo Rakosy, Annette Ehrhardt-Renken, and Marco Zimmermann for their assistance in sampling and valuable exchange of information and to Dietrich Mossakowski for providing us with an additional specimen from Romania. For technical assistance in the lab we thank Marianne Peters, Beatrice Dames, and Gunnar Siedenschnur. We cordially thank Anna Papadopoulou at the Natural History Museum London, who helped with mtDNA sequencing and analysis. We also thank an anonymous reviewer for valuable comments on the manuscript. A. Matern was funded by the German Federal Environmental Foundation (DBU).

References

Assmann T, Weber F (1997) On the allozyme differentiation of *Carabus punctatoauratus* Germar (Coleoptera, Carabidae). J Zool Syst Evol Res 35:33–43

Avise JC (1996) Toward a regional conservation genetics perspective: phylogeography of faunas in the Southeastern United States. In: Avise JC, Hamrick JL (eds) Conservation genetics: case histories from nature. Chapman & Hall, New York, pp 431–470

Avise JC (2000) Phylogeography: the history and formation of species. Harvard University press, Cambridge, MA

Barloy J, Lie P, Prunar F (2004) Inventaire et répartition géographique des espèces du genre *Carabus* (Coleoptera, Carabidae) au Banat roumain. L'Entomologiste 60:63–89

Böhme MU, Schneeweiß N, Fritz U, Schlegel M, Berendonk TU (2007) Small edge populations at risk: genetic diversity of the green lizard (*Lacerta viridis viridis*) in Germany and implications for conservation management. Conserv Genet 8:555–563

Breuning S (1926) Über *Carabus variolosus* Fabr. Koleopterol Rundsch 12:19–25

Brower AVZ (1994) Rapid morphological radiation and convergence among races of the butterfly *Heliconius erato* inferred from patterns of mitochondrial-DNA evolution. Proc Natl Acad Sci USA 91:6491–6495

Casale A, Sturani M, Vigna Taglianti A (1982) Coleoptera, Carabidae. I. Introduzione, Paussinae, Carabinae. Fauna d'Italia 18:1–499

Deffontaine V, Libois R, Kotlík P, Sommer R, Nieberding C, Paradis E, Searle JB, Michaux JR (2005) Beyond the Mediterranean peninsulas: evidence of central European glacial refugia for a temperate forest mammal species, the bank vole (*Clethrionomys glareolus*). Mol Ecol 14:1727–1739

Deuve T (1994) Une classification du genre *Carabus*. Sciences Nat, Venette

Drees C, Matern A, Assmann T (2008) Red light enables direct observation of nocturnal carabid beetles in their habitat. In: Penev L, Erwin TL, Assmann T (eds) Back to the roots and back to the future? Towards a new synthesis between taxonomic, ecological and biogeographical approaches in carabidology. Pensoft, Sofia-Moscow, pp 409–423

Duckett CN, Kjcr KM (2003) Cladistic analysis of the Oedionychines of Southern Brazil (Galerucinae: Alticini) based on two molecular markers. In: Furth DG (ed) Special topics in leaf beetle biology: proceeding of 5th International symposium on the chrysomelidae. Pensoft, Sofia, pp 117–132

Excoffier L, Laval G, Schneider S (2005) Arlequin ver. 3.0: An integrated software package for population genetics data analysis. Evolutionary Bioinformatics Online 1:47–50

Fryxell PA (1962) The "relict species" concept. Acta Biotheoretica 15:105–118

Garnier S, Alibert P, Audiot P, Prieur B, Rasplus J-Y (2004) Isolation by distance and sharp discontinuities in gene frequencies: implications for the phylogeography of an alpine insect species, *Carabus solieri*. Mol Ecol 13:1883–1897

Gersdorf E, Kuntze K (1957) Zur Faunistik der Carabiden Niedersachsens. Ber Naturhist Ges Hannover 103:101–136

Gilpin ME, Soulé ME (1986) Minimum viable populations: processes of species extinction. In: Soulé ME (ed) Conservation biology: the science of scarcity and diversity. Sinauer, Sunderland, MA, pp 19–34

Gröning J, Krause S, Hochkirch A (2007) Habitat preferences of an endangered insect species, Cepero's ground-hopper (*Tetrix ceperoi*). Ecol Res 22:767–773

Hebert PDN, Penton EH, Burns JM, Janzen DH, Hallwachs W (2004) Ten species in one: DNA barcoding reveals cryptic species in the neotropical skipper butterfly *Astraptes fulgerator*. Proc Natl Acad Sci USA 101:14812–14817

Hebert PDN, Ratnasingham S, DeWaard JR (2003) Barcoding animal life: cytochrome c oxidase subunit 1 divergences among closely related species. Proce R Soc Lond B Biol Sci 270:S96–S99

Hewitt GM (1999) Post-glacial re-colonization of European biota. Biol J Linn Soc 68:87–112

Hewitt GM (2001) Speciation, hybrid zones and phylogeography – or seeing genes in space and time. Mol Ecol 10:537–549

Isaac NJB, Mallet J, Mace GM (2004) Taxonomic inflation: its influence on macroecology and conservation. TREE 19:464–469

Jolly MF (1965) Explicit estimates from capture-recapture data both with death and immigration – a stochastic model. Biometrika 52:224–247

Kotlík P, Deffontaine V, Mascheretti S, Zima J, Michaux JR, Searle JB (2006) A Northern glacial refugium for bank voles (*Clethrionomys glareolus*). Proc Natl Acad Sci USA 103:14860–14864

Larsson SG (1939) Entwicklungstypen und Entwicklungszeiten der daenischen Carabiden. Entomol Medd 20:273–562

Magri D, Vendramin GG, Comps B, Dupanloup I, Geburek T, Gömöry D, Latalowa M, Litt T, Paule L, Roure JM, Tantau I, Van der Knaap WO, Petit RJ, De Beaulieu J-L, Jackson ST (2006) A new scenario for the Quaternary history of European beech populations: palaeobotanical evidence and genetic consequences. New Phytol 171:199–221

Matern A, Assmann T (2004) Nationale Verantwortlichkeit und Rote Listen - *Carabus nodulosus* als Fallbeispiel für die Zusammenführung und die damit verbundenen Probleme. Natursch u Biol Vielfalt 8:235–254

Matern A, Drees C, Desender K, Gaublomme E, Paill W, Assmann T (2009) Genetic diversity and population structure of the endangered insect species *Carabus variolosus* in its Western distribution range: implications for conservation. Conserv Genet 10:391–405. doi:10.1007/s10592-008-9606-1

Matern A, Drees C, Kleinwächter M, Assmann T (2007) Habitat modelling for the conservation of the rare ground beetle species *Carabus variolosus* (Coleoptera, Carabidae) in the riparian zones of headwaters. Biol Conserv 136:618–627

Matern A, Drees C, Meyer H, Assmann T (2008) Population ecology of the rare carabid beetle *Carabus variolosus* (Coleoptera: Carabidae) in North-West Germany. J Insect Conserv 12:591–601. doi:10.1007/s10841-007-9096-3

Morati J, Huet M (1995) Présence de *Carabus* (Hygrocarabus) *nodulosus* Creutzer, 1799, dans le Jura français (Coleoptera, Carabidae). Bull Soc Entomol Fr 100:144

Moritz C (1994) Defining "evolutionary significant units" for conservation. TREE 9:373–375

Moritz C (1995) Uses of molecular phylogenies for conservation. Philos Trans R Soc Lond B 349:113–118

Moritz C, Cicero C (2004) DNA barcoding: promise and pitfalls. PLoS Biol 2:e354

Palsbøll PJ, Bérubé M, Allendorf FW (2007) Identification of management units using population genetic data. TREE 22:11–16

Pavicevic D, Mesaros G (1997) Carabini of Jugoslavia and adjacent regions (Coleoptera: Carabidae). Catalogus Faunae Jugoslaviae. Encyclopaedia. Ecolibri-Bionet, Belgrad

Pawlaczyk P, Kepel A, Jaros R, Dzieciolowski R, Wylegala P, Szubert A, Sidlo PO (2004) Natura 2000 Shadow List in Poland - Detailed Analysis of Habitat Directive Implementation, Syntethic Approach to Bird Directive Implementation. Klub Przyrodników, Polish Society for the Protection of Birds, The Polish Society for Nature Protection "Salamandra" & WWF Poland, Warszawa

Perraudin W (1960) Présence de *Hygrocarabus variolosus nodulosus* (Creutzer) en Forêt-Noire. Mitteilung des Badischen Landesvereins für Naturkunde und Naturschutz e.V. 7: 447-450

Pritchard JK, Stephens M, Donnelly P (2000) Inference of population structure using multilocus genotype data. Genetics 155:945–959
Ranius T (2006) Measuring the dispersal of saproxylic insects: a key characteristic for their conservation. Popul Ecol 48:177–188
Raymond M, Rousset F (1995) GENEPOP (version 3.2): population genetics software for exact tests and ecumenicism. J Hered 86:248–249
Reinig WF (1938) Elimination und Selektion - Eine Untersuchung über Merkmalsprogressionen bei Tieren und Pflanzen auf genetisch- und historisch-chorologischer Grundlage. Gustav Fischer, Jena
Ryder OA (1986) Species conservation and systematics: the dilemma of subspecies. TREE 1:9–10
Saitou N, Nei M (1987) The Neighbor-joining method: a new method for reconstructing phylogenetic trees. Mol Biol Evol 4:406–425
Schmeller DS, Merilä J (2007) Demographic and genetic estimates of effective population and breeding size in the amphibian *Rana temporaria*. Conserv Biol 21:142–151
Schwartz MK, Luikart G, Waples RS (2007) Genetic monitoring as a promising tool for conservation and management. TREE 22:25–33
Seber GAF (1965) A note on the multiple-recapture census. Biometrika 52:249–259
Simon C, Frate F, Beckenbach A, Crespi B, Liu H, Flook P (1994) Evolution, weighting and phylogenetic utility of the mitochondrial gene sequences and a compilation of conserved polymerase chain reaction primers. Ann Ent Soc Am 87:651–701
Sturani M (1962) Osservazioni e ricerche biologiche sul genere *Carabus* Linnaeus (Sensu Lato) (Coleoptera Carabidae). Mem Soc Entomol Italiana 41:85–202
Sturani M (1963) Nuove ricerche biologiche e morfologiche sul *Carabus* (Hygrocarabus) *variolosus* Fabricius (Coleoptera Carabidae). Boll Zool Agrar Bachicoltura Serie II:25–34
Tamura K, Dudley J, Nei M, Kumar S (2007) MEGA4: molecular evolutionary genetics analysis (MEGA) software version 4.0. Mol Biol Evol 24:1596–1599
The Council of the European Communities (2004) Council Directive 92/43/EEC of 21 May 1992 on the conservation of natural habitats and of wild fauna and flora – Consolidated Text. Office for Official Publications of the European Communities
Turin H, Penev L, Casale A, Arndt E, Assmann T, Makarov KV, Mossakowski D, Szél G, Weber F (2003) Species accounts. In: Turin H, Penev L, Casale A (eds) The genus *carabus* in Europe – a synthesis. Pensoft Publishers and European Invertebrate Survey, Sofia, pp 151–283
Vogler AP, DeSalle R (1994) Diagnosing units of conservation management. Conserv Biol 8:354–363
Vogler AP, DeSalle R, Assmann T, Knisley CB, Schultz TD (1993) Molecular population genetics of the endangered tiger beetle *Cicindela dorsalis* (Coleoptera: Cicindelidae). Ann Entomol Soc Am 86:142–152
Weber F, Weber I (1966) Wiederentdeckung des Laufkäfers *Carabus variolosus* F. im Arnsberger Wald. Nat Heimat 26:69–70
Woodcock MR, Kippenhan MG, Knisley CB, Foster JA (2007) Molecular genetics of *Cicindela* (Cylindera) *terricola* and elevation of *C. lunalonga* to species level, with comments on its conservation status. Conserv Genet 8:865–877
Zagwijn WH (1992) Migration of vegetation during the Quaternary in Europe. Courier Forschungsinstitut Senckenberg 153:9–20

Peripheral Relict Populations of Widespread Species; Evolutionary Hotspots or Just More of the Same?

Anna Cassel-Lundhagen

Abstract Relicts are species with a former, more widespread distribution range but can also include specific populations in parts of a species distribution (thus called relict populations). Such populations are often found towards the species range limits where conditions may be less optimal than in the more central parts. For peripheral relict populations to be valuable for conservation from a biodiversity point of view, they should be exposed to selection pressures that are different from the more central ones allowing for unique adaptations to arise. I will discuss data from studies of two closely related butterfly species where the aim was to evaluate the uniqueness of marginal relict populations in relation to their more central counterparts. Neutral genetic variation and values of adaptive wing traits among populations of the scarce heath (*Coenonympha hero*) and the pearly heath (*C. arcania*) from three biogeographically different regions were compared representing one central and two peripheral parts of their distributions; one peripheral connected to and one isolated from the main area of their distributions. Individuals from the isolated region were significantly differentiated from the peripheral and central regions in both neutral genetic traits and adaptive wing traits, while the peripheral and central populations showed no signs of differentiation in both trait types. I will discuss the possible mechanisms behind the observed patterns and show that the peripheral, relict populations of also widespread species can contribute with valuable and unique genetic variation. Marginal relict populations constitute ideal study objects of evolution in marginal habitats as they can be used to test how well theory and laboratory experiments reflects more complex natural systems.

1 Introduction

Peripheral populations at the margin of a species' range may be exposed to ecological and evolutionary processes that differ from more central parts of its distribution. If the level of genetic differentiation is high due to exposure to diversifying

A. Cassel-Lundhagen
Department of Ecology, Swedish University of Agricultural Sciences, Ulls väg 16, SE-75007 Uppsala, Sweden
e-mail: anna.lundhagen@ekol.slu.se

J.C. Habel and T. Assmann (eds.), *Relict Species: Phylogeography and Conservation Biology*, DOI 10.1007/978-3-540-92160-8_15,

processes, like random genetic drift and selection, this could eventually lead to speciation (Garcia-Ramos and Kirkpatrick 1997; Mayr 1954, 1963). Thus, such peripheral populations may be unique and important from a conservation point of view (Channell and Lomolino 2000; Lesica and Allendorf 1995).

Relicts are defined as species with a former more widespread distribution range but can also include specific populations in parts of a species distribution (thus called relict populations). Such populations are often found towards the species range limits where conditions may be less optimal than in more central parts (i.e. Gullberg et al. 1998). One could argue that peripheral relict populations of globally widespread species should not be prioritized for conservation as they constitute only a marginal fraction of the species in the global biodiversity. However, both evolutionary history and local selection pressures may differ along a species distribution range and this can lead to unique genotypes and adaptations that are not present elsewhere (Eckert et al. 2008). Especially, peripheral populations of species with limited dispersal capacity are likely to originate from populations that have gone through recurrent bottlenecks during colonization and differentiation can then be rapid and significant if founders are few (Chakraborty and Nei 1977). If the newly established populations lack subsequent gene flow from their sources such effects are expected to remain and accentuate in the region. In addition to the influence of history, genetic drift is likely to be more pronounced towards the range periphery due to a greater influence of stochastic processes. Reduced habitat density as well as climatic constraints will cause size fluctuations of local populations that could cause significant influence of genetic drift with changed allele frequencies as a consequence.

Random effects due to founder events and population fluctuations will lead to changes in the neutral or nearly neutral traits. However, for peripheral relict populations to be valuable for conservation, they should also be exposed to selection pressures that are different from the ones of more central positions (Lesica and Allendorf 1995). If conditions are equal in all parts of a species range or selection pressure is identical, then these populations could differ significantly in neutral markers, while being identical in selected characters. If conditions instead are diverging, selection will favor genotypes that are different from the best adapted ones in the more central regions (Helgadóttir and Snaydon 1986) causing increased global biodiversity.

Diverging selection pressures may be due to a number of factors and can include differences in local or regional climate, species compositions or landscape structures. In the case of landscape structure, for example, selection pressures can act in favor of increased dispersal capacity (Clare and Hughes 2007) as well as against it (Dempster 1991). With increased isolation among patches and a matrix of low quality (lacking for example resting and foraging sites) dispersers may experience increased mortality leading to selection acting against dispersive genotypes (Cody and Overton 1996; den Boer 1990; Heino and Hanski 2001; Schtickzelle et al. 2006). Such altered behavior is likely reflected also in the individuals' phenotype (Berwaerts et al. 2006, 2002; Davis 1984; Dempster 1991; Fric and Konvička 2002; Merckx and Van Dyck 2006; Thomas et al. 1998). For example, short and broad wings are expected to favor slow, agile flight and appear to be more common among forms that fly in restricted spaces and dense vegetation (Betts and Wootton 1988).

However, if gene flow from central parts to the periphery is substantial, diversifying selection pressures may be hidden due to swamping (Garcia-Ramos and Kirkpatrick 1997). This can occur even if the individual dispersal capacity is equal along the whole range due to a larger net inflow of immigrants into a region with smaller population sizes. Such confounding factors will make discrimination between homogenizing gene flow and lack of diversifying selection problematic and a study setup is needed to allow discrimination among these forces. Studies on peripheral populations should, therefore, be performed including those connected to the main area of the species distribution in some regions and isolated from more central parts in others.

2 Examples From Two Heath Butterfly Species

I will discuss data from studies of two closely related butterfly species where the aim was to evaluate the uniqueness of marginal relict populations in relation to more central counterparts. Patterns in neutral genetic markers and values of adaptive wing traits among populations of the scarce heath species (*Coenonympha hero*, Linnaeus, 1761) and the pearly heath species(*C. arcania*, Linnaeus, 1761) from three biogeographically different regions were compared; one central (central Russia for *C. hero* and central Europe for *C. arcania*), one peripheral but connected to the main area of distribution (the Baltic states for both species and termed *peripheral region* in the following text) and one peripheral and isolated (Sweden for both species and termed *isolated region* in the following text) (Fig. 1a,b).

The Scandinavian populations of *C. hero* are relicts of a former more widespread distribution within the Scandinavian peninsula (Berglind 1990, 2007) and the species is therefore classified as "near threatened" in the Swedish Red Data Book (Gärdenfors 2005). On a global scale *C. hero* is widespread, although locally threatened, with a distribution ranging from Japan in the East through central Russia to central Europe in the West. The distribution is fragmentary in Western and Northern Europe and the species is considered threatened in most countries (Eliasson 2005; Kudrna 2002; Tolman and Lewington 1998).

C. arcania has a similar distribution as *C. hero* with peripheral populations in the Baltic states and Scandinavia, with the difference that it is widely distributed in Europe, from the Northern Mediterranean to South-central Scandinavia and central Estonia (Kesküla 1992; Kudrna 2002). In the East, it is found in Turkey, in Southern Russia and central parts of the Ural Mts. and Transcaucasia (Kudrna 2002; Tolman and Lewington 1998). Although the Swedish populations of *C. arcania* represent no typical relict due to their distribution being more stable in Scandinavia, they can serve as an illustrating example and a replicate within the described study frame.

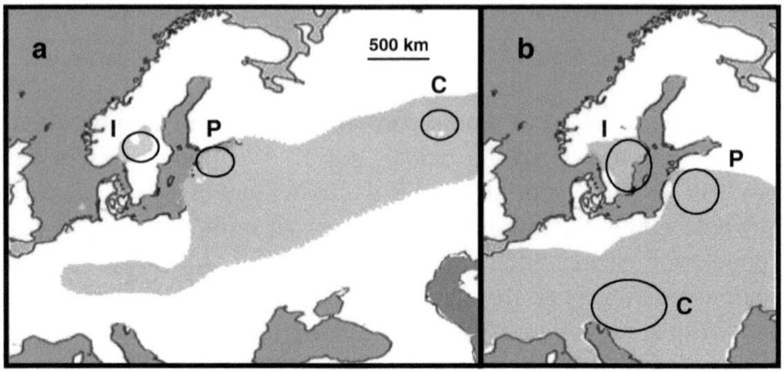

Fig. 1 Distribution range (marked in *light grey*) of (**a**) the scarce heath butterfly (*Coenonympha hero*) and (**b**) the pearly heath butterfly (*Coenonympha arcania*) in Europe and western Asia. *Dark grey circles* indicate sampling sites of both species representing central (C), peripheral (P) and isolated (I) parts of their distributions

In this study design the aim was to evaluate the patterns found not only in the light of the relative geographic locations of the populations, but also on their history inferred from neutral genetic marker. I discuss results obtained from selectively neutral traits, i.e. allozymes (Besold et al. 2008; Cassel and Tammaru 2003), and traits under selection represented by traits describing wing morphology (Cassel-Lundhagen et al. 2009). The allozyme data presented here for *C. arcania* are based on a subset of populations included in Besold et al. (2008) to make the data sets in each region comparable.

3 Gene Flow

Interestingly there were no indications of gradually reduced levels of genetic variability and gradually increased levels of differentiation from the centre via the peripheral to the isolated region. Instead, the populations in the isolated region had considerably lower levels of variation, while populations in the peripheral region did not differ from the more central ones (Table 1, Fig. 2a,b). Further, the populations in the peripheral populations were only marginally differentiated from the central ones, while the populations in the isolated region were markedly differentiated from both the peripheral and central ones (Table 2). This strongly suggests that the observed pattern is due to a combination of the peripheral and isolated populations' post-glacial colonization histories and their current position. Gene flow among central and peripheral parts appears to be significant enough to maintain a high level of genetic diversity and low differentiation, while lack of gene flow will cause both reductions in variability as well as differentiation. Thus, having a peripheral position per se does not necessarily lead to genetic loss and dfferentiation.

Table 1 Comparison of allozyme diversity measures for *C. hero* and *C. arcania* in a central, a peripheral and an isolated region of the two species distributions. Measures include percentage of polymorphic loci where the frequency of the most common allele does not exceed 0.95 ($P_{95\%}$ loci), allelic richness, observed heterozygosity (H_{obs}) and expected heterozygosity (H_{exp})

C. hero				
Regions	$P_{95\%}$ loci	Allelic richness	H_{obs}	H_{exp}
Isolated	16.7	1.20	0.03	0.03
Peripheral	20.0	1.43	0.09	0.12
Central	35.0	1.58	0.09	0.11
Differences among regions[a]				
Isolated/peripheral	n.s.	n.s.	**	*
Isolated/central	**	**	**	**
Peripheral/central	n.s	n.s.	n.s.	n.s.
C. arcania				
Regions	$P_{93\%}$ loci	Allelic richness	H_{obs}	H_{exp}
Isolated	20.4	1.31	0.08	0.09
Peripheral	54.6	1.93	0.18	0.21
Central	45.8	2.06	0.18	0.20
Differences among regions				
Isolated/peripheral	**	***	***	***
Isolated/central	***	***	***	***
Peripheral/central	n.s.	n.s.	n.s.	n.s.

[a]n.s. = non significant, *$p<0.05$, **$p<0.01$, ***$p<0.001$

4 Selection Vs. Gene Flow

The adaptive traits representing overall wing size and shape followed the same pattern as the neutral markers. Individuals from the isolated region were significantly differentiated from the peripheral and central while the peripheral and central populations showed no signs of differentiation. If predicting that selection pressures are equal in the peripheral and isolated region, low or lacking levels of differentiation in both neutral and selected traits in the peripheral populations in combination with a significant differentiation in both marker types in the isolated populations suggest that gene swamping is causing the observed pattern. This is supported by the fact that the peripheral and isolated regions are found on the same latitude while the central populations are found on an order of magnitude more distant, suggesting that environmental conditions should be more equal in the peripheral regions rather than the peripheral and central ones.

However, I cannot from current data exclude that there are actually different selection pressures in the two peripheral regions. For at least two reasons both species experience a more fragmented landscape in Sweden than in the Baltic states. Firstly, Southern Sweden has a history of great changes in landscape configuration due to an intensive afforestation over the last century (Nilsson 1990), while the

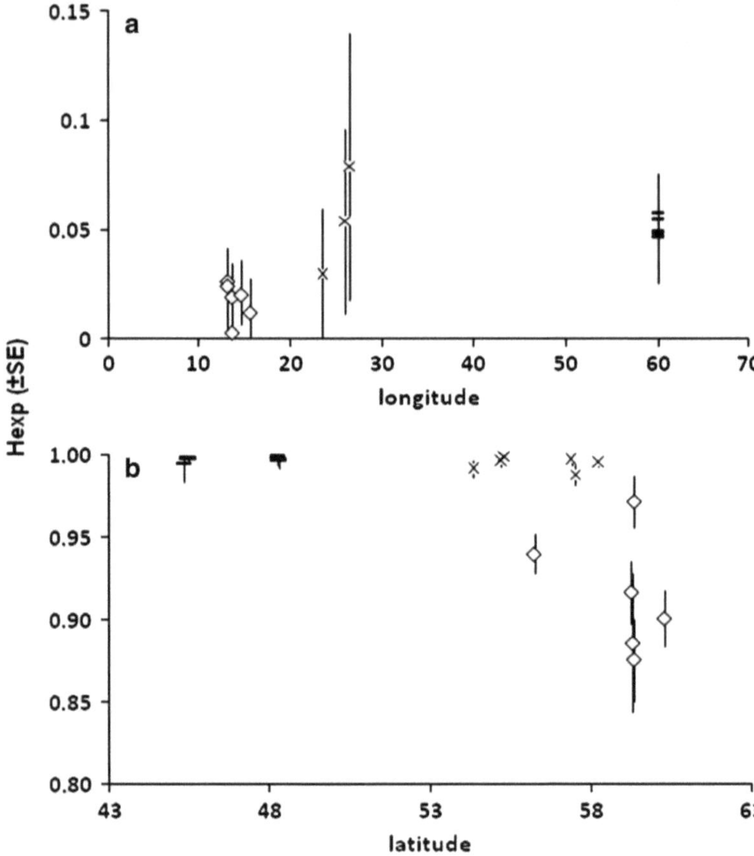

Fig. 2 Expected heterozygosity (± S.E.) for (**a**) 10 allozyme loci plotted against longitudinal position for 13 populations of *C. hero* and (**b**) 18 allozyme loci plotted against latidudinal position for 16 populations of *C. arcania*. Populations originate from three parts of the species' distributions; one core (Ural for C. hero and central Europe for *C. arcania*, population means indicated as straight line), one peripheral (Estonia for both, population means indicated as stars) and one isolated (Sweden for both, population means indicated as rombs)

Table 2 Fixation index (F_{RT}) and their significances among three regions of the two species *C. hero* and *C. arcania* within their distribution range

Regions	*C. hero*	*C. arcania*
Isolated/peripheral	0.26**	0.06***
Isolated/central	0.12**	0.04**
Peripheral/central	0.07**	0.01**

**$p < 0.01$, ***$p < 0.001$

See text for definition of regions

landscape structure has not changed considerably in, for example, Estonia during the same time period (Palang et al. 1998). Secondly, due to a significant decline in low intensive agriculture in Sweden over the past century (Dahlström et al. 2006), the proportion of semi-natural grasslands (primary habitats of *Coenonympha* species (Eliasson et al. 2005) is definitely lower in Sweden compared to the Baltic states. Both these processes have lead to lower amounts of available habitat for the Swedish butterflies. As a consequence, both *Coenonympha* species may occur to a larger extent as locally isolated populations in Sweden, whereas the distribution approaches a continuous one in the peripheral Baltic (states) region.

5 Additional Threats to Peripheral Relict Populations

As described above, peripheral relict populations can also experience reduced levels of genetic variation which may reduce their potential to adapt to the environment and/or cope with stress (Bulmer 1971; Roff 1997). Therefore, even if peripheral relict populations are genetically unique their peripheral position can also cause an additional constraint to their survival capacity, where isolation may threaten their existence if the loss of variation adversely affects their current performance and ability to cope with changes in living conditions. In addition, if peripheral conditions select for less dispersive phenotypes, as appears to be the case in the examples described here, this will adversely affect their ability to react to improved climatic conditions. Even if a new habitat is created beyond the species current distribution range, the species may have lost their ability to track the newly created habitat if selection has removed good dispersers from the gene pool. If the differences observed are mirrored in flight performance, I could then expect that these peripheral populations can differ in reactivity to climate change (Haynes and Cronin 2006). Thus, individuals of the species found at the same range margin may differ in their reaction to changing climatic conditions and this should be taken into account when predicting the effects of climate change on species communities.

6 Conclusions

Irrespective of the exact processes involved in the populations described above, it shows that peripheral, relict populations of also widespread species can add valuable and unique genetic variation. However, these diversifying processes need to act in isolation as homogenizing gene flow from more central parts can significantly counteract both random processes like genetic drift as well as selective forces.

Despite a study setup that aimed to discriminate between the influence of gene flow and selection in these peripheral and isolated populations additional experiments are needed to draw convincing conclusions regarding the opposing alternatives of swamping gene flow and differing selection pressures. If landscape structure is the cause for different phenotypes I would expect a correlation between habitat density and matrix configuration with wing sizes and shapes. If homogenizing gene flow, on the other hand, is causing the observed pattern I would expect the isolated phenotypes to be more successful than the central phenotypes in performance experiments in the peripheral region. These hypotheses can be tested and will add important insights into the evolution of peripheral populations and their reactivity to climate change. Further, marginal relict populations constitute ideal study objects of evolution in marginal habitats as they can be used to test how well theory and laboratory experiments reflects more complex natural systems (Kawecki, 2008).

Acknowledgements I thank all co authors and collaborators involved in various parts of the projects. Thanks to Atte Komonen, Ane Laugen and Thomas Schmitt for valuable comments on earlier versions of this manuscript. Part of the data presented here originates from work that was done at the Department of Conservation Biology and Genetics of Uppsala University, Sweden, and at the Department of Biogeography of University of Trier, Germany.

References

Berglind S-Å (1990) Meadow butterflies about to be replaced by spruce plants. Värmlandsnatur 4:12–17 in Swedish

Berglind S-Å (2007) *Coenonympha hero* Linné 1761, Scarce Heath. Species fact sheet (in Swedish). In: *Rödlistade arter i Sverige - The 200 Swedish Red List of Swedish Species* (ed. Gärdenfors U), p. http://www.artdata.slu.se. Swedish Species Information Centre, SLU, Uppsala

Berwaerts K, Aerts P, Van Dyck H (2006) On the sex-specific mechanisms of butterfly flight: flight performance relative to flight morphology, wing kinematics, and sex in *Pararge aegeria*. Biol J Lin Soc (Lond) 89:675–687

Berwaerts K, Van Dyck H, Aerts P (2002) Does flight morphology relate to flight performance? An experimental test with the butterfly *Pararge aegeria*. Funct Ecol 16:484–491

Besold J, Schmitt T, Tammaru T, Cassel-Lundhagen A (2008) Strong genetic impoverishment from the centre of distribution in Southern Europe to peripheral Baltic and isolated Scandinavian populations. Journal of Biogeography 35:2090–2101

Betts CR, Wootton RJ (1988) Wing shape and flight behaviour in butterflies (Lepidoptera: Papilionoidea and Hesperioidea): a preliminary analysis. J Exp Biol 138:271–288

Bulmer MG (1971) The effect of selection on genetic variability. Am Nat 105:201

Cassel-Lundhagen A, Tammaru T, Windig J, N R, S N (2009) Are peripheral populations special? Congruent patterns in two butterfly species. Ecography (online)

Cassel A, Tammaru T (2003) Allozyme variability in central, peripheral and isolated populations of the scarce heath (*Coenonympha hero*: Lepidoptera, Nymphalidae); implications for conservation. Conserv Genet 4:83–93

Chakraborty R, Nei M (1977) Bottleneck effects on average heterozygosity and genetic distance with stepwise mutation model. Evolution 31:347–356

Channell R, Lomolino MV (2000) Dynamic biogeography and conservation of endangered species. Nature 403:84–86

Clare L, Hughes CDJKH (2007) Modelling and analysing evolution of dispersal in populations at expanding range boundaries. Ecol Entomol 32:437–445

Cody ML, Overton JM (1996) Short-term evolution of reduced dispersal in island plant populations. J Ecol 84:53–61

Dahlström A, Cousins SAO, Eriksson O (2006) The history (1620–2003) of land use, people and livestock, and the relationship to present plant species diversity in a rural landscape in Sweden. Environ Hist 12:191–212

Davis M (1984) The flight and migration ecology of the red milkweed beetle (*Tetraopes teraphthalmus*). Ecology 64:230–234

Dempster JP (1991) Fragmentation, isolation and mobility of insect populations. In: Collins N, Thomas J) (eds) The conservation of insects and their habitats. Academic Press Ltd, London, pp 143–153

den Boer PJ (1990) The survival value of dispersal in terrestrial arthropods. Biol Conserv 54:175–192

Eckert CG, Samis KE, Lougheed SC (2008) Genetic variation across species' geographical ranges: the central-marginal hypothesis and beyond. Mol Ecol 17:1170–1188

Eliasson C (2005) *Coenonympha arcania* Pärlgräsfjäril. In: Nationalnyckeln till Sveriges flora och fauna. Fjärilar: Dagfjärilar. Hesperidae - Nymphalidae, Artdatabanken, SLU, Uppsala, pp 331–332

Eliasson C, Ryrholm N, Holmer M, Jilg K, Gärdenfors U (2005) Nationalnyckeln till Sveriges flora och fauna. Dagfjärilar. Hesperidae - Nymphalidae Artdatabanken, SLU, Uppsala, Fjärilar

Fric Z, Konvička M (2002) Generations of the polyphenic butterfly *Araschnia levana* differ in body design. Evol Ecol Res 4:1017–1032

Garcia-Ramos G, Kirkpatrick M (1997) Genetic models of adaptation and gene flow in peripheral populations. Evolution 51:21–28

Gullberg A, Olsson M, Tegelström H (1998) Colonization, genetic diversity, and evolution in the Swedish sand lizard, Lacerta agilis(Reptilia, Squamata). Biol J Linn Soc 65:257–277

Gärdenfors U (2005) The 2005 Red List of Swedish Species. Swedish Species Information Centre, Swedish University of Agricultural Sciences, Uppsala, Sweden, Uppsala

Haynes K, Cronin J (2006) Interpatch movement and edge effects: the role of behavioral responses to the landscape matrix. OIKOS 113:43–54

Heino M, Hanski I (2001) Evolution of migration rate in a spatially realistic metapopulation model. Am Nat 157:495–511

Helgadóttir Á, Snaydon RW (1986) Patterns of genetic variation among populations of *Poa pratensis* L. and *Poa capillaris* L. from Britain and Iceland. J Appl Ecol 23:703–719

Kawecki TJ (2008) Adaptation to marginal habitats. Annu Rev Ecol Evol Syst 39:321–342

Kesküla T (1992) Distribution maps of Estonian butterflies (Lepidoptera: Hesperioidea, Papilionoidea). In: Acta Musei Zoologici Universitatis Tartuensis, Tartu

Kudrna O (2002) The distribution atlas of European butterflies. Oedippus 20:1–342

Lesica P, Allendorf FW (1995) When are peripheral populations valuable for conservation? Conserv Biol 9:753–760

Mayr E (1954) Change of genetic environment and evolution. In: Huxley J, Hardy A, Ford E (eds) Evolution as a process. Allen and Unwin, London, pp 157–180

Mayr E (1963) Animal species and evolution. Harvard University Press, Cambridge, MA

Merckx T, Van Dyck H (2006) Landscape structure and phenotypic plasticity in flight morphology in the butterfly pararge aegeria. OIKOS 113:226–232

Nilsson N-E (1990) National atlas of Sweden The forest (in Swedish). Skogsstyrelsen, Jönköping

Palang H, Mander Ü, Luud A (1998) Landscape diversity changes in Estonia. Landsc Urban Plan 41:163–169

Roff DA (1997) Evolutionary quantitative genetics. Chapman and Hall, New York

Schtickzelle N, Mennechez G, Baguette M (2006) Dispersal depression with habitat fragmentation in the bog fritillary butterfly. Ecology 87:1057–1065

Thomas CD, Hill JK, Lewis OT (1998) Evolutionary consequences of habitat fragmentation in a localized butterfly. J Anim Ecol 67:485–497

Tolman T, Lewington R (1998) Die Tagfalter Europas und Nordwestafrikas. Franckh-Kosmos Verlag, Stuttgart

Genetic Differentiation Between and Among Refugia

Gabriel Nève and Régine Verlaque

Abstract

1. Since the Last Glacial Maximum, the distribution of terrestrial plant and animal taxa in Europe has shifted from their refugia, Northwards in latitude and upwards in altitude.
2. The populations from refugia, mainly the mountain massifs of S Europe (Pyrenees and Alps, and also the Carpathian and Balkan Mountains), harbor a higher genetic diversity than the populations of N Europe.
3. The isolation by distance effect between populations tends to decrease with the increasing distance from the glacial refugia.
4. In areas of admixture from colonization originating in different refugia the heterozygosity may be higher than near the refugia.
5. Numerous thermophilous species have survived in their Mediterranean refugia at low altitude (particularly in islands).
6. Diploid populations or taxa often remained restricted at mid and high altitudes in S Europe, while their polyploid relatives spread in the lowlands.
7. For cold-adapted species, the refugia are mainly located in the S European mountains, and sometimes in N Eurasia, such as in Andøya island or in Taymyr peninsula.
8. The relict populations retaining a high allelic diversity and the endemic diploid taxa are now key elements in the long term conservation.

1 Introduction

Species which are widely distributed do not generally form one panmictic population. The degree of genetic variability varies within the distribution range. The first to formally recognize and search for genetic variability was the Russian scientist

G. Nève (✉) and R. Verlaque
Institut Méditerranéen d'Ecologie et Paléoécologie, Université de Provence, UMR CNRS 6116, Case 36, 3 Place Victor Hugo, F-13331 Marseille Cedex 3, France
e-mails: gabriel.neve@univ-provence.fr; regine.verlaque@univ-provence.fr

J.C. Habel and T. Assmann (eds.), *Relict Species: Phylogeography and Conservation Biology*, DOI 10.1007/978-3-540-92160-8_16, © Springer-Verlag Berlin Heidelberg 2010

Nicolaï Ivanovitch Vavilov (1885–1943),[1] who set the foundations of research on the origin of cereals (in particular wheat). He demonstrated that their diploid wild relatives, in their original area, had a much higher variability than their widely cultivated generally polyploid relatives (e.g., Vavilov 1950). These principles were later followed by biologists studying the intraspecific variation of wild plants and animals. This same general principle may be applied to glacial relicts, with the question of the identification of the original area for each concerned species.

Since Forbes (1846) we know that the distribution of plants and animals changes with time, in particular during the Quaternary climatic oscillations. Haeckel formally recognised this when he wrote:

> As the glaciation encroached from Northern Europe towards our Alpine chains, the polar inhabitants retreating before it - gentian, saxifrage, polar foxes, and polar hares - must have peopled Germany, in fact all Central Europe. When the temperature again increased, only a portion of these Arctic inhabitants returned with the retreating ice to the Arctic zones. Another portion of them climbed up the mountains of the Alpine chain instead, and there found the cold climate suited to them. The problem is thus solved in a most simple manner (Haeckel 1876, Chap. 14, p. 367).

The ca. 100,000 year cycle of glacial – interglacial periods has had a major impact on the distribution of organisms in the North hemisphere, particularly in Europe because of the East–west orientation of its main mountain massifs (Pyrenees, Alps, Carpathians), leading to a corresponding range shifts (Huntley and Webb 1989). For many cold-adapted species, the latitudinal Northward migration was accompanied by an upward altitudinal migration. These species then ended up with disjunct distribution, often called boreo-mountain or arctic-alpine species, depending on the altitude/latitude of their distribution (de Lattin 1967; Udvardy 1969).

From a genetic point of view, the key factors influencing the genetic structure of the populations of the species with such disjunct distributions are the age of the discontinuity in the distribution and the population size within each of the different distribution patches. For most European species distributed in both Scandinavia and the S European mountains (mostly in the Alps and the Pyrenees), it is assumed that the populations now found in, e.g., the Alps and the ones found in Scandinavia became separated ca. 10,000 years ago, i.e., at the end of the Würm glaciation. However, other scenarios are possible; N Europe may have been colonized from a refugium in Siberia, while Alpine populations were separated from these at a previous colonization cycle (Hewitt 1999). Furthermore, numerous thermophilous species remained at low altitude in the Mediterranean area during glaciations; afterwards some of them remained there while others spread Northwards.

The aim of the present chapter is to briefly discuss: (1) what are the long term consequences of range shifts in terms of endemic and cytotype distributions, and (2) how range shifts influence the genetic structure of populations.

[1]The date of death of N.I. Vavilov is usually given as 1942 (e.g. Harland 1954). Details of his tragic death became known only after the demise of the Soviet Union (Azmanov 2002) and its date is 25 January 1943.

2 Karyology as a Marker of Genetic Variation

Chromosomal data are key variables in the genetic makeup of many plants and animals (Dobigny et al. 2004; Mabble 2004). These data accumulated over the last 80 years constitute a valuable material to look for a general framework about the relict cytotype distribution and the genetic differentiation within and among the refugia.

2.1 *Endemism in the Mediterranean Basin*

The Mediterranean Basin is characterized by its floristic richness (25,000 species) and high rates of endemism (around 12,500 species, i.e., 50%), while there is an obvious South–North gradient of decreasing endemism in Europe (Favarger 1972). The ten Mediterranean hot-spots of plant biodiversity represent privileged endemism areas and the most important refugia: mainly mountains and islands (Médail and Quézel 1997; Médail and Diadema 2008). About two thirds of endemics are perennial taxa that show a clear ability to colonize harsh habitats (coasts, screes, cliffs, rocky places, etc.), at mid and high altitudes (Médail and Verlaque 1997; Verlaque et al. 1997).

The remaining third are lowland endemics and constitute a threatened patrimonial flora (Verlaque 1999). For example, the diploid *Lysimachia minoricensis* of Minorca is now extinct in situ and the plants cultivated in the botanical gardens do not show any genetic variation (Calero et al. 1999). Except in such an extreme case, it is not possible to establish a general pattern of genetic variation according to the restricted or widespread status of taxa (Gitzendanner and Soltis 2000, Thompson 2005). Many rare species show a lower diversity than more common relatives, as theoretically expected, but there are many reverse cases. In the diploid endemic *Cyclamen balearicum*, the threatened populations of S. France (Languedoc) are genetically more differentiated than the healthy and numerous populations of Balearic Islands (Affre et al. 1997). Life history traits (life-form, age, mode of formation etc.; cf. Stearns 1992) and causes of rarity of endemic taxa play a major role in explaining these phenomena (Gitzendanner and Soltis 2000). It should be borne in mind that endemism may have two opposite causes: a small area may be the relict of a former widespread range or the result of a recent speciation. In the latter case, the new taxa may remain restricted neo-endemics, despite some colonization capacities, if they become isolated on an island or at high altitude, or if they are ecologically very specialized in rare habitats, as is the rule for most of the endemics.

In this context, a fourfold classification has been suggested based on the analysis of endemic taxa characters compared to those of their closely related species (Favarger and Contandriopoulos 1961; summarized by Thompson 2005). Proportions of endemic classes calculated in several islands and mainland areas indicate a general Mediterranean framework (Verlaque et al. 1997). The paleo-endemics (relict species morphologically isolated from other taxa) are mainly thermophilous taxa located in the lowlands, and are mostly present on islands. The patro-endemics (generally diploid cytotype progenitors of polyploid relative taxa) occur at mid to high altitudes and are more present in mainland areas than in islands. The schizo-endemics (vicariant taxa with the same chromosome number) are generally diploid taxa living at high altitude. Lastly, the apo- or neo-endemics (polyploid derived cytotypes, younger than its relatives) mainly occur at low and mid altitudes.

Such a pattern allows the identification of two distinct processes: (1) the "passive" or relictual endemism (including paleo- and patro-endemics) and (2) the "active" or recent endemism (including schizo- and apo-endemics). The relictual endemism matches the refugia, and amounts only to 32% of endemics in Corsica, and 28% in Provence and Balearic Islands (Verlaque et al. 1997). As far as the recent endemism is concerned, the schizo-endemics are dominant (49–55% in W Mediterranean) and can be investigated mainly with genetic markers.

2.2 *Polyploidy (i.e. Genome Duplication) in Europe*

The occurrence of polyploidy into a group generally corresponds to an expansion phase (Goldblatt 1980; Lewis 1980; Thompson and Lumaret 1992). Conversely to the pattern found in endemism, by analysis of some lowland native floras, Reese (1958) shows a significant South–North gradient of increasing polyploidy, from the S Mediterranean area to N Europe [Table 1]. Brochmann et al. (2004) confirm this phenomenon and sum up the main historical hypotheses. In Europe, the intensity of Quaternary climatic changes has highly favored this polyploidy, due to a combination of phenomena: (1) the successive movements of floras (Huntley and Webb 1989), (2) their intermingling leading to numerous hybrids, and (3) the recurrent colonization of large virgin areas by the neo-polyploids during periods of glacial retreat. To these ecological factors, one must add (4) the fundamental role of genes controlling the meiosis, which can induce 2n-gametes (i.e., diplogametes) production when exposed to temperature changes and other stresses (Lexer and Fay 2005; Wang et al. 2004; Ravi et al. 2008). Thus, the Arctic region, the most affected by the glaciations, was re-colonized by a young flora very rich in allopolyploid taxa with a high heterozygosity (Brochmann et al. 2004).

The colonization power of polyploids can be explained by their wide ecological amplitude and tolerance: they are much more efficient competitors than their diploid

Table 1 Proportions of polyploid perennial taxa from the native lowland flora

Area	Percentages	Latitude
North Sahara, Algeria	42.3	31° N
East Czechoslovakia	56.2	50° N
Sjaelland, Denmark	57.2	56° N
North Sweden	65.5	66° N
Svalbard	78.2	79° N

From Reese (1958)

Table 2 Proportions of diploid species in the perennial taxa from SE France (Provence)

Total native flora	42%
Endemic taxa	52%
Rare non endemic taxa	60%
Rare endemic taxa	70%

progenitors, particularly in unstable conditions (Stebbins 1971; Ehrendorfer 1980; Levin 1983; Thompson and Lumaret 1992; Lumaret et al. 1997, Bretagnolle et al. 1998). Thus in France, the maximum percentage of polyploids is found among alien invaders (77%, Verlaque et al. 2002). Conversely, in native flora the higher rates of diploids reaches 70% in rare endemic taxa (Table 2) and 75% in critically endangered segetal plants (Verlaque and Filosa 1997). There is a strong correlation between the abundance of taxa and their ploidy level: polyploids correspond to the most common taxa, while the rare or extinct species are mainly diploids (Hodgson 1987).

However, the study of polyploids is often very difficult because of their various possible origins: auto- or allo-polyploids from different populations (Levin 2002; Brochmann et al. 2004; Eidesen et al. 2007). A morphological and karyological approach is needed apart from DNA data in studies of plant phylogeny and "it seems advisable to investigate only the diploids (at least at first)" (Stace 2005). Polyploidy is generally a recurrent process followed by quick karyotype reshuffling, leading to a one-way process (Soltis and Soltis 1999; Crawford and Mort 2003). Genome duplication may also alter sexual reproduction and interspecific barriers; thus polyploids often become autogamous or apomictic (Richards 1997), or inter-hybridize (Breton-Sintès 1974; Gadella 1987). The intensive reticulation appears to be the main cause of phylogenetic incongruence because homogenization (of ITS sequences), via concerted evolution, may occur in various descendant lineages (Mansion et al. 2005). Lastly, the redundant structure of the genome in highly polyploid taxa often leads to a buffer effect against arising mutations.

During the glaciation maxima, the thermophilous species were pushed southward, and only the very dynamic taxa and cytotypes could (re)colonize successfully. In the Pterydophytes genus *Asplenium*, the 22 rupicolous diploidized species ($2n=72$) remained mostly confined in the Mediterranean refugia after the Würm glaciation, only five diploid taxa colonized central Europe, the British Isles or Scandinavia. In contrast, the ten widespread taxa up to N Europe are polyploid ($2n=144$). Furthermore, the diploid populations of *A. ruta-muraria* from Gorges du Verdon in SE France show the highest allelic variability known in ferns (Vogel et al. 1999). In Angiosperms, two examples could be cited: (1) the segetal EuroSiberian *Camelina sativa* includes one diploid relict population ($2n=12$) in the SE France (Monts de Vaucluse, Verlaque and Filosa 1992), now critically endangered, one tetraploid in Bulgaria ($2n=26$) and a majority of hexaploid populations ($2n=40$) in Iceland, Poland, China and Russia; (2) the stenoMediterranean *Calicotome spinosa* is tetraploid, except in SE France (Var: Gapeau Valley) where a relict diploid cytotype still survives at low altitude (Aboucaya 1989, Fig. 1).

During the glaciations some diploid species already adapted to cold conditions persisted in refugia, especially in the S European mountains (Favarger 1967, 1972),

Fig. 1 Distribution of *Calicotome spinosa*. Tetraploid populations (2n=48, 50, 52) are found from Spain to Italy, while only one population in Var (France) was found to be diploid (*Diamond*: 2n=24). Distribution map compiled from Aboucaya (1989) and de Bolós and Vigo (1984)

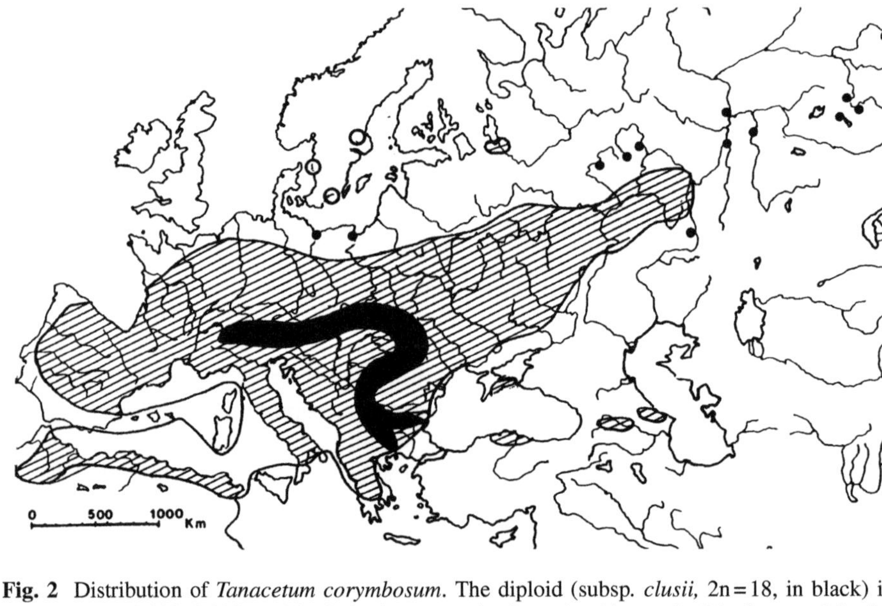

Fig. 2 Distribution of *Tanacetum corymbosum*. The diploid (subsp. *clusii*, 2n=18, in black) is found only at 1,000–2,000 m altitude in the mountains from the Alps to the Rhodopes, while the tetraploid (subsp. *corymbosum* 2n=36, hached) is found in lowlands from Spain to Russia

while only polyploid cytotypes spread widely during the interglacial stages. In *Tanacetum corymbosum*, the diploid subsp. *clusii* is found from the Alps to the Rhodopes (1,000–2,000 m), while the tetraploid subsp. *corymbosum* is widely distributed in the lowlands from N Africa to Russia (Seidenbinder 1989; Verlaque and Contandriopoulos 1990; Fig. 2). In the orophyte *Ranunculus kuepferi* of Corsica and Alps, the diploid sexual relict subsp. *kuepferi* is restricted to the SW margin

(Alpes-Maritimes, Var, Isère), whereas the tetraploid apomictic subsp. *orientalis* is widespread in the rest of the species range (Küpfer 1974).

3 Genetic Structure of Populations

The genetic structure of populations may be studied at different scales. The combination of genetic information from populations through a species' range leads to a knowledge of the spatial aspects of population genetics (Fig. 3). While investigating the population genetics of wild species, several indices may be used (Box 1). The most common for each population are the expected heterozygosity and the allelic richness. When the investigation concerns the structure of the populations, then *F* statistics and their derivatives are the most common indices (Wallis 1994), but the application of geostatistical techniques indicates that these traditional values might not be the best indicators of population differentiation (Joost et al. 2007, 2008).

Each of these genetic indices has its own objectives. The allelic richness is very sensitive to population bottlenecks and to population admixture. In the diploid beech tree *Fagus sylvatica* (W & C Europe, 0–2,000 m), the area of the highest heterozygosity in Europe corresponds to the area which was colonized between 4,000 and 2,000 years ago, which corresponds to the area of admixture of colonization fronts coming from different glacial refugia (Comps et al. 2001). On the other hand, the highest allelic richness is still found near refugia, i.e., the Balkan Peninsula and S Italy.

The process of colonization leads thus to a decrease of allelic richness, due to multiple foundation events, while colonization from different areas may lead to high heterozygosities, because of the admixture from refugia with different alleles (Widmer and Lexer 2001; Petit et al. 2001). For species which have colonized N Europe from a Southern refugium, their heterozygosity decreases with increasing

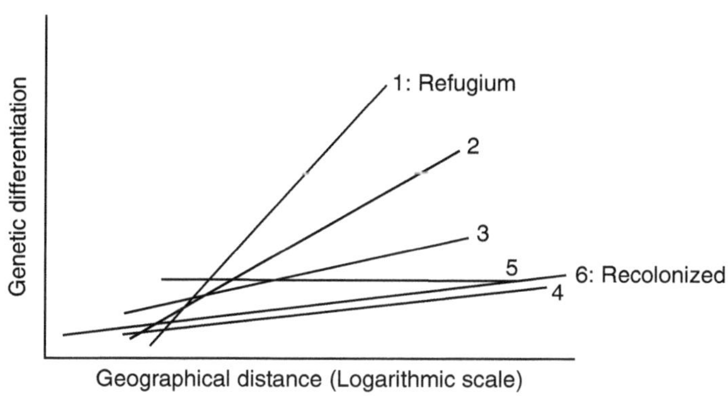

Fig. 3 Isolation by distance effect in *Carabus solieri*.The slope depends on the time since colonization. The northernmost zones (4–6), i.e., the most recently colonized, show a more level slope than the areas close to the refugia (1), with a gradient of slopes for in between zones. Redrawn from Garnier et al. (2004)

Box 1 Indices Used in Population Genetics

H Expected Heterozygosity Probability that two alleles sampled at random from a population are different. It is a function of the allelic frequencies. The expected heterozygosity is basically calculated for each locus, and a mean (or sometimes a median) value is presented for a number of loci. This index is particularly sensitive to bottlenecks having occurred in the population previously. This index is also called the Nei's gene diversity.

Allelic Richness The number of alleles per locus is a good index of diversity. It is highly sensitive to past population bottlenecks (Luikart et al. 1998) and investigated loci. As it depends on the number of sampled individuals, this index has to be normalised to a given sample size (El Mousadik and Petit 1996).

Fixation Index F_{ST} is an index of the way two or more populations are differentiated from each other. It depends on the difference of allele frequencies among the populations. It is usually estimated from allele frequencies in the concerned populations, as follows:

$$F_{ST} = (H_T - \bar{H}_s)/H_T,$$

with $\bar{H}_s$ the mean expected heterozygosity of an individual in each population mating randomly, and H_T the expected heterozygosity of an individual in a total population mating randomly (e.g., Hartl and Clark 1989).

Isolation by distance The strength of any isolation by distance effect is measured as the slope of the regression of an index of genetic distance as a function of the geographic distance, both axes being expressed in logarithmic scale. The genetic distance is expressed either as F_{ST} or as $F_{ST}/(1-F_{ST})$ for each pair of populations. The steeper the slope is, the stronger the isolation by distance effect would be.

Neighborhood Size Individuals often do not reproduce at the location of their parents. The genetic neighborhood has been defined as the area within which most of the individual will reproduce. Dispersal is the movement from the birth or parents' locations to the reproduction location. Wright (1943) defined the genetic neighborhood of an individual as the population of the area within the variance (σ^2) of the dispersal distance. With d the density of individuals, the neighborhood size is given by $N_e = 4\pi d\sigma^2$. As the dispersal distance is usually difficult to measure, the neighborhood size is

(continued)

Box 1 (continued)

often estimated from the differentiation between populations, using F_{ST} values estimated by pairs of populations to estimate the gene flow

$$M_{ST} \approx \frac{1}{4}(\frac{1}{F_{ST}} - 1) .$$

The y-intercept of the log-log plot of the estimated M_{ST} against geographic distance gives an approximation of the local effective population size (Slatkin 1993; example in Nève et al. 2008).

distances from their refugia. This has been called the "leading edge" effect, where the front (here North) part of the distribution moves, and the northernmost populations are the results of multiple colonization events (Hewitt 1993, 1996; Nichols and Hewitt 1994). The slope of the decrease of this heterozygosity depends on the rate of the loss of genetic diversity through the stepping stone colonization process.

During shifts of species range, the newly occupied area is the result of multiple colonization events. If this colonization is the result of rare events involving individuals only at the front edge of the distribution, a strong isolation by distance will follow together with a reduction of allelic diversity due to the repetitive colonization effect (Ibrahim et al. 1996). This resulted in a series of genetic hotspots distributed in the S Europe. Climate warming led to a gradual upward range shift, with little loss of genetic diversity, while the Northwards range shift involved long distance migration and multiple colonization bottleneck events. In the *Frangula alnus*, Eurosiberian diploid (0–1,400 m), the Iberian Peninsula and Turkey still harbor by far most of the genetic diversity. These probably correspond to the two main refugia for this species, and the colonization of the plains of central Europe was possible only by crossing severe barriers to dispersal (and hence gene flow) caused by the Pyrenees and the Black Sea (Hampe et al. 2003). In the arctic-alpine diploid species *Ranunculus glacialis* (1,700–4,000 m), populations from N Europe show a much lower genetic diversity than Alpine populations, as they resulted from a colonization from E Alps, where populations have retained a high genetic diversity (Schönswetter et al. 2003). In *Gentiana ligustica*, tetraploid endemic to the Ligurian Alps, the Shannon's Diversity index, computed on AFLP fragments, decreases with increasing distance from putative Würm refugia (Diadema 2006).

The retreating range margin is also subject to important changes in the population structure. As the area becomes generally unsuitable to the species of interest, its local suitable habitat patches become scarce and far between. The populations which used to be part of large networks of populations exchanging individuals with their neighbors become more isolated. These "trailing edge" populations are then reduced in their effective size, and are subject to drift, independently of other populations (Hampe and Petit 2005). The populations within the different mountain massifs of S Europe have been present in these areas often for several glacial cycles, resulting in levels of population exchanges to have occurred roughly according to the geographic distance. As at the continental scale the population exchanges have been rare, and as this process ended up being leveled over a very long time span (Slatkin 1993), the populations now found in the S European mountains often end up showing a very strong isolation by distance effect. If the North of Europe has been colonized from an Eastern refugium ("*Chorthippus* pattern," Hewitt 1999), the populations of the mountains of S Europe may have been isolated from the Northern populations at least since the Riss glaciation (ca. 130–240 ky BP). In this case their long isolation resulted in a strong differentiation. The Capercaillie *Tetrao urogallus* of the Cordillera Cantabrica and the Pyrenees is an old lineage different from the rest of the species' distribution, from the Alps and Scandinavia into Siberia (Duriez et al. 2007), which is the result of populations from the Iberian refugium not colonizing the lowlands North of the Pyrenees.

Within a given mountain massif, populations may also show a high differentiation, as they are separated by strong barriers such as valleys and rivers. If populations survived in nunataks during the last glaciation, the differentiation between these is the result of a pre-glacial isolation. In such cases, the populations now found at high altitudes of the Alps are the result of in situ survival during the Würm glaciation rather than post-glacial recolonization from peripheral areas. The perennial tetraploid cushion plant *Eritrichium nanum* most probably survived in at least three nunatak areas of the Alps (above 2,500 m), as these populations are the most variable (high number of private AFLP fragments) and the whole data set form three distinct clades (Stehlik et al. 2001). In other cases such as the artic-alpine diploid *Saxifraga oppositifolia*, genetics studies do not allow the identification of nunatak survival in the Alps, presumably because any surviving nunatak population would have been swamped by colonization from the main refugia (Holderegger et al. 2002). Sometimes a single pre-glacial population may end up being split: the tetraploid SW Alpine endemic *Senecio halleri* survived in two nunataks in the South part of its distribution, SW and NE of the Aosta valley, Italy. These two populations are not significantly different from each other, resulting probably from a single pre-glaciation colonization event (Bettin et al. 2007). Survival in and around mountain massifs during the last glaciation favoured the development of genetic differentiation between populations. In the widespread diploid *Arabis alpina*, the high among population variation in the mountains of S Europe (Alps, Tatra and Carpathians) suggests several refugia for this species, whereas all over the North of its distribution (Scandinavia, Iceland, Greenland and Newfoundland) it shows very low genetic variation, as these populations probably result from the recolonization from a single refugium closely following the retreating glaciers (Ehrich et al. 2007). Notably the populations in the East African mountains show high inter-populations variation, an evidence of their old age.

The carabid beetle *Carabus solieri* is an endemic species to the SW Alps near the Italian–French border. Garnier et al. (2004) sampled the whole distribution of this species, and performed genotypic identification of ten microsatellite loci in 1,080 individuals. Globally these populations show a strong isolation by distance, which is not surprising for a wingless species living in the mountain forests, in an area with many barriers to dispersal such as non-forested areas, roads, rivers etc. The 41 studied populations were grouped into three clusters, from South to North. Each cluster was then cut into two parts to have a series of six subclusters. Analysing these clusters separately for isolation by distance effect yielded a strong cline in the slope of the relationship between genetic distance and geographic distance. The slope is the steepest nearest to the refugium, and the most gradual in the recolonized area, i.e., northernmost set of populations (Fig. 3).

Other patterns of postglacial colonization have been described, from the refugia in central Asia or N Europe. The rare Alpine populations of the tetraploid *Carex atrofusca* originate from a central Asian refugium, while Siberia and Greenland populations form a sister clade (Schönswetter et al. 2006a). The diploid *Ranunculus pygmaeus* provides even a more extreme case as the Alpine populations probably originate from a refugium on the Taymyr Peninsula, East of the last glacial ice sheet (Schönswetter et al. 2006b). In Norway, the high nucleotide diversity of the populations of the root vole *Microtus oeconomus* from the island of Andøya strongly suggests in situ glacial survival for this species (Brunhoff et al. 2006). These examples show that the recolonization of the European ice sheet, after the last glacial maximum, took part from all directions, even if the contribution of the Southern refugia was prevalent (Fig. 4).

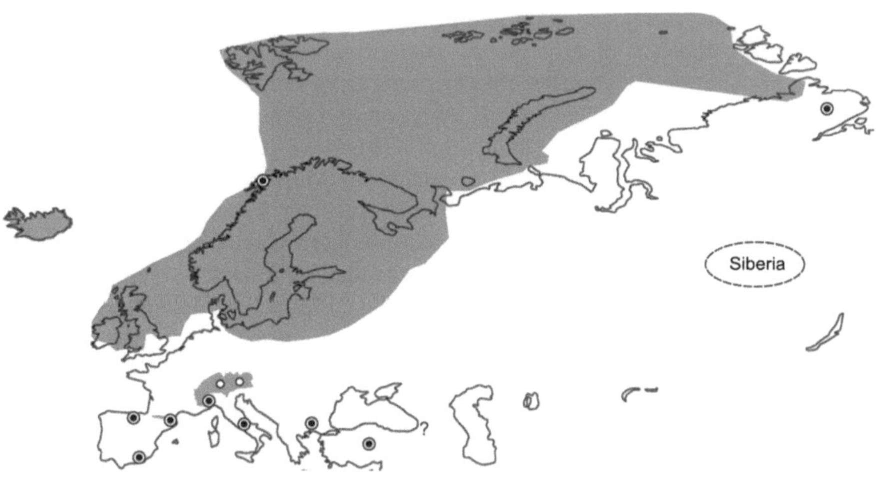

Fig. 4 Map of Eurasia showing the Last Glacial Maximum (LGM; ca. 20 ky BP), with the main identified refugia of European taxa (LGM from Svendsen et al. 2004; Ivy-Ochs et al. 2008). Small glaciated areas in Mediterranean mountains (not shown here) are discussed in Hughes et al. (2006). *Filled circles*: main refugia; *Open circles*: nunatak refugia

4 Isolation by Distance

4.1 Differentiation Among Populations Within a Region

Differentiation among populations within each region is the result of the way individuals move within the region. All species are regionally distributed in an aggregative way: the species is found in its habitat, while it is missing or shows a low density outside its optimal habitats. Some species have a high dispersal power, and their populations are not differentiated within a region. This is the case of many migrating species, where dispersal between the birth location and the reproductive location is common. Many birds and insects (e.g., *Colias* and *Aglais urticae* butterflies; Vandewoestijne et al. 1999; Watt et al. 2003) follow this pattern. Wind pollinated plants also show genetic admixture in large areas, at least as far as the nuclear genome is concerned (e.g., *Quercus* spp., Petit et al. 2002).

The geographical scale of the isolation by distance effect varies with the dispersal scale of the species. In fixed organisms (i.e., plants), the scale of interest may be from a few meters (the distance between two individuals) to the extremes of a species distribution. Depending on the scale and on the dispersal capability, the relationship between geographic and genetic distance may or may not be significant. In the case of the Lycaenid butterfly *Euphilotes enoptes* there is no isolation by distance for distances of less than 30 km, but well in the range 30–400 km (Peterson 1995, 1996). The lack of correlation at small distances was related to the ecological heterogeneity at small scale, such as local topography, which has a profound impact on the species phenology. Isolation by distance at greater distances was attributed to a stepping stone dispersal process. Furthermore, the isolation by distance effect is mainly the result of long-distance migration events, while genetic neighborhood depends chiefly on the rate of movement between neighboring patches. In the butterfly *Proclossiana eunomia* neighborhood size is higher in the mountains than in the lowlands, while isolation by distance is higher in the mountains than in the lowlands (Nève et al. 2008). This shows that dispersal kernels may vary greatly between regions, and that population differentiation patterns may consequently also vary in different ways for short and long distances.

Differences in isolation by distance processes may also go the other way, with significance at the small scale distances of the spectrum, and no significance at a larger scale. In a series of 17 sampled sites for the diploid plant *Rhododendron ferrugineum* in the Alps and Pyrenees, no isolation by distance could be detected for site distances ranging from 4 to 1,000 km while, when a single site was studied in detail, the isolation by distance effect was significant for distances between 10 and 5000 m (Wolf et al. 2004). In this case, short distance isolation by distance was the result of short time scale bio-ecological factors, such as allogamy, pollination and seed dispersal, while the structure of populations on the Alps as a whole resulted from disequilibrium between dispersal and genetic drift, as very long distance dispersal does probably not follow a simple decreasing monotonous curve for this species.

4.2 Differentiation Among Regions and Long Distance Gene Flow

Species which were widely distributed during the Würm glaciation may end up having a now disjunct distribution. This may be the result of different processes: the populations of N Europe may have found themselves separated from the populations moving up the S European mountains or from a long distance colonization of newly available habitats such as islands in the high Arctic Ocean.

In arctic-alpine species, colonization may occur at very long distance. Isolated localities such as Bear Island and the Svalbard Archipelago had to be colonized over hundreds of kilometers of sea. Consequently, there is little isolation by distance in species showing such high colonization power. A recent study of nine species found on Svalbard showed that this archipelago was mainly colonized from NW Siberia (Alsos et al. 2007), rather than from the geographically closer Scandinavia and Greenland, because the main polar wind, from East, favored a colonization of high Arctic islands from Siberia (Coulson et al. 2002), and E Siberia (the Beringian area) was not covered by an ice sheet (Brochmann et al. 2004).

Among all the populations of the diploid mountain avens *Dryas octopetala*, Svalbard populations were most closely related to the Siberian ones. This arctic-alpine species displays an East–west differentiation, with five different groups in the Holarctic region (Skrede et al. 2006). The refugium of the European group probably laid in the tundra of C Europe during the Würm glaciation. As the ice retreated, this population split between the parts which migrated northwards to Scandinavia, and the populations which migrated up the mountains of the Alps. Among the European group, the Alpine populations show a much reduced variability ($H<0.077$) compared to the Arctic ones ($0.101<H<0.144$). This unusual pattern is probably due to the widespread distribution of *D. octopetala* in C Europe during the Würm glaciation, which led to a latitudinal range shift involving several large refugia, while the colonization of the Alps involved fewer ones leading to a strong bottleneck effect.

The mating mode of organisms may increase the effect of isolation by distance. The perennial plant *Draba fladnizensis* mainly reproduces by selfing (as numerous arctic plants: Richards 1997). Morphologically, the individuals are homogenous throughout the artic-alpine range, from N Europe to Alaska. Crossing experiments between specimens from different parts of the range mainly resulted in either no seed, or sterile seeds, while intra-population crossing yielded numerous viable seeds (Grundt et al. 2006). In this diploid species, the differentiation between populations proceeded rapidly, leading to cryptic speciation, even within a limited geographic area: Svalbard archipelago hosts at least three such sibling species. This speciation process (by minor translocation or mutation events, Brochmann et al. 2004) probably occurred in situ during the ca. 10,000 years since the height of the last glaciation was when Svalbard was completely covered by ice. A similar sibling speciation has been found in *Nigella* (Strid 1969) and *Erysimum* (Snogerup 1967) endemic taxa living in harsh habitats in the Aegean islands.

5 Conclusions

Processes of population differentiation had different consequences, according to the spatial and temporal scales at which they occurred. The differentiation between mountain massifs or between North and South Europe is a process due to the occurrence of long distance colonization or lack thereof. Some trends may be suggested: high arctic species tend to have a high colonization power, as their present range had to be efficiently colonized since the Würm glaciation, and also periodically during previous colonization cycles (Huntley and Webb 1989). For genetically spatially structured species, the focus should be on the most variable populations and on the ones which host many private alleles (Diadema et al. 2005). The knowledge of Evolutionary Significant Units (Crandall et al. 2000) taking into account both aspects, would be a valuable tool for conservation.

The postglacial colonization processes have favoured taxa with high vagility of propagules and a great competitive ability. In the latter respect, polyploids generally perform better than their diploid progenitors. However, the conservation of diploids is a key to the future evolution of the flora, their loss would be irreplaceable, as the families with the highest proportion of diploids are generally the ones with the most derived characters. Most of diploids have a better evolutionary potential than polyploids because minor modifications of their genome may alter the species more quickly and deeply. Lastly, harsh habitats, species rarity and endemism are key factors influencing the proportion of diploid in the native flora. Nowadays, disturbances caused by human activities have similar effects as glaciations, with the demise of many diploid stenotopic taxa, and the expansion of a few eurytopic polyploids, leading to an impoverishment and a banalization of the lowland flora.

References

Aboucaya A (1989) La flore des îles d'Hyères: étude des rapports phytogéographiques et biosystématiques avec les Maures et la Corse. Thèse de Doctorat Aix-Marseille III

Affre L, Thompson JD, Debussche M (1997) Genetic structure of continental and island populations of the Mediterranean endemic *Cyclamen balearicum* (Primulaceae). Am J Bot 84:437–451

Alsos IG, Eidesen PB, Ehrich D, Skrede I, Westergaard K, Jacobsen GH, Landvik JY, Taberlet P, Brochmann C (2007) Frequent long-distance plant colonization in the changing Arctic. Science 316:1606–1609

Azmanov I (2002) Vavilov in Ithaca, NY, 1932. In: Buican D, Thieffry D (eds) Proceedings of the XXth International Congress of History of Science (Liège, 20-26 July 1997), vol XI, Biological and medical sciences. Brepols, Turnhout, pp 303–332

Bettin O, Cornejo C, Edwards PJ, Holderegger R (2007) Phylogeography of the high alpine plant *Senecio halleri* (Asteraceae) in the European Alps: in situ glacial survival with postglacial stepwise dispersal into peripheral areas. Mol Ecol 16:2517–2524

Bretagnolle F, Felber F, Calame FG, Küpfer P (1998) La polyploïdie chez les plantes. Bot Helv 5:5–37

Breton-Sintès S (1974) Etude Biosystématique de genre *Knautia* (Dipsacaceae) dans le Massif-Central. II. Analyse morphologique et cytogénétique d'hybrides expérimentaux. Ann Sci Nat Bot Biol Vég 15:277–320

Brochmann C, Brysting AK, Alsos IG, Borgen L, Grundt HH, Scheen AC, Elven R (2004) Polyploidy in arctic plants. Biol J Linn Soc 82:521–536

Brunhoff C, Yoccoz NG, Ims RA, Jaarola M (2006) Glacial survival or late glacial colonization? Phylogeography of the root vole (*Microtus oeconomus*) in North-West Norway. J Biogeogr 33:2136–2144

Calero C, Ibanez O, Mayol M, Rossello JA (1999) Random amplified polymorphic DNA(RAPD) markers detect a single phenotype in *Lysimachia minoricensis* J.J.Rodr. (Primulaceae), a wild extinct plant. Mol Ecol 8:2133–2136

Comps B, Gömöry D, Letouzey J, Thiébaut B, Petit RJ (2001) Diverging trends between heterozygosity and allelic richness during postglacial colonization in the European beech. Genetics 157:389–397

Coulson SJ, Hodkinson ID, Webb NR, Mikkola K, Harrison JA, Pedgley DE (2002) Aerial colonization of high Arctic islands by invertebrates: the diamondback moth *Plutella xylostella* (Lepidoptera: Yponomeutidae) as a potential indicator species. Divers Distrib 8:327–334

Crandall KA, Bininda-Emonds OR, Mace GM, Wayne RK (2000) Considering evolutionary processes in conservation biology. Trends Ecol Evol 15:290–295

Crawford DJ, Mort ME (2003) Polyploidy: some things old to go with the new. Taxon 52:411–413

de Bolós O, Vigo J (1984) Flora dels Països Catalans, vol 1. Barcino, Barcelona

de Lattin G (1967) Grundriss der Zoogeographie. Gustav Fischer, Jena

Diadema K (2006) Apport de la phylogéographie, de la dynamique et de la structure des populations pour la conservation de végétaux endémiques méditerranéens. Thèse de doctorat Aix-Marseille III

Diadema K, Bretagnolle F, Affre L, Yuan YM, Médail F (2005) Geographic structure of molecular variation of *Gentiana ligustica* (Gentianaceae) in the Maritime and Ligurian regional hotspot, inferred from ITS sequences. Taxon 54:887–894

Dobigny G, Ducroz JF, Robinson TJ, Volobouev V (2004) Cytogenetics and cladistics Syst Biol 53:470–484

Duriez O, Sachet JM, Menoni E, Pidancier N, Miquel C, Taberlet P (2007) Phylogeography of the Capercaillie in Eurasia: what is the conservation status in the Pyrenees and Cantabrian Mounts? Conserv Genetics 8:513–526

Ehrendorfer F (1980) Polyploidy and distribution. In: Lewis WH (ed) Polyploidy: biological relevance. Plenum Press, New York, pp 45–60

Ehrich D, Gaudeul M, Assefa A, Koch MA, Mummenhoff K, Nemomissa S, Intrabiodiv Consortium (2007) Genetic consequences of Pleistocene range shifts: contrast between the Arctic, the Alps and the East African mountains. Mol Ecol 16:2542–2559

Eidesen PB, Alsos IG, Popp M, Stensrud O, Suda J, Brochmann C (2007) Nuclear vs. plastid data: complex Pleistocene history of a circumpolar key species. Mol Ecol 16:3902–3925

El Mousadik A, Petit RJ (1996) High level of genetic differentiation for allelic richness among populations of the argan tree [*Argania spinosa* (L) Skeels] endemic to Morocco. Theor Appl Genet 92:832–839

Favarger C (1967) Cytologie et distribution des plantes. Biol Rev 42:163–206

Favarger C (1972) Endemism in the Montane Floras of Europe. In: Valentine DH (ed) Taxonomy phytogeography and evolution. Academic, London, pp 191–204

Favarger C, Contandriopoulos J (1961) Essai sur l'endémisme. Bull Soc Bot Suisse 71:384–408

Forbes E (1846) On the connexion between the distribution of the existing fauna and flora of the British Isles, and the geological changes which have affected their area, especially during the epoch of the Northern Drift. Mem Geol Surv Great Brit 1:336–432

Gadella TWJ (1987) Pattern and process in biosystematic and evolutionary studies. In: Hovenkamp P et al (eds) Systematics and evolution: a matter of diversity. Utrecht University, Utrecht, The Netherlands, pp 35–54

Garnier S, Alibert P, Audiot P, Prieur B, Rasplus JY (2004) Isolation by distance and sharp discontinuities in gene frequencies: implications for the phylogeography of an alpine insect species, *Carabus solieri*. Mol Ecol 13:1883–1897

Gitzendanner MA, Soltis PS (2000) Patterns of genetic variation in rare and widespread plant congeners. Am J Bot 87:783–792

Goldblatt P (1980) Polyploidy in Angiosperms: Monocotyledons. In: Lewis WH (ed) Polyploidy: biological relevance. Plenum Press, New York, pp 219–239

Grundt HH, Kjolner S, Borgen L, Rieseberg LH, Brochmann C (2006) High biological species diversity in the arctic flora. Proc Natl Acad Sci USA 103:972–975

Haeckel E (1876) The History of creation: or the development of the Earth and its inhabitants by the action of natural causes. Appleton and Company, New York

Hampe A, Petit RJ (2005) Conserving biodiversity under climate change: the rear edge matters. Ecol Letters 8:461–467

Hampe A, Arroyo J, Jordano P, Petit RJ (2003) Rangewide phylogeography of a bird-dispersed Eurasian shrub: contrasting Mediterranean and temperate glacial refugia. Mol Ecol 12:3415–3426

Harland SC (1954) Nicolai Ivanovitch Vavilov. Obituary Not Fellows R Soc 9:259–264

Hartl DL, Clark AG (1989) Principles of population genetics, 2nd edn. Sinauer, Sunderland, MA

Hewitt GM (1993) Postglacial distribution and species substructure: lessons from pollen, insects and hybrid zones. In: Lees DR, Edwards D (eds) Linnean society symposium 14. Evolutionary patterns and processes. Academic, London, pp 98–123

Hewitt GM (1996) Some genetic consequences of ice ages, and their role in divergence and speciation. Biol J Linn Soc 58:247–276

Hewitt GM (1999) Post-glacial re-colonization of European biota. Biol J Linn Soc 68:87–112

Hodgson JG (1987) Why do so few plant species exploit productive habitats? An investigation into cytology, plant strategies and abundance within a local flora. Funct Ecol 1:243–250

Holderegger R, Stehlik I, Abbott RJ (2002) Molecular analysis of the Pleistocene history of *Saxifraga oppositifolia* in the Alps. Mol Ecol 11:1409–1418

Hughes PD, Woodward JC, Gibbard PL (2006) Quaternary glacial history of the Mediterranean mountains. Prog Phys Geogr 30:334–364

Huntley B, Webb T III (1989) Migration: Species' response to climatic variations caused by changes in the earth's orbit. J Biogeogr 16:5–19

Ibrahim KM, Nichols RA, Hewitt GM (1996) Spatial patterns of genetic variation generated by different forms of dispersal during range expansion. Heredity 77:282–291

Ivy-Ochs S, Kerschner H, Reuther A, Preusser F, Heine K, Maisch M, Kubik PW, Schluchter C (2008) Chronology of the last glacial cycle in the European Alps. J Quat Sci 23:559–573

Joost S, Bonin A, Bruford MW, Després L, Conord C, Erhardt G, Taberlet P (2007) A spatial analysis method (SAM) to detect candidate loci for selection: towards a landscape genomics approach to adaptation. Mol Ecol 16:3955–3969

Joost S, Bonin A, Taberlet P, Caloz R (2008) Un rôle pour la science de l'information géographique en écologie moléculaire: la détection de régions du génome soumises à la sélection naturelle. Rev Int Géomatique 18:215–237

Küpfer P (1974) Recherches sur les liens de parenté entre la flore orophile des Alpes et celle des Pyrénées. Boissiera 23:1–322

Levin DA (1983) Polyploidy and novelty in flowering plants. Am Nat 122:1–25

Levin DA (2002) The role of chromosomal change in plant evolution. Oxford University Press, Oxford

Lewis WH (1980) Polyploidy: biological relevance. Plenum Press, New York

Lexer C, Fay MF (2005) Adaptation to environmental stress: a rare or frequent driver of speciation? J Evol Biol 18:893–900

Luikart G, Allendorf FW, Cornuet JM, Sherwin WB (1998) Distortion of allele frequency distributions provides a test for recent population bottlenecks. J Hered 89:238–247

Lumaret R, Guillerm JL, Maillet J, Verlaque R (1997) Plant species diversity and polyploidy in islands of natural vegetation isolated in extensive cultivated lands. Biodiv Conserv 6:591–613

Mabble BK (2004) 'Why polyploidy is rarer in animals than in plants': myths and mechanisms. Biol J Linn Soc 82:453–466

Mansion G, Zeltner L, Bretagnolle F (2005) Phylogenetic patterns and polyploid evolution within the Mediterranean genus *Centaurium* (Gentianaceae-Chironieae). Taxon 54:931–950

Médail F, Diadema K (2008) Glacial refugia influence plant diversity patterns in the Mediterranean basin. J Biogeogr 36:1333–1345
Médail F, Quézel P (1997) Hot-spots analysis for conservation of plant biodiversity in the Mediterranean Basin. Ann Mo Bot Gard 84:112–127
Médail F, Verlaque R (1997) Ecological characteristics and rarity of endemic plants from Southeast France and Corsica. Implications for biodiversity conservation. Biol Conserv 80:269–281
Nève G, Barascud B, Descimon H, Baguette M (2008) Gene flow rise with habitat fragmentation in the bog fritillary butterfly (Lepidoptera : Nymphalidae). BMC Evol Biol 8:e84
Nichols RA, Hewitt GM (1994) The genetic consequences of long distance dispersal during colonization. Heredity 72:312–317
Peterson MA (1995) Phenological isolation, gene flow and developmental differences among low-elevation and high-elevation populations of *Euphilotes enoptes* (Lepidoptera, Lycaenidae). Evolution 49:446–455
Peterson MA (1996) Long-distance gene flow in the sedentary butterfly, *Euphilotes enoptes* (Lepidoptera: Lycaenidae). Evolution 50:1990–1999
Petit RJ, Bialozyt R, Brewer S, Cheddadi R, Comps B (2001) From spatial patterns of genetic diversity to postglacial migration processes in forest trees. In: Silvertown J, Antonovics J (eds) Integrating ecology and evolution in a spatial context. Blackwell, Oxford, pp 295–318
Petit RJ, Csaikl UM, Bordacs S, Burg K, Coart E, Cottrell J, van Dam B, Deans JD, Dumolin-Lapegue S, Fineschi S, Finkeldey R, Gillies A, Glaz I, Goicoechea PG, Jensen JS, Konig AO, Lowe AJ, Madsen SF, Matyas G, Munro RC, Olalde M, Pemonge MH, Popescu F, Slade D, Tabbener H, Taurchini D, de Vries SGM, Ziegenhagen B, Kremer A (2002) Chloroplast DNA variation in European white oaks – Phylogeography and patterns of diversity based on data from over 2600 populations. Forest Ecol Manage 156:5–26
Ravi M, Marimuthu MPA, Siddiqi I (2008) Gamete formation without meiosis in *Arabidopsis*. Nature 451:1121–1125
Reese G (1958) Polyploidie und Verbreitung. Zeitsch Bot 46:339–354
Richards AJ (1997) Plant breeding systems. Chapman and Hall, London
Schönswetter P, Paun O, Tribsch A, Niklfeld H (2003) Out of the Alps: Colonization of Northern Europe by East Alpine populations of the Glacier Buttercup *Ranunculus glacialis* L. (Ranunculaceae). Mol Ecol 12:3373–3381
Schönswetter P, Popp M, Brochmann C (2006a) Central Asian origin of and strong genetic differentiation among populations of the rare and disjunct *Carex atrofusca* (Cyperaceae) in the Alps. J Biogeogr 33:948–956
Schönswetter P, Popp M, Brochmann C (2006b) Rare arctic-alpine plants of the European Alps have different immigration histories: the snow bed species *Minuartia biflora* and *Ranunculus pygmaeus*. Mol Ecol 15:709–720
Seidenbinder M (1989) Différenciation et adaptation de certaines Angiospermes en région méditerranéenne. Thèse de doctorat, Aix-Marseille I
Skrede I, Eidesen PB, Portela RP, Brochmann C (2006) Refugia, differentiation and post-glacial migration in arctic-alpine Eurasia, exemplified by the mountain avens (*Dryas octopetala* L.). Mol Ecol 15:1827–1840
Slatkin M (1993) Isolation by distance in equilibrium and non-equilibrium populations. Evolution 47:264–279
Snogerup S (1967) Studies in the Aegean flora IX. Erysimum sect. Cheiranthus. B. Variation and evolution in the small population system. Opera Bot 14:5–86
Soltis DE, Soltis PS (1999) Polyploidy: recurrent formation and genome evolution. Trends Ecol Evol 14:348–352
Stace CA (2005) Plant taxonomy and biosystematics – does DNA provide all the answers ? Taxon 54:999–1007
Stearns SC (1992) The evolution of life histories. Oxford University Press, Oxford
Stebbins GL (1971) Chromosome evolution in higher plants. Edward Arnold, London
Stehlik I, Schneller JJ, Bachmann K (2001) Resistance or emigration: response of the high-alpine plant *Eritrichium nanum* (L.) Gaudin to the ice age within the Central Alps. Mol Ecol 10:357–370

Strid A (1969) Evolutionary trends in the breeding system of *Nigella* (Ranunculaceae). Bot Not 122:380–397

Svendsen JI, Alexanderson H, Astakhov VI, Demidov I, Dowdeswell JA, Funder S, Gataullin V, Henriksen M, Hjort C, Houmark-Nielsen M, Hubberten HW, Ingólfsson O, Jakobsson M, Kjær KH, Larsen E, Lokrantz H, Lunkka J-P, Lyså A, Mangerud J, Matioushkov A, Murray A, Möller P, Niessen F, Nikolskaya O, Polyak L, Saarnisto M, Siegert C, Siegert MJ, Spielhagen RF, Stein R (2004) Late Quaternary ice sheet history of Eurasia. Quat Sci Rev 23:1229–1271

Thompson JD (2005) Plant evolution in the Mediterranean. Oxford University Press, Oxford

Thompson JD, Lumaret R (1992) The evolutionary dynamics of polyploid plants: Origins, Establishment and Persistence. Trends Ecol Evol 7:302–307

Udvardy MDF (1969) Dynamic zoogeography. Van Nostrand Reinhold, New York

Vandewoestijne S, Nève G, Baguette M (1999) Spatial and temporal population structure of the butterfly *Aglais urticae* (Lepidoptera Nymphalidae). Mol Ecol 8:1639–1644

Vavilov NI (1950) The origin, variation, immunity and breeding of cultivated plants. Chron Bot 13:1–364

Verlaque R (1999) L'endémisme en Méditerranée: caractéristiques et menaces. Biosystema 17:45–52

Verlaque R, Contandriopoulos J (1990) Analyse des variations chromosomiques en région méditerranéenne: Polyploïdie, différenciation et adaptation. Ecol Medit 16:93–112

Verlaque R, Filosa D (1992) Mediterranean chromosome number reports – 2. Flora Medit 2:264–272

Verlaque R, Filosa D (1997) Caryologie et Biogéographie des messicoles menacées du Sud-Est de la France; Comparaison avec les autres mauvaises herbes. In: Gap-Charence (eds) AFCEV, BRG, Conserv Bot. Nat. Faut-il sauver les mauvaises herbes? Louis-Jean, Gap, pp 105-124

Verlaque R, Médail F, Quézel P, Babinot JF (1997) Endémisme végétal et Paléogéographie dans le Bassin Méditerranéen. Geobios 21:159–166

Verlaque R, Aboucaya A, Fridlender A (2002) Les xénophytes envahissants en France: ecologie, types biologiques et polyploïdie. Bot Helv 112:121–136

Vogel JC, Rumsey FJ, Schneller JJ, Barrett JA, Gibby M (1999) Where are the glacial refugia in Europe? Evidence from Pteridophytes. Biol J Linn Soc 66:23–37

Wallis GP (1994) Population genetics and conservation in New-Zealand – a hierarchical synthesis and recommendations for the 1990s. J Roy Soc N Z 24:143–160

Wang Y, Magnard J-L, McCormick S, Yang M (2004) Progression through meiosis I and meiosis II in *Arabidopsis* anthers is regulated by an A-type Cyclin predominately expressed in prophase I. Plant Physiol 136:412–4135

Watt WB, Wheat CW, Meyer EH, Martin JF (2003) Adaptation at specific loci. VII. Natural selection, dispersal and the diversity of molecular-functional variation patterns among butterfly species complexes (*Colias*: Lepidoptera, Pieridae). Mol Ecol 12:1265–1275

Widmer A, Lexer C (2001) Glacial refugia: sanctuaries for allelic richness, but not for gene diversity. Trends Ecol Evol 16:267–269

Wolf PG, Doche B, Gielly L, Taberlet P (2004) Genetic structure of *Rhododendron ferrugineum* at a wide range of spatial scales. J Hered 95:301–308

Wright S (1943) Isolation by distance. Genetics 28:114–138

Population Genetic Structure of Two Threatened Dragonfly Species (Odonata: Anisoptera) as Revealed by RAPD Analysis

Mónica A. Amorín, Marta Vila, and Adolfo C. Rivera

Please note the erratum to the authors' names at the end of the book.

Abstract The riverine odonates *Macromia splendens* and *Oxygastra curtisii* were included in the European Habitats directive as taxa of special concern. Nevertheless, there is almost no genetic information about them. We assessed the genetic diversity and population structuring among several Northwest Iberian locations where these species occur. For this, we examined the genetic pattern revealed by RAPD markers in four locations of *M. splendens* and five locations of *O. curtisii*. The former showed strong population structuring, whereas gene flow between different river systems may be the reason for the lower structuring inferred for *O. curtisii*. Based on these results, we support the need of special management for *M. splendens* in Northwest Iberia.

1 Introduction

From the conservation point of view, the order Odonata is among the most thoroughly studied insect groups. This is because odonates are widely acknowledged as biological indicators (Foote and Rice 2005; Rouquette and Thompson 2005). Accordingly, many Odonata (and Lepidoptera) are flag species for conservation in some countries. This chapter evaluates the genetic diversity and population structuring of two species of Odonata (Insecta) occurring in Northwest Iberia: *Macromia splendens* (Pictet 1843) and *Oxygastra curtisii* (Dale 1834). These species are classified as endangered by different administrations

M.A. Amorín and A.C. Rivera (✉)
Department of Ecology and Animal Biology, University of Vigo, EUET Forestal, Campus Universitario, E-36006 Pontevedra, Spain
e-mail: adolfo.cordero@uvigo.es

M. Vila
Department of Molecular and Cell Biology, University of La Coruña, Faculty of Sciences, Campus Zapateira, E-15008 La Coruña, Spain

J.C. Habel and T. Assmann (eds.), *Relict Species: Phylogeography and Conservation Biology*,
DOI 10.1007/978-3-540-92160-8_17,

(Sahlén et al. 2004), but data on their genetic diversity and population structure supporting such a categorization are lacking.

Conservation genetics of insects is a burgeoning scientific field (e.g., Legge et al. 1996; Gadeberg and Boomsma 1997; Rasplus et al. 2001; Jonsson et al. 2003; Lai and Pullin 2004). However, there are a few conservation genetic surveys specifically addressing dragonflies and damselflies (e.g., Watts et al. 2004, 2005a; 2006, 2007; Hadrys et al. 2006; Thompson and Watts 2006). As in most insect groups, the intensive development of highly variable molecular markers (e.g., Hadrys et al. 2005; Keat et al. 2005; Giere and Hadrys 2006; Lorenzo Carballa et al. 2007) and non-invasive techniques (Watts et al. 2005b) will soon provide a great amount of valuable data for nature conservation.

This work is the first population genetics survey on *M. splendens* and *O. curtisii*, two protected Odonata species. We aimed at (1) describing their genetic diversity and (2) estimating their degree of population structure. To achieve this, we screened the fine-scale genetic pattern showed by these species in several Northwest Iberian sampling sites. We used the Random Amplified Polymorphic DNA–Polymerase Chain Reaction (RAPD–PCR) markers.

1.1 The Studied Species

M. splendens is globally acknowledged as being vulnerable (Boudot 2006). This large anisopteran belonging to the Corduliidae family is the only European *Macromia* species. It occurs in Southeast France, Spain, and Portugal (Dommanget and Grand 1996) and inhabits highly localized biotopes (Cordero Rivera 2000). It appears to be distributed in three main groups of locations: France, Northwest Iberia, and Southwest Iberia. Other congeneric species are highly abundant and show more continuous distributions. Such a pattern led Lieftinck (1965) to suggest that *M. splendens* is a Tertiary relict. However, its rareness might also be due to the fact that its habitat has been quite inaccessible to researchers. Pollution and habitat destruction are the main problems for the conservation of this species (Sahlén et al. 2004; Ocharan et al. 2006, and references therein), also in NW Iberia (Cordero Rivera 2000).

O. curtisii is classified as Nearly Threatened by the IUCN Red List (Boudot et al. 2006). This member of the Corduliidae family mainly occurs in the Iberian Peninsula, and in the Western and Southern France. In addition, there are records from North and West Morocco (Jacquemin and Boudot 1999) as well as from Northern France, the Netherlands, Belgium, Western Germany, Switzerland, Northern and central Italy. This species is now extinct in Great Britain (d'Aguilar and Dommanget 1998). *O. curtisii* usually inhabits large rivers, preferring rapid waters. Interestingly, it is abundant in almost all Galician rivers (Northwest Spain, Azpilicueta Amorín et al. 2007). Therefore, a certain level of gene flow is expected.

2 Materials and Methods

2.1 Sampling

The two studied species (*M. splendens* and *O. curtisii)* were sampled between years 2000 and 2002. Sample sites and sizes (Table 1, Fig. 1) were selected according to the restricted distributional range of these taxa in NW Iberia. UTM coordinates were: Lérez 29TNH3501, Cabe 29TPH1201, Tea 29TNG4075, Albarellos 29TNG6492, Deza 29TNH6134, and Arnoia 29TPG0575. The minimum distance between two localities was 26.50 km (Tea–Lérez), whereas the maximum was 76.78 km (Arnoia–Deza). These five localities represented different ecosystems. Thus, sampling areas by rivers Lérez and Cabe represented slow and sunny stretches of deep rivers. Localities by rivers Tea and Deza represented smaller slow rivers. All river banks showed abundant vegetation. Interestingly, the vegetation type was Atlantic for Lérez, Tea, and Deza river basins, but Mediterranean for Cabe and Arnoia. Albarellos is a reservoir collecting water from river Avia. The total river length ranges from 50 km (Tea) to 84.5 km (Arnoia). Only river Lérez is not a tributary, running into the Atlantic Ocean. All sampled locations, except Albarellos and Arnoia, occur at Sites of Community Importance (SCI) within the Natura 2000 network.

Sampled populations of *O. curtisii* were large (about several thousand individuals per generation), whereas populations of *M. splendens* were smaller (hundreds of individuals, Cordero Rivera 2000; Cordero Rivera et al. 1999).

Adults were hand netted. We took one hind leg from each captured individual. Only captured males of *M. splendens* were sampled. Legs were stored in individual 1.5 mL tubes and frozen until analysis.

Table 1 Sample sizes and genetic diversity obtained for two Odonata species in several NW Iberian locations

Species	Population	N	No. (and %) of polymorphic bands	Nei's gene diversity	Shannon's diversity index
Macromia splendens	Lérez	6	10 (40)	0.16±0.21*	0.24±0.30
	Cabe	11	17 (68)	0.26±0.20	0.39±0.29
	Tea	18	25 (100)	0.35±0.12	0.53±0.13
	Albarellos	4	10 (40)	0.16±0.21	0.24±0.30
	Overall	39	25 (100)	0.38±0.11	0.57±0.12
Oxygastra curtisii	Lérez	37	33 (100)	0.44±0.07	0.63±0.08
	Cabe	18	29 (87.9)	0.28±0.17	0.43±0.22
	Tea	20	30 (90.9)	0.33±0.16	0.49±0.22
	Deza	14	33 (100)	0.40±0.09	0.59±0.10
	Arnoia	10	26 (78.8)	0.32±0.19	0.46±0.27
	Overall	99	33 (100)	0.47±0.04	0.66±0.04

*average ± standard deviation

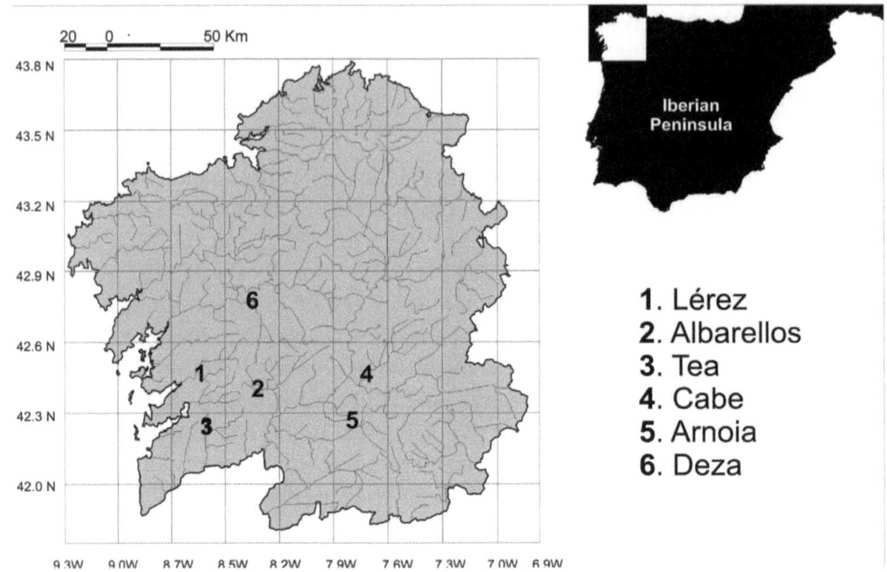

Fig. 1 Map of Galicia (NW Iberia) displaying the sampling locations for this genetic survey on two Odonata species: *Macromia splendens* and *Oxygastra curtisii*

2.2 DNA Extraction and RAPD Profiling

Whole genomic DNA was extracted by grinding the whole leg in liquid nitrogen. The obtained fine powder was homogenized in 500 μL buffer (100 mM Tris-HCl (pH 8.0), 20 mM EDTA (pH 8.0), 0.5% SDS, and 50 mM DTT). The homogenate was incubated at 40°C for four hours in the presence of RNAse (1 unit), and proteinase K (0.25 mg). Proteins were removed by extraction, first with phenol-chloroform and then with chloroform-isoamyl alcohol (2%). DNA was concentrated by ethanol precipitation (Sambrook et al. 1989). DNA samples were stored in TE 1X at a standardized concentration of 50 ng/μL.

RAPD–PCR markers have been successfully used to detect genetic polymorphism in other Odonata species (Andrés et al. 2000). We tested a total of 60 decameric random oligonucleotides (OPA01-OPA20, OPD01-OPD20, OPE01-OPE20, Operon Technologies, Inc.) in order to find polymorphism in a subsample of three individuals per species and from different localities. Eight of the primers (four per species, Table 2) proved to be highly polymorphic and were used to screen the entire collection.

Amplification reactions were performed in a final volume of 25 μL containing 1 μL 10X reaction buffer (Sigma), 1.5 mM $MgCl_2$, 10 mM dNTPs, 2 pmol primer, 1 U Taq (DNA polymerase (Sigma)), and 50 ng genomic DNA. The thermocycler program used for amplification was: initial denaturation step (94°C for 5 min), 45 cycles (94°C for 1 min, 35°C for 1 min, 72°C for 2 min), final extension (72°C for 5 min), and cooling at 4°C. A negative control was always included to monitor any

Table 2 RAPD primers (Operon Technologies Inc.) and summary of RAPD markers obtained for each studied species

Primer name	Primer sequence (5′-3′)	*Macromia splendens*	*Oxygastra curtisii*
OPA–01	CAGGCCCTTC	12, 50–1,000, 6*	–
OPA–02	TGCCGAGCTG	15, 50–750, 7	–
OPA–04	AATCGGGCTG	11, 50–750, 8	–
OPA–08	GTGACGTAGG	–	13, 50–750, 5
OPC–09	CTCACCGTCC	–	17, 50–750, 11
OPA-09	GGGTAACGCC	10, 50–1,000, 4	–
OPE-11	GAGTCTCAGG	–	9, 50–1,000, 6
OPD-01	ACCGCGAAGG	–	10, 150–500, 11

*The three values provided account for (1) number of scored bands, (2) size range of the marker (in base pairs), and (3) number of polymorphic bands (shared by <99% individuals)

possible DNA contamination. Fragments were resolved on 1.5% agarose gels and sized against a one Kilobase ladder (Sigma). Gels were run for 1.75 h at 120 V/400 A. DNA bands, stained with ethidium bromide, were visualized under UV light and photographed. All amplifications were repeated twice to prove the reproducibility of the RAPD–PCR bands. Only well-amplified and reproducible bands were considered for further analyses. RAPD bands were scored as present (1) or absent (0), and a binary matrix of different RAPD phenotypes was so assembled. RAPD bands shared by 99% or more individuals of the same species across the whole dataset were regarded as monomorphic and excluded from genetic analyses.

2.3 Statistical Analyses

We used three estimators to assess genetic diversity: (1) percentage of polymorphic loci (*P*), (2) unbiased Nei's gene diversity for dominant markers, and (3) Shannon's index, which does not assume Hardy-Weinberg equilibrium. We used TFPGA v. 1.3 (Miller 1997) to calculate *P* and Nei's genetic diversity, whereas Shannon's index was calculated using POPGENE v. 1.32 (Yeh and Boyle 1997).

We assessed the extent of genetic structuring for each species using Weir and Cockerham's (1984) unbiased estimator, θ, and its 95% confidence intervals (10,000 bootstrap iterations). We performed exact tests for population differentiation (Raymond and Rousset, 1995). These two analyses were calculated using TFPGA v. 1.3 (Miller 1997). The parameters of the Markov Chain Monte Carlo approach were: 1,000 dememorization steps, 10 batches, and 2,000 permutations per batch.

We tested for evidence of population subdivision using an analysis of molecular variance (AMOVA) as implemented in ARLEQUIN v.3.01 (Excoffier et al. 2005). We also used ARLEQUIN to test the genetic differentiation between pairs of populations using F_{ST} analogs. Statistical significance of covariance components (and

Φ-statistics) from AMOVAs and pairwise Φ_{ST}'s was determined on the basis of distribution of values obtained from 10,000 permutations of data.

We used the Mantel test, as implemented in TFGPA, to estimate the correlation between genetic and geographic distances. A UPGMA tree of populations was constructed with TFPGA, using Nei's unbiased genetic distance. Branch support was estimated by bootstrapping across loci 10,000 times.

An individual-based rather than population-level analysis may provide a more accurate view of the current gene flow. This is because standard approaches like measures of population differentiation by F_{ST} rely on equilibrium assumptions and may, therefore, not be applicable to situations where equilibrium has not yet been attained. We used the program STRUCTURE 2.1 (Pritchard et al. 2000; Falush et al. 2003) for model-based Bayesian clustering of individuals without using prior population definitions. Following Pritchard et al. (2000), this approach was followed only for data not fitting the isolation-by-distance model (i.e., *M. splendens*). Simulations were run under standard settings in the no-admixture model, assuming correlated allele frequencies, without considering the population of origin of the samples. We used a burn-in of 250,000 iterations followed by 1,000,000 iterations for parameter estimation. Each simulation was run 10 times, exploring values for K (the total number of clusters to be constructed in a given simulation) ranging from one to four. To select the most appropriate number of groups, we followed Evanno et al. (2005).

3 Results

We obtained a total of 25 (*M. splendens*) and 33 (*O. curtisi*) polymorphic bands (Tables 1 and 2). Every sample produced a different haplotype. The average gene diversity values ranged from 0.38 in *M. splendens* to 0.47 in *O. curtisii*. (Shannon's index = 0.57 and 0.66, respectively). At a population level, both gene diversity and Shannon's index were lowest for *M. splendens* from Lérez/Albarellos and highest for *O. curtisii* from Lérez (Table 1). Sample size was positively correlated with both Nei's genetic diversity and Shannon's index (Pearson's $r=0.81$, and 0.80, respectively). The degree of population differentiation, as revealed by significant Φ_{ST} values, varied from moderate (0.12, *O. curtisii* Lérez-Deza) to high (0.48, *M. splendens* Lérez-Cabe) (Tables 3 and 4). Within-population variation accounted for most of the total variation in both species ($P<0.001$, Table 5), regardless of grouping criteria such as ecosystem or phylogeny.

3.1 *M. splendens*

We screened the genetic diversity at four Northwest Iberian locations, i.e., three rivers and one reservoir (Albarellos). The most diverse locality was Tea (Table 2). Overall, most of the variation (71.39%) was due to within-population genetic diversity

Table 3 Matrix of population pairwise Φ_{ST}-values obtained for *Macromia splendens*

M. splendens	Lérez	Cabe	Tea	Albarellos
Lérez	–	***	***	NS
Cabe	**0.48**	–	**	***
Tea	**0.31**	**0.13**	–	NS
Albarellos	0.37	0.47	0.25	–

Data in bold remained significant after sequential Bonferroni correction (P-values estimated after 10,000 permutations). Above diagonal: combined probabilities for each pairwise population differentiation (exact differentiation test). * $P>0.05$, ** $P<0.01$, *** $P<0.001$

NS non-significant

Table 4 Matrix of population pairwise Φ_{ST}-values obtained for *Oxygastra curtisii*

O. curtisii	Lérez	Cabe	Tea	Deza	Arnoia
Lérez	–	***	***	***	***
Cabe	**0.21**	–	***	***	***
Tea	**0.15**	**0.40**	–	***	***
Deza	**0.12**	**0.31**	**0.24**	–	***
Arnoia	**0.16**	**0.21**	**0.34**	**0.26**	–

Data in bold remained significant after sequential Bonferroni correction (P-values estimated after 10,000 permutations). Above diagonal: combined probabilities for each pairwise population differentiation (exact differentiation test). * $P>0.05$, ** $P<0.01$, *** $P < 0.001$

NS Non-significant

(Table 5). There was significant population structuring according to sampling location, as suggested by both the test of exact differentiation ($\chi^2=213.77$, d.f. $=50$, $P<0.001$) and F-statistics ($\theta=0.29$, CI $=(0.22$–$0.35)$). Regarding population-wise comparisons, the extent of differentiation ranged from moderate (Tea-Cabe, $\Phi_{ST}=0.13$) to strong (Lérez-Cabe, $\Phi_{ST} = 0.48$). Albarellos was the only locality not significantly differentiated from any other (Table 3). No grouping was significantly differentiated from AMOVA analysis (Table 5), although river Cabe clearly separated from the rest of samples, as revealed by the UPGMA tree (Fig. 2a). Geographic and genetic distances were not correlated (Mantel test, $r=0.38$, $P=0.26$). Results from no-admixture clustering with STRUCTURE 2.1 indicated that the most likely grouping of *Macromia* individuals was $K=2$. The first two sampled localities (Lérez and Albarellos) appeared as purely belonging to a first population (POP1) (*white*, Fig. 3). The third sampled site (Tea) contained (1) 50% of individuals directly assigned to a second population, POP2 (*grey*, Fig. 3), (2) 11% of individuals directly assigned to POP1, and (3) 39% of individuals whose genetic pattern was intermediate between POP1 and POP2, but with a marked trend towards POP2. The fourth sampled locality (Cabe) was almost totally composed of individuals belonging to POP2.

Table 5 Analysis of molecular variance (AMOVA) across a range of putative population groupings

			Percentage of variation accounted for (Φ)		
Species	Grouping criterion	Population grouping	Among groups (Φ_{CT})	Among populations within groups (Φ_{SC})	Within-populations (Φ_{ST})
Macromia splendens	–	All in one group	–	28.61 (*0.29*)***	71.39
	River basin	[Cabe/Tea/ Albarellos] [Deza] [Lérez]	16.01 (0.16)	19.61 (0.23)***	64.38 (0.36)***
	Reservoir vs. river	[Albarellos] [all remaining POPs]	8.64 (0.08)	24.45 (0.27)***	66.91 (0.33)***
	STUCTURE results	[Albarellos/ Lérez] [Tea/Cabe]	21.87 (0.22)	13.51 (0.17)***	64.62 (0.35)***
	UPGMA tree	[Cabe] [all remaining POPs]	−4.69 (−0.05)	31.96 (0.30)***	72.73 (0.27)***
Oxygastra curtisii	–	All in one group	–	22.17 (*0.22*)***	77.83
	River basin	[Cabe/Tea/ Arnoia] [Deza] [Lérez]	−15.05 (−0.15)	34.9 (0.30)***	80.17 (0.19)***
	UPGMA tree	[Cabe/Arnoia] [Tea/Deza/ Lérez]	11.02 (0.11)*	15.13 (0.17)***	73.86 (0.26)***
	UPGMA tree	[Cabe/Arnoia] [Tea] [Deza/ Lérez]	10.09 (0.10)**	13.56 (0.15)***	76.35 (0.24)***

* $P=0.1$, ** $P=0.061$, *** $P<0.001$. Values in italics correspond to Φ_{ST}

3.2 *O. curtisii*

We surveyed the genetic diversity of this species in five Northwest Iberian riverine localities. The most diverse locality was Lérez (Table 1). The within-population genetic diversity accounted for most of the variation (77.83%). There was significant population differentiation between sampling localities, as suggested by both the test of exact differentiation ($\chi^2=446.69$, d.f. $=66$, $P<0.001$) and F-statistics ($\theta=0.16$, CI $=(0.12$–$0.20)$). Regarding population-wise comparisons, the extent of differentiation ranged from moderate (Lérez-Deza, $\Phi_{ST}=0.12$) to strong (Tea-Cabe, $\Phi_{ST}=0.40$) (Table 4). Geographic distance and genetic divergence were significantly

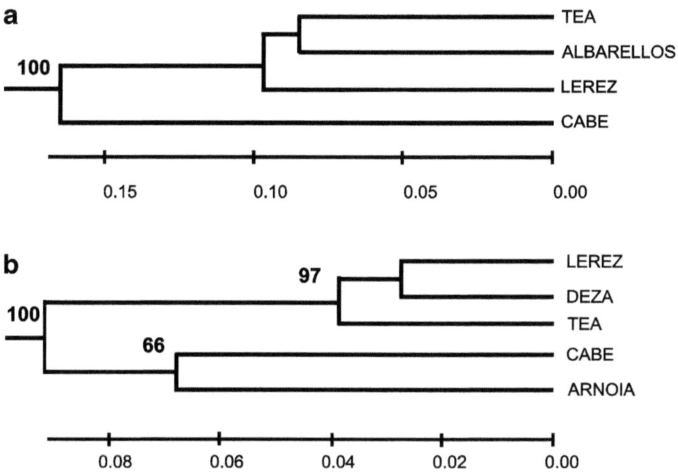

Fig. 2 UPGMA dendrogram based on Nei's unbiased genetic distance. Numbers at nodes are bootstrap values <50% (10,000 pseudoreplicates). (**a**) *Macromia splendens*, (**b**) *Oxygastra curtisii*

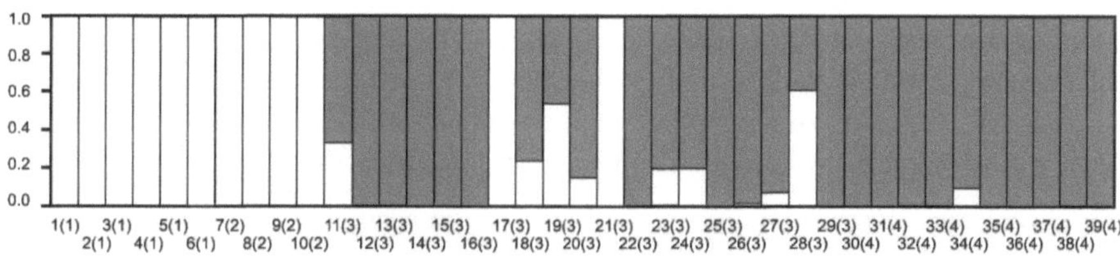

Fig. 3 Results of STRUCTURE no-admixture clustering for *Macromia splendens*. Each individual is represented by a vertical column and the proportion of each of the two colors indicated the putative population they derive from. The most likely number of population ($K=2$) was determined following Evanno et al. (2005). Numbers in the X axis identify each analyzed specimen ($n=38$), giving its geographical origin in parentheses: (1) Lérez, (2) Albarellos, (3) Tea, and (4) Cabe

correlated in the area studied (Mantel test, $r=0.61$, $P<0.05$). The phylogenetic reconstruction showed two differentiated clades (Fig. 2b), in agreement with the geographic location.

4 Discussion

4.1 *M. splendens*

There was a high population structuring according to the sampling location. Regarding significant population-wise comparisons, the extent of differentiation ranged from moderate (Tea-Cabe) to strong (Lérez-Cabe). Lérez, Albarellos, and

Cabe were the poorest localities in terms of genetic diversity. An explanation for this might be that they are at the edge of the species' range. Gene flow may be more restricted between two groups of localities, as revealed by STRUCTURE: POP1 (Lérez-Albarellos) and POP2 (Tea-Cabe). This hypothesis requires further investigation using highly variable co-dominant markers.

Albarellos may have a high conservation value, as (1) it is the only studied locality not significantly differentiated from the others (Table 3) and (2) its geographic condition is intermediate between other sites inhabited by *M. splendens*. The reservoir Albarellos (336 ha, built in 1971) is the only sampled locality not included as SCI in the Natura-2000 network. Such an intermediate site may be important for a species with less dispersal and higher structuring than *O. curtisii*. Low dispersal in *M. splendens* is due to its ecological requirements rather than its actual dispersal ability, as anisopterans are usually good fliers. These facts point to the importance of preserving appropriate habitats and connectivity between them for *M. splendens*, a species less flexible in ecological demands, and clearly linked to remnants of autochthonous forests (Cordero Rivera 2006).

Seven out of the nine known Northwest Iberian populations of *M. splendens* were found in natural rivers. The remaining two sites were located in man-made hydroelectric reservoirs, where aquatic/riverine vegetation is lacking. Despite the negative effect of reservoirs on some insect groups due to habitat destruction (Turner 2007) and fragmentation (Watanabe and Omura 2007), it will be interesting to determine whether Albarellos resulted from a recent colonization from other river basins (e.g., Tea) or descended from individuals previously present in the river.

4.2 *O. curtisii*

There was a significant population structuring according to the sampling location. Regarding population-wise comparisons, the extent of differentiation ranged from moderate (Lérez-Deza) to strong (Tea-Cabe). AMOVA results (Table 5) suggested the presence of phylogeographic structure in our data. Clustering of *O. curtisii* populations, according to the branching pattern showed by the UPGMA tree (Fig. 2b), yielded a nearly significant value ($\Phi_{CT}=0.101$, $P=0.06$). This putative structure is congruent with the isolation-by-distance pattern inferred through Mantel's test. Accordingly, *O. curtisii* appears more widely distributed across the region studied than *M. splendens*. The former shows farther dispersal than the latter (Azpilicueta Amorín et al. 2007).

Arnoia was the only sampled locality not classified as SCI. We recommend this locality to be assigned a certain level of protection. This is based on the fact that Arnoia ($n=10$) showed higher genetic variability than Cabe ($n=18$). Both populations are geographically and phylogenetically close (Figs. 1 and 2). However, gene flow between them is expected to be low (Table 3). It was somewhat surprising that the highest level of genetic diversity was found in Lérez (Table 1). This is because

Lérez is probably the studied location suffering the most harmful human activity, e.g., intensive forestry with eucalypt plantations all over the basin, leisure facilities, and small dams.

4.3 Final Remarks

Different species with contrasting ecological demands react rather differently on the same landscape pattern (e.g., Louy et al. 2007). Our results indicated a higher structure for *M. splendens* than for *O. curtisii*, which agrees with the different dispersal of both species. However, these results should be taken as preliminary, due to the small sample size of some localities. For instance, Lérez was the locality where *M. splendens* ($n=6$) was the least genetically diverse, but the one with highest gene diversity and Shannon's index for *O. curtisii* ($n=37$). According to Aagaard et al. (1998), sample size might also explain the higher structuring of the two studied species compared to other Odonata (also surveyed using RAPDs) at similar geographical scale. Andrés et al. (2000) determined a low degree of structuring for *Ischnura graellsii* ($\theta=0.061$), whereas Andrés et al. (2002) found a moderate differentiation for *Ceriagrion tenellum* ($\theta=0.142$).

Different types of molecular markers may differ in the level of population differentiation they reveal (Volis et al. 2005). Therefore, it would be inappropriate to compare the extent of population structuring (as revealed by *F* analogs) we obtained for *M. splendens* and *O. curtisii*, with none or very little structuring found in *Nehalennia irene* (microsatellites; Wong et al. 2003), *Calopteryx splendens* (AFLPs; Svensson et al. 2004), *Lestes viridis* (allozymes, De Block et al. 2005), or *Ischnura elegans* (AFLPs; Abbott et al. 2008). In addition, comparisons with other studies should also include the geographical scale. The implications of a high population structuring ($F_{ST}=0.25$–0.48) at fine scale (0–26 km), as shown by *Lestes viridis* (allozymes, Geenen et al. 2000) and *Coenagrium mercuriale* (microsatellites, Watts et al. 2004, 2006), might mislead the reconstruction of those processes leading to differentiation of populations as ours separated by 26.5–76.78 km.

To conclude, the present study determined that *M. splendens* is likely to be strongly structured in NW Iberia, whereas gene flow between different river basins may be the reason for the lower differentiation inferred for *O. curtisii*. Based on these results, we agree with the need of special management for *M. splendens*. Estimates of population size and assessment of putative connectivity networks between localities are needed for a better comprehension of the conservation status of *O. curtisii*.

Acknowledgements M. Pimentel and one anonymous referee improved the earlier versions of the manuscript with their comments. R. Sánchez-Guillén kindly helped in the laboratory. This work is part of MAA's PhD thesis. D. Romero designed Fig. 1. Funding was provided by grants from Spanish Ministry of Science (BOS2001-3642) and Xunta de Galicia (PGIDT01MAM37101PR) to ACR. MV was financed by the *Isidro Parga Pondal* program and the University of A Coruna.

References

Aagaard JE, Krutovskii KV, Strauss SH (1998) RAPDs and allozymes exhibit similar levels of diversity and differentiation among populations and races of Douglas-fir. Heredity 81:69–78

Abbott JK, Bensch S, Gosden TP, Svensson EI (2008) Patterns of differentiation in a colour polymorphism and in neutral markers reveal rapid genetic changes in natural damselfly populations. Mol Ecol 17(6):1597–1604. doi:10.1111/j.1365-294X.2007.03641.x

Andrés JA, Sánchez-Guillén RA, Cordero Rivera A (2000) Molecular evidence for selection of female color polymorphis in the damselfly *Ischnura graellsii*. Evolution 54:2156–2161

Andrés JA, Sánchez-Guillén RA, Cordero Rivera A (2002) Evolution of female colour polymorphism in damselflies: testing the hypotheses. Anim Behav 63:677–685

Azpilicueta Amorín M, Rey Rañó C, Docampo Barrueco F, Rey Muñiz XC, Cordero Rivera A (2007) A preliminary study of biodiversity hotspots for odonates in Galicia, NW Spain. Odonatologica 36:1–12

Boudot J-P (2006) *Macromia splendens*. In IUCN 2007. *2007 IUCN Red List of Threatened Species*. Available at http://www.iucnredlist.org. Cited 6 Feb 2008

Boudot J-P, Riservato E, Hardersen S (2006) *Oxygastra curtisii*. In IUCN 2007. *2007 IUCN Red List of Threatened Species*. Available at http://www.iucnredlist.org. Cited 6 Feb 2008

Cordero Rivera A (2000) Distribution, habitat requirements and conservation of *Macromia splendens* Pictet (Odonata: Corduliidae) in Galicia (NW Spain). Int J Odonatol 3:73–83

Cordero Rivera A (ed) (2006) Forests and dragonflies. Pensoft publishers, Sofia, Moscow

Cordero Rivera A, Utzeri C, Santolamazza Carbone S (1999) Emergence and adult behaviour of *Macromia splendens* (Pictet) in Galicia, Northwestern Spain (Anisoptera. Corduliidae). Odonatologica 28:333–342

d'Aguilar JD, Dommanget JL (1998) Guide des Libellules d'Europe et d'Afrique du Nord. Delachaux et Niestlé, Lausanne

De Block M, Geenen S, Jordaens K, Backeljau T, Stoks T (2005) Spatiotemporal allozyme variation in the damselfly, *Lestes viridis* (Odonata: Zygoptera): gene flow among permanent and temporary ponds. Genetica 124:137–144

Dommanget JL, Grand D (1996) *Macromia splendens* (Pictet, 1843). In: Van Helsdingen PJ, Willemse L, Speight MCD (eds) Background information on invertebrates of the habitats directive and the bern convention. Part II – Mantodea, odonata, orthoptera and arachnida. Council of Europe, Strassbourg, pp 315–323

Evanno G, Regnaut S, Goudet J (2005) Detecting the number of clusters of individuals using the software STRUCTURE: a simulation study. Mol Ecol 14:2611–2620

Excoffier L, Laval G, Schneider S (2005) Arlequin ver. 3.0: an integrated software package for population genetics data analysis. Evol Bioinform Online 1:47–50

Falush D, Stephens M, Pritchard JK (2003) Inference of population structure: extensions to linked loci and correlated frequencies. Genetics 164:1567–1587

Foote AL, Rice CL (2005) Odonates as biological indicators of grazing effects on Canadian prairie wetlands. Ecol Entomol 30:273–283

Gadeberg RME, Boomsma JJ (1997) Genetic population structure of the large blue butterfly *Maculinea* alcon in Denmark. J Insect Conserv 1:99–111

Geenen S, Jordaens K, De Block M, Stoks R, De Bruyn L (2000) Genetic differentiation and dispersal among populations of the damselfly *Lestes viridis* (Odonata). J North Am Benthol Soc 19:321–328

Giere S, Hadrys H (2006) Polymorphic microsatellite loci to study population dynamics in a dragonfly, the libellulid *Trithemis arteriosa* (Burmeister, 1839). Mol Ecol Notes 6:933–935

Hadrys H, Schroth W, Schierwater B, Streit B, Fincke OM (2005) Tree hole odonates as environmental monitors: non-invasive isolation of polymorphic microsatellites from the neotropical damselfly *Megaloprepus caerulatus*. Conserv Genet 6:481–483

Hadrys H, Clausnitzer V, Groenefeld LF (2006) The present role and future promise of conservation genetics for forest Odonates. In: Cordero Rivera A (ed) Forest and dragonflies. 4th WDA

International symposium of odonatology, pontevedra (Spain), July 2005. Pensoft publishers, Sofia, Moscow, pp 279–299

Jacquemin G, Boudot JP (1999) Les libellules (Odonates) du Maroc. Sociêté Française d'Odonatologie, Paris

Jonsson M, Johannesen J, Seitz A (2003) Comparative genetic structure of the threatened tenebrionid beetle *Oplocephala haemorrhoidalis* and its common relative *Bolitophagus reticulatus*. J Insect Conserv 7:111–124

Keat S, Thompson DJ, Kemp SJ, Watts PC (2005) Ten microsatellite loci for the red-eyed damselfly *Erythromma viridulum* (Charpentier). Mol Ecol Notes 5:788–790

Lai B-CG, Pullin AS (2004) Phylogeography, genetic diversity and conservation of the large copper butterfly *Lycaena dispar* in Europe. J Insect Conserv 8:27–36

Legge J, Roush R, Desalle R, Vogler A, May B (1996) Genetic criteria for establishing evolutionarily significant units in Cryan's buckmoth. Conserv Biol 10:85–98

Lieftinck MA (1965) *Macromia splendens* (Pictet, 1843) in Europe with notes on its habits, larva and distribution (Odonata). Tijdschr Entomol 108:41–59

Lorenzo Carballa O, Giere S, Cordero A, Hadrys H (2007) Isolation and characterization of microsatellite loci to study parthenogenesis in the citrine forktail, *Ischnura hastata* (Odonata: Coenagrionidae). Mol Ecol Notes 7:839–841

Louy D, Habel J, Schmitt T, Meyer M, Assmann T, Müller P (2007) Strongly diverging population genetic patterns of three skipper species: isolation, restricted gene flow and panmixis. Conserv Genet 8:671–681

Miller MP (1997) Tools For Population Genetic Analyses (TFPGA) A Windows™ program for the analysis of allozyme and molecular population genetic data. http://www.marksgeneticsoftware.net/tfpga.htm. Cited 20 May 2007

Ocharan FJ, Ferreras Romero M, Ocharan R, Cordero Rivera A (2006) *Macromia splendens* (Pictet, 1843). In: Verdú JR, Galante E (eds) Libro Rojo de los Invertebrados de España. Dirección General de Conservación de la Naturaleza, Madrid, pp 258–259

Pritchard JK, Stephens M, Donnelly P (2000) Inference of population structure using multilocus genotype data. Genetics 155:945–959

Rasplus JY, Garnier S, Meusnier S, Piry S, Mondor G, Audiot P, Cornuet JM (2001) Setting conservation priorities: the case study of *Carabus solieri* (Col. Carabidae). Genet Select Evol 33:S141–S175

Raymond M, Rousset F (1995) An exact test for population differentiation. Evolution 49:1280–1283

Rouquette JR, Thompson DJ (2005) Habitat associations of the endangered damselfly *Coenagrion mercuriale*, in a water meadow ditch system in Southern England. Biol Conserv 123:225–235

Sahlén G, Bernard R, Cordero Rivera A, Ketelaar R, Suhling F (2004) Critical species of Odonata in Europe. Int J Odonatol 7:385–398

Sambrook J, Fritsch EF, Maniatis F (1989) Molecular cloning. A laboratory manual, 2nd edn. Cold Spring Harbor Laboratory Press, New York

Svensson EI, Kristoffersen L, Oskarsson K, Bensch S (2004) Molecular population divergence and sexual selection on morphology in the banded demoiselle (*Calopteryx splendens*). Heredity 93:423–433

Thompson DJ, Watts PC (2006) The structure of the *Coenagrion mercuriale* populations in the New Forest, Southern England. In: Cordero Rivera A (ed) Forest and Dragonflies. 4th WDA International symposium of odonatology, pontevedra (Spain), July 2005. Pensoft publishers, Sofia, Moscow, pp 239–258

Turner CR (2007) Water beetles associated with reservoirs on table mountain, cape town: implications for conservation. J Insect Conserv 11:75–83

Volis S, Yakubov B, Shulgina I, Ward D, Mendlinger S (2005) Distinguishing adaptive from nonadaptive genetic differentiation: comparison of Q_{ST} and F_{ST} at two spatial scales. Heredity 95:466–475

Watanabe K, Omura T (2007) Relationship between reservoir size and genetic differentiation of the stream caddisfly *Stenopsyche marmorata*. Biol Conserv 136:203–211

Watts PC, Rouquette JR, Saccheri IJ, Kemp SJ, Thompson DJ (2004) Molecular and ecological evidence for small-scale isolation by distance in an endangered damselfly, *Coenagrion mercuriale*. Mol Ecol 13:2931–2945

Watts PC, Kemp SJ, Saccheri IJ, Thompson DJ (2005a) Conservation implications of genetic variation between spatially and temporally distinct colonies of the endangered damselfly *Coenagrion mercuriale*. Ecol Entomol 30:541–547

Watts PC, Thompson DJ, Daguet C, Kemp SJ (2005b) Exuviae as a reliable source of DNA for population-genetic analysis of odonates. Odonatologica 34:183–187

Watts PC, Saccheri IJ, Kemp SJ, Thompson DJ (2006) Population structure and the impact of regional and local habitat isolation upon levels of genetic diversity of the endangered damselfly *Coenagrion mercuriale* (Odonata: Zygoptera). Freshw Biol 51:193–205

Watts PC, Rousset F, Saccheri IJ, Leblois R, Kemp SJ, Thompson DJ (2007) Compatible genetic and ecological estimates of dispersal rates in insect (*Coenagrion mercuriale*: Odonata: Zygoptera) populations: analysis of "neighbourhood size" using a more precise estimator. Mol Ecol 16:737–751

Weir BS, Cockerham CC (1984) Estimating *F*-statistics for the analysis of population structure. Evolution 38:1358–1370

Wong A, Smith ML, Forbes MR (2003) Differentiation between subpopulations of a polychromatic damselfly with respect to morph frequencies, but not neutral genetic markers. Mol Ecol 12:3505–3513

Yeh FC, Boyle TJB (1997) Population genetic analysis of co-dominant and dominant markers and quantitative traits. Belg J Bot 129:157

Review
Molecular Methods: Blessing or Curse?

Aline Finger and Charlotte Klank

Abstract Conservation genetic studies make use of molecular methods to obtain valuable information which help optimizing management strategies especially for threatened species. This chapter presents an overview of different molecular markers (microsatellites, AFLPs, RFLPs, RAPDs, mtDNA, allozymes) and their applications in conservation and genetic studies. Microsatellites have shown to be, though expensive, currently the most popular genetic marker as the high degree of polymorphism is ideal to study small geographical scales of species. RFLPs, RAPDs and allozymes still represent useful markers for studies of both, small and larger geographical scales. Low degree of polymorphism, no detection of alleles and low reproducibility characterize some drawbacks. To examine phylogeography MtDNA seems to be the best choice.

1 Introduction

Conservation studies rely increasingly on molecular methods to evaluate species "status quo," historical distributions (biogeographical traits), and to develop management strategies for the restoration of populations. Since the invention of the polymerase chain reaction in the late 1980s (Mullis et al. 1996), further possibilities have opened up to apply genetic tools for diverse biological methods (see Fig. 1). Zhang and Hewitt (2003) revealed that, among the 1,758 primary papers and primer notes published between 1994 and 2003 in the journal of *Molecular Ecology*, 29.8 and 42.5% were indexed with mitochondrial and microsatellite DNA markers, respectively. Nevertheless, many more genetic methods exist to answer different ecological and genetic questions. Scientists, unfamiliar with the most commonly used genetic methods in the "conservation genetics field," tend to get lost in the different molecular techniques as each has its own advantages and disadvantages,

A. Finger (✉) and C. Klank
ETH Zürich, ITES-Ecosystem Management, Universitätstrasse 16, CH-8092
Zürich, Switzerland
e-mail: aline.finger@env.ethz.ch

J.C. Habel and T. Assmann (eds.), *Relict Species: Phylogeography and Conservation Biology*, 309
DOI 10.1007/978-3-540-92160-8_18, © Springer-Verlag Berlin Heidelberg 2010

and can be applied in diverse forms. However, all these molecular techniques aim to detect genetic variation and differences within species, populations, or even individuals. This is generally achieved by comparing special DNA sections "marked" by the individual genetic method.

This chapter aims to present an overview of the most commonly used molecular methods during the last two decades (microsatellites, AFLPs, RFLPs, RAPDs, mtDNA, and allozymes) and the variety of ecological questions which can be answered with each method.

2 Molecular Markers and Their Application

2.1 Microsatellites

Microsatellites consist of short, tandemly repeated sequences of 1–6 base pairs within the nucleus of the cell (Palo et al. 1995; Ashley 1999). They have an elevated rate of mutation due to "slipped-strand mispairing" (Levinson and Gutman 1987; Palo et al. 1995; Eisen 1999), resulting in a high proportion of polymorphism even between closely related lines (Semagn et al. 2006 and references therein). Resulting variations (alleles) are scored through differing banding patterns. This marker is neutral to selection and is inherited co-dominantly as a standard Mendelian trait (Meglécz and Solignac 1998; Ashley 1999; Luikart and England 1999).

Microsatellites are the most commonly applied molecular marker in ecological research (Fig. 1). Their use in ecology and conservation studies is very broad and ranges from the identification of population genetic differentiation, demographic

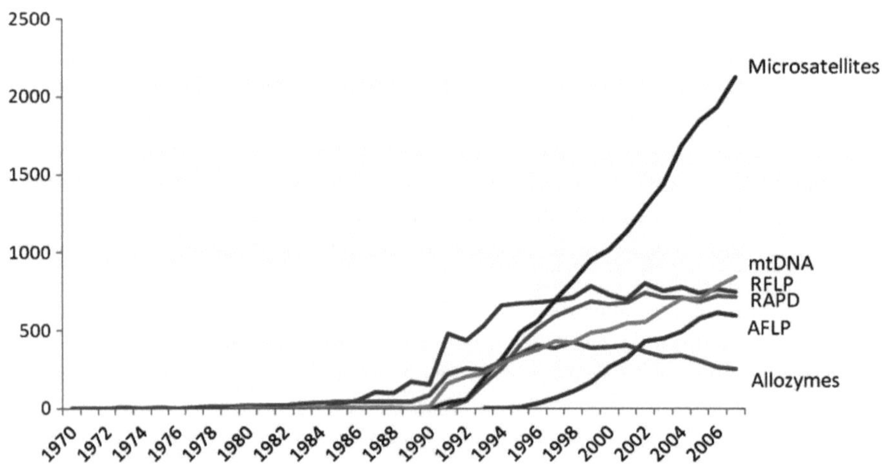

Fig. 1 Number of publications (selected biological subject) between 1970 and 2007 employing mtDNA, Allozymes, Microsatellites, RFLPs, RAPDs and AFLPs found via ISI web of knowledge

changes (e.g., bottlenecks, changes in effective population sizes (Ne), genetic drift), to parentage analysis and the definition of management units (e.g., Jones and Ardren 2003; Wan et al. 2004; Leberg 2005; Olivieri et al. 2008; Orsini et al. 2008). Minimal preserved or dry samples are suitable for the microsatellite technique, allowing the analysis of ancient population patterns and genetic changes over time (Bruford and Wayne 1993; Meglécz and Solignac 1998). However, microsatellites may have limited phylogenetic value due to homoplasy (Goldstein and Pollock 1997; Doyle et al. 1998; Selkoe and Toonen 2006). An informative review about microsatellites and their drawbacks can be found in the study by Selkoe and Toonen (2006).

2.2 *Mitochondrial DNA*

Mitochondrial DNA (mtDNA) is an extra-chromosomal genome in the cell mitochondria that resides outside of the nucleus, and is inherited from mother with no paternal contribution (Avise 1991). The obtained PCR products are sequenced and banding patterns analyzed.

The theory of relatively constant mutation rates (molecular clock) (Lushai et al. 2003) is used to estimate time scales in which populations are split up, allowing the detection of, e.g., species dispersals and dispersal centers (Riddle and Honeycutt 1990; Rocha et al. 2008). Thus, genetic variation can be investigated on a broader geographical scale to unravel the historic (historical or recent) barriers to gene flow (Avise 2000) and genetic basis of speciation and evolution of species and genera. Due to higher evolutionary rates of mtDNA relative to the nuclear genome (Wan et al. 2004 and references therein), this marker is preferred in constructing phylogenies and inferring evolutionary history, and is therefore ideal for within- and between-species comparisons (DeYoung and Honeycutt 2005; Behura 2006). Furthermore, mtDNA is believed to be the best tool for resolving taxonomic problems (Wan et al. 2004), identifying regions of endemism (Proudfoot et al. 2006), and Evolutionary Significant Units (ESUs) (Wan et al. 2004). Drawbacks of mtDNA analyses include hybridization, introgression, and incomplete lineage sorting. Moreover, mtDNA is of little use in investigating the recent loss of genetic variation and any individual-level events such as identity, individual dispersal, and mating systems (Wan et al. 2004).

2.3 *Allozyme*

Allozymes are enzyme variants due to allelic differences and can be visualized through protein electrophoresis. This technique was developed to quantify the genetic and geographic variation in wildlife populations, and it remains a cost-effective and straightforward method (Avise 1994). Genetic variations caused by mutations are expressed as amino acid replacements due to changes in protein compositions, and are resolved as bands (alleles) on electrophoretic gels (DeYoung and Honeycutt 2005).

Allozymes have been important in plant biosystematics (see van der Bank et al. 2001) and are suitable for the detection of genetic variation within and between populations (Steiner and Joslyn 1979; Bartlett 1981; Loxdale et al. 1985). Due to a relatively low average heterozygosity, the application of this marker is suitable for geographically broader scales of extant species (van der Bank et al. 2001; Schmitt 2007). Even the analysis of parentage as single-locus polymorphism is sometimes possible with this marker (Chakraborty and Hedrick 1983; Meagher and Thompson 1986). Similarly, levels of hybridization, introgression, gene flow, and polyploidy can be studied. On the contrary, allozymes have limited phylogenetic value (e.g., Lowe et al. 2004 and references therein), represent phenotypic traits, and might be subject to selection. Another significant drawback is that only fresh material can be used to extract the proteins.

2.4 Multilocus DNA Marker Systems

Three of the commonly used multilocus DNA marker systems in evolutionary, taxonomic, ecological, phylogenetic, and genetic studies are RFLPs, RAPDs, and AFLPs (DeYoung and Honeycutt 2005; Behura 2006; Agarwal et al. 2008). All these markers generate banding patterns that are scored for variation. In all three markers, the detected variation is caused by either point mutation within recognition sequences as well as insertions and/or deletions between the recognition sites, which may lead to an underestimation of genetic variation (DeYoung and Honeycutt 2005). Since none of the described multilocus markers is specific to a certain target organism DNA, there is a risk of false variation generated by contaminations (Sunnucks 2000). Furthermore, dominance of some of the markers (RAPD and AFLP) does not allow for a detection of alleles.

Depending on the sampling strategy, these markers can cover a wide spatial range, allowing for a detailed fine scale analysis of population structure between individuals, especially with AFLPs (Meudt and Clarke 2007), up to taxonomically and spatially coarse studies (e.g., Despres et al. 2003). While RFLP, RAPD, or AFLP are unsuitable to estimate mutation rates or alike, and are thus inappropriate for temporal studies (i.e., evolutionary), they provide a detailed image of the present species or population genetic state.

2.4.1 RFLP: Restriction Fragment Length Polymorphism

Restriction fragment length polymorphisms (Botstein et al. 1980) are highly polymorphic, co-dominantly inherited markers based on the use of restriction enzymes which can be applied as single and multilocus probes with the banding patterns resulting from multilocus probes.

The technique generates highly reproducible banding patterns and is characterized by a high heritability (Lowe et al. 2004; Semagn et al. 2006; Agarwal et al.

2008). It is used in areas such as population and conservation genetics, diversity (e.g., Apostolidis et al. 2008), phylogenetics (e.g., Hu et al. 2008), linkage mapping (e.g., Tanksley et al. 1989), or cultivar identification (e.g., Busti et al. 2004), though their main application is within human genetics (Weising et al. 2005). As RFLPs require relatively large amounts of DNA, they have recently been replaced by PCR-RFLPs or AFLP analyses in most ecological studies.

2.4.2 RAPD: Random Amplified Polymorphic DNA

The dominant marker system RAPD, introduced by Williams et al. (1990), is based on arbitrarily amplifying DNA sequences during PCR, without prior knowledge of the organism sequence, using 10 nucleotide primers (Lowe et al. 2004, Weising et al. 2005). One of the main problems associated with RAPDs is their susceptibility to changes in reaction conditions leading to reproducibility problems (Jones et al. 1997; Agarwal et al. 2008; Assmann et al. 2007). Due to these problems, some peer-reviewed journals (e.g., Molecular Ecology) have recently changed their policy and publish RAPD data only in exceptional cases.

RAPDs have been used in many fields, among them are studies on population and conservation genetics (e.g., Kim et al. 2008), phylogenetics (e.g., Simmons et al. 2007), and linkage mapping (e.g., Sun et al. 2008).

2.4.3 AFLP: Amplified Fragment Length Polymorphism

AFLPs are dominant markers based on a combination of the RFLP and PCR techniques and were developed by Vos et al. (1995). Also, they do not require any previous sequence information and are based on the digestion of DNA by restriction enzymes and adapter ligation (resulting in universal primer binding sites), with two subsequent PCRs using specific primers. AFLPs are highly reproducible (Meudt and Clarke 2007; Agarwal et al. 2008) and primers can be combined to yield a large set of combinations, though this may also lead to a clustering of markers with certain restriction enzymes (Keim et al. 1997; Young et al. 1999; Saal and Wricke 2002).

AFLPs find wide application in studies on population genetics, diversity, and differentiation (e.g., Abbott et al. 2008; Tang et al. 2008), phylogenetics and taxonomy (e.g., Brouat et al. 2004; Schenk et al. 2008), hybridization (e.g., Volkova et al. 2008), linkage, gene, and genome mapping (e.g., Olmstead et al. 2008; Radoev et al. 2008), assignments (e.g., Yang et al. 2008) and kinship (e.g., Hardy et al. 2006). Although they are dominant markers, the large number of loci gives them a high statistical power (Meudt and Clarke 2007). Therefore, they are well suited for intraspecific studies (distinguishing between closely related individuals), where many loci are necessary (i.e., high genomic heterogeneity, low genetic variability), in polyploids, and in systems with hybridization occurring (Meudt and Clarke 2007).

Table 1 Advantages, disadvantages and different features of microsatellites, Allozymes, mtDNA, RFLPs, RAPDs, and AFLPs

	Advantages	Disadvantages	Genomic abundance	DNA amount required	Polymorphism	Inheritance	Single loci
Micro-satellites	Medium costs (1)No sequence information needed (4) High reproducibility & reliability (4)	High start-up costs (1, 4) Species-specific primer pairs (4) PCR-based problems (5)	Medium (1, 2) – high (4)	Low (1, 2)	Medium (2) – high (1,4)	Co-dominant (1, 3)	Yes (3)
Allozymes	Easy & safe protocol (4)Low costs (4, 8) Many enzyme systems (4) Suited for polyploids (4)	Underestimation of genetic variation (4) Not sure whether neutral or not (10) Limited to extant populations (10) Only forzen or fresh samples (10)	–	–	Low (4)	Co-dominat (3)	Yes (3)
mtDNA	Constant mutation rates (11) Available primers Neutral and selective regions (6) Inherited maternally (7)	Little practical value for population genetic studies (8) Expensive	–	–	High (6)	Co-dominant (3)	Yes (3, 8)
RFLP	Detection of nucleotide DNA and organelle DNA polymorphisms (4) Phylogenetic analysis (4) High repeatability (1, 2, 4)	Labor intensive use (1, 4) High start-up costs (1, 4) Comparability between studies difficult (4)	High (1, 2)	High (1, 2, 4)	Medium (1, 2, 4)	Co-dominant (1, 3)	Yes (3)

RAPD	Low costs (1, 4) Easy use (1, 4) No sequence information needed (4)	Low-intermediate reproducibility (1,2) Extensive criticism, see (4)	Very high (1, 2)	Low (1, 2)	Medium (2) – high (1)	Dominant (1, 3)	No (3)
AFLP	Medium costs (1)No sequence information needed (4) High reproducibility & reliability (4)	Initially difficult to set up (1)Technically demanding (4) Relatively high amounts of DNA needed (4)	Very high (1, 2)	Medium (1, 2) – high (4)	Medium (2) – very high (1,4)	Dominant (1, 3)	No (3)

Abbreviations: 1= Semagn et al. 2006; 2= Agarwal et al. 2008; 3= Sunnucks 2000; 4= Lowe et al. 2004; 5= Selkoe and Toonen 2006; 6= Wan et al. 2004; 7= Behura 2006; 8= Zhang and Hewitt 2003; 9= DeYoung and Honeycutt 2005; 10= van der Bank et al. 2C01; 11= Lushai et al. 2003

3 Conclusions

The molecular methods presented in this chapter have question-related advantages and disadvantages (see Table 1). Before choosing a marker, it is essential to evaluate (1) which ecological question ought to be answered, (2) the spatial and temporal scales which ought to be explored, and (3) how exhaustive populations can be sampled (sample design; for detailed sample strategies, see Lowe et al. (2004), and (4) the available financial resources. Furthermore, the popularity of a specific marker might be important for the acceptance in a high ranking peer-reviewed journal, even though many markers could be suitable to answer the same question (Assmann et al. 2007).

In conclusion, we can say that microsatellites are currently one of the most popular genetic markers in ecological studies (see Fig. 1). Especially, the elevated rate of polymorphism is ideal to study small geographical scales of extant species (e.g., Finger et al. 2009). The use of allozyme markers is decreasing since several years (Fig. 1), although the low costs allow a high throughput for studying large geographical scales of extant species (Schmitt 2007). Similarly, AFLPs, RAPDs, and RFLPs can be used to study small to large geographical scales. These markers have their own drawbacks (AFLPs: no detection of alleles, RAPDs: low reproduc-

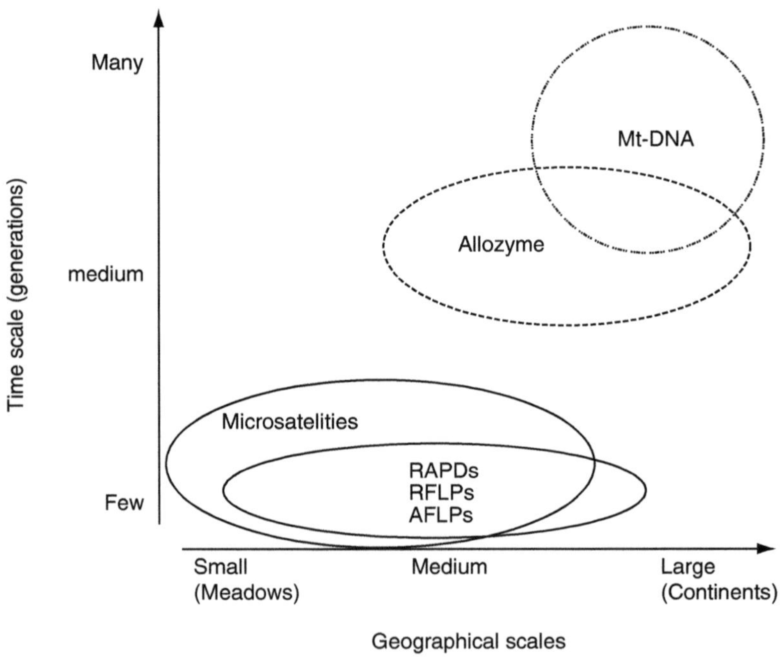

Fig. 2 Different time and geographical scales applicable using mtDNA, Allozymes, Microsatellites, RFLPs, RAPDs, and AFLPs

ibility, RFLPs: high amount of DNA required, see Table 1). Finally, mtDNA is the best choice to study species history and phylogeography on broad geographical scales or deeper timescales (see Fig. 2).

Acknowledgements We would like to thank the professorship of Ecosystem Management, ETH (Jaboury Ghazoul) for funding. Furthermore, we are grateful for comments and corrections from Christopher Kaiser, Chris Kettle, Axel Hochkirch, and Andrea Pluess.

References

Abbott JK, Bensch S, Gosden TP, Svensson EI (2008) Patterns of differentiation in a colour polymorphism and in neutral markers reveal rapid genetic changes in natural damselfly populations. Mol Ecol 17:1597–1604

Agarwal M, Shrivastava N, Padh H (2008) Advances in molecular marker techniques and their applications in plant sciences. Plant Cell Rep 27:617–631

Apostolidis AP, Loukovitis D, Tsigenopoulos CS (2008) Genetic characterization of brown trout (Salmo trutta) populations from the Southern Balkans using mtDNA sequencing and RFLP analysis. Hydrobiologia 600:169–176

Ashley MV (1999) Molecular conservation genetics: tools for assaying the structure of DNA prove valuable in protecting endangered species. Am Sci 87:28–35

Assmann T, Buse J, Drees C, Härdtle HJW, Matern A, von Oheimb G, Schuldt A, Wrase DW (2007) From Latreille to DNA systematics – towards a modern synthesis for carabidology. In: Penev L, Erwin T, Assmann T (eds) Back to the Roots and Back to the Future. Towards a New Synthesis amongst Taxonomic, Ecological and Biogeographical Approaches in Carabidology – Proceedings of the XIII European Carabidologists Meeting, Blagoevgrad, August 20-24. Pensoft Publishers, Sofia, Moscow, pp 41–76

Avise JC (1991) 10 unorthodox perspectives on evolution prompted by comparative population genetic findings on mitochondrial DNA. Annu Rev Genet 25:45–69

Avise JC (1994) Molecular markers, natural history and evolution. Chapman and Hall, New York

Avise JC (2000) Phylogeography: the history and formation of species. Harvard University Press, Cambridge, MA

Bartlett AC (1981) Isozyme polymorphism in populations of the pink-bollworm (Lepidoptera, *Gelechiidae*). Ann Entomol Soc Am 74:9–13

Behura SK (2006) Molecular marker systems in insects: current trends and future avenues. Mol Ecol 15:3087–3113

Botstein D, White RL, Skolnick M, Davis RW (1980) Construction of a genetic-linkage map in man using restriction fragment length polymorphisms. Am J Hum Genet 32:314–331

Brouat C, McKey D, Douzery EJP (2004) Differentiation in a geographical mosaic of plants coevolving with ants: phylogeny of the Leonardoxa africana complex (Fabaceae : Caesalpinioideae) using amplified fragment length polymorphism markers. Mol Ecol 13:1157–1171

Bruford MW, Wayne RK (1993) Microsatellites and their application to population genetic studies. Curr Opin Genet Dev 3:939–943

Busti A, Caceres ME, Calderini O, Arcioni S, Pupilli F (2004) RFLP markers for cultivar identification in tall fescue (Festuca arundinacea Schreb.). Genet Resour Crop Evol 51:443–448

Chakraborty R, Hedrick PW (1983) Paternity exclusion and the paternity index for 2 linked loci. Hum Hered 33:13–23

Despres L, Gielly L, Redoutet W, Taberlet P (2003) Using AFLP to resolve phylogenetic relationships in a morphologically diversified plant species complex when nuclear and chloroplast sequences fail to reveal variability. Mol Phylogenet Evol 27:185–196

DeYoung RW, Honeycutt RL (2005) The molecular toolbox: genetic techniques in wildlife ecology and management. J Wildl Manage 69:1362–1384

Doyle JJ, Morgante M, Tingey SV, Powell W (1998) Size homoplasy in chloroplast microsatellites of wild perennial relatives of soybean (*Glycine* Subgenus *Glycine*). Mol Biol Evol 15:215–218

Eisen J (1999) Mechanistic explanations for variation in microsatellite stability within and between species. In: Goldstein D, Schlötterer C (eds) Microsatellites: evolution and applications. Oxford University Press, Oxford, pp 34–48

Finger A, Schmitt T, Zachos FE, Meyer M, Assmann T, Habel JC (2009) The genetic status of the violet copper *Lycaena helle* – a relict of the cold past in times of global warming. Ecography 32:382–390

Goldstein DB, Pollock DD (1997) Launching microsatellites: a review of mutation processes and methods of phylogenetic inference. J Hered 88:335–342

Hardy OJ, Maggia L, Bandou E, Breyne P, Caron H, Chevallier MH, Doligez A, Dutech C, Kremer A, Latouche-Halle C, Troispoux V, Veron V, Degen B (2006) Fine-scale genetic structure and gene dispersal inferences in 10 Neotropical tree species. Mol Ecol 15:559–571

Hu DC, Zhang QL, Luo ZR (2008) Phylogenetic analysis in some Diospyros spp. (Ebenaceae) and Japanese persimmon using chloroplast DNA PCR-RFLP markers. Sci Hort 117:32–38

Jones AG, Ardren WR (2003) Methods of parentage analysis in natural populations. Mol Ecol 12:2511–2523

Jones CJ, Edwards KJ, Castaglione S, Winfield MO, Sala F, vande Wiel C, Bredemeijer G, Vosman B, Matthes M, Daly A, Brettschneider R, Bettini P, Buiatti M, Maestri E, Malcevschi A, Marmiroli N, Aert R, Volckaert G, Rueda J, Linacero R, Vazquez A, Karp A (1997) Reproducibility testing of RAPD, AFLP and SSR markers in plants by a network of European laboratories. Mol Breed 3:381–390

Keim P, Schupp JM, Travis SE, Clayton K, Zhu T, Shi LA, Ferreira A, Webb DM (1997) A high-density soybean genetic map based on AFLP markers. Crop Sci 37:537–543

Kim C, Na HR, Choi HK (2008) Conservation genetics of endangered Brasenia schreberi based on RAPD and AFLP markers. J Plant Biol 51:260–268

Leberg P (2005) Genetic approaches for estimating the effective size of populations. J Wildl Manage 69:1385–1399

Levinson G, Gutman GA (1987) High frequencies of short frameshifts in poly-CA/TG tandem repeats borne by bacteriophage M13 in *Escherichia coli* K-12. Nucleic Acids Res 15:5323–5338

Lowe A, Harris S, Ashton P (2004) Ecological genetics. Blackwell, Oxford

Loxdale HD, Tarr IJ, Weber CP, Brookes CP, Digby PGN (1985) Electrophoretic study of enzymes from cereal aphid populations.3. Spacial and temporal genetic variation of populations of *Sitobion Avenae* (F.) (Hemiptera: Aphididae). Bull Entomol Res 75:121–141

Luikart G, England PR (1999) Statistical analysis of microsatellite DNA data. Trends Ecol Evol 14:253–256

Lushai G, Smith DAS, Goulson D, Allen JA, Maclean N (2003) Mitochondrial DNA clocks and the phylogeny of Danaus butterflies. Insect Sci Appl 23:309–315

Meagher TR, Thompson E (1986) The relationship between single parent and parent pair genetic likelihoods in genealogy reconstruction. Theor Popul Biol 29:87–106

Meglécz E, Solignac M (1998) Microsatellite loci for *Parnassius mnemosyne* (Lepidoptera). Hereditas 128:179–180

Meudt HM, Clarke HC (2007) Almost forgotten or latest practice? AFLP applications, analyses and advances. Trends Plant Sci 12:106–117

Mullis KF, Caloona F, Scharf S, Saikl R, Horn G, Erlich H (1996) Specific enzymatic amplification of DNA in vitro: the polymerase chain reaction. Cold Spring Harb Symp Quant Biol 51(1):263–273

Olivieri GL, Sousa V, Chikhi L, Radespiel U (2008) From genetic diversity and structure to conservation: genetic signature of recent population declines in three mouse lemur species (*Microcebus* spp.). Biol Conserv 141:1257–1271

Olmstead JW, Sebolt AM, Cabrera A, Sooriyapathirana SS, Hammar S, Iriarte G, Wang D, Chen CY, van der Knaap E, Iezzoni AF (2008) Construction of an intra-specific sweet cherry (*Prunus avium* L.) genetic linkage map and synteny analysis with the Prunus reference map. Tree Genet Genomes 4:897–910

Orsini L, Corander J, Alasentie A, Hanski I (2008) Genetic spatial structure in a butterfly metapopulation correlates better with past than present demographic structure. Mol Ecol 17:2629–2642

Palo J, Varvio SL, Hanski I, Vainola R (1995) Developing microsatellite markers for insect population structure: complex variation in a checkerspot butterfly. Hereditas 123:295–300

Proudfoot GA, Honeycutt RL, Slack RD (2006) Mitochondrial DNA variation and phylogeography of the ferruginous pygmy-owl (Glaucidium brasilianum). Conserv Genet 7:1–12

Radoev M, Becker HC, Ecke W (2008) Genetic analysis of heterosis for yield and yield components in rapeseed (Brassica napus L.) by quantitative trait locus mapping. Genetics 179:1547–1558

Riddle BR, Honeycutt RL (1990) Historical biogeography in North-american arid regions – an approach using mitochondrial-DNA phylogeny in grasshopper mice (genus *Onychomys*). Evolution 44:1–15

Rocha LA, Rocha CR, Robertson DR, Bowen BW (2008) Comparative phylogeography of Atlantic reef fishes indicates both origin and accumulation of diversity in the Caribbean. BMC Evol Biol 8:157

Saal B, Wricke G (2002) Clustering of amplified fragment length polymorphism markers in a linkage map of rye. Plant Breeding 121:117–123

Schenk MF, Thienpont CN, Koopman WJM, Gilissen L, Smulders MJM (2008) Phylogenetic relationships in *Betula* (Betulaceae) based on AFLP markers. Tree Genet Genomes 4:911–924

Schmitt T (2007) Molecular biogeography of Europe: Pleistocene cycles and postglacial trends. Front Zool 4:11

Selkoe KA, Toonen RJ (2006) Microsatellites for ecologists: a practical guide to using and evaluating microsatellite markers. Ecol Lett 9:615–629

Semagn K, Bjornstad A, Ndjiondjop MN (2006) An overview of molecular marker methods for plants. Afr J Biotechnol 5:2540–2568

Simmons MP, Zhang LB, Webb CT, Muller K (2007) A penalty of using anonymous dominant markers (AFLPs, ISSRs, and RAMS) for phylogenetic inference. Mol Phylogenet Evol 42:528–542

Steiner WWM, Joslyn DJ (1979) Electrophoretic techniques for the genetic study of mosquitos. Mosq News 39:35–54

Sun ZN, Liu P, Li J, Meng XH, Zhang XM (2008) Construction of a genetic linkage map in Fenneropenaeus chinensis (Osbeck) using RAPD and SSR markers. Hydrobiologia 596:133–141

Sunnucks P (2000) Efficient genetic markers for population biology. Trends Ecol Evol 15:199–203

Tang SQ, Dai WJ, Li MS, Zhang Y, Geng YP, Wang L, Zhong Y (2008) Genetic diversity of relictual and endangered plant *Abies ziyuanensis* (Pinaceae) revealed by AFLP and SSR markers. Genetica 133:21–30

Tanksley SD, Young ND, Paterson AH, Bonierbale MW (1989) Rflp mapping in plant-breeding – new tools for an old science. Biotechnology 7:257–264

van der Bank H, van der Bank M, van Wyk BE (2001) A review of the use of allozyme electrophoresis in plant systematics. Biochem Syst Ecol 29:469–483

Volkova PA, Shipunov AB, Elven R, Brochmann C (2008) The seashore sedges of the Russian Kola Peninsula: how many species? Flora 203:523–533

Vos P, Hogers R, Bleeker M, Reijans M, Vandelee T, Hornes M, Frijters A, Pot J, Peleman J, Kuiper M, Zabeau M (1995) Aflp – a new technique for DNA-fingerprinting. Nucleic Acids Res 23:4407–4414

Wan QH, Wu H, Fujihara T, Fang SG (2004) Which genetic marker for which conservation genetics issue? Electrophoresis 25:2165–2176

Weising K, Nybom H, Wolff K, Kahl G (2005) DNA fingerprinting in plants – principles, methods, and applications, 2nd edn. CRC Press, Boca Raton, FL

Williams JGK, Kubelik AR, Livak KJ, Rafalski JA, Tingey SV (1990) DNA polymorphisms amplified by arbitrary primers are useful as genetic-markers. Nucleic Acids Res 18:6531–6535

Yang S, Bishop JG, Webster MS (2008) Colonization genetics of an animal-dispersed plant (*Vaccinium membranaceum*) at Mount St Helens, Washington. Mol Ecol 17:731–740

Young WP, Schupp JM, Keim P (1999) DNA methylation and AFLP marker distribution in the soybean genome. Theor Appl Genet 99:785–792

Zhang DX, Hewitt GM (2003) Nuclear DNA analyses in genetic studies of populations: practice, problems and prospects. Mol Ecol 12:563–584

Part V
Conservation of Relict Species

The EU Habitats Directive and the German Natura 2000 Network of Protected Areas as Tool for Implementing the Conservation of Relict Species

Marco Zimmermann, Mareike Vischer-Leopold, Götz Ellwanger, Axel Ssymank, and Eckhard Schröder

Abstract This study analyses whether the Natura 2000 network of Sites of Community Importance (SCIs) is able to protect relict species in the taxonomic groups of higher plants, molluscs, dragonflies and damselflies as well as butterflies (only Rhopalocera) in Germany. Altogether, a total of 157 species from all groups are identified as relict species in Germany. Fourteen of these are included in Annexes II, IV or V of the Habitats Directive.

Most glacial relicts are well covered by an indirect protection regime of the European Union (EU) Habitats Directive as they occur in 46 of Annex I habitat types, and their occurrences are to a large extent covered by Natura 2000 sites (SCIs). For a few relict species and certain relict plant communities a gap remains in the EU protection regime, which can be filled by a national protection regime, for example, in nature reserves. The best way to protect local relict species is to include them in special management plans for their conservation.

1 Introduction

The protection of relict species is a prominent goal in nature conservation (Soulé 1986; Ås et al. 1992). These are taxa with only small recent occurrences and which now show only isolated distribution remaining from a once wider distribution.

M. Zimmermann (✉)
Institute of Forest Botany and Forest Zoology, TU Dresden, Tharandt, Germany
e-mail: marcozimmermann53129@googlemail.com

M. Vischer-Leopold, G. Ellwanger, A. Ssymank and E. Schröder
Department I 2.2 "Habitats Directive/Natura 2000", German Federal Agency for Nature Conservation, Bonn, Germany
e-mail: mareike.vischer-leopold@bfn.de; goetz.ellwanger@bfn.de; axel.ssymank@bfn.de; eckhard.schroeder@bfn.de

J.C. Habel and T. Assmann (eds.), *Relict Species: Phylogeography and Conservation Biology*, DOI 10.1007/978-3-540-92160-8_19, © Springer-Verlag Berlin Heidelberg 2010

There may be very different reasons for their decline in the past. The species may be a relict of former stages of historical vegetation and landscape development resulting from past climate changes (glacial and post-glacial periods). Very often, the anthropogenic effects of cultural landscape development interfere or overlap with this, resulting in additional reduction of the distribution area, the density or fragmentation of patches or the population itself.

The scope of this paper is to focus mainly on the first category of relict species (which have developed mainly as a result of natural changes in abiotic conditions), as anthropogenic fragmentation and isolation has accelerated in recent decades and is at least partly documented in Red Data Books.

Most present-day relict species were widely distributed during the last glacial period, when their ecological requirements were well fulfilled; relict species with this type of distribution pattern are called climatic relicts or glacial relicts (e.g. Cox and Moore 2006).Today, these species are not well adapted to the prevailing climate conditions and are thus often very rare and localized, with the result that their occurrences are generally well documented. A prominent example of such glacial relict species is the dwarf birch (*Betula nana*) in bogs and fens in Germany. After the last glacial period, conditions were good for *B. nana* because cold-steppe vegetation covered the entire Central European landscape. Temperature increase resulted in more and more suboptimal conditions for *B. nana*, because other herbs and trees were more competitive and better adapted to the new conditions. *B. nana* was able to survive only in regions with low temperatures and low nutrient input where competition was reduced. Most glacial relicts, both plants and animals, follow this general pattern (Varga 1977). Often these species show arctic-alpine or boreo-alpine distribution patterns and are nowadays widely distributed in Northern Europe and/or remain isolated in cold areas such as the highest mountain regions or peatlands.

Thermophilic relicts represent a second type of relict species. These are the relicts from once warmer times during the Atlantic period, when thermophilic species were well adapted to warm and dry conditions. During this time, these species were widely distributed, but they later became isolated and restricted to favourable habitats due to colder temperatures and an increase of rainfall. A good example of a thermophilic relict is the Heath Bush-cricket *Gampsocleis glabra* in heathland in Northern Germany. In warm and dry years, Mediterranean species were able to expand to the North. When climate changes resulted in lower temperatures, these species were able to survive only in exposed habitats where higher temperatures are usually reached more quickly. This is often the case in stony and dry biotopes like dunes or rocks on Southern slopes. Areas with thermophilic relicts may often be found in the rain shadow of mountain ranges.

The goal of nature conservation is to conserve all these different types of relict species in their natural habitats as a witness of the natural conditions and developments in the past. Often it is quite clear that these species are not able to expand their ranges because current conditions are bad for them. In these often small populations with no contact to neighbour populations, evolution means that speciation and adaptation occur. This makes such populations very interesting objects of evolutionary research.

We have a responsibility to protect these species and their isolated populations. It is an important policy target to conserve a species' entire gene pool (BMU 2007; BFN 2008; Gruttke et al. 2004). The German government pointed out in its national strategy for biological diversity (BMU 2007) that the entire gene pool of species and populations has to be conserved, e.g. populations with local adaptations to regional microclimatic conditions. Fragmented populations often show significant changes in their character states, so if populations are separated for a long period from each other (several 1,000 years) with an interruption of gene flow, gene drift can mean that these populations change character states. The concept of Evolutionary Significant Units (ESUs) takes into consideration that such populations are on their way to an evolutionary unit of their own, leading to a different taxonomic status (Moritz 1994). For Germany, Gruttke et al. (2004) give a summary of current knowledge on this modern field of conservation biology, where glacial relicts are an important topic.

Nature conservation often means nature management. This involves the question of how relict species can be best protected. Since 1992, there has been an established, important and modern concept in nature conservation: The implementation of the EC Habitats Directive. The main goal of this Directive is to establish a network of Sites of Community Importance (SCI) for species listed in Annex II of the Directive. In Germany there are three biogeographic regions: Atlantic, Continental and Alpine (for further details see Balzer et al. 2008).One hundred and thirty-three plant and animal species as well as 91 habitat types are listed in Annexes II and I of the Habitats Directive, respectively. In Germany, a total of 4,622 SCI have been integrated into the EU network of protected areas "Natura 2000" for these species and habitat types. SCIs in Germany cover a terrestrial area of 3,313,083 ha equalling 9.3% of the territory. No scientific analysis has been performed to date to determine whether relict species are represented in the Natura 2000 network of protected sites in Germany. The current work gives an overview of whether, how and to what extent relict species are conserved by the EC Habitats Directive.

2 Method

A data analysis was carried out to determine whether the EU Habitats Directive provides sufficient protection for relict species in taxa of higher plants, molluscs, dragonflies and damselflies, and butterflies (Rhopalocera only) in Germany. For this purpose a database, containing for the most part glacial relict species, was set up based on studies by CLECOM (2002), De Lattin (1967), Holdhaus (1954), Kudrna (2002), Thorn (1960), Reisch (2001), Welk (2001) and Buchholz and Welk (2005). This database was modified for plants: i.e. *Gentiana clusii* was not accepted as a glacial relict species (populations in the Black Forest) as proposed by Thorn (1960) and Reisch (2001), because the species is not autochthonous there (Oberdorfer 1990). On the other hand, the database does not claim to be exhaustive for all potential glacial relicts with occurrences in Germany. Species on the edge of their distribution area and endemics (other than relicts) in Germany were not included in the data analysis. Assessment was carried out on two levels:

1. Direct species protection was checked, i.e. whether the species is covered by Annex II (site protection within Natura 2000), Annex IV (strict species protection) or Annex V (limited protection of used or managed plants) of the Habitats Directive.
2. Protection of the habitat of the relict species was checked, i.e. whether the species is indirectly protected by the Natura 2000 sites: This can be assumed if the habitat of the species corresponds to a protected Annex I habitat type. We also assessed whether the occurrences of these glacial relicts are really covered by existing SCI and do not by chance lie outside the Natura 2000 network.

3 Results and Discussion

3.1 Glacial Relicts in Higher Plants and Their Protection by the Habitats Directive

In total, 100 plant species were identified as glacial relics in Germany (excluding *Rubus fruticosus* agg.). Most of the glacial relicts in higher plants belong to the alpine, pre-alpine or arctic-alpine distribution types (Table 1). Species that occur in Germany only in the Alps as part of their main distribution area but have relict populations outside of Germany only were not considered (six species, e.g. *Sibbaldia procumbens* in the Vosges or *Moehringia muscosa* in the Bohemian Forest). Six species are already extinct or presumed extinct in Germany, and another two species are extinct or presumed extinct in the isolated parts of their former natural distribution area outside the Alps (and the Alpine foreland), where relict populations remained until the last century (e.g. *Selaginella helvetica* formerly in the Thuringian Forest). A few other species are extinct in one or more discrete

Table 1 Distribution of the glacial relicts of higher plants in Germany according to www.floraweb.de

Area type	Number of species	Examples in Germany
Arctic-alpine	19	*Carex vaginata, Epilobium alsinifolium, Saxifraga stellaris*
Arctic (5), arctic-North (3), North (3), North-continental (1), North-prealpine (1)	13	*Carex paupercula, Pedicularis sceptrum-carolinum, Rubus chamaemorus*
Central Europe (2), central Europe-endemic (1)	3	*Pulsatilla alpina* ssp. *alba*, *Salix bicolor, Saxifraga oppositifolia* ssp. *amphibia*
Alpine	40	*Adenostyles alliariae, Carex sempervirens, Kernera saxatilis*
Prealpine	25	*Anthriscus nitidus, Homogyne alpina, Soldanella montana*
Total	100	–

Table 2 Relict plant species in Germany listed in the Habitats Directive

Species	Habitats Directive Annex	Conservation status (National Report Art. 17, Hab. Dir.; Nationaler Bericht 2007) Biogeographic region: Atlantic	Continental	Alpine
Marsh saxifrage *Saxifraga hirculus*	II, IV	Extinct in Germany		
Yellow gentian *Gentiana lutea*	V	–	Favourable	Favourable
Alpine clubmoss *Diphasiastrum alpinum*	V	–	Unfavourable - inadequate	Favourable
Myosotis rehsteineri	II, IV	–	Unfavourable - bad	–

parts of their former natural distribution area (e.g. *Gnaphalium norvegicum* in the Erzgebirge), but they still have recent relict populations in other parts of their distribution area in Germany. The 92 species that currently exist in relict populations in Germany were analysed in detail.

Direct protection by species annexes of the Habitats Directive is/was given only for four species (Table 2). This includes *Myosotis rehsteineri*, which is a glacial relict species according to Brackel (2001), but is not mentioned by the authors listed in the Methods section above. The species occurs at larger lakes at the Northwestern and Southern edges of the Alps and is classified as endemic in this region. Because the species is listed in Annexes II and IV of the Habitats Directive, we considered it in Table 2. *Saxifraga hirculus* is listed in Annex II; however, it is presumed extinct in its last known German locality in the Murnauer Moos, where Natura 2000 "came too late." No other glacial relict is listed in Annex IV, and only two species (*Gentiana lutea* and *Diphasiastrum alpinum*) are covered by Annex V. Thus, the Habitats Directive affords very little direct species protection of relict species.

Table 3 shows the extent of indirect protection afforded by Annex I habitat types.

Only three glacial relicts (*Carex brunnescens*, *Epilobium alsinifolium* and *E. nutans*) grow outside Annex I habitat types and could not be attributed to one or more habitat types. These species are definitely not protected by the Habitats Directive.

As expected, the main occurrence of glacial relicts is concentrated on a few Annex I habitat types, which serve as refuges. These are hydrophilous tall herb fringe communities (Habitat type 6430), especially in their montane to subalpine subtypes of the class Betulo-Adenostyletea, which host 26 glacial relicts (28.3%); alpine and subalpine calcareous grasslands (habitat type 6170) which host 15 species (16.3%); calcareous rocky slopes with their chasmophytic vegetation (habitat type 8210) with 13 species (14.1%); and the calcareous and calcshist screes of the montane to alpine level (habitat type 8120) which host nine species (9.8%). On siliceous substrates, the situation is similar, with the siliceous alpine and boreal grasslands (habitat type 6150) hosting 20 glacial relicts (21.7%) and the species-rich *Nardus* grasslands in mountain areas (habitat type 6230) hosting 16 glacial relicts (17.4%). All these habitat types have in common the fact that during glacial

Table 3 Coverage of relict species by Annex I habitat types (data base 92 glacial relict species)

Formation/group of habitat types	Number of species per habitat group	Annex I habitatcode	Habitat type	Number of species per habitat type	in %
Freshwater habitats	6	3220	Alpine rivers and the herbaceous vegetation along their banks	5	5.4
Heath and scrub formations	12	4060	Alpine and boreal heaths	5	5.4
		4070	Bushes with *Pinus mugo* and *Rhododendron hirsutum* (Mugo-Rhododendretum hirsuti)	8	8.7
Grassland and tall herb communities	65	6150	Siliceous alpine and boreal grasslands	20	21.7
		6170	Alpine and subalpine calcareous grasslands	15	16.3
		6210	Semi-natural dry grasslands and scrubland facies on calcareous substrates (Festuco-Brometalia)	6	6.5
		*6230	Species-rich *Nardus* grasslands, on silicious substrates in mountain areas (and submountain areas in Continental Europe)	16	17.4
		6410	Molinia meadows on calcareous, peaty or clayey-silt-laden soils (Molinion caeruleae)	3	3.3
		6430	Hydrophilous tall herb fringe communities of plains and of the montane to alpine levels	26	28.3
		6520	Mountain hay meadows	7	7.6
Bogs, transition mires and fens	16	*7110	Active raised bogs	5	5.4
		7120	Degraded raised bogs still capable of natural regeneration	3	3.3
		7140	Transition mires and quaking bogs	7	7.6
		7150	Depressions on peat substrates of the Rhynchosporion	1	1.1
		*7220	Petrifying springs with tufa formation (Cratoneurion)	4	4.3
		7230	Alkaline fens	6	6.5

(continued)

Table 3 (continued)

Formation/group of habitat types	Number of species per habitat group	Annex I habitatcode	Habitat type	Number of species per habitat type	in %
Rock and scree habitats	24	8110	Siliceous scree of the mountains to snow levels (Androsacetalia alpinae and Galeopsietalia ladani)	2	2.2
		8120	Calcareous and calcshist screes of the montane to alpine levels (Thlaspietea rotundifolii)	9	9.8
		8150	Medio-European upland siliceous screes	1	1.1
		*8160	Medio-European calcareous scree of hill and montane levels	1	1.1
		8210	Calcareous rocky slopes with chasmophytic vegetation	13	14.1
		8220	Siliceous rocky slopes with chasmophytic vegetation	6	6.5
		8230	Siliceous rock with pioneer vegetation of the Sedo-Scleranthion or of the Sedo albi-Veronicion dillenii	2	2.2
Forests	9	9140	Medio-European subalpine beech woods with *Acer* and *Rumex arifolius*	5	5.4
		*91D0	Bog woodland	2	2.2
		*91E0	Alluvial forests with *Alnus glutinosa* and *Fraxinus excelsior* (Alno-Padion, Alnion incanae, Salicion albae)	1	1.1
		9410	Acidophilous Picea forests of the montane to alpine levels (Vaccinio-Piceetea)	5	5.4

Double entries of glacial relicts occurring in more than one habitat types of Annex I are possible.
*indicates priority habitat type

and post-glacial times they had a wider distribution between the two main European ice shields in the North and in the Alps. This distribution then slowly diminished and became fragmented during the post-glacial warming-up. This development was aggravated and accelerated by anthropogenic influence which resulted in further fragmentation and extinction in some of the refuges in the lower mountain ranges (e.g. Dörr 2000[1]). As the formations of mountain grasslands, tall herbs, rocky habitats and screes and, to a lesser degree, bogs are rich in glacial relicts and at the same time well represented in Annex I of the Habitats Directive, most of the species theoretically enjoy good indirect protection by the EU nature Directives. As this protection regime is valid only within accepted SCIs of the Natura 2000 network, it was necessary to verify coverage of the occurrences of glacial relicts in SCIs in order to assess the real protection regime. As not every occurrence could be checked individually, the results were classed in four different groups:

(a) Occurrences (almost) completely within SCIs: 43 plant species
(b) Sufficiently represented in SCIs, with a minor proportion of the occurrences outside: 47 species
(c) Unclear, precise verification not possible: 1 species (*Salix myrtilloides*)
(d) Not covered by SCIs: 1 species (*Minuartia stricta*).

Category (b) includes at least seven species, which are probably not or not sufficiently covered in every part of their fragmented occurences, but the species are not in danger of extinction in Germany (e.g. *Hieracium bupleuroides* in the Franconian Alb). Furthermore, three species occur within Natura 2000 without, however, belonging to an Annex I habitat type, so they might not benefit from the EU protection regime.

Not only plant species but also whole plant communities (syntaxa) can be glacial relicts. These communities also need special conservation attention, not only because they are likely to host a number of glacial relicts in their fauna which has not yet been well studied. Rock-crevice communities of *Asplenietea trichomanis* are especially rich in relict species or paleo-endemics, as has been shown for *Kernera saxatilis*, *Draba aoizoides* and *Saxifraga paniculata* in the Swabian Alb by Wilmanns and Rupp (1966) and Wilmanns (1993). Spring formations of the class Montia-Cardaminetea, vegetation of snow-beds (Salicetea herbacea) and fens of the Scheuchzerio-Caricetea nigrae also have relict associations with the higher mountain ranges in central and Southern Germany. Examples of this are the Bryo-Philonotidetum seriatae (acid subalpine springs rich in mosses and *Saxifraga* species) on the Feldberg above 1,300 m altitude (Black Forest) with the nearest occurrences to the Alps, as well as the Bartsio-Caricetum fusca and possibly the Caricetum frigidae. A typical example of snow bed communities is the Nardo-Gnaphalietum supini on the Feldberg. Good examples of glacial relict communities are the Nardus grasslands of the higher mountains such as Leontodonto-Nardetum in the Black Forest, a vicariant association of the Violo-Nardetum (with *Pulsatilla alba* and *Viola lutea*) in the Vosges, the Lycopodio-Nardetum in the Bavarian and Bohemian forest and, finally, the Pulsatillo-Nardetum on the Brocken in the Harz mountains. Today, the subalpine tall herb communities of the Betulo-Adenostyletea have a nordic-alpine distribution with relicts in the lower mountain ranges, such as the Alnetum viridis, the

[1] This article includes some species as glacial relicts which were not listed as glacial relict species by Thorn (1960) and/or Reisch (2000) and were not considered in our study.

Adenostylo-Cicerbitetum and the Salicetum appendiculatae in the Black Forest, and the tall grass vegetation of the Sorbo-Calamagrostietum arundinaceae in both the Black Forest and the Vosges. These relict populations are of outstanding scientific and conservation interest, as their primary stands host species which were the ancestors of many grassland species of our cultivated meadows. The associations of the fluvial gravel deposits of the Salicion eleagni such as the Salici-Myricarietum and the Salicetum eleagni with *Hippophae rhamnoides* ssp. *fluviatile* can also be regarded as relict vegetation, formerly widespread in the fluvioglacial plains of the ice ages. Some pine forests of the Erico-Pinion alliance, such as the Coronillo-Pinetum sylvestris, as a post-glacial relict of the warm period in the Swiss Jurassic (Richard 1972) and the Swabian Alb (Müller 1980) are examples of relict forests.

Of the relict syntaxa listed, most plant associations are covered by Annex I of the Habitats Directive, with two major exceptions: the acidic spring vegetation (Montio-Cardaminetea) and the Erico-Pinion relict pine forests. These merit special protection under national legislation in Germany, as long as they are not covered by the Habitats Directive.

3.2 *Molluscs*

A total of 342 species of non-marine molluscs were recorded from Germany in the CLECOM list (Falkner et al. 2001; CLECOM 2002) (the subterranean freshwater molluscs are not included here), of which 70 are freshwater and 236 terrestrial species. Twenty freshwater species are considered relicts (glacial or post-glacial isolated). The terrestrial molluscs are represented by 18 relicts.

The relict species include mountain molluscs as well as specialist species at low altitudes, for example in mires, fens and calcareous grasslands. Most relict populations suffer severely from bad environmental conditions. This situation has worsened in recent years because of an intensification of anthropogenic land use.

Table 4 Relict molluscs species in Germany listed in the Habitats Directive

Species	Habitats Directive	Conservation status (National Report Art. 17, Hab. Dir.; Nationaler Bericht 2007) Biogeographic region		
	Annex	Atlantic	Continental	Alpine
Snail, *Theodoxus transversalis*	II, IV	–	Unknown	–
Little Whirlpool Ramshorn Snail, *Anisus vorticulus*	II, IV	Unfavourable - bad	Unfavourable - bad	–
Freshwater Pearl Mussel, *Margaritifera margaritifera*	II, V	Unfavourable - bad	Unfavourable - bad	–
Geyer's whorl snail, *Vertigo geyeri*	II	–	Unfavourable - bad	Unfavourable - inadequate
Round-mouthed whorl snail, *Vertigo genesii*	II	Not evaluated 2007 because thought to be extinct; found again in 2008		

In Annexes II, IV and V of the EU Habitats Directive, 5 of these 18 relicts were represented (*Theodoxus transversalis, Anisus vorticulus, Margaritifera margaritifera, Vertigo geyeri, V. genesii*), all of which are represented in the German Natura 2000 network (Table 4). The remaining 13 mollusc relicts mainly live in the following habitat types of Annex I of the Habitats Directive:

- 3110 oligotrophic waters containing very few minerals of sandy plains (Littorelletalia uniflorae)
- 3130 oligotrophic to mesotrophic standing waters with vegetation of the Littorelletea uniflorae and/or of the Isoeto-Nanojuncetea
- 3140 hard oligo-mesotrophic waters with benthic vegetation of *Chara* spp.
- 3150 natural euthrophic lakes with Magnopotamion or Hydrocharition-type vegetation
- 3260 water courses of plain to montane levels with Ranunculion fluitantis and Callitricho-Batrachion vegetation
- 6210 semi-natural dry grasslands and scrubland facies on calcareous substrates (Festuco Brometalia)
- *6240 sub-Pannonic steppe grassland
- 7140 transition mires and quaking bogs
- 7230 alkaline fens
- 8210 calcareous and calcshist screes of the montane to alpine levels (Thlaspietea rotundifolii)
- 9110 Luzulo-Fagetum beech forests
- 9130 Asperulo-Fagetum beech forests
- *9180 Tilio-Acerion forests of slopes, screes and ravines
- *91E0 alluvial forests with *Alnus glutinosa* and *Fraxinus excelsior* (Alno-Padion, Alnion incanae, Salicion alvae)
- 9410 acidophilous Picea forests of the montane to alpine levels (Vaccinio-Piceetea)

This assortment of habitat types and the German Natura 2000 network protects most of the 38 relict species. Only six species which prefer the potamic region of large streams, and *Vertigo lilljeborgii*, a glacial relict of acid sedge beds, are only partly included in the network. Not only the relict species but also other associated species – such as their specialized predators – should be protected. For example, some local populations of the Pyrenean shell crusher beetle *Carabus pyrenaeus* are so highly adapted to their prey, the relict mollusc *Arianta xatarti*, that these populations already show differences in mandible form compared to populations in the Western Pyrenees (Assmann et al. 2000).

3.3 Dragonflies and Damselflies

Eighty-one dragonflies and damselflies are known to occur in Germany, and two of them are listed as extinct. Only 7 of the 81 species were recognized in this study as relict species (cf. Sternberg 1998). *Coenagrion hylas* is extinct in

Table 5 Relict dragonflies and damselflies species in Germany listed in the Habitats Directive

Species	Habitats Directive	Conservation Status (National Report Art. 17, Hab. Dir.; Nationaler Bericht 2007) Biogeographic region		
	Annex	Atlantic	Continental	Alpine
Damselfly, *Coenagrion hylas*	II	Extinct in Germany		
Siberian Winter Damselfly, *Sympecma paedisca*	IV	Unfavourable - bad	Unfavourable - inadequate	Unknown

Germany (one of the two extinct species in Germany), and five of the remaining six species are specialists of dystrophic habitats. Two species are listed in the Annexes of the Habitats Directive (Table 5). Most of the breeding habitats of these five species (*Nehalennia speciosa*, *Aeshna caerulea*, *A. subarctica*, *Somatochlora alpestris*, *S. arctica*) belong to the following habitat types of the Habitats Directive:

- 7110 active raised bogs
- 7120 degraded raised bogs still capable of natural regeneration
- 7140 transition mires and quaking bogs
- 3160 natural dystrophic lakes and ponds.

Although many of the populations and habitats of the relict species are protected by the network of Natura 2000, their conservation status is not good. The main reason is the unfavourable conservation status of their habitat types. In the Atlantic and the Continental biogeographic regions, all four habitat types are considered to have an unfavourable conservation status. Only in the Alpine biogeographic region is the conservation status mostly favourable. The essential threats to these low-nutrient habitats are the intake of nutrients and the exposure level of air pollutants (even by air), the lowering of ground water levels and recultivation of bogs. As a result, the typical poor nutrient water ecosystems and the water chemistry have to be safeguarded, and buffer zones should be created to minimize the inflow of nutrients from surrounding surfaces. The damselfly *Sympecma paedisca* is endangered in Europe and strongly protected by the EU Habitats Directive Annex IV. It is a relict with a Continental distribution. Beneath moorland habitats such as dystrophic waters more nutrient-rich standing waters were also colonized. Although there are no sites especially protected by the Habitats Directive for this species, many of the populations are found in protected areas in Germany. The conservation status of the species in the Atlantic and Continental region is unfavourable - bad and unfavourable - inadequate, respecitvely, and is unknown in the Alpine region. The number of populations of *S. paedisca* in Germany is decreasing severely, at least in Lower Saxony and Baden-Württemberg. Besides habitat degradation, climate change could also be one of the reasons for the decline (Ellwanger and Mauersberger 2003).

3.4 Butterflies

Currently, 188 butterfly species are established in Germany (Lepidoptera: Rhopalocera). Twelve species from two different groups of relict species were identified as relicts. Butterflies such as *Lycaena helle* represent species that are well adapted to cold-stenothermic conditions. Examples for species in Europe that are adapted to cold conditions, and thus defined as glacial relicts, can be found in the work of Varga (1977) and Weidemann (1995). Other species such as *Chazara briseis* are thermophilic relicts from warmer times.

Only one of these relict species in Germany is listed in Annex II of the Habitats Directive (Table 6). This means that relict species in Lepidoptera (Rhopalocera) are underrepresented in Annex II, because SCIs have only been established for one species in the network of Natura 2000 sites. For the other 11 species, it is not obligatory to integrate their populations in the network. In Annex IV another two species are listed (Table 6). These are protected by law, but no special sites (SCIs) need to be set up. Most of the relict species settle in habitat types listed in Annex I of the Habitats Directive, and SCIs should also be selected for those habitat types. Thus, there is a good chance of saving the locations of Germany's butterfly species with the protection of the Annex I habitat types.

All thermophilic relict species live primarily in the following habitat types of Annex I of the Habitats Directive:

- 5110 stable thermoxerophilic formations with *Buxus sempervirens* on rock slopes (Berberidion p.p.)
- 5130 *Juniperus communis* formations on heaths or calcareous grasslands
- *6110 rupicolous calcareous or basophilic grasslands of the Alysso–Sedion albi
- *6120 xeric sand calcareous grasslands
- 6210 semi-natural dry grasslands and scrubland facies on calcareous substrates (Festuco-Brometalia)
- *6240 sub-pannonic steppe grasslands
- 8210 calcareous rocky slopes with chasmophytic vegetation
- 8220 siliceous rocky slopes with chasmophytic vegetation
- 8230 siliceous rock with pioneer vegetation of the Sedo-Scleranthion or of the Sedo albi-Veronicion dillenii.

Table 6 Relict butterfly species in Germany listed in the Habitats Directive

Species	Habitats Directive	Conservation Status (National Report Art. 17, Hab. Dir.; Nationaler Bericht 2007)		
		Biogeographic region		
	Annex	Atlantic	Continental	Alpine
Scarce Heath, *Coenonympha hero*	IV	Unfavourable - bad	Unfavourable - bad	–
Violet Copper, *Lycaena helle*	II, IV	–	Unfavourable - inadequate	Unfavourable - bad
Apollo, *Parnassius Apollo*	IV	–	Unfavourable - bad	Favourable

No thermophilic relict species is listed in Annex II, and only one (*Parnassius apollo*) is listed in Annex IV. But all four thermophilic relict species in Germany can be found in the habitat types described below and also in 2310 (dry sand heaths with *Calluna* and *Genista*), 2320 (dry sand heaths with *Calluna* and *Empetrum nigrum*), 2330 (inland dunes with open *Corynephorus* and *Agrostis* grasslands) and 4030 (European dry heaths).

Eight cold-stenothermic relict species occur in Germany. One of these is listed in Annex II and IV (*Lycaena helle*) and one only in Annex IV (*Coenonympha hero*).

The cold-stenothermic relict species live primarily in the following habitat types of Annex I of the Habitats Directive:

- 4010 Northern Atlantic wet heaths with *Erica tetralix*
- 4060 alpine and boreal heaths
- *6230 species-rich *Nardus* grasslands, on silicious substrates in mountain areas (and submountain areas in Continental Europe)
- 6410 *Molinia* meadows on calcareous, peaty or clayey-silt-laden soils (Molinion caeruleae)
- 6430 hydrophilous tall herb fringe communities of plains and of the montane to alpine levels; important for feeding habitats)
- *7110 active raised bogs
- 7120 degraded raised bogs still capable of natural regeneration
- 7140 transition mires and quaking bogs
- 7150 depressions on peat substrates of the Rhynchosporion
- 7230 alkaline fens
- *91D0 bog woodland.

The locations of most of the cold-stenothermic relict species are protected through this assortment of habitat types and with the German Natura 2000 network.

3.4.1 The Violet Copper (*Lycaena helle*) in Germany

The distribution area of the Violet Copper ranges from Scandinavia to Russia. It is also distributed over Central Europe and the Alps, and reaches East into Siberia (Region Amur) (Weidemann 1995). The species belongs to the Northern-boreal distribution type and is classified as a cold-stenothermic relict species in Germany.

The distribution map shows a high level of fragmentation in Germany (Fig. 1): In the German low mountain ranges, the species is present only in peninsulas in the Westerwald, Eifel, Rothaargebirge and in an isolated area in the Northern German lowlands near the river Oder in Eastern Germany. The best areas are in the prealpine lowlands of the Alps, where this species shows the highest abundances in Germany and the largest coherent area without interruptions. Nearly all localities in Germany are found in the Continental biogeographic region, with the exception of one population in the German Alpine biogeographic region (near Garmisch-Partenkirchen). The conservation status in the German Continental biogeographic region was assessed as unfavourable - inadequate in the German National Report (Art. 17 Habitats Directive)

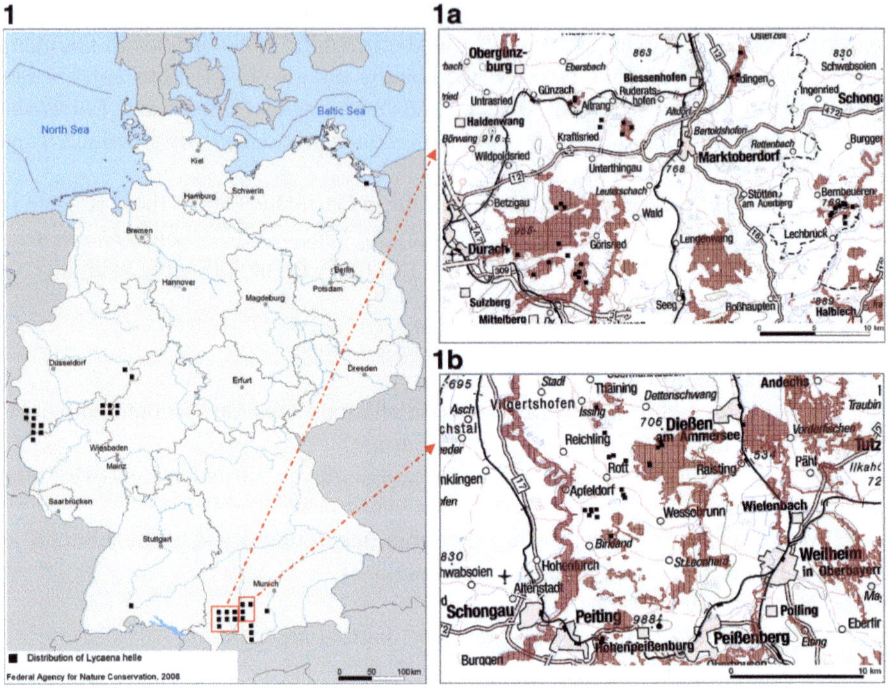

Fig. 1 Distribution of *Lycaena helle* (Source: German National Report according to Art. 17 Hab. Dir.) (**a, b**) Occurrences of *Lycaena helle* (1992–2003, black quadrat) and the Sites of Community Importance (Data as of Dec. 2006, red areas) in Bavaria

in 2007 and in the Alpine biogeographic region as bad (red). Figure 1 shows the current distribution area in Germany.

Imagos of *Lycaena helle* can be found during only a short period from May to June and only in cool climates and wetlands with Calthion (also Cirsietum rivularis and Chaerophyllo-Ranunculetum aconitifolii), where *Polygonum bistorta* occurs. Habitat types with these associations are not well represented in Annex I of the Habitats Directive.

The Violet Copper was submitted to Annexes II and IV during the enlargement of the European Union (EU) to the East in 2004 (Biewald and Nunner 2006). All member states with occurrences of such newly listed species in the Habitats Directive are responsible for their protection. However, the old member states were not urged to install SCIs for the species added to Annex II, but they do fall under the protection of already existing Natura 2000 sites. In Germany, no area has been especially nominated to date for Natura 2000 for *Lycaena helle*, so this species occurs only in a few SCIs where other protected species of Annex II also occur. Only four of the six federal states in Germany where *Lycaena helle* is found have so far completed their duty to update the Standard Data Forms indicating the presence of the species within their sites. Germany has to ensure protection of *Lycaena helle* as it is listed in Annexes II and IV of the Habitats Directive, so a Natura 2000 network will need to be implemented for the protection of this species in the next few years. Figure 1a, b shows the problem that only few habitats of *Lycaena helle* are protected so far in Germany. The best areas should be integrated in the list of SCIs as soon as possible.

Figure 1a shows a small region of Bavaria to the East of Durach, where most areas with populations of the Violet Copper can be found in Natura 2000 sites. In other regions in Bavaria, for example in the West near Lake Ammersee, most remaining populations are located outside Natura 2000 sites (Fig. 1b). Because most settled areas in Western Germany are not represented by Annex I habitat types, there are not enough sites represented in the network of SCIs.

Only one relict butterfly species is listed in Annex II of the Habitats Directive in Germany. Another two species are listed in Annex IV. Most populations of all identified 12 relict species of day-flying butterflies in Germany live in Annex I habitats, so protection for these species is possible in their habitats. The habitats of the Violet Copper as a new species in Annex II are not well represented in the Annex I habitat types. The conservation status of the species in the Continental biogeographic region is unfavourable - inadequate, and in the Alpine biogeographic region it is unfavourable - bad. The specific situation of this species will obviously need additional SCI proposals or enlargement of existing SCIs to fulfil the requirements of the Habitats Directive, and this might be the only means of securing a better conservation status in future. Species-specific protection, site protection and/or management plans are urgently needed.

4 Conclusion

4.1 Habitat Conservation

The habitats of the relict species are mostly covered by Annex I habitat types. This means that even if a species itself is not directly protected, there is an indirect habitat protection in the Natura 2000 network of protected areas. Relict species are sensitive to climate change because they are generally strongly adapted to special conditions. Because of an increase in the anthropogenic causes of climate change, there is a high risk of extinction particularly for glacial relict species, as these represent the largest group of relict species in Germany. The best way to protect these local relict species is to include them in special management conservation plans. For many Natura 2000 sites, management plans should be set up and the relict species in Natura 2000 sites should be integrated as typical species where they occur.

Altogether, relict species from the analysed groups can be found in 46 habitat types of the Directive. At least 60–80% of the total area of most of these habitat types has been included in Natura 2000 sites. The most important habitat types hosting these relict species are bogs and species-rich grasslands, mainly of mountain ranges. In addition, screes represent an important glacial refuge for other relict species groups such as spiders. The importance of bogs has been pointed out in this study for several taxa, but screes were only detected to be important for higher plants among the studied groups. In the latter, the most important habitat types referring to the studied groups were the following:

- 6210 semi-natural dry grasslands and scrubland facies on calcareous substrates (Festuco-Brometalia)
- *7110 active raised bogs
- 7120 degraded raised bogs still capable of natural regeneration
- 7140 transition mires and quaking bogs
- 7230 alkaline fens
- 8210 calcareous rocky slopes with chasmophytic vegetation.

4.2 Species Conservation

One hundred and fifty-seven relict species in the taxonomic groups of higher plants, molluscs, dragonflies, damselflies and butterflies have been analysed in this study. Of these relict species, 143 are not covered by the direct species protection regime or have to be formally integrated into designated sites (not listed in Annexes II and IV of the Habitats Directive). A total of 14 species can be found in Annexes II, IV and V of the Habitats Directive (Table 7). Two of these species are extinct in Germany (*Saxifraga hirculus* and *Coenagrion hylas*), nine species are included in Annex II. The largest group of relict species in Germany directly listed in the Annexes of the Habitats Directive is molluscs (five species). Most relict species have no contact with other populations. Thus, they are isolated without gene flow between the remaining populations, so that they often show criteria that qualifies them as candidates for ESUs; however, more research work is needed on this topic. In accordance with the concept of assessing the responsibility of a region for the worldwide conservation of species (Gruttke et al. 2004), relict species are identified as conservation items of high priority.

Relict species in Germany can mostly be protected by the network of Natura 2000 sites as characteristic or typical species for habitat types of Annex I rather than by direct species protection from Annexes IV and II. Only nine species are listed in Annex II, so SCIs are being established or provided for seven of these nine species, because two species are extinct in the wild in Germany. The refuges of

Table 7 Relict species in four taxonomic groups in Germany

	Number of species in Germany	Relict species	Relict species listed in the annexes of the Habitats Directive
Plants (without mosses and lichens)	ca. 3,500	100	4
Molluscs	342	38	5
Dragonflies and damselflies	81	7	2
Butterflies (Rhopalocera)	188	12	3

most relict species in Germany are at present protected by coverage through the network of Natura 2000 sites conserving the habitats of Annex I, in which the highest number of relict species occur. Because of climate change and their often fragmented and very small survival areas, relict species are subject to a high risk of extinction; so management plans should be provided for their conservation. For a few species, additional sites would be desirable to adequately cover their habitats within Natura 2000. Because the influence of anthropogenic climate change is increasing rapidly, management plans for the protection of relict species should be provided. The urgent need for management plans or species action plans is evident when considering the results of the current conservation status (German National Report, Nationaler Bericht 2007) of relict species listed in the Annexes of the Habitat Directive: two species are extinct and nine of the 14 species are identified as having an unfavourable - inadequate or unfavourable - bad conservation status in Germany (Tables 2–6).

Acknowledgement We would like to thank the LfU Bavaria for providing the distribution data for *Lycaena helle* in Bavaria and Ms. Ersfeld (BfN) for her great support during the production of the maps. Jan Habel (Trier), Petra Dieker (Trier), Aline Finger (Eglisau, Switzerland), Klaus Fischer (Greifswald) and Wolfgang Wipking (Cologne) gave us information about the recent distribution of *Lycaena helle* in Germany. We thank Nora Wuttke, Bonn, for support in English. Finally, we would like to thank one unknown reviewer for helpful comments.

References

Ås S, Bengtsson J, Ebenhard T (1992) Archipelagoes and theories of insularity. In: Hansson L (ed) Ecological principles of nature conservation. Applications in temperate and boreal environments. Elsevier, London, pp 201–251

Assmann T, Schröder E, Terlutter H (2000) Morphometric differentiation in a specialised snail predator: *Carabus pyrenaeus* (Coleoptera, Carabidae). In: Brandmayr P, Casale A, Lövei GL, Vigna Taglianti A (eds) National history and applied ecology of carabid beetles, Pensoft Publishers, Sofia & Moscow, pp 171–178

Balzer S, Ellwanger G, Raths U, Schröder E, Ssymank A (2008) Verfahren und erste Ergebnisse des nationalen Berichts nach Artikel 17 der FFH-Richtlinie. Natur und Landschaft 83(3):111–117

Biewald G, Nunner A (2006) *Lycaena helle* (Denis & Schiffermüller, 1775). In: Petersen B, Ellwanger G Das europäische Schutzgebietssystem Natura 2000. Ökologie und Verbreitung von Arten der FFH-Richtlinie in Deutschland 3: Arten der EU-Osterweiterung. SchrR Landschaftspfl Natursch 69(3):139–153

BMU (Bundesministerium für Umwelt, Naturschutz und Reaktorsicherheit (2007)) Nationale Strategie zur biologischen Vielfalt. BMU, Bonn

Brackel WV (2001) Bodensee-Vergissmeinnicht (*Myosotis rehsteineri*). In: Fartmann T, Gunnemann H, Salm P, Schröder E (eds) Berichtspflichten in Natura 2000-Gebieten. Angew Landschaftsökol 42:119–123

Buchholz A, Welk E (2005) *Minuartia stricta* (Swartz) Hiern (Caryophyllceae): Wiederentdeckung eines in Zentraleuropa verschollen geglaubten Glazialrelikts. Ber Bayer Bot Ges 75:95–108

Bundesamt Für Naturschutz BFN (ed) (2008) Nature data 2008. Landwirtschaftsverlag, Münster

CLECOM (2002) Checklist of species-group taxa of continental mollusca living in Germany (CLECOM Section I). http://www.gnm.se/gnm/clecom/clecom/clecom_D.htm. Cited 25 Feb 2002

Cox CB, Moore PD (2006) Biogeography: an ecological and evolutionary approach. Blackwell, Oxford

De Lattin G (1967) Grundriss der Zoogeographie. VEB Gustav Fischer, Jena

Dörr E (2000) Verbreitung und Rückgang der Glazialrelikte in den Mooren des Allgäuer Raumes. Hoppea 61:567–585

Ellwanger G, Mauersberger R (2003) Sympecma paedisca (Brauer, 1877). In: Petersen B, Ellwanger G, Biewald G, Hauke U, Ludwig G, Pretscher P, Schröder E, Ssymank A Das europäische Schutzgebietssystem Natura 2000. Ökologie und Verbreitung von Arten der FFH-Richtlinie in Deutschland 1: Pflanzen und Wirbellose. SchrR Landschaftspfl Natursch 69/1:611–618

Falkner G, Bank RA, Proschwitz VT (2001) Check-list of the non-marine molluscan species-group taxa of the states of Northern, Atlantic and central Europe (CLECOM 1). Heldia 4(1/2):1–76

Gruttke H, Ludwig G, Schnittler M, Binot-Hafke M, Fritzlar F, Kuhn J, Assmann T, Brunken H, Denz O, Detzel P, Henle K, Kuhlmann M, Laufer H, Matern A, Meinig H, Müller-Motzfeld G, Schütz P, Voith J, Welk E (2004) Memorandum: Verantwortlichkeit Deutschlands für die weltweite Erhaltung von Arten. In: Gruttke H (ed) Ermittlung der Verantwortlichkeit für die Erhaltung mitteleuropäischer Arten. Naturschutz und Biologische Vielfalt 8, Bundesamt für Naturschutz, Bonn-Bad Godesberg, pp 273–280

Holdhaus K (1954) Die Spuren der Eiszeit in der Tierwelt Europas. Abhandlungen der zoologisch-botanischen Gesellschaft in Wien, XVIII. Universitätsverlag Wagner, Innsbruck

Kudrna O (2002) The distribution atlas of European butterflies. Oedippus 20:1–132

Moritz C (1994) Defining "evolutionary significant units" for conservation. Trends Ecol Evol 9:373–375

Müller T (1980) Der Scheidenkronwicken-Föhrenwald (Coronillo-Pinetum) und der Geißklee-Föhrenwald (Cytosi-Pinetum) auf der Schwäbischen Alb. Phytocoenologia 7:392–412

Nationaler Bericht (2007) Nationaler Bericht nach Art. 17 der Fauna-Flora-Habitat-Richtlinie (92/43/EWG). http://www.bfn.de/0316_bericht2007.html. Cited 17 Dec 2008

Oberdorfer E (1990) Pflanzensoziologische Exkursionsflora. Ulmer, Stuttgart

Reisch C (2001) Climatic oscillations and the fragmentation of plant populations – genetic diversity within and among populations of the glacial relict plants Saxifraga paniculata (Saxifragaceae) and Sesleria albicans (Poaceae). Dissertation zur Erlangung des Doktorgrades der Naturwissenschaften der Naturwissenschaftlichen Fakultät III, Biologie und Vorklinische Medizin der Universität Regensburg

Richard J-L (1972) La végétation des Crêtes rocheuses du Jura. Ber Schweiz Bot Ges 82:68–112

Soulé ME (ed) (1986) Conservation biology: the science of scarcity and diversity. Sinauer Associates, Sunderland, MA

Sternberg K (1998) The postglacial colonization of central Europe by dragonflies, with special reference to Southwestern Germany (Insecta, Odonata). J Biogeogr 25:319–337

Thorn K (1960) Bemerkungen zu einer Übersichtskarte vermutlicher Glazialreliktpflanzen Deutschlands. Mitt flor--soziol ArbGem 8:81–85

Varga Z (1977) Das Prinzip der areal-analytischen Methode in der Zoogeographie und die Faunenelemente-Einteilung der europäischen Tagschmetterlinge (Lepidoptera: Diurna). Acta Biol Debrecina 14:223–285

Weidemann HJ (1995) Tagfalter. Beobachten, bestimmen. Naturbuch-Verlag, Augsburg

Welk E (2001) Arealkundliche Analyse und Bewertung der Schutzrelevanz seltener und gefährdeter Gefäßpflanzen Deutschlands. Dissertation zur Erlangung des Doktorgrades vorgelegt an der Mathematisch-Naturwissenschaftlich-Technischen Fakultät der Martin-Luther-Universität Halle-Wittenberg

Wilmanns O (1993) Ökologische Pflanzensoziologie. Quelle & Meyer, Heidelberg

Wilmanns O, Rupp S (1966) Welche Faktoren bestimmen die Verbreitung alpiner Felspflanzengesellschaften auf der Schwäbischen Alb? Veröff Landesst Naturschutz Landschaftspflege Baden-Württemberg 34:62–86

Ecological limits *vis-à-vis* Changing Climate: Relic *Erebia* Butterflies in Insular Sudeten Mountains

Martin Konvička, Jiri Benes, and Thomas Schmitt

Abstract Insular middle mountains of Central Europe, such as Jesenik Mts. (max. altitude 1,495 m) and Krkonose Mts. (1,602 m), both in the Sudeten system, are poor in relic species, but the relic populations are suitable for studying the population responses to recent climatic change. There are two relic butterflies that are native to Jesenik Mts., each traditionally classified as an endemic subspecies: *Erebia sudetica sudetica* (Staudinger, 1851), a species protected by the EU Habitat Directive, and *E. epiphron silesiana* Meyer & Dür, 1852. Whereas Jesenik *E. epiphron* forms continuous populations, restricted to summit *Nardus*-dominated grasslands and not entering forested elevations, the distribution of Jesenik *E. sudetica* forms an archipelago at wet timberline sites and at clearings within mountain forests. Although both species avoid mountain forests, their lower distribution limits seem to be mediated by a limited range of available habitats, rather than of direct ecophysiological constraints. A non-native population of *E. epiphron* in Krkonose, transferred from the Jesenik in the 1930s, descends to ca 1,100 m, colonizing large clearings with cultural hay meadows within the forest belt. There is also behavioral and molecular evidence that *E. epiphron* cannot disperse through continuous forest. On the other hand, *E. sudetica* has a better dispersal power, and recent monitoring of its native Jesenik distribution suggests active colonization of woodland clearings. Thus, both butterflies can survive an altitudinal increase of the timberline, provided open habitats within the forests are secured. This conjecture suggests that these two and other alpine species survived past warmer periods of the Holocene in mountains, such as Jesenik owing to disturbance dynamics that never allowed complete forest closure. In the higher Krkonose, in contrast, they probably lacked suitable open habitats because of the encroachment of such sites by *Pinus mugo* dwarf pine, which is native to Krkonose but non-native to Jesenik.

M. Konvička and J. Benes
Faculty of Sciences, University Southern Bohemia & Institute of Entomology, Czech Academy of Sciences, Branisovska 31, D-37005 Ceske Budejovice, Czech Republic

T. Schmitt (✉)
Department of Biogeography, Trier University, D-54286 Trier, Germany
e-mail: thsh@uni-trier.de

J.C. Habel and T. Assmann (eds.), *Relict Species: Phylogeography and Conservation Biology*,
DOI 10.1007/978-3-540-92160-8_20, © Springer-Verlag Berlin Heidelberg 2010

1 Introduction: Why Study Relics in Middle High Mountains

Isolated mountains of the temperate zone that overtop the timberline in their highest altitudes offer several advantages for studying the biology of alpine relics. In Europe, the subalpine summit habitats of the "middle high mountains" (Jenik 1998) include the uplands of Northern England and Scotland, the French Massive Central, and the highest parts of the Hercynnian system in France (Vosges), Germany (Harz, Schwarzwald, Bayerischer Wald), and the Sudeten Mountains at the Czech/Polish border region (Krkonose Mts. in the West, Kralicky Sneznik and Jesenik Mts. in the East). Due to their small extents of (sub)alpine habitats, studies on alpine relics occurring in such systems can address questions that would not be tractable in the much larger and considerably higher systems of the Alps, Carpathians, or Pyrenees. In particular, the low richness of alpine elements in these small mountain areas reduces the complexity of biotic interactions, their small total areas facilitate research, and the geographic isolation from other mountain systems ensures that local populations do not interact with conspecific populations elsewhere. Relic species inhabiting middle high mountains are particularly vulnerable to habitat loss, aggravated by the ongoing climate change (Hill et al. 2002; Franco et al. 2006).

High mountain insects are subject to ecophysiological constraints, confining their distributions to high altitudes (Hodkinson 2005). Assuming that the outcome of warming climate results in a vertical shift in vegetation belts (e.g., Kullman 2006; Jurasinski and Kreyling 2007), alpine relics occurring above the timberline can be threatened in two ways, either directly by a shift of climatic conditions beyond their physiological limits, or indirectly, via habitat loss (e.g., Grabherr 2003; Holten 2003; Virtanen 2003; Virtanen et al. 2003). Either way, they should respond by uphill shifts and tracking of suitable climate conditions (e.g., Hill et al. 2002; Konvicka et al. 2003; Wilson et al. 2007) and these displacements will eventually hit the upper limits in mountains not high enough to accommodate all the species occurring there now. Still, there had been other periods of relatively warm climate during the Holocene, especially the Atlanticum when the temperature was 1–2 degrees warmer than at present (Prentice et al. 2000), and there is little doubt that all the Northern and alpine relics occurring in European middle high mountains survived these warmer periods in situ, within the mountains where they occur today.

Two relic butterflies inhabit the highest altitudes of the Sudeten mountains: *Erebia epiphron* (Knoch, 1783) and *E. sudetica* Staudinger, 1851 (both Nymphalidae: Satyrinae). Taxonomists treat these populations as an endemic subspecies, *E. epiphron silesiana* (Meyer et Dür, 1852) and *E. s. sudetica* Staudinger, 1851, both belonging to species with considerably wider distributions elsewhere. Therefore, the Sudeten subspecies display extremely restricted native ranges: Exclusively the Jesenik Mts. in the case of *E. epiphron* and Jesenik Mts., Rychlebske Mts. and Kralicky Sneznik Mts. in the case of *E. sudetica* (Kuras et al. 2001c). An artificially established population of *E. epiphron* occurs in the Krkonose Mts., Western Sudetens. The closest native populations of the two species occur 200 km apart in the Carpathian High Tatras,

Slovakia (*E. epiphron*) and 600 and 800 km apart in the Rodna Mts., Eastern Carpathians, Romania and Grindelwald, Berner Oberland, Switzerland, respectively (*E. sudetica*).

Butterflies with such a restricted distribution have long attracted the attention of lepidopterists (e.g., Schwarz 1948; Stiova 1988). A systematic investigation of their ecology, however, began as late as the mid-1990s, when T. Kuras and J. Benes, then from Palacky Univerisity, Olomouc, initiated their mark-recapture surveys. This gradually extended to studies of behavior and population genetics, and to collecting additional information on the ecology of *E. euryale* (Esper, 1805) and *E. ligea* (Linnaeus, 1758), two *Erebia* species inhabiting the Sudeten mountain forests. Inclusion of *E. sudetica* to the EU Habitat Directive recently inspired annual monitoring of its sites. At present, Sudeten *Erebia* is among the most thoroughly studied butterflies in Central Europe.

Here, we review the ecological information available to date, focusing on the prospects of populations *vis a vis* recent climate change. We document that the population structure of *E. epiphron* and *E. sudetica* depends on the spatial distribution of their habitats. We then describe the outcomes of a notable experimental release of *E. epiphron* to the previously unoccupied Krkonose Mts., conducted in the 1930s. Recent surveys of the Krkonose population revealed that under certain conditions, this alpine butterfly can colonize altitudes below its subalpine habitats, and the same applies for *E. sudetica*, which occurs in large numbers at damp openings within the forest belt of Jesenik Mts. These observations suggest that although there exist ecophysiological limits to down-slope distribution of both species, their current distribution within the Sudeten system depends more on habitat availability than on climate.

2 The Biogeography of the Four Mountain *Erebia* Species of the Sudeten Mts

Of the four *Erebia* species occurring in the Sudeten Mts., *E. epiphron* is the most typical alpine species, inhabiting all the large high mountain systems of Europe, and also the Cantabrian range, the Massif Central, the Vosges, the Harz (extinct), mountains of the Western Balkans, and the Jesenik Mts. (Varga and Schmitt 2008). Allozyme studies unravelled the existence of at least five strongly differentiated genetic lineages in Europe: one confined to the Eastern Pyrenees, one in the central Pyrenees and Western Alps, two along the Southern and Southeastern Alps, and one in the Northern Alps and the Jesenik Mts. (Schmitt et al. 2006). The Carpathian populations are morphologically well differentiated from all other populations (Varga 2001). These molecular and morphological characteristics support that *E. epiphron* had several glacial retreats at the foothills of some mountain systems and others between mountain systems, as between the Pyrenees and Western Alps and between the Northern Alps and the Jesenik Mts., thus explaining the strong geographic disjunctions among some of the lineages of the species (Schmitt et al. 2006).

In contrast, *E. sudetica* is missing from most parts of the high mountain systems and is only found in two limited areas of the Western Alps (department Isère, France and near Grindelwald, Switzerland) and in some very localized areas in the Romanian Carpathians, but the species occurs in the Massif Central and in the Eastern Sudeten Mts. (Kudrna 2002). Preliminary allozyme analyses (Haubrich and Schmitt 2007, unpublished data of the authors) show that the populations from the Swiss Alps and the Sudeten Mts. are differentiated; however, the relation between the latter and the Romanian populations is still unknown.

E. euryale is a forest or forest skirt species (Sonderegger 2005) distributed all over the European mountain forest biome (Kudrna 2002). An allozyme study showed that this species is differentiated in at least four well supported genetic lineages: Pyrenees, Western Alps, Eastern Alps, and Southeastern Europe, and thus traces the glacial refugia of the mountain forest biome in Europe (Schmitt and Haubrich 2008).

The boreo-montane *E. ligea* is distributed all over the mountainous areas of central and Eastern Europe, and in the lowlands of Northern Europe (Kudrna 2002). No genetic data over major parts of the distribution of the species are available. However, the North-South disjunction in this species is most probably a postglacial phenomenon (Varga and Schmitt 2008).

3 Jesenik Mts.: Geometry of Habitats Affects Population Structures

With timberline (formed by sparse growths of Norwegian spruce) at about 1,250 m a.s.l. and with the highest summit (Praded Mt., at 1,495 m), Jesenik Mts. do not contain alpine habitats sensu stricto. The flat plateaux above the timberline harbors 12 km^2 of subalpine grasslands, dominated by *Nardus stricta*, *Calamagrostis villosa*, and *Festuca supina*. These Nardeta of the Jesenik plateau represent essentially stabilized habitats not affected by succession (German: Dauergesellschaften). Steeper, wind-shielded valley headwalls affected by periodic avalanches harbor structurally diverse and species-rich tall-herb formations (Jenik 1961; Jenik et al. 1980).

The relic *Erebia* species of this mountain area differ in habitat use. *E. epiphron* occurs in high abundance at these stable subalpine grasslands, but rarely descends to elevations near the timberline, whereas *E. sudetica* is limited to the islets of timberline-adjoining tall-herb formations, which are transitional habitats in a gradient of natural succession (Fig. 1). The two remaining *Erebia* species (*E. euryale* and *E. ligea*) inhabit openings within mountain forests, ranging from ca 700 m to growths of dwarfed spruce at about 1,300 m.

Although these patterns were already noted by Stiova (1988), the first quantitative support originated from a trapping study (Benes et al. 2000). This study used flower-mimicking water-filled traps, an efficient method for capturing Lepidoptera of subalpine habitats. A total of 35 traps exposed for 665 trap-days captured 3,948

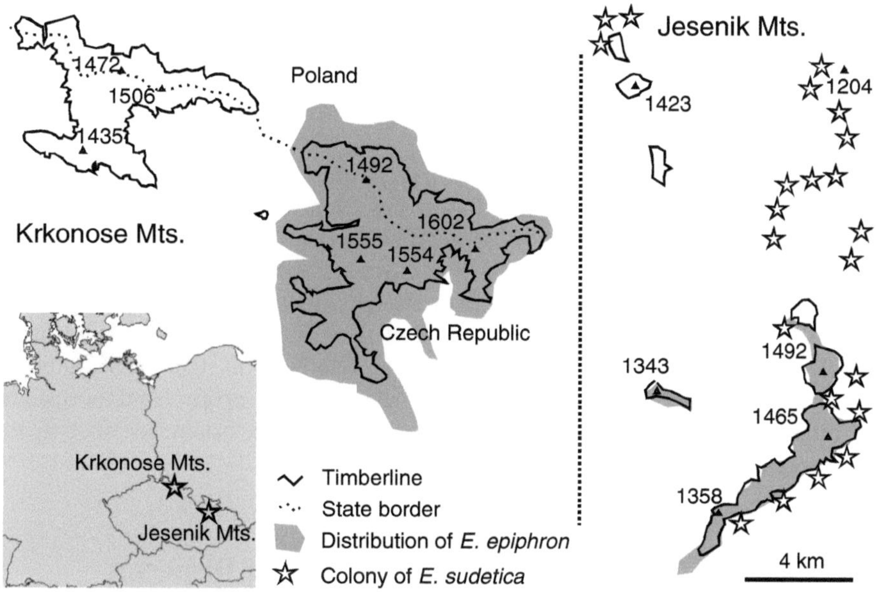

Fig. 1 Map showing the position of Jesenik and Krkonose Mts. in Central Europe (*lower left corner*), extend of subalpine habitats, and the distribution of two alpine butterflies, *Erebia epiphron* and *E. sudetica*. For *E. epiphron*, the entire distribution extend is shown, whereas for *E. sudetica*, the map shows only the larger colonies in Jesenik Mts. The northernmost Jesenik colonies shown are those in Ramzova saddle, which separates Jesenik (South) and Rychlebske Mts. (North)

butterflies and moths, including 3,731 *Erebia* individuals. The tall-herb formations near the timberline yielded the highest species diversity, corroborating patterns known for vegetation (cf. Jenik 1961, 1998). There was a significant habitat separation among woodland-preferring *E. ligea* and *E. euryale*, timberline-preferring *E. sudetica*, and grassland-preferring *E. epiphron* (Kuras et al. 2000).

The summit grasslands form several islets in Jesenik. *E. epiphron* inhabits the two largest units, main ridge (area 10 km^2) and Mravenecnik (area 1.2 km^2), both separated by 4 km of forested saddles (Fig. 1). In contrast, the tall-herb formations inhabited by *E. sudetica* form an archipelago of valley headwalls and springs. Kuras et al. (2003) used the mark-recapture method to illustrate that the differences in spatial configuration of habitats determine the population structures of both species. The main ridge population of *E. epiphron*, studied by mark-recapture in 1996, was really huge: For a total of 4,034 marked individuals, only 236 individuals were recaptured. This was likely due to rather high local density, as the estimated numbers were about 20,000 males for 1 km^2 of habitat, corresponding to hundreds of thousands of individuals for the entire main ridge (female numbers were not estimable), and the unsuitable mountain weather conditions not permitting daily field work. For the lateral Mravenecnik population, marking for 8 days in 1997 yielded 2,542 captured and 440 recaptured individuals, giving an estimate of 4,500 individuals.

In the case of *E. sudetica*, one large and one small population were studied in 1998, the large one yielding 2,171 captured and 893 recaptured individuals, the small one yielding 253 captured and 80 recaptured individuals. The estimates were 5,000 and 300 individuals, respectively, suggesting a population size for the entire main ridge area of around 10,000 individuals.

E. epiphron individuals were considerably poorer dispersers than those of *E. sudetica*. The lifetime probability of moving 1 km was 0.5–1.2% for the former and 2.2–6.4% for the latter, the wide range being due to differences among sexes and populations (Kuras et al. 2003). The difference became even more dramatic over longer distances: the probability of crossing 5 km was 1×10^{-10} for *E. epiphron* and 1×10^{-7}, thousand times higher, for *E. sudetica*. The latter species is thus more mobile, indicating that populations inhabiting relatively unstable environments, such as disturbance-dependent timberline sites, should profit from occasional migrations, while those inhabiting stable and extended environments, such as the summit grasslands, may do without them.

4 Krkonose *E. epiphron*: Novel Biotopes and Dispersal Limitations

Compared to Jesenik, Krkonose Mts. are higher (max. altitude: Snezka Mt., 1602), contain larger subalpine areas (about 50 km^2) and rocky valleys resembling those of much higher mountains. The main difference in vegetation is a native occurrence of *Pinus mugo*, a dwarf pine forming a distinct belt sandwiched between spruce forests and subalpine grasslands.

Krkonose lacked truly alpine *Erebia* until the early 1930s, when the entomologist Josef Soffner sent two parcels, each containing 50 *E. epiphron* and 50 *E. sudetica* females, from a Jesenik site to two Krkonose inn-keepers, asking them to release the butterflies in the wild. The target inns were Lucni bouda (1,410 m.) amidst summit grasslands and Rychorska bouda (1,000 m), a rather uncharacteristic site of pastures below the timberline.

Not much was heard about the releases until the 1960s, when Borkowski (1966) reported *E. epiphron* from Krkonose Northern slopes, about 1 km from the Lucni bouda release site. Soffner (1967) responded by publishing a paper describing the circumstances of the 1930's transfer. Gradually, the butterfly became very abundant within its new range (Liska and Skyva 1997; Cizek et al. 2004). In contrast, neither the released *E. sudetica* nor the other release of *E. epiphron* established a population. It is evident from hindsight that Soffner did not understand the precise habitat preferences of the two species, and selected inappropriate settings for both *E. sudetica* releases and for the Rychorska bouda release of *E. epiphron*.

For decades, there was no systematic interest in the established *E. epiphron* population, despite some fascinating questions brought about by its existence. Specifically, the bottleneck during the transfer could have impaired the adaptive ability of butterflies. On the other hand, the release area contains a broader range of biotopes than the native Jesenik area, which might allow expansion to novel habitats. Last but not least, tracking the newly established population could have allowed discerning possible dispersal barriers.

Cizek et al. (2003) approached these questions in a trapping study, similar to the earlier Jesenik study by Benes et al. (2000). Series of traps were set to all main Krkonose biotopes, from clearings deep in the forest belt up to the rocky summits. The mountains were also systematically surveyed to ascertain the current distribution extent. As in Jesenik, *E. epiphron* abundances peaked at subalpine grasslands near 1,300 m. The species nevertheless reached the uppermost elevations at about 1,600 m and, more surprisingly, has colonized open clearings and hay meadows below the timberline, down to about 1,100 m. Compared to the Jesenik population, the Krkonose population in fact expanded both uphill and downhill.

An important finding concerned the limits of the expansion. Soffner released the butterfly at the Eastern ridge of the Krkonose Mts., which is completely colonized now. The Western ridge, similar to the Eastern ridge in the area of subalpine habitats and maximum elevations, remained uncolonized for over 75 years. The two ridges are separated by about 4 km (aerial distance) of forested elevations, which apparently acted as a dispersal barrier. The butterfly failed to cross the forests – while colonizing the non-forested sites below the timberline.

Another evidence for avoidance of forests originated from a study of the behavior of *E. epiphron* and *E. euryale*, conducted in sympatry at the Jesenik main ridge (Konvička et al. 2002). We recorded the behavior along repeated transects, together with time of day, weather, and vegetation where activities took place. Whereas *E. euryale* utilized woody structures, such as dwarfed spruce, for activities as resting and basking, *E. epiphron* restricted all its activities to open grasslands.

Further evidence that forests restrict *E. epiphron* dispersal originated from a genetic study of the native and introduced populations (Schmitt et al. 2005). The two native populations within Jesenik were significantly differentiated, as were the source population in Jesenik and the introduced population in Krkonose. The genetic distance between the Jesenik source and the Krkonose target sites was less than the distance between samples from northernmost and Southernmost limits of Jesenik main ridge. This revealed isolation by distance along the ridge, consistent with rather limited dispersal. The smaller lateral Jesenik population was strongly genetically impoverished. Its mean heterozygosity was half of the mean heterozygosity of three samples taken along the main ridge (5.5 vs. 10.0%), and similar patterns applied to the total percentages of polymorphic loci, polymorphic loci with the most common allele not exceeding 95%, and the mean number of alleles per locus. No such impoverishment applied to the introduced Krkonose population, indicating that the 50 females transferred by Soffner sufficed to transfer the entire (allozyme) allelic variation of the Jesenik source population.

Perhaps the most intriguing result of the population transfer was the colonization of lower-elevation habitats in Krkonose, as opposed to Jesenik (Cizek et al. 2004). This is explicable by different levels of fragmentation of mountain forests in the two mountains, resulting in different availability of non-wooded habitats within their forested elevations. A peculiar feature of the Krkonose landscape is "cultural" meadows and pastures, created by forest clearance in altitudes of 800–1,200 m (Krahulec et al. 1997). Moreover, much of Krkonose mountain woodlands fell victim to "salvation logging" of the 1980s, which reacted to die-backs caused by atmospheric pollution (cf. Emmer et al. 1998). The forests are thus dissected by a dense network of open habitats. These are practically missing in the forested elevations of Jesenik, making Jesenik mountain forests more restrictive for dispersing *E. epiphron* than Krkonose forests (Fig. 2).

The colonization of elevations near 1,100 m in Krkonose, but not in Jesenik, documents that a lower climatic elevation limit of *E. epiphron* is situated in altitudes suitable for tree growth. The current lower distribution limits are thus not determined solely climatically, but also by the availability of open grasslands below the timberline.

5 Jesenik *E. sudetica*: Disturbance-Dependent Metapopulation

The known distribution of Sudeten *E. sudetica* extends beyond the timberline sites in Jesenik Mts., and the species also inhabits much lower Rychlebske Mts. (highest altitude Smrk, 1,125 m) (cf. Kuras et al. 2001a,c, Maslowski 2005). Based on the then-available information, Kuras et al. (2003) speculated that the populations of the more ephemeral habitats below the timberline are only transitory, depending on colonists originating from the more stable timberline sites colonizing the temporary woodland openings.

Recently (2007), we surveyed as many as possible forest-belt localities of *E. sudetica* in order to ascertain their status and prospects. We spent over 30 person-days visiting such habitats as clearings, open tracks, and meadows along valley streams. Additional surveys covered Rychlebske and Kralicky Sneznik Mts. We failed to confirm the species for the latter two mountains, but we found dozens of extant colonies within the Jesenik forest belt, the lowest-situated ones at 700 m a.s.l. They included boggy hay meadows in saddles and along stream sides, woodland clearings opened up during salvage logging operations in the 1980s and 1990s, and linear open structures such as those below ski lifts. All sites displayed high humidity, typically provided by groundwater springs. Colonies near lower distribution limits, most of them on meadows, tended to be small, hosting just a few individuals. In contrast, colonies in altitudes near 1,000 m, most of them on muddy clearings, were highly abundant. At several sites, we counted several hundreds of individuals per 100 m of transect walk, corresponding to thousands of individuals present. These very large colonies apparently profited from the forest decays in the 1980s and subsequent salvage logging. The conditions were remarkably similar to the clearcuts in Krkonose, with the difference that the Krkonose clearcuts hosted *E. epiphron* that reached them from adjoining subalpine grasslands. Probably because all the Jesenik clearcuts were separated by forested areas from the subalpine sites, they exclusively hosted *E. sudetica*, again demonstrating the good dispersal ability of *E. sudetica* and the poor dispersal ability of *E. epiphron*.

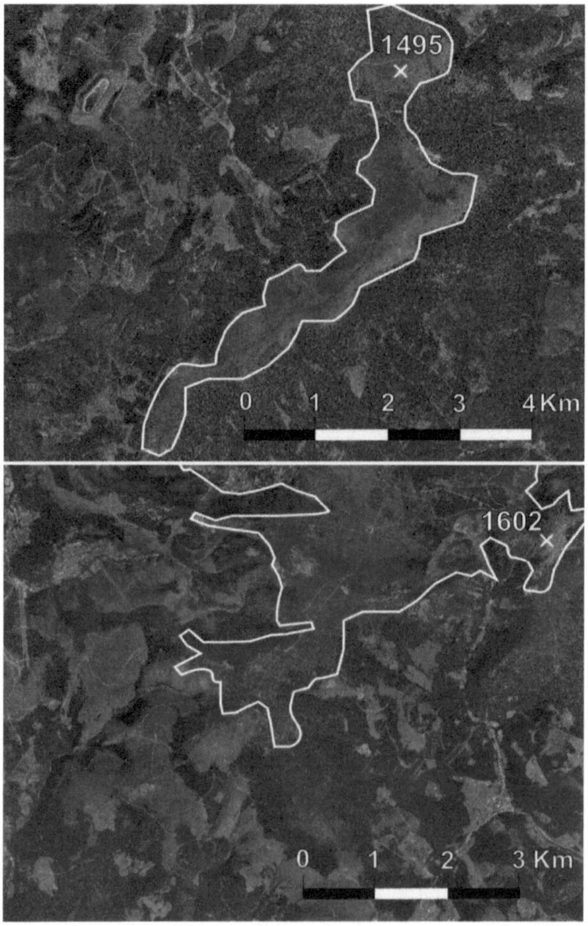

Fig. 2 Aerial photographs of Jesenik main ridge (*above*) and a section of Krkonose Eastern ridge (*below*), with white line showing the position of timberline. Well apparent are the starkly different levels of fragmentation of forest zones of the two mountains

The existence of large forest belt colonies of *E. sudetica* has changed our view of relative conservation importance of these low-elevation sites. It is now clear that the forest belt colonies are scattered over an area of about 200 km^2 and that the total number of butterflies inhabiting these colonies exceeds the number inhabiting the timberline sites. The dynamics of these colonies, however, remains unknown. The forest-belt populations probably form a dynamic metapopulation system, tracking occasional forest openings and forming transient colonies there (see Bergman 2001; Wahlberg et al. 2002, for other examples). The relative role of the timberline vs. forest-elevation sites as source of the colonists remains unknown: Does the entire system depend on the relatively permanent sites near the timberline, or is it essentially self-sustaining, depending on transfers of colonists among forest-elevation patches? Do some "source" sites exist within the forested elevations as well, perhaps at streamside bogs in mountain saddles? Some of the candidate permanent sites could have been heavily altered during the last decades: for example, a pass (800 m) separating Jesenik and Rychlebske Mts. still hosts some colonies at boggy meadows, but the extend of the meadows shrank to tiny remnants during the last few decades as a result of building development, afforestation, and grassland drainage.

However, the ability to occur on open sites below the timberline testifies that similar to *E. epiphron*, *E. sudetica*'s lower limit is habitat-limited rather than climate-limited. Its lowest-elevation populations are small, and we cannot distinguish at this moment whether this is due to the occurrence near a lower climatic limit, or due to the restricted habitat availability. However, in elevations from about 900 m onwards, all clearings and boggy meadows tend to be occupied, usually by high densities of butterfly.

6 Warming Climate, *P. mugo* and Future of Sudeten *Erebia*

The long-term persistence of these two Sudeten species of *Erebia* butterflies will depend on the continuous presence of suitable open habitats in the upper parts of the mountains. The current distribution patterns suggest that the habitats do not necessarily need to be located above the climatic timberline. Also, none of the species is threatened by direct destruction of habitats, as most of their sites already enjoy legal protection, either in the form of reserves within the Jesenik-protected landscape area, or as a core zone of the Krkonose/Karkonosze national parks. The populations are abundant and/or interconnected, further adding to the rather optimistic prospects.

As revealed by pollen and macrofossil records, trees and shrubs ascended to higher elevations, possibly by as much as 100 m, at the height of the Atlanticum period (Jankovská 2004; Treml et al. 2008). This likely diminished the extent of subalpine habitats, relative to the current situation. However, because the specialized alpine biota, including the *Erebia* butterflies in Jesenik, had survived the period in situ, the summit habitats could not be closed-canopy forest, but rather a patchwork of tree growths and open sites. In the same vein, the current diversity of alpine relics might survive the recent warming episode, provided the warming does not get too extreme and the habitats in the uppermost elevations remain rather open. The processes that control tree growth must be adopted by conservationists in order to support the threatened relic species.

An important aspect concerns the species richness in alpine relics in Krkonose and Jesenik. Despite containing much smaller subalpine area, Jesenik hosts approximately as many relic Lepidoptera as do Krkonose (Table 1). The Krkonose fauna is relatively poor, apparently due to the presence of the species-poor vegetation belt of *P. mugo*. The dwarf pine used to cover even larger areas than in present before it was cleared by humans to obtain pastures, and there is no doubt that it expanded to still higher altitudes during the climatic optimum. In this period, relic alpine biota likely retracted to rocky slopes and scree, unsuitable for pine growth. Interestingly, the pine is not native to Jesenik, but had been planted there at the turn of the 20th century. Thus, the existence of *P. mugo* might explain the lack of some relic Lepidoptera associated with grassland formations in the Krkonose Mts., such as both species of *Erebia* and two *Clepsis* tortricids. On the other hand, several species absent from Jesenik are associated with the rocky habitats (*Elophos operaria*,

Table 1 List of all Lepidoptera species associated with high mountain (subalpine and alpine) habitats and recorded from Krkonose (KRK) and Jesenik (JES) Mts., Czech Republic

Species	Family	KRK	JES	Distribution
Incurvaria vetulella (Zetterstedt, 1839)	Incurvariidae	+	+	A, B, C
Lampronia rupella (Dennis et Schiffermüller, 1775)	Prodoxidae	+	+	A, B, C
Argyresthia amiantella (Zeller, 1847)	Yponomeutidae	+	–	A, C
Rhigognostis senilella (Zetterstedt, 1839)	Plutellidae	+	–	A, B, C
Denisia nubilosella (Herrich-Schäffer, 1856)	Oecophoridae	+	+	A, C
Elachista kilmunella Stainton, 1849	Elachistidae	+	+	A, B, C
Chinodes viduella (Fabricius, 1794)	Gelechiidae	+	+	A, B
Sparganothis rubicundana (Herrich-Schäffer, 1856)	Tortricidae	–	+	B
Clepsis steineriana (Hübner, 1799)	Tortricidae	–	+	A
Clepsis rogana (Gueneé, 1845)	Tortricidae	–	+	A, B, C
Phiaris obsoletana (Zetterstedt, 1839)	Tortricidae	+	–	A, B, C
Pseudococcyx mughiana (Zeller, 1839)	Tortricidae	+	–	A, C
Eudonia petrophila (Standfuss, 1848)	Crambidae	+	+	A, C
Eudonia sudetica (Zeller, 1839)	Crambidae	+	+	A, B, C
Catoptria maculalis (Zetterstedt, 1839)	Crambidae	+	–	A, B, C
Catoptria petrificella (Hübner, 1796)	Crambidae	–	+	A, C
Udea alpinalis (Dennis et Schiffermüller, 1775)	Crambidae	+	+	A, C
Erebia epiphron (Knoch, 1783)	Nymphalidae	–[a]	+	A, C
Erebia sudetica Staudinger, 1861	Nymphalidae	–	+	A, C
Isturgia roraria (Fabricius, 1776)	Geometridae	–	+	A, C
Elophos operaria (Hübner, 1813)	Geometridae	+	–	A, C
Psodos quadrifaria (Sulzer, 1776)	Geometridae	+	–	A, C
Glacies alpinata (Scopoli, 1763)	Geometridae	+	+	A, C
Perizoma obsoletata (Herrich-Schäffer, 1838)	Geometridae	+	–	A, C
Eupithecia silenata Assmann, 1848	Geometridae	+	+	A, C
Apamea maillardi (Geyer, 1834)	Noctuidae	+[b]	–	A, B, C
Xestia alpicola (Zetterstedt, 1839)	Noctuidae	+	–	A, B, C
Diarsia mendica borealis Zetterstedt 1839[c]	Noctuidae	–	+	B
Total		20	18	

Based on Liska (2000), with nomenclature modified after Karsholt and van Nieukerken (2004). "Distribution" describes extra-Sudeten distribution of individual species, in which A = occurrence in the Alps and other mountains of W and C Europe; C = occurrence in the Carpathians and other systems of SE Europe; B = occurrence in Northern Europe

[a]Intentionally introduced population: see text for details

[b]A single record, doubted by some authors

[c]A population with diurnal activity, likely distinct from "lowland" *D. mendica*. Its exact status needs clarification

Catoptria maculalis), a habitat mostly absent from the Jesenik Mts., but existing in the Krkonose Mts.

Another important aspect concerns the grasslands. Recently, there is much effort to restore the "cultural" grasslands in the Sudeten Mts. by reinstalling the traditional management, and the prospects are optimistic (e.g., Krahulec et al. 1997; Pavlu et al. 2003; Hejcman et al. 2005, 2006). However, the natural subalpine grasslands had also been managed for centuries by grazing and haymaking. Cessation of activities, and a resulting non-intervention regime, causes a gradual impoverishment of the vegetation, prevalence of a few dominant species, and accelerated trees and shrubs expansion (Klimeš and Klimešová 1991; Hejcman et al. 2006). Effects on *Erebia* species are not yet apparent, but may eventually manifest as direct loss of habitats. The non-intervention management is sometimes advocated as return to a "natural state," which is a risky adventurism in the face of losses of sensitive species, and under the threat of warming-accelerated habitat changes.

E. epiphron requires long-term preservation of open grasslands, preferably interconnected to buffer for its poor dispersal ability. (Recall the decreased genetic variation of the isolated lateral Jesenik population). In Jesenik, where only the subalpine grasslands exist, blocking of tree growth, including the removal of non-native dwarf pine *P. mugo*, represents the only way to support its population. Because Krkonose are higher than Jesenik, the introduced population can be viewed as a genetic reserve, having a higher chance of survival should warming climate eradicate the species in Jesenik. Even here, ascending timberline should be kept in check, and cultural grasslands below the timberline should be managed using approaches mimicking the traditional meadow husbandry.

The situation of *E. sudetica* is more complex, but essentially similar. The structurally rich, "notched" timberline sites with sparse growths of trees may be ascending, but this will not necessarily threaten the populations unless the sites are artificially afforested by spruce or dwarf pine. Practically, all these timberline sites are fairly stable because they are periodically affected by avalanches, which keep them devoid of trees. The fate of the more ephemeral forest-belt populations seem less optimistic, as the vast clearings, now hosting the highest abundance of the species, will gradually succumb to forest re-growth. To support the forest-belt populations, natural disturbance factors opening up canopy, such as windfalls and landslides, should be welcomed rather than abhorred and battled against, and active steps to open the canopy structure at such sites as woodland spring areas and valleys of mountain brooks should be undertaken.

Any predictions on the future of the two butterflies are thwarted by the lack of understanding of ecophysiological limits of the species. So far, we understand that they need cool conditions for development, and it seems likely that the limits operate on requirements of winter larvae. We do not know, however, whether the required condition is sufficiently long allowing uninterrupted winter freezing for uninterrupted hibernation, or whether sufficient height and duration of isolating snow cover is necessary. In the first case, the larvae would be threatened by osmotic stress if the winter gets warmer, while in the latter, they may die from frosts if the snow covers diminish (cf. Hodkinson 2005). The mechanisms of cold (or heat) tolerance of both species are tractable by experimentation and the findings will be critical for forecasting the responses of both species to changing climate.

Acknowledgement We thank O. Cizek, T. Kuras and P. Vrba for inspiring discussions. Field studies of Sudeten *Erebia* were supported by Palacky University Olomouc, and by Jeseniky Landscape Protected Area and Krkonose National Park administrations; particular thanks to conservation officers P. Balaz, V. Kavalcova, and J. Vanek. Molecular study of *E. epiphron* was supported by Grant Agency of the Czech Republic (B6001306/2003). Monitoring of *E. sudetica* is supported by the Czech Conservation Authority, Grant Agency of the Czech Academy of Sciences (600070601) and Czech Ministry of Education (LC06073, 6007665801).

Due to their borderline position and the complex 20th century history, Sudeten Mts. have different names in different languages, which introduces much confusion to biogeographic literature. We present below a simple translation guide for the toponyms appearing in the chapter. They are ordered alphabetically, according to valid names in Czech, and provided, if appropriate, with corresponding toponyms in German [D], English [GB], and Polish [PL]. Jesenik (= Hrubý Jeseník, Jeseníky) [CZ]=Altvater Gebirge (=Hochgesenke) [D]; Kralický Sněžník [CZ]=Glatzer (= Grulicher, Spieglitzer, Großer) Schneeberg [D]=Śnieżnik Kłodzki [PL]; Krkonoše [CZ]=Riesengebirge [D]=Giant Mts. [GB]=Karkonosze [PL]; Lucni bouda [CZ]=Wiesenbaude [D]; Praděd [CZ]=Altvater [D]; Rychlebské Hory [CZ]=Reichensteiner Gebirge [D]=Golden Mts. [GB] = Góry Złote [PL]; Rýchorská bouda [CZ]=Rehornbaude [D]; Sněžka [CZ]=Schneekoppe [D]=Śnieżka [PL]; Šumava [CZ]=Böhmerwald [D]

References

Benes J, Kuras T, Konvicka M (2000) Assemblages of mountainous day-active Lepidoptera studied by the means of yellow pan water traps in the Hruby Jesenik Mountains, Czech Republic. Biologia 55:159–167

Bergman KO (2001) Population dynamics and the importance of habitat management for conservation of the butterfly *Lopinga achine*. J Appl Ecol 38:1303–1313

Borkowski A (1966) *Erebia epiphron silesiana* M.-D. (Lepidoptera, Satyridae) aus dem Riesengebirge. Pol Pismo Entomol 26:85–92

Cizek O, Bakesova A, Kuras T, Benes J, Konvicka M (2003) Vacant niche in alpine habitat: the case of an introduced population of the butterfly *Erebia epiphron* in the Krkonoše Mountains. Acta Oecol 24:15–23

Cizek O, Bakesova A, Vanek J (2004) Mountain Ringlet *Erebia epiphron* in the Krkonose Mts. Six decades of an introduced butterfly population. Acta Mus Reginehrad 30:79–82

Emmer IM, Fanta J, Kobus AT, Kooijman A, Sevink J (1998) Reversing borealization as a means to restore biodiversity in Central-European mountain forests – an example from the Krkonose Mountains, Czech Republic. Biodiv Conserv 7:229–247

Franco AMA, Hill JK, Kitschke C, Collingham YC, Roy DB, Fox R, Huntley B, Thomas CD (2006) Impacts of climate warming and habitat loss on extinctions at species' low-latitude range boundaries. Glob Chang Biol 12:1545–1553

Grabherr G (2003) Alpine vegetation dynamics and climate change – a synthesis of long-term studies and observations. In: Nagy L, Grabherr G, Körner C, Thompson DBA (eds) Alpine biodiversity in Europe. Springer, Berlin, pp 399–410

Haubrich K, Schmitt T (2007) Cryptic differentiation in alpine-endemic, high-altitude butterflies reveals down-slope glacial refugia. Mol Ecol 16:3643–3658

Hejcman M, Auf D, Gaisler J (2005) Year-round cattle grazing as an alternative management of hay meadows in the Giant Mts (Krkonose, Karkonosze), the Czech Republic. Ecol Bratislava 24:419–429

Hejcman M, Pavlu V, Hejcmanova P, Gaisler J, Hakl J, Rauch O (2006) Farmer decision making and its effect on subalpine grassland succession in the Giant Mts., Czech Republic. Acta Soc Bot Pol 75:165–174

Hill JK, Thomas CD, Fox R, Telfer MG, Willis SG, Asher J, Huntley B (2002) Responses of butterflies to twentieth century climate warming: implications for future ranges. Proc R Soc Lond B Biol Sci 269:2163–2171

Hodkinson ID (2005) Terrestrial insects along elevation gradients: species and community responses to altitude. Biol Rev 80:489–513

Holten JI (2003) Altitude ranges and spatial patterns of alpine plants in Northern Europe. In: Nagy L, Grabherr G, Körner C, Thompson DBA (eds) Alpine biodiversity in Europe. Springer, Berlin, pp 173–184

Jankovská V (2004) Krkonoše v dobŽ poledové – vegetace a krajina. [Krkonose in Holocene era – vegetation and landscape]. Opera Corcontica 41:111–123

Jenik J (1961) Alpínská vegetace Krkonoš, Králického SnŽžníku a Hrubého Jeseníku: Teorie anemo-orografických systémŽ [Alpine vegetation of Krkonose, Kralicky Sneznik and Hruby Jesenik: The theory of anemo-orographical systems]. ŽSAV, Praha

Jenik J, Bures L, Buresova Z (1980) Syntaxonomic study of vegetation in Velká Kotlina cirque, the Sudeten Mountains. Folia Geobot Phytotax 15:1–28

Jenik J (1998) Biodiversity of the Hercynian mountains of Central Europe. Pirineos 151–152:83–89

Jurasinski G, Kreyling J (2007) Upward shift of alpine plants increases floristic similarity of mountain summits. J Veg Sci 18:711–718

Karsholt O, van Nieukerken EJ (eds) (2004) Lepidoptera, Moths. Fauna Europaea version 1.1, http://www.faunaeur.org

Klimeš L, Klimešová J (1991) Alpine tundra in the Hrubý Jeseník Mts., the Sudeten, and its development in the 20th century. Preslia 63:245–268

Konvička M, Beneš J, Kuras T (2002) Behaviour and microdistribution of two sympatric alpine butterflies (*Erebia epiphron* and *E. euryale*): relation to weather, vegetation and time of day. Biologia 57:223–233

Konvicka M, Maradova M, Benes J, Fric Z, Kepka P (2003) Uphill shifts in distribution of butterflies in the Czech Republic: effects of changing climate detected on a regional scale. Global Ecol. Biogeogr 12:403–410

Krahulec F, Blažková D, Balátová-Tulácková E, Štursa J, Pechácková S, Fabšicová M (1997) Meadows in the Krkonoše Mts. – plant communities and their dynamics. Opera Corcontica 33:3–250

Kudrna O (2002) The distribution atlas of European butterflies. Oedippus 20:1–342

Kullman L (2006) Long-term geobotanical observations of climate change impacts in the Scandes of West-Central Sweden. Nord J Bot 24:445–467

Kuras T, Beneš J, Konvička M (2000) Differing habitat affinities of four *Erebia* species (Lepidoptera: Nymphalidae, Satyrinae) in the Hrubý Jeseník Mts, Czech Republic. Biologia 55:163–169

Kuras T, Beneš J, Konvička M (2001a) Behaviour and within-habitat distribution of adult Erebia sudetica sudetica, endemic of the Hrubý Jeseník Mts., Czech Republic (Nymphalidae, Satyrinae). Nota Lepid 24:87–101

Kuras T, Konvička M, Beneš J, Čížek O (2001c) *Erebia sudetica* and *Erebia epiphron* (Lepidoptera: Nymphalidae, Satyrinae) in the Czech Republic: Review of present and past distribution, conservation implications. Cas Slez Muz Opava A 50:57–81

Kuras T, Beneš J, Fric Z, Konvicka M (2003) Dispersal patterns of endemic alpine butterflies with contrasting population structures: *Erebia epiphron* and *E. sudetica*. Popul Ecol 45:115–123

Liska J (2000) An attempt at comparing the lepidopteran fauna of subalpine areas of the High Sudetes. Opera Corcontica 37:286–290

Liska J, Skyva J (1997) Historical and recent occurrence of Lepidoptera in mountain sites of the Giant Mountains (Czech Republic). Biologia 52:163–165

Maslowski J (2005) Uwagi o trzech prawnie chronionych gatunkach motyli dziennych (Lepidoptera, papilionoidea) w Sudetach. [Notes on three legally protected butterflies in the Sudetes]. Przyroda Sudetów 8:81–88

Pavlu V, Hejcman M, Pavlu L, Gaisler J (2003) Effect of rotational and continuous grazing on vegetation of an upland grassland in the Jizerske Hory Mts., Czech Republic. Folia Geobot 38:21–34

Prentice IC, Jolly D, BIOME 6000 participants (2000) Mid-Holocene and glacial-maximum vegetation geography of the Northern continents and Africa. J Biogeogr 27:507–519

Schmitt T, Haubrich K (2008) The genetic structure of the mountain forest butterfly *Erebia euryale* unravels the late Pleistocene and postglacial history of the mountain coniferous forest biome in Europe. Mol Ecol 17:2194–2207

Schmitt T, Cizek O, Konvicka M (2005) Genetics of a butterfly relocation: large, small and introduced populations of the mountain endemic *Erebia epiphron silesiana*. Biol Conserv 123:11–18

Schmitt T, Hewitt GM, Müller P (2006) Disjunct distributions during glacial and interglacial periods in mountain butterflies: *Erebia epiphron* as an example. J Evol Biol 19:108–113

Schwarz R (1948) Motýli denní I. [Butterflies]. Vesmír, Praha

Soffner J (1967) *Erebia epiphron silesiana* im Riesengebirge (Lep., Satyridae). Entomol Z Frankfurt 77:125–128

Sonderegger P (2005) Die Erebien der Schweiz. Selbstverlag, Biel/Bienne

Stiova L (1988) Výskyt okáčů rodu *Erebia* (Lep., Satyridae) v Hrubém Jeseníku. [The distribution of Erebia butterflies in Hruby Jesenik Mts.]. Cas Slez Muz Opava A 37:115–133

Štursa J (1998) Research and management of the Giant Mountains' arctic-alpine tundra (Czech Republic). Ambio 27:358–360

Treml V, Jankovska V, Petr L (2008) Holocene dynamics of the alpine timberline in the High Sudetes. Biologia 63:73–80

Varga Z (2001) Die Erebien der Balkanhalbinsel und Karpaten V. Übersicht der subspezifischen Gliederung und der Verbreitung der Erebia Dalman, 1816 Arten (Lepidoptera: Nymphalidae: Satyrinae) in der Balkanhalbinsel und in den Karpaten (II. Teil). Entomol Roman 6:5–39

Varga ZS, Schmitt T (2008) Types of oreal and oreotundral disjunctions in the Western Palearctic. Biol J Linn Soc 93:415–430

Virtanen R (2003) The high mountain vegetation of the Scandes. In: Nagy L, Grabherr G, Körner C, Thompson DBA (eds) Alpine biodiversity in Europe. Springer, Berlin, pp 31–38

Virtanen R, Dirnböck T, Dullinger S, Grabherr G, Pauli H, Staudinger M, Villar L (2003) Patterns in the plant species richness of European high mountain vegetation. In: Nagy L, Grabherr G, Körner C, Thompson DBA (eds) Alpine biodiversity in Europe. Springer, Berlin, pp 149–172

Wahlberg N, Klemetti T, Hanski I (2002) Dynamic populations in a dynamic landscape: the metapopulation structure of the marsh fritillary butterfly. Ecography 25:224–232

Wilson RJ, Gutierrez D, Gutierrez J, Monserrat VJ (2007) An elevational shift in butterfly species richness and composition accompanying recent climate change. Glob Chang Biol 13:1873–1887

Conservation and Management of the Habitats of Two Relict Butterflies in the Belgian Ardenne: *Proclossiana eunomia* and *Lycaena helle*

Philippe Goffart, Nicolas Schtickzelle, and Camille Turlure

Abstract Habitat preservation and restoration are crucial for species conservation and to tackle its main challenges. In many cases, this implies management actions, especially for semi-natural ecosystems. We address here the question of management actions suitable for maintaining viable populations of two relict butterfly species inhabiting wet meadows and peat bogs in the Belgian Ardenne: the bog fritillary *Proclossiana eunomia* and the violet copper *Lycaena helle*. Habitat loss due to natural vegetation succession towards forest can be prevented either by mowing (traditional way of management) or by extensive grazing (with cattle or horses).

Results from field studies and experiments conducted on several sites in the Belgian Ardenne led to two conclusions. Mowing can strongly affect the populations of both species, the impact being particularly pronounced on larval stages due to the removal of grass tussocks and the export of hay containing host plant on which caterpillars feed. Grazing also has a negative effect on both species, with adult density being lower on grazed biotopes compared to abandoned ones. However, the negative impact is more limited for grazing than for mowing.

As both management methods have more or less pronounced negative impacts on populations of these two butterfly species, they should be applied with caution. We suggest that mowing should be applied in long-term rotation, with the preservation of sufficient refuge areas every year, and that grazing should be applied in late summer, every other year, with a low stocking rate (<0.2 LU/ha/year). More particularly, to ensure their effectiveness in maintaining a favorable state, both the biotope and the associated populations using it as their habitat, management actions must be designed according to the species habitat requirements, and adapted to carefully monitor their implementation.

Ph. Goffart (✉)
Département de l'Étude du Milieu naturel et agricde (DEMna), Service Public Wallon (SPW), 23, Avenue Maréchal Juin, B-5030 Gembloux, Belgium
e-mail: philippe.goffart@spw.wallonie.be

N. Schtickzelle and C. Turlure
Biodiversity Research Centre, Université catholique de Louvain, Louvain-la-Neuve, Belgium
e-mail: nicolas.schtickzelle@uclouvain.be; camille.turlure@uclouvain.be

J.C. Habel and T. Assmann (eds.), *Relict Species: Phylogeography and Conservation Biology*, DOI 10.1007/978-3-540-92160-8_21,

1 Introduction

In Europe, and particularly in Belgium, butterfly diversity has strongly decreased during recent decades (Van Swaay et al. 2006; Maes and Van Dyck 2001; Goffart et al. 1992; Fichefet et al. in press); specialist species have been particularly affected. Of prime importance in this context is the conservation of remnant relict butterfly species. The wet meadows and peat bogs of the Belgian Ardenne high plateaus still offer ecological conditions close to those of Northern Europe, allowing populations of relict butterfly species to persist.

However, these populations, at the Southern margin of their distribution range, have a sparse and disjunct distribution (Kudrna 2002). Moreover, they are spatially trapped: they cannot shift their distribution range either to the North, because there is no suitable habitat patch between this region and other populations widespread in Northern Europe (Parmesan et al. 1999; White and Kerr 2006), or uphill because the upper altitudinal limit is reached in the region (Konvicka et al. 2003).

The bog fritillary *Proclossiana eunomia* and the violet copper *Lycaena helle* are two remarkable relict butterflies of the Ardenne massif. Their Belgian populations are amongst the most important of Western Europe and are genetically unique (Meyer 1980, 1981a,b; Barascud and Descimon 1992; Nève et al. 2000; Descimon et al. 2001; Finger et al. 2009). Moreover, they are in decline and threatened (Fichefet et al. 2008), as is the case in other parts of Europe too, where *L. helle* is classified as "vulnerable" (van Swaay and Warren 1999) and recently added to the Appendices II and IV of the European Habitat Directive (92/43/CEE). Their preservation is, therefore, highly desirable.

All remnant populations of relict species in the Ardenne region inhabit pioneer more semi-natural biotopes. Without any disturbance, these biotopes are progressively invaded by shrubs and trees, turn to forest, and ultimately disappear. In extended ecosystems, natural disturbances, like fires or storms, can rejuvenate the biotope by converting it back to open vegetation (Bullock and Pakeman 1997; Svenning 2002). This is no longer the case anywhere in the Western Europe today because of the small area of such remnants. Therefore, human actions are clearly needed to mimic the effects of natural disturbances, through the use of management techniques such as mowing, grazing, and tree cutting (Sutherland 2002). Such management often positively affects the ecological processes that are crucial for biodiversity, by breaking the dominance of highly competitive plants, reducing bush encroachment, or making vegetation structure more varied (Ausden and Treweek 1995). Although these management methods are widely and more or less efficiently implemented on biotopes (WallisDeVries et al. 1998; Krebs et al. 1999), little is known about the detailed effects on butterfly populations and their resources.

In this context, our objectives were to assess the impact of two management techniques, mowing and grazing, on populations of *P. eunomia* and *L. helle*, which inhabit wet meadows and peat bogs in the Belgian Ardenne. Mowing was traditionally applied with non-mechanical means in the Ardenne alluvial meadows (Lambert 1963). Grazing was not used in Ardenne valleys, but is currently becoming popular among nature managers (Delescaille 2002), because of its ease of implementation.

We assessed for both species (1) the impact of mowing on larval stages (through the number of emerging adults) and on the attractiveness of mown biotopes for adults and (2) the impact of three distinct grazing regimes on the adult density. As resources and life history traits of the two species are different, we expected to find contrasted, perhaps even completely different, impacts of both management techniques according to species. Indeed, even if caterpillars of both species feed only on *P. bistorta* leaves, nectar use by adults is different: *P. eunomia* adults only feed on *P. bistorta* flowers, while *L. helle* adults use a wider range of nectar resources, from both herbaceous and woody species. Moreover, overwintering strategies are distinct (Barascud and Descimon 1992): *L. helle* spends the winter as a pupa, while *P. eunomia* spends the winter as a caterpillar. The related applied question is whether there is an efficient management regime for these two relict butterfly species at the same time.

2 The Impact of Habitat Management Via Mowing

2.1 Experiments and Results

Two experiments were conducted in the "Prés du Sommerain" nature reserve (50°10′N 5°47′E) in the Belgian Ardenne to assess the impact of mowing: the first one on the larval stage and the second on the adult stage.

The first experiment aimed at assessing the survival of larval stages after summer and autumn mowing, through the number of emerging adults. Six pairs of 200 m² (25 × 8 m) strips with a dense cover of bistort (i.e., the host plant) were studied in an abandoned wet meadow between 1993 and 1995 (one pair in 1993, two in 1994, three in 1995). In each pair, one strip was mown (in summer between the end of June and the beginning of August, or in autumn in October) and the other strip was left untouched for control. One square meter emergence traps were set up above the vegetation at the end of April, the year after the mowing management. They were arranged in series of four traps, two traps being placed in the mown strip, the other two in the adjacent portion of the control strip. Over a period of three years, a total of 50 traps were devoted to each category (summer control, summer mowing, autumn control, autumn mowing). The presence of emerged butterflies in the trap was then checked daily until the end of summer.

Logistic regression models (implemented using SAS software: www.sas.com) showed no difference in the probability of emergence of adult butterflies between summer and autumn control strips for each species ($\chi_1^2 = 1.43$, $p=0.23$ for *P. eunomia*, $\chi_1^2 = 0$, $p=1$ for *L. helle*) (Fig. 1). Whatever the timing of mowing, summer or autumn, the probability of emergence was higher in the control strips than in the corresponding mown strips for *P. eunomia* (contrast "pooled control strips vs pooled mown strips": $\chi_1^2 = 8.62$, $p=0.0033$; "summer control vs summer mown strips": $\chi_1^2 = 7.19$, $p=0.0073$; "autumn control vs autumn mown strips": $\chi_1^2 = 2.81$, $p=0.0935$); mowing had in fact a very drastic impact on *P. eunomia* as no adult ever

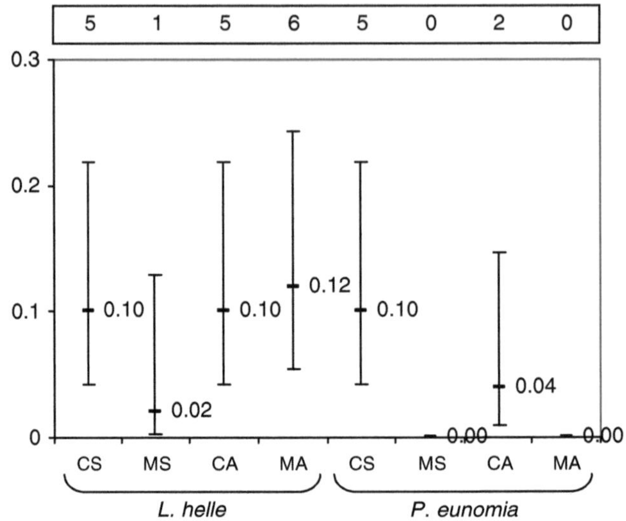

Fig. 1 Impact of mowing treatment on the probability (with 95% confidence intervals) of adult emergence of the two butterfly species. *CS* summer control strips, *MS* summer mown strips, *CA* autumn control strips, *MA* autumn mown strips. Numbers of emerged individuals are presented in the inset

emerged in mown strips (Fig. 1). For *L. helle*, the impact of mowing on larval stages was different according to the timing of mowing (Fig. 1): autumn mowing showed no influence on the probability of emergence of adult butterflies (χ_1^2 = 0.1 p=0.749), but summer mowing seemed harmful (χ_1^2 = 3.08 p=0.079); the difference between summer and autumn mowing was significant (χ_1^2 = 4.23 p=0.039).

The second experiment aimed at comparing how mown and control plots attracted adult butterflies. Three plots of about 500 m^2 (25×20 m) were mown in summer 1994 and the same number of plots were left untouched for control. Several 15 min transect counts were performed during the flight period of both species in summer 1995. There was a difference in the relative attraction exerted by mown and control plots, significant for both species (χ_1^2 = 7.36, p=0.007 for *P. eunomia*; χ_1^2 = 10.98, p<0.01 for *L. helle*), but this response differed between species (χ_1^2 = 16.55, p<0.001). *P. eunomia* adults seemed to avoid – or at least have a lower attraction towards – mown plots, while *L. helle* adults showed a significant attraction towards mown plots (Fig. 2).

2.2 Linking the Impact of Mowing with Habitat Requirements

The impact of mowing was clearly more pronounced on *P. eunomia* than on *L. helle*. The disastrous impact of mowing, both when done in summer and in autumn, on *P. eunomia* larvae is easily explained by the usual location of caterpillars,

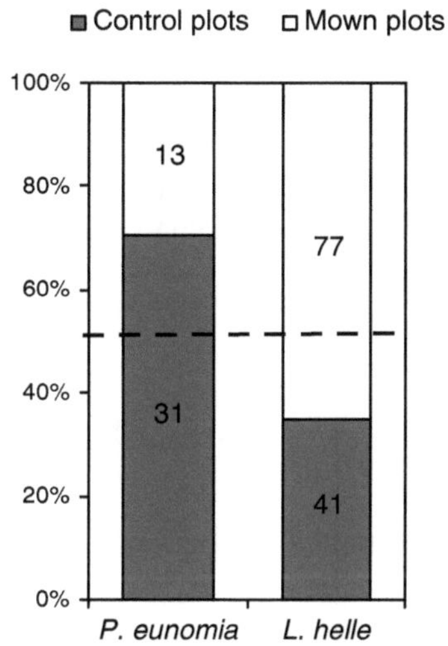

Fig. 2 Impact of mowing treatment on the attractiveness of the area for the two butterfly species. Shown are the numbers and associated percentages of adult butterflies of each species counted in control and mown plots

which find shelter preferentially in grass tussocks, from the beginning of summer until the next spring (Goffart et al. 2001). Indeed, grass tussocks appear to play a crucial role in caterpillar thermoregulation (Turlure *unpublished data*), and probably also allows them to escape from parasitoids, predators, and winter flooding (Joy and Pullin 1997; Severns et al. 2006). These tussocks are cut during mowing, and caterpillars very likely removed with the hay, causing direct mortality. Concerning the avoidance of mown areas – or at least the lower attraction towards the mown areas – displayed by *P. eunomia* adults, it may be related to the change of biotope physiognomy induced by mowing: grass-tussocks are cut and relief becomes almost flat, giving a less attractive vegetation facies for males and females (Turlure et al. 2009). This has been confirmed by another experiment, in which the same plots were manipulated by sequentially placing and removing grass tussocks during a single day, and comparing attractiveness for *P. eunomia* (Goffart unpublished data). Both sexes indeed actively search for tussocks: females find appropriate locations for egg-laying and males find a higher probability to encounter virgin (unmated) females at the precise location of their emergence.

For *L. helle*, the period of mowing proved to be crucial for the survival of larvae. The contrasted responses to summer and autumn mowing can be understood in the light of the butterfly life history: caterpillars, hanging under the leaves in summer, are likely to be carried away with the hay and suffer direct mortality, whereas the overwintering pupae, settling on the top soil in autumn, escape from later mowing. Survival of the pupae (and hence, emergence success) might be even higher in the mown areas, because the shorter vegetation height and associated change in microclimatic conditions (Thomas 1991; Bourn and Thomas 2002) induced by mowing could be more favorable for them. Furthermore, *L. helle* adults seemed to be attracted towards the mown areas. This was clearly linked to the higher density of some nectar flowers like *Cardamine pratensis* (Goffart et al. 2001), as the majority of adult butterflies were observed feeding on them. *L. helle* adults being attracted to mown areas where their offspring suffer from elevated mortality due to mowing actions, an "ecological trapping effect" (Shlaepfer et al. 2002; Battin 2004) might occur if mowing is applied every year in the stipulated period, i.e., in summer.

3 The Impact of Habitat Management Via Grazing

3.1 Experiments and Results

Experiments to assess the impact of light grazing were conducted on several sites in the Belgian Ardenne (Deifelt, Commanster, Shoenberg, Martelange, and Straimont). Within these sites, 14 plots larger than 2 ha were selected, eight lightly grazed (stocking rate <0.2 LU/ha/year) for at least 5 years and six abandoned (control). Three distinct forms of light grazing were distinguished: (1) continuous grazing, cattle being present all the year round; (2) seasonal grazing, cattle being present only during late summer and autumn; and (3) alternate grazing, cattle being present every other year (for 3–9 months). Transect counts of adult butterflies were done during the flight period of both butterfly species in the summer of 2003 and 2004. We analyzed these frequency counts by log linear models with Poisson error, testing the effects of grazing regime (four classes: continuous grazing, seasonal grazing, alternate grazing, and abandoned), sampling year (two classes: 2003 and 2004), sampling effort (transect duration in minutes), and tree/edge density (three classes: some trees and/or edges, intermediate density of trees and/or edges, high density of trees and edges) with AIC-based model selection (Burnham and Anderson 2002; Johnson and Omland 2004; Richards 2005) implemented in SAS. To take into account model selection uncertainty, we report the AIC weight of each explanatory variable, expressing the probability that the variable influences the response (here, the frequency of butterflies), the model-averaged parameter estimates and the confidence limits (Burnham and Anderson 2002).

The frequency of *P. eunomia* adults was highly related to grazing regime, sampling effort, and sampling year, as shown by an AIC weight above 98% (Table 1). *P. eunomia* adults were more numerous in control plots than in grazed plots (Fig. 3a), but much rarer in continuously grazed plots (~75% loss) than in seasonally or alternately grazed plots (~25% loss). More *P. eunomia* butterflies were counted during longer transects, more in 2004 than in 2003. The abundance of *L. helle* adults was highly related to grazing regime and tree/edge density, showing no significant differences between years, and only a marginal increase with sampling effort (Table 1). A significant difference (~50% loss) was detected in the case of seasonal grazing (Fig. 3b). When considering both species together, the impact

Table 1 Factors affecting the adult density of the two butterfly species as observed along transect lines, as modeled using log linear models

		Proclossiana eunomia				*Lycaena helle*			
				95% confidence interval				95% confidence interval	
Variable	Modality	AIC weight (%)	Estimate	Lower	Upper	AIC weight (%)	Estimate	Lower	Upper
Grazing Type	Alternate	100.00	−0.374	−0.661	−0.088	98.57	0.431	−0.145	1.006
	Continuous		−1.041	−1.521	−0.561		−0.185	−1.457	1.086
	Seasonal		−0.510	−0.799	−0.221		−1.239	−2.112	−0.366
	Control		0.000				0.000		
Time	Time	100.00	0.020	0.008	0.032	50.84	0.025	−0.005	0.055
Year	2003	98.97	−1.152	−1.357	−0.947	40.59	0.470	−0.217	1.157
	2004		0.000				0.000		
Tree/Edge	Edge1	43.44	0.242	−0.153	0.636	98.61	−0.714	−1.692	0.264
	Edge2		0.084	−0.212	0.379		0.775	−0.048	1.597
	Edge3		0.000				0.000		

AIC weight expresses the probability that the variable influences the density. Parameter estimates and confidence intervals for each modality were model averaged to take model selection uncertainty into account (Burnham and Anderson 2002). Parameter estimates quantify the difference in the log of the frequency with the last modality of each variable, fixed at 0 by definition

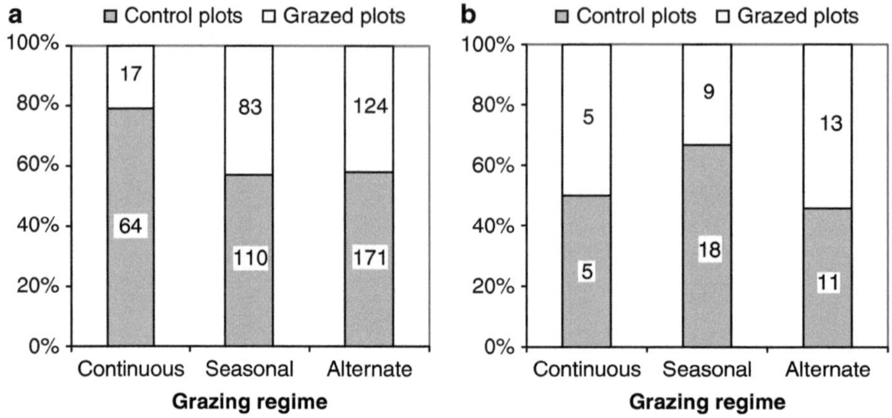

Fig. 3 Impact of grazing treatment on the frequency of the two butterfly species. Shown are the percentages of adult butterflies counted in grazed and neighboring control plots along transects at same dates for same durations, (**a**) for *P. eunomia*, (**b**) for *L. helle*. NB: In the case of alternate grazing (two localities), there was no strict control plot available in one locality (Deifelt); each plot being grazed for 3 months in autumn every second year, the temporarily ungrazed plot from the two adjacent plots was used as a control (in this graph only, for loss estimation – not in the log linear models of Table 1). Mean butterfly densities (measured by the number of individuals counted in 30 min) on these two Deifelt plots were very similar or higher after the start of grazing management than before, in the nineties (*P. eunomia*: 23,9 ind./30 min. after vs 16,9 ind./30 min. before; *L. helle*: 3,7 ind./30 min. after vs 3,8 ind./30 min. before), suggesting no detectable detrimental effect of this grazing regime on these species in this particular site

on populations was the least marked with alternate grazing regime, and higher with seasonal and continuous grazing (Table 1).

3.2 *Linking the Impact of Grazing with Habitat Requirements*

Impact of light grazing on the populations of *P. eunomia* and *L. helle* proved to be non-existent to moderate (<30% loss) or heavy (>50% loss) according to the grazing regime adopted and the species concerned, a periodic application generally appearing less harmful than a continuous one. This could be explained by the negative impact of livestock on (1) nectar flowers during the flight period of the butterflies and on (2) the bistort due to the trampling on rhizomes, and some direct impact on the early stages of the butterflies (Schtickzelle et al. 2007). However, butterfly losses amounting to >70%, as reported by Schtickzelle et al. (2007) on *P. eunomia* in another Ardenne site (Prés de la Lienne), were only recorded on the site with continuous grazing regime (Martelange) in the present study. They were far more reduced on the other sites where periodic grazing was practiced. The regime applied in Prés de la Lienne was also continuous, but with heavier stocking rates (comprised between 0.23 and 0.42 Livestock Units ha^{-1} on an annual basis), even higher during the summer (0.54–0.74 LU ha^{-1}), compared to the sites studied here.

The influence of tree/edge density on *L. helle* counts is readily explained by several aspects of adult life in this species. Indeed, scrubs and trees, from the vicinity of which they are rarely found away (Fisher et al. 1999; Turlure et al. 2009), are used as territorial, nectaring, and roosting places. First, males exhibit a perching territorial behavior along forest edges and bushes, engaging chasing flights after males and waiting for females to mate with. Moreover, both males and females exploit flowering bushes and trees in search of nectar and higher trees for roosting: at the end of the day, they reach the top of the trees, step by step, where they usually spend the night (Goffart and Waeyenbergh 1994; Goffart et al. 2001; Goffart unpublished data). *L. helle* appears to find an optimal habitat in sites with an intermediate density of edge and trees in and around wet grasslands.

4 Options for Management

Open semi-natural habitats, like those inhabited by the two relict butterfly species studied here, must be managed to prevent forest recolonization. The real question of interest to people in charge of the design and implementation of biodiversity conservation plans is how to determine the management protocols that would yield the best conservation results. Mowing and grazing are currently used in many biodiversity conservation and management actions, including Ardenne humid grasslands. However, we showed here that both can have very detrimental effects on target species, if used without caution, and they must therefore not be blindly applied in any case.

Some studies showed that mown grasslands present a reduced butterfly richness and abundance compared with abandoned ones (Erhardt 1985; Erhardt and Thomas 1991; Munguira and Thomas 1992; Feber and Smith 1995, Feber et al. 1996), while other studies showed that hay meadows can harbor a rich and original fauna, including rare species, compared with other biotopes (Swengel 1996; Wettstein and Schmid 1999; Stefanescu et al. 2005). This study and other results of the same experiment (Goffart et al. 2001) suggest that the impact of mowing on populations may be highly negative, dramatically diminishing the butterfly densities (>50% loss). *P. eunomia* appears especially sensitive to mowing, with a real avoidance of mown areas by adults and a disastrous effect on early instars. The likely explanation is that this species is linked to habitats where structural elements like grass tussocks can be found (Goffart et al. 2001; Turlure et al. 2009). Its populations can be affected by a widespread application of cutting management in the presently abandoned humid grasslands. The persistence of populations in the Ardenne valleys at a time when mowing practices were widespread can probably be explained by the far bigger extent of its habitat in the past, allowing for regional persistence of populations despite the decrease in population density and/or extirpation of local populations due to mowing. Indeed, movements of individuals from occupied sites allowed for metapopulation dynamics through rescue effect (immigrants increase population density) and recolonization (Hanski and Gilpin 1997). The non-mechanical way of cutting and its temporal spread might also have had a lower impact than current mowing techniques. Thus, the traditional way of management of these biotopes cannot be retained as an ideal solution for the long-term conservation of *P. eunomia* populations.

Light grazing, being a more "natural" way of management (although not a traditional way) and generating heterogeneity in the vegetation cover (Van Wieren 1991, 1995; WallisDeVries et al. 1998), therefore appears as a prime alternative. Indeed, several studies showed that grazing management can be very efficient to conserve butterfly populations or communities and their habitats, when used with light or moderate stocking level (Dolek and Geyer 1997; Ellingsen et al. 1997; Weiss 1999; WallisDeVries and Raemakers 2001; Ellis 2003; Pöyri et al. 2004; Öckinger et al. 2006). The present study indicates that grazing can also be a management option for *P. eunomia* and *L. helle*, the two threatened species linked to a food plant highly sensible to grazing and trampling, if the following conditions are applied: (1) very light stocking rates are used: <0.2 LU ha^{-1}, as recommended already by others in other contexts (WallisDeVries and Raemakers 2001; Ellis 2003), (2) preferably used in a rotational basis, and (3) avoiding crucial parts of the butterfly life cycle (April to July for *P. eunomia* and *L. helle*).

However, despite its effectiveness in conserving habitats of these target butterflies with relatively limited impacts on their populations, grazing should not be considered as "the" only appropriate and adequate management option in Ardenne humid grasslands. Indeed, light grazing is difficult to apply on small areas, where the risk of overgrazing is higher (Schtickzelle et al. 2007), and this method is better applied in large areas (Bakker and Londo 1998; Ellingsen et al. 1997). Actually, the impact of grazing on butterfly abundance or richness is highly variable and clearly dependent on its intensity and the modalities of its application (Kruess and Tscharntke 2002; Pöyri et al. 2006).Very detrimental effects have been reported at high stocking levels (Wettstein and Schmid 1999; Kruess and Tscharntke 2002; Ellis 2003; Schtickzelle et al. 2007). Mowing may then replace grazing on small areas (< 1 ha), if used with caution, on a rotational basis, with a pluriannual regime.

More generally, the choice of a single management method (i.e., mowing or grazing) is not advisable in a conservation perspective at the landscape scale. The use of different management practices and intensities, including permanent or temporary abandonment, can be fruitful for individual species (Schwarzwälder et al. 1997) as well as entire communities (Swengel 1998; Dolek and Geyer 1997; Wettstein and Schmid 1999; Balmer and Erhardt 2000; Pöyri et al. 2005; Vögel et al. 2007). The particular context of each site, and especially the area of the habitat in the landscape, is clearly an important parameter regarding the choice of the management method aimed at the conservation of viable butterfly populations and of a rich fauna in this kind of biotope.

5 Recommendations for Management and Scientific Monitoring

Applying "general recipes" may be detrimental because the impact of management is not similar in different biotopes, with different species and/or ecological conditions. We must nevertheless remember that every management act has a potential negative impact and we are probably forced to admit some minimal loss in order to achieve management actions that will in the end prove to be positive. To minimize this potential loss, our recommendations follow several major lines:

1. Scientific monitoring of the impact of management should be improved by focusing more precisely on the requirements of species in terms of resources and by considering management impact on the different life stages of the species.
2. Unmanaged areas must be preserved in any implementation of a management procedure. They may serve as a comparison basis for monitoring and assessing the effect of management and also as a refuge for populations (Johst et al. 2006; Swengel and Swengel 2007).
3. Pluriannual rotational regimes (Oates 1995; Morris 2000) that leave some areas unmanaged every year must be clearly preferred over any static management.
4. Implementation of management methods should be decided on the basis of landscape composition. We then urge scientists to take into account the impact of management on the Population Viability Analyses that serve to elaborate conservation guidelines (Schtickzelle and Baguette 2009).

Acknowledgments The authors would like to thank Michel Waeyenbergh, Anne Weiserbs, Gabriel Nève, Isabelle Convié, Isabelle Brasseur, Catherine Burnotte, Julie Choutt, Michel Baguette, Philippe Fayt, and Michel Pirnay who helped during field work. We are very grateful to Philippe Lebrun, Michel Baguette, Hans Van Dyck, Marc Dufrêne, and Pierre Gérard, who provided support at different stages of the work and Thomas Schmitt who commented on an earlier version of the manuscript. Camille Turlure acknowledges a PhD grant from the FRIA institution. Nicolas Schtickzelle is research associate of the Fund for Scientific Research–FNRS.

References

Ausden M, Treweek J (1995) Grasslands. In: Sutherland WJ, Hill DA (eds) Managing habitats for conservation. Cambridge University Press, Cambridge, pp 195–229

Bakker JP, Londo G (1998) Grazing for conservation management in historical perspective. In: WallisDeVries MF (ed) Grazing and conservation management. Kluwer, Dordrecht, pp 23–54

Balmer O, Erhardt A (2000) Consequences of succession on extensively grazed grasslands for central european butterfly communities: rethinking conservation practices. Conserv Biol 14(3):746–757

Barascud B, Descimon H (1992) Deux papillons "reliques glaciaires" en France: *Lycaena helle* (Lycaenidae) et *Proclossiana eunomia* (Nymphalidae): biogéographie, génétique et conservation. Insectes 87(4):5–9

Battin J (2004) When good animals love bad habitats: ecological traps and the conservation of animal populations. Conserv Biol 18(6):1482–1491

Bourn NAD, Thomas JA (2002) The challenge of conserving grassland insects at the margins of their range in Europe. Biol Conserv 104:285–292

Bullock JM, Pakeman RJ (1997) Grazing of lowland heath in England: management methods and their effects on heathland vegetation. Biol Conserv 79:1–13

Burnham KP, Anderson DR (2002) Model selection and multimodel inference: a practical information-theoretic approach, 2nd edn. Springer, New York

Delescaille L-M (2002) Nature conservation and pastoralism in Wallonia. In: Redecker R (ed) Pasture landscapes and nature conservation. Springer, Berlin, pp 39–52

Descimon H, Zimmermann M, Cosson E, Barascud B, Nève G (2001) Diversité génétique, variation géographique et flux géniques chez quelques lépidoptères rhopalocères français. Genet Sel Evol 33(suppl. 1):S223–S249

Dolek M, Geyer A (1997) Influence of management on butterflies of rare grassland ecosystems in Germany. J Insect Conserv 1:125–130

Ellingsen H, Beinlich B, Plachter H (1997) Effects of large-scale cattle grazing on populations of *Coenonympha glycerion* and *Lasiommata megera* (Lepidoptera: Satyridae). J Insect Conserv 1:13–23

Ellis S (2003) Habitat quality and management for the Northern brown argus butterfly *Aricia artaxerxes* (Lepidoptera: Lycaenidae) in North East England. Biol Conserv 113:285–294

Erhardt A (1985) Diurnal Lepidoptera : sensitive indicators of cultivated and abandoned grassland. J Appl Ecol 22:849–861

Erhardt A, Thomas JA (1991) Lepidoptera as indicators of change in semi-natural grasslands of lowland and upland in Europe. In: Collins NM, Thomas J (eds) The conservation of insects and their habitats. Academic, London, pp 213–236

Feber RE, Smith H (1995) Butterfly conservation on arable farmland. In: Pullin AS (ed) Ecology and conservation of butterflies. Chapman and Hall, London, pp 84–97

Feber RE, Smith H, Macdonald DW (1996) The effects on butterfly abundance of the management of uncropped edges of arable fields. J Appl Ecol 33:1191–1205

Fichefet V, Barbier Y, Baugnée J-Y, Dufrêne M, Goffart P, Maes D, Van Dyck H (2008) papillons de jour de Wallonie (1985-2007). Publication du Groupe de Travail Papillons de jour *Lycaena* et du Centre de Recherche de la Nature, des Forêts et du Bois (MRW-DGRNE), Gembloux

Finger A, Schmitt Th, Zachos Fr E, Meyer M, Assmann Th, Habel JC (2009) The genetic status of the violet copper *Lycaena Lelle*–a relict of the cold past in times of global warming. Ecography 32:382–390

Fischer K, Burkhard B, Plachter H (1999) Population structure, mobility and habitat preferences of the Violet Copper *Lycaena helle* (Lepidoptera: Lycaenidae) in Western Germany: implications for conservation. J Insect Conserv 3(1):43–52

Goffart Ph, Waeyenbergh M (1994) Exigences écologiques et gestion des populations de deux papillons des prairies humides ardennaises: le Cuivré et le Nacré de la Bistorte (*Lycaena helle, Proclossiana eunomia*). Cahiers des Réserves Naturelles RNOB 7:21–29

Goffart Ph, Baguette M, De Bast B (1992) La situation des Lépidoptères Rhopalocères en Wallonie ou Que sont nos papillons devenus? Bulletin and Annales de la Société royale belge d'Entomologie 128:355–392

Goffart Ph, Baguette M, Dufrêne M, Mousson L Nève G, Sawchik J, Weiserbs A, Lebrun P (2001) Gestion des milieux semi-naturels et restauration de populations menacées de papillons de jour. Travaux n°25, Publication de la Région wallonne, Division de la Nature et des Forêts, Jambes

Hanski I, Gilpin ME (1997) Metapopulation biology, ecology, genetics, and evolution. Academic, San Diego, CA

Johnson JB, Omland KS (2004) Model selection in ecology and evolution. TREE 19(2):101–108

Johst K, Dreshsler M, Thomas J, Settele J (2006) Influence of mowing on the persistence of two endangered large blue butterfly species. J Appl Ecol 43:333–342

Joy J, Pullin AS (1997) The effects of flooding on the survival and behaviour of overwintering large heath butterfly *Coenonympha tullia* larvae. Biol Conserv 82:61–66
Konvicka M, Maradova M, Benes J, Fric Z, Kepka P (2003) Uphill shifts in distribution of butterflies in the Czech Republic: effects of changing climate detected on a regional scale. Glob Ecol Biogeogr 12(5):403–410
Krebs JR, Wilson JD, Bradbury RB, Siriwardena GM (1999) The second silent spring? Nature 400:611–612
Kruess A, Tscharntke T (2002) Contrasting responses of plant and insect diversity to variation in grazing intensity. Biol Conserv 106:293–302
Kudrna O (2002) The distribution atlas of European butterflies. Oedippus 20:1–342
Lambert J (1963) Recherches phytosociologiques sur les prairies de la moyenne Ardenne. Agricultura 1:1–198 vol. XI, 2ème série
Maes D, Van Dyck H (2001) Butterfly diversity loss in Flanders (North Belgium): Europe's worst case scenario? Biol Conserv 99:263–276
Meyer M (1980) Une sous-espèce méconnue de *Lycaena helle* D. and Schiff., 1775 (Lepidoptera, Lycaenidae). Linneana Belg Pars VIII(3):130–139
Meyer M (1981a) Révision systématique, chorologique et écologique de Lycaena helle D. and Schiff., 1775 (Lepidoptera, Lycaenidae); 1ère partie: la systématique. Linneana Belg Pars VIII(6):238–260
Meyer M (1981b) Révision systématique, chorologique et écologique de Lycaena helle D. and Schiff., 1775 (Lepidoptera, Lycaenidae), 2ème partie: la chorologie. Linneana Belg Pars VIII(8):345–358
Morris MG (2000) The effects of structure and its dynamics on the ecology and conservation of arthropods in British grasslands. Biol Conserv 95:129–142
Munguira ML, Thomas JA (1992) Use of road verges by butterfly and burnet populations, and the effect of roads on adult dispersal and mortality. J Appl Ecol 29:316–329
Nève G, Barascud B, Descimon H, Baguette M (2000) Genetic structure of Proclossiana eunomia populations at the regional scale (Lepidoptera, Nymphalidae). Heredity 84:657–666
Oates MR (1995) Butterfly conservation within the management of grassland habitats. In: Pullin AS (ed) Ecology and conservation of butterflies. Chapman and Hall, London, pp 98–112
Öckinger E, Eriksson AK, Smith HG (2006) Effects of grassland abandonment, restoration and management on butterflies and vascular plants. Biol Conserv 133:291–300
Parmesan C, Ryrholm N, Stefanescu C, Hill JK, Thomas CD, Descimon H, Huntley B, Kaila L, Kullberg J, Tammaru T, Tennent WJ, Thomas JA, Warren M (1999) Poleward shifts in geographical ranges of butterfly species associated with regional warming. Nature 399:579–583
Pöyri J, Lindgren S, Salminen J, Kuussaari M (2004) Restoration of butterfly and moth communities to in semi-natural grasslands by cattle grazing. Ecol Appl 14(6):1656–1670
Pöyri J, Lindgren S, Salminen J, Kuussaari M (2005) Responses of butterfly and moth species to restored cattle grazing in semi-natural grasslands. Biol Conserv 122:465–478
Pöyri J, Luoto M, Paukkunen J, Pykälä J, Raatikainen K, Kuussaari M (2006) Different responses of plants and herbivore insects to a gradient of vegetation height: an indicator of the vertebrate grazing intensity and successional age. Oikos 115:401–412
Richards SA (2005) Testing ecological theory using the information-theoretic approach: examples and cautionary results. Ecology 86(10):2805–2814
Schtickzelle N, Baguette M (2009) (Meta)population viability analysis: a crystal ball for the conservation of endangered butterflies. In: Settele J, Konvicka M, Shreeve T, Van Dyck H (eds), Ecology of butterflies in Europe. Cambridge University Press, Cambridge, chapter 22, pp. 339–352
Schtickzelle N, Turlure C, Baguette M (2007) Grazing management impacts on the viability of the threatened bog fritillary butterfly *Proclossiana eunomia*. Biol Conserv 136(4):651–660
Schwarzwälder B et al (1997) Habitat utilization by the heath fritillary butterfly, *Mellicta athalia* ssp. *celadussa* (Rott.) (Lepidoptera: Nymphalidae) in montane grasslands of different management. Biol Conserv 82:157–165

Severns P, Boldt L, Villegas S (2006) Conserving a wetland butterfly: quantifying early lifestage survival through seasonal flooding, adult nectar, and habitat preference. J Insect Conserv 10(4):361–370

Shlaepfer MA, Runge MC, Sherman PW (2002) Ecological and evolutionary traps. TREE 17(10):474–480

Stefanescu C, Penuelas J, Filella I (2005) Butterflies highlight the conservation value of hay meadows highly threatened by land-use changes in a protected Mediterranean area. Biol Conserv 126:234–246

Sutherland W (2002) Openness in management. Nature 418:834–835

Svenning J-C (2002) A review of natural vegetation openness in North-Western Europe. Biol Conserv 104:133–148

Swengel AB (1996) Effects of fire and hay managements on abundance of prairie butterflies. Biol Conserv 76:73–85

Swengel AB (1998) Effects management on butterfly abundance in tallgrass prairie and pine barrens. Biol Conserv 83(1):77–89

Swengel A, Swengel S (2007) Benefit of permanent non-fire refugia for Lepidoptera conservation in fire-managed sites. J Insect Conserv 11(3):263–279

Thomas JA (1991) Rare species conservation: case studies of European butterflies. In: Spellerberg IF, Goldsmith FB, Morris MG (eds) The Scientific management of temperate communities for conservation, 31st symposium of the British ecological society, Southampton, 1989. Blackwell, Oxford, pp 149–197

Turlure C, Van Dyck H, Schtickzelle N, Baguette M (2009) Resource-based habitat definition, niche overlap and conservation of two sympatric glacial relict butterflies. Oikos 118:950–960

van Swaay C, Warren M (1999) Red data book of European butterflies (Rhopalocera). Nature and environment, n 99. Council of Europe Publishing, Strasbourg

van Swaay C, Warren M, Loïs G (2006) Biotope use and trends of European butterflies. J Insect Conserv 10(2):189–209

Van Wieren SE (1991) The management of populations of large mammals. In: Spellerberg IF, Goldsmith FB, Morris MG (eds) The Scientific management of temperate communities for conservation, 31st symposium of the British ecological society, Southampton, 1989. Blackwell, Oxford, pp 103–127

Van Wieren SE (1995) The potential role of large herbivores in nature conservation and extensive land use in Europe. Biol J Linn Soc 56(suppl.):11–23

Vogel JA, Debinski DM, Koford RR, Miller JR (2007) Butterfly responses to prairie restoration through fire and grazing. Biol Conserv 140(1–2):78–90

WallisDeVries MF, Raemakers I (2001) Does extensive grazing benefit butterflies in coastal dunes? Restor Ecol 9(2):179–188

WallisDeVries MF, Bakker JP, Van Wieren SE (eds) (1998) Grazing and conservation management. Kluwer, Dordrecht

Weiss SB (1999) Cars, cows, and checkerspot butterflies: nitrogen deposition and management of nutrient-poor grasslands for a threatened species. Conserv Biol 13(6):1476–1486

Wettstein W, Schmid B (1999) Conservation of arthropod diversity in montane wetlands: effect of altitude, habitat quality and habitat fragmentation on buterflies and grasshoppers. J Appl Ecol 36:363–373

White P, Kerr JT (2006) Contrasting spatial and temporal global change impacts on butterfly species richness during the 20th century. Ecography 29(6):908–918

Part VI
The Future of Relict Species

Review
Modelling Future Trends of Relict Species

Dennis Rödder and Johannes Dambach

Please note the erratum to the chapter title at the end of the book.

Abstract Distribution patterns of species vary over space and time. This becomes most evident considering the differences between the current and Last Glacial Maximum (LGM, 21,000 BP) distribution patterns of species in the Northern hemisphere (see also Habel et al. 2009, Mini Review 03). Most warm-adapted species experienced reduction and fragmentation of ranges because of intrusion by uninhabitable continental ice sheets causing distributional shifts and fragmentation of primary habitats. On the other hand, cold-adapted species were able to expand their ranges. Today, ranges of those species are restricted to current refugia as can be observed in glacial relict species.

Understanding refugial distributions of species has been a core task in historical bio-geography. Before the 1990s, refugia were preliminarily identified based on disjunctions of species distributions, distribution patterns of sister species, and the fossil records (e.g. Hoffman 1981). More recently, phylogeographic approaches based on intraspecific molecular analyses and spatial modeling approaches based on ecological properties of species have been developed.

1 Climatic Niches and the Spatial Distribution of Species

The observation of ecological properties of species and their areas of distribution being related is not new (Grinnell 1917; James et al. 1984), but the increasing availability of information on the variation of environmental parameters in geographic

D. Rödder (✉)
Department of Herpetology, Zoologisches Forschungsmuseum Alexander Koenig, D-53113 Bonn, Germany
Department of Biogeography, University of Trier, D-54286 Trier, Germany
e-mail: roedder@uni-trier.de

J. Dambach
Section of Lower Arthropods, Zoologisches Forschungsmuseum Alexander Koenig, D-53113 Bonn, Germany

J.C. Habel and T. Assmann (eds.), *Relict Species: Phylogeography and Conservation Biology*, DOI 10.1007/978-3-540-92160-8_22, © Springer-Verlag Berlin Heidelberg 2009

space, species distribution data, and computation capacities during the last decade now allow large scale assessments of relationships between distributions observed and explanatory parameters. Relationships can be assessed by calculating 'environmental' or 'ecological' niches and their subsequent projection into geographic space (Guisan and Zimmermann 2000). Here, GIS-based environmental data offer huge opportunities to assess variations in environmental factors within the species ranges, especially when combined with spatial modeling techniques (Kozak et al. 2008; Waltari et al. 2007). Such techniques were proposed as a useful supplementary tool despite long time/term established techniques for the identification of refuges and potential migration pathways (Waltari et al. 2007).

Model techniques can be classified into two different groups: (1) mechanistic models, which predict the potential distribution of a species based on its physiological tolerances measured in laboratory experiments and (2) spatial model techniques, which derive from the distribution model based on statistical relationships between distribution patterns observed and environmental parameters. In the latter case, Climatic Envelope Models (CEMs) use exclusively climatic variables as predictors whereby a wider range of variables is used in Ecological Niche Models (ENMs) (e.g. soil and vegetation layers or remote sensing data). The development of mechanistic models is just at the beginning (Kearney and Porter 2004; Kearney et al. 2008), but CEMs and ENMs have been applied to predict species' potential distributions (PDs) under current, past, and future climate scenarios for some time now (e.g. Araújo et al. 2004; Araújo and Guisan 2006; Heikkinen et al. 2006; Hijmans and Graham 2006; Pearman et al. 2008; Waltari et al. 2007), invasive species biology (e.g. Peterson 2003; Peterson and Vieglais 2001; Rödder 2009; Rödder et al. 2008), conservation priority setting (e.g. Araújo et al. 2004; Kremen et al. 2008), and ecology and evolutionary biology (e.g. Graham et al. 2004; Kozak et al. 2008; Peterson et al. 1999). CEMs were especially suggested to be useful for identification of potential Pleistocene refugia with high accuracy (Peterson and Nyári 2008; Waltari et al. 2007) and are therefore discussed below in detail.

In CEM approaches, the climate envelope is understood as a part of a species' fundamental or realized niche depending on variables selected and assumptions made (Soberón 2007; Soberón and Peterson 2005). As defined by Hutchinson (1957; 1978) and later extended by Soberón and Peterson (2005), a species' fundamental niche represents the complete set of environmental conditions under which a species can persist, i.e. under which its fitness is greater than or equal to one in the absence of competitors or predators. It's realized niche in environmental space (=realized distribution in geographic space) is a subset of the fundamental niche considering dispersal limitations and biotic interactions, such as food availability, competition, or interactions with pathogens (Fig. 1). Niche variables can be subdivided concerning specific classes regarding the spatial extent in which they operate and if competition may play a role or not. The Grinnellian class is defined by fundamentally non-interactive variables (e.g. climate) (Grinnell 1917) whereby the Eltonian class focuses on biotic interactions and resource-consumer dynamics (Elton 1927). The former operates on a coarse scale and is the main subject in CEM approaches, whereby the latter can principally be measured only at local scales and is commonly not addressed in CEM approaches (Soberón 2007).

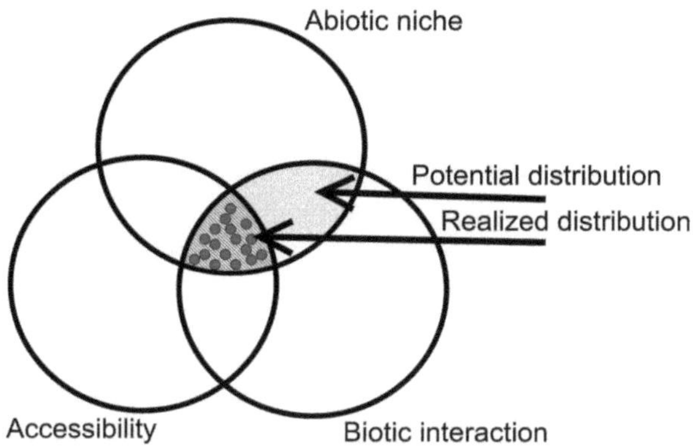

Fig. 1 Relationships between abiotic (=fundamental) niche, biotic interaction and accessibility after Hutchinson (1957) as modified by Soberón and Peterson (2005). The potential distribution is a subset of the abiotic niche considering biotic interactions, whereby the realized distribution is a subset of the potential distribution considering accessibility. *Dots* represent species records

2 How do CEMs Work?

Before CEMs can be computed it is necessary to compile a set of species occurrence records and a set of suitable predictor variables (e.g. GIS layers containing information on climatic parameters). Some examples of freely available climate data sets are given in Kozak et al. 2008; Rödder et al. 2008, Chap. 22. The layers for each climate scenario commonly include the minimum and maximum temperatures and the mean precipitation per month (=36 climatic parameters). Based on these monthly layers, bioclimatic parameters can be generated, e.g. with DIVA-GIS (Hijmans et al. 2001). Bioclimatic parameters are more useful than monthly values, since they are independent of latitudinal variations. This becomes obvious considering that the 'maximum temperature of the warmest month' is more useful than the maximum temperature of a specific month, since not everywhere the same month might also be the warmest.

Multi-collinearity among predictor variables may hamper the analysis of species-environment relationships because ecologically more causal variables may be excluded from models if other correlated variables explain the variation in response variable better in statistical terms (Heikkinen et al. 2006). Therefore, variable selection should be guided by a throughout assessment of the target species' ecology, and a rather minimalist set of predictors should be preferred. Specific adjustment of variable sets according to specific ecological needs of the target species may improve the model output.

Once species records and predictor variables have been compiled, a subsequent step involves the development of a multidimensional view of the climatic niche of a species, which is a considerable challenge given the complex nature of species' niches (Peterson and Vargas 1993). In plain text: in CEM, climatic information for species presence localities are summarized to an 'ideal' climatic niche for the target species (note that this can also be done with reliable absence data) that is afterwards compared to climatic conditions at the query localities, i.e. where the presence/absence of the species is unknown. The results are geographic maps showing the similarity of an area with the 'ideal' climatic niche (Figs. 2a, and 3a). The selection of a suitable algorithm for the computation of the CEM depends on the amount of distribution records available, their quality, and the specific goal of the study (for a brief overview of available algorithms see Jeschke and Strayer 2008 or Rödder et al. 2008, Chap. 22).

3 Evaluation of Results

One of the most frequently used model tests is an assessment of the Area Under the Curve (AUC) statistics, referring to the ROC (Receiver Operation Characteristic) curve using e.g. 25 % of the records as test points and the remaining ones for training (Hanley and McNeil 1982; Phillips et al. 2006). This method is recommended for ecological applications because it is non-parametric (Pearce and Ferrier 2000). Values of AUC range from 0.5 (i.e. random) for models with no predictive ability to 1.0 for models giving perfect predictions. According to the classification of Swets (1988) AUC values >0.9 describe 'very good', >0.8 'good' and >0.7 'useful' discrimination ability. A second possibility is Cohen's kappa statistic of similarity (k) (Fielding and Bell 1997; Pearce and Ferrier 2000). Cohen's kappa yields values ranging from 0.0 (no predictive ability) to 1.0 (perfect predictive ability), whereby k values above 0.7 describe 'very good' discrimination ability (Monserud and Leemans 1992). If only few species records for model building are available (e.g. <25), a jack-knife test can be performed (Pearson et al. 2007). Additionally, in some cases invasive populations of the target species can be used as an independent test (Sax et al. 2008).

It needs to be noted that test statistics, such as AUC or Cohen's kappa scores are a measure of model fit to be compared to the observed distribution of a species and must not be interpreted as a measure in the ecological sense of a model. Profound knowledge of a species' natural history is necessary for an evaluation, when a model meets the ecological requirements of the target species.

4 Strengths of CEMs and Potential 'Pitfalls' When Interpreting Results

CEMs are easy to use and inexpensive. A steadily increasing number of studies have been showing that, when CEMs are properly applied, results can have a high quality (e.g. Peterson 2003; Waltari et al. 2007). They have been a rich source of quantitative projections concerning geographic ranges of species with great value in many areas of both basic and applied ecology and conservation. They allow the

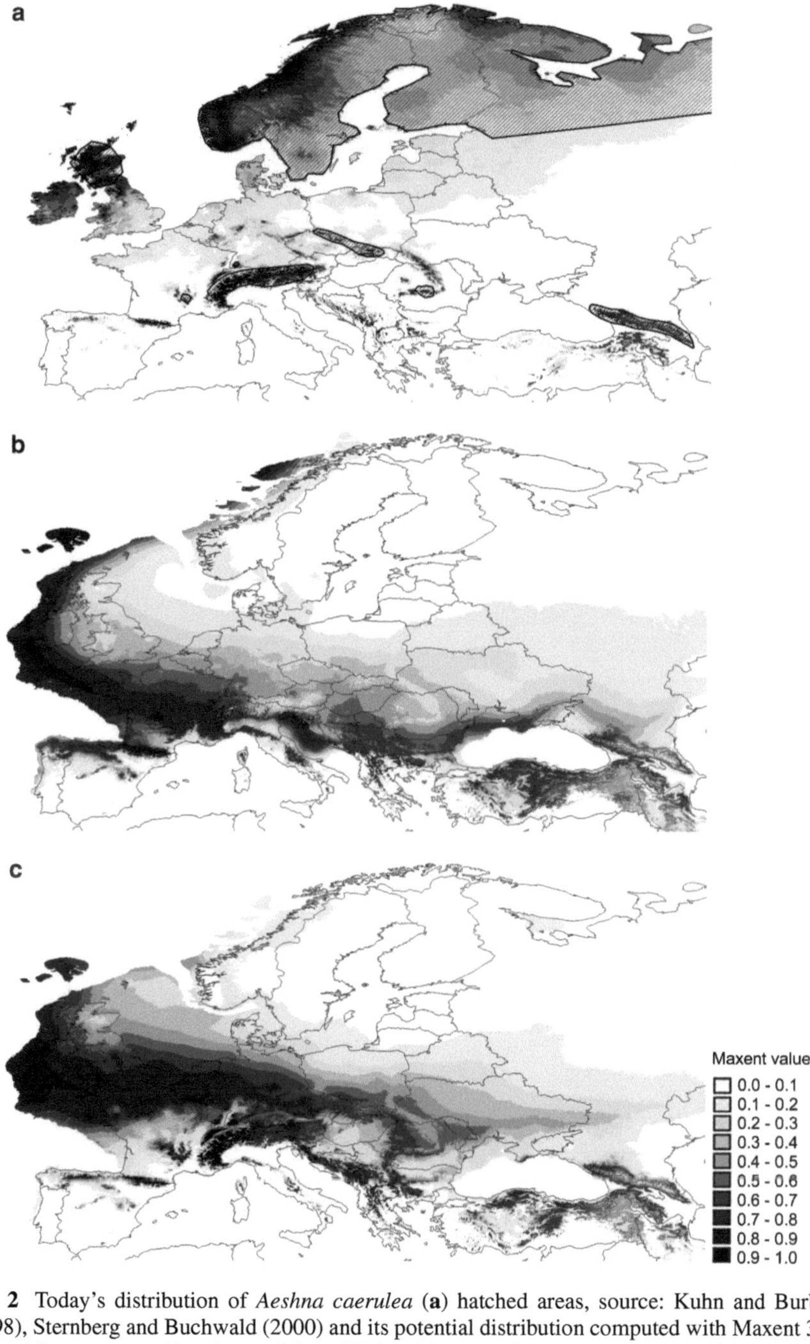

Fig. 2 Today's distribution of *Aeshna caerulea* (**a**) hatched areas, source: Kuhn and Burbach (1998), Sternberg and Buchwald (2000) and its potential distribution computed with Maxent 3.2.1 derived from current climatic conditions. Higher Maxent values suggest higher climatic suitability. Figure 2b,c show potential distributions of *Aeshna caerulea* assuming two different paleoclimatic scenarios depicting climatic conditions as expected for 21,000 BP (**b**) CCSM; 2C: MIROC; for details see Rödder et al. 2008, Chap. 22. Areas of currently known distribution of *A. caerulea* are highly congruent with the proposed potential distribution of the CEM even in small and disjunctive ranges. Projections of the CEM onto palaeoclimatic scenarios suggest potential migration pathways during the LGM connecting most current refugia

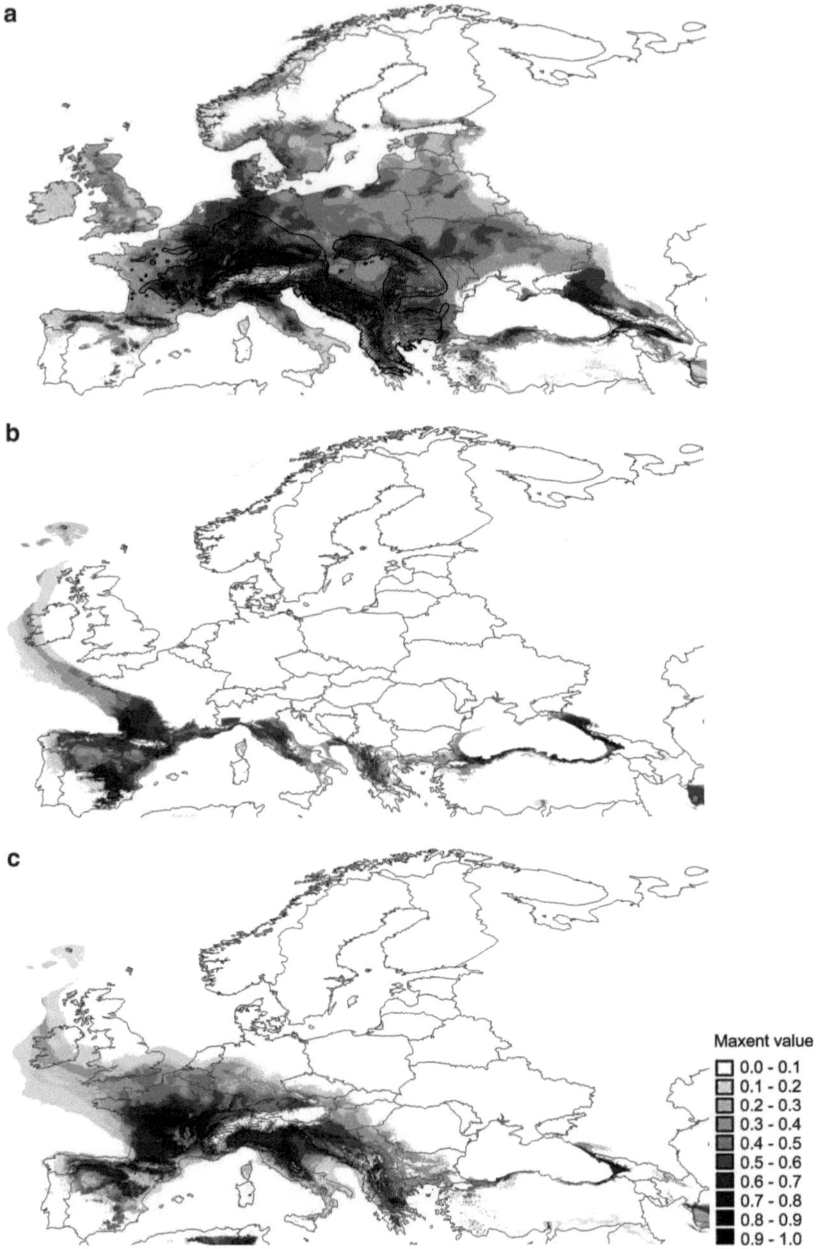

Fig. 3 Today's distribution of *Bombina variegata* (**a**) hatched areas, source: IUCN (2006) and its potential distribution computed with Maxent 3.2.1 derived from current climatic conditions. Higher Maxent values suggest higher climatic suitability. Figure 3b,c show projections of the climate envelope of *B. variegata* assuming two different paleoclimatic scenarios depicting climatic conditions as expected for 21,000 BP (**b**) CCSM; 3C: MIROC; for details see Rödder et al. (2009), Chap. 22. Areas of currently known distribution of *B. variegata* are highly congruent with the proposed potential distribution of the CEM even in small and disjunctive ranges, e.g. in France. Projections of the CEM onto palaeoclimatic scenarios suggest potential refugia during the LGM

identification of possible threats due to climate change (e.g. see Habel et al. 2009, Chap. 25) or past migration pathways (Fig. 2b, c, Rödder 2009, Chap. 22).

For example, CEM projection of the PD of the Azure Hawker (*Aeshna caerulea* (Stroem, 1783)) onto palaeo-climatic scenarios reflecting conditions as expected for the LGM 21,000 BP suggest that climatically suitable areas were widely distributed in Europe throughout the LGM (Fig. 2b,c). The PD would have connected the current refugia (Fig. 2a). It is also possible to reconstruct the past refugia of a currently wide-spread species (Waltari et al. 2007), e.g. as shown for the yellow-bellied toad (Fig. 3). Furthermore, CEMs can be used as a guideline for efficient further sampling, since phylogeographic studies are expensive in terms of both time and resources. Regarding relict species, CEMs may be useful for the identification of potentially suitable areas which may harbor relict populations unknown so far.

However, there are some points that deserve consideration when interpreting results, since a number of potential sources of predictive errors that may be categorized as either 'algorithmic' or 'biotic' do exist (Fielding and Bell 1997). Algorithmic errors are caused by limitations of the modeling technique and data-gathering process. Each algorithm makes specific assumptions and therefore has strengths and weaknesses, e.g. some algorithms are able to account for interactions between predictor variables whereas others are not. This influences their predictive ability and suitability in specific cases. Differences between algorithms need to be evaluated regarding their usefulness depending on the study goal at hand. When projecting CEMs onto past or future climate scenarios it is important to acknowledge that different climate scenarios will reveal different results and no single 'best' model exists. "Strengths and weaknesses" characteristics/qualities of different climate models should be taken into consideration (Beaumont et al. 2008). Evaluation of a variety of scenarios may help to assess output variations. For example, when comparing Figs. 2b,c and 3b,c. the general patterns are similar, but fine scale differences, especially in Northward PD limits, can be identified. These can be traced back to different assumptions made in the palaeo-climatic scenarios used, e.g. concerning CO_2 concentrations.

CEMs are derived from a subset of environmental conditions at species records (= variables selected); hence they capture only a more or less extensive part of the climatic niche depending on the variables selected and other factors influencing the distribution of species (see below). They are generally unable to capture the niche completely, and output maps show regions with similar conditions as the testing records according to the predictor variables selected rather than 'complete' niches.

Quality and spatial properties of distribution records used for model building can significantly influence the results, e.g. when samples do not cover the whole climatic niche space of the target species. Furthermore, some algorithms require that samples are independently distributed of each other, which is seldom the case in biological data. Therefore, spatial autocorrelation – the tendency of neighboring samples to possess similar characteristics – is a potential problem in many niche modeling approaches and several methods to account for this fact have been suggested (Dormann et al. 2007).

When applying CEMs some key assumptions regarding biological aspects are commonly made: (1) species occur at all locations where the climate is favorable,

(2) biotic interactions are unimportant in determining ranges and are constant over space and time and (3) genetic and phenotypic compositions of species are constant over space and time (Jeschke and Strayer 2008). (1) Ideally CEMs highlight all areas that are climatically suitable for the target species reflecting its potential distribution regardless of dispersal limitations and thus accessibility. When interpreting potential distribution maps it is important to evaluate the dispersal properties of the target species. For example, although *Bombina variegata* (Linnaeus, 1758) could find climatically suitable areas in England today (Fig. 3a), the species is unable to colonize the UK due to dispersal limitation by the sea. (2) Biotic interactions are commonly not considered during model building since interactions, such as competition or predation in species assemblages are often too complex to be gathered in a single step during model building. For example, the PD of *B. variegata* is considerably wider than its realized distribution in Germany, the Czech Republic, Slovakia and Hungary (Fig. 3a). Here, climate is not the limiting factor since a hybrid-zone between *B. variegata* and its sister species *Bombina bombina* (Linnaeus 1761) hamper further dispersal (Szymura 1993). Under different climate scenarios, PDs of species can respond spatially in a different manner which can cause strong variations over space and time and result in discrepancies (e.g. as shown by Schweiger et al. (2008) in an ENM approach for a butterfly / host plant system). For simplicity's purpose it may be useful to focus in a first step on the climatic niches of each target species to assess its potential distribution and subsequently compare models for potentially interacting species. (3) An important point when projecting CEMs onto past or future climate scenarios is that fundamental niches can be subject to evolution. In a recent review, it has been shown that, independent of the taxonomic group, the fundamental niche can remain stable for tens of thousands of years or it can shift substantially within only a few generations (Pearman et al. 2007). 'Niche conservatism' of closely related species is a phenomenon that has been observed in several different taxonomic groups (e.g., Peterson et al. 1999; Wiens and Graham 2005), but, on the other hand, niche shifts have also been proposed (e.g. Broennimann et al. 2007; Fitzpatrick et al. 2007; Graham et al. 2004; for a recent review see Pearman et al. 2007). So evidence for the degree of niche shifts or niche conservatism in closely related species is mixed and debated (Pearman et al. 2007; Wiens and Graham 2005). When applying CEMs, it is an important issue if (climatic) niches are rather conservative or not, especially when the aim is an assessment of changes in potential distributions under different climate scenarios. However, in general, there is a considerable lack of knowledge regarding the processes triggering shift in climatic niches as well as in suitable methods to analyze it.

5 Conclusions

Predictions of species distributions derived from correlative models can help to understand the spatial patterns of biodiversity, refugia and identification of possible threats for relict populations. The amount of available data and software is steadily increasing as well as the number of studies applying niche model techniques. However, a discrepancy between increasingly complex studies and the understanding of underlying processes, derivation of valid assumptions, and the development of conceptual backgrounds is still a problem (Jiménez-Valverde et al. 2008). One problem might be that models predicting potential distributions of species are easily computed and that test statistics are sometimes misleading (Lobo et al. 2008) facilitating misinterpretations. However, properly applied and critically evaluated, during each step, from model building to the final interpretation, CEMs can be powerful tools and very helpful to generate and test hypotheses.

Acknowledgements We are indebted to Bernhard Misof and Stefan Lötters for fruitful discussions and Michael Raupach for logistic support. Robert Hijmans helped us with palaeoclimatic scenarios. DR was funded by the 'Graduiertenförderung des Landes Nordrhein-Westfalen' and JD by the 'Deutsche Forschungsgemeinschaft' (DFG).

References

Araújo MB, Guisan A (2006) Five (or so) challenges for species distribution modelling. J Biogeogr 33:1677–1688

Araújo MB, Cabeza M, Thullier W, Hannah L, Williams PH (2004) Would climate change drive species out of reserves? An assessment of existing reserve-selection methods. Glob Chang Biol 10:1618–1626

Beaumont LJ, Hughes L, Pitman AJ (2008) Why is the choice of future climate scenarios for species distribution modelling important? Ecol Lett 11:1135–1146

Broennimann O, Treier UA, Müller-Schärer H, Thuiller W, Peterson AT, Guisan A (2007) Evidence of climatic nich shift during biological invasion. Ecol Lett 10:701–709

Dormann CF, McPherson J, Araújo MB, Bivand R, Bollinger J, Carl G, Davies RG, Hirzel A, Jetz W, Kissling WD, Kühn I, Ohlemüller R, Peres-Neto PR, Reineking B, Schröder B, Schurr FM, Wilson R (2007) Methods to account for spatial autocorrelation in the analysis of species distributional data: a review. Ecography 30:609–628

Elton C (1927) Animal ecology. Sedgwick and Jackson, London

Fielding AH, Bell JF (1997) A review of methods for the assessment of prediction errors in conservation presence/absence models. Environ Conserv 24:38–49

Fitzpatrick MC, Weltzin JF, Sanders NJ, Dunn RR (2007) The biogeography of prediction error: why does the introduced range of the fire ant over-predict its native range? Glob Ecol Biogeogr 16:24–33

Graham CH, Ron SR, Santos JC, Schneider CJ, Moritz C (2004) Integrating phylogenetics and environmental niche models to explore speciation mechanisms in dendrobatid frogs. Evolution 58:1781–1793

Grinnell J (1917) The niche-relationships of the California Thrasher. Auk 34:427–433

Guisan A, Zimmermann N (2000) Predictive habitat distribution models in ecology. Ecol Modell 135:147–186

Habel JC, Augenstein B, Nève G, Rödder D, Assmann T (2009) Population genetics and ecological niche modelling reveal high fragmentation and future extinction of the endangered relict butterfly *Lycaena helle*. Relict Species 417–442

Hanley J, McNeil B (1982) The meaning of the use of the area under a receiver operating characteristic (ROC) curve. Radiology 143:29–36

Heikkinen RK, Luoto M, Araújo MB, Virkkala R, Thuiller W, Sykes MT (2006) Methods and uncertainties in bioclimatic envelope modelling under climate change. Prog Phys Geogr 30:751–777
Hijmans RJ, Graham CH (2006) The ability of climate envelope models to predict the effect of climate change on species distributions. Glob Chang Biol 12:2272–2281
Hijmans RJ, Cruz JM, Rojas E, Guarino L (2001) DIVA-GIS, version 1.4. A geographic information system for the management and analysis of genetic resources data. Manual. International Potato Center and International Plant Genetic Resources Institute, Lima, Peru
Hoffman RS (1981) Different voles for different holes: environmental restrictions on refugial survival of mammals. In: Scudder GGE, Reveal JL (eds) Evolution today proceedings of the second international congress of systematic and evolutionary biology. Hunt Institute for Botanical Documentation, Pittsburgh, pp 25–45
Hutchinson GE (1957) Concluding remarks. Cold Spring Harb Symp Quant Biol 22:415–427
Hutchinson GE (1978) An introduction to population ecology. Yale University Press, New Haven, CT
IUCN, Conservation International, NatureServe (2006) Global amphibian assessment. http://www.globalamphibians.org Cited 15 Oct 2008
James FC, Johnston RF, Warner NO, Niemi G, Boecklen W (1984) The grinnellian niche of the wood thrush. Am Nat 124:17–47
Jeschke JM, Strayer DL (2008) Usefulness of bioclimatic models for studying climate change and invasive species. Ann NY Acad Sci 1134:1–24
Jiménez-Valverde A, Lobo JM, Hortal J (2008) Not as good as they seem: the importance of concepts in species distribution modelling. Divers Distrib . doi:10.111/j.1472-4642.2008.00496.x
Kearney M, Porter WP (2004) Mapping the fundamental niche: physiology, climate, and the distribution of a nocturnal lizard. Ecology 85:3119–3131
Kearney M, Phillips BL, Tracy CR, Christian KA, Betts G, Porter WP (2008) Modelling species distributions without using species distributions: the cane toad in Australia under current and future climates. Ecography 31:423–434
Kozak KH, Graham CH, Wiens JJ (2008) Integrating GIS-based environmental data into evolutionary biology. TREE 23(3):141–148
Kremen C, Cameron A, Moilanen A, Phillips SJ, Thomas CD, Beentje H, Dransfield J, Fisher BL, Glaw F, Good TC, Harper GJ, Hijmans RJ, Lees DC, Louis E, Nussbaum RA, Raxworthy CJ, Razafimpahanana A, Schatz GE, Vences M, Vieites DR, Wright PC, Zjhra ML (2008) Aligning conservation priorities across taxa in Madagascar with high-resolution planning tools. Science 320:222–226
Kuhn K, Burbach K (1998) Libellen in bayern. Ulmer Verlag, Stuttgart
Lobo JM, Jiménez-Valverde A, Real R (2008) AUC: a misleading measure of the performance of predictive distribution models. Glob Ecol Biogeogr 17:145–151
Monserud RA, Leemans R (1992) Comparing global vegetation maps with Kappa statistics. Ecol Modell 62:275–293
Pearce J, Ferrier S (2000) An evaluation of alternative algorithms for fitting species distribution models using logistic regression. Ecol Modell 128:128–147
Pearman PB, Guisan A, Broennimann O, Randin CF (2007) Niche dynamics in space and time. TREE 23:149–158
Pearman PB, Randin CF, Broennimann O, Vittoz P, van der Knaap WO, Engler R, Le Lay G, Zimmermann NE, Guisan A (2008) Prediction of plant species distributions across six millennia. Ecol Lett 11:357–369
Pearson RG, Raxworthy CJ, Nakamura M, Peterson AT (2007) Predicting species distributions from small numbers of occurrence records: a test case using cryptic geckos in Madagascar. J Biogeogr 34:102–117
Peterson AT (2003) Predicting the geography of species' invasions via ecological niche modelling. Q Rev Biol 78:419–433
Peterson AT, Nyári ÁS (2008) Ecological niche conservatism and Pleistocene refugia in the Thrush-like Mourner, *Schiffornis* sp., in the Neotropics. Evolution 62:173–183

Peterson AT, Vargas N (1993) Ecological diversity in scrub jays, *Aphelocoma coerulescens*. In: Ramamoorthy TP, Bye R, Fa J (eds) The biological diversity of Mexico: origins and distribution. Oxford University Press, New York, pp 309–317
Peterson AT, Vieglais DA (2001) Predicting species invasions using ecological niche modelling: new approaches from bioinformatics attack a pressing problem. Bioscience 51:363–371
Peterson AT, Soberón J, Sánchez-Cordero V (1999) Conservation of ecological niches in evolutionary time. Science 285:1265–1267
Phillips SJ, Anderson RP, Schapire RE (2006) Maximum entropy modelling of species geographic distributions. Ecol Modell 190:231–259
Rödder D (2009) 'Sleepless in Hawaii' - does anthropogenic climate change enhance ecological and socioeconomic impacts of the alien invasive *Eleutherodactylus coqui* Thomas, 1966 (Anura: Eleutherodactylidae)? North-West J Zool 5:16–25
Rödder D, Solé M, Böhme W (2008) Predicting the potential distribution of two alien invasive Housegeckos (Gekkonidae: *Hemidactylus frenatus*, *Hemidactylus mabouia*). North-West J Zool 4:236–246
Sax DF, Stachowicz JJ, Brown JH, Bruno JF, Dawson MN, Gaines SD, Grosberg RK, Hastings A, Holt RD, Mayfield MM, O'Connor MI, Rice WR (2008) Ecological and evolutionary insights from species invasions. TREE 22:465–471
Schweiger O, Settele J, Kudrna O, Klotz S, Kühn I (2008) Climate change can cause spatial mismatch of trophically interacting species. Ecology 89:3472–3479
Soberón J (2007) Grinnellian and Eltonian niches and geographic distributions of species. Ecol Lett 10:1115–1123
Soberón J, Peterson AT (2005) Interpretation of models of fundamental ecological niches and species' distributional areas. Biodiv Inform 2:1–10
Sternberg K, Buchwald R (2000) Die Libellen Baden- Württembergs Band 2: Grosslibellen (Anisoptera). Ulmer Verlag, Stuttgart
Swets K (1988) Measuring the accuracy of diagnostic systems. Science 240:1285–1293
Szymura JM (1993) Analysis of hybrid zones with *Bombina*. In: Harrison R (ed) Hybrid zones and the evolutionary process. Oxford University Press, New York, pp 261–289
Waltari E, Hijmans RJ, Peterson AT, Nyári AS, Perkins SL, Guralnick RP (2007) Locating pleistocene refugia: comparing phylogeographic and ecological niche model predictions. PLoS ONE 7:1–11
Wiens JJ, Graham CH (2005) Niche conservatism: integrating evolution, ecology, and conservation biology. Annu Rev Ecol Syst 36:519–539

Conservation of the Grey Bush Cricket *Platycleis albopunctata* (Orthoptera: Tettigoniidae) Under Differing Habitat Conditions: Implications From an Individual-Based Model

Eva M. Griebeler and Eckhard Gottschalk

Abstract Assessing the chance of survival of any species is a great challenge in conservation biology. In this chapter, we analyse the vulnerability of the grey bush cricket *Platycleis albopunctata* in habitats of different food availability under current and increased temperature conditions applying an individual-based model. Our simulations show that populations in warmer habitats with a higher food limitation have a much lower extinction risk than those living in habitats that are less food-limited and colder. An increase in mortalities of life stages severely increases the risk of population extinction, whereas a shift in the termination of egg diapause towards the beginning of the year caused by warmer winters has no significant effect on survival. Thus, *P. albopunctata* may benefit from increasing future temperatures if neither life stage mortalities will increase as the result of simultaneously affected abiotic habitat conditions (due to precipitation and temperature) and/or species interactions nor changes in abiotic conditions will cause a shortage of food. We demonstrate that simulation models are a suitable tool to understand processes influencing population survival of temperature-dependent species under current and future habitat conditions. They are in particular beneficial in conserving relict species that are naturally restricted to patchy and scarce habitats.

1 Introduction

Orthopteran insect species are a major taxonomic group in natural and disturbed ecosystems of temperate and subtropical Eurasia (Sergeev 1998). These species occur in both natural (particularly grasslands) and anthropogenic landscapes

E.M. Griebeler (✉)
Department of Ecology, Zoological Institute, University of Mainz,
P.O. Box 3980, D-55099 Mainz, Germany
e-mail: em.griebeler@uni-mainz.de

E. Gottschalk
Centre for Nature Conservation, University of Göttingen, von-Siebold-Str. 2, D-37075
Göttingen, Germany
e-mail: egottsc1@gwdg.de

J.C. Habel and T. Assmann (eds.), *Relict Species: Phylogeography and Conservation Biology*, 385
DOI 10.1007/978-3-540-92160-8_23, © Springer-Verlag Berlin Heidelberg 2010

(meadows, pastures, agricultural fields, etc.). They are very sensitive to temperature, are especially characteristic of hot and sunny regions and are in general no food specialists. Populations of grasshoppers and bush crickets are declining in Central Europe (Ingrisch and Köhler 1998). More than half of these species are already considered endangered since many of their habitats are rapidly being destroyed, fragmented and changed (Ingrisch and Köhler 1998). However, over the past decade there has been increasing evidence that several orthopterans are able to shift their ranges towards the North as a consequence of present global warming (Kleukers et al. 1996; Cannon 1998; Widgery 2000).

Small population sizes are often the consequence of recent habitat fragmentation or loss. However, they also characterize highly ecologically specialized organisms restricted to scarce habitats such as relict species. Such species show highly localized ranges and are pre-dominantly found in regions where past climate changes have not resulted in major extinction events. Although it is widely acknowledged that population persistence greatly depends on population size and temporal fluctuations, the long historical persistence of many relict and endemic taxa as small populations contradicts this relationship. However, anthropogenic habitat fragmentation, alteration and loss are threats to any species. Thus, assessing the chance of species survival under current and future environmental conditions is especially important for species inhabiting patchy and scarce habitats.

The thermophilic grey bush cricket *Platycleis albopunctata* (Orthoptera: Tettigoniidae) is distributed throughout Europe. Because dry grassland has a patchy distribution in Central Europe, populations of this cricket inhabit habitat islands. During the last century, habitat fragmentation has increased due to changes in land use. The species is considered threatened in Germany (category three, Ingrisch and Köhler 1998) because of the ongoing loss of dry grasslands which results in relict populations. Over the past decades, there has been increasing evidence that *P. albopunctata* is shifting its range towards the North as the result of present global warming (Widgery 2000). Field studies on this cricket have shown that temperature conditions and vegetation density of habitats strongly differ (Gottschalk et al. 2003). It has been suggested that such differences in habitat conditions affect population dynamics of this species and that this may consequently have an impact on its extinction risk (Gottschalk et al. 2003). Although warmer habitats (e.g. dry grasslands) allow for higher reproduction rates and faster development of life stages (Köhler 1999), they have sparser vegetation than colder habitats (e.g. semi-dry grasslands) and a lower rate of primary production which causes food limitation.

In order to better understand the mechanisms that determine the vulnerability of this cricket under current and future thermal conditions in habitats of different food availability, we utilized an individual-based model (IBM) from a previous publication (Griebeler and Gottschalk 2000a). IBMs have been successfully applied to various species (DeAngelis and Mooij 2005) including orthopterans (Wagner and Berger 1996; Griebeler and Gottschalk 2000a,b) to assess their degree of endangering. We test the hypothesis of differing population dynamics generated by differences in temperature conditions and in food availability that was proposed by Gottschalk et al. (2003) based on field observations and laboratory experiments. Therefore, we apply it to the two populations experimentally studied in this paper (in contrast to Griebeler and Gottschalk 2000a,b; Gottschalk et al. 2003), and additionally consider a broader number of habitat conditions by varying temperature profiles and food availability of habitats. Furthermore, we perform a sensitivity analysis for mortalities of life stages and the length of egg diapause, as these may also be altered under global change as a result of simultaneously affected abiotic conditions and/or species interactions.

2 Materials and Methods

2.1 P. albopunctata

The grey bush cricket *P. a. albopunctata* (Orthoptera: Tettigoniidae) is distributed throughout West and Central Europe (Harz 1969). Its sister subspecies *P. a. grisea* extends its distribution area into Asia. *P. a. albopunctata* has an Atlantic-Mediterranean origin, whereas *P. a. grisea* originates from Central Asia (Ingrisch and Köhler 1998).

P. albopunctata is a thermo- and xerothermophilic species (Harz 1957; Ingrisch and Köhler 1998). In Central Europe, it inhabits dry grasslands and similarly structured semi-arid and arid habitats (Harz 1957). Only in habitats with favourable microclimatic conditions is development of instars sufficiently rapid to provide a reproductive output that enables long-term survival (Gottschalk 1996). In Germany, the species is considered threatened due to the loss of its semi-natural habitats (Ingrisch and Köhler 1998; Walter and Gottschalk 1998). Abandonment of grazing as well as afforestation of dry grasslands has led to habitat loss resulting in relict populations within its distribution area. The life cycle of the species is annual with one generation per year (Ingrisch 1986). Growth, development and reproduction of the cricket are primarily affected by temperature (Ingrisch 1986). Under the environmental conditions in our study area (see below), hatching occurs in spring (mainly April). The reproduction period of the cricket lasts from July until temperatures decline in October (Gottschalk et al. 2003). Females produce eggs throughout their adult life. They prefer soil and moss at places with sparse vegetation as oviposition sites. Eggs develop during late summer and in the following spring after an intermediate diapause. The proportion of embryonic development that is completed before hibernation depends on the date of oviposition and ground temperature (Gottschalk et al. 2003). Egg hatching after hibernation needs no frost stimulus, but larvae need higher temperatures to hatch (Ingrisch 1985). The first instars of *P. albopunctata* mainly feed on flowers (anthers, nectar glands, petals). Older nymphs and adults prefer seeds, especially from grasses, and adults also feed on small insects. Both sexes are able to fly.

2.2 Field Studies and Summary of the Field Experiments

Our two study sites are situated about 50 km apart from each other in North Bavaria (southern Germany, 48°N and 10°E). The size of each of the areas is about 1 ha. The "Hammelburg" area is a dry grassland located in the valley of the Franconian Saale on the Southern edge of the Rhön mountains at an altitude of about 280 m. The area is a steep slope that is exposed to the Southwest with an inclination of 35°. The "Prappach" area is a semi-dry grassland in the Main valley at the edge of the Hassberge. This slope is situated at an altitude of 270 m. The area consists of a flat plateau and several steeper embankments with inclinations ranging between 0 and 20°. The area is irregularly grazed by sheep or mowed. Vegetation cover is lower at Hammelburg (median 50%, Gottschalk et al. 2003) than at Prappach (median 80%, Gottschalk et al. 2003). The number of seeds on the ground as a measure for productivity is six-fold higher in Prappach, and the average distance to the next potential food plants is ten-fold lower in Prappach than in Hammelburg (Gottschalk et al. 2003). This suggests that food is more limited at Hammelburg than at Prappach.

In 2002 we performed measurements of ground temperature in both areas between 8th August and 9th September, because this period between time of oviposition and the autumn temperature decline is important for embryonic development (Ingrisch 1979; Gottschalk et al. 2003). The measurements revealed that Hammelburg is on average 4.8°C warmer than Prappach (Gottschalk et al. 2003).

Population abundances estimated from standardized transect walks between 1993 and 1999 showed severely differing population dynamics at both sites. Whereas Prappach exhibited strong fluctuations in population abundance (between 85 (±20) and 3188 (±367) adult animals, mean ± standard deviation, Gottschalk et al. 2003), population sizes at Hammelburg were comparatively constant (between 460 (±78) and 730 (±124) adults, mean ± standard deviation, Gottschalk et al. 2003). The mean population size estimated between 1993 and 1999 was 1,390 adults at Prappach and 614 adults at Hammelburg. However, in 1996 population size at Hammelburg was not estimated because of logistical limitations.

2.3 The Model

A detailed description of the model applied to test our hypothesis (Gottschalk et al. 2003) and to rate the impact of global change on the viability of *P. albopunctata* can be found in Griebeler and Gottschalk (2000a). The following is a short summary of this individual-based day-degree model.

We distinguish three life stages in this model: eggs, larvae and adults. We assume an equal sex ratio. The model is based on the idea that eggs and larvae accumulate temperatures on a daily basis during their development. Both stages need minimum temperatures for daily development ($T_{0,egg}$ = 7.5°C, Ingrisch 1986;

$T_{0,larva}$=9.6°C) and a total temperature input (TS_{egg}=808.5°C days, Ingrisch 1986; TS_{larva}=381.9°C days) to enter the next life stage. The model considers a fixed diapause *D* for eggs between 1st November and 31st March of the following year (Ingrisch 1986). Eggs do not accumulate daily temperatures within these winter months. The model assumes a constant mortality rate for the egg stage (m_{egg}=60%) and for the larva stage (m_{larva}=90%), which summarize all mortalities during the respective life stage. These two rates that determine the final transition of a life stage to the next are randomly applied to each individual before it enters the next life stage. Each adult individual, however, has a constant daily mortality risk (m_{imago}=5.1%) that is independent of its age. The daily oviposition rate of adults depends on temperature ($Eggs(T)=0.1\times(10.1\times T-66.1)$, *T* daily temperature, Waltert et al. 1999) and egg laying requires temperatures higher than $T_{0,egg}$. The total reproductive output of an individual is limited (max_{egg}=100) and adults start reproduction at a minimal age (A_{min}=14 days).

The quality of the area inhabited by the population is modelled by two entities: (1) a constant carrying capacity *K* that mimics food availability and limits the daily number of adults in the population; and (2) a temperature profile that affects all life stages on each simulated day. For density regulation of the population we assume a ceiling model. Surplus adults die if an overflow in capacity occurs. They are randomly removed from the population. We use a long-term meteorological data record from the weather station "Bamberg" to model the effects of daily temperatures on population development (Griebeler and Gottschalk 2000a). This weather station is located in a distance of about 25 km from both study areas. At each simulated day, a temperature value is sequentially taken from a database that contains the daily mean temperatures registered for this meteorological station.

2.4 Computer Simulations

We performed three major simulation studies. In the first study, we tested whether differences in temperature conditions and food limitation account for the observed differences in population development between Prappach and Hammelburg. Therefore, by applying our model we generated population dynamics for each area over the field period between 1993 and 1999. Concerning temperature conditions, we directly applied the daily mean temperatures of the meteorological station Bamberg recorded between 1993 and 1999 for Prappach (Griebeler and Gottschalk 2000a). For Hammelburg, we added on each simulated day 4.8°C to the value obtained from this database reflecting the higher temperatures found in our field studies (Gottschalk et al. 2003). For food limitation, we assumed different carrying capacities. Capacities implementing different levels of food availability were 100, 200, 300, 400, 500, 750, 1,000, 2,000, 3,000, 4,000, 5,000, 7,500 and 10,000 adults per day. All simulations were randomly started for Prappach on 26th July 1993 with

1,218, 1,424, or 1,630 adults or for Hammelburg on 24th July 1993 with 598, 676, or 754 adults. These initial population sizes were chosen according to the population sizes estimated on these dates in the field and the respective estimation errors of sizes (Prappach: ±206, Hammelburg: ±78). For each of the capacities considered per study area, we repeated the Monte Carlo simulation of population size 3,000 (=3 initial sizes × 1,000 simulations) times. In total, we investigated 26 (=2 areas × 13 capacities) simulation scenarios. For each of these simulation scenarios, we annually recorded population sizes in each simulation run.

To derive a more general understanding of how thermal conditions and food availability affect populations of the grey bush cricket and to rate the impact of global warming on this species, we performed a second simulation study. With this study, we aimed to estimate the extinction risk of populations inhabiting areas of different food availability with different temperature profiles. Starting with temperature conditions as observed at Prappach, we added on each simulated day 2, 4, 4.8°C (Hammelburg), 6.8 or 8.8°C to the respective temperature value of the meteorological database. For all temperature profiles we considered 15 carrying capacities (100, 250, 500, 750, 1,000, 2,000, 3,000, 4,000, 5,000, 7,500, 10,000, 15,000, 20,000, 25,000 and 30,000) implementing food availability. All 90 simulation scenarios (=6 temperature profiles × 15 capacities) were started on 1st July. The initial population consisted of adults and its size equalled the respective carrying capacity. For the annual temperature course, we applied mean daily temperatures of the meteorological station Bamberg from 1949 to 1985. For each simulation scenario, we estimated the extinction probability and the mean number of adults based on 1,000 Monte Carlo simulations over 35 years (Griebeler and Gottschalk 2000a). From these estimates, we derived minimum viable population sizes (MVP) for each temperature scenario according to Shaffer (1987). If we assume an extinction probability of 2% for a 35-year time span, this MVP definition conforms to Shaffer's (1987) limit of 5% in 100 years.

In the last simulation study (Table 1), we performed a standard sensitivity analysis for each of the mortalities assumed for life stages (m_{egg}, m_{larva}, m_{adult}) and the assumed end of egg diapause (D_{end}). We aimed to rate their individual impact on both the extinction probability and the mean annual number of adults over 35 years. Changes in habitat temperature conditions expected under global warming are likely to affect mortalities (e.g. by altering the probability at which the cricket is predated) or the termination of diapause (because eggs need no frost as stimulus for hatching and larvae hatch earlier at higher temperatures, Ingrisch 1985). Simulations were performed as in the second simulation study, with the exception that we considered only the temperature conditions of Prappach and Hammelburg. Initial population sizes and capacities were 30,000 adults for Prappach and 3,000 adults for Hammelburg, because we expected low extinction rates for these capacities and default values of mortalities and egg diapause (see Results, Table 1). Again for each simulation scenario, we estimated the extinction probability and the mean annual number of adults based on 1,000 Monte Carlo simulations over 35 years.

Table 1 Mortalities (m_{egg}, m_{larva}, m_{adult}) and ends of egg diapause (D_{end}) studied in the third simulation study

Parameter	
m_{egg}	{0, 0.1, 0.2, 0.3, 0.4, 0.5, **0.6**, 0.7, 0.8, 0.9, 1}
m_{larva}	{0, 0.1, 0.2, 0.3, 0.4, 0.5, 0.6, 0.7, 0.8. **0.9**, 0.925, 0.95, 0.975, 1}
m_{adult}	{0, 0.01, 0.02, 0.03, 0.04, **0.051**, 0.06, 0.07, 0.08, 0.09, 0.1, 0.15, 0.2}
D_{end}	{**0**, 1, 2, 3, 4, 5, 6} weeks before 31st March

Bold values are defaults assumed in the model of Griebeler and Gottschalk (2000a)

3 Results

3.1 Population Dynamics at Prappach and Hammelburg Between 1993 and 1999

The trend in annual population size estimated by our model for Prappach was in overall agreement with field data when carrying capacities between 1,000 and 10,000 adults were assumed (Spearman rank correlation r_s between observed and predicted sizes: $r_s=0.786$, $N=7$, $p=0.05$ for all capacities between 1,000 and 10,000, Fig. 1). Population sizes steadily increased from 1993 until 1995, and the maximum size was reached in 1995. In the following 4 years, the sizes steadily decreased again, but the population never went extinct. The decline between 1995 and 1999 was much weaker in the simulations than observed in the field. Contrary to the natural population, the population frequently went extinct within the simulated time period for carrying capacities (K) below 1,000 adults (extinction probabilities p_{ext}; $K \leq 400$: $p_{ext}=1$, $K=500$: $p_{ext}=0.644$, $K=750$: $p_{ext}=0.216$).

The rather constant population sizes observed at Hammelburg were only reproduced by the model for capacities between 200 and 500 adults (marginal significant Spearman rank correlation r_s between observed and predicted sizes due to the smaller number of years, Sachs 1992: $r_s=0.600$, $N=6$, $p=0.15$ for capacities between 200 and 500 adults, Fig. 1). The simulated course in population size became similar to the course observed at Prappach for capacities higher than 500 adults. For these capacities, a maximal size was observed in 1995 (Fig. 1). However, the population always went extinct for the lowest capacity studied ($K=100$).

3.2 Effect of Temperature Conditions on the Extinction Risk of the Species

Warmer temperature conditions resulted in lower extinction probabilities of populations, higher mean annual numbers of adults and lower capacities for MVP values. The decrease in extinction probability (p_{ext}) with increasing carrying capacities (K) followed Hanski's (1994) model ($p_{ext}(K)=\mathrm{e}K^{-\mathrm{b}}$, where $\mathrm{e} \in [0,1]$ and $\mathrm{b} \in [0,\infty]$, for all temperature profiles: $R^2 \geq 0.972$, $N=15$) for all temperature profiles studied,

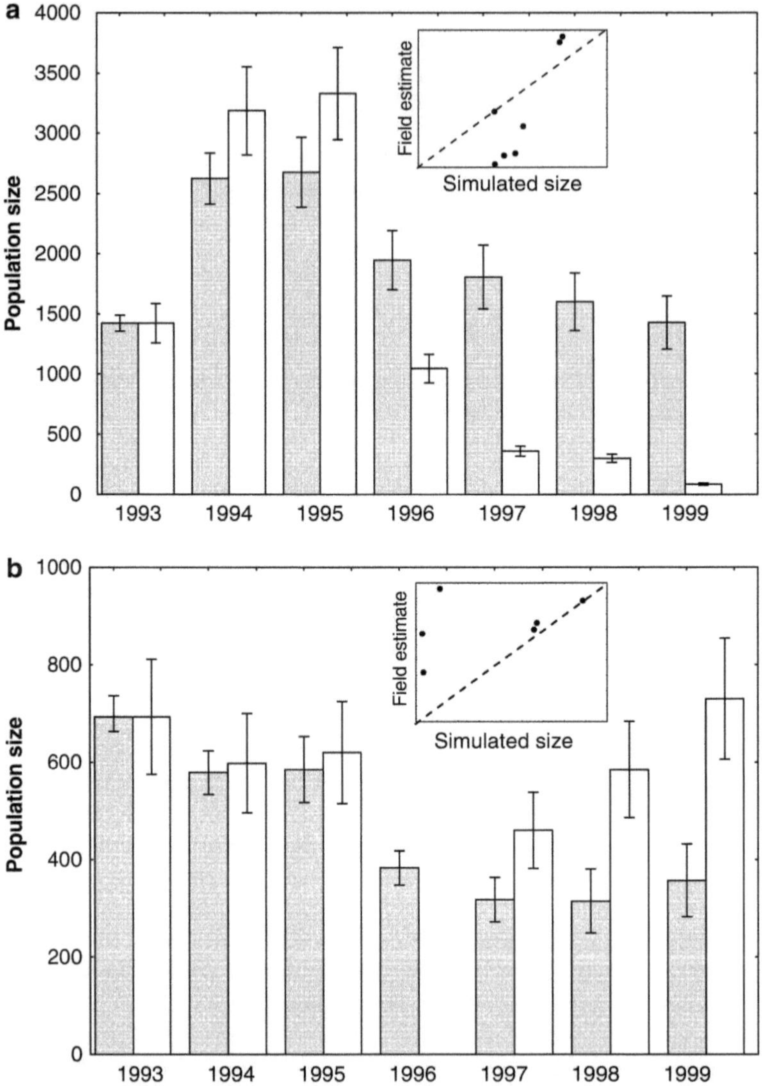

Fig. 1 Simulated population sizes (*dotted bars*) and field estimates (*open bars*) at Prappach (**a**) and at Hammelburg (**b**) as well as correlations between simulated sizes and field estimates (*smaller plots*) for each study area. In 1996, population size at Hammelburg was not estimated because of logistical limitations, but the population was not extinct. The carrying capacities assumed in the simulations were 10,000 adults for Prappach and 300 adults for Hammelburg. The error bars give standard deviations of the mean population size estimated in field and in 3,000 Monte Carlo simulations, respectively

whereas the mean annual number of adults linearly increased with increasing capacities for all temperature profiles (for all temperature profiles: $R^2 \geq 0.999$, $N = 15$). The minimum viable population sizes obtained for each temperature profile (the mean annual number of adults observed for the capacity that resulted in an extinction probability of 2%) exponentially decreased with increasing temperature conditions ($R^2=0.993$, $N=6$). The minimum viable population size was found for $K=30{,}000$ adults (mean annual number of adults = 8,341) under temperature conditions as observed at Prappach and for $K=3{,}000$ adults (mean annual number of adults = 1,828) under temperature conditions as at Hammelburg.

3.3 Effect of Mortalities and Shifts in the Termination of Egg Diapause

Changes in mortalities assumed for eggs, larvae and adults strongly affected extinction probabilities and the mean annual number of adults under both temperature scenarios. For all life stages and both temperature conditions studied, the extinction probability showed a logistic increase with increasing mortalities, whereas mean annual numbers of adults logistically decreased (Fig. 2, Table 2). Under temperature conditions as observed at Prappach, the strength in the shift in the termination of diapause towards the beginning of the year linearly decreased the extinction probabilities, whereas the mean annual population size hyperbolically increased for earlier terminations (Fig. 2, Table 2). In contrast, under conditions as observed at Hammelburg, an earlier termination of diapause had no effect on the extinction probability whereas size decreased in a quadratic manner (Fig. 2, Table 2). However, for both populations the latter effects on population size and extinction probability were very small.

4 Discussion

4.1 Differing Population Dynamics at Prappach and Hammelburg

Based on our simulation results, we were able to show that differences in temperature conditions and food limitation account for the differences in population development observed between Prappach and Hammelburg. This hypothesis had been proposed by Gottschalk et al. (2003) based on field observations and laboratory experiments. The trend observed in population development at Prappach was best explained by the model assuming weak food limitation ($K_{\text{Prappach}} > 1{,}000$, higher carrying capacities that implement food availability). Rather constant population sizes as observed at the warmer Hammelburg area were only obtained under strong food limitation ($200 \leq K_{\text{Hammelburg}} \leq 500$, small carrying capacities).

Population size at Prappach is mainly driven by annual temperature conditions, but density regulation by shortage of food is weak. The more unfavourable thermal conditions allow for lower reproduction than at Hammelburg (Köhler 1999; Waltert et al. 1999) and, thus, in colder years to a poorer compensation of reductions in population size (between 1996 and 1999) than at Hammelburg. Introducing a stronger

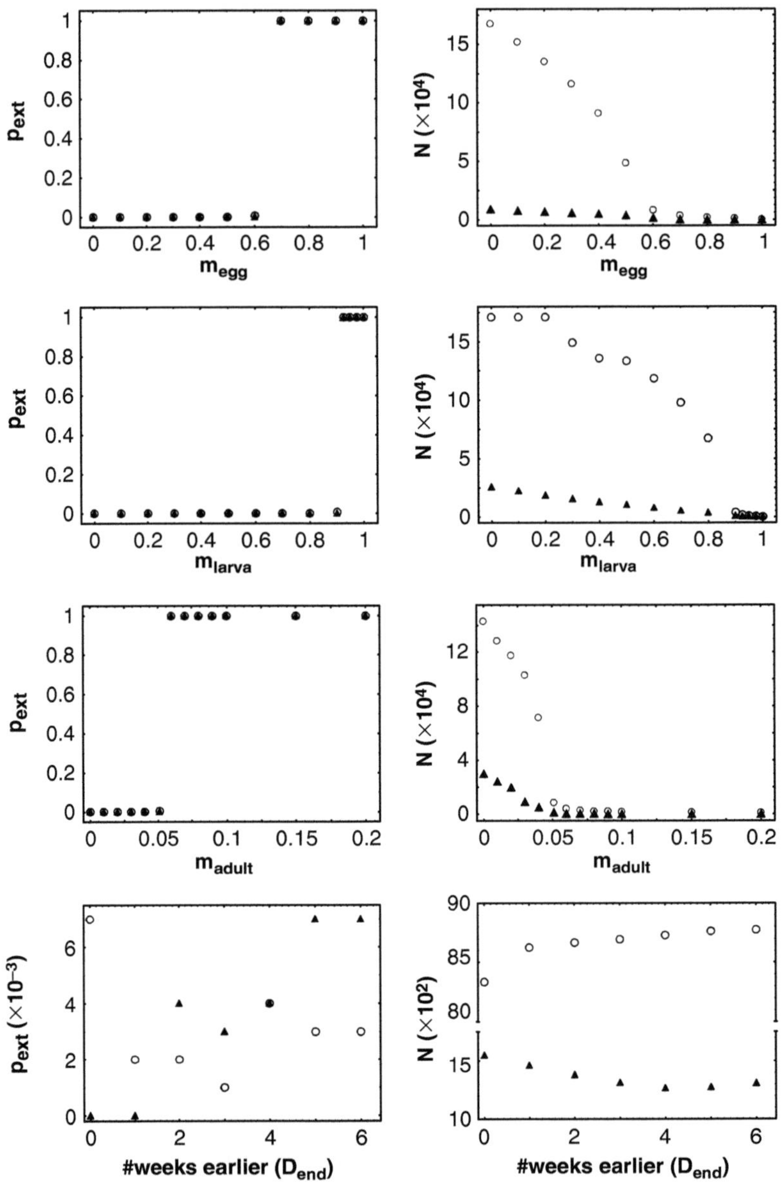

Fig. 2 Results of the sensitivity analysis performed for life stage mortalities (m_{egg}, m_{larva}, m_{adult}) and the assumed end of egg diapause. Extinction probabilities (p_{ext}) and mean annual numbers of adults (N) were estimated for temperature conditions as observed at Prappach (*open dots*) and Hammelburg (*filled triangles*). Capacities and initial population sizes assumed were 3,000 adults for Hammelburg and 30,000 adults for Prappach. Tested model parameter values are summarized in Table 1. Regression functions fitted are given in Table 2

Table 2 Results of regression analysis performed for the sensitivity analysis of life stage mortalities (m_{egg}, m_{larva}, m_{adult}) and the assumed end of egg diapause (D_{end})

Site	Parameter	Estimate	Regression model	N	R^2
Prappach	m_{egg}	p_{ext}	Logistic	11	0.999
	m_{larva}	p_{ext}	Logistic	14	0.999
	m_{adult}	p_{ext}	Logistic	13	0.999
	D_{end}	p_{ext}	Linear	7	0.938
Hammelburg	m_{egg}	p_{ext}	Logistic	11	0.999
	m_{larva}	p_{ext}	Logistic	14	0.806
	m_{adult}	p_{ext}	Logistic	13	0.999
	D_{end}	p_{ext}	Linear	7	n.s.
Prappach	m_{egg}	$\bar{N}$	Logistic	11	0.996
	m_{larva}	$\bar{N}$	Logistic	14	0.983
	m_{adult}	$\bar{N}$	Logistic	13	0.995
	D_{end}	$\bar{N}$	Hyperbolic	7	0.990
Hammelburg	m_{egg}	$\bar{N}$	Logistic	11	0.991
	m_{larva}	$\bar{N}$	Logistic	14	0.993
	m_{adult}	$\bar{N}$	Logistic	13	0.977
	D_{end}	$\bar{N}$	Quadratic	7	0.944

Regression functions applied for extinction probabilities (p_{ext}): logistic: $p_0(x)=\frac{1}{1+\exp(-(a+bx))}$, and linear: $p_0(x)=ax+b$; for mean annual numbers $\bar{N}(x)=\frac{c}{1+\exp(-(a+b(1-x)))}$, hyperbolic: $\bar{N}(x)=\frac{(c-a)x}{b+x}+a$, and quadratic: $\bar{N}(x)=a(x-b)^2+c$; where a, b, c denote parameters of fitted functions and x mortalities or ends of diapause, respectively. Tested parameter values are summarized in Table 1. Capacities and initial population sizes assumed were 3,000 adults for Hammelburg and 30,000 adults for Prappach

food limitation ($K_{Prappach} < 1{,}000$) that increases overall mortalities of juveniles and adults resulted in frequent extinctions of the population due to the lower reproductive output of the population in this colder habitat. However, the population sizes in 1997, 1998 and 1999 were strongly overestimated by the model which might be attributed to increased mortalities of life stages in the natural population due to the cool and wet weather in these years (Richards and Waloff 1954; Ingrisch and Köhler 1998). Such raises in mortalities due to precipitations are not implemented in the model, but have a significant detrimental effect on population size as our third simulation study predicts (Fig. 2, Table 2). Furthermore, indirect effects of the weather such as changes in vegetation that reduce food availability also have a negative influence on population size, which is omitted in the model.

In contrast, population size at Hammelburg is strongly controlled by food. The favourable temperature conditions at this site allow in general for a higher reproduction rate than at Prappach (Waltert et al. 1999). While the reproductive output of the population at Prappach is mainly weather-driven (not limited by food because of high abundance of nutritious seeds, high carrying capacity), the reproductive output at Hammelburg is more constant. It does not strongly increase in warmer years due to low food abundance and does not intensively decrease during colder years due to the more favourable microclimate. The more even and higher production of offspring caused by warmer conditions at Hammelburg stabilizes population dynamics, leads to more constant but lower annual population sizes and allows a faster recovery of population size even in unfavourable years (1997, 1998 and 1999). A reduction in food limitation (higher carrying capacities) increased mean annual population sizes, which were predicted for Hammelburg. In this situation, population development was driven by annual temperature conditions resulting in stronger annual fluctuations of population size. However, conversely to Prappach, our model underestimated population sizes at Hammelburg in the last 3 years. This underestimation is explainable by individual habitat choice of animals in heterogeneous habitats. *P. albopunctata*, like several other orthopteran species (Ehrlinger 1995; Kindvall 1996), prefers areas within habitats with more favourable microclimatic conditions if ambient environmental conditions are worse (Gottschalk 1996). This behaviour increases the individual reproductive output.

Although the rate of detecting density dependence is strongly affected by the length of the time series (Hassel et al. 1989; Woiwood and Hanski 1992; Holyoak 1993), various authors have successfully tested that fluctuations in insect population size can resemble density-independent random walks as observed at Prappach (e.g. Thomas et al. 1994; Whittaker and Tribe 1998) or density-dependent walks as found at Hammelburg (e.g. Alyokhin et al. 2005; Kölzsch et al. 2007). Even for orthopteran species, correlations between temperature and population densities have been reported in the literature (Kemp and Onsager 1986; Haes et al. 1990).

4.2 Extinction Risk of Our Two Natural Populations Under Current Thermal Conditions

Based on our simulations, we can conclude that the survival of none of the two natural populations is secured. The minimum viable size was found for a carrying capacity of 3,000 individuals under temperature conditions as observed at Hammelburg (1,828 adults) and was 10-fold higher under conditions as observed at Prappach (8,341 adults, Griebeler and Gottschalk 2000a). However, the mean population size estimated between 1993 and 1999 was 1,390 adults at Prappach and 614 adults at Hammelburg (Gottschalk et al. 2003), which are both below these limits. Since both study areas are surrounded by a number of additional local populations (Griebeler and Gottschalk 2000b; Gottschalk 1996) and the species is able to capture large distances (maximal observed distance is 2–3 km (Walter and Gottschalk 1998), the protection of these complexes as metapopulations might increase species survival at both study areas. An estimation of the total population size at the Prappach region (6,835 adults) that was calculated by Griebeler and Gottschalk (2000b) is still below the mean annual number of adults derived for a carrying capacity of 30,000 adults (8,341 adults). This suggests that additional conservation measures (grazing or mowing) are needed in this region to enable species persistence. Grazing and mowing improve temperature conditions and cause a shortage of food which stabilizes population dynamics and results in smaller, but safer populations as observed for Hammelburg.

4.3 Survival of P. albopunctata Under Climate Change

Our simulations demonstrate that thermal conditions in the habitat have a severe impact on the extinction risk of the grey bush cricket. The decreasing extinction risk with higher temperatures is caused by the higher reproductive output of adults under improved temperature conditions (Köhler 1999; Waltert et al. 1999). Based on this observation, one might conclude that *P. albopunctata* will be able to successfully survive global environmental warming. However, increasing mortalities, which are realistic for increasing precipitation (Richards and Waloff 1954; Ingrisch and Köhler 1998), will highly influence the population sizes and extinction probabilities as predicted by our model (Fig. 2, Table 2). A slight increase in mortalities assumed for eggs, larvae and adults leads to a rapid extinction of the population under both temperature conditions studied (Fig. 2, Table 2).

How far biotic conditions for thc bush cricket will change in the future is unpredictable, as they range from intra-specific interactions (competition for food) to multi-species interactions (e.g. inter-specific competition, predation). Based on our model, we suggest that if mortalities caused either by abiotic or biotic factors will decrease, this will favour survival of *P. albopunctata* under increasing temperatures whereas an increase in mortalities is detrimental for survival. However, *P. albopunctata* expanded the inhabited area in the Prappach region by colonizing meadows with quite dense vegetation (E. Gottschalk, pers. observation) during warm years. An increase in the potential habitat area under a warmer climate could be predicted based on this observation.

Several studies have demonstrated changes in temporal patterns of species caused by increasing temperatures (Sparks and Carey 1995; Roy and Sparks 2000). Since *P. albopunctata* eggs do not need frost as stimulus for hatching, and larvae hatch earlier at higher temperatures (Ingrisch 1985), increased spring temperatures may shift egg hatching towards the beginning of the year. Our simulation results show that this shift in the phenology has a negligible small effect on both the extinction probability and the size of the two populations if life stage mortalities are not altered. The adult mortality observed at both study areas which we assumed in our model (5.1%) results in about 20 days for mean adult life time. This limitation inhibits that the population expanded its phenology towards the warmer autumn in our simulation, but whether an elongation is possible ultimately depends on future intra- and inter-specific interactions influencing adult mortality.

Future temperatures as well as availability and distribution of rainfall will also affect food availability for the species. Increasing precipitation and temperatures affect vegetation and might reduce food availability. A shortage of food (decreasing *K*) will increase the risk of population extinction independent of temperature conditions assumed as predicted by our model. Whether a detrimental effect of precipitation and temperature on the availability of food might be compensated by the positive effect of increasing temperatures on reproduction needs further empirical and theoretical studies.

5 Conclusions

This study shows that simulation models are able to support conservation biologists in their understanding of processes that affect population extinction under current and future habitat conditions, and that models can also help to develop conservation measurements. Our example of different population dynamics generated by differences in temperature conditions and food availability (vegetation cover) demonstrates a general problem of obtaining estimates for MVPs for any temperature-dependent species. The MVP is not a pure species-specific value; it also severely depends on habitat conditions, which influence population dynamics. Forecasting species survival based on data of any population without adapting the specific local conditions will greatly impair prognosis and might lead to accidental extinction of the species.

Relict species, however, may react more pronouncedly to changes in habitat conditions caused by anthropogenic land use or climate change than our model species *P. albopunctata*. While relict species are also often restricted to patchy habitats, their habitats are scarcer than those of the grey bush cricket and they are often more specialized than the cricket. If the habitats of relict species are altered or destroyed due to global change, populations will be much more endangered than those of any other species that has a broader range distribution, shows a higher ecological and possibly genetic plasticity and may find other suitable habitats by dispersing.

Acknowledgements We are grateful to two anonymous reviewers and Dana Berens who provided helpful comments on the manuscript.

References

Alyokhin A, Drummond FA, Sewell G (2005) Density-dependent regulation in populations of potato-colonizing aphids. Popul Ecol 47:257–266

Cannon RJC (1998) The implications of predicted climate change for insect pests in the UK, with emphasis on non-indigenous species. Glob Chang Biol 4:785–796

DeAngelis DL, Mooij WM (2005) Individual-based modeling of ecological and evolutionary processes. Annu Rev Ecol Evol Syst 36:147–168

Ehrlinger M (1995) Untersuchung der kleinräumigen Wanderung von Heidegrashüpfern (*Stenobothrus lineatus*) zwischen unterschiedlich exponierten Halbtrockenrasen. Artenschutzreport 5:11–15

Gottschalk E (1996) Population vulnerability of the grey bush cricket *Platycleis albopunctata* (GOEZE, 1778) (Ensifera: Tettigoniidae). In: Settele J, Margules CR, Poschlod P, Henle K (eds) Species Survival in Fragmented Landscapes. Kluwer, Doordrecht, pp 275–281

Gottschalk E, Griebeler EM, Waltert M, Mühlenberg M (2003) Population dynamics in the grey bush cricket *Platycleis albopunctata* (Orthoptera: Tettigoniidae) – What causes interpopulation differences? J Ins Conserv 7:45–58

Griebeler EM, Gottschalk E (2000a) The influence of temperature model assumptions on the prognosis accuracy of extinction risk. Ecol Modell 134:343–356

Griebeler EM, Gottschalk E (2000b) An individual based model of the impact of suboptimal habitat on survival of the grey bush cricket, *Platycleis albopunctata* (Orthoptera: Tettigniidae). J Ins Conserv 4:225–237

Haes ECM, Cherrill AJ, Brown VK (1990) Meteorological correlates of wart-biter (Orthoptera: Tettigoniidae) abundance. Entomologist 109:93–99

Hanski I (1994) A practical model of metapopulation dynamics. J Anim Ecol 63:151–162
Harz K (1957) Die Geradflügler Mitteleuropas. Gustav Fischer Verlag, Jena
Harz K (1969) Die Orthoptheren Europas I Ensifera. Junk, The Hague
Hassell MP, Latto J, May RM (1989) Seeing the wood for the trees: detecting density dependence from existing life table studies. J Anim Ecol 58:883–892
Holyoak M (1993) New insights into testing for density dependence. Oecologia 95:435–444
Ingrisch S (1979) Untersuchungen zum Einfluß von Temperatur und Feuchtigkeit auf die Embryogenese einiger mitteleuropäischer Laubheuschrecken (Orthoptera: Tettigoniidae). Zool Beitr N F 25:343–364
Ingrisch S (1985) Effects of hibernation length on termination of diapause in European *Tettigoniidae* (Insecta: Orthoptera). Oecologia 65:376–381
Ingrisch S (1986) The plurennial life cycles of the European *Tettigoniidae*. 1. The effect of temperature on embryonic development and hatching. Oecologia 70:606–616
Ingrisch S, Köhler G (1998) Die Heuschrecken Mitteleuropas. Westarp Wissenschaften, Magdeburg
Kemp WP, Onsager JA (1986) Rangeland grasshoppers (Orthoptera: Acrididae): modelling phenology of natural populations of six species. Environ Entomol 15:924–930
Kindvall O (1996) Habitat heterogeneity and survival in a bush cricket metapopulation. Ecology 77:207–214
Kleukers RMJC, Decleer K, Haes ECM, Kolshorn P, Thomas B (1996) The recent expansion of *Conocephalus discolor* (Thunberg) (Orthoptera: Tettigoniidae) in Western Europe. Entomologist's Gazette 47:37–49
Köhler G (1999) Ökologische Grundlage von Aussterbeprozessen. Fallstudien an Heuschrecken (Caelifera und Ensifera). Laurenti-Verlag, Bochum
Kölzsch A, Sæther SA, Gustafsson A, Fiske P, Höglund J, Kålås JA (2007) Population fluctuations and regulation in great snipe: a time-series analysis. J Anim Ecol 76:740–749
Richards OW, Waloff N (1954) Studies on the biology and population dynamics of British grasshoppers. Anti-Locust Bull 17:1–182
Roy DB, Sparks TH (2000) Phenology of British butterflies and climate change. Glob Chang Biol 6:407–416
Sachs L (1992) Angewandte Statistik. Anwendung statistischer Methoden. Springer, Berlin
Sergeev MG (1998) Conservation of orthopteran biological diversity relative to landscape change in temperate Eurasia. J Insect Conserv 2:247–252
Shaffer M (1987) Minimum viable populations: coping with uncertainty. In: Soulé M (ed) Viable populations for conservation. Cambridge University Press, Cambridge, pp 69–89
Sparks TH, Carey PD (1995) The response of species to climate over the centuries: an analysis of the Marsham phenology record 1736–1947. J Ecol 83:321–329
Thomas JA, Moos D, Pollard E (1994) Increased fluctuations of butterfly populations towards the Northern edges of species' range. Ecogeography 17:215–220
Wagner G, Berger U (1996) A population vulnerability analysis of the red winged grasshopper, *Oedipoda germanica* (Caelifera: Acrididae). In: Settele J, Margules CR, Poschlod P, Henle K (eds) Species survival in fragmented landscapes. Ulmer Verlag, Stuttgart, pp 312–319
Walter R, Gottschalk E (1998) *Platycleis albopunctata* (Goeze 1778) Westliche Beißschrecke. In: Detzel P (ed) Die Heuschrecken Baden-Württembergs. Ulmer Verlag, Stuttgart, pp 253–260
Waltert M, Gottschalk E, Mühlenberg M (1999) Zur Nahrungsökologie der Westlichen Beißschrecke (*Platycleis albopunctata* GOEZE 1778). Articulata 14(1):17–29
Whittaker JB, Tribe NP (1998) Predicting numbers of an insect (*Neophilaenus lineatus*: Homoptera) in a changing climate. J Anim Ecol 67:987–991
Widgery J (2000) Wildlife reports: grasshoppers and relatives. Br Wildl 11:210–211
Woiwood IP, Hanski I (1992) Patterns of density dependence in moths and aphids. J Anim Ecol 61:619–629

Is the 'Lost World' Lost? High Endemism of Aphibians and Reptiles on South American Tepuís in a Changing Climate

Dennis Rödder, Andreas Schlüter, and Stefan Lötters

Abstract The Pantepuí region of the Guyana Shield of Northern South America harbors a unique biota. It is characterized by high local species endemism, especially applicable to its table mountain-like outcrops, the tepuís. For this region, also known as the 'Lost World', different hypotheses have been proposed to explain its Biogeography. We suggest that the amphibian and reptile communities' suit well as target group to study the hypotheses since these vertebrates are relatively well documented through comprehensive species lists and are less mobile than other terrestrial vertebrates. However, records for the majority of endemic amphibian and reptile species naturally numbered only a few. We hence used extant dimensions of the vegetation zones associated to amphibians and reptiles. By running maximum entropy-based Climate Envelope Distribution Models (CEDM), these were projected to climatic scenarios as suggested for the Last Glacial Maximum (LGM, 21,000 years BP). Our results promote the Cool Climate hypothesis for diversity and distribution of amphibians and reptile communities in the Pantepuí region, in which species' geographic ranges were displaced followed by speciation along with the climate change during glacial-interglacial Quaternary cycles. Moreover, we studied the impact of expected future anthropogenic global warming on these vegetation zones. Climate change scenarios used, consider demographic, politico-economic, social and technological futures at different degrees with either more or less environmentally conscious, regionalized solutions to economic, social and environmental sustainability. For the 'Lost World', our CEDM results demonstrated

---

D. Rödder (✉) and S. Lötters
Department of Herpetology, Zoologisches Forschungsmuseum Alexander Koenig, Adenauerallee 160, D-53113 Bonn, Germany
Department of Biogeography, Trier University, Am Wissenschaftspark 25–27, D-54286 Trier, Germany
e-mail: roedder@uni-bonn.de; loetters@uni-trier.de

A. Schlüter
Department of Zoology, Staatliches Museum für Naturkunde Stuttgart, Rosenstein 1, D-70191 Stuttgart, Germany
e-mail: schlueter.smns@naturkundemuseum-bw.de

J.C. Habel and T. Assmann (eds.), *Relict Species: Phylogeography and Conservation Biology*,
DOI 10.1007/978-3-540-92160-8_24,

a strong impact of climate change to vegetation zones resulting in a considerable decrease during future anthropogenic global warming in terms of habitat suitability. For amphibians and reptiles this may mean habitat loss and decline or extinction. However, nothing is known about specific climatic niche breadths (which may be great due to survival under climatic conditions with considerable daily change) or the possibility to adapt over generations (which may be unlikely due to the rapid warming expected) of Pantepuí amphibians and reptiles. Further studies and monitoring efforts are necessary towards a better understanding of the problem and to develop conservation action plans.

1 Towards an Understanding of the Biogeography of the 'Lost World'

One of the most famous fantasy adventure novels has been Sir Arthur Conan Doyle's 1912 'The Lost World'. Purely science fiction, the geological formation of this story finds its paradigm in the real world – the Precambrian Guyana Highlands in South America. This region, which is also known as the Pantepuí harbors an extraordinarily high biodiversity. Pantepuí is situated on the Precambrian Guyana Shield, which had been separated from the African Shield by a continental drift around 80–100 Myr ago (Edmond et al. 1995). Endemism rates are especially remarkable in both plants and animals and at different taxonomic levels (Berry et al. 1995). For examples, in amphibians and reptiles, two well surveyed vertebrate groups in this region, 77.3% (i.e. 75 of 97) and 54.8% (i.e. 34 of 62), respectively, of the known species are endemics (McDiarmid and Donnelly 2005). One reason for this is the unique geomorphology of this region, characterized by isolated table mountains, the tepuís (Fig. 1.). These are remnants of ancient erosion surfaces that have been isolated during the Gondwana breakup, accompanied by the formation of major river systems (Briceno and Schubert 1990; Huber 1994; Edmond et al. 1995). Today, the Pantepuí region constitutes a singular biogeographic province of about 5,000 km^2, which is split into numerous topographical islands, each about 1,500–3,000 m above sea level (Fig. 2; Huber 1994).

Different aspects of the Pantepuí region have been of particular interest to various writers (e.g. Hoogmoed 1979; Duellman 1999; Gorzula and Señaris 1998; McDiarmid and Donnelly 2005; Rull 2005a,b; Señaris and MacCulloch 2005; MacCulloch et al. 2007; Rull and Nogué 2007) and Pantepuí region has been suggested to be well suitable to study biogeographic and evolutionary theories (Rull 2005a,b, 2007). In a recent synopsis, McDiarmid and Donnelly (2005) summarized the different hypotheses trying to explain diversity patterns within the region. The Distance Dispersal hypothesis (1) was derived from observations on the Pantepuí bird fauna by Mayr and Phelps (1967). These authors proposed that some Andean bird species reached the tepuís crossing the valleys of the Orinoco and Negro Rivers and subsequently became established at some of the Western tepuís. Further dispersal through 'island hopping' across the Pantepuí to the Eastern massifs is suggested

Fig. 1 Impressions from the 'Lost World' or Pantepuí region of South America: Cerro Guaiquinima (**a**) and El Venado Tepuí (**b**). Note the strict distinctness in vegetation between basement and summit of these table mountains, here referred to as 'upland' and 'tepuí', respectively. Photos: A. Schlüter

to have taken place, followed by speciation. The Distance Dispersal hypothesis assumes a high mobility of organisms and is therefore only applicable to few organisms. The Habitat Shift hypothesis (2) was introduced by Mayr and Phelps (1967) to explain for more than 36% of the Pantepuí avifauna, and Hoogmoed (1979) further suggested that it may also be applicable to the majority of amphibians and reptiles living at higher elevations. Basically, the Habitat Shift hypothesis assumes that elements of the Pantepuí fauna derived from lowland ancestors that adapted to intermediate and highland habitats and subsequently formed distinct species. Maguire (1970) and Brewer-Carías (1978) suggested that biodiversity, especially the high degree of endemism, in the Guyana Highlands is the result of a long history of evolution in isolation over several geological eras: Lost World hypothesis (3). In a contrasting view, alternating up- and downward displacement during glacial-interglacial Quaternary cycles has been proposed (e.g. Chapman 1931; Tate 1939; Mayr and Phelps 1967; Steyermark and Dunsterville 1980; Rull 2004a,b, 2005a,b). According to this Cool Climate hypothesis (4), the main diversification

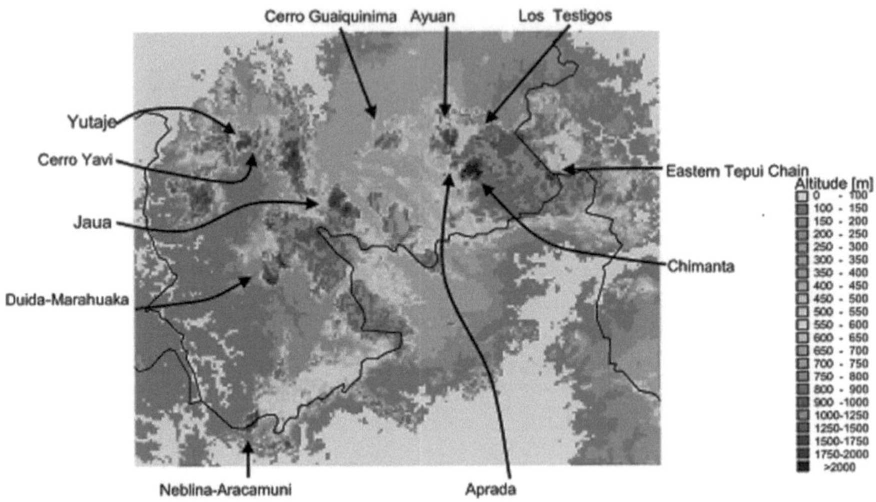

Fig. 2 Altitudinal patterns within the Pantepuí region, and location of massifs mentioned in the text

events occurred during glacial periods, when tepuían biotas migrated down-slope thus escaping from cooling. This allowed for subsequent dispersal through lowlands from one tepuí to another. Later warming again restricted the biota to tepuís and then this resulted in diversification. Palnyological data support that the entire Guyanas were affected by past climatic oscillations (e.g. Huber 1988; Rull 2004a,b), implying past vertical fluctuations of the tepuí flora and fauna.

2 How to Evaluate Hypotheses When Data Availability is Poor?

Selection criteria of appropriate study subjects may involve specific ecological traits of target groups such as niche breadth and dispersal ability. Suitable study subjects are also characterized by data availability, which is especially a problem in remote, poorly known regions. With respect to the Pantepuí region, recent zoogeographic studies have focused on the biogeography of amphibians and reptiles (i.e. herpetofauna), which are comparatively well known (McDiarmid and Donnelly 2005). Such herpetofaunal communities are perhaps better suited for biogeographic analyses than other terrestrial vertebrates since in general these animals are less mobile than birds and most mammals and may be closely tied to specific habitats (e.g. Hoogmoed 1979; McDiarmid and Donnelly 2005; MacCulloch et al. 2007). Sampling efforts largely varies among the tepuís and are hardly comparable between neighbouring tepuís. As a consequence, it may be more useful to compare herpetofaunal communities between different massifs rather than single tepuís (McDiarmid and Donnelly 2005). By doing so, the distinctness of local herpetofaunal

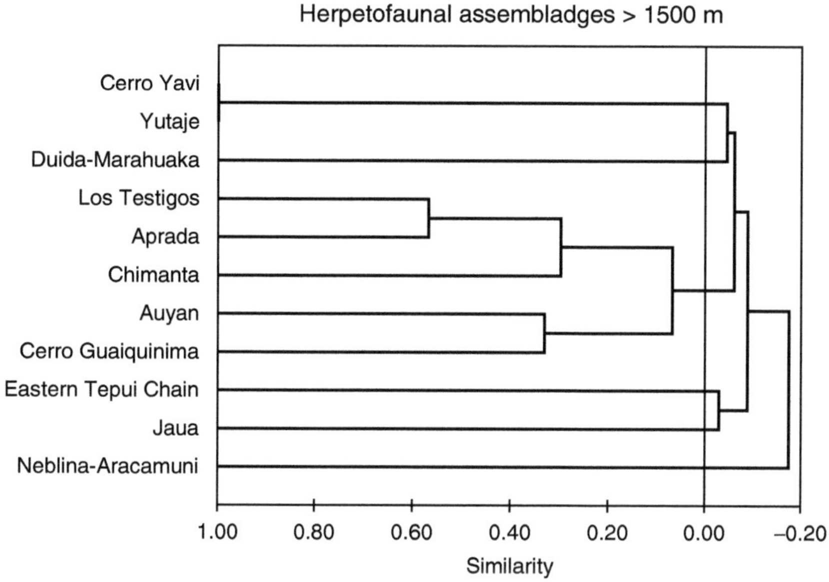

Fig. 3 Comparison of herpetofaunal assemblages at massifs in the Pantepuí region. Most herpetofaunas show limited overlap only, even when compared to neighbouring massifs. Cluster analysis was performed based on a species list of amphibians and reptiles compiled from Donnelly and Myers (1991), Mägdefrau et al. (1991), McDiarmid and Donnelly (2005), MacCulloch et al. (2007), Means and Savage (2007), Schlüter and Rödder (2007), Myers and Donnelly (2008), Rödder and Jungfer (2008)

assemblages as a result of the high degree of local endemism becomes obvious (Fig. 3.). The wide lack of overlapping faunal elements makes broad scale assessments and evaluation of hypotheses difficult. Phylogeographic analyses would be the most suitable tool for an evaluation of the hypotheses formulated above; however, DNA samples, as required, are largely unavailable.

GIS-based Climate Envelope Distribution Models (CEDMs) may provide an easy-to-use-alternative. In recent times, there have been several examples using CEDMs for species or habitat climate envelopes under past, present and future climate scenarios (e.g. Hugall et al. 2002; Bonaccorso et al. 2006; Peterson and Nyári 2008). Such approaches rely on the assumption that climatic tolerances of species are the primary determinants of their current distributions and that specific climatic niches are conservative, at least within an evolutionary short time frames of some hundreds to thousands years (e.g. Wiens and Graham 2005; but see Pearman et al. 2007).

Peterson and Nyári (2008) and Carnaval and Moritz (2008) suggested that spatial modelling of species or habitats under current and palaeoclimate regimes can generate hypotheses about the potential existence and the extent of stable (refugial) regions. On the other hand spatial models may likewise provide information on past – or future – extents of current refuges.

3 Principles of CEDMs

Prior to CEDM performance, it is necessary to compile a list of species records (i.e. geographic coordinates) and of values of climatic variables which are suitable to describe the actual range of the target species. Fortunately, GIS-maps of environmental variables help to find suitable data sets in many regions (e.g. http://www.worldclim.org). A subsequent step involves the development of a multidimensional view of the climate envelope of a species, which is a challenge given the complex nature of a species' ecological niche (Peterson and Vargas 1993). In simple words, in CEDMs as mentioned here, climatic information for the presence of species localities are summarized to an 'ideal' climatic niche for the target species which is afterwards compared to climatic conditions at localities of query, i.e. from which the presence or absence of the target species is unknown. The results are geographic maps showing different degrees of similarity (interpreted as suitability) of a region with the 'ideal' climatic niche. The selection of a moderate algorithm for the computation of a CEDM depends on the amount of distribution records available and their quality.

One of the earlier applied algorithms for the presence of species data is BIOCLIM (Nix 1986; Busby 1991), as implemented in DIVA-GIS 5.4 (Hijmans et al. 2002; http://www.diva-gis.org). It develops CEDMs by intersecting the ranges inhabited by the species along each environmental axis. More sophisticated algorithms are GARP (Stockwell and Noble 1992; Stockwell and Peters 1999), DOMAIN (Carpenter et al. 1993), Maxent (Phillips et al. 2006) and LIVES (Li and Hilbert 2008). The machine-learning Maxent (see below) is often superior to most other methods and hence becomes more and more distributed (e.g. Elith et al. 2006). If records of apparent species absence are available for modeling other algorithms, such as 'artificial neuronal networks', 'classification and regression trees', 'generalized additive models' or 'generalized dissimilarity models' can be applied. These algorithms for example are implemented in the BIOMOD tool (Thuiller 2003).

Unfortunately, the high degree of endemism in the Pantepuí region causes difficulties when using CEDMs at the species level, since a minimum amount of 10–20 distribution records pending on the algorithm applied is at least necessary (e.g. Elith et al. 2006). Looking at the herpetofaunal assemblages, 109 out of 159 (i.e. 97 amphibian and 62 reptile) species are only known from a single tepuí (McDiarmid and Donnelly 2005). As a result, even in this relatively well documented group in the region, for most species CEDMs cannot be seriously generated. Since species distributions are commonly related to specific habitats, one possibility is to perform CEDMs for habitats rather than for a single species using a random set of distribution points within a target habitat. Such an approach was well performed by Carnaval and Moritz (2008) for the identification of climatically stable refugia during the Last Glacial Maximum (LGM), 21,000 years BP, in the Brazilian Atlantic forest.

4 Reconstructing Connectivity Patterns Within Pantepuí Under LGM Climate

For CEDM calculation, MaxEnt 3.2.1 (Phillips et al. 2006; http://www.cs.princeton.edu/~shapire/maxent) was used to assess potential distributions of vegetation zones during LGM. Maxent is a machine-learning algorithm following the principles of maximum entropy (Jaynes 1957). In order to study responses in terms of potential distributions of vegetation zones in the Guyana Highlands, especially the tepuís under different climate scenarios, distribution data points are necessary for CEDM calculation. For this, we generated 10,000 randomly distributed points spanning from latitude 0.3° S to 8.7° N and longitude 68.7° W to 57.6° W with DIVA-GIS 5.4 (Hijmans et al. 2002; http://www.diva-gis.org). This spatial coverage was deliberately made larger than those occupied by the Pantepuí region to ensure that training points later used to generate CEDMs under current-day climate encompassed the full extension of the climate landscape. Altitudinal properties at each of the 10,000 points were extracted and divided into two altitudinal range classes, termed 'upland' and 'tepuí' (see McDiarmid and Donnelly 2005; MacCulloch et al. 2007): i.e. 804 points at 800–1,500 m above sea level and 65 points at ≥1,500 m above sea level. For a better understanding of the discrimination between these two terms see Fig. 1b.

Information on current climate was obtained from the WorldClim database, version 1.4, which is based on weather conditions recorded between 1950 and 2000 with grid cell resolution 2.5 min (Hijmans et al. 2005; http://www.worldclim.org). It was created by interpolation using a thin-plate smoothing spline of observed climate at weather stations, with latitude, longitude and elevation as independent variables (Hutchinson 1995, 2004).

For paleoclimate, General Circulation Model (GCM) simulations from the Community Climate System Model (CCSM) assuming a temperature decrease of approximately 4 °C for the target region were provided by R. J. Hijmans (http://www.ccsm.ucar.edu/; Kiehl and Gent 2004). Original GCM data were downloaded from the PIMP2 website (http://www.pimp2.cnrs-gif.fr/) with spatial resolution of roughly 300×300 km. Surfaces were created at 0.04° spatial resolution via the following procedure: first, the difference between the GCM output for LGM and current-day conditions was calculated. These were then interpolated to the 0.04° resolution grid using the spline function in ESRI ArcInfo with the tension option. Eventually, the interpolated difference was added to the high-resolution current-day climate data set from WorldClim and LGM bioclimatic coverage using DIVA-GIS. This procedure has the dual advantages of producing data at a resolution relevant to the spatial scale of analysis and of calibrating the simulated climate change data to the actual observed climate data.

The layers for each climate scenario included the minimum and maximum temperatures and the mean precipitation per month (=36 climatic parameters) derived from the WorldClim 1.4 interpolation model. Based on these, seven bioclimatic parameters for each climate scenario were generated with DIVA-GIS: 'annual mean temperature', 'temperature seasonality' (standard deviation of monthly mean temperature × 100), 'mean temperature of the warmest quarter', 'mean temperature of the coldest quarter', 'annual precipitation', 'precipitation of the wettest quarter' and 'precipitation of the driest quarter'. These parameters reflect the availability and range of thermal energy and humidity and are suitable for CEDM projections between different climate scenarios (e.g. Carnaval and Moritz 2008). Bioclimatic parameters were exported from DIVA-GIS as *.grd/*.gri files.

CEDMs were calculated for both altitudinal range classes separately and results are shown in Fig. 4. Maxent allows for model testing by calculation of the Area Under the Curve (AUC), referring to the ROC (Receiver Operation Characteristic) curve using e.g. 25% of the records as test points and the remaining ones for training (Hanley and McNeil 1982; Phillips et al. 2006). We acknowledge that currently there has been some doubt raised on the reliability of AUC (Lobo et al. 2008). But for a lack of a better method and because AUC has been recommended and applied for ecological applications by other authors (e.g. Pearce and Ferrier 2000; Elith et al. 2006; Phillips et al. 2006), we hereby continue using it.

AUC values range from 0.5 (i.e. random) for models with no predictive ability to 1.0 for models giving perfect predictions. According to the classification of Swets (1988), AUC values >0.9 describe 'very good', >0.8 'good' and >0.7 'useful' discrimination ability. We received 'very good' AUC values in the models for 'upland' ($AUC_{training} = 0.972$; $AUC_{test} = 0.969$) and 'tepuí' ($AUC_{training} = 0.998$; $AUC_{test} = 0.998$).

Visual comparisons of our CEDM for current conditions and elevation, within the study region confirmed high overlap (Figs. 2 and 4a). CEDMs of 'upland' and 'tepuí' under LGM conditions revealed that the respective areas were apparently much larger supporting the Cool Climate hypothesis of Rull (2005a). Temporal connections between the 'tepuís' during glacial and interglacial oscillations were present and may have allowed exchange of biota. According to our models, at least, the whole upland and the Eastern massifs, such as the Ayuan, Los Testigos, Chimanta, Aprada the Eastern Tepuí Chain the Cerro Yavi and Yutaje massifs (Fig. 2) were connected, whereby the remaining ones exhibited no or limited contact (Fig. 4b). These findings largely coincide with those of Rull and Nogué (2007), who performed a more simple assessment of connectivity patterns assuming a adiabatic temperature gradient and cooling of about 4.1°C.

A comparison of similarities between herpetofaunal assemblages inhabiting different massifs (Fig. 3.) showed that faunas at Cerro Yavi and Yutaje are mostly alike, followed by a clade summarizing Eastern massifs (Los Testigos, Aprada, Chimanta, Auyan and Cerro Guiaquinima). Duida-Marahuaka, Jaua, Neblina-Aracamuni and the Eastern Tepuí Chain are inhabited by more different assemblages. Therefore, we conclude that the connectivity patterns suggested by our palaeoclimate CEDM and hence the Cool Climate hypothesis may explain most of the similarity among herpetofaunal assemblages observed herein, but not entirely. The relatively high similarity between Cerro Guiaquinima and the other Eastern massifs (e.g. Ayuan massif) cannot be well explained with our CEDM results showing

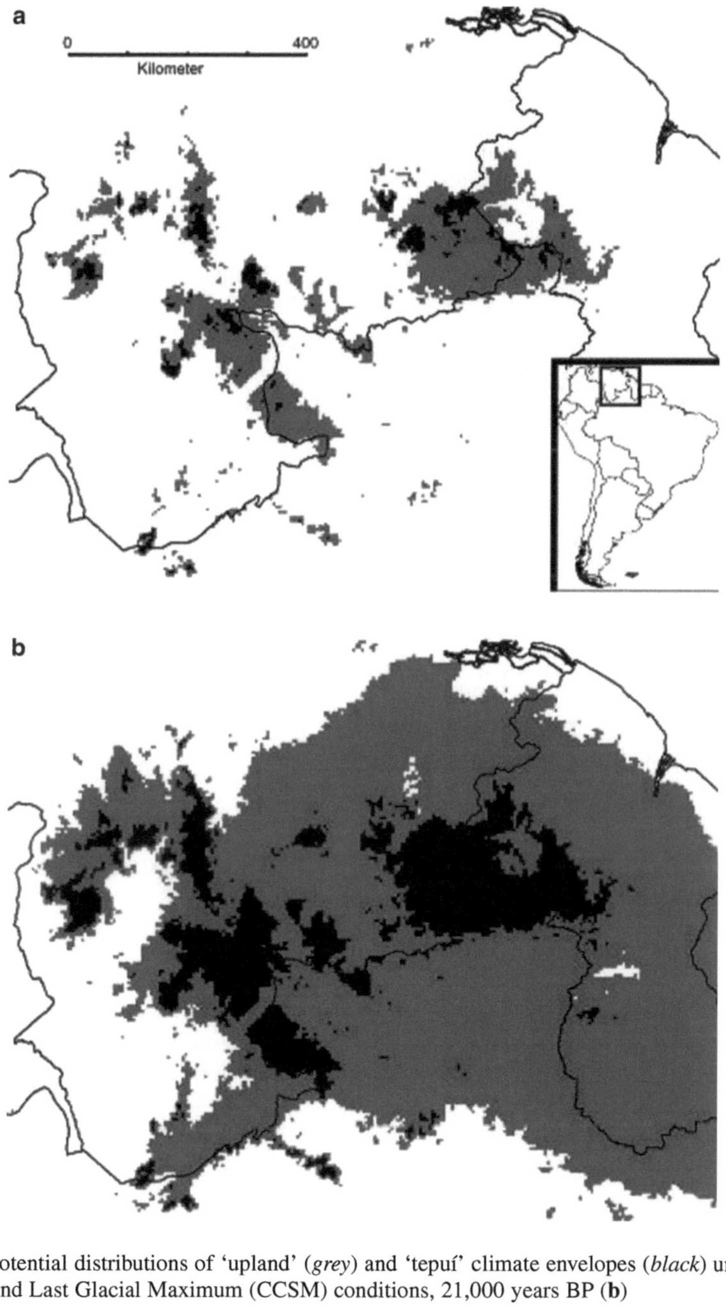

Fig. 4 Potential distributions of 'upland' (*grey*) and 'tepuí' climate envelopes (*black*) under current (**a**) and Last Glacial Maximum (CCSM) conditions, 21,000 years BP (**b**)

no connection between them. Also, the distinctness of the herpetofauna of the Eastern Tepuí Chain, which may have been connected to the other Eastern massifs under LGM conditions, is remarkable.

McDiarmid and Donnelly (2005) argued that the high level of endemic amphibian and reptile species may cast some doubt on the Cool Climate hypothesis. Our results confirm that faunal exchange between most 'tepuís' was restricted, but possible during the LGM within the Eastern massifs at least. This geographic pattern is also reflected when comparing herpetofaunal assemblages, although the high degree of endemism causes remarkable differences even between geographically close massifs (Fig. 3), e.g. the Ayuan, Chimanta and Los Testigos massifs located within a distance of less than 100 km (Fig. 2).

We expect that the extent of faunal exchange during glacial periods depended on the establishment abilities of the invaders in existing faunas. Myers and Donnelly (2001) found that some endemic species have phenotypically and ecologically sibling counterparts on other tepuís. MacCulloch et al. (2007) made similar observations regarding Mt. Roraima and adjacent tepuís. This appears to be especially applicable to frogs in the genera *Anomaloglossus, Oreophrynella* and *Pristimantis* as well as in snakes in the genus *Thamnodynastes*. These species were suggested to comprise monophyletic highland groups rather than separate invasions from lowlands (Myers and Donnelly 2001) and, considering the geological age of the Pantepuí region allowing speciation, many of them may represent sister species. These taxa occupy similar ecological niches (Myers and Donnelly 2001) which would hamper the establishment of one species at tepuís inhabited by a similar species, so much so even if 'tepuís', were connected during the glacial periods, faunal exchange may have been limited, explaining – at least in parts – the distinctness of 'tepuí' herpetofaunas.

5 Assessing Possible Impacts of Future Anthropogenic Climate Change

Next to the interesting opportunity to study relevant biogeographic and evolutionary aspects related to past climate, the unique amount of endemic taxa of Pantepuí may be of certain interest when watching through a window into the future. There is little doubt that within a few decades, the worldwide climate will become warmer due to human impact (IPCC 2007).

Efforts in global biodiversity conservation have underlined the relevance of Pantepuí as a region which deserves protection, as it is listed as one of the WWF/IUCN Neotropical Plant Diversity centres (Pantepuí or SA-2; WWF and IUCN 1997). That is why more than 70 % of the Guyana Highlands is under official protection (Berry et al. 1995). However, as one of the dilemmas of current conservation strategies, land protection alone is not operable to threats caused through future anthropogenic global warming (Rull and Vegas-Vilarrúbia 2006; Rull 2007). Rull and Vegas-Vilarrúbia (2006) tentatively estimated that one tenth to one third of suitable habitat will be subject to loss when temperature increases for 2–4°C over the forthcoming decades. As outlined above, CEDMs provide a powerful tool towards quantification and the response of species to – past and future – climate change in geographic space in terms of habitat suitability.

In order to quantify the expected impact of future anthropogenic global warming to vegetation zones defined above, we computed CDEMs in the manner as described above. For this purpose, we used climate change projections based on the CCCMA, CSIRO and HADCM3 (Flato et al. 2000; Gordon et al. 2000) models and the emission scenarios reported in the Special Report on Emissions Scenarios (SRES) by the Intergovernmental Panel on Climate Change, IPCC (http://www.grida.no/climate/ipcc/emission/). A set of different families of emission scenarios was formulated based on future production of greenhouse gases and aerosol precursor emissions. The SRES scenarios of A2a and B2a were used in this study. Each scenario described one possible demographic, politico-economic, social and technological future. Scenario B2a emphasizes more environmentally conscious, more regionalized solutions to economic, social and environmental sustainability. Compared to B2a, scenario A2a also emphasizes regionalized solutions to economic and social development, but it is less environmentally conscious. For quantitative comparisons between different projections we counted the number of grid cells classified as suitable per vegetation zone with DIVA-GIS.

Our models revealed that under global warming, potential distributions of both 'upland' and 'tepuí' climate envelopes became drastically smaller (Figs. 5 and 6). Applying scenario A2a resulted in a stronger decrease of suitable habitat in both 'upland' and 'tepuí' than scenario B2a. In general, the results of scenarios derived from CCCMA and CSIRO models were roughly similar, suggesting less reduction of potentially suitable habitat than scenarios derived from the HADCM3 model (Table 1). Our findings are coincident with those of Rull and Vegas-Vilarrúbia (2006) who, in a preliminary study which was not subject to CEDM, indicated that 'upland' and 'tepuí' climate envelopes will remarkably shrink under anthropogenic global warming. Different to other Neotropical mountain ecosystems (e.g. Bush 1994), physical circumstances will allow for almost no vertical migration. Now, the question is how will the Pantepuí amphibian and reptile endemics respond to this? Decline and at worst extinction – or is there a chance of survival?

Although it is widely accepted that species' climate envelopes behave conservative, they may have the ability to adapt to climate change over short time periods so that they can compensate potential distribution range loss (e.g. Peterson et al. 1999; Wiens and Graham 2005). This depends on the succession of generations driven by natural selection, ecological potency (i.e. niche breath) and change rates of the environment and species (e.g. Holt and Gomulkiewicz 2004; Thomas et al. 2004; Pearman et al. 2007). However, it is actually unknown how large the climatic niche breadth of Pantepuí amphibian and reptile (or any other) species is. Also, their adaptability over generations remains hidden. But taking the suggested rapid warming, it may be doubted that vertebrates, even with comparatively short generation succession as amphibians and reptiles, are able to succeed. Today, the Guiana Highlands exhibit a mild perhumid climate with mean annual temperature 12–18°C and total annual rainfall 2,500–3,000 mm (Berry et al. 1995), while for the next

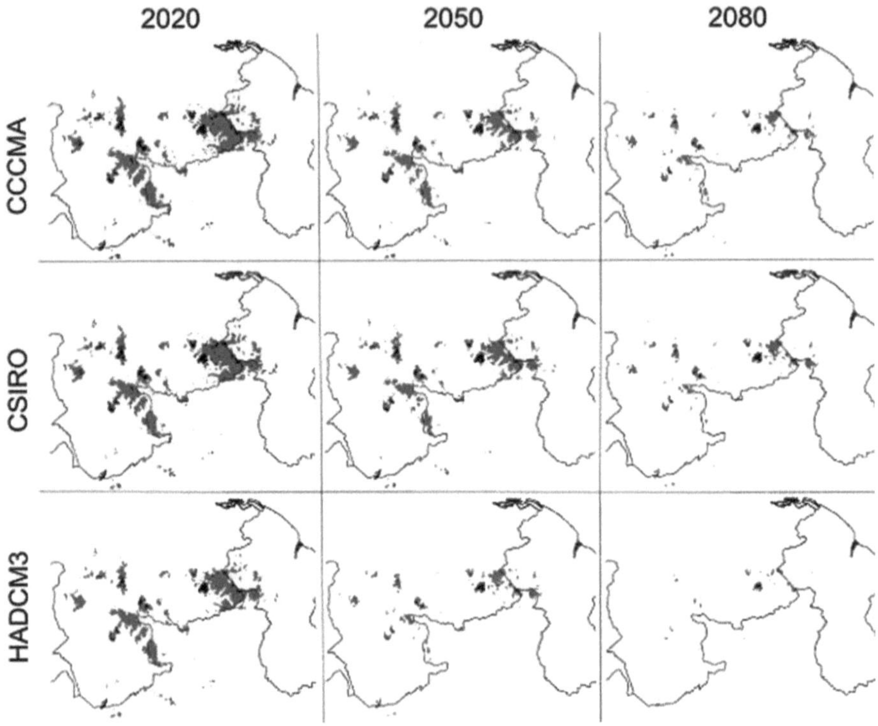

Fig. 5 Potential distributions of 'upland' (*grey*) and 'tepuí' climate envelopes (*black*) under A2a conditions in 2020, 2050, and 2080, according to CCCMA, CSIRO and HADCM3 future climate change (i.e. anthropogenic global warming) scenarios

century there has been predicted a temperature increase of 1.8–5.1°C (IPCC 2007). In contrast, during the last postglacial warming, it has been shown that Guyana Highlands plant species were able to 'handle' a temperature increase of 0.025°C per century (Rull and Vegas-Vilarrúbia 2006) what is a 72–204 times lower rate than the IPCC prediction.

Regarding niche breath, the climate at the tepuí summits is characterized by weak variation between annual minimum and maximum temperatures, but high temperature ranges between day and night (e.g. up to 21°C within 6 h recorded at Chimantá; McDiarmid and Gorzula 1989). This daily temperature range indicates that especially tepuí species are perhaps unaffected by annual climate change even within few generations.

In order to have a better understanding of threats caused by climate change and of adaptation abilities, expected changes in vegetation cover in the Guyana Highlands need to be monitored and quantified spatially over the next decades.

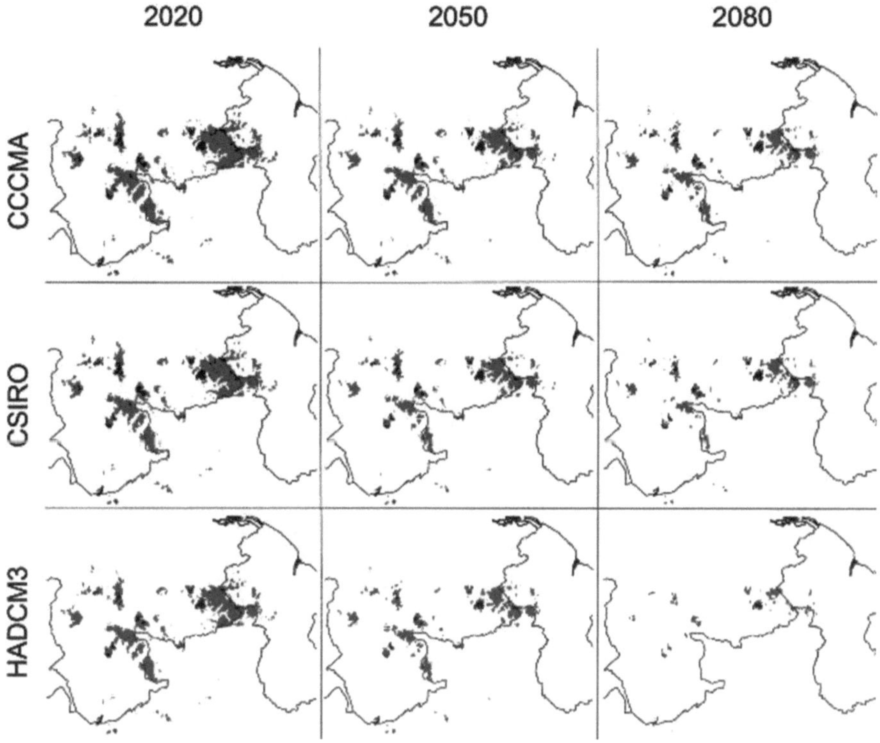

Fig. 6 Potential distributions of 'upland' (*grey*) and 'tepuí' climate envelopes (*black*) under B2a conditions in 2020, 2050, and 2080, according to CCCMA, CSIRO and HADCM3 future climate change (i.e. anthropogenic global warming) scenarios

Table 1 Extension of potential distributions of climate envelopes of 'upland' (800–1,500 m elevation) and 'tepuí' (>1,500 m elevation) relative to current extents [%] in CEDMs for the years 2020, 2050 and 2080

	'Upland'				'Tepuí'		
Future climate change model	2020	2050	2080	Future climate change model	2020	2050	2080
A2a				*A2a*			
HADCM3	54.2	17.7	3.6	HADCM3	27.9	10.4	2.4
CSIRO	63.7	39.3	15.7	CSIRO	49.4	26.3	12.3
CCCMA	67.1	40.3	16.4	CCCMA	51.6	25.5	12.1
B2a				*B2a*			
HADCM3	54.1	32.8	8.9	HADCM3	30.6	15.4	5.6
CSIRO	61.0	39.8	24.6	CSIRO	47.9	26.6	17.5
CCCMA	68.4	48.8	34.0	CCCMA	52.4	30.9	22.0

Acknowledgements The work of DR was funded by the 'Graduiertenförderung des Landes Nordrhein-Westfalen.'

References

Berry PE, Holst BK, Yatskievych K (1995) Flora of the Venezuelan Guayana, vol 1, Introduction. Missouri Botanical Garden Press, St Louis, MO

Bonaccorso E, Koch I, Peterson AT (2006) Pleistocene fragmentation of Amazon species' ranges. Divers Distrib 12:157–164

Brewer-Carías C (1978) La Vegetación del Mundo Perdido. Fundación Eugenio Mendoza, Caracas

Briceno HO, Schubert C (1990) Geomorphology of the Gran Sabana, Guayana Shield, Southeastern Venezuela. Geomorphology 3:125–141

Busby JR (1991) BIOCLIM - a bioclimatic analysis and prediction system. In: Margules CR, Austin MP (eds) Nature conservation: cost effective biological surveys and data analysis. CSIRO, Melbourne, pp 64–68

Bush MB (1994) Amazonian speciation: a necessarily complex model. J Biogeogr 21:5–17

Carnaval AC, Moritz C (2008) Historical climate modelling predicts patterns of current biodiversity in the Brazilian Atlantic forest. J Biogeogr 35:1187–1201

Carpenter G, Gillison AN, Winter J (1993) DOMAIN: a flexible modelling procedure for mapping potential distributions of plants and animals. Biodivers Conserv 2:667–680

Chapman FM (1931) The upper zonal birds of Mt. Auyan-Tepuí, Venezuela. Am Mus Novit 1051:1–5

Donnelly MA, Myers CW (1991) Herpetological results of the 1990 Venezuelan expedition to the summit of Cerro Guaiquinima, with new Tepui reptiles. Am Mus Novit 3017:1–54

Duellman WE (1999) Distribution patterns of amphibians in South America. In: Duellman WE (ed) Patterns of distribution of amphibians. A global perspective. The John Hopkins University Press, Baltimore, pp 255–328

Edmond JM, Palmer MR, Measures CI, Grant B, Stallard RF (1995) The fluvial geochemistry and denudation rate of the Guayana Shield in Venezuela, Colombia, and Brazil. Geochim Cosmochim Acta 59:3301–3325

Elith J, Graham CH, Anderson CH, Dudik M, Ferrier S, Guisan A, Hijmans RJ, Huettmann F, Leathwick JR, Lehmann A, Li J, Lohmann LG, Loiselle BA, Manion G, Moritz C, Nakamura M, Nakazawa Y, Overton JMM, Perterson AT, Phillips SJ, Richardson K, Scachetti-Pereira R, Shapire RE, Soberón J, Williams S, Wisz MS, Zimmermann NE (2006) Novel methods improve prediction of species´ distributions form occurrence data. Ecography 29:129–151

Flato GM, Boer GJ, Lee WG, McFarlane NA, Ramsden D, Reader MC, Weaver AJ (2000) The Canadian centre for climate modelling and analysis global coupled model and its climate. Clim Dyn 16:451–467

Gordon C, Cooper C, Senior CA, Banks HT, Gregory JM, Johns TC, Mitchell JFB, Wood RA (2000) The simulation of SST, sea ice extents and ocean heat transports in a version of the Hadley Centre coupled model without flux adjustments. Clim Dyn 16:147–168

Gorzula S, Señaris JC (1998) Contribution to the herpetofauna of the Venezuelan Guyana I. A data base. Scientia Guaianae 17:1–269

Hanley J, McNeil B (1982) The meaning of the use of the area under a receiver operating characteristic (ROC) curve. Radiology 143:29–36

Hijmans RJ, Guarino L, Rojas E (2002) DIVA-GIS A geographic information system for the analysis of biodiversity data. Manual – International Potato Center, Lima

Hijmans RJ, Cameron SE, Parra JL, Jones PG, Jarvis A (2005) Very high resolution interpolated climate surfaces for global land areas. Int J Climatol 25:1965–1978

Holt RD, Gomulkiewicz R (2004) Conservation implication of niche conservatism and evolution in heterogeneous environments. Cambridge University Press, Cambridge, pp 244–264 Chapter 13 Evolutionary conservation biology
Hoogmoed MS (1979) The herpetofauna of the Guianan Region. In: Duellman WE (ed) The South American herpetofauna: its origin, evolution, and dispersal. Monogr Mus Nat Hist Univ Kansas, Kansas, pp 241–279
Huber O (1988) Guayana highlands versus Guayana lowlands; a reappraisal. Taxon 37:595–614
Huber O (1994) Recent advances in the phytogeography of the Guayana region. South Am Mém Soc Biogéogr 4:53–63
Hugall A, Moritz C, Moussalli A, Stanisic J (2002) Reconciling paleodistribution models and comparative phylogeography in the Wet Tropics rainforest land snail *Gnarosophia bellendenkerensis* Brazier 1875. Proc Natl Acad Sci USA 99:6112–6117
Hutchinson MF (1995) Interpolating mean rainfall using thin plate smoothing splines. Int J Geogr Inform Syst 9:385–403
Hutchinson MF (2004) Anusplin version 4.3. Centre for Resource and Environment Studies. The Australian National University, Canberra
IPCC (2007) Summary for policymakers. The physical science basis. Contribution of Working Group I to the fourth assessment report of the Intergovernmental Panel on Climate Change. Cambridge University Press, Cambridge, pp 1–18
Jaynes ET (1957) Information theory and statistical mechanics. Phys Rev 106:620–630
Kiehl JT, Gent PR (2004) The community climate system model, version 2. J Clim 17:1666–1669
Li J, Hilbert DW (2008) LIVES: a new habitat modelling technique for predicting the distribution of species' occurrences using presence-only data based on limiting factor theory. Biodivers Conserv 17:3079–3095. doi:10.1007/s10531-007-9270-7
Lobo JM, Jiménez-Valverde A, Real R (2008) AUC: a misleading measure of the performance of predictive distribution models. Glob Ecol Biogeogr 17:145–151
MacCulloch RD, Lathrop A, Reynolds RP, Señaris JC, Schneider GE (2007) Herpetofauna of Mount Roraima, Guiana shield region, Northeastern South America. Herpetol Rev 38:24–30
Mägdefrau H, Mägdefrau K, Schlüter A (1991) Herpetologische Daten vom Guaiquinima-Tepui, Venezuela. Herpetofauna 13:13–26
Maguire B (1970) On the flora of the Guayana Highland. Biotropica 2:85–100
Mayr E, Phelps HW Jr (1967) The origin of the bird fauna of the South Venezuelan highlands. Bull Am Mus Nat Hist 136:269–328
McDiarmid RW, Donnelly MA (2005) The herpetofauna of the Guayana Highlands: amphibians and reptiles of the Lost World. In: Donnelly MA, Crother BI, Guyer C, Wake MH, White ME (eds) Ecology and evolution in the tropics: a herpetological perspective. University of Chicago Press, Chicago, pp 461–560 Chapter 18
McDiarmid RW, Gorzula S (1989) Aspects of the reproductive ecology and behaviour of the tepuí toads, genus *Oreophrynella* (Anura: Bufonidae). Copeia 2:445–451
Means DB, Savage JM (2007) Three new malodorous rainfrogs of the genus *Pristimantis* (Anura: Brachycephalidae) from the Wokomung Massif in West-central Guyana, South America. Zootaxa 1658:39–55
Myers CW, Donnelly MA (2001) Herpetofauna of the Yutajé-Corocoro massif Venezuela: second report from the Robert G. Goelet American Museum- TERRAMAR Expedition to the NorthWestern tepuís. Bull Am Mus Nat Hist 261:1–85
Myers CW, Donnelly MA (2008) The summit herpetofaunal of Ayuantepuí Venezuela: report from the Robert G. Goelet American Museum-TERRAMAR Expedition. Bull Am Mus Nat Hist 308:1–147
Nix H (1986) A biogeographic analysis of Australian elapid snakes. In: Longmore R (ed) Atlas of elapid snakes of Australia. Bureau of Flora and Fauna, Canberra, pp 4–15
Pearce J, Ferrier S (2000) An evaluation of alternative algorithms for fitting species distribution models using logistic regression. Ecol Modell 128:127–147

Pearman PB, Guisan A, Broennimann O, Randin CF (2007) Niche dynamics in space and time. Trends Ecol Evol 23:149–158

Peterson AT, Nyári ÁS (2008) Ecological niche conservatism and Pleistocene refugia in the Thrush-like Mourner, *Shifforinis* sp., in the Neotropics. Evolution 62:173–183

Peterson AT, Vargas N (1993) Ecological diversity in scrub jays, *Aphelocoma coerulescens*. In: Ramamoorthy TP, Bye R, Fa J (eds) The biological diversity of Mexico: origins and distribution. New York, Oxford University Press, pp 309–317

Peterson AT, Soberón J, Sánchez-Cordero V (1999) Conservation of ecological niches in evolutionary time. Science 285:1265–1267

Phillips SJ, Anderson RP, Schapire RE (2006) Maximum entropy modelling of species geographic distributions. Ecol Modell 190:231–259

Rödder D, Jungfer K-H (2008) A new *Pristimantis* (Anura: Strabomantidae) from Yuruaní-tepuí, Venezuela. Zootaxa 1814:58–68

Rull V (2004a) An evaluation of the Lost World and Vertical Displacement hypotheses in the Chimantá Massif, Venezuelan Guayana. Global Ecol Biogeogr 13:141–148

Rull V (2004b) Biogeography of the 'Lost World': a paleoecological perspective. Earth-Sci Rev 67:125–137

Rull V (2005a) Biotic diversifications in the Guayana Highlands: a proposal. J Biogeogr 32:921–927

Rull V (2005b) Vegetation and environmental constancy in the Neotropical Guayana Highlands during the last 6000 years? Rev Palaeobot Palynol 135:205–222

Rull V (2007) The Guayana Highlands: a promised but threatened. land for ecological and evolutionary science. Biotropica 39:31–34

Rull V, Nogué S (2007) Potential migration routes and barriers for vascular plants of the Neotropical Guyana Highlands during the Quaternary. J Biogeogr 34:1327–1341

Rull V, Vegas-Vilarrúbia T (2006) Unexpected biodiversity loss under global warming in the Neotropical Guayana Highlands: a preliminary appraisal. Global Chang Biol 12:1–9

Schlüter A, Rödder D (2007) Three new frogs of the genus *Eleutherodactylus* (Amphibia, Leptodactylidae) from Guaiquinima table mountain, Bolívar, Venezuela. Herpetotropicos 3:88–99

Señaris JC, MacCulloch RD (2005) Amphibians. In: Hollowell T, Reynolds RP (eds) Checklist of the terrestrial Vertebrates of the Guiana Shield. Bulletin of the Biological Society of Washington, Washington, DC, pp 9–24

Steyermark JA, Dunsterville GCK (1980) The lowland flora element of the summit of Cerro Guiaquinima and other cerros of the Guayana Highlands of Venezuela. J Biogeogr 7:285–303

Stockwell DRB, Noble IR (1992) Introduction of sets of rules from animal distribution data: a robust and informative method of data analysis. Math Comput Simul 33:385–390

Stockwell DRB, Peters D (1999) The GARP modelling system: Problems and solutions to automated spatial prediction. Int J Geogr Inform Sci 13:143–1158

Swets K (1988) Measuring the accuracy of diagnostic systems. Science 240:1285–1293

Tate GH (1939) The mammals of the Guiana region. Bull Am Mus Nat Hist 76:151–229

Thomas CD, Cameron A, Green RE, Bakkenes M, Beaumont LJ, Collingham YC, Erasmus BFN, Siqueira MF, Grainger A, Hannah L, Hughes L, Huntley B, van Jaarsveld AS, Midgley GF, MIles L, Ortega-Huerta MA, Peterson AT, Phillips OL, Williams SE (2004) Extinction risk from climate change. Nature 427:145–148

Thuiller W (2003) BIOMOD – optimizing predictions of species distributions and projecting potential future shifts under global change. Glob Chang Biol 9:1353–1362

Wiens JJ, Graham CH (2005) Niche conservatism: integrating evolution, ecology, and conservation biology. Annu Rev Ecol Syst 36:519–539

WWF IUCN (1997) Centres of plant diversity, a guide and strategy for their conservation, vol 3, The Americas. IUCN Publication Unit Cambridge, Cambridge

Population Genetics and Ecological Niche Modelling Reveal High Fragmentation and Potential Future Extinction of the Endangered Relict Butterfly *Lycaena helle*

Jan C. Habel, Bettina Augenstein, Marc Meyer, Gabriel Nève, Dennis Rödder, and Thorsten Assmann

Abstract During the post-glacial warming, cold-adapted species shifted their distribution to higher latitudes and altitudes and became widely extinct over the European lowlands. The butterfly *Lycaena helle* shows this feature, and is currently distributed in highly isolated habitat remnants restricted to higher elevations over Central Europe. We analysed five polymorphic microsatellite loci and applied Climate Envelope Modelling. We detected strong genetic differentiation coinciding with the orographic structures of seven distinct mountain regions. This picture become underlined by deflecting levels of genetic diversity and the presence of private alleles, endemic for each single mountain area. Furthermore, genetic differentiation among populations within these mountain groups were detectable and reveal interrupted geneflow on a regional level. This genetic picture of a fragmented distribution coincides with the obtained pattern of potential suitable habitats given by a Climate Envelope Model. A scenario of further climate warming predicts a loss of the major parts of these areas and rising fragmentation of the remainings. The predicted extinction of some populations will cause the loss of unique alleles, which are recently restricted to the given populations.

J.C. Habel(✉) and M. Meyer
Musée national d'histoire naturelle Luxembourg, Section Zoologie des Invertébrés, 25, rue Münster, L-2160 Luxembourg, Luxembourg
e-mail: Janchristianhabel@gmx.de

J.C. Habel, B. Augenstein, and D. Rödder
Department of Biogeography, Trier University, D-54286 Trier, Germany

G. Nève
Institut Méditerranéen d'Ecologie et Paléoécologie, UMR CNRS 6116, Case 36, Université de Provence, 3 Place Victor Hugo, F-13331 Marseille Cedex 3, France

D. Rödder
Department of Herpetology, Zoologisches Forschungsmuseum Alexander Koenig, D-53113 Bonn, Germany

T. Assmann
Institute of Ecology and Environmental Chemistry, Leuphana University Lüneburg, D-21335 Lüneburg, Germany

J.C. Habel and T. Assmann (eds.), *Relict Species: Phylogeography and Conservation Biology*,
DOI 10.1007/978-3-540-92160-8_25, © Springer-Verlag Berlin Heidelberg 2010

1 Introduction

Flora and fauna are affected by both historical and recent climatic oscillations (Hewitt 2001; Schmitt 2007). Most species respond to these climatic modifications through changes in their distribution patterns (Hewitt 1996). The nature of a species' reaction depends on its biotic and abiotic requirements: Thermophilic species become extinct over major parts of Europe during the last glacial period, but survived in Southern refugia (Habel et al. 2005). In contrast, continental species mostly survived in extra-Mediterranean and Eastern refugia and expanded their distribution range westwards (Schmitt 2007), during the postglacial, and cold-adapted species migrated to higher latitudes and altitudes, following the cold climate (Huntley and Webb 1989; Peterson et al. 1999; Parmesan and Yohe 2003; Root et al. 2003; Pearman et al. 2008; Varga and Schmitt 2008). In contrast to the situation in the Alps, where most species can still move and escape uphill, many of the species occurring in the Middle mountains of Europe have reached the altitudinal limit of these mountain areas and cannot compensate further climate warming by further uphill shifts. This means that the few remaining potential suitable habitat areas are highly fragmented, a situation which is aggravated by severe land-use changes. Living in isolation enhances processes of population dynamics (such as population fluctuations, Lesica and Allendorf 1995) and population stochasticity (Melbourne and Hastings 2008), resulting in the loss of genetic diversity in local sites through drift, which cannot be compensated by immigration from neighbouring populations (Hanski 1999). Such genetically impoverished populations often suffer from decreased fitness and viability due to inbreeding, which causes the accumulation of weakly deleterious alleles (Reed and Frankham 2003; Allendorf and Luikart 2006). The effects of this development have been analysed in theoretical and experimental studies (Frankham et al. 2002).

To investigate the effect of post-glacial and anthropogenic habitat fragmentation of boreo-montane species, we chose the endangered Violet Copper butterfly *Lycaena helle*. This species is adapted to cold and moist habitats (Bachelard and Descimon 1999; Steiner et al. 2006) complemented by specific habitat characteristics such as high abundance of its larval food plant *Polygonum bistorta* and habitat structures (shrubs and trees) (Turlure 2006). During the climatic warming of the postglacial period, the species escaped to higher altitudes and northwards (Habel et al. 2009a); today, most of the Central European populations are restricted to higher elevations and thus highly fragmented.

Genetic samples of butterfly specimens were collected on seven Western mountain areas (Pyrenees, Massif Central, Jura, Vosges, Madeleine mountains, Ardennes-Eifel and the Westerwald) to assess the genetic structure and effects of isolation. A Climate Envelope Model (CEM) (Phillips et al. 2006) was performed in order to identify and quantify climatically suitable habitats and possible connections between them (for further details see Rödder et al., 2009). Projection of the climatic envelope into the geographic space allowed us to assess the potential distribution of the butterfly under the current climate and a future climate change scenario (cf. Araujo and Whittaker 2005). The combination of molecular analysis and niche modelling provided evidence of the species' distribution situation and the genetic response of highly fragmented suitable habitats. Both methods revealed an *extinction process* at habitat and species level.

2 Material and Methods

2.1 Study Species

The Violet Copper butterfly, *Lycaena helle* (Denis and Schiffermüller, 1775) is a boreo-montane species of the Palaearctic distributed over Central Europe, Fennoscandia, throughout Russia to the Amur region, China and Mongolia (Bozano 2001), and is present in cool and moist meadows. Essential habitat requirements are the food-plant *Polygonum bistorta* (in Europe), typical vegetation structures (Goffart et al. 2009) and specific moist and cold climatic conditions (Bachelard and Descimon 1999; Fischer et al. 1999; Turlure 2006). The species is classified as a generally sedentary butterfly (Bink 1992), but occasional movements among neighbouring sites or even rare long-distance movements have also been recorded (Bachelard and Descimon 1999). The recent situation which has involved habitat succession, industrialisation of land-use and afforestation has led to severe habitat fragmentation of this species, with severe genetic impact (Finger et al. 2009). The species is classified as one of the most endangered butterflies of Europe, listed in the European Red Data Book (Van Swaay and Warren 1999) and Appendices II and IV of the Habitat Directive (EEC, 92/43/EWG) of the European Union (Drews and Pretscher 2003).

2.2 Genetic Analysis

A total of 560 individuals of *L. helle* were sampled at 30 localities scattered over seven mountain groups in its Western European distribution area (Pyrenees, Massif Central, Jura, Madeleine mountains, Vosges, Ardennes-Eifel and the Westerwald) (Fig. 1, Table 1). Sampling was carried out from the beginning of May to the end of July in 2005–2007. The individuals were netted in the field and one leg per individual was removed and stored in 100% ethanol until analysis. To avoid recaptures of individuals, all sampled butterflies were marked on the wing with a permanent pen before they were released.

We selected five polymorphic microsatellites (LheF12, LheB06, LheE12, Lhe03, and Lhe14), which were developed for this species (Habel et al. 2008). The forward primer of each pair was 5′ end-labelled with the fluorescent phosphoramidites FAM (LheE12 and LheB06), HEX (Lhe03) and TET (Lhe14 and LheF12). Details of the analytical procedures are described in Habel et al. (2008) and Finger et al. (2009).

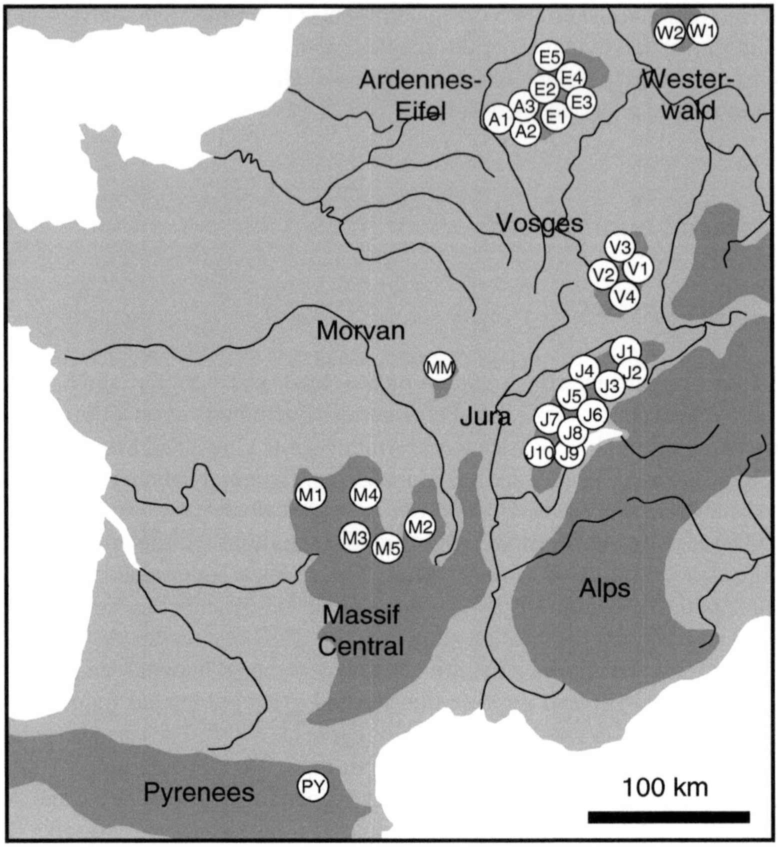

Fig. 1 Geographic locations of the 30 *Lycaena helle* sample sites in the Western European study area. Numbers correspond to localities given in Table 1. *Dark grey* areas mark elevations over 500 m

Potential distortion of the results through stutter bands, large allele drop-out or null alleles (cf. Kimberly and Selkoe 2006; Selkoe and Toonen 2006) was tested using the programme Micro-Checker (Van Oosterhout et al. 2004). Allelic richness was calculated using FSTAT (Goudet 1995). Hierarchical genetic variance analyses (AMOVAs), observed and expected heterozygosities and tests on Hardy Weinberg equilibrium (HWE) and linkage disequilibrium were calculated with Arlequin 3.1 (Excoffier et al. 2005). AMOVAs were carried out using the microsatellite-specific analogous *R*-statistics based on allele lengths and the stepwise mutation model. This statistical approach makes use of the fact that alleles similar in length are more likely to share a common ancestor (i.e. have a higher probability of being identical by descent). The analyses comprised three hierarchical levels: within populations, among populations within regions, and among regions. A population phenogram

Table 1 Sampling localities and number of sampled *Lycaena helle*

Area	Locality	Abbrev.	Altitude	Area (km^2)	Individuals
Pyrenees (PY)	F-Porté-Puymorens	P1	1,038	13,497 (12,809)	18
MassifCentral (M)	F-Complains	M1	1,261	17,172 (4,097)	20
	F-Picheland	M2	1,186		20
	F-Col Chaumoune	M3	1,156		20
	F- Mareuge	M4	1,389		20
	F-La Godivelle	M5	1,240		15
Jura (J)	F-Bellelay	J1	1,130	13,575 (8,476)	9
	CH-Tramelan	J2	1,086		24
	F-Le BémontI	J3	959		20
	F-Le BémontII	J4	1,021		9
	F-Le Mémont	J5	484		20
	F-Le Bélbrieu	J6	902		20
	F-La-R.-Drugeon	J7	809		20
	F-La Vacronnaz	J8	642		20
	F-Bois d'Amont	J9	1,070		25
	F-Tourbière	J10	886		10
Madeleine Mts. (MM)	F-La Verrerie	MM	642	(73515)	7
Vosges (V)	F-Gérardmer	V1	784	11,464 (3,691)	8
	F-Les Hauts Viaux	V2	1,077		25
	F-Xonrupt	V3	959		10
	F-Retournemer	V4	1,019		20
Ardennes (A) and Eifel (E)	B-Witry	A1	410	19,127 (8,429)	26
	B-Pisserotte	A2	493		24
	D-Klengelbachtal	A3	452		21
	D-Rohrvenn	E1	533		30
	D-Kolvenderbach	E2	485		21
	D-Frauenkron1	E3	568		23
	D-Mooshaus	E4	578		19
Westerwald (WW)	D-Lippe	W1	594	18,674 (11,432)	21
	D-Hof	W2	215		15

Abbreviations: *B* Belgium, *D* Germany, *F* France. Numbers of sites are as in Fig. 1. Size of area showing potential suitable habitats obtained from CEM (see text for further details); first value greater than a Maxent value of 0.50, second value greater than a Maxent value of 0.75 in paranthesis

using the neighbour joining algorithm (Saitou and Nei 1987) based on pairwise Cavalli-Sforza and Edwards (1967) distances was constructed with TREEMAKER. Node support was assessed by means of 1,000 bootstrap replicates. Furthermore, STRUCTURE (Pritchard et al. 2000) was used to cluster populations into groups to test the eight geographically isolated areas. Burn-in and simulation lengths were 100,000 and 500,000, respectively.

Population differentiation was tested using a *G* test based on allele frequencies, using Genepop (Raymond and Rousset 1995, Rousset 2008). Isolation by distance was tested by means of a Spearman test performed on half-matrices of pairwise geographic distances (logarithmic scale) and the ratio $F_{ST}/(1-F_{ST})$, using F_{ST} between pairs of populations (Rousset 1997). These computations were performed with Genepop007 (Rousset 2008), on both the total data set and on a partial data set with only the three loci, which did not show heterozygote deficiency.

Alleles which exclusively occurred in one mountain group were identified. These alleles were neither present in any other population of our study area nor in populations of the complete European distribution area analysed (Habel et al. unpublished). Thus, the detected 'private' alleles are endemic for one obtained genetic group (and mountain region) and not for a single population sensu Slatkin (1985).

2.3 Climate Envelope Model

As shown in Waltari et al. (2007), Climate Envelope Models (CEM) provide a useful complement to genetic studies to predict ecological patterns of potential habitat distribution. In this study, 458 presence localities over the distributional range of *Lycaena helle* provided the basis for CEM computation. We used the CEM approach in order to assess the number and spatial distribution of recent and future climatically suitable habitats (cf. Hijmans and Graham 2006).

In order to assess potential suitable habitats, we used bioclimatic variables obtained from the WorldClim database (Hijmans et al. 2005; http://www.worldclim.org/bioclim.htm). Six variables were chosen: 'annual mean temperature', 'maximum temperature of the warmest month', 'minimum temperature of the coldest month', 'annual precipitation', 'precipitation of the wettest month' and 'precipitation of the driest month'. These variables describe the climatic tolerance with respect to mean, minimum and maximum values of temperature and precipitation. They are likely to determine the distribution of our target species. We decided to extend the set of variables to six rather than applying three as used by Huntley et al. (2007) to model bird distribution, since these six variables reflect the ecological requirements of *Lycaena helle* more closely. This was shown by Bachelard and Descimon (1999) and our own data.

In order to assess the future potential distribution of *Lycaena helle*, we used a future anthropogenic climate change scenario developed by the National Center for Atmospheric Research (NCAR) Community Climate Model Version 3 (CCM3) as described by Govindasamy et al. (2003) and Duffy et al. (2003). This climate change scenario assumes a doubling of the CO_2 concentration by the year 2100 (Govindasamy et al. 2003; Duffy et al. 2003). All climatic data were downscaled to a 2.5' spatial resolution, corresponding to a spatial solution of about 3×3 km at 50°N.

To compute the CEM, we used Maxent 3.2.1 (Phillips et al. 2006, http://www.cs.princeton.edu/~schapire/maxent/), a machine-learning algorithm, which was shown to reveal better results than most other algorithms (e.g. Elith et al. 2006).

The default settings were used for the convergence threshold (10^{-5}) with a maximum number of iterations (10,000). A logistic output format was used, whereby Maxent values range from 0 (not suitable) to 1 (very suitable) in a linear relationship. The software automatically selects suitable regularisation values to reduce overfitting (Pearson et al. 2006). For model testing, 75% of the occurrence points were used for model computation, and 25% were used as test points. The Area Under the Curve (AUC) was used to evaluate the model, based on the Receiver Operation Characteristics (ROC) curve as suggested by Phillips et al. (2006).

3 Results

3.1 Genetic Analyses

The five microsatellite loci yielded between 16 and 54 alleles (mean: 32.2 ± 8.8 sd), with a total number of 152 alleles over all loci and populations. No linkage disequilibrium was observed for any pair of loci after Bonferroni correction. Therefore, further analyses were performed on multi-locus data from all five microsatellites. After Bonferroni correction, significant deviations from HWE due to a heterozygote deficit were detected for the loci LheB06 and Lhe14. Using Micro-Checker (Van Oosterhout et al. 2004), we detected a strong signal for null-alleles for these two loci, but there was no indication of errors due to stutter bands or large allele drop-out for any locus. Thus, the HWE deviations at the loci LheB06 and Lhe14 could (at least partly) be due to the presence of null alleles. Values for allelic richness, observed and expected heterozygosity are given in Table 2.

The overall molecular variance among populations was 0.264 ($p<0.0001$) and 0.204 were found within populations (R_{IS}). All populations were subdivided into geographic groups following the orographic structure of the study area (for classification see Table 1). The genetic differentiation among the seven entities was strong (R_{CT}: 0.203, F_{CT}: 0.146, $p<0.0001$ in both cases) all data are given in Table 3. A neighbour joining phenogram based on genetic distances (Cavalli-Sforza and Edvards 1967) revealed separate clusters corresponding with the respective mountain areas (Massif Central, Jura South, Jura North, Vosges, Ardennes-Eifel and Westerwald); the single populations from the Pyrenees and the Madeleine mountains are as strongly distinguished from all other samples as these clusters. Bayesian structure analysis of the data obtained for *Lycaena helle* individuals clearly discriminate the individuals into the same disjunct genetic groups mirroring the mountain areas. Within these groups, the degree of differentiation varies considerably. The Jura splits into two genetic groups – Southern and Northern – both of which were supported by Bayesian structure analysis with only few outliers. These two groups are also clearly discriminated by the neighbour joining phenogram (Fig. 2).

Table 2 Parameters of genetic diversity for all analysed populations of *Lycaena helle* in the West European study area: mean number of alleles (*A*), allelic richeness (*AR*), percentage of expected heterozygosity (*He*), observed heterozygostiy (H_o) and private alleles (a: frequency, b: numbers)

Area	Abbrev.	*A*	*AR*	H_e [%]	H_o [%]	Private alleles[a][%]	Private alleles[b]
Pyrenees (PY)	P1	6.2	4.28	0.76	0.56	0.52	3.28
Massif Central (M)	M1	7.4	4.28	0.71	0.52	0.40	4.60
	M2	6.8	4.04	0.68	0.53	0.21	3.28
	M3	7.0	4.17	0.70	0.48	0.24	2.63
	M4	6.2	4.32	0.74	0.50	0.36	1.31
	M5	7.2	4.86	0.79	0.57	0.20	3.95
Mean (± s.d.)		6.92 (±0.46)	4.33 (±0.31)	0.73 (±0.04)	0.53 (±0.03)	0.38 (±0.09)	3.2 (±1.26)
Northern Jura (J)	J1	4.8	3.93	0.69	0.53	0	0
	J2	7.6	4.39	0.71	0.54	0.03	1.32
	J3	7.2	4.56	0.75	0.51	0.02	0.65
	J4	5.8	4.56	0.71	0.61	0	0
	J5	5.0	3.49	0.62	0.39	0	0
	J6	6.8	4.40	0.72	0.42	0	0
Mean (± s.d.)		6.2 (±1.17)	4.22 (±0.43)	0.7 (±0.04)	0.5 (±0.08)	0.03 (±0.01)	0.99 (±0.47)
Southern Jura (J)	J6	6.8	4.40	0.72	0.42	0	0
	J7	10	5.67	0.83	0.56	0.13	3.28
	J8	7.0	4.51	0.82	0.62	0.29	1.97
	J9	6.0	3.87	0.68	0.52	0.59	2.63
	J10	5.2	3.88	0.75	0.60	0.23	3.95
Mean (± s.d.)		7.00 1.82	4.46 0.73	0.76 0.06	0.54 0.08	0.32 (±0.19)	2.97 (±0.84)
Madeleine Mts. (MM)	MM	5.0	3.66	0.68	0.49	0.09	1.31

Vosges (V)	V1	3.2	2.78	0.49	0.36	0	0
	V2	3.6	2.80	0.51	0.39	0	0
	V3	3.4	2.87	0.52	0.42	0	0
	V4	3.6	2.91	0.58	0.43	0	0
Mean (± s.d.)		3.45 (±0.19)	2.84 (±0.06)	0.53 (±0.04)	0.40 (±0.03)	0 (±0.0)	0 (±0.0)
Ardennes (A) and Eifel (E)	A1	10.8	5.40	0.79	0.71	0.16	3.28
	A2	7.0	4.35	0.74	0.60	0.25	1.97
	A3	7.0	4.18	0.73	0.67	0.22	1.97
	E1	8.0	4.83	0.77	0.75	0.12	1.32
	E2	8.2	4.82	0.77	0.58	0.33	2.63
	E3	8.0	4.66	0.74	0.57	0.20	2.63
	E4	6.0	4.16	0.72	0.59	0.07	1.31
Mean (± s.d.)		7.86 (±1.51)	4.63 (±0.44)	0.76 (±0.03)	0.65 (±0.07)	0.19 (±0.08)	2.16 (±0.73)
Westerwald (WW)	W1	6.4	4.11	0.70372	0.55		
	W2	5.2	3.79	0.675212	0.64	0.09	1.32
Mean (± s.d.)		5.80 (±0.85)	3.95 (±0.23)	0.69 (±0.02)	0.59 (±0.06)	0.09	1.32

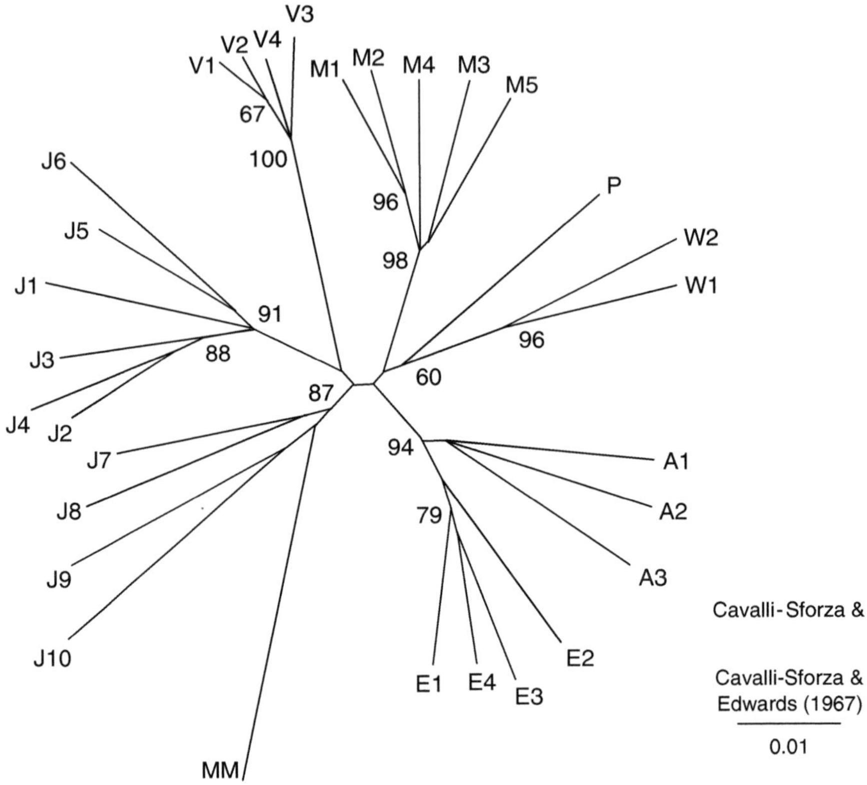

Fig. 2 Neighbour-joining phenogram based on Cavalli-Sforza and Edwards (1967) distances with bootstrap values (derived from 1,000 replicates) of 30 *Lycaena helle* populations from our study area. Abbreviations of localities are as in Table 1 and Fig. 1

The parameters of genetic diversity differ strongly among all classified groups. ANOVAs show significant differences among the regions for the mean number of alleles *A* ($p=0.014$), allelic richness *AR* ($p=0.018$), percentage of expected heterozygosity H_e ($p=0.016$) and observed heterozygosity H_o ($p=0.007$). No significant correlation was found between geographical altitude or latitude and the parameters of genetic diversity. We detected private alleles (see Material and Methods), which were found exclusively within any given region, for each mountain area except for the Vosges. In general, these alleles represent a small part of the complete gene pool. The frequencies varied between 0.02 and 0.35 within a population from one group. While populations from the Northern Jura, the Madeleine mountains and the Westerwald showed low mean frequencies of private alleles (0.03 and 0.09, respectively), the other regions showed higher frequencies (Ardennes-Eifel: 0.19, Southern Jura: 0.32, and Massif Central: 0.38). This was especially true of the Pyrenees, where private alleles exhibit about half of the total allele frequencies of the five studied microsatellite loci (0.52).

Table 3 Non-hierarchical variance analysis of *Lycaena helle* in five regions. All $p<0.001$

Region	F_{ST}	R_{ST}	F_{IS}	R_{IS}
Ardennes-Eifel	0.0854	0.0684	0.1342	0.2424
Massif Central	0.0467	0.1009	0.2657	0.3092
Northern Jura	0.0553	0.1065	0.2794	0.2975
Southern Jura	0.1078	0.2858	0.2700	0.2417
Vosges	0.0276	n.s.	0.1951	0.1195
Westerwald	0.0538	0.0761	0.1396	n.s.
All populations	0.1872	0.2643	0.1701	0.2040

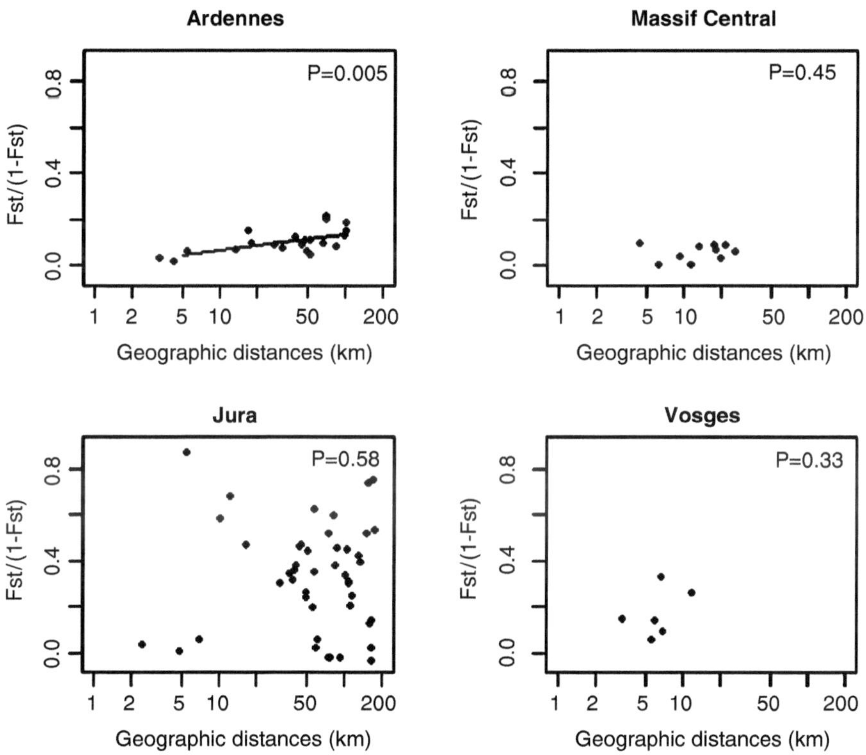

Fig. 3 Isolation by distance in four regions studied. The ratio $F_{ST}/(1-F_{ST})$ as a function of the distances between populations (logarithmic scale)

Within each regional population, all subpopulations were significantly differentiated from each other ($p<0.001$). Isolation by distance was found to be significant only in the Ardennes–Eifel region (Fig. 3). No significant relationship between the number of private alleles or mean frequencies and area of potential suitable habitats (areas with Maxent values greater than 0.50 and 0.75) was detected.

3.2 Climate Envelope Model

The CEM suggests isolated potential distribution areas over the Western range of *Lycaena helle* under current climatic conditions. The areas with a predicted suitability of 0.75 or more (optimal regions) are mostly restricted to higher elevations (Fig. 4b). The areas of predicted suitability of at least 50% surround these mountainous areas. The areas of a predicted suitability higher than 75% are actually separated from each other by pessimal areas. The habitats which are potentially suitable from a climatic point of view also extend across major parts of the Northern and Southern Alps and most of the Balkan region. Distinct areas in Italy were also identified as potentially suitable. However, many of these areas (Italy, Balkans and the Southern and Western Alps) have not recently been colonised by *Lycaena helle*.

Applying the CCM3 scenario, the CEM suggests a strong decline of potentially suitable habitats, especially at the Western edge of the species´ distribution area. Most of the recent areas of predicted suitability may disappear (Fig. 4c). Areas with a predicted suitability of 0.75 exclusively remain over parts of the Jura and the Alps. Areas with a predicted suitability of at least 0.50 remained in parts of the

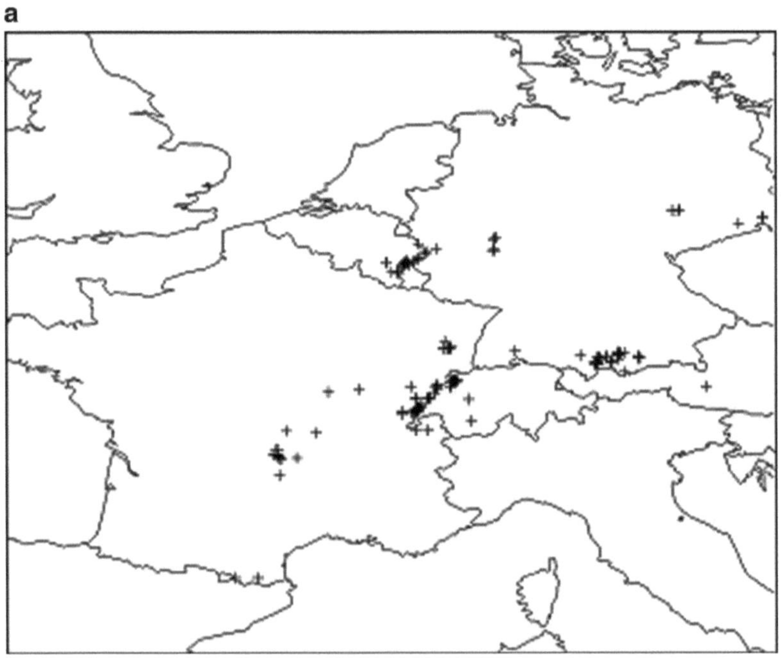

Fig. 4 Recent *L. helle* populations included in the models (**a**) modeled recent (**b**) and future distribution (**c**) of potential suitable habitats of *L. helle* in the Western European study area using Maxent Vers. 3.2.1. Presence of *L. helle* is shown by black crosses (Fig. 4a). Areas with a Maxent value of at least 0.50 are displayed in *grey*, areas with a predicted suitability of 75% or more are displayed in *black*

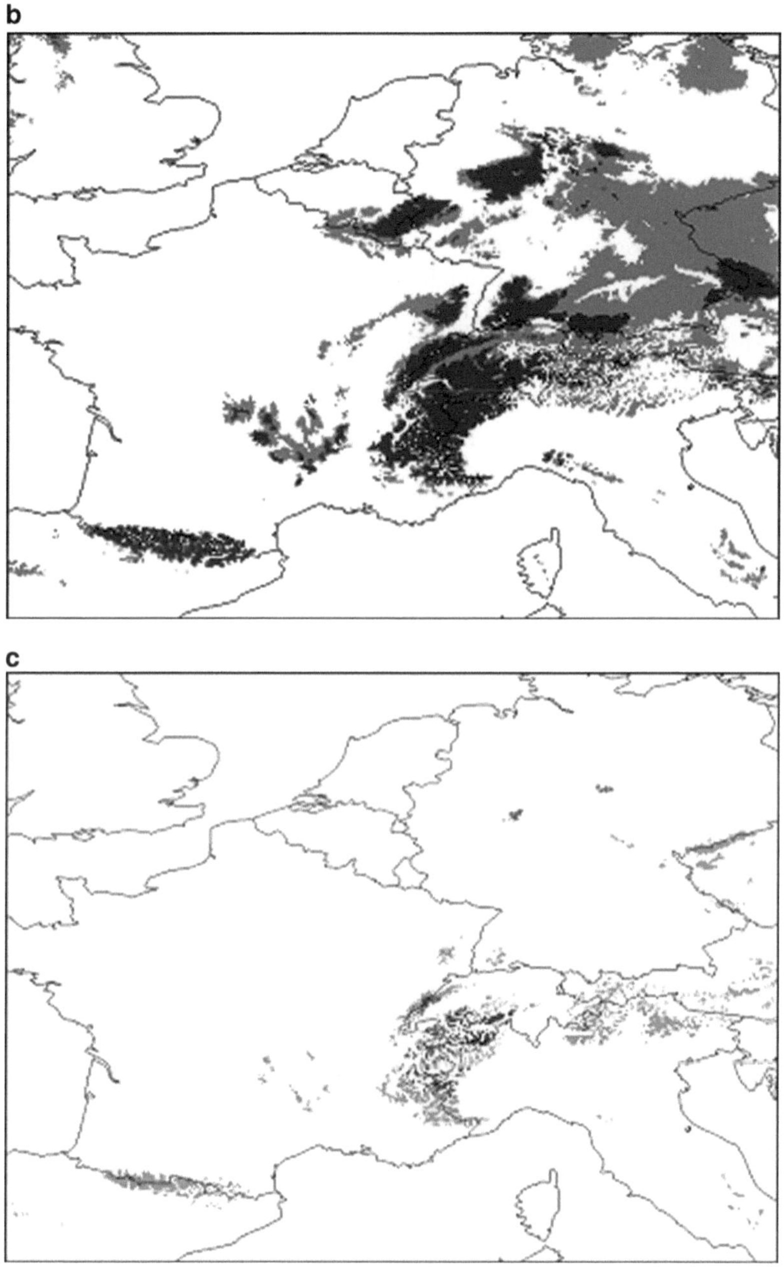

Fig. 4 (continued)

Massif Central, the Pyrenees and Vosges. Even over major parts of the Alps, potentially suitable habitats may largely disappear.

4 Discussion

The genetic data show: (1) eight genetic groups mostly corresponding with the orographic structures of higher elevations over our study area, (2) a high genetic differentiation among populations within these groups, (3) different levels of genetic diversities among the mountain regions and (4) private alleles, endemic for the mountain areas. The fact that the isolation by distance effect is significant only in the Ardennes region suggests that the other areas were subject to more complex colonisation and movement patterns than may be explained by distance *per se*; this pattern is opposite to the one shown for *Proclossiana eunomia* (also, a species feeding exclusively on *Polygonum bistorta*). In this species, the areas of high elevation (Pyrenees and Cantabrian mountains) show a high isolation by distance effect compared with areas of lower elevation as the Ardennes (Nève et al. 2008).

4.1 Historical Range Modifications

The present distribution pattern of *Lycaena helle* reflects demands for cool climatic conditions. From this we can assume a historically wide distribution over the major lowlands of Central Europe. We hypothesise two possible scenarios for the species' historical range modifications: (1) a colonisation of Central Europe from extra-Mediterranean and Eastern European refugia or (2) a wide distribution even during the glacial period. In both scenarios, a restriction of a former distribution range takes place. The first scenario would be in accordance with the butterfly *Proclossiana eunomia*, which represents similar ecological requirements (Nève 1996). However, the second scenario cannot be ruled out as fossil records (e.g. from pollen, charcoal fragments and the elytra of beetles, Huntley and Birks 1983; Coope 1994; Willis and van Andel 2004) support the idea that climatic conditions and basic habitat requirements for *Lycaena helle* (such as the larval food plant) were present during this period, and a CEM suggests a larger area of potential suitable habitats for *Lycaena helle* during the glacial period (Habel et al. unpublished data).

4.2 Patchily Distributed and Genetically Differentiated

In most of the populations analysed we detect a high genetic variability, which might reflect a former wide distribution for this species – or at least interconnected populations within each mountain area. In general, isolated population remnants show reduced genetic diversity as a result of genetic erosion (Lesica and Allendorf 1995), as shown for other butterflies like *Parnassius apollo* (Habel et al. 2009), *Parnassius smintheus* (Keyghobadi et al. 2005), *Speyeria idalia* (Keyghobadi et al. 2006), *Proclossiana eunomia* (Nève et al. in press), and the ground beetle *Carabus auronitens* (Drees et al. 2008). Obviously, the present mountainous retreats of Central Europe left sufficient space for intact metapopulations of *Lycaena helle* to maintain this high genetic diversity to the present day. The phenomenon of high genetic diversity despite long-term isolation was also shown for other relict organisms, which maintained their genetic variability through intact population networks within isolated areas (cf. Baali-Cherif and Besnard 2005). As the detected private alleles do not appear in further populations over the entire European distribution area (Habel et al. unpublished data), their evolution should have taken place in most cases after the species colonised these mountain areas during the post-glacial warming.

The strong genetic differentiation among mountain areas coincides with the orographic structure of the study species and with the optimal areas (potential suitable habitats surpassing a Maxent value of 0.75) of the CEM. Less suitable regions (potential suitable habitats with a Maxent value of at least 0.50) enclose these hotspots. Most of the classified mountain areas are not linked with neighbouring areas by habitats of potential suitability (not even areas with a Maxent value of at least 0.50). As many studies reveal the importance of habitat size to maintain genetic information, we analysed all genetic diversities in respect to area size of suitable habitats obtained from the Ecological Niche Model and could not find any correlations.

Even within each mountain area, we detected a strong genetic differentiation among neighbouring populations. Detected private alleles are often restricted to a single population of a mountain group, which underpins restricted or absent geneflow among populations of *Lycaena helle* within mountain massifs. Within the Ardennes-Eifel, we detected a significant correlation between the level of genetic differentiation among populations and the geographical distance between them, which suggests a restricted geneflow among populations between these two adjoining areas, clustering into two genetic groups with only a few outliers (Finger et al. 2009). For the Jura massif the statistic analysis revealed a significant split into a Southern and Northern group and is in accordance with two morphologically characterised subspecies (Meyer 1982). These two morphological and genetic character sets give results which allow us to conclude that the two groups of populations inhabiting the Jura massif represent two distinct Evolutionarily Significant Units (ESU) sensu Moritz (1994). However, these deep structures are probably not the result of recent habitat isolation but of historic species separation into distinct refugia, as described for the snail *Trochulus villosus* in the same area (Dépraz et al. 2008).

The CEM suggests strongly fragmented potential suitable areas for *Lycaena helle*, corresponding with the higher elevations of our study area. The obtained pattern coincides with the recent genetic and distribution situation of this species (Drews and Pretscher 2003). The CEM projection onto a future climate change scenario suggests a severe decline of most potential suitable habitats over Western

Central Europe, restrictions to higher elevations of the Jura and Alps and an increase in the habitat fragmentation of the remaining areas. Similar changes in distribution patterns have been suggested for birds occurring in the same climatic zone as *Lycaena helle*, such as Hazel Grouse (*Bonasa bonasia*) or the Pygmy owl (*Glaucidium passerinum*) (Huntley et al. 2007).

4.3 Future Habitat Decline and Loss of Genetic Uniqueness

Projecting the CEM onto the CCM3 climate change scenario, our model suggests a complete loss of climatically suitable habitats for *Lycaena helle* on its Western distribution border, except for a few exclaves of potentially suitable areas over the Pyrenees, the Massif Central, the Jura and parts of the Alps. This implies two effects: (1) an immediate range loss involving both population numbers and their genetic diversity and (2) more frequent local extinctions of populations due to stochastic processes by increasing habitat fragmentation (cf. Hanski 2005). The direct loss of distinct population groups would imply the loss of a large part of the genetic diversity, including numerous private alleles, which are exclusively present in these populations and will not be preserved in any other populations over the entire European range (Habel et al. unpublished data).

The interpretation of the CEM results merits some theoretical considerations. Although it is widely accepted that species' climatic envelopes behave conservatively especially over relatively short time periods (e.g. Peterson et al. 1999; Wiens and Graham 2005), species theoretically have the *ability* to adapt to climate change and thus compensate range loss in terms of their potential distribution. This depends on the succession of generations driven by natural selection, ecological potency and change rates of the environment (e.g. Holt and Gomulkiewicz 2004). Furthermore, species inhabiting large areas are *less* vulnerable than species with smaller ranges because of a greater genetic diversity and hence a greater probability of successful adaptations can be expected (e.g. Holt and Gomulkiewicz 2004, Parmesan 2006). In our study, the remaining habitats will become even more isolated from each other than they are today. This situation *may* enforce negative effects, such as genetic drift and the loss of allelic richness and uniqueness in the near future.

5 Conclusions

Combining the advantages of both methods – molecular analysis and Climate Envelope Modelling – creates a complete picture of the status of a species. While genetic data give direct evidence of a species' history and its recent situation over different spatial scales, the data of a CEM show the present distribution and its future tendencies. Thus, both methods give information on different spatial and temporal scales and form a complete picture of a species, and in our case of a species which is representative of a large number of cold-adapted species.

Acknowledgement We acknowledge a grant from the Ministry of Culture, Education and Sciences Luxemburg (grant number BFR-05/118), the Natural History Museum Luxembourg. Thanks go to all lepidopterologists who helped us to collect the species and/or showed us suitable sites. We also thank Aline Finger (Zürich, Switzerland) for the laboratory work and Thomas Schmitt (Trier, Germany) for critical comments on a draft version of this manuscript.

Frequencies of private alleles for distinct mountain regions analysed (no private allele was found for the Vosges); mean frequency/ population, mean frequency/group, mean number/population, mean number/group and total frequency of loci over all populations are given as percentages

Gene/alleles	Pyrenees	Massif Central					Jura						Madeleine Mts.	Ardennes/Eifel							Wester-wald	
	P	M1	M2	M3	M4	M5	J10	J9	J8	J7	J2	J3	MM	A1	A2	A3	E2	E3	E1	E4	W2	Total
He03														1.9			4.8	2.2				0.4
E12														5.8	4.5	4.8						0.6
									2.6													0.1
	32.4																					1.0
He12																		7.1				0.3
																2.5						0.1
																					10.7	0.3
											2.2											0.1
											2.2											0.1
														8.0	2.3		19.0		1.9	2.6		1.4
													7.1									0.1
		25.0	15.0	5.3	35.0	3.3																3.0
						3.3																0.1
							5.0			5.0												0.3
		2.5																				0.1
																	11.9					0.5
	5.6																					0.2
						3.3																0.1
				5.3		3.3																0.3
				2.6		6.7																0.3

F12																	2.2				0.1
												7.1									0.1
															27.3	14.7	19.6	16.1	7.9		3.7
								47.6	25.0												2.8
										2.8											0.1
								2.4													0.1
													2.1								0.1
He14	2.9																				0.1
		2.6																			0.1
			2.5																		0.1
		5.3																			0.2
							5.0														0.1
							5.0														0.1
							5.0														0.1
			2.5																		0.1
										5.9											0.2
	35.3																				1.2
									17.6	2.9											0.7
							10.0	36.8		2.9											1.7
								2.6													0.1
													6.5	31.6							1.5
		18.4	10.0	23.5	21.1	10.7															3.0
																				3.6	0.1
											3.3										0.1

(continued)

(continued)

Gene/alleles	Pyrenees	Massif Central					Jura						Madeleine Mts.	Ardennes/Eifel							Wester-wald	
B06	2.9																					0.1
		2.6																				0.1
			2.5																			0.1
		5.3					5.0															0.2
Mean frequency/ population	0.52	0.40	0.21	0.24	0.36	0.20	0.23	0.59	0.29	0.13	0.03	0.02	0.09	0.16	0.25	0.22	0.33	0.20	0.12	0.07	0.09	0.1
Mean frequency / group	0.52	0.38					0.32				0.03		0.09	0.19							0.09	
Mean number/ population	3.28	4.60	3.28	2.63	1.31	3.95	3.95	2.63	1.97	3.28	1.32		1.31	3.28	1.97	1.97	2.63	2.63	1.32	1.31	1.32	
Mean number / group	3.28	3.2					2.97				0.99		1.31	2.16							1.32	

References

Allendorf FW, Luikart G (2006) Conservation and the genetics of populations. Blackwell, Oxford

Araujo M, Whittaker R (2005) Reducing uncertainty in projections of extinction risk from climate change. Glob Ecol Biogeogr 14:529–538

Baali-Cherif D, Besnard G (2005) High genetic diversity and clonal growth in relict populations of *Olea europaea* subsp. *laperrinei* (Oleaceae) from Hoggar, Algeria. Ann Bot 96:823–830

Bachelard P, Descimon H (1999) *Lycaena helle* (Denis & Schiffermüller, 1775) dans le Massif Central (France): une analyse écogéographique (Lepidoptera: Lycaenidae). Linneana Belg 17:23–41

Bink FA (1992) Ecologische Atlas van de Dagvlinders van Noordwest-Europa. Schuyt, Haarlem

Bozano GC (2001) Guide to the butterflies of the palearctic region. Lycaenidae. 1. Lycaeninae. Omnes Artes, Milano

Cavalli-Sforza LL, Edwards AWF (1967) Phylogenetic analysis: models and estimation procedures. Evolution 21:550–570

Coope GR (1994) The response of insect faunas to glacial-interglacial climatic fluctuations. Phil Trans R Soc Lond B 344:19–26

Dépraz M, Cordellier M, Hausser J, Pfenninger M (2008) Postglacial recolonization at a snail's pace (*Trochulus villosus*): confronting competing refugia hypotheses using model selection. Mol Ecol 17:2449–2462

Drees C, Matern A, Rasplus J-Y, Terlutter H, Assmann T, Weber F (2008) Microsatellites and allozymes as the genetic memory of habitat fragmentation and defragmentation in populations of the ground beetle *Carabus auronitens* (Col., Carabidae). J Biogeogr 35:1937–1949

Drews M, Pretscher P (2003) Schmetterlinge (Insecta, Lepidoptera) der FFH-Richtlinie. In: Das europäische Schutzgebietssystem Natura 2000, Ökologie und Verbreitung von Arten der FFH-Richtlinie in Deutschland. Schriftenr Landschaftspfl Naturschutz 69:445–448

Duffy PB, Govindasamy B, Lorio JP (2003) High resolution simulations of global climate, Part 1. Clim Dyn 21:371–390

Elith J, Graham CH, Anderson RP, Dudik M, Ferrier S, Guisan A, Hijmans RJ, Huettmann F, Leathwick JR, Lehmann A, Li J, Lohmann LG, Loiselle BA, Manion G, Moritz C, Nakamura M, Nakazawa Y, Overton JMM, Perterson AT, Phillips SJ, Richardson K, Scachetti-Pereira R, Shapire RE, Soberón J, Williams S, Wisz MS, Zimmermann NE (2006) Novel methods improve prediction of species´ distributions from occurrence data. Ecography 29:129–151

Excoffier L, Laval G, Schneider S (2005) Arlequin ver. 3.0: an integrated software package for population genetics data analysis. Evol Bioinform Online 1:47–50

Finger A, Zachos FE, Schmitt T, Meyer M, Assmann T, Habel JC (2009) The genetic status of the Violet Copper *Lycaena helle* – a relict of the cold past in times of global warming. Ecography 32:382–390

Fischer K, Beinlich B, Plachter H (1999) Population structure, mobility and habitat preferences of the violet copper *Lycaena helle* (Lepidoptera: Lycaenidae) in Western Germany: implications for conservation. J Insect Conserv 3:43–52

Frankham R, Ballou JD, Briscoe DA (2002) Introduction to conservation genetics. Cambridge University Press, Cambridge

Goffart P, Schtickzelle N, Turlure C (2009) Conservation and management of the habitats of two relict butterflies in the Belgian Ardenne: *Proclossiana eunomia* and *Lycaena helle*. In: Habel JC, Assmann T (eds) Relict species – phylogeography and conservation biology. Springer, Heidelberg

Goudet J (1995) FSTATversion 1.2: a computer programm to calculate F-statistics. Heredity 86:485–486

Govindasamy B, Duffy PB, Coquard J (2003) High-resolution simulations of global climate, part 2: effects of increased greenhouse cases. Clim Dyn 21:391–404

Habel JC, Meyer M, Schmitt T, Assmann T (2008) Polymorphic microsatellite loci in the endangered butterfly *Lycaena helle* (Lepidoptera: Lycaenidae). Eur J Entomol 105:361–362

Habel JC, Finger A, Meyer M, Louy D, Zachos F, Assmann T, Schmitt T (2009) Unprecedented long-term genetic monomorphism in an endangered relict butterfly species. Conserv Genet (in press)

Habel JC, Schmitt T, Müller P (2005) The fourth paradigm pattern of postglacial range expansion of European terrestrial species: the phylogeography of the Marbled White butterfly (Satyrinae, Lepidoptera). J Biogeogr 32:1489–1497

Hanski I (1999) Metapopulation ecology. Oxford University Press, Oxford

Hanski I (2005) The shrinking world: ecological consequences of habitat loss. International Ecology Institute, Oldendorf

Hewitt GM (1996) Some genetic consequences of ice ages, and their role in divergence and speciation. Biol J Linn Soc 58:247–276

Hewitt GM (2001) Speciation, hybrid zones and phylogeography – or seeing genes in space and time. Mol Ecol 10:537–549

Hijmans RJ, Cameron SE, Parra JL, Lones PG, Jarvis A (2005) Very high resolution interpolated climate surfaces for global land areas. Int J Climatol 25:1965–1978

Hijmans RJ, Graham CH (2006) The ability of climate envelope models to predict the effect of climate change on species distributions. Global Chang Biol 12:2272–2281

Holt RD, Gomulkiewicz R (2004) Conservation implication of niche conservatism and evolution in heterogeneous environments. In: Ferrière R, Dieckmann U, Couvet D (eds) Evolutionary conservation biology. Cambridge University Press, Cambridge

Huntley B, Birks HJB (1983) An atlas of past and present pollen maps of Europe: 0–130000 years ago. Cambridge University Press, Cambridge

Huntley B, Green RH, Collingham YC, Willis SG (2007) A climatic atlas of European breeding birds. Durham University, the RSPB and Lynx, Barcelona

Huntley B, Webb T (1989) Migration: Species' response to climatic variations caused by changes in the earth's orbit. J Biogeogr 16:5–19

Keyghobadi N, Roland J, Strobeck C (2005) Genetic differntiation and gene flow among populations of the alpine butterfly *Parnassius smintheus*, vary with landscape connectivity. Mol Ecol 14:1897–1909

Keyghobadi N, Unger KP, Weintraub JD, Fonseca DM (2006) Remnant populations of the Regal Fritillary (*Speyeria idalia*) in Pennsylvania: local genetic structure in a high gene flow species. Conserv Genet 2006:309–313

Kimberly A, Selkoe RJT (2006) Microsatellites for ecologists: a practical guide to using and evaluating microsatellite markers. Ecol Lett 9:615–629

Lesica P, Allendorf FW (1995) When peripheral populations are valuable for conservation. Conserv Biol 9:753–760

Melbourne BA, Hastings A (2008) Extinction risk depends strongly on factors contributing to stochasticity. Nature 454:100–103

Meyer M (1982) Révision systématique, chorologique et écologique de *Lycaena helle* Denis & Schiffermüller, 1775 (Lycaenidae). 3° partie: l'écologie. Linneana Belg 8:451–466

Moritz C (1994) Defining evolutionarily significant units for conservation. Trends Ecol Evol 9:373–375

Nève G, Pavlíc A, Konvicka M (2009) Loss of genetic diversity through spontaneous colonisation in the Bog Fritillary butterfly (*Proclossiana eunomia*) in the Czech Republic (Lepidoptera: Nymphalidae). Eur J Entomol 106:11–19

Nève G (1996) Dispersion chez une espèce à habitat fragmenté : *Proclossiana eunomia* (Lepidoptera, Nymphalidae). PhD Thesis, Louvain-la-Neuve

Nève G, Barascud B, Descimon H, Baguette M (2008) Gene flow rise with habitat fragmentation in the bog fritillary butterfly (Lepidoptera: Nymphalidae). BMC Evol Biol 8:84

Parmesan C (2006) Ecological and evolutionary responses to recent climate change. Ann Rev Ecol Evol 37:637–669

Parmesan C, Yohe G (2003) A globally coherent fingerprint of climate change impacts across natural systems. Nature 421:37–42

Pearman PB, Guisan A, Broennimann O, Randin CF (2008) Niche dynamics in space and time. Trends Ecol Evol 23:149–158

Pearson RG, Thuiller W, Araújo MB, Martinez-Meyer E, Brotons L, McClean C, Miles L, Segurado P, Dawson TP, Lees DC (2006) Model-based uncertainty in species range prediction. J Biogeogr 33:1704–1711

Peterson AT, Soberon J, Sanchez-Cordero V (1999) Conservatism of ecological niches in evolutionary time. Science 285:1265

Phillips SJ, Anderson RP, Schapire RE (2006) Maximum entropy modeling of species geographic distributions. Ecol Modell 190:231–259

Pritchard JK, Stephens M, Donnelly P (2000) Inference of population structure using multilocus genotype data. Genetics 155:945–959

Raymond M, Rousset F (1995) An exact test for population differentiation. Evolution 49:1280–1283

Reed DH, Frankham R (2003) Correlation between fitness and genetic diversity. Conserv Biol 17:230–237

Rödder D, Schlüter A, Lötters S (2009) Is the 'Lost World' lost? High endemism of South American tepuís in a changing climate. In: Habel JC, Assmann T (eds) Relict species – phylogeography and conservation biology. Springer, Heidelberg

Root TL, Price JT, Hall KR, Schneider SH, Rosenzweig C, Pounds JA (2003) Fingerprints of global warming on wild animals and plants. Nature 421:57–60

Rousset F (1997) Genetic differentiation and estimation of gene flow from F-statistics under isolation by distance. Genetics 145:1219–1228

Rousset F (2008) GENEPOP'007: a complete re-implementation of the GENEPOP software for Windows and Linux. Mol Ecol Res 8:103–106

Saitou N, Nei M (1987) The neighbor-joining method: A new method for reconstructing phylogenetic trees. Mol Biol Evol 4:406–425

Schmitt T (2007) Molecular biogeography of Europe: pleistocene cycles and postglacial trends. Front Zool 4:11

Selkoe T, Toonen RJ (2006) Microsatellites for ecologists: a practical guide to using and evaluating microsatellite markers. Ecol Lett 9:615–629

Slatkin M (1985) Rare alleles as indicator of gene flow. Evolution 39:53–65

Steiner R, Trautner J, Grandchamp A-C (2006) Larvalhabitate des blauschillernden Feuerfalters (*Lycaena helle*) am schweizerischen Alpennordrand unter Berücksichtigung des Einflusses von Beweidung. Abhandlungen aus dem Westfälischen Museum für Naturkunde 68:135–151

Turlure C (2006) Deux Lépidoptères qui utilisent la même plante-hôte occupent-ils nécessairement le même habitat? Cas du Nacré et du Cuivré de la bistorte, Diploma-thesis, Université catholique de Louvain, Belgium

Van Oosterhout C, Hutchinson WF, Wills DPM, Shipley P (2004) MICRO-CHECKER (Version 2.2.3): software for identifying and correcting genotyping errors in microsatellite data. Mol Ecol Notes 4:535–538

Van Swaay CAM, Warren M (1999) Red data book of European butterflies (Rhopalocera). Nature and environment 99. Council of Europe Publishing, Strasbourg

Varga Z, Schmitt T (2008) Types of areal and oreotundral disjunctions in the Western Palearctic. Biol J Linn Soc 93:415–430

Waltari E, Hijmans RJ, Peterson AT, Nyari ÁS, Perkins SL, Guralnick RP (2007) Locating pleistocene refugia: comparing phylogeographic and ecological niche model predictions. PLoS ONE 2:e563

Wiens JJ, Graham CH (2005) Niche conservatism: integrating evolution ecology and conservation biology. Annu Rev Ecol Syst 36:519–539

Willis KJ, van Andel TH (2004) Trees or no trees? The environments of central and Eastern Europe during the Last Glaciation. Quat Sci Rev 23:2369–2387

Relict Species Research: Some Concluding Remarks

Jan C. Habel, Thomas Schmitt, and Thorsten Assmann

Relict species or populations mostly occur in small and isolated habitats and are remnants of wider distributions in the past. The recent situation of such populations, which have often been scattered along the periphery of the distribution range over long time periods, has led to independent evolutionary processes. These are reflected today in morphologically and genetically unique characteristics that occur exclusively in these populations. Such evolutionary lineages caused by long-term isolation are often described as "evolutionary significant units"; the theory emphasises the distinctiveness of locally restricted phenological, morphological, and/or genetic units. Furthermore, these differentiation processes represent an important evolutionary potential, which distinguishes relict populations from the populations of the main distribution. This evolution in relict populations may in some cases be of high value for the future adaptiveness of the species concerned. On the other hand, very isolated relict populations often also suffer from repeated bottlenecks and the negative effects of genetic drift which enforce the evolution of local uniqueness, but cause losses of genetic diversity on the taxon and intraspecific level. Thus, ecological selection acts much more strongly and therefore more quickly in small and isolated remnant populations than in core populations. Furthermore, populations at the edge of the species' distribution are often exposed to ecological conditions that are different from those in the core area. This selection under different habitat conditions can lead to new allele combinations essential to survival under conditions which differ from those of the core area; they may therefore be relevant for responses to the challenge of future climate changes. Following the contribu-

J.C. Habel (✉)
Musée national d'histoire naturelle, Section Zoologie des Invertébrés, 25, rue Münster, L-2160 Luxembourg, Luxembourg
e-mail: Janchristianhabel@gmx.de

T. Schmitt
Department of Biogeography, Trier University, D-54286 Trier, Germany

T. Assmann
Institute of Ecology and Environmental Chemistry, Leuphana University Lüneburg, Scharnhorststr 1, D-21335 Lüneburg, Germany

J.C. Habel and T. Assmann (eds.), *Relict Species: Phylogeography and Conservation Biology*,
DOI 10.1007/978-3-540-92160-8_26,

tions presented in this book, we conclude that peripheric populations are of particular importance for conserving the complete evolutionary and adaptive potential of species and their intraspecific variability in terms of the Convention on Biological Diversity.

We hope that this compilation of contributions on relict species and populations is of interest for conservationists and biologists, as many aspects of the fields conservation biology, evolutionary biology, ecology, paleontology, modelling, and conservation politics are highlighted using a variety of methods, including molecular genetics, morphometry, and habitat, as well as climate envelope modelling, to explain the complexity of relict species against the background of climate modifications in the past, present, and future.

Index

Erratum to:

Population Genetic Structure of Two Threatened Dragonfly Species (Odonata: Anisoptera) as Revealed by RAPD Analysis

Mónica A. Amorín, Marta Vila, and Adolfo C. Rivera

M.A. Amorín and A.C. Rivera (✉)
Department of Ecology and Animal Biology, University of Vigo, EUET Forestal, Campus Universitario, E-36006 Pontevedra, Spain
e-mail: adolfo.cordero@uvigo.es

M. Vila
Department of Molecular and Cell Biology, University of La Coruña, Faculty of Sciences, Campus Zapateira, E-15008 La Coruña, Spain

J.C. Habel and T. Assmann (eds.), *Relict Species: Phylogeography and Conservation Biology*, DOI 10.1007/978-3-540-92160-8_17,

DOI: 10.1007/978-3-540-92160-8_27

The correct spelling of the authors' names is:

Mónica Azpilicueta Amorín, Marta Vila, and Adolfo Cordero Rivera

M. Azpilicueta Amorín and A. Cordero Rivera (✉)
Department of Ecology and Animal Biology, University of Vigo, EUET Forestal, Campus Universitario, E-36006 Pontevedra, Spain
e-mail: adolfo.cordero@uvigo.es

M. Vila
Department of Molecular and Cell Biology, University of La Coruña, Faculty of Sciences, Campus Zapateira, E-15008 La Coruña, Spain

---

The original online version for this chapter can be found at
http://dx.doi.org/10.1007/978-3-540-92160-8_17

Erratum to:

Review
Modelling Future Trends of Relict Species

Dennis Rödder and Johannes Dambach

J.C. Habel and T. Assmann (eds.), *Relict Species: Phylogeography and Conservation Biology*,
DOI 10.1007/978-3-540-92160-8_22, © Springer-Verlag Berlin Heidelberg 2010

DOI: 10.1007/978-3-540-92160-8_28

The correct chapter title is:

Bioclimatic Models as Predictive GIS Tools for the Identification of Potential Refuges and Possible Migration Pathways

The original online version for this chapter can be found at
http://dx.doi.org/10.1007/978-3-540-92160-8_22

André Platzer

Logical Analysis of Hybrid Systems

Proving Theorems for Complex Dynamics

Springer

Dr. André Platzer
Carnegie Mellon University
School of Computer Science
5000 Forbes Ave.
Pittsburgh PA 15213
USA
aplatzer@cs.cmu.edu

ISBN 978-3-642-14508-7 e-ISBN 978-3-642-14509-4
DOI 10.1007/978-3-642-14509-4
Springer Heidelberg Dordrecht London New York

Library of Congress Control Number: 2010934645

ACM Computing Classification (1998): F.4.1, F.3, D.2.4, I.2.3, G.1.7, I.2.8

Cover design: KuenkelLopka GmbH, Heidelberg

Printed on acid-free paper

Springer is part of Springer Science+Business Media (www.springer.com)

Foreword

Hybrid Systems are notoriously hard to analyze and verify. So far, techniques based on either explicit or implicit state reachability have failed to scale with the size of such systems. Statistical Model Checking may prove useful, but sacrifices absolute certainty about the correctness of the answer obtained. In both cases, numerical error may change the truth-value of the result from False to True or vice versa. An alternative is to use a combination of decision procedures for real arithmetic and interactive theorem proving. Andre Platzer's Ph.D. thesis explores this alternative approach in great depth. He proposes a logic called Differential Dynamic Logic for specifying properties of Hybrid Systems, investigates the meta-theory of the logic, and gives inference rules for it. He has developed an extremely impressive graphical interface for the resulting tool KeYmaera, which is based on the KeY Prover for verifying Java programs, developed at the University of Karlsruhe, Chalmers, and Koblenz. Particularly noteworthy are the use of Differential Invariants for reasoning about complex Hybrid Systems and the examples that he is able to do: "The European Train Control System" and a curved flight roundabout maneuver for aircraft collision avoidance. Both examples are beyond the scope of current Hybrid System Model Checking tools. I believe that his verification tool is unique – there is no other one like it. I heartily recommend his book and theorem prover for those who need to verify complex cyber-physical systems.

Pittsburgh, February 2010 *Edmund M. Clarke*

Preface

The design of complex systems is essential to much of engineering and science. Equally essential is the effort to fully understand these systems and to develop tools and techniques that can steer us away from unsafe or incorrect designs. In civil engineering, for example, well-understood principles like statics can be used to analyse buildings before they are built, and refined architectural models can be used to predict whether a building will be safe or whether it might collapse during an earthquake. Similarly, in auto body design, wind tunnels and corresponding computer models based on computational fluid dynamics help engineers to gain an understanding of aerodynamic forces and wind resistance for energy efficiency before constructing the actual car and its chassis. Models and their analysis also play an important role in chip design and are used extensively in the semiconductor industry to prevent expensive bugs in hardware. Modelling and model analysis is thus an integral part of science and engineering and is used very effectively in many areas to ensure high-quality system designs, saving replacement cost and preventing dangerous side effects of malfunctioning designs.

Hybrid systems is an emergent area of growing importance, emphasising a systematic understanding of systems that combine discrete (e.g., digital) and continuous (e.g., analog or physical) effects. In fact, it is foreseeable that hybrid systems and the closely related notion of cyber-physical systems will soon play a ubiquitous role in engineering. Combinations of computation and control can lead to very powerful system designs, and computational aspects are being integrated into classical physical, mechanical, and chemical process controls on a routine basis today. The number of systems where both computational and physical aspects are important for really understanding them grows exponentially with modern technological advances. Hybrid systems occur frequently in automotive industries, aviation, railway applications, factory automation, process control, medical devices, mobile robotics, and mixed analog–digital chip design.

Despite the growing relevance in complex system designs, hybrid systems is an area where analytic approaches are still in their infancy. Hybrid systems occur ubiquitously and their analysis faces inherent complexity challenges. Hence, there is probably no other area where the gap is more noticeable between the tremendous complexity of the systems we can build and the modest size of systems that we can analyse. Mankind can build systems that are significantly more complicated than people can understand analytically. This book presents an approach with logical analysis techniques that are intended to help overcome these difficulties and bridge the gap between design demand and analysis power.

In light of this growing interest in the field, the purpose of this book is to provide an introduction to hybrid systems analysis and, in particular, to present a coherent logical analysis approach for hybrid systems. One of the highly successful techniques used for analysing finite-state models in chip designs today is *model checking*, which was pioneered in 1981 by the 2007 ACM Turing Award Laureates Edmund M. Clarke, Allen Emerson, and Joseph Sifakis. Nowadays, model checking is used routinely in the semiconductor industry. Model checking is one of the inspirations for this work. Another area that is strongly related is interactive and *automated theorem proving*, which is also used in advanced industrial settings. Model checking and automated theorem proving complement each other to tackle various aspects of formal system verification. While both areas are ultimately rooted in logic, the basic operating principles are somewhat different. Model checking is based on systematically exploring the state space of a system in a clever way. Model checking searches for counterexamples, i.e., traces of a system that lead to a bug and that serve as a falsification of a correctness property. Impressive results have been demonstrated for finite-state systems where model checking is decidable. In theorem proving, in contrast, the notion of a proof is fundamental and represents a verification of a correctness property. In particular, a proof is a reason and explanation for *why* a system works. Automated theorem proving techniques that construct proofs automatically are another deep source of inspiration for the work presented here. In fact, several of the proof procedures presented in this book are inspired by theorem proving principles that have been used successfully for conventional object-oriented programs.

One important new aspect in hybrid systems is the cardinality and structure of the state space. In (sufficiently small) finite state spaces, for instance, exhaustive state exploration is still feasible, but becomes inherently impossible for the uncountable continuous state spaces of hybrid systems, especially with respect to their complicated interacting discrete and continuous dynamics. Most notably, the continuous dynamics of hybrid systems that is commonly described by differential equations poses significant new challenges compared to classical settings. Thus, verification techniques for differential equations are one very important part of hybrid systems analysis.

Outline

This book is intended as an introduction to hybrid systems and advanced analysis techniques for their dynamics. It covers basic and advanced notions of hybrid systems, specification languages for hybrid systems, verification approaches for hybrid systems, and application scenarios for hybrid systems verification. Starting from a basic background in mathematics and computer science, this book develops all notions required for understanding and analysing hybrid systems. It also provides background material about logic and differential equations in the appendix.

This book presents a coherent logical foundation for hybrid systems analysis that will help the reader understand how behavioural properties of hybrid systems can be analysed successfully. The foundation developed here serves as a basis for advanced hybrid system analysis techniques. The hybrid systems analysis approach has also been implemented in the verification tool KeYmaera for hybrid systems, which is available for download at the book's Web page.

Part I describes specification and verification logics for hybrid systems that are the basis of hybrid systems analysis. It also covers constructive proof calculi that can be used to analyse and verify hybrid systems, including approaches for handling real arithmetic and differential equations. The chapters in Part I show a series of logical systems that are each presented in terms of their syntax, semantics, axiomatics, proof theory, and pragmatics. The syntax defines what can be said about hybrid system behaviour. The semantics gives meaning to the symbolic formulas and shows what we are ultimately interested in: truth, or, more precisely, what is true about the behaviour of the particular dynamics of a hybrid system. In the axiomatic parts, this book develops formal proof techniques that can be used by a human or machine to establish truth by proof. After all, truth that we do not know about is less helpful than truth that we can justify by giving a proof. The development of proof theory connects the semantic notion of truth in the real world with the syntactic device of formal proofs and shows that, in a sense of relative completeness, we can prove all true facts about hybrid systems from elementary properties of differential equations. Finally, this book shows the pragmatics of using verification procedures for analysing hybrid system scenarios. This includes both practical, algorithmic considerations of developing system analysis tools and various examples, application scenarios, and case studies that can be proven with the logics developed in Part I.

Part II focuses on the practical and algorithmic questions of how to turn the theoretical foundation from Part I into automated theorem proving procedures. This part also shows techniques for generating invariants and differential invariants of hybrid systems that are crucial for proving correctness, and shows how to overcome complexity challenges in real arithmetic verification. Part III shows how safety-critical properties of more advanced applications of hybrid systems in railway and aircraft control can be proven with the approach presented in Parts I and II. This part includes a study of collision avoidance in the European Train Control System (ETCS) and roundabout collision avoidance manoeuvres in air traffic control. Numerous examples, illustrations, and proofs throughout the text will also help the reader develop an intuition about hybrid systems behaviour and master the intricacies of the more subtle aspects in hybrid systems analysis.

How to Read This Book

The basic suggested reading sequence is linear (with additional consultation of the appendices for background information as needed). Except for the foundation of this work that is laid out in Part I, however, the chapters are mostly kept self-contained so that they can also be studied independently. The following figure shows the reading order dependencies among the chapters (solid lines) and the partial dependencies of suggested reading sequences that hold for the advanced material of the respective chapters (dashed lines).

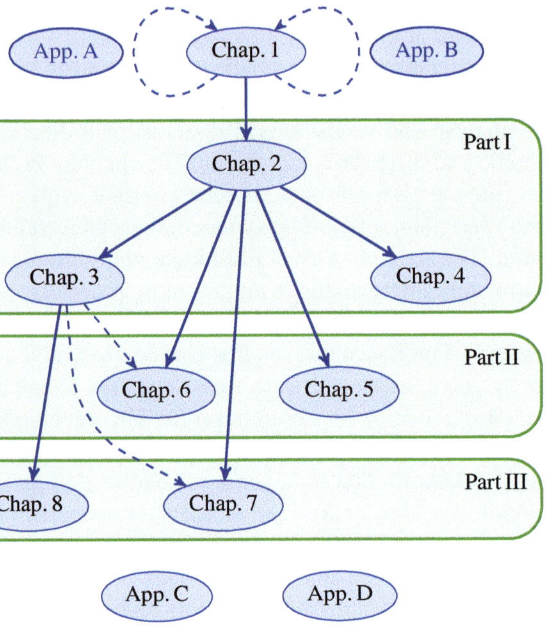

For background on classical first-order logic, we recommend you review App. A as needed. Depending on your interest, field of study, and preference, we recommend you either study the background information in App. A on first-order logic before reading Part I or use the material in App. A as a background reference book on demand while reading the main part of this book. Similarly, we recommend you review the background on ordinary differential equations in App. B either before or during the study of the main part. An intuitive approach to understanding differential equations and formal definitions of their semantics will be given throughout the text. Logic itself is also explained and illustrated intuitively during the main chapters, but some readers may also find it helpful to refresh, update, or learn about the basics of first-order logic from App. A before proceeding to the main part.

While there is a lot of flexibility in the reading sequence of the chapters, we strongly recommend you study the logical foundations of hybrid systems analysis in Chap. 2 of Part I before reading any other chapter of Parts I–III. Some more advanced sections in the applications in Part III also depend on the theory of differential invariants that is developed together with other extensions in Chap. 3.

Appendix C shows a formal relation of hybrid automata with hybrid programs. Appendix D gives more detail on the implementation of the approach put forth in this book in the verification tool KeYmaera. It also presents a survey of computational techniques for handling real arithmetic. Both App. C and D can be read as needed, after studying the introductory material and notions in Chap. 2. The most important formation rules for the logic and proof rules for the calculi are summarised at the end of the book.

Online Material for This Book

The Web page for this book provides online material, including the verification tool KeYmaera that implements our logical analysis approach for hybrid systems. We also provide slide material for parts of this book, an online tutorial for KeYmaera, and several KeYmaera problem files for examples from this book, including train and air traffic control studies. The book Web page is at the following URL:

`http://symbolaris.com/lahs/`

Acknowledgements

This book is based on my Ph.D. thesis and would not have been possible without the support of the PIs and collaborators on the projects that I have been working on. My sincere thanks go to Prof. Ernst-Rüdiger Olderog for his excellent advice and support, and for giving me the opportunity to work in one of the most fascinating areas of science in a group with a friendly and productive atmosphere. My advisor, Prof. Olderog, and the Director of AVACS, Prof. Werner Damm, both deserve my highest gratitude, not only for their continuous support and for their faith, but also for allowing me the freedom to pursue my own research ambitions in the stimulating context of the AVACS project ("Automatic Verification and Analysis of Complex Systems"). Ultimately, this made it possible for me to develop the logic and verification approach presented in this book.

I want to thank the external referees of my Ph.D. thesis, Prof. Tobias Nipkow from the Technical University of Munich and Prof. George J. Pappas from the University of Pennsylvania. It is an honour for me that they were willing to invest their valuable time and effort in the careful reviewing of my thesis. In fact, I am thankful to all members of my Ph.D. committee, Werner Damm, Ernst-Rüdiger Olderog, George J. Pappas, Tobias Nipkow, and Hardi Hungar for fruitful discussions and for the highest support they offered for my work.

I am especially grateful to Prof. Edmund M. Clarke, who invited me to Carnegie Mellon University several times, for his support, interest, and collaboration, and for sharing with me parts of his huge knowledge in all areas of formal methods. I further want to acknowledge the help by Prof. Peter H. Schmitt from the University of

Karlsruhe (TH), Profs. Bernhard Beckert and Ulrich Furbach from the University of Koblenz-Landau, Prof. Reiner Hähnle from the Chalmers University of Technology, Gothenburg, Sweden, Profs. Edmund M. Clarke and Frank Pfenning from Carnegie Mellon University, and Prof. Rajeev Goré from the Australian National University, Canberra, at various stages of my career.

I want to thank the program committee of the TABLEAUX 2007 conference for selecting my first paper on differential dynamic logic for the Best Paper Award, the first award at any TABLEAUX conference. This recognition has encouraged me to continue pursuing my research direction, which ultimately led to the results described in this book. I also thank the program committee of the FM 2009 conference for selecting my paper on formal verification of curved flight collision avoidance maneuvers for the Best Paper Award. I am very grateful to the ACM Doctoral Dissertation Award committee for honoring my Ph.D. thesis with the 2009 ACM Doctoral Dissertation Honorable Mention Award.

I am truly thankful to my colleagues at Carnegie Mellon University for their encouraging feedback about my work and for the friendly and constructive atmosphere at CMU. For many fruitful discussions I thank my colleagues and friends from Oldenburg, Ingo Brückner, Henning Dierks, Johannes Faber, Sibylle Fröschle, Jochen Hoenicke, Stephanie Kemper, Roland Meyer, Michael Möller, Jan-David Quesel, Tim Strazny, and especially my office mate Andreas Schäfer. Ernst-Rüdiger Olderog, Johannes Faber, Ingo Brückner, Roland Meyer, Henning Dierks, Silke Wagner, Nicole Betz, Alex Donzé, and especially Andreas Schäfer also deserve credit for proofreading some of my earlier papers, which formed the basis for this book. I also acknowledge Andreas Schäfer's helpful feedback from proofreading parts of this book. I appreciate the feedback of my students on this book.

Furthermore, I thank Jan-David Quesel for writing a Master's thesis under my supervision and for his invaluable support with the implementation of the verification tool KeYmaera based on the techniques that I present in this book and in prior publications. I also thank him for help with the experiments and ETCS. I am also thankful for indispensable and reliable help from Richard Bubel and Philipp Rümmer with the implementation internals of the KeY basis. I thank the whole KeY team for providing the impressive Java verification tool KeY as a basis for our implementation of KeYmaera.

For help with the book process, I thank Ronan Nugent from Springer.

Especially, I thank my parents, Rudolf and Brigitte Platzer, and my sister, Julia, for their continuous support and encouragement, and I thank my wife, Nicole, for her true faith in me. She also deserves credit for her invaluable help with some of the illustrations in this book.

Funding

This research was partly supported by the German Research Council (DFG) under grant SFB/TR 14 AVACS ("Automatic Verification and Analysis of Complex Systems", see `http://www.avacs.org`); a Transregional Collaborative Research

Center of the Max Planck Institute and the Universities of Oldenburg, Saarbrücken, and Freiburg in Germany, with associated cooperations with the University of Pennsylvania, ETH Zürich, and the Academy of Sciences of the Czech Republic. It was further supported partly by a research fellowship of the German Academic Exchange Service (DAAD) and by a research award of the Floyd und Lili Biava Stiftung. Some part of this work was also supported by the National Science Foundation under grant nos. CNS-0931985 and CNS-0926181, including the NSF Expedition on Computational Modeling and Analysis of Complex Systems (CMACS); see `http://cmacs.cs.cmu.edu` for more information.

The views and conclusions contained in this book are those of the author and should not be interpreted as representing the official policies, either expressed or implied, of any sponsoring institution or government.

Further Sources

This book is based on several sources, most notably the author's Ph.D. thesis [236]. Chapter 2 is an extended version of an article in the Journal of Automated Reasoning [235] and also covers some material from previous work at TABLEAUX [231] and HSCC [232]. Chapter 3 is an extended version of an article in the Journal of Logic and Computation [237], to which we now add a relative completeness argument and prove that DAL is a conservative extension of the sublogic $d\mathcal{L}$. We further combine the solution-based techniques from Chap. 2 with differential induction-based techniques from Chap. 3 by introducing the new extension of differential monotonicity relaxations. Chapter 4 is a substantially extended version of a previous paper at LFCS [233], to which we now add a complete and more elegant calculus and provide a modular relative completeness proof.

In Chap. 5, we extend a previous paper at VERIFY [230] with more details on iterative background closure strategies, including experimental evaluation, and complement this proof technique with a new iterative inflation strategy. Chapter 6 is based on joint work with Edmund M. Clarke at CAV [239] and in Formal Methods in System Design [240].

Chapter 7 is a substantially revised and improved version of joint work with Jan-David Quesel at HSCC [243] with extensions from follow-up work [244]. Chapter 8 is a significantly improved and detailed case study developed on the basis of joint work with Edmund M. Clarke at HSCC [238] and CAV [239] with subsequent extensions at FM [241].

Appendix B summarises classical results from the theory of differential equations from the literature [297]. Finally, App. D uses a few excerpts from joint work with Jan-David Quesel at IJCAR [242], adding an overall discussion of the KeYmaera verification tool that implements the approach presented in this book. Appendix D also adds a thorough description of computational back-ends for real arithmetic, with extensions from joint work with Philipp Rümmer and Jan-David Quesel [246].

Pittsburgh, February 2010 *André Platzer*

Contents

List of Figures

List of Tables

List of Theorems

Chapter 1
Introduction

Time is defined so that motion looks simple [209, p. 23]

Ensuring correct functioning of complex physical systems is among the most challenging and most important problems in computer science, mathematics, and engineering. In addition to nontrivial underlying physical system dynamics, the behaviour of complex systems is determined increasingly by computerised control and automatic analog or digital decision-making, e.g., in aviation, railway, and automotive applications. At the same time, correct decisions and control of these systems are becoming increasingly important, because more and more safety-critical processes are regulated using automatic or semiautomatic controllers, including the European Train Control System [117], collision avoidance manoeuvres in air traffic control [293, 196, 104, 238, 129, 171], car platooning technology for highways following the California PATH project [166], recent driverless vehicle technology [64], and biomedical applications like automatic glucose regulation for diabetes patients [223]. As a more general phenomenon of complex physical systems that are exemplified in these scenarios, correct system behaviour depends on correct functioning of the *interaction* of control with physical system dynamics and is not just an isolated property of only the control logic or only the physical system dynamics. These interactions of computation and control are more difficult to understand and get right than isolated systems. Even if the control software does not crash, the system may still malfunction, because the control software could issue unsafe control actions to the physical process. And even if the pure physics of the process is well understood, an attempt to control the process may still become unsafe, e.g., when the controller reacts to situation changes too slowly because computations take too long, or when sensor values are already outdated once the control actions finally take effect. It is the *interaction* of computation and control that must be taken into account. Systems with such an interaction of discrete dynamics and continuous dynamics are called *hybrid dynamical systems*, or just *hybrid systems* for short.

To illustrate typical aspects and effects in application areas, we take a look at two examples in more detail, which will serve as running examples and case studies throughout this book. In high-speed trains like the ICE (InterCityExpress), TGV (train à grande vitesse), Shinkansen, and the upcoming California High-Speed Rail, whose high mass (1,000–3,000 tons) and high speed (320km/h) cause them to re-

A. Platzer, Logical Analysis of Hybrid Systems,
DOI 10.1007/978-3-642-14509-4_1, © Springer-Verlag Berlin Heidelberg 2010

quire fairly long braking distances (more than 3.8 km), safe driving is impossible just based on sight without automatic technical means that enforce a safe minimum distance between trains. The *European Train Control System* (ETCS), which is currently being developed and installed in Europe [117], regulates and protects train movement according to movement authorities (MAs). A movement authority represents permission for the train to move up to a certain point on the track (the end of the movement authority) and is negotiated dynamically in rapid succession by wireless communication with decentralised radio block controllers (RBCs); see Fig. 1.1.

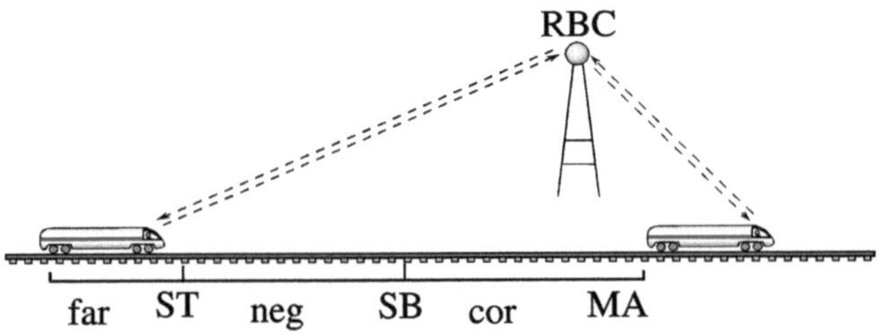

Fig. 1.1 European Train Control System

With the next generation development of ETCS, Level 3, all classical fixed trackside signalling and fixed track segment partitioning with physical separation of the train segments will become obsolete, thereby advancing to a fully autonomous operation of ETCS in order to achieve its performance goals of maximum speed and density on the track. Yet, a safe operation of ETCS requires that—while achieving these performance goals—the train controllers still always respect their local movement authorities and that the radio block controllers only grant compatible movement authorities to each of the trains. Even in emergency situations, the overall train control system must always ensure that the trains cannot crash into one another. To determine correct functioning of these controllers it has to be shown that the train positions, which evolve dynamically over time, are always safely separated. For this, however, we need to be able to analyse the interaction of the train control logic and the wireless ETCS cooperation protocol with a model of the actual physical train dynamics, because collision freedom is not an isolated property of only the discrete cooperation-layer control protocol, only the local train control decision process, or only the continuous train dynamics, but a joint property of their superposition or combination.

Train dynamics is a typical example of what is known as a *hybrid system*: it combines instantaneous discrete jump dynamics with continuous evolution. See Fig. 1.2 for a typical evolution of the acceleration a, which changes instantaneously by discrete control decisions at some points in time, and of the train velocity v and position z, which evolve continuously over time. The continuous dynamics can be described by a differential equation system like $z' = v, v' = a$, saying that the time derivat-

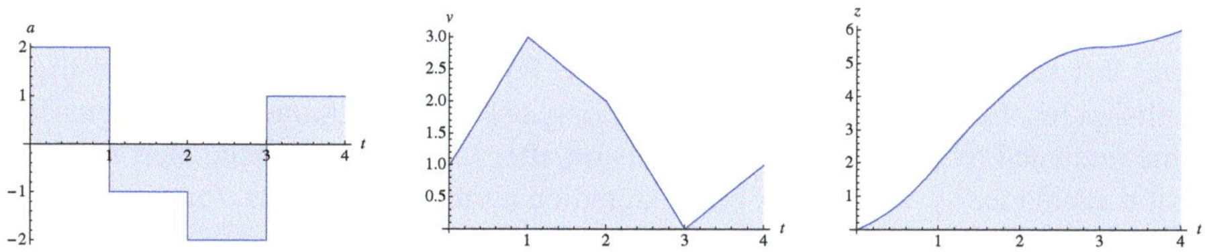

Fig. 1.2 ETCS: discrete evolution of acceleration a, continuous evolution of velocity v and of position z over time t

ive z' of the position equals the velocity v and the time derivative v' of the velocity equals the chosen acceleration a. The discrete dynamics can be described by an assignment such as $a := -2$ together with the conditions of when and under which circumstances this discrete jump will be executed. For example, the discrete dynamics can be described by the operations of the switching control logic, possibly as a small control program that describes under which circumstances the acceleration a switches to braking (negative acceleration), and under which circumstances it switches to positive acceleration. By combining the continuous differential equations and discrete control logic appropriately, we obtain a description of the interacting hybrid dynamics of the hybrid (dynamical) system. Yet the effects of such a combination also need to be well-defined, including a faithful model defining unambiguously how discrete and continuous dynamics work together. Finally, note that full formal verification and proof that the system operates correctly is indeed quite important and of particular practical relevance for safety-critical complex physical systems like ETCS. Despite careful development and testing, safety violations have recently been reported in ETCS [143] even at its moderate currently deployed level.

In air traffic control, collision avoidance manoeuvres [293, 196, 104, 238, 129, 171] are used to help pilots resolve conflicting flight paths that arise during arbitrary free flight of the aircraft. See Fig. 1.3 for an overview of different collision avoidance approaches. These manoeuvres are last-resort means for resolving air traffic

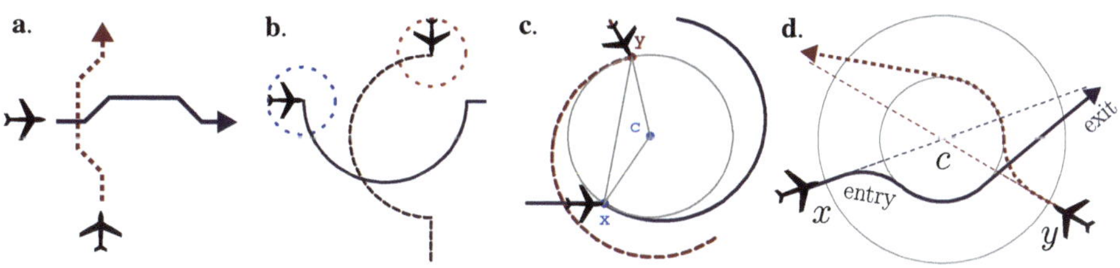

Fig. 1.3 Collision avoidance manoeuvres in air traffic control

conflicts that could otherwise lead to collisions and have not been detected before by the pilots during free flight or by the flight directors of the Air Route Traffic Control Centres. Consequently, complicated online trajectory prediction, trajectory evaluation, or lengthy manoeuvre negotiation may no longer be feasible in the short time that remains for resolving the conflict. For example, in the tragic mid-flight collision in Überlingen [43] in July 2002, only less than one minute of manoeuvring time remained to try to prevent the collision after the on-board traffic alert and collision avoidance system TCAS [196] signalled a traffic alert. Thus, for safe aircraft control we need particularly reliable instant reactions with manoeuvres whose correctness has been established previously by a thorough offline analysis. To ensure correct functioning of aircraft collision avoidance manoeuvres under all circumstances, the temporal evolution of the aircraft in space must be analysed carefully together with the effects that manoeuvring control decisions have on their dynamics, giving again a superposition or combination of physical system dynamics with control, and thus a hybrid system.

The continuous dynamics in air traffic control can again be described by differential equations for the flight dynamics. The relevant system variables for which differential equations need to be found include the two- or three-dimensional position in space and orientation in space for each of the aircraft at positions $x = (x_1, x_2, x_3) \in \mathbb{R}^3$ and $y = (y_1, y_2, y_3) \in \mathbb{R}^3$, for instance. Because rails are essentially a one-dimensional space (except for the track topology) but airspace is three-dimensional, the resulting differential equations for flight are more involved than for trains. The discrete dynamics in air traffic control comes, e.g., from the decisions of when and how to initiate a collision avoidance manoeuvre, and from the discrete change in direction or the angular velocity of the aircraft.

These examples are prototypical and similar phenomena occur in many other application scenarios of hybrid systems. The continuous dynamics often comes from physical movement or physical processes. Discrete dynamics often (but not always) comes from control decisions or digital control implementations.

1.1 Technical Context

In this section we survey the technical context of this book, briefly summarise some important technical concepts in the domain, and preview some technical highlights of this book informally. We also give a short overview of related work.

1.1.1 Hybrid Systems

As a common mathematical model for complex physical systems, *hybrid dynamical systems*, or just *hybrid systems* for short [289, 9, 218, 8, 56, 156, 11, 58, 189, 97, 228, 90, 194], are dynamical systems [227, 177, 279] where the system state evolves over time according to interacting laws of discrete and continuous dynamics, the idea being to capture the superposition or combination of physical system dynamics with control at a natural modelling level. For discrete transitions, the hybrid system changes state instantaneously and possibly discontinuously. During continuous transitions, the system state is a continuous function of continuous time and varies according to a differential equation, which is possibly subject to domain restrictions or algebraic relations resulting from physical circumstances or the interaction of continuous dynamics with discrete control. Continuous dynamics results, for example, from the continuous movement of a train along the track (train position z evolves with velocity v along the differential equation $z' = v$ where z' is the time derivative of z) or from the continuous variation of its velocity over time ($v' = a$ with acceleration a); see evolutions in Fig. 1.2. Other behaviour can be modelled more naturally by discrete dynamics, for example, the instantaneous change of control variables like the acceleration (e.g., the changing of a by setting $a := -b$ with braking force $b > 0$) or the change of status information in discrete controllers; see Fig. 1.2 again. Both kinds of dynamics interact, e.g., when measurements of the continuous state affect decisions of discrete controllers (the train switches to braking mode when v is too high). Likewise, they interact when the resulting control choices take effect by changing the control variables of the continuous dynamics (e.g., changing control variable a in $z'' = a$). The superposition of continuous dynamics with analog or discrete control causes complex system behaviour, which can be analysed neither by purely continuous reasoning (because of the discontinuities caused by discrete transitions) nor by considering discrete change in isolation (because safety depends on continuous states).

Among several other models for hybrid systems [69, 100, 58, 270, 272, 40, 183], the model of *hybrid automata* [156, 8] is one of the more widely used notations. Hybrid automata specify discrete and continuous dynamics in a graph. Even though hybrid automata are a fairly common notation for hybrid systems, there are numerous slightly different notions of hybrid automata or automata-based models for hybrid systems [289, 9, 218, 8, 56, 156, 11, 58, 189, 97, 228, 90]. We consider a number of examples to introduce hybrid automata and illustrate simple hybrid systems informally.

Example 1.1 (Simplistic train control). See Fig. 1.4 for a (much too) simple train control example written as a hybrid automaton. The nodes of the hybrid automaton

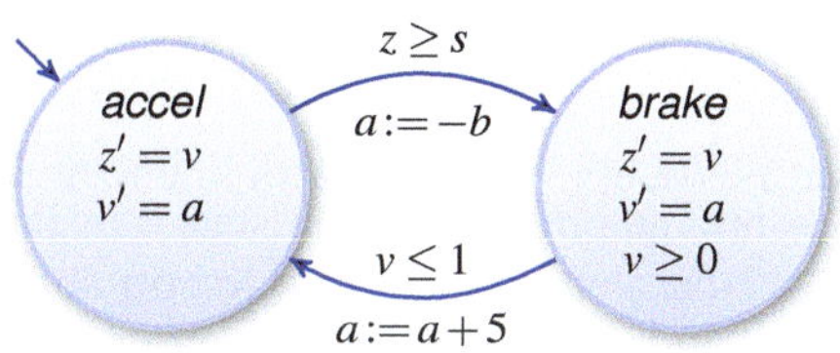

Fig. 1.4 Hybrid automaton for an (overly) simplified train control system

specify the continuous dynamics of the system and the edges specify the discrete switching behaviour between the various modes of continuous dynamics.

Each node of the graph structure corresponds to a continuous dynamical system and is annotated with its differential equations and an evolution domain specifying the maximum domain of evolution. The basic differential equation in both nodes of the train automaton here is $z' = v, v' = a$, because the time derivative of the position z is the velocity v, and the time derivative of the velocity v is the acceleration a. Yet both nodes have different choices for the acceleration a. By specifying a maximum domain of evolution in addition to the differential equations in a node, we restrict the circumstances under which the system is allowed to stay in a node, and enforce when it has to transition to another node instead. The system cannot stay in a node outside this maximum domain of evolution, and is then forced to switch to another node. The system is not required to stay as long as possible in a node until it reaches the border of its maximum domain of evolution, however. Instead, the system is allowed to leave a node earlier if one of the outgoing edges can be used at any time. For example, in the node *brake* of Fig. 1.4, the differential equations $z' = v, v' = a$ only apply within the evolution domain $v \geq 0$ (because the train does not move backwards when braking). Thus, the system is allowed to stay in node *brake* arbitrarily long, but it has to leave before the evolution domain $v \geq 0$ is violated. Edges specify the discrete switching behaviour between the respective modes of continuous evolution. They can be annotated with conditions (*guards*) that need to hold for the state when the system follows the edge, and with discrete state transformations (*jumps*) that take instantaneous effect and transform the continuous state when the system follows the edge. For example, the automaton in Fig. 1.4 can take an edge to leave node *accel* when train position z passes point s (i.e., when the current state satisfies $z \geq s$), which sets the acceleration to braking by changing the state variables according to $a := -b$, and then enter node *brake*. The edge from node *brake* to node *accel* can be taken only when the guard $v \leq 1$ is true for the current continuous state; the transition will then increase the value of the acceleration by $a := a + 5$ and the system will enter node *accel*. The short edge pointing to *accel* from nowhere in the top-left corner indicates that the system will initially start in the *accel* node. In addition, we also need to know what values the continuous variables (z, v, a, b, s) can have in the initial state.

To identify in which state a hybrid automaton is at some point in time, we need to know real number values for all the continuous state variables (z, v, a and the parameters b and s) and we need to know in which node the automaton currently is (either *accel* or *brake*). Over time, the state evolves according to the discrete or continuous dynamics. The state either evolves continuously according to the differential equation (if the state stays inside the evolution domain restriction), or follows one of the discrete transitions of an edge (when the guard is satisfied, thereby transforming the state according to the jump relation). For reference, we formally define hybrid automata and their state transitions in App. C.

Note that an edge of a hybrid automaton can only be taken if the guard holds true and, after updating the state according to the jump, the state is in the evolution domain of the target node. For instance, the automaton starts in *accel*, and may stay some time in *accel* while following the differential equations $z' = v, v' = a$. Then the system could switch to *brake* at some point where the guard $z \geq s$ is satisfied for the current state, thereby setting the acceleration to $a := -b$. Next the system can stay in node *brake*, following the differential equation $z' = v, v' = a$ (now with a different value for a than before), but only while staying inside the permitted evolution domain $v \geq 0$. In particular, the system has to leave node *brake* and switch back to *accel* before the velocity is negative. Because the only outgoing edge from *brake* has a guard $v \leq 1$, the system can switch back to *accel* if and only if the current velocity satisfies $0 \leq v \leq 1$. The system can switch at any point in time where this is satisfied. Suppose the system waits in *brake* until the speed is, say, 0.5. Then the current speed satisfies $v \leq 1$ and the automaton can switch to *accel* again by updating $a := a + 5$.

Note, however, that the resulting acceleration does not have to be positive! If $b > 5$, then the increase $a := a + 5$ for switching to *accel* is less than the braking reset $a := -b$; hence, the value of a after switching from *accel* to *brake* and back to *accel* again will still be negative. Suppose the position still satisfies the guard $z \geq s$ of the outgoing edge back to *brake*. Then, the system might not be allowed to follow this edge, even though $z \geq s$ holds, because the maximum evolution domain of *brake* requires $v \geq 0$ to be true all the time, including when entering the state. That is, in such a run, where the resulting velocity is negative, the automaton is stuck and can never do any other transitions again. In summary, we can see that the operational effect of an edge in an automaton depends not only on the guard and jump of the edge, but also on the evolution domain restrictions in the nodes.

How can we find out if the train model always stays within the safe region on the track? How do we find out for which choices of the parameters s and b the model works as expected? How can we determine if the train is always able to make progress and is never stuck? We will revisit these questions in Chaps. 2 and 7. □

Example 1.2 (Bouncing ball). Another very intuitive example of a hybrid system is the bouncing ball [110]; see Fig. 1.5. The bouncing ball is let loose in the air and is

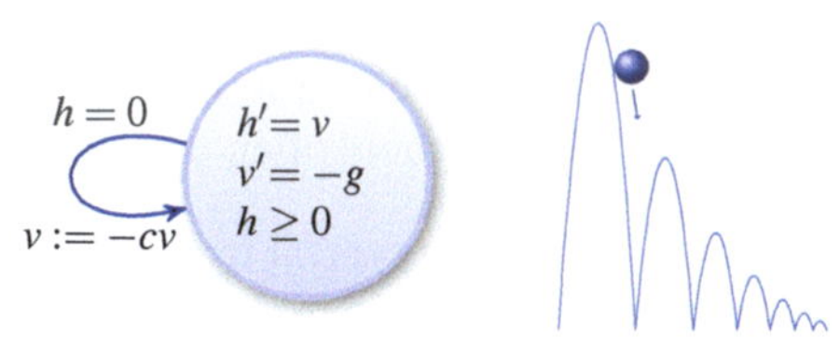

$$\begin{aligned}&(h' = v, v' = -g \,\&\, h \geq 0;\\ &\quad \text{if } (h = 0) \text{ then}\\ &\quad\quad v := -cv\\ &\quad \text{fi})^*\end{aligned}$$

Fig. 1.5 Hybrid automaton and hybrid program of a simple bouncing ball

falling towards the ground. When it hits the ground, the ball bounces back up and climbs until gravity wins and it starts to fall again. The bouncing ball follows the continuous dynamics of physical movement by gravity. It can be understood naturally as a hybrid system, because its continuous movement switches from falling to climbing by reversing its velocity whenever the ball hits the ground and bounces back. Let us denote the height of the ball by h and the current velocity of the ball by v. The bouncing ball is affected by gravity of force $g > 0$, so its height follows the differential equation $h'' = -g$, i.e., the second time derivative of height equals the negative gravity force. The ball bounces back from the ground (which is at height $h = 0$) after an elastic deformation. At every bounce, the ball loses energy according to a damping factor $0 \le c < 1$. Figure 1.5 depicts a hybrid automaton, an illustration of the system dynamics, and a representation of the same system in a program notation for hybrid systems. The hybrid automaton has only one node: falling along the differential equation system $h' = v, v' = -g$ (which is equivalent to $h'' = -g$) restricted to the evolution domain $h \ge 0$, above the floor. In particular, the bouncing ball can never fall through the floor. The hybrid automaton also has only one jump edge: on the ground (when $h = 0$), it can reset the velocity v to $-cv$ and continue in the same node. This jump will change the direction from falling (the velocity v was negative before) to climbing (the velocity $-cv$ is nonnegative again) after dampening the velocity v by c.

The program on the right of Fig. 1.5 represents the same bouncing ball system textually (we will call these textual representations hybrid programs and examine them in detail in Chap. 2). The first line of the program describes the continuous dynamics along the differential equation $h' = v, v' = -g$ restricted to (written &) the maximum evolution domain region $h \ge 0$. After the sequential composition (;), an if-then statement is executed, which resets velocity v to $-cv$ by assignment $v := -cv$ if $h = 0$ holds at the current state. Finally, the sequence of continuous and discrete statements can be repeated arbitrarily often, as indicated by the regular-expression-style repetition operator (*) at the end.

Note one strange phenomenon in the bouncing ball automaton and program. It seems like the bouncing ball will bounce over and over again, switching its direction in shorter and shorter periods of time as indicated in Fig. 1.5 (unless $c = 0$, which means that the ball will just lie flat right away). Even worse, the ball will end up switching directions infinitely often in a short amount of time. This controversial phenomenon is called Zeno behaviour, which we discuss in more detail in later parts of this book.

How can we see that the bouncing ball never bounces higher than its initial height? How can we ensure that the bouncing ball fits closer to the physical reality that it stops after some time, instead of bouncing infinitely often in finite time? □

Example 1.3 (Switching damped oscillators). Another typical example of a hybrid system is a switching damped oscillator with continuous variables x, y; see Fig. 1.6. This system consists of two continuous modes of damped oscillators with different

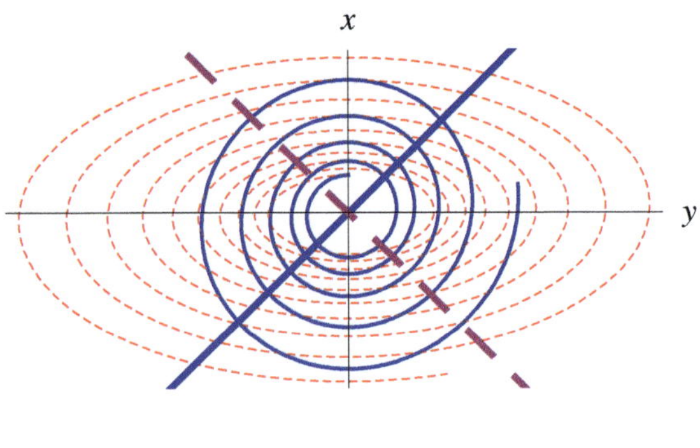

Fig. 1.6 Switching between two damped oscillators

settings. The solid spiral in Fig. 1.6 illustrates the one continuous mode, and the dashed spiral the other continuous mode. The overall system is a hybrid system that switches between those two modes at the switching surfaces depicted by the diagonal lines in Fig. 1.6. Also see a hybrid automaton model of this system in Fig. 1.7.

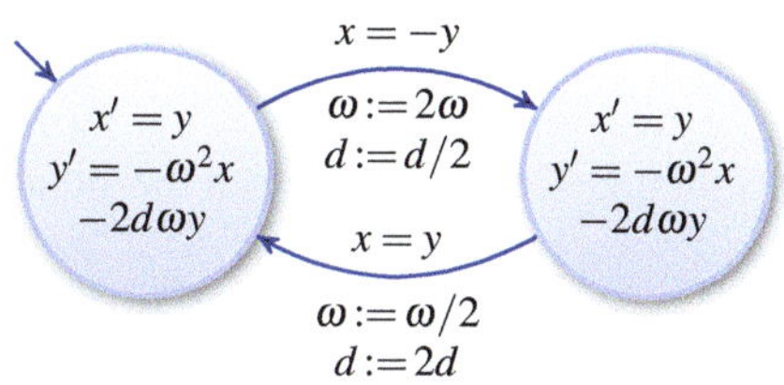

Fig. 1.7 Hybrid automaton for switching damped oscillators

The system starts with the solid dynamics in the node on the left. When passing the dashed secondary diagonal ($x = -y$), it switches to following the dashed continuous mode (right node in Fig. 1.7). When passing the solid main diagonal ($x = y$), the hybrid system switches to the solid continuous dynamics (left node). An example of

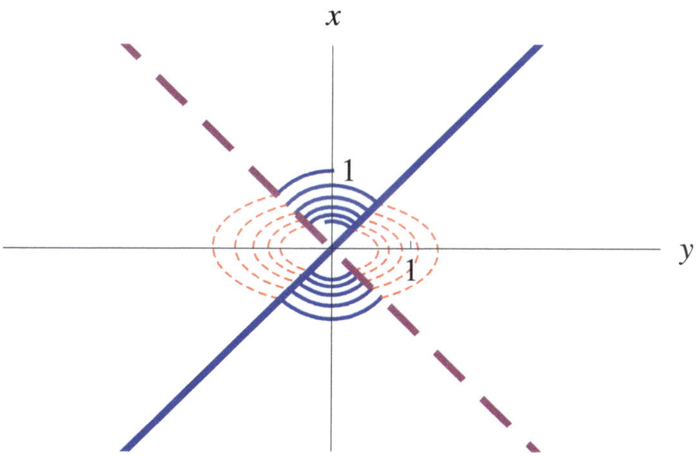

Fig. 1.8 Stable trajectory switching between two damped oscillators

an evolution of this hybrid system that switches between both continuous modes according to these switching conditions is shown in Fig. 1.8. The differential equations for the damped oscillators are of the form

$$x' = y, y' = -\omega^2 x - 2d\omega y$$

with the parameters ω (for the undamped angular frequency) and d (for the damping ratio). These parameters will be adjusted whenever the hybrid system follows an edge in the automaton to a different continuous evolution mode. The solid parts of the curves in Fig. 1.8 correspond to the parts where the system is in the left node of the automaton in Fig. 1.7, and thus follow the solid spiral in Fig. 1.6 The dashed parts of the curves in Fig. 1.8 correspond to the parts where the system has switched to the right node of the automaton in Fig. 1.7, and thus follow the dashed spiral in Fig. 1.6. The switching happens at the diagonals.

The evolution in Fig. 1.8 is *stable*, i.e., it converges to the origin $x = 0, y = 0$. Yet if we tilt the switching axes only slightly and switch a little later —for instance use switching condition $x = -0.5y$ instead of $x = -y$ on the edge from the left to the right node in Fig. 1.7— then the situation is quite different; see Fig. 1.9. Under these

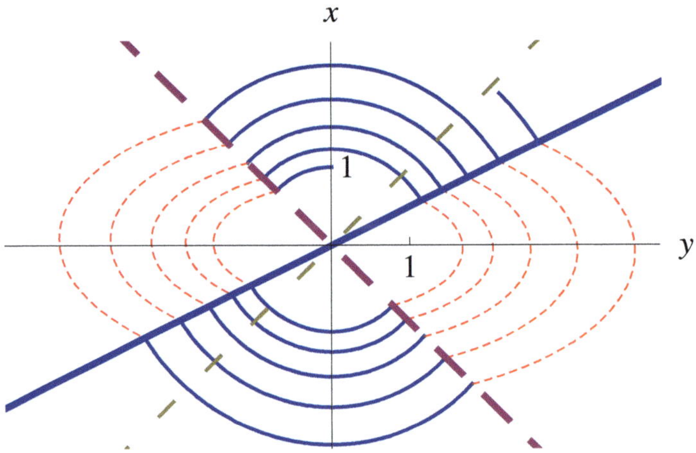

Fig. 1.9 Instable trajectory switching between two damped oscillators

circumstances, the evolution from the same initial point $x = 1$, $y = 0$ is suddenly *instable*, i.e., it does not converge to the origin $x = 0$, $y = 0$. Instead, the evolution diverges and x, y grow infinitely large. This is quite a remarkable difference caused by a seemingly minor variation in the switching conditions of the hybrid systems. In fact, this instability is caused only by the hybrid system switching, because the continuous dynamics alone is stable in both evolution modes (i.e., if both spirals in Fig. 1.6 are considered separately). How do we find out which parameters for the switching surfaces make the system stable, and which parameter choices render it instable? How can we establish whether safety and/or stability depend on the parameters ω and d?

Finally note that the evolutions in Figs. 1.8 and 1.9 both switch nodes in the automaton eagerly, i.e., as soon as the guards of outgoing transitions are true. This is certainly one choice, but hybrid systems do not restrict the evolution to follow permitted transitions as early as possible. Instead, the hybrid automaton in Fig. 1.7 would allow the system to skip a switching possibility and wait for the next one to come along. Nondeterminism like this occurs quite frequently when modelling natural phenomena in complex physical systems, especially in the presence of external control decisions or input. How does this flexibility in switching affect the behaviour of the system? How do we establish for which choices of switching diagonals the overall system behaviour is well behaved. We will come back to these questions about switching damped oscillators in Sect. 3.5.6. □

Example 1.4 (Simple water tank). As a further illustration of a hybrid system, consider a simple water tank model, where a controller tries to keep the water level x in a tank between a lower bound of 1 and an upper bound of 10; see Fig. 1.10. The

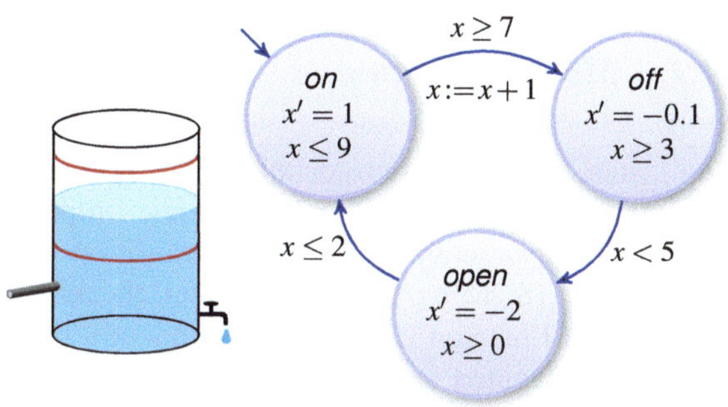

Fig. 1.10 Simple water tank system

water tank can be filled with a pump (node *on*), or the pump can be turned off (node *off*). For a twist, let us assume that, before activating the pump again, the valve has to be opened completely, so that water pours out quickly (node *open*) until the pump is turned *on* again. While the pump is *on*, water enters the tank, increasing the water level continuously by the differential equation $x' = 1$. The tank has a water level sensor that shuts off the pump before the water level exceeds 9; hence, the continuous evolution in mode *on* is restricted to the maximum evolution domain $x \leq 9$. When the water level is $x \geq 7$, the controller is allowed to shut the pump *off* by taking the edge from node *on* to node *off*, but during the shutdown of the valve, more water pours in. That is, the water level increases from x to $x+1$ by the discrete change $x := x+1$ on the edge. Since the evolution in node *on* is restricted to the evolution domain $x \leq 9$, but the transition from node *on* to node *off* requires guard $x \geq 7$, the controller can switch from *on* to *off* no earlier than when $x \geq 7$, but no later than when the water level still is $x \leq 9$. Any point between the water levels 7 and 9 (including 7 and 9) is a permitted switching point.

In node *off*, the pump is off and the valve closed, but, nevertheless, water leaks slowly by the continuous dynamics $x' = -0.1$. The controller knows about this water leakage, so it stays in *off* at most as long as $x \geq 3$. Whenever $x < 5$, the controller can switch along the edge from *off* to *open* for opening the valve. Again, because of the evolution domain restriction $x \geq 3$ in *off* and the guard condition $x < 5$ on the edge, the switch may happen at any water level between 3 (inclusive) and 5 (exclusive). In node *open*, where the valve is open, water drains quickly following the differential equation $x' = -2$, but restricted to the evolution domain $x \geq 0$, because water stops pouring out when there is no water left. Finally, when the water level satisfies guard $x \leq 2$, the system can take the edge from node *open* back to node *on*. It is easy to see that the controller in Fig. 1.10 does not guarantee keeping the water level within the target range of 1 to 10 always, because it can stay in node *open* too long until $x = 0$.

How can we detect this problem and how can we find more subtle problems as automatically as possible? If we found problems, how can we find suggestions for fixes? And, once there are no more bugs, how do we prove that the system always works as intended under all circumstances? □

1.1.2 Model Checking

As a standard verification technique, model checking [75, 114, 78] has been used successfully for verifying temporal logic properties [247, 114, 115, 6, 7, 284] of finite-state abstractions of automata transition structures by exhaustive state space exploration [8, 159, 156, 217]. Model checking was pioneered [76] in 1981 by Clarke and Emerson [75] and independently by Queille and Sifakis [255]. It was originally invented for finite-state systems and has since been extended to handle finite-state abstractions of infinite-state systems. The working principle of model checking is to successively explore all state transitions of a transition system from the set of initial states until an unsafe state is found; see Fig. 1.11 for an illustration. One of the biggest advantages of model checking is this search for concrete

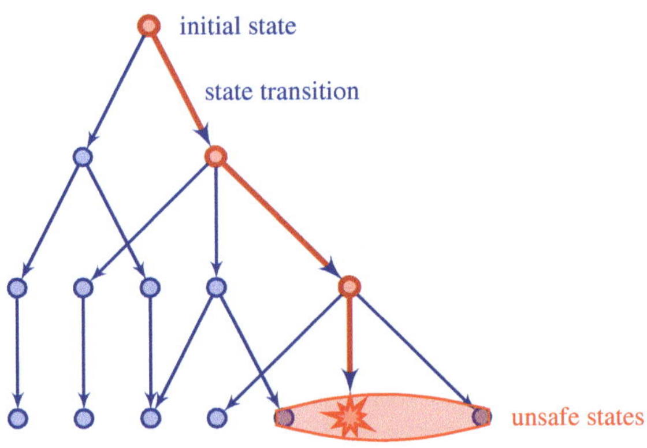

Fig. 1.11 Successive state space exploration in finite-state model checking

counterexamples, because such a sequence of transitions to an unsafe state can be very informative for debugging the system. If no unsafe state can be found this way, however, then model checkers can stop only if they have made sure that they have explored all possibilities. In finite-state systems, this exploration terminates trivially, because the search space is finite. Still, numerous successful optimisations and refinements have been developed to ensure faster termination and avoid exhaustive memory usage, including symbolic model checking with BDDs [66, 65], bounded model checking with propositional SAT solvers [47, 74], and model checking based on Craig interpolants [173, 174]. See the 2007 ACM Turing Award lecture by Edmund M. Clarke, Allen Emerson, and Joseph Sifakis [76] for a detailed survey of model checking, its successes, and its history.

For infinite-state systems the situation is more difficult than for finite automata because no model checker can actually explore an infinite set of states, so various abstractions have been developed that group sets of states into equivalence classes. If a model checker finds an unsafe reachable state in an infinite-state system, this counterexample is still extremely useful. Yet in case it does not find a counterexample after finitely many state explorations, it becomes more difficult to make an argument why no unsafe state could ever be reached by further exploration of the—still infinite—remaining state space. After all, finite exploration of an infinite state space still leaves infinitely many possibilities to consider. For example, timed automata (finite automata with clocks that measure the progress of time) can be abstracted with only finitely many equivalence classes that need to be considered, so that model checking is still guaranteed to terminate by considering one representative of each of these finitely many equivalent behaviours [6, 7]. These timed automata [10, 5, 86] can be understood as hybrid automata in which all differential equations are of the form $x' = 1$, all guards are of a form such as $x \leq c$ or $x - y \leq c$ for numbers $c \in \mathbb{Q}$, and all jumps (if any) are of the form $x := 0$. But even seemingly small extensions like stopwatch automata, which are timed automata with clocks that can be stopped (differential equation $x' = 0$) and resumed ($x' = 1$), have no equivalent finite-state abstractions anymore, so that model checking becomes undecidable [67].

The more general continuous state spaces of hybrid automata are inherently infinite, and do not admit equivalent finite-state abstractions either [156]. Because of this, model checkers for hybrid automata use various approximations [159, 8, 156, 70, 125, 15, 77, 21, 291, 217] and are still more successful in falsification than in verification. Furthermore, for hybrid systems with symbolic parameters in the dynamics, correctness crucially depends on the free parameters (e.g., b and s in Fig. 1.4). It is, however, quite difficult to determine corresponding symbolic parameter constraints from concrete values of a counterexample trace produced by a model checker, especially if they rely on nonstructural state splitting [70, 77, 21, 126]. Finally, in hybrid systems with nontrivial interaction of discrete and continuous dynamics, parameters also have a nontrivial impact on the system behaviour, leading to nonlinear parameter constraints and nonlinearities in the discrete and continuous dynamics. For instance, nonlinear constraints of the form $s \geq \frac{v^2}{2b} + \dots$ will turn out to be important for the safety of train systems such as the one in Fig. 1.4. Thus, the standard model checking approaches [156, 8, 158, 70, 126, 127] cannot be used, because they require linear dynamics in the discrete transitions and various forms of linear approximations in the continuous dynamics. Even in simpler timed automata, in which clocks are the only source of continuous dynamics, parameters immediately make the model checking problem undecidable [13, 61]. Still, parameters occur frequently in practical applications.

1.1.3 Deductive Verification

Deductive approaches [35, 37, 170, 149, 148, 306, 95, 97] have been used for verifying systems by proofs instead of by state space exploration and, thus, do not require finite-state abstractions. Deduction and theorem proving have already been used very successfully for program verification [35, 23, 170], in which proofs are used to show that a conventional program such as a Java program operates as specified. Davoren and Nerode [97] have also argued that deductive methods support formulas with free parameters better than, e.g., the standard model checking problem does. First-order logic, for instance, has widely proven its power and flexibility in handling symbolic parameters as free or quantified logical variables. However, first-order logic has no built-in means for referring to state transitions, which are crucial for verifying dynamical systems where states change over time.

In *temporal logics* [247, 114, 115, 6, 284], state transitions can be referred to using modal operators. For instance, the temporal formula $\Box\phi$ would hold for a system if formula ϕ is true always in all runs of the system. A temporal formula $\Box\phi$ that holds for all systems would be called valid. In deductive approaches, temporal logics have been used to prove the validity of formulas in calculi [97, 306]. Valid formulas of temporal logic, however, only express generic facts that are true for all systems, regardless of their actual behaviour. But these are not exactly the formulas we are interested in. We want to know the special properties of the actual system at hand, not those generic properties shared by all possible systems at once (including the broken implementations).

Hence, the behaviour of a specific hybrid system would need to be characterised declaratively with temporal formulas to obtain meaningful results. It is quite difficult, however, (and generally impossible for reasons of lack of expressive power), to characterise the behaviour of a hybrid system just with a set of temporal formulas such as $\Box x \geq 0$, because there are often myriads of systems satisfying the same formulas. Most temporal logics are not expressive enough to characterise the dynamics of a single system this way. Furthermore, the equivalence of declarative temporal representations and actual system operations would still need to be proven separately using other techniques. Finally even for finite-state systems, direct temporal characterisations can become computationally infeasible. In discrete finite-state systems, for example, direct temporal characterisations of the operational behaviour transform the linear-time model checking problem of the temporal logic CTL [75] into an EXPTIME-complete satisfiability problem [113].

Dynamic logic (DL) [253, 148, 149] is a successful approach for verifying infinite-state discrete systems deductively [35, 37, 170, 149, 148]. Like model checking, DL does not need declarative characterisations of system behaviour but can express and analyse the transition behaviour of actual operational system models directly. Yet, operational models are fully *internalised* within DL-formulas, and DL is closed under logical operators. That is, correctness statements about systems can be combined into bigger formulas with arbitrary propositional operators or quantifiers, and even into nestings of formulas. Within a single specification and verification language, DL combines operational system models with means to talk about the states that are reachable by system transitions. DL provides parametrised modal operators $[\alpha]$ and $\langle\alpha\rangle$ that refer to the states reachable by system α and can be placed in front of any formula. The formula $[\alpha]\phi$ expresses that all states reachable by system α satisfy formula ϕ. Likewise, $\langle\alpha\rangle\phi$ expresses that there is at least one state reachable by α for which ϕ holds. These modalities can be used to express necessary or possible properties of the transition behaviour of α in a natural way. They can be nested or combined propositionally. In first-order dynamic logic [253], where also quantifiers are allowed, $\exists p\,[\alpha]\langle\beta\rangle\phi$ says that there is a choice of parameter p (expressed by $\exists p$) such that for all possible behaviours of system α (expressed by $[\alpha]$) there is a reaction of system β (i.e., $\langle\beta\rangle$) that ensures ϕ. Likewise, $\exists p\,([\alpha]\phi \wedge [\beta]\psi)$ says that there is a choice of parameter p that makes both $[\alpha]\phi$ and $[\beta]\psi$ true, simultaneously, i.e., that makes the conjunction $[\alpha]\phi \wedge [\beta]\psi$ true, saying that formula ϕ holds for all states reachable by α executions and, independently, ψ holds after all β executions. The logical operator $\wedge$ is for conjunction ("and").

On the basis of first-order logic over the reals [288], which we use to describe safe regions of hybrid systems and to quantify over parameter choices, we introduce a first-order dynamic logic over the reals with modalities that directly quantify over the possible transition behaviour of hybrid systems. Since hybrid systems are subject to both continuous evolution and discrete state change, we generalise dynamic logic so that operational models α of hybrid systems can be used in modal formulas such as $[\alpha]\phi$.

With our extension of dynamic logic we can specify desirable properties of hybrid systems. Suppose *train* denotes the hybrid system in Fig. 1.4; then the formula $[train](z \leq s+10)$ expresses that the train always ($[train]$) stays in the region $z \leq s+10$, i.e., it never passes the position $s+10$. This formula cannot be true for fast trains, however, because the model in Fig. 1.4 starts braking at position $z = s$ at the earliest, but there may not be sufficient braking distance to $s+10$ for high velocities v. For some other values of the variables and parameters, instead, the formula may be true, but how can we find out if it is? And how do we determine corresponding parameter constraints that characterise under which circumstances such a formula is true? If we knew these parameter constraints, we may be able to choose s wisely in practise, so that the train model always works as intended.

As a further example, let *ball* be the hybrid system for the bouncing ball from Fig. 1.5. Then the formula $[ball](0 \leq h \leq H)$ expresses that the bouncing ball always ($[ball]$) stays between the ground at height 0 and maximum height H. The answer to the question about whether this formula is true or not depends on the value of H and on the initial height and velocity of the ball (the initial values of h, v) as well as the value of the gravity constant g. For $H = -5$, the formula is certainly false, because the ball can never be lower than the ground ($h < 0$). But for $h = v = 0$ and $H > 1$ it is definitely true, because the ball lies still without moving then. The most interesting cases, however, are those that are neither trivially true nor trivially false, but those where the ball actually bounces for a certain period of time and only the dynamics and relations of parameters and state variables determine the truth of the formula $[ball](0 \leq h \leq H)$.

Finally, suppose *wctrl* is the hybrid system for water control from Fig. 1.10. Then the formula $[wctrl](1 \leq x \leq 10)$ expresses that the water level always ($[wctrl]$) stays between 1 and 10. This formula may be true or false. In fact, it is not true under all circumstances for the model in Fig. 1.10, because the system could still stay in node *open* too long so that $1 \leq x$ no longer holds true. Instead, the dual formula $\langle wctrl\rangle x < 1$ is true, because there is a run of the water tank model ($\langle wctrl\rangle$) that reaches a state where the water level is less than 1 ($x < 1$), which, in fact, holds for any initial water level.

1.1.4 Compositional Verification

As a verification technology for our dynamic logic for hybrid systems, we devise a compositional proof calculus for verifying properties of a hybrid system by proving properties of its parts. Compositional verification is a very powerful proof technique with interesting scalability properties compared to monolithic approaches. It is, in essence, a divide-and-conquer technique for verification. Compositional proof calculi prove properties of a complex system by reducing them to subproperties of its subsystems recursively. Compositional approaches can have the advantage that properties for subsystems are often easier to prove, because they are simpler. For that purpose, our calculus recursively decomposes formula $[\alpha]\phi$ symbolically into an equivalent formula, for instance, $[\alpha_1]\phi_1 \wedge [\alpha_2]\phi_2$ about subsystems α_i of α and subproperties ϕ_i of ϕ. With this, formula $[\alpha]\phi$ can simply be verified by proving the formulas $[\alpha_i]\phi_i$ separately—using the same symbolic decomposition principles recursively—and combining the results conjunctively. In particular, synthesised parameter constraints that are required for the desired property to hold carry over from the latter formula to the former formula just by conjunction. Indeed, our compositional reasoning techniques turn out to be be quite successful in reducing the complexity of more complicated hybrid systems by decomposing them into their submodules.

Unfortunately, hybrid automata are not suitably compositional for this purpose. Their graph structures cannot be decomposed into subgraphs α_i such that the formula $[\alpha_1]\phi_1 \wedge [\alpha_2]\phi_2$ is equivalent to $[\alpha]\phi$, because of the dangling edges between the subgraphs α_i. For instance, the automaton in Fig. 1.4 cannot simply be verified by proving $[accel]\phi \wedge [brake]\phi$, because the effects of edges between the nodes need to be taken into account. Figure 1.12 shows an unsuccessful decomposition attempt of the hybrid automaton for train control from Fig. 1.4. Suppose we had a logic for hybrid automata[1] in which we could formulate logical formulas like $[\alpha](v \leq 8)$, where α is the hybrid automaton from Fig. 1.4. This graphical "formula" is depicted at the bottom of Fig. 1.12. The problem is that this formula at the bottom cannot

[1] The syntax and semantics of such a dynamic logic for hybrid automata can actually be derived from the presentation in Part I quite easily. It is the development of a compositional proof calculus for such a logic for hybrid automata that is technically more involved than the constructions for the logic we present in this book.

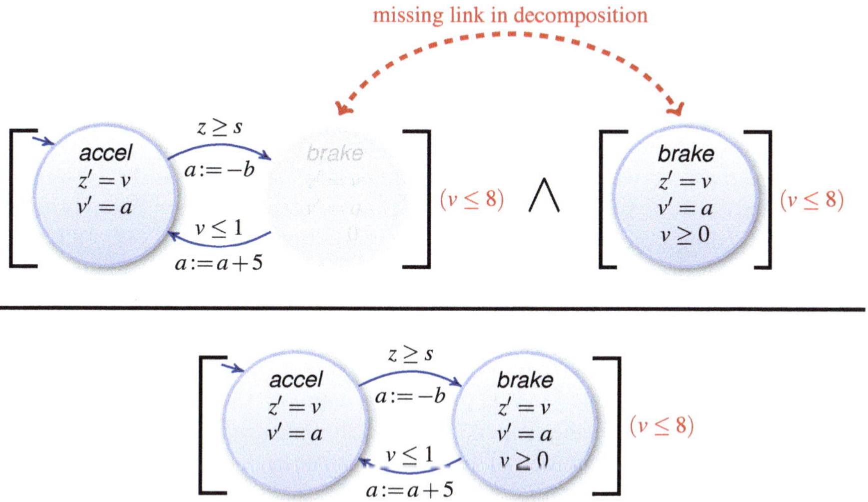

Fig. 1.12 Failed hybrid automaton decomposition attempt

just be proven by decomposing it into a proof of $[accel](v \leq 8) \wedge [brake](v \leq 8)$, because the discrete transition edges are important for the behaviour of the automaton and cannot be ignored. For several reasons, the formula at the bottom of Fig. 1.12 cannot be proven by decomposing it into the conjunction of the two graphical formulas at the top of Fig. 1.12 (the top-left formula results from the bottom formula by isolating node *brake* into the top-right formula, indicated by the grey colour).

One reason why this decomposition fails is that the automaton in the top left formula, which results from the automaton in the bottom by splitting off the *brake* phase, is no longer a hybrid automaton: it now has dangling edges pointing into nowhere. But what is the meaning and behaviour of such a broken, incomplete automaton fragment supposed to be? Remember that node *brake* is not present in the top-left formula, because we wanted to decompose the system in the first place. What should the dynamics look like when the automaton makes a transition into the undefined void? Another part that is missing in the failed decomposition attempt in Fig. 1.12 is the missing (dashed) link from the transitions leaving the automaton in the left formula to the automaton in the right formula and a link that says "if the operations of the automaton in the right formula stop, then the system jumps back to the automaton in the left formula, resuming action with a transition to *accel*." This missing link is pretty strange. It would be a link between the operational behaviour of parts of multiple different formulas. What is the meaning of the left formula supposed to be then if we want to consider it and reason about it alone, without the context of the right formula?

In other words, these decomposition attempts of hybrid automata do not lead to a compositional approach where we can reason separately about the fragments of the system and give a compositional and denotational [277] meaning to all formulas and system models, i.e., a semantics of the whole, solely defined in terms of the semantics of its parts, and not in terms of exterior formulas. The semantics of hybrid automata is just not sufficiently compositional for our purposes. It is not impossible to follow such a graphical approach, but it is technically quite involved and makes decompositions of the hybrid system unnecessarily hard, which is an obstacle to scalable verification, traceability, and tangibility.

For these reasons, we do not impose an automaton structure on the system. Instead, we introduce *hybrid programs* as a textual program notation for hybrid systems by extending conventional discrete program notations [253, 148, 182, 149]. The program on the right of Fig. 1.5 on p. 7, for instance, would be a hybrid program notation for the hybrid automaton on the left of Fig. 1.5. Hybrid programs allow for flexible programmatic combinations of elementary discrete or continuous transitions by structured control programs with a perfectly compositional semantics: The semantics of a compound hybrid program is a simple function of the semantics of its parts and does not further depend on any automata graph structures.

This compositional semantics of hybrid programs and a logic for hybrid programs has significant advantages. Figure 1.13, for instance, shows how the analysis of a property of the hybrid program $accel \cup brake$ that can choose between the operations of *accel* and of *brake* by a nondeterministic choice ($\cup$) can be decomposed directly into the formula $[accel](v \leq 8) \land [brake](v \leq 8)$ on the top. By decomposing the property at the bottom about a more complicated hybrid program into independent conjunctions—$[accel](v \leq 8)$ and $[brake](v \leq 8)$—of properties of simpler hybrid programs, we can recursively make the hybrid program simpler by decomposing it until we can prove the remaining properties of subprograms. As a special case, these decompositions also include cases that define switching conditions between acceleration and braking, but no special non-compositional handling of these is required.

Fig. 1.13 Successful hybrid program decomposition

$$\frac{[accel](v \leq 8) \land [brake](v \leq 8)}{[accel \cup brake](v \leq 8)}$$

In this book we study systematic analysis principles that can be used to decompose properties of complicated hybrid programs into simpler properties as in Fig. 1.13 and that handle the remaining discrete dynamics (discrete assignments), continuous dynamics (differential equations), and their hybrid interactions. We introduce a first-order dynamic logic for hybrid programs that makes these informal ideas rigorous. The resulting logic, which we call *differential dynamic logic* ($d\mathcal{L}$), constitutes a natural specification and verification logic for hybrid systems. With the goal of developing a solid theoretical, practical, and applicable foundation for deductive verification of hybrid systems by automated theorem proving, the focus of this book is a thorough analysis of the logic $d\mathcal{L}$, its calculus, extensions, verification procedures, and applications.

1.1.5 Lifting Quantifier Elimination

When proving formulas of differential dynamic logic (d$\mathcal{L}$), interacting hybrid dynamics cause interactions of arithmetic quantifiers and dynamic modalities, which both affect the values of symbols. For continuous evolutions, we have to prove formulas like $\forall t\,[\alpha]x\geq 0$, expressing that, for all durations t of some continuous evolution in α, the property $x \geq 0$ holds after all runs ($[\alpha]$) of system α. The standard techniques for handling quantifiers in first-order logic [122, 147, 123] are incomplete for handling these situations, because they are based on instantiation or unification (see App. A), which is already insufficient for proving the tautology $\forall z\,(z^2 \geq 0)$ of real arithmetic. Unfortunately, decision procedures for real arithmetic, such as real *quantifier elimination* [288, 81], cannot handle the real quantifier $\forall t$ in $\forall t\,[\alpha]x\geq 0$ either, because of the modality $[\alpha]$. The dynamics of the hybrid program α may depend on the value of t, but, at the same time, the constraints on t depend on the effect of α. Quantifier elimination in real-closed fields [288, 81], however, is only defined for first-order logic, not for first-order differential dynamic logic with modalities for hybrid programs. Indeed, the actual algebraic constraints on the quantified real variable t still depend on how the system variables evolve along the dynamics of α. This effect inherently results from the interacting dynamics of hybrid systems, where the duration t of a continuous evolution determines the resulting state and, hence, affects all subsequent discrete or continuous evolutions in α. Thus, the effect of α first needs to be analysed with respect to the arithmetical constraints it imposes on t for $x \geq 0$ to hold, before the quantifier $\forall t$ can be handled.

In this book, we present a calculus that is suitable for automation and combines deductive and arithmetical quantifier reasoning within a single compositional proof calculus. It introduces *real-valued free variables* and *real Skolem terms* to postpone quantifier elimination and continue reasoning beyond the occurrence of a real quantifier in front of a modality. Quite unlike classical proof techniques, however, our calculus later reintroduces a corresponding quantifier into the proof when its algebraic constraints have been discovered completely. For $\forall t\,[\alpha]x \geq 0$, our calculus will, for instance, continue with the unquantified kernel $[\alpha]x \geq 0$ after replacing variable t with a Skolem term $s(x)$. Once all arithmetical constraints on $s(x)$ are known, a quantifier for $s(x)$ will be reintroduced and handled by real quantifier elimination [288, 81]. The Skolem term $s(x)$ can be thought of as a new name for the universally quantified variable t. In a similar manner, our calculus combines quantifier elimination with deduction for handling existential real quantifiers using real-valued free variables.

In Chap. 2, we introduce a calculus that makes this intuition formally precise. Crucially, we exploit the relationship of Skolem terms and free variables in order to keep track of the lost quantifier nesting to prohibit unsound rearrangements of quantifiers when they are reintroduced. After all, it would be disastrous if we could start out proving a property $\exists x \forall y\,[\alpha]\phi$, turn the quantifiers for x and y into free variables and Skolem terms, and accidentally reintroduce quantifiers in a swapped order to prove the weaker statement $\forall y \exists x \ldots$ only. Clearly, it would not be sufficient to find a different x for each value of y, if, what we started looking for in the first place is one common x for all the y such that $[\alpha]\phi$ holds. The corresponding calculus rules that we introduce, which comply with all soundness requirements, are perfectly natural and coincide nicely with the prerequisites for quantifier elimination over the reals. Further, the d$\mathcal{L}$ semantics and calculus are fully compositional so that properties of a hybrid program can be proven by the reduction to properties of its parts following a structural symbolic decomposition within the compositional d$\mathcal{L}$ proof calculus.

1.1.6 Differential Induction and Differential Strengthening

In Chap. 3, we extend our logic d$\mathcal{L}$ to the *differential-algebraic dynamic logic* DAL, which is the logic of general hybrid change. DAL provides *differential-algebraic programs* (DA-programs) as general models for hybrid systems by allowing for propositional operators and quantifiers in discrete and continuous transitions. DA-programs have a very natural semantics of simultaneous change using just conjunctions ($\wedge$) as logical operators. For instance, the conjunctive *differential-algebraic constraint* $x' = y \wedge y' = -x$, in which each conjunct needs to hold during continuous evolutions, gives a natural semantics to differential equation systems $x' = y, y' = -x$, such that $x' = y, y' = -x$ is a notational variant of $x' = y \wedge y' = -x$. Disjunctions, in contrast, are a natural way to express switching conditions and finite nondeterminism in the dynamics. The disjunctive differential-algebraic constraint $x' = y \vee x' = -y$, for instance, allows x to evolve with velocity y or with velocity $-y$ and even switch between both cases, because only one of the two disjuncts needs to hold at any time. Quantifiers in the dynamics of DA-programs are a very expressive extension with which differential-algebraic equations and differential inequalities can be represented directly, and with which quantified disturbance in the dynamics can be characterised in an elegant and uniform way. The quantified differential-algebraic constraint $\exists y\,(x' = y \wedge y^2 < 5)$, for example, allows *all* evolutions of x with *some* velocity y (i.e., $\exists y$) whose square is less than 5. This quantified nondeterminism makes DA-programs quite expressive for characterising more involved system dynamics.

The standard approach to dealing with continuous dynamics for hybrid systems is to use symbolic or numerical solutions of their respective differential equations. Unfortunately, the range of systems that is amenable to these techniques is fairly limited, because even solutions of simple linear differential equations quickly fall into undecidable classes of arithmetic. For instance, the solutions of the linear differential equation system $s' = c, c' = -s$ are trigonometric functions like sin and cos. But first-order arithmetic with trigonometric functions is undecidable by a simple corollary to Gödel's famous incompleteness theorem [137]. As a means for verifying hybrid systems with challenging continuous dynamics without having to solve their differential equations, we complement discrete induction for loops and repetitions with a new form of *differential induction* for differential equations. Differential induction is a natural induction technique for differential equations. It is based on the local dynamics of the (right-hand side of the) differential equations themselves and does not need closed-form solutions for the differential equations. Because differential equations are simpler than their solutions (which is part of the representational power of differential equations), differential induction techniques working with the differential equations themselves are more scalable than techniques that need solutions of differential equations. Our differential induction techniques even generalise to differential-algebraic constraints with differential inequalities or quantifiers in the dynamics.

To further increase the verification power, we add differential strengthening or differential cuts as a powerful proof technique for refining the system dynamics with auxiliary invariants that can simplify the proof of the original property significantly. The basic insight is that auxiliary properties that are provable invariants of the dynamics can help prove the original property even if it was not provable before. This phenomenon is unique to differential equations and does not happen in classical discrete systems. Overall, our combination of compositional proof calculi with differential induction and differential strengthening turns out to be very powerful for the analysis of advanced hybrid systems, including air traffic control applications.

1.2 Related Work

In this section, we briefly discuss related approaches to verification.

Model Checking of Hybrid Automata

Model checking approaches work by state space exploration and—due to their undecidable reachability problem—require [156] various abstractions or approximations [159, 8, 156, 125, 15, 77, 291, 217] for hybrid automata, including numerical approximations [70, 21].

Beyond standard approaches [8, 156, 126] for linear automata with constant dynamics, the seminal work of Lafferriere et al. [189, 188, 190] presented a decision procedure for o-minimal hybrid automata and classes of linear dynamics with a homogeneous eigenstructure. They have analysed the discrete and continuous dynamics independently, which requires completely decoupled dynamics with forgetful jumps, i.e., where the outcome of a jump is completely independent of the continuous state. Unfortunately, actual systems rarely forget about states completely. For modelling train dynamics accurately, for example, it is important that the train position z and velocity v stays the same when switching from acceleration to braking mode in Fig. 1.4, and these variables do not just change arbitrarily and independently of their prior values.

Chutinan and Krogh [70] presented polyhedral approximations of hybrid automata with polyhedral discrete dynamics, invariants, and initial state sets. The relevant discrete dynamics, initial state regions, and invariants for our train and aircraft applications are nonlinear and thus cannot be described accurately by polyhedra. Fränzle [125] showed that reachability is decidable for specific classes of robust polynomial hybrid automata, where the safe and unsafe states are sufficiently separate and the safe region is bounded. Asarin et al. [21] used piecewise linear numerical approximations in an approximate reachability algorithm for continuous systems with known Lipschitz bounds. Mysore et al. [217] showed decidability of bounded-time and bounded switching reachability prefixes of semialgebraic hybrid automata. Because there seldom is a known bound on the number of transitions that a system can make, nor a (small) bound on the lifetime of a system, we are mostly interested in unbounded horizon properties, which cannot be obtained with bounded model checking.

Model checking tools like HyTech [157], PHAVer [126, 127], d/dt [22], and CheckMate [70] are already quite successful. Still these or other approaches [156, 125, 228] cannot handle our train and aircraft applications with nonlinear switching, nonlinear discrete and continuous dynamics, and high-dimensional state spaces (more than 30 dimensions).

Because hybrid systems do not admit equivalent finite-state abstractions [156] and due to general limits of numerical approximation [238], model checkers are still more successful in falsification than in verification. To obtain a sound verification approach and for improved handling of free parameters [97], we follow a symbolic logic-based approach and support $d\mathcal{L}$ as a significantly more expressive specification language. Finally, we introduce hybrid programs as a more uniform model for hybrid systems that is amenable to compositional symbolic verification.

Logics for Real-time Systems

Logics for real-time systems [159, 275] are not expressive enough to capture the dynamics of hybrid systems, particularly their differential equations, which are the main focus of this book. For instance, Schobbens et al. [275] give complete axiomatisations of two decidable dense time propositional linear temporal logics. Unfortunately, in these propositional logics one cannot express that relevant separation properties hold always during the flight of aircraft guided by specific flight controllers. A crucial property, for instance, is that the system never leaves region $(x_1-y_1)^2+(x_2-y_2)^2 \geq p^2$, which expresses that the distance of the aircraft at (x_1,x_2) and the aircraft at (y_1,y_2) is at least the protected zone p.

Successful model checking approaches for timed automata, which are a real-time restriction of hybrid automata, have been developed before, including the tools KRONOS [98] and UPPAAL [193]. Unlike for hybrid systems, model checking for timed automata is decidable [10, 5, 86]. But the expressive power of timed automata is fairly limited. The only continuous variables are clocks t following the differential equation $t'=1$. Positions or velocities evolving according to other laws of continuous dynamics are not allowed. In fact, even very minor extensions of timed automata make the verification problem undecidable, including stopwatches [67], which are clocks that can be stopped (dynamics $t'=0$) and resumed (dynamics $t'=0$). For a general survey of real-time verification approaches, we refer the reader to the book by Olderog and Dierks [221].

Logics for Hybrid Systems

Zhou et al. [306] extended the duration calculus [305] with mathematical expressions in derivatives of state variables. They use a multitude of calculus rules and a non-constructive oracle that requires external mathematical reasoning about the notions of derivatives and continuity. This undirected reasoning about mathematics, especially derivatives and continuity, is not suitable for automatic verification.

Davoren and Nerode [95, 97] presented a semantics of modal μ-calculus [181] in hybrid systems and examined topological aspects. They provided Hilbert-style calculi to prove formulas that are valid for all hybrid systems simultaneously. With this, however, only limited information can be obtained about a particular system: In propositional modal logics, system behaviour needs to be axiomatised declaratively in terms of abstract actions a,b,c of unknown effect. With unknown effects, however, we cannot characterise, say, a train control system or the flight dynamics for air traffic control, and not even the dynamics of a bouncing ball.

The strength of our logic primarily is that it is an expressive *first-order* dynamic logic: It handles actual operational models of hybrid systems such as $z'' = a$ instead of abstract propositional actions of unknown effect. The advantage of our calculus in comparison to others [306, 95, 97] is that it provides a constructive modular combination of arithmetic reasoning with reasoning about hybrid transitions and works by structural symbolic decomposition. With this, our calculus can be used easily for verifying actual operational hybrid system models, including railway and air traffic control systems, which are of considerable practical interest [156, 58, 70, 77, 217, 90, 238, 91]. Our proof calculus supports free parameters and first-order definable flows, which are well suited for verifying the coordination of train dynamics. First-order approximations of more general flows can be used according to [15, 238, 227]. More general dynamics can also be verified with the technique of differential induction that we introduce in Chap. 3.

Specification Languages for Hybrid Systems

Inspired by a proposal by He [175], Zhou et al. [69] presented a hybrid variant of CSP [162] as a language for describing hybrid systems. They gave a semantics in the extended duration calculus [306]. Zhou et al. did not yet give an actual verification approach for hybrid CSP.

Rönkkö et al. [270] extended guarded command programs with differential relations and gave a weakest-precondition semantics in higher-order logic with built-in derivatives. Without providing a means for verification of this higher-order logic, this approach is limited to providing a notational variant of standard mathematics.

Rounds and Song [272] have developed a hybrid version of the π-calculus [208] as a modelling language for embedded systems. Later, Rounds [271] gave a semantics in a rich set theory for an abstract spatial logic for the hybrid π-calculus, which is also called Φ-calculus. In the hybrid π-calculus, processes interact with a continuously changing environment, but cannot themselves evolve continuously, which would be crucial to capture the physical movement of traffic agents. From the semantics alone, however, no verification is possible in these approaches, except perhaps by manual semantic reasoning.

Other process-algebraic approaches, such as χ [40], have been developed for modelling and simulation purposes [39]. At present, verification is still limited to small fragments that can be translated directly to other verification tools such as PHAVer [127, 126] when converted to hybrid automata or to UPPAAL [193] when converted to timed automata.

Modelling languages for hybrid systems further include SHIFT [100] for networks of hybrid automata, and R-Charon for reconfigurable systems [183]. These approaches focus more on simulation and compilation [100] or the development of a semantics [183], so no verification or formal analysis is possible yet.

Uses of Deduction for Hybrid Systems

Manna et al. [201, 178] and Ábrahám et al. [1] used theorem provers for checking invariants of hybrid automata in STeP [201] and PVS [1], respectively. Their working principle is, however, quite different from ours. Given a hybrid automaton and a global system invariant, they compile, in a single step, a verification condition expressing that the invariant is preserved under all transitions of the hybrid automaton. Hence, hybrid aspects and transition structure vanish completely before the proof starts. All that remains is a flat quantified mathematical formula. Which hybrid systems can be verified with this approach in practise strongly depends on the general mathematical proving capabilities of STeP and PVS, which, because of their general-purpose focus, typically requires user interaction.

In contrast, we follow a fully symbolic approach using a genuine specification and verification logic for hybrid systems. Our dynamic logic for hybrid systems works deductively by symbolic decomposition and preserves the transition structure during the proof, which simplifies the traceability of results considerably. Further, the structure in this symbolic decomposition can be exploited for deriving invariants or parametric constraints. Consequently, invariants do not necessarily need to be given beforehand in our approach. Moreover, in practise, guiding quantifier elimination procedures along natural splitting possibilities of the structural decomposition performed by the our proof calculus turns out to be important for successful automatic proof strategies (Chaps. 5 and 6).

1.3 Contributions

Our main conceptual contribution is a series of differential dynamic logics for hybrid systems (these logics are called $d\mathcal{L}$, DAL, and dTL), which capture the logical quintessence of the dynamics of hybrid systems succinctly. Our logics provide a uniform semantics and concise language for specifying and verifying correctness properties of general hybrid systems with sophisticated dynamics. Our main practical contribution is a concise *free-variable calculus* that axiomatises the transition behaviour of hybrid systems relative to differential equation solving. With our generalisation of free-variable calculi to dynamic logic over the reals, the calculus is suitable for automated theorem proving and for verifying hybrid superpositions of interacting discrete and continuous dynamics compositionally.

Our main theoretical contribution is that we prove our calculi to be sound and complete relative to the handling of differential equations. To the best of our knowledge, this is the first relative completeness proof for a logic of hybrid systems, and even the first formal notion of hybrid completeness. Our results fully align hybrid and continuous reasoning proof-theoretically and show that hybrid systems with interacting repetitive discrete and continuous evolutions can be verified whenever differential equations can.

We further contribute a verification calculus that includes uniform proof rules for differential induction along differential equations or more general differential-algebraic constraints, using a combination of differential invariants, differential variants, and differential strengthening for verifying hybrid systems without having to solve their differential constraints. Based on these calculi, we develop a fixed-point verification algorithm that computes the required invariants and differential invariants for a formula and refines the underlying system dynamics as needed during the proof.

As applied contributions, we demonstrate the capabilities of our logics, calculi, and algorithms by verifying collision avoidance in realistic train control applications and challenging air traffic control manoeuvres. Overall, our logical analysis approach for hybrid systems can successfully verify realistic applications that were out of the scope of other approaches, for both theoretical and scalability reasons.

1.4 Structure of This Book

This book consists of three parts that basically correspond to the theory, practise, and applications, respectively, of the logical analysis of hybrid systems. You are now reading the introduction.

Logics and Calculi

In Part I, which is the core of this book, we introduce novel logics and proof calculi that form the new conceptual, formal, and technical basis for the logical analysis of hybrid systems. In Chap. 2, we introduce the differential dynamic logic $\mathsf{d}\mathcal{L}$ as a variant of dynamic logic that is suitable for specifying and verifying properties of hybrid systems. It generalises dynamic logic to dynamic logic over the reals in the presence of hybrid dynamics with discrete state transitions and with continuous state evolutions along differential equations. As a verification technique, we present a new compositional sequent calculus for $\mathsf{d}\mathcal{L}$ that is suitable for automation and integrates handling of real quantifiers by generalising Skolemisation and free variables to the reals. In Chap. 2, we also prove completeness relative to differential equations as the most fundamental theoretical result in this book.

In Chap. 3, we introduce the differential-algebraic logic DAL that extends the class of hybrid system models by allowing more general differential-algebraic equations, differential inequalities, and quantified nondeterminism. Further, we present a uniform theory of differential induction, differential invariants, differential variants, and differential strengthening as central symbolic verification techniques for handling challenging continuous dynamics in hybrid systems without having to solve their differential equations.

In Chap. 4, we address the handling of temporal properties and introduce the differential temporal dynamic logic dTL along with a calculus that reduces temporal properties to $\mathsf{d}\mathcal{L}$ properties. The extensions of $\mathsf{d}\mathcal{L}$ that we present in Chaps. 3 and 4 are complementary and compatible. Their direct modular combination immediately defines the differential-algebraic temporal dynamic logic DATL.

Automated Theorem Proving

In Part II, we focus on the practical aspects of implementing the verification calculi from Part I. The calculi in Part I have already been designed for the practical needs of automated theorem proving, most notably for the free-variable and Skolemisation techniques from Chap. 2 and the compositional proof calculi from Part I. Immediate implementations of the proof calculi from Part I in automated theorem provers can prove examples of medium complexity directly. Yet, more complex case studies require additional algorithmic techniques for achieving high degree automation and good scalability properties. In Chap. 5, we refine the calculi from Part I to tableau procedures that are suitable for automated theorem proving (ATP) and present proof strategies that navigate through their nondeterminisms to help overcome the complexity issues of integrating real quantifier elimination as a decision procedure for real arithmetic.

In Chap. 6, which is based on joint work with Edmund M. Clarke [239], we introduce the "differential invariants as fixed points" (DIFP) paradigm. We refine the differential induction techniques from Chap. 3 to a fully automatic verification algorithm for computing the required discrete and differential invariants of a hybrid system locally in a logic-based fixed-point loop.

Applications

In Part III, we shift our attention to application scenarios for our logical analysis approach for hybrid systems. Extending smaller hybrid systems which have served as running examples throughout this book, we show full verification case studies of the European Train Control System (ETCS) in Chap. 7, which is based on joint work with Jan-David Quesel [243]. We also extend and show verification case studies for aircraft collision avoidance manoeuvres in air traffic control (ATC) in Chap. 8, which is based on joint work with Edmund M. Clarke [238, 239].

Finally, Chap. 9 concludes this work with a discussion of the results and perspectives for future research.

Appendices

In Part IV, we provide the background in logic and differential equations that we need for the course of this book. In App. A, we give an introduction to basic first-order logic (FOL), its syntax, semantics, and proof techniques. For reference, App. B summarises some classical results about ordinary differential equations (ODEs) that we need as background for this book. In App. C, we formally investigate the relationship between hybrid automata [156] and hybrid programs by embedding hybrid automata into hybrid programs. In App. D, we briefly characterise the verification tool KeYmaera that implements the logics and automated theorem proving techniques presented in this book and that has been implemented in joint work with Jan-David Quesel [242]. We also survey various techniques that can be used to verify real arithmetic.

Online Material

At the Website for this book, we provide the KeYmaera verification tool for download and webstart. KeYmaera has been developed in joint work with Jan-David Quesel [242] and implements the logical analysis approach presented in this book. Slide material, an online tutorial, and KeYmaera problem files for several examples, including the case studies of train and air traffic control, can also be found on the Web.

The Web page for this book is at the following URL:

`http://symbolaris.com/lahs/`

Suggested Reading Sequence

The basic suggested reading sequence in this book is linear (with additional consultation of the appendices for background). Except for the foundation of this work that is laid out in Part I, however, the chapters in this book are mostly self-contained so that they can also be studied independently. Figure 1.14 shows the reading order dependencies among the chapters (solid lines) and the partial dependencies of suggested reading sequences that holds for the advanced material of the respective chapters (dashed lines).

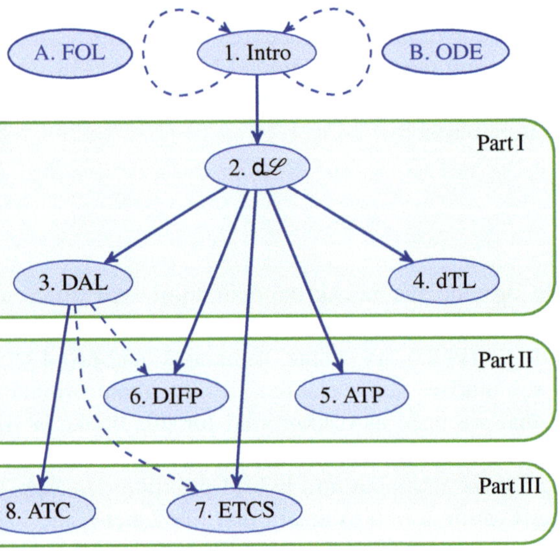

Fig. 1.14 Dependencies and suggested reading sequence of chapters and appendices

For a background in classical first-order logic (FOL), we recommend you review App. A. Depending on the interest, field of study, and preference of the reader, we recommend he either study the background information in App. A on first-order logic before proceeding to Part I or use the material in App. A as a background reference book while reading the main part of this book. Similarly, we recommend he review the background on ordinary differential equations (ODEs) in App. B either before or during the study of this book. An intuitive approach to understanding differential equations and formal definitions of their semantics will be given in the main parts of this book. Logic itself is also explained and illustrated intuitively during the main part of this book, but some readers may also find it helpful to refresh or learn about the basics of first-order logic from App. A before proceeding to the main part.

While there is a lot of flexibility in the reading sequence of the chapters, we strongly recommend you study the logical foundations of hybrid systems analysis and differential dynamic logic (d$\mathcal{L}$) in Chap. 2 of Part I before reading any other chapter of Parts I–III. Chapter 2 develops the logical foundations, the system model of hybrid programs, the differential dynamic logic ($d\mathcal{L}$) for expressing correctness properties, and the proof calculus for verifying these properties that will be needed in the remainder of this book.

Some more advanced sections in the applications in Part III also depend on the theory of differential invariants and the differential algebraic dynamic logic (DAL) that is developed together with other extensions in Chap. 3. We recommend you read Chap. 3 on DAL before studying the air traffic control (ATC) verification in Chap. 8. While most of Chap. 7 on the European Train Control System (ETCS) verification can be read with the foundation in $d\mathcal{L}$ from Chap. 2, some advanced parts also use results from Chap. 3 on DAL. We also recommend you study the theoretical foundations on DAL in Chap. 3 before reading the automation approach "differential invariants as fixed points" (DIFP) in Chap. 6. Still, some level of understanding of the DIFP automation approach in Chap. 6 can also be gained without your having read the full theoretical background on DAL in Chap. 3.

Part I
Logics and Proof Calculi for Hybrid Systems

Overview In this part, which is the core part of this book, we introduce novel logics and proof calculi that form the new conceptual, formal, and technical basis for the logical analysis of hybrid systems. In Chap. 2, we introduce the differential dynamic logic $\mathsf{d}\mathcal{L}$ as a variant of dynamic logic that is suitable for specifying and verifying properties of hybrid systems. It generalises classical dynamic logic to dynamic logic over the reals in the presence of hybrid dynamics with interacting discrete state transitions and continuous state evolutions along differential equations. As a verification technique, we present a new compositional proof calculus for $\mathsf{d}\mathcal{L}$ that is suitable for automation and integrates handling of real quantifiers by generalising Skolemisation and free variables to the reals. In Chap. 2, we also prove completeness relative to differential equations as the most fundamental theoretical result in this book.

In Chap. 3, we introduce the differential-algebraic logic DAL that extends the class of hybrid system models by allowing more general differential-algebraic equations, differential inequalities, and quantified nondeterminism. Further, we present a uniform theory of differential induction, differential invariants, differential variants, and differential strengthening as central symbolic verification techniques for handling challenging continuous dynamics in hybrid systems without having to solve their differential equations.

In Chap. 4, we address the handling of temporal properties and introduce the differential temporal dynamic logic dTL along with a calculus that reduces temporal properties to $\mathsf{d}\mathcal{L}$ properties. The extensions of $\mathsf{d}\mathcal{L}$ that we present in Chap. 3 and Chap. 4 are complementary and compatible. Their direct modular combination immediately defines the differential-algebraic temporal dynamic logic DATL.

The logics and proof techniques developed in this part will form the basis for the automation techniques developed in Part II. They also form the foundation for the formal verification tool KeYmaera. We will also use the differential dynamic logics to formalise safety-critical properties of the train and aircraft control studies in Part III and prove them with the proof techniques we develop in Part I.

Chapter 2
Differential Dynamic Logic $d\mathcal{L}$

Contents

A. Platzer, Logical Analysis of Hybrid Systems,
DOI 10.1007/978-3-642-14509-4_2,

Synopsis Hybrid systems are models for complex physical systems and are defined as dynamical systems with interacting discrete transitions and continuous evolutions along differential equations. With the goal of developing a theoretical and practical foundation for deductive verification of hybrid systems, we introduce a dynamic logic for hybrid programs, which is a program notation for hybrid systems. As a verification technique that is suitable for automation, we introduce a free-variable proof calculus with a novel combination of real-valued free variables and Skolemisation for lifting quantifier elimination for real arithmetic to dynamic logic. The calculus is compositional, i.e., it reduces properties of hybrid programs to properties of their parts. Our main result proves that this calculus axiomatises the transition behaviour of hybrid systems completely relative to differential equations. In a study with cooperating traffic agents of the European Train Control System, we further show that our calculus is well suited for verifying realistic hybrid systems with parametric system dynamics.

2.1 Introduction

In this chapter, we introduce the differential dynamic logic $\mathsf{d}\mathcal{L}$, its syntax, semantics, and proof calculus. It forms the core of this book and is the basis for the extensions, algorithmic refinements, and applications in subsequent chapters of this book.

Contributions

Our main conceptual contribution in this chapter is the differential dynamic logic $\mathsf{d}\mathcal{L}$ for hybrid programs, which captures the logical quintessence of the dynamics of hybrid systems succinctly. Our main practical contribution is a concise free-variable calculus for $\mathsf{d}\mathcal{L}$ that axiomatises the transition behaviour of hybrid systems relative to differential equation solving. It is suitable for automated theorem proving and for verifying hybrid interacting discrete and continuous dynamics compositionally. Our main theoretical contribution is that we prove the $\mathsf{d}\mathcal{L}$ calculus to be sound and complete relative to the handling of differential equations. To the best of our knowledge, this is the first relative completeness proof for a logic of hybrid systems, and even the first formal notion of hybrid completeness. Our results fully align hybrid and continuous reasoning proof-theoretically and show that hybrid systems with interacting repetitive discrete and continuous evolutions can be verified whenever differential equations can. As an applied contribution, we further demonstrate that our logic and calculus can be used successfully for verifying collision avoidance in realistic train control applications.

2.1.1 Structure of This Chapter

After introducing syntax and semantics of the differential dynamic logic d$\mathcal{L}$ in Sects. 2.2 and 2.3, we introduce a free-variable sequent calculus for d$\mathcal{L}$ in Sect. 2.5 and prove soundness and relative completeness in Sects. 2.6 and 2.7, respectively. We present relatively semidecidable fragments of d$\mathcal{L}$ in Sect. 2.8. In Sect. 2.9, we use our calculus to prove an inductive safety property of the train control system that we present in Sect. 2.4. We draw conclusions and discuss future work in Sect. 2.10.

2.2 Syntax of Differential Dynamic Logic

In this section, we introduce the *differential dynamic logic* d$\mathcal{L}$ in which operational models of hybrid systems are internalised as first-class citizens, so that correctness statements about the transition behaviour of hybrid systems can be expressed as formulas. As a basis, d$\mathcal{L}$ includes (nonlinear) real arithmetic for describing concepts like safe regions of the state space. Further, d$\mathcal{L}$ supports real-valued quantifiers for quantifying over the possible values of system parameters or durations of continuous evolutions. For talking about the transition behaviour of hybrid systems, d$\mathcal{L}$ provides modal operators such as $[\alpha]$ or $\langle\alpha\rangle$ that refer to the states reachable by following the transitions of hybrid system α.

The logic d$\mathcal{L}$ is a first-order dynamic logic over the reals for hybrid programs, which is a compositional program notation for hybrid systems. Hybrid programs provide the following constructs.

Discrete jump sets. Discrete transitions are represented as instantaneous assignments of values to state variables, which are, essentially, difference equations. They can express resets like $a := -b$ or adjustments of control variables like $a := a+5$, as occurring in the discrete transformations attached to edges in hybrid automata; see Fig. 2.1. Likewise, implicit discrete state changes such as the changing of evolution modes from one node of an automaton to the other can be expressed uniformly as, e.g., $q := brake$, where variable q remembers the current node. To handle simultaneous changes of multiple variables, discrete jumps can be combined to sets of jumps with simultaneous effect following corresponding techniques in the discrete case [37]. For instance, the discrete jump set $a := a+5, A := 2a^2$ expresses that a is increased by 5 and, simultaneously, variable A is set to $2a^2$, which is evaluated *before* a receives its new value $a+5$.

Differential equation systems. Continuous variation in system dynamics is represented using differential equation systems as evolution constraints. For example the (second-order) differential equation $z'' = -b$ describes deceleration with braking force b and $z' = v, v' = -b \,\&\, v \geq 0$ expresses that the evolution only applies as long as the speed is $v \geq 0$, which represents mode *brake* of Fig. 2.1. This is an evolution along the differential equation system $z' = v, v' = -b$ that is restricted (written &) to remain within the evolution domain region $v \geq 0$, i.e., to stop braking before $v < 0$. Such an evolution can stop at any time within $v \geq 0$, it could even continue with transient grazing along the border $v = 0$, but it is never allowed to enter $v < 0$. The second-order differential equation $z'' = -b$ itself is equivalent to the first-order differential equation system $z' = v, v' = -b$, in which the velocity v is explicit. In this chapter, we separate the respective differential equations in a differential equation system by a comma (,) and separate the evolution domain region (if any) from the differential equations by an ampersand (&). We choose this notation for this chapter to make it easier to identify the evolution domain region. In Chap. 3, we will see that both (,) and (&) can be understood more uniformly as conjunctions.

Control structure. Discrete and continuous transitions—represented as difference or differential equations, respectively—can be combined to form a hybrid program with interacting hybrid dynamics using regular expression operators ($\cup, ^*, ;$) of regular programs [149] as control structure. For example, the hybrid program $q:=accel \cup z'' = -b$ describes a train controller that can choose to either switch to acceleration mode ($q:=accel$) or brake by the differential equation $z'' = -b$, by a nondeterministic choice ($\cup$). The nondeterministic choice $q:=accel \cup z'' = -b$ expresses that either $q:=accel$ or $z'' = -b$ happens. The sequential composition $q:=accel; z'' = -b$, instead, expresses that first $q:=accel$, and then $z'' = -b$ happens. In conjunction with other regular combinations, control constraints can be expressed using tests like $?z \geq s$ as guards for the system state. This test will succeed if, indeed, the current state of the system satisfies $z \geq s$; otherwise the test will fail and execution cannot proceed. In that respect, a test is like an assert statement in conventional programs and cuts the system run if the test is not successful.

Example 2.1 (Embedding hybrid automata). With these operations, hybrid systems can be represented naturally as hybrid programs. For example, the right of Fig. 2.1 depicts a hybrid program rendition of the hybrid automaton on the left, which repeats the automaton from Fig. 1.4 on p. 5. Line 1 represents that, in the beginning,

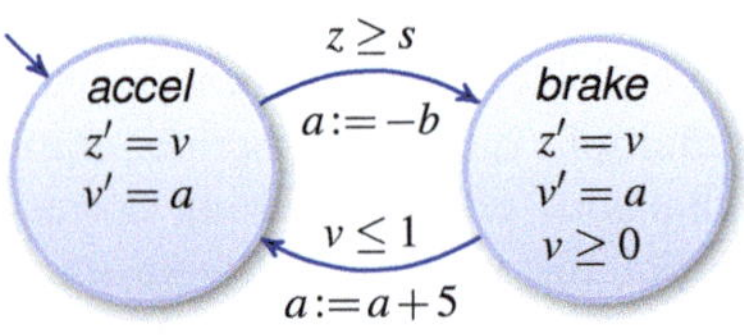

$q:=accel;$ /* *initial mode is node accel* */
$($ $(?q = accel;\ z' = v, v' = a)$
$\cup\ (?q = accel \land z \geq s;\ a:=-b;\ q:=brake;\ ?v \geq 0)$
$\cup\ (?q = brake;\ z' = v, v' = a \,\&\, v \geq 0)$
$\cup\ (?q = brake \land v \leq 1;\ a:=a+5;\ q:=accel))^*$

Fig. 2.1 Hybrid program rendition of hybrid automaton for (overly) simplified train control

the current node q of the system is the initial node *accel*. We represent each discrete and continuous transition of the automaton as a sequence of statements with a nondeterministic choice ($\cup$) between these transitions. Line 4 represents a continuous transition of the automaton. It tests if the current node q is *brake*, and then (i.e., if the test was successful) follows the differential equation system $z' = v, v' = a$ restricted to the evolution domain $v \geq 0$. Line 3 characterises a discrete transition of the automaton. It tests the guard $z \geq s$ when in node *accel*, and, if successful, resets $a := -b$ and then switches q to node *brake*. By the semantics of hybrid automata [8, 156], an automaton in node *accel* is only allowed to make a transition to node *brake* if the evolution domain restriction of *brake* is true when entering the node, which is expressed by the additional test $?v \geq 0$ at the end of line 3. Observe that this test of the evolution domain region generally needs to be checked as the last operation after the guard and reset, because a reset like $v := v - 1$ could affect the outcome of the evolution domain region test. In order to obtain a fully compositional model, hybrid programs make all these implicit side conditions explicit. Line 2 represents the continuous transition when staying in node *accel* and following the differential equation system $z' = v, v' = a$. Line 5 represents the discrete transition from node *brake* of the automaton to node *accel*.

Lines 2–5 cannot be executed unless their tests succeed. In particular, at any state, the nondeterministic choice ($\cup$) among lines 2–5 reduces de facto to a nondeterministic choice between either lines 2–3 or between lines 4–5. At any state, q can have value either *accel* or *brake* (assuming these are different constants), not both. Consequently, when $q = brake$, a nondeterministic choice of lines 2–3 would immediately fail the tests in the beginning and not execute any further. The only remaining choices that have a chance to succeed are lines 4–5 then. In fact, only the single successful choice of line 4 would remain if the second conjunct $v \leq 1$ of the test in line 5 does not hold for the current state. Note that, still, all four choices in lines 2–5 are available, but at least two of these nondeterministic choices will always be unsuccessful. Finally, the repetition operator (*) at the end of Fig. 2.1 expresses that the transitions of a hybrid automaton, as represented by lines 2–5, can repeat indefinitely, possibly taking different nondeterministic choices between lines 2–5 at every repetition. □

2.2.1 Terms

The construction of the logic d$\mathcal{L}$ starts with a set V of logical variables and a *signature* Σ, which is the set of names (called *symbols*) of all entities nameable in a certain context. The signature Σ and set V form the vocabulary or alphabet of signs from which well-formed formulas can be built. For d$\mathcal{L}$ we assume all variables in V are interpreted over the reals and that Σ is a (finite) set of real-valued function and predicate symbols, with the usual function and predicate symbols for real arithmetic, such as $0, 1, +, -, \cdot, /, =, \leq, <, \geq, >$, where $+$ is addition, $\cdot$ is multiplication, $/$ is division and so on. For each function and predicate symbol, we are given the number of arguments that it expects, which is called *arity*, and is a natural number. The arity can be zero, in which case the function or predicate symbol does not have any arguments. The function symbols for the numbers 0,1 have arity zero, because they do not need arguments. The binary arithmetic operators $+, -, \cdot, /$ have arity

2, because they expect two arguments. The binary predicate symbols $=, \leq, <, \geq, >$ also have arity 2, because they need two arguments to compare.

The difference between function and predicate symbols is that function symbols stand for functions that take the values of arguments and give back a function value. Predicate symbols, on the other hand, are interpreted either as true for a vector of arguments or as false. That is, they take the values of arguments and give either the truth-value "*true*" or the truth-value "*false*". No other result is permitted for predicate symbols. For instance, the predicate symbol $\geq$ will be understood such that $\geq(x,y)$ is true if and only if the value of x is greater than or equal to the value of y. For real arithmetic, we use standard notation and standard semantics. In particular, we write $x \geq y$ instead of $\geq(x,y)$. We fix the semantics of $\cdot$ to be multiplication, i.e., the value of $\cdot(x,y)$ is always meant to be the product of the value of x and the value of y. Again, we use the standard notation $x \cdot y$, or just xy if no confusion arises, instead of $\cdot(x,y)$. The denotation of a function symbol could also be, e.g., the function that takes an argument and gives back its cube. A predicate, in contrast, cannot give back any value other than "*true*" or "*false*", but could hold, say, for all real numbers larger than 5. Or it could be the relation that holds for all pairs where the second element of the pair is larger than the square of the first element of the pair. Function symbols are often written as f, g, h, a, b, c and predicate symbols are often written as p, q, r.

State variables of hybrid systems, such as positions, velocities, and accelerations, are represented as real-valued function symbols of Σ of arity 0. Unlike fixed symbols like the number 1, state variables are *flexible*, i.e., their interpretation can change from state to state during the execution of a hybrid program. Flexibility of symbols will be used to represent the progression of system values along states over time during a hybrid evolution. Symbols like 1, on the other hand, are *rigid*, i.e., they have the same value at all states. The symbols of real arithmetic like 1 and $+, \cdot$ are rigid, because we do not want them to change their meaning at any time. State variables like velocity v, in contrast, are flexible, because they can change their value depending on the state. While the velocity v may have been 0 in the beginning, the train could increase its velocity to 10 and then decrease it again later when approaching another train.

Note that there is no need to distinguish between discrete and continuous variables in $d\mathcal{L}$. The distinction between logical variables in V, which can be quantified universally or existentially, and state variables in Σ, which can change their value by discrete jumps and differential equations of hybrid programs in modalities, is not strictly required either. For instance, universal and existential quantification of state variables is definable using auxiliary logical variables. The distinction makes the semantics and soundness proof less subtle, though. Our calculus assumes that V contains sufficiently many variables and Σ contains additional Skolem function symbols, which are reserved for use by the calculus.

Terms

Well-formed arguments to function symbols and predicate symbols are called terms. Logical variables are well-formed terms, and functions applied to the appropriate number of terms as arguments are well-formed terms. The set $\mathrm{Trm}(\Sigma, V)$ of *terms* is defined as in classical first-order logic, yielding polynomial (or rational) expressions over V and over additional Skolem terms $s(t_1,\dots,t_n)$ with terms t_i. Our calculus actually only uses Skolem terms $s(X_1,\dots,X_n)$ with logical variables $X_i \in V$ as arguments.

Definition 2.1 (Terms). $\mathrm{Trm}(\Sigma, V)$ is the set of all *terms*, which is the smallest set such that:

- If $x \in V$, then $x \in \mathrm{Trm}(\Sigma, V)$.
- If $f \in \Sigma$ is a function symbol of arity $n \geq 0$ and, for $1 \leq i \leq n$, $\theta_i \in \mathrm{Trm}(\Sigma, V)$, then $f(\theta_1,\dots,\theta_n) \in \mathrm{Trm}(\Sigma, V)$. The case $n = 0$ is permitted (e.g., for state variables).

More succinctly, we also say that the terms of $\mathsf{d}\mathscr{L}$ are defined by the following grammar (where $\theta_1,\dots,\theta_n$ are terms, f a function symbol of arity n, and $x \in V$ is a logical variable):

$$\theta \ ::= \ x \mid f(\theta_1,\dots,\theta_n).$$

Example 2.2. (Well-formed) terms of $\mathsf{d}\mathscr{L}$ include:

- Logical variables $X \in V$
- State variables $x \in \Sigma$ that may change their value during system evolution
- Expressions of nonlinear polynomial real arithmetic like $x + 5y \cdot (x - 3y + za)$, which we consider as a short notation for $x + 5 \cdot y \cdot (x - 3 \cdot y + z \cdot a)$. Here we assume that $x, y, z, a \in \Sigma$ are state variables. In principle, we also have to mention that the number symbols $5, 3 \in \Sigma$ are (rigid) function symbols without arguments. Yet these number symbols are what we assume as given throughout this book. Note that we could just as well have assumed that $x, y \in \Sigma$ are state variables, $a \in \Sigma$ is a rigid function symbol of arity 0, and $z \in V$ is a logical variable. Then $x + 5y \cdot (x - 3y + za)$ is still a term for this different signature and variables set. For terms, all ways of declaring symbols as state variables, rigid function symbols of arity 0, or logical variables are essentially equivalent. There are many ways to say the same thing. The differences only play a role later for quantification and state change.
- Expressions with Skolem function terms like $x + 5s(X_1, X_2) \cdot (x - 3y + z \cdot t(X_2))$. Here we assume that $x, y \in \Sigma$ are state variables, that $s, t \in \Sigma$ are rigid function symbols of arity 2 and 1, respectively, and that $X_1, X_2 \in V$ are logical variables.
- Real arithmetic expressions with integer powers like $8x^2 + 2x^3(y - a^2bc)$ that can easily be rewritten as $8 \cdot x \cdot x + 2 \cdot x \cdot x \cdot x \cdot (y - a \cdot a \cdot b \cdot c)$. Again, we assume that x, y, a, b, c are symbols in Σ or V.

The following, however, are *no terms* with respect to Σ and V:

- $1+x^y$, because the exponential function x^y cannot be rewritten as a finite product, quite unlike $x^3 = x \cdot x \cdot x$ or $x^4 = x \cdot x \cdot x \cdot x$. In fact, the logical properties of the exponential function are a very exciting and a challenging object of study in recent model theory [107, 206, 44, 108, 45, 2].
- $y^2 - \pi$, unless the transcendental number $\pi = 3.1415926\ldots$ is explicitly added to Σ, because unlike rational numbers, the transcendental number cannot be characterised exactly with a finite combination of sums, products, and 0, 1. Arbitrarily precise approximations of π, instead, can be defined; see Example 2.3. □

First-Order Formulas

Meaningful propositions that are either true or false in a context are called (well-formed) formulas. The well-formed formulas of a logic form a formal language over the alphabet $\Sigma \cup V$ of symbols. The formulas consist of all words that can be built by recursively combining symbols of the signature with logical operator symbols appropriately. We first define only the fragment of first-order logic, then the syntax of hybrid programs, and define the actual formulas of differential dynamic logic afterwards.

The set of formulas of *first-order logic* is defined as usual (cf. App. A), giving first-order real arithmetic [288] augmented with Skolem terms. We will show the precise relationship to standard first-order real arithmetic without Skolem terms in Lemma 2.5 of Sect. 2.5.3.2.

Definition 2.2 (First-order formulas). The set $\mathrm{Fml}_{\mathrm{FOL}}(\Sigma,V)$ of *formulas of first-order logic* is the smallest set with:

- If $p \in \Sigma$ is a predicate symbol of arity $n \geq 0$ and $\theta_i \in \mathrm{Trm}(\Sigma,V)$ for $1 \leq i \leq n$, then $p(\theta_1,\ldots,\theta_n) \in \mathrm{Fml}_{\mathrm{FOL}}(\Sigma,V)$.
- If $\phi, \psi \in \mathrm{Fml}_{\mathrm{FOL}}(\Sigma,V)$, then $\neg\phi, (\phi \wedge \psi), (\phi \vee \psi), (\phi \rightarrow \psi) \in \mathrm{Fml}_{\mathrm{FOL}}(\Sigma,V)$.
- If $\phi \in \mathrm{Fml}_{\mathrm{FOL}}(\Sigma,V)$ and $x \in V$, then $(\forall x\, \phi), (\exists x\, \phi) \in \mathrm{Fml}_{\mathrm{FOL}}(\Sigma,V)$.

More succinctly, we also say that first-order formulas are defined by the following grammar (where ϕ, ψ are first-order formulas, θ_i are terms, p is a predicate symbol of arity n, and $x \in V$ is a logical variable):

$$\phi, \psi ::= p(\theta_1,\ldots,\theta_n) \mid \neg\phi \mid \phi \wedge \psi \mid \phi \vee \psi \mid \phi \rightarrow \psi \mid \forall x\, \phi \mid \exists x\, \phi .$$

Example 2.3. (Well-formed) first-order formulas in our context include:

- $v \cdot v \leq 2b \cdot (m - z)$. Again we assume that v, b, m, z are symbols in the vocabulary Σ or V. For instance, we could assume that $z, v \in \Sigma$ are (flexible) state variables and $b, m \in \Sigma$ are rigid function symbols of arity 0. The rationale for this classification would be that z and v are meant to represent the position and velocity of a train, which of course can change over time (flexible). The symbols b and m are meant to represent the braking force and movement authority of a train, which we assume not to change in this formula (rigid). We could just as well assume that $z, v, b, m \in \Sigma$ are (flexible) state variables.

- $v > 0 \rightarrow v \cdot v \leq 2b \cdot (m - z) \vee b = 0$
- $\forall x \forall y\, (x > y \leftrightarrow x - y > 0)$. Here we assume $x, y \in V$ are logical variables; otherwise the syntax would not allow us to quantify over x, y.
- $x > 0 \wedge \forall y \exists z\, (x > z^2 + y \cdot z - 5)$. Here we assume $y, z \in V$ are logical variables and we could either assume $x \in \Sigma$ to be a state variable, or a rigid function symbol of arity zero, or a logical variable $x \in V$. All those choices are reasonable for this formula, and, in fact, it does not make a real difference here, because they will essentially have the same meaning. These distinctions become somewhat more important for $\mathsf{d}\mathcal{L}$ formulas later.
- Formulas with divisions like $b < x/y$, which can easily be defined in terms of multiplication $(b \cdot y < x \wedge y > 0) \vee (b \cdot y > x \wedge y < 0)$.
- Formulas with rational constants like $a > \frac{2}{3}x^2 + 3.1415x \cdot y^4$, which can easily be defined in terms of successive addition and inverses, say,

$$a > (1+1)/(1+1+1) \cdot x^2 + 31415/10000 \cdot x \cdot y^4.$$

- Arithmetic expressions with roots like $x^4 - y\sqrt{2z} > 0$, which can easily be defined in terms of their characteristic polynomials as $\exists r\, (r^2 = 2z \wedge r \geq 0 \wedge x^4 - y \cdot r > 0)$.

□

2.2.2 Hybrid Programs

As uniform compositional models for hybrid systems, hybrid programs can combine discrete and continuous transitions to structured control programs using the regular-expression-style operators of Kleene algebras [182].

Definition 2.3 (Hybrid programs). The set $\mathrm{HP}(\Sigma, V)$ of *hybrid programs*, with typical elements α, β, is defined inductively as the smallest set such that

1. If $x_i \in \Sigma$ is a state variable and $\theta_i \in \mathrm{Trm}(\Sigma, V)$ for $1 \leq i \leq n$, then the *discrete jump set* $(x_1 := \theta_1, \ldots, x_n := \theta_n) \in \mathrm{HP}(\Sigma, V)$ is a hybrid program. We assume that the $x_1, \ldots, x_n$ are pairwise different state variables.
2. If $x_i \in \Sigma$ is a state variable and $\theta_i \in \mathrm{Trm}(\Sigma, V)$ for $1 \leq i \leq n$, then $x_i' = \theta_i$ is a *differential equation* in which x_i' represents the time derivative of variable x_i. If χ is a first-order formula, then $(x_1' = \theta_1, \ldots, x_n' = \theta_n \,\&\, \chi) \in \mathrm{HP}(\Sigma, V)$. We assume that the $x_1, \ldots, x_n$ are pairwise different state variables.
3. If χ is a first-order formula, then $(?\chi) \in \mathrm{HP}(\Sigma, V)$.
4. If $\alpha, \beta \in \mathrm{HP}(\Sigma, V)$, then $(\alpha \cup \beta) \in \mathrm{HP}(\Sigma, V)$.
5. If $\alpha, \beta \in \mathrm{HP}(\Sigma, V)$, then $(\alpha; \beta) \in \mathrm{HP}(\Sigma, V)$.
6. If $\alpha \in \mathrm{HP}(\Sigma, V)$, then $(\alpha^*) \in \mathrm{HP}(\Sigma, V)$.

Table 2.1 summarises the statements and (informal) effects of hybrid programs. More succinctly, hybrid programs are defined by the following grammar (α, β are

Table 2.1 Statements and (informal) effects of hybrid programs (HPs)

HP Notation	Operation	Effect
$x_1 := \theta_1, \dots, x_n := \theta_n$	discrete jump set	simultaneously assigns terms θ_i to variables x_i
$x_1' = \theta_1, \dots, x_n' = \theta_n \,\&\, \chi$	continuous evolution	differential equations for x_i with terms θ_i within first-order constraint χ (evolution domain)
$?\chi$	state test / check	test first-order formula χ at current state
$\alpha; \beta$	seq. composition	HP β starts after HP α finishes
$\alpha \cup \beta$	nondet. choice	choice between alternatives HP α or HP β
α^*	nondet. repetition	repeats HP α n-times for any $n \in \mathbb{N}$

hybrid programs, θ_i are terms, $x_i \in \Sigma$ are state variables, and χ is a formula of first-order logic):

$$\alpha, \beta ::= x_1 := \theta_1, \dots, x_n := \theta_n \mid x_1' = \theta_1, \dots, x_n' = \theta_n \,\&\, \chi \mid ?\chi \mid \alpha \cup \beta \mid \alpha; \beta \mid \alpha^*.$$

The effect of the discrete jump set $x_1 := \theta_1, \dots, x_n := \theta_n$ is to simultaneously change the interpretations of the x_i to the respective θ_i by performing a discrete jump in the state space. In particular, the new values θ_i are evaluated before changing the value of any variable x_j. The effect of $x_1' = \theta_1, \dots, x_n' = \theta_n \,\&\, \chi$ is an ongoing continuous evolution respecting the differential equation system $x_1' = \theta_1, \dots, x_n' = \theta_n$ that is restricted to remain within the evolution domain region χ. The evolution is allowed to stop at any point in χ. It is, however, required to stop before it leaves χ. For unconstrained evolutions, we write $x' = \theta$ in place of $x' = \theta \,\&\, true$. For structural reasons, we expect both difference equations (discrete jump sets) and differential equations to be given in explicit form, i.e., with the affected variable on the left (we allow more general implicit forms in Chap. 3). The d$\mathcal{L}$ semantics allows arbitrary differential equations. To retain feasible arithmetic, some of our calculus rules in this chapter assume that, as in [8, 125, 217, 156], the differential equations have first-order definable flows or approximations. We assume that standard techniques are used to determine corresponding solutions or approximations, e.g., [15, 189, 238, 227, 297]. We consider verification techniques for more advanced differential equations in Chap. 3.

The test action or state check $?\chi$ is used to define conditions. Its semantics is that of a no-op if the formula χ is true in the current state; otherwise, like *abort*, it allows no transitions. That is, if the test succeeds because formula χ holds in the current state, then the state does not change, and the system execution continues normally. If the test fails because formula χ does not hold in the current state, then the system execution cannot even continue. Thus, the effect of a test action is similar to an *assert* statement in Java. Note that, according to Definition 2.3, we have only allowed first-order formulas as tests. Instead, we could actually allow *rich tests*, i.e., arbitrary d$\mathcal{L}$ formulas χ with nested modalities as tests $?\chi$ inside hybrid programs (and even in evolution domains χ of differential equations). The calculus and our meta-results, including soundness and relative completeness, directly carry over to this rich test version of d$\mathcal{L}$. To simplify the presentation, however, we refrain from allowing arbitrary d$\mathcal{L}$ formulas as tests, because that requires simultaneous inductive handling of hybrid programs and d$\mathcal{L}$ formulas in syntax, semantics, and completeness proofs, because d$\mathcal{L}$ formulas would then be allowed to occur in hybrid programs, and vice versa.

The nondeterministic choice $\alpha \cup \beta$, sequential composition $\alpha;\beta$, and nondeterministic repetition α^* of programs are as in regular expressions but generalised to a semantics in hybrid systems. Choices $\alpha \cup \beta$ are used to express behavioural alternatives between the transitions of α and β. That is, the hybrid program $\alpha \cup \beta$ can choose nondeterministically to follow the transitions of the hybrid program α, or, instead, to follow the transitions of the hybrid program β. The sequential composition $\alpha;\beta$ says that the hybrid program β starts executing after α has finished (β never starts if α does not terminate). In $\alpha;\beta$, the transitions of α take effect first, until α terminates (if it does), and then β continues. Observe that, like repetitions, continuous evolutions within α can take more or less time, which causes uncountable nondeterminism. This nondeterminism is inherent in hybrid systems, because they can operate in so many different ways, and as such reflected in hybrid programs. Repetition α^* is used to express that the hybrid process α repeats any number of times, including zero times. When following α^*, the transitions of hybrid program α can be repeated over and over again, any nondeterministic number of times ($\geq$0). Hybrid programs form a regular-expression-style Kleene algebra with tests [182].

Example 2.4 (Simplistic train). The differential equation $z' = v, v' = a$ expresses continuous movement of position z with velocity v and acceleration a. A very simple (in fact much too simplistic) train controller could be the following hybrid program:

$$((a := -b) \cup (?v < 8; a := A));\ z' = v, v' = a.$$

By a nondeterministic choice ($\cup$), the system either chooses to set the acceleration a to the braking force $-b$ by executing $a := -b$, or the system tries to pass the test $?v < 8$ instead. If the system tries the second choice and the latter test succeeds, i.e., the current velocity is indeed less than 8, then the system sets the acceleration a to A. Otherwise, if it tries the second choice but the test fails, then nothing happens as this execution is blocked and cannot continue. Afterwards (after executing the first part of the sequential composition, which is the nondeterministic choice), the system follows the second part of the sequential composition, which is the differential equation $z' = v, v' = a$ with the previously chosen acceleration. The system then follows this differential equation for a certain (unspecified) period of time.

This controller leaves open too many aspects to be useful, but already illustrates a very simple hybrid program. One of the problems is that the controller can only take a control action for choosing the acceleration a once, at the beginning of the system evolution, and then follows the differential equation for an arbitrarily long time. But the above controller can never react to situation changes and change its mind with a different choice of a when necessary. To improve this issue, the following hybrid program allows repetitive choices by the repetition operator *:

$$(((a:=-b)\cup(?v<8;a:=A));\ z'=v,v'=a)^*. \tag{2.1}$$

Unlike the previous hybrid program, the hybrid program in (2.1) contains a repetition, which can change the acceleration repeatedly over and over again after following the continuous evolution for a certain period of time. While already an improvement over the last controller, this hybrid program has shortcomings. For one thing, the differential equation does not say when it stops. It has no evolution domain restriction and would thus be allowed to evolve as long or as short as it pleases. This may be unsafe if the differential equation would continue indefinitely without giving the controller for the acceleration a chance to react to situation changes. Furthermore, a velocity of 8 may not be a safe choice for the switching condition between acceleration and braking. We will see in Sect. 2.4 how a reasonable train controller can be designed as a hybrid program from first principles and elaborate on train control further in Chap. 7 in full detail. □

Definable Operations

The control flow operations of choice, sequential composition, and repetition in hybrid programs can be combined with $?\chi$ to form all other control structures [149]. All classical discrete control structures can be defined in terms of the basic hybrid program operators (it is easy to see that hybrid programs are Turing-complete). See Table 2.2 for a selection of control structures and statements that are definable as a hybrid program. For instance, $(?\chi;\alpha)^*;?\neg\chi$ corresponds to a while loop that re-

Table 2.2 Statements and control structures definable with hybrid programs

HP Notation	Operation	Effect
$x:=*$	nondet. assignment	assigns any real value to x equivalently definable, see Chap. 3
if χ then α else β	if-then-else	executes HP α if χ holds, otherwise HP β equivalent to $(?\chi;\alpha)\cup(?\neg\chi;\beta)$
if χ then α	if-then	executes HP α if χ holds, otherwise no effect equivalent to $(?\chi;\alpha)\cup(?\neg\chi)$
while χ do α	while loop	repeats α if χ holds, only stops if $\neg\chi$ holds at end equivalent to $(?\chi;\alpha)^*;?\neg\chi$
repeat α until χ	repeat until	repeats HP α until χ holds at end (at least once) equivalent to $\alpha;(?\neg\chi;\alpha)^*;?\chi$
skip	do nothing	has no effect and does not change the state space equivalent to $?true$
abort	aborts execution	blocks current execution and allows no transition equivalent to $?false$

peats α while χ holds and only stops when χ ceases to hold after α. Because the *-operator can repeat arbitrarily often, the subprogram $(?\chi;\alpha)^*$ can repeat α any number of times, but a repetition can only be successful if the test $?\chi$ succeeds. Hence the repetitions have to stop, at the latest, when the test $?\chi$ fails. Now the subprogram $(?\chi;\alpha)^*$ can repeat any number of times and is allowed to stop even if the test $?\chi$ is successful and the loop could be repeated again. But the subsequent sequential composition with the test $?\neg\chi$ makes sure that $(?\chi;\alpha)^*$ can only stop repeating when χ actually ceases to hold. Overall, the hybrid program $(?\chi;\alpha)^*;?\neg\chi$ executes α if χ holds and repeats α again exactly as often as χ still holds after executing α.

If-then-else can be defined with nondeterministic choices and tests. The corresponding hybrid program $(?\chi;\alpha)\cup(?\neg\chi;\beta)$ in Table 2.2 makes a nondeterministic choice between $?\chi;\alpha$ and $?\neg\chi;\beta$. While this choice is nondeterministic, at any state only one of the subsequent tests in the two cases can succeed, because they are complementary. Consequently, hybrid program α will be executed if and only if the test $?\chi$ succeeds because χ is true at the current state. Likewise, hybrid program β will be executed if and only if the dual test $?\neg\chi$ succeeds because $\neg\chi$ is true, i.e., χ is false at the current state. The nondeterministic assignment $x:=*$ that assigns an arbitrary real number to state variable x is definable also. While it is possible define nondeterministic assignments in hybrid programs already, we will come back to this in Chap. 3, where the definition is easier to see.

Example 2.5 (Parametric bouncing ball). As a classical example from the hybrid systems literature [110], consider the bouncing ball. We will describe the bouncing ball as a hybrid program, using the definable hybrid program operations from Table 2.2. Figure 2.2 depicts a hybrid automaton, an illustration of the bouncing ball dynamics, and a representation of the system as a hybrid program.

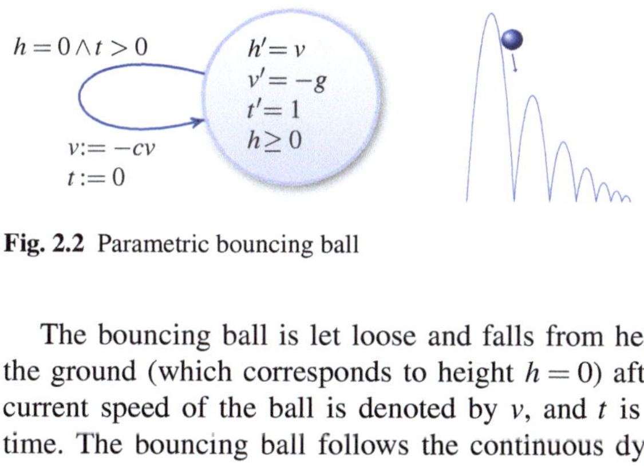

$ball \equiv ($
  $h' = v, v' = -g, t' = 1 \,\&\, h \geq 0;$
  if $(h = 0 \wedge t > 0)$ then
    $c := *;\ ?(0 \leq c < 1);$ *// extra*
    $v := -cv;\ t := 0$
  $\text{fi})^*$

Fig. 2.2 Parametric bouncing ball

The bouncing ball is let loose and falls from height h, but bounces back from the ground (which corresponds to height $h = 0$) after an elastic deformation. The current speed of the ball is denoted by v, and t is a clock measuring the falling time. The bouncing ball follows the continuous dynamics of physical movement by gravity. The ball is affected by gravity of force g, so its height follows the differential equation $h'' = -g$. This second-order differential equation is equivalent to the first-order differential equation system $h' = v, v' = -g$, with an explicit velocity v. Simultaneously, clock t evolves according to the differential equation $t' = 1$. Finally, the ball always stays above the ground and cannot fall through, thus its evolution domain is restricted to $h \geq 0$. Altogether, this gives the continuous evolution $h' = v, v' = -g, t' = 1 \,\&\, h \geq 0$ in the beginning of the hybrid program in Fig. 2.2.

At the ground (which is at height $h = 0$), the ball bounces back after losing energy in an elastic deformation according to a damping factor $0 \leq c < 1$. That is, if the ball is on the ground ($h=0$) and it actually fell (so time has passed, $t>0$), then the ball changes its direction and bounces back into the air by reflecting its current velocity v by a discrete jump $v:=-cv$ and resetting the falling time by $t:=0$.

Now for illustration purposes we have added an extra twist to the hybrid program in Fig. 2.2 that is not in the hybrid automaton. The automaton still enforces infinite bouncing so that the ball can never stop (unless $c=0$, where it stops immediately). In reality, the ball bounces a couple of times and can then come to a standstill when its remaining kinetic energy is insufficient. To model this phenomenon without the need to have a precise physical model for all physical forces and frictions, we allow for the damping factor c to change at each bounce. Line 4 of the hybrid program in Fig. 2.2 represents a corresponding uncountably infinite nondeterministic choice for c as a nondeterministic assignment. The subsequent test $?(0 \le c < 1)$ restricts the arbitrary choices for c to choices in the half-open interval $[0,1)$ and discards all other choices.

For comparison, Fig. 2.3 shows an equivalent hybrid program for the same bouncing, now with all abbreviations for extended statements resolved according to Table 2.2. Note that it is fairly easy to see that height h and clock t always stay nonnegative if they start nonnegative. For that reason, the last test $?(h \neq 0 \vee t \le 0)$ in Fig. 2.3 could even be replaced equivalently by $?(h>0 \vee t=0)$. □

Fig. 2.3 Parametric bouncing ball (with abbreviations resolved)

$$
\begin{aligned}
&ball \equiv (\\
&\quad h' = v, v' = -g, t' = 1 \,\&\, h \ge 0;\\
&\quad (?(h=0 \wedge t>0);\\
&\qquad\quad (c' = 1 \cup c' = -1);\\
&\qquad\quad ?(0 \le c < 1);\\
&\qquad\quad v := -cv;\ t := 0\\
&\quad) \cup ?(h \neq 0 \ \vee\ t \le 0)\\
&)^*
\end{aligned}
$$

Classification of Hybrid Programs

Hybrid programs are designed as a minimal extension of conventional discrete programs. They characterise hybrid systems succinctly by adding continuous evolution along differential equations as the only additional primitive operation to a regular basis of conventional discrete programs. To yield hybrid systems, their operations are interpreted over the domain of real numbers. This gives rise to an elegant syntactic hierarchy of discrete, continuous, and hybrid systems. Hybrid automata [156] can be represented as hybrid programs using a straightforward generalisation of standard program encodings of automata; see App. C for formal details. The fragment of hybrid programs without differential equations corresponds to conventional discrete programs generalised over the reals or to discrete-time dynamical systems [56]. The fragment without discrete jumps corresponds to switched continuous systems [56, 58], whereas the fragment of differential equations gives purely con-

Table 2.3 Operators and (informal) meaning in differential dynamic logic ($d\mathcal{L}$)

$d\mathcal{L}$ Notation	Operator	Meaning
$p(\theta_1,\dots,\theta_n)$	atomic formula	true iff predicate p holds for $(\theta_1,\dots,\theta_n)$
$\neg\phi$	negation / not	true if ϕ is false
$\phi \wedge \psi$	conjunction / and	true if both ϕ and ψ are true
$\phi \vee \psi$	disjunction / or	true if ϕ is true or if ψ is true
$\phi \rightarrow \psi$	implication / implies	true if ϕ is false or ψ is true
$\phi \leftrightarrow \psi$	bi-implication / equivalent	true if ϕ and ψ are both true or both false
$\forall x\, \phi$	universal quantifier / for all	true if ϕ is true for all values of variable x
$\exists x\, \phi$	existential quantifier / exists	true if ϕ is true for some values of variable x
$[\alpha]\phi$	$[\cdot]$ modality / box	true if ϕ is true after all runs of HP α
$\langle\alpha\rangle\phi$	$\langle\cdot\rangle$ modality / diamond	true if ϕ is true after at least one run of HP α

tinuous dynamical systems [279]. Only the composition of mixed discrete jumps and continuous evolutions gives rise to truly hybrid behaviour.

2.2.3 Formulas of Differential Dynamic Logic

The formulas of the differential dynamic logic $d\mathcal{L}$ are defined as in first-order dynamic logic [253, 148, 149] but with real arithmetic as a semantic domain and with hybrid programs as system models. That is, they are built using propositional connectives $\neg,\wedge,\vee,\rightarrow,\leftrightarrow$ and quantifiers $\forall,\exists$ over the reals (first-order part). In addition, if ϕ is a $d\mathcal{L}$ formula and α a hybrid program, then $[\alpha]\phi, \langle\alpha\rangle\phi$ are formulas (dynamic part).

Definition 2.4 ($d\mathcal{L}$ formulas). The set $\mathrm{Fml}(\Sigma,V)$ of *formulas* of $d\mathcal{L}$, with typical elements ϕ,ψ, is the smallest set such that

1. If p is a predicate symbol of arity $n \geq 0$ and $\theta_i \in \mathrm{Trm}(\Sigma,V)$ for $1 \leq i \leq n$, then $p(\theta_1,\dots,\theta_n) \in \mathrm{Fml}(\Sigma,V)$.
2. If $\phi,\psi \in \mathrm{Fml}(\Sigma,V)$, then $\neg\phi, (\phi\wedge\psi), (\phi\vee\psi), (\phi\rightarrow\psi) \in \mathrm{Fml}(\Sigma,V)$.
3. If $\phi \in \mathrm{Fml}(\Sigma,V)$ and $x \in V$, then $\forall x\,\phi, \exists x\,\phi \in \mathrm{Fml}(\Sigma,V)$.
4. If $\phi \in \mathrm{Fml}(\Sigma,V)$ and $\alpha \in \mathrm{HP}(\Sigma,V)$, then $[\alpha]\phi, \langle\alpha\rangle\phi \in \mathrm{Fml}(\Sigma,V)$.

For reference, the logical operators of differential dynamic logic are summarised in Table 2.3. More succinctly, we also say that the formulas of $d\mathcal{L}$ are defined by the following grammar (where ϕ,ψ are $d\mathcal{L}$ formulas, θ_i are terms, p a predicate symbol of arity n, $x \in V$ is a logical variable, and α is a hybrid program):

$$\phi,\psi ::= p(\theta_1,\dots,\theta_n) \mid \neg\phi \mid \phi\wedge\psi \mid \phi\vee\psi \mid \phi\rightarrow\psi \mid \forall x\,\phi \mid \exists x\,\phi \mid [\alpha]\phi \mid \langle\alpha\rangle\phi.$$

We consider the bi-implication or equivalence $\phi \leftrightarrow \psi$ as an abbreviation for $(\phi\rightarrow\psi)\wedge(\psi\rightarrow\phi)$ to simplify the calculus. We often leave out superfluous brackets and use binding priorities instead in order to improve readability. Quantifiers and modalities bind strongly, i.e., their scope only extends to the formula immediately after. Unary operators (negation $\neg$), quantifiers ($\forall, \exists$), and modalities ($[\alpha], \langle\alpha\rangle$) bind stronger than binary operators. Further, conjunction $\wedge$ and disjunction $\vee$ bind stronger than implication $\rightarrow$ and bi-implication $\leftrightarrow$. Thus

$$\phi_0 \wedge \langle\alpha\rangle\phi_1 \wedge \forall x\,\phi_2 \wedge \phi_3 \rightarrow \neg\phi_4 \vee [\alpha]\phi_5 \vee \phi_6$$

is taken to mean

$$\big(\phi_0 \wedge (\langle\alpha\rangle\phi_1) \wedge (\forall x\,\phi_2) \wedge \phi_3\big) \rightarrow \big((\neg\phi_4) \vee ([\alpha]\phi_5) \vee \phi_6\big)$$

and does not mean

$$\phi_0 \wedge \Big(\langle\alpha\rangle\big(\phi_1 \wedge \forall x\,(\phi_2 \wedge \phi_3)\big) \rightarrow \neg\big(\phi_4 \vee [\alpha](\phi_5 \vee \phi_6)\big)\Big).$$

Example 2.6 (Train control). When *train* denotes the hybrid program in Fig. 2.1 or the hybrid program in Example 2.4, or, in fact, any other hybrid program model for a train system, then the following $d\mathcal{L}$ formula expresses that this train is able ($\langle train\rangle$) to enter region $z \geq m$, thereby leaving region $z < m$ when it starts in region $z < m$ with nonnegative initial velocity $v \geq 0$:

$$v \geq 0 \wedge z < m \rightarrow \langle train\rangle\, z \geq m. \tag{2.2}$$

Dually, the following $d\mathcal{L}$ formula expresses that the train will always ($[train]$) stay inside the region $z < m$ when it starts inside it with an initial nonnegative velocity less than 5:

$$v \geq 0 \wedge v < 5 \wedge z < m \rightarrow [train]\, z < m.$$

For most train models *train*, the latter safety property will only be true for additional constraints on the initial state and on the internal parameter choices, including, e.g., braking forces, reaction times, and start braking points. □

Example 2.7 (Parametric bouncing ball). Let *ball* denote the hybrid program for the bouncing ball from Example 2.5. The ball loses energy at every bounce, thus the ball never bounces higher than the initial height. This can be expressed by the property $0 \leq h \leq H$, where H denotes the initial energy level (which corresponds to the initial height if $v = 0$ initially). Then, for instance, the following $d\mathcal{L}$ formula expresses that (under a list of assumptions on the free variables h, v, t and H, g, c) the ball always stays in the region $0 \leq h \leq H$:

$$(v^2 \leq 2g(H-h) \wedge h \geq 0 \wedge g > 0 \wedge H \geq 0 \wedge 1 > c \geq 0) \rightarrow [ball](0 \leq h \leq H). \tag{2.3}$$

This $d\mathcal{L}$ formula follows the pattern of Hoare triples [161]. It expresses that the bouncing ball, when started in an initial state satisfying the precondition on the left of the implication ($\rightarrow$), always respects the postcondition $0 \leq h \leq H$ of the dynamic modality $[ball]$, i.e., all runs of the bouncing ball stay in the region $0 \leq h \leq H$. □

A $\mathsf{d}\mathcal{L}$ formulas of the form $\psi \rightarrow [\alpha]\phi$ corresponds to Hoare triples [161], generalised for hybrid systems. They occur quite frequently, because they specify that system α, when starting in a state satisfying the *precondition* ψ, always respects the *postcondition* ϕ. That is, when started in a state satisfying ψ, all states reachable by α satisfy ϕ. There are several other relevant shapes of $\mathsf{d}\mathcal{L}$ formulas in practical systems verification; see Part III.

Note that, according to Definition 2.4, hybrid programs are not additional external objects but fully internalised [48] as first-class citizens within the logic $\mathsf{d}\mathcal{L}$ itself, and the logic is closed. That is, modalities can be combined propositionally, by quantifiers, or nested. For instance, $[\alpha]\langle\beta\rangle x \leq c$ says that, whatever hybrid program α is doing, hybrid program β can react in some way to reach a controlled state where x is less than some critical value c. That is, for all α actions, there is a β (re)action such that $x \leq c$ holds. Dually, $\langle\beta\rangle[\alpha]x \leq c$ expresses that hybrid program β can stabilise $x \leq c$, i.e., behave in such a way that $x \leq c$ remains true no matter how hybrid program α reacts. That is, there is a β action such that all α actions maintain $x \leq c$. Accordingly, $\exists p\,[\alpha]x \leq c$ says that there is a choice of parameter p such that α remains in $x \leq c$. Nesting modalities and quantifiers in this way can be quite useful for describing interactions of a hybrid program α with an environment β, or for describing the impact of parameter choices on properties of the system behaviour.

During our analysis, we assume differential equations and discrete transitions to be well-defined. In particular, we assume that all divisions p/q are guarded by conditions that ensure $q \neq 0$ as, otherwise, the system behaviour is not well-defined due to an undefined value at a singularity. It is simple but tedious to augment the semantics and the calculus with corresponding side conditions to show that this is respected. For instance, we assume that $x := p/q$ is guarded by $?q \neq 0$ and that continuous evolutions are restricted such that the differential equations are well-defined as $x' = p/q \,\&\, q \neq 0$. Also see our joint work with Beckert [37] for techniques of how such exceptional behaviour can be handled by program transformation while avoiding partial valuations of undefinedness in the semantics. In logical formulas, partiality can be avoided altogether by writing $p = c \cdot q \wedge q \neq 0$ rather than $p/q = c$, and writing $(p > c \cdot q \wedge q > 0) \vee (p < c \cdot q \wedge q < 0)$ rather than $p/q > c$.

2.3 Semantics of Differential Dynamic Logic

We define the semantics of $\mathsf{d}\mathcal{L}$ as a possible world Kripke semantics [185] with worlds representing the possible system states and with reachability along the hybrid transitions of the system representing accessibility relations between worlds. The interpretations of $\mathsf{d}\mathcal{L}$ consist of states (worlds) that are essentially first-order structures over the reals. In particular, real values are assigned to state variables, possibly different values in each state. A potential behaviour of a hybrid system corresponds to a succession of states that contain the observable values of system variables during its hybrid evolution.

2.3.1 Valuation of Terms

Symbols in the logic d$\mathcal{L}$ come from three different syntactic categories that we decided to distinguish in Sect. 2.2.1:

1. rigid symbols in Σ that cannot change their value, e.g., $0, 1, +, \cdot$;
2. flexible symbols in Σ, which are the state variables, whose value can change depending on the current state of the system;
3. logical variables in V that cannot change their value over time by running hybrid programs, but which can be quantified over universally or existentially.

All of those symbols need to be interpreted to give meaning to terms in which they occur. We associate values with rigid symbols by what we call an interpretation I, associate values with flexible symbols by a state ν, and associate values with logical variables by an assignment η. Recall that there is some leeway in declaring symbols as either rigid or flexible symbols or as logical variables. The semantics is unambiguous for each choice, though.

An *interpretation* I assigns functions and relations over the reals to the respective rigid symbols in Σ. The function and predicate symbols of real arithmetic are interpreted as usual by I. Especially, the interpretation $I(+)$ is addition and $I(\cdot)$ is multiplication of real numbers. A *state* is a map $\nu : \Sigma_{fl} \to \mathbb{R}$; the set of all states is denoted by $\mathrm{Sta}(\Sigma)$. Here, Σ_{fl} denotes the set of (flexible) state variables in Σ (they have arity 0, thus take no arguments). Finally, an *assignment for logical variables* is a map $\eta : V \to \mathbb{R}$. It contains the values for logical variables, which are not subject to change by modalities but only by quantification. Observe that flexible symbols (which represent state variables) are allowed to assume different interpretations in different states. Logical variable symbols, however, are rigid in the sense that their value is determined by η alone and does not depend on the state ν.

The *valuation* $val_{I,\eta}(\nu,\cdot)$ of terms is defined as usual [123, 149] with an extra distinction of rigid and flexible functions [37]. It is defined inductively by recursion on the structure of the term, based on the interpretation that assignment η assigns to logical variables, that interpretation I assigns to rigid function symbols, and that state ν assigns to flexible state variables. The semantics of terms is compositional and denotational [277], that is, the semantics of a complex term is defined as a combination of the semantics of its subterms.

Definition 2.5 (Valuation of terms). The *valuation of terms* with respect to interpretation I, assignment η, and state ν is defined by

1. $val_{I,\eta}(\nu,x) = \eta(x)$ if $x \in V$ is a logical variable.
2. $val_{I,\eta}(\nu,a) = \nu(a)$ if $a \in \Sigma$ is a state variable (flexible function symbol of arity 0).
3. $val_{I,\eta}(\nu,f(\theta_1,\dots,\theta_n)) = I(f)\big(val_{I,\eta}(\nu,\theta_1),\dots,val_{I,\eta}(\nu,\theta_n)\big)$ when $f \in \Sigma$ is a rigid function symbol of arity $n \geq 0$.

Example 2.8. Let interpretation I interpret the constant function symbol $b \in \Sigma$ as $I(b) = 2.14$, and interpret the unary function symbol $c \in \Sigma$ as the cubic function

$d \mapsto d^3$, i.e., $\big(I(c)\big)(d) = d^3$. Let the assignment η interpret the logical variable $X \in V$ as $\eta(X) = 4.2$. Finally let the state ν interpret the state variables $x, y, z \in \Sigma$ as $\nu(x) = 3$, $\nu(y) = -5.01$, $\nu(z) = 0$. Throughout this book we assume that the interpretation of $0, 1, +, -, \cdot$ always is as usual in real arithmetic, that is:

$$
\begin{aligned}
I(0) &= 0 \\
I(1) &= 1 \\
\big(I(+)\big)(d,e) &= d+e && \text{(addition)} \\
\big(I(-)\big)(d,e) &= d-e && \text{(subtraction)} \\
\big(I(\cdot)\big)(d,e) &= d\cdot e && \text{(multiplication)}
\end{aligned}
$$

With this we can valuate terms recursively with respect to I, η, ν as follows:

$$
\begin{aligned}
val_{I,\eta}(\nu, x+y) &= val_{I,\eta}(\nu, x) + val_{I,\eta}(\nu, y) = \nu(x) + \nu(y) \\
&= 3 + (-5.01) = -2.01, \\
val_{I,\eta}(\nu, y + 2\cdot X) &= val_{I,\eta}(\nu, y) + val_{I,\eta}(\nu, 2)\cdot val_{I,\eta}(\nu, X) \\
&= \nu(y) + I(2)\cdot \eta(X) = -5.01 + 2*4.2 = 3.39, \\
val_{I,\eta}(\nu, X + b\cdot(x + y\cdot X)) &= \eta(X) + I(b)\cdot(\nu(x) + \nu(z)\cdot\eta(X)) \\
&= 4.2 + 2.14\cdot(3 + 0\cdot 4.2) = 10.62, \\
val_{I,\eta}(\nu, c(x+X) - x) &= val_{I,\eta}(\nu, c(x+X)) - val_{I,\eta}(\nu, x) \\
&= I(c)\big(val_{I,\eta}(\nu, x+X)\big) - \nu(x) \\
&= I(c)\big(\nu(x) + \eta(X)\big) \;-\; \nu(x) \\
&= (3+4.2)^3 - 3 = 370.248.
\end{aligned}
$$

Note here, that the decision about which symbols we consider as rigid function symbols, which ones we consider as flexible function symbols (state variables), and which ones we consider as logical variables is somewhat arbitrary in this example. This decision only becomes relevant when we add quantifiers (for only logical variables can be quantified over) or hybrid programs (for only state variables can be assigned to in hybrid programs). Overall, the syntactic category of symbols is not crucial, as there are often many equivalent ways to assign symbols to syntactic categories. But if we fix a choice of symbols, the semantics becomes less subtle, so we assume a choice has been made for every formula. □

2.3.2 Valuation of Formulas

The *valuation* $val_{I,\eta}(\nu, \cdot)$ of formulas is defined as usual for first-order modal logic [123, 149] with a distinction of rigid and flexible functions [37]. Modalities parametrised by a hybrid program α follow the accessibility relation spanned by the respective hybrid state transition relation $\rho_{I,\eta}(\alpha)$, which is simultaneously inductively defined in Definition 2.7.

The valuation of formulas is defined inductively by recursion on the structure of formulas, based on the interpretation of the terms occurring in it. The semantics of formulas is compositional and denotational, that is, the semantics of a complex formula is defined as a simple function of the semantics of its subformulas. We will use $\eta[x \mapsto d]$ to denote the *modification* of an assignment η that agrees with η except for the interpretation of the logical variable $x \in V$, which is assigned $d \in \mathbb{R}$ in $\eta[x \mapsto d]$.

Definition 2.6 (Valuation of $d\mathcal{L}$ formulas). The *valuation* $val_{I,\eta}(\nu, \cdot)$ *of formulas* with respect to interpretation I, assignment η, and state ν is defined as

1. $val_{I,\eta}(\nu, p(\theta_1, \dots, \theta_n)) = I(p)\big(val_{I,\eta}(\nu, \theta_1), \dots, val_{I,\eta}(\nu, \theta_n)\big)$.
2. $val_{I,\eta}(\nu, \phi \wedge \psi) = true$ iff $val_{I,\eta}(\nu, \phi) = true$ and $val_{I,\eta}(\nu, \psi) = true$.
3. $val_{I,\eta}(\nu, \phi \vee \psi) = true$ iff $val_{I,\eta}(\nu, \phi) = true$ or $val_{I,\eta}(\nu, \psi) = true$.
4. $val_{I,\eta}(\nu, \neg\phi) = true$ iff $val_{I,\eta}(\nu, \phi) \neq true$.
5. $val_{I,\eta}(\nu, \phi \rightarrow \psi) = true$ iff $val_{I,\eta}(\nu, \phi) \neq true$ or $val_{I,\eta}(\nu, \psi) = true$.
6. $val_{I,\eta}(\nu, \forall x\, \phi) = true$ iff $val_{I,\eta[x \mapsto d]}(\nu, \phi) = true$ for all $d \in \mathbb{R}$.
7. $val_{I,\eta}(\nu, \exists x\, \phi) = true$ iff $val_{I,\eta[x \mapsto d]}(\nu, \phi) = true$ for some $d \in \mathbb{R}$.
8. $val_{I,\eta}(\nu, [\alpha]\phi) = true$ iff $val_{I,\eta}(\omega, \phi) = true$ for all states ω for which the transition relation satisfies $(\nu, \omega) \in \rho_{I,\eta}(\alpha)$.
9. $val_{I,\eta}(\nu, \langle\alpha\rangle\phi) = true$ iff $val_{I,\eta}(\omega, \phi) = true$ for some state ω for which the transition relation satisfies $(\nu, \omega) \in \rho_{I,\eta}(\alpha)$.

Following the usual notation, we also write $I, \eta, \nu \models \phi$ iff $val_{I,\eta}(\nu, \phi) = true$. We then say that ϕ is *satisfied* in I, η, ν or holds in I, η, ν. We also say that I, η, ν is a *model* of ϕ. Dually, we write $I, \eta, \nu \not\models \phi$ iff $val_{I,\eta}(\nu, \phi) \neq true$. If ϕ is satisfied for at least one I, η, ν, then ϕ is called *satisfiable*. Occasionally, we write just $\models \phi$ iff $I, \eta, \nu \models \phi$ holds for all I, η, ν. Formula ϕ is then called *valid*, i.e., true in all I, η, ν.

The semantics of modal formulas $[\alpha]\phi$ and $\langle\alpha\rangle\phi$ in $d\mathcal{L}$ is illustrated in Fig. 2.4, showing how the truth of ϕ at (all or some) states ω_i reachable by α relates to the truth of $[\alpha]\phi$ or $\langle\alpha\rangle\phi$ at state ν.

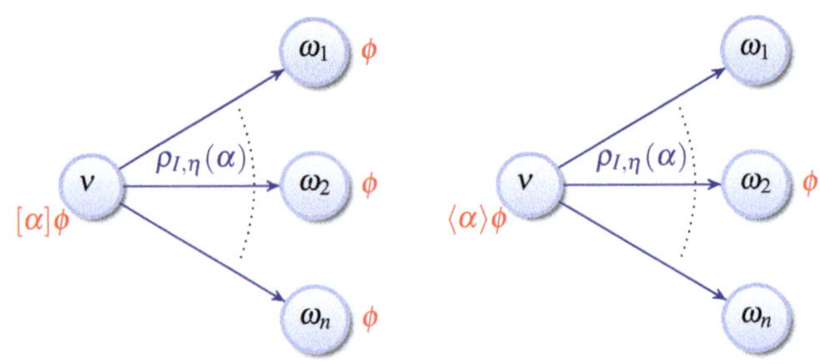

Fig. 2.4 Transition semantics of modalities in $d\mathcal{L}$ formulas

Example 2.9. Consider the following formula that we want to evaluate:

$$v > 0 \rightarrow v \cdot v \leq 2b \cdot (m - z) \vee b = 0. \tag{2.4}$$

First we have to declare which syntactic category the symbols are meant to come from. Suppose $z, v \in \Sigma$ are state variables (flexible function symbols of arity 0), because they represent position and velocity of the train, which are intended to be able to change over time from state to state. Further suppose that $m, b \in \Sigma$ are rigid function symbols, because, for the moment, movement authority and braking force are not allowed to change over time. Note that we could just as well have chosen all symbols z, v, m, b to be flexible state variables. The only notable difference is that if b is a flexible symbol, we would have to prove that a particular hybrid program never changes the value of b to know that it denotes the same value in every part of the program. Otherwise, if b is a rigid symbol, we already know that it cannot possibly change its value by running a hybrid program, because b then is a constant, and only flexible symbols are syntactically allowed to be assigned to or have differential equations in Definition 2.3. While it is certainly not crucial to make this distinction, it can make some things easier to see syntactically.

Now let interpretation I interpret rigid symbol $m \in \Sigma$ as $I(m) = 20$ and interpret $b \in \Sigma$ as $I(b) = 2.2$. Let state ω interpret state variables $v, z \in \Sigma$ as $\omega(v) = 10$, $\omega(z) = 0$. In formula (2.4), suppose we do not have any free logical variables, so that the assignment η of logical variables does not matter. Then we can evaluate formula (2.4) with respect to I, η, ω:

$$\begin{aligned}
&val_{I,\eta}(\omega, v > 0 \rightarrow v \cdot v \leq 2b \cdot (m - z) \vee b = 0) = \mathit{true} \text{ iff}\\
&val_{I,\eta}(\omega, v > 0) \neq \mathit{true}, \text{ or}\\
&val_{I,\eta}(\omega, v \cdot v \leq 2b \cdot (m - z)) = \mathit{true}, \text{ or}\\
&val_{I,\eta}(\omega, b = 0) = \mathit{true}.
\end{aligned}$$

Let us evaluate the terms to determine if the subformulas are true or not:

$$\begin{aligned}
val_{I,\eta}(\omega, v > 0) &= (val_{I,\eta}(\omega, v) \overset{?}{>} val_{I,\eta}(\omega, 0)) = (\omega(v) \overset{?}{>} I(0))\\
&= (10 \overset{?}{>} 0) = \mathit{true},\\
val_{I,\eta}(\omega, v \cdot v \leq 2b \cdot (m - z))&(val_{I,\eta}(\omega, v \cdot v) \overset{?}{\leq} val_{I,\eta}(\omega, 2b \cdot (m - z)))\\
&= \big(\omega(v) \cdot \omega(v) \overset{?}{\leq} 2I(b) \cdot (I(m) - \omega(z))\big)\\
&= \big(10 \cdot 10 \overset{?}{\leq} 2 \cdot 2.2 \cdot (20 - 0)\big) = (100 \overset{?}{\leq} 88) = \mathit{false},\\
val_{I,\eta}(\omega, b = 0) &= (I(b) \overset{?}{=} 0) = (2.2 \overset{?}{=} 0) = \mathit{false}.
\end{aligned}$$

Consequently the formula (2.4) evaluates to *false*. For a different state ν with slower speed $\nu(v) = 8$ and the same position $\nu(z) = 0$, we instead evaluate (2.4) to *true*:

$$val_{I,\eta}(\nu, v > 0 \rightarrow v \cdot v \leq 2b \cdot (m - z) \vee b = 0) = true.$$

Also for a different interpretation J with $J(m) = 20$ and another braking force $J(b) = 4$, but the same original state ω, we evaluate (2.4) to *true*:

$$val_{J,\eta}(\omega, v > 0 \rightarrow v \cdot v \leq 2b \cdot (m - z) \vee b = 0) = true.$$

So we see that the truth-value of formula (2.4) depends on I, η, ω. For some choices of I, η, ω, it evaluates to *true*, for others it evaluates to *false*. Thus, formula (2.4) is not valid, because

$$I, \eta, \omega \models v > 0 \rightarrow v \cdot v \leq 2b \cdot (m - z) \vee b = 0$$

does not hold for all I, η, ω. Still, the formula (2.4) is at least satisfiable, because it holds for some I, η, ω. □

Example 2.10. Consider the assignment η with $\eta(Z) = -2$ and the state ν with $\nu(x) = -4$. Then we can evaluate

$$val_{I,\eta}(\nu, x > -5 \wedge \forall y\, (y^2 + Z > x)) = true$$

because $\nu(x) > -5$ and all squares are greater than or equal zero, so that for all $d \in \mathbb{R}$:

$$\begin{aligned} val_{I,\eta[y \mapsto d]}(\nu, y^2 + Z > x) &= ((\eta[y \mapsto d](y))^2 + \eta[y \mapsto d](Z) \overset{?}{>} \nu(x) \\ &= (d^2 + (-2) \overset{?}{>} -4) = true. \end{aligned}$$

□

Note, that we have not yet explained how to evaluate formulas with modalities like $x > 0 \rightarrow [ctrl; drive^*]\, z \leq m$ in any I, η, ν, because we first have to define the transition semantics $\rho_{I,\eta}(\alpha)$ of hybrid programs, which we will do next.

2.3.3 Transition Semantics of Hybrid Programs

Now we define the transition semantics, $\rho_{I,\eta}(\alpha)$, of hybrid program α. The semantics of a hybrid program is captured by its hybrid state transition relation. For discrete jumps this transition relation holds for pairs of states that respect the discrete jump set. For continuous evolutions, the transition relation holds for pairs of states that can be interconnected by a continuous flow respecting the differential equations and evolution domain restriction throughout the evolution.

The transition semantics of hybrid programs is defined by induction based on the structure of the programs. The semantics of hybrid programs is compositional, that is, the semantics of a complex program is defined as a simple function of the transition semantics of its parts. We will use $\nu[x \mapsto d]$ to denote the *modification* of a state ν that agrees with ν except for the interpretation of the symbol $x \in \Sigma_{fl}$, which is changed to $d \in \mathbb{R}$ in $\nu[x \mapsto d]$.

Definition 2.7 (Transition semantics of hybrid programs). The *valuation of a hybrid program* α, denoted by $\rho_{I,\eta}(\alpha)$, is a *transition relation* on states. It specifies which state ω is reachable from a state ν by operations of the hybrid program α and is defined as follows

1. $(\nu,\omega) \in \rho_{I,\eta}(x_1 := \theta_1, \dots, x_n := \theta_n)$ iff the state ω equals the state obtained by semantic modification of state ν as $\nu[x_1 \mapsto val_{I,\eta}(\nu,\theta_1)]\dots[x_n \mapsto val_{I,\eta}(\nu,\theta_n)]$. Particularly, the values of other variables $z \notin \{x_1,\dots,x_n\}$ remain constant, i.e., $val_{I,\eta}(\omega,z) = val_{I,\eta}(\nu,z)$, and the x_i receive their new values simultaneously, i.e., $val_{I,\eta}(\omega,x_i) = val_{I,\eta}(\nu,\theta_i)$.
2. $(\nu,\omega) \in \rho_{I,\eta}(x_1' = \theta_1, \dots, x_n' = \theta_n \,\&\, \chi)$ iff there is a *flow* f of some duration $r \geq 0$ from state ν to state ω along $x_1' = \theta_1, \dots, x_n' = \theta_n \,\&\, \chi$, i.e., a function $f : [0,r] \to \mathrm{Sta}(\Sigma)$ such that:
   - $f(0) = \nu, f(r) = \omega$;
   - f respects the differential equations: For each variable x_i, the valuation $val_{I,\eta}(f(\zeta),x_i) = f(\zeta)(x_i)$ of x_i at state $f(\zeta)$ is continuous in ζ on $[0,r]$ and has a derivative of value $val_{I,\eta}(f(\zeta),\theta_i)$ at each time $\zeta \in (0,r)$;
   - the value of other variables $z \notin \{x_1,\dots,x_n\}$ remains constant, that is, we have $val_{I,\eta}(f(\zeta),z) = val_{I,\eta}(\nu,z)$ for all $\zeta \in [0,r]$;
   - and f respects the invariant: $val_{I,\eta}(f(\zeta),\chi) = \mathit{true}$ for each $\zeta \in [0,r]$.
3. $\rho_{I,\eta}(?\chi) = \{(\nu,\nu) \;:\; val_{I,\eta}(\nu,\chi) = \mathit{true}\}$
4. $\rho_{I,\eta}(\alpha \cup \beta) = \rho_{I,\eta}(\alpha) \cup \rho_{I,\eta}(\beta)$
5. $\rho_{I,\eta}(\alpha;\beta) = \{(\nu,\omega) : (\nu,\mu) \in \rho_{I,\eta}(\alpha), (\mu,\omega) \in \rho_{I,\eta}(\beta) \text{ for a state } \mu\}$
6. $(\nu,\omega) \in \rho_{I,\eta}(\alpha^*)$ iff there is an $n \in \mathbb{N}$ and states $\nu = \nu_0, \dots, \nu_n = \omega$ such that $(\nu_i,\nu_{i+1}) \in \rho_{I,\eta}(\alpha)$ for all $0 \leq i < n$.

For graphical illustrations of the transition semantics of hybrid programs and example dynamics, see Fig. 2.5. On the left of Fig. 2.5, we illustrate the generic shape of the transition structure $\rho_{I,\eta}(\alpha)$ for transitions along various cases of hybrid programs α from state ν to state ω. On the right of Fig. 2.5, we show examples of how the value of a variable x may evolve over time t when following the dynamics of the respective hybrid program α. The shape of the transition structure of a discrete jump $x := \theta$ (row 1) and of a differential equation $x' = \theta \,\&\, \chi$ (row 2) is an elementary one-step transition from ν to ω. For discrete jumps, however, the transition is an instant jump in the state space (row 1 on the right), while the transition for a differential equation is a continuous evolution in the state space (row 2 on the right). Note that the modifications of a discrete jump set $x_1 := \theta_1, .., x_n := \theta_n$ are executed simultaneously in Definition 2.7 in the sense that all terms θ_i are evaluated in the initial state ν. For simplicity, we assume the x_i to be different, and refer to previous work [37] for a compatible semantics and calculus handling concurrent modifications of the same x_i.

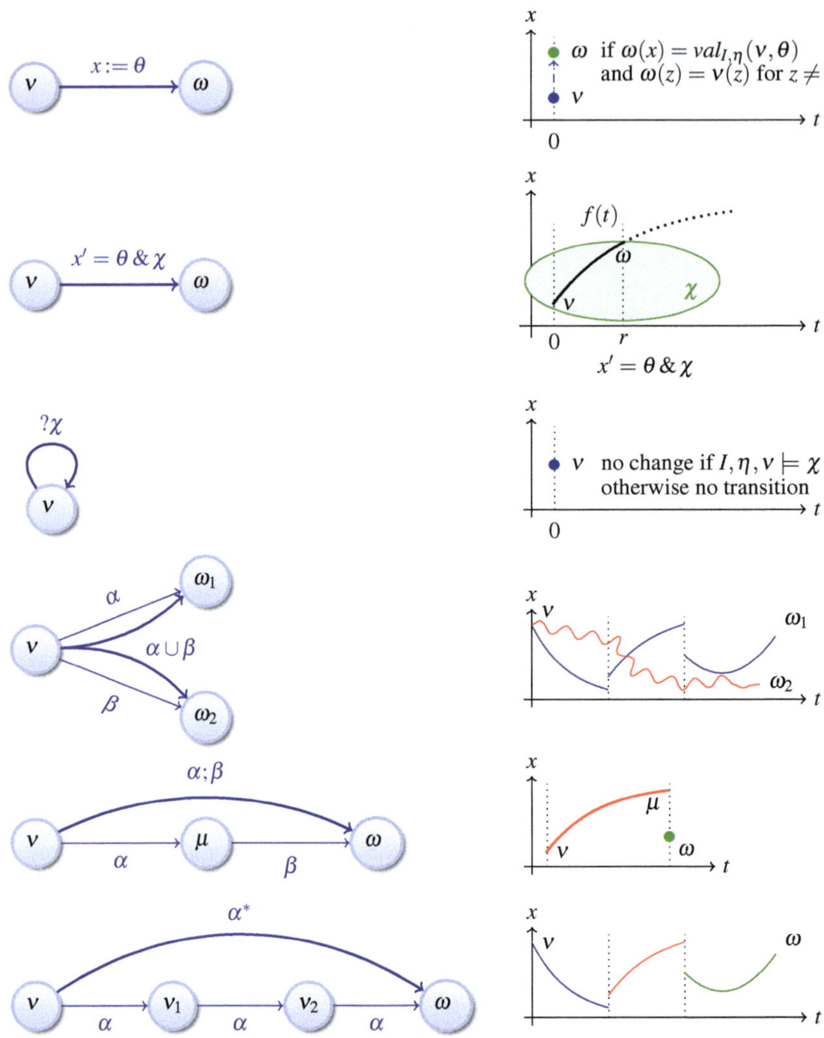

Fig. 2.5 Transition semantics (left) and example dynamics (right) of hybrid programs

For test $?\chi$ (row 3), the only possible transitions in the transition structure are self-loops that do not change the state ν, but even those transitions are only possible if the test succeeds, i.e., $I,\eta,\nu \models \chi$; see Fig. 2.5. The transition structure for choice $\alpha \cup \beta$ (row 4) is a choice between any transition of α and any transition of β. Thus, in the example on the right of row 4, the system can choose between either an evolution like the hybrid evolution (consisting, in this example, of 3 continuous flows and 2 intermediate jumps) leading to ω_1 or the squiggly evolution from ν to ω_2. The transition structure for sequential composition $\alpha;\beta$ (row 5) is that of any α transition to an intermediate state μ, followed by any β transition to ω. In the example evolution on the right of row 5, the system first follows a continuous evolution (which would come from α in this example) and then a discrete jump (which would come from β). The transition structure of a repetition α^* (row 6) repeats any number of α transitions to go from ν to ω via some number of intermediate states ν_i. In the example on the right, the system follows a sequence of various continuous evolutions and discrete jumps, giving truly hybrid behaviour.

For differential equations like $x' = \theta$, Definition 2.7 characterises transitions along a continuous evolution respecting the differential equation; see Fig. 2.6a. A continuous transition along $x' = \theta$ is possible from state ν to state ω whenever there is a continuous flow f of some duration $r \geq 0$ connecting state ν with ω such that f gives a solution of the differential equation $x' = \theta$. That is, its value is continuous on the closed interval $[0,r]$ and differentiable with the value of θ as derivative on the open interval $(0,r)$. Further, only variables subject to a differential equation change during such a continuous transition. Similarly, the continuous transitions of $x' = \theta \,\&\, \chi$ with evolution domain χ are those where f always resides within χ during the whole evolution; see Fig. 2.6b. The evolutions of $x' = \theta \,\&\, \chi$ may still stop at any point in time, but they are no longer allowed to leave χ and have to stop at an arbitrary point in time before that happens; see Fig. 2.6c.

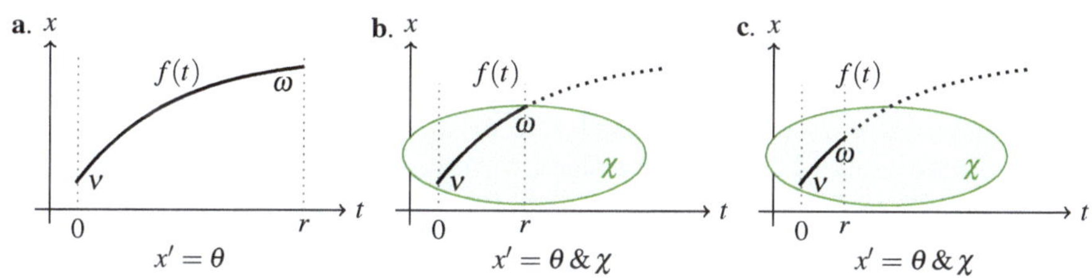

Fig. 2.6 Continuous flow along differential equation $x' = \theta$ over time t

For the semantics of differential equations, derivatives are well defined on the open interval $(0,r)$, because the set $\mathrm{Sta}(\Sigma)$ of states is isomorphic to some finite-dimensional metric real vector space spanned by the variables of the differential equations (derivatives are not defined on the closed interval $[0,r]$ if $r = 0$). For the purpose of a differential equation system, states are fully determined by an assignment of a real value to each occurring variable, which are finitely many. Furthermore, the terms of $\mathrm{d}\mathcal{L}$ are continuously differentiable on the open domain where divisors are nonzero, because the zero set of divisors is closed. Hence, solutions in $\mathrm{d}\mathcal{L}$ are unique:

Lemma 2.1 (Uniqueness). *Differential equations of* $\mathrm{d}\mathcal{L}$ *have unique solutions, i.e., for each differential equation system, each state* ν*, and each duration* $r \geq 0$*, there is at most one flow* $f\colon [0,r] \to \mathrm{Sta}(\Sigma)$ *satisfying the conditions of Case 2 of Definition 2.7.*

Proof. Let $x_1' = \theta_1, \ldots, x_n' = \theta_n \,\&\, \chi$ be a differential equation system with evolution domain χ. Using simple computations in the field of rational fractions, we can assume the right-hand sides θ_i of the differential equations to be of the form p_i/q_i for polynomials p_i, q_i. The set of points in real space where $q_i = 0$ holds is closed. As a finite union of closed sets, the set where $q_1 = 0 \vee \cdots \vee q_n = 0$ holds is closed. Hence, the valuations of the θ_i are continuously differentiable on the complement of the latter set, which is open. Thus, as a consequence of Picard-Lindelöf's theorem, a.k.a. the Cauchy-Lipschitz theorem (Theorem B.2), the solutions are unique on each connected component of this open domain. Consequently, solutions are unique when restricted to χ, which, by assumption, entails $q_1 \neq 0 \wedge \cdots \wedge q_n \neq 0$.

□

Example 2.11 (Evaluation of formula and transition semantics). Recall the following hybrid program from Example 2.4 that models an (overly) simplistic train controller:

$$train \equiv (((a:=-b) \cup (?v < 8; a:=A));\ z' = v, v' = a)^* \tag{2.1*}$$

Recall the $d\mathcal{L}$ formula (2.2) from p. 48 that claims that this simplistic train model can leave the movement authority region m:

$$v \geq 0 \wedge z < m \rightarrow \langle train \rangle z \geq m \tag{2.2*}$$

Let us evaluate this $d\mathcal{L}$ formula. Consider the interpretation I that interprets rigid symbol $m \in \Sigma$ as $I(m) = 20$ and interprets $b \in \Sigma$ as $I(b) = 2$ and $I(A) = 1$. Let state ν interpret state variables $v, z \in \Sigma$ as $\nu(v) = 9$, $\nu(z) = 0$. Then the assumptions $v \geq 0 \wedge z < m$ from the left-hand side of the implication of (2.2) are satisfied, so for formula (2.2) to be evaluated to true, the right-hand side of the implication needs to evaluate to $val_{I,\eta}(\nu, \langle train \rangle z \geq m) = true$. To find out if this is the case, the semantics of $\langle train \rangle$ in Definition 2.6 requires us to find a transition $(\nu, \omega) \in \rho_{I,\eta}(train)$ of the hybrid program $train$ from ν to some state ω, according to Definition 2.7, after which $val_{I,\eta}(\omega, z \geq m)$ holds true. Let us try to find such a state ω by following the transition structure $\rho_{I,\eta}(train)$ depicted in Fig. 2.7a. Essentially, we obtain the transition structure in Fig. 2.7a by gluing the elementary transition patterns from Fig. 2.5 together according to the structure of hybrid program (2.1). We will find a path in the transition structure Fig. 2.7a from ν to ω along the transitions illustrated in Fig. 2.7b, as we explain in the following.

The top-level statement in $train$ is a repetition (corresponding to the outer loop in Fig. 2.7a). We are allowed to execute the repetition twice as illustrated in the double unrolling in Fig. 2.7b (we could also repeat it any other number of times, but two times is sufficient). Thus, we hope to find an intermediate state σ_2 such that both

$$(\nu, \sigma_2) \in \rho_{I,\eta}(((a:=-b) \cup (?v < 8; a:=A));\ z' = v, v' = a) \tag{2.5}$$

and

$$(\sigma_2, \omega) \in \rho_{I,\eta}(((a:=-b) \cup (?v < 8; a:=A));\ z' = v, v' = a) \tag{2.6}$$

In the first transition (2.5), the top-level statement is a sequential composition (;) with a nondeterministic choice ($\cup$) as its first action. This nondeterministic choice can choose either side, indicated as upper and lower choices on the left of Fig. 2.7b.

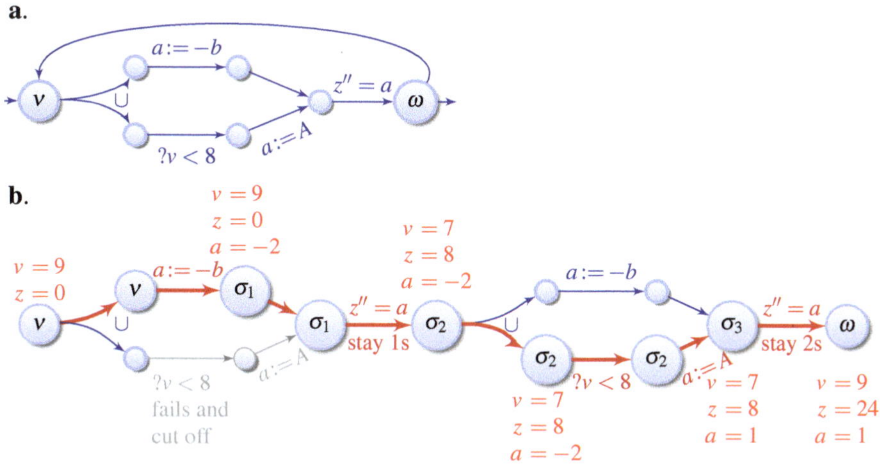

Fig. 2.7 Transition structure and transition example in (overly) simple train control

If it chooses to try to run the second (lower) choice ($?v<8;a:=A$), however, the hybrid program cannot run successfully, because the test $?v<8$ will fail and abort the transition as a dead end, since this test evaluates to $val_{I,\eta}(\nu,v<8)=false$ at state ν. Hence, the hybrid program can only choose the first (upper) option ($a:=-b$) to a state σ_1 whose only difference with ν is that $\sigma_1(a)=-2=val_{I,\eta}(\nu,-b)$. For this state, we have $(\nu,\sigma_1)\in\rho_{I,\eta}(a:=-b)$. Next, the differential equation will run from σ_1 as the second part of the sequential composition. Because $\sigma_1(a)<0$, it will brake and the velocity v will decrease over time. If we just follow this differential equation long enough, say for one second, then the velocity at the end of it will be less than 8. Indeed, after staying in the differential equation for one second, we reach a state σ_2 with $(\sigma_1,\sigma_2)\in\rho_{I,\eta}(z'=v,v'=a)$ and $\sigma_2(v)=7$ and $\sigma_2(z)=8$, because $z(t):=\frac{-2}{2}t^2+9t+0$ and $v(t):=-2t+9$ is the solution of the differential equation when starting in state σ_1 with the interpretation I and staying for t time units. Thus, by the semantics of sequential composition and nondeterministic choice, Definition 2.7, we have that relation (2.5) holds for this state σ_2, and all we need to do is make sure that relation (2.6) holds as well.

For the transition (2.6), we can choose the second (lower) part of the nondeterministic choice, because, unlike before, the test $?v<8$ succeeds in the new state now: $val_{I,\eta}(\sigma_2,v<8)=true$. Hence, we follow $?v<8;a:=A$ to the state σ_3 that is like σ_2 except that we now have $\sigma_3(a)=I(A)=1$. From state σ_3, we can stay with and follow the subsequent differential equation as long as we want to, because there is no evolution domain restriction on it. From this particular initial state σ_3, the solution of the differential equation is $z(t):=\frac{1}{2}t^2+7t+8$ and $v(t):=1t+7$ when staying for t time units. If now we just follow the continuous evolution along this differential equation for long enough, we will eventually reach a state ω with $(\sigma_3,\omega)\in\rho_{I,\eta}(z'=v,v'=a)$ such that $val_{I,\eta}(\omega,z\geq m)=true$. In fact, the minimum time for this to happen is 1.544 time units, after which z has a value greater or equal $I(m) = 20$. But any longer period of time will do too. For instance, after 2 time units, we would have $\omega(z) = 24 \geq 20$ and $\omega(v) = 9$. Thus we have shown $I, \eta, \omega \models z \geq m$ and

$$(\nu, \omega) \in \rho_{I,\eta}((((a:=-b) \cup (?v < 8; a:=A)); z' = v, v' = a)^*) \ ,$$

which implies that formula (2.2) holds for I, η, ν.

So far, we have shown by semantic reasoning that formula (2.2) is true for I, η, ν. Yet formula (2.2) does not evaluate to *true* under all interpretations and states. For a different interpretation J with $J(m) = 1{,}000$, braking force $J(b) = 4$, and (now negative) acceleration $J(A) = -2$, but the same original state ν, we evaluate (2.4) to *false*. The reason is that, no matter which choice the hybrid program uses, the train always brakes, either with braking acceleration $-J(b) = -4$ or with negative acceleration $J(A) = -2$. Either way, the initial velocity $\nu(v) = 9$ is not high enough to reach $I(m) = 1{,}000$ from the initial position $\nu(z) = 0$. Eventually, the train velocity will be 0 and it cannot move forward anymore.

Another trivial example to show that formula (2.2) can evaluate to *false* for some J, η, ω is the interpretation J with $J(m) = 20$, $J(b) = 2$, and $I(A) = 0$ along with the state ω that interprets $\omega(v) = 0$, $\omega(z) = 0$. Then, the train stands still in the beginning and cannot move forward to $I(m) = 20$ at all. In particular, formula (2.2) is not valid, because it does not evaluate to *true* for all I, η, ν. □

What we notice in this example is that it is quite difficult and cumbersome to reason about d$\mathcal{L}$ formulas and the dynamics of hybrid systems on the level of semantics. Especially, we have only analysed particular behaviours starting at specific initial values for all the variables. For validity, we are interested in analysing all possible initial values, all numbers of repetitions, and arbitrary durations of staying in the continuous evolution modes. To do this in an elegant and coherent way, we introduce a proof calculus for d$\mathcal{L}$ in Sect. 2.5. After all, the semantics gives meaning to formulas and captures the intended meaning and behaviour in real systems. The semantics is intended to be intuitive to relate to the real world, not necessarily for being easy to use in meta-reasoning. Supporting simple analysis and proofs is the task of the proof calculus for d$\mathcal{L}$ that we develop in Sect. 2.5. Still, a good semantics like the one we chose is compositional, which makes the proof calculus simpler.

Further note that, for control-feedback loops α with a discrete controller regulating a continuous plant, transition structures involve all safety-critical states; hence, $\psi \to [\alpha]\phi$ is a natural rendition of the safety property that ϕ holds at all states reachable by α from initial states that satisfy ψ. Otherwise, d$\mathcal{L}$ can be augmented with temporal operators to refer to intermediate states or nonterminating traces. The corresponding calculus is compatible and reduces temporal properties to nontemporal properties at intermediate states of the hybrid program, as we illustrate in Chap. 4.

2.4 Collision Avoidance in Train Control

As a case study to illustrate how $d\mathcal{L}$ can be used for specifying and verifying hybrid systems, we examine a scenario of cooperating traffic agents in the European Train Control System (ETCS) [91]. The purpose of ETCS is to ensure that trains cannot crash into other trains or pass open gates. Its secondary objective is to maximise throughput and velocity without endangering safety. To achieve these objectives, ETCS discards the static partitioning of the track into fixed segments of mutually exclusive and physically separated access by trains, which has been used traditionally. Instead, permission to move is granted dynamically by decentralised Radio Block Controllers (RBCs) depending on the current track situation and movement of other traffic agents within the region of responsibility of the RBC; see Fig. 2.8.

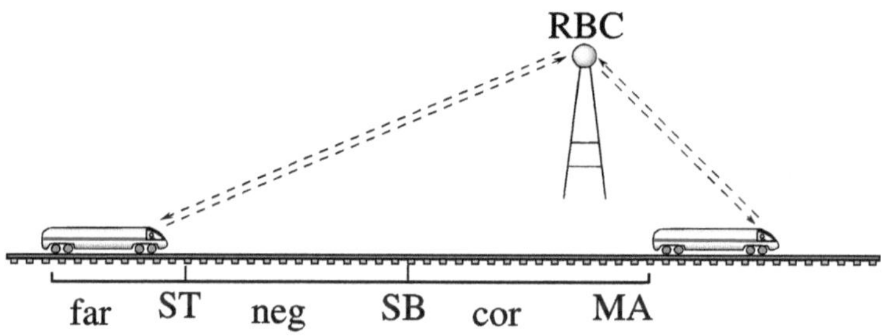

Fig. 2.8 ETCS train coordination protocol using dynamic movement authorities

Movement Authorities

This moving block principle is achieved by dynamically giving a *movement authority* (MA) to each traffic agent, within which it is obliged to remain. Before a train moves into a part of the track for which it does not have MA, it asks the RBC for an MA extension (during the negotiation phase indicated *neg* in Fig. 2.8). Depending on the MA that the RBC has currently given to other traffic agents or gates, the RBC will grant this extension and the train can move on. If the requested MA extension is still in the possession of another train that could possibly occupy the same part of the track, or if the MA is still consumed by an open gate, the RBC will deny the MA extension such that the requesting train needs to reduce speed or start braking in order to safely remain within its old MA. This is the correction phase *cor* in Fig. 2.8, which has to happen at the point SB (for start braking) at the latest. As the negotiation process with the RBC can take time because of possibly unreliable wireless communication and negotiation of the RBC with other agents, the train initiates negotiation well before reaching the end of its MA. This negotiation phase *neg* starts at the start talking point ST at the latest. Only if the train has a very large distance to the end of its MA (phase *far* in Fig. 2.8) is it safe to drive freely and not yet necessary to request MA extensions. When the rear end of a train has safely left a part of a track, the train can give that part of its MA back to RBC control such that it can be used by other traffic agents, including trains or gates.

In addition to increased flexibility and throughput of this moving block principle, the underlying technical concept of movement authorities can be exploited for verifying ETCS. It can be shown that a system of arbitrarily many trains, gates, and RBCs, which communicate in the aforementioned manner, safely avoids collisions if each traffic agent always resides within its MA under all circumstances, provided that the RBCs grant MAs mutually exclusively so that the MAs dynamically partition the track (Chap. 7). This way, verification of a system of unboundedly many traffic agents can be reduced to an analysis of individual agents with respect to their specific MA.

Train Control Model

In trains, speed supervision and automatic train protection are responsible for locally controlling the movement of a train such that it always respects its MA [90]. Depending on the current driving situation, the train controller determines a point SB (for start braking) up to which driving is safe, and adjusts its acceleration a in accordance with SB. Before SB, speed can be regulated freely (to keep the desired speed and throughput of a track profile). Beyond SB (correcting phase *cor* in Fig. 2.8), the train starts braking in order to make sure it remains within its MA if the RBC does not grant an extension in time.

We assume that an MA has been granted up to some track position, which we call m, and the train is located at position z, heading with current speed v towards m. We represent the point SB as the safety distance s relative to the end m of the MA (i.e., $m - s = \text{SB}$). In this situation, d$\mathcal{L}$ can analyse the following crucial safety property of ETCS, which we state as a d$\mathcal{L}$ formula:

$$\begin{aligned} \psi &\rightarrow [(ctrl; drive)^*]\, z \leq m \qquad (2.7)\\ \text{where } ctrl &\equiv (?m - z \leq s; a := -b) \cup (?m - z \geq s; a := A),\\ drive &\equiv \tau := 0; (z' = v, v' = a, \tau' = 1 \,\&\, v \geq 0 \land \tau \leq \varepsilon). \end{aligned}$$

It expresses that a train always ($[(ctrl; drive)^*]$) remains within its MA ($z \leq m$), assuming some constraint ψ for its parameters. The operational system model is a control-feedback loop of the digital controller *ctrl* and the plant *drive*. In *ctrl*, the train controller corrects its acceleration or brakes on the basis of the remaining distance ($m - z$). As a fail-safe recovery manoeuvre [90], it applies brakes with force b if the remaining MA is less than or equal to s. Otherwise, speed is regulated freely. The controller *ctrl* has a nondeterministic choice ($\cup$) where the left option starts with test $?m - z \leq s$ and the right option starts with test $?m - z \geq s$. The controller can try both options, but the left test will only succeed if $m - z \leq s$ holds for the current state, and the right test will only succeed if $m - z \geq s$ holds. In particular, if

$m-z<s$ holds the controller can only choose the left option, leading to braking by the assignment $a:=-b$. If $m-z>s$ holds the controller can only choose the right option, leading to acceleration by the assignment $a:=A$. If both tests could succeed, i.e., $m-z=s$, then either choice can be taken, nondeterministically. For simplicity, we assume the train uses a fixed acceleration A before passing s and does not choose any other accelerations than full braking b and full acceleration A (bang-bang control). The verification is quite similar when the controller can dynamically choose any acceleration $a \leq A$ instead, as we illustrate in Chap. 7.

After acceleration a has been set in *ctrl*, the second half of the sequential composition *ctrl*;*drive* executes, and the train continues moving in *drive*. There, the position z of the train evolves according to the differential equation system $z'=v, v'=a$ (i.e., $z''=a$). The evolution in *drive* stops when the speed v drops below zero (or earlier), because the train would not drive backwards just by braking. Thus, $v \geq 0$ is in the maximum evolution domain of *drive*. Simultaneously, clock τ measures the duration of the current *drive* phase before the controllers react to situation changes again. Clock τ is reset to zero by $\tau:=0$ when entering *drive*, constantly evolves along $\tau'=1$ together with the differential equations $z'=v, v'=a$, and is restricted by the evolution domain $\tau \leq \varepsilon$. Hence, the system can only follow *drive* for up to ε time units and at most as long as $v \geq 0$. The effect is that a *drive* phase is interrupted for reassessing the driving situation after at most ε seconds, and the *ctrl*;*drive* feedback loop repeats by the repetition operator (*). In particular, the continuous evolution cannot just be followed indefinitely without giving the controller *ctrl* a chance to react to situation changes. The corresponding transition structure $\rho_{I,\eta}((ctrl;drive)^*)$ is depicted in Fig. 2.9a. Essentially, we obtain the transition structure in Fig. 2.9a by gluing the elementary transition patterns from Fig. 2.5 together according to the structure of the hybrid program in (2.7).

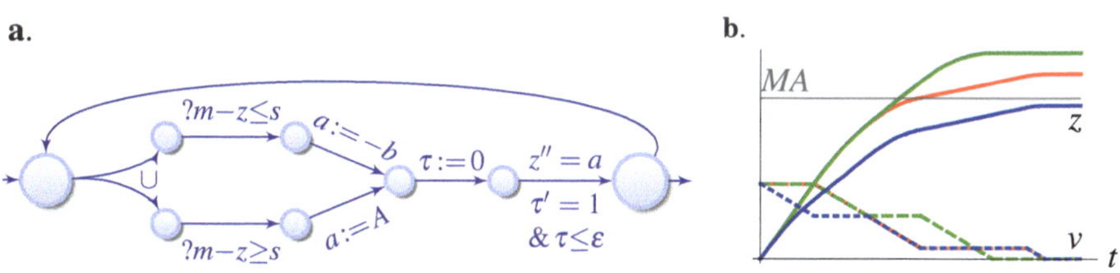

Fig. 2.9 ETCS transition structure and various choices of speed regulation for train speed control

Figure 2.9b shows possible runs of the train where speed regulation successively decreases velocity v because its MA has not been extended in time. Figure 2.9b shows three different runs (three upper position curves and three lower, partially overlapping velocity curves) which correspond to different choices of parameter s, where only the lowest velocity choice is safe. Finally, observe that the evolution domain $v \geq 0 \land \tau \leq \varepsilon$ needs to be true at *all times* during continuous evolutions of *drive*; otherwise there is no corresponding transition in $\rho_{I,\eta}(drive)$. This not only restricts the maximum duration of *drive*, but also imposes a constraint on permitted initial states: The arithmetic constraint $v \geq 0$ expresses that the differential equation only applies for nonnegative speed. Hence, as in a test $?v \geq 0$, program *drive* allows no transitions at all when v is initially less than 0. In that case, $\rho_{I,\eta}((ctrl;drive)^*)$ collapses to the trivial identity transition where only zero repetitions are possible.

Discussion

Here, we explicitly take into account possibly delayed controller reactions to bridge the gap of continuous-time models and discrete-time control design. To get meaningful results, we need to assume a maximum reaction delay ε, because safety cannot otherwise be guaranteed (the system would not be safe if the controllers can never execute). Polling cycles of sensors and digital controllers as well as latencies of actuators such as brakes contribute to ε. Instead of using specific estimates for ε for a particular train, we accept ε as a fully symbolic parameter. Further, instead of manually choosing specific values for the free parameters of (2.7) as in model checking approaches [91], we will use our calculus to synthesise constraints on the relationship of parameters that are required for safe operation of train control. We do not model weather conditions, slope of track, wheel friction, or train mass, because these are less relevant for the cooperation layer of train control [90].

Because of its nonlinear behaviour and nontrivial reset relations, system (2.7) is beyond the modelling capabilities of linear hybrid automata [8, 156, 126] and beyond o-minimal automata [189]. Previous approaches need linear flows [8, 156], do not support the coupled dynamics caused by nontrivial resets [189], require polyhedral initial sets and discrete dynamics [70], only handle robust systems with bounded regions [125] although parametric systems are not robust uniformly for all parameter choices, or handle only bounded-time safety for systems with bounded switching [217]. Finally, in addition to general numerical limits [238], numerical approaches [70, 21] quickly become intractable due to the exponential impact of the number of variables (curse of dimensionality).

2.5 Free-Variable Proof Calculus for Differential Dynamic Logic

In this section, we introduce a sequent calculus for formally verifying hybrid systems by proving validity of corresponding d$\mathcal{L}$ formulas. The basic idea is to symbolically compute the effects of hybrid programs and successively transform them into logical formulas describing these effects by structural symbolic decomposition. The calculus consists of standard propositional rules, rules for dynamic modalities that are generalised to hybrid programs, and novel quantifier rules that integrate real quantifier elimination (or, in fact, any other quantifier elimination procedure) into the modal calculus using free variables and Skolemisation.

2.5.1 Substitution

The d$\mathscr{L}$ calculus uses substitutions that take effect within formulas and programs. The result of applying to a d$\mathscr{L}$ formula ϕ the *substitution* that simultaneously replaces variable y_i by term θ_i (for $1 \leq i \leq m$) is defined as usual. Figure 2.10 shows

$$
\begin{array}{rcll}
\sigma(y_i) &=& \theta_i & \text{for } 1 \leq i \leq n \\
\sigma(z) &=& y & \text{if } z \notin \{y_1,\dots,y_m\} \text{ is a variable} \\
\sigma(f(\theta_1,\dots,\theta_n)) &=& f(\sigma(\theta_1),\dots,\sigma(\theta_n)) & \text{if } f \text{ is a function symbol} \\
\\
\sigma(p(\theta_1,\dots,\theta_n)) &=& p(\sigma(\theta_1),\dots,\sigma(\theta_n)) & \text{if } p \text{ is a predicate symbol} \\
\sigma(\neg\phi) &=& \neg\sigma(\phi) & \\
\sigma(\phi \wedge \psi) &=& \sigma(\phi) \wedge \sigma(\psi) & \\
\sigma(\phi \vee \psi) &=& \sigma(\phi) \vee \sigma(\psi) & \\
\sigma(\phi \rightarrow \psi) &=& \sigma(\phi) \rightarrow \sigma(\psi) & \\
\sigma(\forall x\,\phi) &=& \forall x\,\sigma(\phi) & \text{if admissible} \\
\sigma(\exists x\,\phi) &=& \exists x\,\sigma(\phi) & \text{if admissible} \\
\sigma([\alpha]\phi) &=& [\sigma(\alpha)]\sigma(\phi) & \text{if admissible} \\
\sigma(\langle\alpha\rangle\phi) &=& \langle\sigma(\alpha)\rangle\sigma(\phi) & \text{if admissible} \\
\\
\sigma(x_1 := \theta_1,..,x_n := \theta_n) &=& x_1 := \sigma(\theta_1),..,x_n := \sigma(\theta_n) & \text{if admissible} \\
\sigma(x_1' = \theta_1,..,x_n' = \theta_n \,\&\, \chi) &=& x_1' = \sigma(\theta_1),..,x_n' = \sigma(\theta_n) \,\&\, \sigma(\chi) & \text{if admissible} \\
\sigma(?\chi) &=& ?\sigma(\chi) & \\
\sigma(\alpha;\beta) &=& \sigma(\alpha);\sigma(\beta) & \\
\sigma(\alpha \cup \beta) &=& \sigma(\alpha) \cup \sigma(\beta) & \\
\sigma(\alpha^*) &=& (\sigma(\alpha))^* &
\end{array}
$$

Fig. 2.10 Application of substitution σ that simultaneously replaces variable y_i by term θ_i (for $1 \leq i \leq m$)

how the substitution σ that replaces variable y_i by term θ_i (for each $1 \leq i \leq m$) can be applied to a term, d$\mathscr{L}$ formula, or hybrid program, respectively. The first line in Fig. 2.10 represents that the substitution σ matches on the replaced (logical or state) variables y_i and replaces them by θ_i, respectively. The second line represents that no logical or state variable z other than $y_1,\dots,y_n$ are affected by σ. The third line maps the substitution σ homomorphically over function applications by applying σ recursively to all argument terms. Similarly, the next block of cases in Fig. 2.10 maps substitutions homomorphically over all subformulas. Yet for quantifiers ($\forall,\exists$) and modalities ($[\alpha],\langle\alpha\rangle$), the substitution is only applicable if admissible (as defined below) so that the bound variable x of the quantifier does not interfere with the substitution. We assume *bound variable renaming* (also known as α conversion) for renaming as needed: bound variables can be renamed to resolve conflicts, e.g., $\forall x\,\phi(x) \equiv \forall z\,\phi(z)$. Likewise, for applying the substitution homomorphically to hybrid programs (last block in Fig. 2.10) admissibility of the substitution is crucial in all cases. Admissibility implies, for instance, that the variables y_i replaced by the substitution are different from the changed variables x_j on the left-hand sides of the assignments or differential equations of the hybrid program.

Definition 2.8 (Admissible substitution). An application of a substitution σ is *admissible* if no variable x that σ replaces by $\sigma(x)$ occurs in the scope of a quantifier or modality binding x or a (logical or state) variable of the replacement $\sigma(x)$. A modality *binds* a state variable x iff it contains a discrete jump set assigning to x (such as $x:=\theta$) or a differential equation containing x' (such as $x'=\theta$).

In this book, only admissible substitutions are applicable, which is crucial for soundness. Admissible substitutions are denotation-preserving: They ensure that symbols still denote the same values after a substitution when they did so before.

Example 2.12 (Non-admissible substitution). It is important that only admissible substitutions are applicable. For the following formula, ϕ,

$$x=z \rightarrow \langle z:=z+1\rangle(z \geq x+1),$$

the substitution σ that replaces all occurrences of x by z is not admissible. This is due to the fact that for when we try to apply σ to ϕ forming

$$z=z \rightarrow \langle z:=z+1\rangle(z \geq z+1),$$

the substitution replaces x in postcondition $z \geq x+1$ by z, which is bound by modality $\langle z:=z+1\rangle$. Hence, within the scope of the modality, symbol z denotes a different value than outside the modality, thereby destroying the property of the occurrences of x—or, after the substitution, those of z—to share the same value throughout the formula. Instead, a substitution σ_2 of x by $y+1$ in ϕ to form $\sigma_2(\phi)$ is admissible for other symbols y, giving the formula $\sigma_2(\phi)$:

$$y+1=z \rightarrow \langle z:=z+1\rangle(z \geq y+1+1).$$

□

More succinctly, we abbreviate the result of applying to ϕ the substitution σ that replaces variable y_i with term θ_i (for $1 \leq i \leq m$) by $\phi_{y_1}^{\theta_1}\cdots_{y_m}^{\theta_m}$. Thus $\phi_{y_1}^{\theta_1}\cdots_{y_m}^{\theta_m}$ is an abbreviation for $\sigma(\phi)$ defined according to Fig. 2.10. When no confusion arises, we also use implicit notation for substitutions to improve readability. Let $\phi(z)$ be a formula with a free variable z. Then for any term θ, we use $\phi(\theta)$ as an abbreviation for the formula $\phi(z)_z^\theta$ that results from $\phi(z)$ by substituting θ for z.

Example 2.13 (Admissible versus non-admissible substitutions). Consider the (valid) d$\mathcal{L}$ formula ϕ defined as

$$\phi \equiv x>0 \wedge y>1 \wedge z \geq x \rightarrow [z:=z+xy]z>x.$$

Now the substitution that replaces x by $5a+x^2-y$ is admissible for ϕ, giving the result $\phi_x^{5a+x^2-y}$:

$$5a+x^2-y>0\land y>1\land z\geq 5a+x^2-y \;\rightarrow\; [z:=z+(5a+x^2-y)y]\,z>5a+x^2-y.$$

This formula $\phi_x^{5a+x^2-y}$, which results by an admissible substitution from ϕ, is valid, just like ϕ.

However, the substitution that replaces x with the term az is not admissible for ϕ, because variable z occurs in the replacement az but is bound in ϕ, and could thus have a different value at its various occurrences. So we cannot apply this substitution to ϕ. Yet if we choose a fresh variable u and use bound variable renaming to rename all occurrences of bound variable z to u, we obtain the formula

$$\tilde{\phi} \equiv x>0\land y>1\land z\geq x \;\rightarrow\; [u:=z+xy]\,u>x.$$

This variant $\tilde{\phi}$ is equivalent to ϕ, because only bound variables have been renamed. After this bound variable renaming, the substitution replacing x by az becomes admissible and we obtain

$$\tilde{\phi}_x^{az} \equiv az>0\land y>1\land z\geq az \;\rightarrow\; [u:=z+(az)y]\,u>az.$$

This formula is valid (just like ϕ and $\tilde{\phi}$). But it is quite different from the formula we would obtain if we had just naïvely replaced every occurrence of x (admissible or not) by az, instead of using more careful admissible substitutions:

$$az>0\land y>1\land z\geq az \;\rightarrow\; [z:=z+(az)y]\,z>az.$$

The latter formula is clearly false for all I,η,ν with $val_{I,\eta}(\nu,a)=1$, because z cannot possibly be greater than az then. Contrast this with the validity of the original formula ϕ and its (admissible) substitution instance $\tilde{\phi}_x^{az}$.

Similarly, the substitution that replaces z with ax is not admissible for ϕ, because the replaced variable z is bound in ϕ, and could thus have a different value at its various occurrences. So we cannot apply this substitution. Yet if we again choose a fresh variable u and use bound variable renaming to rename all occurrences of bound variable z to u, we obtain the formula $\tilde{\phi}$ above. After this bound variable renaming, the substitution replacing z with ax becomes admissible and we obtain

$$\tilde{\phi}_z^{ax} \equiv x>0\land y>1\land ax\geq x \;\rightarrow\; [u:=ax+xy]\,u>x.$$

Again, this formula is valid and quite different from the formula we would obtain if we had just naïvely replaced every occurrence of z (admissible or not) by ax:

$$x>0\land y>1\land ax\geq x \;\rightarrow\; [z:=ax+xy]\,ax>x.$$

The latter formula is again false for all I,η,ν with $val_{I,\eta}(\nu,a)=1$, because x cannot possibly be greater than ax then. Contrast this with the validity of the original formula ϕ and the admissible substitution instance $\tilde{\phi}_x^{ax}$.

Thus, there is a close connection between the formula ϕ and its various substitution instances (if admissible!), which we will identify in the next lemma. As part of that, we will show that, since ϕ is valid, all of its (admissible) substitution instances are valid. This close connection (and every other similarity) breaks when we naïvely replace terms in ϕ instead of obeying the requirements of admissible substitutions. □

Example 2.14 (Non-admissibility in repetitions). The last example is prototypical for several d$\mathcal{L}$ formulas and works similarly for all d$\mathcal{L}$ formulas without repetitions or differential equations. Yet repetitions and differential equations themselves are more involved. Consider a (valid) d$\mathcal{L}$ formula with a repetition:

$$\psi \equiv x>0 \land y>1 \land z>x \rightarrow [(z:=z+xy)^*]z>x. \tag{2.8}$$

As with the formula ϕ from the last example, the substitution that replaces x by the term $5a+x^2-y$ is admissible for ψ, giving the result $\psi_x^{5a+x^2-y}$:

$$5a+x^2-y>0 \land y>1 \land z>5a+x^2-y \rightarrow [(z:=z+(5a+x^2-y)y)^*]z>5a+x^2-y$$

This formula $\psi_x^{5a+x^2-y}$, which results by an admissible substitution from ψ is valid, just like ψ.

Again, the substitution that replaces x by the term az is not admissible for ψ, because variable z occurs in the replacement az but is bound in ψ, and could have different values at its occurrences. Hence, we cannot apply this substitution. However, for repetitions, it is not so easy to do bound variable renaming to get around this! We cannot simply replace all bound occurrences of z by one fresh variable u, which would give

$$x>0 \land y>1 \land z>x \rightarrow [(u:=\check{u}+xy)^*]u>x.$$

But here the connection of u with the input z has been lost and the formula is no longer valid. The reason is that the occurrence of z on the right-hand side $z+xy$ of the jump (which corresponds to the occurrence of u we marked ˇ in the last formula) is neither just free nor just bound. During the first iteration of the repetition, it would be free (because it receives its value from outside); during subsequent iterations, however, it would be bound (because it receives its value from the last assignment). The formula we would obtain if we had just naïvely replaced every occurrence of x (admissible or not) by az is also quite different and not valid:

$$x>0 \land y>1 \land z>x \rightarrow [(z:=z+(a\check{z})y)^*]z>az.$$

The reason is that the occurrence marked with ˇ is neither just free nor just bound, because it depends on the number of iterations of the loop.

Likewise, the substitution that replaces z by ax is not admissible for ψ and cannot be applied, because the replaced variable z is bound in ψ. We thus cannot apply this substitution. Once more, it is not so easy to do bound variable renaming to get around this and we cannot just rename z to a fresh variable u to resolve this issue. The formula we would obtain if we had just naïvely replaced every occurrence of x (admissible or not) by az is also quite different and not valid:

$$x > 0 \land y > 1 \land ax > x \to [(z := \check{a}x + xy)^*]\, ax > x.$$

The reason is again that the occurrence of z (prior to replacing) at the position marked ˇ is neither just free nor just bound. While it would be perfectly alright to replace the first dynamic occurrence of z (in the sequential execution order) by ax, subsequent occurrences (including those in repetitions) have a different operational value and cannot be replaced.

In these two cases, the substitutions are just not admissible for ψ and cannot be applied, because the modalities of ψ bind relevant replaced variables or variables in the replacements. Our proof calculus in Sect. 2.5.2 will use other ways that do not need substitution to prove formulas with repetitions like these. □

Example 2.15 (Non-admissibility in differential equations). The situation with differential equations is quite similar. In the d$\mathcal{L}$ formula

$$\psi \equiv x > 0 \land y > 1 \land z > x \to [z' = z + xy]\, z > x \tag{2.9}$$

the occurrences of z in the differential equation are neither just free nor just bound: The value z affects the initial value z of the differential equation, but the value of z also evolves over time when following the differential equation to a new value. Thus, z is both a free initial value and bounded or updated during the evolution. The substitution that replaces x with $5a^2 + x^2 - y$ is still admissible for ψ, giving $\psi_x^{5a^2+x^2-y}$:

$$5a^2 + x^2 - y > 0 \land y > 1 \land z > 5a^2 + x^2 - y \to [z' = z + (5a^2 + x^2 - y)y] z > 5a^2 + x^2 - y$$

But we cannot substitute x with az, because the substitution is not admissible for ψ as bound variable z occurs in the replacement az. Nor can we substitute z with ax, because this substitution is not admissible for ψ either, as the replaced variable z is bound in ψ. The formula we would obtain if we had just naïvely replaced every occurrence of x (admissible or not) with az, is different and not valid:

$$az > 0 \land y > 1 \land z > x \to [z' = z + (\check{a}z)y]\, z > az.$$

The formula we would obtain, instead, if we had just naïvely replaced every occurrence of z (admissible or not) by ax, is also different and not valid:

$$x > 0 \land y > 1 \land ax > x \to [z' = \check{a}x + xy]\, ax > x.$$

In both cases, we marked the positions where the occurrences have been neither free nor bound with ˇ once again.

In these two cases, the substitutions are not admissible for ψ and cannot be applied, because the modalities of ψ bind relevant replaced variables or variables in the replacements. Our proof calculus in Sect. 2.5.2 will prove such properties of differential equations differently. □

Example 2.16 (Bound variable renaming for repetitions and differential equations). On a side note, it would not be impossible to define bound variable renaming for repetitions and differential equations. We decide not to use these extensions in our approach, because they are technically more involved and not necessary for our proof calculus. The purpose of this example is to show how bound variable renaming could be extended appropriately, nevertheless. When we add an extra discrete jump, we could define the following extended bound variable renaming variant of formula (2.8):

$$x > 0 \land y > 1 \land z > x \to [u := z;\ (u := u + xy)^*]\, u > x.$$

This formula separates the initial value assignment from the loop. Similarly when we add an extra discrete jump, we could define the following extended bound variable renaming variant of formula (2.9):

$$x > 0 \land y > 1 \land z > x \to [u := z;\ u' = u + xy]\, u > x.$$

Again, this formula separates the initial value assignment from the differential equation. For both variants, the substitution replacing x with az is admissible, and so is the substitution replacing z with ax. Essentially, the above two variants retain the initial value z explicitly before the repetition or differential equation. We have chosen not to use these extended bound variable renamings in this book and, instead, follow our choice that non-admissible substitutions are not applicable at all. □

There is a direct connection between a formula ϕ and its substitution instance $\sigma(\phi)$, provided that the substitution σ is admissible for ϕ. In fact, the valuation of ϕ and $\sigma(\phi)$ coincide if only we change the interpretation of the replaced symbols appropriately when evaluating ϕ. That is, semantically evaluating ϕ (after modifying the interpretation of the symbols replaced by σ in I, η, ν) is the same as semantically evaluating ϕ in the original I, η, ν after applying the substitution (resulting in $\sigma(\phi)$). Stated differently, we can show that, for admissible substitutions, syntactic substitution in the formula and semantic modification of I, η, ν have the same effect:

Lemma 2.2 (Substitution Lemma). *Let σ be an admissible substitution for the (term or) formula ϕ and let σ replace only logical variables; then*

$$\text{for each } I, \eta, \nu:\ val_{I,\eta}(\nu, \sigma(\phi)) = val_{I,\sigma^*(\eta)}(\nu, \phi),$$

where the semantic modification $\sigma^*(\eta)$ of assignment η is adjoint *to σ, i.e., $\sigma^*(\eta)$ is identical to η, except that $\sigma^*(\eta)(x) = val_{I,\eta}(\nu, \sigma(x))$ for all logical variables $x \in V$.*

Proof. In essence, the proof of this lemma is a simple corollary to the fact that both substitution and valuation are homomorphisms defined inductively on formulas from their effect on atomic symbols. The application of an admissible substitution σ is a homomorphic continuation of its effect on atomic symbols to all $d\mathcal{L}$ formulas by way of Fig. 2.10. That is, the effect of an admissible(!) substitution on a compound formula is just defined by applying the substitution recursively to all subformulas. Likewise, the valuation is a homomorphic continuation of the interpretation I, state ν, and assignment η on atomic symbols to all d$\mathcal{L}$ formulas by way of Definition 2.6. That is, the valuation of a compound formula is just defined by using the valuation on all subformulas.

First we prove the substitution lemma applied to terms θ:

$$\text{for each } I,\eta,\nu:\ val_{I,\eta}(\nu,\sigma(\theta)) = val_{I,\sigma^*(\eta)}(\nu,\theta).$$

The proof is by induction on the structure of the term θ.

1. If θ is a logical variable $x \in V$, then, by definition of $\sigma^*(\eta)$:

$$val_{I,\eta}(\nu,\sigma(x)) = \sigma^*(\eta)(x) = val_{I,\sigma^*(\eta)}(\nu,x).$$

2. If θ is a state variable $x \in \Sigma$, then it is different from replaced logical variables $u \in V$ and $\sigma(x) = x$. Hence

$$val_{I,\eta}(\nu,\sigma(x)) = val_{I,\eta}(\nu,x) = \nu(x) = val_{I,\sigma^*(\eta)}(\nu,x).$$

3. If θ is of the form $f(\theta_1,\dots,\theta_n)$ for a function symbol f of arity $n \geq 1$, then

$$\begin{aligned}
&val_{I,\eta}(\nu,\sigma(f(\theta_1,\dots,\theta_n)))\\
&= val_{I,\eta}(\nu,f(\sigma(\theta_1),\dots,\sigma(\theta_n)))\\
&= I(f)\big(val_{I,\eta}(\nu,\sigma(\theta_1)),\dots,val_{I,\eta}(\nu,\sigma(\theta_n))\big)\\
&= I(f)\big(val_{I,\sigma^*(\eta)}(\nu,\theta_1),\dots,val_{I,\sigma^*(\eta)}(\nu,\theta_n)\big)\\
&= val_{I,\sigma^*(\eta)}(\nu,f(\theta_1,\dots,\theta_n))
\end{aligned}$$

because the θ_i are simpler than $f(\theta_1,\dots,\theta_n)$ so that, by induction hypothesis, we have for each i:

$$val_{I,\eta}(\nu,\sigma(\theta_i)) = val_{I,\sigma^*(\eta)}(\nu,\theta_i).$$

Next we prove the substitution lemma applied to d$\mathcal{L}$ formulas ϕ:

$$\text{for each } I,\eta,\nu:\ val_{I,\eta}(\nu,\sigma(\phi)) = val_{I,\sigma^*(\eta)}(\nu,\phi).$$

The proof is by induction on the structure of the formula ϕ.

1. If ϕ is of the form $p(\theta_1,\dots,\theta_n)$ for a predicate symbol p of arity $n \geq 1$, then the proof is almost identical to that for function symbols above.
2. If ϕ is of the form $\phi_1 \vee \phi_2$, then we use the induction hypothesis on ϕ_1 and ϕ_2 to conclude

$$\begin{aligned}
&val_{I,\eta}(\nu,\sigma(\phi_1 \vee \phi_2))\\
&= val_{I,\eta}(\nu,\sigma(\phi_1) \vee \sigma(\phi_2)) = \mathit{true}\\
&\text{iff } val_{I,\eta}(\nu,\sigma(\phi_1)) = \mathit{true} \text{ or } val_{I,\eta}(\nu,\sigma(\phi_2)) = \mathit{true}
\end{aligned}$$

$$\text{iff } val_{I,\sigma^*(\eta)}(\nu,\phi_1) = true \text{ or } val_{I,\sigma^*(\eta)}(\nu,\phi_2) = true$$
$$\text{iff } val_{I,\sigma^*(\eta)}(\nu,\phi_1 \vee \phi_2) = true$$

If ϕ is of the form $\phi_1 \wedge \phi_2$ or of the form $\phi_1 \to \phi_2$ or $\neg\phi_1$, then the proof is similar.

3. If ϕ is of the form $\exists x\,\psi$, then we use that σ was assumed to be admissible for ϕ. In particular (by bound variable renaming), x is not one of the replaced variables u and x does not occur in any of the replacements $\sigma(u)$. We use the induction hypothesis on ψ to conclude

$$\begin{aligned}
&val_{I,\eta}(\nu,\sigma(\exists x\,\psi)) = val_{I,\eta}(\nu,\exists x\,\sigma(\psi)) = true\\
&\text{iff there is a } d \text{ such that } val_{I,\eta[x\mapsto d]}(\nu,\sigma(\psi)) = true\\
&\text{iff there is a } d \text{ such that } val_{I,\sigma^*(\eta[x\mapsto d])}(\nu,\psi) = true\\
&\text{iff there is a } d \text{ such that } val_{I,\sigma^*(\eta)[x\mapsto d]}(\nu,\psi) = true\\
&\text{iff } val_{I,\sigma^*(\eta)}(\nu,\exists x\,\psi) = true.
\end{aligned}$$

Note that $\sigma^*(\eta[x\mapsto d]) = \sigma^*(\eta)[x\mapsto d]$, because x is not affected by the substitution σ (since admissible); hence x is not affected by adjoint assignments. If ϕ is of the form $\forall x\,\psi$, the proof is similar.

4. If ϕ is of the form $[\alpha]\psi$, then we use that the substitution σ is admissible by assumption. Hence, α does not bind any of the replaced variables nor any of the variables that occur in any of the replacements $\sigma(u)$. We use the induction hypothesis on ψ to conclude

$$\begin{aligned}
&val_{I,\eta}(\nu,\sigma([\alpha]\psi)) = val_{I,\eta}(\nu,[\sigma(\alpha)]\sigma(\psi)) = true\\
&\text{iff for all } \omega \text{ with } (\nu,\omega)\in\rho_{I,\eta}(\sigma(\alpha)) : val_{I,\eta}(\omega,\sigma(\psi)) = true\\
&\text{iff for all } \omega \text{ with } (\nu,\omega)\in\rho_{I,\eta}(\sigma(\alpha)) : val_{I,\sigma^*(\eta)}(\omega,\psi) = true\\
&\text{iff}^\star \text{ for all } \omega \text{ with } (\nu,\omega)\in\rho_{I,\sigma^*(\eta)}(\alpha) : val_{I,\sigma^*(\eta)}(\omega,\psi) = true\\
&\text{iff } val_{I,\sigma^*(\eta)}(\nu,[\alpha]\psi) = true.
\end{aligned}$$

For the middle step marked with $\star$, we still have to prove the substitution lemma for hybrid programs:

$$\rho_{I,\eta}(\sigma(\alpha)) = \rho_{I,\sigma^*(\eta)}(\alpha). \tag{2.10}$$

If α is of the form $\langle\alpha\rangle\psi$ then the proof is similar.

Finally we prove the substitution lemma for hybrid programs α as formulated in (2.10). The proof is by induction on the structure of hybrid program α.

1. If α is of the form $x_1 := \theta_1, \ldots x_n := \theta_n$, then we use the substitution lemma on the terms θ_i to show

$$
\begin{aligned}
&(\nu,\omega) \in \rho_{I,\eta}(\sigma(x_1 := \theta_1, \dots x_n := \theta_n)) = \rho_{I,\eta}(x_1 := \sigma(\theta_1), \dots x_n := \sigma(\theta_n))\\
&\text{iff } \nu[x_1 \mapsto val_{I,\eta}(\nu,\sigma(\theta_1))] \dots [x_n \mapsto val_{I,\eta}(\nu,\sigma(\theta_n))] = \omega\\
&\text{iff } \nu[x_1 \mapsto val_{I,\sigma^*(\eta)}(\nu,\theta_1)] \dots [x_n \mapsto val_{I,\sigma^*(\eta)}(\nu,\theta_n)] = \omega\\
&\text{iff } (\nu,\omega) \in \rho_{I,\sigma^*(\eta)}(x_1 := \theta_1, \dots x_n := \theta_n).
\end{aligned}
$$

2. If α is of the form $?\chi$ for a (first-order) $\mathsf{d}\mathcal{L}$ formula χ, then we use the substitution lemma on the (simpler and even first-order) $\mathsf{d}\mathcal{L}$ formula χ:

$$
\begin{aligned}
&(\nu,\omega) \in \rho_{I,\eta}(\sigma(?\chi)) = \rho_{I,\eta}(?\sigma(\chi))\\
&\text{iff } \nu = \omega \text{ and } val_{I,\eta}(\nu,\sigma(\chi)) = true\\
&\text{iff } \nu = \omega \text{ and } val_{I,\sigma^*(\eta)}(\nu,\chi) = true\\
&\text{iff } (\nu,\omega) \in \rho_{I,\sigma^*(\eta)}(?\chi).
\end{aligned}
$$

3. If α is of the form $x_1' = \theta_1, \dots x_n' = \theta_n \,\&\, \chi$, then we use the substitution lemma on terms and on the (first-order) $\mathsf{d}\mathcal{L}$ formula χ to conclude:

$$
\begin{aligned}
&(\nu,\omega) \in \rho_{I,\eta}(\sigma(x_1' = \theta_1, \dots x_n' = \theta_n \,\&\, \chi))\\
&\quad = \rho_{I,\eta}(x_1' = \sigma(\theta_1), \dots x_n' = \sigma(\theta_n) \,\&\, \sigma(\chi)),
\end{aligned}
$$

which holds if and only if there is a continuous flow function $f\colon [0,r] \to \mathrm{Sta}(\Sigma)$ with $f(0) = \nu, f(r) = \omega$ and $val_{I,\eta}(f(\zeta),z) = val_{I,\eta}(\nu,z)$ for all $\zeta \in [0,r]$ and all $z \notin \{x_1,\dots,x_n\}$ such that:

- for each x_i, $val_{I,\eta}(f(\zeta),x_i) = f(\zeta)(x_i)$ is continuous in ζ on $[0,r]$ and has a derivative of value $val_{I,\eta}(f(\zeta),\sigma(\theta_i))$ at each time $\zeta \in (0,r)$,
- and $val_{I,\eta}(f(\zeta),\sigma(\chi)) = true$ for each $\zeta \in [0,r]$.

By the substitution lemma for terms and formulas, respectively, these conditions are equivalent to

- for each x_i, $val_{I,\sigma^*(\eta)}(f(\zeta),x_i) = f(\zeta)(x_i)$ is continuous in ζ on $[0,r]$ and has a derivative of value $val_{I,\sigma^*(\eta)}(f(\zeta),\theta_i)$ at each time $\zeta \in (0,r)$,
- and $val_{I,\sigma^*(\eta)}(f(\zeta),\chi) = true$ for each $\zeta \in [0,r]$,

which hold if and only if

$$(\nu,\omega) \in \rho_{I,\sigma^*(\eta)}(x_1' = \theta_1, \dots x_n' = \theta_n \,\&\, \chi).$$

4. If α is of the form $\beta \cup \gamma$, then we can use the induction hypothesis on β and γ to conclude

$$
\begin{aligned}
\rho_{I,\eta}(\sigma(\beta \cup \gamma)) &= \rho_{I,\eta}(\sigma(\beta) \cup \sigma(\gamma)) = \rho_{I,\eta}(\sigma(\beta)) \cup \rho_{I,\eta}(\sigma(\gamma))\\
&= \rho_{I,\sigma^*(\eta)}(\beta) \cup \rho_{I,\sigma^*(\eta)}(\gamma) = \rho_{I,\sigma^*(\eta)}(\beta \cup \gamma).
\end{aligned}
$$

5. If α is of the form $\beta;\gamma$, then we use the induction hypothesis on β and on γ to conclude

$$\begin{aligned}&(\nu,\omega)\in\rho_{I,\eta}(\sigma(\beta;\gamma))=\rho_{I,\eta}(\sigma(\beta);\sigma(\gamma))\\ &\text{iff there is a } \mu \text{ with } (\nu,\mu)\in\rho_{I,\eta}(\sigma(\beta)) \text{ and } (\mu,\omega)\in\rho_{I,\eta}(\sigma(\gamma))\\ &\text{iff there is a } \mu \text{ with } (\nu,\mu)\in\rho_{I,\sigma^*(\eta)}(\beta) \text{ and } (\mu,\omega)\in\rho_{I,\sigma^*(\eta)}(\gamma)\\ &\text{iff } (\nu,\omega)\in\rho_{I,\sigma^*(\eta)}(\beta;\gamma).\end{aligned}$$

6. The case where α is of the form β^* is similar, again using admissibility:

$$\begin{aligned}&(\nu,\omega)\in\rho_{I,\eta}(\sigma(\beta^*))=\rho_{I,\eta}(\sigma(\beta)^*)\\ &\text{iff there are } n\in\mathbb{N},\mu_0=\nu,\mu_1,\dots,\mu_n=\omega:(\mu_i,\mu_{i+1})\in\rho_{I,\eta}(\sigma(\beta))\\ &\text{iff there are } n\in\mathbb{N},\mu_0=\nu,\mu_1,\dots,\mu_n=\omega:(\mu_i,\mu_{i+1})\in\rho_{I,\sigma^*(\eta)}(\beta)\\ &\text{iff } (\nu,\omega)\in\rho_{I,\sigma^*(\eta)}(\beta^*).\end{aligned}$$

□

The substitution lemma implies a simple corollary for substituting program variables instead of (or in addition to) logical variables. The proof is an immediate consequence of a double application of the substitution lemma, so that, in the remainder of this book, we do not distinguish between Lemma 2.2 and the following corollary.

Corollary 2.1. *Let σ be an admissible substitution for the (term or) formula ϕ; then*

$$\text{for each } I,\eta,\nu:\ val_{I,\eta}(\nu,\sigma(\phi))=val_{I,\sigma^*(\eta)}(\sigma^*(\nu),\phi),$$

where the semantic modification $\sigma^*(\nu)$ of state ν is adjoint *to σ. The adjoint $\sigma^*(\nu)$ is identical to ν, except that $\sigma^*(\nu)(x)=val_{I,\eta}(\nu,\sigma(x))$ for all state variables $x\in\Sigma$. The adjoint $\sigma^*(\eta)$ is defined as in Lemma 2.2.*

Proof. The proof is a simple corollary to Lemma 2.2, using fresh logical variables z_i to relate $\sigma(\phi)$ with ϕ for gluing two uses of Lemma 2.2 together. To simplify notation, assume that σ only replaces a single state variable x by θ and let us denote the result of applying this substitution to ϕ by ϕ_x^θ. Let z be a fresh logical variable. Since the substitution σ is admissible for ϕ, the replaced variable x and all variables in its replacement θ are not bound in ϕ. Thus, ϕ is of the form ψ_z^x for the formula ψ, which is like ϕ except that it has z in place of x everywhere. Now abbreviate $val_{I,\eta}(\nu,\theta)$ as e, and abbreviate $val_{I,\eta}(\nu[x\mapsto e],x)$ as d. Then, we use Lemma 2.2 at the positions indicated $\star$ to conclude:

$$\begin{aligned}val_{I,\eta}(\nu,\phi_x^\theta)&=val_{I,\eta}(\nu,\psi_{z\,x}^{x\,\theta})=val_{I,\eta}(\nu,\psi_z^\theta)\stackrel{\star}{=}val_{I,\eta[z\mapsto e]}(\nu,\psi)\\ &=val_{I,\eta[z\mapsto d]}(\nu[x\mapsto e],\psi)\stackrel{\star}{=}val_{I,\eta}(\nu[x\mapsto e],\psi_z^x)=val_{I,\eta}(\nu[x\mapsto e],\phi).\end{aligned}$$

Note that the two lines are equal because the value of state variable x in the state does not matter for ψ, where x does not occur, and because $d=val_{I,\eta}(\nu[x\mapsto e],x)=e$. □

Example 2.17. Again consider the formula ϕ, and an instance $\phi_x^{5a^2+b}$ under an admissible substitution:

$$\phi \equiv x = z \to \langle z := z+1 \rangle (z \geq x+1),$$
$$\phi_x^{5a^2+b} \equiv 5a^2+b = z \to \langle z := z+1 \rangle (z \geq 5a^2+b+1).$$

Using the substitution lemma, we can conclude that with respect to any I, η, ν, the formula ϕ and its instance $\phi_x^{5a^2+b}$ evaluate to the same truth-value when adapting the value of x appropriately. That is, let σ be the substitution that replaces x with $5a^2+b$, i.e., $\sigma(\phi) \equiv \phi_x^{5a^2+b}$; then (the corollary to) Lemma 2.2 implies:

$$val_{I,\eta}(\nu, \phi_x^{5a^2+b}) = val_{I,\sigma^*(\eta)}(\sigma^*(\nu), \phi).$$

Let us abbreviate the value $val_{I,\eta}(\nu, 5a^2+b)$ of the replacement $5a^2+b$ of x by e. Then if $x \in V$ is a logical variable, then $\sigma^*(\nu) = \nu$ and $\sigma^*(\eta) = \eta[x \mapsto e]$; hence

$$val_{I,\eta}(\nu, \phi_x^{5a^2+b}) = val_{I,\eta[x \mapsto e]}(\nu, \phi).$$

If, instead, $x \in \Sigma$ is a state variable, then $\sigma^*(\eta) = \eta$ and $\sigma^*(\nu) = \nu[x \mapsto e]$; hence

$$val_{I,\eta}(\nu, \phi_x^{5a^2+b}) = val_{I,\eta}(\nu[x \mapsto e], \phi).$$

In either case (either $x \in V$ or $x \in \Sigma$), if the value of x and its replacement $5a^2+b$ agree in the original I, η, ν already, i.e., if $val_{I,\eta}(\nu, x) = val_{I,\eta}(\nu, 5a^2+b)$, then their valuations agree according to the substitution lemma:

$$val_{I,\eta}(\nu, \phi_x^{5a^2+b}) = val_{I,\eta}(\nu, \phi).$$

□

The substitution lemma is a very powerful tool, because, among other things, we can use it to replace equals for equals without changing the valuation (substitution property). If we know that x and θ have the same value in I, η, ν, then we can substitute θ for x in a formula ϕ (if admissible) without changing the truth-value of ϕ, that is:

Lemma 2.3 (Substitution property). *If $I, \eta, \nu \models x = \theta$, then $I, \eta, \nu \models \phi \leftrightarrow \phi_x^\theta$ for any formula ϕ for which the substitution replacing x with θ is admissible.*

Proof. Consider any I, η, ν with $I, \eta, \nu \models x = \theta$. First, note that this assumption is equivalent to $val_{I,\eta}(\nu, x) = val_{I,\eta}(\nu, \theta)$. We have to show $I, \eta, \nu \models \phi \leftrightarrow \phi_x^\theta$, or, equivalently, $val_{I,\eta}(\nu, \phi) = val_{I,\eta}(\nu, \phi_x^\theta)$. This follows from the Substitution Lemma 2.2 when we choose σ to be the substitution that replaces x by θ since

$$val_{I,\eta}(\nu, \phi_x^\theta) = val_{I,\sigma^*(\eta)}(\sigma^*(\nu), \phi) = val_{I,\eta}(\nu, \phi).$$

The last step follows from the fact that I,η,ν equals $I,\sigma^*(\eta),\sigma^*(\nu)$, respectively, because the substitution σ only replaces x by θ, which already have the same value to begin with, as we assumed $val_{I,\eta}(\nu,x) = val_{I,\eta}(\nu,\theta)$. □

In addition, whenever a formula ϕ is valid (ϕ is true in all I,η,ν), the substitution lemma implies that all of its (admissible) substitution instances $\sigma(\phi)$ are valid too for any substitution σ that is admissible for ϕ.

Lemma 2.4 (Substitutions preserve validity). *If* $\vDash \phi$, *i.e.,* ϕ *is valid, then* $\vDash \sigma(\phi)$ *for any substitution* σ *that is admissible for* ϕ.

Proof. Let ϕ be valid, i.e., $I,\eta,\nu \vDash \phi$ for all I,η,ν. Consider any I,η,ν and any substitution σ that is admissible for ϕ. Now the Substitution Lemma 2.2 implies

$$val_{I,\eta}(\nu,\sigma(\phi)) = val_{I,\sigma^*(\eta)}(\sigma^*(\nu),\phi) = true.$$

The last step holds because ϕ is valid and, in particular, holds for $I,\sigma^*(\eta),\sigma^*(\nu)$. □

Observe that, for soundness, the notion of bound variables in Definition 2.8 could in fact be *any* overapproximation of the set of variables that possibly change their value during a hybrid program. In vacuous identity changes like $x:=x$ or $x' = 0$, variable x will not really change its value, but we still consider x as a bound variable for simplicity. For a hybrid program α, we denote by $\forall^\alpha \phi$ the *universal closure* of formula ϕ with respect to all state variables bound in α. Quantification over state variable x is definable as $\forall X\,[x:=X]\Phi$ using an auxiliary logical variable X.

2.5.2 *Rules of the Calculus for Differential Dynamic Logic*

We present a proof calculus for d$\mathcal{L}$ as a Gentzen-style sequent calculus [133]. Sequents are essentially a standard form for logical formulas that is convenient for proving. A *sequent* is of the form $\Gamma \vdash \Delta$, where the *antecedent* Γ and *succedent* Δ are finite sets of formulas. The semantics of $\Gamma \vdash \Delta$ is that of the formula $\bigwedge_{\phi\in\Gamma}\phi \to \bigvee_{\psi\in\Delta}\psi$. For quantifier elimination rules, we make use of this fact by considering sequent $\Gamma \vdash \Delta$ as an abbreviation for the latter formula. The antecedent Γ can be thought of as the formulas we assume to be true, whereas the succedent Δ can be understood as formulas for which we want to show that at least one of them is true assuming all formulas of Γ are true. So for proving a sequent $\Gamma \vdash \Delta$, we assume all Γ and want to show that one of the Δ is true. For some simple sequents like $\Gamma,\phi \vdash \phi,\Delta$, we directly know that they are valid, because we can certainly show ϕ if we assume ϕ (in fact, we will use this as an axiom). For other sequents, it is more difficult to see whether they are valid (true under all circumstances) and it is the purpose of a proof calculus to provide a means to find out.

For handling quantifiers in the d$\mathcal{L}$ calculus, we cannot use the standard proof rules [147, 122, 123], because these are for uninterpreted first-order logic and (ultimately) work by instantiating quantifiers, either eagerly as in ground tableaux or lazily by unification as in free-variable tableaux [147, 122, 123]. Also, see App. A for an exposition of proving in uninterpreted first-order logic. The basis of d$\mathcal{L}$, in contrast, is first-order logic interpreted over the reals or in the theory of real-closed fields [287, 288]. A formula like $\exists a \forall x\,(x^2 + a > 0)$ cannot be proven by instantiation-based quantifier rules but is valid in the theory of real-closed fields. Unfortunately, quantifier elimination (QE) over the reals [81, 288], which is the standard decision procedure for real arithmetic, cannot be applied to formulas with modalities either. Hence, we introduce new quantifier rules that integrate quantifier elimination in a way that is compatible with dynamic modalities (as we illustrate in Sect. 2.5.3).

Definition 2.9 (Quantifier elimination). A first-order theory admits *quantifier elimination* if, with each formula ϕ, a quantifier-free formula $\mathrm{QE}(\phi)$ can be associated effectively that is equivalent (i.e., $\phi \leftrightarrow \mathrm{QE}(\phi)$ is valid) and has no additional free variables or function symbols. The operation QE is further assumed to evaluate ground formulas (i.e., without variables), yielding a decision procedure for closed formulas of this theory (i.e., formulas without free variables).

Example 2.18. Quantifier elimination uses the special structure of real arithmetic to express quantified arithmetic formulas equivalently without quantifiers and without using more free variables. For instance, QE yields the following equivalence:

$$\mathrm{QE}(\exists x\,(ax^2 + bx + c = 0)) \;\equiv\; (a \neq 0 \wedge b^2 - 4ac \geq 0) \vee (a = 0 \wedge (b = 0 \rightarrow c = 0)).$$

In this particular case, the equivalence can be found by using the generic condition for solvability of quadratic equations over the reals plus special cases when coefficients are zero. For details on quantifier elimination in real-closed fields and an overview of decision procedures for real arithmetic, also see App. D.2. □

As usual in sequent calculus rules—although the direction of entailment in the proof rules is from *premises* (above rule bar) to *conclusion* (below)—the order of reasoning is *goal-directed*: Rules are applied backwards, i.e., starting from the desired conclusion at the bottom (*goal*) to the resulting premises (*subgoals*). To highlight the logical essence of the d$\mathcal{L}$ calculus, Fig. 2.11 provides *rule schemata* with which the following definition associates the calculus rules that are applicable in d$\mathcal{L}$ proofs. The calculus consists of propositional rules ($\neg$r–*cut*), first-order quantifier rules ($\forall$r–i$\exists$), rules for dynamic modalities ($\langle;\rangle$–[$'$]), and global rules ([]*gen*–*con*). All substitutions in the rules in Fig. 2.11 need to be admissible for the rules to be applicable, including the substitution that inserts $s(X_1,..,X_n)$ into $\phi(s(X_1,..,X_n))$. Proof schemata come in three kinds with which the following definition associates proof rules: 1) sequent proof schemata that mention the sequent symbol $\vdash$, 2) symmetric proof schemata that do not mention the sequent symbol $\vdash$ and can be applied on either side of the sequent, 3) the special proof schema i$\exists$ that merges multiple branches.

Definition 2.10 (Rules). The *rule schemata* in Fig. 2.11—in which *all* substitutions need to be admissible for the rules to be applicable, including the substitution that inserts $s(X_1,..,X_n)$ into $\phi(s(X_1,..,X_n))$—induce *calculus rules* by:

1. If
$$\frac{\Phi_1 \vdash \Psi_1 \quad \dots \quad \Phi_n \vdash \Psi_n}{\Phi_0 \vdash \Psi_0} \tag{2.11}$$
is an instance of a rule schema in Fig. 2.11 (rules ∀r–∀l, i∀, and the propositional and global rule schemata have this form), then
$$\frac{\Gamma, \langle \mathcal{J} \rangle \Phi_1 \vdash \langle \mathcal{J} \rangle \Psi_1, \Delta \quad \dots \quad \Gamma, \langle \mathcal{J} \rangle \Phi_n \vdash \langle \mathcal{J} \rangle \Psi_n, \Delta}{\Gamma, \langle \mathcal{J} \rangle \Phi_0 \vdash \langle \mathcal{J} \rangle \Psi_0, \Delta}$$
can be applied as a proof rule of the $d\mathcal{L}$ calculus, where Γ, Δ are arbitrary finite sets of additional context formulas (including empty sets) and $\mathcal{J}$ is a discrete jump set (including the empty set). Hence, the rule context Γ, Δ and prefix $\langle \mathcal{J} \rangle$ remain unchanged during rule applications; only the formulas mentioned in (2.11) are affected.
2. Symmetric schemata can be applied on either side of the sequent: If
$$\frac{\phi_1}{\phi_0}$$
is an instance of one of the symmetric rule schemata (the dynamic rules) in Fig. 2.11, then
$$\frac{\Gamma \vdash \langle \mathcal{J} \rangle \phi_1, \Delta}{\Gamma \vdash \langle \mathcal{J} \rangle \phi_0, \Delta} \quad \text{and} \quad \frac{\Gamma, \langle \mathcal{J} \rangle \phi_1 \vdash \Delta}{\Gamma, \langle \mathcal{J} \rangle \phi_0 \vdash \Delta}$$
can both be applied as proof rules of the $d\mathcal{L}$ calculus, where Γ, Δ are arbitrary finite sets of context formulas (including the empty set) and $\mathcal{J}$ is a discrete jump set (including empty sets). In particular, symmetric schemata yield equivalence transformations, because the same rule applies in the antecedent as in the succedent.
3. Schema i∃ applies to *all* goals containing X at once: If $\Phi_1 \vdash \Psi_1, .., \Phi_n \vdash \Psi_n$ is the list of all open goals of the proof that contain free variable X, then an instance
$$\frac{\vdash \mathrm{QE}(\exists X \bigwedge_i (\Phi_i \vdash \Psi_i))}{\Phi_1 \vdash \Psi_1 \quad \dots \quad \Phi_n \vdash \Psi_n}$$
of rule schema i∃ can be applied as a proof rule of the $d\mathcal{L}$ calculus.

Propositional Rules

For propositional logic, standard propositional rules ¬r–*cut* with the cut rule are listed in the first block of Fig. 2.11. They decompose the propositional structure of formulas. Rules ¬r and ¬l use simple dualities caused by the implicative semantics of sequents. Essentially, instead of showing $\neg\phi$ in the succedent, we assume the contrary ϕ in the antecedent with rule ¬r. In rule ¬l, instead of assuming $\neg\phi$ in

$$(\neg r)\ \frac{\phi \vdash}{\vdash \neg\phi} \quad (\vee r)\ \frac{\vdash \phi, \psi}{\vdash \phi \vee \psi} \quad (\wedge r)\ \frac{\vdash \phi \quad \vdash \psi}{\vdash \phi \wedge \psi} \quad (\to r)\ \frac{\phi \vdash \psi}{\vdash \phi \to \psi} \quad (ax)\ \frac{}{\phi \vdash \phi}$$

$$(\neg l)\ \frac{\vdash \phi}{\neg\phi \vdash} \quad (\vee l)\ \frac{\phi \vdash \quad \psi \vdash}{\phi \vee \psi \vdash} \quad (\wedge l)\ \frac{\phi, \psi \vdash}{\phi \wedge \psi \vdash} \quad (\to l)\ \frac{\vdash \phi \quad \psi \vdash}{\phi \to \psi \vdash} \quad (cut)\ \frac{\vdash \phi \quad \phi \vdash}{\vdash}$$

$$(\langle;\rangle)\ \frac{\langle\alpha\rangle\langle\beta\rangle\phi}{\langle\alpha;\beta\rangle\phi} \quad (\langle{*^n}\rangle)\ \frac{\phi \vee \langle\alpha\rangle\langle\alpha^*\rangle\phi}{\langle\alpha^*\rangle\phi} \quad (\langle:=\rangle)\ \frac{\phi_{x_1}^{\theta_1}\cdots{}_{x_n}^{\theta_n}}{\langle x_1 := \theta_1, .., x_n := \theta_n\rangle\phi}$$

$$([;])\ \frac{[\alpha][\beta]\phi}{[\alpha;\beta]\phi} \quad ([{*^n}])\ \frac{\phi \wedge [\alpha][\alpha^*]\phi}{[\alpha^*]\phi} \quad ([:=])\ \frac{\langle x_1 := \theta_1, .., x_n := \theta_n\rangle\phi}{[x_1 := \theta_1, .., x_n := \theta_n]\phi}$$

$$(\langle\cup\rangle)\ \frac{\langle\alpha\rangle\phi \vee \langle\beta\rangle\phi}{\langle\alpha\cup\beta\rangle\phi} \quad (\langle?\rangle)\ \frac{\chi \wedge \psi}{\langle ?\chi\rangle\psi} \quad (\langle'\rangle)\ \frac{\exists t{\ge}0\,\big((\forall 0{\le}\tilde{t}{\le}t\,\langle\mathcal{S}_{\tilde{t}}\rangle\chi) \wedge \langle\mathcal{S}_t\rangle\phi\big)}{\langle x_1' = \theta_1, .., x_n' = \theta_n \,\&\, \chi\rangle\phi}\ ^1$$

$$([\cup])\ \frac{[\alpha]\phi \wedge [\beta]\phi}{[\alpha\cup\beta]\phi} \quad ([?])\ \frac{\chi \to \psi}{[?\chi]\psi} \quad (['])\ \frac{\forall t{\ge}0\,\big((\forall 0{\le}\tilde{t}{\le}t\,\langle\mathcal{S}_{\tilde{t}}\rangle\chi) \to \langle\mathcal{S}_t\rangle\phi\big)}{[x_1' = \theta_1, .., x_n' = \theta_n \,\&\, \chi]\phi}\ ^1$$

$$(\forall r)\ \frac{\vdash \phi(s(X_1, .., X_n))}{\vdash \forall x\,\phi(x)}\ ^2 \qquad (\exists r)\ \frac{\vdash \phi(X)}{\vdash \exists x\,\phi(x)}\ ^4$$

$$(\exists l)\ \frac{\phi(s(X_1, .., X_n)) \vdash}{\exists x\,\phi(x) \vdash}\ ^2 \qquad (\forall l)\ \frac{\phi(X) \vdash}{\forall x\,\phi(x) \vdash}\ ^4$$

$$(i\forall)\ \frac{\vdash \mathrm{QE}(\forall X\,(\Phi(X) \vdash \Psi(X)))}{\Phi(s(X_1, .., X_n)) \vdash \Psi(s(X_1, .., X_n))}\ ^3 \qquad (i\exists)\ \frac{\vdash \mathrm{QE}(\exists X \bigwedge_i(\Phi_i \vdash \Psi_i))}{\Phi_1 \vdash \Psi_1 \quad \dots \quad \Phi_n \vdash \Psi_n}\ ^5$$

$$([]gen)\ \frac{\vdash \forall^\alpha(\phi \to \psi)}{[\alpha]\phi \vdash [\alpha]\psi} \qquad (\langle\rangle gen)\ \frac{\vdash \forall^\alpha(\phi \to \psi)}{\langle\alpha\rangle\phi \vdash \langle\alpha\rangle\psi}$$

$$(ind)\ \frac{\vdash \forall^\alpha(\phi \to [\alpha]\phi)}{\phi \vdash [\alpha^*]\phi} \qquad (con)\ \frac{\vdash \forall^\alpha \forall v{>}0\,(\varphi(v) \to \langle\alpha\rangle\varphi(v-1))}{\exists v\,\varphi(v) \vdash \langle\alpha^*\rangle\exists v{\le}0\,\varphi(v)}\ ^6$$

[1] t and $\tilde{t}$ are fresh logical variables and $\langle\mathcal{S}_t\rangle$ is the jump set $\langle x_1 := y_1(t), .., x_n := y_n(t)\rangle$ with simultaneous solutions $y_1, .., y_n$ of the respective differential equations with constant symbols x_i as symbolic initial values.

[2] s is a new (Skolem) function symbol and $X_1, .., X_n$ are all free logical variables of $\forall x\,\phi(x)$.

[3] X is a new logical variable. Further, QE needs to be defined for the formula in the premise.

[4] X is a new logical variable.

[5] Among all open branches, free logical variable X only occurs in the branches $\Phi_i \vdash \Psi_i$. Further, QE needs to be defined for the formula in the premise, especially, no Skolem dependencies on X can occur.

[6] Logical variable v does not occur in α.

Fig. 2.11 Rule schemata of the free-variable proof calculus for differential dynamic logic

the antecedent, we show the contrary ϕ in the succedent. Rule $\vee$r uses the fact that formulas are combined disjunctively in succedents, rule $\wedge$l that they are conjunctive in antecedents. The comma between formulas in an antecedent has the same effect as a conjunction, and the comma between formulas in the succedent has the same effect as a disjunction. Rules $\vee$l and $\wedge$r split the proof into two cases, because conjuncts in the succedent can be proven separately ($\wedge$r) and, dually, disjuncts of the antecedent can be assumed separately ($\vee$l). For $\wedge$r we want to show conjunction $\phi \wedge \psi$, so in the left branch we proceed to show $\Gamma \vdash \phi, \Delta$ and, in addition, in the right branch we show $\Gamma \vdash \psi, \Delta$, which, together, entail $\Gamma \vdash \phi \wedge \psi, \Delta$. If, as in rule $\vee$l, we assume disjunction $\phi \vee \psi$ as part of the antecedent, then we do not know if we can assume ϕ to hold or if we can assume ψ to hold in the antecedent, but know only that one of them holds. Hence, as in a case distinction, $\vee$l considers both cases, the case where we assume ϕ in the antecedent, and the case where we assume ψ. If both subgoals can be proven, this entails $\Gamma, \phi \vee \psi \vdash \Delta$. Rules $\rightarrow$r and $\rightarrow$l can be derived from the equivalence of $\phi \rightarrow \psi$ and $\neg\phi \vee \psi$. Rule $\rightarrow$r uses the fact that implication $\rightarrow$ has the same meaning as the sequent arrow $\vdash$ of a sequent. Intuitively, to show implication $\phi \rightarrow \psi$, rule $\rightarrow$r assumes ϕ (in the antecedent) and shows ψ (in the succedent). Rule $\rightarrow$l assumes an implication $\phi \rightarrow \psi$ to hold in the antecedent, but we do not know if this implication holds because ϕ is false, or because ψ is true, so $\rightarrow$l splits into those two branches.

The axiom rule *ax* closes a goal (there are no further subgoals, which we sometimes mark $*$ explicitly), because assumption ϕ in the antecedent trivially entails ϕ in the succedent (sequent $\Gamma, \phi \vdash \phi, \Delta$ is a simple syntactic tautology). Rule *cut* is the *cut* rule that can be used for case distinctions: The right subgoal assumes any additional formula ϕ in the antecedent that the left subgoal shows in the succedent. Dually: regardless of whether ϕ is actually true or false, both cases are covered by proof branches. We only use cuts in an orderly fashion to derive simple rule dualities and to simplify meta-proofs. In practical applications, cuts are not usually needed and we conjecture that this is no coincidence.

According to the definition in Definition 2.10, all propositional rules can be applied with an additional context Γ, Δ. In particular, rules *ax* and *cut* can also be applied as:

$$\text{ax}\frac{}{\Gamma, \phi \vdash \phi, \Delta} \quad \text{and} \quad \text{cut}\frac{\Gamma \vdash \phi, \Delta \quad \Gamma, \phi \vdash \Delta}{\Gamma \vdash \Delta}$$

First-Order Quantifier Rules

The quantifier rules $\forall$r,$\exists$l,$\exists$r,$\forall$l,i$\forall$,i$\exists$ constitute a purely modular interface to arithmetic mathematical reasoning. They can use any theory that admits quantifier elimination and has a decidable ground theory (formulas without quantifiers or variables), including the theory of real arithmetic or real-closed fields [288, 81]. Rules $\forall$r,$\exists$l,$\exists$r,$\forall$l handle quantifiers and replace quantified variables by Skolem function terms ($\forall$r,$\exists$l) or free logical variables ($\exists$r,$\forall$l), respectively. Later in the proof, rules i$\forall$,i$\exists$ can reintroduce quantifiers for these previously quantified symbols and apply quantifier elimination in real-closed fields once the remaining formulas are first-order in the relevant symbols.

Rule $\exists$l, with which we want to show $\exists x\, \phi(x)$ in the succedent, introduces a new free logical variable X for an existentially quantified variable x. Essentially, free variable X can be thought of as a variable for which an appropriate value still needs to be found for the proof to close. This makes sense, because at the time of applying proof rule $\exists$l, it is mostly impossible to know which particular instance to choose for X that will help. But once we find such an X that proves the subgoal $\Gamma \vdash \phi(X), \Delta$

later, we have also proven the goal $\Gamma \vdash \exists x\, \phi(x), \Delta$, because X will be a witness for the existence. We have also proven the goal if, later during the proof, we prove the existence of an X satisfying the constraints indirectly, without directly instantiating a witness. This is what rule i$\exists$ is for.

The dual rule $\forall$l, which assumes $\forall x\, \phi(x)$ in the antecedent, introduces a new free logical variable X for the universally quantified variable x in the antecedent. If, later, we have found an instance of X that proves subgoal $\Gamma, \phi(X) \vdash \Delta$, then we have also proven goal $\Gamma, \forall x\, \phi(x) \vdash \Delta$, because if we can prove the subgoal just from assuming the particular $\phi(X)$ in the antecedent, then the goal also holds where we even assume $\phi(x)$ holds for all x. While this reasoning is perfectly good if it works, it is somewhat surprising why this should always work for all cases. Why should one instance be enough? Why should it not be necessary to assume two different instances $\phi(X)$ and $\phi(Y)$ during the proof? The fact that this is not necessary comes from proof rule i$\exists$, which can reintroduce quantifiers and eliminate them equivalently.

Rule $\forall$r, with which we want to show $\forall x\, \phi(x)$ in the succedent, introduces a new (Skolem) function symbol s for the previously quantified variable x and replaces x by a (Skolem) term $s(X_1, \ldots, X_n)$ where $X_1, \ldots, X_n$ are all the free logical variables of the original formula $\forall x\, \phi(x)$. This works like a proof in mathematics, where we want so show $\forall x\, \phi(x)$ in the succedent and do so by choosing a fresh symbol s for which we prove that $\phi(s(X_1, \ldots, X_n))$ holds. Because s was arbitrary and we did not assume anything special about the value of s, this implies that $\forall x\, \phi(x)$ holds. The free variables $X_1, \ldots, X_n$ of the Skolem terms keep track of the dependencies of the symbols for nested quantifiers. Having all free logical variables $X_1, \ldots, X_n$ in the Skolem term is important for soundness in order to prevent unsound rearrangements of quantifiers, as we elaborate in Sect. 2.5.3.

The dual rule $\exists$l is similar. When we assume $\exists x\, \phi(x)$ in the antecedent, then we only know that such an x exists, not what value it has. Hence, $\exists$l introduces a new name for this object in the form of a new (Skolem) function symbol s and replaces x by a (Skolem) term $s(X_1, \ldots, X_n)$ where $X_1, \ldots, X_n$ are all the free logical variables of the original formula $\exists x\, \phi(x)$. If we can prove the subgoal, the subgoal entails the goal, because we did not assume anything special about s. Having all free logical variables $X_1, \ldots, X_n$ in the Skolem term to track the dependencies of the symbols is again important for soundness to prevent unsound rearrangements of quantifiers. Intuitively, for a formula like $\forall x\, \exists y\, \phi(x,y)$ in the antecedent—which will yield $\exists y\, \phi(X,y)$ after applying $\forall$l—we need to track the dependency of y on X, which yields $\phi(s(X),X)$ when applying $\exists$l. We need to remember that the choice for s may depend on X, because the choice of y may depend on x.

With the rule i$\forall$, we can reintroduce a universal quantifier for a Skolem term $s(X_1, \ldots, X_n)$, which corresponds to a previously universally quantified variable in the succedent or a previously existentially quantified variable in the antecedent. The point of reintroducing the quantifier is that this makes sense when the remaining formulas are first-order in the quantified variable so that they can be handled equivalently by quantifier elimination in real-closed fields. When we have proven the subgoal (with for all X) then this entails the goal for the particular $s(X_1, \ldots, {}_n)$. In particular, when we remove a quantifier with $\forall$r,$\exists$l to obtain a Skolem term, we can continue with other proof rules to handle the dynamic modalities and then reintroduce the quantifier for the Skolem term with i$\forall$ once quantifier elimination for real arithmetic becomes applicable.

The dual rule i$\exists$ can reintroduce an existential quantifier for a free logical variable that was previously existentially quantified in the succedent or previously universally quantified in the antecedent. Again, this makes sense when the resulting formula in the premise is first-order in the quantified variable X so that quantifier elimination can eliminate the quantifier equivalently. When we remove a quantifier with $\exists$r,$\forall$l to obtain a free logical variable, we can continue using other proof rules to handle the dynamic modalities and then reintroduce the quantifier for the free logical variable with i$\exists$ once quantifier elimination is applicable.

The quantifier rules $\forall$r and $\exists$l correspond to the liberalised δ^+-rule of Hähnle and Schmitt [147]. Rules $\exists$r and $\forall$l resemble the usual γ-rule but, unlike in [122, 123, 147, 134], they cannot be applied twice because the original formula is removed ($\exists x\, \phi(x)$ in $\exists$r). The calculus still has a complete handling of quantifiers due to i$\forall$ and i$\exists$, which can reconstruct and eliminate quantifiers once QE is applicable as the remaining constraints are first-order in the respective variables. In the premise of i$\forall$ and i$\exists$, we again consider sequents $\Phi \vdash \Psi$ as abbreviations for formulas. For closed formulas, we do not need other arithmetic rules. We defer illustrations and further discussion of quantifier rules to Sect. 2.5.3. For comparison, App. A gives a summary of the standard γ-rules and δ^+-rules that are used for handling quantifiers in uninterpreted first-order logic. In Sect. 3.5.5, we show an alternative way of handling real arithmetic in a modular way using deduction modulo by side deductions.

Dynamic Rules

The dynamic modality rules transform a hybrid program into structurally simpler logical formulas by symbolic decomposition. Rules $\langle ; \rangle$,$[;]$,$\langle \cup \rangle$,$[\cup]$,$\langle ^{*n} \rangle$,$[^{*n}]$,$\langle ? \rangle$,$[?]$ are as in discrete dynamic logic [149, 37]. Also, see Fig. 2.12 for an illustration of the correspondence of a representative set of proof rules for dynamic modalities to the transition semantics of hybrid programs (from Definition 2.7).

Nondeterministic choices split into their alternatives ($\langle \cup \rangle$,$[\cup]$). For rule $[\cup]$: If all α transitions lead to states satisfying ϕ (i.e., $[\alpha]\phi$ holds) and all β transitions lead to states satisfying ϕ (i.e., $[\beta]\phi$ holds), then, all transitions of program $\alpha \cup \beta$ that choose between following α and following β also lead to states satisfying ϕ (i.e., $[\alpha \cup \beta]\phi$ holds). Dually for rule $\langle \cup \rangle$, if there is an α transition to a ϕ state ($\langle \alpha \rangle \phi$) or a β-transition to a ϕ state ($\langle \beta \rangle \phi$), then, in either case, there is a transition of $\alpha \cup \beta$ to ϕ ($\langle \alpha \cup \beta \rangle \phi$ holds), because $\alpha \cup \beta$ can choose which of those transitions to follow. A general principle behind the $d\mathcal{L}$ proof rules that is most noticeable in $\langle \cup \rangle$,$[\cup]$ is that these proof rules symbolically decompose the reasoning into two separate parts and analyse the fragments α and β separately, which is good for scalability. For these symbolic structural decompositions, it is very helpful that $d\mathcal{L}$ is a full logic that is closed under all logical operators, including disjunction and conjunction, for then the premises in $[\cup]$,$\langle \cup \rangle$ are $d\mathcal{L}$ formulas again (unlike in Hoare logic [161]).

$$([:=])\ \frac{\phi_x^\theta}{[x:=\theta]\phi}$$

$$(\langle'\rangle)\ \frac{\exists t\geq 0\,\langle x:=y_x(t)\rangle\phi}{\langle x'=\theta\rangle\phi}$$

$$(\langle'\rangle)\ \frac{\exists t\geq 0\,(\hat{\chi}\wedge\langle x:=y_x(t)\rangle\phi)}{\langle x'=\theta\,\&\,\chi\rangle\phi}$$

$$\hat{\chi}\equiv\forall 0\leq s\leq t\,\langle x:=y_x(s)\rangle\chi$$

$$([\cup])\ \frac{[\alpha]\phi\wedge[\beta]\phi}{[\alpha\cup\beta]\phi}$$

$$([;])\ \frac{[\alpha][\beta]\phi}{[\alpha;\beta]\phi}$$

$$(ind)\ \frac{\vdash\forall^\alpha(\phi\to[\alpha]\phi)}{\phi\vdash[\alpha^*]\phi}$$

$$(con)\ \frac{\vdash\forall^\alpha\forall v>0\,(\varphi(v)\to\langle\alpha\rangle\varphi(v-1))}{\exists v\,\varphi(v)\vdash\langle\alpha^*\rangle\exists v\leq 0\,\varphi(v)}$$

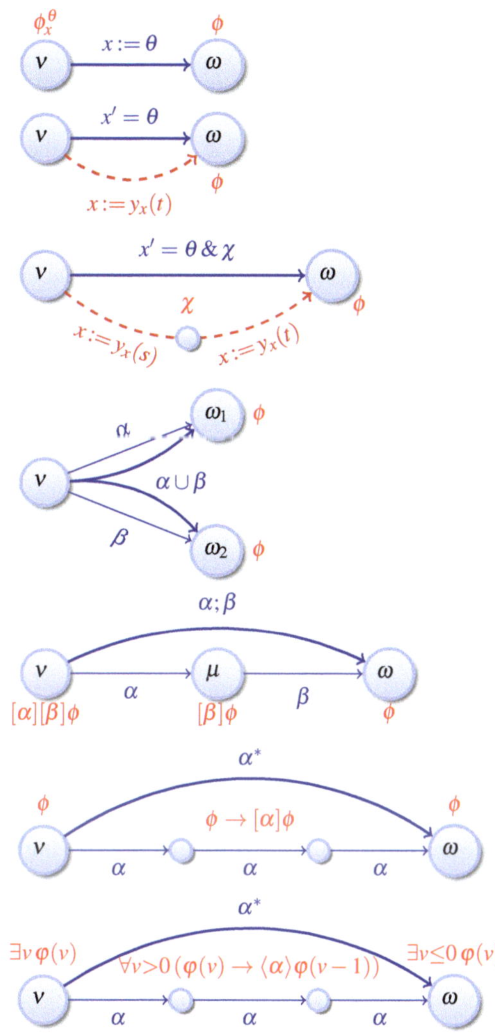

Fig. 2.12 Correspondence of dynamic proof rules and transition semantics

Sequential compositions are proven using nested modalities ($\langle;\rangle$,$[;]$). For rule $[;]$: If after all α-transitions, all β-transitions lead to states satisfying ϕ (i.e., $[\alpha][\beta]\phi$ holds), then also all transitions of the sequential composition $\alpha;\beta$ lead to states satisfying ϕ (i.e., $[\alpha;\beta]\phi$ holds). See, again, Fig. 2.12 for a graphical illustration of this proof principle. The dual rule $\langle;\rangle$ uses the fact that if there is an α-transition, after which there is a β-transition leading to ϕ (i.e., $\langle\alpha\rangle\langle\beta\rangle\phi$), then there is a transition of $\alpha;\beta$ leading to ϕ (that is, $\langle\alpha;\beta\rangle\phi$), because the transitions of $\alpha;\beta$ are just those that first do any α-transition, followed by any β-transition (Definition 2.7).

Rules $\langle^{*n}\rangle$,$[^{*n}]$ are the usual iteration rules, which partially unwind loops. Rule $\langle^{*n}\rangle$ uses the fact that ϕ holds after repeating α (i.e., $\langle\alpha^*\rangle\phi$), if ϕ holds at the beginning (for ϕ holds after zero repetitions then), or if, after one execution of α, ϕ holds after any number of repetitions of α, including zero repetitions (i.e., $\langle\alpha\rangle\langle\alpha^*\rangle\phi$). So rule $\langle^{*n}\rangle$ expresses that for $\langle\alpha^*\rangle\phi$ to hold, ϕ must hold either immediately or after one or more repetitions of α. Rule $[^{*n}]$ is the dual rule expressing that ϕ must hold after all of those combinations for $[\alpha^*]\phi$ to hold.

Tests are proven by showing (with a conjunction in rule $\langle ?\rangle$) or assuming (with an implication in rule $[?]$) that the test succeeds, because test $?\chi$ can only make a transition when condition χ actually holds true (Definition 2.7). Thus, for $d\mathcal{L}$ formula $\langle ?\chi\rangle\phi$ rule $\langle ?\rangle$ is used to prove that χ holds true (otherwise there is no transition and thus the reachability property is false) and that ϕ holds after the resulting no-op. Rule $[?]$ for $d\mathcal{L}$ formula $[?\chi]\phi$, in contrast, assumes that χ holds true (otherwise there is no transition and thus nothing to show) and that ϕ holds after the resulting no-op.

Rule $\langle :=\rangle$ uses simultaneous substitutions from Fig. 2.10 for handling discrete jump sets. To show that ϕ is true after a discrete jump, $\langle :=\rangle$ shows that ϕ has been true before, when replacing the affected variables x_i with their new values θ_i in ϕ by an admissible substitution (Definition 2.8). Alternatively, the discrete jump set can also remain an unchanged prefix ($\mathcal{J}$ in Definition 2.10) for other $d\mathcal{L}$ rules applied to ϕ, until the substitution for rule $\langle :=\rangle$ becomes admissible. This is what our proof calculus uses instead of what we have shown in Example 2.16. Rule $[:=]$ uses the fact that discrete jump sets characterise a unique deterministic transition. Hence, its premise and conclusion are actually equivalent, because there is exactly one terminating transition for each discrete jump set. Assuming the presence of vacuous identity jumps $a := a$ for variables a that do not otherwise change (vacuous identity jumps can be added as they do not change state), we can further use rule $\langle :=\rangle$ to *merge* subsequent discrete jumps into a single discrete jump set (see previous results [37] for a compatible calculus detailing jump set merging, which works without the need to add vacuous identity jumps $a := a$):

$$\cfrac{\cfrac{\cfrac{\vdash \langle z := -\frac{b}{2}t^2 + Vt, v := V+1, a := -b\rangle[\beta]\phi}{\vdash \langle a := -b, v := V\rangle\langle z := \frac{a}{2}t^2 + vt, v := v+1, a := a\rangle[\beta]\phi}\langle :=\rangle}{\vdash \langle a := -b, v := V\rangle[z := \frac{a}{2}t^2 + vt, v := v+1, a := a][\beta]\phi}[:=]}{\vdash \langle a := -b, v := V\rangle[z := \frac{a}{2}t^2 + vt, v := v+1, a := a;\beta]\phi}[;]$$

More generally, $\langle x_1 := \theta_1, \ldots, x_n := \theta_n\rangle\langle x_1 := \vartheta_1, \ldots, x_n := \vartheta_n\rangle\phi$ can be merged by $\langle :=\rangle$ to $\langle x_1 := \vartheta_1{}^{\theta_1}_{x_1}\cdots{}^{\theta_n}_{x_n}, \ldots, x_n := \vartheta_n{}^{\theta_1}_{x_1}\cdots{}^{\theta_n}_{x_n}\rangle\phi$. Also see previous work [37] for more advanced and optimised merging techniques for state changes.

Given first-order definable flows for their differential equations, proof rules $\langle '\rangle$,$[']$ handle continuous evolutions (see [15, 189, 238] and App. B for flow approximation and solution techniques). These flows are combined in the discrete jump set $\mathcal{S}_t$. Given a solution $\mathcal{S}_t$ for the differential equation system with symbolic initial values $x_1, \ldots, x_n$, continuous evolution along differential equations can be replaced by a discrete jump $\langle\mathcal{S}_t\rangle$ with an additional quantifier for the evolution time t. The effect of the constraint on χ is to restrict the continuous evolution such that its solution $\mathcal{S}_{\tilde{t}}$ remains in the evolution domain χ at all intermediate times $\tilde{t} \leq t$. This constraint simplifies to *true* if the evolution domain restriction χ is *true*, which makes sense, because there are no special constraints on the evolution (other than the differential equations) if the evolution domain region is described by *true*, hence the full space $\mathbb{R}^n$. A notable special case of rules $[']$ and $\langle ' \rangle$ is when the evolution domain χ is *true*:

$$\frac{\forall t{\geq}0\, \langle \mathcal{S}_t \rangle \phi}{[x_1' = \theta_1,..,x_n' = \theta_n]\phi} \qquad \frac{\exists t{\geq}0\, \langle \mathcal{S}_t \rangle \phi}{\langle x_1' = \theta_1,..,x_n' = \theta_n \rangle \phi} \tag{2.12}$$

Similar simplifications can be made for convex invariant conditions (Sects. 2.9 and 3.8).

Global Rules

The last block of rules $[]gen$,$\langle\rangle gen$,*ind*,*con* are *global rules*. They depend on the truth of their premises in all states reachable by hybrid program α, which is ensured by the universal closure $\forall^\alpha$ with respect to all bound state variables (Definition 2.8) of the respective hybrid program α. This universal closure overapproximates all possible change caused by α, because it comprises all bound variables. This universal closure is required for soundness in the presence of contexts Γ, Δ (Definition 2.10) or free variables. The global rules are given in a form that best displays their underlying logical principles. The general pattern for applying global rules to prove that the succedent of their conclusion holds is to prove that both their premise and the antecedent of their conclusion hold. In particular, the antecedent can be thought of as holding in the current state, whereas the premise can be thought of as holding in all reachable states because of the universal closure.

Rules $[]gen$,$\langle\rangle gen$ are generalisation rules and can be used to strengthen postconditions: antecedent $[\alpha]\phi$ is sufficient for proving succedent $[\alpha]\psi$ when postcondition ϕ entails ψ in all relevant states reachable by α, which are overapproximated by the universal closure $\forall^\alpha$ with respect to the bound variables of α. Clearly, for rule $[]gen$, if all states reachable by α satisfy ϕ ($[\alpha]\phi$) and ϕ implies ψ in all these states ($\forall^\alpha \phi \to \psi$), then ψ also holds in all states reachable by α ($[\alpha]\psi$). Similarly, for rule $\langle\rangle gen$, if some state reachable by α satisfies ϕ ($\langle\alpha\rangle\phi$) and ϕ implies ψ in all reachable states ($\forall^\alpha \phi \to \psi$), then ψ also holds in some state reachable by α ($\langle\alpha\rangle\psi$).

Rule *ind* is an induction schema with *inductive invariant* ϕ. Similarly, *con* is a generalisation of Harel's convergence rule [149] to the hybrid case with decreasing *variant* φ. Both rules are given in a form that best displays their underlying logical principles and similarity. Rule *ind* says that ϕ holds after any number of repetitions of α if it holds initially (antecedent) and, for all reachable states (as overapproximated by $\forall^\alpha$), invariant ϕ remains true after one iteration of α (premise). If ϕ is true after executing α whenever ϕ has been true before, then, if ϕ holds in the beginning, ϕ will continue to hold, no matter how often we repeat α in $[\alpha^*]\phi$; again, see Fig. 2.12 for an illustration. Rule *con* expresses that the variant $\varphi(v)$ holds for some real number $v \leq 0$ after repeating α sufficiently often if $\varphi(v)$ holds for some real number at all in the beginning (antecedent) and, by premise, $\varphi(v)$ decreases after every execution of α by 1 (or another positive real constant). This rule can be used to show positive progress (by 1) with respect to $\varphi(v)$ by executing α.

For practical verification, rules *ind* or *con* can be combined with generalisation ($[]gen$, $\langle\rangle gen$) to prove a postcondition ψ of a loop α^* by showing that (*a*) the antecedents of the respective goals of *ind* and *con*, which represent the induction start, holds initially (*b*) their subgoals, which represent the induction step, hold and (*c*) the postcondition of the succedentd in their goals entails ψ. The corresponding variants of *ind* and *con* are *derived rules*. That is, these rules are non-essential, because they can be derived easily by chaining the proof rules from Fig. 2.11 together in an appropriate way.

$$(ind')\ \frac{\vdash \phi \qquad \vdash \forall^\alpha(\phi \to [\alpha]\phi) \qquad \vdash \forall^\alpha(\phi \to \psi)}{\vdash [\alpha^*]\psi}$$

$$(con')\ \frac{\vdash \exists v\, \varphi(v) \qquad \vdash \forall^\alpha \forall v{>}0\,(\varphi(v) \to \langle\alpha\rangle \varphi(v-1)) \qquad \vdash \forall^\alpha(\exists v{\leq}0\, \varphi(v) \to \psi)}{\vdash \langle\alpha^*\rangle\psi}$$

For example, using a cut with $\phi \to [\alpha^*]\phi$, rule *ind'* can be derived from *ind* and $[]gen$ as follows:

$$cut\ \dfrac{\to r\ \dfrac{ind\ \dfrac{\vdash \forall^\alpha(\phi \to [\alpha]\phi)}{\phi \vdash [\alpha^*]\phi}}{\vdash \phi \to [\alpha^*]\phi} \qquad \to l\ \dfrac{\vdash \phi \qquad []gen\ \dfrac{\vdash \forall^\alpha(\phi \to \psi)}{[\alpha^*]\phi \vdash [\alpha^*]\psi}}{\phi \to [\alpha^*]\phi \vdash [\alpha^*]\psi}}{\vdash [\alpha^*]\psi}$$

These derived rules are not necessary in theory, but still useful in practise.

Derivability and Proofs

We call any formula ϕ provable or derivable (in the $d\mathcal{L}$ calculus) if we can find a $d\mathcal{L}$ proof for it that starts with axioms (rule *ax*) at the leaves and ends with a sequent $\vdash \phi$ at the bottom. While constructing proofs, however, we would start with the desired goal $\vdash \phi$ at the bottom and work our way backwards to the subgoals until they can be proven to be valid as axioms (*ax*). Once all subgoals have been proven to be valid axioms, they entail their consequences, which, recursively, entail the original goal $\vdash \phi$. This property of preserving truth or preserving entailment, which we prove in Sect. 2.6, is called soundness. Thus, while constructing proofs, we work bottom-up from the goal. When we have found a proof, we justify formulas from the axioms top-down to the original goal.

The notions of derivations and proofs for the $d\mathcal{L}$ calculus are standard, except that $i\exists$ produces multiple conclusions. Hence, we define derivations as finite acyclic graphs instead of trees. We want proofs to be acyclic and not accept a formula that is used to prove itself.

Definition 2.11 (Provability). A *derivation* is a finite acyclic graph labelled with sequents such that, for every node, the (set of) labels of its children must be the (set of) premises of an instance of one of the calculus rules (Definition 2.10) and the (set of) labels of the parents of these children must be the (set of) conclusions of that rule instance. A formula ψ is *provable* from a set Φ of formulas, denoted by $\Phi \vdash_{\mathsf{d}\mathcal{L}} \psi$, iff there is a finite subset $\Phi_0 \subseteq \Phi$ for which the sequent $\Phi_0 \vdash \psi$ is derivable, i.e., there is a derivation with a single root (i.e., node without parents) labelled $\Phi_0 \vdash \psi$.

Example 2.19. A very simple (in fact essentially propositional) proof of the formula

$$v^2 \leq 10 \wedge b > 0 \rightarrow b > 0 \wedge (\neg(v \geq 0) \vee v^2 \leq 10) \tag{2.13}$$

is shown in Fig. 2.13. The proof starts with the proof goal as a sequent at the bottom:

$$\vdash v^2 \leq 10 \wedge b > 0 \rightarrow b > 0 \wedge (\neg(v \geq 0) \vee v^2 \leq 10).$$

$$\rightarrow\!\mathrm{r}\,\dfrac{\wedge\mathrm{r}\,\dfrac{\wedge\mathrm{l}\,\dfrac{ax\,\dfrac{*}{v^2 \leq 10, b > 0 \vdash b > 0}}{v^2 \leq 10 \wedge b > 0 \vdash b > 0} \qquad \vee\mathrm{r}\,\dfrac{\wedge\mathrm{l}\,\dfrac{ax\,\dfrac{*}{v^2 \leq 10, b > 0 \vdash \neg(v \geq 0), v^2 \leq 10}}{v^2 \leq 10 \wedge b > 0 \vdash \neg(v \geq 0), v^2 \leq 10}}{v^2 \leq 10 \wedge b > 0 \vdash \neg(v \geq 0) \vee v^2 \leq 10}}{v^2 \leq 10 \wedge b > 0 \vdash b > 0 \wedge (\neg(v \geq 0) \vee v^2 \leq 10)}}{\vdash v^2 \leq 10 \wedge b > 0 \rightarrow b > 0 \wedge (\neg(v \geq 0) \vee v^2 \leq 10)}$$

Fig. 2.13 Simple propositional example proof

The first (i.e., bottom most) proof step applies proof rule $\rightarrow$r to turn the implication ($\rightarrow$) to the sequent level by moving the assumption into the antecedent. The next proof step applies rule $\wedge$r to split the proof into the left branch for showing that conjunct $b > 0$ follows from the assumptions in the antecedent and into the right branch for showing that conjunct $\neg(v \geq 0) \vee v^2 \leq 10$ follows from the antecedent also. On the left branch, the proof closes with an axiom *ax* after splitting the conjunction $\wedge$ on the antecedent with rule $\wedge$l. We mark closed proof goals with $*$. The right branch closes with an axiom *ax* after splitting the disjunction ($\vee$) in the succedent with rule $\vee$r and then splitting the conjunction ($\wedge$) in the antecedent with rule $\wedge$l. Now that all branches of the proof have closed (with *ax*), we know that all leaves at the top are valid, and, hence, since the premises are valid, each application of a proof rule ensures that their respective conclusions are valid also. By recursively following this derivation from the leaves at the top to the original root at the bottom, we see that the original goal is valid and formula (2.13) is, indeed, true under all circumstances (valid).

While this proof does not show anything particularly exciting, because it only uses propositional rules, it shows how a proof can be build systematically in the $\mathsf{d}\mathcal{L}$ calculus and gives an intuition about how validity is inherited from the premises to the conclusions. □

2.5.3 *Deduction Modulo with Invertible Quantifiers and Real Quantifier Elimination*

The first-order quantifier rules in Fig. 2.11 lift quantifier elimination to $d\mathcal{L}$ by following a generalised deduction modulo approach. They integrate decision procedures, e.g., for real quantifier elimination as a background prover [32], into the deductive proof system. Yet, unlike in the approaches of Dowek et al. [103] and Tinelli [290], the information given to the background prover is not restricted to ground formulas [290] or atomic formulas [103]. Further, real quantifier elimination is different from uninterpreted logic [147, 122, 134] in that the resulting formulas are not obtained by instantiation but by intricate arithmetic recombination. The quantifier rules can use any theory that admits quantifier elimination (see Definition 2.9) and has a decidable ground theory, for instance, the first-order theory of real arithmetic (which is equivalent to the theory of real-closed fields [288, 81]). A *formula of real arithmetic* is a first-order formula with $+,-,\cdot,/,=,\leq,<,\geq,>$ as the only function or predicate symbols besides constant symbols of Σ and logical variables of V. Also see App. D.2.

Integrating quantifier elimination to deal with statements about real quantities is quite challenging in the presence of modalities that influence the values of flexible symbols. In principle, quantifier elimination can be used to handle quantified constraints such as those arising for continuous evolutions. In $d\mathcal{L}$, however, real quantifiers interact with modalities containing further discrete or continuous transitions, which is an effect nherent in the interacting nature of hybrid systems. A hybrid formula like $\exists z\,\langle z''=-b; ?m-z\geq s; z''=0\rangle\, m-z<s$ is not first-order; hence quantifier elimination cannot be applied. Even more so, the effect of a modality depends on the solutions of the differential equations contained therein. The dynamics of a hybrid program depends on the values of its parameters (z,b,m,s in the above case), but, at the same time, the constraints on a quantified variable like z depend on the effect of the hybrid program. For instance, it is hard to know in advance, which first-order constraints need to be solved by QE for the above formula. To find out how z evolves from quantifier $\exists z$ to postcondition $m-z<s$, the system dynamics in the modality needs to be taken into account (as for repetitions). Hence, our calculus first unwraps the first-order structure before applying QE to the resulting arithmetic formulas.

2.5.3.1 Lifting Quantifier Elimination by Invertible Quantifier Rules

The purpose of the quantifier rules in Fig. 2.11 is to postpone QE until the actual arithmetic constraints become apparent. The idea is that ∀r,∃l,∃r, and ∀l temporarily remove quantifiers by introducing new auxiliary symbols for quantified variables such that the proof can be continued beyond the occurrence of the quantifier to further analyse the modalities contained therein. Later, when the actual first-order constraints for the auxiliary symbol have been discovered, the corresponding quantifier

$$
\dfrac{
  \dfrac{
    \dfrac{
      v\geq 0, z<m \vdash T\geq 0
      \qquad
      \dfrac{
        \dfrac{
          \dfrac{v\geq 0, z<m \vdash v^2>2b(m-z)}{\vdash v\geq 0\land z<m\to v^2>2b(m-z)}\;{\scriptstyle\to r,\land l}
        }{v\geq 0, z<m \vdash -\frac{b}{2}T^2+vT+z>m}
      }{v\geq 0, z<m \vdash \langle z:=-\frac{b}{2}T^2+vT+z\rangle z>m}\;{\scriptstyle\langle:=\rangle}
    }{v\geq 0, z<m \vdash T\geq 0\land\langle z:=-\frac{b}{2}T^2+vT+z\rangle z>m}\;{\scriptstyle\land r}
  }{v\geq 0, z<m \vdash \exists t{\geq}0\,\langle z:=-\frac{b}{2}t^2+vt+z\rangle z>m}\;{\scriptstyle\exists r}
}{
  \dfrac{v\geq 0, z<m \vdash \langle z'=v, v'=-b\rangle z>m}{\vdash v\geq 0\land z<m\to\langle z'=v, v'=-b\rangle z>m}\;{\scriptstyle\to r,\land l}
}\;{\scriptstyle\langle'\rangle}
$$

(The rule i∃ is applied to the two branches $v\geq 0, z<m \vdash T\geq 0$ and $v\geq 0, z<m \vdash -\frac{b}{2}T^2+vT+z>m$.)

Fig. 2.14 Deduction modulo for analysis of MA violation in braking mode

can be reintroduced (i∀, i∃) and quantifier elimination QE is applied to reduce the sequents equivalently to a simpler formula with less (distinct) symbols. In ∃r,∀l,i∃, the respective auxiliary symbols are free logical variables. In ∀r,∃l,i∀, Skolem function terms are used instead for reasons that are crucial for soundness and will be illustrated in the remainder of this section. In this context, we think of *free* logical variables as being introduced by γ-rules (∃r and ∀l), and hence implicitly existentially quantified.

To illustrate how quantifier and dynamic rules of $d\mathcal{L}$ interact to combine arithmetic with dynamic reasoning in hybrid systems, we analyse the braking behaviour in train control. The proof in Fig. 2.14 can be used to analyse whether a train can violate its MA although it is braking. That is, if the train position z can leave m ($z>m$) although it starts inside ($z<m$) and is braking will full braking force all the time:

$$v\geq 0\land z<m\to\langle z'=v, v'=-b\rangle z>m.$$

As the proof reveals, the answer depends on the initial velocity v. The proof starts with the conjecture at the bottom and applies propositional transformation rules →r,∧l to obtain a decomposed sequent form. Then it uses rule ⟨′⟩ to replace the differential equation with a quantified formula about its solution. For notational convenience, we use the simplified ⟨′⟩ rule from (2.12), as the differential equation is not restricted to an evolution domain. Now we have a quantified modal formula, $\exists t{\geq}0\,\langle z:=-\frac{b}{2}t^2+vt+z\rangle z>m$, which, unfortunately, cannot be handled by quantifier elimination in real-closed fields, because it is not first-order. Using rule ∃r, however, the proof can continue by introducing a new free variable T for the quantified variable t and postpone QE. After introducing T, the proof can continue by splitting a conjunction in the succedent into two branches (rule ∧r) and applying the assignment with a substitution on the right branch (rule ⟨:=⟩). Finally, the previously quantified free variable T only occurs in first-order formulas on all open goals. Then rule i∃ can be applied in Fig. 2.14 to merge all open proof goals mentioning T, reintroduce the quantifier for T, and apply quantifier elimination. The conjunction of the two goals can be handled by QE and simplification, yielding the resulting subgoal:

$$\text{QE}\left(\exists T\left((v \geq 0 \land z < m \to T \geq 0) \land (v \geq 0 \land z < m \to -\frac{b}{2}T^2 + vT + z > m)\right)\right)$$
$$\equiv v \geq 0 \land z < m \to v^2 > 2b(m-z).$$

After applying rules $\to$r,$\land$l for structural reasons again, the open branch with this formula reveals the speed limit and can be used to synthesise a corresponding parameter constraint. When $v^2 > 2b(m-z)$ holds initially, m can eventually be violated even in braking mode, as the velocity exceeds the braking force.

Similarly, the dual constraint $v^2 \leq 2b(m-z)$ guarantees that m can be respected by appropriate braking. The constraint so discovered thus forms a *controllability constraint* of ETCS, i.e., a constraint that characterises from which states control choices exist that guarantee safety. It is essentially equivalent to $[z'' = -b]z \leq m$ and $\exists a\,(-b \leq a \leq A \land [z'' = a]z \leq m)$. The resulting controllable region of the state space of ETCS is illustrated in Fig. 2.15.

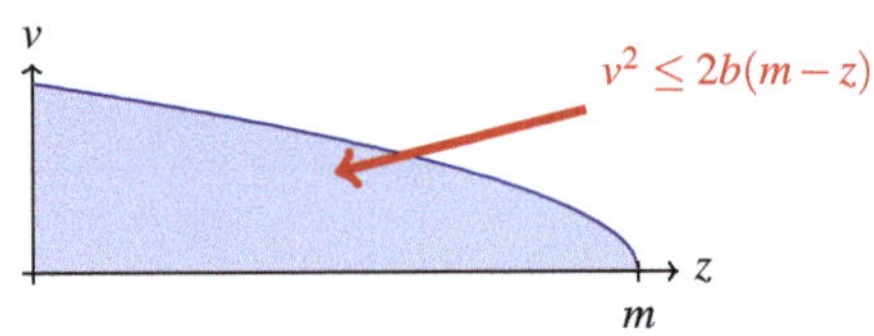

Fig. 2.15 Controllable region of ETCS dynamics

For comparison, the dual formula $v \geq 0 \land z < m \to [z' = v, v' = -b]\,z \leq m$ can be analysed as shown in Fig. 2.16 to study under which circumstances the MA is always respected ($[z'' = -b]z \leq m$) rather than under which it can fail ($\langle z'' = -b\rangle z > m$). The outcome again discovers the controllability constraint. The difference of the

$$\begin{array}{ll}
 & v \geq 0, z < m \vdash v^2 \leq 2b(m-z) \\
\hline
\to\text{r},\land\text{l} & \vdash v \geq 0 \land z < m \to v^2 \leq 2b(m-z) \\
\hline
\text{i}\forall & v \geq 0, z < m, s \geq 0 \vdash -\frac{b}{2}s^2 + vs + z \leq m \\
\hline
\langle:=\rangle & v \geq 0, z < m, s \geq 0 \vdash \langle z := -\frac{b}{2}s^2 + vs + z\rangle z \leq m \\
\hline
[:=] & v \geq 0, z < m, s \geq 0 \vdash [z := -\frac{b}{2}s^2 + vs + z]z \leq m \\
\hline
\to\text{r} & v \geq 0, z < m \vdash s \geq 0 \to [z := -\frac{b}{2}s^2 + vs + z]z \leq m \\
\hline
\forall\text{r} & v \geq 0, z < m \vdash \forall t{\geq}0\,[z := -\frac{b}{2}t^2 + vt + z]z \leq m \\
\hline
['] & v \geq 0, z < m \vdash [z' = v, v' = -b]z \leq m \\
\hline
\to\text{r},\land\text{l} & \vdash v \geq 0 \land z < m \to [z' = v, v' = -b]z \leq m
\end{array}$$

Fig. 2.16 Deduction modulo for analysis of MA-safety in braking mode

deduction in Fig. 2.16 compared to that in Fig. 2.14 is that we now use rule $[']$, which gives a universal quantifier for time t. With rule $\forall$r, the quantifier can be turned into a Skolem constant term s, which does not have any arguments, because no free logical variables occur. After applying the solution of the differential equation with

$[:=],\langle:=\rangle$, the resulting formula is first-order in the Skolem term s. Then rule i$\forall$ can be used to reintroduce a universal quantifier for the previously quantified variable, and to apply quantifier elimination:

$$\begin{aligned}&\mathrm{QE}\left(\forall s\,(v \geq 0 \wedge z < m \wedge s \geq 0 \rightarrow -\frac{b}{2}s^2 + vs + z \leq m)\right)\\ &\equiv v \geq 0 \wedge z < m \rightarrow v^2 \leq 2b(m-z).\end{aligned}$$

2.5.3.2 Admissibility in Invertible Quantifier Rules

The requirement that substitutions in i$\forall$ are admissible implies that no occurrence of $s(X_1,\dots,X_n)$ is within the scope of a quantifier for any of these X_i. Admissibility makes sense, because variables in $s(X_1,\dots,X_n)$ would otherwise be captured by quantifiers when substituting. The admissibility condition prevents i$\forall$ from rearranging the order of quantifiers from $\exists X_i \forall s$ to the weaker $\forall s\,\exists X_i$. Such a rearrangement would be unsound, because it is not sufficient to show the weak subgoal $\forall s\,\exists X_i$ (each s has an X_i) in order to prove the strong statement $\exists X_i \forall s$ saying that the same X_i works for all s. Because this is an important part of soundness, we illustrate in detail why unsound rearrangements are prevented.

$$\begin{array}{ll} & \text{i}\forall \text{ is not applicable}\\ \hline & \vdash \mathrm{QE}(\exists X\,(2X+1 < s(X)))\\ \hline \text{i}\exists & \vdash 2X+1 < s(X)\\ \hline \langle:=\rangle & \vdash \langle x:=2X+1\rangle(x < s(X))\\ \hline \forall\text{r} & \vdash \forall y\,\langle x:=2X+1\rangle(x<y)\\ \hline \exists\text{r} & \vdash \exists x \forall y\,\langle x:=2x+1\rangle(x<y)\end{array}$$

Fig. 2.17a Wrong rearrangement with deduction modulo by invertible quantifiers

$$\begin{array}{ll} & \overbrace{\vdash \mathrm{QE}\,(\exists X\,\mathrm{QE}(\forall s\,(2X+1 < s)))}^{\text{false}}\\ \hline \text{i}\exists & \vdash \mathrm{QE}(\forall s\,(2X+1 < s))\\ \hline \text{i}\forall & \vdash 2X+1 < s(X)\\ \hline \langle:=\rangle & \vdash \langle x:=2X+1\rangle(x < s(X))\\ \hline \forall\text{r} & \vdash \forall y\,\langle x:=2X+1\rangle(x<y)\\ \hline \exists\text{r} & \vdash \exists x \forall y\,\langle x:=2x+1\rangle(x<y)\end{array}$$

Fig. 2.17b Correct reintroduction order with deduction modulo by invertible quantifiers

For the moment, suppose the rules did not contain QE. The requirement for admissible substitutions (Definition 2.8) ensures that the proof attempt of an invalid formula in Fig. 2.17a cannot close in the d$\mathcal{L}$ calculus. At the indicated position at the top, i$\forall$, which would unsoundly invert the quantifier order to $\forall S\,\exists X$, cannot be applied: In i$\forall$, the substitution inserting $s(X)$ gives $\exists Y\,(2Y+1 < s(X))$ by bound variable renaming instead of $\exists X\,(2X+1 < s(X))$, because the substitution would not otherwise be admissible. Thus, i$\forall$ is not applicable, because the quantified formula is not of the form $\Psi(s(X))$.

Now, we consider what happens in the presence of QE. The purpose of QE is to (equivalently) remove quantifiers like $\exists X$. Thus it is no longer obvious that the admissibility argument applies, because the blocking variable X would have disappeared after successful quantifier elimination. However, quantifier elimination over the reals is defined in the first-order theory of real arithmetic [288, 81]. Yet, when eliminating X in Fig. 2.17a, the Skolem term $s(X)$ is no term of real arithmetic, as, unlike that of $+$, the interpretation of the Skolem function s is arbitrary. The truth-value of $\exists X\,(2X+1<s(X))$ depends on the interpretation of s. If $I(s)$ happens to be a constant function, the formula is true, if $I(s)(a)=2a$, however, it is false. In general, such cases cannot be distinguished without quantifiers, because two functions cannot be shown to be identical by evaluating them at finitely many points. Thus, in the presence of uninterpreted function terms, real arithmetic does not generally admit quantifier elimination. Consequently, i$\exists$ and i$\forall$ are only applicable if QE is defined. Yet, we show that QE can be lifted to formulas with Skolem functions, nevertheless, when these are instances of real arithmetic formulas:

Lemma 2.5 (Quantifier elimination lifting). *Quantifier elimination can be lifted to instances of formulas of first-order theories that admit quantifier elimination, i.e., to formulas that result from the base theory by substitution.*

Proof. Let formula ϕ be an instance of ψ, with ψ being a formula of the base theory, i.e., ϕ is $\psi_{z_1}^{\theta_1}\ldots{}_{z_n}^{\theta_n}$ for some variables z_i and arbitrary terms θ_i. As QE is defined for the base theory, let $\mathrm{QE}(\psi)$ be the quantifier-free formula belonging to ψ according to Definition 2.9. Then $\mathrm{QE}\,(\psi)_{z_1}^{\theta_1}\ldots{}_{z_n}^{\theta_n}$ satisfies the requirements of Definition 2.9 for ϕ, because $\vDash \psi_{z_1}^{\theta_1}\ldots{}_{z_n}^{\theta_n} \leftrightarrow \mathrm{QE}\,(\psi)_{z_1}^{\theta_1}\ldots{}_{z_n}^{\theta_n}$: For F defined as $\psi \leftrightarrow \mathrm{QE}(\psi)$, we have that $\vDash F$ implies $\vDash F_{z_1}^{\theta_1}\ldots{}_{z_n}^{\theta_n}$ by a standard consequence of the Substitution Lemma 2.2. And $\psi \leftrightarrow \mathrm{QE}(\psi)$ is indeed valid, by the properties of QE; see Definition 2.9. □

With this, consider again the example in Fig. 2.17a. By Lemma 2.5, QE is defined in the presence of Skolem terms that do not depend on quantified variables, e.g., for $\exists X\,(2X+1<t(Y,Z))$, which is an instance of the form $(\exists X\,(2X+1<z))_z^{t(Y,Z)}$. However, QE is *not defined* in the premise of i$\exists$ when Skolem dependencies on X occur. In Fig. 2.17a, $\exists X\,(2X+1<s(X))$ is no instance of first-order real arithmetic, because, by bound variable renaming $(\exists X\,(2X+1<z))_z^{s(X)}$ yields a different formula $\exists Y\,(2Y+1<s(X))$. An occurrence of $s(X)$, which corresponds to a quantifier nesting of $\exists X\,\forall s$, thus requires $s(X)$ to be eliminated by i$\forall$ before i$\exists$ can eliminate X; see Fig. 2.17b. Hence, inner universal quantifiers are enforced to be handled first and unsound quantifier rearrangements are prevented even in the presence of QE.

Finally, observe that i$\forall$ and i$\exists$ do not require quantifiers to be eliminated in the exact same order in which they occurred in the original formula. The elimination order within homogeneous quantifier blocks like $\forall x_1\,\forall x_2$ is not restricted as there are no Skolem dependencies among the corresponding auxiliary Skolem terms. Yet, eliminating such a quantifier block is sound in any order (accordingly for $\exists x_1\,\exists x_2$). Similarly, i$\exists$ and i$\forall$ could interchange the order of $\forall x\,\exists y$ to the stronger $\exists y\,\forall x$, because the resulting Skolem term s for x in the former formula does not depend on y. In this direction, however, the interchange is sound, as it amounts to proving a stronger statement. This quantifier rearrangement is not necessarily wise, because it requires proving a stronger formula, but it is at least sound.

2.5.3.3 Quantifier Elimination and Modalities

Quantifier elimination over the first-order theory of reals cannot handle modal formulas. Hence, the $d\mathcal{L}$ calculus first reduces modalities to first-order constraints before applying QE. Yet, this is not necessary for all modalities. The modal subformula in the following example does not impose any constraints on X, but its truth-value only determines which first-order constraints are imposed on X:

$$\mathrm{QE}(\exists X\,(X<0\wedge((\langle y:=2y+1\rangle y>0)\rightarrow X>y)))\equiv((\langle y:=2y+1\rangle y>0)\rightarrow y<0.$$

Modal formulas not containing elimination variable X can be handled by propositional abstraction in QE and remain unchanged. Syntactically, the reason for this is that $d\mathcal{L}$ rule applications on modal formulas that do not contain X will never produce formulas which do. The semantical reason for the same fact is a generalisation of the coincidence lemma to $d\mathcal{L}$, which says that values of variables that do not occur will neither affect the transition structure of a hybrid program nor the truth-value of formulas.

Lemma 2.6 (Coincidence lemma). *If the interpretations (and assignments and states, respectively) I,η,ν and J,ε,ω agree on all symbols that occur freely in the formula ϕ, then $val_{I,\eta}(\nu,\phi)=val_{J,\varepsilon}(\omega,\phi)$.*

Proof. The proof is by a simple structural induction using the definitions of valuation $val_{I,\eta}(\nu,\cdot)$ and $\rho_{I,\eta}(\cdot)$ in Definitions 2.5–2.7. □

2.5.3.4 Global Invertible Quantifier Rules

Rules i$\forall$ and i$\exists$ display an asymmetry. While i$\forall$ works locally on a branch, i$\exists$ needs to inspect all branches that contain X. The reason for this is that branches are implicitly combined conjunctively in sequent calculus, as all branches have to close simultaneously for a proof to succeed (Definition 2.11). Universal quantifiers can be handled separately for conjunctions by $\forall x\,(\phi\wedge\psi)\equiv\forall x\,\phi\wedge\forall x\,\psi$. Existential quantifiers, however, can only be dealt with separately for disjunctions but not for conjunctions: $\exists x\,(\phi\vee\psi)\equiv\exists x\,\phi\vee\exists x\,\psi$. In calculi with a disjunctive proof structure, the roles of i$\forall$ and i$\exists$ would be interchanged but the phenomenon remains.

Rule i$\exists$ can be applied to the full proof (i.e., all open goals) as a global closing substitution in the free-variable tableau calculus [122]; cf. App. A. By Lemma 2.6, however, rule i$\exists$ only needs to consider the set of all open goals $\Phi_i\vdash\Psi_i$ that actually contain X. Rule i$\exists$ resembles global closing substitutions in uninterpreted free-variable tableaux [134]. Both avoid the backtracking over closing substitutions that local closing substitutions require. Unlike closing substitutions, however, rule i$\exists$ uses the fixed semantics of function and predicate symbols of real arithmetic such that variables can be eliminated equivalently by QE before the proof completes. Applying i$\forall$ or i$\exists$ early does not necessarily close the proof. Instead, equivalent constraints on the remaining variables will be revealed, which can simplify the proof or help in deriving parametric constraints or invariants.

2.5.4 Verification Example

As a simple example to prove, recall the bouncing ball system (*ball*) from Example 2.5 on p. 45 and its d$\mathcal{L}$ specification from Example 2.7 on p. 48. Consider the intuitive property that the bouncing ball never bounces higher than initial height H when the precondition of property (2.3) holds initially:

$$(v^2 \leq 2g(H-h) \wedge h \geq 0 \wedge g > 0 \wedge H \geq 0 \wedge 1 > c \geq 0) \rightarrow [ball](0 \leq h \leq H). \quad (2.3^*)$$

The bouncing ball is very simple, but shows some interesting aspects of proofs. In order to simplify the proof notation, let us discard clock variable τ. Clock τ is not necessary for the property, and only used to ensure natural switching during the bounce to prevent the ball from bouncing multiple times while still on the ground (which would be *superdense* switching with multiple discrete switches at the exact same point in time).

For the proof, we define some abbreviations. Let ψ denote the general assumptions in the precondition about parameters that do not change during bouncing ball runs, and let ϕ denote the state-dependent part of the precondition, that is:

$$\psi \equiv g > 0 \wedge H \geq 0 \wedge 1 > c \geq 0,$$
$$\phi \equiv v^2 \leq 2g(H-h) \wedge h \geq 0.$$

The d$\mathcal{L}$ proof for the bouncing ball property (2.3) is shown in Fig. 2.18.

The proof starts with the property (2.3) at the bottom (goal). After normalising to sequent form with rules $\rightarrow$r,$\wedge$l, the proof follows an induction (using the rule *ind'* from p. 86) with invariant ϕ. Rule *ind'* produces two other proof subgoals that are not shown in Fig. 2.18: the proof goal that the precondition $\psi \wedge \phi$ implies the invariant ϕ (i.e., $\psi, \phi \vdash \phi$) and the proof goal that the invariant ϕ implies the postcondition, which gives $\psi, \phi \vdash \forall h \forall v (\phi \rightarrow 0{\leq}h{\leq}H)$. Both goals are trivial to prove by *ax* and $\forall$r,i$\forall$, respectively. The quantifiers $\forall h \forall v$ in the latter goal result from the universal closure $\forall^\alpha$ in rule *ind'*. Universal closures are not strictly necessary in this particular proof, because the premise only contains invariant ϕ and formula ψ about symbols that do not change during the hybrid program runs. Thus, the universal closure immediately disappears after applying $\forall$r. In general, however, universal closures in *ind* and the other global proof rules are critical for soundness; see Fig. 2.19a versus Fig. 2.19b.

After splitting the sequential composition by d$\mathcal{L}$ rule [;], the proof uses rule $[']$ with the solution $\langle h{:=}h+vt-\frac{g}{2}t^2, v{:=}v-gt\rangle$ of the differential equation system $h' = v, v' = -g$. We abbreviate this solution by $\langle \mathcal{S}_t \rangle$. Again, we use the simplified $[']$ rule from (2.12). QE cannot be applied to the result quantifier $\forall t{\geq}0 \langle \mathcal{S}_t \rangle \ldots$, because the quantified variable t occurs in modalities to which QE is not applicable. Thus the proof uses $\forall$r to introduce a Skolem function s for the previously quantified variable t. But unlike for the proof in Fig. 2.14, we do not directly apply the resulting solution $\langle \mathcal{S}_s \rangle$ by rule $\langle{:=}\rangle$. The reason is the different system structure of the bouncing ball. In the bouncing ball program, the differential equation comes

$$
\begin{array}{ll}
 & * \\
\text{i}\forall & \psi,\phi,s{\geq}0,h+vs-\tfrac{g}{2}s^2=0 \vdash (-c(v-gs))^2 \leq 2g(H-(h+vs-\tfrac{g}{2}s^2)) \land h+vs-\tfrac{g}{2}s^2 \geq 0 \\
\langle:=\rangle & \psi,\phi,s{\geq}0,\langle \mathscr{S}_s\rangle h=0 \vdash \langle \mathscr{S}_s\rangle\langle v:=-cv\rangle\phi \\
[:=] & \psi,\phi,s{\geq}0,\langle \mathscr{S}_s\rangle h=0 \vdash \langle \mathscr{S}_s\rangle[v:=-cv]\phi \\
\rightarrow\text{r} & \psi,\phi,s{\geq}0 \vdash \langle \mathscr{S}_s\rangle(h=0 \rightarrow [v:=-cv]\phi) \\
[?] & \psi,\phi,s{\geq}0 \vdash \langle \mathscr{S}_s\rangle[?h=0][v:=-cv]\phi \\
[;] & \psi,\phi,s{\geq}0 \vdash \langle \mathscr{S}_s\rangle[?h=0;v:=-cv]\phi
\end{array}
$$

$$
\begin{array}{ll}
 & * \\
\text{i}\forall & \psi,\phi,s{\geq}0,h+vs-\tfrac{g}{2}s^2>0 \vdash (v-gs)^2 \leq 2g(H-(h+vs-\tfrac{g}{2}s^2)) \land h+vs-\tfrac{g}{2}s^2 \geq 0 \\
\langle:=\rangle & \psi,\phi,s{\geq}0,\langle \mathscr{S}_s\rangle h>0 \vdash \langle \mathscr{S}_s\rangle\phi \\
\rightarrow\text{r} & \psi,\phi,s{\geq}0 \vdash \langle \mathscr{S}_s\rangle(h>0 \rightarrow \phi) \\
[?] & \psi,\phi,s{\geq}0 \vdash \langle \mathscr{S}_s\rangle[?h>0]\phi
\end{array}
$$

$$
\begin{array}{ll}
 & \dfrac{\dots}{\psi,\phi,s{\geq}0 \vdash \langle \mathscr{S}_s\rangle[?h>0]\phi} \qquad \dfrac{\dots}{\psi,\phi,s{\geq}0 \vdash \langle \mathscr{S}_s\rangle[?h=0;v:=-cv]\phi} \\
\land\text{r} & \psi,\phi,s{\geq}0 \vdash \langle \mathscr{S}_s\rangle([?h>0]\phi \land [?h=0;v:=-cv]\phi) \\
[\cup] & \psi,\phi,s{\geq}0 \vdash \langle \mathscr{S}_s\rangle[?h>0 \cup (?h=0;v:=-cv)]\phi \\
\rightarrow\text{r} & \psi,\phi \vdash s{\geq}0 \rightarrow \langle \mathscr{S}_s\rangle[?h>0 \cup (?h=0;v:=-cv)]\phi \\
\forall\text{r} & \psi,\phi \vdash \forall t{\geq}0\, \langle \mathscr{S}_t\rangle[?h>0 \cup (?h=0;v:=-cv)]\phi \\
['] & \psi,\phi \vdash [h''=-g][?h>0 \cup (?h=0;v:=-cv)]\phi \\
[;] & \psi,\phi \vdash [h''=-g;(?h>0 \cup (?h=0;v:=-cv))]\phi \\
\textit{ind}' & \psi,\phi \vdash [(h''=-g;(?h>0 \cup (?h=0;v:=-cv)))^*](0{\leq}h{\leq}H) \\
\rightarrow\text{r},\land\text{l} & \vdash \psi\land\phi \rightarrow [(h''=-g;(?h>0 \cup (?h=0;v:=-cv)))^*](0{\leq}h{\leq}H)
\end{array}
$$

Fig. 2.18 Bouncing ball proof (no evolution domain)

$$
\begin{array}{l}
\text{unsound} \\
x \leq 0, x \leq 1 \vdash x+1 \leq 1 \\
x \leq 0, x \leq 1 \vdash [x:=x+1]x \leq 1 \\
x \leq 0 \vdash x \leq 1 \rightarrow [x:=x+1]x \leq 1 \\
x \leq 0 \vdash [(x:=x+1)^*]x \leq 1
\end{array}
$$

Fig. 2.19a Unsound attempt of induction without universal closure $\forall^\alpha$

$$
\begin{array}{ll}
 & \text{not provable} \\
 & x \leq 0, y \leq 1 \vdash y+1 \leq 1 \\
[:=],\langle:=\rangle & x \leq 0, y \leq 1 \vdash [y:=y+1]y \leq 1 \\
\rightarrow\text{r} & x \leq 0 \vdash y \leq 1 \rightarrow [y:=y+1]y \leq 1 \\
\forall\text{r} & x \leq 0 \vdash \forall x\,(x \leq 1 \rightarrow [x:=x+1]x \leq 1) \\
\textit{ind} & x \leq 0 \vdash [(x:=x+1)^*]x \leq 1
\end{array}
$$

Fig. 2.19b Correct use of induction with universal closure $\forall^\alpha$, i.e., $\forall x$

first, and the discrete control equations are executed after that. Thus we keep the discrete jump set for the solution $\langle \mathcal{S}_s \rangle$ as an unmodified discrete jump prefix ($\langle \mathcal{J} \rangle$ in Definition 2.10) for the following rule applications and only apply the assignment with rule $\langle := \rangle$ to first-order formulas at the end of the proof.

On a side note: we could, in fact, just as well have used rule $\langle := \rangle$ here right away and substituted v, h inside the hybrid programs immediately, because there are no remaining loops or differential equations. That would clutter the notation, though, and we want to illustrate how discrete jump sets can be used as unmodified proof rule prefixes in Fig. 2.18.

Leaving prefix $\langle \mathcal{S}_s \rangle$ unchanged, the proof in Fig. 2.18 continues by splitting the choice between $h > 0$ and $h = 0$ with rule $[\cup]$ into two conjuncts, which split into two branches by rule $\wedge$r. On both branches, which are continued as indicated by the arrows, the test statements are turned into implications by rule $[?]$ and, ultimately, the accumulated discrete jump sets are applied (with rules $[:=], \langle := \rangle$) when the remaining formulas are simple. The last step of the proof is to reintroduce quantifiers for Skolem term s by rule i$\forall$ and apply quantifier elimination to the resulting first-order formulas on the left and right branches respectively:

$$\begin{aligned}\mathrm{QE}\Big(\forall s\,\big(\psi \wedge \phi \wedge s{\geq}0 \wedge h + vs - \frac{g}{2}s^2 > 0 \\ \to (v - gs)^2 \leq 2g(H - (h + vs - \frac{g}{2}s^2)) \wedge h + vs - \frac{g}{2}s^2 \geq 0\big)\Big) \equiv \mathit{true};\end{aligned}$$

$$\begin{aligned}\mathrm{QE}\Big(\forall s\,\big(\psi \wedge \phi \wedge s{\geq}0 \wedge h + vs - \frac{g}{2}s^2 = 0 \\ \to (-c(v - gs))^2 \leq 2g(H - (h + vs - \frac{g}{2}s^2)) \wedge h + vs - \frac{g}{2}s^2 \geq 0\big)\Big) \equiv \mathit{true}.\end{aligned}$$

In the proof of Fig. 2.18, we have used a bouncing ball without an evolution domain restriction. The bouncing ball property can also be proven with its evolution domain restricted to $h \geq 0$ on the differential equation system as in Fig. 2.2; see Fig. 2.20 for a proof. The proof is slightly more involved compared to Fig. 2.18, because of the extra constraints from the non-simplified rule $[']$. This time, for a change, we simply choose the full precondition as invariant, although the part marked in grey is still unaffected by the dynamics:

$$\phi \equiv v^2 \leq 2g(H - h) \wedge h \geq 0 \wedge g > 0 \wedge H \geq 0 \wedge 1 > c \geq 0.$$

Note, in particular, that there are many invariants that can be used to prove the same property. We just need to find one invariant that works.

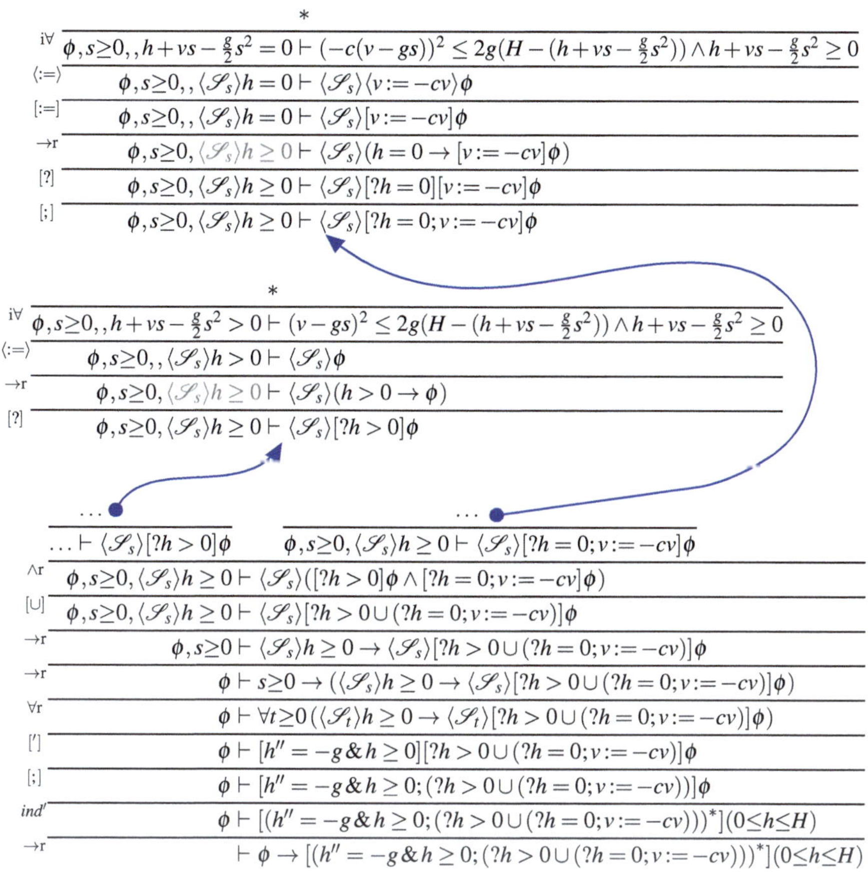

Fig. 2.20 Bouncing ball proof (with evolution domain)

2.6 Soundness

In this section, we prove that the d$\mathcal{L}$ calculus is a sound axiomatisation of the transition behaviour of hybrid systems. Whatever we can prove in the d$\mathcal{L}$ calculus is actually true.

The proof calculus for d$\mathcal{L}$ in Fig. 2.11 needs to fit to the semantics of differential dynamic logic from Sect. 2.3; otherwise, the proof rules would not be meaningful. Fortunately, every differential dynamic logic formula that can be derived in the d$\mathcal{L}$ calculus from Fig. 2.11 really is a valid formula! This property of the calculus is called *soundness* and is crucial, because it would be disastrous if a formula would be called "proven" when it is actually not valid, since we could not trust our proofs then. A calculus is sound iff every formula that can be derived in the calculus is also valid according to the semantics.

We prove that a successful deduction in the d$\mathscr{L}$ calculus always produces correct verification results about hybrid systems: The d$\mathscr{L}$ calculus is *sound*, i.e., all provable (closed) formulas are valid in all states of all interpretations. We can restrict our attention to closed formulas, i.e., formulas without free variables to begin with, because we can start with the universal closure of the formula for validity just as well. To reflect the interaction of free variables and Skolem terms, we adapt the notion of soundness for the liberalised δ^+-rule in free-variable tableau calculi [147] to sequent calculus.

A formula ϕ *has a model* [147] if there is an interpretation I and a state ν such that *for all* variable assignments η we have $I,\eta,\nu \models \phi$. Closed tableaux prove the unsatisfiability of the negated goal [147]. Sequent calculi work dually and show validity of the original proof goal. Consequently, we use the dual notion and say that formula ψ is a *consequence* of ϕ iff, for every I,ν *there is* an assignment η such that $I,\eta,\nu \models \psi$ provided that, for every I,ν, *there is* an assignment η such that $I,\eta,\nu \models \phi$. A proof rule that concludes Ψ from the premises Φ is *sound* if Ψ is, indeed, a consequence of Φ in the sense just defined. As usual, multiple branches in Ψ or Φ are combined conjunctively.

In this context, we think of free logical variables as being introduced by γ-rules, i.e., $\exists$r and $\forall$l (hence the implicit existential quantification of free logical variables by η). For closed formulas (without free logical variables), validity corresponds to being a consequence from an empty set of open goals. Hence, closed formulas that are provable with a sound deduction are *valid* (true in all states of all interpretations).

Theorem 2.1 (Soundness of d$\mathscr{L}$). *The* d$\mathscr{L}$ *calculus is sound.*

Proof. The calculus is sound if each rule instance is sound. All rules of the d$\mathscr{L}$ calculus except $\forall$r,$\exists$l and i$\exists$ are also *locally sound*, i.e., their conclusion is true at I,η,ν if all its premises are true in I,η,ν, which implies soundness. It is also easy to show that locally sound rules remain sound when adding contexts $\Gamma,\Delta,\langle\mathscr{J}\rangle$ as in Definition 2.10, since a discrete jump set $\langle\mathscr{J}\rangle$ characterises a unique state transition. Local soundness proofs of $\langle;\rangle$,$[;]$,$\langle\cup\rangle$,$[\cup]$,$\langle^{*n}\rangle$,$[^{*n}]$,$\langle?\rangle$,$[?]$ and propositional rules are as usual. Note that, for symmetric rules, local soundness implies that the premise and conclusion are *equivalent*, i.e., true in the same states. For an illustration of the dynamics behind the dynamic proof rules, we recall Fig. 2.12 from p. 83.

$\langle:=\rangle$ The rule $\langle:=\rangle$ is locally sound. Assume that the premise holds in I,η,ν, i.e., $I,\eta,\nu \models \phi_{x_1}^{\theta_1}\cdots{}_{x_n}^{\theta_n}$. We have to show that $I,\eta,\nu \models \langle x_1:=\theta_1,..,x_n:=\theta_n\rangle\phi$, i.e., $I,\eta,\omega \models \phi$ for a state ω with $(\nu,\omega)\in\rho_{I,\eta}(x_1:=\theta_1,..,x_n:=\theta_n)$. This follows directly from the Substitution Lemma 2.2 for admissible substitutions (Definition 2.8). The proof for rule $[:=]$ uses the fact that discrete jumps are deterministic.

$\langle;\rangle$ Rule $\langle;\rangle$ is locally sound. Assume that the premise holds in I,η,ν, i.e., $I,\eta,\nu \models \langle\alpha\rangle\langle\beta\rangle\phi$. We have to show that the conclusion holds in I,η,ν, i.e., $I,\eta,\nu \models \langle\alpha;\beta\rangle\phi$. By premise, $I,\eta,\nu \models \langle\alpha\rangle\langle\beta\rangle\phi$, we know that there is a state μ such that $(\nu,\mu)\in\rho_{I,\eta}(\alpha)$ and $I,\eta,\mu \models \langle\beta\rangle\phi$. Hence, there is a state ω such that $(\mu,\omega)\in\rho_{I,\eta}(\beta)$ and $I,\eta,\omega \models \phi$. Now, by the semantics of $\alpha;\beta$ (Definition 2.7), there is a transition from ν to ω (via intermediate state μ) along $\alpha;\beta$. Thus, $(\nu,\omega) \in \rho_{I,\eta}(\alpha;\beta)$ and $I,\eta,\omega \models \phi$, which implies $I,\eta,\nu \models \langle\alpha;\beta\rangle\phi$. The converse direction can be proven similarly to show equivalence and the local soundness of the dual rule [;].

$\langle\cup\rangle$ Rule $\langle\cup\rangle$ is locally sound. Assume that the premise holds in I,η,ν, i.e., $I,\eta,\nu \models \langle\alpha\rangle\phi \vee \langle\beta\rangle\phi$. We have to show that the conclusion holds in I,η,ν, i.e., $I,\eta,\nu \models \langle\alpha\cup\beta\rangle\phi$. If the disjunction in the premise is true, then one of its disjuncts must hold in I,η,ν. Consider the case where $I,\eta,\nu \models \langle\alpha\rangle\phi$. Then there is a state ω such that $(\nu,\omega) \in \rho_{I,\eta}(\alpha)$ and $I,\eta,\omega \models \phi$. By the semantics of $\alpha\cup\beta$ in Definition 2.7, every transition of α is a transition of $\alpha\cup\beta$. Hence $(\nu,\omega) \in \rho_{I,\eta}(\alpha\cup\beta)$ and $I,\eta,\omega \models \phi$, which imply $I,\eta,\nu \models \langle\alpha\cup\beta\rangle\phi$. If, instead, the second disjunct $I,\eta,\nu \models \langle\beta\rangle\phi$ holds, then the proof is similar. Either way, we have $I,\eta,\nu \models \langle\alpha\cup\beta\rangle\phi$. The converse direction can be proven accordingly to show equivalence and the local soundness of the dual rule $[\cup]$.

$\langle{}^{*n}\rangle$ Rule $\langle{}^{*n}\rangle$ is locally sound. Assume that the premise holds in I,η,ν, i.e., assume $I,\eta,\nu \models \phi \vee \langle\alpha\rangle\langle\alpha^*\rangle\phi$. We have to show that the conclusion holds in I,η,ν, i.e., $I,\eta,\nu \models \langle\alpha^*\rangle\phi$. The disjunction in the premise holds; hence, one of the disjuncts holds. Consider the case where $I,\eta,\nu \models \phi$; then $I,\eta,\nu \models \langle\alpha^*\rangle\phi$ already holds with zero repetitions α^* for ϕ is true in the beginning. Consider the case where $I,\eta,\nu \models \langle\alpha\rangle\langle\alpha^*\rangle\phi$. Thus, there is an α-transition to a state μ such that $(\nu,\mu) \in \rho_{I,\eta}(\alpha)$ with $I,\eta,\mu \models \langle\alpha^*\rangle\phi$. Consequently, there is an α^*-transition to a state ω with $(\mu,\omega) \in \rho_{I,\eta}(\alpha^*)$ and $I,\eta,\omega \models \phi$. Obviously, every α-transition also is an α^*-transition, because repetitions may choose to repeat only once. In particular, by chaining the α-transition $(\nu,\mu) \in \rho_{I,\eta}(\alpha) \subset \rho_{I,\eta}(\alpha^*)$ with the α^*-transition $(\mu,\omega) \in \rho_{I,\eta}(\alpha^*)$, we obtain a longer α^*-transition $(\nu,\omega) \in \rho_{I,\eta}(\alpha^*)$ by the transition semantics in Definition 2.7. Hence, in either case, we conclude $I,\eta,\nu \models \langle\alpha^*\rangle\phi$. The converse direction can be proven accordingly to show equivalence and the local soundness of the dual rule $[{}^{*n}]$.

$\langle?\rangle$ Rule $\langle?\rangle$ is locally sound. Assume that the premise holds in I,η,ν, i.e., $I,\eta,\nu \models \chi\wedge\phi$. We have to show that the conclusion holds in I,η,ν, i.e., $I,\eta,\nu \models \langle?\chi\rangle\phi$. We have to show that there is a transition along $?\chi$ to a state where ϕ holds. By the semantics in Definition 2.7, there is only a transition along hybrid program $?\chi$ if $I,\eta,\nu \models \chi$ and the state is not changed by $?\chi$ transitions. Now the premise implies $I,\eta,\nu \models \chi$ and $I,\eta,\nu \models \phi$, which, together, imply $I,\eta,\nu \models \langle?\chi\rangle\phi$. Since this is the only case where $?\chi$ can make a transition to a state satisfying ϕ, it shows equivalence. Local soundness of the dual rule [?] follows from this.

$\langle'\rangle$ The rule $\langle'\rangle$ is locally sound. Let $y_1,\dots,y_n$ be a solution for the differential equation system $x_1' = \theta_1,\dots,x_n' = \theta_n$ with symbolic initial values $x_1,\dots,x_n$. Let further $\langle\mathscr{S}_t\rangle$ be the jump set $\langle x_1 := y_1(t),\dots,x_n := y_n(t)\rangle$. Assume I,η,ν are such that the premise is true: $I,\eta,\nu \models \exists t{\geq}0\,(\bar{\chi}\wedge\langle\mathscr{S}_t\rangle\phi)$ with $\forall 0{\leq}\tilde{t}{\leq}t\,\langle\mathscr{S}_{\tilde{t}}\rangle\chi$ abbreviated as $\bar{\chi}$. For any $\zeta\in\mathbb{R}$, we denote by η^ζ the assignment that agrees with η except that it assigns ζ to t. Then, by assumption, there is a real value $r \geq 0$ such that $I, \eta^r, \nu \models \bar{\chi} \wedge \langle \mathcal{S}_t \rangle \phi$. Abbreviate $x_1' = \theta_1, .., x_n' = \theta_n \,\&\, \chi$ by $\mathcal{D}$. We have to show that $I, \eta, \nu \models \langle \mathcal{D} \rangle \phi$. Equivalently, by Lemma 2.6, we show $I, \eta^r, \nu \models \langle \mathcal{D} \rangle \phi$, because t is a fresh variable that does not occur in $\mathcal{D}$ or ϕ. Let function $f: [0,r] \to \mathrm{Sta}(\Sigma)$ be defined such that $(\nu, f(\zeta)) \in \rho_{I,\eta^\zeta}(\mathcal{S}_t)$ for all $\zeta \in [0,r]$. By premise, $f(0)$ is identical to ν and ϕ holds at $f(r)$. Thus it only remains to be shown that f respects the constraints of Definition 2.7 for $\mathcal{D}$. In fact, f obeys the continuity and differentiability properties of Definition 2.7 by the corresponding properties of the y_i. Moreover, $val_{I,\eta^r}(f(\zeta), x_i) = val_{I,\eta^r}(\nu, y_i(t))$ has a derivative of value $val_{I,\eta^r}(f(\zeta), \theta_i)$, because y_i is a solution of the differential equation $x_i' = \theta_i$ with corresponding initial value $\nu(x_i)$. Further, it can be shown that the evolution domain χ is respected along f as follows: By premise, $I, \eta^r, \nu \models \bar{\chi}$ holds for the initial state ν; thus $val_{I,\eta^r}(f(\zeta), \chi) = \mathit{true}$ for all $\zeta \in [0,r]$. Combining these results, we can conclude that f is a witness for $I, \eta, \nu \models \langle \mathcal{D} \rangle \phi$. The converse direction can be shown accordingly to prove the dual rule $[']$ using Lemma 2.1.

∀r The proof is a sequent calculus adaptation of that in [147]. By contraposition, assume that there are I, ν such that for all η it is the case that $I, \eta, \nu \not\models \forall x \, \phi(x)$; hence $I, \eta, \nu \models \exists x \neg \phi(x)$. We construct an interpretation I' that agrees with I except for the new function symbol s. Let $b_1, \ldots, b_n \in \mathbb{R}$ be arbitrary elements and let η^b assign b_i to the respective X_i for $1 \leq i \leq n$. As $I, \eta, \nu \models \exists x \neg \phi(x)$ holds for all η, we pick a witness d for $I, \eta^b, \nu \models \exists x \neg \phi(x)$ and choose $I'(s)(b_1, \ldots, b_n) = d$. For this interpretation I' and state ν we have $I', \eta, \nu \not\models \phi(s(X_1, \ldots, X_n))$ for all assignments η by Lemma 2.6, as $X_1, .., X_n$ are all free variables determining the truth-value of $\phi(s(X_1, \ldots, X_n))$. To see that the contexts Γ, Δ of Definition 2.10 can be added to instantiate this rule, consider the following. Since s is new and does not occur in the context Γ, Δ, the latter do not change their truth-value by passing from I to I'. Likewise, s is rigid so that it does not change its value by adding jump prefix $\langle \mathcal{J} \rangle$ which concludes the proof. The proof of ∃l is dual.

i∀ i∀ is locally sound. Assume that $I, \eta, \nu \models \mathrm{QE}(\forall X \, (\Phi(X) \vdash \Psi(X)))$. Since QE yields an equivalence, we can conclude $I, \eta, \nu \models \forall X \, (\Phi(X) \vdash \Psi(X))$. Then if the antecedent of the conclusion is true, $I, \eta, \nu \models \Phi(s(X_1, \ldots, X_n))$, we conclude $I, \eta, \nu \models \Psi(s(X_1, \ldots, X_n))$ by choosing $val_{I,\eta}(\nu, s(X_1, \ldots, X_n))$ for X in the premise. By admissibility of substitutions, variables $X_1, \ldots, X_n$ are free at all occurrences of $s(X_1, \ldots, X_n)$, and hence their value is the same in all occurrences.

∃r ∃r is locally sound by a simplified version of the proof in [147]. For any I, η, ν with $I, \eta, \nu \models \phi(X)$ we can conclude $I, \eta, \nu \models \exists x \, \phi(x)$ according to the witness $\eta(X)$. The proof of ∀l is dual.

i∃ For any I, ν let η be such that $I, \eta, \nu \models \mathrm{QE}(\exists X \bigwedge_i (\Phi_i \vdash \Psi_i))$. Again, this implies $I, \eta, \nu \models \exists X \bigwedge_i (\Phi_i \vdash \Psi_i)$, because quantifier elimination yields an equivalence. We pick a witness $d \in \mathbb{R}$ for this existential quantifier. As X does not occur anywhere else in the proof, it disappears from all open premises of the proof by applying i$\exists$. Hence, by the Coincidence Lemma 2.6, the value of X does not change the truth-value of the premise of i$\exists$. Consequently, η can be extended to η' by changing the interpretation of X to the witness d such that $I,\eta',\nu \models \bigwedge_i(\Phi_i \vdash \Psi_i)$. Thus, η' extends I,η,ν to a simultaneous model of all conclusions.

$\langle\rangle gen$ Rules $[]gen$–con are locally sound by a variation of the usual proofs [149] using universal closures for local soundness. $[]gen$, $\langle\rangle gen$ are simple refinements of Lemma 2.6 using the fact that the universal closure $\forall^\alpha$ comprises all variables that change in α. Let $I,\eta,\nu \models \langle\alpha\rangle\phi$, i.e., let $(\nu,\nu') \in \rho_{I,\eta}(\alpha)$ with $I,\eta,\nu' \models \phi$. As α can only change its bound variables, which are quantified universally in the universal closure $\forall^\alpha$, the premise implies $I,\eta,\nu' \models \phi \to \psi$; thus $I,\eta,\nu' \models \psi$ and $I,\eta,\nu \models \langle\alpha\rangle\psi$. The proof of $[]gen$ is similar.

ind For any I,η,ν with $I,\eta,\nu \models \forall^\alpha(\phi \to [\alpha]\phi)$, we know $I,\eta,\nu' \models \phi \to [\alpha]\phi$ for all ν' with $(\nu,\nu') \in \rho_{I,\eta}(\alpha)$. As these share the same η, we can further conclude $I,\eta,\nu \models \phi \to [\alpha^*]\phi$ by induction along the series of states ν' reached from ν by repeating α. The universal closure is necessary as, otherwise, the premise may yield different η in different states ν'.

con Assume that the antecedent and premise hold in I,η,ν. By premise, we have $I,\eta[v \mapsto d],\nu' \models v > 0 \land \varphi(v) \to \langle\alpha\rangle\varphi(v-1)$ for all $d \in \mathbb{R}$ and all states ν' that are reachable by α^* from ν, because $\forall^\alpha$ comprises all variables that are bound by α, which are the same as those bound by α^*. By antecedent, there is a $d \in \mathbb{R}$ such that $I,\eta[v \mapsto d],\nu \models \varphi(v)$. Now, the proof is a well-founded induction on d. If $d \le 0$, we directly have $I,\eta,\nu \models \langle\alpha^*\rangle\exists v{\le}0\,\varphi(v)$ for zero repetitions. Otherwise, if $d > 0$, we have, by premise, that

$$I,\eta[v \mapsto d],\nu \models v > 0 \land \varphi(v) \to \langle\alpha\rangle\varphi(v-1).$$

As $v > 0 \land \varphi(v)$ holds true at $I,\eta[v \mapsto d],\nu$, we have for some ν' with $(\nu,\nu') \in \rho_{I,\eta[v\mapsto d]}(\alpha)$ that $I,\eta[v \mapsto d],\nu' \models \varphi(v-1)$. In particular, we can conclude that $I,\eta[v \mapsto d-1],\nu' \models \varphi(v)$ satisfies the induction hypothesis for a smaller d and a reachable ν', because $(\nu,\nu') \in \rho_{I,\eta}(\alpha)$ as v does not occur in α. The induction is well-founded, because d decreases by 1 up to the base case $d \le 0$. □

With this soundness theorem, we now know that everything we prove in the d$\mathcal{L}$ calculus accurately reflects reality, because the syntactic proofs built with Fig. 2.11 fit to the semantics defined in Sect. 2.3.

2.7 Completeness

In this section, we prove that the d$\mathcal{L}$ calculus is a sound and complete axiomatisation of the transition behaviour of hybrid systems relative to differential equations.

With Soundness Theorem 2.1, we have shown that all provable formulas are valid. So we know that will never prove something that does not even hold (is not valid). The converse question is whether all valid formulas are also provable, i.e., whether we will always be able to prove all formulas that are "true" (valid). Have we just been lucky with the successful proofs that we managed to show so far? Or is there a deeper reason for which we can know that, in principle, we could also find proofs for all other valid formulas?

2.7.1 Incompleteness

Theorem 2.1 shows that all provable closed $d\mathcal{L}$ formulas are valid. The converse question is whether the $d\mathcal{L}$ calculus is *complete*, i.e., all valid $d\mathcal{L}$ formulas are provable. Combining completeness for first-order logic [147] and decidability of real arithmetic [81], it is easy to see that our calculus is complete for closed formulas of first-order real arithmetic by chaining the quantifier rules ∀r,∃l,∃r,∀l with the respective inverse rules i∀,i∃, using propositional rules as needed to unfold the propositional structure. In the presence of modalities, however, $d\mathcal{L}$ is not axiomatisable and, unlike its basis of first-order *real* arithmetic, $d\mathcal{L}$ is undecidable. Both unbounded repetition in the discrete fragment and unbounded evolution in the continuous fragment cause incompleteness. Beyond hybrid dynamics, where reachability is known to be undecidable [156], we show that even the purely discrete and purely continuous parts of $d\mathcal{L}$ are not effectively axiomatisable. Hence, valid $d\mathcal{L}$ formulas are not always provable.

Theorem 2.2 (Incompleteness of $d\mathcal{L}$). *Both the discrete fragment and the continuous fragment of* $d\mathcal{L}$ *are* not effectively axiomatisable, *i.e., they have no sound and complete effective calculus, because natural numbers are definable in both fragments.*

Proof. We prove that natural numbers are definable among the real numbers of $d\mathcal{L}$ interpretations in both fragments. Then these fragments extend first-order *integer* arithmetic such that the incompleteness theorem of Gödel [137] applies. Gödel's incompleteness theorem shows that no logic extending first-order integer arithmetic can have a sound and complete effective calculus. Natural numbers are definable in the discrete fragment without continuous evolutions using repetitive additions:

$$nat(n) \leftrightarrow \langle x:=0; (x:=x+1)^*\rangle\, x=n.$$

In the continuous fragment, an isomorphic copy of the natural numbers is definable using linear differential equations:

$$nat(n) \leftrightarrow \exists s\, \exists c\, \exists \tau\, (s=0 \land c=1 \land \tau=0 \land \langle s'=c, c'=-s, \tau'=1\rangle (s=0 \land \tau=n)).$$

These differential equations characterise sin and cos as unique solutions for s and c,

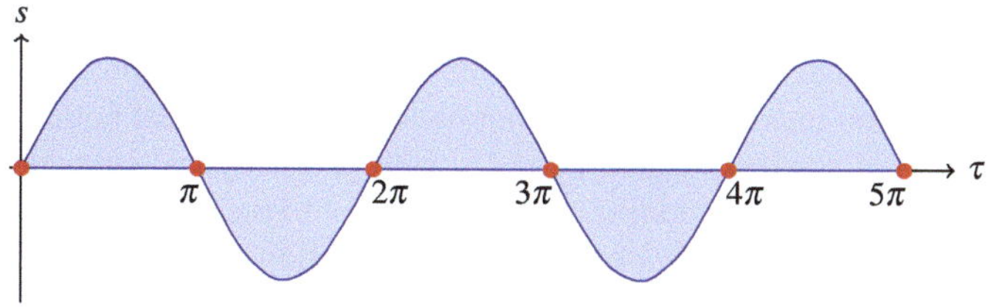

Fig. 2.21 Characterisation of ℕ as zeros of solutions of differential equations

respectively. Their zeros, as detected by τ, correspond to an isomorphic copy of natural numbers, scaled by π, i.e., $nat(n)$ holds iff n is of the form $k\pi$ for a $k \in \mathbb{N}$; see Fig. 2.21. The initial values for s and c prevent the trivial solution identical to 0. □

In this context, note that hybrid programs contain a computationally complete sublanguage and that reachability of hybrid systems is undecidable [156].

2.7.2 Relative Completeness

The standard approach for showing adequacy of a calculus when its logic is not effectively axiomatisable is to analyse the deductive power of the calculus relative to a base logic or to an ineffective oracle rule for the base logic [87, 148, 149]. In calculi for discrete programs, completeness is proven relative to the handling of data [87, 148, 149]. For hybrid systems, this is inadequate: By Theorem 2.2, no sound calculus for $d\mathcal{L}$ can be complete relative to its data (the reals), because its basis, first-order real arithmetic, is a perfectly decidable and axiomatisable theory [288]. If the $d\mathcal{L}$ calculus itself would be complete relative to the data of first-order real arithmetic, then, since this is a decidable logic, the $d\mathcal{L}$ calculus would be complete altogether, which would contradict Theorem 2.2. Thus, we need a different basis for a relative completeness argument. Unlike in classical discrete programs, the data is not where the complexity comes from. In hybrid dynamical systems, the complexity truly originates from the actual dynamics.

According to Theorem 2.2, both continuous evolutions and repetitive discrete transitions, as well as their interaction, cause non-axiomatisability of $d\mathcal{L}$. Discrete transitions and repetition do not supersede the complexity of continuous transitions. Even relative to an oracle for handling properties of discrete jumps and repetition, the $d\mathcal{L}$ calculus is not complete, simply because not all differential equations have solutions that are definable in first-order arithmetic so that rule [′] can be used. For instance, the solutions of $s' = c, c' = -s$ are trigonometric functions (like sin and cos), which are not first-order definable. The question is whether the converse is true, i.e., whether hybrid programs can be verified given that all required differential equations can be handled.

To calibrate the deductive power of the $d\mathcal{L}$ calculus in light of its inherent incompleteness, we analyse the quotient of reasoning about hybrid systems modulo differential equation handling. Using generalisations of the usual notions of relative completeness for discrete systems [87, 148, 149] to the hybrid case, we show that the $d\mathcal{L}$ calculus completely axiomatises $d\mathcal{L}$ relative to one single additional axiom about valid first-order properties of differential equations. Essentially, we drop the effectiveness requirement for one oracle axiom and show that the resulting $d\mathcal{L}$ calculus is sound and complete. We thus show that the $d\mathcal{L}$ calculus would be complete if only we had a complete replacement for $[']$,$\langle'\rangle$. Although repetitions and interactions of hybrid programs are more involved than purely continuous systems, this results emphasises the importance of studying approximations of this continuous oracle for the analysis of hybrid systems, as we do in Chap. 3.

As a basis, we define FOD as the *first-order logic of differential equations*, i.e., first-order real arithmetic augmented with formulas expressing properties of differential equations, that is, $d\mathcal{L}$ formulas of the form $[x_1' = \theta_1, \dots, x_n' = \theta_n]F$ with a first-order formula F. Dually, the diamond formula $\langle x_1' = \theta_1, \dots, x_n' = \theta_n\rangle F$ is expressible as $\neg[x_1' = \theta_1, \dots, x_n' = \theta_n]\neg F$.

Theorem 2.3 (Relative completeness of $d\mathcal{L}$). *The $d\mathcal{L}$ calculus is* complete relative to *FOD, i.e., every valid $d\mathcal{L}$ formula can be derived from FOD tautologies.*

Proof (Outline). The (constructive) proof, which, in full, is contained in the remainder of this section, adapts the techniques of Cook [87] and Harel [148, 149] to the hybrid case. The decisive step is to show that every valid property of a repetition α^* can be proven by rules *ind* or *con*, respectively, with a sufficiently strong invariant or variant that is expressible in $d\mathcal{L}$. For this, we show that $d\mathcal{L}$ formulas can be expressed equivalently in FOD, and that valid $d\mathcal{L}$ formulas can be derived from corresponding FOD axioms in the $d\mathcal{L}$ calculus. In turn, the crucial step is to construct a finite FOD formula that characterises the effect of unboundedly many repetitive hybrid transitions and just uses finitely many real variables. □

This main result completely aligns hybrid and continuous verification proof-theoretically. It gives a formal justification that reasoning about hybrid systems is possible to *exactly* the same extent to which it is possible to show properties of solutions of differential equations. Theorem 2.3 shows that superpositions or combinations of discrete jumps, continuous evolutions, and repetitions of hybrid processes can be verified whenever corresponding (intermediate) properties of differential equations are provable. Moreover, in a proof-theoretical sense, our calculus completely lifts all verification techniques for dynamical systems to hybrid systems perfectly. Summarising Theorems 2.1 and 2.3:

The $d\mathcal{L}$ calculus axiomatises the transition behaviour of hybrid systems completely relative to the handling of differential equations!

In the following subsections, we present a fully constructive proof of Theorem 2.3, which generalises the techniques of Harel [148, 149] and Cook [87] to the hybrid case. It shows that for every valid d$\mathcal{L}$ formula, there is a finite set of valid FOD formulas from which it can be derived in the d$\mathcal{L}$ calculus. Recall the proof outline of Theorem 2.3 for a road map of the proof.

Natural numbers are definable in FOD by Theorem 2.2. In this section, we abbreviate quantifiers over natural numbers, e.g., $\forall x\,(nat(x) \rightarrow \phi)$ by $\forall x\!:\!\mathbb{N}\ \phi$ and $\exists x\,(nat(x) \wedge \phi)$ by $\exists x\!:\!\mathbb{N}\ \phi$. Likewise, we abbreviate quantifiers over integers, e.g., $\forall x\,((nat(x) \vee nat(-x)) \rightarrow \phi)$ by $\forall x\!:\!\mathbb{Z}\ \phi$.

2.7.3 Characterising Real Gödel Encodings

As the central device for constructing a FOD formula that captures the effect of unboundedly many repetitive hybrid transitions and just uses finitely many real variables, we prove that a real version of Gödel encoding is definable in FOD. That is, we give a FOD formula that reversibly packs finite sequences of real values into a single real number.

Observe that a single differential equation system is *not* sufficient for defining these pairing functions as their solutions are differentiable, and yet, as a consequence of Morayne's theorem [213], there is no differentiable surjection $\mathbb{R} \rightarrow \mathbb{R}^2$, nor to any part of $\mathbb{R}^2$ of positive measure. We show that real sequences can be encoded nevertheless by chaining the effects of solutions of multiple differential equations and quantifiers.

Lemma 2.7 ($\mathbb{R}$-Gödel encoding). *The formula $at(Z,n,j,z)$, which holds iff Z is a real number that represents a Gödel encoding of a sequence of n real numbers with real value z at position j (for $1 \leq j \leq m$), is definable in FOD. For a formula $\phi(z)$ we abbreviate $\exists z\,(\mathrm{at}(Z,n,j,z) \wedge \phi(z))$ by $\phi(Z_j^{(n)})$.*

$$\sum_{i=0}^{\infty} \frac{a_i}{2^i} = a_0.a_1a_2\ldots \qquad \sum_{i=0}^{\infty} \frac{b_i}{2^i} = b_0.b_1b_2\ldots$$

$$\longrightarrow \quad \sum_{i=0}^{\infty}\left(\frac{a_i}{2^{2i-1}} + \frac{b_i}{2^{2i}}\right) = a_0b_0.a_1b_1a_2b_2\ldots$$

Fig. 2.22 Fractional encoding principle of $\mathbb{R}$-Gödel encoding by bit interleaving

Proof. The basic idea of the $\mathbb{R}$-Gödel encoding is to interleave the bits of real numbers as depicted in Fig. 2.22 (for a pairing of $n = 2$ numbers a and b). For defining $\mathrm{at}(Z,n,j,z)$, we use several auxiliary functions to improve readability; see Fig. 2.23. Note that these definitions need no recursion. Hence, as in the notation $\phi(Z_j^{(n)})$, we can consider occurrences of the function symbols as syntactic abbreviations for quantified variables satisfying the respective definitions.

$$
\begin{aligned}
\mathrm{at}(Z,n,j,z) &\leftrightarrow \forall i\!:\!\mathbb{Z}\ \mathrm{digit}(z,i) = \mathrm{digit}(Z,n(i-1)+j) \land nat(n) \land nat(j) \land n>0\\
\mathrm{digit}(a,i) &= \mathrm{intpart}(2\,\mathrm{frac}(2^{i-1}a))\\
\mathrm{intpart}(a) &= a - \mathrm{frac}(a)\\
\mathrm{frac}(a) = z &\leftrightarrow \exists i\!:\!\mathbb{Z}\ z = a - i \land -1 < z \land z < 1 \land az \geq 0\\
2^i = z &\leftrightarrow i \geq 0 \land \exists x \exists t\,(x=1 \land t=0 \land \langle x' = x\ln 2, t' = 1\rangle (t = i \land x = z))\\
&\quad \lor\ i < 0 \land \exists x \exists t\,(x=1 \land t=0 \land \langle x' = -x\ln 2, t' = -1\rangle (t = i \land x = z))\\
\ln 2 = z &\leftrightarrow \exists x \exists t\,(x=1 \land t=0 \land \langle x' = x, t' = 1\rangle (x = 2 \land t = z))
\end{aligned}
$$

Fig. 2.23 FOD definition characterising Gödel encoding of $\mathbb{R}$-sequences in one real number

The function symbol $\mathrm{digit}(a,i)$ gives the ith bit of $a \in \mathbb{R}$ when represented with basis 2. For $i > 0$, $\mathrm{digit}(a,i)$ yields fractional bits, and, for $i \leq 0$, it yields bits of the integer part. For instance, $\mathrm{digit}(a,1)$ yields the first fractional bit, $\mathrm{digit}(a,0)$ is the least-significant bit of the integer part of a. The function $\mathrm{intpart}(a)$ represents the integer part of $a \in \mathbb{R}$. The function $\mathrm{frac}(a)$ represents the fractional part of $a \in \mathbb{R}$, which drops all integer bits. The last constraint in its definition implies that $\mathrm{frac}(a)$ keeps the sign of a (or 0). Consequently, $\mathrm{intpart}(a)$ and $\mathrm{digit}(a,i)$ also keep the sign of a (or 0). Exponentiation 2^i is definable using differential equations, using an auxiliary characterisation of the natural logarithm $\ln 2$. The definition of 2^i splits into the case of exponential growth when $i \geq 0$ and a symmetric case of exponential decay when $i < 0$. □

2.7.4 Expressibility and Rendition of Hybrid Program Semantics

In order to show that d$\mathscr{L}$ is sufficiently expressive to state the invariants and variants that are needed for proving valid statements about loops with rules *ind* and *con*, we prove an expressibility result. We give a constructive proof that the state transition relation of hybrid programs is definable in FOD, i.e., there is a FOD formula $\mathscr{S}_\alpha(\vec{x},\vec{v})$ characterising the state transitions of hybrid program α from the state characterised by the vector $\vec{x}$ of variables to the state characterised by vector $\vec{v}$.

For this, we need to characterise hybrid processes equivalently by differential equations in FOD. Observe that the existence of such characterisations does *not* follow from results embedding Turing machines into differential equations [57, 140], because, unlike Turing machines, hybrid processes are not restricted to discrete values on a grid (such as $\mathbb{N}^k$) but work with continuous real values. Furthermore, Turing machines only have repetitions of discrete transitions on discrete data (e.g., $\mathbb{N}$). For hybrid programs, in contrast, we have to characterise repetitive interactions of discrete and continuous transitions in continuous space (some $\mathbb{R}^k$).

Lemma 2.8 (Hybrid program rendition). *For every hybrid program α with variables among $\vec{x} = x_1,\dots,x_k$, there is a FOD formula $\mathscr{S}_\alpha(\vec{x},\vec{v})$ with variables among the $2k$ distinct variables $\vec{x} = x_1,\dots,x_k$ and $\vec{v} = v_1,\dots,v_k$ such that*

$$\vDash \mathscr{S}_\alpha(\vec{x},\vec{v}) \leftrightarrow \langle\alpha\rangle \vec{x} = \vec{v}$$

or, equivalently, for every I,η,ν,

$$I,\eta,\nu \models \mathscr{S}_\alpha(\vec{x},\vec{v}) \textit{ iff } (\nu,\nu[\vec{x}\mapsto val_{I,\eta}(\nu,\vec{v})]) \in \rho_{I,\eta}(\alpha).$$

$$\begin{aligned}
\mathscr{S}_{x_1:=\theta_1,\ldots,x_k:=\theta_k}(\vec{x},\vec{v}) &\equiv \bigwedge_{i=1}^{k}(v_i=\theta_i)\\
\mathscr{S}_{x_1'=\theta_1,\ldots,x_k'=\theta_k}(\vec{x},\vec{v}) &\equiv \langle x_1'=\theta_1,\ldots,x_k'=\theta_k\rangle \vec{v}=\vec{x}\\
\mathscr{S}_{x_1'=\theta_1,\ldots,x_k'=\theta_k\,\&\,\chi}(\vec{x},\vec{v}) &\equiv \exists t\,\big(t=0\wedge\langle x_1'=\theta_1,\ldots,x_k'=\theta_k,t'=1\rangle\big(\vec{v}=\vec{x}\\
&\qquad \wedge[x_1'=-\theta_1,\ldots,x_k'=-\theta_k,t'=-1](t\geq 0\to\chi)\big)\big)\\
\mathscr{S}_{?\chi}(\vec{x},\vec{v}) &\equiv \vec{v}=\vec{x}\wedge\chi\\
\mathscr{S}_{\beta\cup\gamma}(\vec{x},\vec{v}) &\equiv \mathscr{S}_\beta(\vec{x},\vec{v})\vee\mathscr{S}_\gamma(\vec{x},\vec{v})\\
\mathscr{S}_{\beta;\gamma}(\vec{x},\vec{v}) &\equiv \exists\vec{z}\,(\mathscr{S}_\beta(\vec{x},\vec{z})\wedge\mathscr{S}_\gamma(\vec{z},\vec{v}))\\
\mathscr{S}_{\beta^*}(\vec{x},\vec{v}) &\equiv \exists Z\,\exists n{:}\mathbb{N}\,\big(Z_1^{(n)}=\vec{x}\wedge Z_n^{(n)}=\vec{v}\\
&\qquad \wedge\forall i{:}\mathbb{N}\,(1\leq i<n\to\mathscr{S}_\beta(Z_i^{(n)},Z_{i+1}^{(n)}))\big)
\end{aligned}$$

Fig. 2.24 Explicit rendition of hybrid program transition semantics in FOD

Proof. By Lemma 2.6, interpretations of the vectors $\vec{x}$ and $\vec{v}$ characterise the input and output states, respectively, as far as α is concerned. These vectors are finite because α is finite. Vectorial equalities like $\vec{x}=\vec{v}$ or quantifiers $\exists\vec{v}$ are to be understood componentwise. The program rendition is defined inductively in Fig. 2.24. To simplify the notation, we assume that all variables $x_1,\ldots,x_k$ are affected in discrete jumps and differential equations by adding vacuous $x_i:=x_i$, or $x_i'=0$ if x_i does not change in the respective statement.

Differential equations give FOD formulas; no further reduction is needed. Evolution along differential equations with evolution domain restrictions is definable by following the unique flow (Lemma 2.1) backwards. Continuous evolution is reversible, i.e., the transitions of $x_i'=-\theta$ are inverse to those of $x_i'=\theta$. Consequently, with an auxiliary variable t, all evolutions of $[x_1'=-\theta_1,\ldots,x_k'=-\theta_k,t'=-1]$ follow the same flow as $\langle x_1'=\theta_1,\ldots,x_k'=\theta_k,t'=1\rangle$, but backwards. By also reversing clock t, we ensure that, along the reverse flow, χ has been true at all times (because of the box modality) until starting time $t=0$; see Fig. 2.25.

To show reversibility, let $(\nu,\omega)\in\rho_{I,\eta}(x_1'=\theta_1,\ldots,x_k'=\theta_k)$, that is, let $f:[0,r]\to \mathrm{Sta}(\Sigma)$ be a solution of $x_1'=\theta_1,\ldots,x_k'=\theta_k$ starting in state ν and ending in ω. Then $g:[0,r]\to\mathrm{Sta}(\Sigma)$, defined as $g(\zeta)=f(r-\zeta)$, starts in ω and ends in ν. Thus, it only remains to show that g is a solution of $x_1'=-\theta_1,\ldots,x_k'=-\theta_k$, which can be seen for $1\leq i\leq k$ as follows:

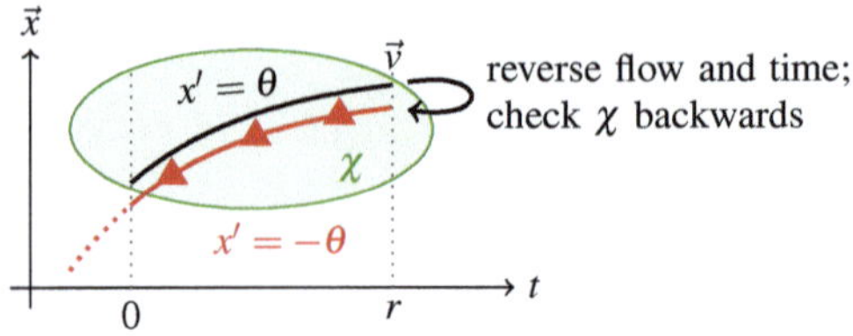

Fig. 2.25 Evolution domain checks along backwards flow over time t

$$\frac{\mathrm{d}g(t)(x_i)}{\mathrm{d}t}(\zeta) = \frac{\mathrm{d}f(r-t)(x_i)}{\mathrm{d}t}(\zeta) = \frac{\mathrm{d}f(u)(x_i)}{\mathrm{d}u}\frac{\mathrm{d}(r-t)}{\mathrm{d}t}(\zeta) = -\frac{\mathrm{d}f(u)(x_i)}{\mathrm{d}u}(\zeta)$$
$$= -\mathit{val}_{I,\eta}(f(\zeta),\theta_i) = \mathit{val}_{I,\eta}(f(\zeta),-\theta_i).$$

Unlike all other cases, case $\mathcal{S}_{x_1'=\theta_1,..,x_k'=\theta_k\,\&\,\chi}(\vec{x},\vec{v})$ in Fig. 2.24 uses nested FOD modalities. Nested modalities can be avoided in $\mathcal{S}_\alpha(\vec{x},\vec{v})$ using an equivalent FOD formula without them; see Fig. 2.25:

$$\exists t\,\exists r\,\big(t=0 \land \langle x_1'=\theta_1,..,x_k'=\theta_k, t'=1\rangle(\vec{v}=\vec{x}\land r=t)\land$$
$$\forall\vec{x}\,\forall t\,(\vec{x}=\vec{v}\land t=r \to [x_1'=-\theta_1,..,x_k'=-\theta_k,t'=-1](t\geq 0\to\chi))\big).$$

With a finite formula, the characterisation of repetition $\mathcal{S}_{\beta^*}(\vec{x},\vec{v})$ in FOD needs to capture arbitrarily long sequences of intermediate real-valued states and the correct transition between successive states of such a sequence. To achieve this with first-order quantifiers, we use the real Gödel encoding from Lemma 2.7 in Fig. 2.24 to map unbounded sequences of real-valued states reversibly to a single real number Z, which can be quantified over in first-order logic. □

Using the program rendition from Lemma 2.8 to characterise modalities, we prove that every d$\mathcal{L}$ formula can be expressed equivalently in FOD by structural induction.

Lemma 2.9 (d$\mathcal{L}$ Expressibility). *Logic* d$\mathcal{L}$ *is* expressible *in FOD: for all* d$\mathcal{L}$ *formulas* $\phi \in \mathrm{Fml}(\Sigma,V)$ *there is a FOD formula* $\phi^\# \in \mathrm{Fml}_{FOD}(\Sigma,V)$ *that is equivalent, i.e.,* $\vDash \phi \leftrightarrow \phi^\#$. *The converse holds trivially.*

Proof. The proof follows an induction on the structure of formula ϕ for which it is imperative to find an equivalent $\phi^\#$ in FOD. Observe that the construction of $\phi^\#$ from ϕ is effective.

0. If ϕ is a first-order formula, then $\phi^\# := \phi$ already is a FOD formula such that nothing has to be shown.
1. If ϕ is of the form $\varphi\lor\psi$, then by the induction hypothesis there are FOD formulas $\varphi^\#, \psi^\#$ such that $\vDash \varphi\leftrightarrow\varphi^\#$ and $\vDash\psi\leftrightarrow\psi^\#$, from which we can conclude by congruence that $\vDash(\varphi\lor\psi)\leftrightarrow(\varphi^\#\lor\psi^\#)$, giving $\vDash\phi\leftrightarrow\phi^\#$ by choosing $\varphi^\#\lor\psi^\#$ for $\phi^\#$. Similar reasoning addresses the other propositional connectives or quantifiers.
2. The case where ϕ is of the form $\langle\alpha\rangle\psi$ is a consequence of the characterisation of the semantics of hybrid programs in FOD. The expressibility conjecture holds

by the induction hypothesis using the equivalence of explicit hybrid program renditions from Lemma 2.8:

$$\vDash \ \langle\alpha\rangle\psi \leftrightarrow \exists\vec{v}\,(\mathscr{S}_\alpha(\vec{x},\vec{v}) \wedge \psi^{\#}{}_{\vec{x}}^{\vec{v}}).$$

3. The case where ϕ is $[\alpha]\psi$ is again a consequence of Lemma 2.8:

$$\vDash \ [\alpha]\psi \leftrightarrow \forall\vec{v}\,(\mathscr{S}_\alpha(\vec{x},\vec{v}) \rightarrow \psi^{\#}{}_{\vec{x}}^{\vec{v}})$$

□

The above proofs directly carry over to rich test d$\mathscr{L}$, i.e., the logic where d$\mathscr{L}$ formulas are allowed in tests $?\chi$ of hybrid programs and evolution domain restrictions χ of differential equations, when using $\chi^{\#}$ in place of χ in Fig. 2.24. Accordingly, nested modalities can be avoided in FOD by using the following formula for $\mathscr{S}_{x_1'=\theta_1,..,x_k'=\theta_k\,\&\,\chi}(\vec{x},\vec{v})$:

$$\begin{aligned}&\exists t\,\exists r\,\big(t=0 \wedge \langle x_1'=\theta_1,..,x_k'=\theta_k, t'=1\rangle(\vec{v}=\vec{x}\wedge r=t)\wedge\\ &\quad \forall\vec{z}\,\big(\exists\vec{x}\,\exists t\,(\vec{x}=\vec{v}\wedge t=r\wedge\langle x_1'=-\theta_1,..,x_k'=-\theta_k, t'=-1\rangle(t\geq 0\wedge\vec{z}=\vec{x}))\\ &\qquad \rightarrow \chi^{\#}{}_{\vec{x}}^{\vec{z}})\big).\end{aligned}$$

2.7.5 Relative Completeness of First-Order Assertions

As special cases of Theorem 2.3, we first prove relative completeness for first-order assertions about hybrid programs. These first-order cases constitute the basis for the general completeness proof for arbitrary formulas of differential dynamic logic.

In the following relative completeness proofs, we use the notation $\vdash_{\mathscr{D}} \phi$ to indicate that a d$\mathscr{L}$ formula ϕ is derivable (Definition 2.11) from a set of FOD tautologies, which is equivalent to saying that ϕ is derivable in the d$\mathscr{L}$ calculus augmented with a single *oracle axiom* $\mathscr{D}$ that gives all valid FOD instances. Likewise, we use the notation $\Gamma \vdash_{\mathscr{D}} \Delta$ to indicate that the sequent $\Gamma \vdash \Delta$ is derivable from $\mathscr{D}$.

For the completeness proof, we use several simplifications. For uniform proofs, we assume formulas to use a simplified vocabulary. A formula ϕ is valid iff it is true in *all* I,η,ν. In particular, we can assume valid ϕ to use Skolem constants (or state variables) instead of free logical variables. Existential quantifiers can be represented as modalities: $\exists x\,\phi \equiv \langle x'=1\rangle\phi \vee \langle x'=-1\rangle\phi$. For simplicity, we use cut (*cut*) and weakening to glue together subproofs propositionally. Weakening (i.e., from $\phi \vdash \psi$ infer $\phi_1,\phi \vdash \psi,\psi_1$) can be emulated using contexts Γ,Δ from Definition 2.10, and we use it implicitly together with rule *cut* in the following. Derivability of sequents and derivability of corresponding formulas are equivalent by the following lemma.

Lemma 2.10 (Derivability of sequents). $\vdash_{\mathscr{D}} \phi \rightarrow \psi$ *iff* $\phi \vdash_{\mathscr{D}} \psi$.

Proof. When we consider sequents as abbreviations for formulas, both sides are identical. Otherwise, let $\vdash_{\mathcal{D}} \phi \to \psi$ be derivable from $\mathcal{D}$. Using *cut* (and weakening) with $\phi \to \psi$, this derivation can be extended to one of $\phi \vdash_{\mathcal{D}} \psi$:

$$\text{cut}\,\dfrac{\dfrac{*}{\phi \vdash \phi \to \psi, \psi} \qquad \to\text{l}\,\dfrac{\text{ax}\,\dfrac{*}{\phi \vdash \phi, \psi} \qquad \text{ax}\,\dfrac{*}{\psi, \phi \vdash \psi}}{\phi, \phi \to \psi \vdash \psi}}{\phi \vdash \psi}$$

The converse direction is by an application of $\to$r. □

Lemma 2.11 (Generalisation). *If $\vdash_{\mathcal{D}} \phi$ is provable without free logical variables, then so are $\vdash_{\mathcal{D}} \forall x\, \phi$ and $\vdash_{\mathcal{D}} \langle x_1 := \theta_1, \ldots x_n := \theta_n \rangle \phi$.*

Proof. For the second conjecture, let $\langle \mathscr{A} \rangle$ abbreviate $\langle x_1 := \theta_1, \ldots x_n := \theta_n \rangle$. We prefix each formula in the proof of ϕ with $\langle \mathscr{A} \rangle$ and show that this gives a proof of $\langle \mathscr{A} \rangle \phi$. i$\exists$ is not needed in the proof due to the absence of free logical variables. As an intermediate step, we first show that prefixing with $\langle \mathscr{A} \rangle$ gives an (extended) proof with rule applications generalised to allow for nested jump prefixes $\langle \mathscr{A} \rangle \langle \mathscr{J} \rangle$: By the argument in Theorem 4.1, it is easy to see for discrete jump sets $\langle \mathscr{A} \rangle$ and $\langle \mathscr{J} \rangle$ that the $d\mathcal{L}$ rules remain sound with nested jump prefix $\langle \mathscr{A} \rangle \langle \mathscr{J} \rangle$ in place of only a single prefix $\langle \mathscr{J} \rangle$ from Definition 2.10. Applicability conditions of rules do not depend on jump prefixes, as Definition 2.10 allows adding *any* jump prefix. Thus, we obtain a sound (extended) proof of $\langle \mathscr{A} \rangle \phi$ when replacing—with arbitrary unchanged context $\Gamma, \Delta, \langle \mathscr{J} \rangle$—every rule application of the form

$$\frac{\Gamma, \langle \mathscr{J} \rangle \Phi_1 \vdash \langle \mathscr{J} \rangle \Psi_1, \Delta \quad \ldots \quad \Gamma, \langle \mathscr{J} \rangle \Phi_n \vdash \langle \mathscr{J} \rangle \Psi_n, \Delta}{\Gamma, \langle \mathscr{J} \rangle \Phi_0 \vdash \langle \mathscr{J} \rangle \Psi_0, \Delta}$$

in the proof of ϕ by a rule application with the additional unchanged prefix $\langle \mathscr{A} \rangle$ for corresponding $\Gamma, \Delta, \langle \mathscr{J} \rangle$:

$$\frac{\Gamma, \langle \mathscr{A} \rangle \langle \mathscr{J} \rangle \Phi_1 \vdash \langle \mathscr{A} \rangle \langle \mathscr{J} \rangle \Psi_1, \Delta \quad \ldots \quad \Gamma, \langle \mathscr{A} \rangle \langle \mathscr{J} \rangle \Phi_n \vdash \langle \mathscr{A} \rangle \langle \mathscr{J} \rangle \Psi_n, \Delta}{\Gamma, \langle \mathscr{A} \rangle \langle \mathscr{J} \rangle \Phi_0 \vdash \langle \mathscr{A} \rangle \langle \mathscr{J} \rangle \Psi_0, \Delta} \tag{2.14}$$

Next, we show that these nested jump prefixes can be reduced to a single jump prefix as Definition 2.10 allows: Let $\langle \mathscr{A}\, \mathscr{J} \rangle$ denote the discrete jump set obtained by merging $\langle \mathscr{A} \rangle$ and $\langle \mathscr{J} \rangle$ using $\langle := \rangle$ as in Sect. 2.5.2. We replace each rule application (with nested prefixes) of the form (2.14) by the following derivation with only a single prefix (assuming $n = 1$ for notational convenience):

$$\langle:=\rangle,\langle:=\rangle\,\dfrac{\text{cut}\,\dfrac{\dfrac{\ldots}{\Gamma, \langle \mathscr{A} \rangle \langle \mathscr{J} \rangle \Phi_1 \vdash \langle \mathscr{A}\, \mathscr{J} \rangle \Psi_1, \Delta} \qquad \langle:=\rangle\,\dfrac{\text{ax}\,\dfrac{*}{\Gamma, \langle \mathscr{A}\, \mathscr{J} \rangle \Phi_1 \vdash \langle \mathscr{A}\, \mathscr{J} \rangle \Phi_1, \Delta}}{\Gamma, \langle \mathscr{A}\, \mathscr{J} \rangle \Phi_1 \vdash \langle \mathscr{A} \rangle \langle \mathscr{J} \rangle \Phi_1, \Delta}}{\dfrac{\Gamma, \langle \mathscr{A}\, \mathscr{J} \rangle \Phi_1 \vdash \langle \mathscr{A}\, \mathscr{J} \rangle \Psi_1, \Delta}{\Gamma, \langle \mathscr{A}\, \mathscr{J} \rangle \Phi_0 \vdash \langle \mathscr{A}\, \mathscr{J} \rangle \Psi_0, \Delta}}}{\Gamma, \langle \mathscr{A} \rangle \langle \mathscr{J} \rangle \Phi_0 \vdash \langle \mathscr{A} \rangle \langle \mathscr{J} \rangle \Psi_0, \Delta}$$

The bottom most $\langle:=\rangle$ applications merge $\langle\mathscr{A}\rangle$ into $\langle\mathscr{J}\rangle$ in the antecedent and succedent, respectively. The unmarked rule applies the same rule that has been used in (2.14), which is applicable on $\Phi_0 \vdash \Psi_0$ for *any* context by Definition 2.10, including $\Gamma, \Delta, \langle\mathscr{A}\,\mathscr{J}\rangle$. The subsequent cut with $\langle\mathscr{A}\rangle\langle\mathscr{J}\rangle\Phi_1$ restores the form of the premise in (2.14). The left branch continues using a dual argument to turn succedent $\langle\mathscr{A}\,\mathscr{J}\rangle\Psi_1$ into $\langle\mathscr{A}\rangle\langle\mathscr{J}\rangle\Psi_1$, thereby yielding a set of non-extended rule applications with the same conclusions and premises as the extended rule application (2.14):

$$\text{cut}\,\dfrac{\Gamma, \langle\mathscr{A}\rangle\langle\mathscr{J}\rangle\Phi_1 \vdash \langle\mathscr{A}\rangle\langle\mathscr{J}\rangle\Psi_1, \Delta \qquad \langle:=\rangle\dfrac{\text{ax}\,\dfrac{*}{\Gamma, \langle\mathscr{A}\,\mathscr{J}\rangle\Psi_1 \vdash \langle\mathscr{A}\,\mathscr{J}\rangle\Psi_1, \Delta}}{\Gamma, \langle\mathscr{A}\rangle\langle\mathscr{J}\rangle\Psi_1 \vdash \langle\mathscr{A}\,\mathscr{J}\rangle\Psi_1, \Delta}}{\Gamma, \langle\mathscr{A}\rangle\langle\mathscr{J}\rangle\Phi_1 \vdash \langle\mathscr{A}\,\mathscr{J}\rangle\Psi_1, \Delta}$$

For reducing the first conjecture of this lemma to the second, let s be a Skolem constant for state variable x. By the above proof, we derive $\vdash_{\mathscr{D}} \langle x := s\rangle\phi$. Using $\forall$r, we continue this derivation to a proof of $\forall X\, \langle x := X\rangle\phi$, which we abbreviate as $\forall x\, \phi$ (see the text below Definition 2.8). Rule $\forall$r is applicable for Skolem constant s as no free logical variables occur in the proof. □

Now we prove two special cases of Theorem 2.3 for formulas of a special form.

Proposition 2.1 (Relative completeness of first-order safety). *For every hybrid program* $\alpha \in \mathrm{HP}(\Sigma, V)$ *and all FOD formulas* $F, G \in \mathrm{Fml}_{FOD}(\Sigma, V)$

$$\vDash F \to [\alpha]G \textit{ implies } \vdash_{\mathscr{D}} F \to [\alpha]G \textit{ (and } F \vdash_{\mathscr{D}} [\alpha]G \textit{ by Lemma 2.10).}$$

Proof. We generalise the relative completeness proof by Cook [87] to d$\mathscr{L}$ and follow an induction on the structure of program α. In the following, *IH* is short for the induction hypothesis.

1. The cases where α is of the form $x_1 := \theta_1, \ldots, x_n := \theta_n$, $?\chi$, $\beta \cup \gamma$, or $\beta;\gamma$ are consequences of the soundness of the symmetric rules $[;]$,$[\cup]$,$[?]$,$\langle:=\rangle$,$[:=]$. Since these rules are symmetric, they perform equivalent transformations. Consequently, whenever their conclusion is valid, their premise is valid and of smaller complexity (the programs get simpler), and hence derivable by IH. Thus, we can derive $F \to [\alpha]G$ by applying the respective rule. We explicitly show the proof for $\beta;\gamma$ as it contains an extra twist.
2. $\vDash F \to [\beta;\gamma]G$, which implies $\vDash F \to [\beta][\gamma]G$. By Lemma 2.9, there is a FOD formula $G^\#$ such that $\vDash G^\# \leftrightarrow [\gamma]G$. From the validity of $\vDash F \to [\beta]G^\#$, we can conclude by IH that $F \vdash_{\mathscr{D}} [\beta]G^\#$ is derivable. Similarly, due to $\vDash G^\# \to [\gamma]G$, we conclude $\vdash_{\mathscr{D}} G^\# \to [\gamma]G$ by IH. Using Lemma 2.11, we conclude that also $\vdash_{\mathscr{D}} \forall^\beta(G^\# \to [\gamma]G)$. With an application of $[]gen$, the latter derivation can be extended to a derivation of $[\beta]G^\# \vdash_{\mathscr{D}} [\beta][\gamma]G$. Combining the above derivations propositionally by a cut with $[\beta]G^\#$, we can derive $F \vdash_{\mathscr{D}} [\beta][\gamma]G$, from which $[;]$ yields $F \vdash_{\mathscr{D}} [\beta;\gamma]G$ as desired (and Lemma 2.10 or $\to$r yield $\vdash_{\mathscr{D}} F \to [\beta;\gamma]G$).

3. $\vDash F \to [x_1' = \theta_1, \dots, x_n' = \theta_n]G$ is a FOD formula and hence derivable as a $\mathcal{D}$ axiom. Continuous evolution $x_1' = \theta_1, \dots, x_n' = \theta_n \,\&\, \chi$ with evolution domain restrictions is definable in FOD by Lemma 2.8, which we consider as an abbreviation in this proof.
4. $\vDash F \to [\beta^*]G$ can be derived by induction. For this, we define the invariant as a FOD encoding of the statement that all potential post-states of β^* satisfy G according to Lemma 2.9:
$$\phi \equiv ([\beta^*]G)^\# \equiv \forall \vec{v}\,(\mathcal{S}_{\beta^*}(\vec{x},\vec{v}) \to G_{\vec{x}}^{\vec{v}}).$$
Since $F \to \phi$ and $\phi \to G$ are valid FOD formulas, they are derivable by $\mathcal{D}$; so is $F \vdash_{\mathcal{D}} \phi$ derivable by Lemma 2.10. By Lemma 2.11 and $[]gen$, $[\beta^*]\phi \vdash_{\mathcal{D}} [\beta^*]G$ is derivable. Likewise, $\phi \to [\beta]\phi$ is valid according to the semantics of repetition, and thus derivable by IH, since β is less complex. Using Lemma 2.11, we can derive $\vdash_{\mathcal{D}} \forall^\beta (\phi \to [\beta]\phi)$, from which *ind* yields $\phi \vdash_{\mathcal{D}} [\beta^*]\phi$. Combining the above derivations propositionally by a cut with $[\beta^*]\phi$ and ϕ yields $F \vdash_{\mathcal{D}} [\beta^*]G$. □

Proposition 2.2 (Relative completeness of first-order liveness). *For each hybrid program $\alpha \in \mathrm{HP}(\Sigma, V)$ and all FOD formulas $F, G \in \mathrm{Fml}_{FOD}(\Sigma, V)$*
$$\vDash F \to \langle\alpha\rangle G \textit{ implies } \vdash_{\mathcal{D}} F \to \langle\alpha\rangle G \textit{ (and } F \vdash_{\mathcal{D}} \langle\alpha\rangle G \textit{ by Lemma 2.10)}.$$

Proof. We generalise the arithmetic completeness proof by Harel [148] to the hybrid case. Most cases of the proof are simple adaptations of the corresponding cases in Proposition 2.1. What remains to be shown is the case of repetitions. Assume that $\vDash F \to \langle\beta^*\rangle G$. To derive this formula by *con*, we use a FOD formula $\varphi(n)$ as a variant expressing that, after n iterations, β can lead to a state satisfying G. This formula is obtained from Lemmas 2.8 and 2.9 as $(\langle\beta^*\rangle G)^\# \equiv \exists \vec{v}\,(\mathcal{S}_{\beta^*}(\vec{x},\vec{v}) \wedge G_{\vec{x}}^{\vec{v}})$, *except* that the quantifier on the repetition count n is removed such that n becomes a free variable (plus index shifting to count repetitions):
$$\varphi(n-1) \equiv \exists \vec{v}\,\exists Z\,(Z_1^{(n)} = \vec{x} \wedge Z_n^{(n)} = \vec{v} \wedge \forall i\!:\!\mathbb{N}\,(1 \le i < n \to \mathcal{S}_\beta(Z_i^{(n)}, Z_{i+1}^{(n)})) \wedge G_{\vec{x}}^{\vec{v}}).$$
By Lemma 2.7, $\varphi(n)$ can only hold true if n is a natural number.

According to the loop semantics, $\vDash n > 0 \wedge \varphi(n) \to \langle\beta\rangle\varphi(n-1)$ is valid by construction: If $n > 0$ is a natural number then so is $n-1$, and if β reaches G after n repetitions, then, after executing β once, $n-1$ repetitions of β reach G. By IH, this formula is derivable, since β contains less loops. By Lemma 2.11, we extend this derivation to $\vdash_{\mathcal{D}} \forall^\beta \forall n{>}0\,(\varphi(n) \to \langle\beta\rangle\varphi(n-1))$. Thus $\exists v\,\varphi(v) \vdash_{\mathcal{D}} \langle\beta^*\rangle \exists v{\le}0\,\varphi(v)$ by *con*. It only remains to show that the antecedent is derivable from F and $\langle\beta^*\rangle G$ is derivable from the succedent. From our assumption, we conclude that the following are valid FOD formulas, hence $\mathcal{D}$ axioms:

- $\vDash F \to \exists v\,\varphi(v)$, because $\vDash F \to \langle\beta^*\rangle G$, and

- $\vDash (\exists v{\le}0\, \varphi(v)) \to G$, because $v \le 0$, and the fact, that by Lemma 2.7, $\varphi(v)$ only holds true for natural numbers, imply $\varphi(0)$. Further, $\varphi(0)$ entails G, because zero repetitions of β have no effect.

From the latter we derive $\vdash_{\mathscr{D}} \forall^{\beta} (\exists v{\le}0\, \varphi(v) \to G)$ by Lemma 2.11 and extend the derivation to $\langle \beta^* \rangle \exists v{\le}0\, \varphi(v) \vdash_{\mathscr{D}} \langle \beta^* \rangle G$ by $\langle\rangle gen$. From $\vdash_{\mathscr{D}} F \to \exists v\, \varphi(v)$ we conclude $F \vdash_{\mathscr{D}} \exists v\, \varphi(v)$ by Lemma 2.10. Now, the above derivations can be combined propositionally by a cut with $\langle \beta^* \rangle \exists v{\le}0\, \varphi(v)$ and with $\exists v\, \varphi(v)$ to yield $F \vdash_{\mathscr{D}} \langle \beta^* \rangle G$. □

2.7.6 Relative Completeness of the Differential Logic Calculus

Having succeeded with the proofs of the above statements we can finish the proof of Theorem 2.3, which is the central theoretical result of this chapter.

Proof (of Theorem 2.3). The proof follows a basic structure analogous to that of Harel's proof for the discrete case [148, Theorem 3.1]. We have to show that every valid d$\mathscr{L}$ formula ϕ can be proven from FOD axioms within the d$\mathscr{L}$ calculus: from $\vDash \phi$ we have to prove $\vdash_{\mathscr{D}} \phi$. The proof proceeds as follows: By propositional recombination, we inductively identify fragments of ϕ that correspond to $\phi_1 \to [\alpha]\phi_2$ or $\phi_1 \to \langle \alpha \rangle \phi_2$ logically. Next, we express subformulas ϕ_i equivalently in FOD by Lemma 2.9, and use Propositions 2.1 and 2.2 to resolve these first-order safety or liveness assertions. Finally, we prove that the original d$\mathscr{L}$ formula can be re-derived from the subproofs.

We can assume ϕ to be given in conjunctive normal form by appropriate propositional reasoning. In particular, we assume that negations are pushed inside over modalities using the dualities $\neg[\alpha]\phi \equiv \langle \alpha \rangle \neg \phi$ and $\neg \langle \alpha \rangle \phi \equiv [\alpha] \neg \phi$. The remainder of the proof follows an induction on a measure $|\phi|$ defined as the number of modalities in ϕ. For a simple and uniform proof, we assume quantifiers to be abbreviations for modal formulas: $\exists x\, \phi \equiv \langle x' = 1 \rangle \phi \vee \langle x' = -1 \rangle \phi$ and $\forall x\, \phi \equiv [x' = 1]\phi \wedge [x' = -1]\phi$.

0. $|\phi| = 0$; then ϕ is a first-order formula; hence derivable by $\mathscr{D}$.
1. ϕ is of the form $\neg \phi_1$; then ϕ_1 is first-order, as we assumed negations to be pushed inside. Hence, $|\phi| = 0$ and Case 0 applies.
2. ϕ is of the form $\phi_1 \wedge \phi_2$, then individually deduce the simpler proofs for $\vdash_{\mathscr{D}} \phi_1$ and $\vdash_{\mathscr{D}} \phi_2$ by IH, which can be combined by rule $\wedge$r.
3. ϕ is a disjunction and—without loss of generality—has one of the following forms (otherwise use associativity and commutativity to select a different order for the disjunction):

$$\phi_1 \vee [\alpha]\phi_2$$
$$\phi_1 \vee \langle \alpha \rangle \phi_2$$

As a unified notation for those cases we use $\phi_1 \vee \langle\!|\alpha|\!\rangle \phi_2$. Then, $|\phi_2| < |\phi|$, since ϕ_2 has less modalities. Likewise, $|\phi_1| < |\phi|$ because $\langle\!|\alpha|\!\rangle \phi_2$ contributes one modality to $|\phi|$ that is not part of ϕ_1.
According to Lemma 2.9 there are FOD formulas $\phi_1^\#, \phi_2^\#$ with $\vDash \phi_i \leftrightarrow \phi_i^\#$ for $i = 1,2$. By congruence, the validity $\vDash \phi$ yields $\vDash \phi_1^\# \vee \langle\!|\alpha|\!\rangle \phi_2^\#$, which directly implies $\vDash \neg\phi_1^\# \rightarrow \langle\!|\alpha|\!\rangle \phi_2^\#$. Then by Propositions 2.1 or 2.2, respectively, we can derive

$$\neg\phi_1^\# \vdash_{\mathscr{D}} \langle\!|\alpha|\!\rangle \phi_2^\#. \tag{2.15}$$

Further $\vDash \phi_1 \leftrightarrow \phi_1^\#$ implies $\vDash \neg\phi_1 \rightarrow \neg\phi_1^\#$, which is derivable by IH, because $|\phi_1| < |\phi|$. By Lemma 2.10, we obtain $\neg\phi_1 \vdash_{\mathscr{D}} \neg\phi_1^\#$, which we combine with (2.15) by a cut with $\neg\phi_1^\#$ to

$$\neg\phi_1 \vdash_{\mathscr{D}} \langle\!|\alpha|\!\rangle \phi_2^\#. \tag{2.16}$$

Likewise $\vDash \phi_2 \leftrightarrow \phi_2^\#$ implies $\vDash \phi_2^\# \rightarrow \phi_2$, which is derivable by IH, as $|\phi_2| < |\phi|$. We can extend the derivation of $\vdash_{\mathscr{D}} \phi_2^\# \rightarrow \phi_2$ to one of $\vdash_{\mathscr{D}} \forall^\alpha(\phi_2^\# \rightarrow \phi_2)$ by Lemma 2.11 and conclude $\langle\!|\alpha|\!\rangle \phi_2^\# \vdash_{\mathscr{D}} \langle\!|\alpha|\!\rangle \phi_2$ by $[]gen$–$\langle\rangle gen$. Finally we combine the latter derivation propositionally with (2.16) by a cut with $\langle\!|\alpha|\!\rangle \phi_2^\#$ to derive $\neg\phi_1 \vdash_{\mathscr{D}} \langle\!|\alpha|\!\rangle \phi_2$, from which $\vdash_{\mathscr{D}} \phi_1 \vee \langle\!|\alpha|\!\rangle \phi_2$ can be obtained, again using *cut*, to complete the proof. □

This concludes the main theoretical proof of relative completeness of the d$\mathscr{L}$ calculus, i.e., of Theorem 2.3.

2.8 Relatively Semidecidable Fragments

To strengthen the completeness result from Theorem 2.3, we consider fragments of d$\mathscr{L}$ where the required FOD tautologies are sufficiently simple as differential equations have first-order definable flows and the required loop invariants (or variants) are expressible in first-order logic over the reals. In these fragments, the only difficulty is to find the required invariants and variants for the proof. Relative to an (ineffective) oracle that provides first-order invariants and variants for repetitions, the d$\mathscr{L}$ calculus can be used as a semidecision procedure. That is, when we assume the oracle to provide suitable (in)variants, validity of formulas can be proven in the d$\mathscr{L}$ calculus. If an imperfect oracle chooses inadequate (in)variants, applying the d$\mathscr{L}$ calculus rules results in goals that are not valid, which is again decidable by quantifier elimination in the d$\mathscr{L}$ calculus.

Theorem 2.4 (Relatively semidecidable fragment). *Relative to an oracle generating first-order invariants and variants, the d$\mathscr{L}$ calculus gives a backtracking-free semidecision procedure for (closed) d$\mathscr{L}$ formulas with differential equations having first-order definable flows.*

Proof (Outline). The (constructive) proof, which, in full, can be found in the remainder of this section, shows that there are always applicable d$\mathcal{L}$ rules that transform the formulas equivalently and that formulas in this d$\mathcal{L}$ proof descend along a well-founded order. For loops, we assume that suitable (in)variants are obtained from the oracle and we can guarantee termination when these (in)variants are first-order (or contain fewer loops). □

As a consequence, enumerating first-order invariants or variants gives a semidecision procedure for the fragment of Theorem 2.4. As a corollary to Theorems 2.2 and 2.4, there are valid d$\mathcal{L}$ formulas that need proper d$\mathcal{L}$ (or FOD) invariants to be provable and cannot be proven just using (in)variants of first-order real arithmetic. Similarly, the fragment with first-order definable flows and bounded loops is decidable: When loops α^* are annotated with natural numbers indicating the maximum number of repetitions of α, an effective oracle for Theorem 2.4 can be obtained by unrolling, e.g., by rule $\langle{*^n}\rangle$.

As an auxiliary result for proving Theorem 2.4, we show that, in d$\mathcal{L}$ proofs, Skolem symbols occur in a uniform way, i.e., a Skolem symbol s always occurs with the same list of arguments.

Lemma 2.12 (Uniform Skolem symbols). *Let ϕ be a d$\mathcal{L}$ formula without Skolem symbols. In any derivation of ϕ, Skolem symbols only occur with a unique list of free logical variables as arguments, provided that the formulas in cuts (rule cut) obey this restriction.*

Proof. The proof is by induction on the structure of proofs in the d$\mathcal{L}$ calculus. For derivations of length zero, the conjecture holds, because ϕ does not contain Skolem symbols. We show that the conjectured Skolem occurrence property is preserved in all subgoals when applying a rule to a goal that satisfies the conjecture.

- ∀r The symbols $s(X_1,\dots,X_n)$ introduced by rules ∀r,∃l are of the required form as the X_i are precisely the free logical variables. In addition, the symbol $s(X_1,\dots,X_n)$ does not occur nested in other Skolem terms, because, by the induction hypothesis, the bound variable x does not occur in Skolem terms of the goal.
- i∀ Rules i∀ and i∃ are only applicable to instances of first-order real arithmetic (Lemma 2.5), for which the equivalence transformations of quantifier elimination preserve the Skolem occurrence property, because they never introduce quantifiers to bind free variables.
- $\langle ' \rangle$ Rule $\langle ' \rangle$ preserves the property, as it only substitutes state variables $x_i \in \Sigma$, not logical variables $X_i \in V$.
- *cut* Cuts preserve the Skolem occurrence property, as we assumed the formulas that *cut* introduces to adhere to the Skolem occurrence property.
- – The other rules of the d$\mathcal{L}$ calculus preserve the property as they never replace arguments of Skolem function symbols (which are free variables by induction hypothesis). □

Proof (of Theorem 2.4). The proof is by well-founded induction. We prove that there is a well-founded strict partial order $\prec$ such that:

IH: For all non-atomic formulas occurring in the sequents during a proof, there is an applicable series of $d\mathcal{L}$ rules such that all resulting subgoals are simpler with respect to $\prec$ and have no additional free variables or function symbols, and their conjunction is equivalent to the conclusion (for suitable oracle choices).

By applying these $d\mathcal{L}$ rules exhaustively, we obtain a decision procedure relative to the oracle, because the subgoals descend along the well-founded order $\prec$, which has no infinite descending chain. Finally, validity of the remaining sequents with atomic formulas is decidable by evaluating ground instances (Definition 2.9), because, by IH, the resulting formulas have no free variables when the initial formula is closed (open formulas, in contrast, yield equivalent parameter constraints as results). We use the derived rules *ind′* and *con′* from p. 86 in place of *ind* and *con*; see Sect. 2.5.2. To obtain a backtracking-free procedure, we remove rules $\langle *^n\rangle$,$[*^n]$, $[]$*gen*,$\langle\rangle$*gen*,*ind*,*con* and *cut* from the calculus: If a calculus with less rules gives a decision procedure, then so does the full calculus.

We define the order $\prec$ as the lexicographical order of, respectively, the number of: loops, differential equations, sequential compositions, choices, modalities, quantifiers, number of different variables and Skolem function symbols, and the number of logical connectives. As a lexicographical order of natural numbers, $\prec$ is well-founded [99]. It lifts to sequents in rule applications (Definition 2.10) when all subgoals of all rule schemata are simpler than their goals with respect to $\prec$, which can be shown to retain well-foundedness as a multiset ordering [99].

Now the proof of IH is by induction along $\prec$. Let ϕ be a non-atomic formula of a sequent in an open branch of the proof. We assume ϕ to occur in the succedent; the respective proofs for the antecedent are dual. Hence, we consider the sequent to be of the form $\Gamma \vdash \phi, \Delta$.

1. If ϕ is of the form $\psi_1 \wedge \psi_2$, then rule $\wedge$r is applicable, yielding smaller sequents (with less logical connectives) that are equivalent. Other logical connectives are handled likewise using rules $\neg$r,$\vee$r,$\wedge$r,$\rightarrow$r, respectively.
2. If ϕ is of the form $[\alpha]\psi$ or $\langle\alpha\rangle\psi$ and α is of the form $?\chi$, $\beta;\gamma$ or $\beta\cup\gamma$, the corresponding rule $\langle;\rangle$,$[;]$,$\langle\cup\rangle$,$[\cup]$ or $\langle?\rangle$,$[?]$ is applicable, yielding a simpler yet equivalent formula.
3. If ϕ is of the form $[x_1' = \theta_1,\dots,x_n' = \theta_n \,\&\, \chi]\psi$, then rule $[']$ is applicable, as we assumed differential equations to have first-order definable flows. The resulting formula is equivalent and simpler, because it contains fewer differential equations. It involves additional bound variables but not free variables. Case $\langle x_1' = \theta_1,\dots,x_n' = \theta_n \,\&\, \chi\rangle\psi$ is similar, by rule $\langle'\rangle$.
4. If ϕ is of the form $[\alpha^*]\psi$, then rule *ind′* is applicable with a first-order invariant F obtained from the oracle. The resulting subgoals are simpler according to $\prec$, because they contain less loops (F does not contain loops). The resulting subgoals do not have additional free variables as all bound variables of α^* remain bound by the universal closure $\forall^\alpha$ in the respective premises. Finally, we assume the oracle to give an invariant such that the conjunction of the resulting subgoals is equivalent to the goal (otherwise we have nothing to show for

inadequate choices by the oracle). The case $\langle\alpha^*\rangle\psi$ is similar, using rule *con'* instead.

5. If ϕ is of the form $\langle x_1 := \theta_1, \dots, x_n := \theta_n\rangle\psi$, there are two cases. If rule $\langle:=\rangle$ is applicable, it yields equivalent simpler sequents. Otherwise, we have

$$\psi \prec \langle x_1 := \theta_1, \dots, x_n := \theta_n\rangle\psi.$$

Thus, by IH, there is a finite sequence of rule applications on ψ yielding equivalent sequents with atomic formulas. Prefixing the resulting proof with $\langle x_1 := \theta_1, \dots, x_n := \theta_n\rangle$ yields a corresponding proof for deriving $\Gamma \vdash \phi, \Delta$ by Lemma 2.11. The formulas of the open branches of this proof resulting from ϕ are of the form $\langle x_1 := \theta_1, \dots, x_n := \theta_n\rangle G$ for atomic formulas G, where, at the latest, rule $\langle:=\rangle$ is applicable, as substitutions are admissible on atomic formulas. Case $[x_1 := \theta_1, \dots, x_n := \theta_n]\psi$ is similar, using rule $[:=]$ first.

6. If ϕ is of the form $\forall x\, \psi(x)$, we can apply rule $\forall$r giving $\psi(s(X_1, \dots, X_n))$. Now, we have $\psi(s(X_1, \dots, X_n)) \prec \forall x\, \psi(x)$; hence, by IH, $\psi(s(X_1, \dots, X_n))$ can be transformed equivalently to a set of sequents of the form

$$\Phi_i(s(X_1, \dots, X_n)) \vdash \Psi_i(s(X_1, \dots, X_n))$$

with atomic formulas (without loss of generality, we can assume $s(X_1, \dots, X_n)$ to occur in all branches). Hence, QE is defined for these atomic formulas and rule i$\forall$ can be applied on each branch, yielding $\mathrm{QE}(\forall s\, (\Phi_i(s) \vdash \Psi_i(s)))$. Consequently, the original sequent $\Gamma \vdash \forall x\, \psi(x), \Delta$ is equivalent to

$$\bigwedge_i \mathrm{QE}(\forall s\, (\Phi_i(s) \vdash \Psi_i(s)))$$

for the following reason: $\Gamma \vdash \psi(s(X_1, \dots, X_n)), \Delta$ is equivalent to

$$\bigwedge_i (\Phi_i(s(X_1, \dots, X_n)) \vdash \Psi_i(s(X_1, \dots, X_n)))$$

by IH, using the equivalence $\mathrm{QE}(\forall s\, (F \wedge G)) \equiv \mathrm{QE}(\forall s\, F) \wedge \mathrm{QE}(\forall s\, G)$ and the fact that s does not occur in Γ, Δ. After applying rule i$\forall$, the result has no additional free symbols, although intermediate formulas do.

7. If ϕ is of the form $\exists x\, \psi(x)$, then rule $\exists$r is applicable giving $\psi(X)$ for a fresh logical variable X. Then $\psi(X) \prec \exists x\, \psi(x)$; hence, by IH, $\psi(X)$ can be transformed equivalently to a set of sequents $\Phi_i \vdash \Psi_i$ with atomic formulas. If no Skolem dependency on X occurs in $\Phi_i \vdash \Psi_i$, then QE is defined and rule i$\exists$ applicable, giving $\mathrm{QE}(\exists X \bigwedge_i (\Phi_i \vdash \Psi_i))$, which is equivalent to $\exists X \bigwedge_i (\Phi_i \vdash \Psi_i)$. By IH, this is equivalent to $\Gamma \vdash \exists X\, \psi(X), \Delta$, because X does not occur in Γ, Δ. Otherwise, if a Skolem term $s(X_1, \dots, X, \dots, X_n)$ occurs in a $\Phi_i \vdash \Psi_i$, then, by IH, the Skolem function s already occurred in $\psi(X)$. By Lemma 2.12, the Skolem term $s(X_1, \dots, X, \dots, X_n)$ itself must already have occurred in $\psi(X)$, which contradicts the fact that X is fresh and that bound variable x does not occur in Skolem

terms of $\exists x\, \psi(x)$, again by Lemma 2.12. After applying rule i$\exists$ the additional free variable X disappears. □

This completes the proof of Theorem 2.4, showing that the $d\mathcal{L}$ calculus can be used as a semidecision procedure for a particular set of (in)variants provided by an oracle. Consequently, these results show that, in a certain sense, finding (in)variants is the only challenge in hybrid systems' verification, because the $d\mathcal{L}$ calculus takes care of everything else. In Chap. 3 we revisit and strengthen this result, because we show that properties of differential equations can be proven by appropriate generalisations of (in)variants that we call differential invariants. Furthermore, we turn to the challenge of finding these (differential) invariants in Chap. 6.

2.9 Train Control Verification

In this section, we verify collision avoidance of the train control system presented in Sect. 2.4. Especially, we identify the constraints required for the free parameters of the system and discover the preconditions for safe driving.

2.9.1 Finding Inductive Candidates

Recall the $d\mathcal{L}$ formula from Sect. 2.4 that expresses that the simplified ETCS train control system ensures that trains always stay inside their movement authority m to ensure collision-freedom:

$$\psi \rightarrow [(ctrl; drive)^*]\, z \le m \tag{2.7*}$$

We want to prove safety statement (2.7) of the simplified version of the European Train Control System. Note that this is a significantly simplified version showing only the true essentials of ETCS. We consider the ETCS cooperation protocol in more detail in Chap. 7.

Using parametric extraction techniques, we identify both the requirement ψ for safe driving and the induction hypothesis ϕ that is required for the proof. Similar to the proof in Fig. 2.16, which is dual to the proof in Fig. 2.14, an unwinding of the loop in (2.7) by rule $[^{*n}]$ can be used to extract a candidate for a parametric inductive hypothesis. It expresses that there is sufficient braking distance at current speed v, which basically corresponds to the controllability constraint for ETCS (as illustrated in Fig. 2.15 on p. 90):

$$\phi \equiv v^2 \le 2b(m-z) \wedge b > 0 \wedge A \ge 0\ . \tag{2.17}$$

2.9.2 Inductive Verification

Using proof rule *ind* to prove d$\mathcal{L}$ formula (2.7) by induction, we show that (*a*) invariant ϕ holds initially, i.e., $\psi \vdash \phi$ (implying the antecedent of the conclusion of *ind*), (*b*) the invariant is sustained after each execution of $ctrl;drive$, and (*c*) invariant ϕ implies postcondition $z \leq m$. Case (*c*) holds by QE, as $0 \leq v^2 \leq 2b(m-z)$ and $b > 0$. The induction start (*a*) will be examined after the full proof, since we want to identify the prerequisite ψ for safe driving by proof analysis. In the proof of the induction step $\phi \to [ctrl;drive]\phi$, we omit condition $m-z \leq s$ from *ctrl*, because it is not used in the proof (braking remains safe with respect to $z \leq m$). The induction is provable in d$\mathcal{L}$ as follows (for notational convenience, we assume rule ∀r calls the Skolem constant for *m* again *m*, and so on, as there are no free logical variables):

$$
\dfrac{\dfrac{\dfrac{\dfrac{\dfrac{\dfrac{\dots}{\phi \vdash \langle a:=-b\rangle[drive]\phi} \qquad {\scriptstyle [?],\to r}\,\dfrac{\dfrac{\dots}{\phi, m-z \geq s \vdash \langle a:=A\rangle[drive]\phi}}{\phi \vdash [?m-z \geq s; a:=A][drive]\phi}}{\phi \vdash [ctrl][drive]\phi}\,{\scriptstyle [\cup],\wedge r}}{\phi \vdash [ctrl;drive]\phi}\,{\scriptstyle [;]}}{\vdash \phi \to [ctrl;drive]\phi}\,{\scriptstyle \to r}}{\vdash \forall^{\alpha}(\phi \to [ctrl;drive]\phi)}\,{\scriptstyle \forall r}}{\phi \vdash [(ctrl;drive)^*]\phi}\,{\scriptstyle ind}
$$

The differential equation system in *drive* is linear with a constant coefficient matrix M. Its solution can be obtained by symbolically computing the exponential series $e^{Mt}\eta$ with symbolic initial value $\eta = (z,v)$ and similar symbolic integration of the inhomogeneous part [297, §18.VI]; also see App. B.4. We abbreviate the solution $\langle z:=-\frac{b}{2}t^2+vt+z, v:=-bt+v\rangle$ thus obtained by $\langle \mathscr{S}_t\rangle$. See Example B.3 in App. B for an explanation of why this is a solution of the differential equations. In this example, the evolution domain restrictions are convex; hence the constraint $\forall 0{\leq}\tilde{t}{\leq}t\, \langle \mathscr{S}_{\tilde{t}}\rangle\chi$ of rule $[']$ can be simplified to $\langle \mathscr{S}_t\rangle\chi$ as in (2.12) to save space. Further, we leave out conditions which are unnecessary for closing the above proof. In the left branch, the constrained evolution of clock τ is irrelevant and will be left out to save space (braking is the safest operation and can be continued indefinitely without extra risk). The left branch closes (marked $*$):

$$
\begin{array}{ll}
 & * \\
\hline
{\scriptstyle \langle:=\rangle,i\forall} & \phi, t \geq 0, -bt+v \geq 0 \vdash \langle \mathscr{S}_t\rangle\phi \\
\hline
{\scriptstyle \langle:=\rangle} & \phi, t \geq 0, \langle v:=-bt+v\rangle v \geq 0 \vdash \langle \mathscr{S}_t\rangle\phi \\
\hline
{\scriptstyle \to r,\to r} & \phi \vdash t \geq 0 \to (\langle v:=-bt+v\rangle v \geq 0 \to \langle \mathscr{S}_t\rangle\phi) \\
\hline
{\scriptstyle \forall r} & \phi \vdash \forall t{\geq}0\,(\langle v:=-bt+v\rangle v \geq 0 \to \langle \mathscr{S}_t\rangle\phi) \\
\hline
{\scriptstyle [']} & \phi \vdash [z'=v, v'=-b \,\&\, v \geq 0]\phi \\
\hline
{\scriptstyle \langle:=\rangle} & \phi \vdash \langle a:=-b\rangle[drive]\phi \\
\hline
{\scriptstyle [:=]} & \phi \vdash [a:=-b][drive]\phi
\end{array}
$$

The right branch does not need the evolution domain constraint $v \geq 0$, because v does not decrease when accelerating. We again use $\langle \mathcal{S}_t \rangle$ as an abbreviation for the solution $\langle z := \frac{A}{2}t^2 + vt + z, v := At + v \rangle$.

$$
\begin{array}{rl}
 & \dots \\ \hline
 & \phi, m-z \geq s \vdash s \geq \frac{v^2}{2b} + \left(\frac{A}{b}+1\right)\left(\frac{A}{2}\varepsilon^2 + \varepsilon v\right) \\ \hline
\langle:=\rangle,\mathrm{i}\forall & \phi, m-z \geq s, 0 \leq t \leq \varepsilon \vdash \langle \mathcal{S}_t \rangle \phi \\ \hline
\rightarrow\mathrm{r},\langle:=\rangle & \phi, m-z \geq s \vdash t \geq 0 \rightarrow (\langle \tau := t \rangle \tau \leq \varepsilon \rightarrow \langle \mathcal{S}_t \rangle \phi) \\ \hline
\forall\mathrm{r} & \phi, m-z \geq s \vdash \forall t{\geq}0 (\langle \tau := t \rangle \tau \leq \varepsilon \rightarrow \langle \mathcal{S}_t \rangle \phi) \\ \hline
\langle:=\rangle & \phi, m-z \geq s \vdash \langle \tau := 0 \rangle \forall t{\geq}0 (\langle \tau := t + \tau \rangle \tau \leq \varepsilon \rightarrow \langle \mathcal{S}_t \rangle \phi) \\ \hline
[{}'] & \phi, m-z \geq s \vdash \langle \tau := 0 \rangle [z' = v, v' = A, \tau' = 1 \,\&\, \tau \leq \varepsilon] \phi \\ \hline
[:=] & \phi, m-z \geq s \vdash [\tau := 0][z' = v, v' = A, \tau' = 1 \,\&\, \tau \leq \varepsilon] \phi \\ \hline
\langle:=\rangle & \phi, m-z \geq s \vdash \langle a := A \rangle [\tau := 0][z' = v, v' = a, \tau' = 1 \,\&\, \tau \leq \varepsilon] \phi \\ \hline
[;] & \phi, m-z \geq s \vdash \langle a := A \rangle [drive] \phi \\ \hline
[:=] & \phi, m-z \geq s \vdash [a := A][drive] \phi
\end{array}
$$

2.9.3 Parameter Constraint Discovery

The right branch only closes when the succedent of its open goal is guaranteed. That formula expresses that there will still be sufficient braking distance even after accelerating by $\leq A$ for up to ε seconds:

$$
s \geq \frac{v^2}{2b} + \left(\frac{A}{b} + 1\right)\left(\frac{A}{2}\varepsilon^2 + \varepsilon v\right). \tag{2.18}
$$

This constraint can be discovered automatically in the above proof by the indicated application of rule i$\forall$ using quantifier elimination with some simplifications. Constraint (2.18) is required to make sure invariant (2.17) still holds after accelerating. In fact, augmenting the case study with (2.18) makes the argument inductive, and the whole proof of the safety statement (2.7) closes when ψ is chosen identical to ϕ. Here, the conditions of ψ cannot be removed without leaving the proof open due to a counterexample, as the invariant (2.17) is a controllability constraint; see Sect. 2.5.3.1.

Quite unlike in the acceleration-free case [231], constraint (2.18) needs to be enforced dynamically as the affected variables change over time. That is, at the beginning of each *ctrl* cycle, s needs to be updated in accordance with (2.18), which admits complex behaviour as in Fig. 2.9b on p. 63. Further, this constraint can be used to find out how densely a track can be packed with trains in order to maximise ETCS throughput without compromising safety. The resulting provably safe train control system can be summarised as follows:

$$
v^2 \leq 2b(m-z) \wedge b > 0 \wedge A \geq 0 \rightarrow [(ctrl;drive)^*]\, z \leq m \tag{2.19}
$$

$$\begin{aligned}\text{where } ctrl &\equiv s := \frac{v^2}{2b} + \left(\frac{A}{b}+1\right)\left(\frac{A}{2}\varepsilon^2 + \varepsilon v\right);\\ &\quad (?m - z \le s; a := -b) \cup (?m - z \ge s; a := A)\\ drive &\equiv \tau := 0; (z' = v, v' = a, \tau' = 1 \,\&\, v \ge 0 \wedge \tau \le \varepsilon).\end{aligned}$$

Using the $d\mathcal{L}$ calculus, similar constraints can be derived (Sect. 4.8) to find out how early a train needs to start negotiation in order to minimise the risk of having to reduce speed when the MA is not extensible in time, which is the ST parameter of Fig. 2.8.

For the resulting ETCS system, liveness can be proven in the $d\mathcal{L}$ calculus by showing that the train can pass every point p by an appropriate choice of m by the RBC:

$$z = z_0 \wedge v = v_0 > 0 \wedge \varepsilon > 0 \wedge b > 0 \wedge A \ge 0 \rightarrow \forall p \exists m \langle (ctrl; drive)^* \rangle z \ge p. \quad (2.20)$$

For $A = 0$, the proof of property (2.20) uses the variant $\varphi(n) \equiv z + n\varepsilon v_0 \ge p \wedge v = v_0$ for rule *con*, which expresses that the speed does not decrease (until $n < 0$) and that the remaining distance from z to target p can be covered after at most n iteration cycles. This directly proves the property even when $A = 0$ for appropriate acceleration choices. For general $A \ge 0$, the following variant proves property (2.20) by *con*:

$$\varphi(n) \equiv ((z + n\varepsilon v_0 \ge p \wedge z_0 \le z \wedge v^2 \le v_0^2 + 2A(z - z_0) \wedge v \ge v_0 \wedge z \le p) \vee z \ge p) \wedge v \ge 0. \quad (2.21)$$

It expresses that, when $z \le p$, the remaining distance can be covered after at most n iterations while the train position and velocity increase, yet the velocity is bounded depending on the initial velocity v_0, acceleration A, and distance $z - z_0$. The appropriate choice of m to prove property (2.20) with this variant is

$$m \ge p + \frac{v_o^2 + 2A(p - z_0)}{2b} + \left(\frac{A}{b}+1\right)\left(\frac{A}{2}\varepsilon^2\right) + \varepsilon\sqrt{v_0^2 + 2A(p - z_0)},$$

which can be obtained by overapproximating braking condition (2.18) with the speed limit $v^2 \le v_0^2 + 2A(z - z_o)$ from the variant. We will analyse ETCS in more detail in Chap. 7.

In this example, we can see the effect of the $d\mathcal{L}$ calculus. It takes a specification of a hybrid system and successively identifies constraints on the parameters which are needed for correctness. These constraints can then be handled in a purely modular way by rules i$\forall$ and i$\exists$. As a typical characteristic of hybrid systems, further observe that intermediate formulas are significantly more complex than the original proof obligation, which can be expressed succinctly in the expressive language of $d\mathcal{L}$. This reflects the fact that the actual complexity of hybrid systems originates from hybrid interaction, not from a single transition. Still, using appropriate proof strategies (Chap. 5) for the $d\mathcal{L}$ calculus, the safety statement (2.7) with invariant (2.17) can be verified automatically in a theorem prover that invokes Mathematica for rules $\langle ' \rangle$,$[']$, i$\forall$, and i$\exists$. In fact, using the invariant computation techniques that we introduce in Chap. 6, the overall safety property (2.7) can be verified fully automatically even without providing an invariant manually.

2.10 Summary

We have introduced a first-order dynamic logic for hybrid programs, which are uniform operational models for hybrid systems with interacting discrete jumps and continuous evolutions along differential equations. For this differential dynamic logic, $d\mathcal{L}$, we have presented a concise generalised free-variable proof calculus over the reals.

Our sequent calculus for $d\mathcal{L}$ is a generalisation of classical calculi for discrete dynamic logic [35, 37, 149, 148] to the hybrid case. It is a compositional verification calculus for verifying properties of hybrid programs by decomposing them into properties of their parts. In order to handle interacting hybrid dynamics, we lift real quantifier elimination to the deductive calculus in a new modular way that is suitable for automation, using real-valued free variables, Skolem terms, and invertible quantifier rules over the reals.

As a fundamental result aligning hybrid and continuous reasoning proof-theoretically, we have proven our calculus to axiomatise the transition behaviour of hybrid systems completely relative to the handling of differential equations. To the best of our knowledge, this is the first relative completeness result for hybrid systems' verification. Moreover, we have demonstrated that our calculus is well suited for practical automatic verification in a realistic case study of a fully parametric version of the European Train Control System.

Dynamic logic can be augmented [37] to support reasoning about dynamically reconfiguring system structures, which we want to extend to hybrid systems in future work. While the $d\mathcal{L}$ calculus is complete relative to the continuous fragment, it is a subtle open problem whether a converse calculus can exist that is complete relative to various discrete fragments.

Chapter 3
Differential-Algebraic Dynamic Logic DAL

Contents

A. Platzer, Logical Analysis of Hybrid Systems,
DOI 10.1007/978-3-642-14509-4_3,

Synopsis We generalise dynamic logic to a logic for differential-algebraic programs, i.e., discrete programs augmented with first-order differential-algebraic formulas as continuous evolution constraints in addition to first-order discrete jump formulas. These programs characterise interacting discrete and continuous dynamics of hybrid systems elegantly and uniformly, including systems with disturbance and differential-algebraic equations. For our logic, we introduce a calculus over real arithmetic with discrete induction and a new *differential induction* with which differential-algebraic programs can be verified by exploiting their differential constraints algebraically without having to solve them. We develop the theory of differential induction and differential refinement and analyse their deductive power. As an example, we present parametric tangential roundabout manoeuvres in air traffic control and prove collision avoidance in our calculus.

3.1 Introduction

In Chap. 2, we have shown a verification logic for hybrid programs and proof rules for differential equations that are solvable in polynomial arithmetic. While the verification approach is, of course, more general, the primary application that we have seen so far is train control. The same proof techniques also work for many other systems, including car control where the continuous dynamics can be solved. Now we consider air traffic control scenarios as a motivating example for hybrid systems where differential equations can no longer be solved in decidable arithmetic, so that more advanced verification techniques for differential equations are needed.

Verification of Hybrid Systems

Flight manoeuvres in air traffic control [293, 196, 203, 104, 92, 238, 129, 171] give rise to hybrid systems with challenging dynamics. There the continuous dynamics results from continuous movement of aircraft in space, and the discrete dynamics is caused by the instantaneous switching of manoeuvring modes or by discrete aircraft controllers that decide when and how to initiate flight manoeuvres. Proper functioning of these systems is highly safety-critical for the spatial separation of aircraft during all flight manoeuvres, especially collision avoidance manoeuvres. Their analysis, however, is challenging due to the superposition of involved continuous flight dynamics with nontrivial discrete control, causing hybrid systems like these to be amenable neither to mere continuous reasoning nor to verification techniques for purely discrete systems. Since, especially in the presence of parameters, hybrid systems cannot be verified numerically [238, 85], we present a purely symbolic approach using combined deductive and algebraic verification techniques.

In practise, correctness of hybrid systems further depends on the choice of parameters that arise naturally from the degrees of freedom of how a part of the system can be instantiated or how a controller can respond to input [293, 97, 91, 238, 171]. For instance, correct angular velocities, proper timing, and compatible manoeuvre points are equally required for safe air traffic control [293, 238]. Additionally, relevant correctness properties for hybrid systems include safety, liveness, and mixed properties like reactivity (see Chap. 7), all of which can possibly involve (alternating) quantifiers or free variables for parameters. As a uniform approach for specifying and verifying these heterogeneous properties of hybrid systems with symbolic parameters, we introduce an extension of our first-order logic and dynamic logic for handling correctness statements about hybrid systems.

Logic for Hybrid Systems

The aim of this chapter is to present logical analysis techniques with which general hybrid systems with interacting discrete and continuous dynamics can be specified and verified in a coherent logical framework. To this end, we introduce the *differential-algebraic dynamic logic* (DA-logic or DAL for short) as the logic of general hybrid change. As an elegant and uniform operational model for hybrid systems in DAL, we introduce *differential-algebraic programs* (DA-programs). These programs combine *first-order discrete jump constraints* (DJ-constraints) to characterise discrete transitions with support for *first-order differential-algebraic constraints* (DA-constraints) to characterise continuous transitions. DA-constraints provide a convenient way for expressing continuous system evolution constraints and give a uniform semantics to differential evolutions, systems of differential equations [297], switched systems [55], invariant constraints [156, 97], triggers [55], differential-algebraic equations [132, 187], and differential inequalities [153, 297]. In DJ-constraints and DA-constraints, first-order quantifiers further give a natural and semantically well-founded way of expressing unbounded discrete or continuous nondeterminism in the dynamics, including nondeterminism resulting from internal choices or external disturbances. In interaction with appropriate control structure, DJ-constraints and DA-constraints can be combined to form DA-programs as uniform operational models for hybrid systems. With this, DA-programs are a generalised program notation for hybrid systems.

As a specification and verification logic for hybrid systems that are given as DA-programs, we design the first-order dynamic logic DAL as an extension of $\mathsf{d}\mathcal{L}$. In particular, we generalise discrete dynamic logic [149] to hybrid control and support DA-programs as actions of a first-order multi-modal logic [123], such that its modalities can be used to specify and verify correctness properties of more advanced hybrid systems. As in Chap. 2, the DAL formula $[\alpha]\phi$ expresses that all runs of DA-program α lead to states satisfying the DAL formula ϕ. Likewise, $\langle\alpha\rangle\phi$ says that there is at least one state reachable by DA-program α which satisfies ϕ. Similarly, $\exists p\,[\alpha]\langle\beta\rangle\phi$ says that there is a choice of parameter p such that for all possible behaviour of DA-program α there is a reaction of DA-program β that ensures ϕ.

Deductive Verification and Differential Induction

As a means for verifying hybrid systems by proving corresponding DAL formulas, we introduce a sequent calculus. It uses side deductions as a simple and concise, yet constructive, modular technique to integrate real quantifier elimination with calculus rules for modalities. For handling discrete transitions, we present a first-order generalisation of standard proof rules [149, 37]. Interacting continuous transitions are more involved. Formulas with very simple differential equations can be verified by using their solutions as in Chap. 2: Linear differential equations with nilpotent constant coefficients (i.e., $x' = Ax$ for a matrix A with $A^n = 0$ for some n) have polynomial solutions so that arithmetic formulas about these solutions can be verified by quantifier elimination [81]; see Apps. B and D.2 for details. This approach, however, does not scale to hybrid systems with more sophisticated differential constraints because their solutions do not support quantifier elimination (e.g., when they involve transcendental functions), cannot be given in closed form [297], are not computable [250], or do not even exist [297, 179]. As part of the descriptive power of differential equations, solutions of differential equations are much more complicated than the original equations and can become transcendental even for simple linear differential equations like $x' = -y, y' = x$, where the solutions will be trigonometric functions.

Instead, as a logical analysis technique for verifying DA-programs with more general differential-algebraic constraints, we introduce *first-order differential induction* as a fully algebraic form of proving logical statements about DA-constraints using their differential-algebraic constraints in a differential induction step instead of using their solutions in a reachability computation. Unlike in discrete induction, the invariant is a *differential invariant*, i.e., a property that is closed under total differentiation with respect to the differential constraints. There, the basic idea for showing invariance of a property F is to show that F holds initially and its total derivative F' holds always along the dynamics (with generalisations of total differentials to logical formulas and corresponding generalisations for quantified DA-constraints). This analysis considers all non-Zeno executions, i.e., where the system cannot switch its mode infinitely often in finite time. In addition, we introduce *differential strengthening* or differential cuts as a technique for refining the system dynamics by differential invariants until the property becomes provable for the refined dynamics, which we show to be crucial in practical applications.

Comparison

In Chap. 2 we have introduced a logic and calculus for verifying *hybrid programs*, which is the quantifier-free subclass of DA-programs without propositional connectives (see Table 3.1 for examples). Further, we have proven this calculus to be complete *relative to* the handling of differential equations (Theorem 2.3). Complementarily, in this chapter, we address the question how sophisticated differential constraints themselves can be specified and verified in a way that lifts to hybrid sys-

Table 3.1 Comparison of DAL with DA-programs versus $d\mathcal{L}$ with hybrid programs

	$d\mathcal{L}$/hybrid programs	DAL/DA-programs
expressive power	single assignments $x:=1$ differential equations $x_1'=d_1, x_2'=d_2$	propositional/quantified DJ-constraints $x>0 \rightarrow \exists a\,(a<5 \wedge x:=a^2+1)$ propositional/quantified DA-constraints $\exists \omega \leq 1\,(d_1'=-\omega d_2 \wedge d_2'=\omega d_1) \vee d_1' \leq d_2' \leq 2d_1$
verification technology	substitutions *polynomial* solutions	quantifier elimination and substitutions first-order differential induction
quantifier integration	real-valued free variables, Skolemisation	side deductions
scope of applications	nilpotent dynamics, e.g., trains in $\mathbb{R}^1$	algebraic dynamics and polynomial differential constraints, e.g., curved aircraft flight

tems, and how these techniques can be integrated seamlessly into a logic. In this chapter, we thus show how properties of systems with more complicated dynamics can be expressed and proven.

To this end, we design differential-algebraic programs as the *first-order completion* of hybrid programs, and we augment both the logic and the calculus with means for handling DA-constraints. In particular, we extend our logic $d\mathcal{L}$ to the logic DAL with general first-order differential constraints plus first-order jump formulas and introduce differential induction for verifying differential-algebraic programs. Specifically, the continuous evolutions which can be handled by differential induction are strictly more expressive than those that previous calculi [306, 270, 97] or the $d\mathcal{L}$ calculus are able to handle. DAL even supports differential-algebraic equations [132, 187]. Consequently, the DAL calculus can verify much more general scenarios, including the dynamics of aircraft manoeuvres or the dynamics of systems with disturbances, which were out of scope for approaches that require polynomial solutions [125, 228]. Table 3.1 summarises the differences in syntactic expressiveness, discrete and continuous verification technology, arithmetic quantifier integration approach, and overall scope of applicability. The DAL extensions presented in this chapter are both complementary to and compatible with our $d\mathcal{L}$ calculus extensions for integrating arithmetic as presented in Chap. 2.

Contributions

The first contribution of this chapter is the generalised differential-algebraic dynamic logic DAL for differential-algebraic programs as the first-order completion of hybrid programs. DAL provides a uniform semantics and a concise language for specifying and verifying correctness properties of general hybrid systems with sophisticated (possibly quantified) first-order dynamics. The main contribution is a verification calculus for DAL including uniform proof rules for differential induction along first-order differential-algebraic constraints with differential invariants, differential variants, and differential strengthening. Our main theoretical contribution is our analysis of the deductive power of differential induction for classes of differential invariants. As an applied contribution, we introduce a generalised tangential roundabout manoeuvre in air traffic control and demonstrate the capabilities of our approach by verifying collision avoidance in the DAL calculus. To the best of our knowledge, this is the first formal proof for safety of the hybrid dynamics of an aircraft manoeuvre with curved flight dynamics and the first sound verification result for collision avoidance with curved aircraft dynamics.

3.1.1 Related Work

Most verification approaches for hybrid systems follow the paradigm of model checking for hybrid automata and use approximations or abstraction refinement, e.g., [156, 77, 21], because reachability is undecidable for hybrid automata [156]. We have shown in previous work [238] that even reachability problems for fairly restricted classes of single continuous transitions are not decidable using numerical computations. Thus, we follow a purely symbolic approach in this book.

Invariants of Hybrid Systems

Several authors [274, 269, 251, 252] argue that invariant techniques scale to more general dynamics than explicit reach-set computations or techniques that require solutions of the differential equations [125, 228, 231, 233]. Among them, there are model checking approaches [274, 269] that use equational polynomial invariants based on Gröbner basis computations. Still, the approach of Rodríguez-Carbonell and Tiwari [269] requires closed-form solutions and is restricted to linear dynamics. The major limitation of these approaches [274, 269], however, is that they only work for equational invariants of fully equation-definable hybrid systems, including equational initial sets and switching surfaces. Yet, this assumes highly regular systems without tolerances and only works for null sets. In practise, the set of initial states usually does not have measure zero, though. A thorough analysis of collision avoidance manoeuvres, for instance, should consider all initial flight paths in free flight instead of just a single restricted position corridor. In train control, the relevant constraints and initial regions are non-equational.

Prajna et al. [251, 252] have generalised Lyapunov functions to barrier certificates, i.e., a function B decreasing along the dynamics whose zero set separates initial from unsafe states. Further, they focus on stochastic extensions. DAL provides barrier certificates as a special case using $B \leq 0$ as a differential invariant. In a similar vein, criticality functions [91] generalise Lyapunov-functions from stability to safety, which DAL provides as a special case of differential invariants.

We generalise purely equational invariants [274, 269] and single polynomial expressions [274, 251, 252, 91] to general differential induction with real arithmetic formulas. In practise, such more general differential invariants are needed for verifying sophisticated hybrid systems, including aircraft manoeuvres. Further, unlike other approaches [274, 269, 251, 252], DAL leverages the full deductive power of logic, combining differential induction with discrete induction to lift these proof techniques uniformly to hybrid systems. In addition, dynamic logic can be used to prove sophisticated statements involving quantifier and modality alternations for parametric verification [231]. Finally, the DAL calculus supports combinations with differential variants for liveness properties and combinations with differential strengthening, which we show to be crucial in verifying realistic aircraft manoeuvres.

Air Traffic Control Verification

In air traffic control, Tomlin et al. [293] analyse competitive aircraft manoeuvres game-theoretically using Hamilton-Jacobi-Isaacs partial differential equations. They derive saddle solutions for purely angular or purely linear control actions. They propose roundabout manoeuvres and give bounded-time verification results for trapezoidal straight-line approximations. Our symbolic techniques avoid exponential state space discretisations that are required for complicated PDEs and are thus more scalable for automation. Further, we handle fully parametric cases, even for more complicated curved flight dynamics.

Hwang et al. [171] have presented a straight-line aircraft conflict avoidance manoeuvre that involves optimisation over complicated trigonometric computations, and validate it on random numerical simulation. They show examples where the decisions of the manoeuvre change only slightly for small perturbations. Hwang et al. do not, however, prove that their proposed manoeuvre is safe with respect to actual hybrid flight dynamics.

Dowek et al. [104] and Galdino et al. [129] consider straight-line manoeuvres and formalise geometrical proofs in PVS. As in the work of Hwang et al. [171], they do not, however, consider curved flight paths nor verify actual hybrid dynamics but work with geometrical meta-level reasoning instead.

In all these approaches [104, 129, 171], it remains to be proven separately that the geometrical meta-level considerations actually fit the hybrid dynamics and flight equations. In contrast, our approach directly works for the hybrid flight dynamics and we verify roundabout manoeuvres with curves instead of straight-line manoeuvres with non-flyable instant turns only. A few approaches [92, 203] have been undertaken to model check discretisations of roundabout manoeuvres, which indicate avoidance of orthogonal collisions. However, the counterexamples found by our model checker in previous work [238] show for these manoeuvres that collision avoidance does not extend to other initial flight paths.

3.1.2 Structure of This Chapter

In Sects. 3.2 and 3.3, we introduce syntax and semantics of the differential-algebraic logic DAL. In Sect. 3.4, we introduce tangential roundabout manoeuvres in air traffic control as a case study and running example. Further, we introduce a sequent calculus with differential induction for DAL in Sect. 3.5 and prove soundness in Sect. 3.6. We show extensions of differential induction techniques in Sect. 3.7. We exploit differential induction techniques for differential monotonicity relaxations in Sect. 3.8. We prove relative completeness of the DAL calculus in Sect. 3.9 and compare the deductive strength of differential invariants in Sect. 3.10. Using the DAL calculus, we prove, in Sect. 3.11, safety of the tangential roundabout manoeuvre in air traffic control. Finally, we draw conclusions and discuss future work in Sect. 3.12.

3.2 Syntax of Differential-Algebraic Logic

In this section, we introduce the *differential-algebraic logic* (DAL) as a specification and verification logic for *differential-algebraic programs* (DA-programs). DA-programs constitute an elegant and uniform model for hybrid systems with expressive dynamics, including differential-algebraic equations, differential inequalities, and disturbance in the dynamics. We start with an informal introduction that motivates the definitions to come. DA-programs have three basic characteristics, as follows.

Discrete jump constraints. Discrete transitions, which can possibly lead to discontinuous change, are represented as discrete jump constraints (*DJ-constraints*), i.e., first-order formulas with instantaneous assignments of values to state variables as additional atomic formulas. DJ-constraints specify what new values the respective state variables assume by an instant change. For instance, $d_1 := -d_2$ specifies that the value of variable d_1 is changed to the value of $-d_2$. Multiple discrete changes can be combined conjunctively ($\wedge$) with simultaneous effect, for instance, $d_1 := -d_2 \wedge d_2 := d_1$, which assigns the previous value of $-d_2$ to d_1 and, simultaneously, the previous value of d_1 to d_2. This operation instantly rotates the vector $d = (d_1, d_2)$ by $\pi/2$ to the left. Using $d := d^{\perp}$ as a short vectorial notation for this jump, the DJ-constraint $(d_1 > 0 \rightarrow d := d^{\perp}) \wedge (d_1 \leq 0 \rightarrow d := -d^{\perp})$ specifies that the direction of the rotation depends on the initial value of d_1. Finally, the DJ-constraint $\exists a\, (\omega := a^2 \wedge a < 5)$ assigns the square of *some* number a less than 5 to ω.

Differential-algebraic constraints. Continuous dynamics is represented with differential-algebraic constraints (*DA-constraints*) as evolution constraints, i.e., first-order formulas with differential symbols x', e.g., in differential equations or inequalities. DA-constraints specify how state variables change continuously over time. For instance, $x_1' = d_1 \wedge x_2' = d_2$ says that the system evolves continuously by moving the vector $x = (x_1, x_2)$ in direction $d = (d_1, d_2)$ along the differential equation system $(x_1' = d_1, x_2' = d_2)$. Likewise, $d_1' = -\omega d_2 \wedge d_2' = \omega d_1 \wedge d_1 \geq 0$ specifies that the vector d is rotating continuously with angular velocity ω, so that (in conjunction with $x_1' = d_1 \wedge x_2' = d_2$), the direction in which point x is heading changes over time. By adding $d_1 \geq 0$ conjunctively to the DA-constraint, we express that the curving will only be able to continue while $d_1 \geq 0$. This evolution will have to stop before $d_1 < 0$. The evolution is impossible altogether if $d_1 \geq 0$ already fails to hold initially. DA-constraints are first-order and can have quantifiers.The quantified DA-constraint $\exists \omega \, (d_1' = -\omega d_2 \wedge d_2' = \omega d_1 \wedge -1 \leq \omega \leq 1)$ characterises rotation with *some* angular velocity $-1 \leq \omega \leq 1$, which may even change over time, in contrast to $d_1' = -\omega d_2 \wedge d_2' = \omega d_1 \wedge -1 \leq \omega \leq 1 \wedge \omega' = 0$ or constraint $d_1' = -\omega d_2 \wedge d_2' = \omega$, where ω is not allowed to change.

Differential-algebraic programs. As an operational model for hybrid systems, DJ-constraints and DA-constraints, which represent general discrete and continuous transitions, respectively, can be combined to form a DA program using regular expression operators ($\cup, ^*, ;$) of regular discrete dynamic logic [149] as control structure. For example, $\omega := 1 \cup \omega := -1$ describes a controller that can choose to set angular velocity ω either to a left or to a right curve by a nondeterministic choice ($\cup$). Similarly, sequential composition $\omega := \omega + 1; d_1' = -\omega d_2 \wedge d_2' = \omega d_1$ says that the system first increases its angular velocity ω by a discrete transition and then switches to a mode in which it follows a continuous rotation with this angular velocity.

Discussion

Not all constraints involving $x := \theta$ or x' qualify as reasonable ways of characterising elementary system transitions. Unlike positive occurrences, negative occurrences of assignments such as in $\neg(x := 5)$ are somewhat pointless, because they impose no meaningful transition constraints on what new value x actually assumes (but only on what value is *not* assigned to x). Likewise, negative occurrences of differential constraints as in $\neg(x' = 5)$ would be pointless as they do not constrain the overall evolution but allow arbitrary transitions.

Further, we disallow duplicate constraints that constrain the same variable in incompatible ways at the same time as, e.g., in $x := 2 \wedge x := 3$ or $x' = 2 \wedge x' = 3$. At any state during a system evolution, variable x can only assume one value at a time, not both 2 and 3 at once. Similarly, variables cannot evolve with contradictory slopes at the same time for any positive duration.

Finally, $\forall a \, (x := a)$ would be equivalent to false, because it is impossible to assign all possible choices for a (hence all reals) simultaneously to x, which can only assume one value at a time. Likewise, no interesting evolutions are possible along $\forall a \, (x' = a)$, because x' can only equal one real value at a time. Dually, $\exists a \, (a := \theta)$ is equivalent to true, because the DJ-constraint imposes no constraints, nor has any visible effects (the scope of the quantified a ends with the DJ-constraint). The situation with $\exists a \, (a' = \theta)$ is similar.

Even though semantics and proof rules for all these cases could still be defined, the respective transitions are degenerate and their technical handling is not very illuminating. Hence, we define DJ-constraints and DA-constraints to avoid these insignificant cases altogether. Note that the syntactical restrictions are non-essential but simplify the presentation by allowing us to focus on the interesting cases instead.

3.2.1 Terms

To simplify the presentation, we use side deduction rules for quantifiers in this chapter. The free-variable calculus rules from Chap. 2 are compatible with the findings in this chapter, but we refrain from using them formally in this chapter to keep things simple. Consequently, we do not need to distinguish between free logical variables from V and free state variables from Σ. Thus, we do not distinguish Σ and V here at all.

The formulas of DAL are built over a signature Σ of real-valued function and predicate symbols. The signature Σ contains the usual function and predicate symbols for real arithmetic, $+,-,\cdot,/,=,\leq,<,\geq,>$, and number symbols such as $0,1$. State variables are represented as real-valued function symbols of arity zero (constants) in Σ. These state variables are *flexible* [37], i.e., their interpretation can change from state to state while following the transitions of a DA-program. Observe that there is no need to distinguish between discrete and continuous variables in DAL. The set $\mathrm{Term}(\Sigma)$ of *terms* is defined as in classical first-order logic, yielding rational expressions over the reals. The set of formulas of first-order logic is defined as usual (Definition 2.2 and App. A), giving first-order real arithmetic.

Although we are primarily interested in polynomial cases, our techniques generalise to the presence of division. Yet to avoid partiality in the semantics, we only allow use of p/q when $q \neq 0$ is present or ensured. Essentially, we assume constraints ϕ containing a term of the form p/q to mean an appropriate constraint like $\phi \wedge \neg(q=0)$. Note that, in a certain sense, divisions cause less difficulties for the calculus than for the semantics. Particularly, our calculus uses indirect means of differential induction to conclude properties of solutions of DA-constraints, thereby avoiding the need to handle singularities in these solutions explicitly, as caused by divisions by zero.

3.2.2 Differential-Algebraic Programs

We build DA-programs with first-order discrete jumps and first-order differential-algebraic constraints as primitive operations, which interact using regular control structure by regular-expression-style operators ($;\,\cup,{}^*$). Reflecting the discussion before Sect. 3.2.1, we characterise reasonable occurrences for changes like $x:=\theta$ or x'

in these constraints as follows. We call a formula G an *affirmative subformula* of a first-order formula F iff:

1. G is a positive subformula of F, i.e., it occurs with an even number of negations, and
2. no variable y that occurs in G is in the scope of a universal quantifier $\forall y$ of a positive subformula of F (or in the scope of an existential quantifier $\exists y$ of a negative subformula of F).

Discrete Jump Constraints

Definition 3.1 (Discrete jump constraint). A *discrete jump constraint (or DJ-constraint)* is a formula $\mathscr{J}$ of first-order real arithmetic over Σ with additional atomic formulas of the form $x := \theta$ where $x \in \Sigma$, $\theta \in \mathrm{Term}(\Sigma)$. The latter are called assignments and are only allowed in affirmative subformulas of DJ-constraints that are not in the scope of a quantifier for x of $\mathscr{J}$. A DJ-constraint without assignments is called *jump-free*. A variable x is (possibly) *changed* in $\mathscr{J}$ iff an assignment of the form $x := \theta$ occurs in $\mathscr{J}$.

The effect of $(x_1 := \theta_1 \wedge .. \wedge x_n := \theta_n \wedge x_1 > 0) \vee (x_1 := \vartheta_1 \wedge .. \wedge x_n := \vartheta_n \wedge x_1 < 0)$ is to simultaneously change the interpretations of the variables x_i to the respective θ_i if $x_1 > 0$, and to change the x_i to ϑ_i if $x_1 < 0$. If neither case applies ($x_1 = 0$), the DJ-constraint evaluates to false as no disjunct applies, so no jump is possible at all, which will prevent the system from continuing any further. In particular, a jump-free DJ-constraint such as $x \geq y$ corresponds to a test. It completes without changing the state if, in fact, $x \geq y$ holds true in the current state, and it aborts system evolution otherwise (deadlock). Especially, unlike the assignment $x := \theta$, which changes the value of x to that of θ, the test $x = \theta$ fails by aborting the system evolution if x does not already happen to have the value θ. If cases overlap, as in $(x := x - 1 \wedge x \geq 0) \vee x := 0$, either disjunct can be chosen to take effect by a nondeterministic choice.

Quantifiers within DJ-constraints express *unbounded discrete nondeterministic choices*. For instance, the following quantified DJ-constraint assigns *some* vector $u \in \mathbb{R}^2$ to e such that the rays spanned by $d = (d_1, d_2)$ and $u = (u_1, u_2)$ intersect:

$$\exists u_1 \exists u_2\, (e_1 := u_1 \wedge e_2 := u_2 \wedge \exists \lambda{>}0\, \exists \mu{>}0\, (\lambda d_1 = \mu u_1 \wedge \lambda d_2 = \mu u_2)).$$

We informally use *vectorial notation* when no confusion arises. Using vectorial quantifiers, equations, arithmetic, and assignments, the latter DJ-constraint simplifies to:

$$\exists u\, (e := u \wedge \exists \lambda{>}0\, \exists \mu{>}0\, \lambda d = \mu u).$$

Example 3.1 (DJ-constraints). Examples of DJ-constraints include:

- $d_1 := -d_2$

- $d_1 := -d_2 \wedge d_2 := d_1$ — simultaneous effect is an instant left rotation by $\frac{\pi}{2}$
- $(d_1 > 0 \rightarrow d_1 := -d_2 \wedge d_2 := d_1) \wedge (d_1 \leq 0 \rightarrow d_1 := d_2 \wedge d_2 := -d_1)$ — effect depends on the sign of d_1
- $\exists a\,(\omega := a^2 \wedge a < 5)$ — quantified nondeterministic effect
- $d_1 > 0 \rightarrow \exists a\,(a < 5 \wedge d_1 := a^2 + 1)$ — but this DJ-constraint is inhomogeneous, it does not specify what happens if $d_1 \leq 0$.

The following cases are not allowed as DJ-constraints:

- $d_1 := -d_2 \wedge d_1 := 0$ — incompatible jump, because d_1 cannot assume both values at once
- $\neg(d_1 := 5)$ — does not specify what really *is* assigned to d_1, only what is not
- $\forall a\,(\omega := a^2)$ — ω cannot assume all those values a^2 at once for all a
- $\exists a\,(a := d_1)$ — is just equivalent to *true*, because it has no visible effects or constraints (the scope of the quantified a ends with the DJ-constraint)

It is not impossible to give a reasonable semantics to these disallowed cases and handle them in proof rules. Yet it would complicate the technical treatment unnecessarily and would distract from the important aspects. □

Differential-Algebraic Constraints

Definition 3.2 (Differential-algebraic constraints). A *differential-algebraic constraint (DA-constraint)* is a formula $\mathscr{D}$ of first-order real arithmetic over $\Sigma \cup \Sigma'$ in which symbols of Σ' only occur in affirmative subformulas that are not in the scope of a quantifier of $\mathscr{D}$ for that symbol. Here, Σ' is the set of all *differential symbols* $x^{(n)}$ with $n \in \mathbb{N}$ for state variables $x \in \Sigma$. A DA-constraint without differential symbols $x^{(n)}$ for $n \geq 1$ is called *non-differential.* A variable x is (possibly) *changed* in $\mathscr{D}$ iff $x^{(n)}$ occurs in $\mathscr{D}$ for an $n \geq 1$.

Syntactically, $x^{(n)}$ is like an ordinary function symbol of arity 0, but only allowed to occur within DA-constraints, and not in any other formula. The intended semantics of a differential symbol $x^{(n)}$ is to denote the nth time derivative of x, which is used to form differential equations (or differential inequalities). We write x' for $x^{(1)}$ and x'' for $x^{(2)}$ and, sometimes, $x^{(0)}$ for the non-differential symbol x. The (partial) *order* $\text{ord}_x \mathscr{D}$ of a DA-constraint $\mathscr{D}$ *in* x is the highest order $n \in \mathbb{N}$ of a differential symbol $x^{(n)}$ occurring in $\mathscr{D}$, or is not defined if no such $x^{(n)}$ occurs. The notion of order is accordingly for terms.

The effect of a DA-constraint $\mathscr{D}$ is an ongoing continuous evolution respecting the differential and non-differential constraints of $\mathscr{D}$ during the whole evolution. For instance, the effect of $(x' = \theta \wedge x > 0) \vee (x' = -x^2 \wedge x < 0)$ is that the system evolves along $x' = \theta$ while $x > 0$, yet evolves along $x' = -x^2$ when $x < 0$. This evolution can stop at any time but is never allowed to enter the region where neither case applies anymore ($x = 0$). There also is no transition at all if $x = 0$ holds in the beginning, because no disjunct applies in the initial state then.

More generally, the differential constraints of $\mathscr{D}$ describe how the valuations of the respective state variables change continuously over time while following $\mathscr{D}$. The non-differential constraints of $\mathscr{D}$ can be understood to express evolution domain restrictions or invariant regions of these evolutions for which the differential equations apply or within which the evolution resides. For instance, in the DA-constraint $d_1' = -\omega d_2 \land d_2' = \omega d_1 \land d_1 \geq 0$, differential equations $d_1' = -\omega d_2$ and $d_2' = \omega d_1$ describe the change and $d_1 \geq 0$ describes the invariance region or maximal domain of evolution. Overlapping cases are resolved as in DJ-constraints, i.e., by nondeterministic choice. Likewise, a DA-constraint where no case applies aborts the system evolution as it does not satisfy the DA-constraint. Hence, non-differential DA-constraints and jump-free DJ-constraints are both equivalent to pure tests ($?\chi$ from Chap. 2). Except for such tests, we need to distinguish DA-constraints from DJ-constraints: Only DA-constraints can have evolutions of nonzero duration and only DJ-constraints can lead to discontinuous changes.

Quantifiers within DA-constraints express *continuous nondeterministic choices*. For example, constraint $\exists u\,(d_1' = -(\omega+u)d_2 \land d_2' = (\omega+u)d_1 \land -0.1 \leq u \leq 0.1)$ expresses that the system follows a continuous evolution in which, at each time, the differential equations are respected for *some* choice of u in $-0.1 \leq u \leq 0.1$. In particular, the choice of u can be different at each time so that u amounts to a bounded nondeterministic disturbance during the rotation in the above DA-constraint.

Example 3.2 (DA-constraints). Examples of DA-constraints include:

- $x_1' = d_1 \land x_2' = d_2$
- $x_1' = d_1 \land x_2' = d_2 \land d_1' = -\omega d_2 \land d_2' = \omega d_1$
- $d_1' = -\omega d_2 \land d_2' = \omega d_1 \land d_1 \geq 0$
- $(d_1 > 0 \to d_1' = -d_2 \land d_2' = d_1) \land (d_1 \leq 0 \to d_1' = d_2 \land d_2' = -d_1)$
- $x_1' \leq d_1 \land x_2' \leq d_2$
- $x_1' = d_1 \land x_2' \leq d_2 \land d_1' < 2 \land d_2' \geq 0$
- $\exists \omega\,(d_1' = -\omega d_2 \land d_2' = \omega d_1 \land -1 \leq \omega \leq 1)$
- $\exists \omega\,(d_1' = -\omega d_2 \land d_2' = \omega d_1 \land -1 \leq \omega \leq 1) \lor (d_1' \leq d_2' \leq 2d_1)$
- $d_1 > 0 \to x_1' = d_1 \land x_2' = d_2 \land d_1' = -\omega d_2$ — but this DA-constraint is inhomogeneous, it does not specify what happens if $d_1 \leq 0$

The following cases are not allowed as DA-constraints:

- $d_1' = -d_2 \land d_1' = 1$ — incompatible slope, because d_1 cannot evolve with two different slopes at once
- $\neg(d_1' = 5)$ — does not specify what the slope of d_1 really *is*, only what it is not
- $\forall \omega\,(d_1' = -\omega d_2 \land d_2' = \omega d_1)$ — d_1 cannot have all those slopes at once
- $\exists a\,(a' = d_1)$ — is just equivalent to *true*, because it has no visible effects or constraints (the scope of the quantified a ends with the DA-constraint)

It is not impossible to give a reasonable semantics to these disallowed cases and handle them in proof rules. But it would complicate the technical treatment unnecessarily. □

When using constraint formulas to characterise system transitions, we face the usual frame problem: Typically, one does not expect variables to change their values unless the respective constraint explicitly specifies how. In this chapter, we indicate constant variables explicitly so that no confusion arises. In practical applications, however, it can be quite cumbersome to have to specify $z := z$ or $z' = 0$ explicitly for all variables z that are not supposed to change in a DJ-constraint or DA-constraint, respectively. To account for that, we will define the DAL semantics so that variables that are not changed by a DJ-constraint or DA-constraint keep their value. Since free nondeterministic change of variable y is expressible using $\exists a\,(y := a)$ or $\exists a\,(y' = a)$, respectively, we expect the changes of all changed variables to be specified explicitly in all cases of the constraints to improve readability:

Definition 3.3 (Homogeneous constraints). A DA-constraint or DJ-constraint $\mathscr{C}$ is called *homogeneous* iff, in each of the disjuncts of a disjunctive normal form of $\mathscr{C}$, every changed variable of $\mathscr{C}$ is changed exactly once.

Note that Lemma 3.3 from Sect. 3.5.2 will show that DA-constraints are equivalent to their disjunctive normal forms and, hence, the notion of homogeneity is well-defined. Throughout the chapter, we assume that all DA-constraints and DJ-constraints are homogeneous, thereby ensuring that all changed variables receive a new value in all cases of the respective constraint (or stay constant because they are changed nowhere in the constraint) and that no change conflicts occur.

Hence, variable y does not change during the DA-constraint $x' = -x \land x \geq y$ but works as a constant lower bound for the evolution of x, because no differential symbol $y^{(n)}$ with $n \geq 1$ occurs so that $y' = 0$ is assumed implicitly. If y is intended to vary, but its variation is not specified by a differential equation, because y varies according to some algebraic relation with x, then quantified DA-constraints can be used to represent such *differential-algebraic equations* [132]. For instance, the differential-algebraic equation $x' = -x, y^2 = x$, in which $y^2 = x$ is an algebraic variational constraint specifying how y changes over time, is expressible as the DA-constraint $x' = -x \land \exists u\,(y' = u \land y^2 = x)$. There, the quantified differential constraint on y essentially says that y can change arbitrarily (with arbitrary disturbance u as slope), but only so that it always respects the relation $y^2 = x$.

Differential-Algebraic Programs

Now we can define DA-programs as regular combinations of DJ-constraints and DA-constraints.

Definition 3.4 (Differential-algebraic programs). The set DA-program(Σ) of *differential-algebraic programs*, with typical elements α, β, is inductively defined as the smallest set such that:

- If $\mathscr{J}$ is a DJ-constraint over Σ, then $\mathscr{J} \in$ DA-program(Σ).
- If $\mathscr{D}$ is a DA-constraint over $\Sigma \cup \Sigma'$, then $\mathscr{D} \in$ DA-program(Σ).
- If $\alpha, \beta \in$ DA-program(Σ) then $(\alpha \cup \beta) \in$ DA-program(Σ).

Table 3.2 Statements and (informal) effects of differential-algebraic programs

DA-program	Operation	Effect
$\mathscr{J}$	discrete jump	jump constraint with assignments holds for discrete jump
$\mathscr{D}$	diff.-alg. flow	differential-algebraic constraint holds during continuous flow
$\alpha;\beta$	seq. composition	DA-program β starts after DA-program α finishes
$\alpha\cup\beta$	nondet. choice	choice between alternative DA-programs α or β
α^*	nondet. repetition	repeats DA-program α n-times for any $n\in\mathbb{N}$

- If $\alpha,\beta\in\text{DA-program}(\Sigma)$ then $(\alpha;\beta)\in\text{DA-program}(\Sigma)$.
- If $\alpha\in\text{DA-program}(\Sigma)$ then $(\alpha^*)\in\text{DA-program}(\Sigma)$.

Choices $\alpha\cup\beta$ are used to express behavioural alternatives between α and β, i.e., the system follows either α or β. In particular, the difference between the DA-constraint $\mathscr{D}\vee\mathscr{E}$ and the DA-program $\mathscr{D}\cup\mathscr{E}$ is that the system has to commit to one choice of either $\mathscr{D}$ or $\mathscr{E}$ in $\mathscr{D}\cup\mathscr{E}$, but it can switch back and forth multiple times between disjunct $\mathscr{D}$ and $\mathscr{E}$ in $\mathscr{D}\vee\mathscr{E}$. The sequential composition $\alpha;\beta$ says that DA-program β starts executing after α has finished (β never starts if α does not terminate, e.g., due to a failed test in α). Observe that, as with repetitions, continuous evolutions within α can take more or less time. This nondeterminism is inherent in hybrid systems and as such reflected in DA-programs. Additional restrictions on the permitted duration of evolutions can simply be specified using auxiliary clocks, i.e., variables of derivative $\tau'=1$. For instance, $\tau:=0; x'=-x^2\wedge\tau'=1\wedge\tau\leq 5; ?\tau\geq 2$ specifies that the system only follows those evolutions along $x'=-x^2$ that take at most five (conjunct $\tau\leq 5$) but at least two time units (subsequent test $?\tau\geq 2$). Repetition α^* is used to express that the hybrid process α repeats any number of times, including zero.

Table 3.2 summarises the statements and (informal) effects of DA-programs. DA-programs still form a Kleene algebra [182] like hybrid programs from Chap. 2, but the primitive programs are more powerful. Further tests $?\chi$ are not necessary, because they can be defined by a jump-free DJ-constraint χ or by a non-differential DA-constraint χ.

Example 3.3 (Train control with disturbance). Recall the hybrid program for a train control system from d$\mathcal{L}$ formula (2.7) in Sect. 2.4:

$$\begin{aligned} &\psi \rightarrow [(ctrl;drive)^*]z\leq m \qquad (2.7^*)\\ \text{where } ctrl &\equiv (?m-z\leq s; a:=-b)\cup(?m-z\geq s; a:=A),\\ drive &\equiv \tau:=0; (z'=v, v'=a, \tau'=1\,\&\, v\geq 0\wedge\tau\leq\varepsilon). \end{aligned}$$

With a minor notational variation, this hybrid program is a DA-program:

$$\begin{aligned} &\psi \rightarrow [(ctrl;drive)^*]z\leq m\\ \text{where } ctrl &\equiv (m-z\leq s\wedge a:=-b)\cup(m-z\geq s\wedge a:=A),\\ drive &\equiv \tau:=0; (z'=v\wedge v'=a\wedge\tau'=1\wedge v\geq 0\wedge\tau\leq\varepsilon). \end{aligned}$$

Note also that we have decided to merge the test $?m-z \leq s$ with the subsequent assignment $a:=-b$ here to form the DJ-constraint $m-z \leq s \wedge a:=-b$. This simplification is non-essential and just used to show the expressiveness of DA-programs. In fact, we could also replace *ctrl* by a single DJ-constraint with the same semantics:

$$(m-z \leq s \wedge a:=-b) \vee (m-z \geq s \wedge a:=A).$$

The simple program assumes ideal-world dynamics with perfect control of the acceleration a, which takes effect exactly by the differential equation $z'=v, v'=a$. Yet this is not quite realistic. In reality, there are more aspects to train control that have an important impact on the dynamics, including wind, track conditions, slope of the track, mass, etc. In order to take these effects into account without having to build a full physical model of all parts, we replace the differential equation by a differential inequality, which allows for minor discrepancies. We thus replace the differential equation DA-constraint $z'=v \wedge v'=a \wedge v \geq 0$ with the differential inequality DA-constraint

$$z'=v \wedge a-l \leq v' \leq a+u \wedge v \geq 0 \tag{3.1}$$

for bounds $l>0$ and $u \geq 0$ on how much the actual acceleration may deviate from the set acceleration a. See Fig. 3.1 for an illustration of how the dynamics can deviate from the ideal-world dynamics in the presence of a disturbance that can change over time. With this generalisation the program can only be represented as a DA-

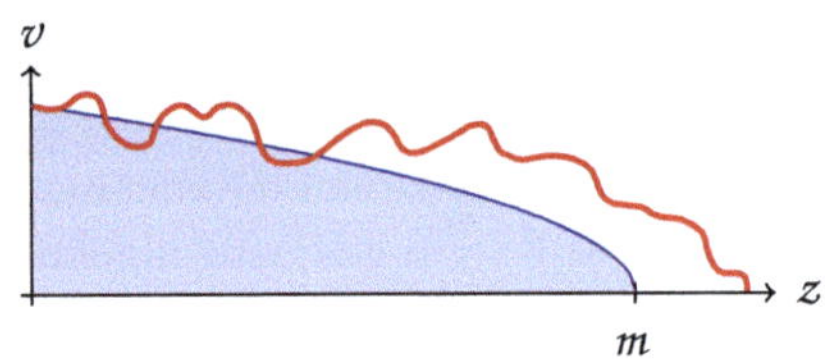

Fig. 3.1 Controllability violated in the presence of disturbance

program, and no longer as a hybrid program. Now, the formula from (2.7) will no longer be a d$\mathcal{L}$ formula, because DA-programs are not allowed in d$\mathcal{L}$. Instead, we will see that it is a well-formed DAL formula in Sect. 3.2.3.

Let us continue this line and replace the original differential equation by a quantified DA-constraint with a quantified disturbance as input to the dynamics:

$$z'=v \wedge \exists d\,(v'=d \wedge a-l \leq d \wedge d \leq a+u) \wedge v \geq 0. \tag{3.2}$$

This DA-constraint expresses that, at every point in time, there is a disturbance d affecting the actual dynamics. Yet this disturbance is bounded at any point by $a-l$ and $a+u$. Again, we obtain a DA-program instead of a hybrid program and again the resulting formula is no longer a d$\mathcal{L}$ formula. One question is what the relation is of those two different ways of generalising the ideal-world dynamics. It will turn out in Sect. 3.5.3 that the DA-constraints (3.1) and (3.2) are, in fact, equivalent, and so are the resulting DA-programs. □

Classification of Differential-Algebraic Programs

DA-programs give rise to an elegant syntactic hierarchy of discrete, continuous, and hybrid systems; see Table 3.3. Purely conjunctive DA-constraints correspond to continuous dynamical systems [279]. DA-constraints with disjunctions correspond to switched continuous dynamical systems [55]. DA-programs without DA-constraints correspond to discrete dynamical systems or, when restricted to domain $\mathbb{N}$ (which is definable in DAL), to discrete while programs [149]. Regular combinations of DJ-constraints form a complete basis of discrete programs [149]. Finally, general DA-programs correspond to (first-order generalisations of) hybrid dynamical systems [55, 156, 97].

Table 3.3 Classification of differential-algebraic programs and correspondence to dynamical systems

DA-program class	Dynamical systems class
conjunctive DA-constraints	continuous dynamical systems
DA-constraints	switched continuous dynamical systems
no DA-constraints	discrete dynamical systems
no DA-constraints, over $\mathbb{N}$	discrete while programs
general DA-programs	hybrid dynamical systems
	+ (generalised to first-order dynamics)

3.2.3 Formulas of Differential-Algebraic Logic

The set of *formulas* of DAL is defined as common in first-order dynamic logic [149]. They are built using propositional connectives and, in addition, if α is a DA-program and ϕ is a DAL formula, then $[\alpha]\phi, \langle\alpha\rangle\phi$ are DAL formulas. The intuitive reading of $[\alpha]\phi$ is that every run of DA-program α leads to states satisfying ϕ. Dually, $\langle\alpha\rangle\phi$ expresses that there is at least one run of DA-program α leading to such a state.

Definition 3.5 (DAL formulas). The set $\mathrm{Fml}(\Sigma)$ of DAL *formulas*, with typical elements ϕ, ψ, is inductively defined as the smallest set with:

- If $\theta_1, \theta_2 \in \mathrm{Term}(\Sigma)$ are terms, then $(\theta_1 \geq \theta_2) \in \mathrm{Fml}(\Sigma)$. The definition is accordingly for $=, \leq, <, >$.
- If $\phi, \psi \in \mathrm{Fml}(\Sigma)$, then $\neg\phi, (\phi \wedge \psi), (\phi \vee \psi), (\phi \to \psi) \in \mathrm{Fml}(\Sigma)$.
- If $\phi \in \mathrm{Fml}(\Sigma)$ and $\alpha \in \text{DA-program}(\Sigma)$, then $[\alpha]\phi, \langle\alpha\rangle\phi \in \mathrm{Fml}(\Sigma)$.

Quantifiers in DAL formulas are definable in terms of DA-constraints or quantified DJ-constraints. We consider quantifiers as abbreviations:

$$\forall x\,\phi \equiv [\exists a\, x{:=}a]\phi \equiv [x' = 1 \vee x' = -1]\phi$$
$$\exists x\,\phi, \equiv \langle\exists a\, x{:=}a\rangle\phi \equiv \langle x' = 1 \vee x' = -1\rangle\phi.$$

Table 3.4 Operators and meaning in differential-algebraic dynamic logic (DAL)

DAL	Operator	Meaning
$\theta_1 \geq \theta_2$	comparison	true iff denotation of θ_1 is greater or equal that of θ_2
$\neg\phi$	negation / not	true if ϕ is false
$\phi \wedge \psi$	conjunction / and	true if both ϕ and ψ are true
$\phi \vee \psi$	disjunction / or	true if ϕ is true or if ψ is true
$\phi \rightarrow \psi$	implication / implies	true if ϕ is false or ψ is true
$\phi \leftrightarrow \psi$	bi-implication / equivalent	true if ϕ and ψ are both true or both false
$\forall x\, \phi$	universal quantifier / for all	true if ϕ is true for all values of variable x
$\exists x\, \phi$	existential quantifier / exists	true if ϕ is true for some values of variable x
$[\alpha]\phi$	$[\cdot]$ modality / box	true if ϕ is true after all runs of DA-program α
$\langle\alpha\rangle\phi$	$\langle\cdot\rangle$ modality / diamond	true if ϕ is true after at least one run of DA-program α

The DAL formula $[\exists a\, x:=a]\phi$ considers *all* possibilities of assigning *some* value a to x, which amounts to universal quantification. Likewise, $\langle\exists a\, x:=a\rangle\phi$ considers some such choice, which is existential quantification. Similarly, the indeterminate continuous evolution $x' = 1 \vee x' = -1$ reaches all values, which amounts to the respective quantifier when combined with the appropriate modality. Note here that we can define quantifiers in terms of both DJ-constraints and DA-constraints although the former are discrete and the latter are continuous. The reason for this is that the time that passed during the continuous evolution is not observable in the system as no clock variable is included in the DA-constraint that could measure the progress of time.

With these abbreviations, we summarise the operators of differential-algebraic dynamic logic in Table 3.4. The structure of the logic DAL is mostly identical to the structure of $\mathsf{d}\mathcal{L}$ except for the significantly more expressive DA-programs inside modalities.

Example 3.4 (Train control with disturbance). In Example 3.3 we have replaced the differential equation of the train example by more general DA-constraints: the differential inequality (3.1) and the quantified DA-constraint (3.2), respectively. The generalised counterpart of the $\mathsf{d}\mathcal{L}$ formula (2.7) for its hybrid program is the following DAL formula for its DA-program:

$$\psi \rightarrow [(ctrl_d;drive_d)^*]z \leq m \tag{3.3}$$

$$\begin{aligned} \text{with } ctrl_d &\equiv (?m-z \leq s; a:=-b) \cup (?m-z \geq s; a:=A) \\ drive_d &\equiv \tau:=0; \\ &\quad (z' = v \wedge \exists d\,(v' = d \wedge a-l \leq d \wedge d \leq a+u) \wedge \tau' = 1 \wedge v \geq 0 \wedge \tau \leq \varepsilon). \end{aligned}$$

DAL formula (3.3) is more general than $\mathsf{d}\mathcal{L}$ formula (2.7), because it makes a safety claim about a more general dynamics, with nondeterministic quantified input d as disturbance. If we prove DAL formula (3.3) then this implies validity of $\mathsf{d}\mathcal{L}$ formula (2.7), because all behaviour of the hybrid program is a special case of the behaviour of the DA-program (with constant disturbance $d = 0$). Yet, how do we prove the more general DAL formula to be valid? Certainly, we can no longer solve its differential equations, because the DA-constraints now depend on quantified input or differential inequalities. □

One common pattern for representing safety statements about hybrid control loops is to use DAL formulas of the form $\phi \to [(controller; plant)^*]\psi$ for specifying that the system satisfies property ψ whenever the initial state satisfies ϕ. There, the system repeats a controller plant feedback loop, with a DA-constraint *plant* describing the continuous plant dynamics and a discrete DA-program *controller* describing the control decisions. The controller plant interaction repeats as indicated by the repetition star. Still, more general forms of systems and properties can be formulated and verified in DAL as well.

3.3 Semantics of Differential-Algebraic Logic

The semantics of DAL is a Kripke semantics with possible states of a hybrid system as possible worlds, where the accessibility relation between worlds is generated by the discrete or continuous transitions of DA-programs. A potential behaviour of a hybrid system corresponds to a succession of states that contain the observable values of system variables during its hybrid evolution.

3.3.1 Transition Semantics of Differential-Algebraic Programs

Since in this chapter we do not distinguish between free logical variables and constants, the semantics does not need to distinguish states and variable assignments.

A *state* is a map $\nu : \Sigma \to \mathbb{R}$; the set of all states is denoted by $\mathrm{State}(\Sigma)$. The function and predicate symbols of real arithmetic are interpreted as usual.

Definition 3.6 (Valuation of terms). The *valuation* $val(\nu,\cdot)$ *of terms* with respect to state ν is defined by

1. $val(\nu,x) = \nu(x)$ if $x \in \Sigma$ is a variable.
2. $val(\nu,\theta_1+\theta_2) = val(\nu,\theta_1)+val(\nu,\theta_2)$.
3. $val(\nu,\theta_1-\theta_2) = val(\nu,\theta_1)-val(\nu,\theta_2)$.
4. $val(\nu,\theta_1\cdot\theta_2) = val(\nu,\theta_1)\cdot val(\nu,\theta_2)$.
5. $val(\nu,\theta_1/\theta_2) = val(\nu,\theta_1)/val(\nu,\theta_2)$ if $val(\nu,\theta_2)\neq 0$.

Note that we do not need the semantics of θ_1/θ_2 for $val(\nu,\theta_2)=0$, because we have assumed the presence of constraints ensuring $\neg(\theta_2=0)$ for divisions.

Discrete Jump Constraints

The interpretation of discrete jump constraints is defined as in first-order real arithmetic, with the addition of an interpretation for assignment formulas.

Definition 3.7 (Interpretation of discrete jump constraints). The *interpretation* $(\nu,\omega) \models \mathcal{J}$ of DJ-constraint $\mathcal{J}$ for the pair of states (ν,ω) is defined as follows, where $\omega(z) = val(\omega,z) = val(\nu,z) = \nu(z)$ for all variables z that are not changed in $\mathcal{J}$:

1. $(\nu,\omega) \models x:=\theta$ iff $val(\omega,x) = val(\nu,\theta)$.
2. $(\nu,\omega) \models \theta_1 \geq \theta_2$ iff $val(\nu,\theta_1) \geq val(\nu,\theta_2)$, and accordingly for $=,\leq,<,>$.
3. $(\nu,\omega) \models \phi \wedge \psi$ iff $(\nu,\omega) \models \phi$ and $(\nu,\omega) \models \psi$. Accordingly for $\neg,\vee,\rightarrow$.
4. $(\nu,\omega) \models \phi \vee \psi$ iff $(\nu,\omega) \models \phi$ or $(\nu,\omega) \models \psi$.
5. $(\nu,\omega) \models \neg\phi$ iff it is not the case that $(\nu,\omega) \models \phi$.
6. $(\nu,\omega) \models \phi \rightarrow \psi$ iff it is not the case that $(\nu,\omega) \models \phi$ or it is the case that $(\nu,\omega) \models \psi$.
7. $(\nu,\omega) \models \forall x\,\phi$ iff $(\nu_x,\omega) \models \phi$ for all states ν_x that agree with ν except for the value of x.
8. $(\nu,\omega) \models \exists x\,\phi$ iff $(\nu_x,\omega) \models \phi$ for some state ν_x that agrees with ν except for the value of x.

Differential-Algebraic Constraints

To give a semantics to DA-constraints, differential symbols $x' \in \Sigma'$ must get a meaning. However, a DA-constraint like $d_1' = -\omega d_2 \wedge d_2' = \omega d_1$ cannot be interpreted in a single state ν, because derivatives are not defined in isolated points. Instead, DA-constraints are constraints that have to hold for an evolution of states over time. Along such a *flow* function $\varphi : [0,r] \to \text{State}(\Sigma)$, DA-constraints can again be interpreted locally by assigning to the formal differential symbol d_1' the analytic time derivative of the value of d_1 along φ at the respective points in time. As we assumed DA-constraints to avoid zero divisions, analytic derivatives are well-defined for $r > 0$ as $\text{State}(\Sigma)$ is isomorphic to a finite-dimensional real space with respect to the finitely many differential symbols occurring in the DA-constraint. We give a uniform definition for all durations $r \geq 0$ and defer the discussion of the understanding for $r = 0$ until the DA-constraint semantics has been presented in full. The philosophy behind hybrid systems is to isolate discontinuities in discrete transitions. Thus we assume that state variables (and their differential symbols, if present) always vary continuously along continuous evolutions over time.

Definition 3.8 (Differential state flow). A function $\varphi : [0,r] \to \text{State}(\Sigma)$ is called *state flow* of duration $r \geq 0$ if φ is componentwise continuous on $[0,r]$, i.e., for all $x \in \Sigma$, $\varphi(\zeta)(x)$ is continuous in ζ. Then, the *differentially augmented state* $\bar{\varphi}(\zeta)$ of φ at $\zeta \in [0,r]$ agrees with $\varphi(\zeta)$ except that it further assigns values to some of the differential symbols $x^{(n)} \in \Sigma'$: If $\varphi(t)(x)$ is n times continuously differentiable in t at ζ, then $\bar{\varphi}(\zeta)$ assigns the nth time derivative $\frac{\mathrm{d}^n \varphi(t)(x)}{\mathrm{d}t^n}(\zeta)$ of x at ζ to differential symbol $x^{(n)} \in \Sigma'$; otherwise the value of $x^{(n)} \in \Sigma'$ is not defined in $\bar{\varphi}(\zeta)$.

For a DA-constraint $\mathcal{D}$, a state flow φ of duration r is called *state flow of the order of $\mathcal{D}$* iff the value of each differential symbol occurring in $\mathcal{D}$ is defined on $[0,r]$, i.e., $\varphi(\zeta)(x)$ is n times continuously differentiable in ζ on $[0,r]$ for $n = \text{ord}_x \mathcal{D}$.

Example 3.5 (Differential state flow). We want to give a semantics to the DA-constraint $x' = x^3$, for which we have to specify where the dynamics of this DA-constraint evolves to when evolution starts in a state ν. Clearly, we can say what the semantics of the right-hand side of the differential equation is, and how the term x^3 evaluates in any state ν. But what should x' evaluate to? And how do we define when the terms x' and x^3 are equal. Remember: we also have to define when one term is less than the other for giving a semantics to differential inequalities like $x' < x^3$ or $x' \leq x^3$. At a single state ν, which is just an isolated point, derivatives are not defined, so that we cannot really give a meaning to the DA-constraints $x' = x^3$ or $x' \leq x^3$ when looking only at a single point.

Yet along a continuous state flow φ, we can make sense of x'. At any time ζ during the continuous state flow φ, we have values for all the variables, and can thus evaluate $val(\bar{\varphi}(\zeta), x^3)$ based on the value of variable $\bar{\varphi}(\zeta)(x)$ and of any other variables that occur. At time ζ along the flow, however, we can also give a value to x' as the value of the derivative of the value $val(\bar{\varphi}(t), x)$ of x by time t at the point in time ζ; see illustration in Fig. 3.2. We assign this value of the derivative at ζ to

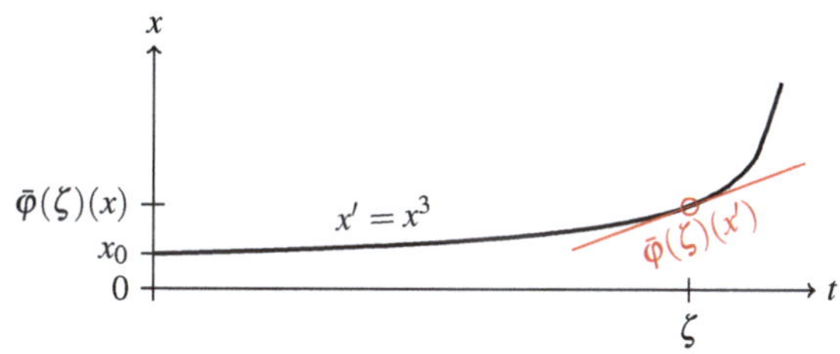

Fig. 3.2 Differential state flow

$\bar{\varphi}(\zeta)(x')$ in the differentially augmented state $\bar{\varphi}$ at time ζ. Intuitively, $\bar{\varphi}(\zeta)(x')$ is determined by considering how the value $val(\bar{\varphi}(\zeta), x)$ of x changes along the flow φ when we change time ζ "only a little bit". With this definition of the differentially augmented states $\bar{\varphi}$, we can easily evaluate DA-constraints like $x' = x^3$ or $x' \leq x^3$ at every point in time ζ by checking if $\bar{\varphi}(\zeta) \models x' = x^3$ or $\bar{\varphi}(\zeta) \models x' \leq x^3$, respectively, hold in the standard valuation of first-order logic, which is the same as checking if $val(\bar{\varphi}(\zeta), x') = val(\bar{\varphi}(\zeta), x^3)$ or $val(\bar{\varphi}(\zeta), x') \leq val(\bar{\varphi}(\zeta), x^3)$, respectively. The same principle can be used to give a semantics to more complicated DA-constraints, just by evaluating the first-order logic (including quantifiers) locally at every point in time along the differentially augmented state flow $\bar{\varphi}(\zeta)$. □

Definition 3.9 (Interpretation of differential-algebraic constraints). The *interpretation of DA-constraint $\mathscr{D}$* with respect to a state flow φ of the order of $\mathscr{D}$ and duration $r \geq 0$ is defined by: $\varphi \models \mathscr{D}$ iff, for all $\zeta \in [0, r]$,

1. $\bar{\varphi}(\zeta) \models_{\mathbb{R}} \mathscr{D}$ using the standard semantics $\models_{\mathbb{R}}$ of first-order real arithmetic (App. A), and
2. $val(\bar{\varphi}(\zeta), z) = val(\bar{\varphi}(0), z)$ for all variables z that are not changed by $\mathscr{D}$.

Observe that, along the state flows for a DA-constraint $\mathscr{D}$, only those variables whose differential symbols occur in $\mathscr{D}$ have to be continuously differentiable to the appropriate order. Quantified variables can change more arbitrarily (even discontinuously) during the evolution. The reason is that the semantics does not directly relate the value of a quantified variable like u in $\exists u x' = u^2$ at time ζ with the values that u assumes at later times. In particular, the value chosen for the quantified variable u can be different at every point in time, thereby giving a semantics of disturbance. Quantified variables may be constrained indirectly by their relations, though: In $\exists u x' = u^2$, the value of u^2 (but not that of u) also varies continuously over time, because x' varies continuously.

As a consequence of Picard-Lindelöf's theorem, a.k.a. the Cauchy-Lipschitz theorem (Theorem B.2), and using the fact that DAL terms are continuously differentiable on the open domain where divisors are nonzero, the flows of explicit quantifier-free, conjunctive DA-constraints of the form $x_1' = \theta_1 \land \cdots \land x_n' = \theta_n \land \chi$ with non-differential constraint χ are unique (as long as they exist): For each duration and initial value, there is at most one state flow φ respecting the DA-constraint (see Lemma 2.1). Yet, this is *not* the case for disjunctive DA-constraints, differential inequalities, quantified DA-constraints, or DA-constraints in implicit form such as $x'^2 - 1 = 0$, which has solutions $x(t) = x(0) + t$ and $x(t) = x(0) - t$. Finally, a non-differential constraint χ imposes no change but only tests whether χ holds. Hence, without differential constraints, a non-differential DA-constraint χ itself only has constant flows (if any), i.e., $\varphi(\zeta) = \varphi(0)$ for all ζ.

Restrictions of differential state flows to a prefix are again state flows. In particular, for all differential equations, the restriction to the point interval $[0,0]$ yields a trivial flow of no effect. For such point duration $r = 0$, however, derivatives and differentiability are not defined. To admit trivial flows nevertheless, the understanding of a DA-constraint is that its differential terms take no effect for flows of zero duration. That is, for trivial flows, atomic formulas with differential symbols are defined to evaluate to *true* as they occur only positively in DA-constraints. Thus, only the non-differential constraints of $\mathscr{D}$ impose constraints for trivial flows. A state flow of duration zero satisfying $\mathscr{D}$ and starting in some state ν exists iff ν satisfies the non-differential part of $\mathscr{D}$, which acts as a test condition.

Differential-Algebraic Programs

Based on the semantics of DJ-constraints and DA-constraints that we have defined above, we can now define the transition semantics of DA-programs. We define the transition semantics, $\rho(\alpha)$, of a DA-program α, compositionally and denotationally in terms of the semantics of its parts. The semantics of a DA-program is captured by the discrete or continuous transitions that are possible by following this DA-program. For DJ-constraints this transition relation holds for pairs of states that satisfy the jump constraints. For DA-constraints, the transition relation holds for pairs of states that can be interconnected by a differential state flow respecting the DA-constraint.

Definition 3.10 (Transition semantics of differential-algebraic programs). The *valuation, $\rho(\alpha)$ of a DA-program α* is a *transition relation* on states. It specifies which state ω is reachable from a state ν by operations of the hybrid system α and is defined as:

1. $(\nu,\omega) \in \rho(\mathscr{J})$ iff $(\nu,\omega) \models \mathscr{J}$ according to Definition 3.7 when $\mathscr{J}$ is a DJ-constraint.
2. $\rho(\mathscr{D}) = \{(\varphi(0),\varphi(r)) : \varphi$ is a differential state flow of the order of $\mathscr{D}$ and some duration $r \geq 0$ such that $\varphi \models \mathscr{D}\}$ when $\mathscr{D}$ is a DA-constraint; see Definition 3.9.
3. $\rho(\alpha \cup \beta) = \rho(\alpha) \cup \rho(\beta)$.
4. $\rho(\alpha;\beta) = \{(\nu,\omega) : (\nu,\mu) \in \rho(\alpha), (\mu,\omega) \in \rho(\beta)$ for some state $\mu\}$.
5. $(\nu,\omega) \in \rho(\alpha^*)$ iff there is $n \in \mathbb{N}$ and there are $\nu = \nu_0,\dots,\nu_n = \omega$ such that $(\nu_i,\nu_{i+1}) \in \rho(\alpha)$ for all $0 \leq i < n$.

3.3.2 Valuation of Formulas

Now, the interpretation of DAL formulas is defined as usual for first-order modal logic [123, 149], with the transition semantics, $\rho(\alpha)$, of DA-programs for modalities. The semantics of formulas is compositional and denotational, that is, the semantics of a complex formula is defined as a simple function of the semantics of its subformulas. The definition is, in fact, an equivalent reformulation of the definition of the semantics of $\mathsf{d}\mathcal{L}$ in Sect. 2.3, yet using the semantics of DA-programs instead of that of hybrid programs.

Definition 3.11 (Interpretation of DAL formulas). The *interpretation* $\models$ of DAL formulas with respect to state ν is defined as

1. $\nu \models \theta_1 \geq \theta_2$ iff $val(\nu,\theta_1) \geq val(\nu,\theta_2)$, and accordingly for $=,\leq,<,>$.
2. $\nu \models \phi \wedge \psi$ iff $\nu \models \phi$ and $\nu \models \psi$.
3. $\nu \models \phi \vee \psi$ iff $\nu \models \phi$ or $\nu \models \psi$.
4. $\nu \models \neg\phi$ iff it is not the case that $\nu \models \phi$.
5. $\nu \models \phi \rightarrow \psi$ iff it is not the case that $\nu \models \phi$ or it is the case that $\nu \models \psi$.
6. $\nu \models [\alpha]\phi$ iff $\omega \models \phi$ for all states ω with $(\nu,\omega) \in \rho(\alpha)$.
7. $\nu \models \langle\alpha\rangle\phi$ iff $\omega \models \phi$ for some state ω with $(\nu,\omega) \in \rho(\alpha)$.

The semantics of quantifiers is defined, because we considered them as abbreviations. In particular:

5. $\nu \models \forall x\,\phi$ iff $\omega \models \phi$ for all states ω that agree with ν except for the value of x.
6. $\nu \models \exists x\,\phi$ iff $\omega \models \phi$ for some ω that agrees with ν except for the value of x.

3.3.3 Time Anomalies

Hybrid systems evolve along piecewise continuous trajectories, which consist of a sequence of continuous flows interrupted by (possibly discontinuous) discrete jumps. A common phenomenon in hybrid system models is that their semantics and analysis are more controversial when discrete and continuous behaviour are allowed to interact without certain regularity assumptions [176, 270, 97, 156]. Zeno-anomalies occur when the hybrid system is allowed to take infinitely many discrete transitions in finite time.

Consider the DA-program $(a' = -1 \wedge d \leq a; d := d/2)^*$ starting in a state where $a > d > 0$ and a and d progress towards goal 0. The (inverse) clock variable a decreases continuously, yet d bounds the maximum duration of each continuous evolution phase. At the latest when $a = d$, variable d decreases by a discrete transition. This Zeno system generates infinitely many transitions in finite time and it is impossible for clock a to finally reach 0, because $a \geq d > 0$ will always remain true; see the dynamics in Fig. 3.3. Yet this behaviour is, in a certain sense,

Fig. 3.3 Zeno system run

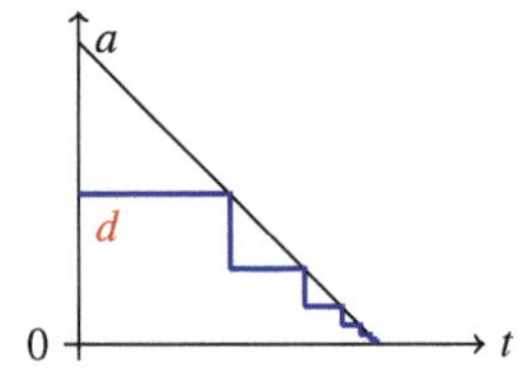

counterfactual, because it fails to obey divergence of time: Real time diverges, whereas clock a converges to 0. Further, systems with Zeno-anomalies cannot be realised [176, 270, 97, 156] so that corresponding regularity assumptions can be justified for practical purposes.

Another example for a Zeno system is the bouncing ball from Fig. 2.2 on p. 45. The bouncing ball will bounce infinitely often in finite time unless the damping coefficient c decreases to 0 at some point.

To avoid pitfalls of time anomalies, we define the DAL semantics so that it only refers to well-defined system behaviour with finitely many transitions in finite time: We restrict the semantics of DA-constraints and disallow infinite numbers of switches between differential equations in bounded time. With DA-constraint $\mathscr{D}$ defined as, say, $(x \geq 0 \rightarrow x'' = -1) \wedge (x < 0 \rightarrow x'' = 1) \wedge y' = 1$, the DAL formula

$$\exists e \, \langle \mathscr{D} \rangle [\mathscr{D}](y > e \rightarrow x \leq d)$$

expresses that, after some time, the system can stabilise such that it always remains within the region $x \leq d$ when $y > e$ for some choice of e. For such a stability property, we do not analyse what happens *after* there have been infinitely many switches from $x'' = 1$ to $x'' = -1$ within the first second. Instead, our semantics is such that our calculus reveals what happens for *any finite* number of switches. Accordingly, we restrict the semantics of DA-constraints to only accept non-Zeno evolutions:

Definition 3.12. A differential state flow φ for a DA-constraint $\mathscr{D}$ is called *non-Zeno* if there only is a finite number of points in time where some variable needs to obey another differential constraint of $\mathscr{D}$ than those before the respective point in time: Let $\mathscr{D}_1 \vee \cdots \vee \mathscr{D}_n$ be a disjunctive normal form of $\mathscr{D}$; then flow

$\varphi : [0,r] \to \text{State}(\Sigma)$ is non-Zeno iff there are $m \in \mathbb{N}$ and $0 = \zeta_0 < \zeta_1 < \cdots < \zeta_m = r$ and indices $i_1, \dots, i_m \in \{1, \dots, n\}$ such that φ respects $\mathscr{D}_{i_k}$ on the interval $[\zeta_{k-1}, \zeta_k]$, i.e., $\varphi|_{[\zeta_{k-1},\zeta_k]} \models \mathscr{D}_{i_k}$ for all $k \in \{1, \dots, m\}$.

The semantics of DA-programs entails that runs with non-Zeno state flows are non-Zeno, because α^* does not accept infinitely many switches.

3.3.4 Conservative Extension

The following result shows that $\mathsf{d}\mathscr{L}$ formulas with hybrid programs can be embedded syntactically into the extension of DAL by DA-programs without changing the meaning of the $\mathsf{d}\mathscr{L}$ formulas. That is, the semantics of DAL formulas given in Definition 3.11 and 3.10 is equivalent to the semantics given in Definitions 2.6 and 2.7 for the sublogic $\mathsf{d}\mathscr{L}$ of dTL using the syntactic embedding of hybrid programs into DA-programs from Table 3.5.

Table 3.5 Embedding hybrid programs as DA-programs

Hybrid program	DA-program
assignment / discrete jump set	DA-constraint
$x_1 := \theta_1, \dots, x_n := \theta_n$	$x_1 := \theta_1 \wedge \cdots \wedge x_n := \theta_n$
differential equation system	DA-constraint
$x_1' = \theta_1, \dots, x_n' = \theta_n \,\&\, \chi$	$x_1' = \theta_1 \wedge \cdots \wedge x_n' = \theta_n \wedge \chi$
test	DJ-constraint / DA-constraint
$?\chi$	χ

Proposition 3.1 (Conservative extension). *The logic* DAL *is a* conservative extension *of* $\mathsf{d}\mathscr{L}$*, i.e., the set of valid* $\mathsf{d}\mathscr{L}$ *formulas is the same with respect to the transition semantics of hybrid programs (Definition 2.7) as with respect to the transition semantics of DA-programs (Definition 3.10).*

Proof. The valuation of formulas of $\mathsf{d}\mathscr{L}$ and DAL is directly compatible (Definition 3.11 and 2.6, respectively). By comparing Definition 2.7 with Definition 3.10, it is easy to see that we only need to show that differential equations generate the same transitions. Using vectorial notation, let $x' = \theta \,\&\, \chi$ be a differential equation with evolution domain restriction χ. Let $(v, \omega) \in \rho(x' = \theta \,\&\, \chi)$ according to a flow φ of duration r as a witness due to Definition 2.7. Then φ is a differential state flow of the order of $x' = \theta$ and $\varphi(0) = v, \varphi(r) = \omega$ and $\varphi \models \chi$ and the value of variables z other than x remains constant. Assume $r > 0$ as there is nothing else to show otherwise. By Definition 2.7, we know that $\varphi \models x' = \theta$ holds on the interval $(0,r)$ and have to show that there is a continuation of φ so that $\varphi \models x' = \theta$ holds on $[0,r]$.

The right-hand side θ of the differential equation assumes values that are defined along φ, because $\varphi \models \chi$ and χ guards against zeros of denominators. Hence, the image of φ remains in the domain of definition of θ. Further, φ is continuous on $[0,r]$;

hence, as a compact image of a continuous map, its image is compact. Thus, by the continuation theorem for solutions of differential equations (Proposition B.1), φ can be continued to a solution of $x' = \theta$ on $[0,r]$.

Conversely, it is easy to see that $(\nu,\omega) \in \rho(x' = \theta \wedge \chi)$ according to Definition 3.10 directly implies $(\nu,\omega) \in \rho(x' = \theta \,\&\, \chi)$ according to Definition 2.7. □

Clearly, the DAL calculus will *not* be a conservative extension of the d$\mathscr{L}$ calculus, because it contains more powerful proof rules for verifying properties of differential equations. We will see that there are d$\mathscr{L}$ formulas that can be proven in the DAL calculus but not in the d$\mathscr{L}$ calculus. With our differential induction techniques in Sects. 3.5.6 and 3.5.7, the DAL calculus has better handling of differential equations than the solution-based proof rules presented in Chap. 2.

3.4 Collision Avoidance in Air Traffic Control

As a case study, which will serve as a running example, we show how succinctly collision avoidance manoeuvres in air traffic control can be described in DAL. In Sect. 3.11, we will verify such manoeuvres in the DAL calculus.

3.4.1 Flight Dynamics

Assuming, for simplicity, aircraft remain at the same altitude, an aircraft is described by its planar position $x = (x_1, x_2) \in \mathbb{R}^2$ and angular orientation ϑ. The dynamics of an aircraft is determined by its linear velocity $v \in \mathbb{R}$ and angular velocity ω; see Fig. 3.4 (depicted with $\vartheta = 0$ in the illustration). When neglecting wind, gravitation,

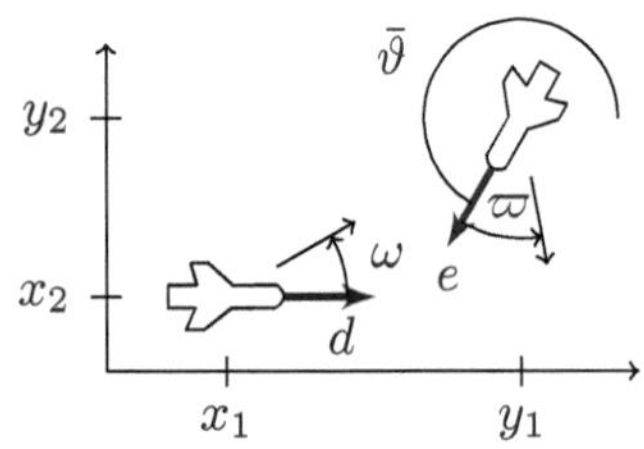

Fig. 3.4 Aircraft dynamics

and so on, which is appropriate for analysing cooperation in air traffic control [293, 196, 203, 92, 238], the in-flight dynamics of an aircraft at x can be described by the following differential equation system; see [293] for details:

$$x_1' = v\cos\vartheta \qquad x_2' = v\sin\vartheta \qquad \vartheta' = \omega. \tag{3.4}$$

That is, the linear velocity v of the aircraft changes both positions x_1 and x_2 in the (planar) direction corresponding to the orientation ϑ the aircraft is currently heading toward. Further, the angular velocity ω of the aircraft changes the orientation ϑ of the aircraft.

3.4.2 Differential Axiomatisation

Unlike for straight-line flight ($\omega = 0$), the nonlinear dynamics in (3.4) is difficult to analyse [293, 196, 203, 92, 238] for curved flight ($\omega \neq 0$), especially due to the trigonometric expressions which are generally undecidable. Solving (3.4) requires the Floquet theory of differential equations with periodic coefficients [297, Theorem 18.X] and yields mixed polynomial expressions with multiple trigonometric functions. A true challenge, however, is the need to verify properties of the states that the aircraft reach by following these solutions, which requires proving that complicated formulas with mixed polynomial arithmetic and trigonometric functions hold true for all values of state variables and all possible evolution durations. However, quantified arithmetic with trigonometric functions is undecidable: By Gödel's incompleteness theorem [137], the resulting first-order real arithmetic with trigonometric functions is not semidecidable, because the roots of sin characterise an isomorphic copy of natural numbers (Theorem 2.2).

To obtain polynomial dynamics, we axiomatise the trigonometric functions in the dynamics differentially and reparametrise the state correspondingly. Instead of angular orientation ϑ and linear velocity v, we use the linear speed vector

$$d = (d_1, d_2) := (v \cos\vartheta, v \sin\vartheta) \in \mathbb{R}^2$$

which describes both the linear speed $\|d\| := \sqrt{d_1^2 + d_2^2} = v$ and the orientation of the aircraft in space; see Figs. 3.4 and 3.5. Substituting this coordinate change into

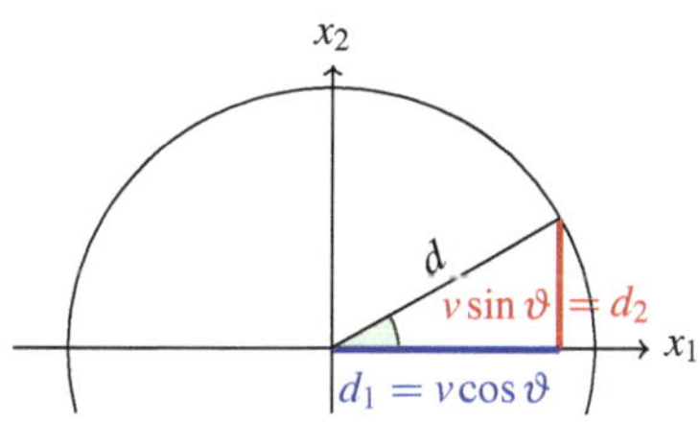

Fig. 3.5 Reparametrise for differential axiomatisation

differential equations (3.4), we immediately have $x_1' = d_1$ and $x_2' = d_2$. With the coordinate change, we further obtain differential equations for d_1, d_2 from differential equation system (3.4) by simple symbolic differentiation:

$$d_1' = (v\cos\vartheta)' = v'\cos\vartheta + v(-\sin\vartheta)\vartheta' = -(v\sin\vartheta)\omega = -\omega d_2,$$

$$d_2' = (v \sin \vartheta)' = v' \sin \vartheta + v(\cos \vartheta)\vartheta' \quad = \quad (v \cos \vartheta)\omega = \quad \omega d_1 .$$

The middle equality holds for constant linear velocity ($v' = 0$), which we assume, because only limited variations in linear speed are possible and cost-effective during the flight [293, 196] so that angular velocity ω is the primary control parameter in air traffic control. Hence, equations (3.4) can be restated as the following DA-constraint $\mathscr{F}(\omega)$:

$$x_1' = d_1 \wedge x_2' = d_2 \wedge d_1' = -\omega d_2 \wedge d_2' = \omega d_1 \qquad (\mathscr{F}(\omega))$$

$$y_1' = e_1 \wedge y_2' = e_2 \wedge e_1' = -\varpi e_2 \wedge e_2' = \varpi e_1 \qquad (\mathscr{G}(\varpi))$$

DA-constraint $\mathscr{F}(\omega)$ expresses that position $x = (x_1, x_2)$ changes according to the linear speed vector $d = (d_1, d_2)$, which in turn rotates according to ω. Simultaneous movement together with a second aircraft at $y \in \mathbb{R}^2$ having linear speed $e \in \mathbb{R}^2$ (also indicated with angle $\bar{\vartheta}$ in Fig. 3.4) and angular velocity ϖ corresponds to the DA-constraint $\mathscr{F}(\omega) \wedge \mathscr{G}(\varpi)$. DA-constraints capture simultaneous dynamics of multiple traffic agents succinctly using conjunction.

By this *differential axiomatisation*, we thus obtain polynomial differential equations. Note, however, that their solutions still involve the same complicated nonlinear trigonometric expressions so that solutions still give undecidable arithmetic (see Example B.4 in App. B). Our proof calculus in this chapter works with the differential equations themselves and not with their solutions, so that differential axiomatisation helps. Since the solutions involve trigonometric functions, previous approaches [306, 156, 125, 270, 97, 228] were not able to handle such dynamics.

3.4.3 Aircraft Collision Avoidance Manoeuvres

Due to possible turbulence or collisions, a flight configuration is unsafe if another aircraft is within a protected zone of radius p, i.e., $\|x - y\|^2 < p^2$. Guiding aircraft by collision avoidance manoeuvres to automatically resolve conflicting flight paths that would lead to possible loss of separation, is a major challenge for both air traffic control and verification [293, 196, 203, 104, 92, 238, 129, 171]. Several different classes of collision avoidance manoeuvres for air traffic control have been suggested [293, 196, 203, 104, 129, 171]. The classical traffic alert and collision avoidance system (TCAS) [196] directs one aircraft on climbing routes and the other on descending routes to resolve conflicts at different altitudes but keeps otherwise unmodified straight-line flight paths. While the simplistic TCAS manoeuvre has several benefits, it does not scale up easily to multiple aircraft or dense traffic situations near airports. As a more scalable alternative, Tomlin et al. [293] suggested roundabout manoeuvres on circular paths (see Fig. 3.6a), where, even at the same altitude, several aircraft can participate in collision avoidance manoeuvres. Because the continuous dynamics of curved flights with $\omega \neq 0$ is quite intricate, Tomlin et al. [293] and Massink and De Francesco [203] have analysed trapezoidal straight-line ($\omega = 0$)

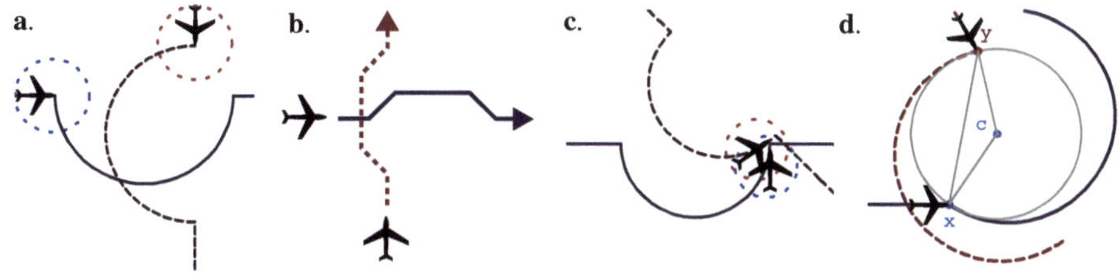

Fig. 3.6 Flight manoeuvres for collision avoidance in air traffic control

approximations of roundabouts instead, which consist only of a series of two to five straight line segments connected by several instant turns (Fig. 3.6b). Unfortunately, the discontinuities in instant turns are not flyable by aircraft.

As a more realistic model, we investigate curved roundabout manoeuvres proposed by Tomlin et al. [293]. Roundabouts have proper flight curves with nonzero angular velocities ω (Fig. 3.6a). We have shown previously [238] that classical roundabout manoeuvres with fixed turns [293, 196, 203, 92] are unsafe for non-orthogonal initial flight paths (see Fig. 3.6c for a counterexample that our model checker found), and we have proposed a tangential roundabout manoeuvre [238] with position-dependent evasive actions to overcome these deficiencies. However, because of general limits of numerical approximation techniques [238, 85], we could not actually verify the tangential roundabout manoeuvre numerically.

In this chapter, we introduce a generalised class of tangential roundabout manoeuvres with curved flight paths (Fig. 3.6d) and formally verify separation properties of this manoeuvre in the purely symbolic DAL calculus. Our main motivation for studying roundabouts are their curved flight paths, which constitute a substantial challenge for verification of hybrid systems with nontrivial dynamics and an important part of realistic flight manoeuvres.

3.4.4 Tangential Roundabout Manoeuvre

In the tangential roundabout manoeuvre, sketched in Fig. 3.6d, the idea is that the aircraft agree on some common angular velocity ω and common centre c around which both can circle safely without coming closer to each other (their linear velocities can differ, though, to compensate for different cruise speeds). Note that neither c nor ω needs to be discovered by complicated online trajectory predictions. Instead, we present a simple characterisation of safe choices for the parameters of the tangential roundabout manoeuvre in Sect. 3.11 and determine safety of the resulting flight paths using formal proofs in the DAL calculus.

In Fig. 3.7, we introduce the DAL model for the tangential roundabout manoeuvre, which is a simplified and more uniform generalisation of our previous work [238]. Observe how concisely complicated aircraft manoeuvres can be specified in DAL. There, safety property ψ for aircraft manoeuvres expresses that pro-

$$\begin{aligned}
\psi &\equiv \phi \to [\mathit{trm}^*]\phi \\
\phi &\equiv \|x-y\|^2 \geq p^2 \equiv (x_1-y_1)^2+(x_2-y_2)^2 \geq p^2 \\
\mathit{trm} &\equiv \mathit{free};\ \mathit{tang};\ \mathscr{F}(\omega) \wedge \mathscr{G}(\omega) \\
\mathit{free} &\equiv \exists\omega\, \mathscr{F}(\omega) \wedge \exists\varpi\, \mathscr{G}(\varpi) \wedge \phi \\
\mathit{tang} &\equiv \text{will be derived in Sect. 3.11}
\end{aligned}$$

Fig. 3.7 Flight control with tangential roundabout collision avoidance manoeuvres

tected zones are respected during the flight (specified by the separation property ϕ). The flight controller (trm^*) performs collision avoidance manoeuvres by tangential roundabouts and repeats these manoeuvres any number of times, as needed, as indicated by the * repetition operator. During each *trm* phase, the aircraft first perform arbitrary free flight (*free*) by (repeatedly) independently adjusting their angular velocities ω and ϖ (by $\exists\omega$ and $\exists\varpi$) while the aircraft are safely separated. This is expressed by conjunct ϕ of the DA-constraint. Observe that, unlike in $\exists u\,(\omega:=u);\mathscr{F}(\omega)$, angular velocities can be (re)adjusted continuously during free flight in $\exists\omega\,\mathscr{F}(\omega)$, rather than just once. In particular, *free* includes piecewise constant choices as in $\big(\exists u\,(\omega:=u) \wedge \exists u\,(\varpi:=u);\,\mathscr{F}(\omega) \wedge \mathscr{G}(\varpi)\big)^*$. Due to evolution domain ϕ of *free*, the tangential roundabout manoeuvre must be initiated (by a tangential initiation controller *tang*) before the flight paths become unsafe. Then, the tangential roundabout manoeuvre itself is carried out by the DA-constraint $\mathscr{F}(\omega) \wedge \mathscr{G}(\omega)$ according to some common angular velocity ω determined by *tang*. Finally, the collision avoidance roundabouts can be left again by repeating the loop trm^* and entering arbitrary free flight at any time. When further conflicts occur during free flight, the controller in Fig. 3.7 again enters roundabout conflict resolution manoeuvres.

In summary, property ψ of Fig. 3.7 expresses that the aircraft remain safe during the flight, especially during evasive roundabout manoeuvres. In Sect. 3.11, we will determine a constraint on the parameter adjustment by *tang* such that the roundabout manoeuvre is safe, and we give a simple choice for *tang* respecting this parameter constraint.

3.5 Proof Calculus for Differential-Algebraic Logic

In this section, we introduce a sequent calculus for proving DAL formulas. The basic idea is to symbolically compute the effects of DA-programs and successively transform them into simpler logical formulas describing their effects by symbolic decomposition. The calculus consists of standard propositional rules, dedicated rules for handling DA-program modalities, including differential induction rules for sophisticated differential constraints, and side deduction rules for integrating real quantifier elimination.

For our calculus, recall the definition of substitutions: The result of applying to ϕ the substitution that replaces x by θ is defined as usual (Sect. 2.5.1); it is denoted by ϕ_x^θ. Likewise, in a simultaneous substitution $\phi_{x_1}^{\theta_1} \dots {}_{x_n}^{\theta_n}$ the x_i are replaced simultaneously by the respective θ_i.

3.5.1 Motivation

DA-constraints are a very expressive formalism and can represent rich forms of continuous dynamics, including differential equations, differential-algebraic equations, differential inequalities, and differential equations with quantified nondeterminism and disturbance input. Even for quite simple differential equations, formal verification of properties of their behaviour is challenging and, in general, undecidable (Theorem 2.2). Even the linear differential equations $x' = a_1x + a_2y \land y' = a_3x + a_4y$ have trigonometric solutions that fall outside decidable classes of arithmetic for appropriate coefficients $a_i \in \mathbb{R}$.

Working with solutions of differential equations as in $d\mathcal{L}$ rules $\langle'\rangle$,$[']$ of Chap. 2 is, thus, not a very promising verification approach. Instead, we are looking for a verification principle that works implicitly based on the local dynamics and the differential equation itself, not on its solution. Solutions of differential equations are usually much more complicated than the differential equations themselves, which makes differential equations representationally powerful—and even more so for the more expressive DA-constraints.

Intuitively, the global solution of a differential equation, while helpful, is not really required for deciding a property. Consider Fig. 3.8, which shows the vector fields of a differential equation system (intuitively, at each point it shows the vector of the right-hand side, i.e., the direction into which the differential equations dynamics points locally). Geometrically, a solution is obtained by following the

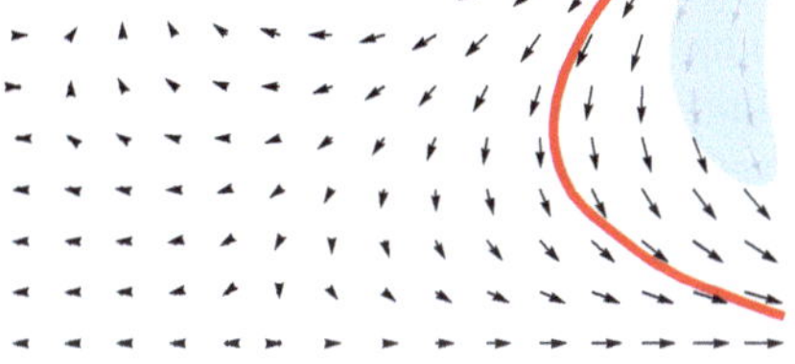

Fig. 3.8 Vector field and a solution of a differential equation

direction of the dynamics "at every point" along the vector field. The solid curve shows one global solution starting at a particular point as initial value. Suppose the two shaded transparent regions represent the set of unsafe states for a system. In principle, verification could proceed by considering each global solution starting at any initial point (typically uncountably many) and show that no such solution enters the unsafe region at any time (uncountably many points in time). Of course, literally following this enumeration of solutions and enumerating all points in time will not work because there are uncountably many initial states and points in time. This principle is what proof rules $\langle ' \rangle$,$[']$ from the $\mathsf{d}\mathcal{L}$ calculus in Fig. 2.11 on p. 79 make formally rigorous and sound using symbolic polynomial solutions and quantifier elimination in real-closed fields. It is generally not very scalable and it can be quite difficult to ensure soundness for more general dynamics.

The point is, however, that proving does not need to use solutions of differential equations! In Fig. 3.8 the question about whether the system can ever enter an unsafe state (in the two transparent shaded regions) can be answered by inspection of the local dynamics of the vector field alone. If the vector field never points from a safe state into the unsafe region, then, intuitively, the system can never end unsafe if it starts safe and always follows the local dynamics of the vector field. Similarly, for a reachability property, we do not necessarily need a solution for the differential equations to decide if the system is able to reach a target region. Intuitively, it can also be answered in the positive if the dynamics is always making good progress towards the target region locally.

The challenge is how to turn this geometrical intuition into a sound proof principle. In fact, it turns out throughout the course of this chapter that this is a surprisingly subtle matter and easy to get wrong. In this section, we develop a sound reasoning framework and a proof calculus for DAL that follows these principles of proving by local dynamics without leading astray into the unsound. We first develop appropriate notions from differential algebra that will be invaluable for our formal development of sound proof rules.

3.5.2 Derivations and Differentiation

As a purely algebraic device for proving properties about continuous evolutions in our calculus, we define syntactic derivations of terms and show that their valuation coincides with analytic differentiation (the *total differential*). With this, we can build proof rules for verifying DA-programs fully algebraically by a differential form of induction without the need to carry out analytic reasoning about analytic limits or similar concepts that would require higher-order logic. The advantage of our syntactic notions that are inspired from differential algebra is that syntactic algebraic reasoning can be carried out with a calculus that is still suitable for automatic theorem proving.

Derivations

We define a syntactic total derivation and prove that its valuation along differential flows coincides with analytic differentiation.

Definition 3.13 (Derivation). The operator $D : \mathrm{Term}(\Sigma \cup \Sigma') \to \mathrm{Term}(\Sigma \cup \Sigma')$ that is defined as follows is called *syntactic (total) derivation*:

$$D(r) = 0 \qquad \text{if } r \in \mathbb{Q} \text{ is a rational number} \tag{3.5a}$$

$$D(x^{(n)}) = x^{(n+1)} \qquad \text{if } x \in \Sigma \text{ is a state variable}, n \geq 0 \tag{3.5b}$$

$$D(a+b) = D(a) + D(b) \tag{3.5c}$$

$$D(a-b) = D(a) - D(b) \tag{3.5d}$$

$$D(a \cdot b) = D(a) \cdot b + a \cdot D(b) \tag{3.5e}$$

$$D(a/b) = (D(a) \cdot b - a \cdot D(b))/b^2 \tag{3.5f}$$

For a first-order formula F, we define the following abbreviations:

$$D(F) \equiv \bigwedge_{i=1}^{m} D(F_i) \qquad \text{where } \{F_1, \ldots, F_m\} \text{ is the set of all literals of } F;$$

$$D(a \geq b) \equiv D(a) \geq D(b) \quad \text{and accordingly for } <, >, \leq, = \text{ or negative literals.}$$

Recall that a *literal* is a logical formula with no logical operators other than $\neg$. A literal is negative if it has an odd number of $\neg$ operators, and is positive if it has an even number. More than one $\neg$ is not needed because $\neg\neg F_i \equiv F_i$.

To illustrate the naturalness of this definition, we briefly align it in terms of the algebraic structures from differential algebra [179]. Case (3.5a) defines number symbols as *differential constants*, which do not change during continuous evolution. Their total derivative is zero. Equation (3.5c) and the *Leibniz* or *product rule* (3.5e) are defining conditions for *derivation operators on rings*. The derivative of a sum is the sum of the derivatives (additivity or a homomorphic property with respect to addition) according to equation (3.5c). Furthermore, the derivative of a product is the derivative of one factor times the other factor plus the one factor times the derivative of the other factor as in (3.5e). Equation (3.5d) is a derived rule for subtraction according to $a - b = a + (-1) \cdot b$ and again expresses a homomorphic property, now with respect to subtraction. In addition, equation (3.5b) uniquely defines operator D on the *differential polynomial algebra* spanned by the *differential indeterminates* $x \in \Sigma$. It says that we understand the higher-order differential symbol $x^{(n+1)}$ as the derivative of the symbol $x^{(n)}$ for all state variables $x \in \Sigma$ and orders $n \geq 0$. Equation (3.5f) canonically extends D to the *differential field of quotients* by the usual *quotient rule*. As the base field $\mathbb{R}$ has no zero divisors, the right-hand side of (3.5f) is defined whenever the original division a/b can be carried out, which, as we assumed, is guarded by $b \neq 0$. The resulting structure $\mathrm{Term}(\Sigma \cup \Sigma')$, together with the derivation D, corresponds to the *differential field of rational fractions* with state variables as *differential indeterminates* over $\mathbb{R}$ and with rational numbers as *differential constants*.

The conjunctive definition of the formula $D(F)$ in Definition 3.13 corresponds to the joint total derivative of all atomic subformulas of F and will be an important tool for differential induction rules of our calculus.

Example 3.6 (Total differential of aircraft separation). Consider the separation property ϕ for aircraft manoeuvres from Fig. 3.7 on p. 152, which expresses that two aircraft at positions $x = (x_1, x_2)$ and $y = (y_1, y_2)$, respectively, have at least distance p:

$$F \equiv (x_1 - y_1)^2 + (x_2 - y_2)^2 \geq p^2.$$

The total differential of formula F is formed by applying the syntactic total derivation $D(\cdot)$ on the terms of each literal, which gives:

$$\begin{aligned} D(F) &\equiv D((x_1 - y_1)^2) + D((x_2 - y_2)^2) \geq D(p^2) \\ &\equiv 2(x_1 - y_1)D(x_1 - y_1) + 2(x_2 - y_2)D(x_2 - y_2) \geq 2pD(p) \\ &\equiv 2(x_1 - y_1)(D(x_1) - D(y_1)) + 2(x_2 - y_2)(D(x_2) - D(y_2)) \geq 2pD(p) \\ &\equiv 2(x_1 - y_1)(x_1' - y_1') + 2(x_2 - y_2)(x_2' - y_2') \geq 2pp'. \end{aligned}$$

An interesting question that we will answer in the following is, what can we conclude about the truth-value of F along the evolution of the system by evaluating its total derivative $D(F)$? □

The following central lemma, which is the differential counterpart of the substitution lemma, establishes the connection between syntactic derivation of terms and semantic differentiation as an analytic operation to obtain analytic derivatives of valuations along differential state flows. It will allow us to draw analytic conclusions about the behaviour of a system along differential equations from the truth of purely algebraic formulas obtained by syntactic derivation. In a nutshell, the following lemma shows that, along a flow, analytic derivatives of valuations coincide with valuations of syntactic derivations. For comparison, the classical substitution lemma (Lemma 2.2) shows that the valuation $val(\sigma^* \nu, \phi)$ in the semantically modified state $\sigma^* \nu$ equals valuation of the syntactic substitution $val(\nu, \phi_x^\theta)$, where $\sigma^* \nu$ is like ν except that x is interpreted as $val(\nu, \theta)$. The derivation lemma has the same form, using derivation instead of substitution.

Lemma 3.1 (Derivation lemma). *The valuation of* DAL *terms is a* differential homomorphism*: Let $\theta \in$ Term(Σ) and let $\varphi : [0, r] \to$ State(Σ) be any state flow of the order of $D(\theta)$ and of duration $r > 0$ along which the value of θ is defined (as no divisions by zero occur). Then we have for all $\zeta \in [0, r]$ that*

$$\frac{\mathrm{d}\, val(\varphi(t), \theta)}{\mathrm{d}t}(\zeta) = val(\bar{\varphi}(\zeta), D(\theta)).$$

In particular, $val(\varphi(t), \theta)$ is continuously differentiable (where θ is defined) and its derivative exists on $[0, r]$.

Proof. The proof is an inductive consequence of the correspondence of the semantics of differential symbols and analytic derivatives in state flows (Definition 3.8). It uses the assumption that the flow φ remains within the domain of definition of θ and is continuously differentiable in all variables of θ. In particular, all denominators are nonzero during φ.

- If θ is a variable x, the conjecture holds immediately by Definition 3.8:

$$\frac{\mathrm{d}\,val(\varphi(t),x)}{\mathrm{d}t}(\zeta) = \frac{\mathrm{d}\,\varphi(t)(x)}{\mathrm{d}t}(\zeta) = \bar{\varphi}(\zeta)(x') = val(\bar{\varphi}(\zeta),D(x)).$$

  There, the derivative exists because the state flow is of order 1 in x and, thus, (continuously) differentiable for x.
- If θ is of the form $a+b$, the desired result can be obtained by using the properties of derivatives, derivations (Definition 3.13), and valuations (Definition 3.6):

$$\begin{aligned}
&\frac{\mathrm{d}}{\mathrm{d}t}(val(\varphi(t),a+b))(\zeta) && \\
&= \frac{\mathrm{d}}{\mathrm{d}t}(val(\varphi(t),a)+val(\varphi(t),b))(\zeta) && val(\nu,\cdot) \text{ homomorphic for } + \\
&= \frac{\mathrm{d}}{\mathrm{d}t}(val(\varphi(t),a))(\zeta)+\frac{\mathrm{d}}{\mathrm{d}t}(val(\varphi(t),b))(\zeta) && \frac{\mathrm{d}}{\mathrm{d}t} \text{ is a (linear) derivation} \\
&= val(\bar{\varphi}(\zeta),D(a))+val(\bar{\varphi}(\zeta),D(b)) && \text{by induction hypothesis} \\
&= val(\bar{\varphi}(\zeta),D(a)+D(b)) && val(\nu,\cdot) \text{ homomorphic for } + \\
&= val(\bar{\varphi}(\zeta),D(a+b)) && D(\cdot) \text{ is a syntactic derivation}
\end{aligned}$$

- The case where θ is of the form $a \cdot b$ or $a - b$ is similar, using Leibniz product rule (3.5e) or subtractivity (3.5d) of Definition 3.13, respectively.
- The case where θ is of the form a/b uses (3.5f) of Definition 3.13 and further depends on the assumption that $b \neq 0$ along φ. This holds as the value of θ is assumed to be defined all along state flow φ.
- The values of numbers $r \in \mathbb{Q}$ do not change during a state flow (in fact, they are not affected by the state at all); hence their derivative is $D(r) = 0$. □

Differential Transformations

The substitution property Lemma 2.3 can be lifted to differential equations, i.e., differential equations can be used for equivalent substitutions along differential state flows respecting the corresponding differential constraints. In a nutshell, the following lemma can be used to substitute right-hand sides of differential equations for the left-hand side derivatives for flows along which these differential equations hold. For comparison, the classical substitution property says that equals can be substituted for equals, i.e., left-hand sides of equations can be substituted by right-hand sides of equations within formulas in which the equations hold.

Lemma 3.2 (Differential substitution property). *If φ is a state flow satisfying $\varphi \models x'_1 = \theta_1 \wedge \cdots \wedge x'_n = \theta_n \wedge \chi$, then $\varphi \models \mathscr{D} \leftrightarrow (\chi \to \mathscr{D}_{x'_1}^{\theta_1} \cdots {}_{x'_n}^{\theta_n})$ holds for all DA-constraints $\mathscr{D}$.*

Proof. The proof is by using the Substitution Lemma 2.2 for first-order logic on the basis of $val(\bar{\varphi}(\zeta), x'_i) = val(\bar{\varphi}(\zeta), \theta_i)$ and $\bar{\varphi}(\zeta) \models \chi$ at each time ζ in the domain of φ. □

Example 3.7 (Differential substitution for aircraft dynamics). Continuing the aircraft scenario from Example 3.6, recall the differential equations ($\mathscr{F}(\omega)$) for flight:

$$x'_1 = d_1 \wedge x'_2 = d_2 \wedge d'_1 = -\omega d_2 \wedge d'_2 = \omega d_1 \qquad (\mathscr{F}(\omega)^*)$$

While the aircraft follow this differential equation, all corresponding differential state flows φ satisfy

$$\varphi \models x'_1 = d_1 \wedge x'_2 = d_2 \wedge d'_1 = -\omega d_2 \wedge d'_2 = \omega d_1$$

Consequently, by Lemma 3.2, along φ we can substitute in the right-hand sides of these differential equations for the left-hand sides. When we perform this substitution for $D(F)$ from Example 3.6, we get

$$D(F)_{x'_1}^{d_1}{}_{x'_2}^{d_2}{}_{d'_1}^{-\omega d_2}{}_{d'_2}^{\omega d_1} \equiv 2(x_1 - y_1)(d_1 - y'_1) + 2(x_2 - y_2)(d_2 - y'_2) \geq 2pp'.$$

If we also substitute in the differential equations ($\mathscr{G}(\varpi)$) for the second aircraft and assume $p' = 0$, we obtain

$$D(F)_{x'_1}^{d_1}{}_{x'_2}^{d_2}{}_{d'_1}^{-\omega d_2}{}_{d'_2}^{\omega d_1}{}_{y'_1}^{e_1}{}_{y'_2}^{e_2}{}_{y'_1}^{-\varpi e_2}{}_{e'_2}^{\varpi e_1}{}_{p'}^{0} \equiv 2(x_1 - y_1)(d_1 - e_1) + 2(x_2 - y_2)(d_2 - e_2) \geq 0.$$

Expressions like these will play a very important role in our proof calculus, because we will use them to show (in)variance of properties without the need for knowing solutions of differential equations. From the truth of such a differential substitution of a total derivative, we will be able to conclude invariance of F by the derivation lemma. □

The following lemma captures that the semantics of DA-constraints is not sensitive to how the DA-constraint is presented. It also plays its part in the soundness proof of our calculus, because it immediately makes all implicational and equivalence transformations of real arithmetic available for DA-constraints.

Lemma 3.3 (Differential transformation principle). *Let $\mathscr{D}$ and $\mathscr{E}$ be two DA-constraints (with the same changed variables). If $\mathscr{D} \to \mathscr{E}$ is a tautology of (non-differential) first-order real arithmetic (that is, when considering $x^{(n)}$ as a new variable independent of x), then $\rho(\mathscr{D}) \subseteq \rho(\mathscr{E})$.*

Proof. Let the first-order formulas ϕ and ψ be obtained from $\mathscr{D}$ and $\mathscr{E}$, respectively, by replacing all x' with new variable symbols X (accordingly for higher-order differential symbols $x^{(n)}$). Using vectorial notation, we write $\phi_X^{x'}$ for the formula obtained from ϕ by substituting all variables X by x'. Thus, $\phi_X^{x'}$ is $\mathscr{D}$ and $\psi_X^{x'}$ is $\mathscr{E}$. Let $\phi \rightarrow \psi$ be valid in (non-differential) real arithmetic. Let $(\nu, \omega) \in \rho(\mathscr{D})$ according to a state flow φ. Then φ also is a state flow for $\mathscr{E}$ that justifies $(\nu, \omega) \in \rho(\mathscr{E})$: For any $\zeta \in [0, r]$, we have $\bar{\varphi}(\zeta) \models \mathscr{D}$. Hence $\bar{\varphi}(\zeta) \models \mathscr{E}$, because $\bar{\varphi}(\zeta) \models \phi_X^{x'}$ immediately implies $\bar{\varphi}(\zeta) \models \psi_X^{x'}$ by validity of $\phi \rightarrow \psi$. The assumption of $\mathscr{D}$ and $\mathscr{E}$ having the same set of changed variables is only required for compatibility with condition 2 of Definition 3.9, which enforces that unchanged variables z remain constant. It can be established easily by adding constraints of the form $z' = 0$ as required. □

DA-constraints $\mathscr{D}$ and $\mathscr{E}$ are *equivalent* iff $\rho(\mathscr{D}) = \rho(\mathscr{E})$. In particular, the semantics of DA-programs is preserved when replacing a DA-constraint by another DA-constraint that is equivalent in non-differential first-order real arithmetic (similarly for DJ-constraints).

Example 3.8. Lemma 3.3 is a very powerful tool for analysing DA-constraints, because it lifts equivalence or implication-preserving transformations from first-order real arithmetic to DA-constraints. In general, it may be difficult to determine the relationship of the reachability relations of differential equations or DA-constraints. With Lemma 3.3, however, we can use the first-order arithmetic structure to find out. For example, the dynamics of DA-constraint $x' = 5x \wedge y' = 1 - x \wedge x > 0$ can be overapproximated safely by the dynamics of $x' = 5x \wedge y' \leq 1 \wedge x > 0$, because the following implication is valid in first-order logic, where we consider $x^{(1)}$ and $y^{(1)}$ as new variables independent of x and y:

$$x^{(1)} = 5x \wedge y^{(1)} = 1 - x \wedge x > 0 \rightarrow x^{(1)} = 5x \wedge y^{(1)} \leq 1 \wedge x > 0.$$

□

Example 3.9 (Disturbance in the train dynamics). Similarly, we have a formally precise way to prove why the transition relation $\rho(\cdot)$ of train dynamics $z' = v \wedge v' = a$ from Chap. 2 is contained in the transition relation of the differential inequality $z' = v \wedge v' \geq a - l \wedge v' \leq a + u$ for bounds $l > 0$ and $u \geq 0$, which, in turn, is contained in the transition relation of $z' = v \wedge v' \geq a - l$. The first differential equation characterises ideal-world dynamics, where the chosen acceleration a takes effect exactly. The second differential inequality characterises dynamics with a bounded disturbance on the actual acceleration a. By the nature of the differential inequality, the actual disturbance may vary quite arbitrarily over time, but only inside the bounds $-l$ and u. The last differential inequality is a variant that is only bounded in one direction.

The proof of inclusion of the above three DA-constraints only amounts to checking the validity of the following formulas in first-order real arithmetic, again considering $z^{(1)}$ and $v^{(1)}$ as new variables:

$$z^{(1)} = v \wedge v^{(1)} = a \rightarrow z^{(1)} = v \wedge v^{(1)} \geq a - l \wedge v^{(1)} \leq a + u,$$

$$z^{(1)} = v \land v^{(1)} \geq a - l \land v^{(1)} \leq a + u \rightarrow z^{(1)} = v \land v^{(1)} \geq a - l.$$

Similarly, we prove that the differential inequality $z' = v \land v' \geq a - l \land v' \leq a + u$ and the following quantified DA-constraint have equivalent reachability relations:

$$z' = v \land \exists d\,(v' = d \land a - l \leq d \land d \leq a + u) \land v \geq 0. \qquad (3.2^*)$$

By Lemma 3.3, we only need to check equivalence of the corresponding formulas in first-order real arithmetic, considering $z^{(1)}$ and $v^{(1)}$ as new variables:

$$(z^{(1)} = v \land v^{(1)} \geq a - l \land v^{(1)} \leq a + u) \\ \leftrightarrow (z^{(1)} = v \land \exists d\,(v^{(1)} = d \land a - l \leq d \land d \leq a + u) \land v \geq 0).$$

□

Counterexample 3.10 (Same changed variables). For correctness of Lemma 3.3, we have to assume that $\mathcal{D}$ is not allowed to have more differential variables than $\mathcal{E}$, because of our implicit assumption that variables without differential constraints are constant. At first sight, it may look like the transition relation $z' = v \land v' = a \land t' = 1$ may be overapproximated by $z' = v \land t' = 1$, because the following arithmetic formula is valid:

$$z^{(1)} = v \land v^{(1)} = a \land t^{(1)} = 1 \rightarrow z^{(1)} = v \land t^{(1)} = 1.$$

But this does not work, because of our convention that assumes variables do not change if they are not mentioned in a differential equation. Thus in $z' = v \land t' = 1$, we silently assume $v' = 0$, because no differential constraint for v is mentioned. Then the transition relations are very different, however, because v may change in the former but will stay constant in the latter. If, instead, we make the implicit assumption $v' = 0$ explicit in the DA-constraint, so that both DA-constraints have the same changed variables, then this subtlety does not happen, because the following formula is *not* valid (unless $a = 0$, in which case both DA-constraints are, indeed, equivalent):

$$z^{(1)} = v \land v^{(1)} = a \land t^{(1)} = 1 \rightarrow z^{(1)} = v \land t^{(1)} = 1 \land v^{(1)} = 0.$$

□

3.5.3 Differential Reduction and Differential Elimination

Using the expressive power of DA-constraints, several reductions can be performed to simplify the syntactic form of DA-constraints. With quantified DA-constraints, we can reduce differential inequalities to quantified differential equations equivalently:

Lemma 3.4 (Differential inequality elimination). *DA-constraints admit* differential inequality elimination, *i.e., with each DA-constraint $\mathscr{D}$, an equivalent DA-constraint without differential inequalities can be effectively associated that has no other free variables.*

Proof. Let $\mathscr{E}$ be obtained from $\mathscr{D}$ by replacing all differential inequalities $\theta_1 \leq \theta_2$ by a quantified differential equation $\exists u\,(\theta_1 = \theta_2 - u \wedge u \geq 0)$ with a new variable u for the quantified disturbance (accordingly for $\geq, >, <$). By Lemma 3.3, the equivalence of $\mathscr{D}$ and $\mathscr{E}$ is a simple consequence of the corresponding equivalences in first-order real arithmetic. □

Example 3.11. Lemma 3.4 can turn differential inequalities to DA-constraints with quantifiers and equations in place of the differential inequalities. Let $l > 0$. We can transform the differential inequality $z' = v \wedge v' \geq a - l \wedge v \geq 0$ to an equivalent quantified DA-constraint

$$\exists d\,(z' = v \wedge v' = a - d \wedge d \leq l \wedge v \geq 0).$$

Both DA-constraints are equivalent. The latter DA-constraint has a structural advantage: we only have differential *equations* and quantifiers, while the inequalities are isolated in the non-differential evolution domain constraints $d \leq l \wedge v \geq 0$. Thus we can handle differential inequalities if we can handle quantified disturbance. In fact, we could even go on and replace *all* inequalities (differential or not) by equivalent equations and quantifiers with Lemma 3.4:

$$\exists d\,(z' = v \wedge v' = a - d \wedge \exists u_1\, d = l - u_1^2 \wedge \exists u_2\, v = u_2^2).$$

But, unlike differential inequalities, those inequalities that only occur in the non-differential part $d \leq l \wedge v \geq 0$ are not complicated for proofs, because they can be handled easily by quantifier elimination in real-closed fields [288, 81]; also see App. D. □

In this book, we assume this transformation has been applied such that we can focus on DA-constraints with differential equations, i.e., where differential symbols only occur in differential equations, and where inequalities do not contain differential symbols. Yet, the DA-constraint resulting from Lemma 3.4 could become inhomogeneous when multiple differential equations are produced for the same variable from multiple differential inequalities. For instance, $\theta_1 \leq x' \leq \theta_2$ produces $\exists u\,\exists v\,(x' = \theta_1 + u \wedge x' = \theta_2 - v \wedge u \geq 0 \wedge v \geq 0)$, which has two differential equations for x. To homogenise this DA-constraint again, we use the following:

Lemma 3.5 (Differential equation normalisation). *DA-constraints admit* differential equation normalisation, *i.e., with each DA-constraint $\mathscr{D}$, an equivalent DA-constraint with at most one differential equation for each differential symbol can be effectively associated that has no other free variables. Furthermore, this differential equation is* explicit, *i.e., of the form $x^{(n)} = \theta$ where* $\mathrm{ord}_x\,\theta < n$.

Proof. For each differential symbol $x^{(n)} \in \Sigma'$ occurring in $\mathscr{D}$, we introduce a new non-differential variable $X_n \in \Sigma$. Let $\mathscr{D}_{x^{(n)}}^{X_n}$ denote the result of substituting X_n for $x^{(n)}$ in $\mathscr{D}$. By Lemma 3.3, the equivalence of $\mathscr{D}$ and $\exists X_n\,(x^{(n)} = X_n \wedge \mathscr{D}_{x^{(n)}}^{X_n})$ is a simple consequence of the corresponding equivalence in first-order logic. Proceeding inductively for all such $x^{(n)} \in \Sigma'$ in $\mathscr{D}$ gives the desired result. □

Example 3.12. Directly using the construction in Lemma 3.4 on $\theta_1 \leq x' \leq \theta_2$ we obtain an equivalent DA-constraint $\exists u\,\exists v\,(x' = \theta_1 + u \wedge x' = \theta_2 - v \wedge u \geq 0 \wedge v \geq 0)$ without differential inequalities but with inhomogeneous parts (two separate differential equations for x). After normalisation with Lemma 3.5, we obtain the equivalent DA-constraint $\exists X_1\,\exists u\,\exists v\,(x' = X_1 \wedge X_1 = \theta_1 + u \wedge X_1 = \theta_2 - v \wedge u \geq 0 \wedge v \geq 0)$.

□

Similarly, higher-order differential constraints reduce to first-order constraints by introducing new non-differential auxiliary variables X_n for each of the higher-order differential symbols $x^{(n)}$. For $1 \leq \mathrm{ord}_x\,\theta < n$, we can replace a higher-order differential equation $x^{(n)} = \theta$ with:

$$x' = X_1 \wedge X_1' = X_2 \wedge \ldots \wedge X_{n-2}' = X_{n-1} \wedge X_{n-1}' = \theta_{x'}^{X_1} \cdots {}_{x^{(n-1)}}^{X_{n-1}}\,{}_{x^{(n)}}^{X_{n-1}'}\,.$$

3.5.4 Rules of the Calculus for Differential-Algebraic Logic

Sequents and substitutions are defined as in Sect. 2.5.2. We again assume bound variable renaming as needed. In the DAL calculus, only admissible substitutions are applicable, which is crucial for soundness.

Definition 3.14 (Admissible substitution). An application of a substitution σ is *admissible* if no replaced variable x occurs in the scope of a quantifier or modality binding x or a variable of the replacement $\sigma(x)$. A modality *binds* x if its DA-program (possibly) changes x, i.e., if it contains a DJ-constraint containing $x := \theta$ or a DA-constraint containing $x^{(n)} \in \Sigma'$ for an $n \geq 1$.

As usual in sequent calculus—although the direction of entailment is from *premises* (above rule bar) to *conclusion* (below)—the order of reasoning and reading is *goal-directed* in practise: Rules are applied backwards, that is, starting from the desired conclusion at the bottom (goal) to the resulting premises (subgoals). To highlight the logical essence of the DAL calculus, Fig. 3.9 provides rule *schemata* with which the following definition associates the calculus rules that are applicable during a DAL proof. The calculus inherits the propositional rules from Fig. 2.11 on p. 79 and further consists of first-order quantifier rules (r∀,l∀,r∃,l∃), rules for dynamic modalities (⟨;⟩–*DS*), and global rules ([]*gen*,⟨⟩*gen*,*ind*,*con*,*DI*,*DV*). The DAL rules that have the same name as corresponding d$\mathscr{L}$ rules are actually identical, except for minor syntactic variations and generalisations. Rule ⟨;⟩, for instance, is identical in the d$\mathscr{L}$ calculus in Fig. 2.11 and the DAL calculus in Fig. 3.9. We just repeat it here for convenience to have a comprehensive representation of the DAL calculus.

The definition of rules is a simplified version of that in Definition 2.10, with side deductions in place of the free-variable quantifier rules from Chap. 2. Further, we can simplify the presentation by avoiding update prefixes (which would also be sound here when allowing conjunctive DA-constraints as prefixes). Note that this choice generally requires more complicated invariants and variants than in the $d\mathcal{L}$ calculus of Chap. 2, where discrete jump set prefixes are allowed for rule applications so that more information about the pre-state is retained automatically.

Definition 3.15 (Rules). The *rule schemata* in Fig. 3.9—in which *all* substitutions need to be admissible—induce *calculus rules* by:

1. If
$$\frac{\Phi_1 \vdash \Psi_1 \quad \dots \quad \Phi_n \vdash \Psi_n}{\Phi_0 \vdash \Psi_0}$$
is an instance of one of the rule schemata in Fig. 3.9, then
$$\frac{\Gamma, \Phi_1 \vdash \Psi_1, \Delta \quad \dots \quad \Gamma, \Phi_n \vdash \Psi_n, \Delta}{\Gamma, \Phi_0 \vdash \Psi_0, \Delta}$$
can be applied as a proof rule of the DAL calculus, where Γ, Δ are arbitrary finite sets of context formulas (including empty sets).
2. Symmetric schemata can be applied on either side of the sequent. If
$$\frac{\phi_1}{\phi_0}$$
is an instance of one of the symmetric rule schemata (dynamic rules $\langle;\rangle$–$[D]$) in Fig. 3.9, then
$$\frac{\Gamma \vdash \phi_1, \Delta}{\Gamma \vdash \phi_0, \Delta} \qquad \text{and} \qquad \frac{\Gamma, \phi_1 \vdash \Delta}{\Gamma, \phi_0 \vdash \Delta}$$
can both be applied as proof rules of the DAL calculus, where Γ, Δ are arbitrary finite sets of context formulas (including empty sets).

Propositional Rules

For propositional logic, we reuse the standard propositional rules $\neg$r–*cut* from the $d\mathcal{L}$ calculus in Fig. 2.11.

First-Order Quantifier Rules

Unlike in uninterpreted first-order logic [122, 123], quantifier rules have to respect the specific semantics of real arithmetic. Thus, our rules handle real quantifiers

$$(\mathrm{r}\forall)\ \frac{\mathrm{QE}(\forall x \bigwedge_i (\Gamma_i \vdash \Delta_i))}{\Gamma \vdash \Delta, \forall x \phi}\,{}^{1} \qquad (\mathrm{r}\exists)\ \frac{\mathrm{QE}(\exists x \bigwedge_i (\Gamma_i \vdash \Delta_i))}{\Gamma \vdash \Delta, \exists x \phi}\,{}^{1}$$

$$(\mathrm{l}\forall)\ \frac{\mathrm{QE}(\exists x \bigwedge_i (\Gamma_i \vdash \Delta_i))}{\Gamma, \forall x \phi \vdash \Delta}\,{}^{1} \qquad (\mathrm{l}\exists)\ \frac{\mathrm{QE}(\forall x \bigwedge_i (\Gamma_i \vdash \Delta_i))}{\Gamma, \exists x \phi \vdash \Delta}\,{}^{1}$$

$$(\langle;\rangle)\ \frac{\langle\alpha\rangle\langle\beta\rangle\phi}{\langle\alpha;\beta\rangle\phi} \qquad (\langle\exists\rangle)\ \frac{\exists x \langle\mathscr{J}\rangle\phi}{\langle\exists x\, \mathscr{J}\rangle\phi} \qquad (\langle:=\rangle)\ \frac{\chi \wedge \phi_{x_1}^{\theta_1} \cdots {}_{x_n}^{\theta_n}}{\langle x_1 := \theta_1 \wedge .. \wedge x_n := \theta_n \wedge \chi\rangle\phi}\,{}^{3}$$

$$([;])\ \frac{[\alpha][\beta]\phi}{[\alpha;\beta]\phi} \qquad ([\exists])\ \frac{\forall x [\mathscr{J}]\phi}{[\exists x\, \mathscr{J}]\phi} \qquad ([:=])\ \frac{\chi \to \phi_{x_1}^{\theta_1} \cdots {}_{x_n}^{\theta_n}}{[x_1 := \theta_1 \wedge .. \wedge x_n := \theta_n \wedge \chi]\phi}\,{}^{3}$$

$$(\langle\cup\rangle)\ \frac{\langle\alpha\rangle\phi \vee \langle\beta\rangle\phi}{\langle\alpha \cup \beta\rangle\phi} \qquad (\langle J\rangle)\ \frac{\langle\mathscr{J}_1 \cup \ldots \cup \mathscr{J}_n\rangle\phi}{\langle\mathscr{J}\rangle\phi}\,{}^{2} \qquad (\langle D\rangle)\ \frac{\langle(\mathscr{D}_1 \cup \ldots \cup \mathscr{D}_n)^*\rangle\phi}{\langle\mathscr{D}\rangle\phi}\,{}^{4}$$

$$([\cup])\ \frac{[\alpha]\phi \wedge [\beta]\phi}{[\alpha \cup \beta]\phi} \qquad ([J])\ \frac{[\mathscr{J}_1 \cup \ldots \cup \mathscr{J}_n]\phi}{[\mathscr{J}]\phi}\,{}^{2} \qquad ([D])\ \frac{[(\mathscr{D}_1 \cup \ldots \cup \mathscr{D}_n)^*]\phi}{[\mathscr{D}]\phi}\,{}^{4}$$

$$([DR])\ \frac{\vdash [\mathscr{E}]\phi}{\vdash [\mathscr{D}]\phi}\,{}^{5} \qquad (\langle DR\rangle)\ \frac{\vdash \langle\mathscr{D}\rangle\phi}{\vdash \langle\mathscr{E}\rangle\phi}\,{}^{5} \qquad (DS)\ \frac{\vdash [\mathscr{D}]\chi \quad \vdash [\mathscr{D} \wedge \chi]\phi}{\vdash [\mathscr{D}]\phi}$$

$$([]gen)\ \frac{\vdash \forall^\alpha(\phi \to \psi)}{[\alpha]\phi \vdash [\alpha]\psi} \qquad (\langle\rangle gen)\ \frac{\vdash \forall^\alpha(\phi \to \psi)}{\langle\alpha\rangle\phi \vdash \langle\alpha\rangle\psi}$$

$$(ind)\ \frac{\vdash \forall^\alpha(\phi \to [\alpha]\phi)}{\phi \vdash [\alpha^*]\phi} \qquad (con)\ \frac{\vdash \forall^\alpha \forall v{>}0\, (\varphi(v) \to \langle\alpha\rangle\varphi(v-1))}{\exists v\, \varphi(v) \vdash \langle\alpha^*\rangle \exists v{\le}0\, \varphi(v)}\,{}^{6}$$

$$(DI)\ \frac{\vdash \forall^\alpha \forall y_1 .. \forall y_k\, (\chi \to F'{}_{x_1'}^{\theta_1} \cdots {}_{x_n'}^{\theta_n})}{[\exists y_1 .. \exists y_k\, \chi]F \vdash [\exists y_1 .. \exists y_k\, (x_1' = \theta_1 \wedge .. \wedge x_n' = \theta_n \wedge \chi)]F}\,{}^{7}$$

$$(DV)\ \frac{\vdash \exists \varepsilon{>}0\, \forall^\alpha \forall y_1 .. y_k\, (\neg F \wedge \chi \to (F' \ge \varepsilon)_{x_1'}^{\theta_1} \cdots {}_{x_n'}^{\theta_n})}{[\exists y_1 .. y_k\, (x_1' = \theta_1 \wedge .. \wedge x_n' = \theta_n \wedge {\sim}F)]\chi \vdash \langle\exists y_1 .. y_k\, (x_1' = \theta_1 \wedge .. \wedge x_n' = \theta_n \wedge \chi)\rangle F}\,{}^{8}$$

---

[1] $\Gamma_i \vdash \Delta_i$ are obtained from the subgoals of side deduction $(\star)$ in Fig. 3.10, in which x is assumed to occur in first-order formulas only, as QE is then applicable. The side deduction starts from goal $\Gamma \vdash \Delta, \phi$ at the bottom (or $\Gamma, \phi \vdash \Delta$ for l$\forall$ and l$\exists$), where x does not occur in Γ, Δ using renaming.

[2] $\mathscr{J}_1 \vee \cdots \vee \mathscr{J}_n$ is a disjunctive normal form of the DJ-constraint $\mathscr{J}$.

[3] Rule applicable for any reordering of the conjuncts of the DJ-constraint where χ is jump-free.

[4] $\mathscr{D}_1 \vee \cdots \vee \mathscr{D}_n$ is a disjunctive normal form of the DA-constraint $\mathscr{D}$.

[5] $\mathscr{D}$ implies $\mathscr{E}$, i.e., satisfies the assumptions of Lemma 3.3.

[6] Logical variable v does not occur in α.

[7] Applicable for any reordering of the conjuncts where χ is non-differential. F is first-order without negative equalities, and F' abbreviates $D(F)$, with z' replaced with 0 for unchanged variables.

[8] Like *DI*, but F contains no equalities and the differential equations are Lipschitz continuous.

Fig. 3.9 Rule schemata of the proof calculus for differential-algebraic dynamic logic

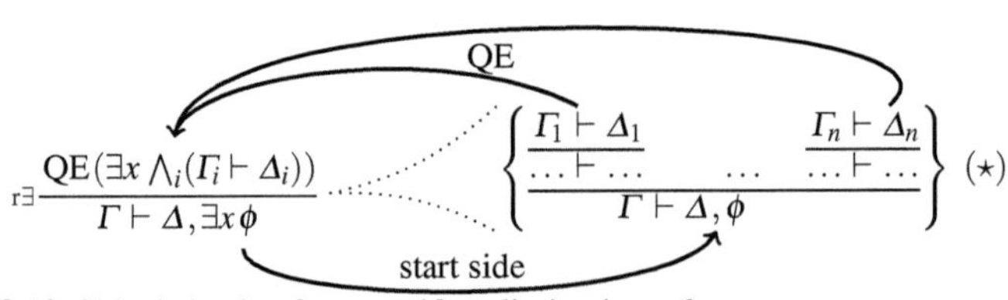

Fig. 3.10 Side deduction for quantifier elimination rules

using quantifier elimination (QE) over the reals [81]. Unfortunately, QE is only defined in first-order real arithmetic and cannot handle DAL modalities, where variables evolve along hybrid trajectories over time. We establish compatibility with dynamic modalities using (cut-like) side deductions for the quantifier rules, as illustrated in Sect. 3.5.5. Alternatively, the quantifier rules in Fig. 3.9 can be replaced by the quantifier rules of Chap. 2, which generalise free variables, Skolemisation, and Deskolemisation to real arithmetic for integrating quantifier elimination with modal rules. Instead, here, we use side deductions that we have introduced in previous work [231] as a very intuitive and simple approach for handling real quantifiers.

Dynamic Rules

The dynamic rules transform a DA-program into simpler logical formulas. Rules $\langle;\rangle$,$[;]$,$\langle\cup\rangle$,$[\cup]$ are as in discrete dynamic logic [149, 37] and in the $\mathsf{d}\mathcal{L}$ calculus in Fig. 2.11. The new rules $\langle J\rangle$ and $[J]$ normalise DJ-constraints to their disjunctive normal form such that the jump alternatives can be read off easily. They also turn the disjuncts of the disjunctive normal form into nondeterministic choices so that $\langle\cup\rangle$,$[\cup]$ will reason separately for each case. Rules $\langle\exists\rangle$ and $[\exists]$ lift quantified choices in DJ-constraints to DAL quantifiers, which are, in turn, handled by quantifier rules. They use the fact that *all choices of* assigning *some value* to x correspond to the universal quantifier (rule $[\exists]$), whereas *some choice of* assigning *some value* to x corresponds to the existential quantifier (rule $\langle\exists\rangle$).

Rules $\langle:=\rangle$ and $[:=]$ use generalised simultaneous substitutions for handling discrete change by substituting the respective new values θ_i for all the affected variables x_i at once (for all $i = 1,\dots,n$) and checking the jump-free constraint χ. In fact, this check or assumption of the jump-free constraint χ is the only essential difference in rules $\langle:=\rangle$,$[:=]$ in Fig. 3.9 compared to those in Fig. 2.11. Notice however, that as in rules $\langle?\rangle$,$[?]$ from Fig. 2.11, the presence of χ already makes $[x:=\theta\wedge\chi]\phi$ and $\langle x:=\theta\wedge\chi\rangle\phi$ non-equivalent. Thus, for $\langle x:=\theta\wedge\chi\rangle\phi$ rule $\langle:=\rangle$ is used to prove that χ holds true (otherwise there is no transition and thus the reachability property is false) and that ϕ holds after replacing x with its new value θ. Rule $[:=]$ for $[x:=\theta\wedge\chi]\phi$ instead assumes that χ holds true (otherwise there is no transition and thus nothing to show) and that ϕ holds after replacing x with its new value θ.

Similarly to rules $\langle J\rangle$,$[J]$, rules $\langle D\rangle$,$[D]$ normalise DA-constraints to a form where their differential evolution alternatives are readily identifiable. They turn a DA-constraints into its disjunctive normal form and split disjuncts into nondeterministic choices. Unlike for $\langle J\rangle$,$[J]$, however, continuous evolutions take time so that the system can switch back and forth (repeatedly) between the various cases of the DA-constraint during one continuous evolution, and hence the repetition. Observe that finitely many repetitions are sufficient for non-Zeno flows (Definition 3.12), which can only switch finitely often in finite time.

Rules $[DR]$,$\langle DR\rangle$, and DS are differential weakening, differential refinement, and differential strengthening rules for DA-constraints, respectively. In rules $[DR]$ and $\langle DR\rangle$, DA-constraint $\mathscr{D}$ implies $\mathscr{E}$ in real arithmetic according to Lemma 3.3, which is easy to decide by QE in practise. Note that $[DR]$ and $\langle DR \rangle$ are sound for *any* such combination of $\mathscr{D}$ and $\mathscr{E}$. Their primary practical purpose is to use $[DR]$ for overapproximating individual variable evolutions by a weaker version (*differential weakening*) and $\langle DR \rangle$ for refining nondeterministic variable evolutions to specific differential equations (*differential refinement*). For instance, rule $[DR]$ can weaken $[x' = 5x \wedge y' = 1 - x \wedge x > 0]\phi$ to $[x' = 5x \wedge y' \leq 1 \wedge x > 0]\phi$, because the former DA-constraint implies the latter DA-constraint according to Lemma 3.3. Such overapproximations have the advantage of decoupling differential equations. In particular, we use $[DR]$ onto project conjunctive differential constraints $\mathscr{D}$ to their non-differential constraints. As we illustrate in Sect. 3.11, this gives a powerful verification technique in combination with differential strengthening (*DS*), which can refine the system dynamics by auxiliary constraints. Rule $\langle DR \rangle$ can refine nondeterministic variable evolutions into specific cases, e.g., $\langle z' = v \wedge v' \geq -b \rangle \phi$ to the borderline case $\langle z' = v \wedge v' = -b \rangle \phi$. The latter refines the former, because it is a special case (the latter DA-constraint implies the former DA-constraint in the sense of Lemma 3.3).

While rule $[DR]$ can be used to remove information from a DA-constraint, the differential strengthening rule *DS* does the opposite and can be used to add information. But, of course, *DS* can only add constraints χ (to the right premise) that have been proven to be invariant (in the left premise); otherwise it would alter the reachable set incorrectly. With the left premise proven, rule *DS* changes the dynamics of the DA-program and restricts the DA-constraint $\mathscr{D}$ to remain within evolution domain χ. But this restriction is a pseudo restriction, because the left premise shows that χ actually is an invariant of the previous dynamics $\mathscr{D}$. In fact, assuming the left premise $[\mathscr{D}]\chi$ to be valid, the right premise $\vdash [\mathscr{D} \wedge \chi]\phi$ and conclusion $\vdash [\mathscr{D}]\phi$ are actually equivalent (as we will show in Proposition 6.2). The differential strengthening rule *DS* is essentially a *differential cut*. We address the problem of automatically determining the respective strengthenings χ that actually help the proof in Chap. 6, where we derive automatic verification algorithms from the results presented in this chapter. Furthermore, $[DR]$ and $\langle DR \rangle$ make all equivalence transformations on DA-constraints from Sect. 3.5.3 available as proof rules, including index reduction techniques for differential-algebraic equations [132].

Note that DAL does not need rules for handling negation in DA-constraints or DJ-constraints, as—possibly after applying $\langle J \rangle$,$[J]$ or $\langle D \rangle$,$[D]$, respectively—negations only occur in jump-free or non-differential parts, because assignments and differential symbols only occur positively by Definitions 3.1 and 3.2. Similarly, no rules for universal quantifiers within DA-constraints or DJ-constraints are needed. Like other propositional operators or quantifiers, negation and universal quantifiers are allowed without restriction in non-differential or jump-free χ and are then handled by $\langle := \rangle$,$[:=]$ or *DI*,*DV*.

Global Rules

The global rules $[]gen$,$\langle\rangle gen$,*ind*,*con*,*DI*,*DV* depend on the truth of their premises in all states, which is ensured by the universal closure with respect to all bound variables of the respective DA-program α (see Definition 3.14). In particular, the rules $[]gen$,$\langle\rangle gen$,*ind*,*con* are as in the $d\mathcal{L}$ calculus of Fig. 2.11. The differential induction rules *DI* and *DV* are new.

Rule *DI* is a rule for *differential induction*, which is a continuous form of induction along differential constraints. The induction rules *ind* and *DI* (or *con* and *DV* respectively) differ in the way the invariant is sustained (or in the way the variant makes progress). Rule *ind* uses the inductive nature of repetition and, ultimately, follows an induction on the number of repetitions. Rule *DI*, in contrast, uses continuity of evolution and the differential equation for a continuous induction step with the *differential invariant* F: If F holds initially (antecedent of conclusion) and its total differential F' satisfies the same relations when taking into account the differential constraints (premise), then F itself is sustained differentially (succedent of conclusion). Formula F' abbreviates $D(F)$ (see Definition 3.13) with z' replaced with 0 for all variables z that are unchanged by the DA-constraint, i.e., that are distinct from $\{x_1,\dots,x_n\}$, because these are assumed constant in the semantics. By bound variable renaming, the y_i do not occur in F. Thus, an important difference between the operating principles of rules *ind* and *DI* is that *ind* uses induction over natural numbers for repetitions of a loop, whereas *DI* uses induction in the continuous domain along the vector fields of differential equations, and is based on differential algebra (Sects. 3.5.2 and 3.5.3). A notable special instance of rule *DI* is the following case for quantifier-free DA-constraints:

$$\frac{\vdash \forall^{\alpha}(\chi \rightarrow F'{}^{\theta_1}_{x_1'}\cdots{}^{\theta_n}_{x_n'})}{[\chi]F \vdash [x_1' = \theta_1 \land \ldots \land x_n' = \theta_n \land \chi]F}$$

Rule *DV* is a *differential variant* rule where the variant F is finally reached differentially (with some minimal progress ε), rather than sustained as in *DI*. Differential induction, the requirement of the differential equations for *DV* to be Lipschitz continuous, and the notations $F' \geq \varepsilon$ and $\sim F$ will be illustrated in more detail in Sects. 3.5.6 and 3.5.7 after side deductions for quantifiers have been explained in Sect. 3.5.5. Finally, global rules can be combined with generalisation ($[]gen$,$\langle\rangle gen$) to strengthen postconditions as needed, similarly to rules *ind'* and *con'* from p. 86. A notable special instance of rule *DV* is the following case for quantifier-free DA-constraints:

$$\frac{\vdash \exists \varepsilon{>}0\,\forall^{\alpha}(\neg F \land \chi \rightarrow (F' \geq \varepsilon)^{\theta_1}_{x_1'}\cdots{}^{\theta_n}_{x_n'})}{[x_1' = \theta_1 \land \ldots \land x_n' = \theta_n \land {\sim}F]\chi \vdash \langle x_1' = \theta_1 \land \ldots \land x_n' = \theta_n \land \chi\rangle F}$$

Derivability and Proofs

Provability can be defined as a simplified version of Definition 2.11, because all DAL rules have only one conclusion; so a proof will be inductively defined as a tree, not as an acyclic graph (as was necessary for the quantifier rules from Chap. 2).

Definition 3.16 (Provability). A formula ψ is *provable* from a set Φ of formulas, denoted by $\Phi \vdash_{DAL} \psi$, iff there is a finite set $\Phi_0 \subseteq \Phi$ for which the sequent $\Phi_0 \vdash \psi$ is derivable. Derivability is inductively defined so that a sequent $\Phi \vdash \Psi$ is *derivable* iff there is a proof rule of the DAL calculus (Definition 3.15) with conclusion $\Phi \vdash \Psi$ such that all premises of the rule are derivable.

3.5.5 Deduction Modulo by Side Deduction

The quantifier rules constitute a purely modular interface to mathematical reasoning. They can use any theory that admits quantifier elimination and has a decidable ground theory, e.g., the theory of first-order real arithmetic, which is equivalent to the theory of real-closed fields [81]. Unlike in deduction modulo approaches of Dowek et al. [103] and Tinelli [290], the information given to the background prover is not restricted to ground formulas [290] or atomic formulas [103], and the effect of modalities has to be taken into account.

Integrating quantifier elimination to deal with statements about real quantities is quite challenging in the presence of modalities that influence the values of variables and terms. Real quantifier elimination cannot be applied to formulas with mixed quantifiers and modalities such as $\exists x\,[x' = -x; x := 2x]x \leq 5$. To find out which first-order constraints are actually imposed on x by this DAL formula, we have to take into account how x evolves from $\exists x$ to $x \leq 5$ along the hybrid system dynamics. Hence, our calculus first unveils the first-order constraints on x before applying QE. To achieve this in a concise and simple way, we use side deductions that we have introduced in previous work [231].

The effect of a side deduction is as follows. First, the DAL calculus discovers all relevant first-order constraints from modal formulas using a side deduction. Secondly, these constraints are re-imported into the main proof and equivalently reduced using QE, and then the main proof continues. For instance, an application of r$\exists$ to a sequent $\Gamma \vdash \Delta, \exists x\,\phi$ starts a side deduction (marked $(\star)$ in Fig. 3.10) with the unquantified kernel $\Gamma \vdash \Delta, \phi$ as a goal at the bottom. This side deduction is carried out in the DAL calculus until x no longer occurs within modal formulas of the remaining open branches $\Gamma_i \vdash \Delta_i$ of $(\star)$. Once all occurrences of x are in first-order formulas, the resulting subgoals $\Gamma_i \vdash \Delta_i$ of $(\star)$ are copied back to the main proof and QE is applied to eliminate x altogether (which determines the resulting subgoal of rule r$\exists$ on the upper left side of Fig. 3.10). The remaining modal formulas not containing x can be considered as atoms for this purpose, as they do not impose constraints on x. Formally, this can be proven using the coincidence lemma 2.6. When

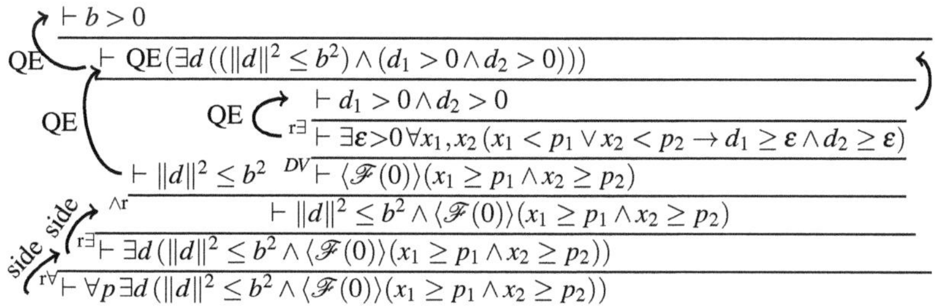

Fig. 3.11 Nested side deductions and differential variants for progress property

several quantifiers are nested, side deductions will be nested in a cascade, as they can again spawn further side deductions. According to the applicability conditions of quantifier rules, inner nested side deductions need to be completed by QE before outer deductions continue. For instance, further side deductions started within $(\star)$ of Fig. 3.10 will be completed before $(\star)$ continues and the quantifier elimination result of $(\star)$ is returned to the main r∃ application.

Example 3.13 (Aircraft progress). To illustrate how our calculus combines arithmetic with dynamic reasoning using side deductions, we look at an aircraft example. Using the notation from Sect. 3.4 where $\mathscr{F}(\omega)$ denotes the flight equation with angular velocity ω, the following DAL formula expresses a simple progress property about aircraft: The aircraft at x can finally fly beyond any point $p \in \mathbb{R}^2$ by adjusting its speed vector d appropriately, using only speed vectors $d \in \mathbb{R}^2$ of bounded speed $\|d\| \le b$, i.e., $\|d\|^2 \le b^2 \equiv d_1^2 + d_2^2 \le b^2$:

$$\forall p \exists d\,(\|d\|^2 \le b^2 \wedge \langle \mathscr{F}(0)\rangle(x_1 \ge p_1 \wedge x_2 \ge p_2)). \tag{3.6}$$

There, point p is constant during the evolution, i.e., $p_1' = p_2' = 0$ and $b' = 0$. The DAL proof in Fig. 3.11 proves this property using nested side deductions for nested quantifiers and differential variant induction *DV*. Applying r∀ in the main branch yields a side deduction for quantifier $\forall p$, which, in turn, yields another side deduction by applying r∃ for the nested quantifier $\exists d$. □

These nested side deductions in Fig. 3.11 are inlined and indicated by indenting the side deductions, with arrows marked "side" pointing to the start of the respective inner side deduction and arrows marked "QE" pointing back to the continuation of the outer deduction (as in Fig. 3.10). The two branches for the side deduction for r∃ recombine conjunctively and, after quantifiers are re-added, quantifier elimination yields $b > 0$, which reveals the parameter constraint on the speed bound b. Consequently, property (3.6) holds true and the proof closes for all positive speed bounds $b > 0$. The right branch of the inner r∃ side deduction uses differential variant induction *DV*, as will be illustrated in Sect. 3.5.7. There, the quantifiers for x_1, x_2 result from the universal closure $\forall^\alpha$ in *DV*. The subsequent innermost r∃ side deduction can be abbreviated by directly applying QE, because the affected formula already is first-order. That is, the side deduction for $\exists\varepsilon$ is the identity proof with zero rule applications.

As with the other aircraft examples in this chapter, formula (3.6) is provable in our theorem prover KeYmaera within a few seconds, despite the complicated underlying aircraft dynamics.

3.5.6 Differential Induction with Differential Invariants

The purpose of *DI* and *DV* is to prove properties about continuous evolutions by differential induction using differential invariants or differential variants, respectively. They work with the differential constraints directly instead of with the complicated (possibly undecidable) arithmetic of their solutions. Unlike approaches using solutions [125, 228, 231, 233, 235], differential induction can even be used to verify systems with nondeterministic quantified input, which would otherwise cause quantified higher-order functions for the time-dependent input of the solutions. Solutions of differential equations lead to a quantifier over time (see $\mathsf{d}\mathcal{L}$ rules $[']$,$\langle'\rangle$ in Fig. 2.11). In addition to their dependency on time, solutions of quantified differential equations or differential equations with input disturbance depend on a function $u : [0,\infty) \to \mathbb{R}$ that specifies how this input $u(t)$ changes over time t. When generalising solution-based proof rules for these situations, quantification over disturbances would thus lead to quantifiers over functions in higher-order logic, as in $\forall u : \mathbb{R} \to \mathbb{R} \forall t{\geq}0 \, \langle x{:=}y_u(t)\rangle\phi$, where $y_u(t)$ is the solution at time t depending on the input disturbance function u. With the theory of differential induction, we avoid these higher-order quantifiers in the proof calculus. Further, unlike in discrete induction, differential induction proof rules exploit continuity of evolution and knowledge of the differential constraints for a continuous induction step. We demonstrate the capabilities and the necessity of the requirements of differential induction rules in a series of examples and counterexamples.

Differential Invariants

Rule *DI* uses differential induction to prove that F is a *differential invariant*, i.e., F is closed under total differentiation (Definition 3.13) relative to the differential constraints. For this, the premise of *DI* shows that the total differential F'—i.e., $D(F)$ with z' replaced by 0 for unchanged variables z— holds within evolution domain χ when substituting the differential equations into F'. Now, if F holds initially (antecedent of conclusion), then F itself is sustained (succedent of conclusion). Intuitively, the premise expresses that, within χ, the total derivative F' along the differential constraints is pointing inwards or transversally to F but never outwards to $\neg F$; see Fig. 3.12. Intuitively, if we start in F and, as indicated by F', the local dynamics never points outside F, then the system always stays in F when following the dynamics. At this point, it is important to note that, even though meta-proofs about

Fig. 3.12 Differential invariants

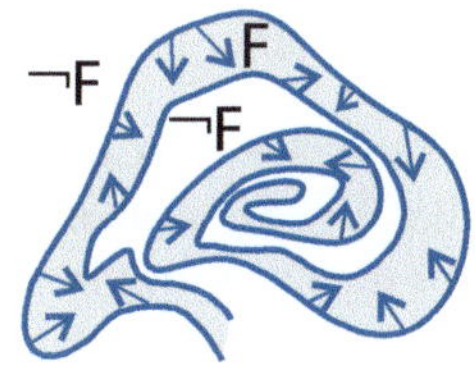

DAL involve analytic reasoning, proofs within the DAL calculus are fully algebraic, including the handling of differential constraints by *DI*. Further observe that the premise of *DI* is a well-formed DAL formula, because all differential symbols are replaced by non-differential terms when forming $F'^{\theta_1}_{x'_1}\dots{}^{\theta_n}_{x'_n}$.

Example 3.14 (Quartic dynamics). As a first simple example, consider the differential equation $x' = x^4$. It is not so easy to see the solution of this differential equation. Still, with implicit means of differential induction, we can establish easily that the solution always stays above $\frac{1}{4}$ whenever the dynamics initially starts above $\frac{1}{4}$:

$$\dfrac{\dfrac{*}{\vdash \forall x\,(x^4 \geq 0)}\,\mathrm{r}\forall}{x \geq \frac{1}{4} \vdash [x' = x^4]x \geq \frac{1}{4}}\,DI$$

This deduction proves the invariance of $x \geq \frac{1}{4}$ along the differential equation $x' = x^4$ by differential induction and without having to solve the differential equation. To apply the differential induction rule *DI*, we form the total derivative of the differential invariant $F \equiv x \geq \frac{1}{4}$ and obtain the differential expression $F' \equiv D(x \geq \frac{1}{4}) \equiv x' \geq 0$. Now, the differential induction rule *DI* takes into account that the derivative of state variable x along the dynamics is known (the trick, of course, is to show why this intuitive reasoning is sound, which we will prove in Sect. 3.6). Substituting the differential equation $x' = x^4$ into the inequality above yields $F'^{x^4}_{x'} \equiv x^4 \geq 0$, which is a valid formula and closes by quantifier elimination with r∀. Observe how elegantly differential induction establishes the desired result indirectly by working with the differential equation itself in an algebraic way instead of requiring its solution.

Even more so, for the differential equations $x' = x^2 + x^4$ or $x' = x^2 - 4x + 6$, solutions are hard to obtain both symbolically and numerically. With differential induction, however, we directly establish the following result about their dynamics:

$$\dfrac{\dfrac{*}{\vdash \forall x\,(3(x^2 + x^4) \geq 0)}\,\mathrm{r}\forall}{3x \geq \frac{1}{4} \vdash [x' = x^2 + x^4]3x \geq \frac{1}{4}}\,DI$$

$$\dfrac{\dfrac{*}{\vdash \forall x\,(3(x^2 - 4x + 6) \geq 0)}\,\mathrm{r}\forall}{3x \geq \frac{1}{4} \vdash [x' = x^2 - 4x + 6]3x \geq \frac{1}{4}}\,DI$$

For the latter proof, note that $3(x^2-4x+6)=3((x-2)^2+2)\geq 0$. □

Example 3.15 (Cubic dynamics). Similarly, differential induction can easily prove that $\frac{1}{3}\leq 5x^2$ is an invariant of the cubic dynamics $x'=x^3$; see the proof in Fig. 3.13a for the dynamics in Fig. 3.13b. To apply the differential induction rule *DI*, we again

$$\dfrac{\dfrac{*}{\vdash \forall x\,(0\leq 5\cdot 2x(x^3))}\,\text{r}\forall}{\frac{1}{3}\leq 5x^2\vdash [x'=x^3]\frac{1}{3}\leq 5x^2}\,DI$$

Fig. 3.13a Cubic dynamics proof

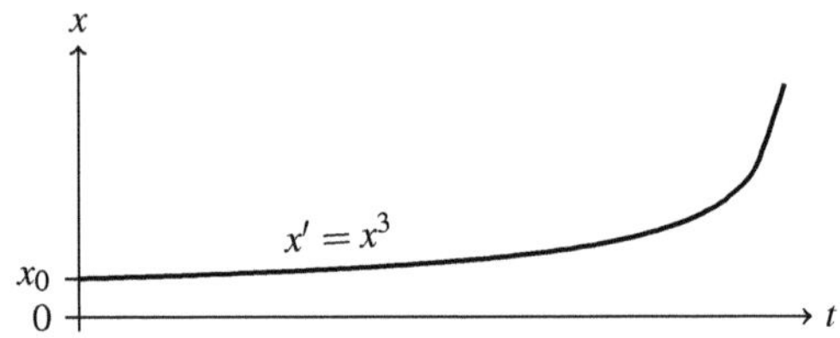

Fig. 3.13b Cubic dynamics

form the total derivative of the differential invariant $F\equiv\frac{1}{3}\leq 5x^2$, which gives the differential expression $F'\equiv D(\frac{1}{3}\leq 5x^2)\equiv 0\leq 5\cdot 2xx'$. Now, the differential induction rule *DI* takes into account that the derivative of state variable x along the dynamics is known. Substituting the differential equation $x'=x^3$ into the inequality yields $F'^{x^3}_{x'}\equiv 0\leq 5\cdot 2xx^3$, which is a valid formula and closes by quantifier elimination with r∀. □

Example 3.16 (Linear versus angular speed). Consider the following simple proof, which shows that the speed v of the aircraft with position x is maintained even when it changes its angular velocity ω nondeterministically during the flight (as in mode *free* of Fig. 3.7). Again, recall the flight equation with angular velocity ω:

$$x_1'=d_1\wedge x_2'=d_2\wedge d_1'=-\omega d_2\wedge d_2'=\omega d_1 \qquad (\mathscr{F}(\omega)^*)$$

$$\dfrac{\dfrac{\dfrac{\dfrac{\dfrac{*}{\vdash \text{QE}(\forall x_1,x_2\,\forall d_1,d_2\,\forall\omega\,(2d_1(-\omega d_2)+2d_2\omega d_1=0))}\,\text{QE}}{\vdash \forall x_1,x_2\,\forall d_1,d_2\,\forall\omega\,(2d_1(-\omega d_2)+2d_2\omega d_1=0)}\,\text{r}\forall}{d_1^2+d_2^2=v^2\vdash [\exists\omega\,\mathscr{F}(\omega)]\,d_1^2+d_2^2=v^2}\,DI}{\vdash d_1^2+d_2^2=v^2\rightarrow[\exists\omega\,\mathscr{F}(\omega)]\,d_1^2+d_2^2=v^2}\,\rightarrow\text{r}}{\vdash \forall v\,(d_1^2+d_2^2=v^2\rightarrow[\exists\omega\,\mathscr{F}(\omega)]\,d_1^2+d_2^2=v^2)}\,\text{r}\forall$$

(QE arrows; side)

The total derivative of the property $F\equiv d_1^2+d_2^2=v^2$ for differential induction with *DI* is $F'\equiv D(d_1^2+d_2^2=v^2)\equiv 2d_1d_1'+2d_2d_2'=2vv'$. Substituting the differential equations $\mathscr{F}(\omega)$ of flight yields $F'^{-\omega d_2\ \omega d_1\ 0}_{d_1'\ \ \ d_2'\ \ \ v'}\equiv 2d_1(-\omega d_2)+2d_2\omega d_1=0$, which is valid and closes by quantifier elimination. This example shows the difference of differential continuous evolution (of d_1,d_2) and nondeterministic continuous evolution (of ω). The DA-constraint specifies how the d_i evolve along differential equations;

hence d_i' is substituted in F'. For ω, in contrast, the DA-constraint is nondeterministic ($\exists\omega$) and does not specify how ω changes precisely. In particular, there is no equation for ω' that could be used for substitution. Yet such an equation is not even needed for forming the premise of *DI*, because, after bound variable renaming, ω cannot occur in F, since the scope of $\exists\omega$ ends with the DA-constraint and does not extend to postcondition F. In the proof, the quantifiers for x_i and d_i result from the universal closure $\forall^\alpha$ in *DI*. The quantifier for ω is introduced by *DV* and ensures that *all* possible evolutions of ω are taken into account as there is no specific equation for ω'. After all, ω is a quantified input and we cannot know a specific value for its slope, but have to expect any ($\forall\omega$) choice. Finally, note that in such cases without existential variables, side deductions can be inlined; see Chap. 2 for formal details on proof rules with Skolem function terms. □

Requirements of Differential Invariants

Next, we illustrate why the requirements formulated for the proof rule *DI* are necessary in general.

Counterexample 3.17. For soundness of differential induction, it is crucial that Definition 3.13 defines the total derivative $D(F \vee G)$ of a disjunction conjunctively as $D(F) \wedge D(G)$ instead of as $D(F) \vee D(G)$. From an initial state ν which satisfies $\nu \models F$, and hence $\nu \models F \vee G$, the formula $F \vee G$ only is sustained differentially if F itself is a differential invariant, not if G is. For instance, $x_1 \geq 0 \vee d_1^2 + d_2^2 = v^2$ is no differential invariant of $\exists\omega\,\mathscr{F}(\omega)$, because $x_1 \geq 0$ can be invalidated by appropriate curved flights along $\mathscr{F}(\omega)$; see formula (3.6). In practise, splitting differential induction proofs over disjunctions can be useful.

Note, however, that $D(F \vee G) \equiv D(F) \wedge D(G)$ is not the only definition that works. We could just as well define $D(F \vee G) \equiv (F \wedge D(F)) \vee (G \wedge D(G))$ instead. It is a simple exercise to show soundness of this modification. □

Counterexample 3.18 (Restricting differential invariance). It may be tempting to suspect that in *DI* the differential invariant F only needs to be differentially inductive at the states where F actually holds true. After all, the rule is used to prove that F stays true all the time, and in discrete loop induction (*ind*), the invariant ϕ can also be assumed to hold when showing the induction step in the premise. So why did we not assume F when proving F' in the premise of rule *DI*? See Fig. 3.14 for a tempting attempt of what a proof rule could be. But the differential induction needs

Fig. 3.14 Unsound restriction of differential invariance

$$\frac{\vdash \forall^\alpha(F \wedge \chi \rightarrow F'^{\theta_1}_{x_1'}\cdots^{\theta_n}_{x_n'})}{[\chi]F \vdash [x_1' = \theta_1 \wedge .. \wedge x_n' = \theta_n \wedge \chi]F}$$

to hold in a neighbourhood, such that adding F (or even just the border of F) to the assumptions in the premise of *DI* would be unsound! Consider the counterexample in Fig. 3.15a, where the differential invariance restricted to F would seem to indicate the region $x^2 \le 0$ was never left when following the dynamics $x' = 1$. This is, of course, completely counterfactual as the dynamics in Fig. 3.15b shows, where region $x^2 \le 0$ is actually left immediately when following the dynamics $x' = 1$.

$$\dfrac{\dfrac{* \text{ (unsound)}}{\vdash \forall x\, (x^2 \le 0 \to 2x \le 0)}}{x^2 \le 0 \vdash [x' = 1]x^2 \le 0}$$

Fig. 3.15a Restricting differential invariance

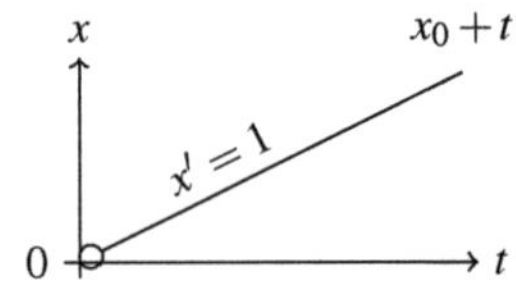

Fig. 3.15b Linear dynamics

Thus, although restricting the domain where proof rules require differential invariance appears to be a tempting idea, we have to be very careful to ensure the proof rules are sound in this subtle domain. In fact, the same counterexample in Fig. 3.15a also demonstrates unsoundness of other approaches that have been proposed to handle nontrivial differential equations [251, 145]. With a minor variation (replacing assumption $x^2 \le 0$ in the premise by $x^2 = 0$), Fig. 3.15a also shows that it is not sufficient to restrict the differential to the border of F. The problem causing this unsoundness is circular reasoning and the fact that derivatives are only defined in domains with non-empty interior. Strictly speaking, in the beginning we only know invariant F to hold at a single point (antecedent of conclusion of *DI*). Now if we assume F to hold in some domain for the induction step (premise of *DI*), then, initially, we only know that F holds in a region with an empty interior. This is not sufficient to conclude anything based on following derivatives, because these are not defined on regions with an empty interior. Thus, the reasoning in Fig. 3.15a assumes more than it has proven already, which explains the unsoundness, and the unsoundness of the rule in Fig. 3.14. □

This counterexample illustrates that differential invariance is a powerful but also subtle matter and that we have to prove soundness of the proof rules very carefully. We prove soundness of the DAL calculus and its differential induction principles in Sect. 3.6. In particular, we cannot generally use differential invariants for proving their continuous induction step. If, however, F describes an open set (e.g., F only involves strict inequalities), then *DI* is sound even when adding F to the assumptions of the premise as in Fig. 3.14. Likewise, F can be added to the assumptions of the premise when strengthening F' to strict inequalities. We will prove both refinements in Sect. 3.7. Furthermore, differential strengthening (*DS*) can be an extraordinarily successful proof technique for successively enriching evolution domain restrictions by derived invariants until F itself becomes differentially inductive, as we illustrate in Sects. 3.10 and 3.11. Also if polynomial solutions exist, they can be used as differential invariants.

Counterexample 3.19 (Negative equations). It is crucial for soundness of differential induction that F be not allowed to contain negative equations. In the following counterexample, variable x can reach $x = 0$ without its derivative ever being 0; again, see Fig. 3.15b for the dynamics.

$$\cfrac{\cfrac{*\ \text{(unsound)}}{\vdash \forall x\,(1 \neq 0)}}{x \neq 0 \vdash [x' = 1]x \neq 0}$$

If, instead, both inequalities $x < 0$ and $x > 0$ are differential invariants of a system (e.g., of the differential equation $x' = x$), then $x \neq 0$ can be proven indirectly by representing it equivalently as $x < 0 \lor x > 0$. We pursue this further in Example 3.29 in Sect. 3.7. □

Differential Weakening

A useful special case of the differential refinement rule $[DR]$ is the following derived weakening rule, with which we can assume the evolution domain constraint χ to hold for all reachable states.

Lemma 3.6 (Differential weakening). *The following is a derived proof rule:*

$$([DR'])\ \frac{\vdash \forall^\alpha \forall y_1 \ldots \forall y_k\,(\chi \to \phi)}{\vdash [\exists y_1 \ldots \exists y_k\,(x_1' = \theta_1 \land \cdots \land x_n' = \theta_n \land \chi)]\phi}$$

Proof. Rule $[DR']$ is sound, because χ is true along all state flows of the DA-constraint and ϕ is a consequence of χ in all reachable states (as overapproximated by $\forall^\alpha$) by premise. Rule $[DR']$ can be derived as follows:

$$\begin{array}{ll}
 & \vdash \mathrm{QE}(\forall y_1 \ldots \forall y_k \forall d_1 \ldots \forall d_n\,(\chi_{x_1}^{d_1}\cdots{}_{x_n}^{d_n} \to \phi_{x_1}^{d_1}\cdots{}_{x_n}^{d_n})) \\ \hline
\mathrm{QE} & \vdash \chi_{x_1}^{d_1}\cdots{}_{x_n}^{d_n} \to \phi_{x_1}^{d_1}\cdots{}_{x_n}^{d_n} \\ \hline
[:=],\ \text{side} & \vdash [x_1 := d_1 \land \cdots \land x_n := d_n \land \chi_{x_1}^{d_1}\cdots{}_{x_n}^{d_n}]\phi \\ \hline
\mathrm{r}\forall & \vdash \forall y_1 \ldots \forall y_k \forall d_1 \ldots \forall d_n\,([x_1 := d_1 \land \cdots \land x_n := d_n \land \chi_{x_1}^{d_1}\cdots{}_{x_n}^{d_n}]\phi) \\ \hline
[\exists] & \vdash [\exists y_1 \ldots \exists y_k \exists d_1 \ldots \exists d_n\,(x_1 := d_1 \land \cdots \land x_n := d_n \land \chi_{x_1}^{d_1}\cdots{}_{x_n}^{d_n})]\phi \\ \hline
[DR] & \vdash [\exists y_1 \ldots \exists y_k \exists d_1 \ldots \exists d_n\,(x_1' = d_1 \land \cdots \land x_n' = d_n \land \chi)]\phi \\ \hline
[DR] & \vdash [\exists y_1 \ldots \exists y_k\,(x_1' = \theta_1 \land \cdots \land x_n' = \theta_n \land \chi)]\phi
\end{array}$$

The second application of $[DR]$ uses the fact that fully nondeterministic continuous state change is equivalent to fully nondeterministic discrete state change, as they generate the same transitions. Finally, $\chi \to \phi$ can be obtained by bound variable renaming. □

Differential Invariants for Disturbance

So far, we have used differential induction rule *DI* for various differential equations. Yet proof rule *DI* and the technique of differential induction is more general than that. The rule also works for DA-constraints with quantifiers, differential inequalities, or systems with disturbance in the dynamics.

Example 3.20 (Disturbance in the train dynamics). Differential induction can be used to handle differential inequalities and disturbance in differential equations. Continuing Example 3.9 from p. 159, let us abbreviate the differential inequality

$$z' = v \wedge a - l \leq v' \leq a + u \wedge v \geq 0 \tag{3.1*}$$

succinctly by $z''{\approx}a$. This differential inequality characterises that the train at position z with velocity v follows approximately the chosen acceleration a, with a deviation from a that is bounded by lower bound $-l \leq 0$ and upper bound $u \geq 0$. Now, for the choice of $a := -b$ for braking, we can prove that the train that we considered in Sect. 2.5.3 always stays inside its movement authority m, despite the differential inequality disturbance in the dynamics, when starting in an initial state that satisfies:

$$\phi \equiv v^2 \leq 2(b-u)(m-z) \wedge b > u \geq 0 \wedge l \geq 0.$$

See Fig. 3.16 for a proof. The proof uses differential strengthening rule *DS* to aug-

$$\mathrm{{\to}r}\,\dfrac{DS\,\dfrac{[DR]\,\dfrac{DI\,\dfrac{\mathrm{r}\forall\,\dfrac{*}{\phi \vdash \forall^\alpha \forall d\,(-l \leq d \leq u \wedge v{\geq}0 \to 2v(d-b) \leq -2v(b-u))}}{\phi \vdash [\exists d\,(-l \leq d \leq u \wedge z'' = -b + d \wedge v \geq 0)]\phi}}{\phi \vdash [z''{\approx}-b \wedge v \geq 0]\phi} \qquad [DR']\,\dfrac{\mathrm{r}\forall\,\dfrac{*}{\vdash \forall^\alpha (v{\geq}0 \wedge \phi \to z \leq m)}}{\vdash [z''{\approx}-b \wedge v{\geq}0 \wedge \phi]z{\leq}m}}{\phi \vdash [z''{\approx}-b \wedge v \geq 0]z \leq m}}{\vdash \phi \to [z''{\approx}-b \wedge v \geq 0]z \leq m}$$

Fig. 3.16 Proof of MA-safety in braking mode with disturbance

ment the differential inequality $z''{\approx}-b$ with ϕ as an auxiliary invariant. The resulting right branch where *DS* added ϕ to the DA-constraint can then be proven easily using differential weakening $[DR']$ to show that the newly augmented evolution domain region $v \geq 0 \wedge \phi$ implies the postcondition $z \leq m$. In the left branch, we use differential induction rule *DI* to prove that the auxiliary invariant ϕ that we assumed in the right branch is actually an invariant of the dynamics and can thus be added soundly to the evolution domain restriction in the right branch, because ϕ is a pseudo restriction. Before the differential induction *DI* on the left branch, we use differential refinement $[DR]$ for an equivalence transformation that replaces the differential inequality $z''{\approx}-b$ from (3.1) with a quantified DA-constraint with a quantifier for the disturbance d that is restricted to the domain $-l \leq d \leq u$:

$$\exists d\,(-l \leq d \leq u \wedge z' = v \wedge v' = -b + d \wedge v \geq 0). \tag{3.7}$$

The equivalence of the quantified DA-constraint (3.7) with disturbance and the differential inequality (3.1) that is required for applying $[DR]$ can be established easily using Lemma 3.3. More detailed examples for properties of dynamics with disturbance, even equivalence properties, can be found in Sect. 7.4. □

Differential Invariants and Oscillation

Differential invariants are an interesting proof technique also for proving properties of dynamical systems with oscillation. In these systems, solutions of the differential equations cannot (really) be used for verification, because the solutions of oscillating processes involve trigonometric functions of undecidable classes of arithmetic.

Example 3.21 (Damped oscillator). Recall the damped oscillator from Example 1.3 on p. 8, which corresponds to the DA-constraint

$$x' = y \land y' = -\omega^2 x - 2d\omega y \tag{3.8}$$

with parameters $\omega \geq 0$ (for the undamped angular frequency) and $d \geq 0$ (for the damping ratio). See Fig. 3.17 for one example of an evolution along this continuous dynamics. Figure 3.17 shows a trajectory in the x, y space on the left, and an evolution of x over time t on the right. For the damped oscillator dynamics, we can easily prove the following invariance property:

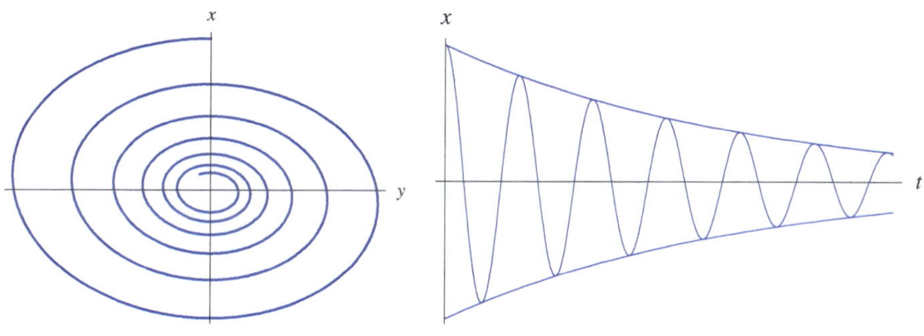

Fig. 3.17 Trajectory and evolution of a damped oscillator

$$\dfrac{\dfrac{*}{\omega \geq 0, d \geq 0 \vdash \forall x \forall y\,(2\omega^2 xy - 2\omega^2 xy - 4d\omega y^2 \leq 0)}\,\text{r}\forall}{\omega \geq 0, d \geq 0, \omega^2x^2 + y^2 \leq c^2 \vdash [x' = y \land y' = -\omega^2 x - 2d\omega y]\,\omega^2 x^2 + y^2 \leq c^2}\,DI$$

The total derivative of the property $F \equiv \omega^2 x^2 + y^2 \leq c^2$ for differential induction with DI is $F' \equiv D(\omega^2x^2 + y^2 \leq c^2) \equiv 2\omega^2 xx' + 2\omega\omega' x^2 + 2yy' \leq 0$. Substituting with the differential equations (3.8) gives $F'^{\,y}_{x'}{}^{-\omega^2x-2d\omega y}_{y'}{}^{0}_{\omega'}{}^{0}_{d'}{}^{0}_{c'}$, which is equivalent to:

$$2\omega^2 xy + 2y(-\omega^2 x - 2d\omega y) \le 0 \equiv 2\omega^2 xy - 2\omega^2 xy - 4d\omega y^2 \le 0 \equiv -4d\omega y^2 \le 0.$$

This formula is valid and closes by quantifier elimination (r∀), because $d\omega \ge 0$. □

Example 3.22 (Switched damped oscillators). Recall the switched damped oscillator from Example 1.3, which switches between two different damped oscillators of the form (3.8) for different choices of ω and d. The switched damped oscillator switches from one damped oscillator to the other at the diagonals of the state space. See Fig. 3.18 for an illustration of one possible evolution of the system. The hybrid sys-

Fig. 3.18 Trajectory switching between two damped oscillators

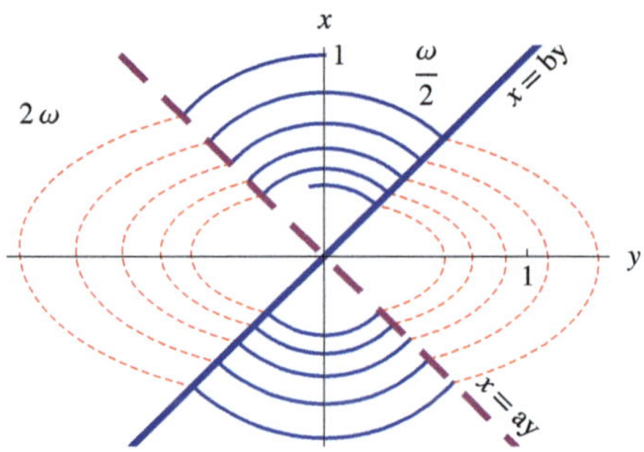

tem switches between two different damped oscillators at the switching surfaces depicted by the diagonal lines in Fig. 1.6 on p. 8 from one mode (shown in solid curves, with smaller ω) to the other mode (shown in dashed curves, with greater ω) and back again. At the solid diagonal, the system switches to the dynamics shown by a solid curve (with a smaller ω value), and at the dashed diagonal, it switches to the dynamics shown by a dashed curve (with greater ω value). We model a generalised parametric version of a corresponding switched damped oscillator in the first line of Fig. 3.19, with subsequent abbreviations. More generally, we assume the system

$$\phi \rightarrow [(\mathscr{D}; (?T \cup (?x = ay; \omega := 2\omega ..) \cup (?x = by; \omega := \frac{\omega}{2} ..)))^*]\phi$$

$$\text{where} \quad \phi \equiv \omega \ge 0 \land d \ge 0 \land \omega^2 x^2 + y^2 \le c$$

$$\mathscr{D} \equiv x' = y \land y' = -\omega^2 x - 2d\omega y$$

$$?T \equiv ?true$$

$$(?x = ay; \omega := 2\omega ..) \equiv ?x = ay; \omega := 2\omega;\ d := \frac{d}{2}; c := c\frac{(2\omega)^2 + 1^2}{\omega^2 + 1^2}$$

$$(?x = by; \omega := \frac{\omega}{2} ..) \equiv ?x = by; \omega := \frac{\omega}{2};\ d := 2d; c := c\frac{\omega^2 + 1^2}{(2\omega)^2 + 1^2}$$

Fig. 3.19 Parametric switched damped oscillator system

switches on a diagonal of the form $x = ay$ (dashed diagonal) to the faster angular frequency ($\omega := 2\omega$), and switches to the slower angular frequency($\omega := \frac{\omega}{2}$) at a diagonal of the form $x = by$ (solid diagonal). When the undamped angular frequency ω changes, the damping factor d is changed conversely and the range coefficient c adapts to the current ω.

Note that the system in Fig. 3.19 does not force the system to switch modes eagerly (there is no evolution domain restriction on the differential equations). Yet, while the system does not force switching, it only allows switching at the respective diagonals, when the tests $?x = ay$ or $?x = by$ are successful, which is all that we need. The intuition why we do not need to enforce switching to obtain a safe model is that the previous example, Example 3.21, showed that the continuous DA-constraint is safe and only switching could make it unsafe. But comparing the safe (and stable) evolution in Fig. 3.18 with the unsafe (and even divergent) evolution in Fig. 3.20, we find that the choice of a and b for the switching surfaces is crucial,

Fig. 3.20 Instable trajectory switching between two damped oscillators

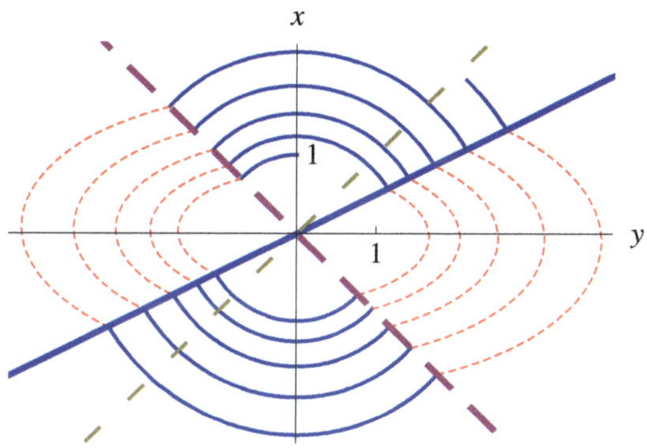

which is the only difference between both evolutions. How do we find out which choices of parameters a and b are safe?

With differential induction (and differential strengthening and differential weakening, respectively), we can analyse the safety property in Fig. 3.19 of the switched damped oscillators, which conjectures that the system always stays within the region $\omega^2x^2 + y^2 \leq c$. We analyse this property in the DAL proof in Fig. 3.21, using the abbreviations from Fig. 3.19.

The proof follows an induction by proof rule *ind* with invariant ϕ and then uses differential strengthening *DS* to show that ϕ is an auxiliary invariant of the DA-constraint $\mathscr{D}$, which can be proven by differential induction *DI* in the left branch immediately. Consequently, differential strengthening rule *DS* adds the auxiliary invariant ϕ. This rule adds auxiliary invariant ϕ into the evolution domain restrictions on the right branch after it has been proven $\in$ on the left branch to be invariant under the dynamics by *DI*. The addition of ϕ on the right branch then is a pseudo restriction of the dynamics, because ϕ is an invariant. The right branch now uses the (derived) differential weakening rule $[DR']$ to overapproximate the dynamics to the new evolution domain restriction ϕ. From ϕ, the remaining switching dynamics of the DA-program can be proven by splitting it into the three cases resulting from

$$\frac{\qquad\qquad\qquad\qquad\qquad\qquad\qquad\qquad\qquad\qquad *}{\qquad\qquad\qquad\qquad\qquad\qquad\qquad\qquad\qquad ^{ax}\ \phi \vdash \phi \qquad \phi \vdash -2 \le a \le 2 \qquad\qquad \phi \vdash b^2 \ge \tfrac{1}{3}}$$

$$\frac{\qquad\qquad\qquad\qquad\qquad\qquad\qquad\qquad\qquad ^{[?]}\ \phi \vdash [?T]\phi \quad \phi \vdash [(?x=ay;\omega:=2\omega..)]\phi \land [(?x=by;\omega:=\tfrac{\omega}{2}..)]\phi}{\wedge r\ \ \phi \vdash [?T]\phi \land [(?x=ay;\omega:=2\omega..)]\phi \land [(?x=by;\omega:=\tfrac{\omega}{2}..)]\phi}$$

$$\frac{}{[\cup]\ \ \phi \vdash [(?T \cup (?x=ay;\omega:=2\omega..) \cup (?x=by;\omega:=\tfrac{\omega}{2}..))]\phi}$$

$$\frac{}{\to r\ \ \phi \vdash \phi \to [(?T \cup (?x=ay;\omega:=2\omega..) \cup (?x=by;\omega:=\tfrac{\omega}{2}..))]\phi}$$

$$\frac{\qquad *}{^{DI}\ \phi \vdash [\mathscr{D}]\phi} \qquad \frac{\forall r\ \ \phi \vdash \forall^\alpha(\phi \to [(?T \cup (?x=ay;\omega:=2\omega..) \cup (?x=by;\omega:=\tfrac{\omega}{2}..))]\phi)}{[DR']\ \ \phi \vdash [\mathscr{D} \land \phi][(?T \cup (?x=ay;\omega:=2\omega..) \cup (?x=by;\omega:=\tfrac{\omega}{2}..))]\phi}$$

$$\frac{}{DS\ \ \phi \vdash [\mathscr{D}][(?T \cup (?x=ay;\omega:=2\omega..) \cup (?x=by;\omega:=\tfrac{\omega}{2}..))]\phi}$$

$$\frac{}{[;]\ \ \phi \vdash [\mathscr{D};(?T \cup (?x=ay;\omega:=2\omega..) \cup (?x=by;\omega:=\tfrac{\omega}{2}..))]\phi}$$

$$\frac{}{ind\ \ \phi \vdash [(\mathscr{D};(?T \cup (?x=ay;\omega:=2\omega..) \cup (?x=by;\omega:=\tfrac{\omega}{2}..)))^*]\phi}$$

$$\frac{}{\to r\ \ \vdash \phi \to [(\mathscr{D};(?T \cup (?x=ay;\omega:=2\omega..) \cup (?x=by;\omega:=\tfrac{\omega}{2}..)))^*]\phi}$$

Fig. 3.21 Parametric switched damped oscillator proof

the three cases of the nondeterministic choice by rules $[\cup]$,$\wedge$r. The leftmost of those branches is provable by the axiom rule *ax* directly, the rightmost two cases need a few more rule applications, including quantifier elimination, and then lead to the two branches indicated on the top right:

$$\phi \vdash -2 \le a \le 2 \qquad \text{and} \qquad \phi \vdash b^2 \ge \frac{1}{3}$$

The left case corresponds to the choice $(?x = ay; \omega := 2\omega..)$, and the right case to the choice $(?x = by; \omega := \frac{\omega}{2}..)$. From this proof, we can thus identify the constraints on the switching regions a and b that we need for the safety property to hold — quite similarly to our approach in Sect. 2.9 and, later, in Sect. 3.11. The resulting constraints we identify by combining the missing requirements are:

$$-2 \le a \le 2 \land b^2 \ge \frac{1}{3}. \tag{3.9}$$

In particular, when we add these assumptions to ϕ, the proof in Fig. 3.21 closes successfully, showing that the switched damped oscillator does not diverge but stays inside its region under these constraints.

Observe that, even though the system switches between two safe and stable continuous modes (damped oscillators), it is not at all evident that the switched damped oscillator is safe. In fact, the counterexample in Fig. 3.20 shows that surprisingly small variations of a and/or b make the switched system unsafe and instable, even make it diverge to infinity. The discovered constraints (3.9) characterise the different situations. For the evolution in Fig. 3.18, we have chosen the switching conditions $a = -1, b = 1$, and for Fig. 3.20 we have chosen $a = -1, b = 0.5$. □

Structural Properties of Differential Invariants

Differential invariants enjoy structural closure properties. They are closed under conjunction (because of the conjunctive definition in Definition 3.13) and the next lemma shows that they are closed under differentiation, which we summarise as:

$$F, G \text{ differential invariants then } F \wedge G \text{ differential invariant (of same system)}$$
$$F \text{ differential invariant then } D(F) \text{ differential invariant (of same system)}$$

Lemma 3.7 (Closure properties of differential invariants). *Differential invariants are closed under differentiation: The total derivative of a differential invariant is an invariant of the same DA-constraint.*

Proof. Let F be a differential invariant, i.e., satisfy DI for some DA-constraint of the form $\exists y\,(x' = \theta \wedge \chi)$, using vectorial notation for x and y. Hence, the premise of DI is provable: $\forall x \forall y\,(\chi \rightarrow F'^{\theta}_{x'})$ where the quantifier for x results from the universal closure $\forall^{\alpha}$. We have to show that the derivative $F'^{\theta}_{x'}$ is invariant and extend the proof to a proof of $[\exists y\,(x' = \theta \wedge \chi)]F'^{\theta}_{x'}$ by weakening (Lemma 3.6):

$$\cfrac{\cfrac{*}{\vdash \mathrm{QE}(\forall x \forall y\,(\chi \rightarrow F'^{\theta}_{x'}))}\;\mathrm{r}\forall}{\vdash [\exists y\,(x' = \theta \wedge \chi)]F'^{\theta}_{x'}}\;[DR']$$

□

3.5.7 *Differential Induction with Differential Variants*

The differential induction techniques with differential invariants (rule DI) can prove invariance properties. There is a dual proof technique of differential induction with differential *variants* (rule DV) to prove reachability or attractor properties.

Differential Variants

Unlike the differential induction rule DI for differential invariants, rule DV uses differential induction to prove that F is a *differential variant*, which is reached differentially as an attractor region rather than sustained differentially as in DI. The essential difference between DV and DI thus is the progress condition $F' \geq \varepsilon$ in the premise, saying that the total differential of F along the DA-constraint is positive and at least some $\varepsilon > 0$. There, $F' \geq \varepsilon$ is a mnemonic notation for replacing all occurrences of inequalities $a \geq b$ in F' with $a \geq b + \varepsilon$ and $a > b$ by $a > b + \varepsilon$ (accordingly for $\leq, >, <$). Intuitively, the premise expresses that, wherever χ holds but F does not yet hold, the total derivative is pointing towards F; see Fig. 3.22.

Fig. 3.22 Differential variants

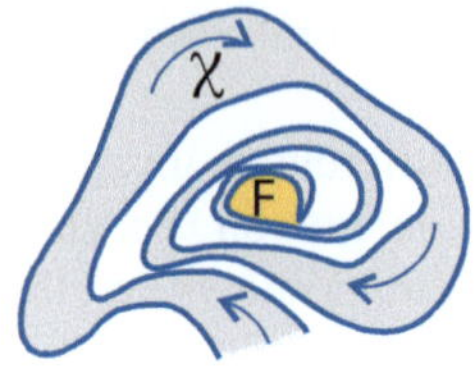

Especially, $F' \geq \varepsilon$ guarantees a minimum progress rate of ε towards F along the dynamics. To further ensure that the continuous evolution towards F remains within χ, the antecedent of the conclusion shows that χ holds *until* F is attained, which can again be proven using *DI*. Overall, the premise of rule *DV* shows that the dynamics makes progress (at least ε) toward F, and the antecedent shows that the dynamics does not leave evolution domain restriction χ on the way to F. In this context, $\sim\! F$ is a shorthand notation for *weak negation*, i.e., the operation that behaves like $\neg$, except that $\sim\!(a \geq b) \equiv b \geq a$ and $\sim\!(a > b) \equiv a \leq b$. Unlike negation, weak negation retains the border of F, which is required in *DV* as χ needs to continue to hold (including the border of F) until F is reached. Especially, for rule *DV*, invariant χ is not required to hold after F has been reached successfully. The operations $F' \geq \varepsilon$ and $\sim\! F$ are defined accordingly for other inequalities (in rule *DV*, we do not permit F to contain equalities; see Counterexample 3.25 below). Again, we demonstrate differential induction and the necessity of its prerequisites in a series of examples.

Example 3.23. As a very simple example for using differential induction with differential variants, consider the property where we want to prove that $\langle x' = a \rangle x \geq b$, i.e., we can finally reach region $x \geq b$, when we follow the dynamics $x' = a$ long enough. We analyse this DAL formula in the following DAL proof:

$$
\cfrac{\cfrac{\vdash a > 0}{\vdash \exists \varepsilon{>}0\, \forall x\, (x \leq b \rightarrow a \geq \varepsilon)}\ \mathrm{r}\exists}{\vdash \langle x' = a \rangle x \geq b}\ DV
$$

As the proof reveals, the property is valid if only $a > 0$, which makes sense, because the system dynamics is then evolving towards $x \geq b$; otherwise it is evolving away from $x \geq b$ (if $a < 0$) or is constant ($a = 0$). For the above proof, we do not need to solve the differential equations. Solving the differential equation would be trivial here for a constant a, but is more involved when a is an arbitrary term. Thus, instead, we just form the total differential of $F \equiv x \geq b$, which gives $F' = x' \geq b'$. When we substitute in the differential equations $x' = a$ and assume that $a' = 0, b' = 0$, we obtain $F'^{a\,0\,0}_{x'\,a'\,b'} \equiv a \geq 0$. Consequently, $(F' \geq \varepsilon)^{a\,0\,0}_{x'\,a'\,b'}$ gives $a \geq \varepsilon$. If we can prove $a \geq \varepsilon$ holds for one common minimum progress $\varepsilon > 0$, then the system makes some minimum progress towards the goal and will reach it in finite time. This even holds if we restrict the progress condition to all x that have not yet reached $x \geq b$ or are on the border of $x \geq b$, which is the assumption $x \leq b$ in the premise. □

Example 3.24 (Aircraft progress). Recall the aircraft progress property that we proved in Fig. 3.11 on p. 169. In the rightmost side deduction, *DV* is used to prove that $F \equiv x_1 \geq p_1 \wedge x_2 \geq p_2$ is finally reached. There, the total derivative is $F' \equiv x_1' \geq 0 \wedge x_2' \geq 0$, which yields $d_1 \geq 0 \wedge d_2 \geq 0$ when substituting the flight equations $\mathscr{F}(\omega)$, because $x_1' = d_1, x_2' = d_2, p_1' = p_2' = 0$. Consequently, for $\omega = 0$, $(F' \geq \varepsilon)^{d_1\, d_2\, -\omega d_2\, \omega d_1\, 0\; 0}_{x_1'\, x_2'\, d_1' \quad d_2'\; p_1'\, p_2'}$ is identical to $(F' \geq \varepsilon)^{d_1\, d_2\, 0\; 0}_{x_1'\, x_2'\, p_1'\, p_2'}$, giving $d_1 \geq \varepsilon \wedge d_2 \geq \varepsilon$. Similarly, the proof for formula (3.6) can be generalised to differential inequalities, again assuming $d_1' = d_2' = p_1' = p_2' = 0$ and $b' = 0$:

$$\forall p\, \exists d\, (\|d\|^2 \leq b^2 \wedge \langle x_1' \geq d_1 \wedge x_2' \geq d_2 \rangle (x_1 \geq p_1 \wedge x_2 \geq p_2)).$$

Using differential refinement rule $\langle DR \rangle$ and Lemma 3.4, the differential inequalities, which express lower bounds on the evolution of x_1 and x_2, can be reduced equivalently to differential equations with quantified disturbance $u \in \mathbb{R}^2$:

$$\forall p\, \exists d\, ..\langle \exists u\, (x_1' = d_1 + u_1 \wedge x_2' = d_2 + u_2 \wedge u_1 \geq 0 \wedge u_2 \geq 0) \rangle (x_1 \geq p_1 \wedge x_2 \geq p_2).$$

The proof for this DAL formula is identical to that in Fig. 3.11, except that *DV* now yields $\forall x\, \forall u\, ((x_1 < p_1 \vee x_2 < p_2) \wedge u_1 \geq 0 \wedge u_2 \geq 0 \rightarrow d_1 + u_1 \geq \varepsilon \wedge d_2 + u_2 \geq \varepsilon)$. □

Requirements of Differential Variants

Like the proof rule *DI* for differential invariants, the proof rule *DV* for differential variants has requirements on the formulas. We illustrate why the requirements formulated for the proof rule *DV* are necessary in general.

Counterexample 3.25 (Equational differential variants). Rule *DV* is not applicable for equations like $x = y$. Even though $x = y$ can be encoded as $F \equiv x \leq y \wedge x \geq y$, the corresponding $F' \geq \varepsilon \equiv x' + \varepsilon \leq y' \wedge x' \geq y' + \varepsilon$ is equivalent to false for $\varepsilon > 0$. Indeed, assuming $a' = b' = 0$, the validity of a formula like $\langle x' = a \wedge y' = b \rangle x = y$ depends on a more involved relationship of the initial values of x and y and the constants a and b: It is true, iff $(x - y)(a - b) < 0 \vee x = y$ holds initially.

More generally, differential variants cannot (directly) verify conjunctive equations as in $\langle x' = a \wedge y' = b \rangle (x = 0 \wedge y = 0)$ because differential variants guarantee that a target region F will be reached, but not when precisely. In particular, differential variants cannot guarantee that $x = 0$ and $y = 0$ would be reached simultaneously. In fact, for $a, b \neq 0$, the above reachability property is only valid iff $bx = ay \wedge ax < 0$ initially. □

Counterexample 3.26 (Minimal progress requirement). Unlike in discrete domains, strictly monotonic sequences can converge in $\mathbb{R}$. Thus, the premise $F' \geq \varepsilon$ for an $\varepsilon > 0$ of *DV* cannot be weakened to $F' > 0$. To see why, consider the counterexample in Fig. 3.23a, in which x converges monotonically to 0 along the dynamics shown in Fig. 3.23b. This counterexample illustrates that differential variance is a

$$\dfrac{\dfrac{*\ \text{(unsound)}}{\vdash \forall x\,(x>0 \to -x<0)}}{\vdash \langle x'=-x\rangle x\leq 0}$$

Fig. 3.23a Monotonically decreasing convergent counterexample

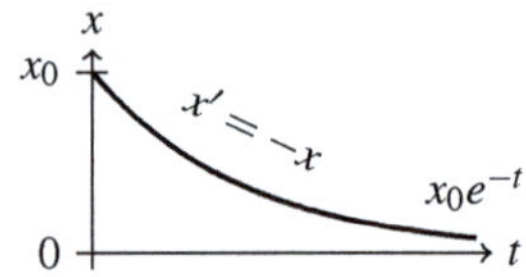

Fig. 3.23b Convergent descent dynamics

$$\dfrac{\dfrac{\mathit{false}}{\vdash \forall x\,(-x\geq 0)}\ \text{r}\forall}{x\geq 0 \vdash [x'=-x]x\geq 0}\ DI$$

Fig. 3.23c Non-inductive property in convergent descent

matter that is no less subtle than differential invariance, and we have to prove soundness of the proof rules very carefully. Moreover, this example demonstrates that, in the presence of convergent dynamics, a property like $x \geq 0$ can be invariant even though it is not differentially invariant; see Fig. 3.23c. □

Counterexample 3.27 (Lipschitz continuity requirement). As the counterexample in Fig. 3.24a shows, Lipschitz continuity (or at least the existence of a solution of sufficient duration) is, in fact, a necessary prerequisite for *DV*. For $x = y = 0$ initially, the solution of the differential equations in Fig. 3.24a is $x(t) = t$ and $y(t) = \tan t$. In explosive examples like the corresponding dynamics in Fig. 3.24b, where solution y grows unbounded in finite time, the duration of existence of solutions is limited so that the target region $x \geq 6$ is physically unreachable. More precisely, the dynamics

$$\dfrac{\dfrac{*\ \text{(unsound)}}{\vdash \exists \varepsilon{>}0\,\forall x\forall y\,(x<6 \to 1\geq \varepsilon)}}{\vdash \langle x'=1 \land y'=1+y^2\rangle x\geq 6}$$

Fig. 3.24a Counterexample of unbounded dynamics without Lipschitz continuity

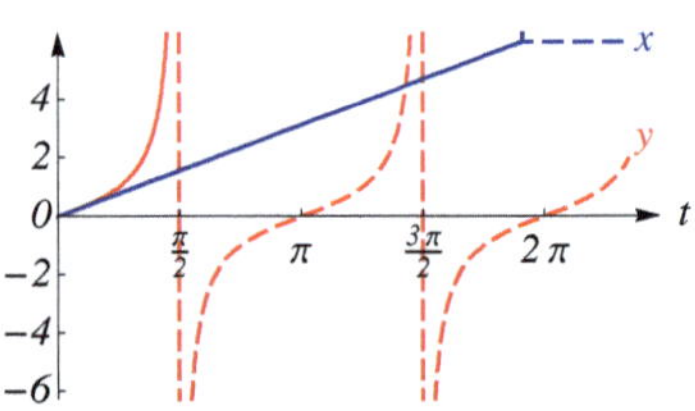

Fig. 3.24b Explosive dynamics with limited duration of solutions

is not well-posed beyond the explosive point of unbounded growth at the singularity $\frac{\pi}{2}$ and is non-physical beyond that singularity. The reason this can happen is that the differential equation is not (globally) Lipschitz continuous but only locally Lipschitz continuous and disobeys divergence of time (Sect. 3.3). The condition of (global) Lipschitz continuity is directly expressible as a formula for *DV*:

$$\exists L \forall y_1 \mathinner{..} \forall y_k \forall x_1 \mathinner{..} \forall x_n \forall \tilde{y}_1 \mathinner{..} \forall \tilde{y}_k \forall \tilde{x}_1 \mathinner{..} \forall \tilde{x}_n$$
$$(\theta_1 - \tilde{\theta}_1)^2 + \cdots + (\theta_n - \tilde{\theta}_n)^2 \le L^2((x_1 - \tilde{x}_1)^2 + \cdots + (x_n - \tilde{x}_n)^2)$$

where $\tilde{\theta}_i$ denotes the result of substituting all x_j in θ_i with the corresponding $\tilde{x}_j$, and the y_j with $\tilde{y}_j$. Observe that, besides Lipschitz continuity, any other condition can be used for rule *DV* that ensures the existence of a solution of sufficient duration. □

3.6 Soundness

In this section we prove that verification with the DAL calculus always produces correct results about DA-programs, i.e., the DAL calculus is sound. In light of the various subtleties and sources of unsoundness we have pointed out in the previous sections, we will devote our attention to a very careful soundness proof.

Theorem 3.1 (Soundness of DAL). *The* DAL *calculus is* sound, *i.e., every* DAL *formula that can be derived in the* DAL *calculus is valid (true in all states).*

Proof. The calculus is sound if each rule instance is sound. The rules of the DAL calculus are even *locally sound*, i.e., their conclusion is true in state ν if all its premises are true in ν. Local soundness implies soundness. The local soundness proofs of $\langle;\rangle$,$[;]$,$\langle\cup\rangle$,$[\cup]$ and the propositional rules are as in Theorem 2.1. Similarly, rules *ind* and *con* are local versions of induction schemes, and the proof is as in Theorem 2.1; likewise for rules $[]gen$,$\langle\rangle gen$. The local soundness of $\langle:=\rangle$,$[:=]$ is a generalisation of the proofs for update rules [37] to first-order DJ-constraints. The proofs for $\langle\exists\rangle$,$[\exists]$,$\langle J\rangle$,$[J]$ are simple. Finally, our results from Theorem 2.1 can be lifted to show that locally sound rules are closed under addition of the Γ,Δ context in Definition 3.16. Soundness would even be closed under the addition of conjunctive DJ-constraints as rule prefixes as in Definition 2.11. For soundness, however, conjunctive DJ-constraints are crucial here [37, 235] as these are deterministic.

r∃ Rule r∃ is locally sound: Let ν be a state in which the premise is true, i.e.,

$$\nu \models \mathrm{QE}(\exists x \bigwedge_i (\Gamma_i \vdash \Delta_i)).$$

We have to show that the conclusion is true in this state. Using the fact that quantifier elimination (QE) yields an equivalence, we see that state ν also satisfies $\exists x \bigwedge_i (\Gamma_i \vdash \Delta_i)$ prior to the quantifier elimination. Hence, for some state ν_x that agrees with ν except for the value of x, we obtain:

$$\nu_x \models \bigwedge_i (\Gamma_i \vdash \Delta_i).$$

As the side deduction ($\star$) in Fig. 3.10 is inductively shown to be locally sound, we can conclude that $\nu_x \models (\Gamma \vdash \Delta, \phi)$. Hence, $\nu \models \exists x (\Gamma \vdash \Delta, \phi)$.

Now the conjecture can be obtained using standard reasoning with quantifiers and the absence of x in Γ,Δ by rewriting the conclusion with local equivalences:

$$\exists x\,(\Gamma \vdash \Delta,\phi) \equiv \exists x\,(\neg\Gamma \vee \Delta \vee \phi) \equiv \neg\Gamma \vee \Delta \vee \exists x\,\phi \equiv \Gamma \vdash \Delta, \exists x\,\phi\,. \tag{3.10}$$

The soundness proof for r$\forall$ is similar since (3.10) holds for any quantifier, as x does not occur in Γ,Δ. The proofs of l$\exists$ and l$\forall$ can be derived using duality of quantifiers.

[D] By Lemma 3.3, there is an equivalent disjunctive normal form $\mathscr{D}_1 \vee \cdots \vee \mathscr{D}_n$ of $\mathscr{D}$. Thus, it only remains to show that $\rho(\mathscr{D}) \subseteq \rho((\mathscr{D}_1 \cup \ldots \cup \mathscr{D}_n)^*)$ as the converse inclusion is obvious. Let φ be a differential state flow for a transition $(\nu,\omega) \in \rho(\mathscr{D})$. We assume that φ is non-Zeno according to Definition 3.12. Thus, there is a finite number, m, of switches between the disjuncts $\mathscr{D}_i$, say $\mathscr{D}_{i_1}, \mathscr{D}_{i_2}, \ldots, \mathscr{D}_{i_m}$. Then, the transition (ν,ω) belonging to φ can be simulated piecewise by m repetitions of $\mathscr{D}_1 \cup \ldots \cup \mathscr{D}_n$, where each piece selects the respective part $\mathscr{D}_{i_j}$. The proof for $\langle D\rangle$ is similar.

[DR] Local soundness of the rules [DR] and $\langle DR\rangle$ is an immediate consequence of Lemma 3.3 and the respective semantics of modalities.

DS Rule *DS* can be proven locally sound using the fact that the left premise implies that every flow φ that satisfies $\mathscr{D}$ also satisfies χ *all along* the flow. Thus, $\varphi \models \mathscr{D}$ implies $\varphi \models \mathscr{D} \wedge \chi$ so that the right premise entails the conclusion.

DI Let ν satisfy the premise and the antecedent of the conclusion as, otherwise, there is nothing to show. Because $D(F)$ is defined in terms of the literals of F, we can assume F to be in disjunctive normal form (also see Lemma 3.3). Consider any disjunct G of F that is true at ν. In order to show that F is sustained during the continuous evolution, it is sufficient to show that each conjunct of G is. We can assume these conjuncts to be of the form $c \geq 0$ (or $c > 0$ where the proof is similar). Finally, using vectorial notation, we write $x' = \theta$ for the differential equation system and $\exists y$ for the chain of quantifiers. Now let $\varphi : [0,r] \to \mathrm{State}(\Sigma)$ be any state flow with $\varphi \models \exists y\,(x' = \theta \wedge \chi)$ beginning in $\varphi(0) = \nu$. In particular, $\varphi \models \exists y\,\chi$, which, by antecedent, implies $\nu \models F$, i.e., $c \geq 0$ holds at ν. We assume duration $r > 0$, because the other case is immediate ($\nu \models c \geq 0$ already holds). We show that $c \geq 0$ holds all along the flow φ, i.e., $\varphi \models c \geq 0$.

Suppose there was a $\zeta \in [0,r]$ where $\varphi(\zeta) \models c < 0$; this will lead to a contradiction. Then the function $h : [0,r] \to \mathbb{R}$ defined as $h(t) = val(\varphi(t),c)$ satisfies $h(0) \geq 0 > h(\zeta)$, because $\nu \models c \geq 0$ by antecedent. Clearly, φ is of the order of $D(c)$, because: φ is of order 1 for all variables in vector x, and trivially of order ∞ for variables that do not change during the DA-constraint. Further, by bound variable renaming, $D(c)$ cannot contain the quantified variables y; hence, φ is not required to be of any order in y. The value of c is defined all along φ, because we have assumed χ to guard against zeros of denominators. Thus, by Lemma 3.1, h is continuous on $[0,r]$ and differentiable at every $\xi \in (0,r)$. The mean value theorem implies that there is a $\xi \in (0,\zeta)$ such that $\frac{dh(t)}{dt}(\xi)\cdot(\zeta-0) = h(\zeta)-h(0) < 0$. In particular, since $\zeta \geq 0$, we can conclude that $\frac{dh(t)}{dt}(\xi) < 0$. Now Lemma 3.1 implies that $\frac{dh(t)}{dt}(\xi) = val(\bar{\varphi}(\xi), D(c)) < 0$. The latter equals[1] $val(\bar{\varphi}(\xi)_y^u, D(c)_{x'}^{\theta})$ by Lemma 3.2, because $\varphi \models \exists y\,(x' = \theta \wedge \chi)$ so that $\bar{\varphi}(\xi)_y^u \models x' = \theta \wedge \chi$ for some $u \in \mathbb{R}$ and because y' does not occur and $y \notin c$. This, however, is a contradiction, because the premise implies that $\varphi \models \forall y\,(\chi \rightarrow D(c)_{x'}^{\theta} \geq 0)$ as $\forall^{\alpha}$ comprises all variables that change during the flow φ along $x' = \theta$, i.e., the vector x. In particular, as $\bar{\varphi}(\xi)_y^u \models \chi$ holds, we have $\bar{\varphi}(\xi)_y^u \models D(c)_{x'}^{\theta} \geq 0$.

DV First, we consider the quantifier free case, again using vectorial notation. Let ν be any state satisfying the premise and the antecedent of the conclusion. Since ν satisfies the premise and, after bound variable renaming, ε is a fresh variable, we can assume ν to satisfy $\nu \models \forall^{\alpha}(\neg F \wedge \chi \rightarrow (F' \geq \varepsilon)_{x'}^{\theta})$. For *DV*, we required $x' = \theta$ to be Lipschitz continuous so that the global Picard-Lindelöf theorem (Theorem B.2 or its corollary Corollary B.1) ensures the existence of a global solution of arbitrary duration $r \geq 0$, which is all we need here. Let φ be a state flow corresponding to a solution of the differential equation $x' = \theta$ starting in ν of some duration $r \geq 0$. If there is a point in time ζ at which $\varphi(\zeta) \models F$, then by antecedent, until (and including, because $\sim F$ contains the closure of $\neg F$) the first such point, χ holds true during φ. Hence, the restriction of φ to $[0,\zeta]$ is a state flow witnessing $\nu \models \langle x' = \theta \wedge \chi \rangle F$. If, otherwise, there is no such point, then we show that extending φ by choosing a larger r will inevitably make F true. We thus have $\varphi \models \neg F \wedge \chi$ and, by premise, $\varphi \models F'^{\theta}_{x'} \geq \varepsilon$, because $\forall^{\alpha}$ comprises the variables x that change during φ. By Definition 3.13, $F'^{\theta}_{x'} \geq \varepsilon$ is a conjunction. Consider one of its conjuncts, say $c'^{\theta}_{x'} \geq \varepsilon$ belonging to a literal $c \geq 0$ of F (the other cases are similar). Again, φ is of the order of $D(c)$ and the value of c is defined along φ, because $\varphi \models \chi$ and χ is assumed to guard against zeros. Hence, by the mean value theorem, Lemma 3.1, and Lemma 3.2, we conclude for each $\zeta \in [0,r]$ that

$$val(\varphi(\zeta),c) - val(\varphi(0),c) = val(\bar{\varphi}(\xi), c'^{\theta}_{x'})(\zeta - 0) \geq \zeta val(\varphi(0),\varepsilon)$$

for some $\xi \in (0,\zeta)$. Now as $val(\varphi(0),\varepsilon) > 0$ we have for all $\zeta > -\frac{val(\varphi(0),c)}{val(\varphi(0),\varepsilon)}$ that $\varphi(\zeta) \models c \geq 0$ and $\varphi(r) \models c \geq 0$, and even $\varphi(r) \models c > 0$. By enlarging r sufficiently, we have that all literals $c \geq 0$ of one conjunct of F are true, which concludes the proof, because, until F finally holds, $\varphi \models \chi$ is implied by the antecedent as shown earlier.

In the presence of quantifiers ($\exists y$ with vectorial notation), rule *DV* implies a slightly stronger statement, because y is quantified universally in the premise (and antecedent): F can be reached for *all* choices of y that respect χ

[1] For $u \in \mathbb{R}$ let $\bar{\varphi}(\xi)_y^u$ denote the (augmented) state that agrees with $\bar{\varphi}(\xi)$ except that the value of y is u.

(rather than just for one). By antecedent, there is a $u \in \mathbb{R}$ such that $\nu_y^u \models \chi$. Hence, ν_y^u satisfies the assumptions of the above quantifier-free case. Thus, $\nu_y^u \models \langle x' = \theta \land \chi \rangle F$, which entails that $\nu \models \langle \exists y\, (x' = \theta \land \chi) \rangle F$ using u constantly as the value for the quantified variable y during the evolution. □

Consequently, we know that all formulas provable in the DAL calculus are valid and hence reflect true properties.

3.7 Restricting Differential Invariants

Example 3.18 on p. 173 shows that differential invariant F cannot generally be assumed to hold in the premise of *DI* without losing soundness. Nevertheless, we present two corresponding refinements of *DI* that are indeed sound, even though they assume the differential invariant F to hold in the induction step.

Proposition 3.2 (Open differential induction). *Using the notation of the proof rulesDI,DV, the following variations of differential induction rule DI are sound (in DI′, F describes an open set):*

$$(DI')\ \frac{\vdash \forall^\alpha \forall y_1 \dots \forall y_k\, (F \land \chi \to F'^{\theta_1}_{x'_1} \cdots ^{\theta_n}_{x'_n})}{[\exists y_1 \dots \exists y_k\, \chi] F \vdash [\exists y_1 \dots \exists y_k\, (x'_1 = \theta_1 \land \dots \land x'_n = \theta_n \land \chi)] F}$$

$$(DI'')\ \frac{\vdash \forall^\alpha \forall y_1 \dots \forall y_k\, (F \land \chi \to (F' > 0)^{\theta_1}_{x'_1} \cdots ^{\theta_n}_{x'_n})}{[\exists y_1 \dots \exists y_k\, \chi] F \vdash [\exists y_1 \dots \exists y_k\, (x'_1 = \theta_1 \land \dots \land x'_n = \theta_n \land \chi)] F}$$

Proof. The proof that rule *DI′* is sound is similar to the soundness proof for *DI* in Theorem 3.1, except that assuming $\varphi(\zeta) \models \neg F$ only yields $h(0) \geq 0 \geq h(\zeta)$, which does not lead to a contradiction. However, by using the fact that F is open, the distance to the border of F is positive in the initial state $\varphi(0)$, which yields the inequality $h(0) > 0 \geq h(\zeta)$, and the contradiction arises accordingly.

The soundness of rule *DI″* needs more adaptation. Repeating the argument for *DI*, we can assume F to be of the form $c \geq 0$. Suppose there was a $\iota \in [0,r]$ where $\varphi(\iota) \models c < 0$, which will lead to a contradiction. Let $\zeta \in [0,r]$ be the infimum of these ι; hence, $\varphi(\zeta) \models c = 0$ by continuity. Then the function $h : [0,r] \to \mathbb{R}$ defined as $h(t) = val(\varphi(t), c)$ satisfies $h(0) \geq 0 \geq h(\zeta)$, because $\nu \models c \geq 0$ by antecedent. By repeating the argument with Lemma 3.1 as in the proof for *DI*, h is continuous on $[0,r]$ and differentiable at every $\xi \in (0,r)$ with a derivative of $\frac{\mathrm{d}h(t)}{\mathrm{d}t}(\xi) = val(\bar{\varphi}(\xi), D(c))$, which in turn equals $val(\bar{\varphi}(\xi), D(c)^{\theta}_{x'})$, as $\varphi \models x' = \theta$. Now, the mean value theorem implies that there is a $\xi \in (0, \zeta)$ such that

$$\frac{\mathrm{d}h(t)}{\mathrm{d}t}(\xi) \cdot (\zeta - 0) = h(\zeta) - h(0) \leq 0.$$

In particular, as $\zeta \geq 0$, we can conclude that $\frac{\mathrm{d}h(t)}{\mathrm{d}t}(\xi) = val(\bar{\varphi}(\xi), D(c)^{\theta}_{x'}) \leq 0$. This, however, contradicts the fact that the premise implies $\bar{\varphi}(\xi) \models D(c)^{\theta}_{x'} > 0$, as the flow satisfies $\varphi \models \chi$ and $\varphi(\xi) \models c \geq 0$, because $\zeta > \xi$ is the infimum of the counterexamples ι with $\varphi(\iota) \models c < 0$. □

Example 3.28. Consider the differential equation $x' = x^3$. With either rule DI' or rule DI'', we can establish easily that the system stays above $\frac{1}{4}$ whenever the dynamics starts above $\frac{1}{4}$ (refer to Fig. 3.13b on p. 172 for the dynamics):

$$\dfrac{\dfrac{*}{\vdash \forall x\,(x > \frac{1}{4} \to x^3 > 0)}\ \text{r}\forall}{x > \frac{1}{4} \vdash [x' = x^3]x > \frac{1}{4}}\ DI'$$

Observe that this property is not provable with rule DI directly, because $x^3 > 0$ does not hold for all x, but only for those where the invariant $x > \frac{1}{4}$ is true already. □

Example 3.29 (Negative equations splitting). As an example with a negative equality consider the DAL formula $x \neq 0 \to [x' = x]x \neq 0$ with the negative equality $x \neq 0$. In examples like these, differential induction does not work directly; see Counterexample 3.19. Yet when we replace $x \neq 0$ by the equivalent $x < 0 \vee x > 0$, we can prove invariance of this formula along the dynamics $x' = x$ separately by reasoning by cases; see Fig. 3.25. In the left branch, we strengthen the differential

$$\dfrac{\dfrac{\dfrac{\dfrac{*}{\vdash \forall x\,(x<0 \to x<0)}\ \text{r}\forall}{x<0 \vdash [x'=x]x<0}\ DI' \qquad \dfrac{\dfrac{*}{\vdash \forall x\,(x<0 \to x\neq 0)}\ \text{r}\forall}{x<0 \vdash [x'=x \wedge x<0]x \neq 0}\ [DR']}{x<0 \vdash [x'=x]x\neq 0}\ DS \qquad \dfrac{*\ \text{similarly}}{x>0 \vdash [x'=x]x\neq 0}\ DS}{x<0 \vee x>0 \vdash [x'=x]x \neq 0}\ \vee\text{l}$$

Fig. 3.25 Differential induction splitting over disjunctions for negative equations

equations with $x < 0$ by DS and prove that $x < 0$ is a differential invariant (by rule DI' or DI''). In the middle branch, we show that the auxiliary evolution domain $x < 0$ implies the postcondition $x \neq 0$ by differential weakening $[DR']$. On the right branch, there is a corresponding proof for strengthening with $x > 0$ by DS and proving differential invariance by rule DI' and differential weakening by $[DR']$ to show that $x > 0$ also implies $x \neq 0$. Note in particular that two different ways to use differential strengthening are required on both branches here when splitting differential induction over disjunctions. □

3.8 Differential Monotonicity Relaxations

Evolution domain constraints of DA-constraints are helpful for differential induction rule DI, because they provide stronger assumptions for the premise. In fact, the whole purpose of the differential strengthening rule *DS* is to enrich evolution domain constraints of DA-constraints in order to weaken the subgoals of *DI*.

In contrast, evolution domain constraints are more demanding for differential variant induction rule *DV*, because the antecedent of its goal requires that evolution domain region χ be shown to remain true throughout the evolution (after all, evolution domains χ of DA-constraints $\mathscr{D}$ are restrictions that can be used as assumptions for $[\mathscr{D} \wedge \chi]\phi$ but have to be shown to hold true for $\langle \mathscr{D} \wedge \chi \rangle \phi$). Similarly, for evolution rules that are based on solutions of differential equations—rules $\langle ' \rangle$,$[']$ from the d$\mathscr{L}$ calculus in Fig. 2.11 on p. 79—evolution domain regions make the subgoal formulas much more complex (even though they lead to weaker subgoals for rule $[']$). In particular, evolution domain regions increase the number of quantifier alternations in $\langle ' \rangle$,$[']$, which have the predominant influence on the complexity of quantifier elimination [94].

For simplifying non-differential evolution domain region χ from $[\mathscr{D} \wedge \chi]\phi$ with a DA-constraint $\mathscr{D}$, we can simply use the differential weakening rule $[DR]$ to drop χ:

$$[DR]\frac{\vdash [\mathscr{D}]\phi}{\vdash [\mathscr{D} \wedge \chi]\phi}$$

Less conservatively, we can approximate the assumption $\forall 0 \leq \tilde{t} \leq t \, \langle \mathscr{S}_{\tilde{t}} \rangle \chi$ on the solution $\mathscr{S}_t$ in the subgoal of rule $[']$ from Fig. 2.11 by $\langle \mathscr{S}_t \rangle \chi$ (or by $\chi \wedge \langle \mathscr{S}_t \rangle \chi$) in a sound yet incomplete way. If every evolution of the solution $\mathscr{S}_t$ satisfying χ at the end satisfies ϕ, then every evolution along $\mathscr{D}$ satisfying χ all the time must satisfy ϕ even more so. Thus, the following variant of $[']$ is sound (but less complete) where $\mathscr{S}_t$ is the solution of the differential equation as for rule $[']$:

$$(['']) \; \frac{\forall t \geq 0 \, (\chi \rightarrow \langle \mathscr{S}_t \rangle (\chi \rightarrow \phi))}{[x_1' = \theta_1 \wedge .. \wedge x_n' = \theta_n \wedge \chi]\phi}$$

For simplifying non-differential evolution domain region χ from $\langle \mathscr{D} \wedge \chi \rangle \phi$, we can use the dual of the differential strengthening rule *DS* and show that χ remains true throughout the evolution along $\mathscr{D}$ (left subgoal, which can be handled using *DI*) so that only some evolution along $\mathscr{D}$ remains to be found that actually reaches ϕ (right subgoal):

$$DS\frac{\vdash [\mathscr{D}]\chi \quad \vdash \langle \mathscr{D} \rangle \phi}{\vdash \langle \mathscr{D} \wedge \chi \rangle \phi}$$

When using *DI* to prove the left subgoal $[\mathscr{D}]\chi$ by showing validity of a formula of the form $\forall^{\mathscr{D}} \chi'^{\theta}_{x'}$, we actually show invariance of χ along $\mathscr{D}$.

More generally, we can use differential-algebraic techniques similar to differential induction to prove the weaker property of monotonicity instead of invariance of χ.

Definition 3.17 (Monotonicity derivation). Let α be a DA-program. For a first-order formula F, the following formula is called *monotonicity derivation* of F, where the syntactic derivative $D(a)$ is defined according to Definition 3.13:

$$M^\alpha(F) \equiv \bigwedge_{i=1}^{m} M^\alpha(F_i) \quad \text{where } \{F_1,\dots,F_m\} \text{ is the set of all literals of } F;$$

$$M^\alpha(a \sim b) \equiv \forall^\alpha(D(a) \geq D(b)) \vee \forall^\alpha(D(a) \leq D(b)) \quad \text{where} \sim \in \{\leq,\geq,<,>,=\}.$$

Proposition 3.3 (Differential monotonicity). *Let $\langle \mathscr{S}_t \rangle$ be the DJ-constraint for the solution at time t of the symbolic initial value problem for the differential equation $\mathscr{D}$ defined as $x_1' = \theta_1 \wedge \dots \wedge x_n' = \theta_n$ as in rule $\langle ' \rangle$ of Fig. 2.11. Let the non-differential constraint χ be a conjunction of atomic formulas without negative equalities; then the following is a sound proof rule:*

$$(\langle ' \rangle') \frac{\vdash \exists t{\geq}0\,(\chi \wedge \langle \mathscr{S}_t \rangle (\chi \wedge \phi)) \qquad \vdash M^{\mathscr{D}}(\chi)_{x_1'}^{\theta_1} \cdots {}_{x_n'}^{\theta_n}}{\vdash \langle x_1' = \theta_1 \wedge \dots \wedge x_n' = \theta_n \wedge \chi \rangle \phi}$$

Proof. Local soundness is a simple consequence of the well-known fact that, for a differentiable function f, monotonic increasing of f on an interval $[c,d]$ is equivalent to $f'(z) \geq 0$ on (c,d). With this, the right subgoal implies that, for any conjunct $a \geq b$ of χ (likewise for $\leq,<,>,=$), the value of $a-b$ is either monotonically increasing or monotonically decreasing along the flow. Either way, if χ holds in the beginning *and* the end of a flow of some duration t (as implied by the left subgoal), monotonicity implies that χ holds all along the flow, so that the subgoals imply the conclusion as in case $\langle ' \rangle$ of the proof of Theorem 2.1.

Formally, let φ be a state flow of an appropriate duration r following solution $\mathscr{S}_t$ according to the left subgoal as in Theorem 2.1. By the left subgoal we have $\varphi(r) \models \phi$ (when r is the witness for $\exists t{\geq}0$) and we only need to show that $\varphi \models \chi$. Consider a conjunct $a \geq b$ of χ and consider the case where the right subgoal implies $\varphi(0) \models \forall^{\mathscr{D}}(D(a) \leq D(b))_{x'}^{\theta}$, using vectorial notation for x and θ. Then $\varphi \models (a' \leq b')_{x'}^{\theta}$, because the universal closure $\forall^{\mathscr{D}}$ comprises all variables that change during the flow φ along $\mathscr{D}$. Thus $\varphi \models a' \leq b'$ by Lemma 3.2. Let $h : [0,r] \to \mathbb{R}$ be the function defined as $h(t) = val(\varphi(t), a-b)$. Again, φ is of the order of $a'-b'$ (φ is of the order 1 in each x_i and of arbitrary order for other variables) and the value of $a-b$ is defined all along φ, because χ guards against zeros in χ. Thus, Lemma 3.1 is applicable and h is differentiable at every $\xi \in (0,r)$. For any $\zeta \in [0,r]$, we have to show that $\varphi(\zeta) \models \chi$. By the mean value theorem, there is a $\xi \in (\zeta, r)$ such that, when using Lemma 3.1, we have

$$h(r) - h(\zeta) = h'(\xi) \cdot (r-\zeta) = val(\bar{\varphi}(\xi), a'-b') \cdot (r-\zeta) \leq 0$$

because $\bar{\varphi}(\xi) \models a' \leq b'$. Thus, we have $h(\zeta) \geq h(r) \geq 0$, since $\varphi(r) \models \chi$, which implies that $\varphi(r) \models a-b \geq 0$ and $\varphi(r) \models a \geq b$.

The other case where the right subgoal implies $\varphi(0) \models \forall^\alpha(a' \geq b')$ is simpler using the fact that $\varphi(0) \models \chi$ and is, in fact, a direct consequence of the proof of *DI* in Theorem 3.1. The other conjuncts of the form $a \leq b, a < b, a > b$, and $a = b$ are almost identical, because the monotonicity argument for $a-b$ carries over easily, as the respective conjunct holds before and after the continuous evolution. □

Example 3.30 (Monotonic invariants in train control). Consider the differential constraints for train control (equation (2.7) on p. 62 in Sect. 2.4). For the DA-constraint $z' = v \wedge v' = a \wedge \tau' = 1 \wedge v \geq 0 \wedge \tau \leq \varepsilon$, evolution domain region $v \geq 0 \wedge \tau \leq \varepsilon$ can be shown to be monotonic or convex with respect to the dynamics. That is, if it holds in the beginning and at the end of an evolution, the invariant also holds in between. Monotonicity is easy to prove with the above proof rule using the following symbolic computations for the right subgoal (where $\forall^\alpha$ is $\forall z \forall v \forall \tau$):

$$\begin{aligned}
&((\forall^\alpha(v' \geq 0) \vee \forall^\alpha(v' \leq 0)) \wedge (\forall^\alpha(\tau' \geq \varepsilon') \vee \forall^\alpha(\tau' \leq \varepsilon')))_{z'\ v'\ \tau'\ \varepsilon'}^{v\ a\ 1\ 0}\\
&\equiv (\forall^\alpha(a \geq 0) \vee \forall^\alpha(a \leq 0)) \wedge (\forall^\alpha(1 \geq 0) \vee \forall^\alpha(1 \leq 0))\\
&\equiv \mathit{true}.
\end{aligned}$$

Observe that the invariant domain will not be a differential invariant, here, because $v \geq 0$ is only an invariant of $z' = v \wedge v' = a$ for $a \geq 0$. For any a, however, $v \geq 0$ will be either a monotonically increasing (if $a \geq 0$ constantly) or a monotonically decreasing (if $a \leq 0$ constantly) property, one of which is true for every constant a. Thus, if $v \geq 0$ has been true *before and after* an evolution along $z' = v \wedge v' = a$, it must have been true throughout this evolution. Likewise, $\tau \leq \varepsilon$ never is a differential invariant of $\tau' = 1$, because the passing of time along $\tau' = 1$ will inevitably violate $\tau \leq \varepsilon$ sooner or later. Still, it is a monotonically decreasing property. Consequently, the monotonicity relaxation of Proposition 3.3 applies for the train control example, thereby simplifying proofs with evolution domain regions considerably, because the invariant only needs to be checked before and after rather than throughout the evolution. For instance, this simplifies the proof of property (2.20) on p. 121. □

Similarly, the right subgoal of rule $\langle ' \rangle'$ is a sufficient condition to ensure that rule $[']'$ is a complete replacement for rule $[']$.

Counterexample 3.31 (Disjunctive monotonicity). For soundness of the differential monotonicity relaxations, it is crucial that rule $\langle ' \rangle'$ only accept conjunctive evolution domain regions. As the counterexample in Fig. 3.26a with the dynamics in Fig. 3.26b shows, differential monotonicity relaxations do not hold for disjunctive evolution domain regions, because the same disjunct has to hold before and after the evolution for monotonicity arguments to be sound. Let χ abbreviate the disjunctive evolution domain region $x \leq 1 \vee x \geq 2$. Then the differential monotonicity criterion $\forall x\,(1 \leq 0) \vee \forall x\,(1 \geq 0)$ would be fulfilled, but a different disjunct holds at the initial state $x = 0$ than at the target $x \geq 3$ so that monotonicity implies neither that $x \leq 1$ nor that $x \geq 2$ holds in between. □

Counterexample 3.32 (Negative equalities). A similar counterexample shows why rule $\langle ' \rangle'$ does not allow negative equalities. Along the dynamics $x' = 1 \wedge x \neq 2$ we cannot conclude from the truth of $x \neq 2$ before and after the evolution that, on the basis of a condition on the derivative $x' \neq 0$, $x \neq 2$ held true throughout the evolution. A continuous evolution from $x = 0$ to $x = 3$ still leaves $x \neq 2$ in between. □

$$\dfrac{\dfrac{*\text{ (unsound)}}{\vdash \exists t{\geq}0\,(\chi \land \langle x:=x+t\rangle(\chi \land x \geq 3))}}{x=0 \vdash \langle x' = 1 \land (x \leq 1 \lor x \geq 2)\rangle x \geq 3}$$

Fig. 3.26a Counterexample for disjunctive monotonicity

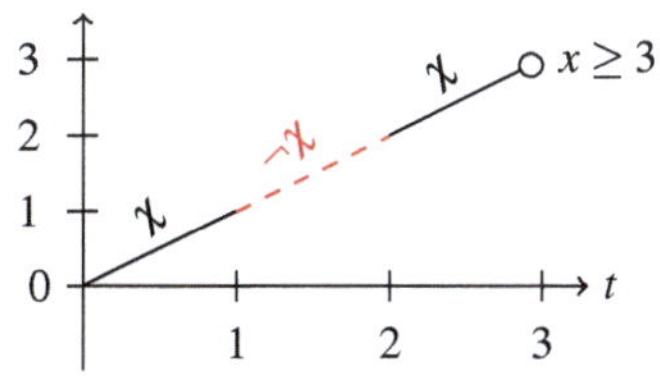

Fig. 3.26b Interrupted dynamics

3.9 Relative Completeness

As a consequence of the Incompleteness Theorem 2.2 for $d\mathcal{L}$ and the fact that DAL is a conservative extension of $d\mathcal{L}$ (Proposition 3.1), the DAL calculus is not effectively axiomatisable (yet even just reachability is undecidable for hybrid systems [156]).

It is easy to see that the relative completeness proof for $d\mathcal{L}$ (Theorem 2.3) generalises to DAL with only minor modifications when using the first-order logic of DA-constraints as a basis in place of FOD (again, nested modalities can be avoided when using quantifiers). The first-order logic of DA-constraints results from FOD by allowing DA-constraints in place of differential equations inside modalities.

Theorem 3.2 (Relative completeness of DAL). *The* DAL *calculus is* complete relative to *DA-constraints, i.e., every valid* DAL *formula can be derived from tautologies of the first-order logic of DA-constraints.*

Proof. The proof is a simple adaptation of the proof of Theorem 2.3 for $d\mathcal{L}$ in Sect. 2.7.2: In the proof of the program rendition Lemma 2.8, we replace all cases for continuous evolutions along differential equations or for discrete jumps by the following cases for DA-constraints $\mathcal{D}$ or DJ-constraints $\mathcal{J}$, respectively:

$$\mathcal{S}_{\mathcal{D}}(\vec{x},\vec{v}) \equiv \langle\mathcal{D}\rangle\vec{v} = \vec{x},$$
$$\mathcal{S}_{\mathcal{J}}(\vec{x},\vec{v}) \equiv \mathcal{J}^{v_i=\theta_i}_{x_i:=\theta_i}.$$

The first-order formula $\mathcal{J}^{v_i=\theta_i}_{x_i:=\theta_i}$ results from $\mathcal{J}$ by replacing all assignments of any form $x_i := \theta_i$ with equations $v_i = \theta_i$. The rest of the relative completeness proof generalises immediately using the fact that the respective rules ($\langle\exists\rangle$,$[:=]$) for DJ-constraints are symmetric, and hence equivalent, and their premises are of smaller complexity. □

Note that, for generalising the relative completeness proof in the most simple way, we formally need to allow update prefixes in DAL proofs as in Definition 2.10, which is easily seen to be sound for deterministic DJ-constraints.

3.10 Deductive Strength of Differential Induction

We analyse the deductive power of differential induction with respect to classes of formulas that are allowed as differential invariants. For purely equational differential invariants, the deductive power is not affected by allowing or disallowing propositional operators in differential invariants:

Proposition 3.4 (Equational deductive power). *The deductive power of differential induction with atomic equations is identical to the deductive power of differential induction with propositional combinations of polynomial equations: Formulas are provable with propositional combinations of equations as differential invariants iff they are provable with only atomic equations as differential invariants.*

Proof. We show that every differential invariant that is a propositional combination ϕ of polynomial equations is expressible as a single atomic polynomial equation (the converse inclusion is obvious). We assume ϕ to be in negation normal form and reduce ϕ inductively using the following transformations:

- If ϕ is of the form $p_1 = p_2 \vee q_1 = q_2$, then ϕ is equivalent to the single equation $(p_1 - p_2)(q_1 - q_2) = 0$. Further, $\phi' \equiv p_1' = p_2' \wedge q_1' = q_2'$ directly implies

$$((p_1 - p_2)(q_1 - q_2))' = 0 \equiv (p_1' - p_2')(q_1 - q_2) + (p_1 - p_2)(q_1' - q_2') = 0.$$

- If ϕ is of the form $p_1 = p_2 \wedge q_1 = q_2$, then ϕ is equivalent to the single equation $(p_1 - p_2)^2 + (q_1 - q_2)^2 = 0$. Further, $\phi' \equiv p_1' = p_2' \wedge q_1' = q_2'$ implies

$$\left((p_1 - p_2)^2 + (q_1 - q_2)^2\right)' = 0 \equiv 2(p_1 - p_2)(p_1' - p_2') + 2(q_1 - q_2)(q_1' - q_2') = 0.$$

- If ϕ is of the form $\neg(p_1 = p_2)$, then ϕ does not qualify as a differential invariant, because it contains a negative equality, which is disallowed for *DI* according to the conditions in Fig. 3.9. □

Observe, however, that the required polynomial degree of atomic equations is larger than for propositional combinations, which can have computational disadvantages for quantifier elimination.

For general differential invariants, where inequalities are allowed, the situation is different: We show that, in general, the deductive power of differential induction depends on which class of formulas is allowed as differential invariants! Some DAL formulas cannot by proven by a differential induction step with only atomic formulas but no propositional operators as differential invariants, while they are provable immediately using unrestricted differential invariants.

Theorem 3.3 (Deductive power). *The deductive power of differential induction with arbitrary formulas exceeds the deductive power of differential induction with atomic formulas: All* DAL *formulas that are provable using atomic differential invariants are provable using general differential invariants, but not vice versa!*

Proof. The inclusion is obvious. Conversely, we have to show that there are DAL formulas that are provable with general differential invariants but not with atomic differential invariants. Consider the following example, which is provable using rule *DI'*, i.e., the variant of *DI* for open sets (Sect. 3.7), with the non-atomic formula $x>0 \land y>0$ as differential invariant:

$$\dfrac{\dfrac{*}{\vdash \forall x \forall y\,(x>0 \land y>0 \rightarrow xy>0 \land xy>0)}\,\text{r}\forall}{x>0 \land y>0 \vdash [x'=xy \land y'=xy](x>0 \land y>0)}\,DI'$$

First, we show that this formula is not provable by a differential induction step with only atomic formulas as differential invariants. Suppose there was a single polynomial $p(x,y)$ in variables x,y such that $p(x,y)>0$ is a differential invariant proving the above formula, which will lead to a contradiction. The conditions for differential invariants (*DI* or *DI'*) imply that the following formulas have to be valid:

1. $x>0 \land y>0 \rightarrow p(x,y)>0$, as differential invariants have to hold in the prestate according to the antecedent of *DI* (or *DI'*).
2. $p(x,y)>0 \rightarrow x>0 \land y>0$, as the differential invariant has to imply the postcondition (when using $[]gen$ to show that the differential invariant implies the postcondition).

In particular, $x>0 \land y>0 \leftrightarrow p(x,y)>0$ is valid, and p is not the zero polynomial. Thus, p enjoys the property:

$$p(x,y) \geq 0 \text{ for } x \geq 0, y \geq 0, \text{ and, otherwise, } p(x,y) \leq 0. \tag{3.11}$$

Assume p has minimal total degree with property (3.11). Now, $p(x,0)$ is a univariate polynomial in x with zeros at all $x>0$; thus $p(x,0)=0$ is the zero polynomial, hence y divides $p(x,y)$. Similarly, $p(0,y)=0$ for all y, and hence x divides $p(x,y)$. Thus, xy divides p. But by comparing the signs (cf. Fig. 3.27), we see that the polynomial $\frac{-p(-x,-y)}{xy}$ also satisfies property (3.11) with a smaller total degree than p, which is a contradiction. In detail: $p(x,y)$ satisfies the sign conditions (3.11) indicated in the outer part of Fig. 3.27. It is divisible by xy, but $\frac{p(x,y)}{xy}$ satisfies different sign conditions (indicated in the middle part of Fig. 3.27). Yet, by flipping the signs, $\frac{-p(-x,-y)}{xy}$ again satisfies the same sign conditions (3.11) as $p(x,y)$ but with a smaller total degree (indicated in the inner part of Fig. 3.27), which is a contradiction.

Similarly, there is no polynomial p such that $x>0 \land y>0 \leftrightarrow p(x,y)=0$, because only the zero polynomial is zero on the full quadrant $(0,\infty)^2$. Finally, property $x>0 \land y>0 \leftrightarrow p(x,y) \geq 0$ is impossible for continuity reasons, which imply that $p(0,0)=0$, which is a contradiction. More generally, the same argument holds for any other sign condition that is supposed to characterise one quadrant of $\mathbb{R}^2$ uniquely.

Observe that, so far, the argument does not depend on the actual dynamics and is, thus, still valid in the presence of arbitrary differential weakening ($[DR]$).

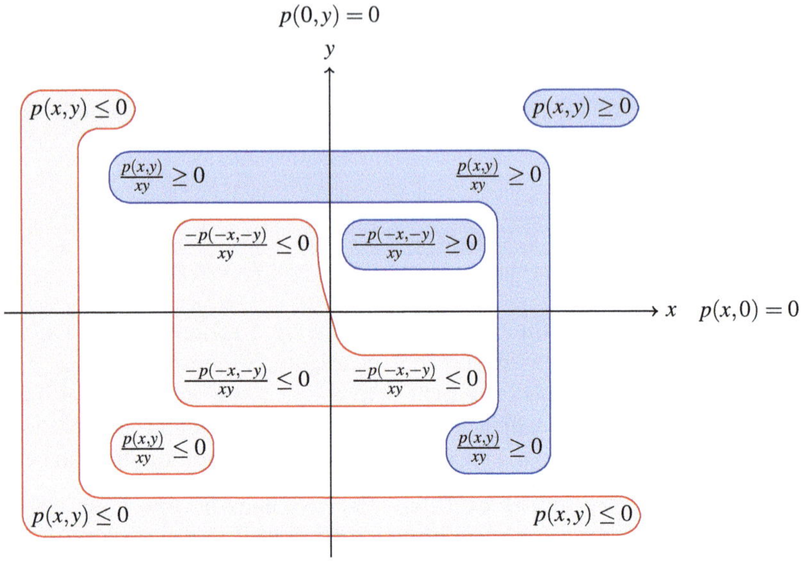

Fig. 3.27 Quadrant sign selection regions of differential invariant

Next, to see that the above example cannot be proven indirectly after differential strengthening (*DS*), we use the fact that, inductively, the strengthening χ itself needs to be a differential invariant: Ultimately, the left subgoal of *DS* can only be shown using differential induction. The above example, however, is built such that, as $x' = xy$ is the differential equation, $xy > 0$ is required for $x > 0$ to be a differential invariant (which thus also requires $y > 0$). Conversely, due to $y' = xy$, formula $xy > 0$ is a prerequisite for the differential invariance of $y > 0$ (which thus also needs $x > 0$). Yet, for differential invariance of $xy > 0$, we have to prove $xy > 0 \rightarrow (y+x)xy > 0$ for DI', because $(xy)'^{xy}_{x'}{}^{xy}_{y'}$ gives $(x'y+yx')^{xy}_{x'}{}^{xy}_{y'}$, i.e., $xyy+yxy$. But the property $xy > 0 \rightarrow (y+x)xy > 0$ is, again, equivalent to $x \geq 0 \lor y \geq 0$, and thus equivalent to $\neg(-x > 0 \land -y > 0)$, which cannot be proven by atomic differential induction (or differential weakening) according to the first part of this proof. Thus, the required atomic differential invariants have

Fig. 3.28 Circular dependencies for differential strengthening

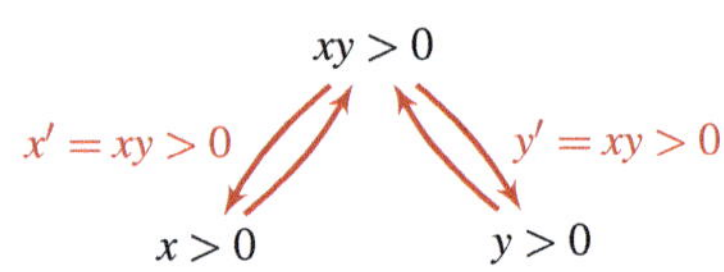

circular dependencies for differential strengthening by $x > 0$, $y > 0$, and $xy > 0$, respectively; see Fig. 3.28. These cannot be resolved in any proof tree without simultaneous differential induction using non-atomic differential invariants, because differential strengthenings have to be ordered totally along each proof branch. □

As a special case, this result implies that differential induction in DAL is deductively stronger than approaches using barrier certificates [251, 252], criticality functions [91], or polynomial invariant equations [274, 269]. On top of that, the DAL calculus adds differential strengthening and weakening techniques, which add further deductive power. The roundabout manoeuvre that we verify in the next section is a practical example where differential induction with mixed non-atomic formulas and successive differential strengthening turns out to be decisive.

3.11 Air Traffic Control Verification

In this section we verify that the tangential roundabout manoeuvre for collision avoidance in air traffic control that we presented in Sect. 3.4 is collision-free. That is, the manoeuvre directs aircraft on flight paths with global minimal distance $p > 0$. We determine a corresponding parameter constraint on the *tang* procedure from the roundabout manoeuvre in Fig. 3.7. Using differential induction and differential strengthening, we can verify the flight manoeuvre despite the complicated hybrid flight dynamics of aircraft whose solutions fall into undecidable classes of arithmetic. Recall the differential flight equations from Sect. 3.4.2:

$$x_1' = d_1 \wedge x_2' = d_2 \wedge d_1' = -\omega d_2 \wedge d_2' = \omega d_1 \qquad (\mathscr{F}(\omega)^*)$$
$$y_1' = e_1 \wedge y_2' = e_2 \wedge e_1' = -\varpi e_2 \wedge e_2' = \varpi e_1 \qquad (\mathscr{G}(\varpi)^*)$$

3.11.1 Characterisation of Safe Roundabout Dynamics

Property ϕ in Fig. 3.7 defines *safe states* as those with separation $\|x-y\| \geq p$. This does *not*, however, characterise the states with *safe dynamics*: Several states that satisfy ϕ will not remain safe when following curved roundabout flight manoeuvres; see Fig. 3.6c on p. 151 for a counterexample violating ϕ after some time. In particular, the angular velocity ω and initial speed vectors d and e must fit to the relative positioning of the aircraft x and y. Otherwise the aircraft dynamics will not remain safe from safely separated initial states. In order to discover the required parametric constraints for safety of the roundabout manoeuvre, we analyse the DAL formula $\phi \to [trm^*]\phi$ in the DAL calculus and identify a corresponding parameter constraint $\mathscr{T}$. For notational convenience, we inline side deductions and slightly simplify the universal closure notation $\forall^\alpha$ by taking free variables as universally quantified here, as with Skolem terms in Chap. 2, because the following DAL proof needs no existential variables.

$$
\dfrac{\dfrac{\dfrac{\dfrac{\dfrac{\dfrac{\dfrac{*}{\phi \vdash \forall x,y,d,e\,(\phi \rightarrow \phi)}\,\text{r}\forall}{\phi \vdash [free]\phi}\,[DR'] \qquad \dfrac{\dfrac{\ldots}{\phi \vdash [tang](\phi \wedge \mathscr{T})} \quad \dfrac{\ldots}{\phi \wedge \mathscr{T} \vdash [\mathscr{F}(\omega) \wedge \mathscr{G}(\omega)]\phi}}{\phi \vdash [tang;\mathscr{F}(\omega) \wedge \mathscr{G}(\omega)]\phi}\,[]gen}{\phi \vdash [free][tang;\mathscr{F}(\omega) \wedge \mathscr{G}(\omega)]\phi}\,[]gen}{\phi \vdash [trm]\phi}\,[;]}{\vdash \forall^{\alpha}(\phi \rightarrow [trm]\phi)}\,\text{r}\forall,\rightarrow\text{r}}{\phi \vdash [trm^*]\phi}\,ind}{\vdash \phi \rightarrow [trm^*]\phi}\,\rightarrow\text{r}
$$

The left branch closes, because postcondition ϕ is the evolution domain restriction in free flight such that its DA-constraint can be weakened by Lemma 3.6. In the other branches, $\mathscr{T}$ is the parameter constraint that *tang* needs to establish in addition to ϕ (middle branch) for the roundabout dynamics to be safe (right branch). Hence condition $\mathscr{T}$ mediates between the middle and right branches. Using successive quantifier elimination on the right branch, we derive the following constraint $\mathscr{T}$ as a prerequisite for ϕ to be differentially inductive. It is the decisive constraint that characterises configurations with safely controllable dynamics in curved roundabout manoeuvres (using vectorial notation and orthogonal complements $d^{\perp}$ from Sect. 3.2):

$$
\begin{aligned}
\mathscr{T} &\equiv d - e = \omega(x-y)^{\perp} \quad (\text{or, equivalently } (d-e)^{\perp} = -\omega(x-y)) \\
&\equiv d_1 - e_1 = -\omega(x_2 - y_2) \wedge d_2 - e_2 = \omega(x_1 - y_1).
\end{aligned} \tag{3.12}
$$

This formula expresses that the relative speed vector $d-e$ is orthogonal to the relative position $x-y$ and compatible with the angular velocity ω and tangential orientation of d and e. Figure 3.29a illustrates the symmetric case of $\mathscr{T}$ with identical linear speed $\|d\| = \|e\|$. Figures 3.29b and 3.29c show asymmetric cases with distinct linear speeds $\|d\| \neq \|e\|$, which is possible as well. Condition $\mathscr{T}$ gives the decisive

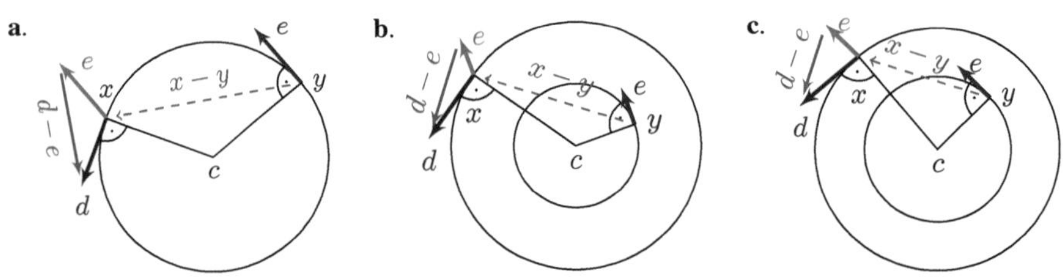

Fig. 3.29 Tangential construction for characteristics $\mathscr{T}$ of roundabout dynamics

handle for an inductive characterisation of safe tangential roundabout configurations: For the right branch of the above proof, we need to show that the tangential configuration $\mathscr{T}$ is sufficient for ϕ to be sustained during curved evasive actions. In the following, we prove that the relative speed vector configuration $\mathscr{T}$ is itself differentially inductive (rule *DI* in left branch). We use differential strengthening with *DS* as a differential cut to augment the dynamics with $\mathscr{T}$ as a derived invariant for proving that the actual safety property ϕ is sustained (right branch), again by differential induction rule *DI*:

$$\dfrac{\dfrac{\dfrac{*}{\vdash \forall^\alpha (\mathscr{T}'_{\mathscr{F}(\omega)\wedge\mathscr{G}(\omega)})}\,\mathrm{r}\forall}{\phi,\mathscr{T} \vdash [\mathscr{F}(\omega)\wedge\mathscr{G}(\omega)]\mathscr{T}}\,DI \qquad \dfrac{\dfrac{*}{\vdash \forall^\alpha (\mathscr{T} \to \phi'_{\mathscr{F}(\omega)\wedge\mathscr{G}(\omega)})}\,\mathrm{r}\forall}{\phi \vdash [\mathscr{F}(\omega)\wedge\mathscr{G}(\omega)\wedge\mathscr{T}]\phi}\,DI}{\dfrac{\phi,\mathscr{T} \vdash [\mathscr{F}(\omega)\wedge\mathscr{G}(\omega)]\phi}{\phi\wedge\mathscr{T} \vdash [\mathscr{F}(\omega)\wedge\mathscr{G}(\omega)]\phi}\,\wedge l}\,DS$$

Observe that differential strengthening by rule *DS* is crucial for the proof, because neither ϕ nor $\mathscr{T}\wedge\phi$ is differentially inductive for $\mathscr{F}(\omega)\wedge\mathscr{G}(\omega)$! Instead, the tangential configuration $\mathscr{T}$ itself is differentially inductive relative to $\mathscr{F}(\omega)\wedge\mathscr{G}(\omega)$ (left branch) and strong enough to make ϕ differentially inductive relative to the augmented DA-constraint $\mathscr{F}(\omega)\wedge\mathscr{G}(\omega)\wedge\mathscr{T}$ (right branch). For readability, we use a slightly weaker rule for differential induction, with ϕ rather than $[\mathscr{T}]\phi$ in the antecedent of the conclusion. This variant can be derived easily using a cut and will again be called *DI*. The differential induction *DI* on the left and right branch close using quantifier elimination in r$\forall$. The arithmetic is also provable by the following algebraic equational reasoning ($\mathscr{T}'_{\mathscr{F}(\omega)\wedge\mathscr{G}(\omega)}$ is a short notation for substituting the differential equations from $\mathscr{F}(\omega)\wedge\mathscr{G}(\omega)$ into $D(\mathscr{T})$; see Lemma 3.2):

$$\begin{aligned}
\mathscr{T}'_{\mathscr{F}(\omega)\wedge\mathscr{G}(\omega)} &\equiv ((d_1-e_1)' = -\omega(x_2-y_2)' \wedge (d_2-e_2)' = \omega(x_1-y_1)')_{\mathscr{F}(\omega)\wedge\mathscr{G}(\omega)}\\
&\equiv (d_1'-e_1' = -\omega(x_2'-y_2') \wedge d_2'-e_2' = \omega(x_1'-y_1'))_{\mathscr{F}(\omega)\wedge\mathscr{G}(\omega)}\\
&\equiv -\omega d_2+\omega e_2 = -\omega(d_2-e_2) \wedge \omega d_1-\omega e_1 = \omega(d_1-e_1) \quad \equiv \mathit{true}\\
\phi'_{\mathscr{F}(\omega)\wedge\mathscr{G}(\omega)} &\equiv (2(x_1-y_1)(x_1-y_1)'+2(x_2-y_2)(x_2-y_2)' \geq 0)_{\mathscr{F}(\omega)\wedge\mathscr{G}(\omega)}\\
&\equiv (2(x_1-y_1)(x_1'-y_1')+2(x_2-y_2)(x_2'-y_2') \geq 0)_{\mathscr{F}(\omega)\wedge\mathscr{G}(\omega)}\\
&\equiv 2(x_1-y_1)(d_1-e_1)+2(x_2-y_2)(d_2-e_2) \geq 0\\
(\text{using } \mathscr{T}) &\equiv 2(x_1-y_1)(-\omega(x_2-y_2))+2(x_2-y_2)\omega(x_1-y_1) = 0 \geq 0 \quad \equiv \mathit{true}.
\end{aligned}$$

Altogether, we have shown that *every* tangential roundabout evasion manoeuvre respecting $\mathscr{T}$ is safe. Further, the middle branch of the above proof reveals the parameter constraint imposed on *tang* for safe roundabouts, which concludes the proof of the following result.

Theorem 3.4 (Safety of tangential roundabout manoeuvre). *For every choice of the tangential entry procedure that satisfies $\phi \to [tang](\phi\wedge\mathscr{T})$, the tangential roundabout flight manoeuvre in Fig. 3.7 safely avoids collisions, i.e., it directs aircraft on flight paths with minimal horizontal aircraft separation at least $p > 0$.*

This result can be proven in our theorem prover [242] in two seconds including user interactions for rules *ind* and *DS*. Its proof does not need rule $[]gen$, which we only used here to shorten the proof presentation. Theorem 3.4 expresses unbounded-time safety for fully parametric tangential roundabouts with arbitrary choices for the free parameters. The proof of Theorem 3.4 generalises to roundabouts entered by more than two participants when ϕ and $\mathscr{T}$ are augmented similarly. For instance, using our automatic proof procedure from Chap. 6, our theorem prover can prove mutual collision avoidance for five aircraft fully automatically; see Chaps. 6 and 8. Likewise, rules *DI* and *DS* can be used to prove that external separation to all other sufficiently far points is maintained during the roundabout manoeuvre. In particular, the manoeuvre only needs bounded space:

Proposition 3.5 (External separation of roundabout manoeuvres). *Separation of aircraft x to all external points $u \in \mathbb{R}^2$ of distance beyond the roundabout diameter $2r$ is maintained, because the following* DAL *formula is provable:*

$$r \geq 0 \wedge (r\omega)^2 = \|d\|^2 \rightarrow \forall u\,(\|x-u\|^2 > (2r+p)^2 \rightarrow [\mathscr{F}(\omega)](\|x-u\|^2 > p^2)).$$

3.11.2 Tangential Entry Procedures

As a simple choice for the tangential initiation procedure *tang* satisfying property $\mathscr{T}$, consider the following operation which chooses an arbitrary angular velocity ω and an arbitrary centre $c \in \mathbb{R}^2$ for the roundabout manoeuvre, and adjusts d and e tangentially:

$$tang \;\equiv\; \exists u\,\omega := u;\; \exists c\,(d := \omega(x-c)^{\perp} \wedge e := \omega(y-c)^{\perp}). \tag{3.13}$$

This formula expresses that the speed vectors d and e of both aircraft at x and y, respectively, are tangential and of the same angular velocity ω relative to the intended centre c of the roundabout, with the same orientation (Fig. 3.29). For this choice, the assumption of Theorem 3.4 can be proven after rule [:=] substitutes the corresponding terms for d and e in $\mathscr{T}$, using rule r∀ (or linearity of $d^{\perp}$):

$$\text{[;],[}\exists\text{],[}\exists\text{]}\;\cfrac{\text{r}\forall\text{,r}\forall\;\cfrac{\text{[:=]}\;\cfrac{\wedge\text{r}\;\cfrac{\text{ax}\;\cfrac{*}{\phi \vdash \phi} \qquad \cfrac{*}{\phi \vdash \omega(x-c)^{\perp} - \omega(y-c)^{\perp} = \omega(x-y)^{\perp}}}{\phi \vdash \phi \wedge \omega(x-c)^{\perp} - \omega(y-c)^{\perp} = \omega(x-y)^{\perp}}}{\phi \vdash [d := \omega(x-c)^{\perp} \wedge e := \omega(y-c)^{\perp}](\phi \wedge \mathscr{T})}}{\phi \vdash \forall \omega \forall c\,[d := \omega(x-c)^{\perp} \wedge e := \omega(y-c)^{\perp}](\phi \wedge \mathscr{T})}}{\phi \vdash [tang](\phi \wedge \mathscr{T})}$$

It can also be shown that $\exists c\,(d = \omega(x-c)^{\perp} \wedge e = \omega(y-c)^{\perp})$ is equivalent to $\mathscr{T}$ for nonzero ω. With choice (3.13), the tangential roundabout manoeuvre in Fig. 3.7 is safe and has been significantly simplified and generalised in comparison to our prior work [238].

3.11.3 Discussion

Our tangential roundabout manoeuvre leaves open the questions of how and when precisely the collision avoidance manoeuvre is initiated or left. For instance, (3.13) does not restrict c and ω but accepts any choice including choices optimising secondary objectives such as fuel consumption. Furthermore, as specified in Fig. 3.7 and proven in this section, the roundabout manoeuvre can be left safely with arbitrary free flight by repeating the loop at any time: The roundabout manoeuvre will simply be initiated again during free flight when necessary. As a special case, this open policy includes free flight, enabling the aircraft to leave the roundabout in their original direction. While the simple choice (3.13) is possibly discontinuous in d and e, it is comparably easy to see that there are fully curved entry and exit procedures that remain safe when the entry procedure is initiated with sufficient distance by using the separation limit of Proposition 3.5. We refine the roundabout collision avoidance manoeuvre and develop a corresponding entry procedure in Chap. 8. Our proof shows that the tangential roundabout manoeuvre is safe for every such entry procedure. In particular, the control parameters c and ω of (3.13) can also be chosen such that the resulting speed vectors d and e are in a bounded range meeting external speed requirements of the aircraft, which can be proven in the DAL calculus easily:

$$\forall v\,(\phi \to \langle tang\rangle(\phi \land \mathscr{T} \land \|d\|^2 = \|e\|^2 = v^2)). \tag{3.14}$$

3.12 Summary

We have introduced a first-order dynamic logic for differential-algebraic programs with interacting first-order discrete jump constraints and first-order differential-algebraic constraints. For this differential-algebraic logic, DAL, we have presented a calculus for verifying hybrid systems given as differential-algebraic programs.

In differential-algebraic programs, both internal choices and disturbances during continuous evolutions and nondeterminism in discrete operations can be described uniformly by quantifiers. Most importantly, we have introduced first-order differential induction with differential invariants and differential variants for proving correctness statements with first-order differential-algebraic constraints purely algebraically, using the differential constraints themselves instead of their solutions. In combination with successive differential strengthening or differential cuts for refining the system dynamics by auxiliary differential invariants, we obtain a powerful verification calculus for systems with challenging dynamics. We have compared the deductive strength for classes of differential invariants and have shown that the deductive power of general differential induction exceeds the deductive power of atomic differential invariants.

We have demonstrated that our calculus can be used successfully for verifying fully parametric roundabout manoeuvres in air traffic control. To the best of our knowledge, this is the first formal proof for unbounded safety of hybrid aircraft dynamics in curved collision avoidance manoeuvres for air traffic control. Moreover, we argue that our fully formal proof about aircraft gives more confidence in flight manoeuvres than informal approaches that do not consider the actual hybrid flight dynamics [171, 104, 129] or results that only prevent orthogonal collisions in discretisations of the system [92, 203]. Our logic DAL is also more convenient, because hybrid systems like the tangential roundabout manoeuvre can be specified and verified uniformly within a single logic. Despite challenging flight dynamics, the DAL formulas about aircraft and roundabout manoeuvres that we presented in this chapter can be proven in our theorem prover KeYmaera within a few seconds.

While this work answers the open issues (1), (3), and (4) raised in the work of Piazza et al. [228], we are interested in extending differential-algebraic methods to address further questions about hybrid systems. In Chap. 6, we investigate algorithms for constructing differential invariants automatically on the basis of our DAL calculus presented here. Interesting future work for the aircraft case study is to find a fully curved manoeuvre that achieves collision avoidance by joint horizontal and vertical evasive actions.

Chapter 4
Differential Temporal Dynamic Logic dTL

Contents

A. Platzer, Logical Analysis of Hybrid Systems,
DOI 10.1007/978-3-642-14509-4_4, © Springer-Verlag Berlin Heidelberg 2010

Synopsis We combine first-order dynamic logic for reasoning about the possible behaviour of hybrid systems with temporal logic for reasoning about the temporal behaviour during their operation. Our logic supports verification of hybrid programs with first-order definable flows and provides a uniform treatment of discrete and continuous evolution. For our combined logic, we generalise the semantics of dynamic modalities to refer to hybrid traces instead of final states. Further, we prove that this gives a conservative extension of our dynamic logic for hybrid systems. On this basis, we provide a modular verification calculus that reduces correctness of temporal behaviour of hybrid systems to nontemporal reasoning, and prove that we obtain a complete axiomatisation relative to the nontemporal base logic. Using this calculus, we analyse safety invariants in a train control system and symbolically synthesise parametric safety constraints.

4.1 Introduction

Correctness of real-time and hybrid systems depends on a safe operation throughout *all* states of all possible trajectories, and the behaviour at intermediate states is highly relevant [90].

Temporal logics (TLs) use temporal operators to talk about intermediate states [247, 114, 115, 6, 284]. In addition to having successful uses in model checking [78, 6, 159, 156, 217], temporal logics have been used in deductive approaches to prove validity of formulas in calculi [97, 96]. Among other shortcomings and difficulties discussed in Chap. 1, the major drawback of TL calculi for our purpose is that TL formulas cannot generally characterise the operations of a specific hybrid system.

Like model checking, *dynamic logic* (DL) [149] can directly analyse the behaviour of actual system models. However, DL only considers the behaviour at the final states, which is insufficient for verifying safety invariants that have to hold all the time, throughout the execution of the system.

We close this gap of expressivity by combining first-order dynamic logic [149] with temporal logic [247, 114, 115]. We use the generalisation of operational system models and semantics to hybrid systems from Chap. 2. In this chapter, we introduce a temporal dynamic logic dTL, which provides modalities for quantifying over traces of hybrid systems based on differential dynamic logic. We equip dTL with temporal operators to state what is true all along a trace or at some point during a trace. In this chapter, we modify the semantics of the dynamic modality $[\alpha]$ to refer to all *traces* of α instead of all final states reachable with α (similarly for $\langle\alpha\rangle$). For instance, the formula $[\alpha]\Box\phi$ expresses that ϕ is true at each state during all traces of the hybrid system α. With this, dTL can also be used to verify temporal statements about the behaviour of α at intermediate states during system runs. As in our nontemporal dynamic logic $d\mathcal{L}$, we use hybrid programs as an operational model for hybrid systems, since they admit a uniform compositional treatment of interacting discrete and continuous evolution in logic.

As a semantical foundation for combined temporal dynamic formulas, we introduce a hybrid trace semantics for dTL. We prove that dTL is a conservative extension of $d\mathcal{L}$, that is, for nontemporal specifications, trace semantics is equivalent to the nontemporal transition semantics of $d\mathcal{L}$ from Chap. 2.

As a means for verification, we introduce a sequent calculus for dTL that successively reduces temporal statements about traces of hybrid programs to nontemporal $d\mathcal{L}$ formulas. In this way, we make the intuition formally precise that temporal safety invariants can be checked by augmenting proofs with appropriate assertions about intermediate states. As in Chap. 2, our calculus works compositionally: It decomposes correctness statements about hybrid programs structurally into corresponding statements about its parts by symbolic transformation. Observe that this is somewhat challenging for hybrid systems, because even a single elementary system operation of continuous evolution exhibits temporal behaviour as it assumes several different states as time passes.

Contributions

Our approach combines the advantages of dynamic logic in reasoning about the behaviour of (multiple and parametric) operational system models with those of temporal logic to verify temporal statements about traces. Our first contribution is the logic dTL, which provides a coherent foundation for reasoning about the temporal behaviour of operational models of hybrid systems with symbolic parameters. The main contribution in this chapter is our calculus for deductively verifying temporal statements about hybrid systems, which is a complete axiomatisation relative to nontemporal $d\mathcal{L}$.

4.1.1 Related Work

Based on [254], Beckert and Schlager [38] added separate trace modalities to dynamic logic and presented a relatively complete calculus for discrete while programs. Their approach only handles discrete state spaces. In contrast, dTL works for hybrid programs with continuous state spaces. There, a particular challenge is that invariants may change their truth-value multiple times during a single continuous evolution; hence relevant temporal behaviour even occurs during single transitions.

Davoren and Nerode [97] extended the propositional modal μ-calculus with a semantics in hybrid systems and examine topological aspects. In [96], Davoren et al. gave a semantics in general flow systems for a generalisation of CTL* [115]. In both cases, the authors of [97] and [96] provided Hilbert-style calculi to prove formulas that are valid for all systems simultaneously using abstract actions.

As discussed in Sect. 1.2, the strength of our logic primarily is that it is a first-order dynamic logic and handles actual hybrid programs like $x:=x+1;x'=2y$ rather than only abstract actions of unknown effect.

4.1.2 Structure of This Chapter

After introducing syntax and semantics of the differential temporal dynamic logic dTL in Sects. 4.2 and 4.3, we introduce a modular sequent calculus for dTL in Sect. 4.5 that extends our previous calculi with temporal proof rules in a completely modular way. We prove soundness and relative completeness in Sects. 4.6 and 4.7, respectively. In Sect. 4.8, we use our calculus to analyse safety invariants in the train control system from Sect. 4.4. We further present extensions for quantifier alternation and liveness in Sect. 4.9. We draw conclusions and discuss future work in Sect. 4.10.

4.2 Syntax of Temporal Dynamic Logic for Hybrid Systems

The *temporal differential dynamic logic* dTL extends dynamic logic [149] with three concepts for verifying temporal specifications of hybrid systems:

Hybrid programs. The behaviour of hybrid systems can be described by hybrid programs (Sect. 2.2.2), which generalise real-time programs [159] to hybrid change. The distinguishing feature of hybrid programs in this context is that they provide uniform discrete jumps and continuous evolutions along differential equations, which can be combined by regular control operations. While hybrid automata [156] can be embedded, program structures are more amenable to compositional symbolic processing by calculus rules.

Modal operators. Modalities of dynamic logic express statements about all possible behaviour ($[\alpha]\pi$) of a system α, or about the existence of a trace ($\langle\alpha\rangle\pi$), satisfying condition π. Unlike in standard dynamic logic, α is a modal of a hybrid system. We use hybrid programs to describe α as in Chap. 2. Yet, unlike in standard dynamic logic [149] or d$\mathcal{L}$, π is a *trace formula* in dTL, and π can refer to all states that occur *during* a trace using temporal operators.

Temporal operators. For dTL, the temporal trace formula $\Box\phi$ expresses that the formula ϕ holds all along a trace selected by $[\alpha]$ or $\langle\alpha\rangle$. For instance, the state formula $\langle\alpha\rangle\Box\phi$ says that the state formula ϕ holds at every state along at least one trace of α. Dually, the trace formula $\Diamond\phi$ expresses that ϕ holds at some point during such a trace. It can occur in a state formula $\langle\alpha\rangle\Diamond\phi$ to express that there is such a state in some trace of α, or as $[\alpha]\Diamond\phi$ to say that along each trace there is a state satisfying ϕ. In this chapter, the primary focus of attention is on homogeneous combinations of path and trace quantifiers like $[\alpha]\Box\phi$ or $\langle\alpha\rangle\Diamond\phi$.

4.2.1 Hybrid Programs

The formulas of dTL are built from a non-empty set Σ of real-valued variables and function and predicate symbols. Signature Σ is assumed to contain the usual function and predicate symbols for real arithmetic: $0, 1, +, \cdot, =, \leq, <, \geq, >$. For simplicity, we do not distinguish between logical variables in V and state variables from Σ. The set $\mathrm{Trm}(\Sigma)$ of *terms* is defined as in classical first-order logic.

The hybrid programs allowed in dynamic modalities of dTL are the same as those of $\mathsf{d}\mathcal{L}$; see Definition 2.3 in Sect. 2.2.2. They are built from elementary discrete jumps and continuous evolutions using a regular control structure. The set $\mathrm{HP}(\Sigma)$ of hybrid programs with variables in Σ is defined in Definition 2.3. Similarly, differential-algebraic programs from Chap. 3 can be allowed when using DAL as a basis instead of $\mathsf{d}\mathcal{L}$, giving *differential-algebraic temporal dynamic logic* DATL.

4.2.2 State and Trace Formulas

The formulas of dTL are defined similarly to first-order dynamic logic [149]. However, the modalities $[\alpha]$ and $\langle\alpha\rangle$ accept trace formulas that refer to the temporal behaviour of *all* states along a trace. Inspired by CTL and CTL* [114, 115], we distinguish between state formulas, which are true or false in states, and trace formulas, which are true or false for system traces. The sets $\mathrm{Fml}(\Sigma)$ of state formulas and $\mathrm{Fml}_T(\Sigma)$ of trace formulas with variables in Σ are simultaneously inductively defined in Definition 4.1.

Definition 4.1 (dTL formulas). The set $\mathrm{Fml}(\Sigma)$ of *(state) formulas* is simultaneously inductively defined as the smallest set such that:

1. If $p \in \Sigma$ is a predicate of arity $n \geq 0$ and if $\theta_1, \ldots, \theta_n \in \mathrm{Trm}(\Sigma)$ are terms, then $p(\theta_1, \ldots, \theta_n) \in \mathrm{Fml}(\Sigma)$.
2. If $\phi, \psi \in \mathrm{Fml}(\Sigma)$, then $\neg\phi, (\phi \wedge \psi), (\phi \vee \psi), (\phi \to \psi) \in \mathrm{Fml}(\Sigma)$.
3. If $\phi \in \mathrm{Fml}(\Sigma)$ and $x \in \Sigma$, then $\forall x\, \phi, \exists x\, \phi \in \mathrm{Fml}(\Sigma)$.
4. If $\pi \in \mathrm{Fml}_T(\Sigma)$ and $\alpha \in \mathrm{HP}(\Sigma)$, then $[\alpha]\pi, \langle\alpha\rangle\pi \in \mathrm{Fml}(\Sigma)$.

The set $\mathrm{Fml}_T(\Sigma)$ of *trace formulas* is the smallest set with:

1. If $\phi \in \mathrm{Fml}(\Sigma)$, then $\phi \in \mathrm{Fml}_T(\Sigma)$.
2. If $\phi \in \mathrm{Fml}(\Sigma)$, then $\Box\phi, \Diamond\phi \in \mathrm{Fml}_T(\Sigma)$.

Formulas without $\Box$ and $\Diamond$, i.e., without Case 2 of the trace formulas, are *nontemporal* $\mathsf{d}\mathcal{L}$ *formulas* (Chap. 2). Unlike in CTL, state formulas are true on a trace (Case 1) if they hold for the *last* state of a trace, not for the first. Thus, $[\alpha]\phi$ expresses that ϕ is true at the end of each trace of α. In contrast, $[\alpha]\Box\phi$ expresses that ϕ is true all along all states of every trace of α. This combination gives a smooth embedding of nontemporal $\mathsf{d}\mathcal{L}$ into dTL and makes it possible to define

Table 4.1 Operators and meaning in differential temporal dynamic logic (dTL)

dTL Notation	Operator	Meaning
$p(\theta_1,..,\theta_n)$	atomic predicate	true iff predicate p holds for $(\theta_1,\ldots,\theta_n)$
$\neg\phi$	negation / not	true if ϕ is false
$\phi\wedge\psi$	conjunction / and	true if both ϕ and ψ are true
$\phi\vee\psi$	disjunction / or	true if ϕ is true or if ψ is true
$\phi\rightarrow\psi$	implication / implies	true if ϕ is false or ψ is true
$\phi\leftrightarrow\psi$	bi-implication / equivalent	true if ϕ and ψ are both true or both false
$\forall x\,\phi$	universal quantifier / for all	ϕ is true for all values of variable x
$\exists x\,\phi$	existential quantifier / exists	ϕ is true for some values of variable x
$[\alpha]\phi$	$[\cdot]$ modality / box	ϕ is true after all runs of HP α
$\langle\alpha\rangle\phi$	$\langle\cdot\rangle$ modality / diamond	ϕ is true after at least one run of HP α
$[\alpha]\Box\phi$	$[\cdot]\Box$ modality nesting	ϕ is true always during all traces of HP α
$\langle\alpha\rangle\Diamond\phi$	$\langle\cdot\rangle\Diamond$ modality nesting	ϕ is true sometimes during some trace of HP α
$[\alpha]\Diamond\phi$	$[\cdot]\Diamond$ modality nesting	ϕ is true sometimes during all traces of HP α
$\langle\alpha\rangle\Box\phi$	$\langle\cdot\rangle\Box$ modality nesting	ϕ is true always during some trace of HP α

a compositional calculus. Like CTL, dTL allows nesting with a branching time semantics [114], e.g., $[\alpha]\Box(x\geq 2\rightarrow\langle\beta\rangle\Diamond x\leq 0)$.

For reference, the operators of differential temporal dynamic logic and typical operator nestings are summarised in Table 4.1.

Example 4.1 (Train control). Recall the simplified ETCS train control system from Sect. 2.4 with the refinements from Sect. 2.9. In Sect. 2.9, we have proven the following d$\mathcal{L}$ formula that expresses that the train control system ensures that trains stay inside their movement authority m, no matter how long the controller runs:

$$v^2\leq 2b(m-z)\wedge b>0\wedge A\geq 0\;\rightarrow\;[(ctrl;drive)^*]z\leq m. \tag{2.19*}$$

But this d$\mathcal{L}$ formula only refers to the safety of the train position at all final states of the hybrid program $(ctrl;drive)^*$, which corresponds to the rightmost state in the transition structure from Fig. 2.9a on p. 63. Formula (2.19) does not say whether $z\leq m$ is actually also ensured in all intermediate states of the transition structure in Fig. 2.9a. For hybrid system verification, it is often insufficient to prove the safety property $z\leq m$ only at the final states, because the passengers will not like to crash into another train at an intermediate state either.

Now because the dTL modality $[\alpha]$ does not need to be followed by a temporal modality $\Box$ or $\Diamond$, the d$\mathcal{L}$ formula (2.7) also is a dTL formula. In fact, all d$\mathcal{L}$ formulas are dTL formulas (those without temporal modalities). But in dTL, we can do better. We can express that the movement authority m is respected *always* throughout the system run, not just at the final states:

$$v^2\leq 2b(m-z)\wedge b>0\wedge A\geq 0\;\rightarrow\;[(ctrl;drive)^*]\Box z\leq m. \tag{4.1}$$

Unlike the d$\mathcal{L}$ formula (2.19), the dTL formula (4.1) also refers to safety at all intermediate states of the transition structure in Fig. 2.9a. In fact, for this particular system model, both formulas are equivalent, because the variables in the postcondition $z \le m$ only change in the last step *drive* of the hybrid program . Also $z \le m$ must have been true at all intermediate states if it holds at the final state, because z increases monotonically (the train is not allowed to drive backwards). In that sense, the final states of the hybrid program $(ctrl;drive)^*$ included all safety-critical states, because the hybrid program was written as an appropriate controller plant loop.

For other systems, where the variables of the postcondition change in multiple parts of the hybrid program, this is no longer the case, and the temporal modality $\Box$ is, in fact, crucial to express safety properly. As a simple (though somewhat contrived) example, consider the following modification, in which we allow *drive* to also drive backwards by removing the evolution domain restriction $v \ge 0$ from it:

$$v^2 \le 2b(m-z) \wedge b > 0 \wedge A \ge 0 \rightarrow [(a:=0;drive;ctrl;drive)^*]\Box z \le m. \quad (4.2)$$

In the hybrid program of this dTL formula, the position z changes at two places: the two occurrences of *drive*. In the first occurrence, the train always keeps its speed for some time by choosing the acceleration $a:=0$, because the sensor information is not yet available. Only in the second occurrence of *drive* do the controller actions *ctrl* actually take effect. For specifying safety adequately in (4.2), we need the temporal modality $\Box$, because we want the movement authority to be respected at all states of the system, including intermediate states. Especially we would not want the movement authority to be violated at some point during the first *drive* even if the system recovers during the second *drive*, e.g., by driving backwards.

Surprisingly, the temporal modality $\Box$ is still redundant in (4.2), because all possibilities of running its hybrid program especially include the case where the first *drive* is run for an arbitrary time $s \le \varepsilon$ and the second *drive* is run for zero seconds. Thus all relevant safety-critical states are covered by final states and $\Box$ is not needed. But as soon as we modify *drive* to the hybrid program

$$\tau:=0;\ (z' = v, v' = a, \tau' = 1 \,\&\, \tau \le \varepsilon);\ ?(\tau = \varepsilon),$$

the temporal modality $\Box$ becomes truly necessary. The reason is that this modification restricts the continuous evolution to take exactly ε time units (the evolution domain region restricts the evolution to $\tau \le \varepsilon$ and the subsequent test to $?\tau = \varepsilon$) and no intermediate state is visible as a final state anymore. □

This example shows that the question about whether all relevant intermediate safety-critical states are covered by a $d\mathscr{L}$ modality can be surprisingly subtle. In larger systems, in fact, it can become quite difficult to analyse this manually. Instead, temporal dTL modality combinations like $[\alpha]\Box\phi$ can be used to ensure that all states are covered and the hybrid system α satisfies ϕ all the time for all possible executions.

Inspired by CTL* [115], syntactic and semantic extensions from dTL to dTL* are straightforward and amount to allowing propositional combinations of trace formulas. Finding appropriate proof calculi, however, is much more difficult, even for CTL* [248, 261].

4.3 Semantics of Temporal Dynamic Logic for Hybrid Systems

In standard dynamic logic [149], the logic $d\mathcal{L}$ from Chap. 2, and the logic DAL from Chap. 3, modalities only refer to the final states of system runs and the semantics is a reachability relation on states: State ω is reachable from state ν using system α if there is a run of α which terminates in ω when started in ν. For dTL, however, formulas can refer to intermediate states of runs as well. To capture this, we change the semantics of a hybrid system α to be the set of its possible *traces*, i.e., successions of states that occur during the evolution of α. The relation between the initial and the final state alone is not sufficient.

4.3.1 *Trace Semantics of Hybrid Programs*

States contain values of system variables during a hybrid evolution. A *state* is a map $\nu : \Sigma \to \mathbb{R}$. In addition, we distinguish a separate state Λ to denote the failure of a system run when it is *aborted* due to a test $?\chi$ that yields *false*. In particular, Λ can only occur at the end of an aborted system run and marks that no further extension is possible because of a failed test. The set of all states is denoted by $\mathrm{Sta}(\Sigma)$.

Hybrid systems evolve along piecewise continuous traces in multi-dimensional space as time passes. Continuous phases are governed by differential equations, whereas discontinuities are caused by discrete jumps in state space. Unlike in discrete cases [254, 38], traces are not just sequences of states, since hybrid systems pass through uncountably many states even in bounded time. Beyond that, continuous changes are more involved than in pure real time [6, 159], because all variables can evolve along differential equations with different slopes. Generalising the real-time traces of [159], the following definition captures hybrid behaviour by splitting the uncountable succession of states into periods σ_i that are regulated by the same control law. For discrete jumps, some of those periods are point flows of duration 0.

The (trace) semantics of hybrid programs is compositional, that is, the semantics of a complex program is defined as a simple function of the trace semantics of its parts.

Definition 4.2 (Hybrid trace). A *trace* is a (nonempty) finite or infinite sequence $\sigma = (\sigma_0, \sigma_1, \sigma_2, \ldots)$ of functions $\sigma_i : [0, r_i] \to \mathrm{Sta}(\Sigma)$ with their respective durations $r_i \in \mathbb{R}$ (for $i \in \mathbb{N}$). A *position* of σ is a pair (i, ζ) with $i \in \mathbb{N}$ and ζ in the interval $[0, r_i]$; the state of σ at (i, ζ) is $\sigma_i(\zeta)$. Positions of σ are ordered lexicographically by $(i, \zeta) \prec (j, \xi)$ iff either $i < j$, or $i = j$ and $\zeta < \xi$. Further, for a state $\nu \in \mathrm{Sta}(\Sigma)$, $\hat{\nu} : 0 \mapsto \nu$ is the *point flow* at ν with duration 0. A trace *terminates* if it is a finite sequence $(\sigma_0, \sigma_1, \ldots, \sigma_n)$ and $\sigma_n(r_n) \neq \Lambda$. In that case, the last state $\sigma_n(r_n)$ is denoted by $\mathrm{last}\,\sigma$. The first state $\sigma_0(0)$ is denoted by $\mathrm{first}\,\sigma$.

Unlike in [6, 159], the definition of traces also admits finite traces of bounded duration, which is necessary for compositionality of traces in $\alpha; \beta$. The semantics of hybrid programs α as the set $\tau(\alpha)$ of its possible traces depends on valuations $val(\nu, \cdot)$

of formulas and terms at intermediate states ν. The valuation of terms and interpretations of function and predicate symbols are as for real arithmetic (Chap. 2). The valuation of formulas will be defined in Definition 4.4. Again, we use $\nu[x \mapsto d]$ to denote the *modification* that agrees with state ν on all variables except for the symbol x, which is changed to $d \in \mathbb{R}$.

Definition 4.3 (Trace semantics of hybrid programs). The *trace semantics, $\tau(\alpha)$, of a hybrid program α,* is the set of all its possible hybrid traces and is defined inductively as follows:

1. $\tau(x_1 := \theta_1, .., x_n := \theta_n) = \{(\hat{\nu}, \hat{\omega}) : \omega = \nu[x_1 \mapsto val(\nu, \theta_1)]..[x_n \mapsto val(\nu, \theta_n)]$ for $\nu \in \mathrm{Sta}(\Sigma)\}$
2. $\tau(x_1' = \theta_1, \dots, x_n' = \theta_n \,\&\, \chi) = \{(\varphi) : \varphi$ is a state flow of order 1 and some duration $r \geq 0$ such that $\varphi \models x_1' = \theta_1 \wedge \dots \wedge x_n' = \theta_n \wedge \chi$; see Definition 3.9$\}$
3. $\tau(?\chi) = \{(\hat{\nu}) : val(\nu, \chi) = true\} \cup \{(\hat{\nu}, \hat{\Lambda}) : val(\nu, \chi) = false\}$
4. $\tau(\alpha \cup \beta) = \tau(\alpha) \cup \tau(\beta)$
5. $\tau(\alpha; \beta) = \{\sigma \circ \varsigma : \sigma \in \tau(\alpha), \varsigma \in \tau(\beta)$ when $\sigma \circ \varsigma$ is defined$\}$;
   the composition of $\sigma = (\sigma_0, \sigma_1, \sigma_2, \dots)$ and $\varsigma = (\varsigma_0, \varsigma_1, \varsigma_2, \dots)$ is

$$\sigma \circ \varsigma := \begin{cases} (\sigma_0, \dots, \sigma_n, \varsigma_0, \varsigma_1, \dots) & \text{if } \sigma \text{ terminates at } \sigma_n \text{ and last}\,\sigma = \text{first}\,\varsigma \\ \sigma & \text{if } \sigma \text{ does not terminate} \\ \text{not defined} & \text{otherwise} \end{cases}$$

6. $\tau(\alpha^*) = \bigcup_{n \in \mathbb{N}} \tau(\alpha^n)$, where $\alpha^{n+1} := (\alpha^n; \alpha)$ for $n \geq 1$, as well as $\alpha^1 := \alpha$ and $\alpha^0 := (?true)$.

Time passes differently during discrete and continuous change. During continuous evolution, the discrete step index i of positions (i, ζ) remains constant, whereas the continuous duration ζ remains 0 during discrete point flows. This permits multiple discrete state changes to happen at the same (super-dense) continuous time, unlike in other approaches [6].

Example 4.2. For comparing the transition semantics of hybrid programs for $\mathsf{d}\mathcal{L}$ from Definition 2.7 and the trace semantics of hybrid programs for dTL from Definition 4.3, consider the following simple hybrid program α:

$$a := -2a;\ a := a^2.$$

The transition semantics is just the relation between initial and final states:

$$\rho(\alpha) \equiv \{(\nu, \omega) : \omega \text{ is like } \nu \text{ except that } \omega(a) = 4\nu(a)^2\}.$$

In particular, the $\mathsf{d}\mathcal{L}$ formula $[\alpha]a \geq 0$ is valid, because all final states have a square as the value of a. In contrast, the trace semantics of α retains all intermediate states:

$$\tau(\alpha) \equiv \{(\hat{\nu}, \hat{s}, \hat{\omega}) : s \text{ is like } \nu \text{ except } s(a) = -2\nu(a) \\ \text{and } \omega \text{ is like } s \text{ except } \omega(a) = s(a)^2 = 4\nu(a)^2\}.$$

During these traces, $a \geq 0$ does not hold at all states. If the trace starts with a positive value ($\nu \models a > 0$), then it will become negative at the point flow s (where $s \models a < 0$), yet recover to a positive value ($\omega \models a > 0$) at the end. □

Example 4.3. The previous example only had discrete jumps, and, thus, the traces only involved point flows. Now consider the hybrid program β from the train context:

$$a := -b;\ z' = v, v' = a;\ ?v \geq 0;\ a := A;\ z' = v, v' = a.$$

The transition semantics of this program only considers successful runs to completion. In particular, if $A > 0$, the velocity v will always be nonnegative at the end (otherwise the test $?v \geq 0$ in the middle fails and the program aborts), because the last differential equation will accelerate and increase the velocity again. Thus, the position z at the end of the program run will never be smaller than at the beginning.

If, instead, we consider the trace semantics of β, all intermediate states are in the set of traces:

$$\begin{aligned}
\tau(\beta) \equiv \{(\hat{\mu_0}, \hat{\mu_1}, \varphi_1, \hat{\mu_2}, \hat{\mu_3}, \varphi_2) \ :\ & \mu_1 = \mu_0[a \mapsto -\mu_0(b)] \text{ and} \\
& \varphi_1 \text{ is a state flow of some duration } r_1 \geq 0 \text{ with } \varphi_1 \models z' = v \wedge v' = a \\
& \text{starting in } \varphi_1(0) = \mu_1 \text{ and ending in a state with } \varphi_1(r_1)(v) \geq 0 \\
& \text{and } \mu_2 = \varphi_1(r_1), \mu_3 = \varphi_1(r_1)[a \mapsto \varphi_1(r_1)(A)] \text{ and} \\
& \varphi_2 \text{ is a state flow of some duration } r_2 \geq 0 \text{ with } \varphi_2 \models z' = v \wedge v' = a \\
& \text{starting in } \varphi_2(0) = \mu_3 \text{ and ending in state } \varphi_2(r_2)\} \\
\cup \{(\hat{\mu_0}, \hat{\mu_1}, \varphi_1, \hat{\mu_2}, \hat{\Lambda}) \ :\ & \mu_1 = \mu_0[a \mapsto -\mu_0(b)] \text{ and} \\
& \varphi_1 \text{ is a state flow of some duration } r \geq 0 \text{ with } \varphi_1 \models z' = v \wedge v' = a \\
& \text{starting in } \varphi_1(0) = \mu_1 \text{ and ending in a state with } \varphi_1(r)(v) < 0 \\
& \text{further } \mu_2 = \varphi_1(r)\}.
\end{aligned}$$

The first set is the set of traces where the test $?v \geq 0$ in the middle succeeds and the system continues. The second set (after the union) is the set of traces that are aborted with $\hat{\Lambda}$ during their execution, because the middle test fails. Note that the traces in the first set have two continuous flows φ_1, φ_2 and four point flows $\hat{\mu_0}, \hat{\mu_1}, \hat{\mu_2}, \hat{\mu_3}$ in each trace. The traces in the second set have only one continuous flow φ_1 and three point flows $\hat{\mu_0}, \hat{\mu_1}, \hat{\mu_2}$, because the subsequent aborting point flow $\hat{\Lambda}$ does not terminate and aborts all further execution. In the trace semantics, $v < 0$ is possible in the middle of some traces, which is a fact that the transition semantics does not notice. Combining traces for $\alpha \cup \beta$, that is, for

$$(a := -2a;\ a := a^2) \cup (a := -b;\ z' = v, v' = a;\ ?v \geq 0;\ a := A;\ z' = v, v' = a)$$

is just the union $\tau(\alpha) \cup \tau(\beta)$ of the traces $\tau(\alpha)$ and $\tau(\beta)$ from Examples 4.2 and 4.3. Note that $a \leq 0$ will hold at least once during every trace of $\alpha \cup \beta$, either in the beginning, or after setting $a := -2a$ or $a := -b$, respectively, when we assume $b > 0$. □

4.3.2 Valuation of State and Trace Formulas

In the semantics of dTL formulas, the dynamic modalities determine the set of traces according to the trace semantics of hybrid programs, and, independently, the temporal modalities determine at which points in time the respective postcondition needs to hold. The semantics of formulas is compositional and denotational, that is, the semantics of a complex formula is defined as a simple function of the semantics of its subformulas.

Definition 4.4 (Valuation of dTL formulas). For state formulas, the *valuation* $val(\nu,\cdot)$ with respect to state ν is defined inductively as follows:

1. $val(\nu, p(\theta_1,\dots,\theta_n)) = p^\ell\big(val(\nu,\theta_1),\dots,val(\nu,\theta_n)\big)$, where p^ℓ is the relation associated with p by the fixed semantics of real arithmetic.
2. $val(\nu,\phi\wedge\psi) = true$ iff $val(\nu,\phi) = true$ and $val(\nu,\psi) = true$
3. $val(\nu,\phi\vee\psi) = true$ iff $val(\nu,\phi) = true$ or $val(\nu,\psi) = true$
4. $val(\nu,\neg\phi) = true$ iff $val(\nu,\phi) \neq true$
5. $val(\nu,\phi\to\psi) = true$ iff $val(\nu,\phi) \neq true$ or $val(\nu,\psi) = true$
6. $val(\nu,\forall x\,\phi) = true$ iff $val(\nu[x\mapsto d],\phi) = true$ for all $d\in\mathbb{R}$
7. $val(\nu,\exists x\,\phi) = true$ iff $val(\nu[x\mapsto d],\phi) = true$ for some $d\in\mathbb{R}$
8. $val(\nu,[\alpha]\pi) = true$ iff for each trace $\sigma\in\tau(\alpha)$ that starts in $\text{first}\,\sigma=\nu$, if $val(\sigma,\pi)$ is defined, then $val(\sigma,\pi) = true$.
9. $val(\nu,\langle\alpha\rangle\pi) = true$ iff there is a trace $\sigma\in\tau(\alpha)$ starting in $\text{first}\,\sigma=\nu$ such that $val(\sigma,\pi)$ is defined and $val(\sigma,\pi) = true$.

For trace formulas, the *valuation* $val(\sigma,\cdot)$ with respect to trace σ is defined inductively as:

1. If ϕ is a state formula, then $val(\sigma,\phi) = val(\text{last}\,\sigma,\phi)$ if σ terminates, whereas $val(\sigma,\phi)$ is *not defined* if σ does not terminate.
2. $val(\sigma,\Box\phi) = true$ iff $val(\sigma_i(\zeta),\phi) = true$ holds for all positions (i,ζ) of σ with $\sigma_i(\zeta)\neq\Lambda$.
3. $val(\sigma,\Diamond\phi) = true$ iff $val(\sigma_i(\zeta),\phi) = true$ holds for some position (i,ζ) of σ with $\sigma_i(\zeta)\neq\Lambda$.

As usual, a (state) formula is *valid* if it is true in all states. Further for (state) formula ϕ and state ν we write $\nu\models\phi$ iff $val(\nu,\phi) = true$. We write $\nu\not\models\phi$ iff $val(\nu,\phi) = false$. Likewise, for trace formula π and trace σ we write $\sigma\models\pi$ iff $val(\sigma,\pi) = true$ and $\sigma\not\models\pi$ iff $val(\sigma,\pi) = false$. In particular, we only write $\sigma\models\pi$ or $\sigma\not\models\pi$ if $val(\sigma,\pi)$ is defined, which it is not the case if π is a state formula and σ does not terminate. The points where a dTL property ϕ has to hold for the various combinations of temporal and dynamic modalities are illustrated in Fig. 4.1.

Example 4.4. Recall the hybrid programs α and β from Examples 4.2 and 4.3. For these, we see the following difference between the $d\mathcal{L}$ transition semantics and the dTL trace semantics, which gives the same difference between the dTL semantics for nontemporal formulas (which are $d\mathcal{L}$ formulas) and the dTL trace semantics

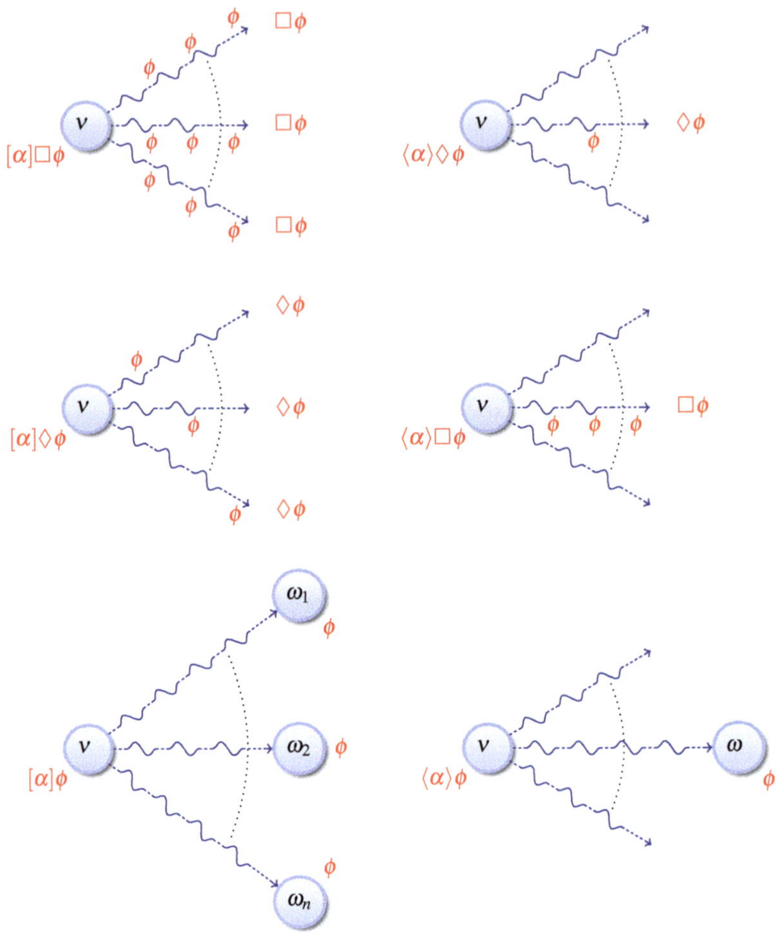

Fig. 4.1 Trace semantics of dTL formulas

for formulas with temporal modalities. The d$\mathcal{L}$ formula $[\alpha]a \geq 0$ is valid (the dTL formula $[\alpha]a \geq 0$ is also valid), because $a \geq 0$ holds at the end of all runs of α. The dTL formula $[\alpha]\Box a \geq 0$, on the other hand, is not valid, because a may be negative after the first step. Likewise, the d$\mathcal{L}$ formula $[\beta]v \geq 0$ is valid (the dTL formula $[\beta]v \geq 0$ is also valid), because $v \geq 0$ holds at the end of all runs of β. The dTL formula $[\beta]\Box v \geq 0$, instead, is not valid, because v may be negative at an intermediate state (those that later fail the middle test $?v \geq 0$ and get aborted). If we add an evolution domain restriction $v \geq 0$ to the differential equation, however, the formula $v \geq 0 \to [\beta]\Box v \geq 0$ is valid.

For the combined hybrid program $\alpha \cup \beta$ from both examples, we find that the dTL formula $[\alpha \cup \beta]\Diamond a \leq 0$ is valid, because a is nonpositive at least once during each of the traces. □

4.3.3 Conservative Temporal Extension

The following result shows that the extension of dTL by temporal operators does not change the meaning of nontemporal formulas. The trace semantics given in Definition 4.4 is equivalent to the final state reachability relation semantics given in Definition 2.6 for the sublogic $d\mathcal{L}$ of dTL.

Proposition 4.1 (Conservative temporal extension). *The logic* dTL *is a* conservative extension *of nontemporal* $d\mathcal{L}$*, i.e., the set of valid* $d\mathcal{L}$ *formulas is the same with respect to transition reachability semantics of* $d\mathcal{L}$ *(Definition 2.6) as with respect to the trace semantics of* dTL *(Definition 4.4).*

The proof of Proposition 4.1 uses the following relationship of reachability and trace semantics of dTL programs, which agree on initial and final states.

Lemma 4.1 (Trace relation). *For hybrid programs* $\alpha \in \mathrm{HP}(\Sigma)$*, we have*

$$\rho(\alpha) = \{(\mathrm{first}\,\sigma, \mathrm{last}\,\sigma) \;:\; \sigma \in \tau(\alpha) \text{ terminates}\}.$$

Proof. The proof follows an induction on the structure of α.

- The cases $x := \theta$, $x' = \theta$, and $\alpha \cup \beta$ are simple comparisons of Definitions 4.3 and 2.7.
- For $?\chi$, the reasoning splits into two directions.

"$\supseteq$" For inclusion "$\supseteq$", assume $\sigma \in \tau(?\chi)$. We distinguish between two cases. If $val(\mathrm{first}\,\sigma, \chi) = true$, then $\sigma = (\hat{v})$ has length one, $\mathrm{last}\,\sigma = \mathrm{first}\,\sigma$, and $(\mathrm{first}\,\sigma, \mathrm{first}\,\sigma) \in \rho(\alpha)$. If, however, $val(\mathrm{first}\,\sigma, \chi) = false$, then $\sigma = (\hat{v}, \hat{\Lambda})$ does not terminate; hence, there is nothing to show.

"$\subseteq$" Conversely, for inclusion "$\subseteq$", assume $(v, v) \in \rho(?\chi)$; then $val(v, \chi) = true$ and $(\hat{v}) \in \tau(\alpha)$ satisfies the conditions on σ.

- For $\alpha;\beta$, the reasoning again splits into the two directions.

"$\supseteq$" For inclusion "$\supseteq$", assume that $\sigma \circ \varsigma \in \tau(\alpha;\beta)$ terminates with $\sigma \in \tau(\alpha)$, $\varsigma \in \tau(\beta)$, and $\mathrm{last}\,\sigma = \mathrm{first}\,\varsigma$. Then, by induction hypothesis, we can assume that $(\mathrm{first}\,\sigma, \mathrm{last}\,\sigma) \in \rho(\alpha)$ and $(\mathrm{first}\,\varsigma, \mathrm{last}\,\varsigma) \in \rho(\beta)$. By the semantics of sequential composition, we have $(\mathrm{first}\,(\sigma \circ \varsigma), \mathrm{last}\,(\sigma \circ \varsigma)) \in \rho(\alpha;\beta)$.

"$\subseteq$" Conversely, for inclusion "$\subseteq$", assume that $(v, w) \in \rho(\alpha;\beta)$. That is, let $(v, z) \in \rho(\alpha)$ and $(z, w) \in \rho(\beta)$. By induction hypothesis, there is a terminating trace $\sigma \in \tau(\alpha)$ with $\mathrm{first}\,\sigma = v$ and $\mathrm{last}\,\sigma = z$. Further, by the induction hypothesis, there is a terminating $\varsigma \in \tau(\beta)$ with $\mathrm{first}\,\varsigma = z$ and $\mathrm{last}\,\varsigma = w$. Hence, $\sigma \circ \varsigma \in \tau(\alpha;\beta)$ terminates with $\mathrm{first}\,(\sigma \circ \varsigma) = v$ and $\mathrm{first}\,(\sigma \circ \varsigma) = w$.

- The case α^* is an inductive consequence of the sequential composition case. □

Proof (of Proposition 4.1). The formulas of $d\mathcal{L}$ are a subset of the dTL formulas. In the course of this proof, we use the notation $val_{d\mathcal{L}}(v, \cdot)$ to indicate that the $d\mathcal{L}$

valuation from Definition 4.4 in Sect. 2.3 is used. For $d\mathcal{L}$ formulas ψ, we show that the valuations with respect to Definitions 4.4 and 2.6 are the same for all states ν:

$$val(\nu,\psi) = val_{d\mathcal{L}}(\nu,\psi) \text{ for all } \nu.$$

We prove this by induction on the structure of ψ. The cases 1–3 of the definition of state formulas in Definition 4.1 are obvious. The other cases are proven as follows.

- If ψ has the form $[\alpha]\phi$, assume that $val(\nu,[\alpha]\phi) = false$. Then there is some terminating trace $\sigma \in \tau(\alpha)$ with $\text{first}\,\sigma = \nu$ such that $val(\text{last}\,\sigma,\phi) = false$. By the induction hypothesis, this implies that $val_{d\mathcal{L}}(\text{last}\,\sigma,\phi) = false$. According to Lemma 4.1, $(\nu,\text{last}\,\sigma) \in \rho(\alpha)$ holds, which implies $val_{d\mathcal{L}}(\nu,[\alpha]\phi) = false$. For the converse direction, assume that $val_{d\mathcal{L}}(\nu,[\alpha]\phi) = false$. Then there is a $(\nu,w) \in \rho(\alpha)$ with $val_{d\mathcal{L}}(w,\phi) = false$. By Lemma 4.1, there is a terminating trace $\sigma \in \tau(\alpha)$ with $\text{first}\,\sigma = \nu$ and $\text{last}\,\sigma = w$. By induction hypothesis, $val(\text{last}\,\sigma,\phi) = false$. Thus, we can conclude that both $val(\sigma,\phi) = false$ and $val(\nu,[\alpha]\phi) = false$.
- The case $\psi = \langle\alpha\rangle\phi$ is proven similarly. □

4.4 Safety Invariants in Train Control

In the European Train Control System (ETCS), trains are coordinated by decentralised Radio Block Centres (RBCs), which grant or deny movement authorities (MAs) dynamically to the individual trains by wireless communication. In emergencies, trains always have to stop within the MA issued by the RBC; see Fig. 4.2. Following

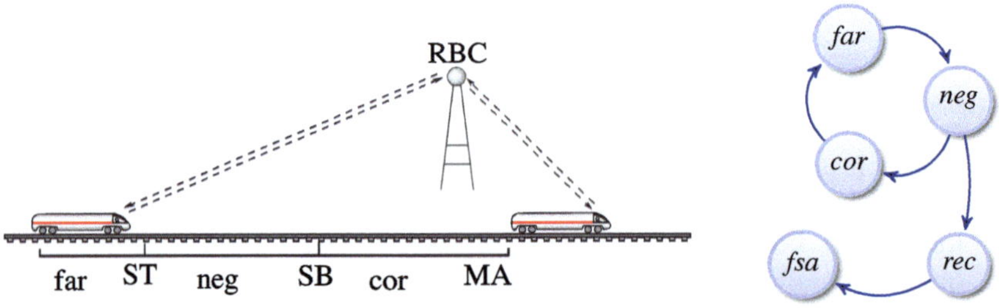

Fig. 4.2 ETCS train coordination protocol phases

the reasoning pattern for traffic agents in [89], each train negotiates with the RBC to extend its MA when approaching the end of its current MA. Since wireless communication takes time, this negotiation is initiated in due time before reaching m. To simplify the presentation, we adopt the assumption of Damm et al. [89] here that trains keep their desired speed (or at least their maximum speed limit) during negotiation. Before entering negotiation at some point ST (for start talking), the train still has sufficient distance to MA (it is in *far* mode) and can regulate its speed freely within the track limits. After the point SB (for start braking), the train has to start applying the brakes as identified in Sect. 2.9.

As a model for train movements, we use the ideal-world model from Sect. 2.4. For a safe operation of multiple traffic agents, it is crucial that the MA be respected at *every* point in time during this protocol, not only at its end. Hence, we need to consider temporal safety invariants. For instance, when the train has entered the negotiation phase at its current position z, dTL can analyse the following safety invariant of a part of the protocol cycle of the train controller:

$$\begin{aligned} \psi &\to [neg; cor; drive]\Box(\ell \leq L \to z < m) \qquad (4.3)\\ \text{where } neg &\equiv z' = v, \ell' = 1 \\ cor &\equiv (?m - z < s; a := -b) \cup (?m - z \geq s; a := \dots) \\ drive &\equiv z' = v, v' = a. \end{aligned}$$

It expresses that—under a sanity condition ψ for parameters—a train will *always* remain within its MA m as long as the accumulated RBC negotiation latency ℓ is at most L. We refer to the work of Faber and Meyer [119] for details on what kind of message passing contributes to ℓ. Like in [89], we model the train to first negotiate while keeping a constant speed ($z' = v$) in *neg*. The differential equation $\ell' = 1$ defines ℓ as a clock that is never reset, but accumulates the time spent in negotiation. Thereafter, in *cor*, the train corrects its acceleration or brakes with force b (as a fail-safe recovery manoeuvre) on the basis of the remaining distance ($m - z$). That is, if the distance of the movement authority and position is less than s (the test $?m - z < s$ succeeds), the acceleration is set to braking by $a := -b$. If, instead, the other test $?m - z \geq s$ succeeds, then acceleration is set to another value (not shown in *cor*). Finally, the train continues moving according to the differential equation system *drive* or, equivalently, $z'' = a$. Instead of manually choosing specific values for the free parameters of (4.3) as in [89, 119], we will use the techniques developed in this book to automatically synthesise constraints on the relationship of parameters that are required for a safe operation of cooperative train control.

4.5 Proof Calculus for Temporal Invariants

In this section, we introduce a sequent calculus for verifying temporal specifications of hybrid systems in differential temporal dynamic logic dTL. With the basic idea being to perform a symbolic decomposition, the calculus transforms hybrid programs successively into simpler logical formulas describing their effects. Statements about the temporal behaviour of a hybrid program are successively reduced to corresponding nontemporal statements about the intermediate states.

$$([\cup]\Box)\ \frac{[\alpha]\pi \wedge [\beta]\pi}{[\alpha\cup\beta]\pi}{}^{1} \qquad (\langle\cup\rangle\Diamond)\ \frac{\langle\alpha\rangle\pi \vee \langle\beta\rangle\pi}{\langle\alpha\cup\beta\rangle\pi}{}^{1}$$

$$([;]\Box)\ \frac{[\alpha]\Box\phi \wedge [\alpha][\beta]\Box\phi}{[\alpha;\beta]\Box\phi} \qquad (\langle;\rangle\Diamond)\ \frac{\langle\alpha\rangle\Diamond\phi \vee \langle\alpha\rangle\langle\beta\rangle\Diamond\phi}{\langle\alpha;\beta\rangle\Diamond\phi}$$

$$([?]\Box)\ \frac{\phi}{[?\chi]\Box\phi} \qquad (\langle ?\rangle\Diamond)\ \frac{\phi}{\langle ?\chi\rangle\Diamond\phi}$$

$$([:=]\Box)\ \frac{\phi \wedge [x:=\theta]\phi}{[x:=\theta]\Box\phi} \qquad (\langle :=\rangle\Diamond)\ \frac{\phi \vee \langle x:=\theta\rangle\phi}{\langle x:=\theta\rangle\Diamond\phi}$$

$$([']\Box)\ \frac{[x'=\theta]\phi}{[x'=\theta]\Box\phi} \qquad (\langle '\rangle\Diamond)\ \frac{\langle x'=\theta\rangle\phi}{\langle x'=\theta\rangle\Diamond\phi}$$

$$([^{*n}]\Box)\ \frac{[\alpha;\alpha^*]\Box\phi}{[\alpha^*]\Box\phi} \qquad (\langle^{*n}\rangle\Diamond)\ \frac{\langle\alpha;\alpha^*\rangle\Diamond\phi}{\langle\alpha^*\rangle\Diamond\phi}$$

$$([^*]\Box)\ \frac{[\alpha^*][\alpha]\Box\phi}{[\alpha^*]\Box\phi} \qquad (\langle^*\rangle\Diamond)\ \frac{\langle\alpha^*\rangle\langle\alpha\rangle\Diamond\phi}{\langle\alpha^*\rangle\Diamond\phi}$$

[1] π is a trace formula and—unlike the state formulas ϕ and ψ—may thus begin with a temporal modality $\Box$ or $\Diamond$.

Fig. 4.3 Rule schemata of the proof calculus for temporal differential dynamic logic

4.5.1 *Proof Rules*

We introduce a proof calculus for differential temporal dynamic logic dTL that inherits the proof rules of d$\mathcal{L}$ from Chap. 2 and adds new proof rules for temporal modalities. We can, in fact, get a corresponding proof calculus when we add the same proof rules for temporal modalities to the DAL proof rules from Chap. 3. In the latter case, the resulting logic is called differential-algebraic temporal dynamic logic DATL.

Inherited Nontemporal Rules

The dTL calculus is presented in Fig. 4.3 and inherits the (nontemporal) d$\mathcal{L}$ proof rules, i.e., the propositional, first-order, dynamic, and global rules from d$\mathcal{L}$. That is, it includes the propositional rules from Fig. 2.11 on p. 79 and either the free-variable quantifier rules from Fig. 2.11 or the simpler quantifier rules from Fig. 3.9 on p. 164 that are based on side deductions. The dynamic rules ($\langle;\rangle$–$[']$) and global rules ($[]$*gen*,$\langle\rangle$*gen*,*ind*,*con*) for handling nontemporal dynamic modalities are also inherited directly from Fig. 2.11. The only possible exception is that $[\cup]$,$\langle\cup\rangle$ can be generalised to apply to formulas of the form $[\alpha\cup\beta]\pi$ where π is an arbitrary trace formula, and not just a state formula as in d$\mathcal{L}$. Thus, π may begin with $\Box$ or $\Diamond$, which is why the rules are repeated in this generalised form as $[\cup]\Box$ and $\langle\cup\rangle\Diamond$ in Fig. 4.3.

$$([:=]\square)\ \frac{\phi \wedge [x:=\theta]\phi}{[x:=\theta]\square\phi}$$

$$(['] \square)\ \frac{[x'=\theta]\phi}{[x'=\theta]\square\phi}$$

$$([;]\square)\ \frac{[\alpha]\square\phi \wedge [\alpha][\beta]\square\phi}{[\alpha;\beta]\square\phi}$$

$$([\cup]\square)\ \frac{[\alpha]\square\phi \wedge [\beta]\square\phi}{[\alpha\cup\beta]\square\phi}$$

$$([^*]\square)\ \frac{[\alpha^*][\alpha]\square\phi}{[\alpha^*]\square\phi}$$

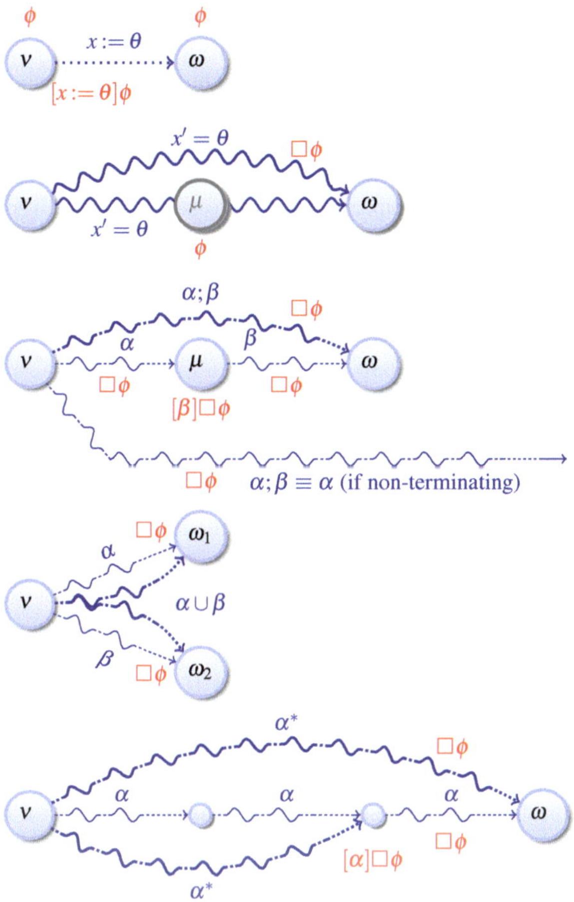

Fig. 4.4 Correspondence of temporal proof rules and trace semantics

Temporal Rules

The new temporal rules in Fig. 4.3 for the dTL calculus successively transform temporal specifications of hybrid programs into nontemporal d$\mathscr{L}$ formulas. The idea underlying this transformation is to decompose hybrid programs and recursively augment intermediate state transitions with appropriate specifications. Also see Fig. 4.4 for an illustration of the correspondence of a representative set of proof rules for temporal modalities to the trace semantics of hybrid programs (Definition 4.3).

Rule $[;]\square$ decomposes invariants of $\alpha;\beta$ (i.e., $[\alpha;\beta]\square\phi$ holds) into an invariant of α (i.e., $[\alpha]\square\phi$) and an invariant of β that holds when β is started in *any* final state of α (i.e., $[\alpha]([\beta]\square\phi)$). Its difference with the d$\mathscr{L}$ rule $[;]$ thus is that the dTL rule $[;]\square$ also checks safety invariant ϕ at the symbolic states in between the execution of α and β, and recursively so because of the temporal modality $\Box$. Again, see Fig. 4.4 for an illustration of this proof principle.

Rule $[:=]\Box$ expresses that invariants of assignments need to hold before and after the discrete change (similarly for $[?]\Box$, except that tests do not lead to a state change, so ϕ holding before the test is all there is to it). Rule $[']\Box$ can directly reduce invariants of continuous evolutions to nontemporal formulas as restrictions of solutions of differential equations are themselves solutions of different duration and thus already included in the evolutions of $x' = \theta$. In particular, observe that the handling of differential equations within hybrid systems is fully encapsulated within the fragment of dynamic rules from Fig. 2.11. The rules $[']\Box$, $\langle'\rangle\Diamond$, $[:=]\Box$, and $\langle:=\rangle\Diamond$ directly generalise to discrete jump sets and systems of differential equations or even to DA-constraints from Chap. 3.

The (optional) iteration rule $[^{*n}]\Box$ can partially unwind loops. It relies on rule $[;]\Box$ and is simpler than d$\mathcal{L}$ rule $[^{*n}]$, because the other rules will inductively produce a premise that ϕ holds in the current state, because of the temporal modality $\Box\phi$. The dual rules $\langle\cup\rangle\Diamond, \langle;\rangle\Diamond, \langle?\rangle\Diamond, \langle:=\rangle\Diamond, \langle'\rangle\Diamond, \langle^{*n}\rangle\Diamond$ work similarly.

In Chaps. 2 and 3, the primary means for handling loops are the invariant induction (*ind*) and variant convergence (*con*) rules. Here, we take a different, completely modular approach for verifying temporal properties of loops based on the d$\mathcal{L}$ capabilities for verifying nontemporal properties of loops. Rules $[^*]\Box$ and $\langle^*\rangle\Diamond$ actually *define* temporal properties of loops inductively. Rule $[^*]\Box$ expresses that ϕ holds at all times during repetitions of α (i.e., $[\alpha^*]\Box\phi$) iff, *after* repeating α any number of times, ϕ holds at all times *during* one execution of α (i.e., $[\alpha^*]([\alpha]\Box\phi)$). See Fig. 4.4 for an illustration. Dually, $\langle^*\rangle\Diamond$ expresses that α holds at some time during repetitions of α (i.e., $\langle\alpha^*\rangle\Diamond\phi$) iff, after some number of repetitions of α, formula ϕ holds at some point during one execution of α (i.e., $\langle\alpha^*\rangle(\langle\alpha\rangle\Diamond\phi)$). In this context, the nontemporal modality $\langle\alpha^*\rangle$ can be thought of as skipping over to the iteration of α during which ϕ actually occurs, as expressed by the nested dTL formula $\langle\alpha\rangle\Diamond\phi$. The inductive definition rules $[^*]\Box$ and $\langle^*\rangle\Diamond$ completely reduce temporal properties of loops to dTL properties of standard nontemporal d$\mathcal{L}$-modalities such that standard induction (*ind*) or convergence rules (*con*) can be used for the outer nontemporal modality of the loop. Hence, after applying the inductive loop definition rules $[^*]\Box$ and $\langle^*\rangle\Diamond$, the standard d$\mathcal{L}$ loop invariant and variant rules can be used for verifying temporal properties of loops without change, except that the postcondition contains temporal modalities.

Rules for handling $[\alpha]\Diamond\phi$ and $\langle\alpha\rangle\Box\phi$ are discussed in Sect. 4.9. Finally, provability in the dTL calculus is denoted by $\Phi \vdash_{\mathrm{dTL}} \psi$, and defined according to Definition 2.11 or Definition 3.16. The notion of a dTL proof directly follows from the previous definitions in Chaps. 2 and 3.

4.5.2 Verification Example

Consider the bouncing ball example from Sect. 2.5.4. The proof in Fig. 2.20 can be generalised easily to a proof of the temporal property

$$v^2 \le 2g(H-h) \land h \ge 0 \land g > 0 \land H \ge 0 \land 1 > c \ge 0$$
$$\to [(h'' = -g \,\&\, h \ge 0; (?h > 0 \cup (?h = 0; v := -cv)))^*]\Box(0 \le h \le H). \quad (4.4)$$

The only aspect of the proof that changes is that the temporal proof rules in Fig. 4.3 are used instead of the dynamic proof rules from Fig. 2.11, and that the resulting extra proof goals for the invariance property at intermediate steps have to be proven.

In contrast, the proof in Fig. 2.18 for the simplified dynamics without evolution domain restriction $h \ge 0$ cannot be generalised to a proof of the temporal property

$$v^2 \le 2g(H-h) \land h \ge 0 \land g > 0 \land H \ge 0 \land 1 > c \ge 0$$
$$\to [(h'' = -g; (?h > 0 \cup (?h = 0; v := -cv)))^*]\Box(0 \le h \le H). \quad (4.5)$$

This difference in provability is for good reasons. The property in (4.4) is valid, but the property in (4.5) is not! While there was no noticeable semantical difference between the nontemporal properties proven in Figs. 2.18 and 2.20, there is a decisive difference between the corresponding temporal properties (4.5) and (4.4). Because there is no evolution domain restriction in (4.5), its hybrid program does not prevent continuous evolution to a negative height under the floor ($h < 0$), for which $0 \le h \le H$ does not hold.

The reason for this discrepancy of the temporal version compared to the nontemporal versions thus is that the nontemporal modalities do not "see" the temporary violation of $0 \le h \le H$. Such a temporary violation of $0 \le h$ during the continuous evolution does not produce a successful run of the hybrid program, because it is blocked by the subsequent tests $?h = 0$ and $?h > 0$. A state with negative height fails both tests. While this behaviour does not give a successful program transition of $(\nu, \omega) \in \rho(ball)$ by Definition 2.7 so that the proof in Fig. 2.18 is correct, the behaviour still gives a valid trace $\sigma \in \tau(ball)$ by Definition 4.3. This trace σ is a partial trace, because it ends in a failure state Λ, but it is still one of the traces that $[ball]\Box(0 \le h \le H)$ quantifies over (quite unlike $[ball](0 \le h \le H)$, which only considers final states of successful traces).

4.6 Soundness

The following result shows that verification with the dTL calculus always produces correct results about the safety of hybrid systems, i.e., the dTL calculus is sound.

Theorem 4.1 (Soundness of dTL). *The* dTL *calculus is* sound*, i.e., derivable (state) formulas are valid.*

Proof. We show that all rules of the dTL calculus are *locally sound*, i.e., for all states ν, the conclusion of a rule is true in state ν when all premises are true in ν. Let ν be any state. For each rule we have to show that the conclusion is true in ν assuming the premises are true in ν. The propositional rules are locally sound by Theorem 2.1. Inductively, the soundness of the dynamic rules follows from Proposition 4.1 and local soundness of the corresponding rules in $d\mathcal{L}$. The proof for the generalisation in $[\cup]$ and $\langle\cup\rangle$ to path formulas π is a straightforward extension. The quantifier rules are sound by Theorems 2.1 and 3.1, respectively.

$[;]\Box$ Assume $\nu \models [\alpha]\Box\phi$ and $\nu \models [\alpha][\beta]\Box\phi$. Let $\sigma \in \tau(\alpha;\beta)$, i.e., $\sigma = \rho \circ \varsigma$ with $\mathrm{first}\,\sigma = \nu$, $\rho \in \tau(\alpha)$, and $\varsigma \in \tau(\beta)$. If ρ does not terminate, then $\sigma = \rho \in \tau(\alpha)$ and $\sigma \models \Box\phi$ by premise. If, instead, ρ terminates with $\mathrm{last}\,\rho = \mathrm{first}\,\varsigma$, then $\rho \models \Box\phi$ by premise. Further, we know $\nu \models [\alpha][\beta]\Box\phi$. In particular for trace $\rho \in \tau(\alpha)$, we have $\mathrm{last}\,\rho \models [\beta]\Box\phi$. Thus, $\varsigma \models \Box\phi$ because $\varsigma \in \tau(\beta)$ starts at $\mathrm{first}\,\varsigma = \mathrm{last}\,\rho$. By composition, $\rho \circ \varsigma \models \Box\phi$. As $\sigma = \rho \circ \varsigma$ was arbitrary, we can conclude $\nu \models [\alpha;\beta]\Box\phi$. The converse direction holds, as all traces of α are prefixes of traces of $\alpha;\beta$. Hence, the assumption $\nu \models [\alpha;\beta]\Box\phi$ directly implies $\nu \models [\alpha]\Box\phi$. Further, all traces of β that begin at a state reachable from ν by α are suffixes of traces of $\alpha;\beta$ starting in ν. Hence, $\nu \models [\alpha][\beta]\Box\phi$ is implied as well.

$[?]\Box$ Soundness of $[?]\Box$ is obvious, since, by premise, we can assume $\nu \models \phi$, and there is nothing to show for Λ states according to Definition 4.4. Conversely, $\hat{\nu}$ is a prefix of all traces in $\tau(?\chi)$ that start in ν.

$[:=]\Box$ Assuming $\nu \models \phi$ and $\nu \models [x := \theta]\phi$, we have to show that $\nu \models [x := \theta]\Box\phi$. Let $\sigma \in \tau(x := \theta)$ be any trace with $\mathrm{first}\,\sigma = \nu$, i.e., $\sigma = (\hat{\nu}, \hat{\omega})$ by Definition 4.3. Hence, the only two states we need to consider are $\sigma_0(0) = \nu$ and $\sigma_1(0) = \omega$. By premise, $\sigma_0(0) = \nu$ yields $\sigma_0(0) \models \phi$. Similarly, for the state $\sigma_1(0) = \mathrm{last}\,\sigma = \omega$, the premise gives $\sigma_1(0) \models \phi$. The converse direction is similar.

$[']\Box$ We prove that $[']\Box$ is locally sound by contraposition. For this, assume that $\nu \not\models [x' = \theta]\Box\phi$; then there is a trace $\sigma = (\varphi) \in \tau(x' = \theta)$ starting in $\mathrm{first}\,\sigma = \nu$ and $\sigma \not\models \Box\phi$. Hence, there is a position $(0,\zeta)$ of σ with $\sigma_0(\zeta) \not\models \phi$. Now φ restricted to the interval $[0,\zeta]$ also solves differential equation $x' = \theta$. Thus, $(\varphi|_{[0,\zeta]}) \not\models \phi$ as $\varphi(\zeta) \not\models \phi$, since the last state is $\varphi(\zeta)$. By consequence, this gives $\nu \not\models [x' = \theta]\phi$. The converse direction is obvious as $\mathrm{last}\,\sigma$ always is a state occurring during σ. Hence $\nu \not\models [x' = \theta]\phi$ immediately implies $\nu \not\models [x' = \theta]\Box\phi$.

$[^{*n}]\Box$ By contraposition, assume that $\nu \not\models [\alpha^*]\Box\phi$. Then there is an $n \in \mathbb{N}$ and a trace $\sigma \in \tau(\alpha^n)$ with $\mathrm{first}\,\sigma = \nu$ such that $\sigma \not\models \Box\phi$. There are two cases. If $n > 0$ then $\sigma \in \tau(\alpha;\alpha^*)$, and thus $\nu \not\models [\alpha;\alpha^*]\Box\phi$. If, however, $n = 0$, then $\sigma = (\hat{\nu})$ and $\nu \not\models \phi$. Hence, all traces $\varsigma \in \tau(\alpha;\alpha^*)$ with $\mathrm{first}\,\varsigma = \nu$ satisfy $\varsigma \not\models \Box\phi$. Finally, it is easy to see that all programs have at least one such trace ς (when V is nonempty) that witnesses $\nu \not\models [\alpha;\alpha^*]\Box\phi$. The converse direction is easy as all behaviour of $\alpha;\alpha^*$ is subsumed by α^*, i.e., $\tau(\alpha;\alpha^*) \subseteq \tau(\alpha^*)$.

$[^*]\square$ Clearly, using the fact that $\tau(\alpha^*) \supseteq \tau(\alpha^*;\alpha)$, the set of states along the traces of α^* at which ϕ needs to be true for the premise is a subset of the corresponding set for the conclusion. Hence, the conclusion entails the premise. Conversely, all states during traces of α^* are also reachable by iterating α sufficiently often to completion and then following a single trace of α. In detail: If $\nu \not\models [\alpha^*]\Box\phi$, then there is a trace $\sigma \in \tau(\alpha^*)$ on which $\neg\phi$ holds true at some state, say, at $\sigma_i(\zeta) \neq \Lambda$. Let $n \geq 0$ be the (maximum) number of complete repetitions of α along σ before discrete step index i. That is, there is some discrete step index $i_n < i$ such that the prefix $\rho = (\sigma_0, \ldots, \sigma_{i_n}) \in \tau(\alpha^n)$ of σ consists of n complete repetitions of α and the suffix $\varsigma = (\sigma_{i_n+1}, \sigma_{i_n+2}, \ldots) \in \tau(\alpha^*)$ starts with a trace of α during which $\neg\phi$ occurs at point $\sigma_i(\zeta)$, namely at relative position $(i-(i_n+1), \zeta)$. Let $\acute{\varsigma} \in \tau(\alpha)$ be this prefix of ς. Consequently, $\acute{\varsigma} \models \langle\alpha\rangle\Diamond\neg\phi$ and the trace $\rho \circ \acute{\varsigma}$ is a witness for $\nu \models \langle\alpha^*\rangle\langle\alpha\rangle\Diamond\neg\phi$.

The proofs for $\langle;\rangle\Diamond$–$\langle^*\rangle\Diamond$ are dual, since $\langle\alpha\rangle\Diamond\phi$ is equivalent to $\neg[\alpha]\Box\neg\phi$ by duality. □

4.7 Completeness

In this section, we show that the strictly modular dTL calculus enables us to lift the Relative Completeness Theorem 2.3 for d$\mathcal{L}$ to dTL.

4.7.1 Incompleteness

The Incompleteness Theorem 2.2 directly generalises to temporal and nontemporal properties of dTL.

Theorem 4.2 (Incompleteness of dTL). *The discrete and continuous fragments of* dTL *are non-axiomatisable for temporal safety ($[\alpha]\Box\phi$) and nontemporal ($[\alpha]\phi$) fragments of* dTL. *Hence, valid* dTL *formulas are not always derivable.*

Proof. We show that the discrete and continuous fragments of the following purely temporal and nontemporal fragments of dTL are non-axiomatisable:

1. the fragment that only contains modalities of the form $[\alpha]\Box\phi$ and $\langle\alpha\rangle\Diamond\phi$
2. the fragment that only contains $[\alpha]\phi$ and $\langle\alpha\rangle\phi$ (d$\mathcal{L}$ fragment).

Case 2 is a consequence of the corresponding incompleteness result (Theorem 2.2) for fragments of the sublogic d$\mathcal{L}$, which carries over to the extension dTL by Proposition 4.1.

For Case 1, we prove that natural numbers are definable amongst the real number domain in both fragments, quite similarly to the proof of Theorem 2.2. Then these fragments extend first-order *integer* arithmetic such that the incompleteness theorem of Gödel applies.

- Natural numbers are definable in the discrete fragment without continuous evolutions $x' = \theta$ using repetitive additions:

$$nat(n) \leftrightarrow \langle x:=0;(x:=x+1)^*\rangle\Diamond x=n.$$

- In the continuous fragment, natural numbers are definable as:

$$nat(n) \leftrightarrow \exists s\,\exists c\,(s=0 \wedge c=1 \wedge \langle s'=c, c'=-s, \tau'=1\rangle\Diamond(s=0 \wedge \tau=n)).$$

  These ODEs have *sin* and *cos* as unique solutions for s and c, respectively. Their zeros characterise an isomorphic copy of natural numbers, scaled by π. □

4.7.2 Relative Completeness

Due to the modular construction of the dTL calculus, we can lift the major relative completeness result Theorem 2.3 from $d\mathcal{L}$ to dTL. By proving dTL completeness relative to Theorem 2.3, we essentially show that dTL is complete relative to $d\mathcal{L}$, which directly implies that dTL is even complete relative to FOD using Theorem 2.3 by a standard argument. Again, we restrict our attention to homogeneous combinations of path and trace quantifiers like $[\alpha]\Box\phi$ or $\langle\alpha\rangle\Diamond\phi$.

Theorem 4.3 (Relative completeness of dTL). *The* dTL *calculus is* complete relative to *FOD, i.e., every valid* dTL *formula can be derived from FOD tautologies.*

Proof (Outline). The proof is a simple extension of the proof of Theorem 2.3, because the dTL calculus successively reduces temporal properties to nontemporal properties and, in particular, handles loops by inductive definition rules in terms of $d\mathcal{L}$ modalities. The temporal rules in Fig. 4.3 transform temporal formulas to simpler formulas, i.e., to where the temporal modalities occur after simpler programs ($[\cup]\Box$, $[^*]\Box$, $\langle\cup\rangle\Diamond$, $\langle^*\rangle\Diamond$) or disappear completely ($[?]\Box$,$[:=]\Box$,$[']\Box$ and $\langle?\rangle\Diamond$,$\langle:=\rangle\Diamond$,$\langle'\rangle\Diamond$). Hence, the inductive relative completeness proof in Sect. 2.7.2 directly generalises to dTL with the following addition: After applying $[^*]\Box$ or $\langle^*\rangle\Diamond$, loops are ultimately handled by the standard $d\mathcal{L}$ rules *ind* and *con*. To show that sufficiently strong invariants and variants exist for the temporal postconditions $[\alpha]\Box\phi$ and $\langle\alpha\rangle\Diamond\phi$, we only have to show that such temporal formulas are expressible in FOD, i.e., Lemma 2.9 generalises to dTL. □

This result, which we prove formally in the remainder of Sect. 4.7.2, gives a formal justification that the dTL calculus reduces temporal properties to nontemporal $d\mathcal{L}$ properties.

4.7.3 Expressibility and Rendition of Hybrid Trace Semantics

The central step for lifting the $d\mathcal{L}$ completeness proof to dTL is the following: To show that, after applying rule $[^*]\Box$ or $\langle^*\rangle\Diamond$, sufficiently strong invariants and variants for dTL postconditions can be expressed in $d\mathcal{L}$ for *ind* or *con* to be able to prove the result, we show that the trace semantics of hybrid programs can be characterised in FOD.

Lemma 4.2 (Hybrid program trace rendition). *For every hybrid program α with variables $\vec{x} = x_1,\dots,x_k$ there is a FOD formula $\mathcal{T}_\alpha(\vec{x},\vec{v})$ with variables among the 2k distinct variables $\vec{x} = x_1,\dots,x_k$ and $\vec{v} = v_1,\dots,v_k$ such that*

$$\vDash \mathcal{T}_\alpha(\vec{x},\vec{v}) \leftrightarrow \langle\alpha\rangle\Diamond\vec{x} = \vec{v}$$

or, equivalently, for every ν, we have that $\nu \vDash \mathcal{T}_\alpha(\vec{x},\vec{v})$ iff

$$\sigma_i(\zeta) = \nu[\vec{x} \mapsto val(\nu,\vec{v})] \text{ for a position } (i,\zeta) \text{ of some trace } \sigma \in \tau(\alpha) \text{ starting in } \nu.$$

$$\begin{aligned}
\mathcal{T}_{x_1:=\theta_1,\dots,x_k:=\theta_k}(\vec{x},\vec{v}) &\equiv \vec{x} = \vec{v} \lor \mathcal{S}_{x_1:=\theta_1,\dots,x_k:=\theta_k}(\vec{x},\vec{v})\\
\mathcal{T}_{x_1'=\theta_1,\dots,x_k'=\theta_k\,\&\,\chi}(\vec{x},\vec{v}) &\equiv \mathcal{S}_{x_1'=\theta_1,\dots,x_k'=\theta_k\,\&\,\chi}(\vec{x},\vec{v})\\
\mathcal{T}_{?\chi}(\vec{x},\vec{v}) &\equiv \mathcal{S}_{?\chi}(\vec{x},\vec{v})\\
\mathcal{T}_{\beta\cup\gamma}(\vec{x},\vec{v}) &\equiv \mathcal{T}_\beta(\vec{x},\vec{v}) \lor \mathcal{T}_\gamma(\vec{x},\vec{v})\\
\mathcal{T}_{\beta;\gamma}(\vec{x},\vec{v}) &\equiv \mathcal{T}_\beta(\vec{x},\vec{v}) \lor \exists\vec{z}\,(\mathcal{S}_\beta(\vec{x},\vec{z}) \land \mathcal{T}_\gamma(\vec{z},\vec{v}))\\
\mathcal{T}_{\beta^*}(\vec{x},\vec{v}) &\equiv \exists\vec{z}\,(\mathcal{S}_{\beta^*}(\vec{x},\vec{z}) \land \mathcal{T}_\beta(\vec{z},\vec{v}))
\end{aligned}$$

Fig. 4.5 Explicit rendition of hybrid program trace semantics in FOD

Proof. The proof is similar to that of Lemma 2.8, yet using the definition in Fig. 4.5. We resort to corresponding characterisations from Lemma 2.8, which simplifies the characterisation of $\mathcal{T}_\alpha(\vec{x},\vec{v})$, because we only have to augment $\mathcal{S}_\alpha(\vec{x},\vec{v})$ by with disjunctions for intermediate states, which can again be defined in terms of $\mathcal{S}_\alpha(\vec{x},\vec{v})$, recursively.

For instance, $\mathcal{T}_{\beta^*}(\vec{x},\vec{v})$ characterises the states reachable during traces of β^* as the states reachable during traces of β that start after running β^* to completion for some number of iterations. □

Using this program rendition to characterise temporal trace modalities, Lemma 2.9 generalises immediately to dTL as follows:

Lemma 4.3 (dTL Expressibility). *Logic* dTL *is* expressible *in FOD: for all* dTL *formulas $\phi \in \text{Fml}(\Sigma)$ there is a FOD formula $\phi^\# \in \text{Fml}_{FOD}(\Sigma)$ that is equivalent, i.e., $\vDash \phi \leftrightarrow \phi^\#$. The converse holds trivially.*

Proof. The proof is by a structural induction identical to that in the proof of Lemma 2.9 with the following additions:

1. The case where ϕ is of the form $\langle\alpha\rangle\Diamond\psi$ is a consequence of Lemma 4.2:

$$\vDash \quad \langle\alpha\rangle\Diamond\psi \leftrightarrow \exists\vec{v}\,(\mathscr{T}_\alpha(\vec{x},\vec{v}) \wedge \psi^{\#}{}_{\vec{x}}^{\vec{v}}).$$

2. The case where ϕ is $[\alpha]\Box\psi$ is again a consequence of Lemma 4.2:

$$\vDash \quad [\alpha]\Box\psi \leftrightarrow \forall\vec{v}\,(\mathscr{T}_\alpha(\vec{x},\vec{v}) \rightarrow \psi^{\#}{}_{\vec{x}}^{\vec{v}}).$$

□

4.7.4 *Modular Relative Completeness Proof for the Differential Temporal Dynamic Logic Calculus*

Now we assemble the proof of Theorem 4.3 from the previous results following a simplified form of the proof of Theorem 2.3, since we can apply Theorem 2.3 for d$\mathscr{L}$ formulas.

Proof (of Theorem 4.3). The proof is a simple extension of the Relative Completeness Theorem 2.3 for d$\mathscr{L}$. Unlike for the rules of the d$\mathscr{L}$ calculus, all new temporal rules are symmetric, hence perform equivalent transformations. Consequently, whenever their conclusion is valid, their premise is valid and of smaller complexity (temporal modalities occur after simpler programs), and hence derivable by induction hypothesis.

For instance, analogously to the induction step for loops in Proposition 2.1, let $\vDash F \rightarrow [\beta^*]\Box G$; then $\vDash F \rightarrow [\beta^*][\beta]\Box G$. By Lemma 4.3, there is a d$\mathscr{L}$ formula or even FOD formula $([\beta]\Box G)^{\#}$ that characterises the temporal postcondition equivalently, i.e., such that $\vDash ([\beta]\Box G)^{\#} \leftrightarrow [\beta]\Box G$. By induction hypothesis, we can derive the simpler formula $\vdash_{\mathscr{D}} ([\beta]\Box G)^{\#} \rightarrow [\beta]\Box G$. Using Lemma 2.11, we conclude $\vdash_{\mathscr{D}} \forall^{\beta}(([\beta]\Box G)^{\#} \rightarrow [\beta]\Box G)$, thus $[\beta^*]([\beta]\Box G)^{\#} \vdash_{\mathscr{D}} [\beta^*][\beta]\Box G$ is derivable by $[]gen$. Furthermore, $\vDash F \rightarrow [\beta^*]([\beta]\Box G)^{\#}$ is a valid d$\mathscr{L}$ formula and, thus, $F \vdash_{\mathscr{D}} [\beta^*]([\beta]\Box G)^{\#}$ is derivable by Theorem 2.3. Combining these derivations by a cut with $[\beta^*]([\beta]\Box G)^{\#}$, we derive $F \vdash_{\mathscr{D}} [\beta^*]\Box G$. Since the temporal rules perform a modular reduction to nontemporal dynamic rules, the case $\vDash F \rightarrow \langle\beta^*\rangle\Diamond G$ is almost identical here, because the differences between variant rule *con* and invariant rule *ind* have already been captured in the proof of Theorem 2.3. □

4.8 Verification of Train Control Safety Invariants

Continuing the ETCS study from Sect. 4.4, we consider a slightly simplified version of equation (4.3) that gives a more concise proof. By a safe abstraction (provable in dTL), we simplify *cor* to permit braking even when $m-z \geq s$, since braking remains safe with respect to $z < m$. Recall (4.3) with additional abbreviations and a simplified *cor*:

$$\psi \to [neg; cor; drive]\Box(\ell \leq L \to z < m) \tag{4.3*}$$

$$\begin{aligned}
\text{where } \psi &\equiv z < m \land v > 0 \land \ell = 0 \land L \geq 0\\
\phi &\equiv \ell \leq L \to z < m\\
neg &\equiv z' = v, \ell' = 1\\
cor &\equiv a := -b \cup (?m - z \geq s; a := \dots)\\
drive &\equiv z' = v, v' = a.
\end{aligned}$$

Within the following proof, $\langle\!|\cdot|\!\rangle$ brackets are used instead of modalities to visually identify the discrete jump set prefix (Definition 2.11). That is, we use $\langle\!|z := \ell v + z|\!\rangle\phi$ to mean $\langle z := \ell v + z\rangle\phi$, just to have more readable bracket grouping. The dTL proof of the safety invariant in (4.3) splits into two cases that correspond to the respective protocol phases:

$$\dfrac{\dfrac{\dfrac{\dots}{\psi \vdash [neg]\Box\phi} \qquad \dfrac{\dots}{\psi \vdash [neg][cor; drive]\Box\phi}}{\psi \vdash [neg; cor; drive]\Box\phi}\,[;]\Box}{\vdash \psi \to [neg; cor; drive]\Box\phi}\,\to\text{r}$$

There, the left branch proves that ϕ holds while negotiating and is as follows:

$$\dfrac{\dfrac{\dfrac{\dfrac{\psi \vdash Lv + z < m}{\psi \vdash \forall l{\geq}0\,(l \leq L \to lv + z < m)}\,\forall\text{r},\text{i}\forall}{\psi \vdash \forall l{\geq}0\,\langle\!|z := lv + z, \ell := l|\!\rangle\phi}\,[:=],\langle:=\rangle}{\psi \vdash [neg]\phi}\,[']}{\psi \vdash [neg]\Box\phi}\,[']\Box$$

The right branch shows that ϕ continues to hold after negotiation has completed when continuing with an adjusted acceleration a in *cor*; *drive*:

$$
\begin{array}{ll}
 & \psi,\ell\geq 0 \vdash v^2 < 2b(m-Lv-z) \wedge Lv+z<m \\ \hline
\text{i}\forall & \psi,\ell\geq 0 \vdash \llparenthesis z:=\ell v+z, a:=-b \rrparenthesis \forall t\geq 0\,(\ell\leq L \rightarrow \frac{a}{2}t^2+vt+z<m) \\ \hline
[:=],\langle:=\rangle & \psi,\ell\geq 0 \vdash \llparenthesis z:=\ell v+z, a:=-b \rrparenthesis \forall t\geq 0\, \llparenthesis z:=\frac{a}{2}t^2+vt+z \rrparenthesis \phi \\ \hline
[']\Box,['] & \psi,\ell\geq 0 \vdash \llparenthesis z:=\ell v+z, a:=-b \rrparenthesis [drive]\Box\phi \qquad \triangleright \\ \hline
[\cup] & \psi,\ell\geq 0 \vdash \llparenthesis z:=\ell v+z \rrparenthesis [cor][drive]\Box\phi \qquad \triangleright \\ \hline
[;]\Box & \psi,\ell\geq 0 \vdash \llparenthesis z:=\ell v+z \rrparenthesis [cor;drive]\Box\phi \\ \hline
\rightarrow\text{r} & \psi \vdash \ell\geq 0 \rightarrow \llparenthesis z:=\ell v+z \rrparenthesis [cor;drive]\Box\phi \\ \hline
\forall\text{r} & \psi \vdash \forall \ell\geq 0\, \llparenthesis z:=\ell v+z \rrparenthesis [cor;drive]\Box\phi \\ \hline
['] & \psi \vdash [neg][cor;drive]\Box\phi
\end{array}
$$

The application of rule $[;]\Box$ in this latter case spawns a third case (marked with $\triangleright$) to show that ϕ holds during *cor*. However, the reasoning in this third case is subsumed by the cases above, since the changes on a in *cor* do not interfere with condition ϕ. Generally, this optimisation of $[;]\Box$ is applicable whenever the modified vocabulary is disjoint from ϕ. Here, proof rules $[']$ and i$\forall$ are implemented in Mathematica to handle evolutions.

The leaves of the proof branches above can even be used to automatically *synthesise parameter constraints* that are necessary to avoid MA violation. The parametric safety constraint obtained by combining the open conditions conjunctively is $Lv+z<m \wedge v^2 < 2b(m-Lv-z)$. It simplifies to $v^2 < 2b(m-Lv-z)$ because $b>0$. This yields bounds for the speed limit and negotiation latency in order to guarantee safe driving and closing of the proof. Similarly, rule $[\cup]$ leads to a branch (marked with $\triangleright$) for the case $[?m-z\geq s; a:=\ldots]$, from which corresponding conditions about the safety envelope s can be derived depending on the particular speed controller, similar to what we have shown in Sect. 2.9.

4.9 Liveness by Quantifier Alternation

Liveness specifications of the form $[\alpha]\Diamond\phi$ or $\langle\alpha\rangle\Box\phi$ are sophisticated (Σ_1^1-hard because they can express infinite occurrence in Turing machines). Beckert and Schlager [38], for instance, note that they failed to find sound rules for a discrete case that corresponds to $[\alpha;\beta]\Diamond\phi$.

For *finitary liveness semantics*, we can still find proof rules. In this section, we modify the meaning of $[\alpha]\Diamond\phi$ to refer to all *terminating* traces of α. Then, the straightforward generalisation $[;]\Diamond$ in Fig. 4.6 is sound, even in the hybrid case. But $[;]\Diamond$ still leads to an incomplete axiomatisation as it does not cover the case where, in some traces, ϕ becomes true at some point during α, and in other traces, ϕ only becomes true during β. To overcome this limitation, we use a program transformation approach. We instrument the hybrid program to monitor the occurrence of ϕ during all changes: In $[\alpha]\Diamond$, $\check{\alpha}$ results from replacing all occurrences of $x:=\theta$ with $x:=\theta; ?(\phi \rightarrow t=1)$ and $x'=\theta$ with $x'=\theta \,\&\, (\phi \rightarrow t=1)$. The latter is a continuous evolution restricted to the region of the state space that satisfies $\phi \rightarrow t=1$.

The effect is that t detects whether ϕ has occurred during any change in α. In particular, t is guaranteed to be 1 after all runs if ϕ occurs at least once along all traces of α. This trick directly works for first-order conditions ϕ. Using the combination presented in [234], nominals can be used as state labels to address the same issue for general ϕ.

$$([;]\lozenge)\ \frac{\vdash [\alpha]\lozenge\phi, [\alpha][\beta]\lozenge\phi}{\vdash [\alpha;\beta]\lozenge\phi} \qquad ([\alpha]\lozenge)\ \frac{\phi \lor \forall t\,[\breve{\alpha}]t = 1}{[\alpha]\lozenge\phi}$$

Fig. 4.6 Transformation rules for alternating temporal path and trace quantifiers

Proposition 4.2 (Local soundness for temporal quantifier alternation). *The rules in Fig. 4.6 are locally sound for finitary liveness semantics.*

Proof. Let ν be any state.

$[;]\lozenge$ Assuming that the premise is true, we need to consider two cases corresponding to the two formulas of its succedent. If $\nu \models [\alpha]\lozenge\phi$, then obviously $\nu \models [\alpha;\beta]\lozenge\phi$, as every trace of $\alpha;\beta$ has a trace of α as prefix, during which ϕ holds at least once. If, however, $\nu \models [\alpha][\beta]\lozenge\phi$, then ϕ occurs at least once during all traces that start in a state reachable from ν by α. Let $\rho \circ \varsigma \in \tau(\alpha;\beta)$ with $\operatorname{first}\rho = \nu$, $\rho \in \tau(\alpha)$ and $\varsigma \in \tau(\beta)$. In finitary liveness semantics, $\rho \circ \varsigma$ can be assumed to terminate (otherwise there is nothing to show). Then, $\operatorname{last}\rho$ is a state reachable from ν by α; hence $\varsigma \models \lozenge\phi$. In particular, $\rho \circ \varsigma \models \lozenge\phi$.

$[\alpha]\lozenge$ For the soundness of $[\alpha]\lozenge$, first observe that the truth of $val(\nu,\phi)$ of ϕ depends on the state ν; hence it can only be affected during state changes. Further, the only actual changes of valuations happen during discrete jumps $x := \theta$ or continuous evolutions $x' = \theta$. All other system actions only cause control flow effects but no elementary state changes. Assume the premise is true in a state ν. If $\nu \models \phi$, the conjecture is obvious. Assume $\nu \models \forall t\,[\breve{\alpha}]t = 1$. Suppose $\nu \not\models [\alpha]\phi$; then there is a trace $\sigma \in \tau(\alpha)$ with $\sigma \not\models \lozenge\phi$. Then, this trace directly corresponds to a trace $\breve{\sigma}$ of $\breve{\alpha}$ in which all $\phi \to t = 1$ conditions are trivially satisfied as ϕ never holds. As there are no changes of the fresh variable t during $\breve{\alpha}$, the value of t remains constant during $\breve{\sigma}$. But then we can conclude that there is a trace, which is essentially the same as $\breve{\sigma}$ except for the constant valuation of the fresh variable t on which no conditions are imposed; hence $t = 0$ is possible. As these traces terminate in finitary liveness semantics, we can conclude $\nu \not\models \forall t\,[\breve{\alpha}]t = 1$, which is a contradiction. Conversely for equivalence of premise and conclusion, assume $\nu \not\models \phi \lor \forall t\,[\breve{\alpha}]t = 1$. Then, the initial state ν does not satisfy ϕ and it is possible for $\breve{\alpha}$ to execute along a terminating trace σ that permits t to be $\neq 1$. Suppose there was a position (i,ζ) of σ at which $\sigma_i(\zeta) \models \phi$. Without loss of generality, we can assume (i,ζ) to be the first such position. Then, the hybrid action which regulates σ_i is accompanied by the immediate condition that $\phi \to t = 1$; hence $t = 1$ holds if σ terminates. Since the fresh variable t is rigid (is never changed during $\breve{\alpha}$) and σ terminates in finitary liveness semantics, we conclude $val(\text{last}\,\sigma, t) = 1$, which is a contradiction. □

4.10 Summary

For reasoning about hybrid systems, we have introduced a temporal dynamic logic, dTL, with modal path quantifiers over traces and temporal quantifiers along the traces. It combines the capabilities of dynamic logic [149] to reason about possible system behaviour with the power of temporal logic [247, 114, 115] in reasoning about the behaviour along traces. Furthermore, we have presented a proof calculus for verifying temporal safety specifications of hybrid programs in dTL.

Our sequent calculus for dTL is a completely modular combination of temporal and nontemporal reasoning. Temporal formulas are handled using rules that augment intermediate state transitions with corresponding sub-specifications. Purely nontemporal $d\mathcal{L}$ rules handle the effects of discrete and continuous evolution. The modular nature of the dTL calculus further enables us to lift the relative completeness result from $d\mathcal{L}$ to dTL. This theoretical result shows that the dTL calculus is a sound and complete axiomatisation of the temporal behaviour of hybrid systems relative to differential equations.

As an example, we demonstrate that our logic is suitable for reasoning about safety invariants in the European Train Control System. Further, we have successfully applied our calculus to automatically synthesise (nonlinear) parametric safety constraints for this system.

Future work includes extending dTL with CTL*-like [115] formulas of the form $[\alpha](\psi \wedge \Box\phi)$ to avoid splitting of the proof into two very similar subproofs for temporal parts $[\alpha]\Box\phi$ and nontemporal parts $[\alpha]\psi$ arising in rule $[;]\Box$. Our combination of temporal logic with dynamic logic is more suitable for this purpose than previous approaches for discrete systems [38], since dTL has uniform modalities and uniform semantics for temporal and nontemporal specifications. This extension will also simplify the treatment of alternating liveness quantifiers conceptually.

Part II
Automated Theorem Proving for Hybrid Systems

Overview After having developed formal specification logics and proof calculi for specifying and verifying safety-critical properties of hybrid systems in Part I, we now turn to practical and algorithmic implementation questions. In this part, we focus on the practical aspects of implementing the proof calculi from Part I. The calculi in Part I have already been designed for the needs of automated theorem proving, most notably with the free-variable and Skolemisation techniques from Chap. 2 and the compositional proof calculi from Part I. Immediate implementations of the proof calculi from Part I in automated theorem provers can prove examples of medium complexity directly. Yet, more complex case studies still require additional algorithmic techniques for achieving high-degree automation and good scalability properties. In Chap. 5, we refine the calculi from Part I to tableau procedures and present proof strategies that navigate among their nondeterminisms to help overcome the complexity issues of integrating real quantifier elimination as a decision procedure for real arithmetic.

In Chap. 6 we introduce the "differential invariants as fixed points" paradigm. We refine the differential induction techniques from Chap. 3 to a fully automatic verification algorithm for computing the required discrete and differential invariants of a hybrid system locally in a logic-based fixed-point loop.

The algorithmic refinement techniques developed in this part of the book add better automation to the proof approach from Part I. These algorithms are crucial for automating the formal verification of properties of complex hybrid systems like the ones we consider in Part III.

Chapter 5
Deduction Modulo Real Algebra and Computer Algebra

Contents

Synopsis We show how deductive, real algebraic, and computer algebraic methods can be combined for verifying hybrid systems in an automated theorem proving approach. In particular, we highlight the interaction of deductive and algebraic reasoning that is used for handling the joint discrete and continuous behaviour of hybrid systems. Systematically, we derive a canonical *tableau procedure modulo* from the calculus of differential dynamic logic. We delineate the nondeterminisms in the tableau procedure carefully and analyse their practical impact in the presence of computationally expensive handling of real algebraic constraints. Based on experience with larger case studies, we analyse proof strategies for dealing with the practical challenges for integrated algebraic and deductive verification of hybrid systems. To overcome the complexity pitfalls of integrating real arithmetic, we propose the *iterative background closure* and *iterative inflation order* strategies, with which we achieve substantial computational improvements.

A. Platzer, Logical Analysis of Hybrid Systems,
DOI 10.1007/978-3-642-14509-4_5, © Springer-Verlag Berlin Heidelberg 2010

5.1 Introduction

While the theoretical background and formal details of our logical analysis approach for hybrid systems can be found in Part I, here we discuss the practical aspects of combining deductive, real algebraic, and computer algebraic prover technologies. In particular, we highlight the principles of how these techniques interact for verifying hybrid systems. We analyse the degrees of freedom in implementing our calculus in terms of the nondeterminisms of its canonical proof procedure. We illustrate the impact that various choices of proof strategies have on the overall performance. For hybrid system verification, we observe that the nondeterminisms in the interaction between deductive and real algebraic reasoning have considerable impact on the practical feasibility. While straightforward combinations are sufficient for verifying examples like those presented in Part I, larger case studies like those that we present in Part III are beyond the capabilities of state-of-the-art decision procedures for real arithmetic. In this chapter, we analyse and explain the causes and consequences of this effect and introduce automatic proof strategies that avoid these complexity pitfalls and work well in practise.

Here we study the modular combination in the $d\mathcal{L}$ calculus (our findings generalise directly to the extensions of the DAL and dTL calculi, so we use $d\mathcal{L}$ interchangeably with DAL and dTL in this chapter). Our observations are of more general interest, though, and we conjecture that similar results hold for other prover combinations of logics with interpreted function symbols that are handled using background decision procedures for computationally expensive theories including real arithmetic, approximations of natural arithmetic, and arrays.

5.1.1 Related Work

As we have pointed out in Sect. 1.2, there are only a few other practical approaches [201, 1] that use deduction for verifying hybrid systems and actually integrate arithmetic reasoning in STeP [201] or PVS [1], respectively. They do not work with a genuine verification logic, however, but only generate flat mathematical verification conditions for hybrid automata with a given invariant. In contrast, the symbolic decompositions in our verification logic preserve the natural problem structure, which enables us to achieve good performance in practise. See Sect. 1.2 for a detailed comparison.

Several other approaches intend to combine deductive and arithmetic reasoning, e.g. [63, 28, 3]. Their focus, however, is on general mathematical reasoning in classes of higher-order logic and is not tailored to verify hybrid systems. Our work, instead, is intended to make practical verification of hybrid systems possible and we aim at automating the verification process.

5.1.2 Structure of This Chapter

In Sect. 5.2, we analyse our calculi for differential dynamic logics from a qualitative perspective and present a corresponding tableau procedure modulo background solvers for handling real algebraic and computer algebraic constraints. We delineate their nondeterminisms carefully in Sect. 5.3 and analyse their practical impact. We present proof procedures for automated theorem proving in differential dynamic logics that navigate through the complexity pitfalls of integrating decision procedures for real arithmetic in Sects. 5.4 and 5.5. In Sect. 5.6, we evaluate their performance in larger case studies.

5.2 Tableau Procedures Modulo

In this section, we derive a canonical tableau procedure modulo background provers from the $d\mathcal{L}$ calculus and analyse the remaining nondeterminisms in the remainder of this chapter.

For the purpose of this chapter, the full details of how the respective quantifier rules of Chaps. 2 and 3 lift quantifier elimination to dynamic logic are not important. What is important to note, however, is that quantifier rules and rules for handling modalities need to interact because the actual constraints on quantified symbols depend on the effect of the hybrid programs within modalities (Sect. 2.5.3). Thus, at some point, after a number of rule applications that handle the dynamic part, quantifier rules will be used to discharge (or at least simplify) proof obligations over real algebraic or semialgebraic constraints by quantifier elimination [80, 81, 288]. The remaining subgoals will be analysed further again using dynamic rules. The quantifier rules constitute the modular interface that combines deduction for handling dynamic reasoning with algebraic constraint techniques for handling continuous reasoning about the reals. Here, we discuss the consequences and principles of this combination and analyse proof strategies.

The principle behind how the $d\mathcal{L}$ calculus in Fig. 2.11 combines deduction technology with methods for handling real algebraic constraints complies with the general background reasoning principles [32, 290, 103]. From an abstract perspective, the $d\mathcal{L}$ calculus selects a set Φ of (quantified) formulas from an open branch (Φ is called the *key*) and hands it over to the quantifier elimination procedure. The resulting formula obtained by applying quantifier elimination (QE) to Φ is then returned to the main sequent prover as a result, and the main proof continues; see Fig. 5.1. Similarly, the $d\mathcal{L}$ calculus triggers symbolic, computer algebraic computations for the rules for differential equations using their solutions (rules $\langle'\rangle$,$[']$ from Fig. 2.11) or total differentials of differential invariants (rule *DI*) or differential variants (rule *DV*) from the DAL calculus in Fig. 3.9.

In this context, the propositional rules, dynamic rules, and global rules of the $d\mathcal{L}$ calculus in Fig. 2.11 or the DAL calculus in Fig. 3.9 constitute the *foreground rules* in the main prover (middle box of Fig. 5.1), except for evolution rules $\langle'\rangle$,$[']$ and

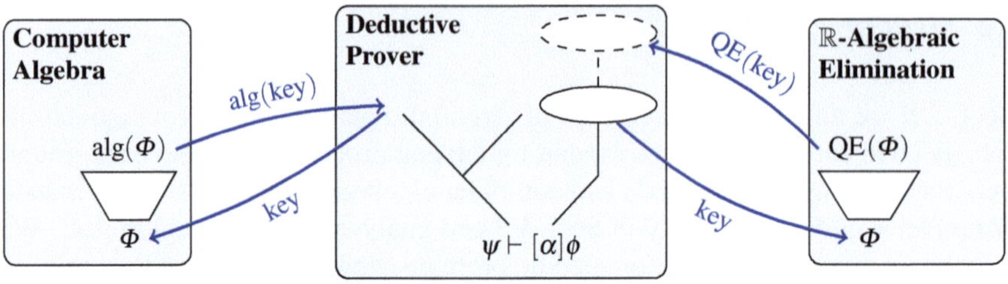

Fig. 5.1 Deductive, real algebraic, and computer algebraic prover combination

differential (in)variant rules *DI*,*DV* that represent the *computer algebraic rules* invoking a computer algebra system as a *background solver* (left box). The arithmetic quantifier rules (in particular i∀ and i∃ from Fig. 2.11 and r∀,l∀,r∃,l∃ from Fig. 3.9) form the set of rules that invoke the *background prover* (right box) for quantifier elimination. Since the primary challenges caused by the proof nondeterminisms in the d$\mathcal{L}$ calculus originate from the interaction of deductive and real algebraic rules, we simplify the presentation in the following and only distinguish between background real arithmetic rules and foreground rules, where computer algebraic rules will simply be considered as foreground rules.

The canonical tableau procedure belonging to the d$\mathcal{L}$ calculus is presented in Fig. 5.2. Observe that the tableau procedure for our d$\mathcal{L}$ calculus has a nonstandard set of nondeterministic steps (indicated by $\mathcal{B}$, $\mathcal{M}$, and $\mathcal{F}$, respectively in Fig. 5.3):

$\mathcal{B}$: *selectBranch*, i.e., select which open branch to choose for further rule applications.

$\mathcal{M}$: *selectMode*, i.e., select whether to apply foreground d$\mathcal{L}$ rules (propositional rules, dynamic rules, and global rules) or background arithmetic rules (i∀, i∃ from Fig. 2.11 or r∀,l∀,r∃,l∃ from Fig. 3.9).

$\mathcal{F}$: *selectFormula*, i.e., which formula(s) to select for rule applications from the current branch in the current mode.

Within the rule applications, there is an additional choice of whether to handle differential equations using their solution (d$\mathcal{L}$ rules $\langle'\rangle$,$[']$ from Fig. 2.11) or by differential induction (DAL rules *DI*,*DV* from Fig. 3.9). We do not follow up on this nondeterminism here but simply choose to use solutions, whenever they are first-order expressible, and fall back to differential induction (Sect. 3.5.6) when no such solution can be found. The computational cost of differential induction is generally less than when working with solutions, but the corresponding differential (in)variants have to be found first, which we handle in Chap. 6. There is a further minor nondeterminism of whether to expand loops using rules $\langle^{*n}\rangle$,$[^{*n}]$ or to go for an induction by rules *ind* and *con*. Yet, as unrolling ($\langle^{*n}\rangle$,$[^{*n}]$) only handles fixed-length reachability prefixes or bounded loops, our proof strategies prefer induction (*ind* and *con*) instead. The other d$\mathcal{L}$ rules do not produce any conflicts once a formula has been selected as they apply to formulas of distinct syntactic structures: The top-level operators uniquely determine which calculus rule to use once a formula has been selected.

```
while tableau T has open branches do
  B := selectBranch(T)          (* 𝓑 nondeterminism *)
  M := selectMode(B)            (* 𝓜 nondeterminism *)
  F := selectFormulas(B,M)      (* 𝓕 nondeterminism *)
  if M = foreground then
    R := result of applying a propositional or
             dynamic or global rule to F in B
    replace branch B by R in tableau T
  else
    send key F to background decision procedure QE
    receive result R from QE
    apply a quantifier rule to T with QE−result R
  end if
end while
```

Fig. 5.2 Tableau procedure for differential dynamic logics

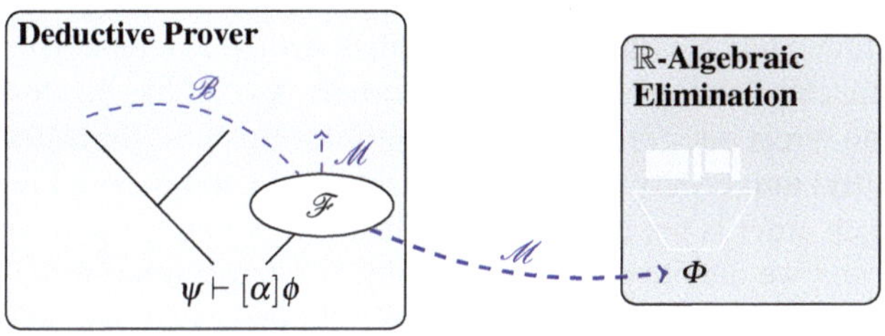

Fig. 5.3 Nondeterminisms in the tableau procedure for differential dynamic logics

At this point, notice that, unlike in the classical tableau procedure [122], we have three rather than four points of nondeterminism, since d𝓛 does not need closing substitutions. The reason for this is that d𝓛 has an interpreted domain. Rather than having to try out instantiations that have been determined by unification or heuristics as in uninterpreted first-order logic [122], we can make use of the structure in the interpreted case of first-order logic over the reals. In particular, arithmetic formulas can be reduced equivalently by QE to simpler formulas in the sense that the quantified symbols no longer occur. As this transformation is an equivalence, there is no loss of information and we do not need to backtrack [122] or simultaneously keep track of multiple local closing instantiations [134].

Despite this seemingly simpler situation, the influence of nondeterminism on the practical prover performance is quite remarkable. Even though the first-order theory of real arithmetic is decidable by quantifier elimination [81], its complexity is *doubly exponential* in the number of quantifier alternations [94]. While more efficient algorithms exist for some linear fragments [198], the practical performance is an issue in nonlinear cases. Hence, the computational cost of individual rule applications is quite different from the linear complexity of applying closing substitutions in uninterpreted tableaux.

5.3 Nondeterminisms in Tableau Modulo

In principle, exhaustive fair application of background rules by the nondeterminisms $\mathscr{M}$ and $\mathscr{F}$ remains complete for appropriate fragments of d$\mathscr{L}$. In practise, however, the complexity of real arithmetic quickly makes this naïve approach infeasible for larger case studies. In the remainder of this chapter, we discuss the consequences of the nondeterminisms and develop proof strategies to tackle the combination and integration challenges.

5.3.1 Nondeterminisms in Branch Selection

In classical uninterpreted tableaux, branch selection has no impact on completeness, and can only have impact on the proving duration, because closing substitutions can sometimes be found much earlier on one branch than on the others. In the interpreted case of d$\mathscr{L}$, branch selection is even less important. As d$\mathscr{L}$ has no closing substitutions, there is no direct interference among multiple branches. Branches with (explicitly or implicitly) universally quantified variables have to be closed independently. Hence the branch order is not important. For instance, when x is a universally quantified variable and we denote the corresponding Skolem symbol by s, the branches in the following proof can be handled separately (branches are implicitly combined by conjunction and universal quantifiers distribute over conjunctions):

$$\dfrac{\dfrac{\dfrac{\mathrm{QE}(\forall x\,(\dots bx^2 \geq 0))}{\Gamma, b>0 \vdash bs^2 \geq 0}\,\mathrm{i}\forall \qquad \dfrac{\mathrm{QE}(\forall x\,(\dots bx^4 + x^2 \geq 0))}{\Gamma, b>0 \vdash bs^4 + s^2 \geq 0}\,\mathrm{i}\forall}{\Gamma, b>0 \vdash bs^2 \geq 0 \land bs^4 + s^2 \geq 0}\,\land\mathrm{r}}{\Gamma, b>0 \vdash \forall x\,(bx^2 \geq 0 \land bx^4 + x^2 \geq 0)}\,\forall\mathrm{r}$$

For existentially quantified variables, the situation is a bit more subtle as multiple branches interfere indirectly in the sense that a simultaneous solution needs to be found for all branches at once. In $\exists v\,(v>0 \land v<0)$, for instance, the two branches resulting from the cases $v>0$ and $v<0$ cannot be handled separately, as the existential quantifier claims the existence of a simultaneous solution for $v>0$ and $v<0$, not two different solutions. Thus, when v is an existentially quantified variable and V its corresponding free existential variable, the branches in the following proof need to synchronise before quantifier elimination is applied:

$$\dfrac{\dfrac{\dfrac{\mathrm{QE}(\exists v\,((b>2 \to b(V-1)>0) \land (b>2 \to (V+1)^2 + b\varepsilon(V+1)>0)))}{b>2 \vdash b(V-1)>0 \qquad\qquad b>2 \vdash (V+1)^2 + b\varepsilon(V+1)>0}}{\mathrm{i}\forall\; b>2 \vdash [v:=V-1]bv>0 \qquad\qquad b>2 \vdash [v:=V+1]v^2 + b\varepsilon v>0}}{\dfrac{b>2 \vdash [v:=V-1]bv>0 \land [v:=V+1]v^2 + b\varepsilon v>0}{b>2 \vdash \exists v\,([v:=v-1]bv>0 \land [v:=v+1]v^2 + b\varepsilon v>0)}\,\forall\mathrm{r}}\,\land\mathrm{r}$$

The order in which the intermediate steps at the two branches are handled, however, has no impact on the proof. Branches like these *synchronise* on an existential free variable V in the sense that all occurrences of V need to be first-order on all branches for quantifier elimination to be applicable. Consequently, the only fairness assumption for $\mathscr{B}$ is that whenever a formula of a branch is selected that is waiting for synchronisation with another branch to become first-order, then it transfers its branch choice to the other branch instead. In the above case the left branch synchronises with the right branch on V. Hence, rule i$\exists$ can only be applied to $b(V-1) > 0$ on the left branch after rule $\langle :=\rangle$ has been applied on the right branch to produce first-order occurrences of V.

Thus, the primary remaining impact of the branch nondeterminism is that closing branches by universally quantified variable reasoning simplifies all subsequent existential variable handling, because fewer branches remain that need to be considered simultaneously. Even for existential variables, universal quantification is sound and can be very helpful; it is just not complete.

5.3.2 Nondeterminisms in Formula Selection

In background proving mode, it turns out that nondeterminism $\mathscr{F}$ is important for practical performance. In practise, when a branch closes or, at least, can be simplified significantly by a quantifier elimination call, then the running time of a single decision procedure call depends strongly on the number of irrelevant formulas that are selected in addition to the relevant ones by $\mathscr{F}$.

Clearly, when Φ is a set of formulas from a sequent that yields a tautology such that applying QE closes a branch, then selecting any superset $\Psi \supseteq \Phi$ from a branch yields the same answer in the end (a sequent forms a disjunction of its formulas; hence it can be closed to true when any subset closes). However, the running time until this result will be found in the larger Ψ can be disturbed strongly by the presence of complicated additional but irrelevant formulas. From our experience with Mathematica [303], decision procedures for full real arithmetic seem to be distracted considerably by such irrelevant additional information.

Example 5.1 (Computational distraction in quantifier elimination). One sequent from the proof of the ETCS kernel in Sects. 2.4 and 2.9 is depicted in Fig. 5.4. It results from the right branch of the proof in Sect. 2.9 by exhaustive splitting. Quantifier elimination, as performed by function Reduce in Mathematica, runs more than 24 hours without producing a result on the formula in Fig. 5.4, which has nine symbolic variables and polynomial degree 2.

The marked constraint in Fig. 5.4 corresponds to the initial state of the system, because it refers to the initial train position z and not to the current train position z_2 in the current induction step. In fact, the induction step does not depend on this part of the initial state information (it *does* depend on the initial $b > 0$, though). When we remove the superfluous constraint on the initial state, the formula in Fig. 5.4 suddenly becomes provable in less than one second, compared to more than a day! The

Fig. 5.4 Computational distraction in quantifier elimination

$$
\begin{aligned}
& t_2 > 0,\ \varepsilon \geq t_2,\ v_2 \geq 0,\ A + 1/t_2 \cdot v_2 \geq 0,\ t_2 \geq 0, \\
& m - z_2 \geq v_2^2/(2b) + (A/b+1)(A/2\varepsilon^2 + \varepsilon v_2), \\
& 2b(m - z_2) \geq v_2^2 \\
& 2b(m - z) \geq v^2, \qquad\qquad /\!*\ \textit{initial state}\ *\!/ \\
& b > 0,\ A \geq 0 \\
\vdash\ & (At_2 + v_2)^2 \leq 2b(m - 1/2(At_2^2 + 2t_2v_2 + 2z_2))
\end{aligned}
$$

d$\mathcal{L}$ calculus presented in Chap. 2 avoids this problem, because of the additional focusing quantifiers introduced by the universal closure $\forall^\alpha$ in global rules and because, later, rule i$\forall$ applies quantifier elimination after reintroducing the quantifier structure. □

Yet, such additional information accumulates in tableaux procedures quite naturally, because the purpose of a proof branch in d$\mathcal{L}$ is to keep track of all that is known about a particular (symbolic) case of the system behaviour. Generally, not all of this knowledge finally turns out to be relevant for that case and only plays a role in other branches. Nevertheless, discarding part of this knowledge arbitrarily would, of course, endanger completeness.

For instance, the train safety statement (2.7) from p. 62 in Sect. 2.4 depends on a constraint on the safety envelope s that regulates braking versus acceleration by the condition $m - z \geq s$ in *ctrl*. A maximal acceleration of A is permitted in case $m - z \geq s$, when adaptively choosing s depending on the current speed v, maximum braking force b, and maximum controller response time ε in accordance with constraint (2.18) as discovered in Sect. 2.9. This constraint is necessary for some but not all cases of the symbolic safety analysis, though. In the case where the braking behaviour of ETCS is analysed, for instance, the constraint on s is irrelevant, because braking is the safest operation that a train can do to avoid crashing into preceding trains. The unnecessary presence of several quite complicated constraints such as (2.18), however, can distract quantifier elimination procedures considerably.

5.3.3 Nondeterminisms in Mode Selection

In its own right, nondeterminism $\mathcal{M}$ has less impact on the prover performance than $\mathcal{F}$. Every part of a branch could be responsible for closing it. The foreground closing rule *ax* of the main prover can only close branches for comparably trivial reasons like $b > 0, \varepsilon > 0 \vdash \varepsilon > 0$. Hence, mode selection has to give a chance to the background procedure every once in a while, following some fair selection strategy. From the observation that some decision procedure calls can run for hours without terminating, however, we can see that realisations of nondeterminism $\mathcal{M}$ need to be devised with considerable care.

As the reason for closing a branch can be hidden in any part of the sequent, some expensive decision procedure calls are superfluous if the branch can be closed by continuing $\mathsf{d}\mathcal{L}$ reasoning on the other parts. For instance, if A is some complicated algebraic constraint, decision procedure calls triggered by nondeterminism $\mathcal{M}$ can lead to nontermination within any feasible time for

$$\ldots, \varepsilon > 0, m - z \geq s \vdash A, [drive]\varepsilon > 0, \ldots$$

If $\mathcal{M}$ chooses foreground rules, then an analysis of $[drive]\varepsilon > 0$ by $\mathsf{d}\mathcal{L}$ rules will quickly discover that the maximum reaction-time ε remains constant while driving. Then, this part of a proof closes without the need to consider constraint A at all. For this reason, proof strategies that *eagerly check* for closing branches by background procedure calls as soon as the formulas are first-order are not successful in practise; see Fig. 5.5. Similarly, for verifying the ETCS case study that we detail in Chap. 7

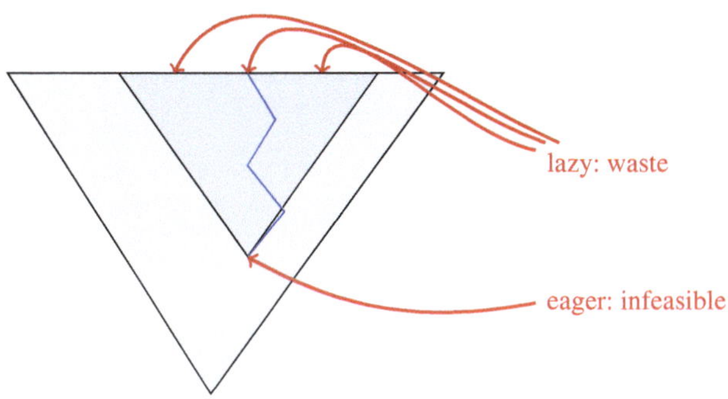

Fig. 5.5 Eager and lazy quantifier elimination in proof search space

we need to prove fairly large subgoals of first-order real arithmetic, which cannot be proven by current quantifier elimination procedures within any feasible amount of time; see Fig. 5.6 for a typical example.

Unfortunately, converse strategies with lazy checks that strongly favour foreground $\mathsf{d}\mathcal{L}$ rule applications in $\mathcal{M}$ are not appropriate either; see Fig. 5.5. The *lazy proof* strategy only uses background solvers if no more foreground proof rules can be used. There, splitting rules like $\wedge$r and $\vee$l can eagerly split the problems into multiple branches without necessarily making them any easier to solve. If this happens, slightly different but similar arithmetic problems of about the same complexity need to be solved repeatedly on multiple branches rather than just one branch, resulting in runtime blowup. The reason this can happen is a substantial syntactic redundancy in the sequent encoding of formulas. For instance, the sets of sequents before and after the following rule application are equivalent:

$$\wedge\text{r},\wedge\text{r}\ \frac{\psi \vdash v^2 \leq 2b(m-z) \quad \psi \vdash \varepsilon > 0 \quad \psi \vdash (z \geq 0 \rightarrow v \leq 0)}{\psi \vdash v^2 \leq 2b(m-z) \wedge \varepsilon > 0 \wedge (z \geq 0 \rightarrow v \leq 0)}$$

Yet, closing the three sequents above the bar by quantifier elimination is not necessarily easier than closing the single sequent below (neither conversely), because

```
  state = 0,
  2 * b * (m - z) >= v ^ 2 - d ^ 2,
  v >= 0, d >= 0, v >= 0, ep > 0, b > 0, A > 0, d >= 0
==>
  v <= vdes
-> \forall R a_3;
     (   a_3 >= 0 & a_3 <= A
      -> (       m - z
              <= (A / b + 1) * ep * v
               + (v ^ 2 - d ^ 2) / (2 * b)
               + (A / b + 1) * A * ep ^ 2 / 2
          -> \forall R t0;
               (   t0 >= 0
                -> \forall R ts0;
                     (0 <= ts0 & ts0 <= t0
                         -> -b * ts0 + v >= 0 & ts0 + 0 <= ep)
                -> 2 * b * (m - 1 / 2 *
                                 (-b * t0 ^ 2 + 2 * t0 * v + 2 * z))
                     >= (-b * t0 + v) ^ 2
                      - d ^ 2
                   & -b * t0 + v >= 0
                   & d >= 0))
        & (     m - z
              >  (A / b + 1) * ep * v
               + (v ^ 2 - d ^ 2) / (2 * b)
               + (A / b + 1) * A * ep ^ 2 / 2
           -> \forall R t2;
               (   t2 >= 0
                -> \forall R ts2;
                     (0 <= ts2 & ts2 <= t2
                         -> a_3 * ts2 + v >= 0 & ts2 + 0 <= ep)
                -> 2 * b * (m - 1 / 2 *
                                  (a_3 * t2 ^ 2 + 2 * t2 * v + 2 * z))
                     >= (a_3 * t2 + v) ^ 2
                      - d ^ 2
                   & a_3 * t2 + v >= 0
                   & d >= 0)))
```

Fig. 5.6 A large subgoal of first-order real arithmetic during ETCS verification

the working principle of arithmetic decision procedures is algebraic and not deductive. Even worse, if the sequents close by applying proof rules to ψ, then similar reasoning has to be repeated for three branches. This threefold reasoning may not even be detected as identical when ψ is again split differently on the three resulting branches.

Further, the representational equivalence in sequents is purely syntactic, i.e., up to permutation, the representations share the same disjunctive normal form. In the uninterpreted case of first-order logic, this syntactic redundancy is exploited by the propositional rules in order to transform sequents into a canonical form with atomic formulas, where partial closing situations are more readily identifiable. In the presence of a background decision procedure, however, reduction to sequents with atomic formulas is no longer necessary as it will be undone when handing the formulas over to the background decision procedure.

Moreover, the logical splitting along the propositional structure does not always help quantifier elimination procedures, because their working principle is not a deductive case analysis but (partial) cylindrical algebraic decomposition [80, 81, 83] based on cell decompositions of polynomials. See App. D.2 for explanations of the basic principle behind quantifier elimination in the background prover, which is quite different from that of symbolic logical case distinctions in the foreground prover. In Sect. 5.4, we will see that the combination of deductive analysis with algebraic arithmetic proving is an extremely important factor for accelerating quantifier elimination, but both have to be applied with care.

Finally, algebraic constraint handling techniques as in the Mathematica function Reduce can come up with a result that is only a restated version of the input if a selected (open) formula cannot be simplified or closed. For instance, the sequent $z < m \vdash v^2 \leq 2b(m-z)$ "reduces" to $\vdash b \geq v^2/(2m-2z) \vee m \leq z$ without any progress. Such arithmetical reformulation cannot even be detected by simple syntactical means and easily leads to infinite proof loops without progress when the outcome is split by $\vee$r and again handled by the background procedures.

5.4 Iterative Background Closure

In this and the following section, we propose strategies to solve the previously addressed computational issues caused by the nondeterminisms of the $d\mathcal{L}$ tableau procedure and its integration with computationally expensive background provers.

Priority-Based Strategies

We propose the priorities for rule applications in Fig. 5.7a (with rules at the top taking precedence over rules at the bottom). In this strategy, algebraic constraints are generally left intact as opposed to being split among multiple branches, because arithmetic rules have a higher priority than propositional splitting rules on first-order constraints. Further, we only accept the result of the background procedure when the number of different variable symbols has decreased to avoid infinite proof loops. We use arithmetic background rules *either* with priority 2 *or* with priority 5.

The effect of using priority 2 is that branches are checked eagerly for closing conditions or variable reductions; see Fig. 5.5. If reasoning about algebraic constraints does not yield any progress (no variables can be eliminated), then $d\mathcal{L}$ rules further analyse the system. For this choice, it is important to work with timeouts to prevent lengthy background decision procedure calls from blocking $d\mathcal{L}$ proof progress.

This problem is reduced significantly when using priority 5 for arithmetic rules instead. The effect of priority 5 is that formulas containing modalities are analysed

1. non-splitting propositional rules ¬r–∨r, ∧l–→r, *ax*
2. *arithmetic rules* if variable eliminated
3. dynamic rules ⟨?⟩–[:=]
4. splitting rules ∧r, ∨l, →l on modalities
5. *arithmetic rules* if variable eliminated
6. (in)variant global rules *ind*, *con*, *DI*, and *DV*
7. splitting rules ∧r, ∨l, →l on first-order formulas

Fig. 5.7a Proof strategy priorities

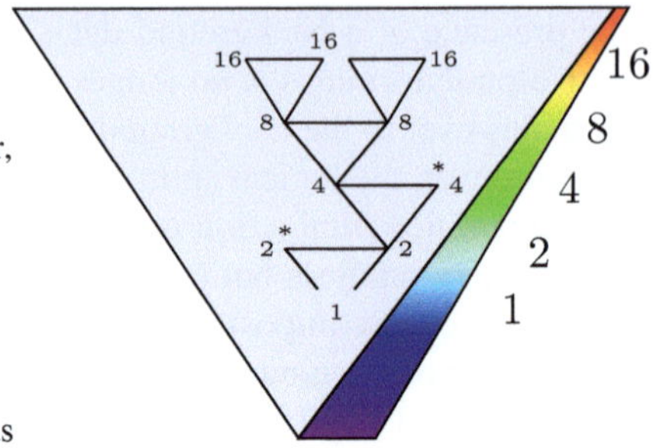

Fig. 5.7b Iterative background closure (IBC) proof strategy

and decomposed as much as possible before arithmetic reasoning is applied to algebraic constraint formulas. Then, however, the prover might consume too much time analysing the effects of programs on branches which would easily close due to simple arithmetic facts like in $\varepsilon > 0, \varepsilon < 0 \vdash [\alpha]\phi$.

A simple compromise is to use a combination of background rules with priority 2 for quick linear arithmetic [198] and expensive quantifier elimination calls for nonlinear arithmetic with priority 5. Another option is to normalise arithmetic formulas in sequents and use polynomial arithmetic to detect simple reasons for closing branches early [273], which has also been implemented in KeYmaera.

Iterative Background Closure

As a more sophisticated control strategy on top of the static priorities in Fig. 5.7a, we have introduced *iterative background closure* (IBC) [230]. There, the idea is to periodically apply arithmetic rules with a timeout T that increases by a factor of 2 after the background decision procedure runs have timed out; see Fig. 5.7b. Thus, background rules interleave with other rule applications (triangles in Fig. 5.7b), and the timeout for the subgoals increases as indicated, until the background procedure successfully eliminates variables on a branch (marked by $*$). The effect is that the prover avoids splitting in the average case but is still able to split cases when combined handling turns out to be prohibitively expensive. As an optimisation, timed-out QE will only be invoked again after the branch has been split (or after a modal formula has disappeared from the sequent) so that QE will work on a different input.

Figure 5.8 shows a proof procedure that implements IBC. Unless the QE call terminated successfully within the current timeout (line 3), IBC applies foreground rules (line 7) until the proof has split. Then, the timeout increases (line 10) and IBC handles the resulting branches recursively (line 11). Our experimental results in Sect. 5.6 show that IBC is surprisingly relevant for handling larger case studies.

```
/* prove validity of the sequent Φ ⊢ Ψ */
function IBC(Φ ⊢ Ψ, timeout):
  if QE(Φ ⊢ Ψ) succeeds within timeout then
    return QE(Φ ⊢ Ψ)
  else
    while foreground rule applicable and proof not split do
      apply foreground rule to current sequent
    end while
    let Φ1 ⊢ Ψ1,...,Φn ⊢ Ψn be the resulting branches
    timeout := 2*timeout
    return IBC(Φ1 ⊢ Ψ1 ,timeout) and ... and IBC(Φn ⊢ Ψn ,timeout)
```

Fig. 5.8 Iterative background closure (IBC) algorithm schema

And/Or Branching

More generally, the d$\mathscr{L}$ calculus gives rise to theorem proving structures with combined and/or-branching. While the branches resulting from one rule application are *and-branches* (all of them have to close for the proof to succeed), the rule alternatives—especially as caused by the tableau procedure nondeterminisms—are *or-branches* (only one of the proof search alternatives has to be successful for the proof); see Fig. 5.9. Most notably, induction rules give rise to or-branches, as there

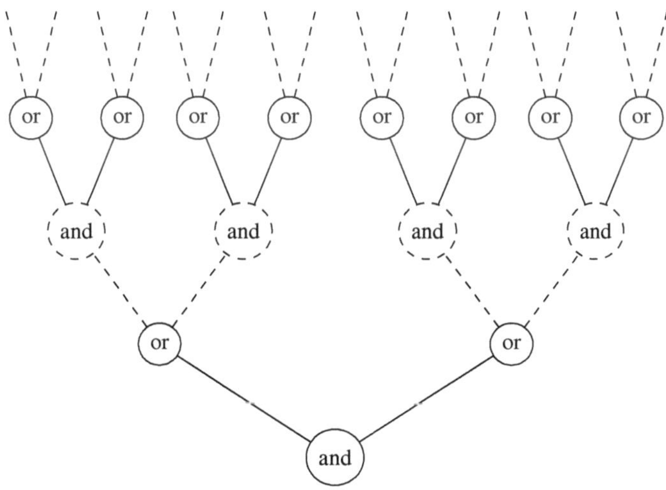

Fig. 5.9 General and/or-branching in proof strategies for differential dynamic logics

are several possible formulas (infinitely many) that could be used as (differential) (in)variants for rules *ind*,*con*,*DI*,*DV*, but one single successful (in)variant is enough for closing the proof. For these or-branch alternatives, any fair parallel exploration scheme or any fair sequential exploration scheme with time interleaving can be used. Iterative background closure is one possible sequential interleaving choice that works well in practise. On parallel computers, distributed clusters, or even multi-core processors, truly parallel exploration schemes would be faster, as they exploit the natural proving parallelism in our calculi from Part I. We will make use of and/or-branching exploration based on iterative background closure in Chap. 6.

5.5 Iterative Inflation

The iterative background closure strategy already has a decisive impact on the feasibility of verifying larger case studies. Yet, there are still cases where the $\mathscr{F}$ nondeterminism has a significant computational impact that limits scalability. These difficulties are primarily caused by superfluous constraints on previous or initial states that accumulate in the sequent; see, e.g., Example 5.1.

A general possible solution for this issue is to iteratively consider more formulas of the sequent and attempt decision procedure calls with fair time interleaving until the respective branch closes. Then, only those additional formulas need to be considered that share variables with any of the other selected formulas. Further, timeouts can be used to discontinue lengthy decision procedure calls and continue along other choices of the nondeterminisms in Fig. 5.2. For complicated cases with a prohibitive complexity, this heuristic process worked well on our examples.

Inflation Order

As a general pattern for building iterative inflation optimisations, we propose an algorithm that selects additional formulas of a sequent successively while a counterexample can be found in bounded time (e.g., using the Mathematica function *FindInstance*) and terminates when a quantifier elimination call yields "true" within the current time bound (increasing timeouts as in IBC).

The *iterative inflation* algorithm schema in Fig. 5.10 proves validity of a disjunction of formulas of first-order real arithmetic that is given as a set S of disjuncts. Validity of a sequent $\Phi \vdash \Psi$ can be proven by $IIO(\{\neg\phi \,:\, \phi \in \Phi\} \cup \{\psi \,:\, \psi \in \Psi\})$. Following the general scheme in Fig. 5.9, the IIO algorithm explores subset candidates of S in parallel by fair time interleaving. Starting from the empty subset of formulas (line 5), the algorithm will try to prove (line 8) or disprove (line 11) each candidate subset ϕ of S. The first successful QE call yielding true within the current timeout shows validity of S (line 9), because a disjunction is valid if any subset is. When a counterexample is found (line 11) or quantifier elimination produced another result than "true" within the time bounds (line 10), candidate ϕ is dropped and any remaining formulas of S are added to ϕ to form new candidates. If no candidate could be proven (line 8) or disproven (lines 10–11) within the current time bounds, the timeout increases (line 14). The algorithm returns that S is not valid if all can-

```
1  /* prove validity of the disjunctive set S */
2  /* of formulas of first-order real arithmetic */
3  function IIO(S):
4    timeout := 1
5    𝒞 := {false}
6    while 𝒞 ≠ ∅ do
7      for ϕ ∈ 𝒞 do
8        if (QE(ϕ) = true) within timeout then
9          return valid
10       else if (QE(ϕ) ≠ true) within timeout
11                 or FindInstance(¬ϕ) within timeout then
12         𝒞 := (𝒞 ∪ {ϕ ∨ A : A ∈ S, A ∉ ϕ}) \ {ϕ}
13     end for
14     timeout := 2*timeout
15   end while
16   return not valid
```

Fig. 5.10 Iterative inflation order (IIO) algorithm schema

didates have counterexamples (or QE yields results other than "true") and no new formulas of S can be added to produce new candidates (line 12).

The IIO algorithm is a schematic algorithm. It allows for several refinements, optimisations, and caching improvements. Most importantly, line 12 can be refined to limit the number of candidates explored in parallel by adding only candidates with computationally promising properties. In general, a *reinclusion ordering* can be used to determine in what order new candidates are added and explored.

Canonical Reinclusion Order

One simple canonical reinclusion ordering is to order formulas by precomputing the overall resulting theoretical complexity based on various complexity results for real quantifier elimination [94, 141, 142, 263, 299, 60, 258, 259, 260], which depend on the number of variables and either the number of quantifiers (practical algorithms) or the number of quantifier alternations (theoretical complexity results). While this has a certain theoretically precise appeal, the disadvantage is that the theoretical complexity measures are rather coarse-grained and do not take into account the structure of the formula but only its asymptotic worst-case elements.

Structural Reinclusion Order

Among all possible orders for reincluding formulas, total orders share the advantage that they prevent the need for parallel exploration of multiple different possibilities of adding mutually incomparable formulas. Yet this can also turn into a disadvantage once the order starts adding computationally problematic high-degree constraints.

As a compromise of the canonical reinclusion order with structural information, we propose ordering formulas for reinclusion according to the lexicographical order of, respectively, relative variable recency, total polynomial degree, number of new variables, and maximum term depth. Favouring more recent variables follows the rationale that formulas that mention variables that have been introduced only recently into the proof are more likely to carry relevant information about the current state than those that only refer to variables from the initial proof obligation, for which more recent state variables have already been introduced during the proof. For instance, in Fig. 5.4, the train position z from the initial state may be less relevant than the current train position z_2 for the induction step. The braking force, b, however, may be as relevant as the recent z_2, because there is no updated copy of the symbolic constant b. The number of new variables that are added to the current candidate $\phi \in \mathscr{C}$ from Fig. 5.10 also has an impact on the complexity. Favouring small polynomial degrees is self-explanatory and follows the observation that the polynomial degree has a substantial impact on overall performance, as we will also see in Chap. 8. The term depth is used as an indicator for the complexity of the formula. To obtain a linear—yet arbitrary—order, this order can be extended easily by breaking ties by lexicographical comparison.

Combining this structural order with more sophisticated combinations of polynomial degree and the number of newly added variables according to the canonical orders may also be a viable alternative to improve on pure lexicographic orders.

5.6 Experimental Results

Tables 5.1–5.4 summarise verification results for our proof strategies for various case studies. Experimental results are from a 2.6 GHz AMD Opteron with 4 GB memory. The timeout for computations was 18000s = 5h and is indicated as ∞. Memory consumption of quantifier elimination is shown, excluding the front end. The dimension of the state space (column Dim) and the number of required proof steps (Steps) are indicated. To isolate effects resulting from our automatic invariant discovery that we describe in Chap. 6, we conduct experiments both with (annotated) user interactions for (differential) invariants and in automatic mode (the number of non-automatic interactions or annotations is indicated in column Int). The respective case studies are based on the examples shown in Part I, the larger case studies in Part III, and some standard examples from Part IV. For the tangential roundabout manoeuvre from Sect. 3.4, the number of participating aircraft is as indicated.

Observe that the performance of the extreme strategies of eager and lazy quantifier elimination depends on the example. For ETCS and larger aircraft roundabout manoeuvres (Tables 5.1 and 5.3), the lazy strategy performs faster, while the eager strategy is faster for the bouncing ball and water tank examples (Tables 5.2 and 5.4), where the lazy approach splits heavily (the number of required proof steps is much higher). As an intermediate strategy, IBC generally shows intermediate perform-

Table 5.1 Experimental results for proof strategies (with standalone QE) I

Case study	Int	Strategy	Time(s)	Memory(MB)	Steps	Dim
ETCS kernel	1	eager	11.1	15.1	44	9
	1	IBC	15.6	14.3	54	
	1	lazy	3.8	14.2	88	
	0	IBC	40.2	25.6	53	
	0	lazy	12.9	24.3	85	
ETCS binary safety	1	eager	9.6	11.4	144	14
	1	IBC	10.7	12.1	148	
	1	lazy	13.4	16.0	413	
	0	IBC	57.3	25.5	148	
	0	lazy	46.7	32.6	293	
ETCS safety	1	eager	∞	∞	∞	14
	1	IBC	26.8	17.6	168	
	1	lazy	25.8	29.1	423	
	0	IBC	2089.2	206.1	171	
	0	lazy	2046.5	203.3	304	
ETCS reactivity	0	eager	∞	∞	∞	14
	0	IBC	1084.1	6.1	34	
	0	lazy	∞	∞	∞	
tangential roundabout	3	eager	1.7	6.8	94	13
	3	IBC	1.5	6.8	93	
	3	lazy	1.6	6.7	139	
	0	IBC	10.3	6.9	114	
	0	lazy	10.2	6.8	197	
tangential roundabout 3	3	eager	∞	∞	∞	18
	3	IBC	75.0	24.7	165	
	3	lazy	52.3	14.9	244	
	0	IBC	1065.5	27.6	186	
	0	lazy	620.2	15.0	342	
tangential roundabout 4	3	eager	4208.2	E	E	23
	3	IBC	513.4	184.4	229	
	3	lazy	57.6	31.4	355	
	0	IBC	10998.1	184.3	256	
	0	lazy	901.7	31.4	520	
tangential roundabout 5	3	eager	7714.1	E	E	28
	3	IBC	2457.9	479.3	317	
	3	lazy	108.9	43.6	502	
	0	IBC	∞	∞	∞	
	0	lazy	3417.5	48.5	735	

Table 5.2 Experimental results for proof strategies (with standalone QE) II

Case study	Int	Strategy	Time(s)	Memory(MB)	Steps	Dim
bouncing ball	1	eager	7.7	13.6	85	7
	1	IBC	8.0	13.6	85	
	1	lazy	∞	∞	∞	
	0	IBC	66.4	19.0	43	
	0	lazy	∞	∞	∞	
water tank	1	eager	3.8	9.1	378	3
	1	IBC	3.9	9.1	375	
	1	lazy	9.2	9.5	1604	

Table 5.3 Experimental results for proof strategies (no standalone QE) I

Case study	Int	Strategy	Time(s)	Memory(MB)	Steps	Dim
ETCS kernel	1	eager	11.3	15.0	42	9
	1	IBC	6.9	14.3	47	
	1	lazy	2.8	14.2	61	
	0	IBC	47.8	25.4	46	
	0	lazy	10.5	24.2	58	
ETCS binary safety	1	eager	9.8	11.3	142	14
	1	IBC	10.4	12.2	148	
	1	lazy	7.2	15.8	235	
	0	IBC	41.0	17.4	144	
	0	lazy	18.6	12.4	204	
ETCS safety	1	eager	∞	∞	∞	14
	1	IBC	26.5	23.1	168	
	1	lazy	18.9	24.2	247	
	0	IBC	2051.3	204.1	163	
	0	lazy	2031.6	203.1	216	
ETCS reactivity	0	eager	∞	∞	∞	14
	0	IBC	104.1	61.7	49	
	0	lazy	∞	∞	∞	
tangential roundabout	3	eager	1.7	6.8	94	13
	3	IBC	1.7	6.8	93	
	3	lazy	1.9	6.7	139	
	0	IBC	10.9	6.9	114	
	0	lazy	9.9	6.8	197	
tangential roundabout 3	3	eager	∞	∞	∞	18
	3	IBC	75.3	24.7	165	
	3	lazy	52.3	15.0	244	
	0	IBC	965.8	32.8	186	
	0	lazy	636.2	15.1	342	
tangential roundabout 4	3	eager	4340.6	E	E	23
	3	IBC	513	185.7	229	
	3	lazy	57	31.4	355	
	0	IBC	3323.5	44.0	247	
	0	lazy	884.9	31.4	520	
tangential roundabout 5	3	eager	7702.3	E	E	28
	3	IBC	2439.7	533.8	317	
	3	lazy	108.9	43.6	503	
	0	IBC	16303.7	827	320	
	0	lazy	3552.6	46.9	735	

Table 5.4 Experimental results for proof strategies (no standalone QE) II

Case study	Int	Strategy	Time(s)	Memory(MB)	Steps	Dim
bouncing ball	1	eager	7.8	13.6	85	7
	1	IBC	7.8	13.6	85	
	1	lazy	20.1	13.7	166	
	0	IBC	65.0	16.9	43	
	0	lazy	78.2	26.0	92	
water tank	1	eager	3.6	9.1	387	3
	1	IBC	3.8	9.1	375	
	1	lazy	6.6	9.3	914	

ance. In our case studies, the eager strategy fails to come up with a result within the timeout fairly often. Furthermore, IBC is the only strategy that is able to prove all case studies. For the reactivity property of the ETCS case study, both lazy and eager proof strategies fail at the same time and only IBC succeeds. A result of "E" indicates that no data is available, because of out of memory errors or other limits in the KeYmaera implementation at the time of performing these experiments. For consistent performance measurements, we have used the same version of KeYmaera for all experiments.

The experiments in Tables 5.1 and 5.2 use standalone quantifier eliminations, while those in Tables 5.3 and 5.4 do not. Standalone quantifier elimination reduces quantifiers of first-order formulas immediately, instead of waiting for the whole sequent to be passed to quantifier elimination at once. By comparing Tables 5.1 and 5.2 with Tables 5.3 and 5.4, we see that the performance is generally better when we do not allow these partial reductions of standalone quantified subformulas of a sequent. The reason for this is that the result of standalone QE-reductions on subformulas (Tables 5.1 and 5.2) can produce several disjunctions which split into further sub-branches, so that the redundancy effects described in Sect. 5.3 and those illustrated in Fig. 5.5 have more impact.

Finally note that, without our range of proof strategies that result from a careful analysis of the nondeterminisms in the tableau procedures for differential dynamic logics, the practical prover performance is significantly worse than all those presented in Tables 5.1–5.4. The small ETCS kernel from Chap. 2 is already provable directly when implementing the $d\mathcal{L}$ calculus naïvely. Without our range of proof-strategic improvements, however, the full ETCS system that we present in Chap. 7 requires as much as 56 user interactions to be provable [256], while our proof strategies and the algorithms that we present in Part II can, in fact, prove ETCS and other complicated systems completely automatically with zero user interaction.

5.7 Summary

From the experience of using our $d\mathcal{L}$ calculus for verifying parametric hybrid systems in traffic applications, we have investigated combinations of deductive, computer algebraic, and real algebraic reasoning from a practical perspective. We have analysed the principles of this prover combination, identified the nondeterminisms that remain in the canonical $d\mathcal{L}$ tableau procedure, and analysed their impact.

We have proposed proof strategies that navigate through these nondeterminisms, including iterative background closure and iterative inflation order strategies. Similarly to the huge importance of subsumption in resolution, background-style tableau proving requires quick techniques to rule out branches closing for simple arithmetic reasons. In our experiments with verifying cooperating traffic agents, the combinations of our proof strategies reduce the number of interactions and the overall running time significantly. In extreme cases, differences can be such that one proof strategy produces no result in a day compared to another that proves immediately, in less than a second.

Chapter 6
Computing Differential Invariants as Fixed Points

Contents

Synopsis We introduce a fixed-point algorithm for verifying safety properties of hybrid systems with differential equations whose right-hand sides are polynomials in the state variables. In order to verify nontrivial systems without solving their differential equations and without numerical errors, we use differential induction as a continuous generalisation of induction, for which our algorithm computes the required *differential invariants*. As a means for combining local differential invariants into global system invariants in a sound way, our fixed-point algorithm works with differential dynamic logic as a compositional verification logic for hybrid systems. To improve the verification power, we further introduce a *saturation procedure* that refines the system dynamics successively by differential cuts with differential invariants until the property becomes provable. By complementing our symbolic verification algorithm with a robust version of numerical falsification, we obtain a fast and sound verification procedure. We verify roundabout manoeuvres in air traffic control and collision avoidance in train control.

A. Platzer, Logical Analysis of Hybrid Systems,
DOI 10.1007/978-3-642-14509-4_6, © Springer-Verlag Berlin Heidelberg 2010

6.1 Introduction

Reachability questions for systems with complex continuous dynamics are among the most challenging problems in verifying embedded systems, in particular for hybrid systems, which are models for these systems with interacting discrete and continuous transitions along differential equations. For simple systems whose differential equations have solutions that are polynomials in the state variables, quantifier elimination [81] can be used for verification as detailed in Chap. 2. Unfortunately, this symbolic approach does not scale to systems with complicated differential equations whose solutions do not support quantifier elimination (e.g., when they are transcendental functions) or cannot be given in closed form.

Numerical or approximation approaches [21, 238, 102] can deal with more general dynamics. However, numerical or approximation errors need to be handled carefully as they easily cause unsoundness, even for principal reasons [238]. More specifically, we have shown previously that even single image computations of fairly restricted classes of hybrid systems are undecidable by numerical computation [238]. Thus, numerical approaches are helpful for falsification but not (ultimately) for verification.

In this chapter, we present a verification algorithm that combines the soundness of symbolic approaches [125, 15, 231, 235] with support for nontrivial dynamics that is classically more dominant in numerical approaches [21, 238, 102]. During continuous transitions, the hybrid system follows a solution of its differential equation. But for nontrivial dynamics, these solutions are much more complicated than the original equations. Solutions quickly become transcendental even if the differential equations are linear. To overcome this, we handle continuous transitions based on their vector fields, which are described by their differential equations. We use *differential induction* (Sect. 3.5.6), a continuous generalisation of induction that works with the differential equations themselves instead of their solutions. For the induction step, we use a condition that can be checked easily based on *differential invariants* (from Sect. 3.5.6), i.e., properties whose derivative holds true in the direction of the vector field of the differential equation. The derivative is a directional derivative in the direction of (the vector field generated by) the differential equation, with derivatives generalised from functions to formulas according to the findings in Chap. 3. For this to work in practise, the most crucial steps are to find sufficiently strong local differential invariants for differential equations and compatible global invariants for the hybrid switching dynamics.

To this end, we introduce a *sound* verification algorithm for hybrid systems that computes the differential invariants and system invariants in a fixed-point loop. Based on the invariants as fixed points paradigm [73], we present the "differential invariants as fixed points" paradigm using $d\mathcal{L}$ as a verification logic that is generalised to hybrid systems accordingly. For combining multiple local differential invariants into a global invariant in a sound way, we exploit the closure properties of the underlying verification logic $d\mathcal{L}$ by forming appropriate logical combinations of multiple safety statements. In addition, we introduce a *differential saturation process* that refines the hybrid dynamics successively by differential cuts with auxiliary differential invariants until the safety statement becomes an invariant of the refined system. Finally, each fixed-point iteration of our algorithm can be combined with numerical falsification to accelerate the overall symbolic verification in a sound way. We validate our algorithm by verifying *aircraft roundabout manoeuvres* [293, 238] and train control applications automatically, continuing the examples in Part I to the full case studies in Part III.

Contributions

The major contribution in this chapter is that we introduce the "differential invariants as fixed points" paradigm and a fixed-point algorithm for computing differential invariants coupled with a differential saturation process. We turn the theoretical and conceptual proof approach from Part I into a practical verification algorithm. We show that this algorithm can verify realistic applications that were out of scope for related invariant approaches [274, 269, 252] or standard model checking approaches [156, 125, 228] based on state-space exploration, for both theoretical and scalability reasons (see discussions in Sects. 2.4 and 3.1.1).

6.1.1 Related Work

Based on successes in discrete programs [268, 267, 266], other authors [251, 274, 269, 252] argued that invariant techniques scale to more general dynamics than explicit reach-set computations or techniques that require solutions for differential equations [125, 228]. However, their techniques are focused on purely equational systems and cannot handle hybrid systems with inequalities in initial sets or switching surfaces [274, 269]. Due to tolerances, inequalities occur in most real applications. *Barrier certificates* [251, 252] only work for single inequalities, but invariants of roundabout manoeuvres require mixed equations and inequalities [237]. Prajna et al. [252] search for barrier certificates of a fixed degree by global optimisation over the set of all possible certificate combinations for the whole system at once, which is infeasible: Even with degree bound 2, this would require solving a 5,848-dimensional optimisation problem for ETCS (Chap. 7) and a 10,005-dimensional problem for roundabouts with 5 aircraft (Chap. 8).

Finally, important distinctions of our work compared to others [251, 274, 269, 252] are: (*i*) we allow arbitrary formulas as differential invariants, which provably improves verification power; (*ii*) we increase the verification power further by nesting differential invariants using differential saturation to refine the system dynamics; and (*iii*) our compositional verification logic allows local generation of differential invariants and natural local existential quantification of formal parameters for local verification subtasks.

Tomlin et al. [293] derive saddle solutions for aircraft manoeuvre games using Hamilton-Jacobi-Isaacs partial differential equations and propose roundabout manoeuvres. Their exponential state space discretisations for PDEs, however, do not scale to larger dimensions (they consider dimension 3) and can be numerically unsound [238]. Differential invariants, in contrast, work for 28-dimensional systems and are sound.

Straight-line aircraft manoeuvres have been analysed by geometrical meta-level reasoning [104, 171]. We directly verify the actual hybrid flight dynamics, including curved roundabout manoeuvres instead of straight-line manoeuvres with non-flyable instant turns. A few approaches [203, 92] have been undertaken to model check if there are *orthogonal* collisions in discretisations of roundabout manoeuvres. However, the counterexamples (see Fig. 3.6c on p. 151) found by our model checker show that *non-orthogonal* collisions can still happen in these classical manoeuvres, which is a problem that we have fixed in Chap. 3.

6.1.2 Structure of This Chapter

In Sect. 6.2, we introduce the "differential invariants as fixed points" paradigm and present an automatic verification algorithm for hybrid systems that is based on the DAL calculus and computes the differential invariants and invariants as fixed points that are required for verification. In Sect. 6.3, we show how soundness of the DAL calculus inherits in a simple and elegant way to soundness of our logic-based verification procedure. We present optimisations of the algorithm in Sect. 6.4. In Sect. 6.5, we present experimental results for the running example of roundabout manoeuvres in air traffic control and conclude in Sect. 6.6.

6.2 Inductive Verification by Combining Local Fixed Points

For verifying safety properties of hybrid systems without having to solve their differential equations, we use differential induction as a continuous form of induction based on our techniques from Chap. 3. In the induction step, we use a condition on directional derivatives in the direction of the vector field generated by the differential equation. The resulting properties are invariants of the differential equation (called *differential invariants* in Chap. 3). The crucial step for verifying discrete systems by induction is to find sufficiently strong invariants (e.g., for loops α^*). Similarly, the crucial step for verifying dynamical systems (which correspond to a single continuous mode of a hybrid system) by induction is to find sufficiently strong invariant properties of the differential equation. Consequently, for verifying hybrid systems inductively, local invariants need to be found for each differential equation and a global system invariant needs to be found that is compatible with all local invariants.

To compute the required invariants and differential invariants, we combine the invariants as in the fixed-points approach from [73] with the lifting of verification logics to hybrid systems from Chaps. 2 and 3. We introduce a verification algorithm that computes invariants of a system as fixed points of safety constraints on subsystems. In order to obtain a local algorithm that works by decomposing global properties of hybrid programs into local properties of subsystems, we exploit the fact that hybrid programs can be decomposed into subsystems and that $\mathsf{d}\mathcal{L}$ can combine safety statements about multiple subsystems simultaneously. Note that the algorithms developed in this chapter apply for the logics $\mathsf{d}\mathcal{L}$, DAL, and dTL from Part I, respectively, even if we mostly refer to $\mathsf{d}\mathcal{L}$ in this chapter.

A simple *safety statement* corresponds to a $\mathsf{d}\mathcal{L}$ formula $\psi \to [\alpha]\phi$ with a hybrid program α, a safety property ϕ about its reachable states, and an arithmetic formula ψ that characterises the set of initial states symbolically. *Validity* of formula $\psi \to [\alpha]\phi$ (i.e., truth in all states) corresponds to ϕ being true in all states reachable by hybrid program α from initial states that satisfy ψ. We describe a verification algorithm that defines the function *prove*$(\psi \to [\alpha]\phi)$ for verifying this safety statement recursively by refining the $\mathsf{d}\mathcal{L}$ and DAL calculi to an automatic verification algorithm. The discrete base cases are discussed in Sect. 6.2.1. In Sect. 6.2.2, we contrast discrete and differential induction. Section 6.2.4 shows the fixed-point algorithm for computing differential invariants for differential equations, and Sect. 6.2.6 for computing loop invariants. In Sect. 6.2.7, we explain the interplay of local and global fixed-point loops of our verification algorithm.

6.2.1 Verification by Symbolic Decomposition

The cases of the verification procedure *prove* where $\mathsf{d}\mathcal{L}$ can verify a property of a hybrid program directly by decomposing it into a property of its parts are shown in Fig. 6.1. They correspond to the cases where the dynamic rules of Fig. 3.9 can

```
function prove(ψ → [x:=θ]φ):
  return prove(ψ∧x̂ = θ → φ_x^x̂)
  where x̂ is a new auxiliary variable
function prove(ψ → [?χ]φ):
  return prove(ψ∧χ → φ)
function prove(ψ → [α∪β]φ):
  return prove(ψ → [α]φ) and prove(ψ → [β]φ) /* i.e. ψ → [α]φ ∧ [β]φ */
function prove(ψ → [α;β]φ):
  return prove(ψ → [α][β]φ)
function prove(ψ → [x:=*]φ):
  return prove(ψ → ∀xφ)
function prove(ψ → φ) where isFirstOrder(φ):
  return QE(ψ → φ)
function prove(ψ → Qxφ) where Qx is ∀x or ∃x:
  return QE(ψ → Qx prove(φ))
```

Fig. 6.1 $\mathsf{d}\mathcal{L}$-based verification by symbolic decomposition be applied without any nondeterminism, and just depending on the top-level program operator. For a concise presentation, the case in line 1 introduces an auxiliary variable $\hat{x}$ to handle discrete assignments by substituting $\hat{x}$ for x in $\phi_x^{\hat{x}}$: E.g., $x \geq 2 \to [x := x-1]x \geq 0$ is shown by proving $x \geq 2 \wedge \hat{x} = x-1 \to \hat{x} \geq 0$. Our implementation in our theorem prover KeYmaera [242] uses proof rule $\langle := \rangle$ and discrete jump sets to avoid auxiliary variables according to Chap. 2. State checks $?\chi$ are shown by assuming the test succeeds, i.e., χ holds true (line 4), nondeterministic choices split into their alternatives (line 6), sequential compositions are proven using nested modalities (line 8), and random assignments are shown by universal quantification (line 10), which is an optimisation of the handling in Chap. 3. Random assignments $[x := *]\phi$ that nondeterministically assign an arbitrary real value to x are definable by $[x' = 1 \cup x' - 1]\phi$ and are equivalent to quantifiers $\forall x\, \phi$.

The base case in line 12, where ϕ is a formula of first-order real arithmetic, can be proven by quantifier elimination (QE) in real-closed fields [81]. Despite its complexity, this can remain feasible, because the formulas resulting from our algorithm do not depend on the solutions of differential equations but only on their right-hand sides. Further, the decompositions in our verification algorithm generally lead to local properties with lower complexities. We present here only a simplified treatment of quantifiers in $d\mathcal{L}$ formulas like $\forall x\, \phi$ that contain modal subformulas: The *prove* function is applied recursively to the unquantified kernel ϕ first and the resulting formula is handled by quantifier elimination (line 14), which resembles the quantifier handling by side deduction in Chap. 3. More generally, quantifier elimination can be lifted to quantifiers in $d\mathcal{L}$ formulas like $\forall x\, \phi$ using the respective Skolemisation and free-variable rules from Chap. 2 . After introducing a fresh Skolem term s for variable x in ϕ, the analysis continues with the unquantified kernel ϕ_x^s. Later on when the formulas resulting from recursive application of *prove* contain no more modalities, the quantifier for s can be reintroduced (Deskolemisation) and eliminated equivalently using quantifier elimination in real-closed fields. Handling $\exists x\, \phi$ for modal formulas ϕ is similar. The crucial part is that Skolem term dependencies can be exploited to prevent unsound quantifier rearrangements; see Chap. 2.

Overall, the algorithm in Fig. 6.1 recursively reduces the safety of hybrid programs to separate properties of continuous evolutions or of repetitions, which we verify in the next sections.

6.2.2 Discrete and Differential Induction, Differential Invariants

In this section, we present algorithms for verifying loops by discrete induction (which corresponds to rule *ind* from the $d\mathcal{L}$/DAL calculi) and continuous evolutions by differential induction, which is a continuous form of induction and corresponds to rule *DI* from the DAL calculus in Fig. 3.9. In either case, we prove that an invariant F holds initially (in the states characterised symbolically by ψ; thus $\psi \to F$ is valid) and finally entails the postcondition ϕ (i.e., $F \to \phi$). The cases differ in their induction step.

Definition 6.1 (Discrete induction). The formula F is a *(discrete) invariant* of $\psi \rightarrow [\alpha^*]\phi$ iff the following formulas are valid:

1. $\psi \rightarrow F$ (induction start), and
2. $F \rightarrow [\alpha]F$ (induction step).

An invariant is *sufficiently strong* if $F \rightarrow \phi$ is valid.

Definition 6.2 (Continuous invariants). Let $\mathscr{D}$ be a differential equation and χ is a first-order formula. Formula F is a *continuous invariant* of $\psi \rightarrow [\mathscr{D} \wedge \chi]\phi$ iff the following formulas are valid:

1. $\psi \wedge \chi \rightarrow F$ (induction start), and
2. $F \rightarrow [\mathscr{D} \wedge \chi]F$ (induction step).

Again, a continuous invariant is *sufficiently strong* if $F \rightarrow \phi$ is valid.

To prove that F is a continuous invariant, it is sufficient by Theorem 3.1 and proof rule *DI* from Fig. 3.9 to check a condition on the directional derivatives of all terms of the formula, which expresses that no atomic subformula of F changes its truth-value along the dynamics of the differential equation. This condition is easier to check than a reachability property ($F \rightarrow [\mathscr{D} \wedge \chi]F$) of a differential equation.

Applications like aircraft manoeuvres need invariants with mixed equations and inequalities. Thus, we use the generalisation of directional derivatives from functions to logical formulas according to the total derivation $D(\cdot)$ from Definition 3.13 on p. 155.

Definition 6.3 (Differential induction). Let the differential equation system $\mathscr{D}$ be $x_1' = \theta_1, \ldots, x_n' = \theta_n$ and χ be a first-order formula. A quantifier-free first-order formula F is a *differential invariant* of $\psi \rightarrow [\mathscr{D} \wedge \chi]\phi$ iff the following formulas are valid:

1. $\psi \wedge \chi \rightarrow F$ (induction start), and
2. $\chi \rightarrow F'_{\mathscr{D}}$ (induction step),

where $F'_{\mathscr{D}}$ is the result of substituting the differential equation $\mathscr{D}$ into the total derivation $D(F)$ of F, which corresponds to the conjunction of all directional derivatives of atomic formulas in F in the direction of the vector field of $\mathscr{D}$ (the partial derivative of b by x_i is $\frac{\partial b}{\partial x_i}$):

$$F'_{\mathscr{D}} \equiv \bigwedge_{(b \sim c) \in F} \left(\left(\sum_{i=1}^{n} \frac{\partial b}{\partial x_i} \theta_i \right) \sim \left(\sum_{i=1}^{n} \frac{\partial c}{\partial x_i} \theta_i \right) \right) \qquad \text{for} \sim \in \{=, \geq, >, \leq, <\}.$$

These partial derivatives of terms are well-defined in the Euclidean space spanned by the variables and can be computed symbolically (Sect. 3.5.2).

The central property of differential invariants for verification purposes is that they can be used to replace infeasible or even impossible reachability analysis with feasible symbolic computation.

Proposition 6.1 (Principle of differential induction). *All differential invariants are continuous invariants.*

Proof. This proposition is a corollary to Theorem 3.1, which shows that the differential induction rule *DI* from the DAL calculus in Fig. 3.9 is sound. □

Example 6.1. Consider the dynamics $x' = x^2, y' = -3$. Differential invariants can show that $3x \geq 4y$ is an invariant for this dynamics without using any state-based reachability verification. We just compute symbolically:

$$(3x \geq 4y)'_{x'=x^2,y'=-3} \equiv \frac{\partial 3x}{\partial x}x^2 + \frac{\partial 3x}{\partial y}(-3) \geq \frac{\partial 4y}{\partial x}x^2 + \frac{\partial 4y}{\partial y}(-3) \equiv 3x^2 \geq -12.$$

Since the latter formula is easily found to be valid, $3x \geq 4y$ is proven to be a differential invariant and thus remains true whenever it holds true initially (case 1 of Definition 6.3). □

We have introduced this proof principle of differential induction in Chap. 3. The above notions form the basis for our development of a systematic algorithm computing the required invariants and differential invariants based on our proof rules from Chap. 3. In Sects. 6.2.4–6.2.6, we present algorithms for finding differential invariants for differential equations, and for finding global invariants for repetitions.

6.2.3 Flight Dynamics in Air Traffic Control

Aircraft collision avoidance manoeuvres resolve conflicting flight paths, e.g., by roundabout manoeuvres [293]; see Fig. 3.6 on p. 151. Their nontrivial dynamics makes safe separation of aircraft difficult to analyse, in particular as good timing and coordination of movement in space are crucial [293, 203, 104, 92, 238, 171]. Correct functioning of these manoeuvres under all circumstances is difficult to guarantee without formal verification, especially in light of the counterexample (Fig. 3.6c) that our model checker discovered [238] for the classical roundabout manoeuvre. Thus, verification is important for aircraft manoeuvres, and, at the same time, aircraft dynamics is a challenge for hybrid systems verification.

For simplicity, we consider planar movement of aircraft (if the aircraft are collision-free in a planar projection, they are collision-free in space). The parameters of two aircraft at (planar) position $x = (x_1, x_2) \in \mathbb{R}^2$ and $y = (y_1, y_2)$ with angular orientation ϑ and ς are illustrated in Fig. 6.2 (with $\vartheta = 0$). Following the work of Tomlin et al. [293], aircraft dynamics is determined by their linear speeds $v, u \in \mathbb{R}$ and angular speeds $\omega, \rho \in \mathbb{R}$, respectively:

$$x_1' = v\cos\vartheta \quad x_2' = v\sin\vartheta \quad \vartheta' = \omega \quad y_1' = u\cos\varsigma \quad y_2' = u\sin\varsigma \quad \varsigma' = \rho \tag{6.1}$$

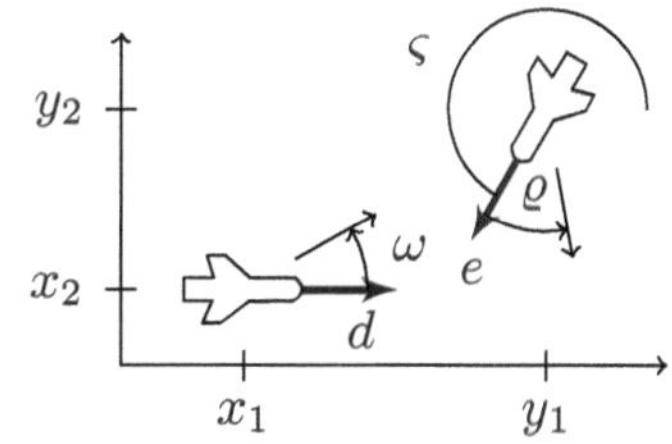

Fig. 6.2 Aircraft dynamics

That is, position x moves with speed v into the direction with angular orientation ϑ, which rotates with angular velocity ω (likewise for a second aircraft at position y with speed u, angular orientation ς, and angular velocity ρ).

In safe flight configurations, aircraft are separated by at least distance p:

$$(x_1 - y_1)^2 + (x_2 - y_2)^2 \geq p^2. \tag{6.2}$$

To handle the transcendental functions in (6.1) without facing the undecidability problems in the arithmetic of trigonometric functions, we axiomatise sin and cos by differential equations and reparametrise the system using a linear velocity vector $d = (d_1, d_2) := (v\cos\vartheta, v\sin\vartheta) \in \mathbb{R}^2$ according to Sect. 3.4.2:

$$\begin{bmatrix} x_1' = d_1 & x_2' = d_2 & d_1' = -\omega d_2 & d_2' = \omega d_1 & t' = 1 \\ y_1' = e_1 & y_2' = e_2 & e_1' = -\rho e_2 & e_2' = \rho e_1 & s' = 1 \end{bmatrix} \tag{$\mathscr{F}$}$$

Using symbolic derivations, it is easy to see that equations ($\mathscr{F}$) and (6.1) are equivalent up to reparametrisation. Here, we add clock variables t, s (with $t' = 1$ and $s' = 1$) that we need for synchronising collision avoidance manoeuvres in Chap. 8.

By a simple computation, $d_1^2 + d_2^2 \geq a^2$ is a differential invariant of ($\mathscr{F}$), thereby showing that the linear speed of aircraft does not drop below some stalling speed a during manoeuvres:

$$\begin{aligned} &(d_1^2 + d_2^2 \geq a^2)'_{\mathscr{F}} \equiv (d_1^2 + d_2^2 \geq a^2)'_{(d_1' = -\omega d_2, d_2' = \omega d_1)} \\ &\equiv \frac{\partial(d_1^2 + d_2^2)}{\partial d_1}(-\omega d_2) + \frac{\partial(d_1^2 + d_2^2)}{\partial d_2}\omega d_1 \geq \frac{\partial a^2}{\partial d_1}(-\omega d_2) + \frac{\partial a^2}{\partial d_2}\omega d_1 \\ &\equiv 2d_1(-\omega d_2) + 2d_2\omega d_1 \geq 0. \end{aligned}$$

As a stronger statement, the equation $d_1^2 + d_2^2 = a^2$ for constant linear speed is a provable differential invariant. Similarly, conjunction $d_1^2 + d_2^2 = a^2 \land e_1^2 + e_2^2 \geq a^2$ can be shown to be a differential invariant of $\mathscr{F}$. The theory of differential invariants shows that this is a general phenomenon: Conjunctions of differential invariants are differential invariants but not conversely so (Sect. 3.5.6). There are examples where only the propositional combination itself is a differential invariant but none of its parts is (Sect. 3.10). Thus, allowing conjunctions in differential invariants is crucial and the verification power of differential invariants is higher than that of other approaches [251, 274, 269, 252], which do not support logical operators.

6.2.4 Local Fixed-Point Computation for Differential Invariants

As with verification with invariants, the central practical question for verification with differential invariants is how to find them. Figure 6.3 depicts our fixed-point algorithm for constructing differential invariants for each continuous evolution $\mathscr{D} \land \chi$ with a differential equation system $\mathscr{D}$ and evolution domain χ. The algorithm in

```
1  function prove(ψ→[𝒟∧χ]φ):
2    if prove(∀^𝒟(χ→φ)) then return true /* property proven */
3    for each F∈Candidates(ψ→[𝒟∧χ]φ, χ) do
4      if prove(ψ∧χ→F) and prove(∀^𝒟(χ→F'_𝒟)) then
5        χ := χ∧F          /* refine by differential invariant */
6        goto 2;           /* repeat fixed-point loop */
7    end for
8    return not provable using candidates
```

Fig. 6.3 Fixed-point algorithm for differential invariants (*Differential Saturation*)

Fig. 6.3 (called *Differential Saturation*) successively refines the evolution domain restriction χ by differential invariants until saturation, i.e., until χ accumulates enough information to become a strong invariant that implies postcondition ϕ (line 2). If evolution domain χ already entails ϕ, then $\psi \to [\mathscr{D} \land \chi]\phi$ is proven (line 2). Otherwise, the algorithm considers candidates F for augmenting χ (line 3). If F is a differential invariant (line 4), then χ can soundly be refined to $\chi \land F$ (line 5) without affecting the states reachable by $\mathscr{D} \land \chi$ (Proposition 6.2 below). Then, the fixed-point loop repeats (line 6). At each iteration of this fixed-point loop, the previous invariant χ can be used to prove the next level of refinement $\chi \land F$ (line 4). Hence, each differential invariant provides more information to simplify subsequent iterations. The refinement of the dynamics at line 5 is correct by the following proposition, using the fact that the conditions in line 4 imply that F is a differential invariant and, thus, a continuous invariant by Proposition 6.1.

Proposition 6.2 (Differential saturation). *If F is a continuous invariant of the formula $\psi \to [\mathscr{D} \land \chi]\phi$, then $\psi \to [\mathscr{D} \land \chi]\phi$ and $\psi \to [\mathscr{D} \land \chi \land F]\phi$ are equivalent.*

Proof. The proof is a stronger version of the soundness of the differential cut rule *DS* by Theorem 3.1: Let F be a continuous invariant, which implies that $\psi \to [\mathscr{D} \land \chi]F$ is valid. Let ν be a state satisfying ψ (otherwise there is nothing to show). Then, $\nu \models [\mathscr{D} \land \chi]F$. Since this means that F is true all along all flows φ of $\mathscr{D} \land \chi$ that start in ν (Definition 3.10), the latter and $\mathscr{D} \land \chi \land F$ have the same dynamics and the same reachable states from ν, i.e., $(\nu, \omega) \in \rho(\mathscr{D} \land \chi)$ holds if and only if $(\nu, \omega) \in \rho(\mathscr{D} \land \chi \land F)$ (Definition 2.7). Thus, we can conclude that $\psi \to [\mathscr{D} \land \chi]\phi$ and $\psi \to [\mathscr{D} \land \chi \land F]\phi$ are equivalent, because their semantics uses the same transition relation. □

This progressive differential saturation turns out to be crucial in practise. For instance, the aircraft separation property (6.2) cannot be proven until ($\mathscr{F}$) has been refined by invariants for d and e, because these determine x' and y'. This makes sense intuitively: Unless we have discovered some invariant about the directions d and e in which the aircraft are flying, we cannot conclude good invariants about their positions x and y, because the evolution of the positions over time depends on the directions.

Function *Candidates* determines candidates for induction (line 3) depending on transitive differential dependencies, as we will explain in Sect. 6.2.5. When these are insufficient for proving $\psi \to [\mathscr{D} \wedge \chi]\phi$, the algorithm fails (line 8, with improvements in subsequent sections). Again, $\forall^\alpha \phi$ denotes the *universal closure* of ϕ. It is required in lines 2 and 4, because the respective formulas need to hold in *all* states (that satisfy χ). Clearly, this set of states is overapproximated conservatively by the universal closure with respect to all variables for which there are differential equations in $\mathscr{D}$. We will reduce the number of quantifiers in Sect. 6.4.

6.2.5 Dependency-Directed Induction Candidates

In this section, we construct likely candidates for differential induction (function *Candidates* in Fig. 6.3). Later, we use the same procedure for finding global loop invariants. We construct two kinds of candidates in an order induced by differential dependencies. By following the effect of hybrid systems symbolically along their decompositions, our verification algorithm enriches precondition ψ successively with more precise information about the symbolic pre-state as obtained by the symbolic decompositions and proof steps in Figs. 6.1 and 6.3. To exploit this, we first look for invariant symbolic state information that accumulated in the respective precondition ψ and postcondition ϕ during the iterative symbolic decomposition by selecting subformulas that are not yet contained in χ. In practise, this gives particularly good candidates for highly parametric hybrid systems. Even if candidates from ψ and ϕ do not yet work in the first iteration, repetitions of outer fixed-point loops (Sect. 6.2.6) may enrich ψ and ϕ so that they become useful in later iterations.

Secondly, we generate parametric invariants. Let $V = \{x_1, \dots, x_n\}$ be a set of relevant variables. We choose fresh names $a^{(l)}_{i_1,\dots,i_n}$ for *formal parameters* of the invariant candidates and build polynomials $p_1, \dots, p_k$ of degree d with variables V using formal parameters as symbolic coefficients:

$$p_l := \sum_{i_1+\cdots+i_n \le d} a^{(l)}_{i_1,\dots,i_n} x_1^{i_1} \dots x_n^{i_n} \quad \text{for } 1 \le l \le k.$$

We define the set of *parametric candidates* (operator $\vee$ is used similarly):

$$\mathit{ParaForm}(k,d,V) := \left\{ \bigwedge_{l=1}^{i} p_l \ge 0 \wedge \bigwedge_{l=i+1}^{k} p_l = 0 \;:\; 0 \le i \le k \right\}.$$

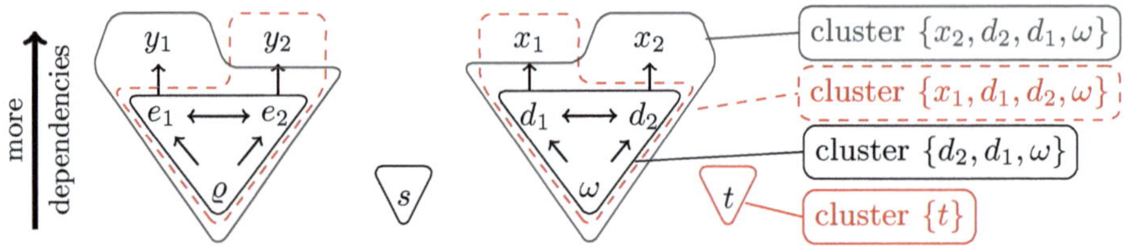

Fig. 6.4 Differential dependencies (arrows) and (triangular) variable clusters of ($\mathscr{F}$)

For instance, the parametric candidate $a_{0,0} + a_{1,0}d_1 + a_{0,1}x_2 = 0$ yields a differential invariant of ($\mathscr{F}$) for the choice $a_{0,0} = 0$, $a_{1,0} = 1$, $a_{0,1} = \omega$. By simple combinatorics, *ParaForm* contains $k+1$ candidates with $k\binom{n+d}{d}$ formal parameters $a^{(l)}_{i_1,\dots,i_n}$, which are existentially quantified. Existence of a common satisfying instantiation for these parameters can be expressed by prefixing the resulting d$\mathscr{L}$ formula that uses the parametric candidate with $\exists a^{(l)}_{i_1,\dots,i_n}$ for each of the formal parameters $a^{(l)}_{i_1,\dots,i_n}$. For this to be feasible, the number of parameters is crucial, which we minimise by respecting (differential) dependencies. We will illustrate the placement of existential quantifiers in Example 6.2 of Sect. 6.2.6 after we have presented the global fixed-point algorithm.

To accelerate the differential saturation process from Sect. 6.2.4, it is crucial to explore candidates in a promising order from simple to complex, because the algorithm in Fig. 6.3 uses successful differential invariants to refine the dynamics, thereby simplifying subsequent proofs: E.g., separation property (6.2) is only provable after the dynamics has been refined with invariants for d and e, because x' and y' depend on the direction d and e. In fact, the safety of roundabouts crucially depends on compatible directions of the aircraft, as the counterexample in Fig. 3.6c shows. We construct candidates in a natural order based on variable occurrence that is consistent with the *differential dependencies* of the differential equations. For a differential equation $\mathscr{D}$, variable x depends on variable y according to the differential equation system $\mathscr{D}$ if y occurs on the right-hand side for x' (or transitively so). The resulting set *depend*($\mathscr{D}$) of dependencies is the transitive closure of

$$\{(x,y) \;:\; (x' = \theta) \in \mathscr{D} \text{ and } y \text{ occurs in } \theta\}.$$

From the differential equation system ($\mathscr{F}$), we determine the differential dependencies indicated as arrows (pointing to the dependent variables x) in Fig. 6.4.

From these dependencies we determine an order on candidates. The idea is that, as the value of x_1 in ($\mathscr{F}$) depends on that of d_1, it makes sense to look for invariant expressions of d_1 first, because refinements with these help differential saturation in proving invariant expressions involving also x_1. Thus, we order variables by differential dependencies, which resembles the back substitution order in Gaussian elimination (if, in triangular form, x_1 depends on d_1, then equations for d_1 must be solved first, except that, in the differential case, variables may remain mutually dependent, e.g., d_1 and d_2). Now we call a set V of variables a *cluster* of the differential equation $\mathcal{D}$ iff V is closed with respect to $depend(\mathcal{D})$, i.e., variables of V only depend on variables in V:

$$x \in V \text{ and } (x,y) \in depend(\mathcal{D}) \text{ then } y \in V.$$

The resulting variable clusters for system ($\mathcal{F}$) are marked as triangular shapes in Fig. 6.4. Finally, we choose candidates from ψ and $ParaForm(k,d,V)$ starting with candidates whose variables lie in small clusters V and cover larger fractions of that cluster. Thus, the differential invariant $d_1^2 + d_2^2 \geq a^2$ of Sect. 6.2.3 within cluster $\{d_2, d_1, \omega\}$ can be discovered before invariants such as $d_1 = -\omega x_2$ that involve x_2, because x_2 depends on d_2. Similarly, $d_1 = -\omega(x_2 - c_2)$ will be discovered before $\|x - y\|^2 \geq p^2$, as the latter lies in a larger cluster with worse coverage percentage of that cluster. The successive differential saturation process along these dependencies further helps to keep the degrees in *ParaForm* small.

6.2.6 Global Fixed-Point Computation for Loop Invariants

With the uniform setup of d$\mathcal{L}$, we can adapt the algorithm in Fig. 6.3 easily to obtain a fixed-point algorithm for loops ($\psi \to [\alpha^*]\phi$) in place of continuous evolutions ($\psi \to [\mathcal{D} \wedge \chi]\phi$): In line 4 of Fig. 6.3, we replace the induction step from Definition 6.3 with the step for loops (Definition 6.1). As an optimisation, we can transfer the reuse of partial differential invariants according to Proposition 6.2 to discrete invariants for loops. Invariants χ of previous iterations can be exploited as refinements of the hybrid system dynamics, similarly to previous differential invariants that can be used in future iterations by refining the dynamics using differential saturation:

Proposition 6.3 (Loop saturation). *If χ is a discrete invariant of $\psi \to [\alpha^*]\phi$, then $\chi \wedge F$ is a discrete invariant iff $\psi \to F$ and $\chi \wedge F \to [\alpha](\chi \to F)$ are valid.*

Proof. Let χ be a discrete invariant of $\psi \to [\alpha^*]\phi$. Let, further, $\chi \wedge F$ be a discrete invariant of $\psi \to [\alpha^*]\phi$. Then $\psi \to \chi \wedge F$ and $\chi \wedge F \to [\alpha](\chi \wedge F)$ are valid by Definition 6.1. Hence, trivially, $\chi \wedge F \to [\alpha](\chi \to F)$ is valid, because all states that satisfy $\chi \wedge F$ also satisfy the weaker property $\chi \to F$. Finally, the validity of $\psi \to \chi \wedge F$ clearly entails $\psi \to F$.

Conversely, let χ be a discrete invariant. Let, further, $\chi \wedge F \to [\alpha](\chi \to F)$ and $\psi \to F$ be valid. For $\chi \wedge F$ to be a discrete invariant, we have to show that F satisfies the induction step of Definition 6.1 (the induction start $\psi \to \chi \wedge F$ is an immediate combination of the validity of $\psi \to \chi$ and $\psi \to F$). Since χ is a discrete invariant, $\chi \to [\alpha]\chi$ is valid, which entails $\chi \wedge F \to [\alpha]\chi$ as a special case. Since $\chi \wedge F \to [\alpha](\chi \to F)$ is valid and $\chi \wedge F \to [\alpha]\chi$ is valid, we conclude that $\chi \wedge F \to [\alpha](\chi \wedge F)$ is valid for the following reason. Let ν be a state satisfying the initial constraints $\chi \wedge F$. Then $\nu \models [\alpha]\chi$ and $\nu \models [\alpha](\chi \to F)$. Hence, all states ω reachable from ν by α satisfy $\omega \models \chi$ and $\omega \models \chi \to F$. Thus, they satisfy $\omega \models \chi \wedge F$, essentially by modus ponens. Consequently, we have shown that

$\chi \wedge F \to [\alpha](\chi \wedge F)$ is valid. and, hence, $\chi \wedge F$ is a discrete invariant of $\psi \to [\alpha^*]\phi$. □

The induction step from Proposition 6.3 can generally be proven faster, because it is a weaker property than that of Definition 6.1.

To adapt our approach from Sect. 6.2.5 to loops, we use discrete data-flow and control-flow dependencies of α. Dependencies can be determined immediately from the syntax of hybrid programs. There is a direct *data-flow dependency* with the value of x depending on y, if $x := \theta$ or $x' = \theta$ occurs in α with a term θ that contains y. Similarly, there is a direct *control-flow dependency* with the value of x depending on y if, for any term θ, $x := \theta$ or $x' = \theta$ occurs in α after a test $?\chi$ or evolution domain restriction $\wedge\chi$ containing y. The respective data-flow and control-flow dependencies are the transitive closures of these relations.

```
function prove(ψ → [α*]φ):
  χ := true     /* currently known invariant of ψ → [α*]φ */
  if prove(∀^α(χ → φ)) then return true /* property proven */
  for each F ∈ IndCandidates(ψ → [α*]φ, χ) do
    if prove(ψ∧χ → F) and prove(∀^α(χ∧F → [α](χ → F))) then
      χ := χ∧F          /* refine by discrete invariant */
      goto 3;           /* repeat fixed−point loop */
  end for
  return not provable using candidates
```

Fig. 6.5 Fixed-point algorithm for discrete loop invariants (loop saturation)

The resulting algorithm in Fig. 6.5 verifies loops. It is a direct adaption of that in Fig. 6.3, except that it uses Proposition 6.3 as an induction step for loops. The algorithm in Fig. 6.5 performs a fixed-point computation for loops and recursively combines the local differential invariants obtained by differential saturation to form a global invariant. It recursively uses *prove* for verifying its subtasks, which handle the discrete switching behaviour according to Fig. 6.1 and infer local differential invariants according to differential saturation by the fixed-point algorithm in Fig. 6.3.

Example 6.2 (Existential parameter quantification). Note that the ability of formulas in differential dynamic logic to have quantifiers in front of reachability modalities is crucial here. To illustrate, consider the simple water tank system from Fig. 6.6, where x denotes the current water level.

The second line in the hybrid program of Fig. 6.6 represents a continuous transition. It tests by $?q = on$ whether the current location q is *on*, and then fills the tank by following the differential equation $x' = 1$ restricted to the evolution domain $x \leq 9$ (i.e., the conjunction $x' = 1 \wedge x \leq 9$). The third line tests the guard $x \geq 5$ when in state *on*, then resets x by a discrete assignment ($x := x - 1$), and then changes location q to *off*. The additional test $?x \leq 9$ in the fourth line is needed, because the hybrid automaton is only allowed to enter mode *on* when its evolution domain re-

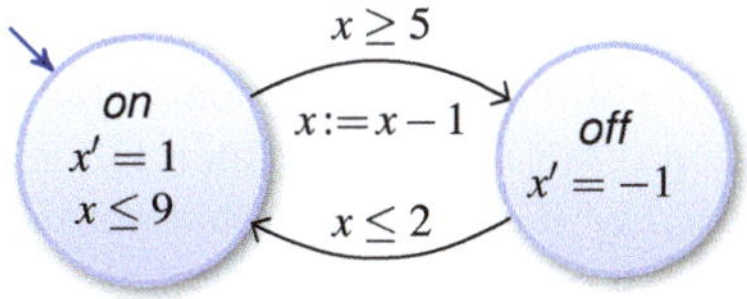

$q := on$; /* initial location is on */
$(\;\; (?q = on;\; x' = 1 \land x \leq 9)$
$\cup\; (?q = on \land x \geq 5;\; x := x-1;\; q := \mathit{off})$
$\cup\; (?q = \mathit{off};\; x' = -1)$
$\cup\; (?q = \mathit{off} \land x \leq 2;\; q := on;\; ?x \leq 9))^*$

Fig. 6.6 Hybrid program rendition of hybrid automaton for simple water tank

gion $x \leq 9$ is satisfied. The * at the end of the hybrid program indicates that the transitions of a hybrid automaton repeat indefinitely.

Let *wctrl* abbreviate the body of the loop of the hybrid program for the water controller from Fig. 6.6 such that $q := on; (wctrl)^*$ corresponds to the full hybrid program from Fig. 6.6. Then the following d$\mathcal{L}$ formula states that the water level is always below 10 when it starts at level $x \leq 3$:

$$x \leq 3 \rightarrow [q := on; (wctrl)^*]\, x < 10. \tag{6.3}$$

During the verification run for property (6.3), we need to show a property of the following form for the loop $wctrl^*$:

$$x \leq 3 \land q = on \rightarrow [(wctrl)^*]\, x < 10.$$

For a parametric candidate F of the form $a_1 x + a_0 \geq 0$, line 5 of the algorithm in Fig. 6.5 produces subtasks for discrete induction (Definition 6.1), which will be handled recursively by the *prove* function:

$$\begin{aligned} &\text{prove}(x \leq 3 \land q = on \land \chi \;\rightarrow\; a_1 x + a_0 \geq 0) \\ \textbf{and}\; &\text{prove}\big(\forall^\alpha (a_1 x + a_0 \geq 0 \land \chi \;\rightarrow\; [wctrl](\chi \rightarrow a_1 x + a_0 \geq 0))\big) \end{aligned}$$

In the first iteration (where χ is still *true*), the combination of these subtasks by conjunction corresponds to proving the following overall d$\mathcal{L}$ formula:

$$\exists a_0 \exists a_1 \big((x \leq 3 \land q = on \rightarrow a_1 x + a_0 \geq 0) \land \forall x\, (a_1 x + a_0 \geq 0 \rightarrow [wctrl]\, a_1 x + a_0 \geq 0)\big). \tag{6.4}$$

The universal quantifier $\forall x$ in (6.4) results from the universal closure $\forall^\alpha$ with respect to all variables changed in *wctrl*, i.e., x. The existential quantifiers for a_0 and a_1 are for formal parameters in the parametric candidate (Sect. 6.2.5). Observe that the outer placement of existential quantifiers is required for the resulting instance of $a_1 x + a_0 \geq 0$ to be a *common solution* implied by the precondition (left conjunct) and inductive for *wctrl* (right conjunct). In particular, the right conjunct requires finding parameter choices for which $a_1 x + a_0 \geq 0$ holds true after executing one iteration *wctrl* of the loop from any initial state if it was true before.

Note that this inherently requires quantifiers around reachability properties! Yet these nestings are naturally expressible in the first-order verification logic d$\mathcal{L}$. The left conjunct expresses the fact that the parameter choices need to make $a_1 x + a_0 \geq 0$ true in the beginning. The conjunction and outer placement of $\exists a_0 \exists a_1$ ensures that the required parameter choices fit together. Equation (6.4) also illustrates the need for the universal closure, because we need the same choice for the formal parameters a_0, a_1 to be able to conclude the induction step (case 2 of Definition 6.1) for all states x ($\forall x$). Different incompatible choices for a_0, a_1 at each state would not yield an inductive argument. To prove the example in Fig. 6.6, our algorithm will discover the parameter combination $a_1 = -1$ and $a_0 = 9.5$, for instance. □

For verification with parametric candidates to be feasible, we exploit the fact that we can keep existential quantifiers as local as possible in d$\mathcal{L}$, which we ensure by the symbolic decompositions in our logic. For instance, quantifiers for the formal parameters of a differential invariant will remain local to the formulas resulting from the algorithm in Fig. 6.3. Contrast this to a computationally more complicated global use of quantifiers around the whole verification problem at once. Minimising the number of parameters according to Sect. 6.2.5 further improves the computational tractability.

6.2.7 Interplay of Local and Global Fixed-Point Loops

Together, our local and global fixed-point algorithms jointly verify correctness properties of hybrid programs. Their interplay needs to be coordinated with fairness, though. If the local fixed-point algorithm in Fig. 6.3 does not converge, stronger invariants may need to be found by the global fixed-point algorithm which iteratively result in stronger preconditions ψ_i for the local fixed-point algorithm; see Fig. 6.7.

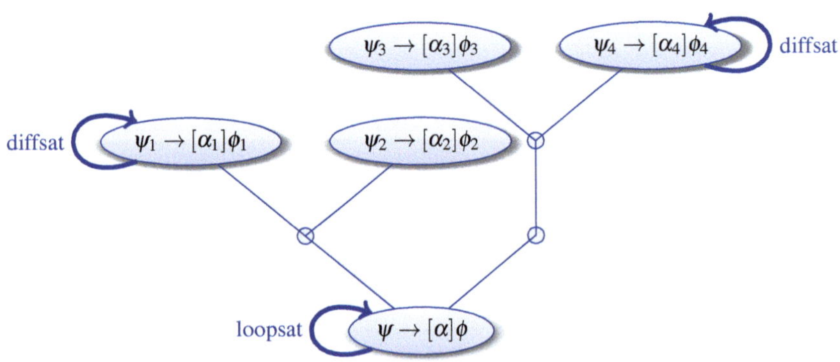

Fig. 6.7 Interplay of local (diffsat) and global (loopsat) fixed-points verification loops during symbolic decomposition

Thus, for fairness reasons, the local fixed-point algorithm should stop when it cannot prove its postcondition, either because of a counterexample or because it runs out of candidates for differential invariants. As in the work of Prajna et al. [252], the degrees of parametric invariants, therefore, need to be bounded and increased iteratively. As in [252], there is no natural measure for how these degrees should be increased. Instead, here, we exploit the fact that the candidates produced by function *Candidates* are independent and we explore them in parallel with fair time interleaving. The interleaving by iterative timeout increase is similar to iterative background closures from Sect. 5.4, except that it works for full subproofs instead of for just one local quantifier elimination call.

During fixed-point computations, wrong choices of candidates are time consuming. Thus, in practise, it is important to discover futile attempts quickly. For this, non-exhaustive sampling with numerical simulations can be used to look for counterexamples. Note that this numerical counterexample search can be performed separately for each local subtask in the decomposition. For instance, a counterexample for a candidate F for node $\psi_4 \rightarrow [\alpha_4]\phi_4$ in Fig. 6.7 will only abort the attempt to prove this particular candidate F, not attempts with other candidates. A counterexample found by local numerical simulation for the reachability property $\psi_4 \rightarrow [\alpha_4]\phi_4$ itself, on the other hand, will terminate all proof search for that subtask by propagating the need for a stronger precondition ψ_4 up the decomposition tree in Fig. 6.7. Likewise, a counterexample found for the original property $\psi \rightarrow [\alpha]\phi$ will abort the whole proof search. We discuss the implications of including numerical procedures in Sect. 6.4.

In well-designed control loop systems, global fixed points are easier to find than local fixed points of differential invariants, because the global invariant often coincides with the controllability region of the system. In these systems, the minimal control objective is to keep the system in the controllable state region. That is, if the system is in a controllable state, i.e., there is a control choice such that the system can remain safe, suitable controllers will pick one such control option such that the system again ends up in a controllable state. Inductively, this controllability region directly corresponds to a global system invariant. We make this intuition more precise in the context of Chap. 7.

6.3 Soundness

Even though the interplay of the fixed-point verification algorithms in this chapter is already quite complicated, we can exploit the fact that the algorithms are always working with formula transformations in the verification logic $d\mathcal{L}$ to ensure soundness quite easily. Since all formula transformations in the algorithm are justified by sound proof rules, soundness is fully captured in the logic and the algorithm can only be incomplete but not unsound:

Theorem 6.1 (Soundness of fixed-point verification algorithm). *The verification algorithm in Sect. 6.2 is sound, i.e., whenever* `prove(`$\psi \rightarrow [\alpha]\phi$`)` *returns "*`true`*", the $d\mathcal{L}$ formula $\psi \rightarrow [\alpha]\phi$ is true in all states, i.e., all states reachable by α from states satisfying ψ satisfy ϕ.*

Proof. Soundness is a direct consequence of Theorem 3.1, as every statement that returns true is justified by a sound proof rule of DAL. To show this in full detail, we prove by induction on the structure of the algorithm that, for any d$\mathcal{L}$ formula ϕ, ϕ is true in every state ν where (the formula returned by) *prove*(ϕ) is true. That is, $\nu \models$ *prove*(ϕ) implies $\nu \models \phi$. In particular, if the algorithm returns "true", the input formula is true in all states.

- In the base case (line 12 of Fig. 6.1), *prove* returns the result of quantifier elimination, which is a sound decision procedure [81]. The result of quantifier elimination is equivalent to its input. Hence one formula is true if and only if the other is.
- If α is of the form $x := \theta$, the algorithm in line 1 of Fig. 6.1 is responsible. Consider any state ν where the formula returned by *prove*($\psi \to [x := \theta]\phi$) is true. That is, *prove*($\psi \wedge \hat{x} = \theta \to \phi_x^{\hat{x}}$) is true at ν. Hence, by the induction hypothesis, $\psi \wedge \hat{x} = \theta \to \phi_x^{\hat{x}}$ is true at ν. Now, because $\hat{x}$ was a fresh variable, the Substitution Lemma 2.2 can be used to show that $\psi \to \phi_x^{\theta}$ and $\psi \to [x := \theta]\phi$ are true at ν.
- If α is of the form $x := *$, the algorithm in line 10 of Fig. 6.1 is responsible. The proof is a direct consequence of the fact that ϕ being true after *all* random assignments to x is equivalent to ϕ being true for all real values of x. Hence, $\psi \to [x := *]\phi$ is true in a state ν if and only if $\psi \to \forall x\, \phi$ is.
- For formulas of the form $\psi \to Qx\, \phi$ where Q is a quantifier, line 14 of Fig. 6.1 is responsible. If the formula returned by *prove*($\psi \to Qx\,\phi$) is true in state ν, then QE($\psi \to Qx$*prove*(ϕ)) is true in ν. As quantifier elimination yields an equivalent formula, this implies that $\psi \to Qx$*prove*(ϕ) itself is true at ν. Assume that ψ holds at ν as there is nothing to show otherwise. Then Qx*prove*(ϕ) is true at ν. Thus for all (or some if Q is $\exists$) states ω that agree with ν except for the value of x, we know that the formula returned by *prove*(ϕ) is true at ω. The formula ϕ is structurally simpler. Hence, by induction hypothesis, ϕ is true at ω. Consequently, both $Qx\,\phi$ and $\psi \to Qx\,\phi$ are true at ν, as ω was arbitrary.
- The other cases of Fig. 6.1 are similar consequences of the soundness of the DAL rules in Fig. 3.9 by Theorem 3.1.
- If α is of the form $\mathcal{D} \wedge \chi$ for a differential equation system $\mathcal{D}$, the algorithm in Fig. 6.3 is responsible. If it returns "true" in line 2 in the first place, then the calls to *prove* in line 2 must have resulted in "true". Hence, by induction hypothesis, χ entails ϕ. Thus, soundness of rule [*DR*] justifies soundness as follows. Postcondition ϕ is true in a subregion of the evolution domain χ. Thus $\psi \to [\mathcal{D} \wedge \chi]\phi$ is valid, trivially, because all evolutions along $\mathcal{D} \wedge \chi$ always satisfy χ and, hence, ϕ. If, however, χ was changed in line 5 during the fixed-point computation, then the calls to *prove* for the properties in line 4 must have returned "true". Thus, by the induction hypothesis, the d$\mathcal{L}$ formulas $\psi \wedge \chi \to F$ and $\forall^{\mathcal{D}}(\chi \to F'_{\mathcal{D}})$ are valid. Hence *DS* justifies soundness as follows. Formula F is a differential invariant of $\psi \to [\mathcal{D} \wedge \chi]\phi$ by Definition 6.3. Consequently, by Proposition 6.1, F also is a continuous invariant (Definition 6.2). Thus, by Proposition 6.2, the d$\mathcal{L}$ formulas $\psi \to [\mathcal{D} \wedge \chi]\phi$ and $\psi \to [\mathcal{D} \wedge \chi \wedge F]\phi$ are equivalent, and we can (soundly) verify the former by proving the latter. Consequently, the modification of the

evolution domain χ to $\chi \wedge F$ in line 5 is sound, because the algorithm will continue proving a refined but equivalent formula for a refined but equivalent system.
- If α is a loop of the form β^*, the proof is similar to the case for differential equations, except that it uses Proposition 6.3 instead of Proposition 6.1. □

Since reachability of hybrid systems is undecidable, our algorithm must be incomplete. It can fail to converge when the required invariants are not expressible in first-order logic (yet, the required invariants are always expressible in $d\mathcal{L}$ by Theorem 2.3). Consequently, a fixed point in $d\mathcal{L}$ always exists but our algorithm may fail to find appropriate (differential) invariants in first-order logic.

6.4 Optimisations

In this section, we consider some optimisations of the fixed-point verification algorithm.

6.4.1 Sound Interleaving with Numerical Simulation

During fixed-point computations, wrong choices of candidates are time consuming. Thus, it is important to discover futile proof attempts quickly. For this, we use non-exhaustive numerical simulation to look for a counterexample for each candidate. To prevent rejecting good candidates due to numerical errors, we discard fragile counterexamples. We consider counterexamples with distance $<\varepsilon$ to safe states as *fragile*, because small numerical perturbations could make it safe (the right x in Fig. 6.8 marks fragile examples or counterexamples). The left mark in

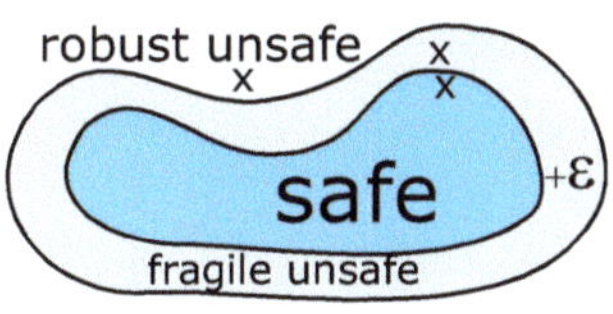

Fig. 6.8 Robustness in counterexamples

Fig. 6.8, instead, is a *robust* counterexample, i.e., only large ($\geq \varepsilon$) perturbations could make it safe. Robust counterexamples can be ensured by replacing, e.g., $a \geq b$ with $a \geq b + \varepsilon$ in the formulas for numerical reachability simulation for some estimate $\varepsilon \geq 0$ of the numerical error. Unlike in other approaches [21, 192, 228, 252, 238], numerical errors are *not* critical for soundness here, because safety is exclusively established by sound symbolic verification. Numerical computations are only used to stop futile proof attempts.

We can further exploit the symbolic decomposition performed by our algorithm in Sect. 6.2 and prefix recursive calls to $prove(\psi \to [\alpha]\phi)$ with a partial simulation of α. Using approximate cylindrical algebraic decomposition [81] in the *FindInstance* function of Mathematica, we often find good samples of states satisfying ψ to start the simulation of α.

6.4.2 Optimisations for the Verification Algorithm

Formulas with variables that do not change in a fragment of a hybrid program are trivial invariants, as their truth-value is unaffected. For instance, $\omega = \rho$ is a trivial invariant of system ($\mathscr{F}$). Hence, it can be used as an invariant without proof. A formula like $\omega^2(d_1^2 + d_2^2) > r^2$ in ψ, in contrast, is not trivially invariant, because d_i changes during ($\mathscr{F}$). Still, it has invariant consequences such as $\omega \neq 0$. To make use of these direct and indirect trivial invariants from ψ , we (soundly) weaken all universal closures of the form $\forall^\alpha \phi$ in lines 2 and 4 of Fig. 6.3 by $\psi \to \forall^\alpha \phi$, which directly reflects the rule context instantiations from Definition 2.10.

6.5 Experimental Results

We analyse aircraft roundabout manoeuvres [293] as a system with nontrivial dynamics. Curved flight as in roundabouts is challenging for verification, because of its transcendental solutions. The manoeuvre from [293] in Fig. 3.6a on p. 151 and the manoeuvre in Fig. 3.6d from [238, 237] are not flyable, because they still involve a few instant turns. A flyable roundabout manoeuvre without instant turns is depicted in Fig. 6.9. We verify safety properties for several (but not yet all) phases

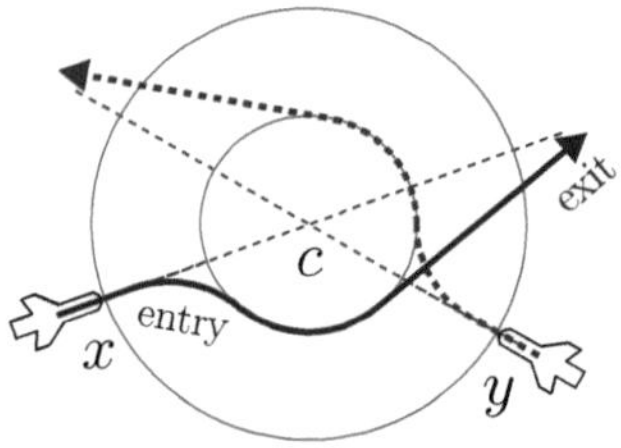

Fig. 6.9 Flyable aircraft roundabout

of Fig. 6.9 fully automatically and provide verification results in Table 6.1. Details on the case studies are presented in Part III. Finally, note that the required invariants found by our approach for the roundabout manoeuvre cannot even be found from Differential Gröbner Bases [202].

Verification results for roundabout aircraft manoeuvres [293, 92, 238, 237] and the European Train Control System (ETCS) are in Table 6.1. Results are from a 2.6 GHz AMD Opteron with 4 GB memory. Memory consumption of quantifier elimination is shown in Table 6.1, excluding the front end. The results are only slightly

Table 6.1 Experimental results for differential invariants as fixed points

Case study	Time(s)	Memory(MB)	Proof steps	Dim
tangential roundabout (2 aircraft)	14	8	117	13
tangential roundabout (3 aircraft)	387	42	182	18
tangential roundabout (4 aircraft)	730	39	234	23
tangential roundabout (5 aircraft)	1964	88	317	28
ETCS kernel safety	41	28	53	9
ETCS binary safety	56	27	147	14
ETCS safety	183	87	169	14

worse on a 1.7 GHz Pentium M laptop with 1 GB memory. The dimension of the continuous state space is indicated (column Dim). Notice that we handle *all* the variables symbolically leading to state spaces up to $\mathbb{R}^{28}$. The experimental results are encouraging, in particular as the memory consumption is fairly moderate. High memory consumption would limit scalability much more than time consumption. For instance, numerical techniques on a rather coarse numerical mesh with only 10 samples per variable would need to consider reachability analysis from 10^{28} initial states for this case. The results in Table 6.1 indicate that scalability of our approach is substantially less limited by the number of variables than in other approaches. Instead, the complexity of the constraints and dependencies among the variables have more impact. In parallel systems, there is usually a smaller fraction of dependencies among continuous variables. Multiple aircraft, e.g., are not coupled physically (see Fig. 6.4) but interact only by discrete dynamics originating from sporadic communication.

6.6 Summary

We have introduced the "differential invariants as fixed points" paradigm and we have presented a *sound* algorithm for verifying hybrid systems with nontrivial dynamics. It handles differential equations using differential invariants instead of requiring solutions of the differential equations, because the latter quickly yield undecidable arithmetic. We compute differential invariants as fixed points using $d\mathcal{L}$, DAL, or dTL as verification logics for hybrid systems. In the logics we can decompose the system for computing local invariants and we obtain sound recombinations into global invariants. Moreover, we introduce a differential saturation procedure that verifies more complicated properties by refining the system dynamics successively in a sound way. We validate our algorithm on challenging *roundabout collision avoidance manoeuvres* for aircraft and on collision avoidance protocols for trains. We give verification results with a sound algorithm for challenging dynamics of curved flight in up to 28-dimensional continuous state spaces.

Our algorithm works particularly well for highly parametric hybrid systems, because their parameter constraints can often be combined faster to find invariants, than for systems with a single initial state, where simulation is more appropriate.

Our decompositional approach exploits locality in system designs. In well-designed systems, subsystems do not generally depend on all other parts of the system but are built according to modularity and locality principles in engineering. In these cases, the d$\mathcal{L}$ calculus achieves good decompositions and the required invariants for one part of the system only need very little information about other parts of the systems, so the fixed-point algorithm terminates faster. Our algorithm probably performs worse for systems that violate locality principles.

Scalability to applications from other domains like chemical process control or biomedical devices remains an interesting topic for future research. An independent question for future research, underlying virtually all hybrid systems' verification approaches, is how to scale real arithmetic handling. Semidefinite programming relaxations [224] may be an interesting direction for future research (see App. D). Differential induction and the logic d$\mathcal{L}$ generalise to liveness properties and to systems with disturbances (Chap. 3). In future work, we want to generalise the synthesis of corresponding differential (in)variants for these cases.

Part III
Case Studies and Applications in Hybrid Systems Verification

Overview In Part I we have developed differential dynamic logics as specification and verification languages for both hybrid programs and DA-programs as operational system models for hybrid systems. We have also developed proof calculi as formal verification techniques for hybrid systems in differential dynamic logic. In Part II, we have studied algorithmic refinements, automation techniques, and invariant generation procedures for the respective proof calculi.

In this part of the book, we now shift our attention to application scenarios for our logical analysis approach for hybrid systems. Extending smaller hybrid systems which have served as running examples throughout this book, we show full case studies of the European Train Control System in Chap. 7 and for aircraft collision avoidance manoeuvres in Chap. 8. We show how complex physical systems can be modeled as hybrid programs (or DA-programs) and how safety-critical properties can be formalised in differential dynamic logics and proven in their proof calculi.

Chapter 7
European Train Control System

Contents

Synopsis Complex physical systems have several degrees of freedom. They only work correctly when their control parameters obey corresponding constraints. Based on the informal specification of the *European Train Control System* (ETCS), we design a controller for its cooperation protocol. For the free parameters of the system, we successively identify constraints that are required to ensure collision freedom. We formally prove the parameter constraints to be sharp by characterising them equivalently in terms of reachability properties of the hybrid system dynamics. We use the calculus of our differential dynamic logic for hybrid systems and formally verify controllability, safety, liveness, and reactivity properties of the ETCS protocol that entail collision freedom. We prove that the ETCS protocol remains correct even in the presence of perturbation by disturbances in the dynamics.

A. Platzer, Logical Analysis of Hybrid Systems,
DOI 10.1007/978-3-642-14509-4_7, © Springer-Verlag Berlin Heidelberg 2010

7.1 Introduction

Complex physical control systems often contain many degrees of freedom, including how specific parameters are instantiated or adjusted [205, 91, 27]. Yet, virtually all of these systems are hybrid systems and only work correctly under certain constraints on these parameters. The *European Train Control System* (ETCS) [117] has a wide range of different possible configurations of trains, track layouts, and different driving circumstances. It is only safe for certain conditions on external parameters, e.g., as long as each train is able to avoid collisions by braking with its specific braking force on the remaining distance to the rear end of the next train. Similarly, internal control design parameters for supervisory speed control and automatic braking triggers need to be adjusted in accordance with the underlying train dynamics. Moreover, parameters must be constrained such that the system remains correct when passing from continuous models with instant reactions to sampled-data discrete-time controllers of hardware implementations. Finally, parameter choices must preserve correctness robustly in the presence of disturbances caused by unforeseen external forces (wind, friction, etc.) or internal modelling inaccuracies of ideal-world dynamics, e.g., when passing from ideal-world dynamics to *proportional-integral* (PI) controller implementations.[1] Yet, determining the range of external parameters and the choice of internal design parameters for which complex control systems like ETCS are safe, is not possible just by looking at the model, and even less so in the presence of disturbance.

Likewise, it is difficult to read off the parameter constraints that are required for correctness from a failed verification attempt of model checkers [156, 217, 126], since concrete numeric values of a counterexample trace cannot simply be translated into a generic constraint on the free parameters of the system, which would have prevented this kind of error. While approaches like counterexample-guided abstraction refinement [72, 126] are highly efficient in undoing automatic abstractions of an abstract hybrid system from spurious counterexamples, they stop when true counterexamples remain in the concrete system. For discovering constraints on free parameters, though, even concrete models will have counterexamples until all required parameter constraints have been identified. Even worse, concrete numeric values of a counterexample trace as produced by a model checker with state splitting cannot simply be translated into a uniform global constraint on the free parameters of the system, which would prevent this kind of error.

Instead, we use our techniques based on symbolic decompositions from Part I to systematically explore the design space of a hybrid system and for discovering correctness constraints on free parameters. We have already shown how some such constraints can be identified for a very simple train control model in the examples from Chap. 2. For a complex physical system, we now show step by step how a control system can be developed that meets its control design goals and desired correctness properties. Starting from a coarse skeleton of the ETCS cooperation

[1] PI is a standard control technique and also used for controlling trains [91]. We refer to joint work with Jan-David Quesel [244] for verification results of ETCS in the presence of PI control.

protocol obtained from its official specification [117], we systematically develop a safe controller and identify the parameter constraints that are required for collision freedom. Although these parameter constraints are safety-critical, they are not stated in the official specification [117]. They only originate quite indirectly from the system dynamics and the overall control objectives of ETCS. For all practical implementations, however, these constraints need to be made explicit in order to find safe parameter range choices. These constraints are also nontrivial, especially those needed to ensure a safe interplay of physics and sampled control implementations. Using the parametric constraints so discovered, we prove correctness properties of the ETCS cooperation protocol that entail collision freedom.

During the course of this case study, we prove rich properties, including safety, controllability, reactivity, and liveness, which are not uniformly expressible and verifiable in other analysis approaches. Moreover, we prove those correctness properties of the parametric ETCS case study almost fully automatically in our verification tool KeYmaera.

Contributions

We show how realistic fully parametric hybrid systems for traffic protocols can be designed and verified using our logical analysis approach. For ETCS, we identify all relevant safety constraints on free parameters, including external system parameters and internal design parameters of controllers. Safe control choices will be important for more than two million passengers in Europe per day. Similar constraints hold for other upcoming train control systems. Our first contribution is that we characterise safe parameter choices equivalently in terms of properties of the train dynamics and that we prove controllability, reactivity, safety, and liveness properties of ETCS. Other issues often arise from verification results for ideal-world dynamics that cease to hold for real-world dynamics. Our second contribution is showing how to extend formal ETCS verification to the presence of disturbances in the dynamics, which account for friction and other discrepancies of the real-world dynamics and simplistic control models. Most notably, the full ETCS model with its rich set of properties is out of scope for other approaches. Nevertheless, we have formally verified all propositions in this chapter with our verification tool KeYmaera [242]. Furthermore, we have proven ETCS to work correctly when using a proportional-integral (PI) controller for speed supervision. In contrast to their routine use in control, giving formal proofs for the correct functioning of PIs has been an essentially unsolved problem. We refer to joint work with Jan-David Quesel [245] for details on the PI verification.

ETCS further illustrates a more general phenomenon in hybrid systems: safely combining dynamics with control requires parameter constraints that are much more complicated than the original dynamics. Safe control of hybrid dynamics requires more than just an understanding of the physics underlying a system and a separate understanding of the computing processes in the system. Rather, safe choices of hybrid systems control usually depend on a joint understanding of how physics and computation interact.

7.1.1 Related Work

Model checkers for hybrid systems, for example HyTech [11] and PHAVer [126], verify by exploring the state space of the system as exhaustively as possible. In contrast to our approach they need concrete numbers for most parameters and cannot verify liveness, such as whether the train controller is guaranteed to make progress, or existential properties, e.g., whether and how a control parameter can be instantiated so that the system is always safe.

Batt et al. [27] give heuristics for splitting regions by linear constraints that can be used to determine parameter constraints. Frehse et al. [128] synthesise parameters for linear hybrid automata. However, realistic systems like ETCS require nonlinear parameter constraints, often have a nonlinear system dynamics, and are out of scope for these approaches.

Tomlin et al. [294] show a game-theoretic semidecision algorithm for hybrid controller synthesis. For systems like ETCS, which are more general than linear [127] or o-minimal hybrid automata [188], they suggest numerical approximations. We give exact results for fully parametric ETCS using symbolic techniques. We also focus on verification and parameter synthesis of parameter constraints for a given controller template, rather than for full controller synthesis. While full controller synthesis is a very interesting and tempting idea, it is less scalable than verification [75, 249]. In addition to theoretical complexity results that separate verification and synthesis in the discrete case, verification and even parameter synthesis are more modular than full synthesis. In verification, several properties can be proven separately for the same model, but in synthesis all properties need to be considered at once. A synthesised controller satisfying a safety property may not be live, e.g., it may be safe, just because it disallows movement altogether, which is perfectly useless. Another synthesised controller satisfying a liveness property, however, may be unsafe. So in synthesis all relevant properties need to be considered at once for the resulting system to make sense. In verification and also in proof-based parameter synthesis, properties can be considered one at a time, thereby gaining scalability advantages.

In independent work, Cimatti et al. [71] have recently analysed consistency of informal requirements of ETCS expressed as temporal properties, including the continuous system dynamics, using an approach based on the combination of temporal logic with regular expressions. Our work is complementary, as we focus on developing and formally verifying an actual hybrid systems controller that can be implemented later on, and not on showing the consistency of the requirement specification properties. The fact that we are able to construct a controller satisfying ETCS requirements also entails consistency of these requirements. Our goal, however, is to develop a verified controller to show how the system can be controlled safely, and not whether the requirements are incompatible.

7.1.2 Structure of This Chapter

We introduce a formal model for parametric ETCS in Sect. 7.2. Using our symbolic logical decomposition approach from Part I, we systematically derive parametric correctness constraints for ETCS and verify several correctness properties of parametric ETCS using these constraints in Sect. 7.3. In Sect. 7.4, we generalise the physical transmission model to the presence of disturbances and verify ETCS despite disturbances. We present experimental results for our verification approach in Sect. 7.5. For details on the experiment and formal KeYmaera proofs of the propositions, we refer the reader to [245].

7.2 Parametric European Train Control System

The *European Train Control System* (*ETCS*) [117, 205] is a standard to ensure safe and collision-free operation as well as high throughput of trains that currently run at speeds up to 320 km/h. Correct functioning of ETCS is highly safety-critical, because the upcoming installation of ETCS Level 3 will replace all previous track-side safety measures in order to achieve its high throughput objectives. In this section, we present a system skeleton, which corresponds to a simple representation of the train dynamics and controller reflecting the informal ETCS cooperation protocol [117]. The system obtained from the informal specification turns out to be unsafe. In Sect. 7.3, we will systematically refine this skeleton with the parameter constraints that are required for safety but not stated in the informal ETCS specification [117], and only indirectly result from the effects of the system dynamics on the safety requirements.

7.2.1 Overview of the ETCS Cooperation Protocol

ETCS Level 3 follows the *moving block principle*, i.e., neither are movement permissions known beforehand nor is the size of movement blocks fixed statically. Rather, permission to move is determined based on the current track situation by a *Radio Block Controller* (RBC). Trains are only allowed to move within their current *movement authority* (MA), which can be updated dynamically by the RBC using wireless communication. Hence the train controller needs to regulate the movement of a train locally such that it always remains within its MA, because there could be

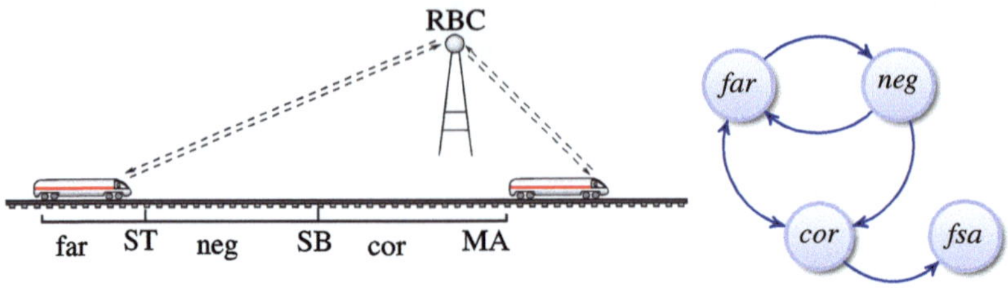

Fig. 7.1 ETCS train cooperation protocol phases and dynamic movement authorities

open gates, other trains, or speed restrictions due to tunnels beyond the end of the MA.

The automatic train protection unit (*atp*) dynamically determines a safety envelope around a train τ, within which it considers driving safe, and adjusts the train acceleration accordingly. Figure 7.1 illustrates the dynamic assignment of MA. The ETCS controller switches according to the protocol pattern in Fig. 7.1, which corresponds to a simplified version of Damm et al. [91]. When approaching the end of its MA the train switches from *far* mode (where speed can be regulated freely) to negotiation (*neg*), which, at the latest, happens at the point indicated by *ST* (for *start talking*). During negotiation the RBC grants or denies MA extensions. If the extension is not granted in time, the train starts braking in the correcting mode (*cor*) and returns to *far* after the situation has cleared. Emergency messages announced by the RBC can also put the controller into *cor* mode. If so, the train may switch to a fail-safe state (*fsa*) after the train has come to a full stop and await manual clearance by the train operator.

Lemma 7.1 (Principle of separation by movement authorities). *If each train stays within its MA and, at any time, MAs issued by the RBC form a disjoint partitioning of the track, then trains can never collide.*

Proof. To simplify notation, we assume trains are points (the proof is a simple extension when each train has some maximal length l). Consider any point in time ζ. For some $n \in \mathbb{N}$, let $z_1, \ldots, z_n$ be the positions of all the trains 1 to n at ζ. Let M_i be the MA range, i.e., the set of positions on the track for which train i has currently been issued MA. Suppose there was a collision at time ζ. Then $z_i = z_j$ at ζ for some $i, j \in \mathbb{N}$. However, by assumption, $z_i \in M_i$ and $z_j \in M_j$ at ζ, thus $M_i \cap M_j \neq \emptyset$; which contradicts the assumption of disjoint MAs. □

Lemma 7.1 effectively reduces the verification of an unbounded number of traffic agents to a finite number. We exploit MAs to decouple reasoning about global collision freedom and reduce it to local cooperation of every traffic agent with its RBC. In particular, we verify correct coordination for a train without having to consider gates or railway switches, because these only communicate via RBC mediation and can be considered as special reasons for denial of MA extensions. We only need to prove that the RBC handles all interaction between the trains by assigning or revoking MA correctly and that the trains actually respect their MA always. Yet, in order to enable the RBC to guarantee disjoint partitioning of the track, ETCS has to rely on properties like appropriate safe rear end computation of the train. Additionally, safe operation of the train plant in conjunction with its environment depends on proper functioning of the gates. As these properties have a more static nature, they are much easier to show once the actual hybrid train dynamics and movements have been proven to be controlled correctly and once the relevant safety constraints have been identified.

As trains are not allowed to drive backwards without manual clearance by track supervision personnel, the relevant part of the safety envelope is the closest distance to the end of its current MA. The point *SB*, for *start braking*, is the latest point where the train needs to start correcting its acceleration (in mode *cor*) to make sure it always stays within the bounds of its MA. In Sect. 7.3, we derive a necessary and sufficient constraint on parameter *SB* that guarantees safe driving.

We generalise the concept of MA to a vector $\mathbf{m} = (d, e, r)$, meaning that, beyond end point $\mathbf{m}.e$, the train is not permitted to have a velocity greater than the desired speed $\mathbf{m}.d$. Additionally, the train should try not to outspeed the *recommended speed* $\mathbf{m}.r$ for the current track segment, but short periods of slightly higher speed are not considered safety-critical. Figure 7.2 shows an example of possible train behaviour in conjunction with the current value of $\mathbf{m}$ that changes over time due to RBC communication.

Fig. 7.2 ETCS track profile

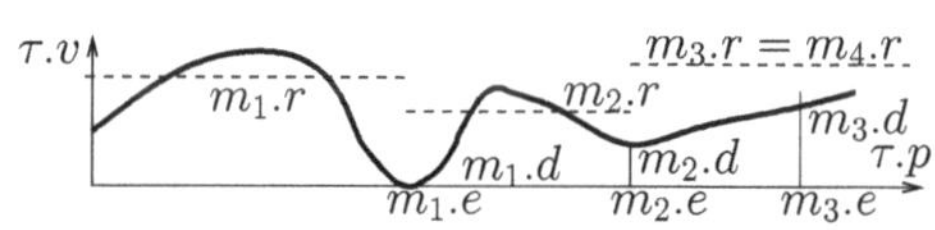

For a train $\tau = (p, v, a)$ at position $\tau.p$ with current velocity $\tau.v$ and acceleration $\tau.a$, we want to determine sufficient conditions that ensure safety and formally verify that $\tau.v$ is always safe with respect to its current MA, satisfying:

$$\tau.p \geq \mathbf{m}.e \rightarrow \tau.v \leq \mathbf{m}.d. \qquad (\mathscr{S})$$

Formula ($\mathscr{S}$) expresses that the train velocity $\tau.v$ does not exceed the strict speed limit $\mathbf{m}.d$ after passing the point $\mathbf{m}.e$ (i.e., $\tau.p \geq \mathbf{m}.e$). Generalised MAs are a uniform composition of two safety-critical features. They are crucial aspects for ensuring collision-free operation in ETCS (Lemma 7.1) and can take into account safety-critical velocity limits due to bridges, tunnels, or passing trains. For example high-speed trains need to reduce their velocity while passing non-airtight or freight trains with a pressure-sensitive load within a tunnel. Our model captures this by reducing the speed component $\mathbf{m}.d$ of MA $\mathbf{m}$ appropriately.

$$
\begin{array}{lll}
ETCS_{\text{skel}} & \equiv & (train \cup rbc)^* \\
train & \equiv & spd;\ atp;\ drive \\
spd & \equiv & \big(?(\tau.v \leq \mathbf{m}.r);\ \tau.a{:=}*;\ ?(-b \leq \tau.a \leq A)\big) \\
 & & \cup \big(?(\tau.v \geq \mathbf{m}.r);\ \tau.a{:=}*;\ ?(-b \leq \tau.a \leq 0)\big) \\
atp & \equiv & \text{if } (\mathbf{m}.e - \tau.p \leq SB \vee rbc.message = emergency) \text{ then } \tau.a := -b \text{ fi} \\
drive & \equiv & t := 0;\ (\tau.p' = \tau.v \wedge \tau.v' = \tau.a \wedge t' = 1 \wedge \tau.v \geq 0 \wedge t \leq \varepsilon) \\
rbc & \equiv & \big(rbc.message := emergency\big) \cup \big(\mathbf{m} := *;\ ?(\mathbf{m}.r > 0)\big)
\end{array}
$$

Fig. 7.3 Formal model of parametric ETCS cooperation protocol (skeleton)

7.2.2 Formal Model of Fully Parametric ETCS

For analysing the proper functioning of ETCS, we have developed a formal model of ETCS as a hybrid program (see Fig. 7.3) that is based on the informal ETCS specification [117]. The RBC and the train are independently distributed components running in parallel. They cooperate by message passing over wireless communication. Since the RBC is a purely digital track-side controller and has no dependent continuous dynamics, we can express parallel composition equivalently by interleaving using nondeterministic choice ($\cup$) and repetition (*): the decisions of the train controller only depend on the point in time when RBC messages arrive at the train, not on the communication latency. Thus, the nondeterministic interleaving in ETCS where either the train or ($\cup$) the RBC chooses to take action faithfully models every possible arrival time without the need for an explicit (delayed) channel model. The * at the end of $ETCS_{\text{skel}}$ indicates that the interleaving of train and RBC repeats arbitrarily often. Successive actions in each component are modelled using sequential composition (;). The train checks for the difference between the current speed and the recommended speed (in speed supervision *spd*) before checking if emergency braking is necessary (in automatic train protection *atp*).

Train Controller

It is quite difficult to use highly detailed models for the train and its mechanical transmission [91] directly in the verification and parameter discovery process. Hence, we first approximate it by a controller with a ranged choice for the effective acceleration $\tau.a$ between its lower bound ($-b < 0$) and upper bound ($A > 0$); we will refine the dynamics in Sect. 7.4. This nondeterministic range controller provides a model that we can use both to derive parameter constraints and to overapproximate the actual choices made by the physical train controller [91] conservatively. For Sects. 7.2 and 7.3, we model the continuous train dynamics by the differential equation system

$$\tau.p' = \tau.v \wedge \tau.v' = \tau.a \wedge t' = 1 \wedge \tau.v \geq 0 \wedge t \leq \varepsilon. \qquad (\mathscr{I})$$

It formalises the ideal-world physical laws for movement, restricted to the evolution domain $\tau.v \geq 0 \wedge t \leq \varepsilon$ in *drive*. Recall from Part I that primed variables stand for the first time derivative of the respective unprimed variable. Therefore, $\tau.p'$ gives the rate with which the position $\tau.p$ of the train τ changes, i.e., the velocity ($\tau.p' = \tau.v$). The velocity $\tau.v$ itself changes continuously according to the acceleration $\tau.a$, i.e., $\tau.v' = \tau.a$. The train speeds up when $\tau.a > 0$ and brakes when $\tau.a < 0$. In particular, for $\tau.a < 0$, the velocity would eventually become negative, which would mean the train is driving backwards. But that is prohibited without manual clearance, so we restrict the evolution domain to nonnegative speed ($\tau.v \geq 0$). Time can be measured by clocks, i.e. variables changing with constant slope 1 (along differential equation $t' = 1$). To further account conservatively for delayed effects of actuators like brakes or for delays caused by cycle times of periodic sensor polling and sampled-data discrete-time controllers, we permit the continuous movement of the train to continue for up to $\varepsilon > 0$ time units until control decisions finally take effect. This is expressed using the evolution domain restriction $t \leq \varepsilon$ on the clock t that is reset by the discrete assignment $t := 0$ before the continuous evolution starts. It is used to keep track of the progress of time advancing with constant slope 1. When the system executes the system of differential equations in *drive*, it can follow a continuous evolution respecting the constraints of ($\mathscr{I}$) all the time.

The speed supervision *spd* has two choices ($\cup$). The first option in Fig. 7.3 can be taken if the test $?(\tau.v \leq \mathbf{m}.r)$ succeeds, the second one if the test $?(\tau.v \geq \mathbf{m}.r)$ is successful. If both tests can succeed, either choice is possible by a nondeterministic choice ($\cup$). The speed supervision *spd* chooses the acceleration $\tau.a$ to keep the recommended speed $\mathbf{m}.r$ by a random assignment $\tau.a := *$, which assigns an arbitrary value to $\tau.a$. By the subsequent test $?(-b \leq \tau.a \leq 0)$ an acceleration is chosen from the interval $[-b, 0]$ if the current speed $\tau.v$ exceeds recommended speed $\mathbf{m}.r$ (otherwise the full range $[-b, A]$ is available.) Our nondeterministic controller includes controllers optimising speed and energy consumption as secondary objectives.

As a supervisory controller, the automatic train protection (*atp* in Fig. 7.3) checks whether the train has passed point *SB* ($\mathbf{m}.e - \tau.p \leq SB$, that is, the remaining distance of $\tau.p$ to $\mathbf{m}.e$ is less than or equal *SB*) or whether a message from the RBC was received notifying of a track-side emergency situation. Both events cause immediate braking with full deceleration $-b$. Thus, *atp* decisions take precedence over the recommended *spd* speed advisory. In the case where $\mathbf{m}.e - \tau.p > SB$ and no emergency message arrived, the decisions made by *spd* take effect. We will identify safety-critical constraints on parameter *SB* in Sect. 7.3.2.

Radio Block Controller

We model the RBC as a controller with two possible choices ($\cup$). It may choose to either demand immediate corrective action by sending emergency messages (*rbc.message* := *emergency*) or the RBC may update the MA by assigning arbitrary new values to its three components ($\mathbf{m} := *$). These nondeterministic changes to $\mathbf{m}$ reflect different real-world effects like extending $\mathbf{m}.e$ and $\mathbf{m}.d$ if the heading train has advanced significantly or notify of a new recommended speed $\mathbf{m}.r$ for a track segment. We will identify safety-critical constraints on MA updates in Sect. 7.3.2.

Figure 7.4 illustrates the transition structure $\rho(ETCS_{\text{skel}})$, as defined in Definition 2.7, that corresponds to the hybrid program of the ETCS skeleton in Fig. 7.3.

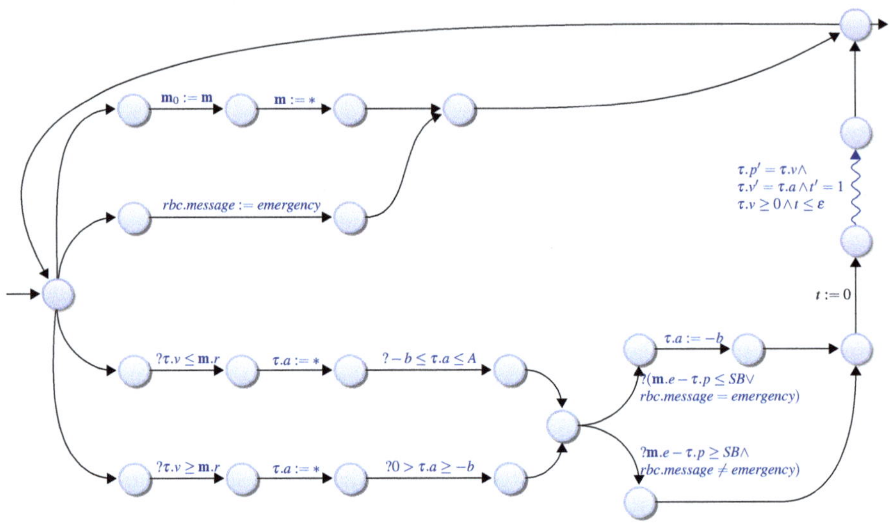

Fig. 7.4 Transition structure of ETCS skeleton

7.3 Parametric Verification of Train Control

The model in Fig. 7.3 that was read off from the informal ETCS specification [117] is unsafe, i.e., it does not always prevent collisions. To correct this we identify constraints on the free parameters of ETCS by analysing increasingly more complex correctness properties of ETCS. Using these constraints we refine the train control model iteratively into a safe model with constraints on design parameter choices and physical prerequisites on external parameters resulting from the safety requirements on ETCS and the behaviour of the train dynamics.

Iterative Refinement Process

For discovering parametric constraints required for system correctness, we follow an *iterative refinement process* that is of more general interest. It can be used to synthesise parameters with the $d\mathcal{L}$ calculus, i.e., successively identify safety-critical parameter constraints for a template controller.

1. *Controllability discovery*: Start with uncontrolled system dynamics. Use structural symbolic decomposition in $d\mathcal{L}$ until a first-order formula is obtained revealing the controllable state region, which specifies for which parameter combinations the system dynamics can actually be controlled safely by any control law.
2. *Control refinement*: Successively add partial control laws to the system while leaving its decision parameters (such as *SB* or **m**) free. Use structural symbolic decomposition to discover parametric constraints which preserve controllability under these control laws.
3. *Safety convergence*: Repeat step 2 until the resulting system is proven safe.
4. *Liveness check*: Prove that the discovered parametric constraints do not over-constrain the system inconsistently by showing that it remains live.

In practise, variants of the controllable domain as discovered by step 1 constitute good candidates for inductive invariants, and the parameter constraints discovered by step 2 ensure that the actual control choices taken by the controller never leave the controllable domain. For step 4, liveness can be verified again by structural symbolic decomposition in $d\mathcal{L}$ and no need for separate verification techniques or different system models arises.

7.3.1 Controllability Discovery in Parametric ETCS

By analysing the uncontrolled train dynamics, we obtain a controllability constraint on the external train parameters, i.e., a formula characterising the parameter combinations for which the train dynamics can be controlled safely by any control law at all. For our analysis we choose the global assumptions

$$\tau.v \geq 0 \land \mathbf{m}.d \geq 0 \land b > 0 \land A \geq 0 \tag{$\mathscr{A}$}$$

stating that the velocity is nonnegative ($\tau.v \geq 0$), the movement authority issued by the RBC does not force the train to drive backwards ($\mathbf{m}.d \geq 0$), and the train has some positive braking force $b > 0$ and some nonnegative maximum acceleration $A \geq 0$. The controllability constraint is now obtained by applying the $d\mathcal{L}$ proof calculus from Chap. 2 to the following $d\mathcal{L}$ (or DAL) formula:

$$(\mathscr{A} \land \tau.p \leq \mathbf{m}.e) \to [\tau.p' = \tau.v \land \tau.v' = -b \land \tau.v \geq 0]\mathscr{S}.$$

This formula expresses that—starting in some state where assumptions ($\mathscr{A}$) hold and the train has not yet passed $\mathbf{m}.e$ (i.e., $\tau.p \leq \mathbf{m}.e$)—every possible evolution of the train system that applies full brakes ($\tau.v' = -b$) is safe, i.e. does not violate ($\mathscr{S}$). By a proof similar to that in Fig. 2.14 on p. 89, we can discover that this $d\mathcal{L}$ formula only holds if $\tau.v^2 - \mathbf{m}.d^2 \leq 2b(\mathbf{m}.e - \tau.p)$ is satisfied in the initial state. We prove that the so-discovered constraint, illustrated in Fig. 7.5, characterises the set of states where the train dynamics can still respect MA by appropriate control choices (expressed by the left-hand side d$\mathcal{L}$ formula).

Fig. 7.5 Controllable region of ETCS

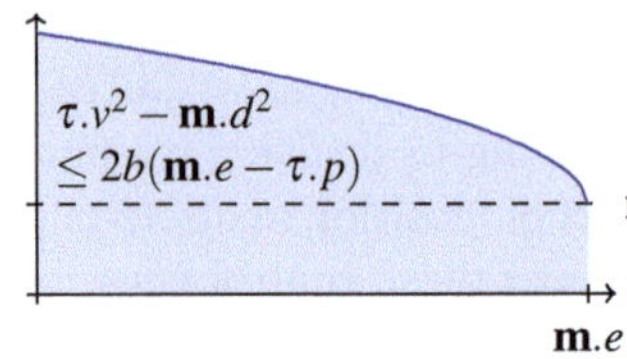

Proposition 7.1 (Controllability). *The constraint* $\tau.v^2 - \mathbf{m}.d^2 \le 2b(\mathbf{m}.e - \tau.p)$ *is a* controllability constraint *for the train* τ *with respect to property* $(\mathcal{S})$ *on page 283, i.e., this constraint retains the ability of the train dynamics to respect the safety property. Formally, with* $\mathcal{A} \wedge \tau.p \le \mathbf{m}.e$ *as regularity assumptions, the following equivalence is a provable* d$\mathcal{L}$ *formula:*

$$\begin{aligned} &[\tau.p' = \tau.v \wedge \tau.v' = -b \wedge \tau.v \ge 0]\,(\tau.p \ge \mathbf{m}.e \to \tau.v \le \mathbf{m}.d) \\ \equiv\, &\tau.v^2 - \mathbf{m}.d^2 \le 2b(\mathbf{m}.e - \tau.p). \end{aligned}$$

This formula expresses that *every run* of a train in braking mode always satisfies $(\mathcal{S})$ if and only if condition $\tau.v^2 - \mathbf{m}.d^2 \le 2b(\mathbf{m}.e - \tau.p)$ holds initially. Observe how the above equivalence reduces a dynamic d$\mathcal{L}$ formula about future controllable train dynamics to a single static constraint on the current state. We use this key reduction step from dynamically safe train dynamics to controllably safe state constraints by analysing whether each part of the ETCS controller preserves train controllability.

Definition 7.1 (Controllable state). A train τ is in a *controllable state*, if the train is always able to stay within its movement authority $\mathbf{m}$ by appropriate control actions, which, by Proposition 7.1, is equivalent to

$$\tau.v^2 - \mathbf{m}.d^2 \le 2b(\mathbf{m}.e - \tau.p) \wedge \mathcal{A}. \qquad (\mathcal{C})$$

ETCS cannot be safe unless trains start and stay in controllable states. Hence we pick $(\mathcal{C})$ as a minimal candidate for an inductive invariant. This invariant will be used to prove safety of the system by induction.

7.3.2 Iterative Control Refinement of ETCS Parameters

Starting from the constraints for controllable trains, we identify constraints for their various control decisions and refine the ETCS model correspondingly, so that the train stays controllable for every control choice.

RBC Control Constraints

For safe functioning of the ETCS it is important that trains always respect their current MA. Consequently, RBCs are not allowed to issue MAs that are physically impossible for the train to obey, such as instantaneous full stops. Rather, RBCs are only allowed to send new MAs that remain within the controllable range of the train dynamics. For technical reasons the RBC does not reliably know the train positions and velocities in its domain of responsibility to a sufficient precision, because the communication with the trains has to be performed wirelessly with possibly high communication delay and message loss. Thus, we give a fail-safe constraint for MA updates which is reliably safe even for loss of position recording communication.

Proposition 7.2 (RBC preserves train controllability). *The constraint*

$$\mathbf{m}_0.d^2 - \mathbf{m}.d^2 \le 2b(\mathbf{m}.e - \mathbf{m}_0.e) \land \mathbf{m}_0.d \ge 0 \land \mathbf{m}.d \ge 0 \tag{$\mathscr{M}$}$$

ensures that the RBC preserves train controllability ($\mathscr{C}$) *when changing the MA from* $\mathbf{m}_0$ *to* $\mathbf{m}$. *That is, the following* d$\mathscr{L}$ *formula is provable:*

$$\forall\tau\Big(\mathscr{C} \to [\mathbf{m}_0 := \mathbf{m};\ rbc]\ (\mathscr{M} \to \mathscr{C})\Big). \tag{7.1}$$

Moreover, RBC controllability is characterised by the following provable d$\mathscr{L}$ *formula:*

$$\mathbf{m}.d \ge 0 \land b > 0\ \to\ [\mathbf{m}_0 := \mathbf{m};\ rbc]\ \Big(\mathscr{M} \leftrightarrow \forall\tau\,\big((\langle \mathbf{m} := \mathbf{m}_0\rangle \mathscr{C}) \to \mathscr{C}\big)\Big). \tag{7.2}$$

Constraint ($\mathscr{M}$) characterises that an extension is safe if it is possible to reduce the speed by braking with deceleration b from the old target speed $\mathbf{m}_0.d$ to the new target speed $\mathbf{m}.d$ within the extension range $\mathbf{m}.e - \mathbf{m}_0.e$, regardless of the current speed of train τ. It imposes constraints on feasible track profiles. Formula (7.1) expresses that, for all trains ($\forall\tau$, i.e., $\forall\tau.v\forall\tau.p$) in a controllable state ($\mathscr{C}$), every RBC change ($[rbc]$) of MA from $\mathbf{m}_0$ to $\mathbf{m}$ that complies with ($\mathscr{M}$) ensures that the train is still in a controllable state ($\mathscr{C}$). Constraint ($\mathscr{M}$) is characterised by the equivalence (7.2), expressing that for every decision of *rbc*, ($\mathscr{M}$) holds for the RBC change from $\mathbf{m}_0$ to $\mathbf{m}$ if and only if *all* trains ($\forall\tau$) that were controllable ($\mathscr{C}$) for the previous MA (set using $\langle \mathbf{m} := \mathbf{m}_0\rangle$) still remain controllable for the new MA $\mathbf{m}$.

Train Control Constraints

Now that we have found constraints characterising when the cooperation of train and RBC is controllable, we need to find out under which circumstances the actual control choices by *spd* and *atp* retain controllability. In particular, the design parameter *SB* (start braking point relative to the end of the movement authority) needs to be chosen appropriately to preserve ($\mathscr{C}$). First we show that *there is* a safe choice of *SB*:

Proposition 7.3 (Reactivity of ETCS). *For all feasible RBC choices and all choices of speed control, there is a choice for SB that makes the train always stay within its MA, i.e., for controllable states, the following d$\mathscr{L}$ formula is provable:*

$$\mathscr{C} \rightarrow [\mathbf{m}_0 := \mathbf{m};\ rbc]\big(\mathscr{M} \rightarrow [spd]\langle SB := *\rangle[atp;\ drive]\mathscr{S}\big).$$

The formula expresses that, starting in a controllable region $\mathscr{C}$, if the RBC updates the MA from $\mathbf{m}_0$ to any $\mathbf{m}$ respecting ($\mathscr{M}$), then after arbitrary *spd* choices, the train controller is still able to find some choice for *SB* ($\langle SB := *\rangle$) such that it always respects the fresh MA when following *atp* and *drive*. Since Proposition 7.3 is provable in KeYmaera we know that *there is a safe solution for ETCS*. On the formula level the assumptions are expressed using implications such that the formula does not make any proposition if either ($\mathscr{C}$) is not initially satisfied or the RBC updates do not respect ($\mathscr{M}$). The train controller in this property is split up into the proposition that for all executions of the speed supervision ($[spd]$) there is a choice for *SB* ($\langle SB := *\rangle$) such that the automatic train protection unit (*atp*) always preserves safety during the train movement in the *drive* phase. For *atp* and *drive* we again make a statement over all possible executions of the components ($[atp; drive]$). Only the choice of *SB* is existentially quantified by an $\langle SB := *\rangle$ modality. We see that alternation of d$\mathscr{L}$ modalities is quite expressive and can capture interaction of system components.

To find a particular constraint on the choice of *SB*, we need to take the maximum reaction latency ε of the train controllers into account. With $\varepsilon > 0$, the point where the train needs to apply brakes to comply with $\mathbf{m}$ is not determined by the physical constraints ($\mathscr{C}$) alone, but needs additional safety margins to compensate for reaction delays in the controller. Therefore, we search for a constraint that characterises that for every possible end of the movement authority ($\forall \mathbf{m}.e$) and every train position ($\forall \tau.v$), train movement with an acceleration of A preserves ($\mathscr{C}$) if it started in a state where ($\mathscr{C}$) holds and the point *SB* has not been passed yet ($\mathbf{m}.e - \tau.p \geq SB \wedge \mathscr{C}$).

Proposition 7.4 (Reactivity constraint). *If the train is in a controllable state, the supervisory ETCS controller reacts appropriately in order to maintain controllability iff SB is chosen according to the following provable equivalence:*

$$\Big(\forall \mathbf{m}.e \forall \tau.p\, \big(\mathbf{m}.e - \tau.p \geq SB \wedge \mathscr{C} \rightarrow [\tau.a := A;\ drive]\mathscr{C}\big)\Big)$$
$$\equiv\ SB \geq \frac{\tau.v^2 - \mathbf{m}.d^2}{2b} + \left(\frac{A}{b} + 1\right)\left(\frac{A}{2}\varepsilon^2 + \varepsilon\, \tau.v\right). \qquad (\mathscr{B})$$

Reactivity constraint ($\mathscr{B}$) on *SB* can be derived using a projection of the train behaviour to the worst-case acceleration A in a state where *SB* has not been passed yet, similarly to the approach from Sect. 2.9. We choose this projection because the train controller needs to ensure that it can drive safely with maximum acceleration A for ε time units even right before passing *SB* in order for an acceleration choice of A to be safe. Constraint ($\mathscr{B}$) is not at all obvious from the original system model in Fig. 7.3.

$$
\begin{aligned}
ETCS_{\mathrm{r}} &\equiv (train_{\mathrm{r}} \cup rbc_{\mathrm{r}})^{*}\\
train_{\mathrm{r}} &\equiv spd;\ atp_{\mathrm{r}};\ drive\\
spd &\equiv \ \big(?(\tau.v \le \mathbf{m}.r);\ \tau.a{:=}{*};\ ?(-b \le \tau.a \le A)\big)\\
&\quad \cup \big(?(\tau.v \ge \mathbf{m}.r);\ \tau.a{:=}{*};\ ?(-b \le \tau.a \le 0)\big)\\
atp_{\mathrm{r}} &\equiv SB := \tfrac{\tau.v^2-\mathbf{m}.d^2}{2b} + \left(\tfrac{A}{b}+1\right)\left(\tfrac{A}{2}\varepsilon^2+\varepsilon\,\tau.v\right);\ atp\\
atp &\equiv \text{if } (\mathbf{m}.e-\tau.p \le SB \lor rbc.message = emergency) \text{ then } \tau.a := -b \text{ fi}\\
drive &\equiv t:=0;\ (\tau.p' = \tau.v \land \tau.v' = \tau.a \land t' = 1 \land \tau.v \ge 0 \land t \le \varepsilon)\\
rbc_{\mathrm{r}} &\equiv \ (rbc.message := emergency)\\
&\quad \cup \big(\mathbf{m}_0 := \mathbf{m};\ \mathbf{m}{:=}{*};\\
&\qquad ?(\mathbf{m}.r \ge 0 \land \mathbf{m}.d \ge 0 \land \mathbf{m}_0.d^2 - \mathbf{m}.d^2 \le 2b(\mathbf{m}.e-\mathbf{m}_0.e))\big)
\end{aligned}
$$

Fig. 7.6 ETCS cooperation protocol refined with parameter constraints

After discovering constraint ($\mathscr{B}$), however, it can be explained in retrospect: It characterises the relative braking distance required to reduce speed from $\tau.v$ to target speed $\mathbf{m}.d$ with braking deceleration b, which corresponds to controllability and is expressed by the term $\frac{\tau.v^2-\mathbf{m}.d^2}{2b}$. In addition, the constraint involves the distance travelled during one reaction cycle of at most ε time units with acceleration A, including the additional distance needed to reduce the speed down to $\tau.v$ again after accelerating at A for ε time units, as expressed by $\left(\frac{A}{b}+1\right)\left(\frac{A}{2}\varepsilon^2+\varepsilon\,\tau.v\right)$. This extra distance results from speed changes and depends on the relation $\frac{A}{b}$ of maximum acceleration A and maximum braking force b.

Propositions 7.1–7.4 prove d$\mathscr{L}$ equivalences. Hence, counterexamples exist that demonstrate unsafety of the ETCS skeleton in Fig. 7.3 whenever the respective parameter constraints are not met. Consequently, these constraints must be obeyed for correctness of *any* model of ETCS controllers, including all technical implementation refinements. It is, thus, important to identify these safety constraints early in the overall design and verification process of ETCS. Similar parameter constraints hold for other upcoming train control systems.

7.3.3 Safety Verification of Refined ETCS

By refining the system from Fig. 7.3 with the parametric constraints obtained from Propositions 7.1–7.4, we synthesise a safe system model completing the ETCS protocol skeleton. In Fig. 7.6, we present the refined model, which bugfixes the model in Fig. 7.3 taken from the informal specification. Parts *spd*, *drive*, and the subprogram *atp* of atp_{r} stay as in Fig. 7.3. The constraints in atp_{r} and rbc_{r} are new, as is the precondition $\mathscr{C}$ on the initial states.

Proposition 7.5 (Safety of ETCS). *Assuming the train starts in a controllable state, the following global and unbounded horizon* d$\mathscr{L}$ *safety formula about the refined ETCS system in Fig. 7.6 is provable:*

$$\mathscr{C} \rightarrow [ETCS_r](\tau.p \ge \mathbf{m}.e \rightarrow \tau.v \le \mathbf{m}.d). \tag{7.3}$$

This provable $d\mathcal{L}$ formula states that, starting in a controllable region ($\mathscr{C}$), the refined ETCS model is safe, i.e., trains always respect their movement authority even when movement authorities change dynamically based on wireless communication with the RBC.

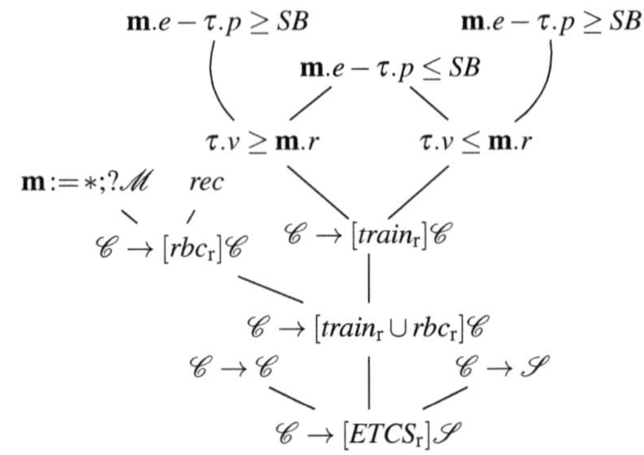

Fig. 7.7 Proof sketch for ETCS safety (Proposition 7.5)

As an example to illustrate the proof structure for the verification of Proposition 7.5, consider the sketch in Fig. 7.7. The proof starts with the conjecture at the bottom and proceeds by decomposition in the $d\mathcal{L}$ calculus to the leafs. Figure 7.7 does not give a full formal proof in the $d\mathcal{L}$ calculus but still shows a sketch of the overall proof structure for Proposition 7.5. We need to prove that the assumption that the train is in a controllable state expressed by ($\mathscr{C}$) implies $[ETCS_r]\mathscr{S}$. As the system consists of a global loop $(train_r \cup rbc_r)^*$, we prove that ($\mathscr{C}$) is an invariant of this loop and strong enough to imply ($\mathscr{S}$) by $d\mathcal{L}$ rule ind' from p. 86, which splits into the three lower branches indicated in Fig. 7.7. Using KeYmaera it can be shown easily that the invariant is initially valid (left branch) and implies the postcondition (right branch). As usual, proving that the invariant is preserved by the loop body is the most challenging part of the proof (middle branch), which splits into two cases by applying $d\mathcal{L}$ rule $[\cup]$, and rule $\wedge r$ from Fig. 2.11 subsequently. For the left case, we have to show that the RBC preserves the invariant, which follows from Proposition 7.2 and splits into the case where the RBC changes the MA according to constraint ($\mathscr{M}$) (in the left branch $\mathbf{m} := *$) and the case where an emergency message requests immediate recovery by emergency braking (denoted by *rec* in the right branch). For the right choice, we show that the train controller preserves the invariant. The proof splits again by rule $[\cup]$ due to the choice in the *spd* component depending on the relation of the current speed $\tau.v$ to the recommended speed $\mathbf{m}.r$. The next split with rule $[\cup]$ on both of these branches depends on the value of *SB* according to *atp*. If the train has passed point *SB* (the shared middle case), we can close this goal using Proposition 7.1, because the invariant describes a controllable state and the *spd* controller applies safety brakes. The outer branches, where the train has not yet passed *SB*, can be closed using Proposition 7.4, because the controllers will react to situation changes after at most ε time units (induction). Overall, the d$\mathscr{L}$ formula in Proposition 7.5 can be proven automatically by our approach.

7.3.4 Liveness Verification of Refined ETCS

In order to show that the discovered parameter constraints do not over-constrain the system inconsistently, we show liveness, i.e., that an ETCS train is able to reach every track position with appropriate RBC permissions.

Proposition 7.6 (Liveness of ETCS). *The refined ETCS system is live. That is, assuming the RBC can safely grant the required MAs because preceding trains are moving on, trains are able to reach any track position P by appropriate RBC choices, which is expressed in the following provable* d$\mathscr{L}$ *formula:*

$$\tau.v \geq 0 \land A > 0 \land \varepsilon > 0 \to \forall P \langle ETCS_r \rangle \, \tau.p \geq P. \tag{7.4}$$

This d$\mathscr{L}$ formula expresses that, starting in a state where the velocity is non-negative and where maximum acceleration and maximum evolution time limit are positive, every point P (that is, $\forall P$) can be reached ($\tau.p \geq P$) by some execution of the ETCS model ($\langle ETCS_\mathrm{r} \rangle$). Here the diamond modality is used to say that not all, but *some* appropriate execution reaches a state where the postcondition ($\tau.p \geq P$) holds. The validity of formula (7.4) means that the train model is always able to reach the postcondition *for all* corresponding values of the variables. Mere satisfiability of (7.4) would correspond to a weaker property saying that the train can move only under some circumstance for appropriate positions (e.g., when $\tau.p \geq P$ trivially holds in the beginning), and not under all circumstances for all values of the free variables, as Proposition 7.6 shows.

Further note that, for proving that the ETCS model is live, a more liberal initial state is possible with regard to the controllability of the train. The only important restrictions on the initial region are those ensuring that the system of differential equations used for modelling the train movement can actually be followed for some positive amount of time $\varepsilon > 0$. As usual, the velocity of the train must be nonnegative ($\tau.v \geq 0$). If either of these assumptions is violated, the train may be unable to move, because the evolution domain ($\tau.v \geq 0 \land t \leq \varepsilon$) of the differential equation system ($\mathscr{I}$) is violated immediately and thus no continuous movement of the model would be allowed at all. Finally, the maximum acceleration A must be positive ($A > 0$) or at least the initial velocity must be $\tau.v > 0$; otherwise the train would be locked to its current position without any effective control choices.

The proof of Proposition 7.6 uses unwinding (d$\mathscr{L}$ rule $\langle ^{*n} \rangle$) and generalisation (rule $\langle\rangle gen$) to split the proof into one part to show that positive speed $\tau.v > 0$ is always reachable, and another part to show that positive speed can be maintained during sufficiently many control cycles with a non-Zeno progress towards the goal position P. The latter part of the proof can be shown using the convergence rule *con* with variant $\varphi(n) \equiv \tau.p + n\varepsilon v_0 \geq P \land \tau.v \geq v_0$, which expresses that goal P is within reach of at most n cycles of duration ε at minimum speed v_0 and that the speed $\tau.v$ does not decrease below the speed v_0 reached after the first unwinding cycle.

Contrast this simple variant with the complex variant (2.21) needed for proving the liveness property of the simple train control subsystem considered in property (2.20) on p. 121 for nonnegative acceleration $A \geq 0$. At first sight, this complexity difference sounds somewhat surprising. After all, the system considered in Chap. 2 is a lot simpler than the full ETCS protocol considered here, so the variant and proof in Chap. 2 should be simpler too. While this was perfectly true for safety properties, the situation is slightly different for liveness. In terms of liveness, the system considered in (2.20) had significantly less flexibility compared to the full ETCS system in Fig. 7.6. For safety, flexibility is harder, because we have to prove more cases and combinations to stay safe. For liveness, flexibility is good, because we have more options to choose from for satisfying the goal. The simple train controller in Chap. 2, for instance, allowed only one choice of a movement authority which then remains constant forever. Consequently, our liveness proof had to establish the existence of an appropriate MA with only one single control shot. For the ETCS system in Fig. 7.6, multiple adjustments of the MA are permitted, with which it is a lot easier to prove liveness, because inappropriate prior MAs can simply be readjusted again by RBC communication. Essentially, the liveness proof in Chap. 2 had to plan ahead and prove without any intermediate adaptation, whereas our liveness proof for Proposition 7.6 is able to use dynamic feedback to adapt to the actual movement (dynamically during the proof).

7.3.5 Full Correctness of ETCS

By collecting Propositions 7.1–7.6 together, we obtain the following main result of this chapter, which demonstrates the feasibility of d$\mathcal{L}$-based parametric discovery and formal verification. It gives important insights into the fully parametric ETCS case study and yields conclusive and fully verified choices for the free parameters in ETCS. By virtue of the parametric formulation, this result applies to all concrete instantiations of the ETCS cooperation protocol (Sect. 7.2), including controllers that further optimise speed or model refinements in hardware implementations. These constraints also apply to related train control systems.

Theorem 7.1 (Correctness of ETCS cooperation protocol). *The refined ETCS system with constraints ($\mathcal{B}$) and ($\mathcal{M}$) is correct as given in Fig. 7.6. Starting in any controllable state respecting ($\mathcal{C}$), trains remain in the controllable region at any time. They safely respect movement authorities issued by the RBC so that ETCS is collision-free. Further, trains can always react safely to all RBC decisions respecting ($\mathcal{M}$). ETCS is live: when tracks become free, trains are able to reach any track position by appropriate RBC actions. Furthermore, the refinements ($\mathcal{C}$) and ($\mathcal{B}$) are necessary and sharp: Every configuration violating ($\mathcal{C}$) or ($\mathcal{B}$), respectively, gives rise to a concrete counterexample violating safety property ($\mathcal{S}$). Finally, every RBC* *choice violating ($\mathscr{M}$) gives rise to a counterexample in the presence of lossy wireless communication channels.*

7.4 Disturbance and the European Train Control System

In Sects. 7.2 and 7.3, we assumed there was direct control of acceleration in ideal-world physics. In reality, actual acceleration is caused by the physical transmission of corresponding forces that depend on the electrical current in the engine and are regulated by PID controllers [91]. Yet even a full model of the physical transmission system is still somewhat incomplete, because wind and friction, slippery rail conditions, and turbulence may have some small impact on the dynamics.[2] As a conservative overapproximation of all these effects, we instead generalise the ETCS model to a model with differential inequalities, where we also take into account disturbances in the physical transmission of forces:

$$\tau.p' = \tau.v \land \tau.a - l \leq \tau.v' \leq \tau.a + u \land t' = 1 \land \tau.v \geq 0 \land t \leq \varepsilon. \qquad (\mathscr{I}_d)$$

This differential inequality tolerates a disturbance within the interval $[-l,u]$. That is, the acceleration $\tau.a$ chosen by the train controller may take effect with an error bounded by $-l$ and u. Especially, the derivative $\tau.v'$ of the velocity will not need to equal $\tau.a$ exactly in ($\mathscr{I}_d$), but $\tau.v'$ can vary arbitrarily between $\tau.a - l$ and $\tau.a + u$ over time. We thus generalise the differential equation ($\mathscr{I}$) in component *train* from Figs. 7.3 and 7.6 by replacing it with the differential inequality ($\mathscr{I}_d$) and denote the result by $train_d$. For the precise semantics of hybrid systems with differential inequalities, we refer to the semantics of DA-programs with DA-constraints for continuous evolution given in Chap. 3; see Examples 3.3 and 3.20 on pp. 137 and 176, respectively. Especially, because of the differential inequalities, the resulting programs are DA-programs and the resulting formulas are, strictly speaking, DAL formulas. See Fig. 7.8 for an illustration of how the original controllability region can be violated, because the disturbed dynamics can deviate from the ideal-world dynamics with a time-dependent, bounded disturbance.

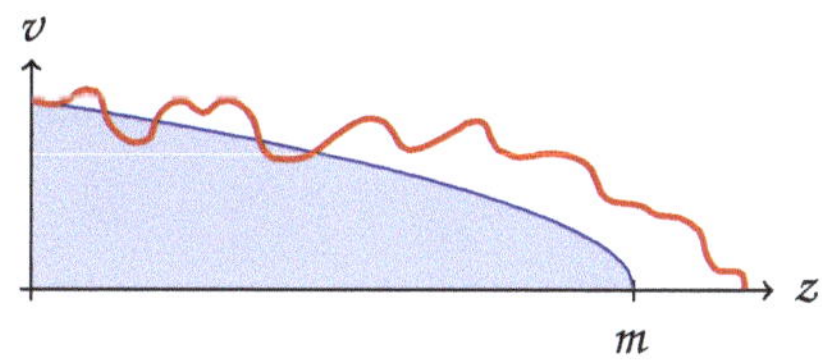

Fig. 7.8 Controllability region changes in the presence of disturbance

[2] Obtaining a fully complete model is thus de facto impossible, probably not even when resolving to the Schrödinger equation of quantum mechanics [276].

Notice that, unlike with the differential equation ($\mathscr{I}$), we cannot simply solve differential inequality ($\mathscr{I}_d$), because its actual solution depends on the precise value of the disturbance, which is a quantity that changes over time. Thus, solutions will only be relative to this disturbance function and a reachability analysis would have to consider all choices of this function, which would require higher-order logic. Instead, we use differential invariants from Chap. 3 as a first-order characterisation for the proofs behind the following propositions.

7.4.1 Controllability in ETCS with Disturbances

The controllability characterisation from Proposition 7.1 carries over to train control with disturbance when taking into account the effect that a maximum disturbance u can have on the maximum braking force b and limits the effective braking force to $(b-u)$:

Proposition 7.7 (Controllability despite disturbance). *The constraint*

$$\tau.v^2 - \mathbf{m}.d^2 \leq 2(b-u)(\mathbf{m}.e - \tau.p) \wedge \mathbf{m}.d \geq 0 \wedge b > u \geq 0 \wedge l \geq 0 \qquad (\mathscr{C}_d)$$

is a controllability constraint *with respect to property* ($\mathscr{S}$) *for the train* τ *with disturbance* ($\mathscr{I}_d$)*, i.e., it retains the ability of the train dynamics to respect the safety property despite disturbance. Formally, with* $\mathscr{A} \wedge \tau.p \leq \mathbf{m}.e \wedge b > u \geq 0 \wedge l \geq 0$ *as regularity assumptions, the following equivalence is provable in the* DAL *calculus:*

$$\begin{aligned}&\left[\tau.p' = \tau.v \wedge -b-l \leq \tau.v' \leq -b+u \wedge t' = 1 \wedge \tau.v \geq 0 \wedge t \leq \varepsilon\right]\mathscr{S}\\ &\equiv \tau.v^2 - \mathbf{m}.d^2 \leq 2(b-u)(\mathbf{m}.e - \tau.p).\end{aligned}$$

Here ($\mathscr{C}_d$) results from ($\mathscr{C}$) by replacing b with $(b-u)$. In worst-case disturbance, the train cannot brake with its chosen deceleration $-b$ but might be off by a disturbance up to u. In order to guarantee that the train is still always able to stay within its MA despite the disturbance, the controller has to assume worst-case guaranteed deceleration $-(b-u)$ when making control decisions.

Since the proof of Proposition 7.7 follows interesting principles, we present a formal DAL proof for Proposition 7.7 in Fig. 7.9, using the DAL calculus from Fig. 3.9. The proof uses the following abbreviations:

$$\begin{aligned}X \approx -b &\equiv -b-l \leq X \wedge X \leq -b+u\\ \psi &\equiv \mathscr{A} \wedge \tau.p \leq \mathbf{m}.e \wedge b > u \geq 0 \wedge l \geq 0\\ \phi &\equiv \tau.v^2 - \mathbf{m}.d^2 \leq 2(b-u)(\mathbf{m}.e - \tau.p)\\ \langle \mathscr{S}_t \rangle &\equiv \langle \tau.p := \frac{-b+u}{2}t^2 + \tau.v\,t + \tau.p \wedge \tau.v := (-b+u)t + \tau.v \rangle.\end{aligned}$$

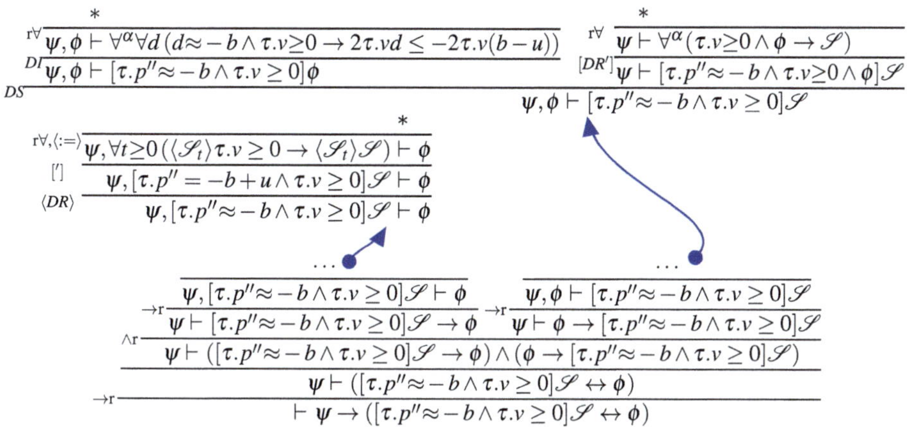

Fig. 7.9 Proof of ETCS controllability despite disturbance (Proposition 7.7)

The bottommost proof block in Fig. 7.9 splits the proof into branches for proving the two directions of the equivalence: the direction for proving necessity (middle proof block) and the direction for proving sufficiency (top proof block).

In the top block, sufficiency can be proven by differential strengthening (DAL rule *DS* from Fig. 3.9), with the controllability constraint $\mathscr{C}_d$ (or its relevant part ϕ) as an auxiliary invariant. On the left branch of the top block, validity of the differential strengthening step can be proven by differential induction rule *DI* from Fig. 3.9 to show that ϕ is actually an invariant, using the differential inequality reduction techniques from Sect. 3.5.3. On the right branch of the top block, differential weakening rule $[DR']$ from p. 175 can be used to conclude $\mathscr{S}$ from the auxiliary invariant ϕ and the evolution domain restriction $\tau.v \geq 0$. Also see Example 3.20 on p. 176 for details on this kind of reasoning.

In the middle block, necessity can be proven by differential refinement (DAL rule $\langle DR\rangle$ from Fig. 3.9). We use differential refinement to replace the differential inequality ($\mathscr{I}_d$)—which we abbreviate as $\tau.p''\approx -b$ in Fig. 7.9—with the differential equation $\tau.p'' = -b+u$. This differential equation represents the worst case with constant maximum disturbance u in the differential inequality ($\mathscr{I}_d$). Intuitively, this replacement is possible, because the differential equation describes one possible special case of the differential inequality and we only need to find one permitted evolution that implies ϕ for proving necessity. Formally, the differential refinement is justified by proof rule $\langle DR\rangle$ and the fact that

$$\tau.p' = \tau.v \land \tau.v' = -b+u \land \tau.v \geq 0$$

entails the following differential inequality (in the sense of Lemma 3.3 from p. 158):

$$\tau.p' = \tau.v \land -b-l \leq \tau.v' \leq -b+u \land \tau.v \geq 0.$$

Note that DAL rule $\langle DR\rangle$ can be applied to the $[]$-formulas in the antecedent, because these are equivalent to $\langle\rangle$-formulas in the succedent.

7.4.2 Iterative Control Refinement of Parameters with Disturbances

When taking into account worst-case effects of disturbance on control, reactivity constraint ($\mathscr{B}$) carries over to the presence of disturbance in the train dynamics.

Proposition 7.8 (Reactivity constraint despite disturbance). *For trains in controllable state, the supervisory ETCS controller reacts appropriately despite disturbance in order to maintain controllability iff SB is chosen according to the following provable equivalence:*

$$\Big(\forall \mathbf{m}.e \forall \tau.p \left((\mathbf{m}.e - \tau.p \geq SB \wedge \tau.v^2 - \mathbf{m}.d^2 \leq 2(b-u)(\mathbf{m}.e - \tau.p)) \rightarrow \right.$$
$$\left. [\tau.a := A;\ drive_d](\tau.v^2 - \mathbf{m}.d^2 \leq 2(b-u)(\mathbf{m}.e - \tau.p))\right)\Big)$$
$$\equiv SB \geq \frac{\tau.v^2 - \mathbf{m}.d^2}{2(b-u)} + \left(\frac{A+u}{b-u} + 1\right)\left(\frac{A+u}{2}\varepsilon^2 + \varepsilon\tau.v\right). \qquad (\mathscr{B}_d)$$

For reactivity constraint ($\mathscr{B}_d$), not only the maximum deceleration but also the maximum acceleration matters—both could be affected by disturbances. Therefore, we need to substitute not only every b with $(b-u)$ but also every A with $(A+u)$, the maximum acceleration under disturbance, to get a (provable) reactivity constraint for the disturbed system.

Accordingly, we adapt the RBC constraint ($\mathscr{M}$) to account for the maximum effect of disturbance by again replacing every b with $(b-u)$:

$$\mathbf{m}_0.d^2 - \mathbf{m}.d^2 \leq 2b(\mathbf{m}.e - \mathbf{m}_0.e) \wedge \mathbf{m}_0.d \geq 0 \wedge \mathbf{m}.d \geq 0. \qquad (\mathscr{M}_d)$$

Constraint ($\mathscr{M}_d$) is fairly easy to obtain from ($\mathscr{M}$), because the RBC itself is not subject to disturbance, and only the train is.

7.4.3 Safety Verification of ETCS with Disturbances

When we refine the ETCS model with the constraints ($\mathscr{B}_d$) and ($\mathscr{M}_d$), ETCS is safe even in the presence of disturbance in the dynamics when starting in a state respecting ($\mathscr{C}_d$).

Proposition 7.9 (Safety despite disturbance). *Assuming the train starts in a controllable state satisfying ($\mathscr{C}_d$), the following global and unbounded horizon safety* *formula about the ETCS system from Fig. 7.10 with disturbance in the dynamics is provable:*

$$\mathscr{C}_d \rightarrow [ETCS_d](\tau.p \geq \mathbf{m}.e \rightarrow \tau.v \leq \mathbf{m}.d).$$

$$\begin{aligned}
ETCS_d &\equiv (train_d \cup rbc_d)^* \\
train_d &\equiv spd;\ atp_d;\ drive_d \\
spd &\equiv \big(?(\tau.v \leq \mathbf{m}.r);\ \tau.a := *;\ ?(-b \leq \tau.a \leq A)\big) \\
&\quad \cup \big(?(\tau.v \geq \mathbf{m}.r);\ \tau.a := *;\ ?(0 > \tau.a \geq -b)\big) \\
atp_d &\equiv SB := \tfrac{\tau.v^2 - \mathbf{m}.d^2}{2(b-u)} + \left(\tfrac{A+u}{b-u}+1\right)\left(\tfrac{A+u}{2}\varepsilon^2 + \varepsilon\tau.v\right); \\
&\quad \text{if } (\mathbf{m}.e - \tau.p \leq SB \lor rbc.message = emergency) \text{ then } \tau.a := -b \text{ fi} \\
drive_d &\equiv t := 0;\ (\tau.p' = \tau.v \land \tau.a - l \leq \tau.v' \leq \tau.a + u \land t' = 1 \land \tau.v \geq 0 \land t \leq \varepsilon) \\
rbc_d &\equiv (rbc.message := emergency) \\
&\quad \cup \big(\mathbf{m}_0 := \mathbf{m}; \mathbf{m} := *; \\
&\qquad ?(\mathbf{m}.r \geq 0 \land \mathbf{m}.d \geq 0 \land \mathbf{m}_0.d^2 - \mathbf{m}.d^2 \leq 2(b-u)(\mathbf{m}.e - \mathbf{m}_0.e))\big)
\end{aligned}$$

Fig. 7.10 Parametric ETCS cooperation protocol with disturbances

The proof sketched in Fig. 7.7 for undisturbed ETCS generalises to ETCS with disturbance when using differential induction (DAL proof rule *DI* from Fig. 3.9) and differential strengthening (*DS*) with a time-dependent version of ($\mathscr{B}_d$) as a differential invariant for the acceleration case:

$$\mathbf{m}.e - \tau.p \geq \frac{\tau.v^2 - \mathbf{m}.d^2}{2(b-u)} + \left(\frac{A+u}{b-u}+1\right)\left(\frac{A+u}{2}(\varepsilon - t)^2 + (\varepsilon - t)\tau.v\right). \qquad (7.5)$$

We also use differential strengthening by this differential invariant in the proof of Proposition 7.8. The full ETCS system with disturbance, where all abbreviations are resolved, is depicted in Fig. 7.11.

7.5 Experimental Results

Experimental results for verifying ETCS in our $d\mathcal{L}$- and DAL-based verification tool KeYmaera are presented in Table 7.1. Experimental results are from a 2.6 GHz AMD Opteron with 4 GB memory. All those correctness properties and parameter constraints of ETCS can be verified with 94% to 100% degree automation. In the universal fragment of $d\mathcal{L}$, user interactions are only needed for supplying invariants (or differential invariants), which, in turn, can be discovered using our iterative refinement process and our *differential invariants as fixed-points* algorithm from Chap. 6. For liveness properties or substantial quantifier alternations beyond the capabilities of currently available decision procedures for real arithmetic, KeYmaera still needs more user guidance. Yet, those properties can nevertheless be verified formally with KeYmaera! Further, we see that the symbolic state dimension (Dim) has more impact on the computational complexity than the number of proof steps in $d\mathcal{L}$ decompositions (Steps). Table 7.1 gives the number of user interactions in the

$$\psi \equiv \tau.v^2 - \mathbf{m}.d^2 \le 2(b-u)(\mathbf{m}.e - \tau.p) \wedge \mathbf{m}.d \ge 0 \wedge b > u \ge 0 \wedge l \ge 0$$
$$\to [ETCS_{\mathrm{d}}](\tau.p \ge \mathbf{m}.e \to \tau.v \le \mathbf{m}.d)$$

$$
\begin{aligned}
ETCS_{\mathrm{d}} \equiv & \Big(\\
\textit{spd}: \quad & \big(((?(\tau.v \le \mathbf{m}.r);\ \tau.a := *;\ ?(-b \le \tau.a \le A)) \\
& \cup (?(\tau.v \ge \mathbf{m}.r);\ \tau.a := *;\ ?(0 > \tau.a \ge -b))); \\
\textit{atp}_d: \quad & SB := \tfrac{\tau.v^2 - \mathbf{m}.d^2}{2(b-u)} + \left(\tfrac{A+u}{b-u} + 1\right)\left(\tfrac{A+u}{2}\varepsilon^2 + \varepsilon\tau.v\right); \\
& \text{if } (\mathbf{m}.e - \tau.p \le SB \vee rbc.message = emergency) \text{ then } \tau.a := -b \text{ fi}; \\
\textit{drive}_d: \quad & t := 0;\ (\tau.p' = \tau.v \wedge \tau.a - l \le \tau.v' \le \tau.a + u \wedge t' = 1 \wedge \tau.v \ge 0 \wedge t \le \varepsilon)\big) \\
& \cup \\
\textit{rbc}: \quad & \big((rbc.message := emergency) \\
& \cup (\mathbf{m}_0.d := \mathbf{m}.d;\ \mathbf{m}_0.e := \mathbf{m}.e;\ \mathbf{m}_0.r := \mathbf{m}.r; \\
& \quad \mathbf{m}.d := *;\ \mathbf{m}.e := *;\ \mathbf{m}.r := *; \\
& \quad ?(\mathbf{m}.r \ge 0 \wedge \mathbf{m}.d \ge 0 \wedge \mathbf{m}_0.d^2 - \mathbf{m}.d^2 \le 2(b-u)(\mathbf{m}.e - \mathbf{m}_0.e)))\big) \Big)^*
\end{aligned}
$$

Fig. 7.11 Parametric ETCS cooperation protocol with disturbances (full instantiation)

column Int, and, for comparison, the total number of applied proof rules in column Steps. The interactions in Propositions 7.8 and 7.9 are for the time-dependent version (7.5) of the constraint on *SB*.

The experimental results in Table 7.1 for Proposition 7.3 can be improved significantly. Most of the user interactions are only required to overcome the current limitations of the preliminary implementation of iterative inflation order proof strategies from Sect. 5.5 in KeYmaera. Finally, contrast our overall experimental results with earlier implementations [256] that required as much as 56 user interactions even for property Proposition 7.5 to be provable, which KeYmaera can prove completely automatically using our new automated theorem proving techniques for differential dynamic logics and our logic-based verification algorithms from Part II.

Table 7.1 Experimental results for the European Train Control System

Case study		Int	Time(s)	Memory(MB)	Steps	Dim
Controllability	Proposition 7.1	0	0.6	6.9	14	5
RBC Control	Proposition 7.2 property (7.1)	0	0.5	6.4	42	12
RBC Control	Proposition 7.2 property (7.2)	0	0.9	6.5	82	12
Reactivity	Proposition 7.3	13	279.1	98.3	265	14
Reactivity	Proposition 7.4	0	103.9	61.7	47	14
Safety	Proposition 7.5	0	2052.4	204.3	153	14
Liveness	Proposition 7.6 kernel	4	35.2	92.2	62	10
Liveness	Proposition 7.6 simplified	6	9.6	23.5	134	13
Controllability	Proposition 7.7 disturbance	0	2.8	8.3	26	7
Reactivity	Proposition 7.8 disturbance	1	23.7	47.6	76	15
Safety	Proposition 7.9 disturbance	1	5805.2	34	218	16

7.6 Summary

As a case study for parametric verification of hybrid systems, we have verified controllability, reactivity, safety, and liveness of the fully parametric cooperation protocol of the European Train Control System. We have demonstrated the feasibility of $\mathsf{d}\mathcal{L}$- and DAL-based verification of parametric hybrid systems and identified parametric constraints that are both sufficient and necessary for a safe collision-free operation of ETCS. We have characterised these constraints on the free parameters of ETCS equivalently in terms of corresponding reachability properties of the underlying train dynamics. We have proven that the system remains correct even when the train dynamics is subject to disturbances caused, e.g., by the physical transmission, friction, or wind. We have also verified a corresponding fully parametric proportional-integral (PI) controller for ETCS [245] using KeYmaera.

We have shown how the properties of train control can be expressed in $\mathsf{d}\mathcal{L}$. We have proven all propositions formally in our logic-based verification tool KeYmaera. Our experimental results with KeYmaera show a scalable approach by combining the power of completely automatic verification procedures with the intuition behind user guidance to tackle even highly parametric hybrid systems and properties with substantial quantifier alternation (reactivity or liveness) or disturbance; also see our report [245] for details on the experiments and KeYmaera proofs.

For future work, we want to generalise ETCS using probabilistic information about sensor and communication accuracy and availability.

Chapter 8
Air Traffic Collision Avoidance

Contents

Synopsis Aircraft collision avoidance manoeuvres are important and complex applications. Curved flight exhibits nontrivial continuous behaviour. In combination with the control choices during air traffic manoeuvres, this results in hybrid systems with challenging interactions of discrete and continuous dynamics. As a case study for demonstrating the scalability of logical analysis for hybrid systems with challenging dynamics, we analyse collision freedom of roundabout manoeuvres in air traffic control, where appropriate curved flight, good timing, and compatible manoeuvring are crucial for guaranteeing safe spatial separation of aircraft throughout their flight. We show that our DAL-based proof techniques can scale to curved flight manoeuvres required in aircraft control applications. Our logical analysis approach can be used successfully to verify collision avoidance of the tangential roundabout manoeuvre automatically, even for five aircraft. Moreover, we introduce a fully flyable variant of the roundabout collision avoidance manoeuvre and verify safety properties by compositional verification in our calculus.

A. Platzer, Logical Analysis of Hybrid Systems,
DOI 10.1007/978-3-642-14509-4_8, © Springer-Verlag Berlin Heidelberg 2010

8.1 Introduction

In air traffic control, collision avoidance manoeuvres [293, 104, 129, 171] are used to resolve conflicting flight paths that arise during free flight. See Fig. 8.1 for a series of increasingly more realistic—yet also more complicated—aircraft collision avoidance manoeuvres. Figures 8.1a and b show successful collision avoid-

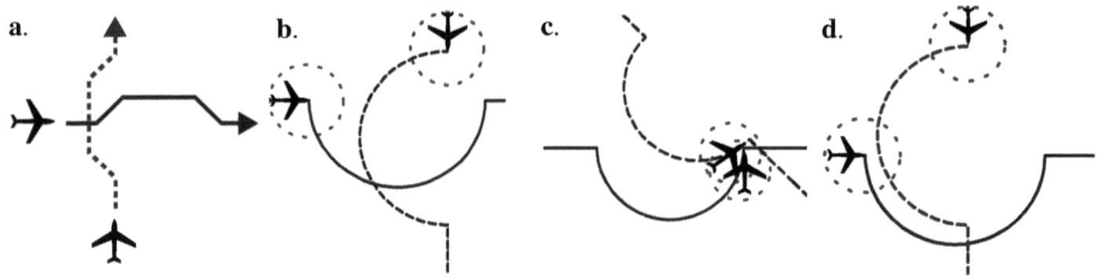

Fig. 8.1 Evolution of collision avoidance manoeuvres in air traffic control

ance manoeuvres. Figure 8.1c shows a malfunctioning collision avoidance attempt, and Fig. 8.1d shows an improved manoeuvre. Collision avoidance manoeuvres are a "last resort" for resolving air traffic conflicts that could lead to collisions. They become important whenever flight route conflicts have not been detected by the pilots during free flight or by the flight directors of the Air Route Traffic Control Centres. Consequently, complicated online trajectory prediction, manoeuvre planning, or negotiation may no longer be feasible in the short time that remains for resolving the conflict. In the tragic 2002 mid-flight collision in Überlingen [43], the aircraft collided tens of seconds after the on-board traffic alert and collision avoidance system TCAS [196] signalled a traffic alert. Thus, for safe aircraft control we need particularly reliable reactions with manoeuvres whose correctness has been established previously by a thorough offline analysis. To ensure correct functioning of aircraft collision avoidance manoeuvres under all circumstances, the temporal evolution of the aircraft in space must be analysed carefully together with the effects that manoeuvring control decisions have on their dynamics. This results in complicated superpositions of physical system dynamics with control, which can be modelled naturally as hybrid systems.

Several numerical [293, 180, 46, 168, 171] or optimisation-based [180, 46, 167, 171] approaches have been proposed for air traffic control. It is difficult to give sound formal verification results for these approaches due to errors in numerical computations or the implicit definition of manoeuvres in terms of complicated optimisation processes. Formal verification is important to avoid collisions and prevent malfunctioning collision avoidance attempts like that in Fig. 8.1c. Formal verification results have been given by geometrical reasoning [104, 129, 295, 296] in PVS. Yet, it still remains to be proven by other techniques that the hybrid dynamics of a flight controller actually follows the geometrical shapes. In contrast, here we verify the hybrid system dynamics directly using a formally sound approach (assuming sound elementary decision procedures), consider curved flight, and achieve better automation.

Control Challenges

Because of the complicated spatio-temporal movement of aircraft, their manoeuvres are challenging for verification. Unlike in ground transportation, braking and waiting is not an option to resolve conflicts, because, in contrast to quadrotors [164], fixed-wing aircraft drop out of the sky if they do not move fast enough for sufficient lift. Consequently, aircraft manoeuvres have to be coordinated such that the aircraft always respect minimal and maximal lateral and angular speed constraints yet always remain safely separated. Further, angular velocity for curving is the primary means of control, because changes in thrust and linear speed are less efficient for aircraft.

Technical Challenges

Complexities in analysing aircraft manoeuvres manifest most prominently in difficulties with analysing hybrid systems for flight equations. General solutions of flight equations involve trigonometric functions that depend on the angular velocity ω and the orientation of the aircraft in space. For straight-line flight (angular velocity $\omega = 0$), the movement in space is just linear, so classical analysis techniques can be used [156]. These include pure straight-line manoeuvres [293, 203, 104, 129, 171]; see, e.g., Fig. 8.1a. They have to assume instant turns for heading changes of the aircraft between multiple straight line segments; see, e.g., Fig. 8.2. Instant turns, however, are impossible in mid-flight, because they are *not flyable*: Aircraft cannot suddenly change their flight direction from 0 to 45 degrees discontinuously. They need to follow a smooth curve instead, in which they slowly steer towards the desired direction by adjusting the angular velocity ω appropriately. Moreover, the

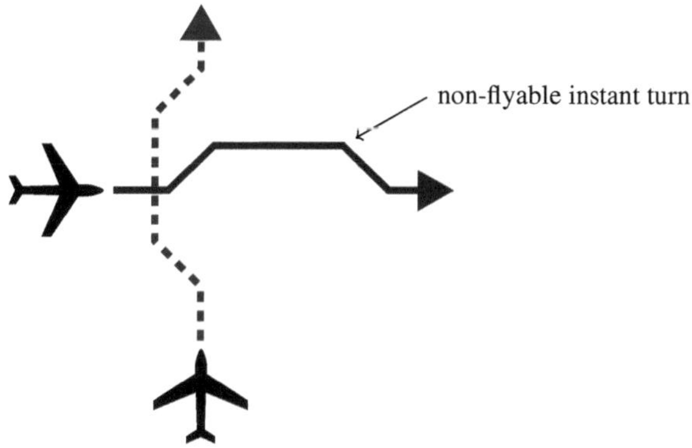

Fig. 8.2 Non-flyable straight-line manoeuvre with instant turns

area of the airspace required by manoeuvres for which instant turns could possibly be understood as adequately close approximations of properly curved flight is prohibitively large. Curved flight is thus an inherent part of real aircraft control. All the manoeuvres in Fig. 8.1 still include some instant turns and are, thus, not practical.

During curved flight, the angular velocity ω is nonzero. For $\omega \neq 0$, however, flight equations have transcendental solutions, which generally fall into undecidable classes of arithmetic. Consequently, manoeuvres with curves, as in Figs. 8.1b and d, are more realistic but also substantially more complicated for verification than straight-line manoeuvres such as that in Fig. 8.1a. In Chap. 6 we have developed a *sound* verification algorithm that works with differential invariants from Chap. 3 instead of with solutions of differential equations to address this arithmetic. Now we show how a fully curved manoeuvre can be verified, i.e., a flyable manoeuvre that only consists of curves and straight lines but *no* instant turns.

In this chapter, we introduce and verify the *fully flyable tangential roundabout manoeuvre (FTRM)*. It refines the non-flyable tangential roundabout manoeuvre (NTRM) from Fig. 8.1d, which still has discontinuities at the entry and exit points of roundabouts, to a fully flyable curved manoeuvre. Unlike most previously proposed manoeuvres [293, 46, 203, 104, 92, 129, 171], FTRM does not have any non-flyable instant turns. It is flyable and smoothly curved. Unlike the authors of other approaches emphasising the importance of flyability [180], we give formal verification results.

As a case study, we show how safety properties of collision avoidance manoeuvres in air traffic management can be verified with the DAL proof calculus from Chap. 3 using our verification algorithm from Chap. 6.

Contribution

Our main contribution is to show that reality in model design and coverage in formal verification are no longer incompatible desires even for applications as complex as aircraft manoeuvres. As a case study illustrating the use of the differential dynamic logics for hybrid systems from Part I of this book, we demonstrate how tricky and nonlinear dynamics can be verified with our verification algorithm from Chap. 6. We introduce a fully curved flight manoeuvre and verify its hybrid dynamics formally. In contrast to previous approaches, we handle curved flight and hybrid dynamics, and produce formal proofs with almost complete automation. Manual effort is still needed to simplify arithmetical complexity and modularise the proof appropriately. We further illustrate the resulting verification conditions for the respective parts of the manoeuvre. Finally, we identify the most difficult steps during the verification and present new transformations to handle the enormous computational complexity. To reduce complexity, we still use some of the simplifications assumed in related work, e.g., synchronous manoeuvring (i.e., aircraft make simultaneous, symmetric manoeuvre choices).

8.1.1 Related Work

Lafferriere et al. [189] gave important decidability results for hybrid systems with some classes of linear continuous dynamics but only random discrete resets. These results do not apply to air traffic manoeuvres, because they have non-trivial resets: the aircraft's position does not just jump randomly when switching modes but, rather, systematically according to the manoeuvre.

Tomlin et al. [293] analyse competitive aircraft manoeuvres game-theoretically using numerical approximations of partial differential equations. As a solution, they propose roundabout manoeuvres (Fig. 8.1b) and give bounded-time verification results for straight-line approximations (Fig. 8.1a). We verify actual curved roundabout manoeuvres with up to 28 variables and use a sound symbolic approach that avoids numerical approximation errors. Our symbolic techniques avoid state space discretisations that are exponential in the number of variables as required for partial differential equations. This enables our logical analysis techniques to scale better and to handle more complicated curved flight dynamics.

Flyability has been identified as one of the major challenges in Košecká et al. [180], where planning based on superposition of potential fields has been used to resolve air traffic conflicts. This planning does not guarantee flyability but, rather, defaults to classical vertical altitude changes whenever a non-flyable path is detected. The resulting manoeuvre has not yet been verified. The planning approach has been pursued further by Bicchi and Pallottino [46] with numerical simulations in two scenarios.

Numerical simulation algorithms approximating discrete-time Markov chain approximations of aircraft behaviour have been proposed by Hu et al. [168]. They approximate bounded-time probabilistic reachable sets for one initial state. We consider hybrid systems combining discrete control choices and continuous dynamics instead of uncontrolled, probabilistic continuous dynamics.

Hwang et al. [171] have presented a straight-line aircraft conflict avoidance manoeuvre that involves optimisation over complicated trigonometric computations, and have validated it using random numerical simulation and informal arguments. They also show examples where the decisions of the manoeuvre change only slightly for small perturbations.

The works of Dowek et al. [104] and Galdino et al. [129] are probably closest to ours. They consider straight-line manoeuvres and formalise geometrical proofs in PVS. As in the work of Hwang et al. [171], they do not, however, consider curved flight paths nor verify actual hybrid dynamics but work with geometrical meta-level reasoning instead.

A few attempts [203, 92] have been undertaken to model check discretisations of roundabout manoeuvres, which indicate avoidance of orthogonal collisions (Fig. 8.1b). However, counterexamples found by our model checker in previous work show that collision avoidance does not extend to other initial flight paths of the classical roundabout; see Fig. 8.1c.

Pallottino et al. [222] have presented a spatially distributed pattern for multiple roundabout circles at different positions. They reason manually about desirable properties of the system and estimate probabilistic results in [168]. Pallottino et al. thus take a view that is complementary to ours: they determine the global compatibility of multiple roundabouts while assuming correct functioning within each local roundabout. We verify that the actual hybrid dynamics of each local roundabout is collision-free. Generalising our approach to a spatial pattern of verified local roundabouts would be interesting future work.

Similarly, the work by Umeno and Lynch [296, 295] is complementary to ours. They consider real-time properties of airport protocols using Timed I/O Automata. We are interested in proving properties of the actual hybrid system dynamics.

Our approach has a very different focus than other complementary work:

- Our manoeuvre directly involves curved flight, unlike [293, 168, 104, 129, 171, 296, 295]. This makes our manoeuvre more realistic but also much more difficult to analyse.
- Unlike [180, 168, 171], we do not give results for a finite (sometimes small) number of initial flight positions in bounded-time horizons (simulation). Instead, we verify correct functioning for uncountably many initial states and give unbounded-time horizon verification results.
- Unlike [293, 180, 46, 168, 167, 171], we use symbolic instead of numerical computation so that numerical and floating-point errors cannot cause soundness problems.
- Unlike [46, 203, 168, 104, 129, 171, 296, 295], we directly analyse the hybrid system dynamics and do not use discrete or geometrical approximations that can be difficult to relate to the actual dynamics.
- Unlike [180, 293, 46, 168, 171, 203, 222] we produce formal, deductive proofs. Further unlike the formal proofs in [104, 129, 296, 295], our verification is much more automatic but also covers a different level of models.
- In [104, 129, 171, 296, 295], it remains to be proven that the hybrid dynamics and flight equations really follow geometrical intuition. Yet geometrical intuition can sometimes be quite misleading for more complicated cases such as the classical roundabout manoeuvre in Fig. 8.1b that looks promising, but fails for the situation in Fig. 8.1c. In contrast, our approach directly works for the hybrid flight dynamics. We illustrate verification results graphically to help understand them, but the figures do not prove anything.
- Unlike [216], we consider collision avoidance manoeuvres, and not just detection.
- Unlike [46, 167], we do not guarantee optimality of the resulting manoeuvre. Instead, we verify safety properties.

8.1.2 Structure of This Chapter

We verify safety-critical properties of curved flight in roundabout manoeuvres in Sect. 8.2. We analyse synchronisation properties of roundabout manoeuvres for multiple aircraft in Sect. 8.3. In Sect. 8.5, we combine the previous results to a safety theorem for fully flyable tangential roundabout manoeuvres following a compositional verification approach in our proof calculi according to Sect. 8.4. We present experimental results in Sect. 8.6 and conclude in Sect. 8.7.

8.2 Curved Flight in Roundabout Manoeuvres

In this section, we introduce and verify the fully flyable roundabout manoeuvre that is depicted in Fig. 8.3. It refines the tangential roundabout manoeuvre from Fig. 8.1d (which we considered in Chap. 3, but which still has discontinuities at the entry and exit points of roundabouts) to a fully flyable curved manoeuvre. Unlike the manoeuvres in Fig. 8.1, the resulting manoeuvre does not contain any instant turns. All of its curves are sufficiently smooth to be flyable.

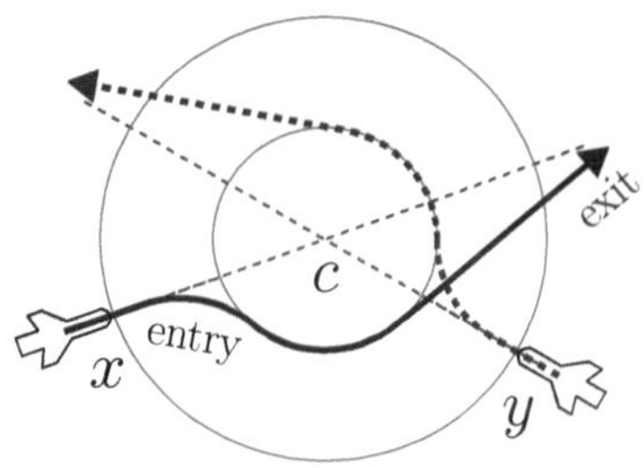

Fig. 8.3 Flyable aircraft roundabout

8.2.1 Flight Dynamics

The parameters of two aircraft at (planar) position $x=(x_1,x_2)$ and $y=(y_1,y_2)$ in $\mathbb{R}^2$ flying in directions $d=(d_1,d_2)\in\mathbb{R}^2$ and $e=(e_1,e_2)$ are illustrated in Fig. 8.4. Their dynamics is determined by their angular speeds $\omega,\varpi\in\mathbb{R}$ and linear velocity

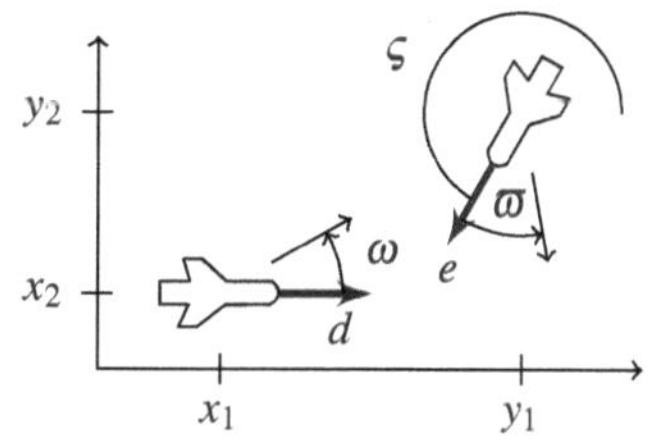

Fig. 8.4 Flight dynamics

vectors d and e, which describe both the linear velocity $\|d\| := \sqrt{d_1^2+d_2^2}$ and the orientation of the aircraft in space. Roundabout manoeuvres are horizontal collision avoidance manoeuvres, so as in [293, 203, 167, 92, 222, 129, 171], we simplify them to planar positions. We denote the flight equations for aircraft at x with angular velocity ω by $\mathscr{F}(\omega)$ and the flight equations for aircraft at y with angular velocity ϖ by $\mathscr{G}(\varpi)$:

$$[x' = d \quad d' = \omega d^{\perp}] \tag{$\mathscr{F}(\omega)$}$$

$$[y' = e \quad e' = \varpi e^{\perp}] \tag{$\mathscr{G}(\varpi)$}$$

This differential equation system is equivalent to the system ($\mathscr{F}(\omega)$) from p. 150 that has been derived from the literature [293]. Its difference with the differential equations from Chap. 3 is that we use vectorial notation in this chapter. The vector $d^{\perp} := (-d_2, d_1)$ is the *orthogonal complement* of vector d. Differential equations $\mathscr{F}(\omega)$ express that position x is moving in direction d, which is rotating with angular velocity ω, i.e., evolves orthogonal to d. Equations $\mathscr{G}(\varpi)$ are similar for y, e and ϖ of the second aircraft. The complexity of differential equations $\mathscr{F}(\omega)$ comes from the fact that, for curved flight with $\omega \neq 0$, their solutions involve trigonometric functions that fall into undecidable classes of arithmetic. See Example B.4 on p. 362 in App. B for an explanation of the solutions of these differential equations.

In safe flight configurations, aircraft respect protected zone p. That is, they are separated by at least distance $p \geq 0$, i.e., the state satisfies formula $\mathscr{S}(p)$:

$$\mathscr{S}(p) \equiv \|x - y\|^2 \geq p^2 \equiv (x_1 - y_1)^2 + (x_2 - y_2)^2 \geq p^2 \quad \text{for } p \in \mathbb{R} \tag{8.1}$$

We treat p, like all other parameters, purely symbolically, without assuming a specific value. In practise, horizontal separation should be $\geq$5 mi, and vertical separation $\geq$1000 ft.

8.2.2 Roundabout Manoeuvre Overview

The *flyable tangential roundabout manoeuvre* (*FTRM*) consists of the phases in the protocol cycle in Fig. 8.5a, which correspond to the marked flight phases in Fig. 8.5b. During free flight, the aircraft move without restriction by repeatedly

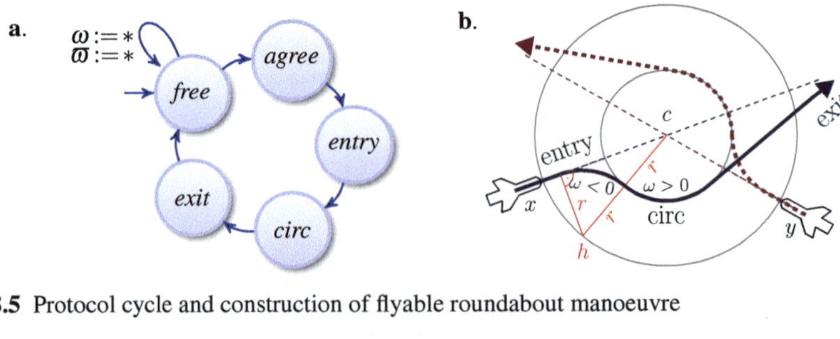

Fig. 8.5 Protocol cycle and construction of flyable roundabout manoeuvre

choosing arbitrary new angular velocities ω and ϖ respectively (as indicated by the self-loop at *free* in Fig. 8.5a). As in Chap. 3, continuous nondeterministic variation of ω and ϖ could be permitted instead of the self-loop by adding $\exists\omega\,\exists\varpi$ to the DA-constraints in *free*. The subsequent proofs carry over to this generalisation immediately.

When the aircraft come too close to one another, they agree on a roundabout manoeuvre by negotiating a compatible roundabout centre $c = (c_1, c_2)$ in coordination phase *agree* by communication. Next, the aircraft approach the roundabout circle in a right curve with $\omega < 0$ (*entry* mode) according to Fig. 8.5b, and reach a tangential position around centre c. During the *circ* mode, the aircraft follow the circular roundabout manoeuvre around the agreed centre c with a left curve of common angular velocity $\omega > 0$. Finally, the aircraft leave the roundabout in cruise mode ($\omega = 0$) in their original direction (*exit*) and enter free flight again (*free*) when they have reached sufficient distance (the protocol cycle repeats as necessary). The collision avoidance manoeuvre is symmetric when exchanging left and right curves, which corresponds to exchaing a value $\omega < 0$ for $\omega > 0$.

8.2.3 Compositional Verification Plan

For verifying safety properties and collision avoidance of FTRM, we decompose the verification problem and pursue the following overall verification plan:

AC1 *Tangential roundabout manoeuvre cycle*: We prove that the protected zones of aircraft are safely separated at all times during the whole manoeuvre (including repetitive collision avoidance manoeuvre initiation and multiple aircraft) with a simplified but not yet flyable entry operation $entry_n$. Subsequently, we refine this verification result to a flyable manoeuvre by verifying that we can replace $entry_n$ with its flyable variant *entry*.

AC2 *Bounded control choices for aircraft velocities*: We show that linear speeds remain unchanged during the whole manoeuvre (the aircraft do not stall).

AC3 *Flyable entry*: We prove that the simplified $entry_n$ procedure can be replaced with a flyable curve *entry* reaching the same configuration around the inner roundabout circle as $entry_n$.

AC4 *Bounded entry duration*: We show why the flyable *entry* procedure succeeds in bounded time, i.e., aircraft reach the roundabout circle in some bounded time $\leq T$.

AC5 *Safe entry separation*: Most importantly, we prove that the protected zones of aircraft are still respected during the flyable entry procedure *entry*.

AC6 *Successful negotiation*: We prove that the negotiation phase (*agree*) satisfies the respective requirements of multiple aircraft simultaneously and identify which constraints are needed for negotiation to succeed.

AC7 *Safe exit separation*: We show that, for its bounded duration, the exit procedure cannot produce collisions and that the initial *far separation* for free flight is reached again, so the FTRM cycle repeats safely.

This plan modularises the proof and allows us to identify the respective safety constraints imposed by the various manoeuvre phases successively. We present details on these verification tasks in this chapter and summarise the respective verification results in a joint safety property of FTRM in Sect. 8.5.

Finally note that to simplify notation we use square roots, vectorial notation, and norms within formulas in this chapter, because those are easily definable. For instance, for vectors $x = (x_1, x_2)$ and $y = (y_1, y_2)$, the formula $\|x-y\| \geq p$ is definable by $(x_1-y_1)^2+(x_2-y_2)^2 \geq p^2$ for $p \geq 0$. Likewise, $\|x-y\| = \sqrt{3}r$ is definable by $\|x-y\|^2 = 3r^2$ for $r \geq 0$.

8.2.4 Tangential Roundabout Manoeuvre Cycles (AC1)

First, we analyse roundabouts with a simplified instant entry procedure and without an exit procedure (property AC1 from Sect. 8.2.3). This corresponds to the non-flyable NTRM depicted in Fig. 8.1d on p. 304. We refine this manoeuvre and its verification to the flyable FTRM afterwards.

Modular Correctness of Tangential Roundabout Cycles

For AC1, we have proven in Sect. 3.11 that, for arbitrary choices of the *entry* manoeuvre that satisfy the prerequisites of Theorem 3.4 from p. 199, the tangential roundabout manoeuvre safely avoids collisions, i.e., the aircraft always maintain a safe distance $\geq p$ during the curved flight in the roundabout circle. In addition, these results show that arbitrary repetitions of the protocol cycle are safe at all times for a simplified choice of the entry manoeuvre. The model and specification for this tangential roundabout, as constructed in Sects. 3.4 and 3.11, are summarised in Fig. 8.6, recalling the separation property $\mathscr{S}(p)$ from equation (8.1).

$$
\begin{aligned}
\psi &\equiv \mathscr{S}(p) \rightarrow [NTRM]\,\mathscr{S}(p)\\
\mathscr{S}(p) &\equiv \|x-y\|^2 \geq p^2 \equiv (x_1-y_1)^2+(x_2-y_2)^2 \geq p^2\\
NTRM &\equiv (\mathit{free};\ \mathit{agree};\ \mathit{entry}_n;\ \mathit{circ})^*\\
\mathit{free} &\equiv (\omega := *;\ \varpi := *;\ \mathscr{F}(\omega) \wedge \mathscr{G}(\varpi) \wedge \mathscr{S}(p))^*\\
\mathit{agree} &\equiv \omega := *;\ c := *\\
\mathit{entry}_n &\equiv d := \omega(x-c)^{\perp};\ e := \omega(y-c)^{\perp}\\
\mathit{circ} &\equiv \mathscr{F}(\omega) \wedge \mathscr{G}(\omega)
\end{aligned}
$$

Fig. 8.6 Non-flyable tangential roundabout collision avoidance manoeuvre NTRM

The simplified flight controller in Fig. 8.6 performs collision avoidance manoeuvres by tangential roundabouts and repeats these manoeuvres any number of times as needed. During each cycle of the loop of *NTRM*, the aircraft first perform arbitrary free flight (*free*) by choosing arbitrary new angular velocities ω and ϖ (repeatedly as indicated by the loop in *free*). Aircraft only fly freely while they are safely separated, which is expressed by the conjunctive constraint $\mathscr{S}(p)$ in the differential equation for *free*, which amounts to an evolution domain restriction. Then the aircraft agree on an arbitrary roundabout centre c and angular velocity ω (*agree*). We model this communication by nondeterministic assignments to the shared variables ω, c. Refinements include all negotiation processes that reach an agreement on common ω, c choices in bounded time. Next, the aircraft perform the simplified non-flyable entry procedure (*entry*$_n$) with instant turns (according to Fig. 8.1d). This operation identifies the goal state that entry procedures need to reach:

$$\mathscr{T} \equiv d = \omega(x-c)^{\perp} \wedge e = \omega(y-c)^{\perp}. \tag{8.2}$$

As identified in equation (3.12) on p. 198, the constraint $\mathscr{T}$ expresses that, at the positions x and y, respectively, the directions d and e of the aircraft are tangential to the roundabout circle with centre c and angular velocity ω; see Fig. 8.7. Finally,

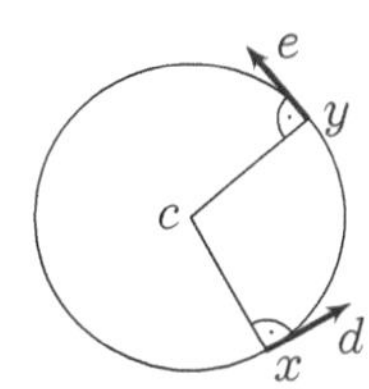

Fig. 8.7 Tangential configuration $\mathscr{T}$

the roundabout manoeuvre itself is carried out in *circ* with the common angular velocity ω that was agreed upon in *agree*. The collision avoidance roundabouts can be left again by repeating the loop and entering arbitrary free flight at any time. When further conflicts occur during free flight, the overall process repeats and the controller in Fig. 8.6 again enters roundabout conflict resolution manoeuvres. The d$\mathscr{L}$ formula ψ in Fig. 8.6 expresses the fact that aircraft following the *NTRM* protocol always stay safely separated by at least the protected zone p if they start safely separated. This formula can be proven similarly to the proof in Sect. 3.11.

Multiple Aircraft

We prove a corresponding collision avoidance property for up to five aircraft, which jointly participate in the roundabout manoeuvre. There, the safety property is mutual collision avoidance, i.e., each of the aircraft has a safe distance $\geq p$ from all the other aircraft, which yields a quadratic number of constraints. This quadratic increase in the property that actually needs to be proven for a safe roundabout of n aircraft and the increased dimension of the underlying continuous state space causes increased verification times for more aircraft in our experiments (Sect. 8.6). For instance, Fig. 8.8 illustrates the (flyable) roundabout manoeuvre with multiple air-

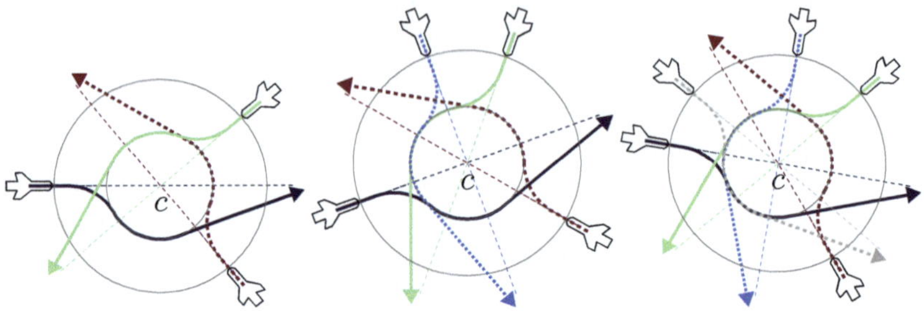

Fig. 8.8 Flyable aircraft roundabout (multiple aircraft)

craft and Fig. 8.9 contains the system and separation property specification for the four aircraft tangential roundabout manoeuvre (still with simplified entry procedure). There, d$\mathscr{L}$ formula ψ expresses that the four aircraft at positions $x, y, z, u \in \mathbb{R}^2$, respectively, keep mutual distance $\geq p$ each.

$$
\begin{aligned}
\psi &\equiv \phi \to [NTRM]\phi \\
\phi &\equiv (x_1-y_1)^2+(x_2-y_2)^2 \geq p^2 \wedge (y_1-z_1)^2+(y_2-z_2)^2 \geq p^2 \\
&\quad \wedge (x_1-z_1)^2+(x_2-z_2)^2 \geq p^2 \wedge (x_1-u_1)^2+(x_2-u_2)^2 \geq p^2 \\
&\quad \wedge (y_1-u_1)^2+(y_2-u_2)^2 \geq p^2 \wedge (z_1-u_1)^2+(z_2-u_2)^2 \geq p^2 \\
NTRM &\equiv (free;\ agree;\ entry_n;\ circ)^* \\
free &\equiv (\omega_x := *;\ \omega_y := *;\ \omega_z := *;\ \omega_u := *; \\
&\quad x_1' = d_1 \wedge x_2' = d_2 \wedge d_1' = -\omega_x d_2 \wedge d_2' = \omega_x d_1 \\
&\quad \wedge y_1' = e_1 \wedge y_2' = e_2 \wedge e_1' = -\omega_y e_2 \wedge e_2' = \omega_y e_1 \\
&\quad \wedge z_1' = f_1 \wedge z_2' = f_2 \wedge f_1' = -\omega_z f_2 \wedge f_2' = \omega_z f_1 \\
&\quad \wedge u_1' = g_1 \wedge u_2' = g_2 \wedge g_1' = -\omega_u g_2 \wedge g_2' = \omega_u g_1 \wedge \phi)^* \\
agree &\equiv \omega := *;\ c := * \\
entry_n &\equiv d_1 := -\omega(x_2-c_2);\ d_2 := \omega(x_1-c_1); \\
&\quad e_1 := -\omega(y_1-c_1);\ e_2 := \omega(y_2-c_2); \\
&\quad f_1 := -\omega(z_1-c_1);\ f_2 := \omega(z_2-c_2); \\
&\quad g_1 := -\omega(u_1-c_1);\ g_2 := \omega(u_2-c_2) \\
circ &\equiv x_1' = d_1 \wedge x_2' = d_2 \wedge d_1' = -\omega_x d_2 \wedge d_2' = \omega_x d_1 \\
&\quad \wedge y_1' = e_1 \wedge y_2' = e_2 \wedge e_1' = -\omega_y e_2 \wedge e_2' = \omega_y e_1 \\
&\quad \wedge z_1' = f_1 \wedge z_2' = f_2 \wedge f_1' = -\omega_z f_2 \wedge f_2' = \omega_z f_1 \\
&\quad \wedge u_1' = g_1 \wedge u_2' = g_2 \wedge g_1' = -\omega_u g_2 \wedge g_2' = \omega_u g_1
\end{aligned}
$$

Fig. 8.9 Tangential roundabout collision avoidance manoeuvre (four aircraft)

8.2.5 Bounded Control Choices (AC2)

For property AC2 from Sect. 8.2.3, we show that bounded speed is sufficient for the choices required for the entry procedure and that the bounded speed is always maintained safely throughout the manoeuvre (no stalling).

Bounded Entry Choices

The NTRM in Fig. 8.6 maintains collision avoidance for all its choices of centre c and angular velocity ω in *agree*. Next, we show that *there always is* a choice respecting external requirements on linear speed (aircraft are safe neither at too high speeds nor when they are travailing too slowly, as they would stall), which corresponds to property AC2. The fact that all external speed requirements can be met is a consequence of the proof of property (3.14) on p. 201. Consequently, the constraints on the entry procedure are feasible with parameter choices respecting *any* bounds for velocities that are imposed by the aircraft.

Bounded Manoeuvre Speed

As a simple consequence of the proof in Example 3.16 on p. 172, it is easy to see that external requirements on the linear speed of aircraft are maintained throughout the whole roundabout manoeuvre. Example 3.16 shows that the linear speed is maintained for arbitrary repeated choices of the angular velocity ω, which, as a special case, includes the respective control choices during the roundabout manoeuvre. Thus, the aircraft do not stall.

8.2.6 Flyable Entry Procedures (AC3)

For property AC3 from Sect. 8.2.3, we generalise the verification results about NTRM with simplified entry procedures (cf. Fig. 8.1d) to FTRM (Fig. 8.5b) by replacing the non-flyable $entry_n$ procedure with flyable curves. The resulting flyable entry procedure is called *entry*. This turns the non flyable NTRM into the flyable FTRM manoeuvre.

Flyable Entry Properties

We construct a flyable entry manoeuvre that follows the smooth entry curve from Fig. 8.5b using the anchor point h indicated in Fig. 8.10. Anchor h is positioned relative to the roundabout centre c and the x position at the start of the entry curve (i.e., with x at the right angle indicated in Fig. 8.10). This anchor h has a distance

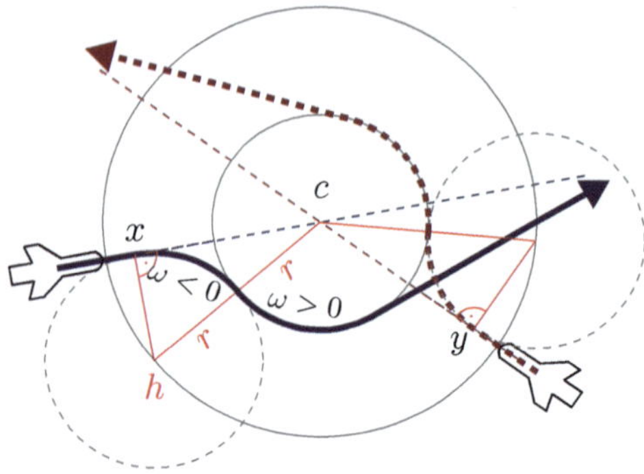

Fig. 8.10 Flyable entry characteristics

of r to x, and distance $2r$ to c, and $x-h$ is orthogonal to the aircraft direction d while taking into account the relative orientation of the roundabout as implied by angular velocity ω in the corresponding specification of the flyable entry procedure depicted in Fig. 8.10. The following property specifies that the tangential configuration of the simple choice for *agree* in Fig. 8.6 is reached by a flyable curve while waiting until aircraft x and centre c have distance r:

$$\begin{aligned}(r\omega)^2 = \|d\|^2 \wedge \|x-c\| = \sqrt{3}r \wedge \exists \lambda \geq 0\,(x+\lambda d = c)\\ \wedge\ \|h-c\| = 2r \wedge d = -\omega(x-h)^{\perp}\\ \rightarrow \left[\mathscr{F}(-\omega)\,\&\,\|x-c\| \geq r\right]\left(\|x-c\| \leq r \rightarrow d = \omega(x-c)^{\perp}\right)\end{aligned} \tag{8.3}$$

The assumptions in line 1 of d$\mathscr{L}$ formula (8.3) express that r is the radius corresponding to speed $\|d\|$ and angular velocity ω (which corresponds to $(r\omega)^2 = \|d\|^2$) and that *entry* starts when *agree* has finished with distance $\sqrt{3}r$ to c and is heading towards c (i.e., $\exists \lambda \geq 0\,(x+\lambda d = c)$). For the construction of the manoeuvre and positioning in space, we use the auxiliary anchor point $h \in \mathbb{R}^2$ identified in Fig. 8.10 and line 2 of (8.3). Anchor h is positioned relative to the roundabout centre c and the x position at the start of the entry curve (i.e., with x at the right angle indicated in Fig. 8.10). The entry curve around h is similar to the roundabout curve around c. It only goes the other way. Formally, h is characterised by distance r to x, and distance $2r$ to c (i.e., $\|h-c\| = 2r$) and, further, vector $x-h$ is orthogonal to d and obeys the relative orientation of the curve belonging to the right curve with angular velocity $-\omega$ (hence $d = -\omega(x-h)^{\perp}$). The formula in (8.3) specifies that the tangential goal configuration (8.2) around c is reached by a flyable curve when waiting until aircraft x and centre c have distance r, because the domain restriction of the dynamics is $\|x-c\| \geq r$ (line 3) and the postcondition assumes $\|x-c\| \leq r$, which, together, imply $\|x-c\| = r$.

The feasibility of choosing anchor point h can be shown by proving the dual existence formula: For AC3, we thus also prove that the anchor point h can always be chosen as illustrated in Fig. 8.10. That is, we show feasibility of the assumptions on h from property (8.3) by the following existence property:

$$
\begin{aligned}
&(r\omega)^2 = \|d\|^2 \land \|x-c\| = \sqrt{3}r \land \exists \lambda \geq 0\,(x+\lambda d = c)\\
\rightarrow &\exists h\,(d = -\omega(x-h)^\perp \land \|h-c\| = 2r)
\end{aligned}
\tag{8.4}
$$

Spatial Symmetry Reduction

The formulas (8.3) and (8.4) can be proven in a simplified version. We use a (new) *spatial symmetry reduction* to simplify these properties computationally. We exploit symmetries to reduce the spatial dimension by fixing variables and conclude safety for states that result from the fixed class of states by symmetry transformations. Without loss of generality, we can recentre the coordinate system to have centre c at position 0 (the manoeuvre is invariant under translation in space). Further, we can assume aircraft x to come from the left by changing the orientation of the coordinate system (the manoeuvre is invariant under planar rotation in space). Finally, we can assume, without loss of generality, the linear speed to be 1 (by rescaling units appropriately). Observe, however, that we cannot fix a value for both the linear speed and the angular velocity, because their units are strictly interdependent. In other words, if we fix the linear speed, we need to consider all angular velocities to verify the manoeuvre for all possible curve radii r for the roundabout manoeuvre. The x position resulting from these symmetry reductions can be determined by the Pythagoras theorem (i.e., $(2r)^2 = r^2 + x_1^2$ for the triangle enclosed by h,x,c in Fig. 8.10) or simple trigonometry, as follows:

$$
x = (2r\cos\frac{\pi}{6}, 0) = (\sqrt{(2r)^2 - r^2}, 0) = (\sqrt{3}r, 0) \quad . \tag{8.5}
$$

To express the square root function using polynomial terms, we can easily use a random assignment for x_1 with a test condition $?(x_1^2 = (\sqrt{3}r)^2 = 3r^2)$. Consequently, we simplify formulas (8.3) and (8.4) by specialising them to the following situation:

$$
\begin{aligned}
&d_1 := 1;\ d_2 := 0;\ \ c_1 := 0;\ c_2 := 0;\\
&x_2 := 0;\\
&r := *;\ \ ?(r > 0);\ \ \omega := 1/r;\\
&x_1 := *;\ \ ?(x_1^2 = 3r^2 \land x_1 \leq 0);
\end{aligned}
$$

Verification results for the resulting entry procedure after symmetry reduction, and a proof of existence of a corresponding anchor point h will be shown in Sect. 8.6.

For proving the feasibility property in (8.4) within reasonable time, it is, in fact, sufficient to specialise the state by symmetry reduction to $c_1 := 0 \land c_2 := 0 \land \omega := 1$. Interestingly, this example shows another surprise of quantifier elimination complexity and the power of symmetry reduction quite impressively: Without the reduction to $\omega := 1$, QE of Mathematica runs for more than 20 days without producing a result on an Intel Xeon X5365 with 3 GHz and 16 GB memory (of which only 3 GB are used), even when adding the state space reduction $r := 1$.

8.2.7 Bounded Entry Duration (AC4)

As the first step for showing that the entry procedure finally succeeds at goal (8.2) and maintains a safe distance all the time, we show that *entry* succeeds in bounded time and cannot take arbitrarily long to succeed (property AC4 from Sect. 8.2.3).

Qualitative Analysis

By a simple consequence of the proof for formula (8.3), the entry procedure follows a circular motion around the anchor point h, which is chosen according to (8.3); see Fig. 8.10. That is, when r is the radius belonging to angular velocity ω and linear speed $\|d\|$, the property $\|x-h\| = r$ is a differential invariant (Sect. 3.5.6) of *entry*: For AC4, we prove constant distance to anchor point h, i.e., that, indeed, $\|x-h\| = r$ is an invariant of *entry*:

$$\begin{aligned}(r\omega)^2 = \|d\|^2 \wedge \|x-c\| = \sqrt{3}r \wedge \exists\lambda\geq 0\,(x+\lambda d = c)&\\ \wedge d = -\omega(x-h)^{\perp} \wedge \|h-c\| = 2r&\\ \rightarrow \left[\mathscr{F}(-\omega)\,\&\,\|x-c\| \geq r\right]\left(\|x-h\| = r\right)& \qquad (8.6)\end{aligned}$$

Indeed, the corresponding formula in (8.6) is provable easily. For the remainder of AC4, we use informal arguments. By AC2, the speed $\|d\|$ is constant during the *entry* procedure. Thus, the aircraft proceeds with nonzero minimum progress rate $\|d\|$ around the circle with radius r. Consequently, the entry manoeuvre ends in bounded time, because the arc length of the circle of radius r around h is bounded ($2\pi r$ for radius r) and every bounded curve will be traversed in bounded time with constant linear speed and orientation.

Quantitative Analysis

To obtain a quantitative bound on the duration of the entry manoeuvre, consider the following. The flight duration for a full circle of radius r around h at constant linear speed $\|d\|$ is $\frac{2\pi r}{\|d\|}$, because its arc length is $2\pi r$. From the trigonometric identities underlying equation (8.5), we can read off that the aircraft do not even have to complete the full circle but only a $\frac{\pi}{3} = 60^\circ$ arc; see Fig. 8.10. Consequently, the maximum duration T of the *entry* procedure can be determined as:

$$T := \frac{1}{6}\cdot\frac{2\pi r}{\|d\|} = \frac{\pi r}{3\|d\|}. \qquad (8.7)$$

Using the standard relation $|\omega| = \frac{v}{r}$ between the angular velocity ω, the corresponding radius r of the circle (of evasive actions), and the linear velocity $v = \|d\|$ of the aircraft, maximum duration T can also be expressed as:

$$T = \frac{1}{6} \cdot \frac{2\pi}{|\omega|}.$$

Note that the constant π in the expression (8.7) for T is not definable in the theory of real-closed fields (the transcendental number π is not even contained in the real-closed field of algebraic numbers, and thus it cannot be definable; see App. D.2). Yet, the maximum duration T only has to be some upper bound on the maximum duration of the *entry* procedure. Consequently, any overapproximation of the maximum duration computed using a rational $P \geq \pi$ will do, for instance, $\frac{3.15r}{3\|d\|}$ when using the approximation 3.15 of $\pi = 3.1415926\ldots$ in (8.7).

8.2.8 Safe Entry Separation (AC5)

In Sect. 8.2.6, we have shown that the simplified $entry_n$ procedure from NTRM can be replaced by a flyable *entry* manoeuvre that meets the requirements of approaching tangentially (according to Theorem 3.4) for each aircraft. Unlike in instant turns ($entry_n$), we have to show in addition that the flyable entry manoeuvres of multiple aircraft do not produce mutually conflicting flight paths, i.e., spatial separation of all aircraft is maintained during the entry of multiple aircraft (property AC5 from Sect. 8.2.3). See Fig. 8.8 on p. 314 for multiple aircraft FTRM where entry separation is important.

Bounded Overapproximation

We show that entry separation is a consequence of the bounded speed (AC2) and bounded duration (AC4) of the flyable entry procedure when initiating the negotiation phase *agree* with sufficient distance. We prove that, when following bounded speed for a bounded duration, aircraft only come closer by a bounded distance. This gives a fairly generic proof that generalises to other entry choices.

Let b denote the overall speed bound during FTRM according to AC2 and let T be the time bound for the duration of the entry procedure due to AC4. We overapproximate the actual behaviour during the *entry* phase by *arbitrary curved flight* (see Fig. 8.11 where angular velocities ω, ϖ are free inputs). As an overapproximation of the actual behaviour during the *entry* phase, the following strong DAL formula expresses that, when the *entry* procedure is initiated with sufficient distance $p + 2bT$, the protected zone will still be respected after the two aircraft follow *any* curved flight (including the actual choices during *entry* and subsequent *circ*) with speed $\leq b$ up to T time units (see Fig. 8.11):

$$\begin{aligned}&\|d\|^2 \leq \|e\|^2 \leq b^2 \land \|x-y\|^2 \geq (p+2bT)^2 \land p \geq 0 \land b \geq 0 \land T \geq 0\\&\quad \to \left[\tau := 0; \exists \omega\, \mathscr{F}(\omega) \land \exists \varpi\, \mathscr{G}(\varpi) \land \tau' = 1 \land \tau \leq T\right](\|x-y\|^2 \geq p^2) \qquad (8.8)\end{aligned}$$

Fig. 8.11 Entry separation by bounded nondeterministic overapproximation

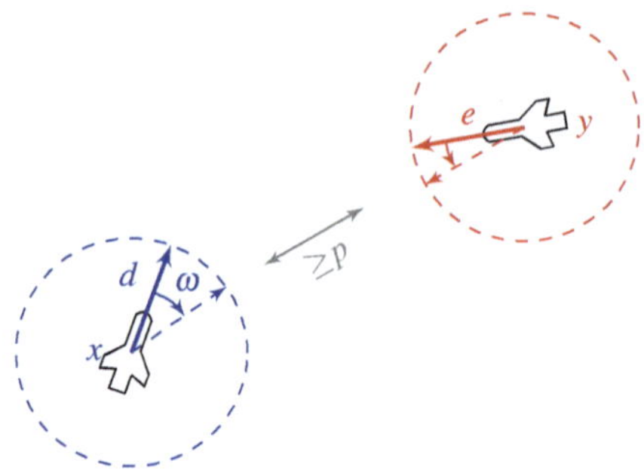

This formula is a consequence of the fact that—regardless of the actual angular velocity choices—aircraft only make limited progress in bounded time when starting from some initial point z with bounded speeds, even when changing angular velocity ω arbitrarily as a nondeterministic quantified input ($\exists\omega$ as in Chap. 3):

$$\|d\|^2 \le b^2 \land b \ge 0 \land x = z \to [\tau := 0;\ \exists\omega\, \mathscr{F}(\omega) \land \tau' = 1]\|x - z\|_\infty \le \tau b \qquad (8.9)$$

The maximum distance $\|x-z\|_\infty$ from initial position z depends on clock τ and speed bound b. To reduce the polynomial degree and the computational verification complexity in the proof, we overapproximate distances from quadratic Euclidean norm $\|\cdot\|$ in terms of linearly definable supremum norm $\|\cdot\|_\infty$, which is definable as:

$$\|x\|_\infty \le c \equiv -c \le x_1 \le c \land -c \le x_2 \le c;$$
$$\|x\|_\infty \ge c \equiv (x_1 \le -c \lor c \le x_1) \lor (x_2 \le -c \lor c \le x_2).$$

The limited progress property (8.9) is provable immediately by differential induction for the postcondition (the antecedent of *DI* is provable as $x = z \vdash \|x-z\|_\infty = 0$):

$$\dfrac{\dfrac{\dfrac{*}{\|d\|^2 \le b^2 \land b \ge 0 \vdash \forall x,y,\tau\,\forall\omega\,(-b \le d_1 \le b \land -b \le d_2 \le b)}\ \forall r}{\|d\|^2 \le b^2 \land b \ge 0 \land x = z \vdash \langle\tau := 0\rangle[\exists\omega\,\mathscr{F}(\omega) \land \tau' = 1]\|x-z\|_\infty \le \tau b}\ DI}{\|d\|^2 \le b^2 \land b \ge 0 \land x = z \vdash [\tau := 0;\ \exists\omega\,\mathscr{F}(\omega) \land \tau' = 1]\|x-z\|_\infty \le \tau b}\ [;],[:=]$$

Cartesian Degree Reduction

Due to the complexity of QE, the proof of the original property (8.8) does not terminate quickly enough. To overcome this issue, we simplify property (8.8) and use the (linearly definable) supremum norm $\|\cdot\|_\infty$ in place of the (quadratically definable) Euclidean 2-norm $\|\cdot\|_2$, thereby yielding the following provable variant of (8.8):

$$\|d\|^2 \le \|e\|^2 \le b^2 \wedge \|x-y\|_\infty \ge (p+2bT) \wedge p \ge 0 \wedge b \ge 0 \wedge T \ge 0$$
$$\to \left[\tau := 0; \exists \omega\, \mathscr{F}(\omega) \wedge \exists \varpi\, \mathscr{G}(\varpi) \wedge \tau' = 1 \wedge \tau \le T\right](\|x-y\|_\infty \ge p) \quad (8.10)$$

It can be shown (and is provable even by QE) that the supremum norm $\|\cdot\|_\infty$ and the standard Euclidean norm $\|\cdot\|_2$ are equivalent, i.e., their values are identical up to constant factors:

$$\forall x\, (\|x\|_\infty \le \|x\|_2 \le \sqrt{n}\|x\|_\infty) \quad (8.11)$$
$$\forall x\, (\frac{1}{\sqrt{n}}\|x\|_2 \le \|x\|_\infty \le \|x\|_2) \quad (8.12)$$

where n is the dimension of the vector space, here 2. From this equivalence of norms, we can conclude that the following variant of (8.10) with 2-norms is valid:

$$\|d\|^2 \le \|e\|^2 \le b^2 \wedge \|x-y\|_2 \ge \sqrt{2}(p+2bT) \wedge p \ge 0 \wedge b \ge 0 \wedge T \ge 0$$
$$\to \left[\tau := 0; \exists \omega\, \mathscr{F}(\omega) \wedge \exists \varpi\, \mathscr{G}(\varpi) \wedge \tau' = 1 \wedge \tau \le T\right](\|x-y\|_2 \ge p) \quad (8.13)$$

The extra factor of $\sqrt{2}$ in the separation requirement results from the relaxation of the 2-norm to the ∞-norm. In the preconditions of formula (8.13), we assume $\|x-y\|_2 \ge \sqrt{2}(p+2bT)$, which implies the assumption $\|x-y\|_\infty \ge (p+2bT)$ in the precondition of (8.10) using (8.12). The postcondition $\|x-y\|_\infty \ge p$ of (8.10) in turn implies postcondition $\|x-y\|_2 \ge p$ of (8.13) using (8.11).

Now, with the bounded duration property AC4 of the entry manoeuvre, it is easy to see that *entry* is a special case of the hybrid program with quantified nondeterministic angular velocities in (8.13). Thus we conclude that the following property is valid:

$$\|x-y\| \ge \sqrt{2}(p+2bT) \wedge p \ge 0 \wedge \|d\|^2 \le \|e\|^2 \le b^2 \wedge b \ge 0 \wedge T \ge 0$$
$$\to [entry]\, (\|x-y\| \ge p) \quad (8.14)$$

Far Separation

By combining the estimation of the entry duration (8.7) at speed $\|d\| = b$ with the entry separation property (8.14), we determine the following magnitude as the *far separation* f, i.e., the initial distance guaranteeing that the *FTRM* protocol can be repeated safely in case new collision avoidance is needed:

$$f := \sqrt{2}(p+2bT) = \sqrt{2}\left(p + \frac{2}{3}\pi r\right). \quad (8.15)$$

Using f as an abbreviation for that term, we can abbreviate property (8.14) as follows:

$$\|d\|^2 \le \|e\|^2 \le b^2 \wedge \|x-y\|_2 \ge f \wedge p \ge 0 \wedge b \ge 0 \to [entry](\|x-y\|_2 \ge p) \quad (8.16)$$

8.3 Synchronisation of Roundabout Manoeuvres

In the previous sections, we have analysed collision freedom and separation properties of the various curved flight phases of FTRM. According to our verification plan in Sect. 8.2.3, we also need to show that the various actions of the respective aircraft are synchronised appropriately to ensure safety of the manoeuvre. We analyse the negotiation phase and compatible exit procedures next.

8.3.1 Successful Negotiation (AC6)

For the negotiation phase to succeed, we have to show that there is a common choice of the roundabout centre c and angular velocity ω (or radius r) so that all participating aircraft can satisfy the requirements of their respective entry procedures simultaneously. That is, we show that the assumptions of formula (8.3) can be satisfied simultaneously for all aircraft at once. We thus have to show property AC6 from Sect. 8.2.3.

Separate Success

First of all, we show that there is always a choice of centre c, radius r, and common angular velocity ω during the *agree* mode such that the antecedent of the formula (8.3) is satisfied, i.e., the *agree* mode is feasible:

$$\langle c:=*;\ r:=*;\ ?\|x-c\| = \sqrt{3}r \wedge r \geq 0; \omega := *;\ ?(r\omega)^2 = d_1^2 + d_2^2\rangle \\ \left(\|x-c\| = \sqrt{3}r \wedge \exists \lambda \geq 0\,(x+\lambda d = c)\right) \quad (8.17)$$

The nondeterministic assignments choose any c, r, ω and the subsequent tests ensure that the *agree* mode succeeds in a state with appropriate distance $\sqrt{3}r$ to the centre c as identified in (8.3). The test $?(r\omega)^2 = d_1^2 + d_2^2$ ensures that ω is an angular velocity that fits to the linear speed $\|d\|$ and roundabout radius r.

The dual property shows that, in fact, *all* choices of *agree* that also satisfy the (hence, feasible) constraint $\exists \lambda \geq 0\,(x+\lambda d = c)$ already satisfy the *entry* requirements:

$$[c:=*; r:=*; ?\|x-c\| = \sqrt{3}r \wedge r \geq 0;\ ?\exists \lambda \geq 0\,(x+\lambda d = c); \\ \omega := *;\ ?(r\omega)^2 = d_1^2 + d_2^2] \\ \left(\|x-c\| = \sqrt{3}r \wedge \exists \lambda \geq 0\,(x+\lambda d = c)\right) \quad (8.18)$$

With constraint $\exists \lambda \geq 0\,(x+\lambda d = c)$ we chose to have the centre c on the straight line from x in its current direction d. This is one option but not the only choice for an entry construction.

Joint Negotiation Success

In addition to the entry requirements working out separately for each aircraft, we prove that all corresponding choices of *agree* satisfy the mutual requirements of multiple aircraft simultaneously. There are many possibilities for simultaneous solutions of negotiation. To simplify the computational complexity, we only consider symmetric situations with identical speed and common distance to the point of potential collision, which we can then choose as the roundabout centre c; see Fig. 8.12. This is a restriction but simplifies the computational proof complexity considerably.

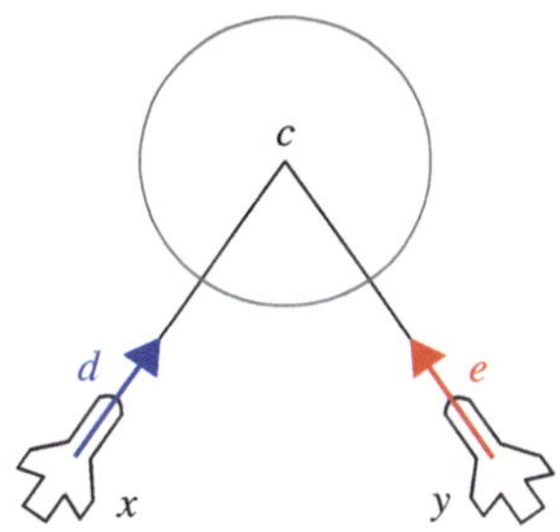

Fig. 8.12 Some mutually agreeable negotiation choices for aircraft

A very simple choice is to use the simultaneous intersection (provided there is an intersection $x+\lambda d = y+\lambda e$ after time λ) of the flight paths of the aircraft at x and y as the roundabout centre c. Then the choices for c, r, ω are compatible for multiple aircraft:

$$\begin{aligned}
&\lambda > 0 \land x+\lambda d = y+\lambda e \land \|d\| = \|e\| \to \\
&[c := x+\lambda d;\ r := *;\ ?\|x-c\| = \sqrt{3}r;\ ?\|y-c\| = \sqrt{3}r;\ \omega := *;\ ?(r\omega)^2 = \|d\|^2] \\
&\quad (\|x-c\| = \sqrt{3}r \land \lambda \geq 0 \land x+\lambda d = c \land \|y-c\| = \sqrt{3}r \land y+\lambda e = c)
\end{aligned} \tag{8.19}$$

The tests in the dynamics again ensure that the *entry* curve starts when x, y, and c have appropriate distance $\sqrt{3}r$ (as identified in Sect. 8.2) and that r is defined to be the radius belonging to angular velocity ω and linear speed $\|d\|$.

Secondly, these choices are feasible, i.e., there always is a mutually agreeable choice, according to the following diamond formula:

$$\begin{aligned}
&\|d\| = \|e\| \land \lambda > 0 \land x+\lambda d = y+\lambda e \to \\
&\langle c := x+\lambda d;\ r := *;\ ?\|x-c\| = \sqrt{3}r;\ ?\|y-c\| = \sqrt{3}r;\ \omega := *;\ ?(r\omega)^2 = d_1^2 + d_2^2\rangle
\end{aligned}$$

$$\left(\|x-c\| = \sqrt{3}r \land \lambda \geq 0 \land x+\lambda d = c \land \|y-c\| = \sqrt{3}r \land \lambda \geq 0 \land y+\lambda e = c\right) \tag{8.20}$$

This property follows from the following provable simplified variant, obtained from formula (8.20) by gluing the equations $\|x-c\| = \sqrt{3}r$ and $\|y-c\| = \sqrt{3}r$ together:

$$\|d\| = \|e\| \land \lambda > 0 \land x+\lambda d = y+\lambda e \rightarrow \langle c := x+\lambda d;\ r := *;\ ?\|x-c\| = \|y-c\| = \sqrt{3}r\rangle \mathit{true} \tag{8.21}$$

The fact that $\|x-c\|$ can be written in the form $\sqrt{3}r$ for some r is easy to see just by division; similarly for the choice of ω.

Far Separation

The *entry* procedure has to be initiated while the aircraft are still sufficiently far apart for safety reasons. Otherwise, there would not be enough airspace for manoeuvring and approaching the roundabout circle safely. Correspondingly, the *agree* procedure will negotiate a roundabout choice while the aircraft have far distance. Thus, the *agree* procedure will have to maintain far separation, i.e., satisfy the property

$$\|x-y\| \geq \sqrt{2}\left(p+\frac{2}{3}\pi r\right) \rightarrow [\mathit{agree}]\left(\|x-y\| \geq \sqrt{2}\left(p+\frac{2}{3}\pi r\right)\right) \tag{8.22}$$

This appears to be a trivial property, because *agree* models the successful completion of the negotiation, so no time elapses during *agree*; hence the positions x and y do not even change. Observe, however, that the far separation distance according to equation (8.15) depends on the protected zone p and the radius r of evasive actions. Unlike p, radius r may change during *agree*, which allows for the flexibility of changing the flight radius r adaptively when repeating the roundabout manoeuvre loop at different positions. Consequently, the far separation distance $\sqrt{2}(p+\frac{2}{3}\pi r)$ is affected when changing r.

To ensure that the new radius r is chosen such that far separation is still maintained, i.e., formula (8.22) is respected, we add a corresponding constraint to *agree*. Thus, changes of r are only accepted as long as they do not compromise far separation. We show that, when adding a corresponding constraint to property (8.19), all choices by *agree* maintain far separation of the aircraft at x and y according to (8.15):

$$\|d\| = \|e\| \land \lambda > 0 \land x+\lambda d = y+\lambda e \rightarrow$$

$$\left[c := x+\lambda d;\ r := *;\ ?\|x-c\| = \sqrt{3}r;\ ?\|y-c\| = \sqrt{3}r;\ ?\|x-y\| \geq \sqrt{2}\left(p+\frac{2}{3}\pi r\right);\right.$$

$$\left.\omega := *;\ ?(r\omega)^2 = d_1^2 + d_2^2\right]$$

$$\left(\|x-c\| = \sqrt{3}r \land \lambda \geq 0 \land x+\lambda d = c \land \|y-c\| = \sqrt{3}r \land \lambda \geq 0 \land y+\lambda e = c\right.$$

$$\land \|x-y\| \geq \sqrt{2}\left(p+\frac{2}{3}\pi r\right)\Big) \quad (8.23)$$

Finally, we analyse when such choices of *agree* are feasible with a diamond formula:

$$\|d\| = \|e\| \land \lambda > 0 \land x + \lambda d = y + \lambda e \rightarrow$$
$$\langle c := x + \lambda d;\ r := *;\ ?\|x-c\| = \|y-c\| = \sqrt{3}r\rangle \left(\|x-y\| \geq \sqrt{2}\left(p+\frac{2}{3}\pi r\right)\right) \quad (8.24)$$

The corresponding distance constraints on x, y and c for *agree*, respectively, are depicted in Fig. 8.13. Using standard trigonometric relations for each half of the tri-

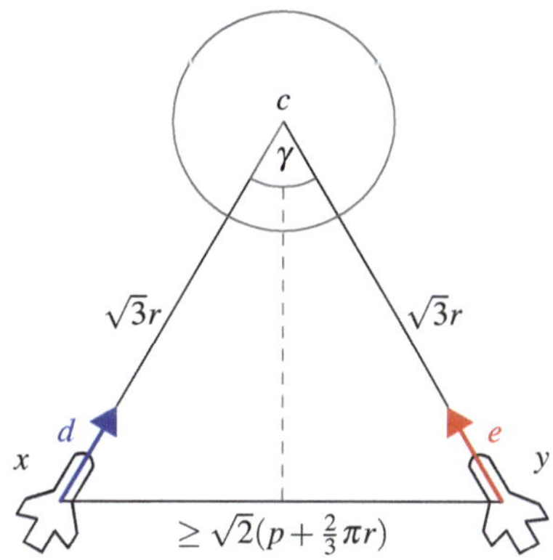

Fig. 8.13 Far separation for mutually agreeable negotiation choices

angle, we compute the resulting distance between x and y as $\|x-y\| = 2\sqrt{3}r\sin\frac{\gamma}{2}$. With quantifier elimination and simple evaluation for the remaining trigonometric expressions, we can determine under which circumstances d$\mathcal{L}$ formula (8.24) holds true, i.e., for all protected zones p there is a radius r satisfying the distance requirements:

$$\mathrm{QE}\left(\forall p \exists r{\geq}0 \left(2\sqrt{3}r\sin\frac{\gamma}{2} \geq \sqrt{2}(p+\frac{2}{3}\pi r)\right)\right) \equiv \sin\frac{\gamma}{2} > \frac{1}{3}\sqrt{\frac{2}{3}}\pi \equiv \gamma > 117.527^\circ$$

Consequently, corresponding choices are feasible for identical speeds and all protected zones with flight paths that do not intersect with narrow collision angles. The constraint on the flight path intersection angle would relax to $\gamma > 74.4^\circ$ when removing the extra factor of $\sqrt{2}$ from (8.15), which only results from our computational simplification of Cartesian degree reduction from Sect. 8.2.8. Likewise, the constraints can be relaxed by allowing different angular velocities for *entry* and *circ*, which would be an interesting extension for future work.

Despite the presence of trigonometric expressions, the above formula is a substitution instance of first-order real arithmetic and can thus be handled by QE using

Lemma 2.5. Note that the primary difference compared to trigonometric expressions occurring in the solutions of flight equations for curved flight—which do not support quantifier elimination—is that the argument $\frac{\gamma}{2}$ of sin is not quantified over. Consequently, we can consider $\sin\frac{\gamma}{2}$ as a variable, apply QE, and then use (quantifier-free) ground evaluation.

8.3.2 *Safe Exit Separation (AC7)*

NTRM (from Fig. 8.1d) does not need an exit procedure for safety, because the manoeuvre can repeat immediately in case further air traffic conflicts arise. For FTRM, on the other hand, we need to show that the exit procedure produces safe flight paths until the aircraft are sufficiently separated: When repeating the FTRM manoeuvre, the *entry* procedure needs far separation (8.15) for manoeuvring, and not just safety distance p; see Fig. 8.5b.

Safe Separation

In order to establish the safe separation of aircraft during their *exit* procedures, we show that, for its bounded duration, the exit procedure does not conflict with other flight curves such that the initial far separation is again maintained as needed by the flyable entry procedure when re-initiating collision avoidance manoeuvres repeatedly. For showing property AC7 of Sect. 8.2.3, we have to show the following d$\mathcal{L}$ formula:

$$\mathcal{T} \land \|x-y\|^2 \geq p^2 \rightarrow \left[x' = d \land y' = e \land e' = e^{\perp}; x' = d \land y' = e\right](\|x-y\|^2 \geq p^2) \tag{8.25}$$

This formula expresses that, when x already exits on a straight line while y keeps following the roundabout for a while until (after the sequential composition) both aircraft exit on straight lines, then the protected zones are respected at any point. Since, in Sect. 8.3.1, we have assumed simultaneous entry and identical speed, the aircraft can also exit simultaneously. With that simplification, property (8.25) simplifies to the following property, saying that aircraft that exit simultaneously (from tangential positions $\mathcal{T}$ of the roundabout circle) always respect their protected zones:

$$\mathcal{T} \land \|x-y\|^2 \geq p^2 \rightarrow [x' = d \land y' = e](\|x-y\|^2 \geq p^2) \tag{8.26}$$

This property expresses that safely separated aircraft that exit simultaneously along straight lines from tangential positions of a roundabout always remain safely separated. To reduce the arithmetical complexity, we overapproximate this property by showing that even the whole exit rays never cross when the aircraft exit the same roundabout tangentially (see Fig. 8.14a; the counterexample in Fig. 8.14b shows that the assumption on identical radii is required for this relaxation):

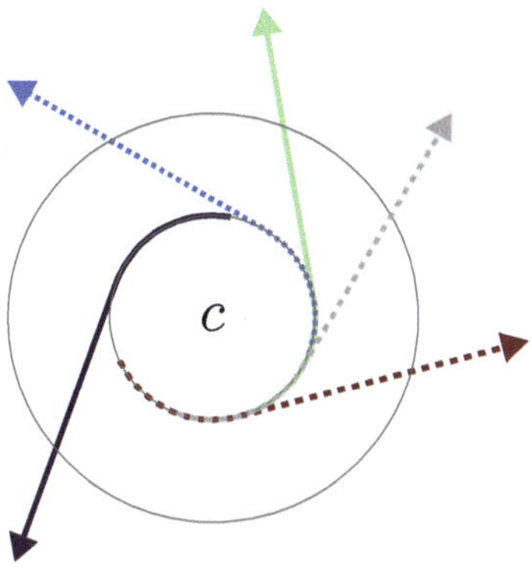

Fig. 8.14a Exit ray separation

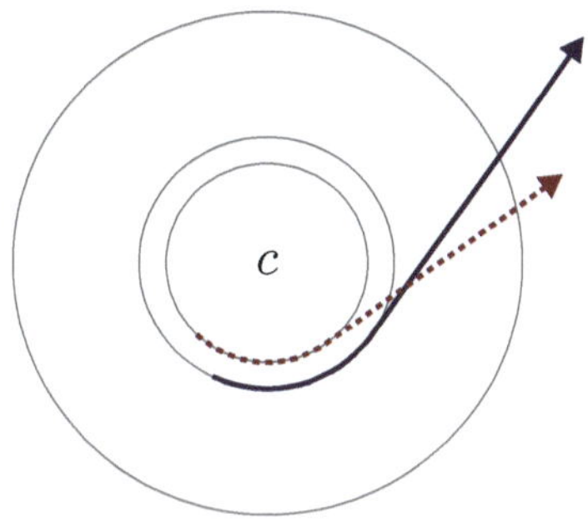

Fig. 8.14b Incompatible exit rays

$$\mathscr{T} \wedge \|x-c\|^2 = \|y-c\|^2 \wedge x \neq y \rightarrow [x' = d; y' = e] x \neq y \tag{8.27}$$

The difference with (8.26) is that, here, the aircraft exit independently by a sequential composition of two separate differential equation systems, rather than by a single differential equation system. Again the computational complexity of proving this property can be simplified substantially by adding $c_1 := 0 \wedge c_2 := 0$ by spatial symmetry reduction (the manoeuvre is invariant under translation in space). From this property, the original separation property follows using the geometric fact that, for linearity reasons, rays that never cross cannot come closer than the original distance p, which can be expressed elegantly in d$\mathscr{L}$ as:

$$\|x-y\|^2 \geq p^2 \wedge [x' = d \wedge y' = e] x \neq y \rightarrow [x' = d \wedge y' = e](\|x-y\|^2 \geq p^2) \tag{8.28}$$

Thus, by combining (8.27) with (8.28) propositionally, and by using the simple fact that the sequential independent ray evolution $x' = d; y' = e$ is an overapproximation of the synchronous evolution $x' = d \wedge y' = e$, we conclude that property (8.26) is valid. The rationale behind this overapproximation is that, if all combinations of two aircraft exiting independently (with a sequential composition) are safely separate, then, in particular, simultaneous exit of the aircraft along one common differential equation system is safely separate.

Far Separation

Aircraft finally reach arbitrary far separation when following the exit procedure long enough. Using the ray overapproximation principle from Fig. 8.14a, we prove that, due to different exit directions $d \neq e$, i.e., $\|d-e\| > 0$, the exit procedure will finally separate the aircraft arbitrarily far (starting from tangential configuration $\mathscr{T}$ of the roundabout):

$$\mathscr{T} \wedge d \neq e \rightarrow [\mathscr{F}(\omega) \wedge \mathscr{G}(\omega)](\|d-e\| > 0) \tag{8.29}$$

Aircraft in a simultaneous roundabout manoeuvre can, indeed, never enter in the same direction: Either they would already have collided (when they are entering in the same direction at the same position) or would crash later (when entering in the same direction with opposing orientations at different positions of the roundabout circle), but we have already shown that the roundabout manoeuvre is collision-free (Theorem 3.4). Thus by formula (8.29), they maintain their different directions while following the roundabout. Again, we can combine (8.29) with the geometric fact that rays into different directions which never cross will be arbitrarily far apart after sufficient time (see Fig. 8.14a):

$$d \neq e \wedge [x' = d \wedge y' = e] x \neq y \rightarrow \forall a \langle x' = d \wedge y' = e \rangle (\|x - y\|^2 > a^2)$$

By combining this geometric fact with (8.29), we obtain the final separation property saying that, due to their different directions, the exit procedure will finally separate the aircraft arbitrarily far:

$$\mathscr{T} \wedge d \neq e \rightarrow \forall a \langle x' = d \wedge y' = e \rangle (\|x - y\|^2 > a^2) \tag{8.30}$$

8.4 Compositional Verification

Subsequently, we want to combine the previous results into a full model of the flyable tangential roundabout manoeuvre. To begin with, we show how verification results for large systems can be obtained compositionally in our proof calculi (Part I of this book) from verification lemmas about small subsystems whenever the large system has been constructed by composition from the corresponding small subsystems.

Assume we have subsystems α_i with lemmas guaranteeing $\psi_i \rightarrow [\alpha_i]\phi_i$, as is the case for the respective phases *agree*, *entry*, and *circ* of FTRM. Then, in order to obtain a verification result for their sequential composition $\alpha_1; \alpha_2; \ldots; \alpha_n$, we have to show that postcondition ϕ_{i-1} of α_{i-1} implies precondition ψ_i of α_i for $i \leq n$. In fact, this is a derived proof rule of the compositional $\mathsf{d}\mathscr{L}$ calculus from Fig. 2.11 or the DAL calculus from Fig. 3.9 and can be obtained directly from the generalisation rule $[]gen$ (with the usual cut rule application *cut* to shift the antecedent of a goal to the succedent of a new subgoal, as for rule *ind'* in Sect. 2.5.2):

$$\dfrac{\vdash \forall^\alpha(\phi_i \rightarrow \phi) \qquad \dfrac{\dfrac{\psi \vdash [\alpha_{i-1}]\psi_i \qquad \vdash \forall^\alpha(\psi_i \rightarrow [\alpha_i]\phi_i)}{\psi \vdash [\alpha_{i-1}][\alpha_i]\phi_i}\,[]gen}{\psi \vdash [\alpha_{i-1}; \alpha_i]\phi_i}\,[;]}{\psi \vdash [\alpha_{i-1}; \alpha_i]\phi}\,[]gen$$

8.5 Flyable Tangential Roundabout Manoeuvre

In this section, we combine the results of this chapter about the individual phases of flyable roundabouts into a full model of the flyable tangential roundabout manoeuvre that inherits safety in a modular way by joining the individual safety properties of the respective phases together. Essentially, we collect the various manoeuvre parts together according to the protocol cycle of Fig. 8.5 and take care to ensure that all safety prerequisites are met that we have identified for the respective flight phases previously. Formally, safety properties about the individual phases will be glued together using the generalisation rule $[]gen$ and the compositional technique from Sect. 8.4.

One possible instance of FTRM is the hybrid program in Fig. 8.15, which is composed of previously illustrated parts of the manoeuvre. The technical construction

$$
\begin{aligned}
\psi &\equiv \|d\| = \|e\| \wedge r > 0 \wedge \mathscr{S}(f) \rightarrow [FTRM]\mathscr{S}(p)\\
\mathscr{C} &\equiv \|x-c\| = \sqrt{3}r \wedge \exists\lambda{\geq}0\,(x+\lambda d = c) \wedge \|y-c\| = \sqrt{3}r \wedge \exists\lambda{\geq}0\,(y+\lambda e = c)\\
FTRM &\equiv \big(free^*;\;\; agree;\;\; \Pi(entry;\, circ;\, exit)\big)^*\\
free &\equiv \omega := *;\;\; \varpi := *;\;\; \mathscr{F}(\omega) \wedge \mathscr{G}(\varpi) \wedge \mathscr{S}(f)\\
agree &\equiv c := *;\;\; r := *;\;\; ?(\mathscr{C} \wedge r > 0);\;\; ?\mathscr{S}(f);\\
&\qquad \omega := *;\;\; ?(r\omega)^2 = \|d\|^2;\;\; x_0 := x; d_0 := d; y_0 := y; e_0 := e\\
entry &\equiv \mathscr{F}(-\omega)\ until\ \|x-c\|^2 = r^2\\
circ &\equiv \mathscr{F}(\omega)\ until\ \exists\lambda{\geq}0\,\exists\mu{>}0\,(x+\lambda d = x_0 + \mu d_0)\\
exit &\equiv \mathscr{F}(0);\;\; ?\mathscr{S}(f)
\end{aligned}
$$

Fig. 8.15 Flight control with flyable tangential roundabout collision avoidance

and protocol cycle of the entry procedure have already been illustrated in Fig. 8.5. The operation $\mathscr{F}(\omega)$ *until* G expresses that the system follows differential equation $\mathscr{F}(\omega)$ until condition G is true. It is defined in terms of evolution domain restrictions and tests to make sure the system leaves mode $\mathscr{F}(\omega)$ neither later nor earlier than G specifies. We define $\mathscr{F}(\omega)$ *until* G as $\mathscr{F}(\omega) \wedge {\sim}G;\ ?G$, using the weak negation ${\sim}G$ from Sect. 3.5.7 to retain the border of G in the evolution domain restriction. For instance, $\mathscr{F}(\omega)$ *until* $x_1 \geq 0$ is $\mathscr{F}(\omega) \wedge x_1 \leq 0; ?x_1 \geq 0$.

Finally, in FTRM, Π denotes the parallel product operator. As in the work of Hwang et al. [171], the FTRM manoeuvre is assumed to operate synchronously, i.e., all aircraft make simultaneous mode changes. Consequently, the parallel product $\Pi(entry; circ; exit)$ simplifies to the conjunction of the respective differential equations in the various modes and can be defined easily as follows (with corresponding simplifications to resolve simultaneous tests):

$$(entry_x \wedge entry_y);\ (circ_x \wedge circ_y);\ (exit_x \wedge exit_y)$$

where $entry_x$ is the entry procedure of the aircraft at position x etc.

To verify this manoeuvre, we split the proof into the modular properties that we have shown previously following the verification plan from Sect. 8.2.3. Formally, we use the generalisation rule ($[]gen$) to split the system at its sequential compositions. The corresponding subproperties are depicted in Fig. 8.16 with the mapping to previous results according to Table 8.1. Formula $\mathscr{T}$ is the characterisation of tangential configurations due to equation (8.2), and the separation formula $\mathscr{S}(p)$ is from (8.1).

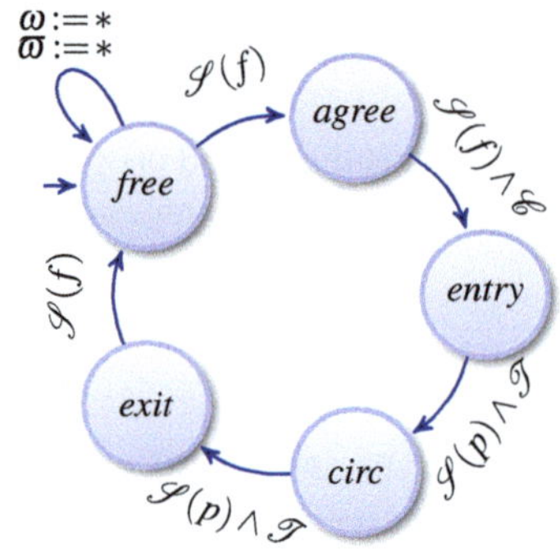

Fig. 8.16 Verification loop for flyable tangential roundabout manoeuvres

Table 8.1 Verification loop properties for flyable tangential roundabout manoeuvres

Property	Consequence of
$\mathscr{S}(f) \to [free]\,\mathscr{S}(f)$	Fig. 8.6
$\mathscr{S}(f) \to [agree](\mathscr{S}(f) \wedge \mathscr{C})$	(8.19), (8.23)
$\mathscr{C} \wedge \mathscr{S}(f) \to [entry]\,\mathscr{S}(p)$	(8.14)
$\mathscr{C} \wedge \mathscr{S}(f) \to [entry]\,\mathscr{T}$	(8.3)
$\mathscr{T} \wedge \mathscr{S}(p) \to [circ](\mathscr{S}(p) \wedge \mathscr{T})$	Fig. 8.6
$\mathscr{T} \wedge \mathscr{S}(p) \to [exit]\,\mathscr{S}(p)$	(8.26)
$\mathscr{T} \wedge \mathscr{S}(p) \to [exit]\,\mathscr{S}(f)$	(8.26), (8.30)

By combining the results about the FTRM flight phases as summarised in Fig. 8.16, we conclude that FTRM avoids collisions safely. Note that the modular proof structure in Fig. 8.16 still holds when replacing any part of the manoeuvre with a different choice that still satisfies the same specification, e.g., for different entry procedures that still succeed in tangential configuration $\mathscr{T}$ within bounded time. This includes roundabouts with *asymmetric positions*, i.e., where the initial distance to c can be different, and with *near conflicts*, where the flight paths do not intersect in one point but in a larger critical region [171]. Most notably, the separation proof in Sect. 8.2.8 is by overapproximation and tolerates asymmetric distances to c (Fig. 8.11).

Theorem 8.1 (Safety property of flyable tangential roundabouts). *FTRM is collision-free, i.e., the collision avoidance property ψ in Fig. 8.15 is valid. Furthermore, any variation of FTRM with a modified entry procedure that safely reaches tangential configuration $\mathcal{T}$ in some bounded time T is safe. That is, if the following formula holds, saying that, until time T the aircraft have safe distance p and will have reached configuration $\mathcal{T}$ at time T, where τ is a clock:*

$$\mathcal{S}(f) \rightarrow [\tau := 0; \mathit{entry} \wedge \tau' = 1]\big((\tau \leq T \rightarrow \mathcal{S}(p)) \wedge (\tau = T \rightarrow \mathcal{T})\big).$$

For reference, a possible instantiation of Fig. 8.15 with abbreviations resolved is shown in Fig. 8.17.

8.6 Experimental Results

Table 8.2 summarises experimental results for the air traffic control case study of flyable tangential roundabout manoeuvres. The rows marked with $*$ indicate a property where simplifications like symmetry reduction have been used to reduce the computational complexity of quantifier elimination. The results in Table 8.2 show that even aircraft manoeuvres with challenging dynamics can be verified with our logical analysis approach for hybrid systems.

Table 8.2 Experimental results for air traffic control (initial timeout = 10s)

Case study		Time(s)	Memory(MB)	Steps	Dim
tangential roundabout	(2 aircraft)	10.5	6.8	197	13
tangential roundabout	(3 aircraft)	636.1	15.1	342	18
tangential roundabout	(4 aircraft)	918.4	31.4	520	23
tangential roundabout	(5 aircraft)	3552.6	46.9	735	28
bounded speed control	(3.14)	19.6	34.4	28	12
bounded manoeuvre speed	Example 3.16	0.3	6.3	14	4
flyable roundabout entry*	(8.3)	10.2	9.6	132	8
flyable entry feasible*	(8.4)	104.4	87.9	16	10
flyable entry circular	(8.6)	2.9	7.6	81	5
limited progress	(8.9)	2	6.5	60	8
entry separation	(8.10), 29 int.	140.1	20.1	512	16
negotiation feasible	(8.17)	4.5	8.4	27	8
negotiation successful	(8.18)	2.7	21.8	34	8
mutual negotiation successful	(8.19)	0.9	6.4	60	12
mutual negotiation feasible	(8.21)	7.5	23.8	21	11
mutual far negotiation	(8.23)	2.4	8.1	67	14
simultaneous exit separation*	(8.27)	4.7	12.9	44	9
different exit directions	(8.29)	3	11.1	42	11

Experimental results are from a 2.6 GHz AMD Opteron with 4 GB memory. Memory consumption of quantifier elimination is shown in Table 8.2, excluding the front end. The dimension of the continuous state space and number of proof steps

$$
\begin{aligned}
\psi &\equiv d_1^2+d_2^2=e_1^2+e_2^2 \wedge r>0 \wedge (x_1-y_1)^2+(x_2-y_2)^2 \geq 2\left(p+\frac{2}{3}\pi r\right)^2 \\
&\quad \rightarrow [FTRM^*]\phi(p) \\
FTRM &\equiv \Bigg(\Big(\omega := *; \varpi := *; \\
\textit{free}:&\quad x_1' = d_1 \wedge x_2' = d_2 \wedge d_1' = -\omega d_2 \wedge d_2' = \omega d_1 \\
&\quad \wedge y_1' = e_1 \wedge y_2' = e_2 \wedge e_1' = -\varpi e_2 \wedge e_2' = \varpi e_1 \\
&\quad \wedge (x_1-y_1)^2+(x_2-y_2)^2 \geq 2\left(p+\frac{2}{3}\pi r\right)^2 \Big)^*; \\
\textit{agree}:&\quad c := *;\ r := *;\ ?(x_1-c_1)^2+(x_2-c_2)^2 = 3r^2; \\
&\quad ?\exists \lambda \geq 0\, (x_1+\lambda d_1 = c_1 \wedge x_2+\lambda d_2 = c_2); \\
&\quad ?(y_1-c_1)^2+(y_2-c_2)^2 = 3r^2; \\
&\quad ?\exists \lambda \geq 0\, (y_1+\lambda e_1 = c_1 \wedge y_2+\lambda e_2 = c_2); \\
&\quad ?(x_1-y_1)^2+(x_2-y_2)^2 \geq 2\left(p+\frac{2}{3}\pi r\right)^2; \\
&\quad \omega := *;\ ?(r\omega)^2 = d_1^2+d_2^2 \\
&\quad x_1^0 := x_1;\ x_2^0 := x_2;\ d_1^0 := d_1;\ d_2^0 := d_2; \\
&\quad y_1^0 := y_1;\ y_2^0 := y_2;\ e_1^0 := e_1;\ e_2^0 := e_2; \\
\textit{entry}_x \wedge \textit{entry}_y:&\quad x_1' = d_1 \wedge x_2' = d_2 \wedge d_1' = -(-\omega) d_2 \wedge d_2' = -\omega d_1 \\
&\quad \wedge y_1' = e_1 \wedge y_2' = e_2 \wedge e_1' = -(-\omega) e_2 \wedge e_2' = -\omega e_1 \\
&\quad \wedge \sim\left((x_1-c_1)^2+(x_2-c_2)^2 = r^2\right); \\
&\quad ?(x_1-c_1)^2+(x_2-c_2)^2 = r^2; \\
\textit{circ}_x \wedge \textit{circ}_y:&\quad x_1' = d_1 \wedge x_2' = d_2 \wedge d_1' = -\omega d_2 \wedge d_2' = \omega d_1 \\
&\quad \wedge y_1' = e_1 \wedge y_2' = e_2 \wedge e_1' = -\omega e_2 \wedge e_2' = \omega e_1 \\
&\quad \wedge \sim\big(\exists \lambda \geq 0\, \exists \mu > 0\, (x_1+\lambda d_1 = x_1^0+\mu d_1^0 \wedge x_2+\lambda d_2 = x_2^0+\mu d_2^0) \\
&\qquad \wedge \exists \lambda \geq 0\, \exists \mu > 0\, (y_1+\lambda e_1 = y_1^0+\mu e_1^0 \wedge y_2+\lambda e_2 = y_2^0+\mu e_2^0)\big); \\
&\quad ?\big(\exists \lambda \geq 0\, \exists \mu > 0\, (x_1+\lambda d_1 = x_1^0+\mu d_1^0 \wedge x_2+\lambda d_2 = x_2^0+\mu d_2^0) \\
&\qquad \wedge \exists \lambda \geq 0\, \exists \mu > 0\, (y_1+\lambda e_1 = y_1^0+\mu e_1^0 \wedge y_2+\lambda e_2 = y_2^0+\mu e_2^0)\big); \\
\textit{exit}_x \wedge \textit{exit}_y:&\quad x_1' = d_1 \wedge x_2' = d_2 \wedge y_1' = e_1 \wedge y_2' = e_2; \\
&\quad ?(x_1-y_1)^2+(x_2-y_2)^2 \geq 2\left(p+\frac{2}{3}\pi r\right)^2 \Bigg)
\end{aligned}
$$

Fig. 8.17 Flight control with FTRM (synchronous instantiation) are indicated as well. For comparison, the results reported in Table 8.2 use settings that are closer to those that we had used in Table 5.3. Using different settings for initial timeouts for differential saturation results in faster performance for larger dimensions, particularly for multiple aircraft; see Table 8.3.

Table 8.3 Experimental results for air traffic control (initial timeout = 4s)

Case study		Time(s)	Memory(MB)	Steps	Dim
tangential roundabout	(2 aircraft)	10.4	6.8	197	13
tangential roundabout	(3 aircraft)	253.6	7.2	342	18
tangential roundabout	(4 aircraft)	382.9	10.2	520	23
tangential roundabout	(5 aircraft)	1882.9	39.1	735	28
bounded speed control	(3.14)	19.5	34.4	28	12
bounded manoeuvre speed	Example 3.16	0.5	6.3	14	4
flyable roundabout entry*	(8.3)	10.1	9.6	132	8
flyable entry feasible*	(8.4)	104.5	87.9	16	10
flyable entry circular	(8.6)	3.2	7.6	81	5
limited progress	(8.9)	1.9	6.5	60	8
entry separation	(8.10), 29 int.	140.1	20.1	512	16
negotiation feasible	(8.17)	4.8	8.4	27	8
negotiation successful	(8.18)	2.6	13.1	34	8
mutual negotiation successful	(8.19)	0.8	6.4	60	12
mutual negotiation feasible	(8.21)	7.5	23.8	21	11
mutual far negotiation	(8.23)	2.4	8.1	67	14
simultaneous exit separation*	(8.27)	4.3	12.9	44	9
different exit directions	(8.29)	3.1	11.1	42	11

The experimental results in Tables 8.2 and 8.3 for property (8.10) can be improved. Currently, they still need as much as 29 simple user interactions to overcome preliminary simplifications in the implementation of our proof strategies in the KeYmaera tool. An improved implementation of iterative inflation order (Sect. 5.5) will reduce the number of required interactions to at most one interaction that specifies the postcondition of the limited progress property (8.9) as an invariant.

8.7 Summary

We have analysed complex air traffic control applications. Real aircraft can only follow sufficiently smooth flyable curves. Hence, mathematical manoeuvres that require instant turns give physically impossible conflict resolution advice. We have developed a new collision avoidance manoeuvre with smooth, fully flyable curves. Despite its complicated dynamics and manoeuvring, we have verified collision avoidance in this flyable tangential roundabout manoeuvre formally using our verification algorithm for a logic of hybrid systems. Because of the intricate spatio-temporal movement of aircraft in curved roundabouts, some of the properties require intricate arithmetic, which we handled by symmetry reduction and degree-based reductions. The proof is automatic except for the modularisation and arithmetical simplifications that are still needed to overcome the computational complexity.

While the flyable roundabout manoeuvre is a highly nontrivial and challenging study, we still use some modelling assumptions from the literature that should be relaxed in future work, e.g., synchronous, symmetric conflict resolution. Further generalisations include asymmetric entry procedures, different varying cruise speeds, disturbances, or new aircraft. The proof structure behind Theorem 8.1 is already sufficiently general, but the computational complexity is quite high. It would be interesting future work to see if the informal robustness studies of Hwang et al. [171] can be carried over to a formal verification result.

Chapter 9
Conclusion

Hybrid systems are an equally important and challenging class of systems, whose numerous occurrences in safety-critical complex physical systems call for formal verification techniques to establish their correct functioning by rigorous mathematical analysis. The characteristic feature of the modelling idea behind hybrid systems is that they admit interacting discrete and continuous dynamics to capture the superposition of physical system dynamics with control at a natural modelling level. With these superpositions, hybrid systems can model challenging system dynamics fairly easily but also require sophisticated analysis techniques.

Correct functioning of hybrid systems plays an important role in many practical applications. This includes small embedded controllers in automotive industries that regulate isolated processes like air bag inflation, and complex physical traffic systems for train or air traffic control. Similar effects occur in small biomedical devices like glucose regulators following, e.g., model-predictive control, and large-scale physical or chemical process control like injection control in combustion engines or full nuclear reactors. In other domains, hybrid effects also become increasingly important, including in robotic applications, where mobile robots have to work reliably in safety-critical environments, e.g., in driverless vehicle technology. Circuit designs also often exhibit relevant hybrid effects, where increasing clock rates lead to mixed analog and digital effects, because larger parts of the chip remain analog. That happens, for instance, when reducing the number of extra latches that stabilise values computed by analog circuits.

Logic for Hybrid Systems

As a general analysis technique for hybrid systems, we have described a systematic logical analysis approach in this book that is based on symbolic and mathematical logic, automated theorem proving, differential algebra, computer algebra, semialgebraic geometry, analysis, and results from the theory of differential equations and dynamical systems; see Fig. 9.1.

A. Platzer, Logical Analysis of Hybrid Systems,
DOI 10.1007/978-3-642-14509-4_9, © Springer-Verlag Berlin Heidelberg 2010

Fig. 9.1 Topics contributing to the logical analysis of hybrid systems

We introduced a series of *differential dynamic logics* ($d\mathcal{L}$, DAL, dTL) as concise languages for specifying correctness properties of hybrid systems along with concise practical proof calculi for verifying hybrid systems. The logic $d\mathcal{L}$ is a first-order dynamic logic for hybrid programs, which extend classical discrete programs to uniform operational models for hybrid systems with interacting discrete jumps and continuous evolutions along differential equations. Its first-order completion, DAL, is a first-order dynamic logic for more general differential-algebraic programs that can combine general first-order discrete jump constraints and first-order differential-algebraic constraints. In a further independent dimension, we augment $d\mathcal{L}$ and DAL to a temporal dynamic logic, dTL, with independent modal path quantifiers over traces and temporal quantifiers along traces, thereby combining the capabilities of dynamic logic to reason about possible system behaviour with the power of temporal logic in reasoning about the temporal behaviour along traces.

Proof Calculi

Our concise proof calculi for differential dynamic logics work compositionally by decomposing properties of hybrid systems symbolically into properties of their parts, which improves both traceability and scalability of results. In order to handle interacting hybrid dynamics, we lift real quantifier elimination to the deductive calculi in a new modular way that is suitable for automation, using a combination of real-valued free variables, Skolem terms, and invertible quantifier rules over the reals.

As a fundamental result aligning hybrid and continuous reasoning proof-theoretically, we have proven our calculi to axiomatise the transition behaviour of hybrid systems sound and complete relative to the handling of differential equations. This result is based on an entirely new notion of hybrid completeness and shows that the calculi are adequate for verification purposes. This is the first relative completeness result for hybrid systems and, in fact, the first notion of hybrid completeness.

Furthermore, we have complemented discrete induction with a new first-order differential induction that uses differential invariants and differential variants for proving correctness statements about first-order differential-algebraic constraints and differential inequalities purely algebraically based on the differential constraints themselves instead of their solutions. This is particularly relevant for formal verification, because solutions of differential equations quickly yield undecidable arithmetic or may not even be expressible in closed form. In combination with successive differential refinement and differential cuts for refining the system dynamics with auxiliary differential invariants, we obtain a powerful verification calculus for hybrid systems with challenging dynamics.

Finally, our sequent calculus for the temporal extension dTL of differential dynamic logic is a completely modular combination of temporal and nontemporal reasoning, where temporal formulas are handled using rules that augment intermediate state transitions with corresponding sub-specifications recursively. We have shown that this gives a sound and complete axiomatisation for temporal properties relative to the nontemporal base logic.

Logic-Based Verification Algorithms

To address practical scalability challenges for larger case studies, we have introduced proof strategies that navigate through the nondeterminisms in the proof calculi for differential dynamic logics, including iterative background closure and iterative inflation order strategies. Further, we have introduced a verification algorithm that computes invariants and differential invariants as fixed points in a proof loop for differential dynamic logic, thereby using the compositional properties of our calculi to exploit locality in system designs for formal verification. Our compositional proof calculi result in a compositional verification approach that is, essentially, a divide-and-conquer approach for hybrid systems verification.

Applications

In a series of examples and case studies from practical application domains, we have demonstrated that our techniques can be used successfully for verifying safety, controllability, and liveness properties in realistic train control applications and challenging case studies for fully curved roundabout manoeuvres in air traffic control.

In a nutshell, we have introduced and developed the theory, practise, and applications of differential dynamic logics, leading to a concise yet complete novel logical analysis approach for hybrid systems with which we can verify applications that were out of scope for other approaches.

Part IV
Appendix

Overview In the appendices, we summarise the background in logic and differential equations that we need in the main parts of this book. In App. A, we give an introduction to basic first-order logic, its syntax, semantics, and proof techniques. For reference, App. B summarises some classical results about differential equations. In App. C, we formally investigate the relationship of hybrid automata and hybrid programs by embedding hybrid automata into hybrid programs. In App. D, we briefly characterise the verification tool KeYmaera that implements the logics and automated theorem proving techniques presented in this book. We also survey various techniques that can be used to prove formulas of real arithmetic.

Appendix A
First-Order Logic and Theorem Proving

Contents

Synopsis In this chapter, we briefly review basic elements of first-order logic, which is the basis for the logics developed in this book. We define the formal syntax of first-order logic, i.e., the language, in which sentences and formulas of this logic can be written. We further define the semantics of first-order logic, which gives meaning to the syntactic expressions. Depending on the interpretation of their elementary symbols, some formulas then express true facts about the world while others do not. Most informative are those formulas that are valid, i.e., true under all interpretations. Finally we review sequent calculi as systematic proof procedures for proving that a formula is valid in first-order logic.

A.1 Overview

In this chapter, we review first-order logic to a level of detail that is used in this book. A comprehensive treatment of first-order logic and details about automatic proof procedures can be found in the book by Fitting [122]. Gentzen-style sequent

A. Platzer, Logical Analysis of Hybrid Systems,
DOI 10.1007/978-3-642-14509-4, © Springer-Verlag Berlin Heidelberg 2010

calculi can be found in the book by Gallier [130]. As a reference for the specifics of dynamic logic for conventional discrete programs, we recommend the book by Harel et al. [149] and the earlier work by Harel [148]. For further background on first-order logic we refer the reader to the literature [282, 42, 169, 17].

Classical logic studies how truth is preserved in reasoning and investigates valid reasoning schemes of formal proofs. In the context of hybrid systems, logic becomes useful on multiple levels: on the system level to describe the local system operations and on the reasoning level to analyse the global system behaviour. What we are most interested in for hybrid systems is what facts are true about a particular hybrid system at hand. We start with an informal introduction to first-order logic and then review its syntax, its semantics, and a proof calculus.

Logical System Modelling

On the system level, logic is very useful for describing the control logic of the system itself, for instance, for describing the conditions under which the system controller switches to a new mode. In a train control system, for instance, typical conditions in the controller are:

> **If** the train runs faster than the recommended speed, **then** the controller switches to braking mode.
> **If** the train runs slower than the recommended speed **and** the distance to the next train is large enough, **then** the controller switches to acceleration mode, **unless** the train received an emergency message.
> **If** the train received an emergency message, **then** the controller follows emergency procedures.
> **If** the controller follows emergency procedures **and** the emergency situation has **not** been cleared, **then** the controller switches to braking mode.

These switching conditions seem to imply the following consequences:

> **If** the train received an emergency message **and** the emergency situation has **not** been cleared, **or if** the train is running faster than the recommended speed, **then** the controller switches to braking mode.
> **If** the train received no emergency message **and** the distance to the next train is large enough, **then** the controller does **not** switch to braking mode **or** the train runs faster than the recommended speed.

Among many other uses of logic, logical reasoning can be used as a systematic way to analyse whether these conditions are really consequences of the above conditions. In order to find out if they are indeed consequences, we try to formalise the above conditions as logical formulas in first-order logic:

$$\begin{aligned}
&\mathit{faster}(v,r) \to \mathit{brake}(v)\\
&\mathit{slower}(v,r) \land \mathit{far}(z) \land \neg EM \to \mathit{accel}(v)\\
&EM \to EP\\
&EP \land \neg EC \to \mathit{brake}(v)
\end{aligned} \tag{A.1}$$

The first formula directly represents the first condition above using a logical implication. If the train with speed v runs faster than the recommend speed r (which we write as $faster(v,r)$), then (expressed by the implication arrow $\rightarrow$) the train brakes. By v we mean the train's velocity, by r the recommended speed. The predicate $faster(v,r)$ is meant to hold of v and r if v is more than r. By $brake(v)$ we mean that the train brakes when it has velocity v, by $accel(v)$ that it accelerates. Formula EM is meant to be true if an emergency message has been received, EP if the train follows emergency procedures. By EC we mean the condition that is true if the emergency situation has been cleared. Note that the meaning we associate with these symbols is quite arbitrary and subject to our interpretation. The logical formulas themselves are not limited to this particular interpretation of the symbols. The truth of logical formulas depends on the interpretation of symbols, which can be arbitrary. What is fixed, however, is the meaning of the logical operators as indicated in Table A.1.

Table A.1 Intuitive meaning of logical operators in first-order logic

Formula	Read	Informal meaning
$\neg\phi$	"not ϕ"	negation, holds if ϕ is false
$\phi \wedge \psi$	"ϕ and ψ"	conjunction, holds if both ϕ and ψ are true
$\phi \vee \psi$	"ϕ or ψ"	disjunction, holds if ϕ is true or ψ is true
$\phi \rightarrow \psi$	"ϕ implies ψ"	implication, holds if ϕ is false or ψ is true
$\phi \leftrightarrow \psi$	"ϕ equivalent to ψ"	equivalence, holds if ϕ and ψ are either both true or both false
$\forall x\, \phi$	"for all x: ϕ"	universal quantifier, holds if ϕ is true for all values of x
$\exists x\, \phi$	"for some x: ϕ"	existential quantifier, holds if ϕ is true for some value of x
(ϕ)		brackets, only used for grouping logical formulas

Next, we can phrase the switching conditions that we suspected to be consequences as logical formulas using the same symbols:

$$(EM \wedge \neg EC) \vee faster(v,r) \rightarrow brake(v) \tag{A.2}$$

$$\neg EM \wedge far(z) \rightarrow \neg brake(v) \vee faster(v,r) \tag{A.3}$$

Logical Consequences

Now, it makes sense to ask if the original switching conditions (A.1) actually imply the suspected consequences (A.2) and (A.3). Thus we ask if the following first-order formula, which has the conjunction of the formulas in (A.1) on the left hand side of an implication ($\rightarrow$ in the last line) and formula (A.2) on the right hand side, is true under all circumstances:

$$\begin{aligned}\Big(\big(faster(v,r) \rightarrow brake(v)\big) \wedge \big(slower(v,r) \wedge far(z) \wedge \neg EM \rightarrow accel(v)\big) \wedge \\ (EM \rightarrow EP) \wedge \big(EP \wedge \neg EC \rightarrow brake(v)\big)\Big) \\ \rightarrow \big((EM \wedge \neg EC) \vee faster(v,r) \rightarrow brake(v)\big)\end{aligned} \tag{A.4}$$

Using a simple proof in first-order logic, this formula turns out to be valid, i.e., true under all interpretations of the symbols. Because it is true under all interpretations, it is also true under the specific train interpretation we had in mind when we started the formalisation.

Let us try to conduct an informal proof of property (A.4). Assume all assumptions are true, because an implication does not say anything unless the assumption on the left of the implication operator ($\rightarrow$) holds. Thus we assume $(EM \wedge \neg EC) \vee faster(v,r)$ to be true. If this disjunction is true, then the left subformula or the right subformula needs to be true. Now if the disjunction is true because the right subformula $faster(v,r)$ is true, then we are done because, together with the first condition $faster(v,r) \rightarrow brake(v)$, this implies that $brake(v)$ must be true (if $faster(v,r)$ and $faster(v,r) \rightarrow brake(v)$ are true then $brake(v)$ is too). If, instead, the first subformula $EM \wedge \neg EC$ is true, then both EM and $\neg EC$ must be true, hence EC false. Thus, the combination with condition $EM \rightarrow EP$ implies that EP must be true. Since EP is true but EC is not, condition $EP \wedge \neg EC \rightarrow brake(v)$ can be used to conclude that $brake(v)$ must be true. Hence we have found out by logical analysis that condition (A.2) is indeed a consequence of conditions (A.1). While successful, this manual reasoning is not very scalable to large systems. Automated theorem proving studies the question how proving can be automated such that formulas such as (A.4) can be proven by a machine. In fact, formula (A.4) can be proven very easily by a computer.

Formula (A.3), however, turns out *not* to be a consequence of formulas (A.1). The reason is that—with our specific train intuition in mind—we had originally associated a special meaning with the symbols $faster, slower, accel$ and $brake$ that is not reflected in the formalisation. First-order logic in its own right, allows an arbitrary interpretation of these symbols, including one where $faster(v,r)$ and $slower(v,r)$ would be true for the same arguments v,r, or where $accel(v)$ and $brake(v)$ would both be true for all v. What we may have had in mind from the names—which are arbitrary though—is that *faster* and *slower* should be opposite notions and that *accel* and *brake* should not hold simultaneously. In order to explain these additional constraints in first-order logic, let us assume that by $faster(v,r)$ we mean exactly the opposite of $slower(v,r)$, that is $faster(v,r)$ is true if and only if $slower(v,r)$ is false. Let us further assume that an accelerating train cannot brake at the same time, and that a braking train cannot accelerate at the same time, while the train could still neither brake nor accelerate when it keeps moving with constant speed. This background knowledge corresponds to the following formulas in first-order logic:

$$\begin{aligned} &slower(v,r) \leftrightarrow \neg faster(v,r)\\ &accel(v) \rightarrow \neg brake(v)\\ &brake(v) \rightarrow \neg accel(v) \end{aligned} \tag{A.5}$$

Observe that logic can be used to show that the last two conditions are actually equivalent! The middle condition says: If the train accelerates, then it does not brake. The last condition says: If the train brakes, then it does not accelerate. But that is already entailed: if the train brakes, then it cannot accelerate for otherwise the middle condition would imply that it does not brake, contradicting the assumption that it indeed brakes. We defer logical reasoning like this and continue to work with all background assumptions, even though some of them turn out to be redundant or superfluous.

Now, with this background knowledge, (A.3) is a consequence of the conjunction of (A.1) and (A.5).

$$\Big(\big(\mathit{faster}(v,r) \to \mathit{brake}(v)\big) \land \big(\mathit{slower}(v,r) \land \mathit{far}(z) \land \neg EM \to \mathit{accel}(v)\big) \land$$
$$\big(EM \to EP\big) \land \big(EP \land \neg EC \to \mathit{brake}(v)\big) \land$$
$$\big(\mathit{slower}(v,r) \leftrightarrow \neg\mathit{faster}(v,r)\big) \land \big(\mathit{accel}(v) \to \neg\mathit{brake}(v)\big) \land \big(\mathit{brake}(v) \to \neg\mathit{accel}(v)\big)\Big)$$
$$\to \big(\neg EM \land \mathit{far}(z) \to \neg\mathit{brake}(v) \lor \mathit{faster}(v,r)\big) \quad \text{(A.6)}$$

The interpretation of *faster* and *slower* is still arbitrary, but now only those interpretations that respect assumption $\mathit{slower}(v,r) \leftrightarrow \neg\mathit{faster}(v,r)$ matter, because the assumptions on the left side of the implication (upper three lines) are not met such that formula (A.6) is true trivially. With assumptions (A.5), we have not characterised the interpretations of *faster* and *slower* uniquely. In fact, there are still many interpretations of what *faster* means, which, indeed, is a somewhat subjective notion. At least, (A.5) characterise just enough background knowledge for our analysis above.

Limitations of First-Order Logic

Our reasoning about the switching conditions of a train in first-order logic has been successful so far. But first-order logic has its limitations that make it difficult and, in general, even impossible to prove all properties of hybrid systems in first-order logic. In reality, variables like velocity v of a train are not restricted to some arbitrary but fixed value (as first-order logic assumes), but, instead, may change its value over time as the train accelerates or brakes. Likewise, the emergency may not have been cleared yet (*EC* is still false), but may get cleared a minute later (*EC* suddenly becomes true). These changes in interpretation over time are not well reflected in first-order logic. In particular, plain first-order logic itself does not support reasoning about the differential equations and real arithmetic domains underlying hybrid systems.

Still, first-order logic is a very successful basis for hybrid systems analysis. The differential dynamic logics presented in Part I of this book are extensions of first-order logic and the proof procedures for differential dynamic logic extend first-order logic theorem proving. In this appendix, we thus review first-order logic.

A.2 Syntax of First-Order Logic

In this section, we introduce the syntax of logical formulas in *first-order logic* (FOL). First-order logic is the logic of function symbols, predicate symbols, propositional logical operators, and quantifiers over objects of the domain of discourse.

The language of first-order logic, in which sentences can be written, is defined by a formal syntax. This section presents the syntactic structure of the language of first-order logic. Given a vocabulary of symbols the language of the logic will be inductively defined by combining them using logical operators.

A.2.1 Terms

The construction of a logic starts with a *signature* Σ, which is the set of names (called *symbols*) of all entities nameable in a certain context. A signature is the vocabulary or alphabet of symbols or signs from which well-formed formulas can be built. Formulas of first-order logic are built over a set V of logical variable symbols and a signature Σ of function and predicate symbols. The difference between function and predicate symbols is that function symbols stand for functions that take the values of arguments and give back a function value. Predicate symbols, in contrast, are either true or false of their arguments. That is, they take the values of arguments and give either the truth-value "*true*" or the truth-value "*false*". No other result is permitted for predicate symbols. Function symbols are often written as f,g,h,a,b,c and predicate symbols are often written as p,q,r.

In addition to declaring which function and predicate symbols are allowed, the signature Σ further declares the arity of each function and predicate symbol, i.e., the number of arguments it expects. Formally, arity can be understood as a mapping $\Sigma \to \mathbb{N}$. Arity 0 means that the function or predicate symbol does not expect (nor admit) any arguments. A common short notation to specify the arities of function and predicate symbols is

$$\Sigma = \{a/0, c/0, f/1, g/2;\ p/1, r/2\}. \tag{A.7}$$

This notation would say that a and c are function symbols without arguments (0), f is a function symbol that expects one argument, and g a function symbol that expects two arguments. Further p would be a predicate symbol with one argument and r a predicate symbol with two arguments. In the above notation in (A.7), predicate symbols are separated from function symbols by a semicolon (;). In addition to function and predicate symbols from the signature Σ, we also allow logical variables from a set V that stand for objects. Logical variables are often written as x,y,z.

Well-formed arguments to function symbols and predicate symbols are called terms. Variables are well-formed terms, and functions applied to the appropriate number of terms as arguments are well-formed terms.

Definition A.1 (Terms). $\mathrm{Trm}(\Sigma)$ is the set of all *terms*, which is the smallest set such that:

- If $x \in V$ is a logical variable, then $x \in \mathrm{Trm}(\Sigma)$.
- If $f \in \Sigma$ is a function symbol of arity $n \geq 0$ and, for $1 \leq i \leq n$, $\theta_i \in \mathrm{Trm}(\Sigma)$, then $f(\theta_1,\ldots,\theta_n) \in \mathrm{Trm}(\Sigma)$. The case $n = 0$ is permitted.

More succinctly, we also say that the terms of first-order logic are defined by the following grammar (where $\theta_1,\ldots,\theta_n$ are terms, f a function symbol of arity n, and $x \in V$ is a logical variable):

$$\theta ::= x \mid f(\theta_1,\ldots,\theta_n).$$

Example A.1. When we assume the signature from (A.7) and the set of variables $V = \{x,y,z\}$, then the following are well-formed terms:

- $f(a)$
- $f(f(g(a,f(x))))$
- $g(f(a),f(c))$
- $f(g(g(a,f(a)),g(c,f(c))))$
- $f(x,g(g(y,a),x))$

The following, however, are no terms with respect to the chosen Σ and V:

- $f(a,c)$ — wrong number of arguments for f; exactly one argument is expected
- $g(c,f)$ — f is not a term but a function symbol that expects two arguments, not zero
- $g(g(f(a)$ — incomplete term
- $f(g(a,b))$ — b is not in V or in Σ according to (A.7). The expression is a term if we add b to V or use $\Sigma \cup \{b/0\}$ instead of Σ. □

A.2.2 First-Order Formulas

The well-formed formulas of a logic form a formal language over the alphabet $\Sigma \cup V$. The formulas consist of all words that can be built by recursively combining symbols of the signature with logical operator symbols. Formulas are defined as follows.

Definition A.2 (First-order formulas). The set $\mathrm{Fml}_{\mathrm{FOL}}(\Sigma)$ of *formulas of first-order logic* is the smallest set with:

- If $p \in \Sigma$ is a predicate symbol of arity $n \geq 0$ and $\theta_i \in \mathrm{Trm}(\Sigma)$ for $1 \leq i \leq n$, then $p(\theta_1,\ldots,\theta_n) \in \mathrm{Fml}_{\mathrm{FOL}}(\Sigma)$.
- If $\phi,\psi \in \mathrm{Fml}_{\mathrm{FOL}}(\Sigma)$, then $\neg\phi, (\phi \wedge \psi), (\phi \vee \psi), (\phi \rightarrow \psi) \in \mathrm{Fml}_{\mathrm{FOL}}(\Sigma)$.
- If $\phi \in \mathrm{Fml}_{\mathrm{FOL}}(\Sigma)$ and $x \in V$, then $(\forall x\, \phi), (\exists x\, \phi) \in \mathrm{Fml}_{\mathrm{FOL}}(\Sigma)$.

See Table A.1 on p. 343 for an overview of operators in logical formulas. More succinctly, we also say that first-order formulas are defined by the following grammar (where ϕ, ψ are first-order formulas, θ_i are terms, p a predicate symbol of arity n, and $x \in V$ is a logical variable):

$$\phi, \psi ::= p(\theta_1, \ldots, \theta_n) \mid \neg\phi \mid \phi \wedge \psi \mid \phi \vee \psi \mid \phi \rightarrow \psi \mid \forall x \phi \mid \exists x \phi.$$

Example A.2. When we again use the signature from (A.7) and the set of variables $V = \{x, y, z\}$, then the following are well-formed formulas:

- $p(a) \rightarrow r(a, f(c))$
- $p(x) \rightarrow (r(x, f(c)) \vee \neg p(c))$
- $\forall x (p(x) \rightarrow r(f(x), f(c)))$
- $\big(p(a) \vee \neg(r(a,a) \wedge r(f(a), f(a)))\big) \leftrightarrow (\neg p(c) \wedge r(c,c))$
- $\forall x \big(p(f(x)) \rightarrow \exists y\, r(y, f(x))\big)$
- $r(a,y) \leftrightarrow \neg \forall x \exists y \forall z \big(r(x,y) \wedge r(y, g(x,z))\big)$

The following, however, are not formulas with respect to the chosen Σ and V:

- $p(a) \rightarrow r(g(a))$ — second argument of r missing; second argument of g missing
- $r(x,y) \wedge \rightarrow p(y)$ — formula missing between operators $\wedge$ and $\rightarrow$
- $p(a) \rightarrow \forall x$ — formula missing after quantifier
- $p(a) \rightarrow \forall r\, r(a, f(a))$ — r is not a variable symbol in V but a predicate symbol in Σ; thus r cannot be quantified over (in first-order logic). Predicate symbol r stands for a predicate, not for an individual object. The logic where quantification over functions and predicates is allowed is called higher-order logic. □

Example A.3 (Trains). The examples in App. A.1 are formulas with respect to $V = \{v, z\}$ and the following signature, where r is the only function symbol:

$$\Sigma = \{r/0;\ EM/0, EP/0, EC/0, slower/2, faster/2, accel/1, brake/1, far/1\}.$$

□

A.3 Semantics of First-Order Logic

The semantics of formulas of first-order logic depends on the interpretation of its symbols. Given such an interpretation of each of the symbols, the truth-value of the full formula is defined from the meaning of its logical operators.

A.3.1 Valuation of Terms

An *interpretation* I chooses some non-empty set D_I as the universe of objects and it assigns functions and relations over D_I to the respective function and predicate symbols in Σ. In particular, if f is a function symbol of arity n, then $I(f)$ is a function $I(f) : D_I^n \to D_I$ with n arguments. Here $D_I^n := \{(d_1,\dots,d_n) \ : \ d_i \in D_I\}$ is the nth crossproduct of D_I, i.e., the set of n-tuples of D_I. Thus $I(f)(d_1,\dots,d_n) \in D_I$ is the value of the function associated with f by I at position $(d_1,\dots,d_n) \in D_I^n$. Further, if p is a predicate symbol of arity n, then $I(p)$ would be a relation, i.e., a subset $I(p) \subseteq D_I^n$. Thus $(d_1,\dots,d_n) \in I(p)$ if and only if the predicate associated with p holds true at position $(d_1,\dots,d_n) \in D_I^n$. Interchangeably, we use a slightly different notation and consider the interpretation $I(p)$ of a predicate symbol to be the characteristic function $I(p) : D_I^n \to \{true, false\}$ of the predicate. Then p holds true at position $(d_1,\dots,d_n) \in D_I^n$ if and only if $I(p)(d_1,\dots,d_n) = true$. The difference between function symbols and predicate symbols then is that the former have values in D_I and the latter have values in $\{true, false\}$.

The meaning of a formula further depends on the interpretation of its variable symbols from V. An *assignment for logical variables* is a map $\eta : V \to D_I$ that assigns an object of the universe of I to each variable.

Given an interpretation I of function (and predicate symbols) and an assignment η of logical variables, we can evaluate terms inductively.

Definition A.3 (Valuation of terms). The *valuation $val_{I,\eta}(\cdot)$ of terms* with respect to interpretation I and assignment η is defined by

1. $val_{I,\eta}(x) = \eta(x)$ if $x \in V$ is a logical variable.
2. $val_{I,\eta}(f(\theta_1,\dots,\theta_n)) = I(f)\big(val_{I,\eta}(\theta_1),\dots,val_{I,\eta}(\theta_n)\big)$ when $f \in \Sigma$ is a function symbol of arity $n \geq 0$.

A.3.2 Valuation of First-Order Formulas

Given an interpretation I of function and predicate symbols and an assignment η of logical variables, we can evaluate first-order formulas. We will use $\eta[x \mapsto d]$ to denote the assignment that agrees with assignment η except on variable $x \in V$, which is assigned $d \in D_I$ instead.

Definition A.4 (Valuation of first-order formulas). The *valuation, $val_{I,\eta}(\cdot)$, of first-order formulas* with respect to interpretation I and assignment η is defined as

1. $val_{I,\eta}(p(\theta_1,\dots,\theta_n)) = I(p)\big(val_{I,\eta}(\theta_1),\dots,val_{I,\eta}(\theta_n)\big)$
2. $val_{I,\eta}(\phi \wedge \psi) = true$ iff $val_{I,\eta}(\phi) = true$ and $val_{I,\eta}(\psi) = true$
3. $val_{I,\eta}(\phi \vee \psi) = true$ iff $val_{I,\eta}(\phi) = true$ or $val_{I,\eta}(\psi) = true$
4. $val_{I,\eta}(\neg\phi) = true$ iff $val_{I,\eta}(\phi) \neq true$
5. $val_{I,\eta}(\phi \to \psi) = true$ iff $val_{I,\eta}(\phi) \neq true$ or $val_{I,\eta}(\psi) = true$

6. $val_{I,\eta}(\forall x\,\phi) = true$ iff $val_{I,\eta[x\mapsto d]}(\phi) = true$ for all $d \in \mathbb{R}$
7. $val_{I,\eta}(\exists x\,\phi) = true$ iff $val_{I,\eta[x\mapsto d]}(\phi) = true$ for some $d \in \mathbb{R}$

We also write $I,\eta \models \phi$ iff $val_{I,\eta}(\phi) = true$ and say that $I,\eta\,\phi$ *satisfies* ϕ. Dually, we write $I,\eta \not\models \phi$ iff $val_{I,\eta}(\phi) \neq true$. We write just $\models \phi$ iff $I,\eta \models \phi$ holds for all I,η. We say that ϕ is *valid* if $\models \phi$ i.e., ϕ is true in all interpretations and under all assignments. The study of valid formulas is an important part of logic, in particular of systematic ways to establish validity by proofs.

Example A.4 (Trains). Continuing Example A.3, consider one particular interpretation I_0 with the universe $D_{I_0} := \{0,1,2,3,\ldots,100\}$ of natural numbers up to 100 and the following interpretations:

$$\begin{aligned}
I_0(r) &= 90,\\
I_0(EM) &= false,\\
I_0(EP) &= false,\\
I_0(EC) &= true,\\
I_0(slower) &= \{(d,e) \in \{0,1,2,\ldots,100\} \;:\; d < e - 10\},\\
I_0(faster) &= \{(d,e) \in \{0,1,2,\ldots,100\} \;:\; d > e\},\\
I_0(accel) &= \{d \in \{0,1,2,\ldots,100\} \;:\; d > 0, d < 50\},\\
I_0(brake) &= \{d \in \{0,1,2,\ldots,100\} \;:\; d > 80\},\\
I_0(far) &= \{d \in \{0,1,2,\ldots,100\} \;:\; d < 20\}.
\end{aligned}$$

Further, consider the particular assignment η_0 with $\eta_0(v) = 85$ and $\eta_0(z) = 10$. Interpretation I_0 and assignment η_0 correspond to one particular situation or system snapshot in train control. In this interpretation, formula (A.4) evaluates to true, as it does in all other interpretations, i.e., (A.4) is valid. In I_0, η_0, however, formula (A.3) evaluates to false, although all formulas in (A.1) evaluate to true. Since we argued that the similar formula (A.6) is valid under all interpretations, but the right-hand side (A.3) of its implication is false, this means that one of the assumptions from A.5 must be false in I_0, η_0. Indeed, $I_0, \eta_0 \not\models slower(v,r) \leftrightarrow \neg faster(v,r)$, because $val_{I_0,\eta_0}(slower(v,r)) = false$ but $val_{I_0,\eta_0}(\neg faster(v,r)) = true$, which, in turn, holds because $val_{I_0,\eta_0}(faster(v,r)) = false$. □

A.4 A Sequent Proof Calculus for First-Order Logic

After having defined syntax and semantics of first-order logic, we know how well-formed formulas look and how the truth-value of a formula is defined for an interpretation of its symbols. The most interesting formulas for us are those that are valid, i.e., true in all interpretations and assignments, because they will hold under all circumstances including any particular interpretation that we could have in mind.

The question is, how can these valid formulas be identified? Given a formula, how can we find out if it is actually valid?

If a formula is not valid, then it has an interpretation (and assignment) where it is false. Such an interpretation (and assignment) then is a witness for the fact that the formula is not valid. This witness is also known as a *counterexample* to the claim that the formula is valid. It can be difficult to find such a counterexample, but once we have it, it is often fairly easy to check. But if a formula is valid, how could we show that? Certainly, enumerating all interpretations and checking if the formula is true in each of these is a hopeless enterprise, because there is an infinite number of interpretations. Nevertheless, clever refinements of enumeration can give a proof procedure following the famous Herbrand theorem [160] and recent techniques based mostly on instantiation [30, 229, 29, 131].

A witness for a formula that is not valid is a counterexample, while a witness for a formula that is valid is a formal proof. In our setting, a formal proof is a derivation of the formula using a set of valid proof rules from axioms or formulas that are obviously valid for simple structural reasons, such as $\phi \to \phi$. Finding a proof can again be a difficult task, but once we have found it, the proof can be checked easily for compliance with the proof rules to establish validity. In fact, in a goal-directed sequent calculus like the one we show, the proof rules are written in a way that helps finding proofs by working in a systematic way to analyse the original formula. A full treatment of how the proving process can be automated for first-order logic is beyond the scope of this book. We refer the reader to the book by Fitting [122] for more details on automated theorem proving in first-order logic.

A.4.1 Proof Rules for First-Order Logic

Sequents are essentially a standard form for logical formulas that are convenient for proving. A *sequent* is of the form $\Gamma \vdash \Delta$, where the *antecedent* Γ and *succedent* Δ are finite sets of formulas. The semantics of sequent $\Gamma \vdash \Delta$ is that of the formula $\bigwedge_{\phi \in \Gamma} \phi \to \bigvee_{\psi \in \Delta} \psi$. The antecedent Γ can be thought of as the formulas we assume to be true, whereas the succedent Δ can be understood as formulas for which we want to show that at least one of them is true assuming all formulas of Γ are true. So for proving a sequent $\Gamma \vdash \Delta$, we assume all Γ and want to show that one of the Δ is true.

Standard propositional rules are listed in Fig. A.1. They decompose the propositional structure of formulas. Rules $\neg$r and $\neg$l use simple dualities caused by the implicative semantics of sequents. Essentially, instead of showing $\neg\phi$, we assume the contrary ϕ in rule $\neg$r. In rule $\neg$l, instead of assuming $\neg\phi$, we show the contrary ϕ. Rule $\vee$r uses the fact that formulas are combined disjunctively in succedents. Rule $\wedge$l uses the fact that they are conjunctive in antecedents. The comma between formulas in an antecedent has the same effect as a conjunction, and the comma between formulas in the succedent has the same effect as a disjunction. Rules $\vee$l and $\wedge$r split the proof into two cases, because conjuncts in the succedent can be

$$(\neg r)\frac{\Gamma,\phi\vdash\Delta}{\Gamma\vdash\neg\phi,\Delta}\quad(\vee r)\frac{\Gamma\vdash\phi,\psi,\Delta}{\Gamma\vdash\phi\vee\psi,\Delta}\quad(\wedge r)\frac{\Gamma\vdash\phi,\Delta\quad\Gamma\vdash\psi,\Delta}{\Gamma\vdash\phi\wedge\psi,\Delta}$$

$$(\neg l)\frac{\Gamma\vdash\phi,\Delta}{\Gamma,\neg\phi\vdash\Delta}\quad(\vee l)\frac{\Gamma,\phi\vdash\Delta\quad\Gamma,\psi\vdash\Delta}{\Gamma,\phi\vee\psi\vdash\Delta}\quad(\wedge l)\frac{\Gamma,\phi,\psi\vdash\Delta}{\Gamma,\phi\wedge\psi\vdash\Delta}$$

$$(\rightarrow r)\frac{\Gamma,\phi\vdash\psi,\Delta}{\Gamma\vdash(\phi\rightarrow\psi),\Delta}\quad(ax)\frac{}{\Gamma,\phi\vdash\phi,\Delta}$$

$$(\rightarrow l)\frac{\Gamma\vdash\phi,\Delta\quad\Gamma,\psi\vdash\Delta}{\Gamma,(\phi\rightarrow\psi)\vdash\Delta}\quad(cut)\frac{\Gamma\vdash\phi,\Delta\quad\Gamma,\phi\vdash\Delta}{\Gamma\vdash\Delta}$$

$$(\exists r)\frac{\Gamma\vdash\phi(\theta),\exists x\,\phi(x),\Delta}{\Gamma\vdash\exists x\,\phi(x),\Delta}{}^{1}\quad(\forall r)\frac{\Gamma\vdash\phi(s(X_1,..,X_n)),\Delta}{\Gamma\vdash\forall x\,\phi(x),\Delta}{}^{2}$$

$$(\forall l)\frac{\Gamma,\phi(\theta),\forall x\,\phi(x)\vdash\Delta}{\Gamma,\forall x\,\phi(x)\vdash\Delta}{}^{1}\quad(\exists l)\frac{\Gamma,\phi(s(X_1,..,X_n))\vdash\Delta}{\Gamma,\exists x\,\phi(x)\vdash\Delta}{}^{2}$$

[1] θ is an arbitrary term, usually a new logical variable X.
[2] s is a new (Skolem) function and $X_1,..,X_n$ are all free logical variables of $\forall x\,\phi(x)$.

Fig. A.1 Rule schemata of the sequent calculus for first-order logic

proven separately ($\wedge$r) and, dually, disjuncts of the antecedent can be assumed separately ($\vee$l). For rule $\wedge$r we want to show conjunction $\phi\wedge\psi$, so in the left branch we proceed to show $\Gamma\vdash\phi,\Delta$, and in the right branch we show $\Gamma\vdash\psi,\Delta$. If, as in rule $\vee$l, we assume disjunction $\phi\vee\psi$ as part of the antecedent, then we do not know whether we can assume ϕ to hold or whether we can assume ψ to hold, but can only assume that one of them holds. Hence, as in a case distinction, rule $\vee$l considers both cases, the case where we assume ϕ in the antecedent, and the case where we assume ψ. If we have proven both subgoals, the subgoal assuming ϕ and the subgoal assuming ψ, then we have also justified the goal assuming only their disjunction $\phi\vee\psi$. Rules $\rightarrow$r and $\rightarrow$l can be derived from the equivalence of $\phi\rightarrow\psi$ and $\neg\phi\vee\psi$. Rule $\rightarrow$r uses that implication $\rightarrow$ has the same meaning as the turnstile arrow $\vdash$ of a sequent. Intuitively, to show implication $\phi\rightarrow\psi$, rule $\rightarrow$r assumes ϕ (in the antecedent) and shows ψ (in the succedent). Rule $\rightarrow$l assumes an implication $\phi\rightarrow\psi$ to hold in the antecedent, but we do not know if this implication holds because ϕ is false or because ψ is true, so rule $\rightarrow$l splits into those two branches.

The axiom rule *ax* closes a goal (there are no further subgoals), because assumption ϕ in the antecedent trivially entails ϕ in the succedent (sequent $\Gamma,\phi\vdash\phi,\Delta$ is a simple syntactic tautology). The *cut* rule can be used for case distinctions: The right subgoal assumes any additional formula ϕ in the antecedent that the left subgoal shows in the succedent. Dually: regardless of whether ϕ is actually true or false, both cases are covered by proof branches.

Rules $\exists$r,$\forall$l,$\forall$r,$\exists$l are standard proof rules for first-order logic. For explaining these quantifier proof rules, let us first assume for a moment there are no free variables $X_1,\ldots,X_n$ (i.e. $n=0$) and use what is known as the ground calculus.

The quantifier proof rules work much as in mathematics. Consider $\forall$r, where we want to show a universally quantified property. When a mathematician wants to show a universally quantified property $\forall x\,\phi(x)$ to hold, he could choose a fresh symbol s (called Skolem function symbol) and prove that $\phi(s)$ holds (for s). Then the mathematician would remember that s was arbitrary and his proof did not assume anything special about the value of s. So he would conclude that $\phi(s)$ must indeed hold for all s, and that hence $\forall x\,\phi(x)$ holds true. For example, to show that the square of all numbers is nonnegative, a mathematician could start out by saying "let s be an arbitrary number", prove $s^2 \geq 0$ for s, and then conclude $\forall x\,(x^2 \geq 0)$, since s was arbitrary. Proof rule $\forall$r essentially makes this reasoning formal. It chooses a *new* (function) symbol s and replaces the universally quantified formula in the succedent by a formula for s (with all free logical variables $X_1,\dots,X_n$ added as arguments, as we explain below). Notice, of course, that it is important to choose a new symbol s that has not been used (in the sequent) before. Otherwise, we would assume special properties about s that may not be justified.

Consider $\exists$r, where we want to show an existentially quantified property. When a mathematician proves $\exists x\,\phi(x)$, he could directly produce any witness θ for this existential property and prove that, indeed, $\phi(\theta)$, for then he would have shown $\exists x\,\phi(x)$ with this witness. For example, to show that there is a number whose cube is less than its square, a mathematician could start by saying "let me choose 0.5 and show the property for 0.5". Then he could prove $0.5^3 < 0.5^2$, because $0.125 < 0.25$, and conclude that there, thus, is such a number, i.e., $\exists x\,(x^3 < x^2)$. Proof rule $\exists$r does that. It allows the choice of *any* term θ for x and accepts a proof of $\phi(\theta)$ as a proof of $\exists x\,\phi(x)$. However note that the claim "θ is a witness" may turn out to be wrong, for example, the choice 2 for x would be a bad start for attempting to show $\exists x\,(x^3 < x^2)$. Consequently, proof rule $\exists$r keeps both options $\phi(\theta)$ and $\exists x\,\phi(x)$ in the succedent. If the proof with θ is successful, the sequent is valid and the part of the proof can be closed successfully. If the proof with θ later turns out to be unsuccessful, another attempt can be used to prove $\exists x\,\phi(x)$, e.g., by applying $\exists$r again with another attempt for a different witness θ_2.

Rules $\forall$l,$\exists$l are dual to $\exists$r,$\forall$l. Consider $\forall$l, where we have a universally quantified formula in the assumptions (antecedent) that we can use, and not in the succedent, which we want to show. In mathematics, when we know a universal fact, we can use this knowledge for any particular instance. If we know that all positive numbers have a square root, then we can also use the fact that 5 has a square, because 5 is a positive number. Hence from assumption $\forall x\,(x > 0 \rightarrow \mathit{hasSqrt}(x))$ in the antecedent, we can also assume instance $5 > 0 \rightarrow \mathit{hasSqrt}(5))$. Rule $\forall$l can produce an instance $\phi(\theta)$ for arbitrary terms θ of the assumption $\forall x\,\phi(x)$. Since we may need the universal fact $\forall x\,\phi(x)$ for multiple instantiations with $\theta_1, \theta_2, \theta_3$ during the proof, rule $\forall$l keeps the assumption $\forall x\,\phi(x)$ in the antecedent so that it can be used repeatedly.

Consider rule $\exists$l in which we can use an existentially quantified formula from the antecedent. In mathematics, if we know an existential fact, then we can give a name to the object that we then know does exist. If we know that there is a smallest integer less than 10 that is a square, we can call it s, but we cannot denote it by a different term like 5, because 5 may be (and in fact is) the wrong answer. Rule $\exists$l gives a fresh name s (with all logical variables $X_1,\dots,X_n$ as arguments) to the object that exists. Since it does not make sense to give a different name for the same existing object later, $\exists x\,\phi(x)$ is removed from the antecedent when adding $\phi(s(X_1,\dots,X_n))$.

There are two ways of using the proof rules in Fig. A.1. One way is to avoid free variables X_i altogether and only choose ground terms without variables for instantiations θ in $\exists$r,$\forall$l. Then the Skolem functions used in $\forall$r,$\exists$l have $n=0$ free logical variables $X_1,\dots,X_n$ as arguments. This case is called a *ground calculus*, because free variables are never used and all term instantiations are ground (no free variables).

The other way is to work with free variables and always use some fresh logical variable X for instantiation of θ every time $\exists$r,$\forall$l are used. This is a free-variable calculus [147, 122, 123] where $\exists$r,$\forall$l are called γ-rules and $\forall$r,$\exists$l are called δ^+-rules [147], which is an improvement of what is known as the δ-rule [122, 123]. This case is called a *free-variable calculus*, because instantiations are with free variables. At the end of the proof, these free variables can be instantiated by a global substitution on the full proof to close all branches. The free variables $X_1,\dots,X_n$ in the Skolem terms keep track of the dependencies of symbols and prevent instantiations where we instantiate X_1 by a term such as $s(X_1,\dots,X_n)$ depending on X_1. The ground calculus and free-variable calculus uses of Fig. A.1 can also be mixed.

A.4.2 Proof Example: Ground Proving Versus Free-Variable Proving

Figure A.2a shows a proof in the FOL calculus from Fig. A.1 in the ground calculus version (without using free logical variables) of formula $\exists y\forall x\,c(x,y)\rightarrow\forall x\exists y\,c(x,y)$. This property says that if there is one common y for all x such that $c(x,y)$, then for each of those x there also is a y such that $c(x,y)$. Proofs are constructed bottom-up starting with the conjecture $\exists y\forall x\,c(x,y)\rightarrow\forall x\exists y\,c(x,y)$ at the bottom, and working backwards to the top by using proof rules from Fig. A.1.

$$
\begin{array}{ll}
 & \qquad\qquad * \\ \hline
\text{ax} & \forall x\,c(x,r), c(s,r) \vdash \exists y\,c(s,y), c(s,t), c(s,r) \\ \hline
\exists\text{r} & \forall x\,c(x,r), c(s,r) \vdash \exists y\,c(s,y), c(s,t) \\ \hline
\forall\text{l} & \forall x\,c(x,r) \vdash \exists y\,c(s,y), c(s,t) \\ \hline
\exists\text{l} & \exists y\forall x\,c(x,y) \vdash \exists y\,c(s,y), c(s,t) \\ \hline
\exists\text{r} & \exists y\forall x\,c(x,y) \vdash \exists y\,c(s,y) \\ \hline
\forall\text{r} & \exists y\forall x\,c(x,y) \vdash \forall x\exists y\,c(x,y)
\end{array}
$$

Fig. A.2a Ground proof example

$$
\begin{array}{ll}
 & \qquad\qquad *[X\mapsto s][Y\mapsto t] \\ \hline
\text{ax} & \forall x\,c(x,t), c(X,t) \vdash \exists y\,c(s,y), c(s,Y) \\ \hline
\forall\text{l} & \forall x\,c(x,t) \vdash \exists y\,c(s,y), c(s,Y) \\ \hline
\exists\text{l} & \exists y\forall x\,c(x,y) \vdash \exists y\,c(s,y), c(s,Y) \\ \hline
\exists\text{r} & \exists y\forall x\,c(x,y) \vdash \exists y\,c(s,y) \\ \hline
\forall\text{r} & \exists y\forall x\,c(x,y) \vdash \forall x\exists y\,c(x,y)
\end{array}
$$

Fig. A.2b Free-variable proof example

One of the practical difficulties with rules $\exists$r,$\forall$l is how to choose the right instantiation θ that will help close the proof. This is also visible in Fig. A.2a, where the first choice for the instantiation of $\exists$r with t is not successful for the proof, because the Skolem function symbol for the subsequent $\exists$l application must be new; hence it is called r. But $\Gamma, c(s,r)\vdash c(s,t),\Delta$ is not an instance of axiom *ax*, so the proof cannot yet be closed unless $\exists$r is used again to instantiate with r. After this, the proof closes successfully by axiom rule *ax* (the notation $*$ marks the successful end of a proof).

Thus, the primary difficulty when using ground calculi is the need to find smart instantiations that will only turn out to work or fail much later in the proof. For automated theorem provers, it is hard to predict which instantiations are likely to work or fail, although very powerful heuristics and even successful methods based entirely on instantiation [30, 229, 29, 131] have been developed already.

Figure A.2b shows a proof of the same first-order formula in the free-variable calculus version of Fig. A.1. In particular, $\exists$r,$\forall$l instantiate the formulas just with new free variables X and Y, respectively, and the actual instantiations for X and Y are chosen only at the end of the proof, once it becomes obvious which choices prove the formula. The proof closes with the choice s for free variable X and the choice t for Y, which can be detected easily because this instance of $\Gamma, c(X,t) \vdash c(s,Y), \Delta$ can be closed by *ax*. We mark this substitution for closing successfully by $*[X \mapsto s][Y \mapsto t]$ at the end of the proof. Notice also that the free-variable calculus proof Fig. A.2b here uses only one application of $\exists$r, not two applications as in Fig. A.2a, because the smart instantiation choice can easily be deferred until the end of the proof.

In contrast, the failed proof attempt in Fig. A.3 does not close, because we would

$$
\begin{array}{ll}
 & \text{not closed} \\
\hline
 & c(X,t(X)), \forall x \exists y\, c(x,y) \vdash c(s(Y),Y), \exists y \forall x\, c(x,y) \\
\hline
\exists\text{l} & \exists y\, c(X,y), \forall x \exists y\, c(x,y) \vdash c(s(Y),Y), \exists y \forall x\, c(x,y) \\
\hline
\forall\text{r} & \exists y\, c(X,y), \forall x \exists y\, c(x,y) \vdash \forall x\, c(x,Y), \exists y \forall x\, c(x,y) \\
\hline
\forall\text{l} & \forall x \exists y\, c(x,y) \vdash \forall x\, c(x,Y), \exists y \forall x\, c(x,y) \\
\hline
\exists\text{r} & \forall x \exists y\, c(x,y) \vdash \exists y \forall x\, c(x,y) \\
\hline
\rightarrow\text{r} & \vdash \forall x \exists y\, c(x,y) \rightarrow \exists y \forall x\, c(x,y)
\end{array}
$$

Fig. A.3 Wrong proof attempt in first-order logic

need to make $c(X,t(X))$ and $c(s(Y),Y)$ identical by substituting terms for X and Y. But if we try to substitute $s(Y)$ for X, then we have to apply this substitution everywhere. In particular, we have to substitute $t(X)$ for Y, which after applying the X substitution corresponds to replacing Y with $t(s(Y))$. This, however, is recursive, as Y occurs inside its own replacement (also known as occurs check). If we try to replace Y by $t(s(Y))$, then we have to apply this replacement everywhere. Thus, we no longer have to substitute $s(Y)$ for X, but now we have to substitute $s(t(s(Y)))$ for X. But then Y needs to be identical to $t(X)$; thus we need to replace Y with $t(s(t(s(Y))))$. This process never terminates! Formally, we say that $c(X,t(X))$ and $s(Y),Y)$ cannot be unified [265].

Yet the fact that we cannot prove the property in Fig. A.3 is actually good news! The formula $\forall x \exists y\, c(x,y) \rightarrow \exists y \forall x\, c(x,y)$ that we were trying to prove in the first place is not valid, so we should never be able to prove it at all. This formula expresses the false statement that, if every x has a y for which $c(x,y)$ holds, then there also is a single y such that for all x the property $c(x,y)$ holds. But being able to choose the same common y for all x is a much stronger property than just being able to choose some (possibly different) y for every x separately. Note that the attempt in Fig. A.3 also shows that it is necessary to keep the dependency on all free variables as arguments in the Skolem terms $s(X)$ and $t(Y)$. If the proof rules $\forall$r and $\exists$l would not add the free variables X and Y as arguments, then nothing would have prevented us from closing the proof in Fig. A.3 by substituting s for X and t for Y. Consequently, the Skolem dependencies, which correspond to the dependencies and orders between the quantifiers, are necessary to make $\forall$r and $\exists$l sound, i.e., ensure that they only prove valid formulas.

A.5 Soundness

The proof calculus for first-order logic in App. A.4 needs to fit the semantics of first-order logic from App. A.3. Fortunately, every first-order logic formula that can be derived in the FOL calculus from Fig. A.1 is a valid formula. This property of the calculus is called *soundness* and is crucial, because it would be disastrous if some formula would be called "proven" when it is actually not valid. For instance, being able to prove the invalid formula in Fig. A.3 would be disastrous. We certainly would not want to rely on safety-critical embedded systems that were built based on any wrong reasoning principles.

A calculus is sound iff every formula that can be derived in the calculus is also valid according to the semantics.

Theorem A.1 (Soundness of FOL). *The calculus for first-order logic is sound.*

The proof is a version of a classical soundness proof for first-order logic [147, 122, 130] and shows that every instance of the proof rules is sound, because valid premises imply that the conclusion is valid. The soundness proof for the $\forall$r,$\exists$l can be found in the article by Hähnle and Schmitt [147].

A.6 Completeness

Soundness shows that all provable formulas are valid. So we know that we will never prove something that does not even hold (is not valid). The converse question is about whether all valid formulas are also provable, i.e., whether we will always be able to prove all formulas that are true (or, more precisely, valid).

Theorem A.2 (Completeness of FOL). *The first-order logic calculus is* complete, *i.e., every valid first-order logic formula can be derived in the first-order calculus.*

Completeness of a first-order logic proof calculus was first proven by Kurt Gödel [135, 136]. The proof technique was simplified considerably by Leon Henkin [155], and such Henkin completeness proofs are used very successfully today. Completeness proofs for modern first-order calculi that are such as the one in Fig. A.1 can be found in the literature [147, 122, 130].

As a corollary to the soundness and completeness results, a proof calculus for first-order logic such as that in Fig. A.1 is "perfect" in the sense that it can successfully prove exactly the valid formulas. Still, numerous improvements and refinements are needed to make first-order logic proving successful in practise.

A.7 Computability Theory and Decidability

In this section, we provide a brief, informal introduction to decidability, semidecidability, and undecidability in the context of logic. We refer the reader to the literature [280] for a formal introduction to computability theory and related notions.

First-order logic has a sound and complete calculus that we have shown in App. A.4. In particular, given any formula of first-order logic, we can use the proof calculus to check whether the formula is valid by systematically generating proofs. The primary challenge for proving formulas is to find appropriate instantiations for quantifiers, but the completeness theorem A.2 shows that appropriate instantiations exist for all valid formulas. Hence, given a first-order formula ϕ, systematically generating proofs will eventually be successful if ϕ is valid, but may go on forever if ϕ is not valid. This makes validity of first-order logic semidecidable, because there is an algorithm that always terminates and answers correctly for valid formulas but may fail to terminate for formulas that are not valid. In fact, it turns out that there is no algorithm for first-order logic that always terminates with the correct answer also for formulas that are not valid.

In general, a *decision problem* is the task to decide for each instance of the problem between "yes" and "no". For instance, in the decision problem of validity in first-order logic, the task is for each formula of first-order logic to decide between "yes, valid" and "no, not valid". A decision problem is called *decidable* iff there is an algorithm that always terminates and produces the correct answer for the decision problem. The problem is called *undecidable* otherwise, i.e., no algorithm exists that is guaranteed to terminate with the correct answer in all cases. A decision problem is called *semidecidable* iff there is an algorithm that always produces the correct answer for the decision problem and always terminates for "yes" instances of the problem, but does not need to terminate for "no" instances.

Validity in first-order logic is semidecidable, but not decidable. Validity in equational first-order logic, i.e., with equality as the only predicate symbol and no function symbols, is decidable, as has been shown by Skolem [281], based on prior work by Löwenheim [199]. Tarski [288] showed that validity in first-order real arithmetic, i.e., first-order logic interpreted over real-closed fields is decidable by quantifier elimination.

Appendix B
Differential Equations

Contents

Synopsis In this chapter, we give an intuition for and examples of differential equations and briefly summarise some classical results about differential equations. We state Peano's existence theorem for solutions of differential equations and the Picard-Lindelöf or Cauchy-Lipschitz existence and uniqueness theorem.

B.1 Ordinary Differential Equations

In this chapter, we briefly introduce differential equations, show examples, and summarise classical results. We refer to the book by Walter [297] for details and proofs about differential equations. For further background on differential equations, we refer you to the literature [153, 257, 116, 172]. For a presentation of differential equations that is closer to the needs in dynamical systems, we refer you to the book by Perko [227]. Differential-algebraic equations are also covered in the literature in more detail [132, 187].

In this chapter, $C^k(D,\mathbb{R}^n)$ denotes the space of k times continuously differentiable functions from domain D to $\mathbb{R}^n$.

An ordinary differential equation in explicit form is an equation $y'(t) = f(t,y)$ where $y'(t)$ is meant to be the derivative of y with respect to t. A solution is a differentiable function Y which satisfies this equation when substituted in the differential equation, i.e., when substituting $Y(t)$ for y and the derivative $Y'(t)$ of Y at t for $y'(t)$.

Definition B.1 (Ordinary differential equation). Let $f : D \to \mathbb{R}^n$ be a function on a domain $D \subseteq \mathbb{R} \times \mathbb{R}^n$. The function $Y : I \to \mathbb{R}^n$ is a *solution* on the interval $I \subseteq \mathbb{R}$

of the *initial value problem*

$$\begin{bmatrix} y'(t) = f(t,y) \\ y(t_0) = y_0 \end{bmatrix} \tag{B.1}$$

with *ordinary differential equation (ODE)* $y' = f(t,y)$, if, for all $t \in I$

1. $(t, Y(t)) \in D$,
2. $Y'(t)$ exists and $Y'(t) = f(t, Y(t))$,
3. $Y(t_0) = y_0$.

If $f : D \to \mathbb{R}^n$ is continuous, then it is easy to see that $Y : I \to \mathbb{R}^n$ is continuously differentiable. The definition is accordingly for higher-order differential equations, i.e., differential equations involving higher-order derivatives $y^{(n)}(t)$ for $n > 1$.

Let us consider the intuition for this definition. A differential equation (system) can be thought of as a vector field such as the one in Fig. B.1, where, at each point, the vector shows in which direction the solution evolves. At every point, the vector would correspond to the right-hand side of the differential equation. A solution of a differential equation adheres to this vector field at every point, i.e., the solution (e.g., the solid line in Fig. B.1) locally follows the direction indicated by the vector of the right-hand side of the differential equation. There are many solutions of the differential equation corresponding to the vector field illustrated in Fig. B.1. For the particular initial value problem, however, a solution also has to start at the position y_0 at time t_0 and then follow the differential equations or vector field from this point. In general, there could still be multiple solutions for the same initial value problem.

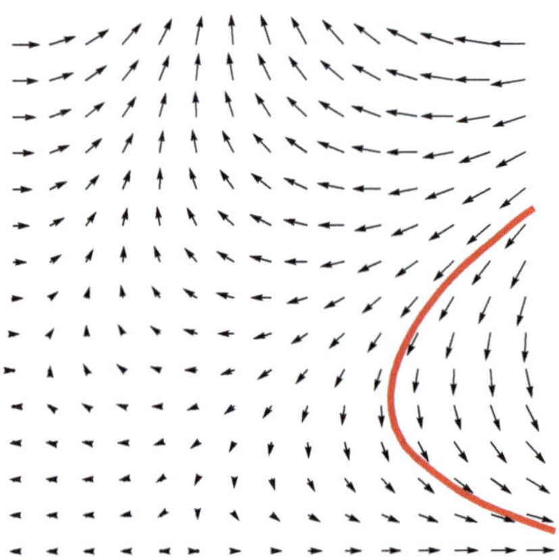

Fig. B.1 Vector field and a solution of a differential equation

Example B.1. Some differential equations are easy to solve. For instance, the initial value problem

$$\begin{bmatrix} x'(t) = 5 \\ x(0) = 2 \end{bmatrix}$$

has a solution $x(t) = 5t+2$. This can be checked easily by inserting the solution into the differential equation and initial value equation:

$$\begin{bmatrix} (x(t))' = (5t+2)' = 5 \\ x(0) = 5 \cdot 0 + 2 = 2 \end{bmatrix}$$

□

Example B.2. Consider the initial value problem

$$\begin{bmatrix} x'(t) = -2x \\ x(1) = 3 \end{bmatrix}$$

which has a solution $x(t) = 3e^{-2(t-1)}$. The test, again, is to insert the solution into the (differential) equations of the initial value problems and check:

$$\begin{bmatrix} (3e^{-2(t-1)})' = -6e^{-2(t-1)} = -2x(t) \\ x(1) = 3e^{-2(1-1)} = 3 \end{bmatrix}$$

□

Example B.3 (Trains). For the train control examples in this book (especially in Chaps. 2 and 7), consider the differential equation system $z' = v, v' = a$ from (2.7) on p. 62 and the initial value problem

$$\begin{bmatrix} z'(t) = v(t) \\ v'(t) = a \\ z(0) = z_0 \\ v(0) = v_0 \end{bmatrix}$$

Note that this initial value problem is a *symbolic initial value problem* with symbols z_0, v_0 as initial values (not specific numbers like 5 and 2.3). Moreover, the differential equation has a constant symbol a, and not a specific number like 0.6, in the differential equation. In vectorial notation, the initial value problem with this differential equation system corresponds to a vectorial system when we denote $y(t) := (z(t), v(t))$, i.e., with dimension $n = 2$ in Definition B.1:

$$\begin{bmatrix} y'(t) = \begin{pmatrix} z \\ v \end{pmatrix}'(t) = \begin{pmatrix} v(t) \\ a \end{pmatrix} \\ y(0) = \begin{pmatrix} z \\ v \end{pmatrix}(0) = \begin{pmatrix} z_0 \\ v_0 \end{pmatrix} \end{bmatrix}$$

The solution of this initial value problem is

$$z(t) = \frac{a}{2}t^2 + v_0 t + z_0$$
$$v(t) = at + v_0$$

We can show that this is the solution by inserting the solution into the (differential) equations of the initial value problems and checking:

$$\begin{bmatrix} (\frac{a}{2}t^2 + v_0t + z_0)' = 2\frac{a}{2}t + v_0 = v(t) \\ (at + v_0)' = a \\ z(0) = \frac{a}{2}0^2 + v_0 0 + z_0 = z_0 \\ v(0) = a0 + v_0 = v_0 \end{bmatrix}$$

□

Example B.4 (Aircraft). For the aircraft control examples in this book (especially in Chaps. 3 and 8), consider the differential equation system ($\mathscr{F}(\omega)$) from p. 150 and the corresponding initial value problem:

$$\begin{bmatrix} x_1'(t) = d_1(t) \\ x_2'(t) = d_2(t) \\ d_1'(t) = -\omega d_2(t) \\ d_2'(t) = \omega d_1(t) \\ x_1(0) = x_{1,0} \\ x_2(0) = x_{2,0} \\ d_1(0) = d_{1,0} \\ d_2(0) = d_{2,0} \end{bmatrix}$$

This initial value problem is again a symbolic initial value problem with variable symbols $x_{1,0}, x_{2,0}, d_{1,0}, d_{2,0}$ as initial values instead of specific numbers. The differential equations are also symbolic with a variable symbol ω instead of a specific number like 2.3. When we assume the particular number $\omega = 0$, which corresponds to straight-line flight (see Fig. 1.3a on p. 3), the solution is a simple linear function:

$$\begin{aligned} x_1(t) &= x_{1,0} + d_{1,0}t \\ x_2(t) &= x_{2,0} + d_{2,0}t \\ d_1(t) &= d_{1,0} \\ d_2(t) &= d_{2,0} \end{aligned}$$

For the general case with $\omega \neq 0$, which corresponds to curved flight (Fig. 1.3b–d), the solution of this initial value problem is much more complicated:

$$\begin{aligned} x_1(t) &= x_{1,0} + \frac{1}{\omega}(d_{2,0}\cos(\omega t) + d_{1,0}\sin(\omega t) - d_{2,0}) \\ x_2(t) &= x_{2,0} - \frac{1}{\omega}(d_{1,0}\cos(\omega t) - d_{2,0}\sin(\omega t) - d_{1,0}) \\ d_1(t) &= d_{1,0}\cos(\omega t) - d_{2,0}\sin(\omega t) \\ d_2(t) &= d_{2,0}\cos(\omega t) + d_{1,0}\sin(\omega t) \end{aligned}$$

We can show that this is the solution by inserting the solution into the (differential) equations of the initial value problems and checking:

$$\begin{bmatrix} x_1'(t) = \frac{1}{\omega}(-\omega d_{2,0}\sin(\omega t) + \omega d_{1,0}\cos(\omega t)) = d_1(t) \\ x_2'(t) = -\frac{1}{\omega}(-\omega d_{1,0}\sin(\omega t) - \omega d_{2,0}\cos(\omega t)) = d_2(t) \\ d_1'(t) = -\omega d_{1,0}\sin(\omega t) - \omega d_{2,0}\cos(\omega t) = -\omega d_2(t) \\ d_2'(t) = -\omega d_{2,0}\sin(\omega t) + \omega d_{1,0}\sin(\omega t) = \omega d_1(t) \\ x_1(0) = x_{1,0} + \frac{1}{\varphi}(d_{2,0}\cos(0) + d_{1,0}\sin(0) - d_{2,0}) = x_{1,0} \\ x_2(0) = x_{2,0} - \frac{1}{\omega}(d_{1,0}\cos(0) - d_{2,0}\sin(0) - d_{1,0}) = x_{2,0} \\ d_1(0) = d_{1,0}\cos(0) - d_{2,0}\sin(0) = d_{1,0} \\ d_2(0) = d_{2,0}\cos(0) + d_{1,0}\sin(0) = d_{2,0} \end{bmatrix}$$

□

As a general phenomenon, observe that solutions of differential equations can be much more involved than the differential equations themselves, which is part of the representational and descriptive power of differential equations.

Often, differential equations are more difficult to solve than the above examples. In a certain sense, "most" differential equations are impossible to solve in that they have no explicit closed-form solution with elementary functions, for instance, $x''(t) = e^{t^2}$; see [304]. Likewise, differential equations like $y'(t) = \frac{2}{t^3}y$ can have non-analytic smooth solutions like:

$$y(t) = e^{-\frac{1}{t^2}}.$$

B.2 Existence Theorems

There are several classical theorems that guarantee existence and/or uniqueness of solutions if differential equations (not necessarily closed-form solutions with elementary functions, though). The existence theorem is due to Peano [225]. A proof can be found in [297, Theorem 10.IX].

Theorem B.1 (Existence theorem of Peano). *Let $f : D \to \mathbb{R}^n$ be a continuous function on an open, connected domain $D \subseteq \mathbb{R} \times \mathbb{R}^n$. Then, the initial value problem* (B.1) *with $(t_0, y_0) \in D$ has a solution. Further, every solution of* (B.1) *can be continued arbitrarily close to the border of D.*

Peano's theorem only proves that a solution exists, not for what duration it exists. Still, it shows that every solution can be *continued arbitrarily close to the border* of the domain D. That is, the closure of the graph of solution φ, when restricted to $[0,0] \times \mathbb{R}^n$, is not a compact subset of D. Especially, there is then a global solution on the interval $[0,\infty)$ if $D = \mathbb{R}^{n+1}$.

Peano's theorem shows the existence of solutions of continuous differential equations on open, connected domains, but there can still be multiple solutions.

Example B.5. The initial value problem with the following continuous differential equation

$$\begin{bmatrix} y' = \sqrt[3]{|y|} \\ y(0) = 0 \end{bmatrix}$$

has multiple solutions:

$$y(t) = 0$$

$$y(t) = \left(\frac{2}{3}t\right)^{\frac{3}{2}}$$

$$y(t) = \begin{cases} 0 & \text{for } t \leq s \\ \left(\frac{2}{3}(t-s)\right)^{\frac{3}{2}} & \text{for } t > s \end{cases}$$

where $s \geq 0$ is any nonnegative real number. □

B.3 Existence and Uniqueness Theorems

If we know that the differential equation (its right-hand side) is continuously differentiable on an open, connected domain, then the Picard-Lindelöf theorem gives a stronger result than Peano's theorem. It shows that there is a unique solution (except, of course, that the restriction of any solution to a sub-interval is again a solution). For this, recall that a function $f : D \to \mathbb{R}^n$ with $D \subseteq \mathbb{R} \times \mathbb{R}^n$ is called *Lipschitz continuous* with respect to y iff there is an $L \in \mathbb{R}$ such that for all $(t,y),(t,\bar{y}) \in D$,

$$\|f(t,y) - f(t,\bar{y})\| \leq L\|y - \bar{y}\|.$$

If, for instance, $\frac{\partial f(t,y)}{\partial y}$ exists and is bounded on D, then f is Lipschitz continuous with $\max_{(t,y)\in D} \|\frac{\partial f(t,y)}{\partial y}\|$ by mean value theorem. Similarly, f is *locally Lipschitz continuous* iff for each $(t,y) \in D$, there is a neighbourhood in which f is Lipschitz continuous. In particular, if f is continuously differentiable, i.e., $f \in C^1(D,\mathbb{R}^n)$, then f is locally Lipschitz continuous.

Most importantly, Picard-Lindelöf's theorem [195], which is also known as the Cauchy-Lipschitz theorem, guarantees existence and uniqueness of solutions. As restrictions of solutions are always solutions, we understand uniqueness up to restrictions. A proof can be found in [297, Theorem 10.VI]

Theorem B.2 (Uniqueness theorem of Picard-Lindelöf). *In addition to the assumptions of Theorem B.1, let f be locally Lipschitz continuous with respect to y (for instance, $f \in C^1(D,\mathbb{R}^n)$ is sufficient). Then, there is a unique solution of the initial value problem* (B.1).

Picard-Lindelöf's theorem does not show the duration of the solution, but shows only that the solution is unique. Under the assumptions of Picard-Lindelöf's theorem, every solution can be extended arbitrarily close to the border of D by Peano's theorem, however. The solution is unique, except that all restrictions of the solution to a sub-interval are also solutions.

The following global uniqueness theorem shows a stronger property when the domain is $[0,a] \times \mathbb{R}^n$. It is a corollary to Theorems B.1 and B.2, but used prominently in the proof of Theorem B.2, and is of independent interest. A direct proof of the following global version of the Picard-Lindelöf theorem can be found in [297, Proposition 10.VII].

Corollary B.1 (Global uniqueness theorem of Picard-Lindelöf). *Let $f : [0,a] \times \mathbb{R}^n \to \mathbb{R}^n$ be a continuous function that is Lipschitz continuous with respect to y. Then, there is a unique solution of the initial value problem* (B.1) *on* $[0,a]$.

The following result is a componentwise generalisation of [297, Proposition 6.VI] to vectorial differential equations and can be used to extend solutions.

Proposition B.1 (Continuation of solutions). *Let $f : D \to \mathbb{R}^n$ be a continuous function on the open, connected domain $D \subseteq \mathbb{R} \times \mathbb{R}^n$. If φ is a solution of differential equation $y' = f(t,y)$ on $[0,b)$ whose image $\varphi([0,b))$ lies within a compact set $A \subseteq D$, then φ can be continued to a solution on $[0,b]$. Furthermore, if φ_1 is a solution of differential equation $y' = f(t,y)$ on $[0,b]$ and φ_2 is a solution of $y' = f(t,y)$ on $[b,c]$ with $\varphi_1(b) = \varphi_2(b)$, then their concatenation*

$$\varphi(t) := \begin{cases} \varphi_1(t) & \text{for } 0 \le t \le b \\ \varphi_2(t) & \text{for } b < t \le c \end{cases}$$

is a solution on $[0,c]$.

B.4 Linear Differential Equations with Constant Coefficients

For linear differential equation systems with constant coefficients there is a well-established constructive theory for obtaining closed-form solutions of initial value problems using classical techniques from linear algebra. A proof and more details can be found in [297, §18.VI].

Proposition B.2 (Linear systems with constant coefficients). *For a constant matrix $A \in \mathbb{R}^{n\times n}$, the initial value problem*

$$\begin{bmatrix} y'(t) = Ay(t) + b(t) \\ y(\tau) = \eta \end{bmatrix} \tag{B.2}$$

has the solution

$$y(t) = e^{A(t-\tau)}\eta + \int_\tau^t e^{A(t-s)} b(s)\, ds$$

where exponentiation of matrices is defined by the usual power series (generalized to matrices):

$$e^{At} = \sum_{n=0}^{\infty} \frac{1}{n!} A^n t^n.$$

In particular, if the matrix A is *nilpotent*, i.e., $A^n = 0$ for some $n \in \mathbb{N}$, and the terms $b(t)$ are polynomials in t, then the solution of the initial value problem is a polynomial function, because the exponential series stops at A^n and is a finite polynomial in t then:

$$e^{At} = \sum_{k=0}^{\infty} \frac{1}{k!} A^k t^k = \sum_{k=0}^{n-1} \frac{1}{k!} A^k t^k.$$

In particular, as products and sums of polynomials are polynomials (polynomials form an algebra) and polynomials in t are closed under integration, the solution identified in Proposition B.2 is a polynomial. Furthermore, this solution is unique by Theorem B.2.

Example B.6 (Trains). In the initial value problem from Example B.3, we guessed the solution of the differential equation system and then checked that it is the right solution by inserting it into the differential equations. But how do we compute the solution constructively in the first place without having to guess? The train differential equations $z' = v, v' = a$ from (2.7) on p. 62 are linear with constant coefficients. When we denote $y(t) := (z(t), v(t))$, we can rewrite (2.7) in the form of (B.2) as follows:

$$\begin{bmatrix} y'(t) = \begin{pmatrix} z'(t) \\ v'(t) \end{pmatrix} = \begin{pmatrix} 0 & 1 \\ 0 & 0 \end{pmatrix} \begin{pmatrix} z(t) \\ v(t) \end{pmatrix} + \begin{pmatrix} 0 \\ a \end{pmatrix} =: Ay(t) + b(t) \\ y(0) = \begin{pmatrix} z(0) \\ v(0) \end{pmatrix} = \begin{pmatrix} z_0 \\ v_0 \end{pmatrix} = \eta \end{bmatrix}$$

This system, as a linear differential equation system with a constant coefficient matrix, has the form required in Proposition B.2. First, we compute the exponential series for the matrix A, which terminates quickly because $A^2 = 0$:

$$\begin{aligned} e^{At} &= \sum_{n=0}^{\infty} \frac{1}{n!} A^n t^n = A^0 + At + \frac{1}{2!} \underbrace{A^2}_{0} t^2 + \underbrace{A^2}_{0} \sum_{n=3}^{\infty} \frac{1}{n!} A^{n-2} t^n \\ &= \begin{pmatrix} 1 & 0 \\ 0 & 1 \end{pmatrix} + \begin{pmatrix} 0 & 1 \\ 0 & 0 \end{pmatrix} t = \begin{pmatrix} 1 & t \\ 0 & 1 \end{pmatrix} \end{aligned}$$

Now we can use Proposition B.2 to compute a solution of this differential equation:

$$\begin{aligned} y(t) &= e^{At}\eta + \int_0^t e^{A(t-s)} b(s)\, ds \\ &= \begin{pmatrix} 1 & t \\ 0 & 1 \end{pmatrix} \begin{pmatrix} z_0 \\ v_0 \end{pmatrix} + \int_0^t \begin{pmatrix} 1 & t-s \\ 0 & 1 \end{pmatrix} \begin{pmatrix} 0 \\ a \end{pmatrix} ds \\ &= \begin{pmatrix} z_0 + v_0 t \\ v_0 \end{pmatrix} + \int_0^t \begin{pmatrix} at - as \\ a \end{pmatrix} ds \\ &= \begin{pmatrix} z_0 + v_0 t \\ v_0 \end{pmatrix} + \begin{pmatrix} \int_0^t (at - as)\, ds \\ \int_0^t a\, ds \end{pmatrix} \\ &= \begin{pmatrix} z_0 + v_0 t \\ v_0 \end{pmatrix} + \begin{pmatrix} ats - \frac{a}{2}s^2 \\ as \end{pmatrix} \Bigg|_{s=0}^{s=t} \end{aligned}$$

$$
\begin{aligned}
&= \begin{pmatrix} z_0 + v_0 t \\ v_0 \end{pmatrix} + \begin{pmatrix} at^2 - \frac{a}{2}t^2 \\ at \end{pmatrix} - \begin{pmatrix} a0^2 - \frac{a}{2}0^2 \\ a0 \end{pmatrix} \\
&= \begin{pmatrix} z_0 + v_0 t + \frac{a}{2}t^2 \\ v_0 + at \end{pmatrix}
\end{aligned}
$$

The last equation is exactly the solution we had guessed and checked in Example B.3. Now we have computed it constructively. □

Appendix C
Hybrid Automata

Contents

Synopsis To formally relate the notions of hybrid programs and hybrid automata, we show that hybrid automata can be embedded canonically into hybrid programs. We further prove that reachability in hybrid automata directly corresponds to satisfying models of associated formulas in differential dynamic logic. Finally, safety corresponds to validity of those formulas in differential dynamic logic. Differential dynamic logics can express more general properties about hybrid systems than plain safety though.

C.1 Syntax and Traces of Hybrid Automata

Among several other models for hybrid systems [69, 100, 58, 270, 272, 40, 183], the model of *hybrid automata* [156, 8] is one of the more widely used notations. Even though hybrid automata are a fairly common notation for hybrid systems, there are several slightly different notions of hybrid automata or automata-based models for hybrid systems [289, 9, 218, 8, 56, 156, 11, 58, 189, 97, 228, 90]. An early mention of a formalism that is related to what has later been called hybrid automata is the work of Tavernini [289]. Even earlier inspirations for hybrid systems can be found in the work by Witsenhausen [302]. We follow the notion of hybrid automata from Henzinger [156] most closely, with corresponding care for actual definability of the relations as in other approaches [125, 189, 228, 238] and with invariants that are required to hold at all times following [8, 90, 238].

Hybrid automata are graph models with two kinds of transitions: discrete jumps in the state space caused by mode switches (edges in the graph), and continuous evolution along continuous flows within a mode (vertices in the graph).

Definition C.1 (Hybrid automata). A *hybrid automaton* A consists of

- a continuous state space $\mathbb{R}^n$;
- a finite directed graph (*control graph*) with vertices Q (as *modes*) and edges E (as *control switches*);
- flow conditions $flow_q \subseteq \mathbb{R}^n \times \mathbb{R}^n$ that determine the relationship of the continuous state $x \in \mathbb{R}^n$ and its time derivative $x' \in \mathbb{R}^n$ during continuous evolution in mode $q \in Q$;
- invariant conditions $inv_q \subseteq \mathbb{R}^n$ or evolution domain restrictions that have to be true while in mode $q \in Q$;
- jump relations $jump_e \subseteq \mathbb{R}^n \times \mathbb{R}^n$ that determine the new value of the continuous state $x \in \mathbb{R}^n$ depending on its old value when following edge $e \in E$;

We assume at least that the relations $jump_e$ and inv_q are definable in first-order real arithmetic [288], but several additional restrictions apply for $flow_q$ depending on the class of hybrid automata.

Typically, the jump relation $jump_e$ is given as a conjunction of transition guards $guard_e \subseteq \mathbb{R}^n$, which determine from which states an edge can be taken, and variable resets $reset_e \subseteq \mathbb{R}^n \times \mathbb{R}^n$, which adjust the state value to its new value. In most cases, the reset relation is specified by a list of assignments $x_1 := \theta_1, \ldots x_n := \theta_n$, which correspond to a discrete jump set of differential dynamic logic (Definition 2.3). Further, flow conditions are usually just specified by a set of differential equations $x_1' = \theta_1, \ldots, x_n' = \theta_n$. See the left part of Fig. C.1 for an example of a hybrid automaton and the right part of Fig. C.1 for a description of the same hybrid system as a hybrid program. We refer you to Sect. 1.1.1 for more examples of hybrid automata.

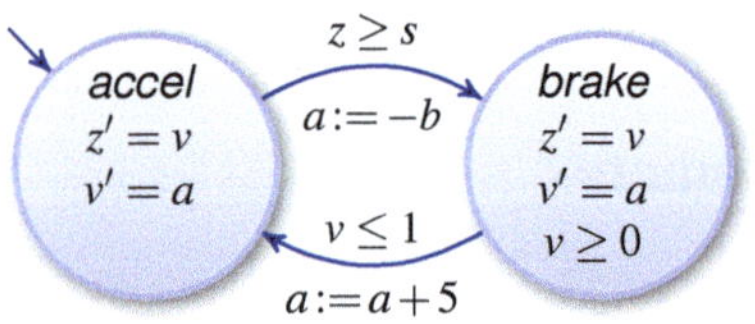

$q := accel;$ /* *initial mode is node accel* */
$(\ (?q = accel;\ z' = v, v' = a)$
$\cup\ (?q = accel \wedge z \geq s;\ a := -b;\ q := brake;\ ?v \geq 0)$
$\cup\ (?q = brake;\ z' = v, v' = a \,\&\, v \geq 0)$
$\cup\ (?q = brake \wedge v \leq 1;\ a := a + 5;\ q := accel))^*$

Fig. C.1 Hybrid automaton and corresponding hybrid program

Although often neglected in the literature, definability of the constituent relations of hybrid automata, e.g., in first-order real arithmetic, is a crucial prerequisite for dealing with any state reachability question. If the parts of a hybrid automaton are not reasonably computational or constructive, then no (interesting) computable analysis can be built on top of it. For instance, if the switching conditions in the $jump_e$ relation are undecidable, then analysis procedures have no way of knowing if the system can jump along an edge. Likewise, in a hybrid automaton that uses an undecidable set like the Mandelbrot set as invariant, it would already be undecidable whether a state $x \in \mathbb{R}^n$ satisfies the invariant of the current mode. These difficulties even arise in the strong computational model of real Turing machines by Blum et al. [49].

Definition C.2 (Transition semantics of hybrid automata). The *transition system* of a hybrid automaton A is a transition relation $\curvearrowright$ defined as follows

- the state space is defined as $S := \{(q,x) \in Q \times \mathbb{R}^n \; : \; x \in inv_q\}$;
- the transition relation $\curvearrowright \subseteq S \times S$ is defined as the union $\bigcup_{e\in E} \overset{e}{\curvearrowright} \cup \bigcup_{q\in Q} \overset{q}{\curvearrowright}$ where
  1. $(q,x) \overset{e}{\curvearrowright} (\tilde{q},\tilde{x})$ iff $e \in E$ is an edge from $q \in Q$ to $\tilde{q} \in Q$ in the hybrid automaton and $(x,\tilde{x}) \in jump_e$ (*discrete transition*).
  2. $(q,x) \overset{q}{\curvearrowright} (q,\tilde{x})$ iff $q \in Q$ and there is a function $f : [0,r] \to \mathbb{R}^n$ that has a time derivative $f' : (0,r) \to \mathbb{R}^n$ such that $f(0) = x, f(r) = \tilde{x}$ and that respects $(f(\zeta), f'(\zeta)) \in flow_q$ at each $\zeta \in (0,r)$. Further, $f(\zeta) \in inv_q$ for each $\zeta \in [0,r]$ (*continuous transition*).

State $\sigma \in S$ is *reachable* from state $\sigma_0 \in S$, denoted by $\sigma_0 \curvearrowright^* \sigma$, iff, for some $n \in \mathbb{N}$, there is a sequence of states $\sigma_1, \sigma_2, \ldots, \sigma_n = \sigma \in S$ such that $\sigma_{i-1} \curvearrowright \sigma_i$ for $1 \leq i \leq n$.

Most often, the semantics of hybrid automata is further restricted to non-Zeno traces with only finitely many transitions in finite time [156, 90, 97].

C.2 Embedding Hybrid Automata into Hybrid Programs

A hybrid automaton as on the left of Fig. C.1 can be represented faithfully as the hybrid program on the right of Fig. C.1. More generally, we show that it is always possible to represent hybrid automata as hybrid programs, thus showing that hybrid automata can be embedded faithfully into differential dynamic logic.

Proposition C.1 (Hybrid automata embedding). *There is an effective mapping ι from hybrid automata to hybrid programs and DA-programs such that the following diagram commutes:*

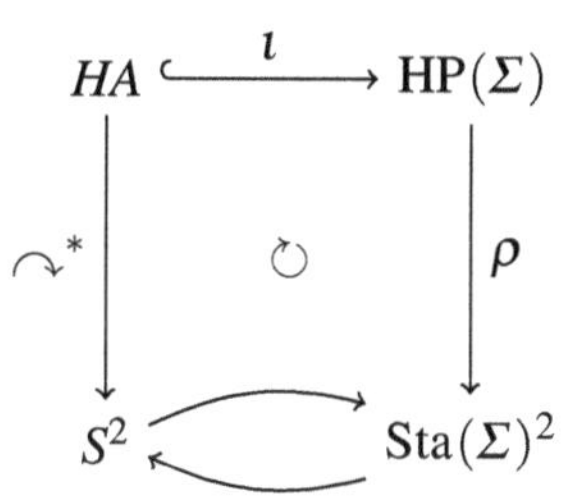

That is, the transition semantics $\rho(\iota(A))$ of the hybrid program $\iota(A)$ corresponding to a hybrid automaton A is identical to the reachability relation $\curvearrowright^*$ corresponding to hybrid automaton A when identifying states of hybrid programs in $\mathrm{Sta}(\Sigma)^2$ *with states of hybrid automata in S by a canonical bijection.*

Proof. Let $\Sigma = \{q, x_1, \ldots, x_n, x_1{}^+, \ldots, x_n{}^+\}$ be the signature. We use vectorial notation x for the vector $(x_1, \ldots, x_n)$ and x^+ for $(x_1{}^+, \ldots, x_n{}^+)$. We define ι as the function that maps hybrid automaton A to the following HP:

$$\begin{aligned}&(?q = q_i; \mathit{flow}_{q_i}(x, x')\,\&\, \mathit{inv}_{q_1}\\ &\cup ?q = q_i;\ (x^+ := *;\ ?\mathit{jump}_e(x, x^+);\ x := x^+);\ ?\mathit{inv}_{q_j};\ q := q_j\\ &\cup \ldots\\ &)^*\end{aligned}$$

The respective lines in this HP are subject to a choice for each mode q_i or each edge e from some state q_i to some state q_j. Let α^* denote this program $\iota(A)$.

States of the hybrid automaton A and states of its HP $\iota(A)$ immediately correspond to each other using the bijection $\Phi : S \to \mathrm{Sta}(\Sigma)$ that maps $(\tilde{q}, \tilde{x}) \in S$ to the state ν that is defined as $\nu(q) = \tilde{q}$ and $\nu(x_i) = \tilde{x}_i$ for $1 \le i \le n$. Observe that Φ is a bijection up to forgetful projections of internal variables x^+. In the following, we use the state identification Φ implicitly to simplify the notation. We have to show that the diagram commutes, that is, $\curvearrowright^* = \rho \circ \iota$ (up to identification of states by Φ, i.e., $\Phi \circ \curvearrowright^* = \rho \circ \iota$).

"$\subseteq$" Let $\sigma_0 \curvearrowright^* \sigma$. Hence, $n \in \mathbb{N}$ and $\sigma_1, \sigma_2, \ldots, \sigma_n = \sigma \in S$ such that $\sigma_{i-1} \curvearrowright \sigma_i$ for all $1 \le i \le n$. The proof is by induction on n.

- IA If $n = 0$ then $(\sigma_0, \sigma) \in \rho(\alpha^*)$ using zero repetitions.
- IS By induction hypothesis, we can assume that $(\sigma_0, \sigma_{n-1}) \in \rho(\alpha^*)$. We have to show $(\sigma_{n-1}, \sigma_n) \in \rho(\alpha)$, thereby implying $(\sigma_0, \sigma_n) \in \rho(\alpha^*)$. Consider the case where the last transition $\sigma_{n-1} \curvearrowright \sigma_n$ is a continuous transition in mode q_i of some duration $r \ge 0$. Then, up to identification by Φ, there is a state flow $\varphi : [0, r] \to \mathrm{Sta}(\Sigma)$ with $\varphi(0) = \sigma_{n-1}$, $\varphi(r) = \sigma_n$ and $\varphi \models \mathit{flow}_{q_i} \wedge \mathit{inv}_{q_i}$. Thus, α can copy the transition as $(\sigma_{n-1}, \sigma_n) \in \rho(\alpha)$ using the choice $?q = q_i; \mathit{flow}_{q_i}(x, x')\,\&\,\mathit{inv}_{q_i}$. The test succeeds, because $\Phi(\sigma_{n-1})(q) = q_i$.
  Consider the case where the last transition $\sigma_{n-1} \curvearrowright \sigma_n$ is a discrete transition from mode q_i to q_j along edge e. Then $(\sigma_{n-1}, \sigma) \in \mathit{jump}_e$. Thus, by choosing the values of σ_n for x^+, we have that $(\sigma_{n-1}, \sigma_n) \in \rho(\alpha)$ by the choice $?q = q_i;\ (x^+ := *;\ ?\mathit{jump}_e(x, x^+);\ x := x^+);\ ?\mathit{inv}_{q_j};\ q := q_j$.

"$\supseteq$" Let $(\sigma_0, \sigma_n) \in \rho(\alpha^*)$ following n repetitions of α: Let $\sigma_1, \ldots, \sigma_{n-1} \in \mathrm{Sta}(\Sigma)$ such that $(\sigma_{i-1}, \sigma_i) \in \rho(\alpha)$ for all $1 \le i \le n$. The proof is by induction on n.

- IA For $n = 0$, there is nothing to show.
- IS By induction hypothesis, we can assume that $\sigma_{i-1} \curvearrowright \sigma_i$ for all $1 \le i < n$. We have to show that $\sigma_{n-1} \curvearrowright \sigma_n$, thereby showing that $\sigma_0 \curvearrowright^* \sigma_n$. If $q_i := \sigma_{n-1}(q) = \sigma_n(q)$, then it is easy to see from the structure of α that $(\sigma_{n-1}, \sigma_n) \in \rho(?q = q_i; \mathit{flow}_{q_i}(x, x')\,\&\,\mathit{inv}_{q_i})$. Thus, $\sigma_{n-1} \overset{q_i}{\curvearrowright} \sigma_n$ by a continuous transition.
  If, however, $q_i := \sigma_{n-1}(q)$, $q_j = \sigma_n(q)$, then it is easy to see that

$$(\sigma_{n-1}, \sigma_n) \in \rho(?q = q_i;\ (x^+ := *;\ ?jump_e(x, x^+); x := x^+);\ ?inv_{q_j}; q := q_j)$$

according to a line of α that originates from some edge e from q_i to q_j. Thus, $(\sigma_{n-1}, \sigma_n) \in jump_e$ and $\sigma_n \models inv_{q_j}$; hence, $\sigma_{n-1} \overset{e}{\curvearrowright} \sigma_n$ by a discrete transition. □

Corollary C.1. *There is an effective mapping from safety properties of hybrid automata to* d$\mathscr{L}$/ DAL *formulas such that the hybrid automaton A, starting in initial mode q_0, safely remains in the region $F \in \mathrm{Fml}_{\mathrm{FOL}}(\Sigma)$ if and only if the corresponding* d$\mathscr{L}$/ DAL *formula is valid.*

Proof. Let α^* be the HP $\iota(A)$ belonging to A according to Proposition C.1. Then, when starting A in initial mode q_0, it safely remains in the region F iff the following formula is valid:

$$q = q_0 \wedge inv_{q_0} \rightarrow [\alpha^*]F.$$

□

Example C.1 (Water tank). Consider the classical simple water tank example, which regulates water level y between 1 and 12 by filling or emptying the water tank. The control in the hybrid automaton of Fig. C.2a further uses a clock variable x to model delayed reactions of pumps or valves. In node *fill*, the pump is activated and the wa-

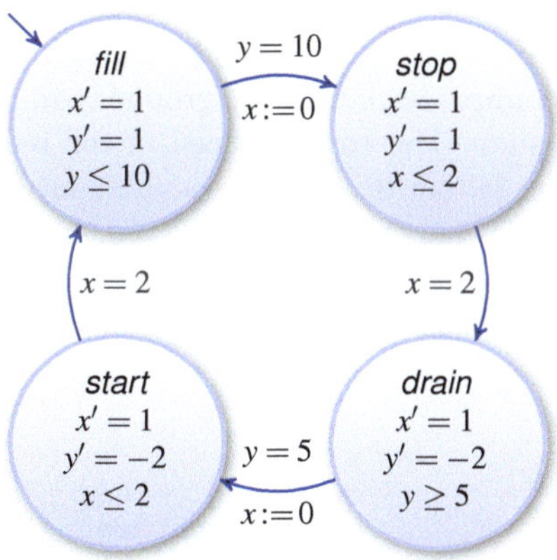

Fig. C.2a Hybrid automaton for water tank

$$
\begin{aligned}
&q = \mathit{fill} \rightarrow [(\\
&\quad (?q = \mathit{fill};\ x' = 1, y' = 1 \,\&\, y \le 10)\\
&\cup (?q = \mathit{fill} \wedge y = 10;\ x := 0;\ q := \mathit{stop})\\
&\cup (?q = \mathit{stop};\ x' = 1, y' = 1 \,\&\, x \le 2)\\
&\cup (?q = \mathit{stop} \wedge x = 2;\ q := \mathit{drain})\\
&\cup (?q = \mathit{drain};\ x' = 1, y' = -2 \,\&\, y \ge 5)\\
&\cup (?q = \mathit{drain} \wedge y = 5; x := 0; q := \mathit{start})\\
&\cup (?q = \mathit{start};\ x' = 1, y' = -2 \,\&\, x \le 2)\\
&\cup (?q = \mathit{start} \wedge x = 2;\ q := \mathit{fill})\\
&)^*](1 \le y \wedge y \le 12)
\end{aligned}
$$

Fig. C.2b Hybrid program for water tank

ter tank is filling slowly with $y' = 1$, while staying in the region $y \le 10$ all the time. At level $y = 10$, it switches to the *stop* mode, which deactivates the pump. Full deactivation of the pump takes two time units, though, such that the water is still pumped into the tank with $y' = 1$ in *stop*. When in node *drain*, water pours out with speed $y' = -2$. Again, to start the pump, the water switches to *start* first. In *start* node, the water continues to pour out with speed $y' = -2$, until, after two seconds exactly, the system switches into the *fill* node. The timing of those transitions is being taken care of deterministically by the clock variable x which has differential equation $x' = 1$ in all nodes. The fact that the evolution domain restrictions or invariant regions in the *start* and *stop* nodes are $x \leq 2$ and the outgoing edges have a guard $x = 2$ ensures that the system can only leave these nodes after two seconds exactly. The automaton cannot switch at any other time than after two seconds when $x = 2$, but it cannot remain in the *start* or *stop* nodes any longer, because the invariant region $x \leq 2$ would cease to hold then. Similarly, the switching from *fill* to *stop* is triggered by water level $y = 10$ precisely, because of the invariant region $y \leq 10$ in *fill* and the guard $y = 10$ at the outgoing edge. Finally, the switching from *drain* to *start* is at $y = 5$ precisely, for the same reasons. Consequently, the hybrid automaton in Fig. C.2b is deterministic and, thus, very easy for simulation.

Figure C.2b shows a corresponding representation of the hybrid automaton in Fig. C.2a as a hybrid program. Each line of the hybrid program corresponds to a discrete or continuous transition of the water tank hybrid automaton. The constants *fill, stop, drain, andstart* are pairwise different. The water tank is provable with the following state-dependent invariant:

$$1 \leq y \leq 12 \wedge (q = start \rightarrow y \geq 5 - 2x) \wedge (q = stop \rightarrow y \leq 10 + x).$$

□

The transformation in Proposition C.1 is a canonical embedding but hybrid programs allow for more flexible programming structures with which more natural characterisations of the system behaviour can be obtained.

Example C.2 (Parametric bouncing ball). Consider the well-known bouncing ball example [110]. A ball falls from height h and bounces back from the ground (which corresponds to $h = 0$) after an elastic deformation. The current speed of the ball is denoted by v, and t is a clock measuring the falling time. We assume an arbitrary positive gravity force g and that the ball loses energy according to a damping factor $0 \leq c < 1$. Figure C.3 depicts the hybrid automaton, an illustration of the system dynamics, and a representation of the system as a hybrid program.

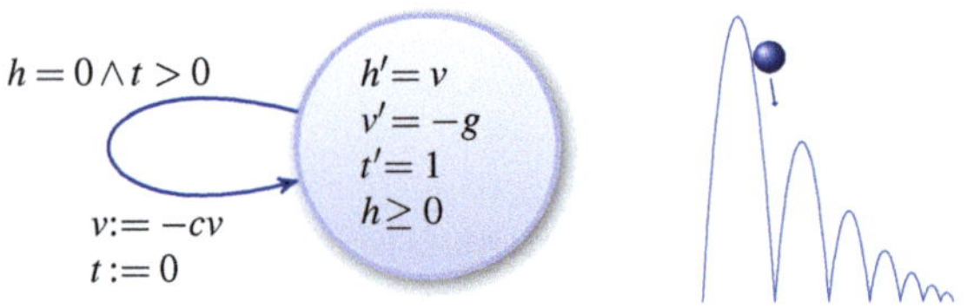

$ball \equiv ($
$\quad h' = v, v' = -g, t' = 1 \,\&\, h \geq 0;$
$\quad \text{if } (h = 0 \wedge t > 0) \text{ then}$
$\quad\quad v := -cv;\ t := 0$
$\quad \text{fi})^*$

Fig. C.3 Parametric bouncing ball

The ball loses energy at every bounce. Thus the ball never bounces higher than the initial height. This can be expressed by the safety property $0 \leq h \leq H$, where H denotes the initial energy level, i.e., the initial height if $v = 0$. For instance, we can prove the following property:

$$(v^2 \leq 2g(H - h) \wedge h \geq 0 \wedge g > 0 \wedge H \geq 0 \wedge 1 > c \geq 0) \rightarrow [ball](0 \leq h \leq H).$$

This specification follows the pattern of Hoare triples. It expresses that the bouncing ball, when started in an initial state satisfying the precondition, always respects the postcondition $0 \leq h \leq H$. □

Appendix D
KeYmaera Implementation

Contents

Synopsis KeYmaera is a hybrid verification tool for hybrid systems that combines deductive, real algebraic, and computer algebraic prover technologies. It is an automated and interactive theorem prover for specification and verification logics for hybrid systems. KeYmaera supports *differential dynamic logic*, which is a real-valued first-order dynamic logic for hybrid systems. For automating the verification process, KeYmaera implements our generalised free-variable sequent proof calculi and automatic proof strategies that decompose the hybrid system specification symbolically. To overcome the complexity of real arithmetic, we integrate real quantifier elimination following our iterative background closure strategy. Our formal verification tool is particularly suitable for verifying parametric hybrid systems and has been used successfully for verifying collision avoidance in case studies from train and air traffic control.

D.1 KeYmaera: A Hybrid Theorem Prover for Hybrid Systems

In the interest of a comprehensive presentation, we briefly characterise the implementation of our logical analysis and verification approach from Parts I and II in our new verification tool KeYmaera [242].[1] KeYmaera [242] is a hybrid theorem prover for hybrid systems that combines the deductive, real algebraic, and computer algebraic prover technologies developed in this book. It is an automated and interactive theorem prover for differential dynamic logics for hybrid systems. KeYmaera implements the proof calculi for differential dynamic logic $d\mathcal{L}$, differential-algebraic dynamic logic DAL, and differential temporal dynamic logic dTL that we introduced in Part I.

KeYmaera has been implemented as a combination of the deductive theorem prover KeY [4, 34, 35] with computer algebraic and real algebraic techniques. The general architecture of KeYmaera is depicted in Fig. D.1. KeYmaera has built-in

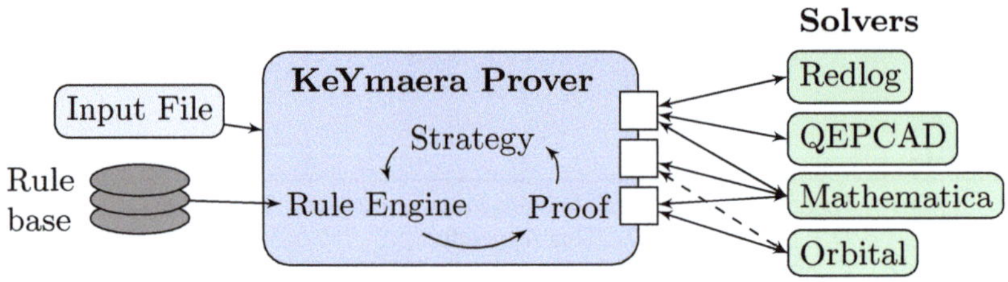

Fig. D.1 Architecture and plug-in structure of the KeYmaera prover

techniques for handling differential equations and real arithmetic, but it can also interface with other tools handling real arithmetic and/or computer algebra for improved performance. KeYmaera can interface with the computer algebra system Mathematica by Wolfram Research [303] for quantifier elimination, differential equation handling, and symbolic computation. Alternatively, it can interface with Redlog [101] or QEPCAD B [59] as quantifier elimination tools for real arithmetic. KeYmaera can also interface with computer algebraic tools and differential equation handling from the Orbital library developed by the author. KeYmaera has been developed on the basis of KeY [4, 34, 35], which is a semi-interactive theorem prover with a user-friendly graphical interface for proving correctness properties of Java programs. KeY has been developed in the group of Peter Schmitt at the University of Karlsruhe, Germany, the group of Reiner Hähnle at Chalmers University in Gothenburg, Sweden, and the group of Bernhard Beckert at the University of Koblenz-Landau, Germany. We have generalised KeY from discrete Java programs to hybrid systems by adding support for the differential dynamic logic $d\mathcal{L}$ (and DAL and dTL, respectively). Figure D.2 shows a screenshot of the graphical user interface of KeYmaera.

In the conventional KeY prover for Java programs, rule applications are comparably fast, but in KeYmaera, proof rules that use decision procedures for real arithmetic can require a substantial amount of time to produce a result. To overcome this, KeYmaera has new automatic proof strategies for the hybrid case that navigate through computationally expensive rule applications, following the strategies we

[1] KeYmaera is available from the web at `http://symbolaris.com/`

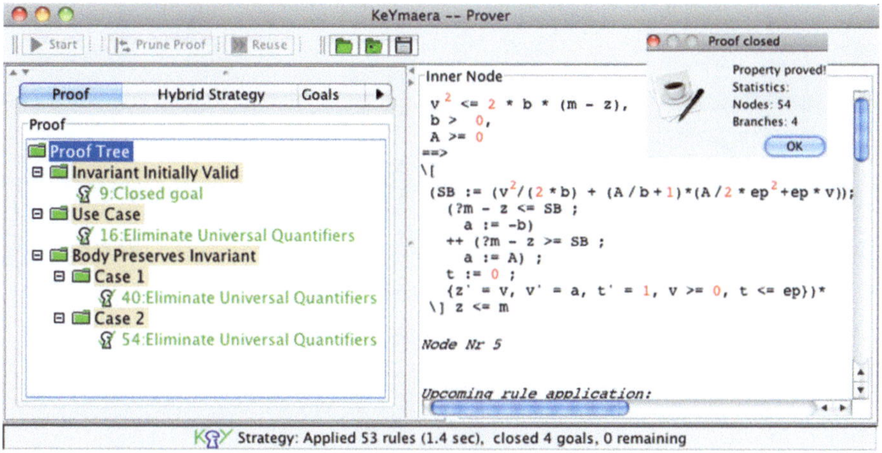

Fig. D.2 Screenshot of the KeYmaera user interface

developed in Chap. 5. KeYmaera also implements the (differential) invariant generation techniques we have developed in Chap. 6.

KeYmaera follows a plug-in architecture for integrating multiple implementations of decision procedures for real arithmetic handling, differential equation solving, and computer algebra. KeYmaera can integrate with a range of tools, some of which are shown in Fig. D.1. It does not need all of these tools to run, but can interface with any of these background solvers. KeYmaera can also run as a standalone tool based on the built-in techniques without any background solvers. Yet, the performance of KeYmaera generally improves by interfacing with additional background solvers, e.g., with Mathematica.

To overcome the complexity pitfalls of quantifier elimination and to scale to real-world application scenarios, KeYmaera implements the *iterative background closure* strategy from Sect. 5.4 that interleaves background solver calls with deductive $d\mathcal{L}$ rules. For performance reasons, the KeYmaera implementation further optimises the order of quantifiers in universal closures $\forall^\alpha \phi$ according to the modification order in the hybrid program α that produced $\forall^\alpha \phi$ (see variable dependency orders from Sects. 6.2.5 and 6.2.6). KeYmaera provides several options for adjusting the prover strategy; see Fig. D.3.

D.1.1 Structure of This Appendix

In App. D.2, we summarise techniques that can be used as back-ends for handling real arithmetic in theorem provers for differential dynamic logics. These alternatives for real arithmetic are available in KeYmaera. We discuss the consequences of the KeYmaera architecture and implementation in Sect. D.3, concerning how soundness of proof calculi is related to soundness of actual implementations. In Sect. D.4, we

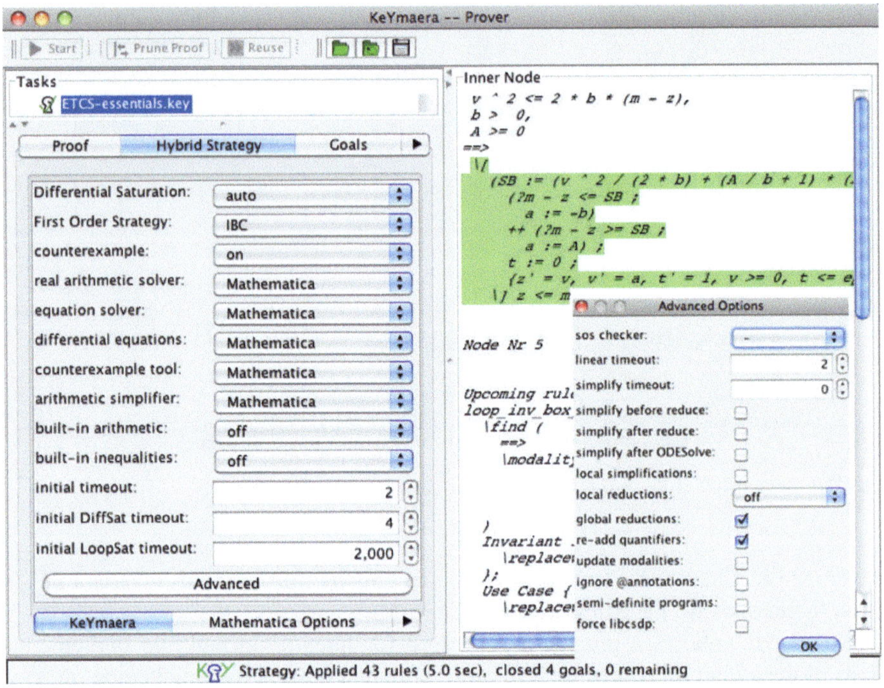

Fig. D.3 KeYmaera proof strategy options

explain the setup for our experimental evaluations and performance measurements in this book.

D.2 Computational Back-ends for Real Arithmetic

In this section, we briefly summarise a range of techniques that we use as computational back-ends for handling real arithmetic in background provers for the foreground prover KeYmaera. These techniques are various alternatives for implementing the QE quantifier elimination procedure for the quantifier rules of Fig. 2.11 or Fig. 3.9, respectively. We also refer to recent work [246], where we have introduced a new decision procedure for real arithmetic based on Gröbner bases [62], semidefinite programming [224, 151], and Stengle's real Nullstellensatz [283] that combines the techniques we present in this section.

D.2.1 Real-Closed Fields

Generalising classical techniques for finding the number of real roots for univariate polynomials due to Sturm [286], Tarski showed that validity of closed formulas in the first-order theory of real arithmetic is equivalent to the theory of real-closed fields and is decidable by quantifier elimination [287, 288]; also see Definition 2.9 on p. 77. Due to its importance, there are numerous refinements of this result that turn it into a more practical decision procedure [278, 264, 197, 79, 184, 93, 80, 19, 20, 81, 82, 84, 83, 94, 142, 299, 68, 198, 24, 285, 292, 26, 60]. A historical overview of results in this area can be found in an article by van den Dries [105]. The complexity of quantifier elimination in real-closed fields is doubly exponential in the number of quantifier alternations and has been analysed carefully [94, 141, 142, 263, 299, 60, 258, 259, 260, 259, 260, 25, 24]. Technical detail about real algebraic geometry can also be found in the book by Bochnak and Coste and Roy [50]. Algorithmic details can be found in the book by Basu, Pollack, and Roy [26].

Real numbers and real variables are omniscient in hybrid systems, because of their connection with the real physical world. Positions, velocities, accelerations, and the like are real quantities. They need to be understood properly for faithful hybrid system models. Like all other infinite structures, the reals, however, cannot be characterised uniquely up to isomorphisms in first-order logic, which is a simple corollary to the Skolem-Löwenheim theorem. Real-closed fields try to capture the properties and axioms of reals as closely as possible. Real-closed fields are an axiomatic generalisation of the field $\mathbb{R}$ of reals and share the same first-order axioms with the actual model of reals.

First we briefly recapitulate some basic notions from algebra [191, 52, 53, 112, 211]. A *ring* R is an algebraic structure that is an Abelian group with respect to addition and a (commutative) semigroup with respect to multiplication where multiplication distributes over addition, i.e., for all $a,b,c \in R$:

$$\begin{aligned} a(b+c) &= ab+ac, \\ (a+b)c &= ac+bc. \end{aligned}$$

Here R has an *Abelian group* structure with respect to addition iff, for all $a,b,c \in R$:

$$\begin{aligned} a+b &\in R, \\ (a+b)+c &= a+(b+c), \\ a+0 &= 0+a=a, \\ a+(-a) &= (-a)+a=0, \\ a+b &= b+a. \end{aligned}$$

Further, R has an *commutative semigroup* structure with respect to multiplication iff, for all $a,b,c \in R$:

$$\begin{aligned} a\cdot b &\in R,\\ (a\cdot b)\cdot c &= a\cdot(b\cdot c),\\ a\cdot 1 &= 1\cdot a = a,\\ a\cdot b &= b\cdot a. \end{aligned}$$

A *field* K is a ring such that $1 \neq 0$ and all elements $x \in K \setminus \{0\}$ have a multiplicative inverse, i.e., a $y \in K$ such that $xy = 1$.

Now we can define formally real fields, and real-closed fields subsequently.

Definition D.1 (Formally real field). A field R is a *(formally) real field* iff any of the following (equivalent) conditions holds (see [26, Theorem 2.7]):

1. -1 is not a sum of squares in R.
2. For every $x_1,\dots,x_n \in R$ we have that $\sum_{i=1}^{n} x_i^2 = 0$ implies $x_1 = \cdots = x_n = 0$.
3. R admits an ordering that makes R an ordered field, i.e., a total order $\leq \in R\times R$ on R that is compatible with the field structure:

   a. $x \leq y$ implies $x+z \leq y+z$ for all $x,y,z \in R$;
   b. $0 \leq x$ and $0 \leq y$ imply $0 \leq xy$ for all $x,y \in R$.

   Recall that the order $\leq$ is a total order iff the following conditions hold:

   a. Transitive: $x \leq y$ and $y \leq z$ imply $x \leq z$ for all $x,y,z \in R$;
   b. Antisymmetric: $x \leq y$ and $y \leq x$ imply $x = y$ for all $x,y \in R$;
   c. Total: for all $x,y \in R$: $x \leq y$ or $y \leq x$.

There are several equivalent characterisations of real-closed fields [26]:

Definition D.2 (Real-closed field). A field R is a *real-closed field* iff any of the following (equivalent) conditions holds (see [26, Theorem 2.11]):

1. R is an ordered field where every positive element is a square and every polynomial in $R[X]$ of odd degree has a root in R (then this order is, in fact, unique).
2. R is not algebraically closed but its field extension $R[\sqrt{-1}] = R[i]/(i^2+1)$ is algebraically closed.
3. R is not algebraically closed but its algebraic closure is a finite extension, i.e., finitely generated over R.
4. R has the *intermediate value property*, i.e., R is an ordered field such that for any polynomial $p \in R[X]$ with $a,b \in R, a < b$ and $p(a)p(b) < 0$, there is a ζ with $a < \zeta < b$ such that $p(\zeta) = 0$.
5. R is a real field such that no proper algebraic extension is a formally real field.

Example D.1 (Quadratic formula). The following sets are real-closed fields:

- Real numbers $\mathbb{R}$, as the prototypical but not the only example.
- Real algebraic numbers $\bar{\mathbb{Q}}\cap\mathbb{R}$, that is, real numbers in the algebraic closure of $\mathbb{Q}$, i.e., real numbers that are roots of a nonzero polynomial with rational or integer coefficients.
- Computable numbers [298], i.e., those real numbers that can be approximated by a computable function up to any desired precision.

- Definable numbers, i.e., those real numbers $a \in \mathbb{R}$ for which there is a first-order formula φ in the language of set theory with one free variable such that a is the unique real number for which φ holds true. □

According to the important Tarski-Seidenberg Principle [288, 278] or the transfer principle for real-closed fields, the first-order theory of real arithmetic is equivalent to the first-order theory of real-closed fields; see [26, Theorems 2.80 and 2.81].

Theorem D.1 (Tarski-Seidenberg principle). *The first-order theory of reals is identical to the first-order theory of real-closed fields, i.e., the set of closed first-order formulas over the signature* $+,\cdot,=,<,0,1$ *that are valid over the reals* $\mathbb{R}$ *is the same as the corresponding set of formulas that are valid in any real-closed field.*

Consequently, the notion of real-closed fields captures exactly the first-order logic point of view of real arithmetic. The technical device exploiting this result for deciding formulas of real arithmetic is quantifier elimination via cylindrical algebraic decomposition, which we examine next.

D.2.2 Semialgebraic Geometry and Cylindrical Algebraic Decomposition

From an algebraic or model-theoretic perspective, a quantifier-free formula of real arithmetic directly corresponds to the set of its satisfying assignments in $\mathbb{R}^n$. Formally, a *semialgebraic set* is a subset of $\mathbb{R}^n$ that is defined by a finite conjunction of polynomial equations and inequalities or any finite union of such sets, which, up to normalisation, is a quantifier-free formula of real arithmetic. From an algebraic or geometrical perspective, a purely equational quantifier-free formula of real arithmetic also corresponds directly to an algebraic variety, which is the basic object of study in algebraic geometry [150, 154, 215].

The most important part about projections in the following central theorem about semialgebraic sets is due to Tarski [288, 278, 163, 50, 106].

Theorem D.2 (Semialgebraic sets). *Semialgebraic sets are closed under finite union, finite intersection, complement, and projection (to linear subspaces).*

It is easy to see that projection of a semialgebraic set in $\mathbb{R}^n$ to a linear subspace (e.g., $\mathbb{R}^{n-1}$) corresponds to elimination of existential quantifiers. The projection of the set of points in $\mathbb{R}^n$ where a formula $\phi(x_1,\dots,x_n)$ holds true corresponds directly to the subspace $\mathbb{R}^{n-1}$ spanned by the variables $x_1,\dots,x_{n-1}$where the formula $\bar{\phi}(x_1,\dots,x_{n-1}) \equiv \exists x_n\, \phi(x_1,\dots,x_n)$ holds true. See Fig. D.4 for an illustration of how the quantifier elimination of a quantified real arithmetic formula $(\exists y\,(y \geq 0 \land 1 - x - 1.83x^2 + 1.66x^3 > y))$ corresponds to projection (to the thick region on the x axis that satisfies $0.75 < x \land x < 0.68 \lor x > 1.17$).

Simply speaking, the basic insight of Tarski [288] with subsequent simplifications by Seidenberg [278], Robinson [264], Łojasiewicz [197], Cohen [79], and

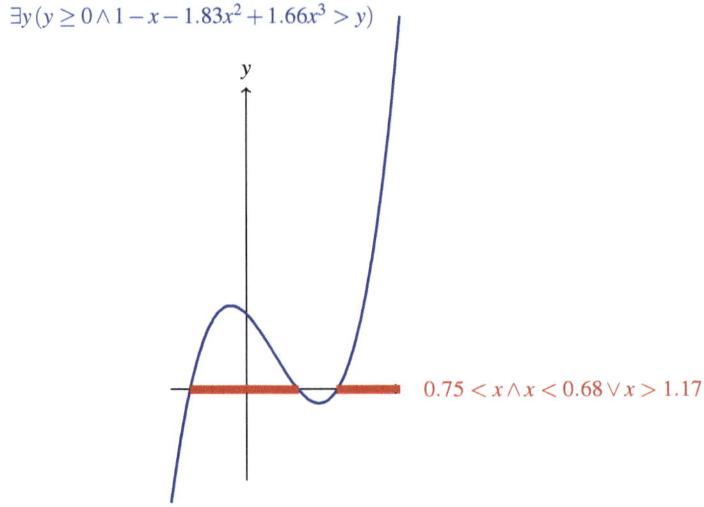

Fig. D.4 Projection of semialgebraic sets and quantifier elimination

Kreisel and Krivine [184], which led to the development of cylindrical algebraic decomposition (CAD) [80] and partial cylindrical algebraic decomposition [81] as decision procedures for real-closed fields, is that the conjunction of a set of polynomial equations and inequalities partitions the real space $\mathbb{R}^n$ into finitely many equivalence classes based on their sign combinations. The basic observation is that each polynomial $p \in K[X_1, \ldots, X_n]$ partitions the space into three equivalence classes based on its sign:

1. $\hat{+} := \{x \in \mathbb{R}^n \; : \; p(x) > 0\}$,
2. $\hat{0} := \{x \in \mathbb{R}^n \; : \; p(x) = 0\}$,
3. $\hat{-} := \{x \in \mathbb{R}^n \; : \; p(x) < 0\}$.

These classes can have multiple (but only finitely many) connected components. Further, Tarski showed that a set of polynomials partitions the real space into finitely many equivalence classes that essentially correspond to the coarsest joint refinement of all these sign relations (and possibly the relations of these polynomials on various connected components). That is, a *(natural) algebraic decomposition* of $\mathbb{R}^n$ is a partitioning of $\mathbb{R}^n$ into maximal connected regions where each of the relevant polynomials has invariant sign, which is used more systematically in the cell construction behind CAD [80]. Now, a satisfying assignment for a quantifier-free arithmetic formula exists if and only if the formula is true in one of those equivalence classes. Consequently, working from inside out, existential quantifiers can be replaced equivalently by a big disjunction, replacing the existentially quantified variable x with any representative point inside each of these finitely many equivalence classes. Let S be a (finite) representative system of the equivalence classes for all polynomials in the quantifier-free formula F; then

$$\text{QE}(\exists x F) \equiv \bigvee_{x \in S} F.$$

The computationally expensive part in the quantifier elimination algorithms is the construction of the representative system S needed for a given formula $\exists x F$.

Example D.2 (Quantifier elimination by interior points and basic virtual substitution). In the following simple example, quantifier elimination can simply insert a range of representative cases into the formula and evaluate the remaining arithmetic:

$$\begin{aligned}
&\text{QE}(\exists x (x > 2 \land x < \frac{17}{3})) && \\
&\equiv (2 > 2 \land 2 < \frac{17}{3}) && \text{border case "}x = 2\text{"} \\
&\lor (\frac{17}{3} > 2 \land \frac{17}{3} < \frac{17}{3}) && \text{border case "}x = \frac{17}{3}\text{"} \\
&\lor (\frac{2+\frac{17}{3}}{2} > 2 \land \frac{2+\frac{17}{3}}{2} < \frac{17}{3}) && \text{intermediate case "}x = \frac{2+\frac{17}{3}}{2}\text{"} \\
&\lor (-\infty > 2 \land -\infty < \frac{17}{3}) && \text{extremal case "}x = -\infty\text{"} \\
&\lor (\infty > 2 \land \infty < \frac{17}{3}) && \text{extremal case "}x = \infty\text{"} \\
&\equiv \textit{true}.
\end{aligned}$$

The basic idea behind this is illustrated in the following real line:

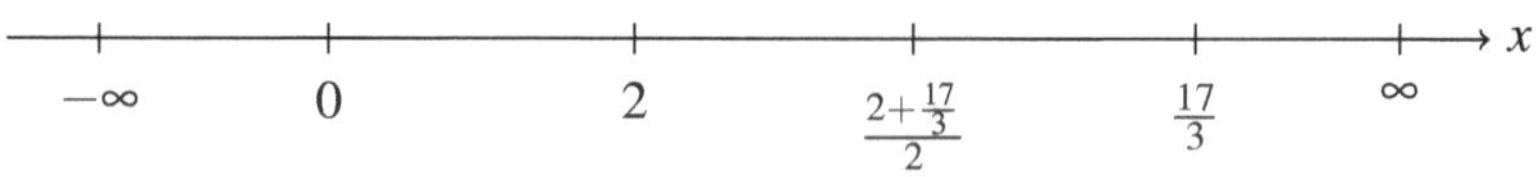

The intuition is to substitute in all the special points mentioned in the formula (2 and $\frac{17}{3}$) and to find some representative of the open interval between these points, say the middle $\frac{2+17/3}{2}$. Any other point in the open interval would be an equally good representative. The other two special cases are representatives for infinitely large (∞) or infinitely small ($-\infty$) points. Considering one disjunct above at a time, the remaining ground formulas (i.e., no variables) can be evaluated based on the numbers to obtain *true* or *false*. Here, only the intermediate case $\frac{2+17/3}{2}$ evaluates to *true*. All other disjuncts evaluate to *false*. A systematic account of quantifier elimination by a more general technique called virtual substitution can be found in the work of Weispfenning [301, 300]. □

Example D.3. A quadratic example is the following equivalence constructed by quantifier elimination:

$$\text{QE}(\exists x\, (ax^2 + bx + c = 0)) \;\equiv\; a \neq 0 \land b^2 - 4ac \geq 0 \lor a = 0 \land (b = 0 \to c = 0).$$

It basically follows the standard determinant condition $b^2 \geq 4ac$. Note however, that for this to be a true equivalence, the special cases of what happens if $a = 0$ or if $b = 0$ also need to be considered by quantifier elimination procedures. □

Universal quantifiers can be handled by duality from the handling of existential quantifiers:

$$\mathrm{QE}(\forall x F) \equiv \neg\,\mathrm{QE}(\exists x \neg F).$$

For multiple quantifiers, this procedure can be used recursively by applying quantifier elimination from inside out, i.e., starting from inner quantifiers. The final result for a closed formula gives a propositional formula without variables, which is decidable by evaluating the remaining arithmetic expressions with concrete numbers. Practical quantifier elimination procedures for real arithmetic improve on this theoretical decision procedure, e.g., by minimising the number of required equivalence classes. Practical decision procedures for quantifier elimination include cylindrical algebraic decomposition [80], partial cylindrical algebraic decomposition [81], and virtual substitution [301, 300].

KeYmaera integrates Mathematica by Wolfram Research [303], Redlog [101], and QEPCAD B [59] as alternative implementations of quantifier elimination procedures. KeYmaera also includes a built-in implementation of an adaptation of John Harrison's version [152] of the Cohen-Hörmander procedure [165] that was developed by David Renshaw. Finally, KeYmaera can also interface directly with Harrison's implementation [152] or a proof-producing quantifier elimination procedure by Sean McLaughlin and John Harrison [204].

D.2.3 Nullstellensatz and Gröbner Bases

Gröbner bases [62, 31, 211, 88] can be used as a sound but incomplete procedure for proving validity of formulas in the universal fragment of equational first-order real arithmetic. This approach is, in fact, not specific to real arithmetic but also applies for other fields. Gröbner bases have been introduced by Bruno Buchberger [62] as a systematic theory and algorithm for symbolic computations in factor rings of multivariate polynomial rings. Gröbner basis algorithms can be considered as a multivariate joint generalisation of the Euclidean algorithm for computing greatest common divisors in univariate polynomial rings and of Gaussian elimination for solving linear equation systems. An analogous concept for local rings, called standard bases, was developed independently by Heisuke Hironaka in 1964.

Preliminaries

First we briefly recapitulate some basic notions from algebra [191, 52, 53, 112, 211]. We do not always give the most general definition of these notions but restrict our attention to what we actually need in our context.

For a field K, the set $K[X_1,\ldots,X_n]$ of *multivariate polynomials* over K forms a ring and is defined as the free commutative and associative algebra over the indeterminates $X_1,\ldots,X_n$. Products $X_1^{i_1},\ldots,X_n^{i_n}$ of the indeterminates are called *monomials*. The polynomials over K are sums of monomials with coefficients in K, of the form

$$\sum_{i_1,\ldots,i_n\in\mathbb{N}} a_{i_1,\ldots,i_n} X_1^{i_1},\ldots,X_n^{i_n},$$

for which only finitely many coefficients $a_{i_1,\ldots,i_n}\in K$ are nonzero. A subset $I\subseteq R$ is an *ideal* of a ring R, denoted as $I\trianglelefteq R$, iff I is a subgroup of the additive group of R (that is $x+y\in I$ for all $x,y\in I$) and

$$rx\in I \text{ for all } x\in I, r\in R.$$

This ideal property is a "magnetic property". The elements of I are magnetic in the sense that they again land in I when multiplied with any element of the full ring R (not just when multiplied with elements in I itself).

The ideal *generated* by a set $G\subseteq R$ is the smallest ideal $I\trianglelefteq R$ containing G. In that case, G is called a *generating system* of I. In particular, if $G\subseteq K[X_1,\ldots,X_n]$ is a set of polynomials that is zero at a particular point $(x_1,\ldots,x_n)\in K^n$, then every element of the ideal (G) is zero at that point, because sums of zero stay zero, and products of zero with any element stay zero.

To start the development of Gröbner bases, we will need to define polynomial reductions. The notions of Gröbner bases and polynomial reductions are relative to an admissible monomial order, which also determines the leading term (largest monomial) in multivariate polynomials. An *admissible monomial order* $\prec$ is a strict well-order (well-founded total order; in particular, every non-empty set has a least element) on monomials such that $uw\prec vw$ for all monomials u,v,w with $u\prec v$. Admissible orders extend canonically to polynomials $K[X_1,\ldots,X_n]$ as a multiset order [62, 99]. The monomial order determines the *leading term* in multivariate polynomials, i.e., the maximal monomial with respect to $\prec$. Here, we simply assume some fixed admissible order and refer to the literature on Gröbner bases [31, 211, 88, 62] for details on monomial orders, their admissibility conditions, and their properties.

As a multivariate generalisation of the Euclidean algorithm we need multivariate polynomial division:

Definition D.3 (Reduction). Let $f,g\in K[X_1,\ldots,X_n]$ be polynomials. We say that *f reduces to g* with respect to a set $G\subset K[X_1,\ldots,X_n]$ of polynomials iff for some $m\in\mathbb{N}$ there are $f_0,f_1,\ldots,f_m$ in $K[X_1,\ldots,X_n]$ with $f_0=f, f_m=g$ such that, for all i,

$$f_{i+1}=f_i-h_ig_i$$

for some $h_i\in K[X_1,\ldots,X_n]$, $g_i\in G$, and $f_{i+1}\prec f_i$. We write $g=\mathrm{red}_G f$ if, in addition, g cannot be reduced further, i.e., there is no $h_{m+1}\in K[X_1,\ldots,X_n]$ and $g_{m+1}\in G$ with $g-h_{m+1}\,g_{m+1}\prec q$. Then polynomial $\mathrm{red}_G f$ is called *reduction* (or *remainder*) of f by G. Furthermore, polynomial f is called *reduced* with respect to G iff $f = \mathrm{red}_G f$.

The most common approach to reducing a polynomial is by following a series of elementary reductions. These elementary reductions are a special case of the above definition. Let $f, g \in K[X_1, \ldots, X_n]$ be polynomials and let l be the leading term of g. If $a_{i_1,\ldots,i_n} X_1^{i_1} \cdots X_n^{i_n}$ is some (e.g., the largest) term of f which is divisible by l, then the following polynomial is called an *elementary reduction* of f by g:

$$f - \frac{a_{i_1,\ldots,i_n} X_1^{i_1} \cdots X_n^{i_n}}{l} g$$

The theory of Gröbner bases characterises exactly under which circumstances reduction produces unique remainders, which is not generally so in multivariate polynomial rings.

Definition D.4 (Gröbner basis). A finite subset G of $I \trianglelefteq K[X_1, \ldots, X_n]$, i.e., an ideal I of a polynomial ring, is called a *Gröbner basis* iff the ideal generated by G is I and any of the following (equivalent) conditions holds:

1. Reduction with respect to G gives 0 for any $p \in I$.
2. $\mathrm{red}_G\, p = 0$ iff $p \in I$.
3. Reduction with respect to G gives a unique remainder.
4. The polynomials that are reduced with respect to G form representatives of the factor ring $K[X_1, \ldots, X_n]/I$.
5. The leading ideal of I, i.e., the ideal generated by leading terms of polynomials of I, is generated by the leading terms of G.

A Gröbner basis G is *reduced* if all $g \in G$ are reduced with respect to $G \setminus \{g\}$.

Because ideals are closed under multiplication by nonzero constants, we can assume Gröbner basis G to contain normalised polynomials only, i.e., where the leading coefficient of their leading terms is 1.

The most important result about Gröbner bases is that every ideal of a polynomial ring $K[X_1, \ldots, X_n]$ over a field K in finitely many variables has a unique reduced Gröbner basis (with leading coefficient 1) that can be computed effectively by the Buchberger algorithm [62] or its subsequent improvements [31, 211, 88, 120, 121]. Thus, Gröbner bases give an effective version of Hilbert's basis theorem:

Theorem D.3 (Hilbert's basis theorem). *Every ideal in the ring $K[X_1, \ldots, X_n]$ of multivariate polynomials over a field K is finitely generated. That is, every ideal $I \trianglelefteq K[X_1, \ldots, X_n]$ has a finite generating system G, i.e., the ideal generated by G is I.*

Using Gröbner Basis Eliminations

Gröbner basis reductions can be used for handling fragments of equational universal arithmetic. Assume we have a sequent of the following form (when rewriting to

(A1) $\dfrac{(f-g)z=1 \vdash}{\vdash f=g}\,^{1}$

(A2) $\dfrac{f-g=z^2 \vdash}{f \geq g \vdash}\,^{1}$

(A3) $\dfrac{(f-g)z^2=1 \vdash}{f > g \vdash}\,^{1}$

(A4) $\dfrac{*}{g_1=\tilde{g}_1,\ldots,g_n=\tilde{g}_n \vdash}\,^{2}$

(A5) $\dfrac{*}{g_1=\tilde{g}_1,\ldots,g_n=\tilde{g}_n \vdash f=h}\,^{3}$

(A6) $\dfrac{\vdash 1+s_1^2+\cdots+s_n^2=0}{\vdash}\,^{4}$

[1] z is a fresh variable.
[2] Applicable if $\mathrm{red}_G\, 1 = 0$ where G is the Gröbner basis of the ideal $(g_1-\tilde{g}_1,\ldots,g_n-\tilde{g}_n)$.
[3] Applicable if $\mathrm{red}_G\, f = h$ where G is the Gröbner basis of the ideal $(g_1-\tilde{g}_1,\ldots,g_n-\tilde{g}_n)$.
[4] Polynomials $s_1,\ldots,s_n$ can be chosen arbitrarily

Fig. D.5 Rule schemata of Gröbner calculus rules

normalise equations, disequations, and inequalities as necessary):

$$\Gamma, g_1=\tilde{g}_1,\ldots,g_n=\tilde{g}_n \vdash f_1=h_1,\ldots,f_e=h_e,\Delta. \tag{D.1}$$

Let G be a Gröbner basis of the ideal generated by the $g_i-\tilde{g}_i$, i.e., of the ideal

$$(g_1-\tilde{g}_1,\ldots,g_n-\tilde{g}_n). \tag{D.2}$$

If the reduction with respect to G of some f_i equals the reduction of the corresponding h_i with respect to G, i.e., $\mathrm{red}_G\, f_i = \mathrm{red}_G\, h_i$, then the sequent (D.1) is valid and can be closed. This most naïve use of Gröbner bases for real arithmetic is summarised in the proof rules A4 and A5 of Fig. D.5. The rule A4 closes a goal (marked by $*$) if the ideal G generated by equations in the antecedent contains 1. At every point where all equations of G are zero, all equations of the ideal generated by G are also zero. If the ideal generated by G contains 1, then $1=0$ holds on the points where all polynomials of G are zero. Hence, these polynomials do not have a common zero (they are contradictory). Similarly, rule A5 closes a goal if the sides f,h of an equation in the succedent have the same remainder modulo the Gröbner basis G of the equations of the antecedent, which means $f-h \in (G)$. Thus, whenever the elements of G are zero, $f-h$ is zero. So far, rules A4 and A5 can prove some but not all properties.

The scope of the rules can be extended further by testing for *radical membership* instead of ideal membership, which can prove problems like $x^2=0 \vdash x=0$ that rule A5 cannot prove. The *radical* of an ideal I is the set

$$\sqrt{I} = \bigcup_{i=1}^{\infty} \{g \in \mathbb{Q}[X_1,\ldots,X_n] \;:\; g^i \in I\} \supseteq I$$

Because the inclusion $I \subseteq \sqrt{I}$ can be strict (e.g., $\sqrt{(x^2)} = (x)$), testing for radical membership is more liberal than testing for ideal membership, while still being sound.

In practise, the rule A1, which is known as *Rabinowitch's trick*, represents a simple way of testing for radical membership. It is based on the observation that $g \in \sqrt{I}$ if and only if $1 \in (I \cup \{gz-1\})$ (where z is a fresh variable). The latter property can be tested by first applying rule A1 and then rule A4.

Finally, inequalities can be translated to equations using proof rules A2 and A3, which exploit the fact that a real number is positive iff it is a square (rule A3 is an optimised version including Rabinowitch's trick). Rules A2, A3, and A5 are similar for inequalities in the succedent. Combined with the rules A4 and A5, this encoding of inequalities is rather weak, and not able to derive simple facts like $a \leq b \wedge b \leq c \rightarrow a \leq c$. It is, however, an important basis for extensions to full decision procedures [246], in particular, in combination with proof rule A6, which we discuss in Sect. D.2.4.

Proposition D.1 (Soundness of Gröbner basis rules). *The rules in Fig. D.5 are sound.*

Proof. The rules are locally sound. For rules A1, A2, and A3, we show satisfiability-equivalence of premise and conclusion, which implies soundness.

A1 Satisfiability-equivalence of A1 is a consequence of the Rabinowitch trick equivalence $x \neq 0 \leftrightarrow \exists z\,(xz = 1)$, using $f-g$ for x. More generally, this holds in fields where nonzero elements are exactly the elements that have some inverse z. By introducing a free new variable z, we obtain that the premise is satisfiable if and only if the conclusion is satisfiable.

A2 Satisfiability-equivalence follows from equivalence $f \geq g \leftrightarrow \exists z\,(f-g=z^2)$ in the domain of reals. More generally, this holds in real-closed fields where squares are exactly the positive numbers. By introducing a free new variable z in the rule, we see that the premise and conclusion are satisfiability-equivalent, i.e., the premise is satisfiable if and only if the conclusion is satisfiable. Using the equivalence and the soundness of ∃l from Theorem 2.1, soundness can also be obtained easily from the following derivation when using the state variable z, which is implicitly quantified universally in the sequent, in place of the Skolem symbol s:

$$\cfrac{\exists\text{l}\;\cfrac{f-g=s^2 \vdash}{\exists z\, f-g=z^2 \vdash}}{f \geq g \vdash}$$

A3 Satisfiability-equivalence follows from the equivalence $x > 0 \leftrightarrow \exists z\,(xz^2 = 1)$ in the reals, using $f-g$ for x.

A4 The soundness of A4 is a special case of the soundness of A5 when assuming the false formula $1=0$ for the succedent $f=h$.

A5 Suppose the conclusion was false in ν: $\nu \models g_1 = \tilde{g}_1 \wedge \cdots \wedge g_n = \tilde{g}_n \wedge f \neq h$. Thus, $\nu \models g = 0$ for all $g \in G$ (using the notions of algebraic geometry, this would correspond to ν being in the algebraic variety of G). Consequently, $\nu \models g = 0$ for all polynomials g in the ideal (G) of G. As a consequence of the applicability condition, we have $\text{red}_G(f-h) = 0$, which, by Definition D.4, implies that $f-h$ is in the ideal of G. In combination, we have $\nu \models f-h=0$, and hence $\nu \models f=h$, which is a contradiction.

A6 Since a sum of squares is nonnegative over $\mathbb{R}$, the value of $1+s_1^2+\cdots+s_n^2$ is strictly positive and $1+s_1^2+\cdots+s_n^2=0$ is a contradiction over the reals. Consequently, if the premise is valid, then so is the conclusion. □

While rules A1, A2, A3, A4, and A5 alone without A6 are incomplete for real arithmetic, the situation is different over the complex numbers. Hilbert's Nullstellensatz shows that rules A4 and A1 yield a decision procedure for universal equational problems in $\mathbb{C}$. For this, we need the following notions. The *algebraic variety* $V(I)$ of a set of polynomials $F \trianglelefteq K[X_1,\dots,X_n]$ is defined as the set of points where all polynomials of F are zero:

$$V(F) := \{x \in K^n \ : \ p(x)=0 \text{ for all } \ p \in F\}.$$

See Fig. D.6 for examples of algebraic varieties $V(p)$ generated by a single polynomial p in two variables x and y (thus, algebraic curves in the Euclidean plain). Conversely, the *vanishing ideal* $I(V)$ of any set $V \subseteq K^n$ is defined as the set of polynomials that are zero at all points of V:

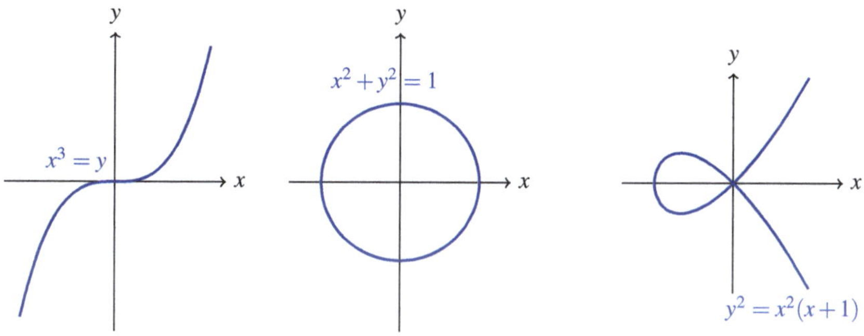

Fig. D.6 Some algebraic varieties generated by one polynomial equation in two variables

$$I(V) := \{p \in K[X_1,\dots,X_n] \ : \ p(x)=0 \text{ for all } \ x \in V\}.$$

The set $I(V)$ is, indeed, an ideal. With those notions from algebraic geometry, Hilbert's Nullstellensatz [191] can be formulated in an elegant way.

Theorem D.4 (Hilbert's Nullstellensatz). *Let K be an algebraically closed field and $I \trianglelefteq K[X_1,\dots,X_n]$ an ideal. Then*

$$I(V(I)) = \sqrt{I}$$

In particular, $V(I)=\emptyset$ iff $\sqrt{I}=K[X_1,\dots,X_n]$.

As a corollary, the sequent $g_1=\tilde{g}_1,\dots,g_n=\tilde{g}_n \vdash$ is valid iff $g_1=\tilde{g}_1 \wedge\cdots\wedge g_n=\tilde{g}_n$ is unsatisfiable over $\mathbb{R}$. Over $\mathbb{C}$, according to Theorem D.4, the sequent is valid, i.e., the equations in the antecedent are unsatisfiable over $\mathbb{C}$ iff 1 is in the radical ideal of (D.2). With Rabinowitch's trick in rule A1, Gröbner bases with rule A4 can be used for this radical ideal membership check, giving a complete procedure for complex arithmetic.

The situation with the field of reals is more difficult. For reals, the Gröbner basis approach with rules A1, A2, A3, A4, and A5 alone gives a sound but incomplete overapproximation. To see why Gröbner bases are incomplete for real arithmetic, consider the following. Gröbner bases are a general approach for polynomial rings over fields (or even rings). They do not take into account the special properties of the reals. For instance, the sequent $x^2 = -1 \vdash$ is valid, i.e., the formula $x^2 = -1$ is unsatisfiable over $\mathbb{R}$, but the Gröbner basis of x^2+1 is $\{x^2+1\}$ and, in fact, $x^2 = -1$ is satisfiable over $\mathbb{C}$ but not over $\mathbb{R}$.

KeYmaera integrates with Mathematica by Wolfram Research [303] and the Orbital library developed by the author as implementations of Gröbner basis algorithms.

D.2.4 Real Nullstellensatz

Gröbner bases and the Nullstellensatz alone are complete for universal properties of complex arithmetic, but incomplete for the domain of real arithmetic. A corresponding completeness result for real-closed fields is Stengle's real Nullstellensatz [283], which extends previous real Nullstellensätze [186, 109, 262]:

Theorem D.5 (Real Nullstellensatz for real-closed fields). *Let R be a real-closed field (e.g., $R = \mathbb{R}$) and let $F \subset R[X_1,\dots,X_n]$ be a finite subset. Then*

$$V(F) = \emptyset$$

if and only if there are polynomials $s_1,\dots,s_m \in R[X_1,\dots,X_n]$ such that

$$1 + s_1^2 + \cdots + s_m^2 \in (F).$$

If, moreover, $F \subseteq \mathbb{Q}[X_1,\dots,X_n]$, then also the polynomials $s_1,\dots,s_m$ can be chosen among the elements of $\mathbb{Q}[X_1,\dots,X_n]$.

This theorem leads to an extremely simple, complete proof method for the universal fragment of real arithmetic: in addition to the rules A1, A2, A3, A4, and A5, we add rule A6 in Fig. D.5 for injecting the equation $1 + s_1^2 + \cdots + s_m^2 = 0$ into a proof goal. The proof rule A6 expresses that an equality $1 + s_1^2 + \cdots + s_n^2 = 0$ can always be added to the succedent at any time, because a sum of squares is nonnegative, and, if we add 1, is nonzero. Consequently, $1 + s_1^2 + \cdots + s_n^2 = 0$ is obviously equivalent to false. Such an equation can be very powerful when using Gröbner basis computations on equations.

Any valid formula in the universal fragment of real-closed field can then be proven as follows:

1. turn inequalities and equations in the succedent, or disequations and inequalities in the antecedent, into equations in the antecedent with rules A1, A2, and A3.
2. add the *witness* $1+s_1^2+\cdots+s_m^2$ due to the real Nullstellensatz to the succedent with rule A6; and
3. prove the resulting sequent with Gröbner Basis computations by rule A5.

Corollary D.1 (Completeness). *Along with propositional rules from Fig. 2.11, the rules in Fig. D.5 are complete for the universal fragment of real arithmetic and in real-closed fields.*

Proof. Completeness follows from Theorem D.5 using the satisfiability-equivalence properties for the transformation by A1, A2, and A3 according to Proposition D.1. □

The main difficulty with this calculus is obvious: it does not provide any guidance for choosing the witness $1+s_1^2+\cdots+s_m^2=0$. One technique to tackle the required search is semidefinite programming, following the work based on Stengle's Positivstellensatz (Sect. D.2.5) in [224, 151]. See joint work with Quesel and Rümmer [246] for details.

Example D.4. In Fig. D.7, we show a proof for the following implication (leaving out propositional reasoning):

$$x\geq y\wedge z\geq 0\rightarrow xz\geq yz. \qquad \text{(D.3)}$$

First, we turn the inequalities $x\geq y$ and $z\geq 0$ into equations using rule A2. Prov-

$$\dfrac{\dfrac{\dfrac{\dfrac{\dfrac{\dfrac{*}{x-y=a^2,z=b^2,(yz-xz)c^2=1\vdash 1+(abc)^2=0}\text{A5}}{x-y=a^2,z=b^2,(yz-xz)c^2=1\vdash}\text{A6}}{x\geq y,z\geq 0,yz>xz\vdash}\text{A2,A3}}{x\geq y,z\geq 0\vdash xz\geq yz}\textit{cut}}{x\geq y\wedge z\geq 0\vdash xz\geq yz}\wedge\text{l}}{\vdash x\geq y\wedge z\geq 0\rightarrow xz\geq yz}\rightarrow\text{r}$$

Fig. D.7 Example proof using the real Nullstellensatz

ing by contradiction (or using propositional rules and the *cut* rule), the conclusion $xz\geq yz$ is considered as an assumption $yz>xz$ and subsequently eliminated with rule A3. Once this is done, we rely on an oracle to tell us the witness $1+(abc)^2$, which is introduced using rule A6. Finally, we close the proof by A5: the set $\{a^2-x+y,\ b^2-z,\ xzc^2-yzc^2+1\}$ is a Gröbner basis representing the equations in the antecedent. This Gröbner basis reduces the term $1+(abc)^2$ to 0 by subtracting multiples of the basis polynomials as indicated:

$$1+a^2b^2c^2 \overset{b^2-z}{\rightsquigarrow} 1+a^2zc^2 \overset{a^2-x+y}{\rightsquigarrow} 1+xzc^2-yzc^2 \rightsquigarrow 0.$$

Consequently, at every point where all equations in the antecedent are zero, all elements of the above Gröbner basis are zero (they are in the same ideal), and, thus, $1+(abc)^2$ is zero at the same point, because the Gröbner basis reduces this to zero. Hence, if the equations in the antecedent hold, the equation in the succedent holds too, thereby proving the property. □

D.2.5 Positivstellensatz and Semidefinite Programming

The Positivstellensatz for real-closed fields [283, 50], which has been introduced by Stengle along with the real Nullstellensatz for real-closed fields [283], can be used as a sound procedure for proving formulas in the universal fragment of first-order real arithmetic. The Positivstellensatz has recently been exploited in combination with relaxations from semidefinite programming [224, 151]. In joint work with Quesel and Rümmer, the author has recently introduced a combination of Stengle's real Nullstellensatz with Gröbner bases and semidefinite programming as a decision procedure for the universal fragment of real arithmetic [246].

A property of the ideal (G) generated by $G \subseteq R[X_1,\dots,X_n]$ is that products of an element that is zero with any element are zero and that sums of zero elements are zero. In particular, if every polynomial of G is zero at a point $x_1,\dots,x_n$, then every polynomial of (H) is zero at that point. The *multiplicative monoid* $\mathrm{mon}(H)$ generated by $H \subseteq R[X_1,\dots,X_n]$ is the set of finite products of elements of H (including the empty product 1). The intuition behind this monoid construction is that a product of nonzero elements remains nonzero. So if we know that all elements of H are nonzero at a point $x_1,\dots,x_n$, then any product in $\mathrm{mon}(H)$ is also nonzero at that point. The *cone* $\mathrm{con}(F)$ generated by a set $F \subseteq R[X_1,\dots,X_n]$ is the subsemiring of $R[X_1,\dots,X_n]$ generated by F and arbitrary squares, i.e., the smallest set containing F and squares s^2 of arbitrary polynomials $s \in R[X_1,\dots,X_n]$ that is closed under addition and multiplication. The idea behind the cone construction is that sums and products of nonnegative elements remain nonnegative and that nonnegative elements remain nonnegative if we add arbitrary squares s^2. In particular, if we assume all elements in F to be nonnegative at a point $x_1,\dots,x_n$,, then all elements of $\mathrm{con}(F)$ are nonnegative at that point as well. For more computational representations of cones and ideals, we refer you to the literature [224, 50].

With these constructions we can present Stengle's Positivstellensatz [283], which relates the feasibility of a set of equations, disequations, and inequalities to a polynomial formed from the ideal of equations, the monoid of disequations, and the cone of inequalities.

Theorem D.6 (Positivstellensatz for real-closed fields). *Let R be a real-closed field (e.g., $R=\mathbb{R}$) and F,G,H finite subsets of $R[X_1,\dots,X_n]$. Then*

$$\{x \in R^n \,:\, f(x) \geq 0 \textit{ for all } f \in F,\ g(x) = 0 \textit{ for all } g \in G,\ h(x) \neq 0 \textit{ for all } h \in H\}$$

is empty iff

$$\text{(A7)}\ \frac{*}{f_1 \geq \tilde{f}_1, \ldots, f_m \geq \tilde{f}_m, g_1 = \tilde{g}_1, \ldots, g_n = \tilde{g}_n, h_1 \neq \tilde{h}_1, \ldots, h_l \neq \tilde{h}_l \vdash}{}^{1}$$

[1] Applicable iff $s+g+m^2=0$ for some polynomial $s \in \mathrm{con}(\{f_1-\tilde{f}_1,\ldots,f_m-\tilde{f}_m\})$, some polynomial $g \in (g_1-\tilde{g}_1,\ldots,g_n-\tilde{g}_n)$, and some polynomial $m \in \mathrm{mon}(\{h_1-\tilde{h}_1,\ldots,h_l-\tilde{h}_l\})$

Fig. D.8 Rule schema of Positivstellensatz calculus rule

there are $s \in \mathrm{con}(F), g \in (G), m \in \mathrm{mon}(H)$ *such that* $s+g+m^2=0$.

If, moreover, $F,G,H \subseteq \mathbb{Q}[X_1,\ldots,X_n]$, *then also the polynomials* s,g,m *can be chosen among the elements of* $\mathbb{Q}[X_1,\ldots,X_n]$.

The polynomials s,g,m are polynomial infeasibility witnesses. For bounded degree, witnesses s,g,m can be searched for using numerical semidefinite programming [224] by parametrising the resulting polynomials. As (theoretical) degree bounds exist for the certificate polynomials s,g,m, the Positivstellensatz yields a theoretical decision procedure. These bounds are, however, at least triply exponential [224]. Thus, the approach advocated by Parrilo [224] is to increase the bound successively and solve the existence of bounded degree witnesses due to the Positivstellensatz by semidefinite programming [54].

As a simple corollary to Theorem D.6 we have that rule A7 in Fig. D.8 is sound.

Corollary D.2. *The rule in Fig. D.8 is sound.*

Example D.5. A proof for the implication (D.3) from p. 393 that uses the Positivstellensatz can be found in Fig. D.9. In contrast to the proof in Fig. D.7, it is now

$$\dfrac{\dfrac{\dfrac{\dfrac{\dfrac{*}{x \geq y, z \geq 0, (yz-xz)c^2 = 1 \vdash}\text{A7}}{x \geq y, z \geq 0, yz > xz \vdash}\text{A3}}{x \geq y, z \geq 0 \vdash xz \geq yz}\textit{cut}}{x \geq y \wedge z \geq 0 \vdash xz \geq yz}\wedge\text{l}}{\vdash x \geq y \wedge z \geq 0 \rightarrow xz \geq yz}\rightarrow\text{r}$$

Fig. D.9 Example proof using the Positivstellensatz

unnecessary to eliminate the inequalities $x \geq y$ and $z \geq 0$ while we still use the proof rule A3 for $xz \geq yz$ (corresponding to $yz > xz$ in the antecedent). A witness for the problem is:

$$\underbrace{c^2 \cdot (x-y) \cdot z}_{s} + \underbrace{(yz-xz)c^2 - 1}_{g} + \underbrace{1}_{m^2} = 0.$$

The terms $x-y$ and z in s result from the inequalities in the sequent, while the term g is derived from the equation. □

KeYmaera supports an integration with CSDP [51] as an implementation of semidefinite programming algorithms and can interface with Harrison's implementation [151] of semidefinite programming for the Positivstellensatz in HOL Light.

D.3 Discussion

As usual, it is a nontrivial task to ensure that the soundness that has been proven for a calculus inherits to soundness of a tool implementing this calculus. Despite our simple and concise calculi from Part I, the tool KeYmaera qualifies as a nontrivial piece of code, because of its numerous features, totalling to around 33,000 SLOC[2] on top of the KeY base system with 170,000 SLOC (not all parts of the base system are used in the context of hybrid systems, though). KeYmaera can also use external background solvers as faster decision procedures. In this section, we discuss the relationship of the soundness of our calculi and the soundness of KeYmaera. We refer to [36] for a general discussion of verification for verification systems.

Deductive Kernels

One advantage of deductive verification approaches is that they can have a comparably small trusted deductive kernel. The basic rationale underlying tools like LCF [207], HOL [138], or Isabelle/HOL [220] is that programming bugs in implementations that affect completeness are less fatal than bugs that would make the tool unsound and "prove" properties that are counterfactual. Essentially, when a verification tool like KeYmaera can only prove formulas by applying proof rules to them, it is sufficient for soundness purposes to ensure the correct implementation of these calculus rules and the proof rule application mechanism. In particular, this removes all sophisticated implementations of proof strategies from the trusted computing base. See, for instance, higher-order logic proof systems like Isabelle/HOL [220] for tools that follow this approach.

In KeYmaera, however, there are some additional pitfalls. Most of the rules of the calculi for our logics from Part I can be written as schematic rules, called *taclets* [33] in the KeY system, so that the careful soundness proofs carry over from the $d\mathcal{L}$/DAL/dTL calculi from Part I to the implementation KeYmaera. Yet even this requires the (long-tested) rule application mechanism of KeY [34, 35] to be correct. Other rules like i$\forall$ and i$\exists$ from Fig. 2.11, however, cannot be written as a KeY taclet but have to be implemented in Java directly. Taclet notations for some of the rules also require meta-operators that are implemented in Java, e.g., for obtaining the solution of differential equations in rules $\langle'\rangle$,$[']$. To improve the quality of these Java implementations, in addition to software engineering techniques like code reviews and testing, we work with runtime assertions to try to ensure that the results

[2] Source lines of code, not counting comments, estimated with SLOCCount from `www.dwheeler.com/sloccount/`

KeYmaera produces during the current verification run are correct. For instance, the Java rules for proof rules i$\forall$ and i$\exists$ can assert at runtime that their goal (conclusion) is actually a substitution instance of the subgoals (premises) that they constructed, thereby ensuring in retrospect that their formula transformation was justified by the quantifier elimination lifting lemma, Lemma 2.5, from p. 92; see further explanations in Sect. 2.5.3.2. Such runtime checks can eliminate most of the corresponding rule implementations in KeYmaera from the trusted basis, leaving essentially only the implementation of substitutions and quantifier rearrangements in the trusted deductive kernel.

External Black Box Procedures

The practical soundness problem is deeper, though, because, for performance reasons, the KeYmaera tool calls external decision procedures following the cooperation scheme in Chap. 5, which use the techniques described in App. D.2. Thus, soundness of KeYmaera generally requires the external decision procedures to be implemented correctly. We discuss the respective external procedures and means for removing them from the required trusted computing base in this appendix.

Differential Equation Handling

For the implementation of differential equation handling rules like $\langle ' \rangle$,$[']$ from the d$\mathcal{L}$ calculus in Fig. 2.11, it is easy to check at runtime by simple symbolic differentiation and symbolic computations in polynomial rings whether a solution produced by a (complicated) procedure for solving differential equations actually is a solution. These checks can eliminate the differential equation solver (Mathematica by Wolfram Research and the Orbital library by the author) from the trusted computing base completely. Similar observations hold for differential induction rules *DI*,*DV* from the DAL calculus in Fig. 3.9, which only require symbolic differentiation and symbolic polynomial computations.

Positivstellensatz and Semidefinite Programming

A pleasant property of the Positivstellensatz Theorem D.6 and the approach in Sect. D.2.5 is that it produces a witness (the polynomials f,g,h) for the validity of a formula (which corresponds to the closing of the branch by rule A7). Once the witness has been found, it is checkable by simple computations in the polynomial ring to determine whether $s+g+m^2=0$ by comparing the coefficients. Thus, complicated numerical semidefinite programming tools [54] do not need to be part of the trusted computing base concerning soundness. Similarly, the approach in Sect. D.2.4 based on the real Nullstellensatz Theorem D.5 gives checkable proof witnesses based on simple polynomial computations.

Semialgebraic Geometry and Cylindrical Algebraic Decomposition

Most quantifier elimination procedures follow a combination of cylindrical algebraic decomposition [80], partial cylindrical algebraic decomposition [81], or virtual substitution [301], which are efficient but quite intricate procedures. Unlike approaches using the Positivstellensatz, general quantifier elimination, unfortunately, does not produce simple checkable certificates.

In order to remove the implementations of quantifier elimination procedures in Mathematica and other background solvers from the trusted computing base, diversification (i.e., cross-validating results by multiple background solvers) only is a partial answer, because the solvers might still agree on the same wrong result. Further, cross-validation limits the overall performance to the worst-of-breed rather than best-of-breed running times. As a more fundamental approach, we can use verified implementations, e.g., the verified quantifier elimination procedure for linear real arithmetic by Nipkow [219] in an executable fragment of Isabelle/HOL. Another approach is to use a non-verified but proof-producing implementation of general quantifier elimination by McLaughlin and Harrison [204], which can be used in KeYmaera.

Unfortunately, the practical performance achieved by those prototype implementations is not yet sufficient for larger case studies. A good compromise is to use the paradigm of re-verification: The proof search procedures IBC and IIO generate several calls to the quantifier elimination procedure to find a proof, but only those in the final proof are soundness-critical; see Chap. 5 for details. Thus, for soundness, it is sufficient to use a fast and possibly untrusted implementation of QE during the proof search and to re-verify the final proof in a proof checker with a verified or proof-producing QE implementation [204, 219]. For this purpose, the iterative background closure strategy from Chap. 5 is especially useful, because it iteratively identifies the sweet spot for using quantifier elimination during the proof search.

Nullstellensatz and Gröbner Bases

While Gröbner basis approaches do not have as simple witnesses as Positivstellensatz approaches, their working principle is strictly based on appropriate symbolic computations in the ring of polynomials. Consequently, these polynomial reductions can be carried out from a small set of rewrite rules within a logic. Based on an implementation for integer arithmetic by Philipp Rümmer in KeY [273], KeYmaera provides a built-in real arithmetic version [246] of Gröbner bases for nonlinear equations and Fourier-Motzkin elimination for linear inequalities [214, 124]. These give fully internalised KeYmaera proofs for the arithmetic.

Simplification

As an optional procedure, KeYmaera can call simplification procedures that simplify mathematical expressions to improve readability. Yet, an external simplification procedure that simplifies x/x to 1 would be unsound for the compactified domain $\mathbb{R} \cup \{-\infty, +\infty\}$, because ∞/∞ is undefined and not identical to 1. Likewise, this simplification could be unsound depending on whether x is allowed to assume 0. In our case, the above simplification is in fact sound for the domain of reals ($\mathbb{R}$) when adopting our convention from Sects. 2.2.3 and 3.2.1 that any formula containing x/x is understood to imply that $x \neq 0$ holds. Generally, however, arbitrary mathematical simplification algorithms might perform non-equivalent transformations and are thus disabled in KeYmaera by default.

Summary

The level of formal assurance of the correct functioning of the implementation of KeYmaera is, unfortunately, not quite comparable to that of systems like LCF [207], HOL [138], or Isabelle/HOL [220]. The built-in proof rule language in KeY allows us to state most proof rules of the $d\mathcal{L}$, DAL, and dTL proof calculi primarily as schematic proof rules in the taclet language [33]. Ensuring a correct implementation of the remaining code is non-trivial. The challenge is simplified substantially by using a deductive kernel with a smaller trusted basis and by using runtime assertions that check for coherence of the current proof rule application with its formal prerequisites. Fortunately, the simplicity of the proof rules of our calculi makes it possible to come up with a correct implementation.

The most tricky part, however, is that, for performance reasons, KeYmaera relies on tuned external decision procedures to ensure scalability to our larger case studies. These procedures can be eliminated from the trusted verification base by checking witnesses when working with solutions of differential equations, the Positivstellensatz, the real Nullstellensatz, and Göbner basis rules. For full quantifier elimination over real-closed fields, verified or proof-generating implementations can be helpful.

D.4 Performance Measurements

Unless otherwise indicated, the measurements in this book have been performed on a Dual-Core AMD Opteron™ 1218 Processor (F2 stepping) running at 2.6 GHz clock speed with 64 kB instruction cache, 64 kB first level data cache, and 1 MB second-level cache per core, sharing 4 GB of DDR2-667 main memory with CAS latency 5. In these measurements, KeYmaera was used on a Java™ SE Virtual Machine, build 1.6.0_06-b02, with a HotSpot™ engine in mixed mode on a Gentoo

Linux with a 2.6.25-g SMP kernel for x86/64 bit. The measurements are further based on the Mathematica 6.0.2 kernel.

Due to several effects including Java dynamic class loading and caching effects, the timing measurements are not necessarily always statistically significant. Rather, the timing information is intended to give a feeling for differences in magnitude. The measurements have been repeated to ensure reproducibility of results to increase confidence. While the Java Virtual Machine itself already contributes to nondeterministic effects that result from dynamic class loading, garbage collection, and caching, the implementation of the KeYmaera system itself is subject to nondeterminism as well, including the proof strategies, timing effects of iterative background closures, nondeterminstic term orderings based on memory references, and so on. Finally, computational back-ends allow for variations in overall performance.

Still, qualitative performance differences can be read off from the measurements. For instance, it makes quite a remarkably dramatic practical difference whether a strategy is able to prove a case study within only 100 seconds or cannot even come up with an answer after five hours, which happened a lot when comparing our proof strategies developed in Part II to simpler procedures.

References

1. Ábrahám-Mumm, E., Steffen, M., Hannemann, U.: Verification of hybrid systems: Formalization and proof rules in PVS. In: Andler and Offutt [16], pp. 48–57. DOI 10.1109/ICECCS.2001.930163
2. Achatz, M., McCallum, S., Weispfenning, V.: Deciding polynomial-exponential problems. In: J.R. Sendra, L. González-Vega (eds.) ISSAC, pp. 215–222. ACM (2008). DOI 10.1145/1390768.1390799
3. Adams, A., Dunstan, M., Gottliebsen, H., Kelsey, T., Martin, U., Owre, S.: Computer algebra meets automated theorem proving: Integrating Maple and PVS. In: R.J. Boulton, P.B. Jackson (eds.) TPHOLs, *LNCS*, vol. 2152, pp. 27–42. Springer (2001). DOI 10.1007/3-540-44755-5_4
4. Ahrendt, W., Baar, T., Beckert, B., Bubel, R., Giese, M., Hähnle, R., Menzel, W., Mostowski, W., Roth, A., Schlager, S., Schmitt, P.H.: The KeY tool. Software and System Modeling **4**, 32–54 (2005). DOI 10.1007/s10270-004-0058-x
5. Alur, R.: Timed automata. In: N. Halbwachs, D. Peled (eds.) CAV, *LNCS*, vol. 1633, pp. 8–22. Springer (1999). DOI 10.1007/3-540-48683-6_3
6. Alur, R., Courcoubetis, C., Dill, D.L.: Model-checking for real-time systems. In: Mitchell [210], pp. 414–425
7. Alur, R., Courcoubetis, C., Dill, D.L.: Model-checking in dense real-time. Inf. Comput. **104**(1), 2–34 (1993). DOI 10.1006/inco.1993.1024
8. Alur, R., Courcoubetis, C., Halbwachs, N., Henzinger, T.A., Ho, P.H., Nicollin, X., Olivero, A., Sifakis, J., Yovine, S.: The algorithmic analysis of hybrid systems. Theor. Comput. Sci. **138**(1), 3–34 (1995). DOI 10.1016/0304-3975(94)00202-T
9. Alur, R., Courcoubetis, C., Henzinger, T.A., Ho, P.H.: Hybrid automata: An algorithmic approach to the specification and verification of hybrid systems. In: Grossman et al. [144], pp. 209–229
10. Alur, R., Dill, D.L.: A theory of timed automata. Theor. Comput. Sci. **126**(2), 183–235 (1994). DOI 10.1016/0304-3975(94)90010-8
11. Alur, R., Henzinger, T.A., Ho, P.H.: Automatic symbolic verification of embedded systems. IEEE T. Software Eng. **22**(3), 181–201 (1996)
12. Alur, R., Henzinger, T.A., Sontag, E.D. (eds.): Hybrid Systems III: Verification and Control, Proceedings of the DIMACS/SYCON Workshop, October 22–25, 1995, Rutgers University, New Brunswick, NJ, USA, *LNCS*, vol. 1066. Springer (1996)
13. Alur, R., Henzinger, T.A., Vardi, M.Y.: Parametric real-time reasoning. In: STOC, pp. 592–601 (1993). DOI 10.1145/167088.167242
14. Alur, R., Pappas, G.J. (eds.): Hybrid Systems: Computation and Control, 7th International Workshop, HSCC 2004, Philadelphia, PA, USA, March 25–27, 2004, Proceedings, *LNCS*, vol. 2993. Springer (2004). DOI 10.1007/b96398

15. Anai, H., Weispfenning, V.: Reach set computations using real quantifier elimination. In: M.D.D. Benedetto, A.L. Sangiovanni-Vincentelli (eds.) HSCC, *LNCS*, vol. 2034, pp. 63–76. Springer (2001). DOI 10.1007/3-540-45351-2_9
16. Andler, S.F., Offutt, J. (eds.): 7th International Conference on Engineering of Complex Computer Systems (ICECCS 2001), 11–13 June 2001, Skövde, Sweden. IEEE Computer Society, Los Alamitos (2001)
17. Andrews, P.B.: An Introduction to Mathematical Logic and Type Theory: To Truth Through Proof, 2 edn. Kluwer (2002)
18. Armando, A., Baumgartner, P., Dowek, G. (eds.): Automated Reasoning, Third International Joint Conference, IJCAR 2008, Sydney, Australia, Proceedings, *LNCS*, vol. 5195. Springer (2008)
19. Arnon, D.S., Collins, G.E., McCallum, S.: Cylindrical algebraic decomposition I: The basic algorithm. SIAM J. Comput. **13**(4), 865–877 (1984)
20. Arnon, D.S., Collins, G.E., McCallum, S.: Cylindrical algebraic decomposition II: An adjacency algorithm for the plane. SIAM J. Comput. **13**(4), 878–889 (1984)
21. Asarin, E., Dang, T., Girard, A.: Reachability analysis of nonlinear systems using conservative approximation. In: Maler and Pnueli [200], pp. 20–35. DOI 10.1007/3-540-36580-X_5
22. Asarin, E., Dang, T., Maler, O.: The d/dt tool for verification of hybrid systems. In: E. Brinksma, K.G. Larsen (eds.) CAV, *LNCS*, vol. 2404, pp. 365–370. Springer (2002). DOI 10.1007/3-540-45657-0_30
23. Barnett, M., Chang, B.Y.E., DeLine, R., Jacobs, B., Leino, K.R.M.: Boogie: A modular reusable verifier for object-oriented programs. In: F.S. de Boer, M.M. Bonsangue, S. Graf, W.P. de Roever (eds.) FMCO, *LNCS*, vol. 4111, pp. 364–387. Springer (2005). DOI 10.1007/11804192_17
24. Basu, S.: New results on quantifier elimination over real closed fields and applications to constraint databases. J. ACM **46**(4), 537–555 (1999). DOI 10.1145/320211.320240
25. Basu, S., Pollack, R., Roy, M.F.: On the combinatorial and algebraic complexity of quantifier elimination. J. ACM **43**(6), 1002–1045 (1996). DOI 10.1145/235809.235813
26. Basu, S., Pollack, R., Roy, M.F.: Algorithms in Real Algebraic Geometry, 2 edn. Springer (2006)
27. Batt, G., Belta, C., Weiss, R.: Model checking genetic regulatory networks with parameter uncertainty. In: Bemporad et al. [41], pp. 61–75. DOI 10.1007/978-3-540-71493-4_8
28. Bauer, A., Clarke, E.M., Zhao, X.: Analytica — an experiment in combining theorem proving and symbolic computation. J. Autom. Reasoning **21**(3), 295–325 (1998). DOI 10.1023/A: 1006079212546
29. Baumgartner, P.: FDPLL — a first order Davis-Putnam-Longeman-Loveland procedure. In: D.A. McAllester (ed.) CADE, *LNCS*, vol. 1831, pp. 200–219. Springer (2000). DOI 10. 1007/10721959_16
30. Baumgartner, P., Tinelli, C.: The model evolution calculus as a first-order DPLL method. Artif. Intell. **172**(4-5), 591–632 (2008). DOI 10.1016/j.artint.2007.09.005
31. Becker, T., Weispfenning, V.: Gröbner Bases, *Springer Graduate Texts in Mathematics*, vol. 141. Springer (1998)
32. Beckert, B.: Equality and other theories. In: M. D'Agostino, D. Gabbay, R. Hähnle, J. Posegga (eds.) Handbook of Tableau Methods. Kluwer (1999)
33. Beckert, B., Giese, M., Habermalz, E., Hähnle, R., Roth, A., Rümmer, P., Schlager, S.: Taclets: A new paradigm for constructing interactive theorem provers. Revista de la Real Academia de Ciencias Exactas, Físicas y Naturales, Serie A: Matemáticas (RACSAM) **98**(1) (2004)
34. Beckert, B., Giese, M., Hähnle, R., Klebanov, V., Rümmer, P., Schlager, S., Schmitt, P.H.: The KeY System 1.0 (deduction component). In: F. Pfenning (ed.) Proceedings, International Conference on Automated Deduction, Bremen, Germany, LNCS. Springer (2007)
35. Beckert, B., Hähnle, R., Schmitt, P.H. (eds.): Verification of Object-Oriented Software: The KeY Approach, *LNCS*, vol. 4334. Springer (2007). DOI 10.1007/978-3-540-69061-0
36. Beckert, B., Klebanov, V.: Must program verification systems and calculi be verified? In: 3rd International Verification Workshop VERIFY'06, at FLoC, Seattle, USA, pp. 34–41 (2006)

37. Beckert, B., Platzer, A.: Dynamic logic with non-rigid functions: A basis for object-oriented program verification. In: U. Furbach, N. Shankar (eds.) IJCAR, *LNCS*, vol. 4130, pp. 266–280. Springer (2006). DOI 10.1007/11814771_23
38. Beckert, B., Schlager, S.: A sequent calculus for first-order dynamic logic with trace modalities. In: Goré et al. [139], pp. 626–641. DOI 10.1007/3-540-45744-5_51
39. van Beek, D.A., Man, K.L., Reniers, M.A., Rooda, J.E., Schiffelers, R.R.H.: Deriving simulators for hybrid Chi models. Intelligent Control, 2006. IEEE International Symposium on pp. 42–49 (2006). DOI 10.1109/CACSD-CCA-ISIC.2006.4776622
40. van Beek, D.A., Man, K.L., Reniers, M.A., Rooda, J.E., Schiffelers, R.R.H.: Syntax and consistent equation semantics of hybrid Chi. J. Log. Algebr. Program. **68**(1-2), 129–210 (2006). DOI 10.1016/j.jlap.2005.10.005
41. Bemporad, A., Bicchi, A., Buttazzo, G. (eds.): Hybrid Systems: Computation and Control, 10th International Conference, HSCC 2007, Pisa, Italy, Proceedings, *LNCS*, vol. 4416. Springer (2007)
42. Ben-Ari, M.: Mathematical Logic for Computer Science, 2 edn. Springer (2003)
43. BFU: Investigation report. Tech. Rep. AX001-1-2/02, German Federal Bureau of Aircraft Accidents Investigation (2004)
44. Bianconi, R.: Nondefinability results for expansions of the field of real numbers by the exponential function and by the restricted sine functions. J. Symb. Log. **62**(4), 1173–1178 (1997)
45. Bianconi, R.: Undefinability results in o-minimal expansions of the real numbers. Ann. Pure Appl. Logic **134**(1), 43–51 (2005). DOI 10.1016/j.apal.2004.06.010
46. Bicchi, A., Pallottino, L.: On optimal cooperative conflict resolution for air traffic management systems. IEEE Trans. Intelligent Transportation Systems **1**(4), 221–231 (2000)
47. Biere, A., Cimatti, A., Clarke, E.M., Zhu, Y.: Symbolic model checking without BDDs. In: R. Cleaveland (ed.) TACAS, *LNCS*, vol. 1579, pp. 193–207. Springer (1999). DOI 10.1007/3-540-49059-0_14
48. Blackburn, P.: Internalizing labelled deduction. J. Log. Comput. **10**(1), 137–168 (2000)
49. Blum, L., Cucker, F., Shub, M., Smale, S.: Complexity and real computation. Springer, Secaucus, NJ, USA (1998)
50. Bochnak, J., Coste, M., Roy, M.F.: Real Algebraic Geometry, *Ergebnisse der Mathematik und ihrer Grenzgebiete*, vol. 36. Springer (1998)
51. Borchers, B.: CSDP, a C library for semidefinite programming. Optimization Methods and Software **11**(1-4), 613–623 (1999). DOI 10.1080/10556789908805765
52. Bourbaki, N.: Algebra I: Chapters 1–3. Elements of mathematics. Springer (1989)
53. Bourbaki, N.: Commutative Algebra. Elements of mathematics. Hermann, Paris (1989). Translated from the French
54. Boyd, S., Vandenberghe, L.: Convex Optimization. Cambridge University Press (2004)
55. Branicky, M.S.: General hybrid dynamical systems: Modeling, analysis, and control. In: Alur et al. [12], pp. 186–200. DOI 10.1007/BFb0020945
56. Branicky, M.S.: Studies in hybrid systems: Modeling, analysis, and control. Ph.D. thesis, Dept. Elec. Eng. and Computer Sci., Massachusetts Inst. Technol., Cambridge, MA (1995)
57. Branicky, M.S.: Universal computation and other capabilities of hybrid and continuous dynamical systems. Theor. Comput. Sci. **138**(1), 67–100 (1995). DOI 10.1016/0304-3975(94)00147-B
58. Branicky, M.S., Borkar, V.S., Mitter, S.K.: A unified framework for hybrid control: Model and optimal control theory. IEEE T. Automat. Contr. **43**(1), 31–45 (1998). DOI 10.1109/9.654885
59. Brown, C.W.: QEPCAD B: A program for computing with semi-algebraic sets using CADs. SIGSAM Bull. **37**(4), 97–108 (2003). DOI 10.1145/968708.968710
60. Brown, C.W., Davenport, J.H.: The complexity of quantifier elimination and cylindrical algebraic decomposition. In: D. Wang (ed.) ISSAC, pp. 54–60. ACM (2007). DOI 10.1145/1277548.1277557
61. Bruyère, V., Raskin, J.F.: Real-time model-checking: Parameters everywhere. Logical Methods in Computer Science **3**(1) (2007). DOI 10.2168/LMCS-3(1:7)2007. Online journal

62. Buchberger, B.: An algorithm for finding the basis elements of the residue class ring of a zero dimensional polynomial ideal. Ph.D. thesis, University of Innsbruck (1965)
63. Buchberger, B., Jebelean, T., Kriftner, F., Marin, M., Tomuta, E., Vasaru, D.: A survey of the Theorema project. In: ISSAC, pp. 384–391 (1997)
64. Buehler, M.: Summary of DGC 2005 results. Journal of Field Robotics **23**, 465–466 (2008). DOI 10.1002/rob.20145
65. Burch, J.R., Clarke, E.M., McMillan, K.L., Dill, D.L., Hwang, L.J.: Symbolic model checking: 10^{20} states and beyond. In: Mitchell [210], pp. 428–439
66. Burch, J.R., Clarke, E.M., McMillan, K.L., Dill, D.L., Hwang, L.J.: Symbolic model checking: 10^{20} states and beyond. Inf. Comput. **98**(2), 142–170 (1992). DOI 10.1016/0890-5401(92)90017-A
67. Cassez, F., Larsen, K.G.: The impressive power of stopwatches. In: CONCUR, pp. 138–152 (2000). DOI 10.1007/3-540-44618-4_12
68. Caviness, B.F., Johnson, J.R. (eds.): Quantifier Elimination and Cylindrical Algebraic Decomposition, Texts and Monographs in Symbolic Computation. Springer (1998)
69. Chaochen, Z., Ji, W., Ravn, A.P.: A formal description of hybrid systems. In: Alur et al. [12], pp. 511–530
70. Chutinan, A., Krogh, B.H.: Computational techniques for hybrid system verification. IEEE T. Automat. Contr. **48**(1), 64–75 (2003). DOI 10.1109/TAC.2002.806655
71. Cimatti, A., Roveri, M., Tonetta, S.: Requirements validation for hybrid systems. In: A. Bouajjani, O. Maler (eds.) CAV, *LNCS*, vol. 5643. Springer (2009). DOI 10.1007/978-3-642-02658-4_17
72. Clarke, E., Grumberg, O., Jha, S., Lu, Y., Veith, H.: Counterexample-guided abstraction refinement for symbolic model checking. J. ACM **50**(5), 752–794 (2003). DOI 10.1145/876638.876643
73. Clarke, E.M.: Program invariants as fixedpoints. Computing **21**(4), 273–294 (1979). DOI 10.1007/BF02248730
74. Clarke, E.M., Biere, A., Raimi, R., Zhu, Y.: Bounded model checking using satisfiability solving. Formal Methods in System Design **19**(1), 7–34 (2001)
75. Clarke, E.M., Emerson, E.A.: Design and synthesis of synchronization skeletons using branching-time temporal logic. In: D. Kozen (ed.) Logic of Programs, *LNCS*, vol. 131, pp. 52–71. Springer (1981)
76. Clarke, E.M., Emerson, E.A., Sifakis, J.: Model checking: algorithmic verification and debugging. Commun. ACM **52**(11), 74–84 (2009). DOI 10.1145/1592761.1592781
77. Clarke, E.M., Fehnker, A., Han, Z., Krogh, B.H., Ouaknine, J., Stursberg, O., Theobald, M.: Abstraction and counterexample-guided refinement in model checking of hybrid systems. Int. J. Found. Comput. Sci. **14**(4), 583–604 (2003). DOI 10.1142/S012905410300190X
78. Clarke, E.M., Grumberg, O., Peled, D.A.: Model Checking. MIT Press, Cambridge, MA, USA (1999)
79. Cohen, P.J.: Decision procedures for real and *p*-adic fields. Communications in Pure and Applied Mathematics **22**, 131–151 (1969)
80. Collins, G.E.: Hauptvortrag: Quantifier elimination for real closed fields by cylindrical algebraic decomposition. In: H. Barkhage (ed.) Automata Theory and Formal Languages, *LNCS*, vol. 33, pp. 134–183. Springer (1975). DOI 10.1007/3-540-07407-4_17
81. Collins, G.E., Hong, H.: Partial cylindrical algebraic decomposition for quantifier elimination. J. Symb. Comput. **12**(3), 299–328 (1991). DOI 10.1016/S0747-7171(08)80152-6
82. Collins, G.E., Johnson, J.R.: Quantifier elimination and the sign variation method for real root isolation. In: ISSAC, pp. 264–271 (1989). DOI 10.1145/74540.74574
83. Collins, G.E., Johnson, J.R., Krandick, W.: Interval arithmetic in cylindrical algebraic decomposition. J. Symb. Comput. **34**(2), 145–157 (2002). DOI 10.1006/jsco.2002.0547
84. Collins, G.E., Johnson, J.R., Küchlin, W.: Parallel real root isolation using the coefficient sign variation method. In: R. Zippel (ed.) CAP, *LNCS*, vol. 584, pp. 71–87. Springer (1990)
85. Collins, P., Lygeros, J.: Computability of finite-time reachable sets for hybrid systems. In: CDC-ECC'05, pp. 4688– 4693. IEEE (2005)

86. Comon, H., Jurski, Y.: Timed automata and the theory of real numbers. In: J.C.M. Baeten, S. Mauw (eds.) CONCUR, *LNCS*, vol. 1664, pp. 242–257. Springer (1999). DOI 10.1007/3-540-48320-9_18
87. Cook, S.A.: Soundness and completeness of an axiom system for program verification. SIAM J. Comput. **7**(1), 70–90 (1978). DOI 10.1137/0207005
88. Cox, D.A., Little, J., O'Shea, D.: Ideals, Varieties and Algorithms: An Introduction to Computational Algebraic Geometry and Commutative Algebra. Undergraduate Texts in Mathematics. Springer, New York (1992)
89. Damm, W., Hungar, H., Olderog, E.R.: On the verification of cooperating traffic agents. In: F.S. de Boer, M.M. Bonsangue, S. Graf, W.P. de Roever (eds.) FMCO, *LNCS*, vol. 3188, pp. 77–110. Springer (2003). DOI 10.1007/b100112
90. Damm, W., Hungar, H., Olderog, E.R.: Verification of cooperating traffic agents. International Journal of Control **79**(5), 395–421 (2006). DOI 10.1080/00207170600587531
91. Damm, W., Mikschl, A., Oehlerking, J., Olderog, E.R., Pang, J., Platzer, A., Segelken, M., Wirtz, B.: Automating verification of cooperation, control, and design in traffic applications. In: C.B. Jones, Z. Liu, J. Woodcock (eds.) Formal Methods and Hybrid Real-Time Systems, *LNCS*, vol. 4700, pp. 115–169. Springer (2007). DOI 10.1007/978-3-540-75221-9_6
92. Damm, W., Pinto, G., Ratschan, S.: Guaranteed termination in the verification of LTL properties of non-linear robust discrete time hybrid systems. In: Peled and Tsay [226], pp. 99–113. DOI 10.1007/11562948_10
93. Dantzig, G.B., Eaves, B.C.: Fourier-Motzkin elimination and its dual. J. Comb. Theory, Ser. A **14**(3), 288–297 (1973)
94. Davenport, J.H., Heintz, J.: Real quantifier elimination is doubly exponential. J. Symb. Comput. **5**(1/2), 29–35 (1988). DOI 10.1016/S0747-7171(88)80004-X
95. Davoren, J.M.: On hybrid systems and the modal μ-calculus. In: P.J. Antsaklis, W. Kohn, M.D. Lemmon, A. Nerode, S. Sastry (eds.) Hybrid Systems, *LNCS*, vol. 1567, pp. 38–69. Springer (1997). DOI 10.1007/3-540-49163-5_3
96. Davoren, J.M., Coulthard, V., Markey, N., Moor, T.: Non-deterministic temporal logics for general flow systems. In: Alur and Pappas [14], pp. 280–295. DOI 10.1007/b96398
97. Davoren, J.M., Nerode, A.: Logics for hybrid systems. IEEE **88**(7), 985–1010 (2000). DOI 10.1109/5.871305
98. Daws, C., Olivero, A., Tripakis, S., Yovine, S.: The tool KRONOS. In: Hybrid Systems III, *LNCS*, vol. 1066, pp. 208–219 (1996)
99. Dershowitz, N., Manna, Z.: Proving termination with multiset orderings. Commun. ACM **22**(8), 465–476 (1979). DOI 10.1145/359138.359142
100. Deshpande, A., Göllü, A., Varaiya, P.: SHIFT: A formalism and a programming language for dynamic networks of hybrid automata. In: P.J. Antsaklis, W. Kohn, A. Nerode, S. Sastry (eds.) Hybrid Systems, *LNCS*, vol. 1273, pp. 113–133. Springer (1996). DOI 10.1007/BFb0031558
101. Dolzmann, A., Sturm, T.: Redlog: Computer algebra meets computer logic. ACM SIGSAM Bull. **31**, 2–9 (1997). DOI 10.1145/261320.261324
102. Donzé, A., Maler, O.: Systematic simulation using sensitivity analysis. In: Bemporad et al. [41], pp. 174–189. DOI 10.1007/978-3-540-71493-4_16
103. Dowek, G., Hardin, T., Kirchner, C.: Theorem proving modulo. J. Autom. Reasoning **31**(1), 33–72 (2003). DOI 10.1023/A:1027357912519
104. Dowek, G., Muñoz, C., Carreño, V.A.: Provably safe coordinated strategy for distributed conflict resolution. In: Proceedings of the AIAA Guidance Navigation, and Control Conference and Exhibit 2005, AIAA-2005-6047 (2005)
105. van den Dries, L.: Alfred Tarski's elimination theory for real closed fields. J. Symb. Log. **53**(1), 7–19 (1988)
106. van den Dries, L.: Tame Topology and O-minimal Structures. Cambridge University Press (1998)
107. van den Dries, L., Miller, C.: On the real exponential field with restricted analytic functions. Israel J. Math. **85**(1-3), 19–56 (1994). DOI 10.1007/BF02758635

108. van den Dries, L., Speissegger, P.: The real field with convergent generalized power series. Trans. Amer. Math. Soc. **350**, 4377–4421 (1998). DOI 10.1090/S0002-9947-98-02105-9
109. Dubois, D.W.: A Nullstellensatz for ordered fields. Ark. Mat. **8**(2), 111–114 (1970). DOI 10.1007/BF02589551
110. Egerstedt, M., Johansson, K.H., Sastry, S., Lygeros, J.: On the regularization of Zeno hybrid automata. Systems and Control Letters **38**, 141–150 (1999)
111. Egerstedt, M., Mishra, B. (eds.): Hybrid Systems: Computation and Control, 11th International Conference, HSCC 2008, St. Louis, MO, USA, Proceedings, *LNCS*, vol. 4981. Springer (2008)
112. Eisenbud, D.: Commutative algebra with a view toward algebraic geometry, *Graduate Texts in Mathematics*, vol. 150, 3 edn. Springer, New York (1999)
113. Emerson, A.: Temporal and modal logic. In: J. van Leeuwen (ed.) Handbook of Theoretical Computer Science, Volume B: Formal Models and Sematics (B), pp. 995–1072. MIT Press (1990)
114. Emerson, E.A., Clarke, E.M.: Using branching time temporal logic to synthesize synchronization skeletons. Sci. Comput. Program. **2**(3), 241–266 (1982)
115. Emerson, E.A., Halpern, J.Y.: "Sometimes" and "Not Never" revisited: on branching versus linear time temporal logic. J. ACM **33**(1), 151–178 (1986). DOI 10.1145/4904.4999
116. Eriksson, K., Estep, D., Hansbo, P., Johnson, C.: Computational Differential Equations. Cambridge University Press (1996)
117. ERTMS User Group: ERTMS/ETCS System requirements specification. `http://www.era.europa.eu` (2002)
118. Etessami, K., Rajamani, S.K. (eds.): Computer Aided Verification, 17th International Conference, CAV 2005, Edinburgh, Scotland, UK, July 6–10, 2005, Proceedings, *LNCS*, vol. 3576. Springer (2005)
119. Faber, J., Meyer, R.: Model checking data-dependent real-time properties of the European Train Control System. In: FMCAD, pp. 76–77. IEEE Computer Society Press (2006). DOI 10.1109/FMCAD.2006.21
120. Faugére, J.C.: A new efficient algorithm for computing Gröbner bases (F4). Journal of Pure and Applied Algebra **139**(1-3), 61–88 (1999). DOI 10.1016/S0022-4049(99)00005-5
121. Faugére, J.C.: A new efficient algorithm for computing Gröbner bases without reduction to zero (F5). In: ISAAC. ACM Press (2002). DOI 10.1145/780506.780516
122. Fitting, M.: First-Order Logic and Automated Theorem Proving, 2 edn. Springer, New York (1996)
123. Fitting, M., Mendelsohn, R.L.: First-Order Modal Logic. Kluwer, Norwell, MA, USA (1999)
124. Fourier, J.B.J.: Solution d'une question particulière du calcul des inégalités. Nouveau Bulletin des Sciences par la Société Philomatique de Paris pp. 99–100 (1823)
125. Fränzle, M.: Analysis of hybrid systems: An ounce of realism can save an infinity of states. In: J. Flum, M. Rodríguez-Artalejo (eds.) CSL, *LNCS*, vol. 1683, pp. 126–140. Springer (1999)
126. Frehse, G.: PHAVer: Algorithmic verification of hybrid systems past HyTech. In: Morari and Thiele [212], pp. 258–273. DOI 10.1007/b106766
127. Frehse, G.: PHAVer: algorithmic verification of hybrid systems past HyTech. STTT **10**(3), 263–279 (2008). DOI 10.1007/s10009-007-0062-x
128. Frehse, G., Jha, S.K., Krogh, B.H.: A counterexample-guided approach to parameter synthesis for linear hybrid automata. In: Egerstedt and Mishra [111], pp. 187–200. DOI 10.1007/978-3-540-78929-1_14
129. Galdino, A.L., Muñoz, C., Ayala-Rincón, M.: Formal verification of an optimal air traffic conflict resolution and recovery algorithm. In: D. Leivant, R. de Queiroz (eds.) WoLLIC, *LNCS*, vol. 4576, pp. 177–188. Springer (2007)
130. Gallier, J.H.: Logic for Computer Science. Longman Higher Education (1986)
131. Ganzinger, H., Korovin, K.: Theory instantiation. In: M. Hermann, A. Voronkov (eds.) LPAR, *LNCS*, vol. 4246, pp. 497–511. Springer (2006). DOI 10.1007/11916277_34
132. Gear, C.W.: Differential-algebraic equations index transformations. SIAM J. Sci. Stat. Comput. **9**(1), 39–47 (1988). DOI 10.1137/0909004

133. Gentzen, G.: Untersuchungen über das logische Schließen. Math. Zeit. **39**, 405–431 (1935). DOI 10.1007/BF01201363
134. Giese, M.: Incremental closure of free variable tableaux. In: Goré et al. [139], pp. 545–560. DOI 10.1007/3-540-45744-5_46
135. Gödel, K.: Über die Vollständigkeit des Logikkalküls. Ph.D. thesis, Universität Wien (1929)
136. Gödel, K.: Die Vollständigkeit der Axiome des logischen Funktionenkalküls. Mon. hefte Math. Phys. **37**, 349–360 (1930). DOI 10.1007/BF01696781
137. Gödel, K.: Über formal unentscheidbare Sätze der Principia Mathematica und verwandter Systeme I. Mon. hefte Math. Phys. **38**, 173–198 (1931). DOI 10.1007/BF01700692
138. Gordon, M.J.C., Melham, T.F.: Introduction to HOL: A Theorem-Proving Environment for Higher-Order Logic. Cambridge Univ. Press (1993)
139. Goré, R., Leitsch, A., Nipkow, T. (eds.): Automated Reasoning, First International Joint Conference, IJCAR 2001, Siena, Italy, June 18–23, 2001, Proceedings, *LNCS*, vol. 2083. Springer (2001)
140. Graça, D.S., Campagnolo, M.L., Buescu, J.: Computability with polynomial differential equations. Advances in Applied Mathematics (2007)
141. Grigoriev, D.: Complexity of deciding Tarski algebra. J. Symb. Comput. **5**(1/2), 65–108 (1988). DOI 10.1016/S0747-7171(88)80006-3
142. Grigoriev, D., Vorobjov, N.: Solving systems of polynomial inequalities in subexponential time. J. Symb. Comput. **5**(1/2), 37–64 (1988). DOI 10.1016/S0747-7171(88)80005-1
143. Gross, J.: Schlussbericht über die Entgleisung von Güterzug 43647 der BLS AG auf der Weiche 34 (Einfahrt Lötschberg-Basisstrecke). Tech. Rep. 07101601, Unfalluntersuchungsstelle Bahnen und Schiffe (2007)
144. Grossman, R.L., Nerode, A., Ravn, A.P., Rischel, H. (eds.): Hybrid Systems, *LNCS*, vol. 736. Springer (1993)
145. Gulwani, S., Tiwari, A.: Constraint-based approach for analysis of hybrid systems. In: Gupta and Malik [146], pp. 190–203. DOI 10.1007/978-3-540-70545-1
146. Gupta, A., Malik, S. (eds.): Computer Aided Verification, CAV 2008, Princeton, NJ, USA, Proceedings, *LNCS*, vol. 5123. Springer (2008)
147. Hähnle, R., Schmitt, P.H.: The liberalized δ-rule in free variable semantic tableaux. J. Autom. Reasoning **13**(2), 211–221 (1994). DOI 10.1007/BF00881956
148. Harel, D.: First-Order Dynamic Logic. Springer, New York (1979)
149. Harel, D., Kozen, D., Tiuryn, J.: Dynamic logic. MIT Press, Cambridge (2000)
150. Harris, J.: Algebraic Geometry: A First Course. Graduate Texts in Mathematics. Springer (1995)
151. Harrison, J.: Verifying nonlinear real formulas via sums of squares. In: K. Schneider, J. Brandt (eds.) TPHOLs, *LNCS*, vol. 4732, pp. 102–118. Springer (2007). DOI 10.1007/978-3-540-74591-4_9
152. Harrison, J.: Handbook of Practical Logic and Automated Reasoning. Cambridge Univ. Press (2009)
153. Hartman, P.: Ordinary Differential Equations. John Wiley (1964)
154. Hartshorne, R.: Algebraic Geometry, *Graduate Texts in Mathematics*, vol. 52. Springer (1977)
155. Henkin, L.: The completeness of the first order functional calculus. J. Symb. Log. **14**(3), 159–166 (1949)
156. Henzinger, T.A.: The theory of hybrid automata. In: LICS, pp. 278–292. IEEE Computer Society, Los Alamitos (1996)
157. Henzinger, T.A., Ho, P.H.: HYTECH: The Cornell HYbrid TECHnology tools. In: P.J. Antsaklis, W. Kohn, A. Nerode, S. Sastry (eds.) Hybrid Systems, *LNCS*, vol. 999, pp. 265–293. Springer (1994)
158. Henzinger, T.A., Ho, P.H., Wong-Toi, H.: HyTech: A model checker for hybrid systems. In: O. Grumberg (ed.) CAV, *LNCS*, vol. 1254, pp. 460–463. Springer (1997)
159. Henzinger, T.A., Nicollin, X., Sifakis, J., Yovine, S.: Symbolic model checking for real-time systems. In: LICS, pp. 394–406. IEEE Computer Society (1992). DOI 10.1006/inco.1994.1045

160. Herbrand, J.: Recherches sur la théorie de la démonstration. Travaux de la Société des Sciences et des Lettres de Varsovie, Class III, Sciences Mathématiques et Physiques **33**, 33–160 (1930)
161. Hoare, C.A.R.: An axiomatic basis for computer programming. Commun. ACM **12**(10), 576–580 (1969). DOI 10.1145/363235.363259
162. Hoare, C.A.R.: Communicating Sequential Processes. Prentice Hall International (1985)
163. Hodges, W.: Model Theory. Cambridge University Press (1993)
164. Hoffmann, G.M., Huang, H., Waslander, S.L., Tomlin, C.J.: Quadrotor helicopter flight dynamics and control: Theory and experiment. In: Proceedings of the AIAA Guidance, Navigation, and Control Conference. Hilton Head, SC (2007). AIAA Paper Number 2007-6461
165. Hörmander, L.: The Analysis of Linear Partial Differential Operators II, *Grundlehren der mathematischen Wissenschaften*, vol. 257. Springer (1983)
166. Hsu, A., Eskafi, F., Sachs, S., Varaiya, P.: Design of platoon maneuver protocols for IVHS. PATH Research Report UCB-ITS-PRR-91-6, Institute of Transportation Studies, University of California, Berkeley (1991)
167. Hu, J., Prandini, M., Sastry, S.: Optimal coordinated motions of multiple agents moving on a plane. SIAM Journal on Control and Optimization **42**, 637–668 (2003)
168. Hu, J., Prandini, M., Sastry, S.: Probabilistic safety analysis in three-dimensional aircraft flight. In: CDC, vol. 5, pp. 5335 – 5340 (2003). DOI 10.1109/CDC.2003.1272485
169. Huth, M., Ryan, M.: Logic in Computer Science: Modelling and Reasoning about Systems, 2 edn. Cambridge Univ. Press (2004)
170. Hutter, D., Langenstein, B., Sengler, C., Siekmann, J.H., Stephan, W., Wolpers, A.: Deduction in the verification support environment (VSE). In: M.C. Gaudel, J. Woodcock (eds.) FME, *LNCS*, vol. 1051, pp. 268–286. Springer (1996)
171. Hwang, I., Kim, J., Tomlin, C.: Protocol-based conflict resolution for air traffic control. Air Traffic Control Quarterly **15**(1), 1–34 (2007)
172. Ilyashenko, Y., Yakovenko, S.: Lectures on Analytic Differential Equations, *Graduate Studies in Mathematics*, vol. 86. AMS (2008)
173. Jhala, R., McMillan, K.L.: Interpolant-based transition relation approximation. In: Etessami and Rajamani [118], pp. 39–51. DOI 10.1007/11513988_6
174. Jhala, R., McMillan, K.L.: Interpolant-based transition relation approximation. Logical Methods in Computer Science **3**(4) (2007). DOI 10.2168/LMCS-3(4:1)2007
175. Jifeng, H.: From CSP to hybrid systems. In: A.W. Roscoe (ed.) A classical mind: essays in honour of C. A. R. Hoare, pp. 171–189. Prentice Hall, Hertfordshire, UK (1994)
176. Johansson, K.H., Sastry, S., Zhang, J., Lygeros, J.: Zeno hybrid systems. Int. J. Robust and Nonlinear Control **11**, 435–451 (2001). DOI 10.1002/rnc.592
177. Katok, A., Hasselblatt, B.: Introduction to the Modern Theory of Dynamical Systems. Cambridge University Press, New York, NY (1996)
178. Kesten, Y., Manna, Z., Pnueli, A.: Verification of clocked and hybrid systems. Acta Inf. **36**(11), 837–912 (2000). DOI 10.1007/s002360050177
179. Kolchin, E.R.: Differential Algebra and Algebraic Groups. Academic Press, New York (1972)
180. Košecká, J., Tomlin, C., Pappas, G., Sastry, S.: 2-1/2D conflict resolution maneuvers for ATMS. In: CDC, vol. 3, pp. 2650–2655. Tampa, FL, USA (1998). DOI 10.1109/CDC.1998.757853
181. Kozen, D.: Results on the propositional μ-calculus. Theor. Comput. Sci. **27**, 333–354 (1983). DOI 10.1016/0304-3975(82)90125-6
182. Kozen, D.: Kleene algebra with tests. ACM Trans. Program. Lang. Syst. **19**(3), 427–443 (1997). DOI 10.1145/256167.256195
183. Kratz, F., Sokolsky, O., Pappas, G.J., Lee, I.: R-Charon, a modeling language for reconfigurable hybrid systems. In: J.P. Hespanha, A. Tiwari (eds.) HSCC, *LNCS*, vol. 3927, pp. 392–406. Springer (2006). DOI 10.1007/11730637_30
184. Kreisel, G., Krivine, J.L.: Elements of mathematical logic: Model Theory, 2 edn. North-Holland (1971)

185. Kripke, S.A.: Semantical considerations on modal logic. Acta Philosophica Fennica **16**, 83–94 (1963)
186. Krivine, J.L.: Anneaux préordonnés. Journal d'analyse mathématique **12**, 307–326 (1964)
187. Kunkel, P., Mehrmann, V.: Differential-Algebraic Equations: Analysis and Numerical Solution. European Mathematical Society (2006)
188. Lafferriere, G., Pappas, G.J., Sastry, S.: O-minimal hybrid systems. Mathematics of Control, Signals, and Systems **13**(1), 1–21 (2000). DOI 10.1007/PL00009858
189. Lafferriere, G., Pappas, G.J., Yovine, S.: A new class of decidable hybrid systems. In: F.W. Vaandrager, J.H. van Schuppen (eds.) HSCC, *LNCS*, vol. 1569, pp. 137–151. Springer (1999). DOI 10.1007/3-540-48983-5_15
190. Lafferriere, G., Pappas, G.J., Yovine, S.: Symbolic reachability computation for families of linear vector fields. J. Symb. Comput. **32**(3), 231–253 (2001). DOI 10.1006/jsco.2001.0472
191. Lang, S.: Algebra. Addison-Wesley Publishing Company (1978)
192. Lanotte, R., Tini, S.: Taylor approximation for hybrid systems. In: Morari and Thiele [212], pp. 402–416. DOI 10.1007/b106766
193. Larsen, K.G., Pettersson, P., Yi, W.: UPPAAL in a nutshell. STTT **1**(1-2), 134–152 (1997)
194. Liberzon, D.: Switching in Systems and Control. Systems and Control: Foundations and Applications. Birkhäuser, Boston, MA (2003)
195. Lindelöf, M.E.: Sur l'application de la méthode des approximations successives aux équations différentielles ordinaires du premier ordre. Comptes rendus hebdomadaires des séances de l'Académie des sciences **114**, 454–457 (1894)
196. Livadas, C., Lygeros, J., Lynch, N.A.: High-level modeling and analysis of TCAS. Proc. IEEE – Special Issue on Hybrid Systems: Theory & Applications **88**(7), 926–947 (2000)
197. Łojasiewicz, S.: Triangulations of semi-analytic sets. Annali della Scuola Normale Superiore di Pisa **18**, 449–474 (1964)
198. Loos, R., Weispfenning, V.: Applying linear quantifier elimination. Comput. J. **36**(5), 450–462 (1993). DOI 10.1093/comjnl/36.5.450
199. Löwenheim, L.: Über Möglichkeiten im Relativkalkül. Mathematische Annalen **76**, 447–470 (1915). DOI 10.1007/BF01458217
200. Maler, O., Pnueli, A. (eds.): Hybrid Systems: Computation and Control, 6th International Workshop, HSCC 2003 Prague, Czech Republic, April 3–5, 2003, Proceedings, *LNCS*, vol. 2623. Springer (2003)
201. Manna, Z., Sipma, H.: Deductive verification of hybrid systems using STeP. In: T.A. Henzinger, S. Sastry (eds.) HSCC, *LNCS*, vol. 1386, pp. 305–318. Springer (1998). DOI 10.1007/3-540-64358-3_47
202. Mansfield, E.L.: Differential Gröbner bases. Ph.D. thesis, University of Sydney (1991)
203. Massink, M., Francesco, N.D.: Modelling free flight with collision avoidance. In: Andler and Offutt [16], pp. 270–280. DOI 10.1109/ICECCS.2001.930186
204. McLaughlin, S., Harrison, J.: A proof-producing decision procedure for real arithmetic. In: R. Nieuwenhuis (ed.) CADE, *LNCS*, vol. 3632, pp. 295–314. Springer (2005). DOI 10.1007/11532231_22
205. Meyer, R., Faber, J., Hoenicke, J., Rybalchenko, A.: Model checking duration calculus: A practical approach. Formal Aspects of Computing pp. 1–25 (2008). DOI 10.1007/s00165-008-0082-7
206. Miller, C.L.: Expansions of the real field with power functions. Ann. Pure Appl. Logic **68**(1), 79–94 (1994)
207. Milner, R.: Logic for computable functions: description of a machine implementation. Tech. rep., Stanford University, Stanford, CA, USA (1972)
208. Milner, R.: Communicating and Mobile Systems: the π-Calculus. Cambridge Univ. Press (1999)
209. Misner, C.W., Thorne, K.S., Wheeler, J.A.: Gravitation. W. H. Freeman, New York (1973)
210. Mitchell, J. (ed.): Proceedings, Fifth Annual IEEE Symposium on Logic in Computer Science, 4–7 June 1990, Philadelphia, Pennsylvania, USA. IEEE Computer Society (1990)
211. Mora, T.: Solving Polynomial Equation Systems II: Macaulay's Paradigm and Gröbner Technology. Cambridge University Press (2005)

212. Morari, M., Thiele, L. (eds.): Hybrid Systems: Computation and Control, 8th International Workshop, HSCC 2005, Zurich, Switzerland, March 9–11, 2005, Proceedings, *LNCS*, vol. 3414. Springer (2005)
213. Morayne, M.: On differentiability of Peano type functions. Colloquium Mathematicum **LIII**, 129–132 (1987)
214. Motzkin, T.S.: Beiträge zur Theorie der Linearen Ungleichungen. Ph.D. thesis, Basel, Jerusalem (1936)
215. Mumford, D.: Algebraic Geometry I: Complex Projective Varieties. Classics in Mathematics. Springer (1995)
216. Muñoz, C., Carreño, V., Dowek, G., Butler, R.W.: Formal verification of conflict detection algorithms. STTT **4**(3), 371–380 (2003). DOI 10.1007/s10009-002-0084-3
217. Mysore, V., Piazza, C., Mishra, B.: Algorithmic algebraic model checking II: Decidability of semi-algebraic model checking and its applications to systems biology. In: Peled and Tsay [226], pp. 217–233. DOI 10.1007/11562948_18
218. Nicollin, X., Olivero, A., Sifakis, J., Yovine, S.: An approach to the description and analysis of hybrid systems. In: Grossman et al. [144], pp. 149–178
219. Nipkow, T.: Linear quantifier elimination. In: Armando et al. [18]
220. Nipkow, T., Paulson, L.C., Wenzel, M.: Isabelle/HOL — A Proof Assistant for Higher-Order Logic, *LNCS*, vol. 2283. Springer (2002)
221. Olderog, E.R., Dierks, H.: Real-Time Systems: Formal Specification and Automatic Verification. Cambridge Univ. Press (2008)
222. Pallottino, L., Scordio, V.G., Frazzoli, E., Bicchi, A.: Decentralized cooperative policy for conflict resolution in multi-vehicle systems. IEEE Trans. on Robotics **23**(6), 1170–1183 (2007)
223. Parker, R.S., Doyle, F.J., Peppas, N.A.: The intravenous route to blood glucose control. IEEE Engineering in Medicine and Biology **20**(1), 65–73 (2001). DOI 10.1109/51.897829
224. Parrilo, P.A.: Semidefinite programming relaxations for semialgebraic problems. Math. Program. **96**(2), 293–320 (2003). DOI 10.1007/s10107-003-0387-5
225. Peano, G.: Demonstration de l'intégrabilité des équations différentielles ordinaires. Mathematische Annalen **37**(2), 182–228 (1890). DOI 10.1007/BF01200235
226. Peled, D., Tsay, Y.K. (eds.): Automated Technology for Verification and Analysis, Third International Symposium, ATVA 2005, Taipei, Taiwan, October 4–7, 2005, Proceedings, *LNCS*, vol. 3707. Springer (2005)
227. Perko, L.: Differential equations and dynamical systems, 3 edn. Springer, New York, NY, USA (2006)
228. Piazza, C., Antoniotti, M., Mysore, V., Policriti, A., Winkler, F., Mishra, B.: Algorithmic algebraic model checking I: Challenges from systems biology. In: Etessami and Rajamani [118], pp. 5–19. DOI 10.1007/11513988_3
229. Plaisted, D.A., Zhu, Y.: Ordered semantic hyper-linking. J. Autom. Reasoning **25**(3), 167–217 (2000)
230. Platzer, A.: Combining deduction and algebraic constraints for hybrid system analysis. In: B. Beckert (ed.) VERIFY'07 at CADE, Bremen, Germany, *CEUR Workshop Proceedings*, vol. 259, pp. 164–178. CEUR-WS.org (2007)
231. Platzer, A.: Differential dynamic logic for verifying parametric hybrid systems. In: N. Olivetti (ed.) TABLEAUX, *LNCS*, vol. 4548, pp. 216–232. Springer (2007). DOI 10.1007/978-3-540-73099-6_17
232. Platzer, A.: Differential logic for reasoning about hybrid systems. In: Bemporad et al. [41], pp. 746–749. DOI 10.1007/978-3-540-71493-4_75
233. Platzer, A.: A temporal dynamic logic for verifying hybrid system invariants. In: S.N. Artëmov, A. Nerode (eds.) LFCS, *LNCS*, vol. 4514, pp. 457–471. Springer (2007). DOI 10.1007/978-3-540-72734-7_32
234. Platzer, A.: Towards a hybrid dynamic logic for hybrid dynamic systems. In: P. Blackburn, T. Bolander, T. Braüner, V. de Paiva, J. Villadsen (eds.) LICS International Workshop on Hybrid Logic, HyLo'06, Seattle, USA, Proceedings, *ENTCS*, vol. 174, pp. 63–77 (2007). DOI 10.1016/j.entcs.2006.11.026

235. Platzer, A.: Differential dynamic logic for hybrid systems. Journal of Automated Reasoning **41**(2), 143–189 (2008). DOI 10.1007/s10817-008-9103-8
236. Platzer, A.: Differential dynamic logics: Automated theorem proving for hybrid systems. Ph.D. thesis, Department of Computing Science, University of Oldenburg (2008)
237. Platzer, A.: Differential-algebraic dynamic logic for differential-algebraic programs. Journal of Logic and Computation **20**(1), 309–352 (2010). DOI 10.1093/logcom/exn070. Advance Access published on November 18, 2008
238. Platzer, A., Clarke, E.M.: The image computation problem in hybrid systems model checking. In: Bemporad et al. [41], pp. 473–486. DOI 10.1007/978-3-540-71493-4_37
239. Platzer, A., Clarke, E.M.: Computing differential invariants of hybrid systems as fixedpoints. In: Gupta and Malik [146], pp. 176–189. DOI 10.1007/978-3-540-70545-1_17
240. Platzer, A., Clarke, E.M.: Computing differential invariants of hybrid systems as fixedpoints. Formal Methods in System Design **35**(1), 98–120 (2009). DOI 10.1007/s10703-009-0079-8
241. Platzer, A., Clarke, E.M.: Formal verification of curved flight collision avoidance maneuvers: A case study. In: A. Cavalcanti, D. Dams (eds.) FM, *LNCS*, vol. 5850, pp. 547–562. Springer (2009). DOI 10.1007/978-3-642-05089-3_35
242. Platzer, A., Quesel, J.D.: KeYmaera: A hybrid theorem prover for hybrid systems. In: Armando et al. [18], pp. 171–178. DOI 10.1007/978-3-540-71070-7_15
243. Platzer, A., Quesel, J.D.: Logical verification and systematic parametric analysis in train control. In: Egerstedt and Mishra [111], pp. 646–649. DOI 10.1007/978-3-540-78929-1_55
244. Platzer, A., Quesel, J.D.: European Train Control System: A case study in formal verification. In: K. Breitman, A. Cavalcanti (eds.) ICFEM, *LNCS*, vol. 5885, pp. 246–265. Springer (2009). DOI 10.1007/978-3-642-10373-5_13
245. Platzer, A., Quesel, J.D.: European Train Control System: A case study in formal verification. Tech. Rep. 54, Reports of SFB/TR 14 AVACS (2009). ISSN: 1860-9821, http://www.avacs.org.
246. Platzer, A., Quesel, J.D., Rümmer, P.: Real world verification. In: R.A. Schmidt (ed.) CADE, *LNCS*, vol. 5663, pp. 485–501. Springer (2009). DOI 10.1007/978-3-642-02959-2_35
247. Pnueli, A.: The temporal logic of programs. In: FOCS, pp. 46–57. IEEE (1977)
248. Pnueli, A., Kesten, Y.: A deductive proof system for CTL*. In: L. Brim, P. Jancar, M. Kretínský, A. Kucera (eds.) CONCUR, *LNCS*, vol. 2421, pp. 24–40. Springer (2002). DOI 10.1007/3-540-45694-5_2
249. Pnueli, A., Rosner, R.: On the synthesis of a reactive module. In: POPL, pp. 179–190 (1989). DOI 10.1145/75277.75293
250. Pour-El, M.B., Richards, I.: A computable ordinary differential equation which possesses no computable solution. Annals of Mathematical Logic **17**, 61–90 (1979). DOI 10.1016/0003-4843(79)90021-4
251. Prajna, S., Jadbabaie, A.: Safety verification of hybrid systems using barrier certificates. In: Alur and Pappas [14], pp. 477–492. DOI 10.1007/b96398
252. Prajna, S., Jadbabaie, A., Pappas, G.J.: A framework for worst-case and stochastic safety verification using barrier certificates. IEEE T. Automat. Contr. **52**(8), 1415–1429 (2007). DOI 10.1109/TAC.2007.902736
253. Pratt, V.R.: Semantical considerations on Floyd-Hoare logic. In: FOCS, pp. 109–121. IEEE (1976)
254. Pratt, V.R.: Process logic. In: POPL, pp. 93–100 (1979). DOI 10.1145/567752.567761
255. Queille, J.P., Sifakis, J.: Specification and verification of concurrent systems in CESAR. In: M. Dezani-Ciancaglini, U. Montanari (eds.) Symposium on Programming, *LNCS*, vol. 137, pp. 337–351. Springer (1982). DOI 10.1007/3-540-11494-7_22
256. Quesel, J.D.: A theorem prover for differential dynamic logic. Master's thesis, University of Oldenburg, Department of Computing Science. Correct System Design Group (2007)
257. Reid, W.T.: Ordinary Differential Equations. John Wiley (1971)
258. Renegar, J.: On the computational complexity and geometry of the first-order theory of the reals, part I: Introduction. Preliminaries. The geometry of semi-algebraic sets. The decision problem for the existential theory of the reals. J. Symb. Comput. **13**(3), 255–300 (1992)

259. Renegar, J.: On the computational complexity and geometry of the first-order theory of the reals, part II: The general decision problem. Preliminaries for quantifier elimination. J. Symb. Comput. **13**(3), 301–328 (1992)
260. Renegar, J.: On the computational complexity and geometry of the first-order theory of the reals, part III: Quantifier elimination. J. Symb. Comput. **13**(3), 329–352 (1992)
261. Reynolds, M.: An axiomatization of PCTL*. Inf. Comput. **201**(1), 72–119 (2005). DOI 10.1016/j.ic.2005.03.005
262. Risler, J.J.: Une caractérisation des idéaux des variétés algébraiques réelles. C.R.A.S. Paris série A **271**, 1171–1173 (1970)
263. Risler, J.J.: Some aspects of complexity in real algebraic geometry. J. Symb. Comput. **5**(1/2), 109–119 (1988). DOI 10.1016/S0747-7171(88)80007-5
264. Robinson, A.: Complete Theories. North-Holland (1956)
265. Robinson, J.A.: A machine-oriented logic based on the resolution principle. J. Assoc. Comput. Mach. **12**, 23–41 (1965). DOI 10.1145/321250.321253
266. Rodríguez-Carbonell, E., Kapur, D.: An abstract interpretation approach for automatic generation of polynomial invariants. In: R. Giacobazzi (ed.) SAS, *Lecture Notes in Computer Science*, vol. 3148, pp. 280–295. Springer (2004). DOI 10.1007/b99688
267. Rodríguez-Carbonell, E., Kapur, D.: Automatic generation of polynomial invariants of bounded degree using abstract interpretation. Sci. Comput. Program. **64**(1), 54–75 (2007). DOI 10.1016/j.scico.2006.03.003
268. Rodríguez-Carbonell, E., Kapur, D.: Generating all polynomial invariants in simple loops. J. Symb. Comput. **42**(4), 443–476 (2007). DOI 10.1016/j.jsc.2007.01.002
269. Rodríguez-Carbonell, E., Tiwari, A.: Generating polynomial invariants for hybrid systems. In: Morari and Thiele [212], pp. 590–605. DOI 10.1007/b106766
270. Rönkkö, M., Ravn, A.P., Sere, K.: Hybrid action systems. Theor. Comput. Sci. **290**(1), 937–973 (2003)
271. Rounds, W.C.: A spatial logic for the hybrid π-calculus. In: Alur and Pappas [14], pp. 508–522. DOI 10.1007/b96398
272. Rounds, W.C., Song, H.: The ϕ-calculus: A language for distributed control of reconfigurable embedded systems. In: HSCC, *LNCS*, vol. 2623, pp. 435–449 (2003). DOI 10.1007/3-540-36580-X_32
273. Rümmer, P.: A sequent calculus for integer arithmetic with counterexample generation. In: B. Beckert (ed.) VERIFY'07 at CADE, Bremen, Germany, *CEUR Workshop Proceedings*, vol. 259, pp. 179–194. CEUR-WS.org (2007)
274. Sankaranarayanan, S., Sipma, H., Manna, Z.: Constructing invariants for hybrid systems. In: Alur and Pappas [14], pp. 539–554. DOI 10.1007/b96398
275. Schobbens, P.Y., Raskin, J.F., Henzinger, T.A.: Axioms for real-time logics. Theor. Comput. Sci. **274**(1-2), 151–182 (2002). DOI 10.1016/S0304-3975(00)00308-X
276. Schrödinger, E.: An undulatory theory of the mechanics of atoms and molecules. Phys. Rev. **28**, 1049–1070 (1926). DOI 10.1103/PhysRev.28.1049
277. Scott, D., Strachey, C.: Toward a mathematical semantics for computer languages? Tech. Rep. PRG-6, Oxford Programming Research Group (1971)
278. Seidenberg, A.: A new decision method for elementary algebra. Annals of Mathematics **60**, 365–374 (1954)
279. Sibirsky, K.S.: Introduction to Topological Dynamics. Noordhoff, Leyden (1975)
280. Sipser, M.: Introduction to the Theory of Computation, 2 edn. Course Technology (2005)
281. Skolem, T.: Logisch-kombinatorische Untersuchungen über die Erfüllbarkeit oder Beweisbarkeit mathematischer Sätze nebst einem Theorem über dichte Mengen. Videnskapsselskapet Skrifter, I. Matematisk-naturvidenskabelig Klasse **6**, 1–36 (1920)
282. Smullyan, R.M.: First-Order Logic. Dover (1995)
283. Stengle, G.: A Nullstellensatz and a Positivstellensatz in semialgebraic geometry. Math. Ann. **207**(2), 87–97 (1973). DOI 10.1007/BF01362149
284. Stirling, C.: Modal and temporal logics. In: Handbook of Logic in Computer Science (vol. 2): Background: Computational Structures, pp. 477–563. Oxford University Press, Inc., New York, NY, USA (1992)

285. Strzebonski, A.W.: Cylindrical algebraic decomposition using validated numerics. J. Symb. Comput. **41**(9), 1021–1038 (2006). DOI 10.1016/j.jsc.2006.06.004
286. Sturm, C.F.: Mémoire sur la résolution des équations numériques. Mémoires des Savants Etrangers **6**, 271–318 (1835)
287. Tarski, A.: Sur les ensembles définissables de nombres réels I. Fundam. Math. **17**, 210–239 (1931)
288. Tarski, A.: A Decision Method for Elementary Algebra and Geometry, 2 edn. University of California Press, Berkeley (1951)
289. Tavernini, L.: Differential automata and their discrete simulators. Non-Linear Anal. **11**(6), 665–683 (1987). DOI 10.1016/0362-546X(87)90034-4
290. Tinelli, C.: Cooperation of background reasoners in theory reasoning by residue sharing. J. Autom. Reasoning **30**(1), 1–31 (2003). DOI 10.1023/A:1022587501759
291. Tiwari, A.: Approximate reachability for linear systems. In: Maler and Pnueli [200], pp. 514–525. DOI 10.1007/3-540-36580-X_37
292. Tiwari, A.: An algebraic approach for the unsatisfiability of nonlinear constraints. In: C.H.L. Ong (ed.) CSL, *LNCS*, vol. 3634, pp. 248–262. Springer (2005). DOI 10.1007/11538363_18
293. Tomlin, C., Pappas, G.J., Sastry, S.: Conflict resolution for air traffic management: a study in multi-agent hybrid systems. IEEE T. Automat. Contr. **43**(4), 509–521 (1998). DOI 10.1109/9.664154
294. Tomlin, C.J., Lygeros, J., Sastry, S.: A game theoretic approach to controller design for hybrid systems. Proc. IEEE **88**(7), 949–970 (2000). DOI 10.1109/5.871303
295. Umeno, S., Lynch, N.A.: Proving safety properties of an aircraft landing protocol using I/O automata and the PVS theorem prover: A case study. In: J. Misra, T. Nipkow, E. Sekerinski (eds.) FM, *LNCS*, vol. 4085, pp. 64–80. Springer (2006). DOI 10.1007/11813040_5
296. Umeno, S., Lynch, N.A.: Safety verification of an aircraft landing protocol: A refinement approach. In: Bemporad et al. [41], pp. 557–572. DOI 10.1007/978-3-540-71493-4_43
297. Walter, W.: Ordinary Differential Equations. Springer (1998)
298. Weihrauch, K.: Computable Analysis. Springer (2005)
299. Weispfenning, V.: The complexity of linear problems in fields. J. Symb. Comput. **5**(1/2), 3–27 (1988). DOI 10.1016/S0747-7171(88)80003-8
300. Weispfenning, V.: Quantifier elimination for real algebra — the cubic case. In: ISSAC, pp. 258–263 (1994). DOI 10.1145/190347.190425
301. Weispfenning, V.: Quantifier elimination for real algebra — the quadratic case and beyond. Appl. Algebra Eng. Commun. Comput. **8**(2), 85–101 (1997). DOI 10.1007/s002000050055
302. Witsenhausen, H.S.: A class of hybrid-state continuous-time dynamic systems. IEEE Trans. Automat. Contr. **11**, 161–167 (1966)
303. Wolfram Research, Inc., Champaign, IL: Mathematica, version 5.2 edn. (2005). `http://www.wolfram.com/`
304. Zeidler, E. (ed.): Teubner-Taschenbuch der Mathematik I. Teubner (2003)
305. Zhou, C., Hansen, M.R.: Duration Calculus: A Formal Approach to Real-Time Systems. Monographs in Theoretical Computer Science. Springer (2004)
306. Zhou, C., Ravn, A.P., Hansen, M.R.: An extended duration calculus for hybrid real-time systems. In: Grossman et al. [144], pp. 36–59

Index

Symbols

A

Table 2.1 Statements and effects of hybrid programs (HPs)

HP Notation	Operation	Effect
$x_1 := \theta_1, \dots, x_n := \theta_n$	discrete jump set	simultaneously assigns terms θ_i to variables x_i
$x_1' = \theta_1, \dots, x_n' = \theta_n \,\&\, \chi$	continuous evolution	differential equations for x_i with terms θ_i within first-order constraint χ (evolution domain)
$?\chi$	state test / check	test first-order formula χ at current state
$\alpha; \beta$	seq. composition	HP β starts after HP α finishes
$\alpha \cup \beta$	nondet. choice	choice between alternatives HP α or HP β
α^*	nondet. repetition	repeats HP α n-times for any $n \in \mathbb{N}$

Tables 2.3, 3.4 and 4.1 Operators of differential dynamic logic ($\mathsf{d}\mathcal{L}$), and additional operators of differential-algebraic dynamic logic (DAL) and differential temporal-dynamic logic (dTL)

$\mathsf{d}\mathcal{L}$ Notation	Operator	Meaning
$p(\theta_1, \dots, \theta_n)$	atomic formula	true iff predicate p holds for $(\theta_1, \dots, \theta_n)$
$\neg\phi$	negation / not	true if ϕ is false
$\phi \wedge \psi$	conjunction / and	true if both ϕ and ψ are true
$\phi \vee \psi$	disjunction / or	true if ϕ is true or if ψ is true
$\phi \rightarrow \psi$	implication / implies	true if ϕ is false or ψ is true
$\phi \leftrightarrow \psi$	bi-implication / equivalent	true if ϕ and ψ are both true or both false
$\forall x\, \phi$	universal quantifier / for all	true if ϕ is true for all values of variable x
$\exists x\, \phi$	existential quantifier / exists	true if ϕ is true for some values of variable x
$[\alpha]\phi$	$[\cdot]$ modality / box	true if ϕ is true after all runs of HP α
$\langle\alpha\rangle\phi$	$\langle\cdot\rangle$ modality / diamond	true if ϕ is true after at least one run of HP α
$[\alpha]\phi$	$[\cdot]$ modality / box (DAL)	true if ϕ is true after all runs of DA-program α
$\langle\alpha\rangle\phi$	$\langle\cdot\rangle$ modality / diamond (DAL)	true if ϕ is true after some run of DA-program α
$[\alpha]\Box\phi$	$[\cdot]\Box$ modality nesting (dTL)	if ϕ is true always during all traces of HP α
$\langle\alpha\rangle\Diamond\phi$	$\langle\cdot\rangle\Diamond$ modality nesting (dTL)	if ϕ is true sometimes during some trace of HP α
$[\alpha]\Diamond\phi$	$[\cdot]\Diamond$ modality nesting (dTL)	if ϕ is true sometimes during all traces of HP α
$\langle\alpha\rangle\Box\phi$	$\langle\cdot\rangle\Box$ modality nesting (dTL)	if ϕ is true always during some trace of HP α

Table 3.2 Statements and effects of differential-algebraic programs

DA-program	Operation	Effect
$\mathcal{J}$	discrete jump	jump constraint with assignments holds for discrete jump
$\mathcal{D}$	diff.-alg. flow	differential-algebraic constraint holds during continuous flow
$\alpha; \beta$	seq. composition	DA-program β starts after DA-program α finishes
$\alpha \cup \beta$	nondet. choice	choice between alternative DA-programs α or β
α^*	nondet. repetition	repeats DA-program α n-times for any $n \in \mathbb{N}$

$$(\neg r)\ \frac{\phi \vdash}{\vdash \neg\phi} \qquad (\vee r)\ \frac{\vdash \phi, \psi}{\vdash \phi \vee \psi} \qquad (\wedge r)\ \frac{\vdash \phi \quad \vdash \psi}{\vdash \phi \wedge \psi} \qquad (\to r)\ \frac{\phi \vdash \psi}{\vdash \phi \to \psi} \qquad (ax)\ \frac{}{\phi \vdash \phi}$$

$$(\neg l)\ \frac{\vdash \phi}{\neg\phi \vdash} \qquad (\vee l)\ \frac{\phi \vdash \quad \psi \vdash}{\phi \vee \psi \vdash} \qquad (\wedge l)\ \frac{\phi, \psi \vdash}{\phi \wedge \psi \vdash} \qquad (\to l)\ \frac{\vdash \phi \quad \psi \vdash}{\phi \to \psi \vdash} \qquad (cut)\ \frac{\vdash \phi \quad \phi \vdash}{\vdash}$$

$$(\langle;\rangle)\ \frac{\langle\alpha\rangle\langle\beta\rangle\phi}{\langle\alpha;\beta\rangle\phi} \qquad (\langle{}^{*n}\rangle)\ \frac{\phi \vee \langle\alpha\rangle\langle\alpha^*\rangle\phi}{\langle\alpha^*\rangle\phi} \qquad (\langle:=\rangle)\ \frac{\phi_{x_1}^{\theta_1} \cdots {}_{x_n}^{\theta_n}}{\langle x_1 := \theta_1, \ldots, x_n := \theta_n\rangle\phi}$$

$$([;])\ \frac{[\alpha][\beta]\phi}{[\alpha;\beta]\phi} \qquad ([{}^{*n}])\ \frac{\phi \wedge [\alpha][\alpha^*]\phi}{[\alpha^*]\phi} \qquad ([:=])\ \frac{\langle x_1 := \theta_1, \ldots, x_n := \theta_n\rangle\phi}{[x_1 := \theta_1, \ldots, x_n := \theta_n]\phi}$$

$$(\langle\cup\rangle)\ \frac{\langle\alpha\rangle\phi \vee \langle\beta\rangle\phi}{\langle\alpha\cup\beta\rangle\phi} \qquad (\langle ?\rangle)\ \frac{\chi \wedge \psi}{\langle ?\chi\rangle\psi} \qquad (\langle'\rangle)\ \frac{\exists t{\geq}0\,\big((\forall 0{\leq}\tilde{t}{\leq}t\,\langle\mathcal{S}_{\tilde{t}}\rangle\chi) \wedge \langle\mathcal{S}_t\rangle\phi\big)}{\langle x_1' = \theta_1, \ldots, x_n' = \theta_n \,\&\, \chi\rangle\phi}$$ [1]

$$([\cup])\ \frac{[\alpha]\phi \wedge [\beta]\phi}{[\alpha\cup\beta]\phi} \qquad ([?])\ \frac{\chi \to \psi}{[?\chi]\psi} \qquad (['])\ \frac{\forall t{\geq}0\,\big((\forall 0{\leq}\tilde{t}{\leq}t\,\langle\mathcal{S}_{\tilde{t}}\rangle\chi) \to \langle\mathcal{S}_t\rangle\phi\big)}{[x_1' = \theta_1, \ldots, x_n' = \theta_n \,\&\, \chi]\phi}$$ [1]

$$(\forall r)\ \frac{\vdash \phi(s(X_1, \ldots, X_n))}{\vdash \forall x\, \phi(x)}$$ [2]

$$(\exists r)\ \frac{\vdash \phi(X)}{\vdash \exists x\, \phi(x)}$$ [4]

$$(\exists l)\ \frac{\phi(s(X_1, \ldots, X_n)) \vdash}{\exists x\, \phi(x) \vdash}$$ [2]

$$(\forall l)\ \frac{\phi(X) \vdash}{\forall x\, \phi(x) \vdash}$$ [4]

$$(i\forall)\ \frac{\vdash \mathrm{QE}(\forall X\, (\Phi(X) \vdash \Psi(X)))}{\Phi(s(X_1, \ldots, X_n)) \vdash \Psi(s(X_1, \ldots, X_n))}$$ [3]

$$(i\exists)\ \frac{\vdash \mathrm{QE}(\exists X \bigwedge_i (\Phi_i \vdash \Psi_i))}{\Phi_1 \vdash \Psi_1 \quad \ldots \quad \Phi_n \vdash \Psi_n}$$ [5]

$$([]gen)\ \frac{\vdash \forall^\alpha(\phi \to \psi)}{[\alpha]\phi \vdash [\alpha]\psi} \qquad (\langle\rangle gen)\ \frac{\vdash \forall^\alpha(\phi \to \psi)}{\langle\alpha\rangle\phi \vdash \langle\alpha\rangle\psi}$$

$$(ind)\ \frac{\vdash \forall^\alpha(\phi \to [\alpha]\phi)}{\phi \vdash [\alpha^*]\phi}$$

$$(con)\ \frac{\vdash \forall^\alpha \forall v{>}0\, (\varphi(v) \to \langle\alpha\rangle\varphi(v-1))}{\exists v\, \varphi(v) \vdash \langle\alpha^*\rangle\exists v{\leq}0\, \varphi(v)}$$ [6]

[1] t and $\tilde{t}$ are fresh logical variables and $\langle\mathcal{S}_t\rangle$ is the jump set $\langle x_1 := y_1(t), \ldots, x_n := y_n(t)\rangle$ with simultaneous solutions $y_1, \ldots, y_n$ of the respective differential equations with constant symbols x_i as symbolic initial values.

[2] s is a new (Skolem) function symbol and $X_1, \ldots, X_n$ are all free logical variables of $\forall x\, \phi(x)$.

[3] X is a new logical variable. Further, QE needs to be defined for the formula in the premise.

[4] X is a new logical variable.

[5] Among all open branches, free logical variable X only occurs in the branches $\Phi_i \vdash \Psi_i$. Further, QE needs to be defined for the formula in the premise, especially, no Skolem dependencies on X can occur.

[6] Logical variable v does not occur in α.

Fig. 2.11 Proof calculus for differential dynamic logic ($d\mathcal{L}$)

$$(\mathrm{r}\forall)\ \frac{\mathrm{QE}(\forall x \bigwedge_i(\Gamma_i \vdash \Delta_i))}{\Gamma \vdash \Delta, \forall x\, \phi}\,{}^1 \qquad (\mathrm{r}\exists)\ \frac{\mathrm{QE}(\exists x \bigwedge_i(\Gamma_i \vdash \Delta_i))}{\Gamma \vdash \Delta, \exists x\, \phi}\,{}^1$$

$$(\mathrm{l}\forall)\ \frac{\mathrm{QE}(\exists x \bigwedge_i(\Gamma_i \vdash \Delta_i))}{\Gamma, \forall x\, \phi \vdash \Delta}\,{}^1 \qquad (\mathrm{l}\exists)\ \frac{\mathrm{QE}(\forall x \bigwedge_i(\Gamma_i \vdash \Delta_i))}{\Gamma, \exists x\, \phi \vdash \Delta}\,{}^1$$

$$(\langle;\rangle)\ \frac{\langle\alpha\rangle\langle\beta\rangle\phi}{\langle\alpha;\beta\rangle\phi} \qquad (\langle\exists\rangle)\ \frac{\exists x\,\langle\mathscr{J}\rangle\phi}{\langle\exists x\,\mathscr{J}\rangle\phi} \qquad (\langle:=\rangle)\ \frac{\chi \wedge \phi_{x_1}^{\theta_1}\cdots{}_{x_n}^{\theta_n}}{\langle x_1 := \theta_1 \wedge .. \wedge x_n := \theta_n \wedge \chi\rangle\phi}\,{}^3$$

$$([;])\ \frac{[\alpha][\beta]\phi}{[\alpha;\beta]\phi} \qquad ([\exists])\ \frac{\forall x\,[\mathscr{J}]\phi}{[\exists x\,\mathscr{J}]\phi} \qquad ([:=])\ \frac{\chi \to \phi_{x_1}^{\theta_1}\cdots{}_{x_n}^{\theta_n}}{[x_1 := \theta_1 \wedge .. \wedge x_n := \theta_n \wedge \chi]\phi}\,{}^3$$

$$(\langle\cup\rangle)\ \frac{\langle\alpha\rangle\phi \vee \langle\beta\rangle\phi}{\langle\alpha\cup\beta\rangle\phi} \qquad (\langle J\rangle)\ \frac{\langle\mathscr{J}_1 \cup \ldots \cup \mathscr{J}_n\rangle\phi}{\langle\mathscr{J}\rangle\phi}\,{}^2 \qquad (\langle D\rangle)\ \frac{\langle(\mathscr{D}_1 \cup \ldots \cup \mathscr{D}_n)^*\rangle\phi}{\langle\mathscr{D}\rangle\phi}\,{}^4$$

$$([\cup])\ \frac{[\alpha]\phi \wedge [\beta]\phi}{[\alpha\cup\beta]\phi} \qquad ([J])\ \frac{[\mathscr{J}_1 \cup \ldots \cup \mathscr{J}_n]\phi}{[\mathscr{J}]\phi}\,{}^2 \qquad ([D])\ \frac{[(\mathscr{D}_1 \cup \ldots \cup \mathscr{D}_n)^*]\phi}{[\mathscr{D}]\phi}\,{}^4$$

$$([DR])\ \frac{\vdash [\mathscr{E}]\phi}{\vdash [\mathscr{D}]\phi}\,{}^5 \qquad (\langle DR\rangle)\ \frac{\vdash \langle\mathscr{D}\rangle\phi}{\vdash \langle\mathscr{E}\rangle\phi}\,{}^5 \qquad (DS)\ \frac{\vdash [\mathscr{D}]\chi \quad \vdash [\mathscr{D} \wedge \chi]\phi}{\vdash [\mathscr{D}]\phi}$$

$$([]gen)\ \frac{\vdash \forall^\alpha(\phi \to \psi)}{[\alpha]\phi \vdash [\alpha]\psi} \qquad (\langle\rangle gen)\ \frac{\vdash \forall^\alpha(\phi \to \psi)}{\langle\alpha\rangle\phi \vdash \langle\alpha\rangle\psi}$$

$$(ind)\ \frac{\vdash \forall^\alpha(\phi \to [\alpha]\phi)}{\phi \vdash [\alpha^*]\phi} \qquad (con)\ \frac{\vdash \forall^\alpha \forall v{>}0\,(\varphi(v) \to \langle\alpha\rangle\varphi(v-1))}{\exists v\,\varphi(v) \vdash \langle\alpha^*\rangle \exists v{\leq}0\,\varphi(v)}\,{}^6$$

$$(DI)\ \frac{\vdash \forall^\alpha \forall y_1 .. \forall y_k\,(\chi \to F'{}_{x'_1}^{\theta_1}\cdots{}_{x'_n}^{\theta_n})}{[\exists y_1 .. \exists y_k\,\chi]F \vdash [\exists y_1 .. \exists y_k\,(x'_1 = \theta_1 \wedge .. \wedge x'_n = \theta_n \wedge \chi)]F}\,{}^7$$

$$(DV)\ \frac{\vdash \exists \varepsilon{>}0\, \forall^\alpha \forall y_1 .. y_k\,(\neg F \wedge \chi \to (F' \geq \varepsilon)_{x'_1}^{\theta_1}\cdots{}_{x'_n}^{\theta_n})}{[\exists y_1 .. y_k\,(x'_1 = \theta_1 \wedge .. \wedge x'_n = \theta_n \wedge {\sim}F)]\chi \vdash \langle\exists y_1 .. y_k\,(x'_1 = \theta_1 \wedge .. \wedge x'_n = \theta_n \wedge \chi)\rangle F}\,{}^8$$

[1] $\Gamma_i \vdash \Delta_i$ are obtained from the subgoals of side deduction ($\star$) in Fig. 3.10, in which x is assumed to occur in first-order formulas only, as QE is then applicable. The side deduction starts from goal $\Gamma \vdash \Delta, \phi$ at the bottom (or $\Gamma, \phi \vdash \Delta$ for l$\forall$ and l$\exists$), where x does not occur in Γ, Δ using renaming.

[2] $\mathscr{J}_1 \vee \cdots \vee \mathscr{J}_n$ is a disjunctive normal form of the DJ-constraint $\mathscr{J}$.

[3] Rule applicable for any reordering of the conjuncts of the DJ-constraint where χ is jump-free.

[4] $\mathscr{D}_1 \vee \cdots \vee \mathscr{D}_n$ is a disjunctive normal form of the DA-constraint $\mathscr{D}$.

[5] $\mathscr{D}$ implies $\mathscr{E}$, i.e., satisfies the assumptions of Lemma 3.3.

[6] Logical variable v does not occur in α.

[7] Applicable for any reordering of the conjuncts where χ is non-differential. F is first-order without negative equalities, and F' abbreviates $D(F)$, with z' replaced with 0 for unchanged variables.

[8] Like DI, but F contains no equalities and the differential equations are Lipschitz continuous.

Fig. 3.9 Proof calculus for differential-algebraic dynamic logic (DAL)

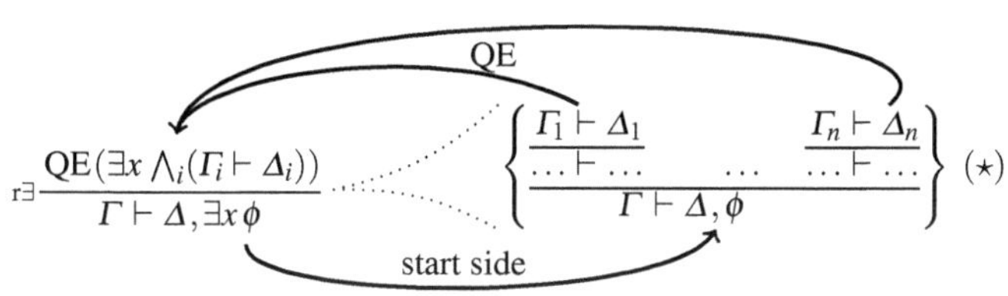

Fig. 3.10 Side deduction

$$([\cup]\Box)\ \frac{[\alpha]\pi \land [\beta]\pi}{[\alpha \cup \beta]\pi}\ ^{1} \qquad (\langle\cup\rangle\Diamond)\ \frac{\langle\alpha\rangle\pi \lor \langle\beta\rangle\pi}{\langle\alpha \cup \beta\rangle\pi}\ ^{1}$$

$$([;]\Box)\ \frac{[\alpha]\Box\phi \land [\alpha][\beta]\Box\phi}{[\alpha;\beta]\Box\phi} \qquad (\langle;\rangle\Diamond)\ \frac{\langle\alpha\rangle\Diamond\phi \lor \langle\alpha\rangle\langle\beta\rangle\Diamond\phi}{\langle\alpha;\beta\rangle\Diamond\phi}$$

$$([?]\Box)\ \frac{\phi}{[?\chi]\Box\phi} \qquad (\langle ? \rangle\Diamond)\ \frac{\phi}{\langle ?\chi\rangle\Diamond\phi}$$

$$([:=]\Box)\ \frac{\phi \land [x:=\theta]\phi}{[x:=\theta]\Box\phi} \qquad (\langle:=\rangle\Diamond)\ \frac{\phi \lor \langle x:=\theta\rangle\phi}{\langle x:=\theta\rangle\Diamond\phi}$$

$$([']\Box)\ \frac{[x'=\theta]\phi}{[x'=\theta]\Box\phi} \qquad (\langle'\rangle\Diamond)\ \frac{\langle x'=\theta\rangle\phi}{\langle x'=\theta\rangle\Diamond\phi}$$

$$([^{*n}]\Box)\ \frac{[\alpha;\alpha^*]\Box\phi}{[\alpha^*]\Box\phi} \qquad (\langle^{*n}\rangle\Diamond)\ \frac{\langle\alpha;\alpha^*\rangle\Diamond\phi}{\langle\alpha^*\rangle\Diamond\phi}$$

$$([^{*}]\Box)\ \frac{[\alpha^*][\alpha]\Box\phi}{[\alpha^*]\Box\phi} \qquad (\langle^{*}\rangle\Diamond)\ \frac{\langle\alpha^*\rangle\langle\alpha\rangle\Diamond\phi}{\langle\alpha^*\rangle\Diamond\phi}$$

[1] π is a trace formula and—unlike the state formulas ϕ and ψ—may thus begin with a temporal modality $\Box$ or $\Diamond$.

Fig. 4.3 Proof calculus for differential temporal dynamic logic (dTL)

Meat Products Preservation

Natural and Non-Conventional Methods

Edited by

Renata Ernlund Freitas de Macedo

Pontifical Catholic University of Parana, Curitiba, Paraná, Brazil

Carlos Adam Conte-Junior

Federal University of Rio de Janeiro, Rio de Janeiro, Rio de Janeiro, Brazil

Humana Press

Editors
Renata Ernlund Freitas de Macedo
Pontifical Catholic University of Parana
Curitiba, Paraná, Brazil

Carlos Adam Conte-Junior
Federal University of Rio de Janeiro
Rio de Janeiro, Rio de Janeiro, Brazil

ISSN 2662-950X ISSN 2662-9518 (electronic)
Methods and Protocols in Food Science
ISBN 978-1-0716-5067-7 ISBN 978-1-0716-5068-4 (eBook)
https://doi.org/10.1007/978-1-0716-5068-4

This Humana imprint is published by the registered company Springer Science+Business Media, LLC, part of Springer Nature.
The registered company address is: 1 New York Plaza, New York, NY 10004, U.S.A.

Preface to the Series

Methods and Protocols in Food Science series is devoted to the publication of research protocols and methodologies in all fields of food science. The series is unique as it includes protocols developed, validated, and used by food and related scientists, and it provides the theoretical basis for each protocol. Aspects related to improvements, adaptations, and further developments of the protocols may also be approached.

The *Methods and Protocols in Food Science* series aims to bring the most recent developments in research protocols in the field and well-established methods. As such the series targets undergraduate and graduate students, as well as researchers in the field of food science and correlated areas. The protocols documented in the series will be highly useful for scientific inquiries in the field of food sciences, and are presented in such a way that the readers will be able to reproduce the experiments in a step-by-step style.

Each protocol will be characterized by a brief introductory section, followed by a short aims section, in which the precise purpose of the protocol is clarified. Then, an in-depth list of materials and reagents required for employing the protocol is presented, followed by a comprehensive and step-by-step procedures on how to perform that experiment. The next section presents the do's and don'ts for carrying out the protocol, followed by the main pitfalls faced and how to troubleshoot them. Finally, template results will be presented and their meaning/conclusions addressed.

The *Methods and Protocols in Food Science* series will fill an important gap by addressing a common complaint about food scientists regarding the difficulties in repeating experiments detailed in scientific papers. With this, the series has a potential to become a reference material in food science laboratories of research centers and universities throughout the world.

Campinas, Brazil ***Anderson S. Sant'Ana***

Preface

Meat and meat products, essential in the global diet as sources of high-biological-value nutrients, face an inherent challenge: high perishability. From slaughter through processing, distribution, and storage, meat undergoes chemical and microbiological deterioration. Oxidation and microbial growth generate undesirable flavors, odors, and compounds, compromising quality and safety while also causing major economic losses.

Given the growing demand for foods combining nutritional value, freshness, and safety, while limiting chemical additives, industry and research are focused on innovative preservation strategies. Traditional technologies, often based on thermal and chemical treatments, may alter nutritional and sensory attributes and raise concerns about toxicity and residues. In this context, natural and unconventional methods, particularly non-thermal technologies, emerge as promising alternatives to extend shelf life without compromising intrinsic product characteristics.

This book provides a comprehensive overview of these methodologies, presenting their applications, mechanisms, and efficacy in preserving meat and fish products. Its 16 chapters address key aspects in advancing preservation technologies.

The volume highlights innovative non-thermal technologies. Chapter 2 details high hydrostatic pressure (HPP) for decontaminating ready-to-eat products like sliced ham and sausages. Chapter 3 discusses ultrasound in fresh beef, optimizing quality attributes. Chapter 4 presents UVC-LED, a mercury-free technology, for decontamination and shelf-life extension of chicken and fish products. Application of Pulsed Light (PL) to reduce microbial populations on meat surfaces is featured in Chapter 5. Chapter 6 explores cold plasma (CP) to inactivate microorganisms and control lipid oxidation in beef, pork, and chicken. Chapter 7 covers microwave applications in fish, addressing drying, heating, and sterilization. Chapter 8 describes supercritical carbon dioxide (sc-CO_2) to remove water and reduce microbial loads in meat and fish.

This book also examines natural compounds and delivery systems. Chapter 1 describes liposomal encapsulation of bioactive compounds using microfluidics to improve health benefits and preservation in ham and sausages. Chapter 9 addresses nanoemulsions of bioactives in sausages and patties. Chapter 12 explores edible coatings with antimicrobial agents to extend shelf life. Chapter 14 considers polyphenols as natural antioxidants to improve oxidative stability in fresh meat. Chapter 15 provides a guide to applying organic acids (lactic, acetic, citric) for carcass and raw meat decontamination.

Biopreservation methods are also explored. Chapter 10 addresses immobilized lactic acid bacteria for dry fermented products, highlighting protective cultures. Chapter 11 discusses antimicrobial metabolites in cooked sausages. Chapter 13 validates bacteriophage use as processing aids against *Escherichia coli* and *Salmonella* in fresh meat.

This book concludes with Chap. 16, exploring combined methods and active packaging.

This volume is intended for researchers, food scientists, industry professionals, and academics, serving as a practical and reliable reference for applying natural and unconventional methods in meat and fish preservation. Its goal is not only to ensure quality, safety, and extended shelf life, but also to align with and advance the continuous evolution of the contemporary meat industry.

The editors express their sincere gratitude to all contributors who generously shared their expertise and knowledge in the methods presented. We hope this volume will inspire new studies and research, contributing to the advancement of preservation technologies and supporting safer, healthier, and more sustainable solutions for the global market.

Curitiba, Paraná, Brazil *Dr. Renata Ernlund Freitas de Macedo*
Rio de Janeiro, Rio de Janeiro, Brazil *Dr. Carlos Adam Conte-Junior*

Contents

Contributors

SARRA AKERMI • *Laboratory of Microbial and Enzymes Biotechnology and Biomolecules (LMEBB), Centre of Biotechnology of Sfax (CBS), University of Sfax, Sfax, Tunisia*

ÉDIPO DA SILVA ALMEIDA • *Center for Food Analysis (NAL), Technological Development Support Laboratory (LADETEC), Federal University of Rio de Janeiro (UFRJ), Cidade Universitária, Rio de Janeiro, RJ, Brazil; Graduate Program in Food Science (PPGCAL), Institute of Chemistry (IQ), Federal University of Rio de Janeiro (UFRJ), Cidade Universitária, Rio de Janeiro, RJ, Brazil*

MARCO ANTONIO TRINDADE • *Department of Food Engineering, Universidade de Sao Paulo Faculdade de Zootecnia e Engenharia de Alimentos, Pirassununga, Brazil*

TERESA AYMERICH • *Food Safety and Functionality, IRTA (Institute of AgriFood Research and Technology), Monells, Girona, Spain*

P. BARCIELA • *Instituto de Agroecoloxía e Alimentación (IAA), Universidade de Vigo, Nutrition and Food Group (NuFoG), Ourense, Spain; Nutrition and Food Group (NuFoG), Galicia Sur Health Research Institute (IIS Galicia Sur), Vigo, Spain*

HELOÍSA VALARINE BATTAGIN • *Department of Food Engineering, Universidade de Sao Paulo Faculdade de Zootecnia e Engenharia de Alimentos, Pirassununga, Brazil*

BOUTHAINA BEN HADJ HMIDA • *Laboratory of Biochemistry and Enzymatic Engineering of Lipases, National Engineering School of Sfax, University of Sfax, Sfax, Tunisia*

YAGO A. A. BERNARDO • *Analytical and Molecular Laboratorial Center (CLAn), Institute of Chemistry (IQ), Federal University of Rio de Janeiro (UFRJ), Cidade Universitária, Rio de Janeiro, RJ, Brazil; Center for Food Analysis (NAL), Technological Development Support Laboratory (LADETEC), Federal University of Rio de Janeiro (UFRJ), Cidade Universitária, Rio de Janeiro, RJ, Brazil; Department of Food Technology, Federal Rural University of Rio de Janeiro, Seropédica, RJ, Brazil; Graduate Program in Food Science (PPGCAL), Institute of Chemistry (IQ), Federal University of Rio de Janeiro (UFRJ), Cidade Universitária, Rio de Janeiro, RJ, Brazil; Graduate Program in Food Science and Technology, Federal Rural University of Rio de Janeiro, Seropédica, RJ, Brazil; Laboratory of Advanced Analysis in Biochemistry and Molecular Biology (LAABBM), Department of Biochemistry, Federal University of Rio de Janeiro (UFRJ), Cidade Universitária, Rio de Janeiro, RJ, Brazil*

FERNANDA CRISTINA KANDALSKI BORTOLOTTO • *Laboratory of Agrifood Research and Innovation, Graduate Program in Animal Science, School of Medicine and Life Sciences, Pontifícia Universidade Católica do Paraná, Curitiba, Brazil*

LOURDES MARIA CORRÊA CABRAL • *Embrapa Food Technology, Rio de Janeiro, Brazil*

MOUFIDA CHAARI • *Laboratory of Microbial and Enzymes Biotechnology and Biomolecules (LMEBB), Centre of Biotechnology of Sfax (CBS), University of Sfax, Sfax, Tunisia*

CARLOS ADAM CONTE-JUNIOR • *Analytical and Molecular Laboratorial Center (CLAn), Institute of Chemistry (IQ), Federal University of Rio de Janeiro (UFRJ), Cidade Universitária, Rio de Janeiro, RJ, Brazil; Center for Food Analysis (NAL), Technological Development Support Laboratory (LADETEC), Federal University of Rio de Janeiro (UFRJ), Cidade Universitária, Rio de Janeiro, RJ, Brazil; Graduate*

Program in Biochemistry (PPGBq), Institute of Chemistry (IQ), Federal University of Rio de Janeiro (UFRJ), Cidade Universitária, Rio de Janeiro, RJ, Brazil; Graduate Program in Chemistry (PGQu), Institute of Chemistry (IQ), Federal University of Rio de Janeiro (UFRJ), Cidade Universitária, Rio de Janeiro, RJ, Brazil; Graduate Program in Food Science (PPGCAL), Institute of Chemistry (IQ), Federal University of Rio de Janeiro (UFRJ), Cidade Universitária, Rio de Janeiro, RJ, Brazil; Graduate Program in Veterinary Hygiene (PPGHV), Faculty of Veterinary Medicine, Fluminense Federal University (UFF), Vital Brazil Filho, Niterói, RJ, Brazil; Laboratory of Advanced Analysis in Biochemistry and Molecular Biology (LAABBM), Department of Biochemistry, Federal University of Rio de Janeiro (UFRJ), Cidade Universitária, Rio de Janeiro, RJ, Brazil

LUIZ CARLOS CORRÊA-FILHO • *Graduate Program in Food Science and Technology, Federal Rural University of Rio de Janeiro - UFRRJ, Seropédica, Brazil*

ANNADA DAS • *Department of Livestock Products Technology, IVSAH, Siksha 'O' Anusandhan University, Bhubaneswar, India; Department of Livestock Products Technology, West Bengal University of Animal and Fishery Sciences, Kolkata, India*

ARUN K. DAS • *Eastern Regional Station, ICAR-Indian Veterinary Research Institute, Kolkata, India*

SEVERINO MATHIAS DE ALENCAR • *Department of Agri-Food Industry, Food and Nutrition, Universidade de Sao Paulo Escola Superior de Agricultura Luiz de Queiroz, Piracicaba, Brazil*

TAMIRES SOUSA DE OLIVEIRA • *Graduate Program in Food Science, Institute of Chemistry, Federal University of Rio de Janeiro – UFRJ, Rio de Janeiro, Brazil*

GONZALO DELGADO-PANDO • *Instituto de Ciencia y Tecnología de Alimentos y Nutrición (ICTAN), Consejo Superior de Investigaciones Científicas (CSIC), Madrid, Spain*

KHAOULA ELAHDEF • *Laboratory of Microbial and Enzymes Biotechnology and Biomolecules (LMEBB), Centre of Biotechnology of Sfax (CBS), University of Sfax, Sfax, Tunisia*

WALID ELFALLEH • *Department of Biology, College of Science, Imam Mohammad Ibn Saud Islamic University (IMSIU), Riyadh, Saudi Arabia*

ALBERTO GONÇALVES EVANGELISTA • *Microbiology Research Laboratory, Biopark Educação, Av. Max Planck - Biopark, Toledo, Brazil*

LUCIMARA APARECIDA FORATO • *Embrapa Instrumentação, São Carlos, SP, Brazil*

VANESSA CRISTINA FRANCISCO • *Embrapa Pecuaria Sudeste, São Carlos, SP, Brazil*

MARIA JOÃO FRAQUEZA • *Associate Laboratory for Animal and Veterinary Sciences (AL4AnimalS), Lisboa, Portugal; CIISA - Centro de Investigação Interdisciplinar em Sanidade Animal, Faculdade de Medicina Veterinária, Universidade de Lisboa, Lisboa, Portugal*

CARMEN SÍLVIA FÁVARO-TRINDADE • *Department of Food Engineering, Universidade de Sao Paulo Faculdade de Zootecnia e Engenharia de Alimentos, Pirassununga, Brazil*

MOHAMMAD GAGAOUA • *PEGASE, INRAE, Institut Agro, Saint-Gilles, France*

LUCIANA RUGGERI MENEZES GOTARDO • *Department of Food Engineering, Universidade de Sao Paulo Faculdade de Zootecnia e Engenharia de Alimentos, Pirassununga, Brazil*

ELAHESADAT HOSSEINI • *Department of Chemical Engineering, Payame Noor University, Tehran, Iran; National Nutrition and Food Technology Research Institute, Faculty of Nutrition Sciences and Food Technology, Shahid Beheshti University of Medical Sciences, Tehran, Iran*

A.O.S. Jorge • *Instituto de Agroecoloxía e Alimentación (IAA), Universidade de Vigo, Nutrition and Food Group (NuFoG), Ourense, Spain; REQUIMTE/LAQV, Department of Chemical Sciences, Faculty of Pharmacy, University of Porto, Porto, Portugal*
Marilia Silva Malvezzi Karwowski • *Laboratory of Agrifood Research and Innovation, Graduate Program in Animal Science, School of Medicine and Life Sciences, Pontifical Catholic University of Parana, Curitiba, Brazil*
Loíse C. S. Lima • *Department of Animal and Food Sciences, University of Kentucky, Lexington, KY, USA; Graduate Program in Food Science, Chemistry Institute, Federal University of Rio de Janeiro, Rio de Janeiro, RJ, Brazil*
Renata Ernlund Freitas de Macedo • *Laboratory of Agrifood Research and Innovation, Graduate Program in Animal Science, School of Medicine and Life Sciences, Pontifical Catholic University of Parana, Curitiba, Brazil*
Amilton de Mello • *Department of Agriculture, Veterinary and Rangeland Sciences, University of Nevada, Reno, NV, USA*
Pramod Kumar Nanda • *Eastern Regional Station, ICAR-Indian Veterinary Research Institute, Kolkata, India*
Renata Tieko Nassu • *Embrapa Pecuaria Sudeste, São Carlos, SP, Brazil*
Rafael Nogueira-Marques • *Instituto de Agroecoloxía e Alimentación (IAA), Universidade de Vigo, Nutrition and Food Group (NuFoG), Ourense, Spain; Nutrition and Food Group (NuFoG), Galicia Sur Health Research Institute (IIS Galicia Sur), Vigo, Spain*
A.G. Pereira • *Instituto de Agroecoloxía e Alimentación (IAA), Universidade de Vigo, Nutrition and Food Group (NuFoG), Ourense, Spain; Nutrition and Food Group (NuFoG), Galicia Sur Health Research Institute (IIS Galicia Sur), Vigo, Spain*
A. Perez-Vazquez • *Instituto de Agroecoloxía e Alimentación (IAA), Universidade de Vigo, Nutrition and Food Group (NuFoG), Ourense, Spain; Nutrition and Food Group (NuFoG), Galicia Sur Health Research Institute (IIS Galicia Sur), Vigo, Spain*
M.A. Prieto • *Instituto de Agroecoloxía e Alimentación (IAA), Universidade de Vigo, Nutrition and Food Group (NuFoG), Ourense, Spain; Nutrition and Food Group (NuFoG), Galicia Sur Health Research Institute (IIS Galicia Sur), Vigo, Spain*
Dannaya Julliethy Gomes Quirino • *Department of Food Engineering, Universidade de Sao Paulo Faculdade de Zootecnia e Engenharia de Alimentos, Pirassununga, Brazil*
Ana Paula A. A. Salim • *Department of Animal and Food Sciences, University of Kentucky, Lexington, KY, USA*
Fabio Santi • *Department of Industrial Engineering, University of Padova, Padova, Italy*
Kaushik Satyaprakash • *Department of Veterinary Public Health and Epidemiology, Faculty of Veterinary and Animal Sciences, Banaras Hindu University, Mirzapur, India*
Sepidar Seyyedi-Mansour • *Instituto de Agroecoloxía e Alimentación (IAA), Universidade de Vigo, Nutrition and Food Group (NuFoG), Ourense, Spain; Nutrition and Food Group (NuFoG), Galicia Sur Health Research Institute (IIS Galicia Sur), Vigo, Spain*
Slim Smaoui • *Laboratory of Microbial and Enzymes Biotechnology and Biomolecules (LMEBB), Centre of Biotechnology of Sfax (CBS), University of Sfax, Sfax, Tunisia*
Cecilia de Souza Cordeiro • *Universidade Estadual Paulista (Unesp), Faculdade de Ciências Farmacêuticas, Araraquara, SP, Brazil*
Sara Spilimbergo • *Department of Industrial Engineering, University of Padova, Padova, Italy*

Alexandros Ch. Stratakos • *Centre for Sustainable Agri-Food and Environment(SAFE), School of Applied Sciences, College of Health, Science and Society, University of the West of England, Bristol, UK*

Surendranath P. Suman • *Department of Animal and Food Sciences, University of Kentucky, Lexington, KY, USA*

Renata Valeriano Tonon • *Embrapa Food Technology, Rio de Janeiro, Brazil*

Alessandro Zambon • *Department of Civil, Chemical, Environmental, and Materials Engineering, University of Bologna, Bologna, Italy*

Riccardo Zulli • *Department of Industrial Engineering, University of Padova, Padova, Italy*

Check for updates

Chapter 1

Liposomal Encapsulation of Bioactive Compounds Using Microfluidics for Application in Ham and Sausage

Gonzalo Delgado-Pando and Alexandros Ch. Stratakos

Abstract

Liposomal formulations that encapsulate bioactive compounds can be used in meat products, such as sausages and ham, to enhance their health benefits and preservation. This chapter describes a method for producing liposomes containing a hydrophobic bioactive compound using a simple, continuous flow process. The process employs a push-pull, pressure-controlled microfluidic system. The liposomes are prepared using 1,2-dimyristoyl-sn-glycero-3-phosphocholine (DSPC) lipid and an optimal lipid-to-cholesterol ratio of 2:1. To facilitate homogeneous distribution into the meat product, the liposomes can be previously mixed into the brine. In addition, they can also be added to active packaging.

Keywords Lipids, Cholesterol, Liposomes, Encapsulation, Bioactive, Meat products

1 Introduction

Bioactive compounds are substances that have biological activity and affect directly a living organism [1]. In the food industry, bioactive compounds are used to enhance food quality thanks to their antimicrobial and antioxidant effects. By inhibiting oxidative processes and suppressing pathogen and spoilage microorganism growth, bioactive compounds help maintain the quality, safety, and sensory properties, thereby reducing spoilage and waste [2]. In the case of meat and meat products, the auto-oxidation and microbial deterioration lead to the formation of undesirable flavors, odors, and toxic compounds, posing a risk to human health [3]. To counter this issue and as an alternative to synthetic preservatives, bioactive compounds, through their clean label and GRAS (Generally Recognized as Safe) status, have been incorporated as ingredients in the meat industry [4]. Bioactive compounds from plant sources are often preferred to be used in meat products due to their antimicrobial properties against pathogens, as well as their ability to extend shelf-life through their antioxidant properties [5]. Despite their potential, the incorporation

Renata Ernlund Freitas de Macedo and Carlos Adam Conte-Junior (eds.), *Meat Products Preservation*, Methods and Protocols in Food Science, https://doi.org/10.1007/978-1-0716-5068-4_1,

of bioactive compounds into meat products has certain limitations, such as poor solubility, low chemical stability, or adverse impacts on food quality, appearance, texture, or flavor [6]. To overcome these challenges, encapsulation technology has emerged as a viable solution. Encapsulation involves encasing bioactive compounds within a protective matrix or coating, which shields them from adverse environmental conditions and enhances their stability. This process not only protects the bioactive compounds from degradation but also allows for their controlled release during storage and consumption. The encapsulated material is known as the active agent, core material, fill, payload phase, or internal phase. The substance that encapsulates is known as the membrane, coating, shell, wall material, carrier material, matrix, or the exterior phase [5]. Some criteria should be considered before selecting the encapsulation method, such as the characteristics and types of the active substance to encapsulate, the carrier material, and its suggested application [7]. The nature of the core material (bioactive compound), the wall material, the type of meat product that the bioactive component is going to be incorporated into and the environmental conditions (pH, temperature, ionic strength, temperature, light, oxygen, ingredient interactions mechanical forces) are the key parameters to take into account [5, 6]. In this chapter, we will use the technique of liposomal encapsulation of a hydrophobic substance by using microfluidics for its application in cooked ham and sausage.

2 Materials

2.1 Liposome Encapsulation and Characterization

1. Bioactive compound (in this case of hydrophobic compound) at 0.4% concentration solution: add 40 μL of the bioactive compound to 9.96 mL of ethanol (*see* **Note 1**).
2. Aqueous phase: phosphate-buffered saline (PBS): dissolve one PBS tablet in 200 mL of deionized water.
3. Lipid phase: add 0.05 g of the phospholipid you plan to use (*see* **Note 2**) and 0.025 g of cholesterol (surfactant) in 10 mL of 0.4% of step 1 solution. Vortex the tube to mix the content.
4. 50 mL Falcon tubes.
5. 10 mL Falcon tubes.
6. UV spectrophotometer.
7. 4.5 × 4.5 mm disposable cuvette.
8. Quartz cuvette.
9. 0.2 μm filter.
10. Disposable syringes.
11. Disposable bottles (10 mL).

12. Disposable Pasteur pipettes.
13. Glass microfluidic chips (e.g., zigzag and Y microchannel pattern).
14. Folded capillary zeta cuvettes.

2.2 Cooked Ham Manufacture

1. Lean pork meat
2. Brine ingredients: water, salt, sodium nitrite (150 ppm), dextrose (0.5%), phosphates (0.3%), and sodium ascorbate (0.05%) (*see* **Note 3**). The brine should be prepared according to the following formula for each of the ingredients:

$$\%\ \text{Ingredient}_{(\text{in brine})} = \frac{\text{wanted}\ \%\ \text{ingredient}_{(\text{in sample})} \times \left(100 + \%\ \text{added brine}\right)}{\%\ \text{added brine}}$$

(The concentration of added brine can range from 20 to 80%)

3. Mixer
4. Refrigerator
5. Knife and cutting table
6. pH meter
7. Mincer with kidney blade or 10–30 mm ones
8. Tumbler
9. Vacuum packer
10. Ham molds
11. Cooking oven
12. Slicer

2.3 Sausage Manufacture

1. Lean pork and beef meat
2. Pork fat
3. Ice and water
4. Additives: salt, phosphates (0.25%), nitrite (150 ppm), sodium ascorbate (0.05%), spices mix (0.7%)
5. Mincer (blades 3–5 mm)
6. Bowl cutter
7. Filler
8. Smoker oven

2.4 Active Packaging

1. Chitosan powder
2. Glacial acetic acid
3. Tween 80
4. Glycerol

5. Homogenizer
6. Glass Petri dishes
7. Environmental chamber

3 Methods

Four distinct procedures must be followed. First, the encapsulation of the bioactive compound, followed by the assessment of encapsulation efficiency and the characterization of the loaded liposomes. Next, the bioactive compound can be incorporated either as an ingredient or as part of bioactive packaging. While the following methodology uses carvacrol as the bioactive compound, it can be applied to others as well. Frankfurters have been chosen as the representative sausage example.

3.1 Preparation of Liposomes Using Microfluidics

1. Switch on the microfluidics device/pressure pump.
2. Connect Falcon tubes containing PBS (aqueous phase) and ethanol (lipid phase) to the pressure pump.
3. Flush the capillary tubes and microfluidic chip channels with PBS and Ethanol to clear the capillary tubes of any impurities or previous residual solutions (*see* **Note 4**).
4. PBS and the lipid solvent containing the bioactive compounds travel through the capillary tubes and are mixed in the microchannel of the microfluidic chip. This causes the liposomes to self-assemble.
5. Maintain the total flow rate (TFR) at 500 μL/min.
6. Use a 3:1 flow rate ratio (FRR) to prepare the liposomes. The flow rate ratio is the volume flow ratio of aqueous and organic phases introduced in the microfluidic device (*see* **Note 5**).
7. While maintaining the FRR, ensure the pressure does not exceed the maximum limit of your microfluidic device (e.g., 2000 mbars).
8. Once the pressure is stable start collecting the loaded liposomes in a 10 mL Falcon tube.
9. The liposomes can be freeze-dried for posterior use.

3.2 Encapsulation Efficiency and Characterization of Loaded Liposomes

1. The liposome solution collected in 3.1.8 is centrifuged at 11,000 × *g* for 20 min.
2. The supernatant is used for determining the encapsulation efficiency (EE) utilizing the following formula:

$$\% \text{ EE} = \frac{\text{initial concentration} - \text{supernatant concentration}}{\text{initial concentration}} \times 100$$

3. The concentration of the bioactive compound is determined using a calibration curve and the determination methodology specific for the selected compound.
4. Liposome size, polydispersity index (PDI), and Zeta potential are liposome characteristics that will be determined using a Dynamic Light Scattering (DLS) analytical instrument (*see* **Note 6**).
5. The pellet obtained from the liposome solution in 3.2.1 is resuspended in 1 mL PBS and centrifuged at 11,000×*g* for 20 min.
6. The pellet obtained is resuspended in PBS and the centrifugation step is repeated. This is done to remove any free compound from the liposomal pellet.
7. 30 μL of the above is diluted with 2970 μL of PBS (the total volume is 3 mL).
8. The liposomes are filtered through a 0.2 μm filter (*see* **Note 7**).
9. Add approximately 1 mL of filtered liposome sample in a 4.5×4.5 mm disposable cuvette. Place the cuvette in the slot of the DLS instrument and close the lid. Create a folder with an appropriate label and select the protocol for determining liposome size. The size and PDI of liposomes are measured simultaneously.
10. Place the capillary cuvette in the appropriate slot in DLS equipment (*see* **Note 8**). Select the proper protocol/program in your device to measure zeta potential.

3.3 Direct Application in Brine-Added Ham

1. Prepare the meat for ham production: cut pork lean meat into small pieces with a knife and mince them in a mincer (*see* **Note 10**).
2. Check the pH of the meat (*see* **Note 11**).
3. Calculate the percentages of all the ingredients in the brine, and then calculate the grams of each depending on the quantity added.
4. Add the liposomes (3.1.8) to the brine in the appropriate amount, and mix thoroughly (*see* **Note 12**).
5. Add the meat to a tumbler and then add the brine.
6. Tumble them for 60 min at 6 rpm on intervals of 15 min work/rest periods under chilling conditions (2–4 °C).

7. Vacuum pack the tumbled meat and store it overnight at 2 ± 1 °C.
8. Put the meat in molds and cook in an oven under steam conditions (100% relative humidity) until the core temperature reaches 72 °C.
9. Apply a chill shower for 30 min and store them for 24 h at 2 ± 1 °C.
10. Slice and package accordingly.

3.4 Direct Application in Cooked Sausage

1. Prepare the meat and fat: mince pork and beef meat in a 3–5 mm blade, vacuum pack, and store at freezing temperatures. Do the same for pork fat.
2. Add the semi-frozen minced meat to the bowl cutter and start cutting at 1500 rpm.
3. Add phosphates and keep cutting at the same speed until tacky consistency is obtained (*see* **Note 9**).
4. Add around 70% of the water as a mix of ice and water followed by the salt.
5. Increase the bowl cutter speed to maximum (at least 4000 rpm) controlling the temperature.
6. When reaching 4 °C add the liposomes (3.1.8) and mix for 1 extra minute (*see* **Note 12**).
7. Add the remaining ice followed by the minced fat and the remaining additives. Keep mixing and controlling the temperature (*see* **Note 9**).
8. Use a filler to fill the batter into casings (*see* **Note 13**).
9. Use the smoker oven at 55°C, relative humidity (RH) 85% for 12 min, then reduce the RH (~40%) and dry at 62°C for 18 min.
10. Smoke the samples at 65°C for 12 min RH 40–60% and dry them for 2–3 min at 60–65°C to fix the smoke.
11. Cook the sausages in steam until 72°C is reached in the core (*see* **Note 14**).
12. Apply chill shower for 20 min and store them for 24 h at 2 ± 1°C.

3.5 Application in Active Edible Packaging Material

1. Chitosan stock solution (2% w/v): 2 g chitosan powder in 100 mL of 1% (w/v) glacial acetic acid. Stir for 20 h at 20–25°C.
2. Add 0.25% (w/v) of Tween 80 and 0.5 mL of glycerol/g chitosan to the solution and stir for 30 min.
3. Add liposomes (3.1.8) at the required quantity and homogenize at 12,000 rpm for 2 min.
4. The bioactive edible packaging is ready.

5. You can coat the piece of ham directly into the solution for 5 min and then dry it in aseptic conditions.
6. You can also create individual chitosan films for the sausages and the ham by casting the solution (3.4.4) into glass Petri dishes (*see* **Note 15**). Then leave them dry in an environmental chamber at controlled conditions: 20–25 °C and 20–30% relative humidity for 48 h.

4 Notes

1. In the case of hydrophilic antimicrobial compounds, the aqueous phase such as phosphate-buffered saline should be used to dissolve the compound.
2. Usually, 1,2-distearoyl-sn-glycero-3-phosphocholine (DSPC) or 1,2-Dimyristoyl-sn-glycero-3-phosphocholine (DMPC). The literature will provide guidance based on the specific application.
3. It is preferable to add the ascorbate last to the mix.
4. If the microfluidic chip is blocked, passing ethanol through may help unblock the chip.
5. The ratio of lipid to aqueous phase is crucial for forming stable liposomes. Experiment with different ratios to find the optimal balance for your specific lipid composition and target liposome size.
6. PDI is a metric that represents the size variation of liposomes in a sample, ranging from 0 to 1. Lower PDI values indicate that the liposomal sample is more monodisperse, meaning the liposomes are more uniform in size. Zeta potential value predicts the stability of the liposomes, which means if the liposomes have like charges, they will tend to repel each other and thus prevent aggregation. Typically, liposomes are considered stable when their zeta potential values are greater than +30 mV or less than −30 mV.
7. Use a 2.5 or 1 mL syringe to filter them.
8. Folded capillary cuvettes are used for measuring Zeta potential. The minimum sample volume required is 0.75 mL. The cuvette exhibits a U shape, therefore make sure no bubbles are present when the liposome sample is introduced. Use a 1 mL syringe to add the liposome sample.
9. You can dissolve the phosphates in water beforehand to ensure even distribution throughout the product. Be mindful of the temperature during the cutting process, making sure it does not exceed 12°C.

10. Use a kidney blade or blade with a diameter between 10 and 20 mm.
11. pH between 5.7 and 6.1 is optimal.
12. This is a crucial step, as the liposomes must be well distributed in the brine, so the bioactive compound is evenly absorbed in the meat while tumbling.
13. Usually, sheep natural casings of 18–22 mm are used.
14. A hot water bath at 75–80°C can be also used.
15. The diameter of your Petri dishes should match the diameter of your ham slices or be large enough to cover your sausages. You should also calculate the exact amount of the 3.4.4 solution needed for each dish to ensure a uniform film across all Petri dishes.

References

1. Smith A, Datta SP, Smith GH et al (2000) Oxford dictionary of biochemistry and molecular biology, Revised edition. Oxford University Press, Oxford, New York
2. Smaoui S, Ben Hlima H, Ben Braïek O et al (2021) Recent advancements in encapsulation of bioactive compounds as a promising technique for meat preservation. Meat Sci 181:108585. https://doi.org/10.1016/j.meatsci.2021.108585
3. Olvera-Aguirre G, Piñeiro-Vázquez ÁT, Sanginés-García JR et al (2023) Using plant-based compounds as preservatives for meat products: a review. Heliyon 9. https://doi.org/10.1016/j.heliyon.2023.e17071
4. Delgado-Pando G, Ekonomou SI, Stratakos AC, Pintado T (2021) Clean label alternatives in meat products. Foods 10:1615. https://doi.org/10.3390/foods10071615
5. Zhang D, Ivane NMA, Haruna SA et al (2022) Recent trends in the micro-encapsulation of plant-derived compounds and their specific application in meat as antioxidants and antimicrobials. Meat Sci 191:108842. https://doi.org/10.1016/j.meatsci.2022.108842
6. Domínguez R, Pateiro M, Munekata PES et al (2021) Encapsulation of bioactive phytochemicals in plant-based matrices and application as additives in meat and meat products. Molecules 26:3984. https://doi.org/10.3390/molecules26133984
7. Dias DR, Botrel DA, Fernandes RVDB, Borges SV (2017) Encapsulation as a tool for bioprocessing of functional foods. Curr Opin Food Sci 13:31–37. https://doi.org/10.1016/j.cofs.2017.02.001

Chapter 2

2-High-Pressure Application in Ready-to-Eat Meat Products (Sliced Ham, Sliced Dry-Fermented Sausages)

Maria João Fraqueza and Teresa Aymerich

Abstract

High hydrostatic pressure (HPP) technology is commercially utilized as a non-thermal decontamination method for ready-to-eat (RTE) meat products. Its well-documented lethal effect on pathogenic and spoilage bacteria improves the microbial safety of meat products while extending their shelf life. HPP is a chemical residue-free processing method that applies high pressure to products using an aseptic pressure system ranging from 100 to 600 MPa, with water or other fluids serving as the pressure-transmitting medium. Beyond being considered a non/thermal pasteurization process, HPP also enhances water binding, reduces the need for additives, and helps to achieve the desired texture, particularly in products with previously denatured proteins, making it a valuable technology for extending shelf life and ensuring both safety and quality in meat processing.

This chapter will explain the principles of high hydrostatic pressure (HHP) and its main effects on ready-to-eat meat products, highlighting both the advantages and disadvantages of its applications. In addition, the HHP application process for meat products will be described, focusing on optimizing the technology for different types of ready-to-eat meat products.

Keywords High hydrostatic pressure, Sliced meat products, Shelf life, Safety, Dry fermented sausages, Ready-to-eat meat products

1 Introduction

High-pressure processing (HPP) is a commercially applied, non-thermal decontamination method used to assure the safety and quality of ready-to-eat (RTE) meat products. This technology inactivates enzymes and damages certain microorganisms, particularly the vegetative forms of Gram-negative bacteria. Its proven efficacy against pathogenic and spoilage microorganisms significantly improves the microbial safety of meat products while extending their shelf life. It is important to highlight that HPP achieves this without causing notable changes to the aroma, flavour, or nutritional profile of RTE-products with already denatured protein, such as fermented meat products, cooked meat products [1, 2].

Renata Ernlund Freitas de Macedo and Carlos Adam Conte-Junior (eds.), *Meat Products Preservation*, Methods and Protocols in Food Science, https://doi.org/10.1007/978-1-0716-5068-4_2,

HPP is a chemical-residue-free process that applies pressures ranging from 100 to 600 MPa using water or another fluid as the pressure-transmitting medium. The process is typically carried out in a steel chamber where pressure is applied uniformly in all directions. This isostatic compression preserves the original shape of solid foods and operates independently of product size and geometry. Products are sealed in flexible packaging to prevent direct contact with the fluid during processing. For meat products, a pasteurizing effect is usually achieved at pressures between 400 and 600 MPa for 3–7 min [3–9], although it depends on the product and microorganisms.

Beyond microbial safety, HPP also improves product quality by enhancing water-binding capacity, reducing the need for additives, and contributing to desirable textural properties. These features make it particularly valuable in the processing of cured meat products cooked and dried [10].

However, the effectiveness of HPP can be influenced by several factors, including the pressure-time combination, temperature, characteristics of the food matrix (e.g., pH, water activity, salt content, additives), as well as the type, resistance, and growth phase of the target microorganisms, along with the preservation conditions [11–13].

High Pressure Processing (HPP) is currently implemented in various countries for an extensive range of ready-to-eat (RTE) meat products, and it is accepted by the Food Safety and Inspection Service (FSIS) as a non-thermal pasteurization technique [14]. These include thermally processed (e.g., ham, sausages, turkey, chicken), dry-cured (e.g., ham, loin), fermented (e.g., salami, sausages), marinated (e.g., beef, pork), and raw (e.g., carpaccio) meat matrices [15]. In fermented meat production, HPP can be applied at various stages: to raw materials and ingredients, before fermentation, or to the final product. Nevertheless, when used in meat batter, texture and other properties may change if not well-controlled conditions are applied [16]. Several studies have explored its application in meat [17, 18], ingredients [19], and various finished products [4, 6, 20–23]. On fresh meat, HPP >200 MPa may provoke myoglobin denaturation, inducing color changes, and making the meat appear cooked or pale, although flavor typically remains unaffected [24–26]. Nevertheless, lower HHP treatments may be used as processing aids for salt diffusion [27] or to increase tenderness [28, 29]. Besides, HPP may positively decrease aminogenesis of fermented sausages when HPP (200 MPa) where applied to meat batter [30, 31].

HPP can eliminate pathogenic microorganisms of major concern to the food industry, such as *Salmonella* spp., *E. coli*, and *Campylobacter jejuni*, which have low infectious doses [32–37]. At the same time, it can preserve the fermentative microbiota essential to the quality of fermented meat products, thereby maintaining the natural microbial competition that contributes to product stability and safety. *Listeria monocytogenes* is one of the major potential hazards associated with delicatessen meat products. Numerous studies have investigated the inactivation of *L. monocytogenes* in meat products using High Hydrostatic Pressure (HHP), concluding that its efficacy depends on the specific processing parameters (e.g., pressure level and holding time) and the physicochemical characteristics of the food matrix (such as pH, fat content, water activity, presence of spices and additives) [2, 4, 36, 38–40]. Additionally, certain additives, such as lactate, have been reported to exert a baroprotective effect on microbial populations of *L. monocytogenes*, potentially reducing HHP efficacy [41–43].

According to Roobab et al. [44], HPP could be combined with natural clean-label ingredients improving the physicochemical quality but also helping to reduce salt or nitrate/nitrite contents in cured meat products.

All the previously mentioned advantages should be weighed against the main drawback of this technology—its high equipment cost.

This chapter will describe the HHP application process for meat products, focusing on optimizing the technology for different types of ready-to-eat meat products according to the binomial condition of pressure and time.

2 Materials

- Vacuum package equipment.
- Vacuum bags.
- Whole meat product or slices of a meat product.
- HHP equipment (JBT Avure Technologies or Hiperbaric), *see* **Note 1**.

3 Method

Ensure operators have proper training. Preventive measures should be implemented to ensure safe products based on the General Food Hygiene Code, from Codex Alimentarius [45]. If HHP is a CCP in the process, the treatment should be monitored and recorded.

a) Meat product pre-package: the meat product or slices of a meat product should be packaged under vacuum (Fig. 1). Meat products or their slices must be hermetically sealed in flexible packaging.

b) Ensure the hermetic sealing of the packaging to prevent ruptures.

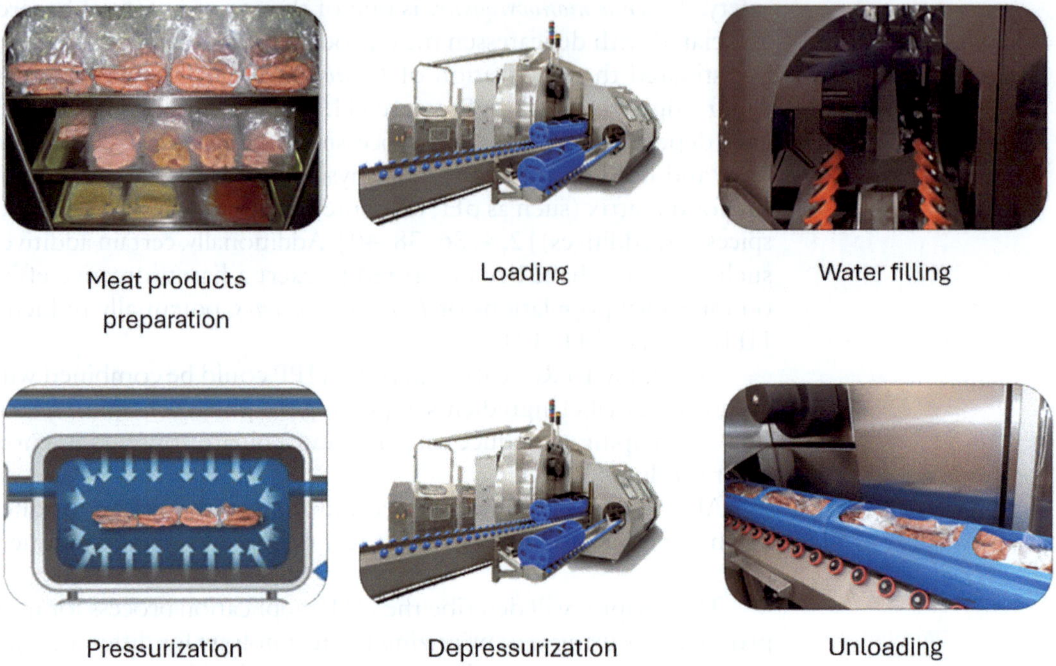

Fig. 1 HHP technology steps applied to whole meat products

c) Group products by weight or size to optimize the process (*see* **Note 2**).

d) Loading: place the packaged products into the HPP baskets (perforated baskets are commonly used for loading and unloading the packages).

e) The baskets are inserted into the pressure chamber. Sealing closures and a locking mechanism ensure the chamber closes.

f) Check seals and sensors before starting the cycle.

g) Water filling: the chamber is filled with potable water, the medium to transmit the pressure. The system ensures the chamber is filled to allow uniform pressure distribution.

h) The HPP equipment includes a chamber with varying volume capacities to hold the product and pressurize fluid.

i) Pressurization (process pressure): a hydraulic pump pressurizes (high-pressure intensifier pump) the water from 2000 to 6000 bar (200–600 MPa). The pressure used to decontaminate meat products usually varies from 400 to 600 MPa (please see Table 1 with examples of time-pressure to be applied and assure decontamination with pathogens, *see* **Note 3**). The time to achieve the pressure applied to the product should be monitored.

Table 1

Conditions of binomial pressure/time applied to different type of cured meat products

Name and type of meat product	Commercial presentation	Conditions of binomial pressure/time	Microbial effect	Physico-chemical effect	Sensory effect	References
Cooked Ham	Sliced	400 MPa/10 min	3.10 log reduction of *Listeria monocytogenes* strains on the control without lactate. All HPP lots stored at 1°C led to counts lower than 100 CFU/g at day 60 (*see* **Note 4**)	Not tested	Not tested	[46]
Dry Ham	Sliced	HPP treatment: pressure, holding time, initial fluid temperature (347–852 MPa; for 2.3–15.75 min; at 7.6–24.4°C)	600 MPa (for 9 min) resulted in an average reduction of *L. monocytogenes* counts of 2.9 logarithmic units (logs), while treatments at 600 MPa (for 3 and 10 min) reduced colony counts of *L. monocytogenes* by more than 7 logs	Not tested	Not tested	[4]

Table 1
continued

Name and type of meat product	Commercial presentation	Conditions of binomial pressure/time	Microbial effect	Physico-chemical effect	Sensory effect	References
Iberian and Serrano dry cured ham		HHP 600 MPa 6 min and storage at 2–3 °C for 5 months	Not tested	No impact of HHP in the physicochemical characteristics, fatty acid profiles was observer. Nevertheless, color was affected, L* increased in both type of hams and b* in Serrano ham	Dynamic sensory analysis reported that HHP increased cured flavour and saltiness. Besides, appearance and texture were not modified unless increased fibrousness and reduced pastiness, especially in Serrano dry cured ham. Whole piece processing compared to other reports may minimize effects	[47]

Table 1
continued

Name and type of meat product	Commercial presentation	Conditions of binomial pressure/time	Microbial effect	Physico-chemical effect	Sensory effect	References
Choriço (Fermented dry and smoked sausage)	whole	HPP treatment pressure and holding time: 200–600 MPa at controlled temperature (10°C) and from 154 to 1800 s	The Gram positive microbiota evaluated on the present study was less affected by HPP treatments than the Gram negative. High pressure (>400 MPa and longer time than 154 s) produced a reduction in fungi, aerobic count and on Gram-negative bacteria, such as *Enterobacteriaceae* and *Pseudomonas*, without negative effect on Gram-positive fermentative microbiota (LAB)	Pressurization at 400 and 600 MPa resulted in significantly higher ($P<0.001$) pH values compared to 200 –260 MPa, regardless of holding time. The nutritional value of the lipid fraction was only affected by the ratio of $n-6/n-3$ FA, and their stability. Total fatty acids and lipid stability were not affected by HPP	No exudates were observed at 400 MPa for 154 s or 960 s. However, prolonged holding times negatively affected product appearance, with greyish-white exudates reported in moderate to high levels by focus group participants	[20]
Iberian dry cured "chorizo"		HHP treatment at 600 MPa 8 min 16°C	High pressure reduces 3.2 log de counts of *L. monocytogenes* either when stores at 4° or 18°C. The mesophilic aerobic plate counts, and yeast and moulds were reduced after treatment and after 30 days	HHP did not affect color and lipid or protein oxidation during the first 60 days but they increment after 120 days		[48]

Table 1
continued

Name and type of meat product	Commercial presentation	Conditions of binomial pressure/time	Microbial effect	Physico-chemical effect	Sensory effect	References
Alheira (Fermented semi dried-smoked sausage)	whole	HPP treatment pressure and holding time: combinations of high pressure (300–600 MPa) and time (154–1800 s)	HPP treatment led to a reduction in *Enterobacteriaceae*, moulds, and yeasts, thereby minimizing the formation of major biogenic amines. However, levels of polyamines (spermine and spermidine) increased, likely due to the absence of microbial metabolic activity. Given the significant reduction of key spoilage microorganisms and the absence of detectable sensory changes, the condition of 600 MPa for 960 s was identified as the optimal treatment for further application	The TBARS values were low in all samples, and thus, the application of HPP did not promote lipid oxidation ($P > 0.05$)	The use of HPP in the processing of this semidried fermented smoked sausage showed promising results in improving its microbial characteristics, with minimal impact on product oxidation, color, and sensory characteristics	[6]
Dry-cured beef "Cecina de Leon"		HHP treatment at 500 MPa 5 min	HHP avoided the growth of *Enterobacteriaceae*, enterococci and *Pseudomonas*, and delayed growth of lactic acid bacteria, *Micrococcaceae* and yeasts and molds after 210 days at 6°C	HHP did not affect the physicochemical characteristics of the product	HHP did not affect the sensory characteristics of the product	[22]

Table 1
continued

Name and type of meat product	Commercial presentation	Conditions of binomial pressure/ time	Microbial effect	Physico-chemical effect	Sensory effect	References
Iberian dry-cured fermented sausages		600 MPa 8 min stored at 4° and 20°C for 180 days	Mesophilic aerobic counts significantly decreased	The HHP affected color, decreasing lightness and redness, and increasing yellowness, lipid oxidation increased and protein oxidation also in one of the products. Packaging and product composition even in the same commercial categories behaved different	Sensory analysis also reported color changes, higher hardness, some unpleasant odours after 180 days storage oh HHP samples. Besides higher saltiness after treatment was also reported	[49]
Low salt cured chicken sausages		200 MPa 10 min		When HHP was applied before cooking, the firmness was increased, more compact and continuous sausages were obtained. Color L, a, b parameters were lower, but color and textures were kept during 40 days of chilled storage	Sensory evaluation reported not negative effect of HHP and texture improved	[50]

j) After attaining the selected pressure, the process hold time according to the selected binomial should be monitored. The process typically lasts 1–6 min, depending on the product and the goal (e.g., elimination of pathogens, yeasts, etc.).

k) The initial product temperature and the temperature after treatment should be monitored. Although the processing temperature will not exceed 10°C, the temperature of the pressurization fluid may still influence the effectiveness of the High Hydrostatic Pressure (HHP) treatment and should therefore be considered during process validation.

l) Monitor water temperature and pressure through the control panel.

m) Depressurization or decompression: after the set time, the pressure is released in a controlled manner. Depending on the system, the water can be cooled or reused in subsequent cycles. A defined decompression period is required before opening the chamber to ensure safety and system stability.

n) Unloading and drying: remove the baskets from the chamber. Products are dried (if necessary) and sent to storage, refrigeration, or distribution.

o) Repeating or maintenance: if more batches are to be processed, repeat the steps.

p) Perform periodic cleaning and preventive maintenance of the equipment as scheduled.

q) Most systems also include filters and fluid recycling mechanisms for the compression medium.

4 Notes

1. Please visualize the sites: (https://www.jbtc.com/foodtech/products-and-solutions/products/high-pressure-processors/) or (https://www.hiperbaric.com/en/).
2. Although in HHP the pressure is applied isostatically, it is better to group products of similar physical and compositional characteristics in the same batch to optimize treatment parameters (time, temperature, pressure), thereby improving process efficiency, ensuring food safety, reducing operational costs, and enabling reliable shelf-life validation and quality control.
3. It is essential to demonstrate the validation of the HHP treatment based on the specific characteristics of the meat product (formulation, technology, pH, water activity, presence of additives).
4. This result was different when lactate was added, with a baro-protective effect to *L. monocytogenes*.

Acknowledgments

This work is supported by FCT, under projects UIDB/00276/2020 (CIISA), and LA/P/0059/2020 (AL4AnimalS) with financial support of the project AL4A-PROJ-LT1.3—Safety and acceptability assessment of Green label cured meat products.

References

1. Yeung C, Huang S (2016) Effects of high-pressure processing technique on the quality and shelf life of Chinese style sausages. J Food Nutr Res 4(7):442–447
2. Cava R, Higuero N, Ladero L (2021) High-pressure processing and storage temperature on *Listeria monocytogenes*, microbial counts and oxidative changes of two traditional dry-cured meat products. Meat Sci 171:108273. https://doi.org/10.1016/j.meatsci.2020.108273
3. Morales P, Calzada J, Nuñez MJ (2006) Effect of high-pressure treatment on the survival of Listeria monocytogenes Scott A in sliced vacuum-packaged Iberian and Serrano cured hams. Food Prot 69(10):2539–2543. https://doi.org/10.4315/0362-028x-69.10.2539
4. Bover-Cid S, Belletti N, Garriga M, Aymerich T (2011) Model for Listeria monocytogenes inactivation on dry-cured ham by high hydrostatic pressure processing. Food Microbiol 28:804–809
5. Bajovic B, Bolumar T, Heinz V (2012) Quality considerations with high pressure processing of fresh and value added meat products. Meat Sci 92(3):280–289
6. Borges AF, Cózar A, Patarata L, Gama LT, Alfaia CM, Fernandes MJ, Fernandes MH, Pérez HV, Fraqueza MJ (2020) Effect of high hydrostatic pressure challenge on biogenic amines, microbiota, and sensory profile in traditional poultry- and pork-based semidried fermented sausage. J Food Sci 85(4):1256–1264. https://doi.org/10.1111/1750-3841.15101
7. Meloni D (2019) High-hydrostatic-pressure (HHP) processing technology as a novel control method for *Listeria monocytogenes* occurrence in Mediterranean-Style Dry-fermented Sausages. Foods 8(12):672. https://doi.org/10.3390/foods8120672
8. Trejo A, Martín MJ, Gómez-Quintana A, Cava R, García-Parra JJ, Ramírez MR (2021) Effect of high-pressure treatment and storage temperature on top-quality (Montanera) Iberian dry-cured pork sausages (chorizo). J Food Sci 86(5):1963–1978. https://doi.org/10.1111/1750-3841.15698
9. Rigdon M, Thippareddi H, Thomas CL, Kumar S, McKee RW, Stelzleni AM (2023) High-pressure processing helps meet the *Escherichia coli* O157:H7 and Shiga toxin–producing *E. coli* (STEC) performance standards for beef summer sausage. MMB 7(1), https://doi.org/10.22175/mmb.16076
10. Liu H, Xu Y, Zu S, Wu X, Shi A, Zhang J, Wang Q, He N (2021) Effects of high hydrostatic pressure on the conformational structure and gel properties of myofibrillar protein and meat quality: a review. Foods 10(8):1872. https://doi.org/10.3390/foods10081872
11. Kameník J, Dušková M, Dorotíková K, Hušáková M, Ježek F (2023) The effect of high-pressure processing on the survival of non-O157 Shiga toxin-producing *Escherichia coli* in steak tartare: the good- or best-case scenario? Microorganisms 11(2):377. https://doi.org/10.3390/microorganisms11020377
12. Hereu A, Bover-Cid S, Garriga M, Aymerich T (2012) High hydrostatic pressure and biopreservation of dry-cured ham to meet the Food Safety Objectives for Listeria monocytogenes. Int J Food Microbiol 154:107–112
13. Hugas M, Garriga M, Monfort JM (2002) New mild technologies in meat processing: high pressure as a model technology. Meat Sci 62:359–371
14. FSIS (2021) Verification activities for high pressure processing, irradiation and microwave tempering. FSIS Directive 5000(15):9. https://www.fsis.usda.gov/sites/default/files/media_file/2021-08/5000.15.pdf
15. Campus M (2010) High pressure processing of meat, meat products and seafood. Food Eng Rev 2:256–273. https://doi.org/10.1007/s12393-010-9028-y
16. Marcos B, Aymerich T, Guardia MD, Garriga M (2007) Assessment of high hydrostatic pressure and starter culture on the quality properties of low-acid fermented sausages. Meat Sci 76(1): 46–53. https://doi.org/10.1016/j.meatsci.2006.09.020
17. Bajovic B, Bolumar T, Heinz V (2012) Quality considerations with high-pressure processing of fresh and value-added meat products. Meat Sci

92(3):280–289. https://doi.org/10.1016/j.meatsci.2012.04.024

18. Souza CM, Boler DD, Clark DL, Kutzler LW, Holmer S, Summerfield JW, Cannon JE, Smit NR, McKeith FK, Killefer J (2011) The effects of high-pressure processing on pork quality, palatability, and further processed products. Meat Sci 87(4):419–427
19. Fraqueza MJ, Martins C, Gama LT, Fernandes MJ, Fernandes MH, Ribeiro MHL, Hernando BR, Barreto AS, Alfaia AJI (2019) High hydrostatic pressure and time effects on hygienic and physical characteristics of natural casings and condiments used in the processing of cured meat sausage. IFSET 58:102242. https://doi.org/10.1016/j.ifset.2019.102242
20. Alfaia A, Alfaia CM, Patarata L, Fernandes MJ, Fernandes MH, Elias M, Ribeiro MH, Fraqueza MJ (2015) Binomial effects of high isostatic pressure and time on the microbiological, sensory characteristics and lipid composition stability of vacuum packed dry fermented sausages "chouriço". IFSET 32:37–44
21. Rubio B, Martinez B, Gonzalez-Fernandez C, Garcıa-Cachan MD, Rovira J, Jaime I (2006) Influence of storage period and packaging method on sliced dry cured beef "Cecina de Leon": effects on microbiological, physicochemical and sensory quality. Meat Sci 74(4):710–717
22. Rubio B, Martinez B, Garcia-Cachan MD, Rovira J, Jaime I (2007) Effect of high pressure preservation on the quality of dry-cured beef "Cecina de Leon". IFSET 8:102–110
23. Black EP, Setlow P, Hocking AD, Stewart CM, Kelly AL, Hoover DG (2007) Response of spores to high-pressure processing. Compr Rev Food Sci Food Saf 6:104–119
24. Knorr D (2007) Effizienz Analyse von Prozess-und Anlagekonzepten zur schonenden Haltbarmachung von Lebensmittel mittels neuartiger Hochdruckverfahren, BMBF Forschungsvorhaben 0330089A. Technische Universität Berlin, Berlin
25. Kruk ZA, Yun H, Rutley DL, Lee EJ, Kim YJ, Jo C (2011) The effect of high pressure on microbial population, meat quality and sensory characteristics of chicken breast fillet. Food Control 22(1):6–12. https://doi.org/10.1016/J.FOODCONT.2010.06.003
26. McArdle R, Marcos B, Kerry JP, Mullen A (2010) Monitoring the effects of high pressure processing and temperature on selected beef quality attributes. Meat Sci 86(3):629–634. https://doi.org/10.1016/j.meatsci.2010.05.001
27. Villacis MF, Rastogi NK, Balasubram VM (2008) Effect of high pressure on moisture and NaCl diffusion into turkey breast. LWT 41(5):836–844. https://doi.org/10.1016/j.lwt.2007.05.018
28. MacFarlane JJ (1973) Pre-rigor pressurization of muscle: effects on pH, shear value and taste panel assessment. J Food Sci 38(2):294–298. https://doi.org/10.1111/j.1365-2621.1973.tb01409.x
29. Suzuki A, Watanabe M, Iwamura K, Ikeuchi Y, Saito M (1990) Effect of high pressure treatment on the ultrastructure and myofibrillar protein of beef skeletal muscle (Food & Nutrition). Agric Biol Chem 54(12):3085–3091. https://doi.org/10.1016/0309-1740(93)90067-R
30. Latorre-Moratalla ML, Bover-Cid S, Aymerich T, Marcos B, Vidal-Carou MC, Garriga M (2007) Aminogenesis control in fermented sausages manufactured with pressurized meat batter and starter culture. Meat Sci 75(3):460–469. https://doi.org/10.1016/j.meatsci.2006.07.020
31. Bajovic B, Bolumar T, Heinz V (2012) Quality considerations with high pressure processing of fresh and value added meat products. Meat Sci 92(3):280–289
32. Chai H-E, Shiowshuh Sheen S (2021) Effect of high pressure processing, allyl isothiocyanate, and acetic acid stresses on Salmonella survivals, storage, and appearance color in raw ground chicken meat. Food Control 123:107784. https://doi.org/10.1016/j.foodcont.2020.107784
33. Chuang S, Sheen S, Sommers CH, Zhou S, Sheen L-Y (2020) Survival evaluation of Salmonella and Listeria monocytogenes on selective and nonselective media in ground chicken meat subjected to high hydrostatic pressure and carvacrol. J Food Prot 83(1):37–44. https://doi.org/10.4315/0362-028X.JFP-19-075
34. Chien SY, Sheen S, Sommers C, Sheen L-Y (2019) Combination effect of high-pressure processing and essential oil (*Melissa officinalis* Extracts) or their constituents for the inactivation of *Escherichia coli* in ground beef. Food Bioprocess Technol 12:359–370. https://doi.org/10.1007/s11947-018-2211-5
35. Aras S, Kabir MN, Chowdhury S, Fouladkhah AC (2020) Augmenting the pressure-based pasteurization of *Listeria monocytogenes* by synergism with nisin and mild heat. Int J Env Res Public Health 17(2):563. https://doi.org/10.3390/ijerph17020563
36. Porto-Fett ACS, Jackson-Davis A, Kassama LS, Daniel M, Oliver M, Jung Y, Luchansky JB (2020) Inactivation of Shiga toxin-producing *Escherichia coli* in refrigerated and frozen meatballs using high pressure processing. Microorganisms 8(3):360. https://doi.org/10.3390/microorganisms8030360

37. Gunther NW, Abdul-Wakeel A, Ramos R, Sheen S (2019) Evaluation of hydrostatic high pressure and cold storage parameters for the reduction of Campylobacter jejuni in chicken livers. J Food Prot 82(6):1039–1044. https://doi.org/10.4315/0362-028X.JFP-18-469
38. Cava R, García-Parra J, Ladero L (2020) Effect of high hydrostatic pressure processing and storage temperature on food safety, microbial counts, colour and oxidative changes of a traditional dry-cured sausage. LWT 128:109462. https://doi.org/10.1016/j.lwt.2020.109462
39. Rubio B, Possas A, Rincón F, García-Gímeno RM, Martínez B (2018) Model for Listeria monocytogenes inactivation by high hydrostatic pressure processing in Spanish chorizo sausage. Food Microbiol 69:18–24. https://doi.org/10.1016/j.fm.2017.07.012
40. Pérez-Baltar A, Serrano A, Montiel R, Medina M (2020) Listeria monocytogenes inactivation in deboned dry-cured hams by high pressure processing. Meat Sci 160:107960. https://doi.org/10.1016/j.meatsci.2019.107960
41. Aymerich T, Jofré A, Garriga M, Hugas M (2005) Inhibition of Listeria monocytogenes and Salmonella by natural antimicrobials and high hydrostatic pressure in sliced cooked ham. J Food Prot 68:173–177
42. Serra-Castello C, Jofre A, Belletti N, Garriga M, Bover-Cid S (2021) Modelling the piezo-protection effect exerted by lactate on the high pressure resistance of Listeria monocytogenes in cooked ham. Food Res Int 140:110003. https://doi.org/10.1016/j.foodres.2020.110003
43. Serra-Castelló C, Ferrocino I, Jofré A, Cocolin L, Bover-Cid S, Rantsiou K (2021) Unravelling the molecular mechanisms underlying the protective effect of lactate on the high-pressure resistance of *Listeria monocytogenes*. Biomolecules 11(5):677. https://doi.org/10.3390/biom11050677
44. Roobab U, Rehan Afzal R, Ranjha MMAN, Zeng X-A, Ahmed Z, Aadil RM (2021) High pressure-based hurdle interventions for raw and processed meat: a clean-label prospective. IJFST 57(2):816–826. https://doi.org/10.1111/ijfs.15499
45. FAO & WHO (2023) General principles of food hygiene. Codex Alimentarius Code of Practice, No. CXC 1-1969. Codex Alimentarius Commission, Rome https://doi.org/10.4060/cc6125en
46. Marcos B, Aymerich T, Monfort JM, Garriga M (2008) High-pressure processing. and antimicrobial biodegradable packaging to control *Listeria monocytogenes* during storage of cooked ham. Food Microbiol 25(1):177–182. https://doi.org/10.1016/j.fm.2007.05.002
47. Lorido L, Estévez M, Ventanas J, Ventanas S (2015) Comparative study between Serrano and Iberian dry-cured hams in relation to the application of high hydrostatic pressure and temporal sensory perceptions. LWT 64(2):1234–1242. https://doi.org/10.1016/j.lwt.2015.07.029
48. Cava R, García-Parra J, Ladero L (2020) Effect of high hydrostatic pressure processing and storage temperature on food safety, microbial counts, colour and oxidative changes of a traditional dry-cured sausage. LWT 128:109462. https://doi.org/10.1016/j.lwt.2020.109462
49. Ramírez R, Trejo A, Delgado-Adámez J, Martín-Mateos MJ, García-Parra J (2022) Effect of high-hydrostatic-pressure processing and storage temperature on sliced Iberian dry-cured sausage ("salchichón") from pigs reared in montanera system. Foods 11(9):1338. https://doi.org/10.3390/foods11091338
50. Zhou Y, Watkins P, Oiseth S, Cochet-Broch M, Sikes AL, Chen C, Buckow R (2021) High pressure processing improves the sensory quality of sodium-reduced chicken sausage formulated with three anion types of potassium salt. Food Control 126:108008. https://doi.org/10.1016/j.foodcont.2021.108008

Chapter 3

Ultrasound Application in Fresh Beef

Ana Paula A. A. Salim, Loíse C. S. Lima, Surendranath P. Suman and Carlos Adam Conte-Junior

Abstract

Ultrasound is an emerging technology applied to improve red meat quality attributes. The determination of optimal ultrasound intensity requires the consideration of equipment specifications and sample characteristics. In this context, we outline the procedural steps and mathematical formulations to determine the ultrasound intensity relative to the sample area and acoustic energy density (AED) relative to sample weight when applied to beef samples utilizing an ultrasonic bath.

Keywords High-intensity ultrasound, Beef, Ultrasonic bath, Calorimetric, Acoustic energy density, Ultrasound intensity

1 Introduction

Ultrasound (US) is an energy produced by a longitudinal mechanical wave with a vibration frequency above the range of human hearing (20 kHz). In a US system, electrical energy is converted into vibrational (mechanical) and transmitted through the sonicated medium. In this process, part of the energy is lost as heat, and the rest can produce cavitation [1, 2].

Cavitation is the growth and collapse of vapor bubbles embedded in a liquid medium during wave compression and expansion [1]. There are many ways to generate acoustic cavitation, such as shock waves or steady state. In a steady state, cavitation occurs when fluid pressure rapidly decreases with the generation of small bubbles. Then, when the fluid pressure increases again, the bubbles implode, and the potential energy obtained is transformed into kinetic energy [3].

Ultrasound frequencies range from 20 kHz to 10 MHz and can be classified into: (1) high-intensity (>5 W/cm^2 or 10–1000 W/cm^2)

Renata Ernlund Freitas de Macedo and Carlos Adam Conte-Junior (eds.), *Meat Products Preservation*, Methods and Protocols in Food Science, https://doi.org/10.1007/978-1-0716-5068-4_3,

and low-frequency (20–100 kHz); (2) medium-intensity and intermediate-frequency (100 kHz–1 MHz); and (3) low-intensity (<1 W/cm^2) and high-frequency (1–10 MHz) [1].

High-intensity ultrasound (HIU) causes changes in the physical, chemical, or mechanical properties of foods [1]. Previous investigations have reported potential uses of high-intensity ultrasound on fresh meat [4–8]. However, some authors report undesirable changes (e.g., discoloration, oxidation) resulting from HIU application depending on frequency, power, or exposure time [1, 9].

Regarding the influence of HIU on beef color, previous investigations have reported conflicting results. Caraveo et al. [5] documented that HIU reduced redness (a^*) and increased lightness (L^*), which represents a negative effect for red meats. Alves et al. [10], however, reported no change in beef color under ultrasonication, whether in a bath (45 kHz, 500 W, 1.8 W/cm^2) or with a probe (20 kHz, 750 W, 26.5 W/cm^2) for 0, 60, 120, or 240 s. Additionally, Lima et al. [7, 8] investigated the influence of ultrasound intensities (control: 0; low: 276.36; moderate: 422.67; and high: 747.79 W/cm^2) on color, oxidative stability and texture of *Longissimus lumborum* steaks from Nellore cattle and reported no adverse effects of HIU on muscle pH, color, texture, lipid, and protein oxidation.

Thus, the influence of HIU on red meat quality depends on intrinsic factors (e.g., post-mortem time, muscle type, packaging, etc.) and equipment settings (intensity, frequency, temperature, time, pulse, amplitude, probe size, etc.) [6, 12, 13]. It is important to highlight that the ultrasound frequency is relatively narrow, typically ranging from 20–40 kHz, whereas ultrasound intensities (from 2.39 to 109.7 W/cm^2) and treatment times (from 1 to 120 min) exhibit greater variability [4].

Three different methods can be used to apply ultrasound to products: (a) direct application, (b) coupling to the device, and (c) immersion in an ultrasound bath [11]. In this chapter, we will outline the procedural steps and mathematical formulations to determine the ultrasound intensity relative to the sample area and acoustic energy density (AED) relative to the sample weight for application in beef samples utilizing an ultrasonic bath.

2 Materials

1. Ultrasound bath: VCX 750 ultrasonic liquid processor, cup horn assembly, and chiller (Sonics & Materials, INC., Connecticut, EUA);
2. Beef samples;
3. Knife;
4. Ruler;
5. Polyethylene bags: 12 μm polyethylene and nylon; moisture vapor transfer rate of 4.91 g^{-2} 24 h^{-1} 1 atm^{-1} (38°C; 90%

relative humidity), O_2 permeability of 150 cm^{-3} m^{-2} 24 h^{-1} 1 atm^{-1} (23°C; dry);

6. Chamber Vacuum Sealer: Modulare CV65 (Conceito Vácuo, São Paulo, Brazil);
7. Deionized water (1000 mL);
8. Liquid-in-glass thermometer;
9. Semi-analytical scale.

3 Methods

3.1 Effective Power Calculation

1. Calculate the ultrasound intensities using the calorimetric method (*see* **Note 1**). The effective power or real power output (*Po*) should be found through the following equation:

$$Po = m \bullet cp \bullet ((Tf - Ti)/(tf - ti))$$

where: *m* is the mass of liquid (g); *cp* is the specific heat constant (J/g °C); *ti* is the initial temperature (°C); *Tf* is the final temperature; *ti*(s) is the initial time; and *tf*(s) is the final time.

2. Fill the cup horn assembly with 1000 mL of deionized water; the volume will be applied as the mass of liquid (g) at the effective power (*Po*) equation (*see* **Note 2**).
3. The deionized water acts as the acoustic medium, and the specific heat constant (*cp*) is considered when calculating the effective power. In this regard, the *cp* will be 4.18 J/g °C, the specific heat constant of water.
4. Measure and take note of the water temperature before turning on the ultrasonic bath. This temperature refers to the initial temperature (*Ti*). Make sure to maintain the thermometer immersed in the water.
5. Turn on the ultrasonic bath and set the amplitude/power to maximum (*see* **Note 3**). In the case of the VCX 750 ultrasonic liquid processor, the amplitude of 100% corresponds to 750 W.
6. The time variation (*tf* – *ti*) to determine the (*Po*) starts at 0 s (initial time; *ti*) and finishes at 180 s (final time; *tf*) (*see* **Note 4**). The time setting can be observed in Fig. 1a.
7. Take note of the water temperature after 180 s (Fig. 1b); this temperature refers to the final temperature (*Tf*).
8. Apply the mass of water, water-specific heat constant, temperature variation, and time variation on the equation to determine the effective power. The results are expressed in watts (W) or joules per second (J/s).

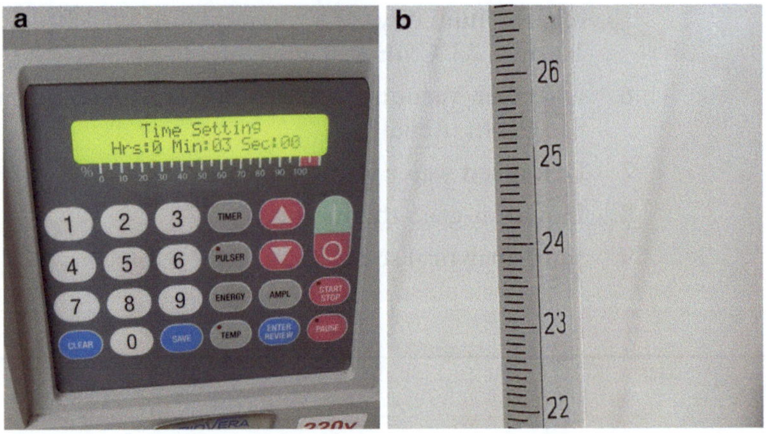

Fig. 1 (**a**) Display of ultrasound equipment with time setting for effective power calculation. The time setting of 3 min corresponds to 180 s (*see* **Note 4**). (**b**) The final temperature of the water bath was measured to calculate the effective power

9. Using the aforementioned ultrasound equipment and considering an amplitude of 25% (187.5 W) (*see* **Note 5**), the effective power would be:

$$Po = 1000 \bullet 4.18 \bullet ((25.1 - 23)/(180 - 0))$$

$$Po = 48.77\ W$$

3.2 Sample Preparation

1. Slice the beef samples into 2.5 cm thick, 7.5 cm length, and 4.5 cm height steaks (*see* **Note 6**). Calculate the sample area to determine the ultrasound intensity (UI).
2. Weight each steak and take notes to determine the acoustic energy density (AED), posteriorly.
3. Place the steaks individually in each polyethylene bag and vacuum package them (Fig. 2).
4. Keep the vacuum-packed samples at 4°C until the ultrasound application starts.

3.3 Ultrasound Intensities Calculation

1. Calculate the sample surface area considering: 7.5 cm length × 4.5 cm height = 33.75 cm^2.
2. After calculating the effective power and the sample surface area, add the results in the following equation to obtain the ultrasound intensities (UI):

$$UI = (Po \bullet t)/A$$

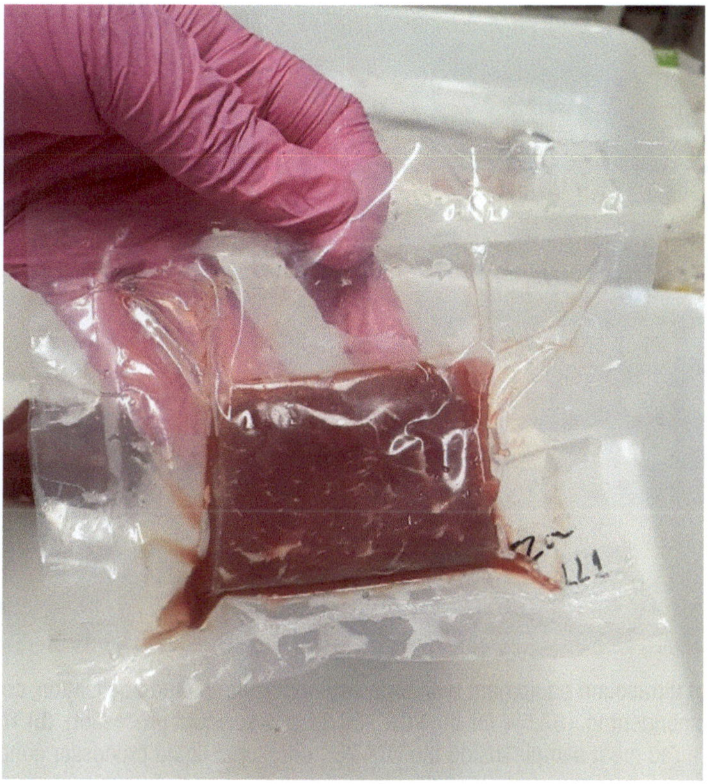

Fig. 2 Vacuum-packed sample of beef *Longissimus lumborum* with 2.5 cm thick, 7.5 cm length, and 4.5 cm height

where: *Po* is the effective power (W); *t* is the duration of the treatment (s); and *A* is the surface area of the sample (cm^2).

3. The duration of the treatment directly influences the ultrasound intensity. In this regard, the intensities could be predicted according to the samples' surface area and time in the ultrasonic bath.
4. The results are expressed in W/cm^2.
5. Considering the sample and a sonication time of 10 min (600 s), the ultrasound intensity would be:

$$UI = (48.77 \bullet 600)/33.75$$

$$UI = 867.02\ W/cm^2$$

3.4 Acoustic Energy Density Calculation

1. After calculating the effective power and weight of the samples, the results will be applied in the following equation to obtain the acoustic energy density (*AED*):

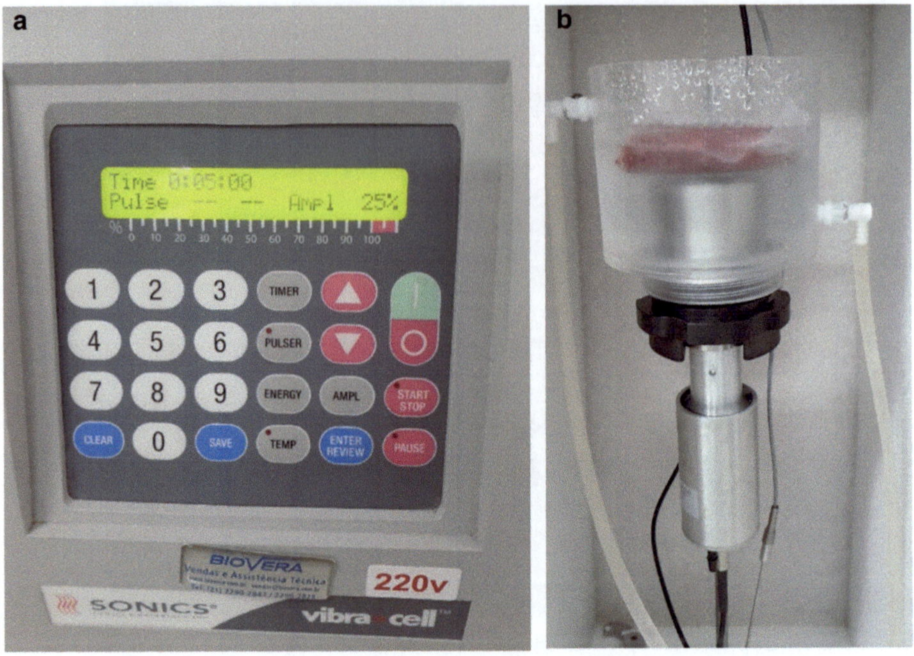

Fig. 3 (**a**) Display of ultrasound equipment with time setting for ultrasound application, considering an amplitude of 25% (correspondent to 187,5W for the VCX 750 ultrasonic liquid processor). (**b**) The ultrasound application occurs with a red meat sample inside the VCX 750 ultrasonic liquid processor coupled to the cup horn assembly and chiller

$$AED = \frac{\left[\frac{Po \cdot t}{m}\right]}{1000}$$

where: *Po* is the effective power (W or KJ/s), *t* is the duration of the treatment (s) and *m* is the sample weight (g).

2. The intensities can be predicted based on the weight of the samples and the duration of their treatment in the ultrasonic bath. The duration of the treatment directly influences the ultrasound intensity and the ultrasound intensity calculation.
3. The results are expressed in KJ/g.
4. Considering the sample weights of 64.85 g and the treatment time to calculate the ultrasound intensity (10 min or 600 s), the AED would be:

$$AE = \frac{\left[\frac{48.77 \cdot 600}{64.85}\right]}{1000}$$

$$AED = 5.01\ KJ/g$$

3.5 Ultrasound Application

1. Set the desired treatment duration for the beef steaks, considering the previous ultrasound intensity or acoustic energy density calculated (Fig. 3a).
2. Turn on the chiller before starting the ultrasound treatment. Set the temperature to 4°C and wait for the water to chill.
3. Add one of the vacuum-packed samples to the ultrasonic bath (Fig. 3b) and check the parameters: Power/Amplitude and time (*see* **Note 7**).
4. Start the ultrasound treatment for each sample individually and repeat steps 1–4 of Sect. 3.5 for each one.

4 Notes

1. The effective power needs to be calculated [14, 15]. According to da Silva et al. [16], the amount of energy dissipated in the medium depends on the volume of the liquid. This energy needs to be computed and presented as a result. This energy uses a volume function to measure the flow of sound energy. Reference [15] is not cited in text. Please check the inserted citation in the text for correctness.
2. Verify the volume of the bath and consider the volume as the mass of the liquid (*m*) in grams to apply the effective power (***Po***) equation.
3. The amplitude (%) of ultrasound equipment refers to the nominal power (W). This relation is reported by Kim et al. [17] and Neves et al. [18].
4. The time of the calorimetric method is determined according to Kang et al. [19].
5. Some ultrasound equipment has adjustable amplitude allowing to test and determine the effective power by adjusting the amplitude/power.
6. The dimensions of the samples can be adapted to the size of the ultrasonic bath and adjusted to achieve specific ultrasound intensities according to the objective of the study.
7. The settings may vary according to the model of ultrasound equipment. Some ultrasonic baths have adjustable power/amplitude (W/%), frequency (kHz), and sonication time (min or s), while others have fixed operating parameters.

References

1. Alarcon-Rojo AD, Carrillo-Lopez LM, Reyes-Villagrana R, Huerta-Jimenez M, Garcia-Galicia IA (2019) Ultrasound and meat quality: a review. Ultrason Sonochem 55:369–382
2. Alarcon-Rojo AD, Janacua H, Rodriguez JC, Paniwnyk L, Mason TJ (2015) Power ultrasound in meat processing. Meat Sci 107:86–93

3. McNamara W, Didenko Y, Suslick K (1999) Sonoluminescence temperatures during multi-bubble cavitation. Nature 401:772–775
4. Got F, Culioli J, Berge P, Vignon X, Astruc T, Quideau JM et al (1999) Effects of high-intensity high-frequency ultrasound on ageing rate, ultrastructure and some physico-chemical properties of beef. Meat Sci 51:35–42
5. Caraveo O, Alarcon-Rojo AD, Renteria A, Santellano E, Paniwnyk L (2015) Physicochemical and microbiological characteristics of beef treated with high-intensity ultrasound and stored at 4°C. J Sci Food Agric 95:2487–2493
6. Carrillo-Lopez LM, Cruz-Garibaldi BY, Huerta-Jimenez M, Garcia-Galicia IA, Alarcon-Rojo AD (2022) The physicochemical, microbiological, and structural changes in beef are dependent on the ultrasound system, time, and one-side exposition. Molecules 27:541
7. Lima LCS, Salim APAA, Trezze IMM, Ferreira MS, Monteiro MLG, Vasconcellos-Junior FJ, Mano SB, Conte-Junior CA (2023a) High-intensity ultrasound improves color and oxidative stability of beef from grain-fed and pasture-fed Nellore cattle. Meat Sci 206:109324
8. Lima LCS, Salim APAA, Trezze I, Ferreira MS, Monteiro MLG, Mano SB, Conte-Junior CA (2023b) Influence of ultrasound intensities on colour, oxidative stability, and texture of longissimus lumborum steaks from Nellore cattle. Int J Food Sci Technol 54:1990–2002
9. Gonzalez-Gonzalez L, Alarcon-Rojo AD, Carrillo-Lopez LM, Garcia-Galicia IA, Huerta-Jimenez M, Paniwnyk L (2020) Does ultrasound equally improve the quality of beef? An insight into Longissimus lumborum, infraspinatus and cleidooccipitalis. Meat Sci 160:107963
10. Alves LL, Rampelotto C, Silva MS, De Moura HC, Durante EC, Mello RO et al (2018) The effect of cold storage on physicochemical and microbiological properties of beef Semitendinosus muscle subjected to ultrasonic treatment in different systems (bath or probe). Int Food Res J 25:504–514
11. Chemat F, Zill-E-Huma, Khan MK (2011) Applications of ultrasound in food technology: processing, preservation and extraction. Ultrason Sonochem 18(4):813–835
12. Alarcon-Rojo AD, Janacua H, Rodriguez JC, Paniwnyk L, Mason TJ (2015) Power ultrasound in meat processing. Meat Sci 107:86–93
13. Carrillo-Lopez LM, Huerta-Jimenez M, Garcia-Galicia IA, Alarcon-Rojo AD (2019) Bacterial control and structural and physicochemical modification of bovine Longissimus dorsi by ultrasound. Ultrason Sonochem 58:104608
14. Barekat S, Soltanizadeh N (2018) Effects of ultrasound on microstructure and enzyme penetration in beef longissimus lumborum muscle. Food Bioprocess Technol 11:680–693
15. Bernardo YAA, do Rosario DKA, Conte-Junior CA (2023) Comment on the paper Ultrasonication induced nano-emulsification of thyme essential oil: optimization and antibacterial mechanism against *Escherichia coli*. Food Control 148:109651
16. da Silva BD, Rosario DKA, Conte-Junior CA (2022) Can droplet size influence antibacterial activity in ultrasound-prepared essential oil nanoemulsions? Crit Rev Food Sci Nutr 63:12567–12577
17. Kim I, Lee I, Jeon SH, Hwang T, Han JI (2015) Hydrodynamic cavitation as a novel pretreatment approach for bioethanol production from reed. Bioresour Technol 192:335–339
18. Neves MIL, Socas-Rodríguez B, Valdés A, Silva EK, Cifuentes A, Meireles MAA, Ibáñez E (2022) Synergic effect of natural deep eutectic solvent and high-intensity ultrasound on obtaining a ready-to-use genipin extract: crosslinking and anti-neurodegenerative properties. Food Chem X 16:100489
19. Kang DC, Zou YH, Cheng YP, Xing LJ, Zhou GH, Zhang WG (2016) Effects of power ultrasound on oxidation and structure of beef proteins during curing processing. Ultrason Sonochem 33:47–53

Check for updates

Chapter 4

UVC-LED Application in Chicken and Fish-Based Products

Yago A. A. Bernardo and Carlos Adam Conte-Junior

Abstract

The ultraviolet-C light-emitting diode (UVC-LED) application has grown for animal-origin product preservation and shelf-life extension. This technology exhibits a high lifetime, mercury absence, and no heating requirements. The mechanism of bacterial inactivation is based on direct DNA damage, affecting transcription and cell replication. However, UVC-LED can lead to quality losses for the treated food product, including undesirable color and texture alterations and oxidative stress. Therefore, we described in this chapter an optimized methodology for applying UVC-LED in chicken breast and fish fillets to improve decontamination and extend shelf-life regarding total aerobic viable bacteria and critical physicochemical characteristics, such as instrumental color, texture, and lipid oxidation.

Keywords Spoilage bacteria, Shelf-life, Non-thermal technology, Food safety, Ultraviolet-C light-emitting diode, Instrumental color, Texture profile analysis

1 Introduction

Applying traditional thermal methods for food preservation results in undesirable nutritional and sensory alterations [1]. Also, these methods have been argued due to the ability of microorganisms to maintain viability after processing [2, 3]. Non-thermal technologies, such as ultraviolet C light (UVC), have been proposed as alternatives, as they are considered safe, environmentally friendly, and can extend the shelf-life of the product [4].

UVC light (200–280 nm) changes microbial DNA's structure by forming cytosine-thymine cross-bridges and producing free radicals via water radiolysis [5]. Recently, the use of light-emitting diode UVC (UVC-LED) has been explored for fish [6–8] and chicken [9–12] decontamination (Table 1) due to the absence of mercury lamps, high flexibility, long shelf-life, and no heating requirement [7].

On the other hand, according to Monteiro et al. [5], despite extending shelf-life from 2 (at low doses) to 6 (at high doses) days by retarding microbial growth, the use of UVC induces changes in the instrumental color, texture profile, and lipid oxidation rates

Renata Ernlund Freitas de Macedo and Carlos Adam Conte-Junior (eds.), *Meat Products Preservation*, Methods and Protocols in Food Science, https://doi.org/10.1007/978-1-0716-5068-4_4,

Table 1
Recent studies on applying UVC-LED for decontamination (spoilage and pathogenic bacteria) of chicken and fish

Food matrix	Conditions	Findings	References
Rainbow trout (*Oncorhynchus mykiss*)	0.97–1.78 mW/cm², 244–3355 s	Increased the shelf-life from 1.4 to 4.7 days	Monteiro et al. [6]
Nile tilapia (*Oreochromis niloticus*)	0.97–1.78 mW/cm², 244–3355 s	Increased the shelf-life in 2 days	Monteiro et al. [7]
Chicken breast	0.041 W/cm², 60–660 s	1.5 log CFU/mL inactivation of *Campylobacter jejuni*	Soro et al. [10]
Chicken breast	0.204–0.408 W min/cm², 180–360 s	UVC-LED mitigated the growth of *C. jejuni* and *Salmonella* Typhimurium during storage	Soro et al. [9]
Chicken breast	0–3.6 J/cm², 60–900 s	Bacterial inactivation (*Salmonella* spp.) is dependent of power and time	Calle et al. [11]
Yellowfin tuna (*Thnunnus albacores*)	0.5 mW/cm², 0–20 s	Reduction (approximately 6 log CFU/cm²) of *S.* Typhimurium, *E. coli* O157:H7, and *L. monocytogenes*	Fan et al. [8]
Chicken breast	1000–4000 mJ/cm², 500–2000 s	Reduction (approximately 5 log CFU/cm²) of *S.* Typhimurium, *E. coli* O157:H7, and *L. monocytogenes*	Wang et al. [12]

CFU, colony-forming units

of the matrices. Therefore, there is a need to apply this technology using optimal conditions of power and time to achieve a high-quality product without undesirable physicochemical alterations.

In this context, we demonstrate in this protocol how UVC-LED can be efficiently applied using optimized power (W/cm²) and time (s) to improve the shelf-life of chicken breasts and fish fillets-based products without causing significant changes in the color, texture, and oxidative parameters. The protocol presented here was prepared considering the findings of Monteiro et al. [6, 7], which optimized the UVC-LED treatment aiming to assess the stability of the product over storage and confirm the efficiency of UVC-LED.

2 Materials

All solutions should be prepared using distilled water and analytical-grade reagents. The waste disposal regulations must be followed when disposing of waste materials.

2.1 Sample Collection and Storage

1. Fresh chicken breast or fish fillet samples.
2. Analytical scale.

3. Styrofoam box with ice (0 ± 0.1 °C).
4. Nylon/polyethylene bags (250 mm × 170 × 170 mm).
5. Vacuum-packaging machine.
6. Biochemical oxygen demand (BOD) incubator (4 ± 0.1 °C).

2.2 UVC-LED

1. 275 nm LED irradiator (Black Box Smart®, BioLambda) equipped with a controller module and irradiation chamber of 163.5 cm^2 beam area, anodized aluminum, and ultra-clear glass, as reported by Monteiro et al. [6, 7].

2.3 Bacteriological Analysis

1. Biosafety cabinet (class II).
2. Stomacher.
3. Sodium chloride solution.
4. Plate count agar (PCA).
5. Autoclave.
6. Sterile Petri dishes.
7. Microcentrifuge tubes.
8. Incubator (36 ± 0.1 °C).
9. BOD incubator (7 ± 0.1 °C).

2.4 Physicochemical Analysis

1. CM-600D portable colorimeter spectrophotometer (Konica Minolta Sensing Americas Inc., Ramsey, NJ, USA).
2. Texture Analyzer (Stable Micro Systems, Surrey, UK), equipped with a 36 mm cylindrical probe (P/36R).
3. Ultra turrax (IKA® T10, China).
4. Trichloroacetic acid (TCA).
5. Thiobarbituric acid (TBA).
6. UV-1900 spectrophotometer (Shimadzu Corporation, Kyoto, Japan).

3 Method

3.1 Sample Preparation

1. Acquire and transport the samples (chicken breast or fish fillets) to the laboratory in the Styrofoam box with ice, maintaining the temperature close to 0 °C (*see* **Note 1**).
2. Use previously acquired samples to prepare the animal-origin product of interest, e.g., hamburger, nuggets, sausage.
3. Cut the samples into pieces of 20 ± 0.5 g each.
4. Pack the pieces using nylon/polyethylene bags and vacuum seal them (*see* **Note 2**).

3.2 UVC-LED Treatment

1. Insert the package containing the samples individually in the 275 nm LED irradiator (*see* **Note 3**).
2. The package must be illuminated on top and bottom simultaneously, at a distance of 15 cm from the LED lamps.
3. Power and time should be set at the equipment interface, considering the optimal values described by Monteiro et al. [6, 7], between 0.97 and 1.46 W/cm^2 and 2407.8–2503.6 s, respectively.
4. The dose (energy density) should be measured according to the Eq. (1), considering the irradiance (P, in W/cm^2), and the time (t, in seconds).

$$Dose\ (mJ/cm^2) = p \,.\, t \tag{1}$$

3.3 Total Mesophilic and Psychrotrophic Counts

Steps 3–9 must be performed under the biosafety cabinet, and sterile gloves must be used to avoid cross-contamination.

1. Prepare the PCA and sodium chloride solution by dissolving 22.5 g and 0.85 g each in 1000 mL water (*see* **Note 4**).
2. Autoclave PCA and sodium chloride solution by 15 min at 121°C.
3. After cooling (50–60°C), pour the PCA into the sterile Petri dishes (15–20 mL).
4. Use the sodium chloride solution to fill the microcentrifuge tubes (0.9 mL).
5. Open the polyethylene bags, collect a 10 g aliquot of the samples, and mix with sodium chloride solution at 1:10 w/v (90 mL) in a Stomacher bag.
6. Homogenize the solution using the Stomacher for 1 min.
7. Transfer 0.1 mL of the homogenized solution to a microcentrifuge tube containing 0.9 mL of sodium chloride 0.85% and homogenize using a vortex.
8. Repeat step 7 serially to generate the serial dilutions (*see* **Note 5**).
9. Plate the serial dilutions (0.1 mL) in the PCA plates and incubate at 36°C/48 h in the incubator for the mesophilic bacteria and at 10°C/7 days in the BOD for the psychrotrophic bacteria (*see* **Note 6**).
10. Count the characteristic colonies in the plates and calculate the colony-forming units per gram of samples (CFU/g), according to Eq. (2).

$$\frac{CFU}{g} = Colony\ count \cdot Dilution\ factor \cdot Ailquote \tag{2}$$

11. Obtain the logarithmic count using the function "$\log_{10}$" using a calculator or statistical software (*see* **Note 7**).

3.4 Color Measurement

1. Open the polyethylene bags and collect a 10 g aliquot of the samples.
2. Perform a blooming of 30 min in air and daylight before color measurements to allow myoglobin (Mb) oxygenation.
3. Record L^* (lightness), a^* (redness), and b^* (yellowness) from at least three different locations randomly on the surface of each sample using the CM-600D portable colorimeter spectrophotometer (*see* **Note 8**).
4. The measurements should be performed using illuminant A, 10° standard observer, and 8 mm diameter aperture.
5. Calculate the total color difference (ΔE) using Eq. (3). The ΔE is applied to compare the color differences between samples before and after UVC-LED treatment and through storage, considering day 0 as the control.

$$\Delta E = \sqrt{(L_2^* - L_1^*)^2 + (a_2^* - a_1^*)^2 + (b_2^* - b_1^*)^2} \quad (3)$$

where ΔE is the total color difference, L^* is the lightness, a^* is the redness, and b^* is the yellowness.

3.5 Texture Measurement

1. Obtain 4 cubic samples ($2 \times 2 \times 2\ cm^3$) from the treated polyethylene bags.
2. Perform the texture profile analysis using the TA.XT plus Texture Analyzer equipped with a 36 mm cylindrical probe (P/36R).
3. The conditions for analysis (i.e., compression force, interval, and test speed) should observe the standards proposed by Bernardo et al. [13] (*see* **Note 9**).
4. Measure the parameters of hardness (N), chewiness (N × mm), cohesiveness (ratio), springiness (ratio), and resilience (ratio).
5. Use the Exponent software package (version 6.1.9.1; Stable Micro Systems) to calculate the texture parameters.

3.6 Lipid Oxidation

1. Collect 5 g of the treated samples to perform the thiobarbituric acid reactive substances (TBARS) assay, according to Yin et al. [14].
2. Prepare a TCA 11% solution using TCA and distilled water.
3. Prepare a TBA 20 mM solution using TCA and distilled water.
4. Homogenize the samples with 22.5 mL of 11% trichloroacetic acid (TCA) solution using the Ultra-turrax for 1 min (*see* **Note 10**).

5. Centrifuge at 15,000 ×*g* for 15 min at 4°C.
6. Remove the supernatant (1 mL) and mix it with 1 mL of the aqueous solution of 20 mM thiobarbituric acid (TBA).
7. Incubate at 25°C for 20 h without light.
8. Read the absorbance values at 532 nm utilizing the UV-1900 spectrophotometer.
9. Express the results in mg malonaldehyde (MDA) per kg of sample. The concentration of MDA is obtained from a standard curve containing seven different MDA concentrations (1–500 μmol) (*see* **Note 11**).

4 Notes

1. The samples (chicken breast or fish fillets) can be frozen and stored (−18°C) before treatments.
2. At least six samples (approximately 20 g each) can be packaged in a single package before the treatments. This procedure should be considered to optimize the experiment and homogenize the treatments.
3. The UVC-LED treatment described above can be performed on novel products, chicken breasts, and fish fillets.
4. The PCA and sodium chloride solution can be stored before and after the experiments at refrigerated conditions (4°C) to avoid contamination and generation of waste.
5. Considering the efficiency of the decontamination process, the microbial dilutions and plating can be performed between 10^0 and 10^{-2}.
6. Other microbial groups, such as molds and yeasts, and pathogen bacteria, are suggested for the microbiological analyses, including *Salmonella* spp., *Escherichia coli*, *Listeria monocytogenes*, and *Vibrio parahaemolyticus*.
7. The samples can be stored (4°C) and analyzed through the storage period to establish the shelf-life of the novel product developed and treated by UVC-LED.
8. For the instrumental color analysis, it is recommended to calibrate the equipment and use a white support for the samples during the measurement.
9. The instrumental texture measurement conditions can be adapted considering the compression force, interval time, and test speed, according to Bernardo et al. [13]. Moreover, the texturometer probe can also suffer slight modification (diameter) considering the geometry of the samples.

10. During the homogenization of the sample for lipid oxidation analysis, the Ultra-turrax is recommended in 30 s intervals (on and off), and a double jacket with ice to protect the samples from overheating.
11. Other quality control parameters can be performed to verify the effects of UVC-LED, such as protein oxidation.

References

1. Bernardo YAA, Do Rosario DKA, Conte-Junior CA (2023) Ultrasound on milk decontamination: potential and limitations against foodborne pathogens and spoilage bacteria. Food Rev Int 39:320–333. https://doi.org/10.1080/87559129.2021.1906696
2. Machado MAM, Castro VS, Monteiro MLG et al (2023) Effect of UVC-LED and ultrasound alone and combined on heat-resistant *Escherichia coli* isolated from pasteurised milk. Int J Dairy Tech 76:74–80. https://doi.org/10.1111/1471-0307.12925
3. Machado MAM, Castro VS, Monteiro MLG et al (2023) Can extraintestinal pathogenic *Escherichia coli* with heat resistance profile overcome nonthermal technologies? Foodborne Pathog Dis fpd.2023:0111. https://doi.org/10.1089/fpd.2023.0111
4. Bernardo YAA, Conte-Junior CA (2024) Modeling and optimization of non-thermal technologies for animal-origin food decontamination. Adv Food Nutr Res 113:65–101
5. Monteiro MLG, Do Rosario DKA, De Carvalho APA, Conte-Junior CA (2021) Application of UV-C light to improve safety and overall quality of fish: a systematic review and meta-analysis. Trends Food Sci Technol 116:279–289. https://doi.org/10.1016/j.tifs.2021.07.027
6. Monteiro MLG, Torres Neto L, Mutz YDS et al (2024) Optimized UVC-LED condition to improve the shelf life of vacuum-packed refrigerated stored rainbow trout (*Oncorhynchus mykiss*) fillets. Food Control 156:110141. https://doi.org/10.1016/j.foodcont.2023.110141
7. Monteiro MLG, Torres Neto L, Mutz YDS et al (2024) Optimizing UVC-LED application to improve the shelf life of vacuum-packed refrigerated stored Nile tilapia (*Oreochromis niloticus*) fillets. J Food Sci 89:552–565. https://doi.org/10.1111/1750-3841.16874
8. Fan L, Liu X, Dong X et al (2021) Effects of UVC light-emitting diodes on microbial safety and quality attributes of raw tuna fillets. LWT 139:110553. https://doi.org/10.1016/j.lwt.2020.110553
9. Soro AB, Ekhlas D, Shokri S et al (2023) The efficiency of UV light-emitting diodes (UV-LED) in decontaminating Campylobacter and Salmonella and natural microbiota in chicken breast, compared to a UV pilot-plant scale device. Food Microbiol 116:104365. https://doi.org/10.1016/j.fm.2023.104365
10. Soro AB, Ekhlas D, Marmion M et al (2023) Investigation of differences in susceptibility of *Campylobacter jejuni* strains to UV light-emitting diode (UV-LED) technology. Sci Rep 13, 9459. https://doi.org/10.1038/s41598-023-35315-0
11. Calle A, Fernandez M, Montoya B et al (2021) UV-C LED irradiation reduces Salmonella on chicken and food contact surfaces. Foods 10:1459. https://doi.org/10.3390/foods10071459
12. Wang W, Zhao D, Li K et al (2021) Effect of UVC light-emitting diodes on pathogenic bacteria and quality attributes of chicken breast. J Food Prot 84:1765–1771. https://doi.org/10.4315/JFP-21-066
13. Bernardo YAA, Do Rosario DKA, Monteiro MLG et al (2022) Texture profile analysis: how parameter settings affect the instrumental texture characteristics of fish fillets stored under refrigeration? Food Anal Methods 15:144–156. https://doi.org/10.1007/s12161-021-02095-0
14. Yin MC, Faustman C, Riesen JW, Williams SN (1993) α-Tocopherol and ascorbate delay oxymyoglobin and phospholipid oxidation in vitro. J Food Sci 58:1273–1276. https://doi.org/10.1111/j.1365-2621.1993.tb06164.x

Chapter 5

5-Pulsed Light Field Application in Meat and Meat Products

Teresa Aymerich and Maria João Fraqueza

Abstract

Pulse light (PL) treatments are considered an emergent, safe, and eco-friendly technology that could have applications on food, packages, or food contact materials since they have been proven capable of reducing microbial populations on their surfaces. This non-thermic technology replaces or could be complementary to chemical and thermic treatments applied to food to decontaminate pathogens and spoilage microorganisms on the surface. PL is based on the emission of short-duration pulses of high-intensity broad-spectrum light (radiations from a range of ultraviolet (UV) to near-infrared), from an excited xenon lamp. The number of pulses and pulse characteristics are important to be reported, but it is mandatory to know the dose of PL treatment applied and quantified by the fluence. The present chapter describes PL foundations and recent research on the potential applications of PL on meat and meat products to assure safety and extend shelf life. It discusses the advantages and disadvantages of PL. The procedure of PL treatment applied to meat and meat products is described with all the essential steps to warrant a successful and comparable application of this technology.

Keywords Pulse light, UV-C, Meat, Meat products, Pathogens, Spoilage microorganisms, Fluence

1 Introduction

Pulsed light (PL) processing is a non-thermal technology that has become increasingly popular in the food industry. The present chapter describes PL foundations and recent research on the potential applications of PL on meat and meat products to assure safety and extend shelf life, since to assess in situ effectiveness PL conditions should be validated for each type of meat product.

This emergent technology enhances food safety and stability by using short pulses of high-intensity light to decontaminate food surfaces without the detrimental effects of heat treatments [1]. It exhibits potential for application in food, packaging, and food contact materials due to its demonstrated efficacy in reducing microbial populations on their surfaces. Although the approval by the Food and Drug Administration since 1996 for food-surface applications (up to 12 J/cm^2) [2], pulsed light (PL) still has not been used on a

Renata Ernlund Freitas de Macedo and Carlos Adam Conte-Junior (eds.), *Meat Products Preservation*, Methods and Protocols in Food Science, https://doi.org/10.1007/978-1-0716-5068-4_5,

large scale by industry [3]. Within the European Union, according to Regulation (EU) 2015/2283 [4], the application of pulsed light (PL) is classified as an emerging food processing technology and constitutes a novel production method. Consequently, any food product rendered novel by this technological intervention is subject to a comprehensive safety evaluation by the European Food Safety Authority (EFSA) as an integral component of the EU-wide novel food authorization procedure.

The most recent developments regarding PL application have been reported in the dairy-free industry, as well as in the plant-based food and beverage sectors [5]. This non-thermic technology, which does not generate unwanted residuals, replaces or could complement chemical and thermal treatments applied to food to eliminate pathogens and spoilage microorganisms.

The underlying technological principle involves storing a high discharge voltage in a capacitor, followed by the rapid release of energy through xenon flash lamps. These intense light flashes generate a broad-spectrum emission ranging from approximately 200 to 1100 nm, with around 25% falling within the UV region, critical for the successful inactivation of microorganisms [6]. This method is more efficient than continuous UV irradiation, as it delivers, high fluence in a very short time [7]. Thus, pulsed light (PL) technology relies on the emission of high-intensity, short-duration pulses of broad-spectrum radiation, ranging from ultraviolet (UV) to near-infrared, generated by an electrically excited xenon flash lamp. These lamps can produce light pulses multiple times per second.

The number of pulses and pulse characteristics are important to be reported, but it is mandatory to know the dose of PL treatment applied and quantified by the fluence. The process variables include the type of gas filled in the lamp, light intensity, the type of reflector cover, the sensitivity of the sample to light, the light transmission medium, the distance between the lamp and the sample, frequency (in the range of 0.5–10 Hz), pulse width, and treatment time [8–12] Reducing the distance between the lamp and the sample increases surface decontamination efficiency (as the sample absorbs more light energy [13]. The following parameters [14] are commonly used to describe a Pulsed Light (PL) treatment:

- Fluence rate: measured in watts per square meter (W/m^2), it indicates the energy the lamp delivers to the sample per unit area each second.
- Fluence: measured in joules per square meter (J/m^2), it represents the total energy received by the sample per unit area throughout the treatment duration.
- Dose: sometimes used interchangeably with fluence.

- Exposure time: the total duration of the treatment, measured in seconds.
- Pulse width: the duration of each pulse, typically a fraction of a second, during which energy is emitted.
- Pulse repetition rate (PRR): the number of pulses delivered per second, expressed in hertz (Hz) or as pulses per second (p/s)
- Peak power: measured in watts (W), it is calculated by dividing the energy of a single pulse by its duration.

Pulsed light conditions applied to a food sample should consider the electric potential difference (Volt, V) and distance (D, cm) of the source of Pulsed light to the sample with fixed pulses. The Fluence should be measured to know the reduction of microbial contamination resulting from this parameter applied to the food.

The microbial inactivation mechanisms of PL involve several actions: photochemical effects that inhibit DNA replication by preventing the formation of new DNA strands and inducing dimer formation; photothermal effects resulting from localized temperature increases; and photophysical damage, which disrupts cell membranes and causes protein leakage [15].

Several factors can influence the efficacy of the PL treatment on microbial inactivation, such as food characteristics (non-uniform surfaces and opacity, pH, aw, salt content), pathogen type (Gram positive more resistant than Gram negative), and PL dose applied [16–18].

The application of LP on meat products should be considered only to slices or surface decontamination of meat products (as the energy penetration is restricted to only a few mm). Sample thickness is a critical limiting factor in microbial inactivation by pulsed light (PL), as the limited penetration of UV light allows overlapping or opaque materials to shield underlying surfaces from effective decontamination [5].

2 Materials

Dry or cooked ham, cured loin, or similar meat products.

Slicer

Vacuum machine or MAP packaging (*see* **Note 1**)

Vacuum or MAP package films (*see* **Note 2**)

Pulsed Light equipment, PL tecum unit, with 15% UV-C light (200–280 nm), 50% visible light (280–780 nm) and 35% proximal IR light (780–1100 nm) (CLARANOR, Manosque, France)

Laser power detector (Joulmeter UP17P connected to MAESTRO-monitor, Gentec-EO, Canada) (*see* **Note 3**)

3 Method

The PL treatment process flow represented in Fig. 1, is described below:

a) Meat products preparation: meat product slicing and vacuum packaging.

b) Product are loaded to the PL treatment chamber (*see* **Note 4**).

c) PL light generation:
   - Power supply unit charges capacitors.
   - Discharge is sent to xenon flash lamps, generating high-intensity broad-spectrum UV pulses (including UV-C).

d) Exposure stage (see Table 1):
   - Product is exposed to short pulses of intense UV light.
   - Parameters control (*see* **Note 5**):
     i. Pulse duration (ms).
     ii. Pulse frequency (Hz).
     iii. Energy per pulse (Fluence J/m^2).
     iv. Total exposure time.
     v. Distance from lamp.

e) Cooling and ventilation system:
   - Heat from lamps is dissipated.
   - Airflow protects components and prevents overheating.

f) Monitoring and control unit:
   - Interfaces for:
     i. Lamp operation.
     ii. Energy control.
     iii. Safety interlocks.
     iv. Dose monitoring.

g) Unloading:
   - Treated product exit (*see* **Note 6**).

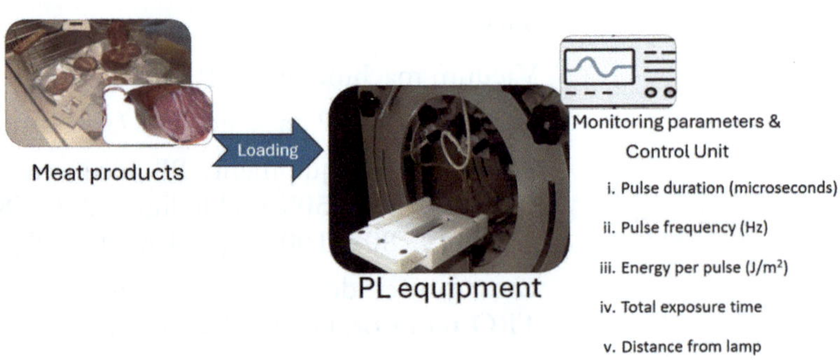

Fig. 1 Pulsed light treatment process

Table 1
Pulse light treatment conditions applied on cured meat products and respective effects on pathogenic or spoilage microorganisms

Type of meat product	Sample preparation	Conditions of PL treatment	Microorganisms	Effects	References
Cooked ham and bologna sausage	Slices (1 mm thickness), vacuum packaged	0.7, 2.1, 4.2 and 8.4 J/cm²	*Listeria monocytogenes*	PL treatment at 8.4 J/cm² reduced *L. monocytogenes* by 1.78 cfu/cm² in cooked ham and by 1.11 cfu/cm² in bologna without changing the quality of the product. The shelf-life of ham was extended up to 49 days under vacuum packaging	[19]
Cooked ham, Frankfurter sausages	Slices ham 1 mm vacuum packaged	0.35 J/cm²–3.6 J/cm²	*Listeria innocua*	Pulsed light treatments with the highest fluence (3.6 J/cm²) reduced *Listeria innocua* on packaged ham by 1 log Pulsed light treatments with the highest fluence (3.6 J/cm²) reduced *Listeria innocua* on packaged frankfurter sausages by 3–4 log. The reduction level of *Listeria innocua* was maintained during storage	[20]
Dry cured Salchichón and loin	Slices 1 mm thick, vacuum packaged	0.7, 2.1, 4.2, 8.4 and 11.9 J/cm²	*L. monocytogenes* and *Salmonella enterica* serovar Typhimurium	1.5 and 1.8 cfu/cm² were obtained for both microorganisms when a fluence of 11.9 J/cm² was applied without changing their sensory properties	[21]
Serrano and Iberian dry ham	Dry ham slices (1 mm thick), vacuum packaged	2.1, 4.2 and 8.4 J/cm²	*Listeria innocua*	The reduction of Listeria was higher in Iberian (2 log cfu/cm²) than in Serrano ham (1 log cfu/cm²) with 8.4 J/cm²	[22]

Table 1
continued

Type of meat product	Sample preparation	Conditions of PL treatment	Microorganisms	Effects	References
Chinese traditional dry cured meat	Slices	(8 J, 300 pulses, 10 cm)	*Micrococcus* and *Staphylococcus* (MSA), yeasts and molds (PDA), *lactic acid bacteria* (MRS)	PL alone was a little less effective than PL-UV equipment that contained xenon and UV lamps. PL-UV was able to reduce 2 log the total counts, *Micrococcus* and *Staphylococcus*, while PL was only able to reduce 1.22 log MSA counts, and 0.45 log PDA counts, while MRS was not affected. The treatment did not negatively contribute to color or sensory changes. Microbial decontamination was correlated with lower POV and effective extension of shelf life. PL was able to extend, but less than PL-UV	[23]
Poultry meat	Poultry breast (4–5 mm) slight vacuum pack	Fluence from 2.82 to 9.67 J/cm^2	*Campylobacter*	PL treatment was able to reduce 1–1.3 log of *Enterobacteriaceae* but was unable to reduce even 1 log of *Campylobacter*.	[17]
			Enterobacteriaceae	Volatile profile changed with treatment, reducing aldehyde content. Concerning color, L* and b* parameters were influenced while a* increased	
Cured and smoked meat product *salpicão*	Slices (3 mm thick), vacuum packaged	Fluence from 1.74 to 16.11 J/cm^2	*L. monocytogenes*	PL treatment was able to inactivate *L. monocytogenes* from the surface of a packaged sliced cured meat product *salpicão*, with a maximum log reduction of 1.58 cfu/g obtained with a fluence of 5.31 J/cm^2. The application of higher PL fluence did not result in a higher reduction of *L. monocytogenes*.	[18]
				Enterobacteriaceae, LAB, and CNS endogenous microbial populations were not influenced by PL treatment	
Mortadella	Slices (2.5 mm thick, 1 g)	Fluence of 5.28 J/cm^2 and an irradiance of 10,469 W/cm^2	Total mesophilic aerobic bacteria and lactic acid bacteria	PL application alone effectively reduced Total mesophilic aerobic bacteria and lactic acid bacteria populations by 1.4–1.55 log CFU/g and 1.0–2.24 log CFU/g, respectively, but negatively affected mortadella's pH, color, and oxidative quality throughout storage	[24]

h) Security measures:
   - Enclosed chamber or shielded system.
   - UV-blocking materials.
   - Emergency shut-off and interlocks.

4 Conclusion

The effect of PL against several food borne pathogens (mainly *Listeria* and *Salmonella*) together with the effect against total mesophilic aerobic bacteria and lactic acid bacteria in several meat products (cooked ham, bologna sausage, frankfurter sausages, dry cured salchichón and loin, serrano and Iberian dry ham, cured and smoked salpicão and mortadella) has been reported and it is shown in Table 1. In summary, the best treatments were between 3.6 and 11.9 J/cm^2, and mainly they were able to achieve a reduction of 1–2 log of the bacterial target previously inoculated in the products. The treatments did not significantly change the quality of some of the products tested, except for mortadella.

5 Notes

1. Packaging of the products may be performed either before or following PL treatment; however, effective exposure of the product surface to the pulsed light is essential. The overlapping of slices may negatively impact treatment efficacy by obstructing light penetration. Additionally, in the case of modified atmosphere packaging, the composition and volume of the headspace could further influence the effectiveness of the PL treatment.
2. The packaging film must be suitable for transmitting pulsed light, including ultraviolet (UV) wavelengths, to ensure the effectiveness of the treatment. Validating the fluence on the product surface through the packaging film is essential to ensure effective treatment.
3. The temperature of the surface and the energy absorbed by the surface should be monitored, namely the (Fluence J/m^2).
4. To optimize product quality, it is very important that the product is well refrigerated and that the environmental control of the processing rooms, including that of the PL equipment, is properly managed and cooled.

5. To ensure comparability of results, the fluence should be standardized, irrespective of the number of pulses, pulse duration, distance from the light source, total exposure time, or food matrix.
6. After processing, the product must be quickly transferred to preservation conditions.

Acknowledgments

This work is supported by FCT, under projects UID/276/2025 (CIISA), and LA/P/0059/2020 (AL4AnimalS) with financial support of the project AL4A-PROJ-LT1.3—Safety and acceptability assessment of Green label cured meat products.

References

1. Cheigh CI, Hwang HJ, Chung MS (2013) Intense pulsed light (IPL) and UV-C treatments for inactivating *Listeria monocytogenes* on solid medium and seafoods. Food Res Int 54:745–752
2. FDA (1996) U.S. Food and Drug Administration. 1996. Code of Federal Regulations. 21 CFR 179.41. https://www.ecfr.gov/current/title-21/chapter-I/subchapter-B/part-179/subpart-B/section-179.41
3. Rowan NJ (2019) Pulsed light as an emerging technology to cause disruption for food and adjacent industries – Quo vadis? Trends Food Sci Technol 88:316–332. https://doi.org/10.1016/j.tifs.2019.03.027
4. Regulation (EU) 2015/2283 Regulation (EU) 2015/2283 of the European Parliament and of the Council on novel foods, amending Regulation (EU) No. 1169/2011 of the European Parliament and of the Council and repealing Regulation (EC) No. 258/97 of the European Parliament and of the Council and Commission Regulation (EC) No. 1852/2001. Official J Eur Union L 327:1–22. eur-lex.europa.eu/homepage.html
5. CLARANOR (2025) Food & beverage pulsed light packaging sterilization. https://www.claranor.com/en/solutions-2/sterilization-of-food-beverage-packaging-by-pulsed-light/ Accessed
6. Kramer B, Muranyi P (2014) Effect of pulsed light on structural and physiological properties of *L. innocua* and *Escherichia coli*. J Appl Microbiol 116:596–611
7. Garvey M, Rowan NJ (2019) Pulsed UV as a potential surface sanitizer in food production processes to ensure consumer safety. Curr Opin Food Sci 26:65–70. https://doi.org/10.1016/j.cofs.2019.03.003
8. Ozer NP, Demirci A (2005) Inactivation of *Escherichia coli* O157:H7 and *Listeria monocytogenes* inoculated on raw salmon fillets by pulsed UV-light treatment. Int J Food Sci Technol 41:354–360
9. Ortega-Rivas E (2012) Pulsed light technology. In: Ortega-Rivas E (Ed.) In non-thermal food engineering operations. Springer, Boston, MA, pp. 263–273
10. Kim AJ, Shin JK (2014) Nonthermal sterilization of dried laver by intense pulsed light with batch system. Korean J Food Sci Technol 46:778–781
11. Park JA (2017) Effect of sterilization as chamber and packaging film materials by intense pulsed light treatment. M.S. thesis, Jeonju Univ, Jeonju, Korea
12. Park JA, Shin JK (2021) Effect of inactivation on *Escherichia coli* as panel materials of treatment chamber used in intense pulsed light treatment. Food Eng Prog 25:240–246
13. Kim A-J, Shin J-K (2015) Nonthermal sterilization of pathogenic *Escherichia coli* by intense pulsed light using a batch system. Korean J Food Sci Technol 47(1):81–86. https://doi.org/10.9721/kjfst.2015.47.1.81
14. IUPAC (2023) Quantities, units, and symbols in physical chemistry, the IUPAC Green Book Abridged Version, 4th ed. RSC Publishing, https://doi.org/10.1039/9781839163180
15. Hwang H-J, Seo J-H, Jeong C, Cheigh C-I, Chung M-S (2019) Analysis of bacterial inactivation by intense pulsed light using a double-Weibull survival model. Innov Food Sci Emerg Technol 56:102185. https://doi.org/10.1016/j.ifset.2019.102185

16. Bryant MT, Degala HL, Mahapatra AK, Gosukonda RM, Kannan G (2021) Inactivation of *Escherichia coli* K12 by pulsed UV light on goat meat and beef: Microbial responses and modelling. Int J Food Sci Technol 56(2):563–572. https://doi.org/10.1111/IJFS.14733
17. Baptista E, Borges A, Aymerich T, Alves SP, Gama LT, Fernandes H, Fernandes MJ, Fraqueza MJ (2022) Pulsed light application for campylobacter control on poultry meat and its effect on color and volatile profile. Foods 11(18):2848. https://doi.org/10.3390/foods11182848
18. Borges A, Baptista E, Aymerich T, Alves SP, Gama LT, Fraqueza MJ (2023) Inactivation of *Listeria monocytogenes* by pulsed light in packaged and sliced salpicão, a ready-to-eat traditional cured smoked meat sausage. LWT 179:114641. https://doi.org/10.1016/j.lwt.2023.114641
19. Hierro E, Barroso E, De la Hoz L, Ordóñez JA, Manzano S, Fernández M (2011) Efficacy of pulsed light for shelf-life extension and inactivation of *Listeria monocytogenes* on ready-to-eat cooked meat products. Innov Food Sci Emerg Technol 12(3):275–281
20. Kramer B, Wunderlich J, Muranyi P (2019) Inactivation of Listeria innocua on packaged meat products by pulsed light. Food Packag Shelf Life 21, https://doi.org/10.1016/j.fpsl.2019.100353
21. Ganan M, Hierro E, Hospital XF, Barroso E, Fernández M (2013) Use of pulsed light to increase the safety of ready-to-eat cured meat products. Food Control 32(2):512–517. https://doi.org/10.1016/J.FOODCONT.2013.01.022
22. Fernández M, Hospital XF, Cabellos C, Hierro E (2020) Effect of pulsed light treatment on Listeria inactivation, sensory quality and oxidation in two varieties of Spanish dry-cured ham. Food Chem 316:126294. https://doi.org/10.1016/J.FOODCHEM.2020.126294
23. Liu N, Zhu Q, Zeng X, Yang B, Liang M, Hu P, He L, Deng L, Liang C, Zhang R, Zhou J (2019) Influences of pulsed light-UV treatment on the storage period of dry-cured meat and shelf life prediction by ASLT method. J Food Sci Technol 56:1744–1756. https://doi.org/10.1007/s13197-019-03603-1
24. Fracari PR, Tomasevic I, Massia AG, Laroque DA, Balzan MM, dos Santos BA, Cichoski AJ, Wagner R, Carciofi BAM, Campagnol PCB (2025) Pulsed light and jabuticaba peel extract for nitrite reduction and quality enhancement in sliced mortadella. Meat Sci 224:109777. https://doi.org/10.1016/j.meatsci.2025.109777

Check for updates

Chapter 6

Cold Plasma Application in Meat and Meat Products

Édipo da Silva Almeida and Carlos Adam Conte-Junior

Abstract

Cold plasma (CP) is an innovative non-thermal technology applied to meat preservation, offering microbial inactivation, control of lipid oxidation, and shelf-life extension without compromising sensory and nutritional attributes. Its mechanism of action relies on the generation of reactive oxygen and nitrogen species (RONS), ultraviolet radiation, and charged particles that interact with food surfaces, affecting microbial cells and biochemical components. The effectiveness of CP depends on operational parameters such as gas composition, exposure time, electrode distance, and applied voltage, which highlights the need for protocol standardization. In this chapter, we present a standardized and experimentally validated protocol for CP application to beef loin, pork loin, and chicken breast, using a dielectric barrier discharge (DBD) system with ambient air. All experimental parameters are described, including sample preparation, treatment conditions, analytical procedures, and critical considerations for reproducibility and safety assessment. This protocol is designed to serve as a technical guide for researchers and food technologists, facilitating the adoption of cold plasma technology as a chemical-free decontamination strategy. Standardizing CP application contributes to food safety improvements, reduces the need for additives, and supports the development of scalable, non-thermal preservation approaches for meat products.

Keywords Cold plasma, Dielectric barrier discharge, Microbial inactivation, Lipid oxidation, Fresh meat

1 Introduction

Cold plasma (CP) is a non-thermal processing technology that has shown high efficacy in microbial inactivation without compromising the physicochemical and sensory attributes of foods [1, 2]. Generated through electrical discharges—such as dielectric barrier discharge (DBD) [3, 4], plasma jets, and plasma-activated water (PAW) [5], cold plasma produces a complex mixture of reactive oxygen and nitrogen species (RONS), charged particles, and ultraviolet photons [6, 7]. These species interact synergistically at the surface of foods, promoting membrane disruption, DNA fragmentation, and microbial inactivation [8].

In fresh meat systems, CP application has demonstrated significant potential for surface decontamination, oxidative stability, and

Renata Ernlund Freitas de Macedo and Carlos Adam Conte-Junior (eds.), *Meat Products Preservation*, Methods and Protocols in Food Science, https://doi.org/10.1007/978-1-0716-5068-4_6,

quality retention [9]. However, the effectiveness of CP treatments depends on several parameters, including the electrical configuration, gas composition, exposure time, electrode gap, and sample characteristics [8, 10]. Standardizing these variables is essential to ensure reproducibility, safety, and scalability.

This chapter provides a detailed and experimentally validated protocol for the application of dielectric barrier discharge (DBD) cold plasma using ambient air to three raw meat matrices: beef loin (*Longissimus dorsi*), pork loin, and chicken breast (*Pectoralis major*). These specific cuts were selected due to their commercial relevance, uniform surface morphology, and frequent appearance in cold plasma literature, which supports the development of a consistent and transferable treatment protocol [11–16].

All methodological aspects are described in a clear, prescriptive manner, including sample preparation, plasma exposure conditions, and post-treatment analyses. By adopting standardized operational parameters and analytical procedures, this chapter aims to provide researchers and food technologists with a reproducible protocol that supports the development of chemical-free decontamination strategies for the meat industry.

2 Materials

2.1 Sample Collection and Storage

Fresh samples of beef loin (*Longissimus dorsi*), pork loin, and chicken breast (*Pectoralis major*) should be obtained from a certified local butcher or a reliable meat supplier on the day of the experiment. Select cuts that are free of visible fat, connective tissue, or surface damage to ensure uniform structure and appearance across all samples.

1. Using sterile instruments, portion the meat into rectangular pieces of approximately 5.0 cm × 5.0 cm × 1.5 cm (length × width × thickness), with a target weight of 40–50 g per sample. Maintain consistency in size and thickness between samples to reduce variability during treatment and analysis.
2. Immediately after portioning, individually vacuum-pack the samples using food-grade polyethylene bags to prevent dehydration and contamination. Store all samples at 4 ± 1 °C, in the dark, for no longer than 24 h prior to cold plasma exposure. Avoid freezing to preserve the structural integrity and original physicochemical properties of fresh meat.
3. All handling steps should be performed under hygienic conditions, using sterile gloves and sanitized surfaces.

2.2 Equipment and Gases

The use of a dielectric barrier discharge (DBD) cold plasma system is strongly recommended for the treatment of meat cuts such as beef loin, pork loin, and chicken breast, due to its ability to generate a uniform, surface-acting plasma over flat or slightly irregular biological surfaces. DBD systems are widely used in food decontamination studies and offer a good balance between microbial inactivation efficiency and preservation of sensory quality. Their relatively simple configuration also facilitates laboratory-scale applications and reproducibility across different sample types.

1. A typical DBD setup consists of two parallel electrodes, at least one of which is covered with a dielectric material such as glass, quartz, or ceramic [8]. The meat sample should be placed on a non-conductive support positioned between the electrodes or just beneath the active electrode, depending on the system design. The discharge gap, defined as the distance between the electrode and the sample surface should be maintained at 5.0 ± 0.5 mm to ensure effective exposure to reactive species while minimizing thermal effects.
2. The system should operate at an output voltage between 60 and 80 kV, with a frequency in the range of 10–40 kHz, as commonly used in food-grade DBD applications [16, 17]. These parameters are sufficient to generate a stable plasma discharge capable of producing reactive oxygen and nitrogen species (RONS), including ozone, singlet oxygen, nitric oxide, and other radicals essential for microbial inactivation.
3. For the working gas, ambient air is the preferred option due to its accessibility, low cost, and the generation of both oxygen- and nitrogen-based reactive species. Alternatively, argon or oxygen may be used to fine-tune the oxidative potential of the discharge, particularly when milder or more specific effects on meat quality are desired. The gas flow should be adjusted to maintain a flow rate of 1–3 L/min if an external gas supply is used.
4. The electrode surface area should cover at least 2 cm beyond the borders of each meat sample to ensure uniform plasma exposure. In multi-sample setups, treat each portion individually or ensure sufficient spacing to avoid shadowing effects.
5. All components of the plasma setup should be grounded and shielded appropriately to prevent electrical hazards. The operation should be conducted in a well-ventilated area or under an exhaust hood due to the possible generation of ozone and nitrogen oxides during treatment.

2.3 Reagents for Microbiological and Physicochemical Analysis

To evaluate the effects of cold plasma treatment on meat quality and safety, the following reagents, solutions, and equipment are required for microbiological and physicochemical analyses:

2.3.1 Total Viable Count of Mesophilic and Psychrotrophic Microorganisms: Culture Media, Diluentes, Materials and Equipments

1. Plate Count Agar (PCA);
2. Sterile 0.1% peptone water;
3. Petri dishes (sterile);
4. Pipettes and micropipettes with sterile tips;
5. Stomacher or blender with sterile bags;
6. Incubators set at 30°C (mesophiles) and 7°C (psychrotrophs);
7. Colony counter or manual counting grid;
8. Laminar flow hood or biosafety cabinet;
9. Autoclave for sterilization of media and instruments.

2.3.2 pH Measurement: Reagents and Equipments

1. None required beyond distilled water for electrode rinsing;
2. Bench-top digital pH meter with meat-penetration electrode (calibrated at pH 4.00 and 7.00);
3. Beakers, gloves, and wipes for cleaning.

2.3.3 Color Measurement (CIE Lab): Equipment and Conditions*

1. Portable or bench-top colorimeter or spectrocolorimeter calibrated to D65 standard illuminant;
2. White calibration tile;
3. Glass or quartz sample holders (if needed);
4. Perform color readings on freshly cut meat surfaces after 30 min of blooming at 4°C.

2.3.4 Lipid Oxidation (TBARS Assay): Reagents and Equipments

1. Trichloroacetic acid (TCA) 10% (w/v);
2. Thiobarbituric acid (TBA) 0.02 M;
3. Butylated hydroxytoluene (BHT) – antioxidant;
4. Malondialdehyde (MDA) standard (optional for calibration);
5. Distilled water;
6. Centrifuge (≥3000 rpm);
7. Water bath (90–100°C);
8. Spectrophotometer set at 532 nm;
9. Glass cuvettes.

2.3.5 Texture Profile Analysis (TPA): Equipments and Conditions

1. Texture analyzer or texturometer with cylindrical probe (e.g., 50–75 mm diameter);
2. Data acquisition software;
3. Perform analysis at room temperature (~22°C);
4. Samples should be standardized in shape and temperature.

2.3.6 Total Volatile Basic Nitrogen (TVB-N): Reagents and Equipments

1. Magnesium oxide (MgO);
2. Boric acid solution (2%);

3. Mixed indicator solution (methyl red and bromocresol green);
4. Hydrochloric acid 0.01 N (for titration);
5. Steam distillation apparatus (e.g., Kjeldahl-type or semi-automatic);
6. Buret and titration setup;
7. Conical flasks and distillation flasks;
8. Heating mantle or distillation unit with temperature control.

All analyses should be performed in triplicate for each sample type. Analytical-grade reagents and freshly prepared solutions are recommended to ensure reliability and reproducibility.

3 Methods

3.1 Sample Preparation

3.1.1 Beef Loin

1. Trim visible fat and connective tissue from the beef loin (*Longissimus dorsi*) using a sterile scalpel or knife. Cut the meat into rectangular portions measuring 5.0 cm × 5.0 cm × 1.5 cm, with a target weight of 40–50 g per sample. Prepare at least three replicates per treatment condition. Gently pat the surface of each sample with sterile paper to remove excess moisture, without compromising the natural hydration of the meat.
2. Place each portion into a separate sterile polyethylene bag, and store at 4 ± 1°C for up to 24 h before plasma treatment. Avoid freezing or excessive handling to preserve surface integrity.

3.1.2 Pork Loin

1. Select lean portions of pork loin and remove any visible fat or membrane. Using a sterile cutting tool, prepare rectangular cuts with dimensions of 5.0 cm × 5.0 cm × 1.5 cm and weight between 40 and 50 g. Maintain consistency in shape and size across all samples to ensure uniform plasma exposure.
2. Transfer each sample to a vacuum-sealable food-grade plastic pouch and seal under mild vacuum. Store at 4 ± 1°C and proceed with plasma treatment within 24 h of preparation.

3.1.3 Chicken Breast

1. Use fresh, skinless chicken breast (Pectoralis major). Remove any residual connective tissue or blood vessels. Cut standardized portions measuring 5.0 cm × 5.0 cm × 1.5 cm, ensuring homogeneous size and weight (40–50 g). Surface-dry each piece with sterile absorbent paper to reduce free surface moisture.

2. Individually place the samples in sterile vacuum bags and store at refrigeration temperature (4 ± 1°C). Perform plasma treatment within 24 h of preparation to ensure sample freshness.

3.2 Cold Plasma Treatment

This protocol standardizes the cold plasma treatment procedure for beef loin, pork loin, and chicken breast using the same DBD system parameters. However, slight variations in composition, fat content, or surface hydration between these meats may affect plasma interactions. Such considerations are addressed in the Notes section and should be reviewed before execution (*see* **Note 1**).

1. Utilize a dielectric barrier discharge (DBD) plasma system equipped with two parallel electrodes, one of which is coated with a dielectric material, such as glass, quartz, or ceramic. Position the meat sample on a non-conductive support placed beneath the discharge electrode. Maintain a discharge gap of 5.0 ± 0.5 mm between the electrode and the upper surface of the meat to ensure adequate exposure while minimizing arcing (*see* **Note 2**).
2. Set the plasma system to operate at an output voltage of 70 kV and a frequency of 20 kHz. These values ensure consistent plasma discharge for microbial inactivation without significant heating of the sample surface (*see* **Note 3**).
3. Use ambient air as the working gas, introduced at a flow rate of 2.0 L/min, directed transversally or perpendicularly across the sample area to ensure the dispersion of reactive species. If using a gas source with pressure regulation, maintain a stable output pressure of 1.0–1.5 bar (*see* **Note 4**).
4. Place only one sample at a time beneath the electrode, ensuring the electrode surface area exceeds the meat sample's area by at least 2 cm on all sides to avoid edge effect and ensure uniform exposure. Do not stack or overlap samples during treatment (*see* **Note 5**).
5. Expose each sample to plasma for 5 min, maintaining constant treatment conditions across all replicates. Use a timer to ensure precise control of exposure time. After each sample, wait at least 2 min before treating the next one to allow the electrodes to cool and avoid energy drift (*see* **Note 6**).
6. During treatment, monitor the sample's surface temperature using an infrared thermometer or thermographic camera. Do not allow the surface temperature to exceed 30°C, as heating may compromise the "cold" nature of the treatment and introduce thermal effects (*see* **Note 7**).
7. Immediately after plasma exposure, transfer each sample to a sterile, labeled polyethylene bag, and store at 4 ± 1°C until analysis. All physicochemical and microbiological analyses should be conducted within 2 h post-treatment to avoid bias from storage degradation (*see* **Note 8**).

3.3 Microbiological Analysis

1. To determine the total viable count (TVC) of mesophilic and psychotropic microorganisms, use Plate Count Agar (PCA) as the culture medium. Prepare the medium according to the manufacturer's instructions, autoclave at 121°C for 15 min, and allow to cool to approximately 45°C before pouring into sterile Petri dishes.
2. Using sterile scissors and tweezers, aseptically transfer 25 g of meat sample into a sterile stomacher bag containing 225 mL of 0.1% sterile peptone water. Homogenize the sample for 2 min using a stomacher or paddle blender at medium speed. This generates the first dilution (10^{-1}).
3. Prepare serial decimal dilutions (e.g., 10^{-2}–10^{-6}) in sterile 0.1% peptone water. For each selected dilution, pipette 1.0 mL into sterile Petri dishes in duplicate, and immediately pour approximately 15 mL of molten PCA (45 ± 1°C) into each plate. Gently swirl to mix and allow the medium to solidify at room temperature.
4. Incubate mesophilic counts at 30 ± 1°C for 48 h and psychotropic counts at 7 ± 1°C for 10 days. After incubation, count plates that present 30–300 colonies to ensure accuracy. If necessary, average the colony counts from duplicate plates.
5. Calculate the result in colony-forming units per gram (CFU/g) using the following formula:

$$CFU/g = \frac{Average\ number\ of\ colonies \times Diluition\ Factor}{Volume\ plated\ (mL)} \quad (1)$$

6. Record all results with one decimal place, and express them as $\log_{10}$ CFU/g for comparative purposes. For samples below detection limits (less than 1.0 log CFU/g), report as "not detected" (*see* **Note 9**).

All microbiological procedures should be performed under aseptic conditions using a laminar flow hood or biosafety cabinet. Clean and disinfect all work surfaces and equipment before and after analysis. More details on this microbiological analysis can be found in the studies carried out by Moutiq et al. [15], Zhuang et al. [13], and Liao et al. [18].

3.4 Color, pH, and Texture Analysis

3.4.1 Color (CIE L*a*b*)

1. Color evaluation should be performed on the fresh surface of each meat sample, following a blooming period of 30 min at 4 ± 1°C to allow oxygenation and pigment stabilization. Use a portable or benchtop colorimeter previously calibrated with

a standard white tile. Set the illuminant to D65 and use an observation angle of 10°.

2. Record values of L^* (lightness), a^* (red-green component), and b^* (yellow-blue component) from three different points on the sample surface. Avoid areas with visible damage, discoloration, or fat inclusions.
3. To assess overall color change before and after plasma treatment, calculate the total color difference (ΔE) using the following equation:

$$\Delta E = \sqrt{(L_1^* - L_2^*)^2 + (a_1^* - a_2^*)^2 + (b_1^* - b_2^*)^2} \quad (2)$$

where:
L_1^*, a_1^*, b_1^* = color values before treatment;
L_2^*, a_2^*, b_2^* = color values after treatment.

4. Color measurements should be conducted in triplicate and reported as mean ± standard deviation.

3.4.2 pH Measurement

1. Measure the pH using a bench-top digital pH meter equipped with a penetration electrode, previously calibrated with buffer solutions at pH 4.00 and 7.00. Insert the electrode directly into the geometric center of each meat sample. Ensure measurements are taken at room temperature (20–22°C) and that the electrode is rinsed with distilled water between each reading to avoid cross-contamination.
2. Perform measurements in triplicate for each sample and report values as mean ± standard deviation.

3.4.3 Texture Profile Analysis (TPA)

1. Perform texture analysis using a texture analyzer fitted with a cylindrical probe (typically 50–75 mm in diameter). Standardize meat samples to a uniform size and conduct the test at room temperature (~22°C) after removing them from refrigeration for 15 min.
2. Use a double compression cycle to 50% of the original sample height at a crosshead speed of 1.0 mm/s. Record the following parameters using the instrument's software:
   - Hardness: peak force of the first compression.
   - Cohesiveness: ratio of areas under the first and second compressions.
   - Springiness: distance recovered between compressions.
   - Chewiness: derived from hardness × cohesiveness × springiness.
3. Conduct each analysis in triplicate and average the values for reporting.

3.5 Lipid Oxidation

1. Lipid oxidation should be assessed using the thiobarbituric acid reactive substances (TBARS) assay, which quantifies the malondialdehyde (MDA) formed as a secondary product of lipid peroxidation. The MDA reacts with thiobarbituric acid (TBA) to form a pink chromogen, which is measurable by spectrophotometry.
2. To begin, homogenize 5 g of meat sample with 15 mL of 10% (w/v) trichloroacetic acid (TCA) containing 0.1% butylated hydroxytoluene (BHT) to prevent artificial oxidation during preparation. Centrifuge the mixture at 3000 × *g* for 10 min and collect the supernatant. Add 2 mL of 0.02 M TBA solution to 2 mL of the supernatant, mix thoroughly, and incubate in a water bath at 95°C for 30 min. Cool the reaction tubes to room temperature under running water.
3. Measure the absorbance at 532 nm using a UV–visible spectrophotometer. Calculate the concentration of MDA using the following equation:

$$MDA\left(\frac{mg}{kg}\right)=\frac{A_{352}\times V\times K}{W} \quad (3)$$

where:
A_{352} = absorbance at 532 nm;
V = final volume of extract (mL)
K = constant related to the molar absorptivity of the MDA–TBA complex
W = weight of the sample (g)

4. In addition to TBARS, protein oxidation may also be evaluated using specific assays such as carbonyl content determination or free thiol group quantification; however, these methods are not included in this protocol (*see* **Note 10**).
5. The value of K depends on the use of a standard curve or specific protocol conditions and should be adjusted accordingly (*see* **Note 11**). Results should be expressed as mg of MDA per kg of sample and reported as mean ± standard deviation of triplicate measurements.

3.6 Total Volatile Basic Nitrogen (TVB-N)

1. The total volatile basic nitrogen (TVB-N) content is an indicator of meat freshness and protein degradation. It quantifies the concentration of volatile amines such as ammonia, trimethylamine, and dimethylamine, released from microbial and enzymatic activity in meat matrices.
2. To perform the analysis, homogenize 10 g of meat sample with 100 mL of distilled water. Alkalinize the mixture by adding 1 g of magnesium oxide (MgO), and immediately distill the volatile bases using a steam distillation apparatus (e.g., Kjeldahl or

semi-automatic distiller). Collect the distillate in 25 mL of 2% boric acid solution containing a mixed indicator (methyl red and bromocresol green).

3. Titrate the boric acid solution with 0.01 N hydrochloric acid (HCl) until the endpoint is reached (color change from green to pink). Record the volume of HCl used. Calculate the concentration of TVB-N using the following equation:

$$Total\ Volatile\ Basic\ Nitrogen\left(\frac{mg}{100g}\right)=\frac{(V\times N\times 14.007)\times 100}{W} \quad (4)$$

where:
V = volume of HCl used in the titration (mL);
N = normality of HCl (mol/L);
14.007 = molar mass of nitrogen (mg/mmol);
W = weight of the meat sample (g).

4. The result should be expressed as mg of nitrogen per 100 g of sample and reported as the mean ± standard deviation of triplicate measurements. Ensure that all reagents are freshly prepared and the apparatus is properly calibrated before analysis (*see* **Note 12**).

4 Notes

1. Although this protocol applies uniformly to beef loin, pork loin, and chicken breast, slight differences in fat content, water activity, and pigment composition may affect the response to cold plasma treatment. Users should interpret results with consideration of each matrix's specific characteristics. For further examples of DBD application to each specific cut, see Jayasena et al. [11], Huang et al. [13], and Abdel-Naeem et al. [16].
2. The discharge gap between the electrode and the meat surface must remain consistent to avoid arc formation or uneven plasma distribution. Always verify the gap with a non-metallic gauge prior to treatment.
3. Output voltage and frequency settings may vary slightly depending on the plasma generator model. Ensure that the device provides a stable discharge and does not exceed the thermal threshold of the sample.
4. When using compressed gas, ensure a stable and regulated flow rate. Excessive flow may disturb reactive species concentration or dry out the sample surface.
5. Plasma exposure is highly surface-dependent. Ensure the electrode area exceeds the sample dimensions by at least 2 cm on all sides to avoid incomplete treatment at the edges.

6. Allow a cooling period of at least 2 min between successive treatments to prevent electrode overheating and ensure consistent plasma activity.
7. Monitor the sample surface temperature during treatment. If it exceeds 30°C, reduce the exposure time or increase rest intervals to preserve non-thermal conditions.
8. Treated samples must be analyzed within 2 h post-treatment to ensure valid physicochemical and microbiological readings. Longer storage may lead to degradation or shifts in oxidative status.
9. For microbial counts below 1.0 log CFU/g, results should be reported as "not detected" or "below detection limit." Ensure appropriate dilution ranges are tested to avoid false negatives.
10. Protein oxidation can be assessed using methods such as carbonyl group quantification (e.g., DNPH assay), sulfhydryl content determination, or FTIR spectroscopy. These methods are not covered in this protocol but can be consulted in specialized literature and standard procedures for muscle protein evaluation [13, 19].
11. The constant K in Eq. (3) may vary depending on laboratory protocols, reagent purity, and whether a calibration curve is used. Users are advised to follow validated methods or commercial TBARS kits for consistent results.
12. Distillation-based TVB-N measurements may vary based on equipment (manual or semi-automated), titration accuracy, and operator experience. Adherence to AOAC or ISO standard methods is recommended for regulatory or comparative studies.

Acknowledgments

The authors gratefully acknowledge the Coordination for the Improvement of Higher Education Personnel (CAPES) for funding this research through the Institutional Postdoctoral Program (PIPD) under grant number 88887.007873/2024-00. We also thank the Food Analysis Center (NAL) and the Technological Development Support Laboratory (LADETEC), both part of the Federal University of Rio de Janeiro (UFRJ), for providing technical support and access to analytical infrastructure.

This protocol was developed based on a critical review of current scientific literature and adapted from validated methodologies published in recent experimental studies involving cold plasma applications in meat systems.

References

1. Harikrishna S, Anil PP, Shams R, Dash KK (2023) Cold plasma as an emerging nonthermal technology for food processing: a comprehensive review. J Agric Food Res 14:100747. https://doi.org/10.1016/j.jafr.2023.100747
2. Ganesan AR, Tiwari U, Ezhilarasi PN, Rajauria G (2021) Application of cold plasma on food matrices: a review on current and future prospects. J Food Process Preserv 45. https://doi.org/10.1111/jfpp.15070
3. Birania S, Attkan AK, Kumar S, Kumar N, Singh VK (2022) Cold plasma in food processing and preservation: a review. J Food Process Eng 45. https://doi.org/10.1111/jfpe.14110
4. Akhtar J, Abrha MG, Teklehaimanot K, Gebrekirstos G (2022) Cold plasma technology: fundamentals and effect on quality of meat and its products. Food Agric Immunol 33:451–478. https://doi.org/10.1080/09540105.2022.2095987
5. Yawut N, Mekwilai T, Vichiansan N, Braspaiboon S, Leksakul K, Boonyawan D (2024) Cold plasma technology: transforming food processing for safety and sustainability. J Agric Food Res 18:101383. https://doi.org/10.1016/j.jafr.2024.101383
6. Perera KY, Prendeville J, Jaiswal AK, Jaiswal S (2022) Cold plasma technology in food packaging. Coatings 12:1896. https://doi.org/10.3390/coatings12121896
7. Farooq S, Dar AH, Dash KK, Srivastava S, Pandey VK, Ayoub WS, Pandiselvam R, Manzoor S, Kaur M (2023) Cold plasma treatment advancements in food processing and impact on the physiochemical characteristics of food products. Food Sci Biotechnol 32:621–638. https://doi.org/10.1007/s10068-023-01266-5
8. Zhang C, Luo Y, Deng Z, Du R, Han M, Wu J, Zhao W, Guo R, Hou Y, Wang S (2025) Recent advances in cold plasma technology for enhancing the safety and quality of meat and meat products: a comprehensive review. Food Res Int 202:115701. https://doi.org/10.1016/j.foodres.2025.115701
9. Misra NN, Jo C (2017) Applications of cold plasma technology for microbiological safety in meat industry. Trends Food Sci Technol 64:74–86. https://doi.org/10.1016/j.tifs.2017.04.005
10. Laroque DA, Seó ST, Valencia GA, Laurindo JB, Carciofi BAM (2022) Cold plasma in food processing: design, mechanisms, and application. J Food Eng 312:110748. https://doi.org/10.1016/j.jfoodeng.2021.110748
11. Jayasena DD, Kim HJ, Yong HI, Park S, Kim K, Choe W, Jo C (2015) Flexible thin-layer dielectric barrier discharge plasma treatment of pork butt and beef loin: effects on pathogen inactivation and meat-quality attributes. Food Microbiol 46:51–57. https://doi.org/10.1016/j.fm.2014.07.009
12. Bauer A, Ni Y, Bauer S, Paulsen P, Modic M, Walsh JL, Smulders FJM (2017) The effects of atmospheric pressure cold plasma treatment on microbiological, physical-chemical and sensory characteristics of vacuum packaged beef loin. Meat Sci 128:77–87. https://doi.org/10.1016/j.meatsci.2017.02.003
13. Huang M, Wang J, Zhuang H, Yan W, Zhao J, Zhang J (2019) Effect of in-package high voltage dielectric barrier discharge on microbiological, color and oxidation properties of pork in modified atmosphere packaging during storage. Meat Sci 149:107–113. https://doi.org/10.1016/j.meatsci.2018.11.016
14. Cui H, Wu J, Li C, Lin L (2017) Promoting anti-listeria activity of lemongrass oil on pork loin by cold nitrogen plasma assist. J Food Saf 37. https://doi.org/10.1111/jfs.12316
15. Moutiq R, Misra NN, Mendonça A, Keener K (2020) In-package decontamination of chicken breast using cold plasma technology: microbial, quality and storage studies. Meat Sci 159:107942. https://doi.org/10.1016/j.meatsci.2019.107942
16. Abdel-Naeem HHS, Ebaid EMSM, Khalel KHM, Imre K, Morar A, Herman V, EL-Nawawi FAM (2022) Decontamination of chicken meat using dielectric barrier discharge cold plasma technology: the effect on microbial quality, physicochemical properties, topographical structure, and sensory attributes. LWT 165:113739. https://doi.org/10.1016/j.lwt.2022.113739
17. Jayasena DD, Kang T, Wijayasekara KN, Jo C (2023) Innovative application of cold plasma technology in meat and its products. Food Sci Anim Resour 43:1087–1110. https://doi.org/10.5851/kosfa.2023.e31
18. Liao X, Xiang Q, Cullen PJ, Su Y, Chen S, Ye X, Liu D, Ding T (2020) Plasma-activated water (PAW) and slightly acidic electrolyzed water (SAEW) as beef thawing media for enhancing microbiological safety. LWT 117:108649. https://doi.org/10.1016/j.lwt.2019.108649
19. Sharifian A, Soltanizadeh N, Abbaszadeh R (2019) Effects of dielectric barrier discharge plasma on the physicochemical and functional properties of myofibrillar proteins. Innov Food Sci Emerging Technol 54:1–8. https://doi.org/10.1016/j.ifset.2019.03.006

Chapter 7

Microwave Applications in Fish

Khaoula Elahdef, Bouthaina Ben Hadj Hmida, Elahesadat Hosseini, Sarra Akermi, Moufida Chaari, Walid Elfalleh and Slim Smaoui

Abstract

Currently, in the food industry, microwave energy is progressively outcome applications in numerous processing techs. A microwave, electromagnetic waves of precise frequencies ranging between 300 MHz and 3000 GHz, is a novel processing method that has been broadly applied in food processing fields. Thereafter, this innovative tool was explored by researchers as an upgrading method and has been effectively applied in blanching, cooking, thawing, pasteurization, drying, frying, extraction, and sterilization. This chapter outlines the principles and characteristics of microwave-assisted fish processing methods, and their impacts on drying, heating, and sterilization. Here, we discussed the impact of microwave-assisted fish processing methods on the quality features, microbial growth, nutritional and physicochemical composition of fish products. In addition, the advantages of microwave processing and a comparison against conventional methods were discussed. Interestingly, microwave processing ensures gaining time, effective yields, and energy with high preservation and could maintain sensory traits of the fish's nutritional value. This chapter will provide evidence about these topics and their advances that could be motivating for food engineering and technology community. Moreover, the chapter will describe methods for microwave application in the fish industry.

Keywords Microwaves applications, Fishing industry, Engineering, Drying, Heating, Sterilization

1 Introduction

Thermal processing, employed in most of the food industry and applied numerous disposable heated techniques, includes electro-thermal engineering viz direct resistance, radio frequency, induction, or by the features of the electric field [1, 2]. The application of high temperatures is awfully mutual progressions within the food industry. In the food field, heat preservation has been a significant dispensation technology since the appreciation process was discovered and could ensure that processed food will be safe and satisfactory to the consumer [3, 4].

Microwaves, a non-ionizing category of electromagnetic radiation, could pervade within matrices and meet the target components.

Renata Ernlund Freitas de Macedo and Carlos Adam Conte-Junior (eds.), *Meat Products Preservation*, Methods and Protocols in Food Science, https://doi.org/10.1007/978-1-0716-5068-4_7,

The interaction, of microwave/biological materials could result in ionic transmission and dipole spin and, therefore, could exercise too much pressure on the cell wall, which provokes it to break and deliver the intracellular components. This phenomenon at a pressure ranging between 30 MHz and 300 GHz, could enhance the matrix's porosity, and amplify the extraction yield [5, 6]. To prevent disturbance with the waves operated in numerous fields (e.g., industry, medical, and research instruments), the Federal Communications Commission (FCC) governs and supervises the microwave frequencies at a national and international level. As an illustration, for industrial scale, scientific, and medical applications, frequencies like 13.56, 27.12, 40.68, 433.92, 915, 2450, 5800, and 24,125 MHz are fixed [7]. For industrial heating operators, the commonly used frequencies were 915 MHz or 2.45 GHz [8]. The familiar industry where microwaves are frequently used is food technology. In this field, microwaves are used for drying, thawing, pasteurization, or sterilization of food products [9]. In several food industries, compared to conventional methods, the microwave application displayed many advantages as the reduction of handling time and costs and preserving the sensory traits without provoking variations in the nutritional value of food products [10]. Up to now, microwaves have been extensively applied to food processing owing to their advantages like negligible energy damage throughout transmission, durable penetrating aptitude, and the ability to transfer adequate energy to heat the food materials. In the chemical industry, microwaves can be applied to a manifold of extraction processes. For instance, the unconventional microwave-assisted extraction (MAE) method has been examined as a conceivably renewable method for recovering oil from fish by-products. From salmon backbones, the MAE allows the recovery of >50% of the total lipid content, heads, and viscera in less than 11 min of extraction [11]. During MAE, microwaves endorse a rise of pressure within the sample cells, triggering the disagreement of their membranes and the consequent release of solutes into the solvent [12]. Over conventional extraction, MAE has some advantages, including shorter time of extraction, abridged solvent consumption, advanced yield, better precision, and adequacy for thermo-labile chemical components [13]. Moreover, MAE increased the bioactivity of fish protein hydrolysates due to enhanced proteolysis and the formation of low microwave (MWs) peptides [8]. However, the heat generated during the extraction process may damage heat-sensitive compounds.

Although microwaves have been utilized and have been well-known by the fish industry for decades, this applied science progresses very quickly, with new knowledge and data performed annually. Therefore, the aim of this chapter is to describe methods for microwave application in the fish industry.

2 Material

2.1 Microwave Drying

- Fish sample: fish fillets (e.g., tilapia, salmon, or cod).
- Microwave oven: microwave with adjustable power settings.
- Balance: precision electronic scale (±0.001 g).
- Thermocouples: fiber-optic temperature sensors for real-time monitoring.
- Moisture analyzer: for initial and final moisture content determination.
- Petri dishes or microwave-safe trays.
- Desiccator: for post-drying stabilization.
- Humidity and temperature probe: for environmental monitoring.
- Gloves: for sample handling.
- Ventilation system: to manage water vapor release.
- Distilled water.

2.2 Microwave Heating

- Microwave oven: specific power 0.3–300 GHz, with wavelengths that span from 1 mm to 1 m [14].
- Dishes: (1) transparent materials: ceramics, quartz, and glass, which allow microwaves to pass through without significant interaction; (2) reflective materials: metals, which reflect microwaves rather than absorb them; and (3) absorbing materials: carbon, water making them more relevant due to their high thermal response [15, 16] (*see* **Note 1**).
- Microwave safe covers: to improve the uniformity of temperature distribution and prevent splattering and moisture loss [17].
- Microwave susceptors: for selective heating or crisping [18].
- Thermometer: check the temperature of the heated fish.
- Gloves: to protect your hands after heating.
- Fish samples: tuna, salmon, mackerel, and sardine.

2.3 Microwave Sterilization

2.3.1 Fish Samples

- Fish fillets (e.g., salmon, cod, tilapia…) were obtained from a local fish market [19].
- Samples were examined for the presence of *Aeromonas hydrophila*, and only those free from contamination were used.

2.3.2 Bacterial Strains and Inoculum Preparation

- The following bacterial strains were used for inoculation:
  - *Salmonella* Enteritidis.
  - *Escherichia coli*.

  - Bacteria of the genus *Enterococcus.*
  - Spores of *Clostridium sporogenes.*
  - *A. hydrophila.*

- Bacterial suspensions were prepared in sterile Ringer's solution from 24-h cultures on nutrient agar.
- Suspension density was standardized at 0.3 McFarland units using a densitometer.
- *C. sporogenes* spore suspensions were obtained by:
  - Culturing in BHI broth at 4°C, then transferring to FTG medium under anaerobic conditions (37°C, 24 h).
  - Further incubation in Duncan-Strong medium (37°C, 7 days, anaerobic conditions).
  - Harvesting spores via centrifugation (3000 rpm, 15 min), washing, and resuspension in 50 mL sterile distilled water to a concentration of 10^7 CFU/mL.

2.3.3 Microwave System

- Microwave treatments were performed using a household microwave oven (Clatronic MWG-748, 2.45 GHz, 850 W) with a rotating glass plate.
- The power settings were adjusted to 100, 180, 300, 450, 600, and 700 W.
- The exposure times were:
  - 2.5 min for bacterial inactivation.
  - 11 min for spore inactivation.

2.3.4 Other Equipment

- Selective media for microbial enumeration:
  - *Salmonella* spp.: buffered peptone water, Rappaport broth, BPLS agar.
  - *E. coli*: MacConkey broth, ENDO agar.
  - *Enterococci*: glucose-azide broth, kanamycin-esculin-azide agar.
  - *C. sporogenes*: DRCM agar for anaerobic culture.
  - *A. hydrophila*: Aeromonas agar with selective supplement.
- Centrifuge (5000 × *g*, 10 min) for bacterial preparation.
- Water bath (80°C, 15 min) for spore inactivation.
- Stomacher for homogenizing fish samples before bacterial enumeration.

3 Methods

3.1 Protocol for Microwave Drying

1. Fillet the fish to a uniform thickness of 3–5 mm.
2. Rinse fillets with distilled water to remove surface contaminants.
3. Blot dry using filter paper to remove excess moisture.
4. Salting: immerse fish fillets in a 3% NaCl solution for 30 min to enhance flavor and texture.
5. Blanching: submerge fillets in hot water (60°C, 1 min) to inactivate enzymes and microbial load.
6. Marination (if required): use vinegar or antioxidant solutions (citric acid) to prevent lipid oxidation (*see* **Note 1**).

3.1.1 Microwave Drying Setup

- Record the initial weight (Wi) of fish fillets using a precision balance.
- Place fish fillets on microwave-safe trays in a single layer.
- Maintain a 1 cm gap between pieces for uniform heating.
- Insert thermocouples to monitor internal temperature.
- Set microwave power level (300–600 W) based on desired drying rate.
- Drying time: 5–15 min, monitored at 1-min intervals.
- Intermittent mode (recommended): 1 min on/30 s off to prevent overheating.
- Adjust power settings if excessive charring or case hardening occurs.
- Weigh the sample at regular intervals until the final weight (Wf) is achieved.
- Calculate the moisture content (*see* **Note 2**).
- Stop drying when the moisture content reaches ≤10% (dry basis).
- Transfer dried fish to a desiccator for 30 min to prevent moisture absorption.
- Store in airtight packaging at 4°C.

4 Protocol for Microwave Heating

1. Place the dish containing the fish in the microwave turntable (*see* **Note 2**).
2. Cover with a microwave-safe lid (*see* **Note 3**).
3. Choose the right microwave power level according to the fish type and thickness:

- Low (300–600 W): gentle cooking to retain texture.
- Medium (600–900 W): general heating setting for fish.
- High (900–1200 W): used for quick heating but may cause uneven cooking.

4. Adapt the heating time based on the thickness and type of fish:
   - Thin fillets (tilapia, sole): 30 s–2 min.
   - Thicker fillets (salmon, cod): 2–4 min.
   - Whole fish: 4–6 min.
5. After heating, use an infrared thermometer to check the internal temperature. The fish is safe for eating when the core temperature hits to 63°C.
6. Let the fish rest for 1–2 min after microwaving to allow heat equilibration.
7. Wear gloves when handling hot containers to prevent burns.
8. If reheating fish, use a lower power setting (30–50%) to prevent drying out (*see* **Note 4**)
9. Ensure proper ventilation when microwaving fish to prevent odor buildup.

4.1 Protocol for Microwave Sterilization

4.1.1 Preparation of Inoculated Samples

- Fish meal samples (7 g each) were inoculated with (Table 1):
  - 1 mL of bacterial suspensions (*S. enteritidis*, *E. coli*, *Enterococcus*).
  - 1 mL of *C. sporogenes* spore suspension.
- The inoculated samples were mixed thoroughly to obtain a dough-like consistency.
- The mixture was dried for 45 min at 37°C before microwave treatment.

4.1.2 Microwave Treatment

- Inoculated fish meal was spread in a thin layer (0.5 g per sample) on ceramic trays (6.25 cm^2 area).
- Trays were subjected to microwave radiation at 100–700 W (*see* **Note 6**) for:
  - 2.5 min for bacterial inactivation.
  - 11 min for spore inactivation.

Table 1
Summary of experimental protocol for microwave sterilization

Step	Procedure	Equipment
1. Sample preparation	Inoculate fish meal and fillets with bacterial suspensions	Pipettes, centrifuge
2. Microwave treatment	Expose fish to microwave radiation (100–700W, 2.5–11 min)	Microwave oven
3. Microbial testing	Plate fish homogenate on selective media, incubate	Incubator, agar plates
4. Physical analysis	Sensory evaluation of color, texture, and odor	Expert panel

4.1.3 Microbiological Analysis

Enumeration of Microorganisms

- The most probable number (MPN) method (3-tube set) was used for microbial quantification.
- Serial dilutions (10^{-1} to 10^{-8}) were prepared in 0.1% peptone water.
  - Selective media incubation conditions:
  - *Salmonella* spp.: buffered peptone water (37°C, 24 h), Rappaport broth (43°C, 24 h), BPLS agar (37°C, 24 h).
  - *E. coli*: MacConkey broth (43°C, 24 h), ENDO agar (43°C, 24 h) [20].
  - *Enterococci*: glucose-azide broth (37°C, 48 h), kanamycin-esculin-azide agar (37°C, 48 h).
  - *C. sporogenes*: DRCM agar (37°C, 3 days, anaerobic conditions).
  - *A. hydrophila*: Aeromonas agar (37°C, 24 h) [21].

Determination of Inactivation Parameters

- Lethal dose (D-value): the regression line slope was used to estimate bacterial elimination rates.
- D90 dose: the dose required to achieve a 90% reduction (1 log CFU/MPN) was calculated.

5 Notes

1. The steps 3–6 are optional
2. Moisture content (%) is calculated using:

$$MC\% = \frac{\text{Wi}-\text{Wf}}{\text{wi}} \times 100$$

where:

- Wi = Initial weight (g)
- Wf = Final weight after drying (g)

3. Avoid using aluminum, which can cause sparking inside the microwave.
4. The fish sample should be in the center of the microwave to ensure uniformity in heating and even temperature distribution.
5. The microwave-safe lid ensures quicker cooking, keeps odors contained, and reduces the chance of contamination while preventing drying and preserving the fish's texture.
6. To prevent overcooking and drying out the fish, it's crucial to reheat it at a reduced power level (30–50%). The fish's texture is preserved, and moisture retention is aided by the slower, kinder heating procedure that the lower power guarantees.
7. Microwave radiation dose (D, kJ/g) was calculated using the equation [19]:

$$\text{D} = \frac{M \times T}{m}$$

where:

- M = microwave power (kW)
- T = exposure time (s)
- m = sample mass (g)

References

1. Atuonwu J, Tassou S (2021) Decarbonisation of food manufacturing by the electrification of heat: a review of developments, technology options and future directions. Trends Food Sci Technol 107:168–182
2. Ali S, Rezende VT, Ullah S, de Paiva EL, Tonin FG, Corassin CH, de Oliveira CAF (2023) Food processing and challenges in the food production and quality: the foodomics approach. Food Biosci 103217. https://doi.org/10.1016/j.fbio.2023.103217
3. Dadhaneeya H, Nema PK, Arora VK (2023) Internet of Things in food processing and its potential in Industry 4.0 era: a review. Trends Food Sci Technol. https://doi.org/10.1016/j.tifs.2023.07.006
4. Hassoun A, Boukid F, Ozogul F, Aït-Kaddour A, Soriano JM, Lorenzo JM, Câmara JS (2023) Creating new opportunities for sustainable food packaging through dimensions of industry 4.0: new insights into the food waste perspective. Trends Food Sci Technol 142:104238
5. Viji P, Rao BM, Debbarma J, Ravishankar CN (2022) Research developments in the applications of microwave energy in fish processing: a review. Trends Food Sci Technol 123:222–232
6. Kutlu N, Pandiselvam R, Saka I, Kamiloglu A, Sahni P, Kothakota A (2022) Impact of different microwave treatments on food texture. J Texture Stud 53(6):709–736

7. Orsat V, Raghavan GSV, Krishnaswamy K (2017) Microwave technology for food processing: an overview of current and future applications. In: The microwave processing of foods. Woodhead Publishing, Amsterdam, p 100
8. Zhang J, Luan D (2024) Microwave ovens: domestic and industrial. Microwave processing of foods: challenges, advances and prospects: microwaves and food. Springer International Publishing, Cham, pp 17–33
9. Principato L, Spigno G (2024) Microwave heating in food processing. Food packaging and preservation. Academic, pp 299–329
10. Ekezie FGC, Sun DW, Han Z, Cheng JH (2017) Microwave-assisted food processing technologies for enhancing product quality and process efficiency: a review of recent developments. Trends Food Sci Technol 67:58–69
11. de la Fuente B, Pinela J, Calhelha RC, Heleno SA, Ferreira IC, Barba FJ, Barros L (2022) Sea bass (Dicentrarchus labrax) and sea bream (Sparus aurata) head oils recovered by microwave-assisted extraction: nutritional quality and biological properties. Food Bioprod Process 136:97–105
12. Li H, Zhao Z, Xiouras C, Stefanidis GD, Li X, Gao X (2019) Fundamentals and applications of microwave heating to chemicals separation processes. Renew Sustain Energy Rev 114:109316. Ling B, Tang J, Kong F, Mitcham E J, Wang S (2015) Kinetics of food quality changes during thermal processing: a review. Food Bioprocess Technol 8:343–358
13. Noor NQIM, Razali RS, Ismail NK, Ramli RA, Razali UHM, Bahauddin AR et al (2021) Application of green technology in gelatin extraction: a review. Processes 9(12):2227
14. Zhang L, Zhang Z, Yu W, Miao Y (2023) Review of the application of microwave heating technology in asphalt pavement self-healing and de-icing. Polymers 15(7):1696
15. Farag S, Sobhy A, Akyel C, Doucet J, Chaouki J (2012) Temperature profile prediction within selected materials heated by microwaves at 2.45 GHz. Appl Thermal Eng 36:360–369
16. Mello PA, Barin JS, Guarnieri RA (2014) Microwave heating. In: Microwave-assisted sample preparation for trace element analysis. Elsevier, Amsterdam, pp 59–75
17. Schiffmann R (2017) Packaging for microwave foods. In: The microwave processing of foods. Woodhead Publishing, Amsterdam, pp 273–299
18. Perry MR, Lentz RR (2020) Susceptors in microwave packaging. In Development of packaging and products for use in microwave ovens. Woodhead Publishing, Amsterdam, pp 261–291
19. Bauza-Kaszewska J, Skowron K, Paluszak Z, Dobrzański Z, Śrutek M (2014) Effect of microwave radiation on microorganisms in fish meals. Ann Anim Sci 14(3):623–636
20. Rougier C, Prorot A, Chazal P, Leveque P, Leprat P (2014) Thermal and nonthermal effects of discontinuous microwave exposure (2.45 gigahertz) on the cell membrane of *Escherichia coli*. Appl Environ Microbiol 80(16):4832–4841
21. Mohamad WS, Megahed AA, El-Ghaiaty HA, Hafez TA (2016) Effect of gamma radiation and microwave cooking on Aeromonas hydrophila in bolti fish fillet. Arab J Nucl Sci Appl 49:100–106

Chapter 8

Supercritical Carbon Dioxide Drying Applied to Meat and Fish Products

Fabio Santi, Riccardo Zulli, Sara Spilimbergo and Alessandro Zambon

Abstract

Supercritical carbon dioxide (sc-CO_2) drying is an innovative, low-temperature technique for removing water from various matrices. This method has gained attention for its ability to reduce moisture content and lower water activity (a_w), thereby extending shelf life and enabling the development of novel products. This chapter presents the operational principles of sc-CO_2 drying, with a focus on its application in meat and fish products, where it has demonstrated significant potential. Studies have shown that sc-CO_2 drying effectively reduces water activity in protein matrices such as codfish ($a_w < 0.815$) and chicken breast ($a_w < 0.50$), while simultaneously inactivating microbial loads, including pathogenic bacteria. Furthermore, the integration of sc-CO_2 drying with High-Power Ultrasound (HPU) has been shown to enhance mass transfer, significantly reducing drying times. This chapter highlights the versatility and efficacy of sc-CO_2 drying as a promising technology for food preservation and innovation.

Keywords Supercritical carbon dioxide, Drying, Innovative technologies, Meat, Fish, Microorganism inactivation

1 Introduction

Carbon dioxide (CO_2) is classified as Generally Recognized as Safe (GRAS) by the FDA and is widely used in the food and beverage industry. It plays a key role in extending the shelf life of food by acting as a bacteriostatic agent, effectively slowing the growth of aerobic microorganisms. High concentrations of CO_2 are commonly employed in Modified Atmosphere Packaging (MAP) to preserve food. In beverages, CO_2 is responsible for the characteristic effervescence. When CO_2 reaches its supercritical (sc) phase—above 31.1°C and 7.38 MPa—it can be used as an innovative extraction technique, to remove solvents and/or water. Specifically, sc-CO_2 drying has been extensively used for the production of aerogels, as it allows the removal of solvents from the gel matrix without compromising its delicate porous structure [1]. In recent years, the technique has been investigated also for the drying of food [2, 3]. Sc-CO_2 drying

Renata Ernlund Freitas de Macedo and Carlos Adam Conte-Junior (eds.), *Meat Products Preservation*, Methods and Protocols in Food Science, https://doi.org/10.1007/978-1-0716-5068-4_8,

has been applied to several food matrices, mainly vegetables and fruits like carrot [4], basil [5], coriander [6, 7], apple [8, 9] peas [10], strawberry [11, 12], persimmon, mango [13], red bell pepper [14], and beetroot [15].

In the meat and fish industry, common drying technologies include hot air drying, sun drying, and lyophilization [15]. These methods are widely used for products such as codfish, salted meat, and beef jerkies. However, while they effectively reduce moisture content, they often compromise the product's appearance and nutritional value, mainly due to the use of high temperatures (>50°C). In contrast, the relatively low temperatures used for sc-CO_2 drying (<50°C) allow a reduction in browning and hardening of the treated food compared with conventional technologies, as reported by Tomic et al. [15, 16].

The optimization of key parameters—temperature, time, pressure, and flow rate—plays a critical role in determining the drying efficiency and the final quality of the product. These four parameters are closely interdependent and are crucial for achieving effective drying, maintaining the sample's appearance, and ensuring microbial inactivation. Several studies have focused on optimizing process conditions, taking into account the specific food matrix [16]. The use of the Design of Experiment (DOE) methodology is highly recommended for conducting process optimization for the drying or new food products. DOE can lead to more efficient optimization, reducing the need for trial-and-error testing and improving both process performance and resource utilization.

In summary, sc-CO_2 drying is a promising method for reducing water content and microbial load in food while maintaining good appearance quality. Although primarily studied at lab and pilot scales, this chapter provides a detailed description of the operative protocol for this innovative method.

As described above, this technique not only removes water but also reduces the microbial load. In fact, at supercritical phase, CO_2 becomes bactericidal as extensively described in [17, 18] and is effective in reducing both naturally present and pathogenic microorganisms. Referring to chicken meat samples [19], pressurization and depressurization alone resulted in a reduction of mesophilic bacteria from 5.67 ± 0.67 log CFU/g to 2.15 ± 0.20 log CFU/g, while molds and yeasts (initially 4.38 ± 0.69 log CFU/g) were completely inactivated [19]. The same study on chicken breast meat also reported a reduction in inoculated *Salmonella enterica* (initially 7.97 ± 0.10 log CFU/g), achieving a reduction of 1.76 ± 0.16 log CFU/g with pressurization and depressurization alone. After 45 min of treatment, *Salmonella enterica* levels were further reduced by 7.71 ± 0.05 log CFU/g. Different microorganisms exhibit different sensitivity to the antimicrobial effect of sc-CO_2. For example, yeasts and molds are the most sensitive to the treatment, as also demonstrated in vegetables products [20], while Gram-negative bacteria are usually more susceptible than Gram-positive ones [17]. Bacterial spores are instead not inactivated by sc-CO_2 and co-solvents, or higher temperatures are needed [7].

To enhance the efficiency of microbial inactivation, vegetative and spores [6], and accelerate dehydration, sc-CO_2drying can be coupled with High Power Ultrasounds (HPU), as reported by Michelino et al. [6]. The HPU system improves mass transfer in the supercritical CO_2 extraction process. Specifically, in experiments using inoculated chicken breast meat with three different strains of *Salmonella*, sc-CO_2 drying process coupled with HPU achieved a significant microbial reduction. After 15 min of treatment, a reduction of 5.33±2.51 log CFU/g was observed, compared to 1.52±0.14 log CFU/g achieved with sc-CO_2 alone. After 45 min, both treatments resulted in the complete inactivation of *Salmonella*. Regarding the improved mass transfer, the synergistic effect of HPU significantly accelerated the drying kinetics [20]. In chicken breast meat, a weight reduction of 75.1% was achieved after 300 min of treatment, leading to a final product with low water activity.

Sc-CO_2 drying is achieved inside a high-pressure vessel where the samples are placed, and the CO_2 is continuously flowed to remove the water [6, 11]. The typical range of process conditions adopted includes the following: temperature between 35 and 50°C, pressure between 6 and 14 MPa, duration of 180–360 min, and a flow up to 30 kg/h.

The process can be divided into three important phases: the pressurization phase, the drying phase, and the depressurization phase [18].

1. **Pressurization phase:** The pump is activated, and the pressure rises from the tank pressure (usually around 5 MPa) to reach the desired pressure level by pumping CO_2 into the drying vessel.
2. **Drying phase:** Once the desired pressure is reached, CO_2 flows continuously through the samples, either in a semi-continuous or fully continuous manner (Sect. 2). As reported by Garcia-Gonzalez et al. [17], the drying principles can be described through the following steps:

   a. Dry supercritical carbon dioxide is continuously introduced into the drying vessel.
   b. Water is dissolved in the sc-CO_2 and exits the product.
   c. The mixture of wet sc-CO_2 (sc-CO_2+extracted water) exits the vessel.

At this stage, two possibilities arise:

- The effluent may be expanded so that the density of sc-CO_2 rapidly decreases until CO_2 becomes gaseous and loses its solvent properties. The effluent then splits into two phases: a gaseous CO_2-rich phase and a liquid water-rich phase (see Sect. 2.1).
- Sc-CO_2 is not expanded but instead enters a second vessel to be recycled. Water can be removed from sc-CO_2 by flowing it through solid water adsorbents or by bubbling it through liquid desiccants (*see* **Note 1**), a process patented by Riera et al. [21] (see Sect. 2.2). The process with recirculation reduces the amount of CO_2 consumed, making it more environmentally friendly [12, 24]

3. Depressurization phase: After the required time to achieve the target moisture content, the CO_2 flow is stopped and the pressure in the drying vessel is reduced until reaching the atmospheric one. It is crucial to carefully determine the timing of depressurization, as it can significantly impact the quality of the samples [17].

2 Materials

2.1 High-Pressure Apparatus Without Recirculation

Sc-CO_2 drying can be achieved using a high-pressure plant where the CO_2 flow is completely discharged after passing through the food samples.

A possible scheme is shown in Fig. 1.

The plant should include the following elements:

- High-pressure vessel: the main section of the plant where the food samples are inserted and the water removal happens. The vessel should be designed and built following different criteria based on the maximum operating conditions (pressure and temperature). In particular, the safety criteria in compliance with the regulatory laws, such as the European Pressure Equipment Directive (PED 2014/68/EU), must be followed. Guidelines for the vessel design can be found in different manuals, such as [22, 23].
  The vessel must be built in stainless steel, usually AISI316 or AISI304. It must be easily opened from the top for the sample insertion. Different designs are possible, such as flanged lids or screw caps.
  The vessel must be provided with a heating system and temperature and pressure probes to check the operating conditions. The heating system could be electric or fluid based. In the

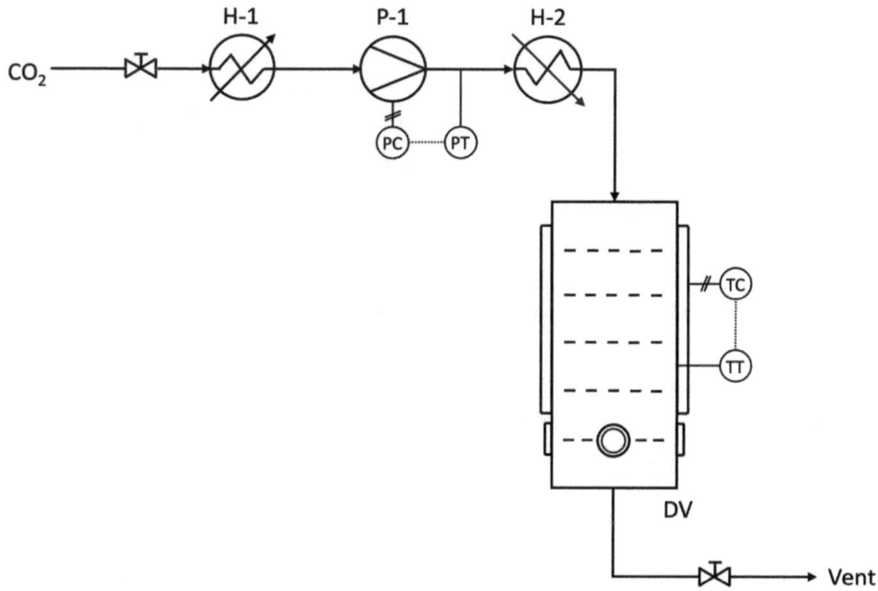

Fig. 1 Process diagram: supercritical carbon dioxide drying without recirculation. H-1 Chiller, H-2 Heater, P Pump, PT Pressure Transmitter, PC Pressure Controller, TT Temperature Transmitter, TC Temperature Controller, DV Drying Vessel

first case, the vessel is surrounded by an electric jacket or coil that must be connected to the temperature probe inside the vessel. An on-off controller or a PID (Proportional-integral-derivative) one is needed. In the second case, the vessel temperature is controlled by a fluid, usually water, that flows in the vessel jacket and is heated in an external system, such as a water bath. A proper temperature is usually addressed by directly controlling the heating fluid temperature, which must be at a few degrees higher than the temperature needed in the vessel. This difference is dependent on the vessel design and dimensions, and the heating system properties.
Moreover, the vessel must include at least one inlet, usually at the top, and one outlet, usually at the bottom.

- CO_2 tank with dip tube, which can be kept at room temperature close to the plant.
- Refrigerating system: a chiller reservoir or a similar system is needed to decrease the temperature of the CO_2 at the pumping system inlet and the pumping system itself. At this stage, indeed, the CO_2 must be in liquid state before entering the pump and avoid cavitations.
- Pumping system: in this case, a pump with pressure and flow rate control is recommended.

- Connecting lines: the different equipment parts are connected by stainless-steel high-pressure pipes. The pipe diameter depends on the wanted CO_2 flow rate. A velocity of liquid CO_2 in the range 1–2 m/s is usually advised.
- Micrometric valve, instressurization rate of the vessel.

This design is quite simple to be built and used but it is advised to be adopted for small vessel capacity (<100 mL) and to address preliminary lab tests on small food samples (<5 g per batch) due to the high amount of consumed CO_2.

2.2 High-Pressure Apparatus with Recirculation

Sc-CO_2 drying can also be achieved using a high-pressure apparatus with CO_2 recirculation. This solution can be potentially adopted for large volumes thanks to the low CO_2 consumption. However, the design, construction and operation are more complicated than the first case, and the plant must include a larger number of controlling systems due to safety reasons. A semi-automation of the system with a controlling panel is strongly recommended. The connection of the plant to a computer with a process control software could be also adopted, even at small scale.

A possible scheme of a CO_2 drying plant with recirculation is shown in Fig. 2.

The plant could be divided into two sections: the feed section and the recycling section.

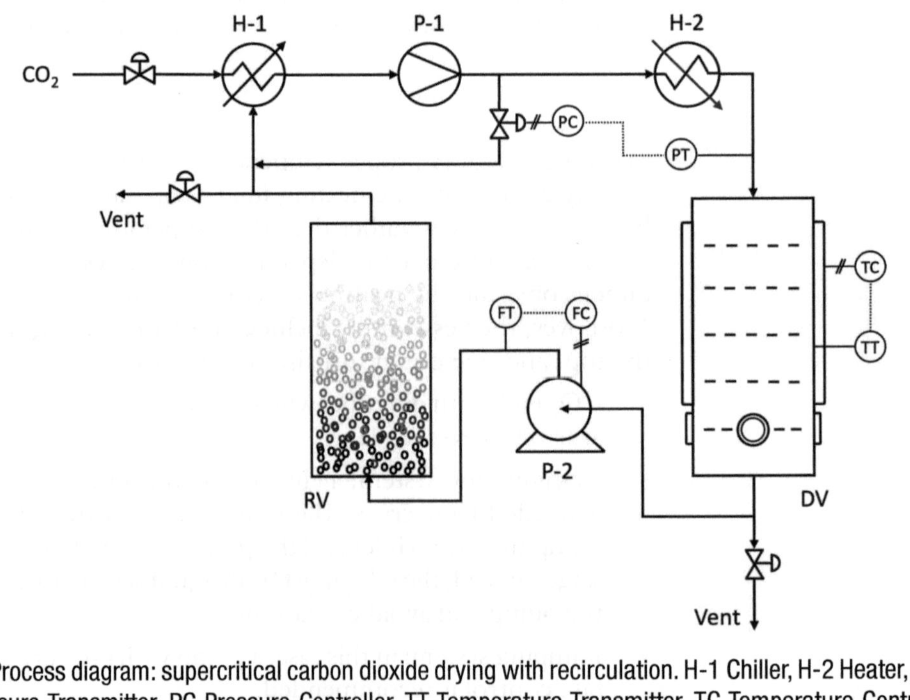

Fig. 2 Process diagram: supercritical carbon dioxide drying with recirculation. H-1 Chiller, H-2 Heater, P Pump, PT Pressure Transmitter, PC Pressure Controller, TT Temperature Transmitter, TC Temperature Controller, FT Flow Transmitter, FC Flow Controller, DV Drying Vessel, RV Regeneration Vessel

The first one includes the CO_2 tank, a high-pressure pumping system and a refrigerating system, similar to the case without recirculation.

The recirculation section is instead usually composed of:

- Drying vessel: similar consideration of the plant without recirculation can be done. However, PID controllers for both temperature and pressure are strongly recommended.
- Ball valve: it allows or not the recirculation of the CO_2 in the loop.
- Recirculation pump: an additional pump, usually centrifugal, is needed to allow the flow of CO_2 in the recirculation loop and control the flow rate. The pump must be surrounded by a refrigeration tube or jacket to ensure the liquid state of CO_2 inside the pump and avoid cavitation. It is recommended to connect it to the same refrigeration system of the feed section to improve energy efficiency and reduce equipment.
- Regeneration vessel: a second high-pressure vessel, including an adsorbent material, is needed to remove the water present in the CO_2. The vessel design must follow the same regulation and safety criteria adopted for the drying vessel. The dimensions depend on the plant capacity and the adsorbent material adopted. A volume at least 25% larger than the drying vessel is recommended.
- Heat exchanger: an electric heat exchanger must be used to increase the CO_2 temperature before entering the vessel.

The five units of the recirculation loop should be connected in the order shown in the list. The feed section, which introduces CO_2 into the system, must be placed between the recirculation pump and the regeneration vessel to ensure that the CO_2 flows through the loop in a controlled manner before being regenerated and recycled. This positioning ensures the precise introduction of CO_2 and improves the overall efficiency of the system, as shown in Fig. 2.

A discharging line must be connected to the recirculation loop to depressurize the plant.

A micrometric valve is recommended to control the depressurization rate.

As stated before, a controlling system is needed for a proper plant operation and for safety reasons and should include the following elements:

- A normally-closed valve between the CO_2 tank and the plant. This allows the control of the start of the operations and shut down the connection between the tank and the plant in the case of anomalies.

- A normally-open valve on the discharge line of the plant. This allows the control of the end of the operation (start of depressurization phase) and depressurize the vessel in the case of anomalies.
- A controller for the heat exchanger.
- A controller for the drying vessel jacket connected to a temperature probe.
- A flow rate controller that modifies the recirculation pump velocity.
- A pressure probe inside the drying vessel that control the pressure value by two possible actions:
  - Opening a back-pressure regulator (BPR) which connects the recirculation loop with the high-pressure pump inlet. It usually acts with a PID controller to pressure overshoot and for small pressure variations.
  - Opening the normally-open discharge valve. It usually acts for larger pressure variation. It is recommended to design the controller to open for a small interval of time if the pressure overshoot is in the order of 1 MPa, or together with the complete shutdown of the plant when values close to the maximum allowed pressure is reached.

2.3 Food Sample Preparation

The following instruments and equipment are needed for the preparation of the food samples before the insertion in the drying plant:

1. Fresh food product for treatments,
2. Precision balance,
3. Thermic bags and eutectic plates,
4. Knives and cutting boards,
5. Ethanol for cleaning and disinfection,
6. Bunsen flame or safety cabinet,
7. Strain of microorganisms (if microbial inactivation studies are needed) and material needed (culture media and microbiology lab materials).

2.4 Samples Treatment

The following elements are instead needed for the operation of the plant:

1. Food-grade CO_2 tank 3.0 (99.9%) or higher with dip tube (*see* **Note 2**).
2. Baskets for sample positioning in the drying vessel, metallic ones are recommended.
3. Aspiration system.
4. Adsorbent material: molecular sieves like zeolites, 3A (pore size of 3 Armstrong) bead-shaped material is recommended to ease the substitution. This is needed only in the case of a plant with recirculation.

3 Methods

3.1 Samples Preparation and Storage

The following steps are suggested for the preparation and the storage of the food samples before the drying processes:

1. Purchase the food product as fresh as possible, to guarantee a good quality. It is also important to transport food (*see* **Note 3**) in the correct storage conditions (fridge or room temperature).
2. If the experimental campaign is addressed in a long interval of time (longer than a few days), it is suggested to use products as homogeneous as possible. A possible way is to divide the product in small portions, freeze it at −80°C (−18°C if not possible) and thaw each portion at 4°C before the drying test. This allows to eliminate, or at least reduce, the risk of the product variability impacts on the process optimization.
3. Cut the samples with the size and weight desiderated (*see* **Note 4**).
4. Measure the weight of each basket that will be used (tare weight).
5. Put the sample in the clean metallic baskets (*see* **Note 5**).
6. Measure the weight of each basket with the samples (gross weight). This will be used to calculate the drying performance.

3.2 Treatment Without Recirculation

In this section, the protocol used to dry food samples with the sc-CO_2 plant without recirculation (Sect. 2.1) is presented.

3.2.1 Preliminary Activities

The following actions should be addressed around 1 h before the starting of trial:

1. Turn on the refrigeration system.
2. Preheat the drying vessel by turning on the heating system at the desired temperature. If the temperature control is addressed by a fluid-based jacket, it is recommended to set the temperature at least a few degrees higher than the one required by the process (*see* **Note 6**).

3.2.2 Sample Loading and Pressurization Phase

1. Turn on the aspiration system (*see* **Note 7**).
2. Put the baskets with the samples in the drying vessel and close it (*see* **Note 10**). The order of the samples in the vessel could influence the process performance.
3. Set the desired flow rate and the pressure of the pump (*see* **Note 8**).
4. Open the CO_2 tank.
5. Verify that the system has a pressure about 5–6 MPa (tank pressure) and eventual leaks (*see* **Note 9**). If there are leaks, close the CO_2 bottle and better close the point where there are leaks.

Moreover, if there are no leaks and the pressure is not between 5 and 6 MPa change the CO_2 tank.

6. Activate the pump to transport the CO_2 from the tank to the vessel.
7. When the system achieves the desired pressure, slightly open the micrometric exit valve to guarantee the desired CO_2 flow rate.
8. Turn on the chronometer to check the time of the drying phase.

3.2.3 Drying Phase

During this stage, it is necessary to monitor the process parameters to make prompt adjustments if necessary. It is possible to act on the heating system to regulate the temperature and on the pump settings and the valve opening to regulate pressure and flow rate.

3.2.4 Depressurization Phase

1. Close the CO_2 exit valve to ensure that the depressurization phase does not begin before the bottle is sealed, allowing for a well-controlled depressurization process.
2. Turn off the pump.
3. Close the CO_2 bottle.
4. Slowly open the CO_2 valve exit. It is recommended to have a slow depressurization with a constant rate (lower than 20 MPa/min) (*see* **Note 11**).
5. When the pressure is equal to the atmospheric one, it is possible to open the vessel.

3.3 Treatment with Recirculation

To process a larger number of samples, a system equipped with CO_2 recirculation is required. As previously mentioned, operating with CO_2 recirculation presents greater challenges, and a high level of process automation is strongly recommended.

The following sections outline a generic operating protocol, detailing the actions required from operators and identifying those that can be either manual or automated. Minor adjustments may be necessary depending on the specific automation system and software employed.

3.3.1 Preliminary Activities

The following actions should be addressed at least one hour before the start of the trials:

1. Turn on the refrigeration system. The chiller temperature must be set at around 2°C in order to cool down the two pumps and the inlet pipes of CO_2 to the high-pressure pump.
2. Preheat the plant. Set the desired temperatures of the heating jacket and the heat exchanger. The time to achieve the temperature depends on the size of the plant, the characteristic of instrumentation, and the desired temperature.

3. Introduce dry adsorbent materials inside the regenerator vessel (*see* **Note 1**).

3.3.2 Sample Loading and Pressurization Phase

When the desired temperature of the refrigeration and heating systems is reached, it is possible to start the following procedure for the start of the experimental run:

1. Set the desired value of pressure, flow rate, and processing time (*see* **Note 8**).
2. Insert the samples inside the drying vessel and carefully close the vessel.
3. Check that all the manual valves are in the correct position. All the discharge valves must be closed. Moreover, the ball valve in the recirculation loop must be closed. This avoids the establishment of CO_2 flows in the opposite way during the pressurization phase.
4. Open the CO_2 bottle and start the pressurization phase. The valve between the bottle and the feed section should automatically open to allow the entrance of CO_2 into all the sections of the plant. If an automatic valve is not available, open it manually.
5. If any leaks are present (*see* **Note 9**), proceed with a quick shutdown of the plant (go to procedure 3.3.4).
6. Activate the high-pressure pump. If the regulation of the pump flow rate is allowed, it is recommended to start the pump with the maximum value.
7. When the pressure is higher than 7 MPa, open the manual ball valve in the recirculation loop.
8. When the pressure is reaching the desired value, adjust the high-pressure pump flow rate to avoid pressure overshoot. At the same stage, the BPR should start working.
9. When the pressure and temperature are stable, the recirculation pump can be activated (manually or automatically). The flow rate controller should control the pump velocity. In this first stage, small adjustments to the pump flow rates and the operating parameters could be needed before stabilization.
10. When all the parameters are stable, the operating time can be started.

3.3.3 Drying Phase

During this phase, the process parameters must be manually and/or automatically checked to allow a stable process and avoid safety issues. The automatic discharge valve should operate when the pressure reaches values much higher than the desired one.

3.3.4 Depressurization Phase

1. Turn off both the pumps.
2. Close the CO_2 bottle.
3. Open the exit valve of CO_2 and regulate the micrometric discharge valve to control the depressurization rate as described in Sect. 3.2.4 (*see* **Note 11**).
4. Open the vessel when the pressure reaches the atmospheric one.

3.4 After-Treatment Operations

After the treatment, the samples must be quickly removed from the drying vessel to be analyzed or stored.

The vessel and the plant must be carefully cleaned and/or sterilized (*see* **Note 12**).

In the only case of samples for microbiological analyses, they should be removed in a sterile manner and inserted in sterile containers right after the removal from the vessel to avoid recontamination. It is not necessary to operate in a sterile or aseptic manner for drying.

If the sample weight is needed, it is recommended to weigh them together with their basket (gross weight) to avoid the loss of sample portion. Other analyses are usually addressed, such as moisture content measures, water activity, pH, rehydration capacity, and nutritional content [16, 19].

Similar analysis can also be addressed during the shelf life of the products. Storage conditions can vary in function of the product properties and use in terms of temperature (usually room or fridge) and packaging methods (under vacuum or modified atmosphere, transparent or not).

Sensorial tests with panelists can be helpful to address the impact of the process on product acceptability and eventually compare the treatments with other drying techniques.

4 Notes

1. Zeolites are water-adsorbent materials. In our system, we use 250 g of zeolites, which is slightly above the amount required to ensure complete water adsorption from the sample. However, we recommend not exceeding this quantity, as an excessive amount of zeolites may reduce the CO_2 flow rate and significantly increase the drying time. After using the zeolites, it is possible to re-use them after drying, heating up for 4 h under a flow of hot air at 150°C. Other water-adsorbent materials can be used; however, a preliminary study is necessary to evaluate their performance.
2. It is necessary to use a bottle with a dip tube because the pump works for liquid CO_2.

3. We find it useful to transport fresh foods using thermal bags with eutectic plates. To avoid variability, it is recommended to supply a sufficient amount of product and then store them at refrigerated temperature for the entire experimental campaign.
4. The size of the samples was determined based on the dimensions of the plant. It is important to consistently maintain these dimensions, as any variations could impact the treatment's performance. When evaluating the process on a new food product, it is particularly recommended to use samples that are uniform in size, weight, and composition. The range of sizes used in our trials for thickness is from 5 mm to 1.0 cm, and for weight, it ranges from 0.5 to 10 g.
5. It is important to clean the metallic basket thoroughly, ensuring that any remnants of previously treated food are removed. Additionally, when evaluating microbial inactivation, we find it useful to immerse the baskets in ethanol and then flame them to ensure a high level of cleanliness. When the samples are inoculated with a high load of microorganisms (e.g., for a challenge test), we prefer to autoclave the metallic baskets.
6. We find it useful to set the temperature 2°C above the desired temperature to guarantee the correct temperature.
7. The aspiration system is essential for safety reasons, as it prevents the release and accumulation of CO_2 in the work environment.
8. The parameters of Temperature, Pressure, Flow rate and time are chosen in base of preliminary optimization on different characteristics such as aw, moisture quantity, color and microbial inactivation.
9. To verify CO_2 leaks, we check the manometer connected to the vessel. Additionally, it can be useful to apply soap to the critical points of the system; leaks are indicated by bubbles.
10. The drying vessel, as shown in Fig. 2, is vertically oriented. Due to the small-scale nature of the setup, the samples are stacked one on top of the other. As a result, the order of the samples becomes an important factor to consider. Experimental tests have shown that the first and last samples tend to have different moisture levels compared to the others. This variation is likely due to the short treatment time. However, if longer treatment times are applied, this issue does not occur.
11. The depressurization rate is an important parameter as it can significantly affect the characteristics of the samples. A slow depressurization rate is typically desired. This can be achieved by carefully opening and closing the exit valve to maintain a

well-controlled rate, allowing for different depressurization profiles as needed. This process can also be automated if the system is equipped with the appropriate software and instruments designed for this application.

12. Cleaning the plant after use is essential, as food residues can accumulate and lead to contamination of the treated food. Additionally, inoculated microorganisms may survive the treatment and proliferate within the plant. Usually, regular cleaning is sufficient when the plant is used for drying. However, if microbial inactivation is being tested, it is recommended to use ethanol, followed by rinsing with demineralized water. After this procedure, run a cycle without samples, using only CO_2.

References

1. Baldino L, Zuppolini S, Cardea S, Diodato L, Borriello A, Reverchon E, Nicolais L (2020) Production of biodegradable superabsorbent aerogels using a supercritical CO_2 assisted drying. J Supercrit Fluids 156:104681
2. Wang W, Rao L, Wu X, Wang Y, Zhao L, Liao X (2021) Supercritical carbon dioxide applications in food processing. Food Eng Rev 13:570–591
3. Li P, Mei J, Xie J (2023) The regulation of carbon dioxide on food microorganisms: a review. Food Res Int 172:113170
4. Brown ZK, Fryer PJ, Norton IT, Bakalis S, Bridson RH (2008) Drying of foods using supercritical carbon dioxide — investigations with carrot. Innov Food Sci Emerg Technol 9(3):280–289
5. Bušić A, Vojvodić A, Komes D, Akkermans C, Belščak-Cvitanović A, Stolk M, Hofland G (2014) Comparative evaluation of CO_2 drying as an alternative drying technique of basil (*Ocimum basilicum* L.)—the effect on bioactive and sensory properties. Food Res Int 64:34–42
6. Michelino F, Zambon A, Vizzotto MT, Cozzi S, Spilimbergo S (2018) High power ultrasound combined with supercritical carbon dioxide for the drying and microbial inactivation of coriander. J CO_2 Utilization 24:516–521
7. Zambon A, Michelino F, Bourdoux S, Devlieghere F, Sut S, Dall'Acqua S et al (2018) Microbial inactivation efficiency of supercritical CO_2 drying process. Dry Technol 36(16):2016–2021
8. Zambon A, Bourdoux S, Pantano MF, Pugno NM, Boldrin F, Hofland G, Spilimbergo S (2021) Supercritical CO_2 for the drying and microbial inactivation of apple's slices. Drying Technol 39(2):259–267
9. Tomic N, Djekic I, Zambon A, Spilimbergo S, Bourdoux S, Holtze E, Rajkovic A (2019) Challenging chemical and quality changes of supercritical CO_2 dried apple during long-term storage. Lwt 110:132–141
10. Cardin M, Zulli R, Santi F, Andrigo P, Lincetti E, Collini D, Zambon A (2024) Exploring novel supercritical CO_2 drying combined with high power ultrasounds: a case on peas and apples. Chem Eng Trans 110:325–330
11. Zambon A, Zulli R, Boldrin F, Spilimbergo S (2022) Microbial inactivation and drying of strawberry slices by supercritical CO_2. J Supercrit Fluids 180:105430
12. Zambon A, Facco P, Morbiato G, Toffoletto M, Poloniato G, Sut S, Andrigo P, Dall'Acqua S, de Bernard M, Spilimbergo S (2022) Promoting the preservation of strawberry by supercritical CO_2 drying. Food Chem 397:133789
13. Braeuer AS, Schuster JJ, Gebrekidan MT, Bahr L, Michelino F, Zambon A, Spilimbergo S (2017) In situ Raman analysis of CO_2-assisted drying of fruit slices. Foods 6:37
14. Zambon A, Tomic N, Djekic I, Hofland G, Rajkovic A, Spilimbergo S (2020) Supercritical CO_2 drying of red bell pepper. Food Bioprocess Technol 13:753–763
15. Tomic N, Djekic I, Hofland G, Smigic N, Udovicki B, Rajkovic A (2020) Comparison of supercritical CO_2-drying, freeze-drying, and frying on sensory properties of beetroot. Foods 9(9):1201
16. Bernardo YA, Zambon A, Cardin M, Zulli R, Andrigo P, Santi F, Conte-Junior CA (2024) Development and optimization of general drying models for cod (*Gadus* spp.) using supercritical carbon dioxide. Innov Food Sci Emerg Technol 94:103688

17. Garcia-Gonzalez L, Geeraerd AH, Spilimbergo S, Elst K, Van Ginneken L, Debevere J, Van Impe JF, Devlieghere F (2007) High pressure carbon dioxide inactivation of microorganisms in foods: the past, the present and the future. Int J Food Microbiol 117:1–28
18. Bourdoux S, Li D, Rajkovic A, Devlieghere F, Uyttendaele M (2016) Performance of drying technologies to ensure microbial safety of dried fruits and vegetables. Compr Rev Food Sci Food Saf 15:1056–1066
19. Morbiato G, Zambon A, Toffoletto M, Poloniato G, Dall'Acqua S, de Bernard M, Spilimbergo S (2019) Supercritical carbon dioxide combined with high power ultrasound as innovate drying process for chicken breast. J Supercrit Fluids 147:24–32
20. Zambon A, Michelino F, Bourdoux S, Devlieghere F, Sut S, Dall'Acqua S, Spilimbergo S (2018) Microbial inactivation efficiency of supercritical CO_2 drying process. Dry Technol 36:2016–2021
21. Riera E, Blanco A, García J, Benedito J, Mulet A, Gallego-Juárez JA, Blasco M (2010) High-power ultrasonic system for the enhancement of mass transfer in supercritical CO_2 extraction processes. Ultrasonics 50:306–309
22. Guarise GB (2007) Lezioni di impianti chimici ad alta pressione. Padova, Italy. CLEUP.
23. Bertucco A, Vetter G (2011) High pressure process technology: fundamentals and applications. Elsevier Science, Amsterdam
24. Bernardo YA, Zulli R, Andrigo P, Santi F, Do Rosario DK, Zambon A, Spilimbergo S, Conte-Junior CA (2025) Modelling and optimization of supercritical CO_2 drying of tuna (Thunnus albacares) fillets: unraveling physicochemical and structural changes. J Supercrit Fluids 223:106621

Chapter 9

Application of Nanoemulsions of Bioactive Compounds in Sausages and Meat Patties

A.G. Pereira, A. Perez-Vazquez, P. Barciela, A.O.S. Jorge, Rafael Nogueira-Marques, Sepidar Seyyedi-Mansour and M.A. Prieto

Abstract

Nanoemulsion-based delivery systems involve the formation of fine oil-in-water dispersions where the compounds of interest are encapsulated within nano-sized droplets stabilized by food-grade emulsifiers. The process typically includes creating a pre-emulsion of the bioactive compound with carrier oils and surfactants, followed by the application of high-energy methods, such as high-shear homogenization, to reduce droplet size to the nanoscale. Nanoemulsions are particularly suitable for encapsulating and protecting sensitive bioactive compounds, such as essential oils, plant extracts, and lipid-soluble vitamins, enabling their uniform incorporation into food matrices. This is especially relevant for perishable products with a complex matrix like meat patties and sausages, where shelf-life extension and functional enhancement are critical. The purpose of using nanoemulsions in these products is to enhance stability, bioavailability, and controlled release of functional compounds while maintaining desirable sensory and technological properties. Nanoemulsion delivery systems help protect bioactive compounds from degradation during processing and storage, and can modulate their release during digestion, contributing to functional or nutritionally enhanced meat products. While several high-energy techniques are available for preparing nanoemulsions, this chapter will focus specifically on high-shear homogenization as a primary method for encapsulating bioactive compounds into sausages and meat patties. This mechanical process is characterized by using intense shear forces to break droplets in the nanometer range, resulting in kinetically stable emulsions. Therefore, the aim of this chapter is to provide a detailed protocol for this nanoemulsion preparation technique with the goal of improving their safety, shelf life, and nutritional value of meat patties and sausages.

Keywords Nanoemulsion preparation, High-shear homogenization, Bioactive compounds, Meat patties, Sausages

1 Introduction

1.1 Nanoencapsulation Methods for Bioactive Compounds Incorporation into Meat Patties and Sausages

The meat industry faces multiple challenges, including the need to meet current product demand while simultaneously responding to consumer demands for healthier and more sustainable foods, without compromising the sensory quality of the products [1]. In this context, the incorporation of bioactive compounds into various

Renata Ernlund Freitas de Macedo and Carlos Adam Conte-Junior (eds.), *Meat Products Preservation*, Methods and Protocols in Food Science, https://doi.org/10.1007/978-1-0716-5068-4_9,

processed meat products (e.g., meat patties, sausages) represents an innovative strategy capable of improving the product's techno-functional properties and exerting potential beneficial effects on consumer health [2]. However, to make this application viable, it is necessary to implement parallel processes that stabilize the bioactive compounds, increase their bioavailability, and prevent potential adverse effects on the final product, such as negative sensory perception [3]. Among these parallel processes, one of the most relevant alternatives in recent years is nanoencapsulation. This process consists of the design of a nanometric-scale aggregate, generally with sizes less than 100 nm, as in the case of nanocapsules (<0.2 μm), microcapsules (0.2–5000 μm) and macrocapsules (>5000 μm). This aggregate integrates the bioactive molecule and acts as a system for the controlled and precise release of said substances [4]. This is particularly advantageous in meat products where sustained action of antioxidants and antimicrobials is required to extend shelf life and guarantee microbiological safety during storage [4, 5]. Furthermore, nanoencapsulation helps to mask undesirable flavors and aromas associated with the addition of certain bioactive compounds, which enables their inclusion in effective concentrations without compromising the organoleptic characteristics of the final product [6]. The general vision of this process is shown in Fig. 1. The incorporation of this type of nanoencapsulation in meat products such as meat patties and sausages can be achieved by using different techniques, such as direct incorporation into the meat formulation, incorporation into edible coatings, or addition in active packaging systems. This variety of techniques allows the use of nanoencapsulated materials in this type of industry to present a great technological adaptability [7, 8]. Regarding the encapsulating materials used, biodegradable naturally derived materials (e.g., polysaccharides, proteins, and lipids) and hybrid materials (e.g., biopolymers) are the most frequently employed [9]. Once the encapsulating material is designated, it is necessary to prepare an initial mixture that allows the bioactive compounds to be dissolved or dispersed in the encapsulating material. Among the most frequently reported strategies are gelation, the design of polymeric nanoparticles through coacervation, the development of nanoemulsions (NE) for lipophilic compounds, the design of liposomes or micelles for amphiphilic compounds, and the development of nanofibers [9, 10]. The selection of the type of material will be determined by the chemical nature of the bioactive compound, the encapsulating material, and the target food matrix. Other factors that must be consider are the type of processing (especially when heat treatments are applied), pH variations, water activity, and specific request of the formulation, such as extending shelf life or adding functional properties to the meat product [11, 12].

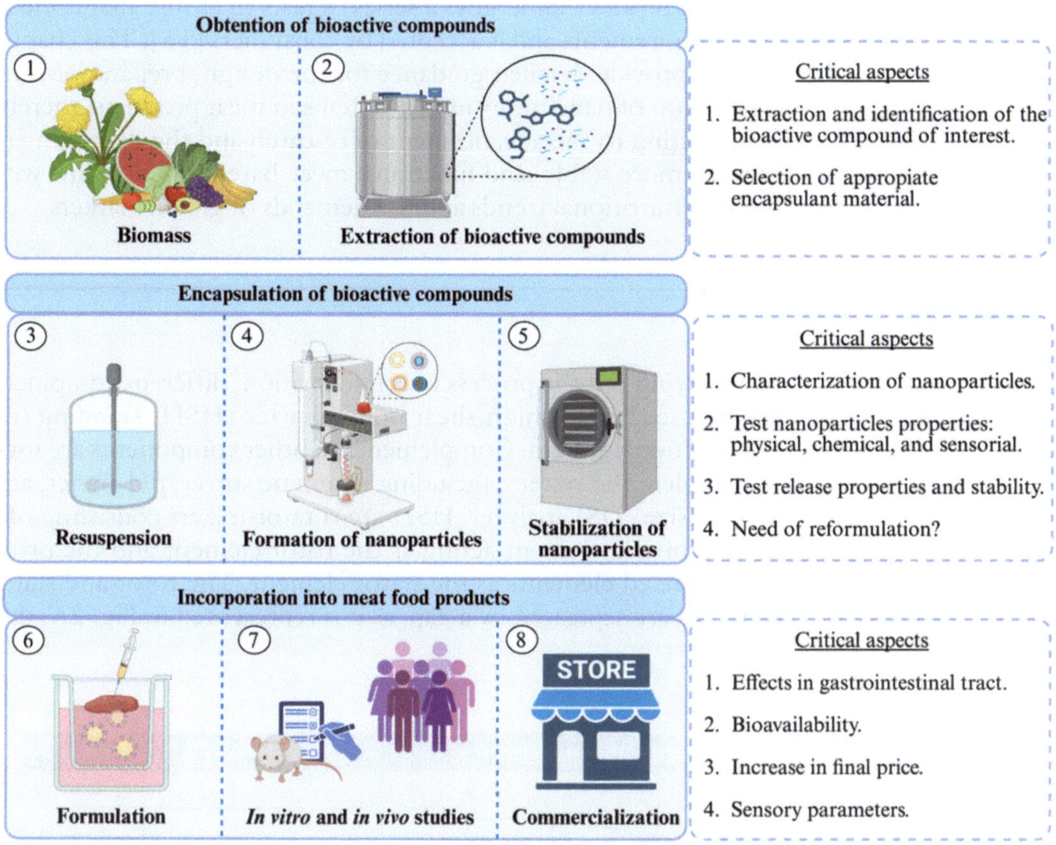

Fig. 1 Steps followed to provide nanoencapsulated bioactive compounds incorporated into meat products

The mixture then undergoes various physical and chemical processes that allow its incorporation into the food matrix [4]. These processes must ensure that the final product is safe, effective, and stable during processing, storage, and distribution. To achieve this, the release of the bioactive compound must remain constant across different batches of the product. This requires conducting kinetic studies of the release of these compounds, as well as validating their effect on the quality and safety of the final food. Furthermore, the final product must have consistent physicochemical and microbiological composition across different production batches. Therefore, it is necessary to develop standardized protocols to ensure compliance with all these requirements [13, 14]. The development of these types of protocols for nanoencapsulation of bioactive compounds for incorporation into meat patties and sausages will provide innovative solutions to the main problems facing the meat industry today. These protocols should explain how to carefully select encapsulation materials and methods, as well as serve as the basis for developing rigorous protocols for their implementation in meat matrices. Finally, the meat product undergoes a series of tests to ensure that it meets legal requirements and is accepted by consumers [15]. This chapter aims to provide detailed guidance for the design, preparation, and application of nanoencapsulated systems in meat products, thereby contributing to the advancement of research and the development of safer, more stable, and functional meat-based foods, in line with modern nutritional trends and the demands of global markets.

2 Materials

For carrying out the process of NE formation, different equipment can be used, with a high-shear homogenizer (HSH) standing out as a functional system. Complementary, other components are used to complete the system, including magnetic stirrer, pH-meter, and droplet size (DS) analyzer. HSH are rotator-mixers consisting of a driven mixing element acting as the rotor element and the proximity of fixed elements as the stator element. The rotor and stator elements are separated by a gap, as it is represented in Fig. 2A, that

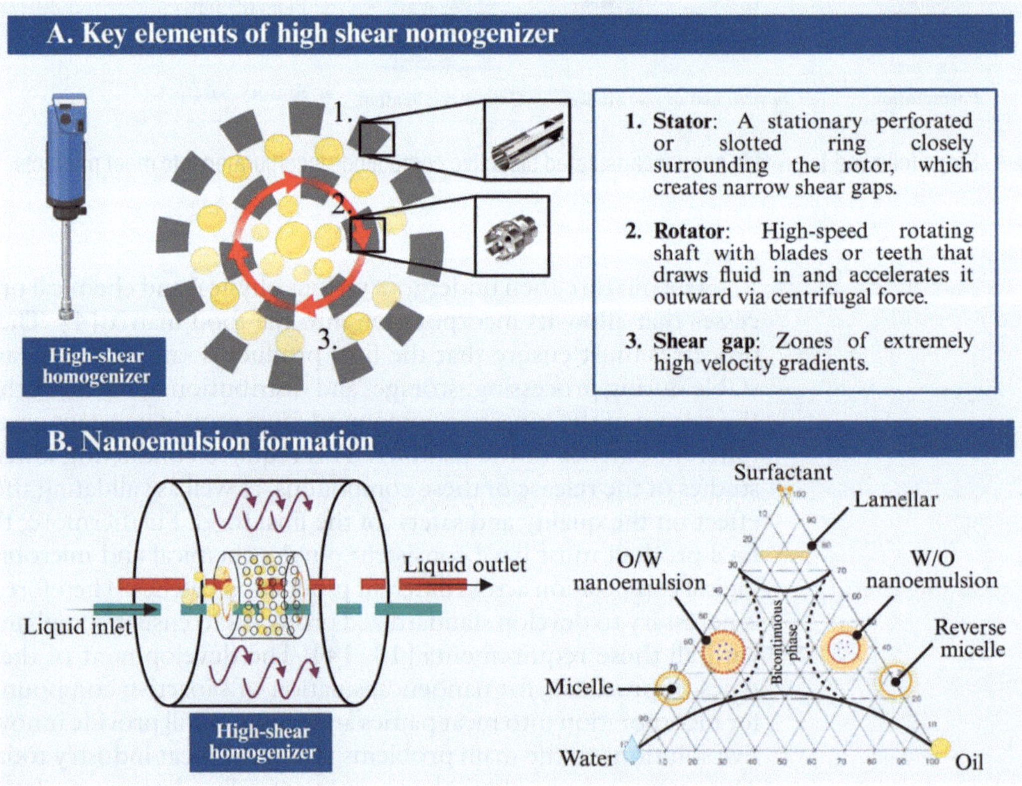

Fig. 2 High-shear homogenizer. **A** Description of the elements forming a high-shear homogenizer. **B** Schematic representation of high-shear homogenizer mode of action with a ternary diagram of the nanoemulsion phases

usually varies between 100 and 3000 μm and operates at shear rates of 20,000–100,000 s^{-1}. The tip speed, which can vary from 10 to 50 m/s, affects the applications of this equipment. Thus, high shear rate extends HS applications to those requiring mixing of immiscible fluids or size reduction with short cycle times, such as homogenization, emulsification, solid-liquid dispersion, micro-mixing, dissolution, and milling [16]. HSH mode of action can be divided into phases. First, the rotor starts to move and, when it reaches critical speed, a low-pressure zone is generated at the tip of the rotor blades. Then, when liquid vapor pressure is reached, cavities (small vapor bubbles) are created and directed to the edge of the tank along with the liquid phase. Finally, the pressure increases in this zone, causing the bubbles to implode and energy to be released. This energy (force) results in droplet breakage, and therefore, emulsification occurs [17] (Fig. 2B).

3 Methods

3.1 Nanoemulsion Preparation and Homogenization by High-Shear Homogenization

NE by HSH often employs a two-stage process to convert oil and water phases into a fine NE. First, a coarse pre-emulsion is formed by mixing heated oil and aqueous phases, and then a high-shear rotor-stator homogenizer is applied to reduce DS into the nanometer range. The oil (lipid) phase, containing the hydrophobic bioactive compounds, and the aqueous phase (with surfactant) are typically heated separately (often to 60–80°C) to lower viscosity and ensure complete dissolution of components [18, 19]. The heated oil is then slowly added into the hot aqueous phase under stirring to form a coarse emulsion [18, 20]. This pre-emulsion is immediately subjected to HSH. Throughout the process, the temperature is maintained above about 60°C during mixing to keep waxy lipids molten. Rapid cooling follows homogenization to lock in the nano-sized droplets [18, 19]. Below, the procedure is divided into the following steps: Pre-emulsion preparation, high-shear homogenization, and post-homogenization cooling and stabilization. Each subsection lists the detailed laboratory steps in order.

3.1.1 Pre-Emulsion Preparation

Weigh and Dissolve Components

Prepare the lipidphase by weighing the oil (e.g., edible oil, wax, or lipid carrier) and dissolving any hydrophobic bioactive compound into it. Prepare the aqueous phase by weighing distilled water and dissolving the chosen surfactant or mixture of surfactants and any hydrophilic bioactive compounds. Typical formulations use on the order of 5–10% oil and 1–5% emulsifier by weight (e.g., 5 wt% oil, 1–2 wt% surfactant, balance water) [19]. Optionally pre-mix co-surfactants (e.g., Poloxamers or Tween) into the oil or water as appropriate.

Heat Phases Separately

Heat the oil phase and aqueous phase separately to about 60–80°C. It is suggested to place each in a beaker on a temperature-controlled hotplate or water bath at ~70–80°C until the lipid is fully molten and the surfactant is fully dissolved [18, 20]. The two phases should be at the same temperature, often between 10 and 20°C above the melting point of the oil components [19]. Gentle magnetic stirring (≈200–500 rpm) during heating helps prevent local overheating and ensures homogeneous mixtures.

Combine Phases to Form a Coarse Emulsion

With both phases hot (~70–80°C), begin stirring the aqueous phase at moderate speed (~300–600 rpm on a stirrer or propeller). Then, slowly add the hot oil phase into the hot water phase in a thin stream [18].

Continuously stir to disperse the oil and the mixture for about 3–10 min after the oil addition to form a coarse pre-emulsion, which is typically a white or opaque emulsion. Throughout this step, keep the temperature above ~60°C to prevent premature solidification of the lipid [18]. As an example, one study formed a coarse emulsion by homogenizing at 15,000 rpm for 4 min in an Ultra-Turrax (T18 IKA) after initial mixing [21]. This step can also be done with lower speed stirring if a homogenizer is not used yet.

Prepare for High-Shear Homogenization

If the pre-emulsion is not already in a suitable container, transfer it to the homogenization vessel (e.g., a double-walled beaker with temperature control, or a jacketed reactor). Ensure the total batch volume (e.g., 50–100 mL for lab scale, or larger as needed) leaves enough headspace above the rotor. The vessel and mixers (stir bars, temperature probe, etc.) should be clean and ready for use. If using a batch homogenizer, place the vessel on a magnetic or overhead stirrer. If using continuous orin-line homogenization, set up the feed pump and homogenizer as described in the following section.

3.1.2 High-Shear Homogenization

Configure the Homogenizer

Use a high-speed rotor-stator homogenizer such as an Ultra-Turrax (IKA) or Silverson device. Ultra-Turrax T25 or T18 equipped with a dispersing tool/probe (e.g., S25N-18G) is commonly used [19, 22]. Alternatively, a Silverson L5M-A lab homogenizer or an in-line rotor-stator unit may be used. Place the disperser probe so that its tip is submerged ~1–2 cm below the emulsion surface (in the aqueous phase) [18]. Secure the homogenizer to avoid vibration. If possible, set the vessel in an ice-water bath or recirculating jacket to remove heat since prolonged shear will further warm the emulsion.

Set Rotation Speed

Start the homogenizer at a moderate speed (4000–6000 rpm) briefly to begin emulsification, then increase to the target high speed. Typical rotor speeds range from about 8000 to 15,000 rpm for laboratory-scale emulsification [19, 22]. For example, Souto et al. obtained ~150 nm droplets by homogenizing at 8000 rpm for 10 min [19]. Other studies have used 15,000 rpm for 5–10 min on an Ultra-Turrax to prepare essential oil NEs [21, 22]. The exact speed and time depend on volume, viscosity, and desired DS. Higher speed and longer time produce smaller droplets but generate more heat.

Homogenize with High Shear

Run the homogenizer at the chosen speed (usually 8–15,000 rpm) for 5–15 min, or until the emulsion appears uniformly milky and no large droplets are visible. For example, 10 min at 8000 rpm was sufficient to reduce 1% lipid droplets to ~150 nm [19], whereas 10 min at 15,000 rpm is also reported for nanoemulsions of essential oils [22]. If processing large batches continuously, pump the pre-emulsion through the rotor-stator at a constant flow. In industrial practice, feed pumps have delivered ~2400 L/h into a multi-stage IKA homogenizer [20]. To prevent overheating, one can operate in pulses (e.g., 2 min on, 1 min off) or run in an ice bath. It is recommended to maintain the vessel temperature above ~50°C during homogenization if waxes or solidifying lipids are present [18].

Monitor Processing

During homogenization, the emulsion should become fine and opaque. Occasionally pause and inspect the mixture. The absence of visible oil pools and smooth consistency indicate completion. It is recommended to perform homogenization entirely at elevated temperature (above 60°C) [18], as cooling too early can lead to coalescence. Typical final DSs with sufficient shear are in the order of 0.1–1 μm or below, depending on formulation.

Finish Homogenization

Once the desired time and appearance are reached, stop the homogenizer. If using a continuous set-up, turn off the feed pump and homogenizer and decouple lines. For batch devices, remove the probe from the vessel. Emulsion at this point is a fine, hot NE.

3.1.3 Post-Homogenization Cooling and Stabilization

Cool Immediately

Promptly reduce the temperature of the NE to ambient temperature or below to stabilize the droplets. Transfer the homogenization vessel to an ice-water bath or apply cooling (e.g., refrigerated circulator) while gently stirring. This rapid cooling maintains the droplets distribution and, for lipid-containing emulsions, solidifies any melted lipid cores [19]. Cooling should continue until the emulsion reaches room temperature, or slightly below.

Gentle Agitation (Optional)

During cooling, it can be useful to stir slowly (e.g., 100–300 rpm) to prevent creaming or sedimentation. If an organic solvent or volatile component is used, the mixture may be stirred gently (for example, at 300 rpm overnight) to allow solvent evaporation [22]. In that case, control the final volume and adjust by adding water if needed, to account for evaporative loss. If no solvent is present, prolonged stirring is not required; simply let the emulsion stand on ice until cooled.

Finalize Nanoemulsion

After cooling, the NE should appear as a homogeneous, milky dispersion. At this point, no further high-shear processing is used. The emulsion may be transferred to storage containers. If desired, measure and record temperature (should be ~20–25°C), pH, and conductivity at this stage. The DS distribution can be measured later by light scattering [18, 19, 22].

Storage and Stability

Label the NE and store it protected from light and cold until use. Typically, such NEs remain physically stable for days to weeks if cooled properly. Any further processing, such as dilution or incorporation into food, is done in later sections not covered here [18, 19, 22].

By following the above protocol, carefully heating, and mixing the oil and water phases, then applying intense rotor-stator homogenization under controlled conditions, a reproducible O/W NE can be produced. Example conditions reported in the literature include heating phases to ≈70–80°C [18], homogenizing at 8000–15,000 rpm for 5–10 min [19, 22], and cooling immediately in an ice bath [19]. These steps ensure that droplet sizes are minimized (often <200 nm) and the final NE remains stable for subsequent use.

3.2 Nanoemulsion Incorporation into Meat Patties and Sausages

3.2.1 Sausages Preparation

For the preparation of the sausages with NEs embedded in their matrix, the preparation steps described in Sects. 3.1.1, 3.1.2, and 3.1.3 will be followed. Once the desired NEs are formed, the sausage will be prepared as described below [22]:

1. *Mincing and mixing.* Mince the lean part and the fat part of the selected meat in the chosen proportions (generally 70% lean part and 28% fat part).
2. *Sausage formation.* Incorporate the rest of the ingredients (salt, ice, and other condiments) for mincing all together, and the NE previously formed so everything is distributed uniformly, favoring emulsification and homogeneous distribution in the sausage. Ensure the mixture does not reach temperatures above 12°C.

3. *Sterilization*. The mixture obtained is subjected to a bath at 75°C for 40 min.
4. *Conservation*. Subsequently, it is allowed to cool and kept refrigerated at 4°C to guarantee its preservation.

3.2.2 Meat Patties Preparation

In the same way as in sausageformulation, the preparation steps described in Sects. 3.1.1, 3.1.2, and 3.1.3 will be followed before meat patties elaboration. Once the desired NE have been formed, the meat patties will be prepared as described below [5]:

1. *Preparation of meat mixture*. Mince the lean part and the fat part of the selected meat in the chosen proportions (generally 70% lean part and 28% fat part).
2. *Homogeneous mass formation*. Mix the lean minced parts with the rest of the ingredients (salt and other condiments) gradually adding water or ice. Incorporate gradually the NEs previously formed and use a vacuum mixer, chopper or similar for mixing and extracting proteins until a homogeneous and sticky mass is formed. Once this mass is formed, incorporate the minced fat and continue mixing. Rest the homogenizer mixture for 1 h ensuring it does not reach temperatures above 12°C.
3. *Patty formation*. Weight portions for each patty and incorporate them into a mold or hand-press. Place the formed patties on trays lined with waxed paper, preventing them from sticking together.
4. *Conservation*. Storage at 4°C for short time preservation or at −18°C for long time preservation.

4 Notes for Nanoemulsion of Bioactive Compounds Incorporated into Meat Patties and Sausages

1. *Emulsifier selection*. The selection of the emulsifier must be considering recognized food compounds, such as soy lecithin, polysorbate 80, and gum arabic [23].
2. *Homogenization conditions*. Cycles of 3–5 min with intermediate cooling are recommended to avoid overheating that degrades the bioactive compounds. Specifically, it is crucial to keep the temperature of the mixture to minimize compounds degradation [18].
3. *Droplet size control*. In meat products, the average suitable size of NEs to ensure homogeneity is between 100 and 200 nm [24].

4. *Meat product incorporation*. In meat products whose base is minced meat, the Nes can be mixed directly with the mixture or wrapped in edible coatings to improve the distribution and activity of the compound [25].
5. *Scalability*. High-shear homogenization allows scalability for industrial production [26].

Acknowledgments

The research leading to these results was supported by MICIU/AEI/10.13039/501100011033, supporting the predoctoral industrial grant for A. Perez-Vazquez (DIN2024-013416) in collaboration with Mercantia Desarrollos Alimentarios SL; by Xunta de Galicia for supporting the post-doctoral grant of A.G. Pereira (IN606B-2024/011), and the pre-doctoral grant of P. Barciela (ED481A-2024-230). The authors are grateful to the National funding by FCT, Foundation for Science and Technology, through the individual research grant of A.O.S. Jorge (2023.00981.BD). The authors thank the EU-FOR A Fellowship Program (EUBA-EFSA-2023-ENREL-01) that supports the work of S. Seyyedi-Mansour (INNOV2SAFETY-GA13). The authors thank the Ibero-American Program on Science and Technology (CYTED— GENOPSYSEN, P222RT0117). Funding for open access charge: Universidade de Vigo/CISUG

References

1. Schulze-Ehlers B, Anders S (2018) Towards consumer-driven meat supply chains: opportunities and challenges for differentiation by taste. Renew Agric Food Syst 33:73–85. https://doi.org/10.1017/S1742170517000059
2. Ayuso P, García-Pérez P, Nieto G (2025) New insights and strategies in the nutritional reformulation of meat products toward healthier foods. Molecules 30. https://doi.org/10.3390/molecules30122565
3. Ribas-Agustí A, Olga M-B, Robert S-F, Elez-Martínez P (2018) Food processing strategies to enhance phenolic compounds bioaccessibility and bioavailability in plant-based foods. Crit Rev Food Sci Nutr 58:2531–2548. https://doi.org/10.1080/10408398.2017.1331200
4. Pateiro M, Gómez B, Munekata PES, Barba FJ, Putnik P, Kovačević DB, Lorenzo JM (2021) Nanoencapsulation of promising bioactive compounds to improve their absorption, stability, functionality and the appearance of the final food products. Molecules 26. https://doi.org/10.3390/molecules26061547
5. Ojeda-Piedra SA, Zambrano-Zaragoza ML, González-Reza RM, García-Betanzos CI, Real-Sandoval SA, Quintanar-Guerrero D (2022) Nano-encapsulated essential oils as a preservation strategy for meat and meat products storage. Molecules 27. https://doi.org/10.3390/molecules27238187
6. Mehta N, Kumar P, Verma AK, Umaraw P, Kumar Y, Malav OP, Sazili AQ, Domínguez R, Lorenzo JM (2022) Microencapsulation as a noble technique for the application of bioactive compounds in the food industry: a comprehensive review. Appl Sci 12. https://doi.org/10.3390/app12031424
7. Xie Q, Guishan L, Zhang Y (2024) Edible films/coatings containing bioactive ingredients with micro/nano encapsulation: a comprehensive review of their fabrications, formulas, multifunctionality and applications in food packaging. Crit Rev Food Sci Nutr

64:5341–5378. https://doi.org/10.1080/10408398.2022.2153794
8. Trajkovska Petkoska A, Daniloski D, Kumar N, Pratibha Broach AT (2021) Active edible packaging: a sustainable way to deliver functional bioactive compounds and nutraceuticals. In: Muthu SS (ed) Sustainable Packaging. Springer Singapore, Singapore, 225–264
9. Mondéjar-López M, García-Simarro MP, Navarro-Simarro P, Gómez-Gómez L, Ahrazem O, Niza E (2024) A Review on the encapsulation of "eco-friendly" compounds in natural polymer-based nanoparticles as next generation nano-agrochemicals for sustainable agriculture and crop management. Int J Biol Macromol 280:136030–https://doi.org/10.1016/j.ijbiomac.2024.136030
10. Ezhilarasi PN, Karthik P, Chhanwal N, Anandharamakrishnan C (2013) Nanoencapsulation techniques for food bioactive components: a review. Food Bioprocess Technol 6:628–647
11. Pisoschi AM, Pop A, Cimpeanu C, Turcuş V, Predoi G, Iordache F (2018) Nanoencapsulation techniques for compounds and products with antioxidant and antimicrobial activity – a critical view. Eur J Med Chem 157:1326–1345. https://doi.org/10.1016/j.ejmech.2018.08.076
12. Jafari SM (2017) An overview of nanoencapsulation techniques and their classification. In: Nanoencapsulation technologies for the food and nutraceutical industries. Academic Press, Oxford
13. Assadpour E, Jafari SM (2020) Chapter One - importance of release and bioavailability studies for nanoencapsulated food ingredients. In: Jafari SM (ed) Release and bioavailability of nanoencapsulated food ingredients. Academic Press, Oxford, p 1–24
14. Witika BA, Makoni PA, Matafwali SK, Chabalenge B, Mwila C, Kalungia AC, Nkanga CI, Bapolisi AM, Walker RB (2020) Biocompatibility of biomaterials for nanoencapsulation: current approaches. Nanomaterials 10. https://doi.org/10.3390/nano10091649
15. Cushen M, Kerry J, Morris M, Cruz-Romero M, Cummins E (2012) Nanotechnologies in the food industry – recent developments, risks and regulation. Trends Food Sci Technol 24:30–46. https://doi.org/10.1016/j.tifs.2011.10.006
16. Soyombo DA (2024) Navigating regulatory challenges and safety considerations in lactation supplement development: a Nigeria and US comparison. World J Biol Pharm Heal Sci 18:001–011. https://doi.org/10.30574/wjbphs.2024.18.1.0165
17. Asaithambi N, Singha P, Dwivedi M, Singh SK (2019) Hydrodynamic cavitation and its application in food and beverage industry: a review. J Food Process Eng 42:e13144. https://doi.org/10.1111/JFPE.13144;PAGEGROUP:STRING:PUBLICATION
18. Evonik Nutrition & Care GmbH TEGO\textregistered\Care APD 18 (2016) Superior PEG-free O/W emulsifier for diverse AP/Deo roll-on systems. Tech. Data Sheet n. B10162
19. Souto EB, Doktorovova S, Zielinska A, Silva AM (2019) Key production parameters for the development of solid lipid nanoparticles by high shear homogenization. Pharm Dev Technol 24:1181–1185. https://doi.org/10.1080/10837450.2019.1647235
20. Silva TM, Cerize NNP, Oliveira AM (2016) The effect of high shear homogenization on physical stability of emulsions. Int J Chem 8:52. https://doi.org/10.5539/ijc.v8n4p52
21. Hashemi H, Shad E, Ghiasi F, Eskandari MH (2024) Savory and peppermint essential oils-loaded emulsions and nanoemulsions effects on *Enterococcus faecium* isolated from vacuum-packed cured sausage. Foods 13:341. https://doi.org/10.3390/FOODS13020341
22. Baldassarre F, Schiavi D, Di Lorenzo V, Biondo F, Vergaro V, Colangelo G, Balestra GM, Ciccarella G (2023) Cellulose nanocrystal-based emulsion of thyme essential oil: preparation and characterisation as sustainable crop protection tool. Molecules 28:7884. https://doi.org/10.3390/molecules28237884
23. Tripathi AD, Sharma R, Agarwal A, Haleem DR (2021) Nanoemulsions based edible coatings with potential food applications. Int J Biobased Plast 3:112–125. https://doi.org/10.1080/24759651.2021.1875615
24. Hashemi H, Shad E, Ghiasi F, Eskandari MH (2024) Savory and peppermint essential oils-loaded emulsions and nanoemulsions effects on *Enterococcus faecium* isolated from vacuum-packed cured sausage. Foods 13:341. https://doi.org/10.3390/FOODS13020341
25. Ujilestari T, Febrisiantosa A, Sholikin MM, Wahyuningsih R, Wahyono T (2023) Nanoemulsion application in meat product and its functionality. Rev J Anim Sci Technol 65:275. https://doi.org/10.5187/JAST.2022.E120
26. Cano-Sarmiento C, Alamilla-Beltrán L, Azuara-Nieto E, Hernández-Sánchez H, Téllez-Medina DI, Jiménez-Martínez C, Gutiérrez-López GF (2015) High shear methods to produce nano-sized food related to dispersed systems. In: Food engineering series. Springer, Cham, p 145–161

Chapter 10

Immobilization of Lactic Acid Bacteria for Application in Dry Fermented Meat Products

Marilia Silva Malvezzi Karwowski and Renata Ernlund Freitas de Macedo

Abstract

Innovative strategies have been developed to enhance consumer acceptance of reformulated meat products, with the incorporation of probiotics emerging as a promising approach. However, this strategy is challenged by the adverse conditions probiotics face during processing, storage, and gastrointestinal (GI) passage. Key obstacles include low water activity, acidic pH, and the presence of curing salts in meat products, which compromise probiotic survival. Additionally, many probiotics are highly sensitive to the low pH of the stomach and the high bile salt concentrations in the proximal intestine. To address these challenges, the immobilization of lactic acid bacteria (LAB) within food matrices has proven effective in enhancing probiotic viability in meat products and during GI transit. Grain-based immobilization offers physical entrapment, providing a protective barrier that shields LAB from the harsh conditions of fermented meat production and digestion. This chapter details a method for immobilizing LAB in grains and its subsequent application in fermented meat products, offering insights into improving probiotic delivery and functionality.

Keywords Entrapment, Grain, Fermented sausage, Probiotic, Viability, Functional

1 Introduction

Fermentation is an ancient method for producing meat products with distinct sensory and functional characteristics [1]. This process relies on naturally occurring bacteria or selected starter cultures under controlled conditions [2, 3]. In the meat industry, starter cultures are essential for standardizing fermentation processes and ensuring the safety and stability of products [4]. For dry-fermented sausages, high populations of starter cultures are critical for predictable fermentation [1]. Typical ingredients in these products include meat (usually pork), animal fat, salt, spices, and additives such as nitrites or nitrates [5].

While meat offers nutritional benefits, its consumption, particularly processed meat, has been associated with increased risks of cardiovascular diseases, obesity, certain cancers, and other

Renata Ernlund Freitas de Macedo and Carlos Adam Conte-Junior (eds.), *Meat Products Preservation*, Methods and Protocols in Food Science, https://doi.org/10.1007/978-1-0716-5068-4_10,

health issues [6]. Reformulation of processed meat products has emerged as a strategy to address these concerns [7]. Reformulation typically follows two approaches: reduction—lowering harmful ingredients such as sodium [8–10] and nitrites [11, 12]—and/or addition—enhancing products with health-promoting ingredients like probiotics [13].

Probiotics are beneficial microorganisms, with common genera in the food industry including *Lactobacillus, Bifidobacterium, Lactococcus, Streptococcus, Enterococcus*, and *Bacillus* [14]. In meat fermentation, lactic acid bacteria (LAB), such as *Lactobacillus casei, Lactiplantibacillus, plantarum, Latilactobacillus curvatus*, and others, play a dominant role. Many of these LAB strains exhibit probiotic properties, offering health benefits beyond basic nutrition. Incorporating probiotics into fermented meat products is a promising strategy for developing functional foods with added health value [15].

However, integrating probiotics into meat products poses significant technological challenges. Probiotics must maintain viability under adverse conditions, including variations in temperature, exposure to acid and bile, osmotic pressure, and oxidative stress during gastrointestinal (GI) transit [16]. These challenges are even greater in dry fermented meats, where low water activity and extended storage periods can further compromise probiotic survival. Consequently, there is a pressing need for techniques that maintain high probiotic concentrations (10^8 CFU/g) under challenging storage and environmental conditions.

Immobilization techniques offer a viable solution to these challenges by enhancing the stability, viability, and functionality of probiotic bacteria. Immobilization involves entrapping cells within a matrix that protects them from harsh processing and physiological conditions, such as stomach acidity and bile salts in the intestine [17, 18]. This approach also supports higher probiotic densities, reduces microbial contamination risks [19], and accelerates ripening rates [20].

Grains rich in polysaccharides, with their prebiotic properties and specific physicochemical behavior, are promising immobilization matrices. They not only enhance probiotic viability but also contribute to the nutritional profile of the final product by providing dietary fiber, proteins, and other nutrients [21]. Grains such as wheat and millet (*Panicum miliaceum* L.) have shown promise as immobilizing supports for probiotics in dry-fermented sausages [22–25].

This chapter outlines a method for immobilizing LAB on grain matrices for application in fermented meat products. The described approach details the preparation of the support matrix and procedural steps, providing a comprehensive guide for readers to understand and implement this technique (Fig. 1).

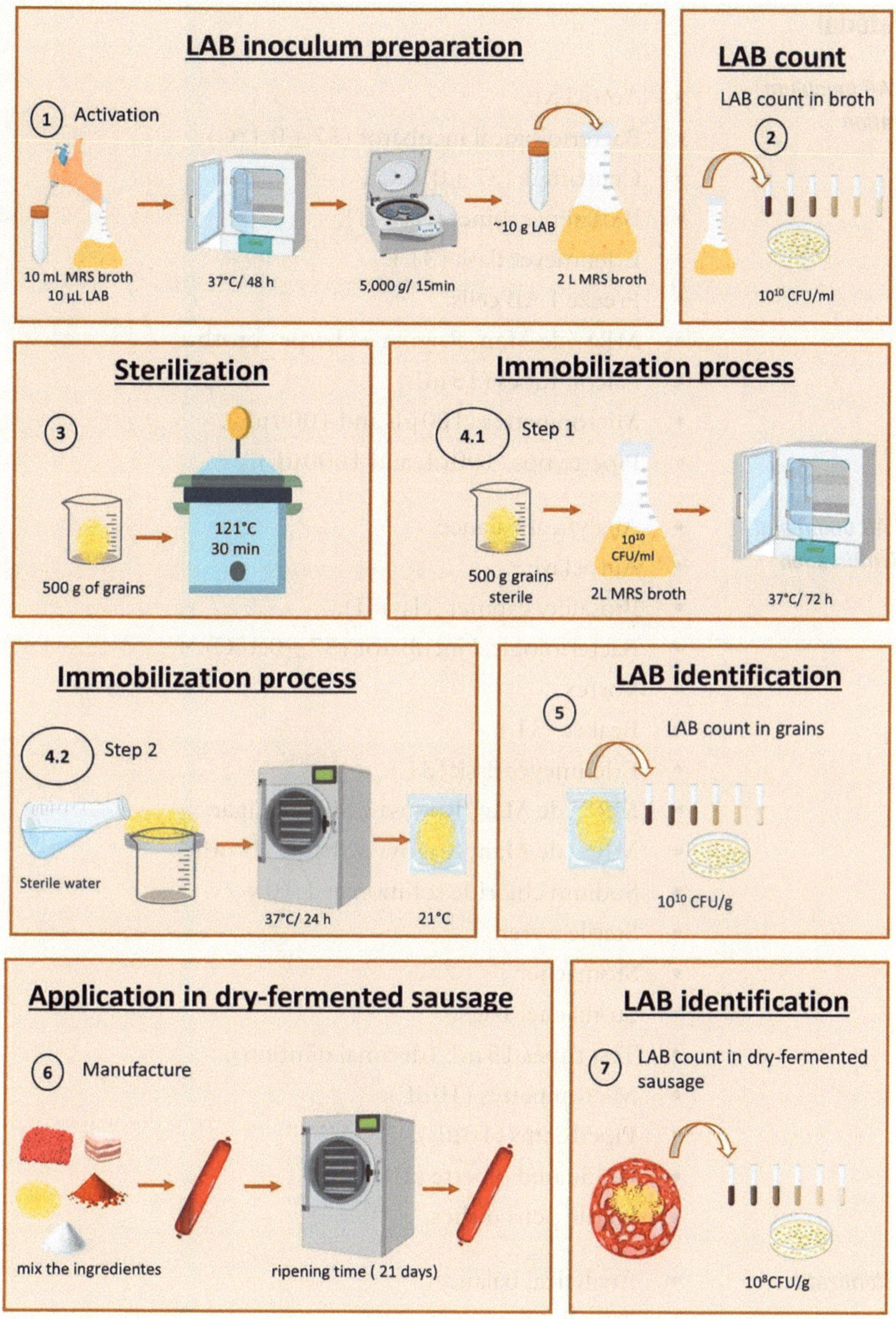

Fig. 1 Procedural steps for immobilizing probiotic bacteria and their application in fermented meat products

2 Material

2.1 LAB Inoculum Preparation

- Autoclave.
- Bacteriological incubator (37 ± 0.1°C).
- Centrifuge (37 ± 0.1°C).
- Biosafety cabinet (class II).
- Erlenmeyer flask (3 L).
- Freeze LAB cells.
- MRS (de Man, Rogosa e Sharpe) broth.
- Falcon tubes (15 mL).
- Micropipettes (100 μL and 1000 μL).
- Pipette tips (100 μL and 1000 μL).

2.2 LAB Count and LAB Identification

- Analytical balance.
- Autoclave.
- Biosafety cabinet (class II).
- Bacteriological incubator (37 ± 0.1°C).
- Vortex.
- Beaker (3 L).
- Erlenmeyer flask (3 L).
- MRS (de Man, Rogosa e Sharpe) agar.
- MRS (de Man, Rogosa e Sharpe) broth.
- Sodium chloride solution at 1:10 w/v.
- Sterile water.
- Stomacher.
- Stomacher bags.
- Test tubes 15 mL (decimal dilution).
- Micropipettes (10 μL).
- Pipette tips (10 μL).
- Graduated pipette (10 mL).
- Sterile petri dishes.

2.3 Sterilization

- Analytical balance.
- Autoclave.
- Beaker (3 L).
- Grains of quinoa (*Chenopodium quinoa Willd*), proso millet (*Panicum miliaceum* L.), amaranth (*Amaranthus hypochondriacus*), or wheat (*Triticum aestivum*).

2.4 Immobilization

- Analytical balance.
- Biosafety cabinet (class II).
- Bacteriological incubator (37 ± 0.1 °C).
- Drying chamber.
- Stomacher.
- Vacuum sealers.
- Vortex.
- Erlenmeyer flask (3 L).
- Sterile water.
- MRS (de Man, Rogosa e Sharpe) agar.
- MRS (de Man, Rogosa e Sharpe) broth.
- Sodium chloride solution at 1:10 w/v.
- Sterile metal sieve.
- Stomacher bags.
- Test tubes 15 mL (decimal dilution).
- Micropipettes (1000 μL).
- Pipette tips (1000 μL).
- Sterile petri dishes.
- Vacuum plastic bags.
- Sterile grains of quinoa (*Chenopodium quinoa Willd*), proso millet (*Panicum miliaceum* L.), amaranth (*Amaranthus hypochondriacus*), or wheat (*Triticum aestivum*).

2.5 Dry-Fermented Sausage Manufacture

- Automatic filler.
- Mixer.
- Meat grinder (disc 8 mm).
- Chamber cabinet.
- pH meter.
- Aqualab CX2 instrument.
- Pork (*biceps femoralis, semimembranosus, and semitendinosus*).
- Beef meat (trimmings).
- Pork back fat.
- Salt.
- Glucose (0.5%) and sucrose (0.5%).
- Additives: ascorbate (0.5%), nitrate (50 ppm), nitrite (150 ppm), sodium ascorbate (0.05%), spices mix (1.0%).

- Collagen casings (45 mm diameter).
- Culture starter *Staphylococcus xylosus* activated (*see* **Note 1**).

3 Methods

Three distinct steps must be followed. First, the immobilization of the LAB in grains, followed by the assessment of immobilization efficiency. Next, the grains immobilized with LAB can be incorporated as an ingredient of fermented meat products. Dry-fermented sausages have been chosen as the representative example.

3.1 LAB Inoculum Preparation

- Prepare the MRS broth following the manufacturer's instructions in an Erlenmeyer flask (3 L).
- Autoclave the MRS broth at 121°C for 15 min.
- Measure 10 mL of MRS broth using a graduated pipette.
- Transfer 10 mL of MRS broth to a Falcon tube (15 mL).
- Transfer 10 μL of freeze-dried LAB probiotic bacteria.
- Grow the culture at 37°C for 48 h.
- Centrifuge at 5000 × *g* for 15 min at 25°C.
- Resuspend the pellets (~10 g) of LAB cells with 2 L of sterile MRS broth in Erlenmeyer flask (3 L).

3.2 LAB Count

3.2.1 LAB Count in MRS Broth

- Weigh the required amount of sodium chloride into a 1000 mL beaker and dissolve in sterile water (typically 1:10 w/v).
- Collect a 10 mL aliquot of MRS broth (obtained in 3.1), and mix with sodium chloride solution at 1:10 w/v (90 mL) in a stomacher bag.
- Homogenize the mixture using the stomacher for 1 min.
- Transfer 1 mL of the homogenized solution to a test tube containing 9 mL of sodium chloride 0.9% and homogenize using a vortex.
- Perform decimal serial dilutions.
- Prepare the MRS agar following the manufacturer's instructions.
- Autoclave MRS agar by 15 min at 121°C.
- Pour plate the serial dilutions (1 mL) into Petri dishes using 15–20 mL of MRS agar.
- Homogenize the mixture.
- Incubate at 37°C/48 h.

- Count the characteristic colonies and calculate the colony-forming units per gram of samples (CFU/g), according to Equation (1).

$$\frac{CFU}{g} = Colony\ count \cdot Dilution\ factor \tag{1}$$

- Obtain the logarithmic count using the function "$\log_{10}$" using a calculator or statistical software (*see* **Note 2**).

3.3 Preparation of Support and Cell Immobilization

3.3.1 Sterilization Process

- Weigh 500 g of grains-Quinoa (*Chenopodium quinoa Willd*), proso millet (*Panicum miliaceum* L.), amaranth (*Amaranthus hypochondriacus*), or wheat (*Triticum aestivum*) in beaker 3 L.
- Sterilize the grains at 121°C for 30 min (*see* **Note 3**).

3.3.2 Immobilization Process

- Add the 2 L of MRS broth with the LAB (Sect. 3.1) into the Erlenmeyer flask containing the sterilized grains (500 g) (Sect. 3.3.1).
- Grow the mixture at 37°C for 72 h, without agitation (*see* **Note 4**).
- Drain the liquid using a sterile metal sieve supported on a beaker.
- Wash the grains twice with sterile water.
- Dry the grains in an air circulation drying chamber at 37°C for 24 h.
- After drying, vacuum store the grains hermetically sealed in plastic bags and stored at room temperature (20°C).

3.4 LAB Count

3.4.1 LAB Enumeration in Grains

- Weigh the required amount of sodium chloride into a 1000 mL beaker and dissolve in sterile water (typically 1:10 w/v).
- Collect 10 g of grains (obtained in 3.3.2), and mix with sodium chloride solution at 1:10 w/v (90 mL) in a stomacher bag.
- Homogenize the mixture using the stomacher for 1 min.
- Transfer 1 mL of the homogenized solution to a test tube containing 9 mL of sodium chloride 0.9% and homogenize using a vortex.
- Perform decimal serial dilutions.
- Prepare the MRS agar following the manufacturer's instructions.
- Autoclave MRS agar by 15 min at 121°C.

- Pour plate the serial dilutions (1 mL) into Petri dishes using 15–20 mL of MRS agar.
- Homogenize the mixture.
- Incubate at 37°C/48 h.
- Count the characteristic colonies and calculate the colony-forming units per gram of samples (CFU/g), according to Equation (1).

$$\frac{CFU}{g} = Colony\ count \cdot Dilution\ factor \tag{1}$$

- Obtain the logarithmic count using the function "$\log_{10}$" using a calculator or statistical software (*see* **Note 5**).

3.5 Application of Immobilized LAB in Dry-Fermented Sausage

3.5.1 Manufacture of Dry-Fermented Sausage

- Calculate the percentages of all the ingredients (*see* **Note 6**).
- Ground the pork and beef meat for dry-fermented sausage production.
- Mix the ingredients in this order: minced pork and beef, pork back fat, salt, glucose, sucrose, additives, and spices.
- Mix the starter culture (*Staphylococcus xylosus*) in a sodium chloride solution (0.9%).
- Mix the grains with immobilized LAB.
- Stuff the batter into collagen casings.
- Place the dry-fermented sausage in a chamber cabinet (*see* **Note 7**).
- Measure the pH during the ripening time. Ensure that the pH of the dry-fermented sausage will be less than 5.5.
- Measure the Aw (water activity). Ensure that the Aw of the dry-fermented sausage will be less than 0.920.
- After the ripening, vacuum pack the sausages in polyethylene bags.
- Store at 21°C ± 1°C and 83 ± 2% RH.

3.5.2 LAB Count in Dry-Fermented Sausage

- Weigh the required amount of sodium chloride into a 1000 mL beaker and dissolve in sterile water (typically 1:10 w/v).
- Collect a 10 g of dry-fermented sausage (obtained in Sect. 3.5.1), and mix with sodium chloride solution at 1:10 w/v (90 mL) in a stomacher bag.

- Homogenize the mixture using the stomacher for 1 min.
- Transfer 1 mL of the homogenized solution to a test tube containing 9 mL of sodium chloride 0.9% and homogenize using a vortex.
- Perform decimal serial dilutions.
- Prepare the MRS agar following the manufacturer's instructions.
- Autoclave MRS agar for 15 min at 121°C.
- Pour plate the serial dilutions (1 mL) into Petri dishes using 15–20 mL of MRS agar.
- Homogenize the mixture.
- Incubate at 37°C/48 h.
- Count the characteristic colonies and calculate the colony-forming units per gram of samples (CFU/g), according to Equation (1).

$$\frac{CFU}{g} = Colony\ count \cdot Dilution\ factor \tag{1}$$

- Obtain the logarithmic count using the function "$\log_{10}$" using a calculator or statistical software (*see* **Note 8**).

After cell enumeration, the identification of the LAB probiotic species can be performed using molecular methods such as RAPD-PCR [26] (*see* **Note 9**).

4 Notes

1. Activate the lyophilized starter *Staphylococcus xylosus* in TBS (Tryptic Soy Broth) broth following the manufacturer's instructions.
2. As the MRS broth will be used for the immobilization process, it is important to ensure that the count of probiotic bacteria in the broth is ~ 10^{10} CFU/mL.
3. Some studies sterilize the food matrix for 15 min, the time will depend on the porosity of the matrices [25, 27, 28].
4. Some studies incubate the grains for 24 or 48 h, the duration will depend on the matrix and the LAB species [23, 27, 28].
5. The LAB count is expected to be around 10^9 and 10^{10} CFU/g.
6. The grain percentage should be calculated according to the LAB count obtained on the grain after immobilization.
7. On the first day of ripening, the temperature should be set at 25°C for fast fermentation, and program it to drop by 1°C every 24 h until reaching 18°C. The relative humidity (RH)

should initially be set at 91% and decrease 1% every 24 h up to 85%. The ripening period is 21 days. Ensure that the final Aw is lower than 0.920 when fermented sausages are considered "ready to eat."

8. The LAB count is expected to be around 10^8 and 10^9 CFU/g.
9. The RAPD procedure follows the methodology described by Oneca et al. [29]. RAPD is performed in a 25 μL volume containing 3 mM MgCl2, 1× Buffer, 0.25 mM dNTP, 1 U Taq and 3 μM primer. Thermal cycling for the RAPD reaction starts with 5 min at 95°C of initial denaturation, followed by 45 cycles of 1 min at 94°C, 1 min at 36°C, and 1 min at 72°C, followed by a final extension of 7 min at 72°C. For RAPD analysis, 10 colonies are randomly selected from the MRS agar plates (obtained in Sect. 3.5.2). The extraction of the DNA follows the protocol described by Pospiech and Neumann [26]. Quantify the DNA extracted from the strains in a Nanodrop apparatus. Amplify and separate the product on a 1.5% agarose gel by electrophoresis at 70 V stained with ethidium bromide solution.

References

1. Pilevar Z, Hosseini H (2017) Effects of starter cultures on the properties of meat products: a review. Annu Res Rev Biol 17(6):1–17
2. Pilevar Z, Nutrition N, Hosseini H (2017) Effects of starter cultures on the properties of meat products: a review. Annu Res Rev Biol 17:1–17
3. Speranza B, Racioppo A, Beneduce L, Bevilacqua A, Sinigaglia M, Corbo MR (2017) Autochthonous lactic acid bacteria with probiotic aptitudes as starter cultures for fish-based products. Food Microbiol 65:244–53. https://doi.org/10.1016/j.fm.2017.03.010
4. Zhang Y, Zhang Y, Zhou X, Wang S, Li P (2022) Salt replacement changed the bacterial community composition and physicochemical characteristics of sodium-reduced fermented sausages during fermentation and ripening. Foods. 2021:10(3):630
5. Lee S, Jo K, Jeong SKC, Jeon H, Choi YS, Jung S (2023) Recent strategies for improving the quality of meat products. J Anim Sci Technol 65(5):895–911
6. Boada LD, Henríquez-Hernández LA, Luzardo OP (2016) The impact of red and processed meat consumption on cancer and other health outcomes: epidemiological evidences. Food Chem Toxicol 92:236–244. https://doi.org/10.1016/j.fct.2016.04.008
7. Hong X, Li C, Wang L, Wang M, Grasso S, Monahan FJ (2023) Consumer preferences for processed meat reformulation strategies: a prototype for sensory evaluation combined with a choice-based conjoint experiment. Agriculture 13(2):id.234
8. Pateiro M, Munekata PE, Cittadini A, Domínguez R, Lorenzo JM (2021) Metallic-based salt substitutes to reduce sodium content in meat products. Curr Opin Food Sci 38:21–31. https://doi.org/10.1016/j.cofs.2020.10.029
9. Kumar S, Yadav S, Pal A, Vaquil (2021) Studies on the impact of partial replacement of sodium chloride with potassium lactate on quality attributes of buffalo calf meat rolls. Asian J Dairy Food Res 40(2):206–12
10. Muchaamba F, Stoffers H, Blase R, von Ah U, Tasara T (2021) Potassium lactate as a strategy for sodium content reduction without compromising salt-associated antimicrobial activity in salami. Foods 10(1):114
11. Perea-Sanz L, López-Díez JJ, Belloch C, Flores M (2020) Counteracting the effect of reducing nitrate/nitrite levels on dry fermented sausage aroma by *Debaryomyces hansenii* inoculation. Meat Sci 164:108103. https://doi.org/10.1016/j.meatsci.2020.108103
12. Zhang Y, Zhang Y, Jia J, Peng H, Qian Q, Pan Z et al (2023) Nitrite and nitrate in meat processing: functions and alternatives. Curr Res Food Sci 6:100470. https://doi.org/10.1016/j.crfs.2023.100470
13. Karwowski MSM, Cavalari CM de A, Oliveira K, Rosset M, Macedo REF de (2023) Millet grains as an immobilizing matrix for probiotics in dry

fermented sausage. Food Bioprocess Technol https://doi.org/10.1007/s11947-023-03003-6
14. Markowiak P, Ślizewska K (2017) Effects of probiotics, prebiotics, and synbiotics on human health. Nutrients 9(9):1021
15. Ojha KS, Kerry JP, Duffy G, Beresford T, Tiwari BK (2015) Technological advances for enhancing quality and safety of fermented meat products. Trends Food Sci Technol 44(1):105–116. https://doi.org/10.1016/j.tifs.2015.03.010
16. Wilkinson MG (2018) Flow cytometry as a potential method of measuring bacterial viability in probiotic products: a review. Trends Food Sci Technol 78:1–10. https://doi.org/10.1016/j.tifs.2018.05.006
17. Kallenberg AI, Van Rantwijk F, Sheldon RA (2005) Immobilization of penicillin G acylase: the key to optimum performance. Adv Synth Catal 347(7–8):905–26
18. Doherty SB, Gee VL, Ross RP, Stanton C, Fitzgerald GF, Brodkorb A (2010) Efficacy of whey protein gel networks as potential viability-enhancing scaffolds for cell immobilization of *Lactobacillus rhamnosus* GG. J Microbiol Methods 80(3):231–41. https://doi.org/10.1016/j.mimet.2009.12.009
19. Mitropoulou G, Nedovic V, Goyal A, Kourkoutas Y (2013) Immobilization technologies in probiotic food production. J Nutr Metab 2013:716861
20. Kourkoutas Y, Xolias V, Kallis M, Bezirtzoglou E, Kanellaki M (2005) *Lactobacillus casei* cell immobilization on fruit pieces for probiotic additive, fermented milk and lactic acid production. Process Biochem 40(1):411–6
21. Kaur R, Sharma M (2019) Cereal polysaccharides as sources of functional ingredient for reformulation of meat products: a review. J Funct Foods. 62:103527. https://www.sciencedirect.com/science/article/pii/S1756464619304517
22. Karwowski MSM, Sayuri Kassuia Y, Turchenski DG, Aust ACC de O, Macedo REF de (2024) Novel immobilization method of *Lactiplantibacillus plantarum* in grains for dry food products. Food Sci Technol Int 31:506–515
23. Sidira M, Kandylis P, Kanellaki M, Kourkoutas Y (2015) Effect of immobilized Lactobacillus casei on volatile compounds of heat treated probiotic dry-fermented sausages. Food Chem 178:201–7
24. Sidira M, Santarmaki V, Kiourtzidis M, Argyri A, Papadopoulou O, Chorianopoulos N et al (2017) Evaluation of immobilized *Lactobacillus plantarum* 2035 on whey protein as adjunct probiotic culture in yoghurt production. LWT Food Sci Technol 75:137–46
25. Sidira M, Kandylis P, Kanellaki M, Kourkoutas Y (2015) Effect of immobilized *Lactobacillus casei* on the evolution of flavor compounds in probiotic dry-fermented sausages during ripening. Meat Sci 100:41–51
26. Pospiech A, Neumann B (1995) A versatile quick-prep of genomic DNA from Gram-positive bacteria. Trends Genet 11(6):217–8
27. Bosnea LA, Kourkoutas Y, Albantaki N, Tzia C, Koutinas AA, Kanellaki M (2009) Functionality of freeze-dried *L. casei* cells immobilized on wheat grains. LWT Food Sci Technol 42(10):1696–1702. https://doi.org/10.1016/j.lwt.2009.05.011
28. Sidira M, Galanis A, Nikolaou A, Kanellaki M, Kourkoutas Y (2014) Evaluation of Lactobacillus casei ATCC 393 protective effect against spoilage of probiotic dry-fermented sausages. Food Control 42:315–20
29. Oneca M, Irigoyen A, Ortigosa M, Torre P (2003) PCR and RAPD identification of *L. plantarum* strains isolated from ovine milk and cheese. Geographical distribution of strains. FEMS Microbiol Lett 227(2):271–7

Chapter 11

Evaluation and Application of Antimicrobial Metabolites (Nisin and ε-Poly-L-lysine) in Cooked Sausage

Fernanda Cristina Kandalski Bortolotto, Alberto Gonçalves Evangelista and Renata Ernlund Freitas de Macedo

Abstract

Cooked sausages are highly susceptible to microbial spoilage, particularly when stored at room temperature. Certain bacterial groups, especially lactic acid bacteria (LAB), are capable of surviving the cooking process and remaining viable in the final product. This persistence not only accelerates microbial spoilage but also reduces shelf life and compromises sensory quality before the expiration date. In line with the growing interest in using natural antimicrobials to control microbial growth in food, antimicrobial metabolites such as nisin (NI) and ε-poly-L-lysine (ε-PL) have emerged as promising agents due to their broad-spectrum activity. NI is a natural antimicrobial peptide produced by *Lactococcus lactis* subsp. *lactis*, and is recognized as Generally Recognized as Safe (GRAS) by the United States Food and Drug Administration. ε-PL is a polyamino acid obtained through the fermentation of Streptomyces albulus, and is likewise classified as GRAS. The combination of these natural antimicrobials can produce synergistic effects, making them suitable for the control of LAB-related spoilage in meat products. This chapter addresses the methods required for the incorporation of NI and ε-PL into cooked sausages, with a focus on the control of spoilage bacteria.

Keywords Natural antimicrobials, Nisin, ε-Poly-L-lysine, Food preservation, Lactic acid bacteria, Sausage

1 Introduction

Antimicrobial metabolites, also referred to as secondary metabolites with antimicrobial properties, are compounds synthesized by microorganisms that inhibit the growth or eliminate other microorganisms, including bacteria, fungi, and yeasts. These substances are typically produced during the secondary phase of microbial metabolism and are widely applied in infection control, as well as in the food industry as natural preservatives [1].

Renata Ernlund Freitas de Macedo and Carlos Adam Conte-Junior (eds.), *Meat Products Preservation*, Methods and Protocols in Food Science, https://doi.org/10.1007/978-1-0716-5068-4_11,

Among these compounds, antimicrobial peptides (AMPs) play a critical role as a first line of defense and are integral components of the innate immune system in a wide range of organisms. AMPs exhibit diverse mechanisms of action, most commonly through disruption of microbial cell membranes, leading to irreversible leakage of intracellular contents and cell death. Bacteria are a rich source of AMPs, which have been extensively studied for applications in food preservation. Bacteriocins, for example, were first described in Gram-negative bacteria, such as colicins produced by *Escherichia coli*, but lactic acid bacteria (LAB), a group of Gram-positive bacteria, have since become a major focus due to their ability to produce bacteriocins with proven efficacy in food systems [2].

Nisin (NI) is a well-characterized bacteriocin produced by *Lactococcus lactis* subsp. *lactis*. It holds Generally Recognized as Safe (GRAS) status by the United States Food and Drug Administration [3]. NI exerts potent bactericidal activity against a broad range of Gram-positive bacteria, including LAB. Its antimicrobial mechanism involves inhibition of cell wall peptidoglycan biosynthesis and pore formation in the cytoplasmic membrane, ultimately resulting in cell lysis [4]. NI's efficacy can be enhanced when used in combination with other hurdles, such as organic acids, essential oils, or other AMPs [5, 6].

ε-Poly-L-lysine (ε-PL) is a cationic polyamino acid produced by *Streptomyces albulus* during fermentation. It exhibits broad-spectrum antimicrobial activity against both Gram-positive and Gram-negative bacteria, including relevant foodborne pathogens such as *E. coli* O157:H7, *Salmonella* Typhimurium, and *Listeria monocytogenes*, as well as against yeasts and molds [7]. ε-PL is approved as a food preservative in Japan and has GRAS status in the United States [8]. Its mode of action is based on electrostatic interaction with negatively charged microbial membranes, leading to membrane destabilization, leakage of intracellular components, and cell death [9].

In food systems, preservatives are applied to extend shelf life by reducing or delaying the growth of spoilage and pathogenic microorganisms, thus maintaining the sensory and microbiological quality of the product [9]. Cooked sausages are particularly susceptible to spoilage due to their high-water activity, relatively high pH (typically >6.0), and rich nutrient composition. LAB, such as *Lactobacillus*, *Leuconostoc*, and *Carnobacterium*, are the predominant spoilage organisms in vacuum-packed cooked sausages, such as calabrese [10]. The use of natural antimicrobials in these products must strike a balance between effective microbial control, minimal impact on sensory attributes, and consumer demand for clean-label solutions. To this end, understanding the efficacy and interaction of antimicrobial compounds is essential.

The evaluation of antimicrobial activity typically comprises two main stages: *in vitro* efficacy testing and application in the food matrix. *In vitro* analysis includes the determination of the Minimum Inhibitory Concentration (MIC), Minimum Bactericidal Concentration (MBC), and, when assessing combinations of antimicrobial compounds, the Fractional Inhibitory Concentration (FIC) and the calculation of the Fractional Inhibitory Concentration Index (FIC_{index}). This chapter outlines the methodologies used to determine MIC via broth microdilution, MBC, and FIC of NI and ε-PL using the checkerboard assay, as well as the interpretation of their interactions through FIC_{index} calculation [11, 12]. Once the *in vitro* methods are established, the chapter proceeds to the application method in a food system, employing direct incorporation, a strategy in which antimicrobial agents are added directly to the product formulation. This approach ensures uniform distribution and immediate interaction with the target microbiota.

2 Material

2.1 Determination of MIC, MBC, FIC, and FIC_{index} of NI and ε-PL Against LAB

2.1.1 Equipment (see Note 1)

- Single-channel micropipettes for handling volumes between 100 and 1000 μL.
- Multichannel micropipette for handling volumes between 100 and 200 μL (*see* **Note 2**).
- Autoclave.
- Biological safety cabinet with appropriate biosafety level (*see* **Note 3**).
- Microbiological incubator set at 37°C.
- Spectrophotometer or equivalent equipment capable of measuring absorbance in the visible light range (*see* **Note 4**).
- Precision analytical balance, if the tested compound is in solid form.
- Microplate reader.

2.1.2 Sterile or Sterilized Items

- De Man, Rogosa and Sharpe broth (MRSb), prepared according to the manufacturer's instructions.
- De Man, Rogosa and Sharpe agar (MRSa), prepared according to the manufacturer's instructions.
- Test tubes or centrifuge tubes.
- 96-well microplates.
- Pipette tips for handling volumes between 100 and 1000 μL (*see* **Note 5**).

2.2 Preparation of Calabrese Sausage Added with Natural Antimicrobials NI and ε-PL

2.2.1 Equipment

- Meat grinder (8 mm plate).
- Mixer.
- Stuffing machine.
- Steam chamber.
- Ice-water bath.
- Vacuum-packing machine.
- Vacuum-packing low-density barrier pouches (permeability of $<30\,cm^3/m^2$/day at 1 atm and 23°C).

2.2.2 Sausage Ingredients (see Note 6)

- Pork shoulder trim (68.45%).
- Pork back fat (15.0%).
- Mechanically deboned poultry meat (10.0%).
- Water (2.0%).
- Texturized soybean protein (2.0%).
- Garlic (0.25%).
- Calabrese pepper (0.15%).
- Sodium tripolyphosphate (0.25%).
- Antioxidant mix (0.25%), containing sucrose 55%, sodium erythorbate 30%, sodium chloride 15%.
- Salt (1.4%).
- Curing salt (0.25%), containing NaCl 90%, $NaNO_2$ 6%, $NaNO_3$ 4%.
- Natural antimicrobials (ε-PL and NI).
- Natural pork casings.

3 Methods

3.1 Determination of MIC, MBC, FIC, and FIC_{index} of NI and ε-PL Against LAB

3.1.1 Execution of the MIC Assay

- The analysis will be conducted following the Clinical and Laboratory Standards Institute guidelines [11], with modifications to accommodate lactic acid bacteria.

- Although MRSa and MRSb are standard media for lactic acid bacteria, they must be validated for bacterial growth before use in the assay.
- To better simulate inhibition within the microbiome of a food matrix, it is recommended to perform the assay using a pool of at least five isolates derived from the matrix under study. Reference strains may also be included.
- Each strain or isolate must first be activated individually in MRSb at 37°C for 24–48 h (*see* **Note 7**). If the culture was stored frozen or lyophilized, activation should be performed at least twice before use in the assay.
- Once activated, the bacteria must be cultured in MRSb until they reach the mid-exponential phase, which is the ideal phase for microbiology testing (*see* **Note 8**).
- At mid-exponential phase, bacterial concentration should be estimated by absorbance measurement using a spectrophotometer or equivalent device (*see* **Note 9**). The population is calculated using a standard growth curve (Fig. 1), with the equation:

$$Population\ (CFU/mL) = a \times Absorbance \pm b$$

- Based on this result, an inoculum of 10^6 CFU/mL total bacteria in MRSb is prepared. For pooled cultures, 10^6 must be divided

1. General equation for estimating bacterial population (CFU/mL) based on absorbance measurements:

Population (CFU/mL) = a × Absorbance ± b

2. For illustrative purposes, let us consider the following hypothetical linear regression curve:

Population (×10^8 CFU/mL) = 18.961 × Absorbance – 1.6297

3. For explanatory purposes, let us consider the following absorbance value obtained after adjusting for the absorbance of the sterile culture medium:

Absorbance = 0.450

4. The calculation is presented as follows:

Population (×10^8 CFU/mL) = 18.961 × Absorbance – 1.6297
Population (×10^8 CFU/mL) = 18.961 × 0.450 – 1.6297
Population (×10^8 CFU/mL) = 6.90275

Population = 6.9×10^8 CFU/mL

Fig. 1 Calculation of bacterial concentration (CFU/mL) from absorbance data by the number of strains in the pool to calculate each strain's contribution. Inoculum volume is calculated using:

$$c_1 \times v_1 = c_2 \times v_2$$

where c_1 is the original concentration, v_1 the volume to be added (parameter to be determined), c_2 the desired concentration, and v_2 the final volume of inoculum.

- So, the total volume of MRSb (v_2) is used to dilute the individual volumes (v_1) of each strain, producing a final concentration of 10^6 CFU/mL (Fig. 2).
- An antimicrobial compound solution is then prepared in MRSb at twice the highest desired concentration. The purity of the compound should be considered and adjusted to 100%. Up to ten concentrations can be tested simultaneously per plate using the microdilution method.
- Using a standard 96-well plate (12 columns × 8 rows), assemble the layout as follows (Fig. 3):
  - Add 100 μL of sterile MRSb to columns 2–11.
  - Add 200 μL of sterile MRSb to column 12.
  - Add 200 μL of the antimicrobial solution to column 1.
  - Perform a 1:2 serial dilution by transferring 100 μL from column 1 to 2, then from 2 to 3, and so on until column 10. Discard 100 μL from the last column to ensure a final volume of 100 μL.

1. Based on the previously established population:

$$\text{Population} = 6.9\times10^8 \text{ CFU/mL}$$

2. The following formula is used for inoculum preparation:

$$c_1 \times v_1 = c_2 \times v_2$$

3. To prepare 10 mL (10,000 μL) of inoculum at a concentration of 10^6 CFU/mL, the following equation is used:

$$6.9\times10^8 \text{ CFU/mL} \times v_1 = 10^6 \text{ CFU/mL} \times 10,000\ \mu\text{L}$$

4. The calculation is presented as follows:

$$6.9\times10^8 \text{ CFU/mL} \times v_1 = 10^6 \text{ CFU/mL} \times 10,000\ \mu\text{L}$$
$$v_1 = (10^6 \text{ CFU/mL} \times 10,000\ \mu\text{L}) \div 6.9\times10^8 \text{ CFU/mL}$$
$$v_1 = 14.49\ \mu L$$

5. Based on the results, to prepare the inoculum, the user should use 10 mL of sterile culture medium, remove 14.49 μL, and add 14.49 μL of the original culture. This procedure will yield an inoculum with a final concentration of 10^6 CFU/mL.

Fig. 2 Preparation of bacterial inoculum based on previously measured absorbance

- Add 100 μL of the bacterial inoculum to wells in columns 1–11.

- With this setup, 8 replicates of the same condition are possible. Column 12 serves as the negative control (no growth expected), column 11 as the positive control (growth without antimicrobial), and columns 1–10 contain decreasing concentrations of the antimicrobial agent.
- In this configuration, due to the new 1:2 dilution performed at the moment of microorganism inoculation, the concentration of the tested antimicrobial compounds is reduced by half, reaching the desired concentration, considering that the stock solution was prepared at twice the target concentration, and the microbial population decreases to 5×10^5 CFU/mL, as recommended by the assay protocol.
- Plates should be sealed and incubated at 37°C for 24 h. Due to slower growth rates of lactic acid bacteria, longer incubation times may be necessary. Regular monitoring of the positive control is recommended to determine the minimum required incubation time.
- After incubation, the MIC is defined as the lowest concentration of the compound that inhibits visible bacterial growth (absence of turbidity in the culture medium) (Fig. 4) (*see* **Note 10**).
- The described method refers to the determination of MIC by microdilution. Other methods may also be employed, such as disk diffusion, agar diffusion, or macrodilution. However, it

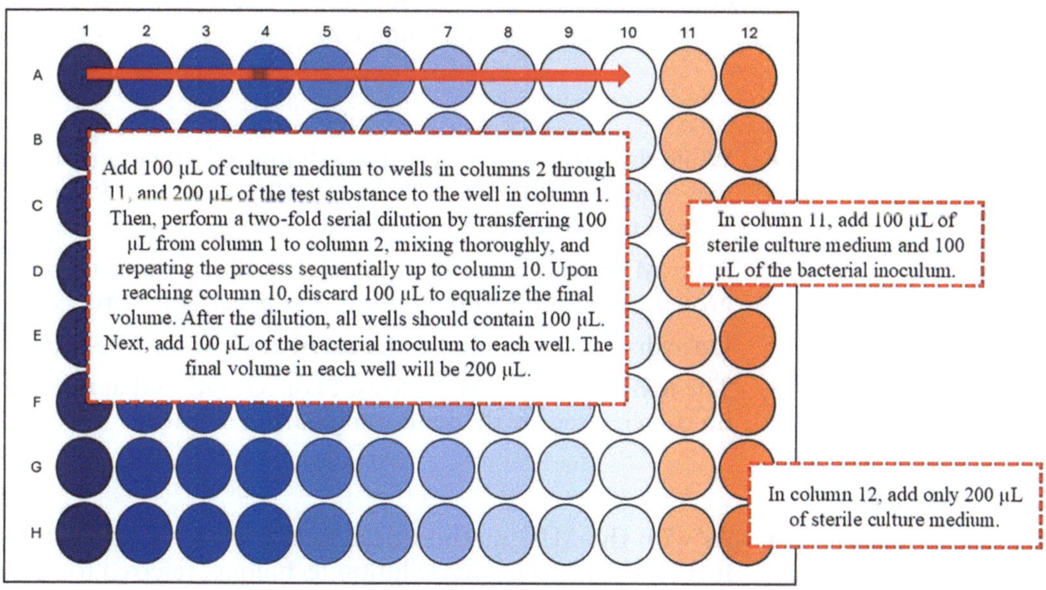

Fig. 3 Setup of a 96-well microtiter plate for minimum inhibitory concentration determination

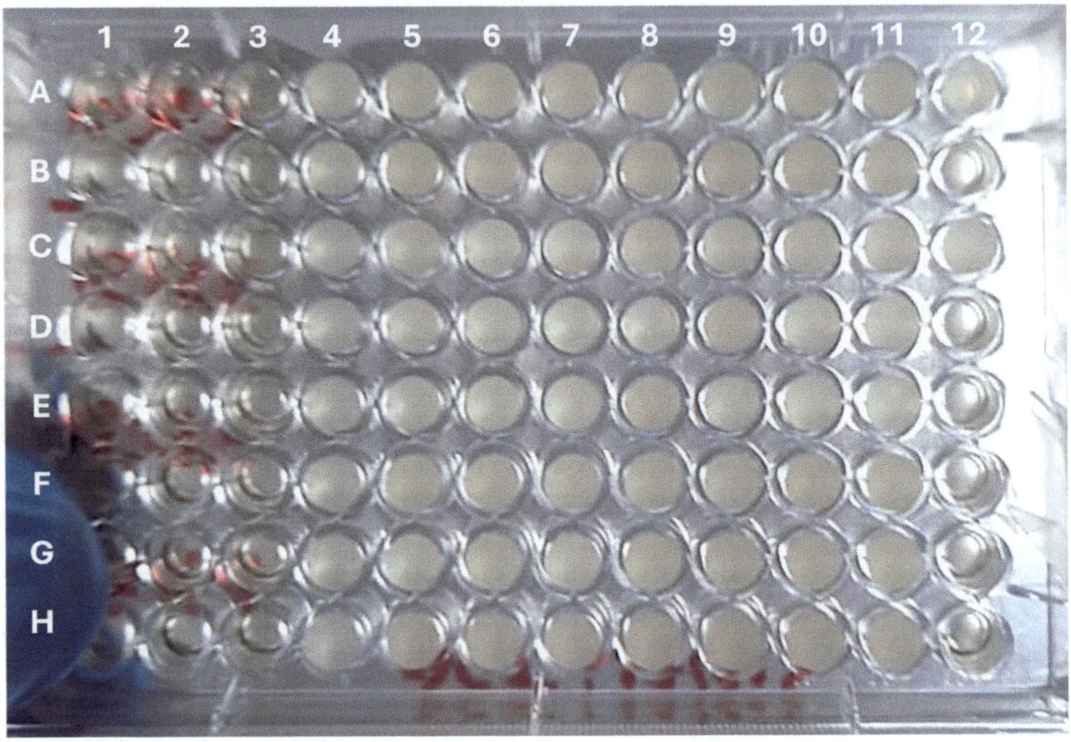

Fig. 4 MIC interpretation. Absence of growth is observed in column 12 (with contamination in well 12C, leading to the exclusion of row C), and appropriate growth is seen in column 11, representing positive control, as well as in columns 4 to 10. The MIC is determined as the concentration present in column 3

is recommended to standardize the use of the microdilution technique due to its advantages in terms of material and time optimization, as well as its capacity to evaluate a greater number of concentrations within a single batch.

3.1.2 Execution of the MBC Assay

- While the MIC assay reveals growth inhibition, target bacteria may remain viable. Therefore, an MBC assay should be performed to assess bactericidal activity.
- After MIC reading, 5 μL from each well showing no visible growth should be spotted onto MRSa plates (*see* **Note 11**).
- Once dry, plates should be incubated at 37°C for 24–48 h.
- The MBC is defined as the lowest concentration that results in no bacterial regrowth, confirming bacterial death.

3.1.3 Execution of the FIC Assay and Calculation of the FIC_{index}

- As with the MIC method described previously, it is recommended to use the microdilution technique, referred to in this context as the checkerboard assay, for FIC determination, due to

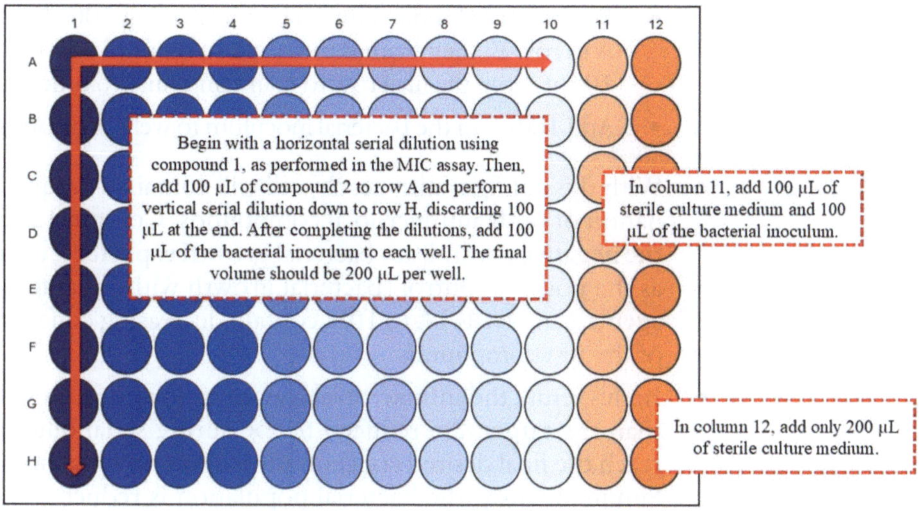

Fig. 5 Setup of a 96-well microtiter plate for fractional inhibitory concentration determination

its optimization of time, resources, and its capacity to evaluate a greater number of concentration combinations. The test will be described as determined in Evangelista et al. [12].

- For this analysis, the bacterium must be reactivated and prepared in advance according to the MIC method.
- The FIC assay can only be performed after prior determination of the MIC values for the compounds under evaluation.
- The preparation of the bacterial inoculum must follow the same protocol used for the MIC assay, with a final concentration of 10^6 CFU/mL of total bacteria.
- Since this is a combinatory analysis involving two antimicrobial compounds, two independent stock solutions must be prepared, each containing a concentration 4× higher than the maximum desired concentration on the microplate. The purity of the compound should be considered and adjusted to 100%.
- Once the solutions are prepared, the 96-well microplate should be set up as follows (considering a 12-column × 8-row layout) (Fig. 5):
  - Add 100 μL of sterile MRSb to columns 2–11.
  - Add 200 μL of sterile MRSb to column 12.
  - Add 200 μL of one of the antimicrobial solutions to column 1.
  - Perform a 1:2 serial dilution by transferring 100 μL from column 1 to column 2, then from column 2–3, and so on until column 10. Discard 100 μL from the final column to maintain a consistent volume of 100 μL.
  - Add 100 μL of the second antimicrobial solution to row A.

- Perform a 1:2 serial dilution by transferring 100 μL from row A to row B, then from B to C, and so on until row H. Discard 100 μL from the final row to maintain a final volume of 100 μL.
- Add 100 μL of the bacterial inoculum to wells in columns 1–11.

- With this configuration, each well will contain a unique combination of both antimicrobial compounds. Column 12 serves as the negative control (no bacterial growth expected), column 11 as the positive control (bacterial growth without antimicrobial agents), and wells 1A to 10H contain decreasing concentrations of the test compounds.
- In this setup, the antimicrobial concentrations, initially 4× higher than the target, are reduced to 2× during serial dilution, and reach the final desired concentration upon bacteria inoculation. Simultaneously, the bacterial population is reduced to 5×10^5 CFU/mL, as recommended by the assay protocol.
- Plates should be sealed and incubated at 37°C for 24 h. Due to the slower growth rate of lactic acid bacteria, longer incubation times may be required. Positive controls should be monitored frequently to determine the optimal incubation period.
- After incubation, the FIC of each compound is defined as the lowest concentration of that compound (in combination) that inhibits visible bacterial growth (absence of turbidity in the culture medium). Each compound will have its own FIC value.
- Once FIC values are determined, they are used along with the previously determined MIC values to calculate the FIC_{index}, using the following formula:

$$FIC_{index} = \left(\frac{FIC_{compound\ A}}{MIC_{compound\ A}} \right) + \left(\frac{FIC_{compound\ B}}{MIC_{compound\ B}} \right)$$

- Based on the calculated FIC_{index}, interactions are interpreted as follows: synergism is indicated when $FIC_{index} \leq 0.5$; additive effect is indicated when $FIC_{index} > 0.5$ and ≤ 2.0; no interaction is indicated when $FIC_{index} > 2.0$ and < 4.0; and antagonism is indicated when $FIC_{index} \geq 4.0$ (Fig. 6) (*see* **Note 12**).

3.2 Preparation of Calabrese Sausage Added with NI and ε-PL

- After determining the *in vitro* inhibitory doses of NI and ε-PL against LAB, it is proposed that concentrations ten times higher than the FIC should be applied directly to the Calabrese sausage, as described by Bortolotto et al. [13] (*see* **Note 13**).

1. Considering the following hypothetical values:

Minimum Inhibitory Concentration of nisin = 10 mg/L
Minimum Inhibitory Concentration of ε-poly-L-lysine = 30 mg/L

Fractional Inhibitory Concentration of nisin = 2 mg/L
Fractional Inhibitory Concentration of ε-poly-L-lysine = 0.5 mg/L

2. The Fractional Inhibitory Concentration Index is calculated as follows:

$$FIC_{index} = \left(\frac{FIC_{nisin}}{MIC_{nisin}}\right) + \left(\frac{FIC_{\varepsilon\text{-poly-L-lysine}}}{MIC_{\varepsilon\text{-poly-L-lysine}}}\right)$$

$$FIC_{index} = \left(\frac{2\ mg/L}{10\ mg/L}\right) + \left(\frac{0.5\ mg/L}{30\ mg/L}\right)$$

$$FIC_{index} = 0.22$$

3. Considering the hypothetical values, the interaction between the compounds results in a **synergistic effect**.

Fig. 6 Calculation of the fractional inhibitory concentration index

- Pork shoulder trim, pork fat, and mechanically deboned poultry meat (maintained at a maximum temperature of 7°C) are to be ground using an electric meat grinder equipped with an 8 mm plate.
- All ground ingredients should then be transferred to a mixer and blended for 2 min.
- Dry ingredients must be added to the ground meat in the following order: texturized soybean protein, salt, garlic, Calabrese pepper, sodium tripolyphosphate, antioxidant blend, and curing salt.
- The mixture should be blended for 2 min.
- Subsequently, the antimicrobial compounds should be dissolved in water and incorporated into the batter during the final stage of mixing. The batter should be mixed for over 2 min before proceeding to stuffing.
- The prepared batter is then stuffed into natural pork casings that have been sanitized by immersion in a 10% lactic acid solution for 30 min.
- The sausages should be thermally processed in a steam chamber at 90°C until the internal temperature reaches 76°C (~2 h), immediately cooled in an ice-water bath until the internal temperature reaches 25°C, and vacuum-packed in low-barrier pouches.
- The sausages are stored at room temperature, not exceeding 25°C, with an estimated shelf life of ~60 days (Fig. 7).

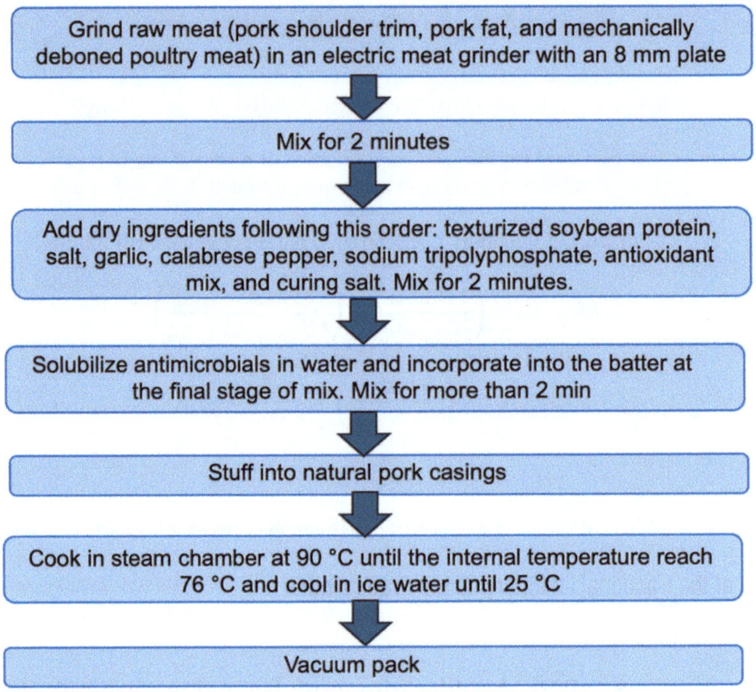

Fig. 7 Flowchart of the production of cooked sausage added with Nisin and ε-Poly-L-lysine.

4 Notes

1. Calibration and certification of all equipment (e.g., autoclaves, pipettes) are essential to ensure proper function and reproducibility.
2. In the absence of a multichannel pipette, single-channel pipettes can be used with extra caution to avoid pipetting errors.
3. Most members of the Lactobacillaceae family fall under biosafety level 1. Confirm the biosafety level of the specific microorganism before selecting equipment.
4. If a spectrophotometer is unavailable, the McFarland scale may be used. However, spectrophotometric methods are preferred for greater standardization and objectivity.
5. Ensure proper tip-pipette compatibility to avoid handling errors, particularly with multichannel pipettes.
6. The formulation of sausages may vary depending on the manufacturer, including the choice of casing type

7. Each strain's growth kinetics must be validated in advance. Lactic acid bacteria may require longer incubation times to reach high populations.
8. The mid-exponential phase should be determined using lab-specific growth curves, including lag, exponential, stationary, and decline phases. Alternatively, visible turbidity may serve as an approximate indicator of this phase. By convention, it is assumed that once turbidity is immediately observed, the microorganisms have reached the mid-exponential phase.
9. To maintain linearity, absorbance values should not exceed 1. Dilutions must be performed if necessary.
10. Visual MIC reading is standard per Clinical and Laboratory Standards Institute guidelines. However, plate readers at 560 nm or colorimetric methods using 0.01% resazurin (blue = non-viable bacteria; pink = viable bacteria) may be used for quantitative assessment. In the resazurin method, the MIC corresponds to the lowest concentration yielding a blue color.
11. MBC analysis should follow visual MIC reading. Complementary methods (e.g., plate reading or dye addition) may compromise sterility and interfere with MBC accuracy. If such analyses are to be performed, the MBC incubation should be carried out beforehand.
12. Some methodologies do not consider the "additive effect" category and interpret FIC_{index} values >0.5 and <4.0 as indicating "no interaction." Several journals provide specific guidelines for interpreting these values. Therefore, it is advisable to consult the methodological recommendations of the selected journal and to consider the interpretation criteria used by other researchers in the field.
13. After selecting the meat product to be tested with natural antimicrobials, it is necessary to define the target microorganism to be controlled. Once the target microorganism is established, the individual MIC of each antimicrobial compound must be determined, followed by FIC tests to assess the nature of their interaction. Subsequently, doses should be extrapolated to ten times the FIC and applied to the meat product in question. It is also important to emphasize that after producing the meat product with the added natural antimicrobials, tests such as microbial counts, physicochemical analyses, and especially sensory evaluations should be conducted to verify whether color and flavor have been altered.

References

1. Riley MA, Wertz JE (2002) Bacteriocin diversity ecological and evolutionary perspectives. Biochemie 84:57–364.
2. Cleveland J, Montville JT, Nes IF, Chikindas Ml (2001) Bacteriocins: safe, natural antimicrobials for food preservation. Int J Food Microbiol 71:1–20
3. Müller-Auffermann K, Grijalva F, Jacob F, Hutzler M (2015) Nisin and its usage in breweries: a review and discussion. J Inst Brew 121(3):309–319
4. Cotter PD, Hill C, Ross RP (2005) Bacteriocins: developing innate immunity for food. Nat Rev Microbiol 3:773–788
5. Shi DL, Shi H (2022) The synergistic antibacterial effect and inhibition of biofilm formation of nisin in combination with terpenes against Listeria monocytogenes. Lett Appl Microbiol 75(3):632–642
6. Liu H, Pei H, Han Z, Feng G, Li D (2015) The antimicrobial effects and synergistic antibacterial mechanism of the combination of Ɛ-Poly-L-lysine and Nisin against *Bacillus subtilis*. Food Control 47:444–450
7. Yaghoubi M, Alirezalu K, Mazloomi SM, Marcinkowska-Lesiak M, Azadmard-Damirchi S, Peighambardoust SH, et al (2024) Enhancing beef sausage packaging with calcium alginate active film infused with nisin and ε-polylysine nanoparticles and beetroot extract. LWT 191:115665
8. Benli H, Sanches-Plata MX, Keeton JT (2011) Efficacy of ε-polylysine, lauric arginate, or acidic calcium sulfate applied sequentially for Salmonella reduction on membrane filters and chicken carcasses. J Food Prot 74:743–750
9. Hyldgaard M, Mygind T, Vad BS, Stenvang M, Otzen DE, Meyer RL (2014) The antimicrobial mechanism of action of epsilon-poly-L-lysine. Appl Env Microbiol 80:7758–7770
10. Bortolotto FCK, Hacke TM, Karwowski MSM, Jede NCK, Poniewas L, Ferreira GA, Macedo REF (2025) Ɛ-poly L-Lysine and Nisin Z as biopreservatives in reduced-sodium Calabresa sausage stored at room temperature. Meat Sci 226:109847
11. Clinical and Laboratory Standards Institute (CLSI) (2022) Performance standards for antimicrobial susceptibility testing. CLSI Supplement M100. CLSI, Wayne, PA
12. Evangelista AG, Janotto L. dS, Matté EHC, Perry CC, Ribeiro FL, Luciano FB (2023) *In vitro* interaction between essential oil compounds and halquinol against zoonotic pathogenic bacteria. Biocatal Agric Biotechnol 50:e102672
13. Bortolotto FCK, Farfan MHR, Jede NCK, Danielski GM, Macedo REF (2021) Nisin and ε-poly-L-lysine as natural antimicrobials towards spoilage-associated *Lactobacillus plantarum*. Ciência Rural 51:e20200423

Check for updates

Chapter 12

Producing Edible Coatings Incorporated with Antimicrobial Compounds for Application on Meat Products

Tamires Sousa de Oliveira, Luiz Carlos Corrêa-Filho, Lourdes Maria Corrêa Cabral and Renata Valeriano Tonon

Abstract

The use of edible coatings with active compounds to enhance meat quality and inhibit microbial growth has been intensified in recent years. Usually, active coatings are designed by blending edible polymers with food-grade additives. These coatings are composed of polymers like polysaccharides, proteins, and lipids, which are biodegradable, edible, and act as carriers for antimicrobial, antioxidant, or flavoring compounds. Incorporating natural active compounds in these coatings extends meat product shelf life and delays the deterioration of sensory quality and their potential as flavorings. This chapter describes and critically evaluates methods for preparing and applying edible active coatings with antimicrobial compounds on meat products, offering a straightforward and replicable protocol.

Keywords Polymers, Active compounds, Food preservation, Edible coatings, Natural products, Meat products

1 Introduction

Meat and meat products are among the best sources of essential nutrients, including high-quality proteins and essential amino acids, with an annual global production exceeding 370 million tons in 2023 [1]. However, being perishable, meat products are susceptible to chemical and microbiological deterioration during processing, distribution, and storage, which compromises their quality, safety, and shelf life. Meat deterioration begins immediately after livestock slaughter. Complex chemical reactions in the meat result in discoloration, off-flavors, and various undesirable chemical and structural changes. The primary deterioration processes include meat darkening and lipid oxidation, which can occur through lipid autoxidation, photooxidation, and enzymatic oxidation [2, 3]. These processes pose risks to consumer health and cause financial losses for producers.

Renata Ernlund Freitas de Macedo and Carlos Adam Conte-Junior (eds.), *Meat Products Preservation*, Methods and Protocols in Food Science, https://doi.org/10.1007/978-1-0716-5068-4_12,

As the population's standard of living continues to rise, there is an increasing demand for high-nutritional-value foods that provide quality, freshness, and safety, while minimizing the use of preservatives and chemical additives [4]. Therefore, food preservation technologies face the challenge of extending the shelf life of meat products. To mitigate these deteriorative processes, incorporating functional agents into edible coatings and packaging has emerged as a promising approach for the sustainable preservation of meat and meat products, thereby reducing the rate of meat deterioration [4, 5].

Edible coatings are thin layers of polymeric material applied directly onto the surface of meat products, providing protection and altering little or no visual change [6]. Moreover, these coatings are designed to preserve the physicochemical quality and extend the shelf life of animal products during storage and distribution. They are effective barriers against water and oxygen, preventing dehydration and reducing lipid oxidation [7].

The use of polymers as a structural matrix to produce edible coatings has become increasingly common, primarily due to their affordability and environmental friendliness. Polysaccharides, proteins, and lipids are among the most prominent polymers used for this purpose. In the production process of edible coatings, polymers are dispersed and dissolved in solvents such as water, alcohol, or their combinations. Adjusting the pH and/or heating the solutions can facilitate the dispersion of specific polymers. Furthermore, some substances can be added to improve the final functional properties. Plasticizers (such as glycerol, sorbitol and fatty acids) can provide more malleability and cross-linking agents (calcium chloride, sodium trimetaphosphate, and citric acid) can improve strength and barrier properties. Additionally, the incorporation of antimicrobial and/or antioxidant agents guarantees the biological and chemical action necessary to extend the shelf life, making the coatings active [8, 9].

Edible coatings are commonly applied using three methods: dipping, spraying, and spreading [10]. In the dipping method, the meat product is immersed directly into the coating solution, allowing a homogeneous layer to form on the surface after natural drying. Alternatively, the spray method is suitable for solutions with low viscosity, applied at low pressure. Finally, the spreading method involves applying the coating solution directly onto the product using a brush. The effectiveness of these coatings depends on the physicochemical characteristics of the materials, the selection and proportion of active agents and the choice of application technique [8, 10].

Polysaccharides and proteins isolated from biomass have been widely studied for use as a polymeric base in edible coatings [11]. Proteins can be of animal (creatine, gelatin, casein) or plant (soy proteins, zein, gluten) origin, with those of animal origin being the most explored in this field [12]. These molecules are polymers or can form clean polymers, making them suitable for use as coatings due to

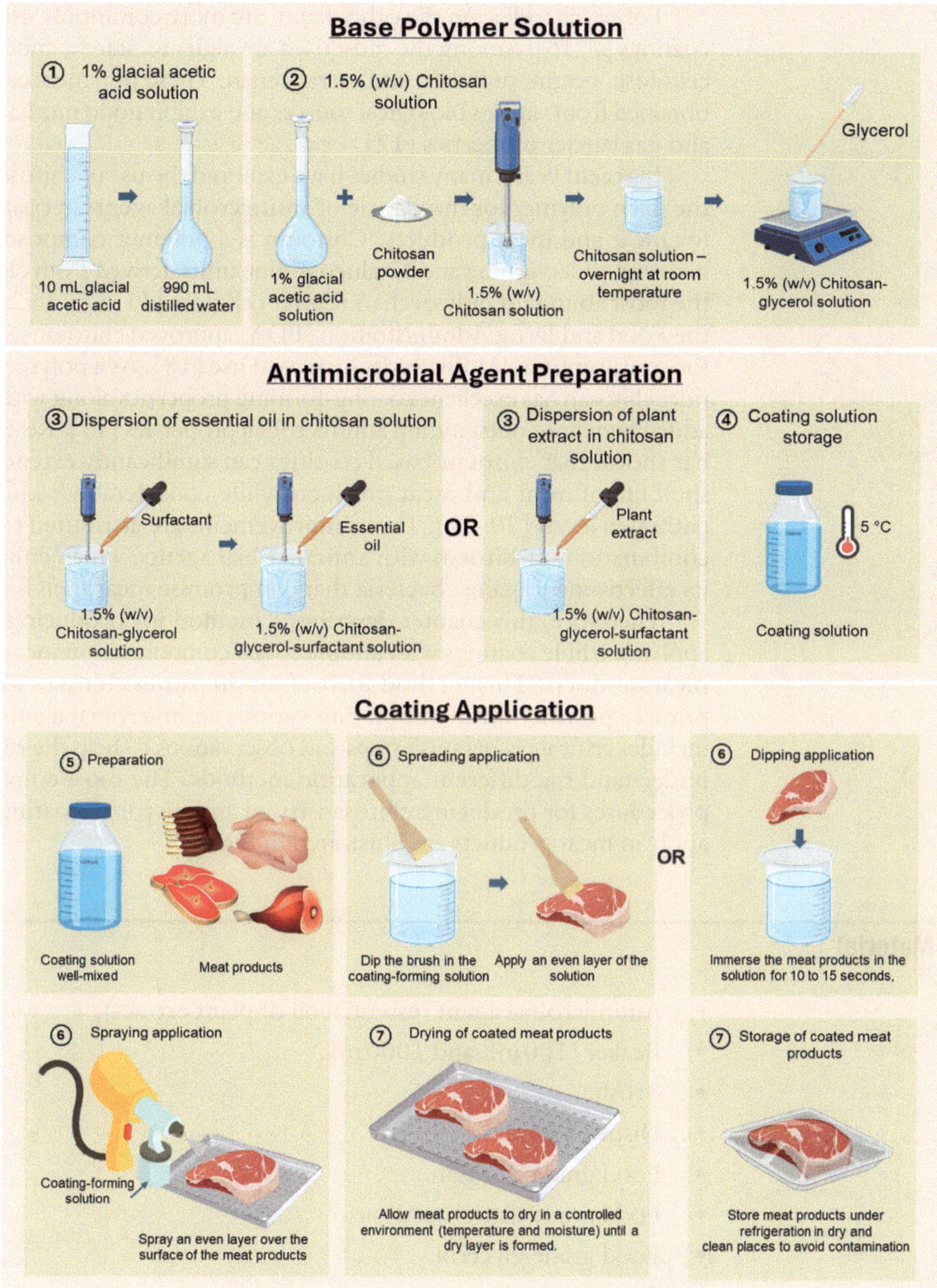

Fig. 1 Preparation of an edible antimicrobial coating for preserving meat products

their strong emulsifying capabilities and excellent barrier properties [13, 14]. However, protein-based coatings can be highly permeable to water vapor [15].

Polysaccharides, on the other hand, are more commonly used as coatings [7, 16]. Among the most used are alginate, starch, chitosan, cellulose, pectin, pullulan, and carrageenan. These polymers can be obtained from various biological sources and exhibit good mechanical and gas barrier properties [17].

In recent years, many studies have explored the use of chitosan as the main polymer for the vehicle of antimicrobial agents in coatings for meat and meat products. Chitosan is a polymer composed of N-acetyl-glucosamine and N-glucosamine units derived from chitin, the most abundant polysaccharide after cellulose [11, 17]. In 2001, the Food and Drug Administration (FDA) approved chitosan as safe for consumption, leading to its increased use [18]. As a polysaccharide, chitosan has excellent coating-forming properties along with the added benefit of outstanding antimicrobial properties [19]. Research has shown that chitosan-based coatings can significantly extend the shelf life of meat and meat products while considerably reducing pathogen levels [20–22]. These improvements are attributed to the combination of chitosan with antimicrobial agents, which enhance its effectiveness against bacteria that compromise meat shelf life.

Therefore, this chapter describes a method for producing and applying edible coatings with antimicrobial compounds on meat and meat products. This method focuses on preparing chitosan as the primary polymer for incorporating various antimicrobial agents. It includes critical preparation steps and observations to help the reader understand the different application methods. The most common procedures for producing chitosan-based antimicrobial coatings to apply in meat products are illustrated in Fig. 1.

2 Material

- Antimicrobial agent (essential oil or plant extract).
- Beaker (100 mL and 1000 mL).
- Brush.
- Distilled water.
- Food grade chitosan.
- Food grade glacial acetic acid.
- Food grade glycerol.
- Graduated cylinder.
- Rotor-stator homogenizer.
- Lab scale.
- Magnetic stirrer and magnetic bars.
- Perforated stainless-steel screens.
- Pipettes.

- Pneumatic sprayer.
- Surfactant (Tween 80).
- Volumetric flask.

3 Methods

3.1 Base Polymer Solution Preparation

3.1.1 1% Glacial Acetic Acid Solution (see Note 1)

- Measure 10 mL of glacial acetic acid using a graduated cylinder.
- Transfer 10 mL of glacial acetic acid to a 1000 mL volumetric flask.
- Fill the volumetric flask to the 1000 mL mark with distilled water.

3.1.2 1.5% (w/v) Chitosan Solution (see Note 2)

- Weigh the required amount of chitosan (typically 1–2% w/v) into a 100 mL beaker.
- Dissolve the chitosan in 1000 mL of 1% glacial acetic acid solution in a 1000 mL beaker.
- Use the rotor-stator homogenizer at 10,000 rpm to facilitate complete solubilization of the polymer (*see* **Note 3**).
- Allow the solution to stand overnight at room temperature to ensure complete solubilization of the polymer.
- Once the chitosan is completely dissolved, weigh the glycerol (typically 30%–50% of the polymer's weight).
- Slowly add the glycerol to the chitosan solution while stirring with a magnetic stirrer for 30 min (*see* **Note 4**).

3.2 Preparation of the Antimicrobial Agent

If the antimicrobial agent is an essential oil, refer to Sect. 3.2.1. If it is an aqueous plant extract, see Sect. 3.2.2.

3.2.1 Dispersion of Essential Oil in Chitosan Solution (see Note 5)

- Weigh the essential oil (typically between 0.1 and 1% w/v, depending on desired antimicrobial efficacy).
- Weigh the surfactant (Tween 80) to achieve a 1:1 ratio with the essential oil.
- Slowly add the surfactant to the chitosan and glycerol solution under constant stirring using a rotor-stator homogenizer at 10,000 rpm for 3 min.
- After 3 min, add the essential oil to the surfactant solution and continue stirring under the same conditions for an additional 5 min to fully incorporate the essential oil into the solution (*see*

Note 6). This creates an antimicrobial coating-forming solution with essential oil for application to meat products.

- Store the coating solution in sterilized, tightly closed glass bottles under refrigeration, and protect them from light. The solution can be kept for up to one week under these conditions (*see* **Note 7**).

3.2.2 *Dispersion of Aqueous Plant Extract in Chitosan Solution* (see **Note 8**)

- Weigh the plant extract (typically between 0.1 and 1% w/v, depending on the desired antimicrobial efficacy).
- Slowly add the plant extract to the chitosan and glycerol solution under constant stirring using a rotor-stator homogenizer at 10,000 rpm for 5 min. This creates an antimicrobial coating-forming solution with plant extracts for application to meat products.
- Store the coating solution in sterilized, tightly closed glass bottles under refrigeration, and protect them from light. The solution can be kept for up to one week under these conditions.

3.3 Coating Application

Coatings are typically applied to meat products through brushing, dipping, or spraying methods. Details for each method are provided below:

3.3.1 *Preparation of Meat Products*

- Ensure meat products' surfaces are clean and ready for application (*see* **Note 9**).
- Ensure the coating solution is homogeneous and ready for use.

3.3.2 *Spreading*

- Dip the brush in the coating-forming solution.
- Apply an even layer of the solution to the surface of the meat products (*see* **Note 10**).
- Place the meat products on perforated stainless-steel screens to dry and remove excess coating.
- If necessary, apply a second layer after the first has dried to ensure full coverage.
- Allow meat products to dry in a controlled environment (temperature and moisture) until a dry layer is formed.

3.3.3 *Dipping* (see **Note 11**)

- Fill a large, clean container with the coating-forming solution.
- Immerse the meat products in the solution for 10–15 s, ensuring complete coverage.
- Remove the meat products from the solution and allow the excess solution to drain.
- Place the meat products on perforated stainless-steel screens to dry and remove excess coating.

- Allow meat products to dry in a controlled environment (temperature and moisture) until a dry layer is formed.
- Ensure all areas of the meat products are covered.

3.3.4 Spraying

- Ensure the sprayer is clean and in good working order.
- Pour the coating-forming solution into the sprayer reservoir (a pneumatic type of sprayer is generally used).
- Support the pieces of meat or meat products on perforated stainless-steel screens.
- Spray an even layer over the surface of the meat products, maintaining an adequate distance to avoid excess (*see* **Note 12**).
- Ensure the applied area is dry and completely covered.
- Carefully turn over the pieces of meat or meat products.
- Repeat the application of the coating-forming solution.
- Allow meat products to dry in a controlled environment (temperature and moisture) until a dry layer is formed.
- Ensure all areas of the meat product are covered.

3.3.5 Storage of Coated Meat Products

- After drying, store meat products under refrigeration in dry and clean places to avoid contamination.
- Conduct periodic analyses to evaluate the quality and safety of the products, typically every 2 days.

4 Notes

1. Chitosan has low solubility in neutral and alkaline pH. Weak organic acids such as acetic, formic, and citric acids can be used to prepare base solutions for solubilizing chitosan. In most studies of coating-forming solutions for meat products, a 1% acetic acid solution has been used [5, 18].
2. When solubilizing chitosan, it is important to consider that other components will be added during the production of the coating-forming solution. Ideally, the initial concentration of chitosan should be slightly higher than the desired concentration in the final solution. Therefore, before starting the protocol, calculate the final volume of the solution and the final concentration of each component separately.
3. Some studies employ heating in combination with a rotor-stator homogenizer to enhance polymer dissolution. In such cases, temperature control is crucial, as temperatures exceeding the glass transition point of chitosan can compromise its physical and chemical properties, including solubility, viscosity,

and film-forming capacity, thereby affecting its applicability. The glass transition temperature of chitosan can vary depending on the technique used to measure it. However, from 85°C onwards, thermal relaxation behavior is observed in the structure [23, 24].

4. Plasticizing agents typically have low molecular density. Depending on the chosen plasticizer, some studies may use a rotor-stator homogenizer to achieve faster and more thorough mixing with the polymeric solution. In such cases, the plasticizer should be homogenized gently at low speeds (up to 2000 rpm), as this ensures effective solvation within the polymeric structure with minimal energy input [25].
5. In some studies, the essential oil is first homogenized with the surfactant before being added to the polymeric solution. This method can facilitate the dispersion of highly hydrophobic essential oils.
6. When dispersing the essential oil directly into the chitosan solution using a rotor-stator homogenizer, a coarse emulsion is initially formed. To enhance the dispersion and stability of this solution, some studies employ low (Phase inversion temperature and spontaneous emulsification) or high-energy (High-pressure and ultrasonic homogenization) methods to reduce droplet size, particularly at the nanoscale. At this scale, increased surface area of the droplets can enhance the physical and biological properties of the antimicrobial agent [26]. When aiming to produce fine emulsions (micro or nanoemulsions), careful attention must be given to the quantities of emulsion components and chitosan solution. Depending on the method used, adjustments to the initial concentrations may be necessary to achieve the desired final concentration.
7. Essential oils consist of complex mixtures of volatile components, each with distinct characteristics. Therefore, the stability of the solution depends on the specific properties of the oil, as well as the type and proportions of the surfactant used. If the solution is not applied within a few days of production, it is advisable to conduct a preliminary stability study on the coating-forming solution to detect early signs of instability.
8. Plant extracts can be classified as hydrophilic or hydrophobic. The dispersion of hydrophilic extracts was described in the Sect. 3.2.2. Hydrophobic extracts require careful attention due to their potential toxicity if organic solvents are used during extraction, necessitating complete evaporation. Enhancing the dispersion and stability of these extracts in chitosan solutions can be facilitated by using surfactants, similar to essential oils. It is also crucial to carefully monitor and adjust the final concentrations as needed.

9. It is important to ensure uniformity in the weight and size of cuts of meat or meat products to facilitate comparison in physical-chemical and sensory analyses between treatments.
10. Avoid excessive brushing of meat cuts or meat products, as this may remove the coating-forming solution from their surface.
11. The dipping method requires careful attention to minimize cross-contamination risks. The coating solution can transfer microorganisms between batches, and the dipping container can become a source of contamination if not properly cleaned and sanitized between uses. Improper handling of products during immersion can also introduce contaminants into the solution. Therefore, it is essential to change the solution regularly, maintain adequate sanitary conditions, and clean and disinfect the container after each use. Additionally, using clean handling equipment, disposable gloves, and following strict hygiene practices are crucial measures.
12. Standardizing the distance and number of sprayer activations for applying the coating-forming solution to meat products is crucial. This practice helps maintain consistency and minimizes potential errors in subsequent analysis results.

References

1. Food and Agriculture Organization of the United Nations (FAO) (2025). MEAT MARKET REVIEW: Overview of global market developments in 2024; FAO: Rome, Italy. https://openknowledge.fao.org/items/b553c447-f6b2-4ff8-825e-024767b27760
2. Song D-H, Hoa VB, Kim HW, Khang SM, Cho S-H, Ham J-S et al (2021) Edible films on meat and meat products. Coatings 11:1344. https://doi.org/10.3390/coatings11111344
3. Priya K, Thirunavookarasu N, Chidanand DV (2023) Recent advances in edible coating of food products and its legislations: a review. J Agric Food Res 12:100623. https://doi.org/10.1016/j.jafr.2023.100623
4. Vlčko T, Golian J, Fikselová M, Rybnikár S (2022) Current overview in the field of application of edible coatings/films (meat products examples). J Microbiol Biotechnol Food Sci 12:e9281. https://doi.org/10.55251/jmbfs.9281
5. Hashemi SMB, Kaveh S, Abedi E, Phimolsiripol Y (2023) Polysaccharide-based edible films/coatings for the preservation of meat and fish products: emphasis on incorporation of lipid-based nanosystems loaded with bioactive compounds. Foods 12:3268. https://doi.org/10.3390/foods12173268
6. Gil M, Rudy M (2023) Innovations in the packaging of meat and meat products—a review. Coatings 13:333. https://doi.org/10.3390/coatings13020333
7. Gagaoua M, Bhattacharya T, Lamri M, Oz F, Dib AL, Oz E et al (2021) Green coating polymers in meat preservation. Coatings 11:1379. https://doi.org/10.3390/coatings11111379
8. Antonino C, Difonzo G, Faccia M, Caponio F (2024) Effect of edible coatings and films enriched with plant extracts and essential oils on the preservation of animal-derived foods. J Food Sci 89:748–772. https://doi.org/10.1111/1750-3841.16894
9. Moghadas HC, Chauhan R, Smith JS (2024) Application of plant oils as functional additives in edible films and coatings for food packaging: a review. Foods 13:997. https://doi.org/10.3390/foods13070997
10. Wang T, Zhai X, Huang X, Li Z, Zhang X, Zou X et al (2023) Effect of different coating methods on coating quality and mango preservation. Food Packag Shelf Life 39:101133. https://doi.org/10.1016/j.fpsl.2023.101133
11. Mohamed SAA, El-Sakhawy M, El-Sakhawy MA-M (2020) Polysaccharides, protein and lipid-based natural edible films in food packaging: a

review. Carbohydr Polym 238:116178. https://doi.org/10.1016/j.carbpol.2020.116178
12. Umaraw P, Munekata PES, Verma AK, Barba FJ, Singh VP, Kumar P et al (2020) Edible films/coating with tailored properties for active packaging of meat, fish and derived products. Trends Food Sci Technol 98:10–24. https://doi.org/10.1016/j.tifs.2020.01.032
13. Chen H, Wang J, Cheng Y, Wang C, Liu H, Bian H et al (2019) Application of protein-based films and coatings for food packaging: a review. Polymers 11:2039. https://doi.org/10.3390/polym11122039
14. Ucpinar Durmaz B, Aytac A (2021) Effects of polyol-based plasticizer types and concentration on the properties of polyvinyl alcohol and casein blend films. J Polym Environ 29:313–322. https://doi.org/10.1007/s10924-020-01881-x
15. Tkaczewska J (2020) Peptides and protein hydrolysates as food preservatives and bioactive components of edible films and coatings – a review. Trends Food Sci Technol 106:298–311. https://doi.org/10.1016/j.tifs.2020.10.022
16. Santos Junior JR dos, Corrêa-Filho LC, Pereira VO, Barboza HTG, Ferreira JCS, Soares AG et al (2024) Application of rosin resin and zinc oxide nanocomposites to chitosan coatings for extending the shelf life of passion fruits. Sustain Food Technol 2:415–425. https://doi.org/10.1039/D3FB00189J
17. Moeini A, Pedram P, Makvandi P, Malinconico M, Gomez d'Ayala G (2020) Wound healing and antimicrobial effect of active secondary metabolites in chitosan-based wound dressings: a review. Carbohydr Polym 233:115839. https://doi.org/10.1016/j.carbpol.2020.115839
18. No HK, Meyers SP, Prinyawiwatkul W, Xu Z (2007) Applications of chitosan for improvement of quality and shelf life of foods: a review. J Food Sci 72:R87–100. https://doi.org/10.1111/j.1750-3841.2007.00383.x
19. Cazón P, Velazquez G, Ramírez JA, Vázquez M (2017) Polysaccharide-based films and coatings for food packaging: a review. Food Hydrocoll 68:136–148. https://doi.org/10.1016/j.foodhyd.2016.09.009
20. Chaparro-Hernández S, Ruiz-Cruz S, Márquez-Ríos E, Ornelas-Paz J de J, Del-Toro-Sánchez CL, Gassos-Ortega LE et al (2019) Effect of chitosan–tomato plant extract edible coating on the quality, shelf life, and antioxidant capacity of pork during refrigerated storage. Coatings 9:827. https://doi.org/10.3390/coatings9120827
21. Lekjing S (2016) A chitosan-based coating with or without clove oil extends the shelf life of cooked pork sausages in refrigerated storage. Meat Sci 111:192–197. https://doi.org/10.1016/j.meatsci.2015.10.003
22. Cheng Y, Hu J, Wu S (2021) Chitosan based coatings extend the shelf-life of beef slices during refrigerated storage. LWT 138:110694. https://doi.org/10.1016/j.lwt.2020.110694
23. Qiao C, Ma X, Wang X, Liu L (2021) Structure and properties of chitosan films: effect of the type of solvent acid. LWT 135:109984. https://doi.org/10.1016/j.lwt.2020.109984
24. Quijada-Garrido I, Laterza B, Mazón-Arechederra JM, Barrales-Rienda JM (2006) Characteristic features of chitosan/glycerol blends dynamics. Macromol Chem Phys 207(19): 1742–1751. https://doi.org/10.1002/macp.200600230
25. Song D-H, Hoa VB, Kim HW, Khang SM, Cho S-H, Ham J-S et al (2021) Edible films on meat and meat products. Coatings 11:1344. https://doi.org/10.3390/coatings11111344.
26. Espitia PJP, Fuenmayor CA, Otoni CG (2019) Nanoemulsions: synthesis, characterization, and application in bio-based active food packaging. Compr Rev Food Sci Food Saf 18:264–285. https://doi.org/10.1111/1541-4337.12405

Check for updates

Chapter 13

Validation of Bacteriophage Applications as Processing Aids in Fresh Meat Processing

Amilton de Mello

Abstract

Bacteriophage solutions are commonly used in the meat industry to reduce the risk of microbial contamination during processing. In this chapter, we describe methods to validate the effects of bacteriophage applications on reducing *E. coli* and *Salmonella* counts in intact and ground fresh meat. The methods encompass all phases, including proper strain selection, inoculum preparation, inoculation, bacteriophage application, bacteria recovery, and enumeration.

Keywords Bacteriophage, Fresh meat, *E. coli*, *Salmonella*

1 Introduction

Bacteriophages are the most abundant microorganisms on earth, with a concentration of 10^{30} in the environment [1, 12]. They are giant viruses with complex structures about 90 nm wide and 200 nm long [5]. Their icosahedral capsid contains a double-stranded DNA genome of around 168 kbp size [3, 7]. Bacteriophages attach to the bacteria wall and inject their DNA into the host, which is incorporated into the host genome, hijacking the bacterium's cellular machinery and forcing the host to produce viral components. Subsequently, components are assembled, generating new phages that burst the cell, causing bacterium death [4]. Bacteriophages usually perform two types of cycles: lysogenic or lytic. During the lysogenic cycle, the infected host genome is latent; it may produce viral components but does not lead to cell death until the bacteria undergo stress conditions. Stressful events trigger the lytic cycle, making the cellular machinery produce viral components and assemble new viruses, leading to bacteria death [6]. Bacteriophages directly adopting the lytic cycle are the most recommended for food safety interventions since there is no need for additional events to activate the cell death mechanism [11].

Renata Ernlund Freitas de Macedo and Carlos Adam Conte-Junior (eds.), *Meat Products Preservation*, Methods and Protocols in Food Science, https://doi.org/10.1007/978-1-0716-5068-4_13,

Bacteriophages are microorganisms that have an affinity for specific hosts since their external structures, such as the tail fibers, identify specific components of the bacteria's cell wall that allow them to attach to it [2]. Because of this natural characteristic, they can be isolated to target bacteria of interest. Shiga-toxin *E. coli* (STEC) and *Salmonella* are pathogens of significant concern in fresh meats. In the United States, while regulatory procedures are required to ensure meat products are contaminated with STEC O157:H7, O145, O121, O111, O103, O45, and O26, significant efforts have also been made to follow regulatory guidelines aiming to decrease the *Salmonella* prevalence in meat and poultry.

Applying bacteriophages for pathogen biocontrol offers an additional hurdle for robust food safety management systems. Since phages are specific and self-limit in the absence of the host, the meat industry has adopted phage solutions as an effective processing aid to decontaminate fresh products. In this chapter, we detail a generic methodology used to validate the antimicrobial effect of bacteriophages in lowering *Salmonela* and *E. coli* loads in fresh meats. Since meat processing is performed in different conditions, further testing scenarios can be found in the literature [8–10, 13, 14].

2 Materials

2.1 Fresh Meat and Poultry Matrices

1. Beef Chuck and Flank, Rose Meat (IM) (IMPS 194).
2. Beef trimmings (IMPS 138).
3. Pork trimmings (IMPS 418).
4. Poultry: breast, thighs, whole birds, skin on/off.
5. Ground beef, ground pork, and ground poultry.

2.2 Recommended Strains

1. *Salmonella*: Infantis (ATCC® 51741™), Heidelberg (ATCC® 8326™), Newport (ATCC® 27869™), Typhimurium (ATCC® BAA-190™).
2. *E. coli*: STECs O26 (ATCC® BAA-2196™), O45 (ATCC® BAA-2193™), O103 (ATCC® BAA-2215™), O111 (ATCC® BAA-2440™), O121 (ATCC® BAA-2219™), O145 (ATCC® BAA-2192™), and O157:H7 (ATCC® 35150™). Surrogates MP-26 (ATCC® BAA-1427™, ATCC® BAA-1428™, ATCC® BAA-1429™, ATCC® BAA-1430™, and ATCC® BAA-1431™).

2.3 Solutions and Media

1. Bacteriophage solution at manufacturer-recommended concentration.
2. LB broth (Miller).

3. 0.1% Buffered Peptone Water (BPW, 10% Peptone, 5% Sodium chloride, 3.5% Sodium phosphate, and 1.5% Potassium phosphate).
4. LB Broth with agar (Lennox).
5. XLD agar.
6. Ethanol 70%.

2.4 Equipment and Disposables

1. Biosafety level 2 cabinet (BSL 2) equipped with a germicidal light.
2. UV spectrophotometer.
3. Vortex mixer.
4. Incubator.
5. Incubator shaker.
6. Bunsen burner.
7. Disposable cuvette compatible with spectrophotometer.
8. Meat grinder with autoclavable parts.
9. Colony counter.
10. Paddle Blender or Stomacher™ circulator.
11. Styrofoam trays 8 1/4"×5 3/4" and food graded sorbent pads.
12. PVC film.
13. Sterile inoculation loops and sterile plastic rods.
14. Petri dishes.
15. Vacuum packaging machine.
16. 50 mL Falcon and 1.5 mL microcentrifuge tubes with snap cap.
17. Erlenmeyer flasks and beakers.
18. Nitrile gloves, sterile gloves, and paper towels.
19. Oxygen permeable film and vacuum bags.
20. Paddle Blender or Stomacher™ bags.
21. Moisturized swabs.

3 Methods

3.1 Media Preparation

3.1.1 LB Broth (Miller)

1. Suspend 25 g in 1 L of distilled water.
2. Autoclave for 15 min at 121°C.

3.1.2 LB Broth with Agar (Lennox) Plates

1. Suspend 17.5 g of LB Lennox powder in 500 mL of distilled water.
2. Heat to boiling and stir to dissolve all ingredients.
3. Autoclave for 15 min at 121°C.
4. Set Petri dishes in the BSL2 cabinet.
5. Gently pour the media into the Petri plates while hot to avoid cooling and solidifying the agar.

3.1.3 Tryptic Soy Broth (TSB)

1. Suspend 15 g of TSB powder in 500 mL of distilled water.
2. Autoclave for 15 min at 121°C.

3.1.4 XLD Plates

1. Suspend 27.72 g pf XLD powder in 500 mL of distilled water.
2. Heat to boiling and stir to dissolve all ingredients.
3. Do not autoclave.
4. Set Petri dishes in the BSL 2 cabinet.
5. Gently pour the media into the Petri plates while hot to avoid cooling and solidifying the agar.

3.2 Inoculum Preparation

3.2.1 Strain Recovering (Glass Vial)

1. Hold the tip of the glass vial within the flame of the Bunsen burner for about 10 s.
2. Open the glass vial aseptically using paper towels lightly dampened with ethanol.
3. Remove the cotton and transfer the freeze-dried bacterial pellet into 10 mL of LB broth (Miller) (*E. coli*) or TSB (*Salmonella*). Mix by vortexing.
4. Incubate tubes overnight in the incubator shaker at 200 RPM and 37°C.
5. With a sterile loop, streak isolate the *E. coli* cultures on LB agar (Lennox) and *Salmonella* on XLD.
6. Incubate plates inverted at 37°C overnight.

3.2.2 Strain Recovering (GC Vial with Aluminum Lid)

1. Peel back the lid's top part to remove the aluminum piece.
2. Remove the rubber piece.
3. Transfer the bacterial pellet into 10 mL of LB broth (Miller) (*E. coli*) or TSB (*Salmonella*). Mix by vortexing.
4. Incubate tubes overnight in the incubator shaker at 200 RPM and 37°C.
5. With a sterile loop, streak isolate the *E. coli* cultures on LB agar (Lennox) and *Salmonella* on XLD.
6. Incubate plates inverted at 37°C overnight.

3.2.3 Obtaining a Master Culture

1. With a sterile loop, pick up one well-isolated colony from the Petri dish and inoculate LB broth (Miller) (*E. coli*) or TSB (*Salmonella*). Mix by vortexing.
2. Incubate tubes overnight in the incubator shaker at 200 RPM and 37°C.

3.2.4 Obtaining the Main Inoculum

1. Transfer 1 mL of the master culture into 20 mL of LB broth (Miller), mix by vortexing, and incubate in the incubator shaker at 200 RPM and 37°C until the desired inoculum concentration is reached.
2. Determine the concentration by reading the inoculum in the UV spectrophotometer. Transfer 1 mL of the inoculum into the cuvette and read the optical density at 600 nm (OD_{600}).
3. Combine individual cultures and dilute the main inoculum in BPW to achieve the desirable concentration.

3.3 Sample Inoculation

3.3.1 Intact Meats and Poultry

1. Fabricate samples to obtain a $10 \times 10\,cm^2$ ($100\,cm^2$) sampling surface.
2. Place samples on styrofoam trays on top of the sorbent pads.
3. Uniformly pipette 500 µL of the inoculum on the sample surface and distribute thoroughly using a sterile plastic rod.

3.3.2 Ground Meats and Poultry

1. Fabricate samples to obtain pieces ~4 in (10.16 cm) in length, 1/2 in (1.27 cm) in width, and 1/8 in (0.32 cm) thick.
2. Uniformly pipette 240 µL of the inoculum on the surface of the pieces and homogenize by mixing
3. Keep samples refrigerated for 30 min.

3.4 Bacteriophage Solution Application and Bacteria Recovery

3.4.1 Intact Meats and Poultry

1. Remove samples from packages and apply 500 µL of the bacteriophage solution at the manufacturer's recommended concentration. Using a sterile plastic rod, ensure the solution is distributed thoroughly.
2. Repackage samples using the desired method (aerobic or vacuum) and store them under refrigeration (5 ± 2°C).
3. After 1 h, remove the samples from the package and thoroughly swab the treated surfaces.

4. Homogenize and vortex the swab's content using the solution in the swab container and serially dilute the homogenate in BPW in microcentrifuge tubes.
5. Spread plate 100 μL of the dilutions using a plastic sterile rod on LB agar plates if testing for *E. coli* or on XLD if testing for *Salmonella*.
6. Incubate plates inverted at 35°C for 24 h.

3.4.2 Ground Meats and Poultry

1. Remove the pieces from the packaging and uniformly pipette 1200 μL of the bacteriophage solution at the concentration recommended by the manufacturer. Using a sterile plastic rod, ensure the solution is distributed thoroughly.
2. Coarsely grind samples using the grinder equipped with a sterile pan, head, auger, blade, plate, and retaining ring. Regrind the samples using the fine plate.
3. Collect a 10 g aliquot of the fine grind sample and homogenize it in 900 mL of BPW for 2 min at 230 rpm using a paddle blender or a Stomacher™ circulator.
4. Collect 40 mL of the homogenate, homogenize it, vortex it, and serially dilute it in BPW in microcentrifuge tubes.
5. Spread plate 100 μL of the dilutions using a plastic sterile rod on LB agar plates if testing for *E. coli* or on XLD if testing for *Salmonella*.
6. Incubate plates inverted at 35°C for 24 h.

3.5 Bacteria Enumeration

1. Using the colony counter, estimate the number of colonies and the final bacteria concentration.

4 Notes

1. Fresh meat and poultry must be screened for *Salmonella* and *E. coli*.
2. The QS1000, Hygiena, Camarillo, CA, United States, is recommended for moisturized swabs.
3. When making media plates, they must be set in the BSL2 cabinet with the "germicidal" feature on.
4. After making plates, they must sit in the cabinet for at least 30 min before being used or stored. Plates must be stored in aseptic conditions for up to one month.
5. Experiments testing bacteriophage solutions must be conducted for individual genera.
6. When recovering strains, use the broth to flush the pellet from the vial.

7. Master cultures can be refrigerated for 72 h and used in subsequent trial replications.
8. Use LB broth (Miller) for the main inoculum's *E. coli* and *Salmonella* strains.
9. Estimated concentrations for *E. coli* and *Salmonella* can be obtained using the Agilent *E. coli* Cell Culture Concentration from OD600 Calculator (Agilent, 2019). Desirable values are from 0.5 to 0.6 McFarland units.
10. The recommended inoculum concentration for bacteriophage validations on meat and poultry is between 1×10^7 and 1×10^8 CFU/mL.
11. Meat and poultry samples must be kept refrigerated at 5 ± 2°C.
12. For testing in aerobic conditions, cover the samples with oxygen-permeable film (aerobic conditions). In vacuum conditions, place the tray inside the vacuum bag and seal it using the vacuum packaging machine.
13. After inoculation, keep samples refrigerated for 30 min before application of bacteriophage solution.
14. Sterile gloves must be used when homogenizing the inoculum by hand into ground samples, and a plastic sterile rod must be used when homogenizing the distribution of inoculum on the sample surface.
15. If the concentration must be adjusted, dilute the bacteriophage stock solution in BPW to achieve the desired concentration before application.

References

1. Chibani-Chennoufi S, Bruttin A, Dillmann ML, Brüssow H (2004) Phage-host interaction: an ecological perspective. J Bacteriol 186(12): 3677–3686. https://doi.org/10.1128/JB.186.12.3677-3686.2004
2. Elois MA, Silva R, da Pilati GVT, Rodríguez-Lázaro D, Fongaro G (2023) Bacteriophages as biotechnological tools. Viruses 15(2):1–25. https://doi.org/10.3390/v15020349
3. Fuller DN, Raymer DM, Kottadiel VI, Rao VB, Smith DE (2007) Single phage T4 DNA packaging motors exhibit large force generation, high velocity, and dynamic variability. Proc Natl Acad Sci USA 104(43):16868–16873. https://doi.org/10.1073/pnas.0704008104
4. Gao Z, Feng Y (2023) Bacteriophage strategies for overcoming host antiviral immunity. Front Microbiol 14:1–16. https://doi.org/10.3389/fmicb.2023.1211793
5. Karimi M, Mirshekari H, Moosavi Basri SM, Bahrami S, Moghoofei M, Hamblin MR (2016) Bacteriophages and phage-inspired nanocarriers for targeted delivery of therapeutic cargos. Adv Drug Deliv Rev 106:45–62. https://doi.org/10.1016/j.addr.2016.03.003
6. Makky S, Dawoud A, Safwat A, Abdelsattar AS, Rezk N, El-Shibiny A (2021) The bacteriophage decides its own tracks: When they are with or against the bacteria. Curr Res Microbial Sci 2:1–9. https://doi.org/10.1016/j.crmicr.2021.100050
7. Miller ES, Kutter E, Mosig G, Arisaka F, Kunisawa T, Rüger W (2003) Bacteriophage T4 genome. Microbiol Mol Biol Rev 67(1):86–156. https://doi.org/10.1128/mmbr.67.1.86-156.2003
8. Shebs EL, Giotto FM, de Mello AS (2022) Effects of MS bacteriophages, ultraviolet light, and organic acid applications on beef

trim contaminated with STEC O157:H7 and the "Big Six" serotypes after a simulated High Event Period Scenario. Meat Sci 188:108783. https://doi.org/10.1016/j.meatsci.2022.108783
9. Shebs-Maurine EL, Giotto FM, Laidler ST, de Mello AS (2021) Effects of bacteriophages and peroxyacetic acid applications on beef contaminated with Salmonella during different grinding stages. Meat Sci 173:1–6. https://doi.org/10.1016/j.meatsci.2020.108407
10. Shebs-Maurine EL, Torres ES, Yeh-Parker Y, de Mello AS (2020) Application of MS bacteriophages on contaminated trimmings reduces Escherichia coli O157 and non-O157 in ground beef. Meat Sci 170:1–5. https://doi.org/10.1016/j.meatsci.2020.108243
11. Sillankorva SM, Oliveira H, Azeredo J (2012) Bacteriophages and their role in food safety. Int J Microbiol 2012:863945. https://doi.org/10.1155/2012/863945
12. Wommack KE, Colwell RR (2000) Virioplankton: viruses in aquatic ecosystems. Microbiol Mol Biol Rev 64(1):69–114. https://journals.asm.org/journal/mmbr
13. Yeh Y, de Moura FH, Van Den Broek K, de Mello AS (2018) Effect of ultraviolet light, organic acids, and bacteriophage on Salmonella populations in ground beef. Meat Sci 139:44–48. https://doi.org/10.1016/j.meatsci.2018.01.007
14. Yeh Y, Purushothaman P, Gupta N, Ragnone M, Verma SC, de Mello AS (2017) Bacteriophage application on red meats and poultry: effects on Salmonella population in final ground products. Meat Sci 127:30–34. https://doi.org/10.1016/j.meatsci.2017.01.001

Chapter 14

Direct Application of Polyphenols (Phenolic Acids and Other Polyphenolics) in Fresh Meat Products: Application of Red Propolis Extract in Fresh Lamb Sausage

Luciana Ruggeri Menezes Gotardo, Heloísa Valarine Battagin, Dannaya Julliethy Gomes Quirino, Carmen Sílvia Fávaro-Trindade, Severino Mathias de Alencar and Marco Antonio Trindade

Abstract

Red propolis in lamb sausages is a healthy option with good prospects for replacing synthetic antioxidants with natural products. The extract is added with the aim of reducing lipid oxidation during storage, showing that natural antioxidants can be a more efficient and healthier option than synthetic ones. This chapter describes the methodologies for preparing the propolis extract, making fresh lamb sausages, and evaluating the oxidative stability of meat products.

Keywords Natural antioxidants, Meat products, Lipid oxidation, Shelf life

1 Introduction

Oxidation causes fresh meat products to lose the functionality of proteins and lipids, lose color and develop a rancid taste, as well as undesirable aromas [1]. Controlling product contact with oxygen, light, and metals, as well as temperature, prevents the formation of free radicals [2].

Among the most common meat products in Brazil are sausages, which are obtained from the processing of meat from butchered animals with or without adipose tissue, ingredients, stuffed in natural or artificial casings, and subjected to the appropriate technological process [3].

Synthetic antioxidants used in meat products have phenolic structures that give up a proton to the free radical in order to interrupt or delay the lipid oxidation reaction [4]. However, their replacement by antioxidants extracted from natural sources is interesting due to their availability in nature and their beneficial effects on health [5].

Renata Ernlund Freitas de Macedo and Carlos Adam Conte-Junior (eds.), *Meat Products Preservation*, Methods and Protocols in Food Science, https://doi.org/10.1007/978-1-0716-5068-4_14,

Propolis extracts have high antioxidant power due to the presence of phenolic compounds, but the levels vary depending on the type of propolis and the region where it was obtained, which makes it difficult to determine its chemical composition [6]. Phenolic compounds play an important role for cells due to their ability to neutralize oxidative stress by capturing reactive oxygen species [7] and are chemically characterized by their aromatic ring with one or more hydroxyls, including their derivatives [8].

Phenolic compounds are divided into flavonoids (catechin, naringin, myricetin, naringenin, hesperetin, kaempferol, and apigenin) and phenolic acids (gallic acid, chlorogenic acid, caffeic acid, coniferous acid, benzoic acid, and trans-cinnamic acid) [9].

Red propolis is rich in flavonoids [10] and the main flavonoids account for 25–30% of the ethanolic extract of propolis [11]. One of the properties of red propolis is its antioxidant power and the antioxidant activity of plant extracts can be analyzed using various methods: Folin–Ciocalteu reagent reduction (total phenolics) [12] and total flavonoids [13]. The methodologies for preparing red propolis extract, evaluating its antioxidant potential and application to fresh lamb sausage and the evaluation of the obtained meat product oxidative stability are presented here.

2 Materials

2.1 Preparation and Characterization of Propolis Extract

1. "Type 13" red propolis, taken from colonies of Apis mellifera bees.
2. Solvent for preparing the propolis extract: 80% ethyl alcohol (v/v) diluted in absolute ethanol with ultrapure water.
3. Blender.
4. Water bath with mechanical stirring at 800 rpm.
5. Rotaevaporator.
6. Refrigerator.
7. Centrifuge.
8. UV spectrophotometer.
9. Glass bottle with hermetic amber closure or wrapped in aluminum foil
10. Gallic acid.
11. Folin–Ciocalteu reagent.
12. Sodium carbonate.
13. 80% alcohol.
14. Aluminum nitrate 10%.
15. Potassium acetate 1M.

2.2 Manufacture of Fresh Lamb Sausage

1. Meat grinder with 8 mm and 6 mm disks.
2. Knives, boards, and trays.
3. Fridge.
4. Fryer.
5. BOD set at 2°C.
6. Vacuum packing machine.
7. Scale.
8. Water bath.
9. Spectrophotometer.
10. Ultra freezer.
11. Lean lamb meat (examples of meat cuts that can be used: eye of round, outside flat, and knuckle) ground into an 8 mm disk (75% of the formulation).
12. Pork back fat ground into a 6 mm disk (20%).
13. Ingredients and additives: Salt (1.8%), onion (0.3%), garlic (0.15%), ground white pepper (0.1%), water (2.6%), sodium erythorbate (0.05%) or red propolis extract (0.18%, *see* **Notes 2 and 3**), and sodium nitrite (0.0075%).
14. Natural pork casing (*see* **Note 4**).

2.3 Determination of Free Fatty Acids

2.3.1 First Stage

1. Chloroform.
2. Methanol.
3. Distilled water.
4. Sodium sulfate.
5. Filter paper.
6. Eppendorfs.
7. Ultra-freezer.
8. Shaker table.
9. Vórtex mixer.
10. Drying oven.
11. Exhaust hood.

2.3.2 In the Second Stage

1. NaOH (Sodium Hydroxide) solution 15%.
2. Thin toilet paper.
3. Hexane P.A.
4. Nitrogen.

5. BF3 (Boron Trifluoride).
6. Hexane HPLC.
7. Na_2SO_4 (Sodium Sulfate).
8. Exhaust hood.
9. Water Bath.
10. Vórtex mixer.
11. Gas chromatography-mass spectrometry.

2.4 Determination of Peroxide Index

1. Acetic acid-chloroform solution (1:1).
2. Potassium iodide.
3. Starch solution.
4. Sodium thiosulphate solution 0.1 N or 0.01 N.
5. Semi-analytical balance.
6. Buret.
7. Magnetic stirrer.

2.5 TBARS

1. Trichloroacetic acid (TCA).
2. EDTA (disodium salt).
3. PG (n-propyl gallate).
4. TBA (thiobarbituric acid) 0.02 M.
5. Turrax.
6. Water bath.
7. Semi-analytical balance.
8. Magnetic stirrer.

2.6 Residual Nitrite

1. Sodium tetraborate decahydrate ($Na_2B_4O_7{\cdot}10H_2O$).
2. Potassium ferrocyanide trihydrate ($K_4Fe(CN)_6{\cdot}3H_2O$).
3. Zinc acetate dihydrate ($Zn(CH_3COO)_2{\cdot}2H_2O$) or zinc sulfate heptahydrate ($ZnSO_4{\cdot}7H_2O$).
4. Glacial acetic acid.
5. Sulfanilamide ($C_6H_8N_2O_2$).
6. Hydrochloric acid.
7. NED - Alpha-naphthylethylenediamine chloride ($C_{12}H_{16}Cl_2N_2$).
8. Sodium nitrite ($NaNO_2$).
9. Semi-analytical balance.
10. Vórtex mixer.

3 Methods

3.1 Preparing the Propolis Extract

1. Grind the red propolis in a blender, only the solid part to reduce the size.
2. Mix the red propolis and alcohol in a ratio of 30 g of propolis to 100 mL of 80% ethanol.
3. Place the ethanolic extract of red propolis in a water bath at 50°C with mechanical stirring at 800 rpm for 30 min.
4. Centrifuge the mixture for 10 min at 4500 ×*g* and 25°C to remove the beeswax.
5. Concentrate the extract to 40% of its initial volume in a rotaevaporator.
6. Store the extract in an airtight glass bottle, protected from light and kept refrigerated until use.

3.2 Characterization of the Red Propolis Extract

3.2.1 Determination of Total Phenolic Content [14]

1. Use gallic acid as a standard and the total phenolic content is determined using the Folin–Ciocalteu reagent.
2. Mix a 0.5 mL aliquot of the ethanolic extract of red propolis with 2.5 mL of Folin–Ciocalteu reagent (1:10 dilution in distilled water) and set aside for 5 min.
3. Add 2 mL of sodium carbonate (7.5%, diluted in distilled water).
4. Place for 15 min in a water bath at 50°C in a dark environment.
5. Measure the absorbance at 740 nm on a spectrophotometer.

3.2.2 Determination of Total Flavonoids [15]

1. Mix a 0.5 mL aliquot of the ethanolic extract of red propolis with 4.3 mL of 80% alcohol, 0.1 mL of 10% aluminum nitrate and 0.1 mL of 1 M potassium acetate.
2. Homogenize and leave to stand for 40 min, protected from light.
3. Measure the absorbance of the samples at 415 nm on a spectrophotometer.

3.3 Sausage Preparation with Propolis Extract Application

1. Defrost the lamb cuts in the refrigerator at 4°C for 24 h (*see* **Note 1**).
2. Chop the lean meat using knives to reduce the size to fit in the grinder.
3. Insert the pieces of lamb meat into the grinder with an 8 mm disk.

4. Place the minced meat in labeled white bowls, cover them with plastic wrap and put them in the fridge to maintain the temperature until all the ingredients are mixed.
5. Cut the pork back fat into smaller pieces and insert into the grinder using the 6 mm disk.
6. Separate bowls labeled with the name of each treatment and formulation.
7. Weigh all the ingredients separately and identify the containers with the name and weight of the respective ingredient used.
8. Weigh out the ground meat and pork back fat and place in the bowls according to treatment and formulation.
9. Spread the powdered ingredients (salt, onion, garlic, chili, sodium nitrite and sodium erythorbate) on top of the weighed meat and back fat and then spread the liquid ingredients (water and propolis extract; *see* **Note 5**).
10. Mix all the ingredients until a homogeneous mass is obtained.
11. Chill the sausage mixture in the refrigerator for 1 h.
12. Stuff the meat mixture into natural pork casings to form the sausages.
13. Portion out and tie the wedges, each weighing an average of 80 g.
14. Pack the sausage in a plastic bag (for instance, polynylon 0.16 mm thick).
15. Vacuum-pack and store at 2°C (*see* **Note 6**).

The methodologies described below are for evaluating sausage lipidic oxidation (TBARS, Peroxide, and Free Fatty Acids), as well as residual nitrite, in order to understand the influence of propolis extract on these parameters.

3.4 Determination of Free Fatty Acids

The sausages free fatty acids profile can be determined in two stages.

The first stage is to extract the lipid fraction [16], as follows:

1. Weigh 20.00 g of the crushed sample into a 250 mL Erlenmeyer flask with a lid.
2. Add 10 mL of chloroform, 20 mL of methanol and 8 mL of distilled water together with the crushed sample.
3. Place this mixture in a shaker and shake for 20 min.
4. Add another 20 or 10 mL of chloroform and 10 mL of sodium sulfate.
5. Shake for a further 10 min.
6. Leave overnight in contact with the solvent.

7. Separate beckers, funnels and filter paper. Pipette the entire volume of the bottom layer of the tube (containing chloroform) and place in the 50 mL falcon tubes (*see* **Note 7**).
8. Add ~1 g of anhydrous sodium sulfate to each of the tubes, cap and vortex for ~2 min.
9. Pour the solutions into the beakers with funnels and wait for them to filter.
10. Pour all the filtrate into the beakers and place in an oven at 85°C for 30 min or until the residue turns yellow.
11. Increase the temperature to 95°C and leave for 4 h (until the solvent evaporates).
12. Then pipette the extracted fat into an Eppendorf, identify and freeze at −80°C in an ultra-freezer.

In the second stage, thaw the prepared samples [17]:

1. Weigh 3.0 g of the fat sample into a 50 mL Falcon tube.
2. Add 5 mL of NaOH solution and vortex the Falcon tube for 15 min, then centrifuge for 10 min at 2500 rpm.
3. Remove all the liquid phase using a glass pipette and discard it.
4. Keep the solid part, dry it on fine toilet paper and place the solid content in another clean 50 mL Falcon tube.
5. Add 3 mL of hexane P.A.
6. Vortex for 15 s and centrifuge for 5 min.
7. Repeat the procedure in item 5 five times, collecting the upper contents and transferring them to a glass.
8. After collecting the contents of the hexane + salts that were placed in the vial five times, dry in nitrogen. The remaining contents will be a viscous liquid.
9. Then add 3 mL of BF_3 and heat in a water bath at 100°C for 2 min.
10. When removed from the water bath, rinse under running water to cool, safeguarding the integrity of the sample.
11. Add 6 mL of Hexane HPLC and vortex.
12. Wait for the phases to separate and remove the supernatant, placing it in another glass.
13. Add Na_2SO_4, shake manually and pipette 1.5 mL of the sample into the GC-MS injection vial.
14. Place the samples in the gas chromatography-mass spectrometry analyzer in an apparatus equipped with a Split/Splitless injector with an automatic injector (model AOC-500). The

stationary phase is a capillary column (100 m × 0.25 mn di × 0.20 μm df), using helium as the carrier gas, at a flow rate of 1.59 mL/min. The samples are injected into an injector with a split ratio of 1:40, a temperature of 250°C and a column pressure of 300 Kpa.

15. After reading on the chromatograph, analyze the spectra by searching the library for the compounds identified in the sample.

3.5 Determination of the Peroxide Index

The methodology used to determine the peroxide index [18] is as follows:

1. Weigh 1 ± 0.05 g of sample into a flask.
2. Add 30 mL of the 1:1 acetic acid-chloroform solution while stirring.
3. Add powdered KI (the tip of a spatula) and leave to stand in the dark for 5 min.
4. After the reaction time, carry out the titration.
5. Add 30 mL of water and 2 mL of starch indicator solution to the flask.
6. Fill the buret with 0.1 N or 0.01 N sodium thiosulphate solution.
7. Start the titration, the Erlenmeyer flask should be shaken constantly until the yellow color has almost disappeared and white or a gray tint appears.
8. Note down the volume of sodium thiosulphate used in the titration and calculate. Peroxide value (mEq/1000 g of sample) = (N * V * 1000)/g

   Where: N = normality of sodium thiosulphate (N = Eq/L)

   V = volume of titrant (mL)

   g = grams of sample.
9. The Peroxide Index is expressed in mEq/1000 g of sample.

3.6 TBARS

The oxidative stability of the samples can be analyzed using the method developed by [19] with modifications. It is carried out in a few stages:

1. After making solutions of TCA (*see* **Note 8**), TBA (*see* **Note 9**), and TEP (*see* **Note 10**), switch on the water bath and wait until the temperature reaches 98°C.
2. For each sample, separate three 250 mL beakers and identify them.
3. Separate the test tubes, taking into account the repetitions + 2 blank tubes for subsequent calibration of the spectrophotometer.
   a. Identify the test tubes.

4. Separate Erlenmeyers with 125 mL, funnels with 9 mm with filter paper for filtering after homogenization—identify beakers.
5. Grind the sausage samples for 2 min in a domestic grinder and weigh the samples on a semi-analytical balance (*see* **Note 11**).
6. For the TBARS analysis, weigh out 5 g of the sample in each beaker.
7. Pipette 25 mL of TCA solution into each of the beakers and into the two blank tubes (*see* **Note 12**).
8. Homogenize at speed 2 for 2 min in a homogenizer with Turrax—wash between each sample.
9. Filter all the samples containing TCA through ordinary filter paper.
10. Make white tubes: 5 mL of TCA solution + 5 mL of TBA solution.
11. From the resulting filtrate, pipette 5mL and place in the corresponding 30 test tubes+ 5 mL of TBA.
12. Homogenize all the test tubes in a tube shaker, including the blanks.
13. After homogenizing, place the tubes in a rack together with the blanks.
14. Place tubes in a water bath at 98°C for 40 min.
15. Switch on the spectrophotometer 30 min before use.
16. Remove the tubes from the water bath and place them in a plastic tray with water—you can leave them under running water for ~15 min to cool down.
17. Take absorbance readings on spectrophotometer at 538 nm (*see* **Notes 13** and **14**).
18. Record the results and calculate the TBAR values.
19. Make the standard curve (*see* **Note 15**).

3.7 Residual Nitrite

The methodology used for residual nitrite determination [20], number 283/IV—Spectrophotometric determination of nitrites.

1. Weigh 10 g of the crushed and homogenized sample into a 200 mL beaker.
2. Add 5mL of sodium tetraborate solution, mix with a glass rod and add about 50 mL of hot water (80°C).
3. Leave in a water bath for 15 min, stirring frequently with a whisk. Proceed in the same way with a reagent blank without adding the sample.
4. Using a glass rod and a funnel, transfer the contents quantitatively to a 200 mL volumetric flask. Rinse the beaker thoroughly with ~50 mL of water.

5. Leave to cool and add 5 mL of potassium ferrocyanide solution and 5 mL of zinc acetate solution (or alternatively zinc sulfate). Shake by rotation after adding each reagent and top up with water. Shake vigorously.
6. Leave to stand for 15 min, shaking vigorously several times during this period.
7. Filter through qualitative filter paper into a 250 mL Erlenmeyer flask.
8. Pipette 10 mL of the reagent blank and samples, respectively, into 50 mL amber volumetric flasks.
9. Add 5 mL of sulfanilamide reagent, leave to react for 5 min and add 3 mL of NED reagent, shaking after each addition. Make up to volume with water and homogenize.
10. Leave to stand for 15 min and read the absorbance at 540 nm against the reagent blank.
11. Draw the curve (*see* **Note 16**).
12. Make the calculations using the formula:

$$\text{Sodium nitrite in mg/Kg} = [(A - b) \times 1000]/p \times a$$

A = sample absorbance

b = linear coefficient of the line obtained from the standard curve

a = absorptivity (angular coefficient of the line obtained from the standard curve)

p = sample mass in grams

1000 = dilution factor.

4 Notes

1. The amounts of red propolis extract added should be calculated based on the antioxidant equivalence of the total phenolic value of erythorbate, which for 500 mg was equivalent to 1800 mg in propolis in previously tested samples.
2. The value of total phenolic content found in the previous study was 182.60 mg AGE/g for red propolis extract and for sodium erythorbate the value found was 695 mg AGE/g. Thus, the equivalence between the antioxidant capacity of the two antioxidants was 3.6 times (695/182.6), so the proposed sausages were: 0.18% (1800 ppm—equivalent to erythorbate) and 0.36% (3600 ppm—twice as much as erythorbate), considering the maximum value of 0.05% (500 ppm) of sodium erythorbate added to the formulations.

3. Leave the natural pork casing immersed in water to remove excess salt.
4. The cuts used must have samples taken and the Total Bacterial Count made to get an idea of the microbiological quality of the raw material, in order to guarantee the initial quality of the raw material.
5. The propolis extract was sprinkled on after all the ingredients had been mixed, it was left on last and the sausage mixture was left in a bowl covered in plastic for 1 h until it was stuffed.
6. The storage temperature of 2°C and the vacuum packaging were determined from the literature and through preliminary tests, taking into account sensory characteristics, color, odor and microbiology stability (standard count).
7. Take care that the solvent does not come with the pipette.
8. Preparation of the 7.5% TCA solution (trichloroacetic acid) with 0.1% EDTA (disodium salt) and 0.1% PG (n-propyl gallate). Weigh 37.5 g TCA (on a semi-analytical balance as it is highly corrosive) + 0.5 g EDTA (analytical balance) + 0.5 g PG (analytical balance). Complete Becker (250 mL) with 200 mL of water and add the reagents that were weighed. Place the magnet and the beaker containing the solution on a magnetic stirrer (medium speed) until completely diluted. After dissolving, transfer the solution from the Becker to a 500 mL volumetric flask (you can use a funnel or not) and top up with distilled water.
9. Preparation TBA (thiobarbituric acid) 0.02 M. Weigh out 0.288 g for 100 mL and 0.72 g of TBA. Pour ~150 mL of distilled water into a 250 mL Becker. Add the TBA little by little (to avoid solidification) to the beaker containing the distilled water already stirred on a magnetic stirrer. Pour the diluted solution into a 250 mL volumetric flask and fill the flask with distilled water up to the meniscus and homogenize the flask.
10. Weigh out 0.2203 g of TEP (Tetraethoxypropane) and dissolve this amount in 100 mL of TCA in a 100 mL volumetric flask. TEP—simulates malanaldehyde and reacts with TBA.
11. TBA generates color by reacting with malonaldehyde, which reacts with TBA to form a pinkish color, so be careful if the sample has the same pinkish color resulting from this reaction. If this happens, you should look in the literature for adaptations to the methodology.
12. Place the TCA solution in a smaller beaker to make it easier to obtain using a 25 mL volumetric pipette.
13. Calibrate with the white solution and the black cuvette alternately until 0 is obtained for the white cuvette and 100 for the black cuvette.

14. Halfway through the readings, take a reading of the white solution to check.
15. To draw the standard curve, prepare solutions of TBA (250 mL) and TCA (500 mL); weigh out 0.2203 g of TEP (Tetraethoxypropane) and dissolve this amount in 100 mL of TCA in a 100 mL volumetric flask (Solution A). Pipette 0.1 mL of Solution A and top up 100 mL with TCA in another 100 mL volumetric flask (Solution B). Identify the 26 tubes and place in the rack [(0.01 mL(10 μL)/0.02 mL(20 μL)/0.05 mL(50 μL)/ 0.1 mL/0.2 mL/0.3 mL/0.4 mL/0.5 mL/1 mL/2 mL/3 mL/ 4 mL/5 mL] of solution B. In each tube, place 4.99 mL/4.98 mL/ 4.95 mL/4.9 mL/4.8 mL/4.7 mL/4.6 mL/4.5 mL/4 mL/ 3 mL/2 mL/1 mL/0 mL of TCA in the 26 tubes opposite those of solution B, making up 5 mL. Add 5 mL of TBA to all the tubes. Make 2 white tubes, adding 5 mL of TCA + 5 mL of TBA. Place the rack in a water bath for 40 min, cool the tubes and take the readings on a spectrophotometer. Record the results and draw the curve.
16. Make the standard curve—Pipette aliquots of 1, 2, 3, 4, 5, 6, and 7 mL of the sodium nitrite working standard solution (8 μg/mL) into 50 mL volumetric flasks. Add 5 mL of sulfanilamide reagent to each, mix by rotation and after 5 min add 3 mL of NED reagent. Complete the volume and homogenize. Allow the color to develop for 15 min and determine the absorbance at 540 nm against the reagent blank. Construct the curve with the absorbance values on the y-axis and the sodium nitrite concentration 0.16, 0.32, 0.48, 0.64, 0.80, 0.96, and 1.12 μg/mL on the x-axis. Calculate the linear and angular coefficients of the line (absorptivity, considering the optical path of the cuvette of 1 cm).

Acknowledgments

This work was carried out with the support of the Coordination for the Improvement of Higher Education Personnel—Brazil (CAPES)—Financing Code 001 and funding from CNPQ (National Council for Scientific and Technological Development—number 43155.

References

1. Munekata PES, Rocchetti G, Pateiro M, Lucini L, Domínguez R, Lorenzo JM (2020) Addition of plant extracts to meat and meat products to extend shelf-life and health-promoting attributes: an overview. Curr Op Food Sci 31:81–87
2. Ribeiro EF, Jorge N (2017) Oxidative stability of soybean oil added to coffee husk extract (*Coffea arabica* l.) under accelerated storage conditions. Food Sci Technol 37:5–10
3. Brasil (2000) Ministério da Agricultura Pecuária e Abastecimento. Secretária de Defesa

Agropecuária (SDA). Instrução Normativa n° 4, de 31 de março de 2000. Regulamento Técnico de Identidade e Qualidade de Carne Mecanicamente Separada, de Mortadela, de Linguiça e de Salsicha

4. Ramalho VC, Jorge N (2006) Antioxidantes utilizados em óleos, gorduras e alimentos gordurosos. Química Nova 29(4):755–760
5. Sarmadi BH, Ismail A (2010) Antioxidative peptides from food proteins: a review. Peptides Nova Iorque 31(10):1949–1956
6. Teixeira EW et al (2010) Seasonal variation, chemical composition and antioxidant activity of Brazilian propolis samples. Evid Bas Complement Altern Med 7(3):307–315
7. Dumitriu D et al (2015) Grape pomace extract improves the in vitro and in vivo antioxidant properties of wines from sun light dried Pedro Ximénez grapes. J Funct Foods 17:380–387
8. Shahidi F, Han XQ (1993) Encapsulation of food ingredients. Crit Rev Food Sci Nutr 33(6):501–547
9. Moniruzzaman M et al (2014) Identification of phenolic acids and flavonoids in monofloral honey from bangladesh by high performance liquid chromatography: determination of antioxidant capacity. BioMed Res Int 2014:737490
10. Daugsch A, Moraes CS, Fort P, Park YK (2008) Brazilian red propolis – chemical composition and botanical origin. Evid Bas Complement Altern Med 5(4):435–441. https://doi.org/10.1093/ecam/nem057
11. Havsteen B (1983) Flavonoids, a class of natural products of high pharmacological potency. Biochem Pharmacol 32(7):1141–1148
12. Woisky RG, Salatino A (1998) Analysis of propolis: some parameters and procedures for chemical quality control. J Apicultural Res 37(2):99–105
13. Miller HE (1971) A simplified method for the evaluation of antioxidants. J Am Oil Chem Soc 48(2):91–91
14. Singleton VL et al (1999) Analysis of total phenols and other oxidation substrates and antioxidants by means of folin-ciocalteu reagent. Methods Enzymol 299:152–178
15. Park Y et al (1995) Survey of some components of propolis which were collected by *Apis mellifera* in Brazil. Web Sci 432:137175
16. Bligh EG, Dyer WJ (1959) A rapid method of total lipid extraction and purification. Can J Biochem Physiol 37(8):911–917. https://doi.org/10.1139/o59-099
17. Vicentini-Polette CM, Ramos PR, Gonçalves Oliveira AL (2021) Determination of free fatty acids in crude vegetable oil samples obtained by high-pressure processes. Food Chem X 12:100166. https://doi.org/10.1016/j.fochx.2021.100166
18. Pearson D (1976) Cap.5p Índice de aceites y enranciamento. In: Técnicas de laboratório para el analisis de alimentos. Acribia, Zaragoza, pp 137–139
19. Vyncke BW (1970) Direct determination of the thiobarbituric acid value in trichloracetic acid extracts of fish as a measure of oxidative rancidity. Fette Seifen Anstrichmittel Leinf 72(12):1084–1087
20. Adolfo Lutz Institute (1985) Analytical standards of the Adolfo Lutz Institute. In: Chemical and physical methods for food analysis, 3rd edn. IMESP, São Paulo, Brazil, p 97–98

Chapter 15

Direct Application of Organic Acids for Decontamination of Carcass and Raw Meat

Annada Das, Kaushik Satyaprakash, Pramod Kumar Nanda, Mohammad Gagaoua and Arun K. Das

Abstract

The meat industry continually seeks to enhance food safety, especially following outbreaks of meat-related illnesses. Effective decontamination of carcasses and raw meat is essential, yet traditional methods that use chemical compounds and heat treatments can compromise the nutritional and sensory properties, raising concerns about the toxicity and safety of meat products. Organic acids, such as lactic, citric, propionic, acetic, and peracetic acids, have emerged as effective alternatives due to their natural antimicrobial properties, significantly reducing bacterial contamination without compromising intrinsic meat quality traits. This chapter offers a comprehensive guide for using organic acids to decontaminate carcasses and meat, aiming to extend shelf life, ensure safety, and protect the health of consumers. Detailed procedures for selecting organic acids, preparing solutions, and applying them through various methods such as spraying/rinsing, dipping/immersing, and brushing/surface application are provided, alongside guidelines for microbial testing and safety precautions.

Keywords Organic acid, Lactic acid, Citric acid, Decontamination, Carcass, Raw meat, Food safety

1 Introduction

The intrinsic quality of meat products, mainly safety, is of paramount importance in the food industry, where contamination by pathogenic microorganisms not only poses serious health risks to consumers but also results in considerable economic losses and food waste [1, 2]. Meat, being a highly perishable product, is susceptible to numerous pathogenic and spoilage microorganisms, which are often associated with foodborne illnesses [3]. According to the World Health Organization (WHO), an estimated 7.69% of the global population (600 million, or nearly 1 in every 10 people) experience foodborne illness each year. Of the total 56 million annual deaths worldwide, 7.5% (420,000) are attributed to foodborne diseases, resulting in

Renata Ernlund Freitas de Macedo and Carlos Adam Conte-Junior (eds.), *Meat Products Preservation*, Methods and Protocols in Food Science, https://doi.org/10.1007/978-1-0716-5068-4_15,

the loss of 33 million healthy life years due to the consumption of contaminated, spoiled, and unsafe food [4].

A huge economic loss amounting to 110 billion US$ occurs towards medical expenses related to foodborne illness in low- and middle-income countries [5]. The Food and Drug Administration (FDA) and the Centers for Disease Control and Prevention (CDC) have also estimated that about 48 million cases of foodborne illness (1 in 6 Americans), 128,000 hospitalizations and 3000 deaths occur in the USA due to unsafe food consumption [6, 7]. In the unabating pursuit of food safety, the meat industry struggles with the challenge of decontaminating carcasses and raw meat, amidst the recent notorious episodes of meat-borne illnesses. Traditionally, decontamination methods using chemical washes, thermal treatments, and irradiation have been widely used to control bacterial contamination in carcass and raw meat products [8]. However, these methods often have some limitations such as altering nutritional and sensory qualities, besides posing toxicity, carcinogenicity, residue, and environmental concerns [9]. Additionally, the health-conscious consumers and regulatory bodies have increasingly demanded the use of more natural and safer methods for carcass decontamination that do not compromise the meat quality.

Organic acids stand out as viable alternatives for raw meat and carcass decontamination due to their inherent antimicrobial properties, effectiveness, and minimum impact on the quality attributes of meat [10, 11]. Organic acids are food-grade and believed to be more potent than inorganic acids of equivalent concentrations [12]. Organic acids such as lactic acid, acetic acid, formic acid, citric acid, and peracetic acid are being effectively used for combating pathogenic and spoilage bacteria in meat, such as *Salmonella* Typhimurium, *Escherichia coli* O157:H7, *Listeria monocytogenes*, *Bacillus cereus*, and *Staphylococcus aureus* [13]. These organic acids have the ability to lower the pH of the meat, creating an unfavorable environment for bacterial growth [10]. Organic acids have attained the generally recognized as safe (GRAS) status by the United States Food and Drug Administration [14, 15] due to past experience with their common use in food [16, 17].

Many researchers have used organic acids, such as acetic, lactic, peracetic, and citric acid as rinses, dips, and sprays to reduce microbial load in raw meat and decontaminate carcasses (Table 1). In a recent study, it was observed that dipping raw poultry wings in peracetic acid alone and in combination with sodium bisulfate significantly reduced the spoilage bacteria counts with a 7-day extension in shelf life at refrigerated storage (4°C) [13]. Another study evaluated the effects of pre- and post-slaughter spray of acetic acid and lactic acid (3% v/v) on beef carcass and found that the two times spray reduced 2–5 log CFU of each bacteria, including *Staphylococcus*, *Shigella*, *Bacillus*, *Escherichia*, and *Salmonella* spp. [18]. Solutions (1%–3%) of lactic and acetic acids are most commonly used as sprays for decontaminating beef and lamb carcasses [19]. Some researchers also used

Table 1
Decontamination of meat surfaces with organic acids and meat preservation

Carcass/ meat	Organic acid	Concentration	Effective antimicrobial activity	Microbial load reduction	Sensory attributes	Shelf-life extension	Reference
Poultry	Lactic acid	2%–5% v/v	Effective against *Salmonella*, *Escherichia coli*	Up to 3 log CFU	Minimal impact on color and odor	7 days	[13]
Beef	Acetic acid	1%–3% v/v	Effective against *Staphylococcus*, *Listeria*	2–5 log CFU	Acceptable color, slight odor changes	5 days	[18]
Pork	Citric acid	1%–3% v/v	Effective against *Bacillus*, *Salmonella*	Significant reduction observed	Slight color changes, odor acceptable	6 days	[20]
Lamb	Peracetic acid	1%–2% v/v	Broad-spectrum activity	Up to 90%	No noticeable changes	7–10 days	[14]
Fresh beef	Citric acid	1%–2% v/v	Effective against *E. coli*, aerobic plate counts, *Salmonella*, and Enterobacteriaceae	0.7–2.0 log CFU	Maintained sensory qualities	7 days	[21]

organic acids up to 6% for carcass decontamination [18]. However, these acids when used at higher concentrations cause discoloration of meat and carcass surface [14]. Several factors like nature and extent of bacterial contamination, type of tissue, degree of bacterial attachment to tissue, carcass condition, type and concentration of organic acid, contact time, and temperature affect the degree or effectiveness of decontamination (Fig. 1). Typically, higher reduction in bacterial counts up to 70%–90% is achieved using organic acid carcass decontaminants, thereby reducing the instances of foodborne illnesses [13], if the procedure is followed.

Considering the above, this chapter outlines the methods for direct application of organic acids for decontamination of carcasses and raw meat, aiming to prolong the shelf life of meat, thereby ensuring food safety, and safeguarding consumer health. Further, this chapter details the stepwise procedures for decontamination of carcass and raw meat using organic acids, including selection of organic acids, preparation of organic acid solutions, and different application methods like spraying, dipping, and brushing Additionally, monitoring through microbial testing, sensory evaluation to assess meat quality, and precautions or safety measures to be taken during the application process are also discussed here.

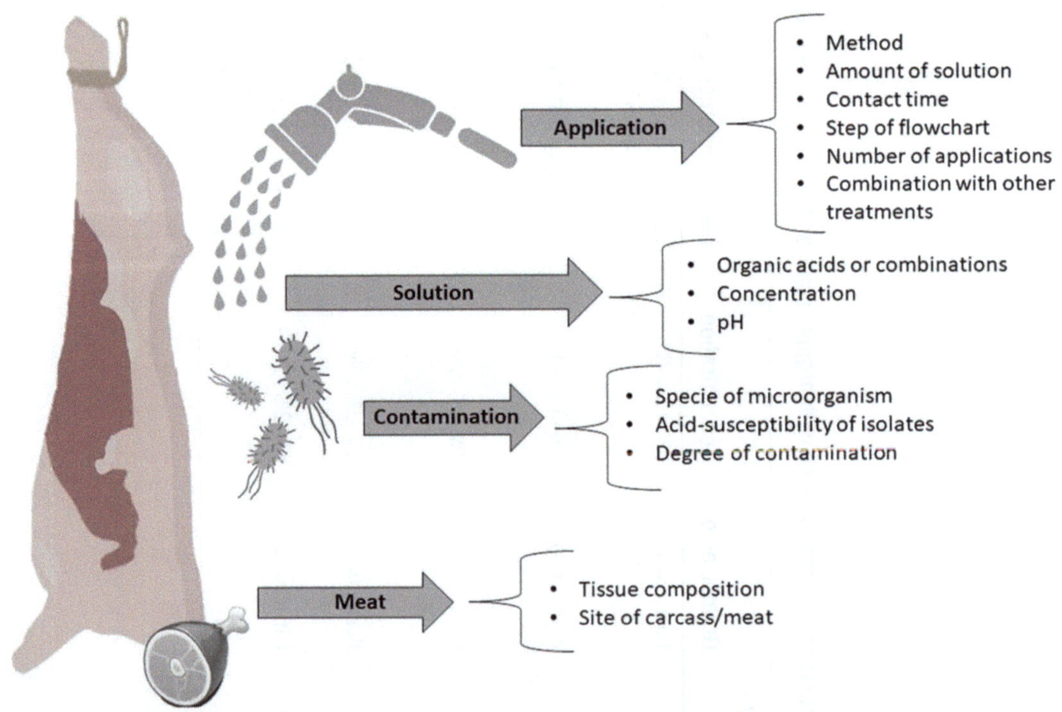

Fig. 1 Factors that affect the effectiveness of the application of organics and peracetic acid in carcasses and/or meat are categorized into four groups (application method, characteristics of the solution, type of contamination, and characteristics of the meat) (Source: [14] from Wiley)

2 Materials

1. Organic acids: food-grade organic acid viz., lactic acid ($C_3H_6O_3$) (2%–5%, v/v concentration) or acetic acid (CH_3COOH) (1%–3%, v/v concentration), or citric acid ($C_6H_8O_7$) (1%–3%, v/v concentration) is used for decontamination of carcass or raw meat.
2. Sterile distilled water: Milli-Q water is recommended for dilution of organic acids and washing of equipment.
3. Glass wares and plastic wares: sterile glass wares, plastic wares like graduated cylinders, pipettes, beakers, bottles, funnels, and conical flasks are required
4. Mechanical stirrers: required for thorough mixing of the acid solutions.
5. pH meter: required for measurement of pH or acidity of the prepared solution and meat samples.
6. Refrigerator and packaging materials: required for storage of meat samples, chemicals, and reagents at 4 ± 1°C, or the required temperature, depending on the need.
7. Application equipment:
    1. High-pressure sprayers: for uniform spraying of organic acid solution under pressure to every part of the carcass.
    2. Dipping tanks or vats: corrosion-resistant tanks or vats containing organic acid solutions as dipping tanks for submerging the carcass or meat cuts.
    3. Acid-resistant brushes: for direct application of organic acids on the meat or carcass surface by brushing manually.
8. Documentation tools:
    1. Labels and markers: for clear identification of samples, acid solutions, equipment, glasswares, etc.
    2. Digital devices, laboratory notebooks, and pens: for recording observations, data, and results.
9. Requirements for microbiological testing:
    1. Sterile swabs, stomacher bags, and tissue homogenizer: for collection and preparation of surface samples before and after decontamination.
    2. Petri plates, spreader, reagents, and media: sterile petri plates, spreaders, buffered peptone water, sterile nutrient broth, general purpose media selective media, and other chemicals for isolation and identification of target pathogenic and spoilage microorganisms.

3. Incubator and anaerobic jar: for maintaining the appropriate temperature required for microbial growth.

10. Sensory evaluation tools: sterile spoons, knives, forks, proforma, etc. for assessing the raw meat color and odor scores.
11. Precautionary safety kits:
    1. Personal protective kits like gloves, masks, face shields or goggles, protective clothing, respirators, and gum boots.
    2. First aid kit for immediate treatment of accidental exposure to concentrated acids.
    3. Emergency eye wash station to wash eyes in case of accidental contact with acids.
    4. Absorbent pads with basic substances (such as sodium bicarbonate, calcium carbonate, sodium carbonate, or combinations thereof) for neutralizing acids, or to manage acid spills.

3 Methods

3.1 Selection of the Appropriate Organic Acid

1. Determine the target pathogens or spoilage microbes present on carcass or meat surfaces, such as *Salmonella* Typhimurium, *Escherichia coli* O157:H7, *Listeria monocytogenes*, *Bacillus cereus*, *Staphylococcus aureus*, etc.
2. Choose the suitable organic acid amongst the commonly used acids viz., lactic acid (2%–5%), acetic acid (1%–3%), or citric acid (1%–3%), that have strong and broad-spectrum antimicrobial activity, are cost-effective and have minimum negative impact on meat quality. Also, consider the meat type (beef, pork, lamb, or poultry), and type of application (spraying, brushing, or dipping) during the selection of decontaminant.

3.2 Preparation of Acid Solutions

1. For preparing the desired concentration of organic acid, proper dilution is essential.
   For example, if 20 L of 3% organic acid solution have to be prepared from the volume of 88% concentrated acid, then the volumes of sterile distilled water and concentrated acid required to prepare desired dilution can be calculated using the formula:

$$\text{Initial volume} \times \text{Initial Concentration} = \text{Final Volume} \times \text{Final Concentration}$$

From the above example, it is known that
Initial concentration = 88%
Final volume = 20 L
Final concentration = 3%

Initial volume = unknown, i.e., to be calculated

$$\text{So initial volume of concentrated acid required} = \frac{20 \times 3}{88} = 0.68 \text{ litres}$$

Now the volume of sterile distilled water required is

$$20\text{ L} - 0.68\text{ L} = 19.32\text{ L}$$

So, adding 19.32 L of sterile distilled water to 0.68 L of concentrated acid and mixing the solution thoroughly will give the desired volume (20 L) of 3% organic acid solution.

2. Measure the required volumes of sterile distilled water/concentrated acid using graduated glass wares/pipettes.
3. Add concentrated acid to sterile distilled water slowly with continuous shaking to ensure uniform mixing.
4. While pouring concentrated organic acid; it is advisable to wear goggles, gloves, and an apron to protect the body.
5. The working organic acid solution should be stored in a corrosion-resistant container with proper labelling, until further use.

3.3 Application Methods of Organic Acid Solutions

3.3.1 Spraying

1. Pour the prepared acid solution into the high-pressure sprayer.
2. Focus on areas prone to contamination, such as rear parts, joints, corners, etc.
3. Spray the organic acid solution onto the carcass or meat surface, with a uniform pressure of 15 psi and a spraying speed of 2 mm/s. The nozzle of the spray should be within 12 inches of the carcass surface to ensure uniform/even spray.
4. Spraying should be done methodically, starting from top to bottom, to ensure that the entire carcass surface is treated.
5. Allow a minimum contact time of 1–5 min to maximize microbial reduction.

3.3.2 Dipping

1. Fill the immersion tank or vat to a level sufficient for submerging the carcass or meat cut.
2. Immerse the carcass or meat cut in the vat containing the organic acid solution, ensuring complete dipping for 2–5 min.
3. Remove the meat sample after adequate contact time and allow the excess acid solution to drain off.

3.3.3 Brushing

1. Dip an acid-resistant brush into the prepared acid solution.

2. Manually brush the acid solution on the meat or carcass surface, ensuring uniform brushing at all areas of the carcass or meat cut.
3. Allow a contact time of 1–5 min for optimum decontamination.

3.4 Monitoring the Effectiveness of Organic Acid Treatment

1. In order to monitor the effectiveness of organic acid treatment, collect representative meat or carcass surface samples (pre- and post-organic acid treatment) using sterile cotton swabs and prepare serial dilution.
2. Inoculate selective or general-purpose agar plates with the collected samples based on the target microbe, incubate plates at specific times and temperatures (37°C for 24–48 h for mesophiles, 4°C for 7 days for psychrophiles) for counting the number of colonies and express as colony-forming units (CFU) by using the dilution factor.
3. Compare the CFU from the pre-treated and treated samples to evaluate the effectiveness of the organic acid application.
4. The microbial counts obtained in organic acid-treated samples can also be compared with those of some standard antimicrobials used as positive control.
5. The data are recorded and analyzed to interpret, and compare the results with the standard microbial limits for raw meat (5–7 CFU/g).

3.5 Sensory Evaluation to Assess Certain Meat Quality Parameters

1. Sensory evaluation of the control and treated meat or carcass samples should be conducted by a panel of 10 semi-trained assessors, focusing on sensory attributes such as color and odor. It is important to adapt the required method following the standards and guidelines available in the literature.
2. The meat samples can be further evaluated for color and odor using a 5-point descriptive scale where for meat color, 1—very undesirable/unattractive/dull, 2—undesirable, 3—moderately desirable, 4—desirable, 5—very desirable or, 1 = pale pink, 2 = pink, 3 = pinkish red, 4 = bright red, 5 = reddish brown; and for meat odor, 1 = very unpleasant, 2 = moderately unpleasant, 3 = moderately pleasant, 4 = pleasant, 5 = very pleasant.
3. The obtained scores for each sample are recorded in a predesigned proforma and analyzed statistically to observe any effect of organic acid treatment on the sensory quality of meat.
4. Additional technological quality features can be assessed based on their extensive application in the literature related to technological and commercial quality traits.

3.6 Precautions or Safety Measures While Using Organic Acids

1. Wear personal protective equipment (hand gloves, aprons, goggles, etc.) to prevent direct contact of the human body with acid.
2. Conduct the experiment in a well-ventilated area and open the concentrated acids under a fume hood to minimize exposure to fumes.
3. Always add acid to water, not water to acid, by continuous mixing to avoid exothermic reactions.
4. Avoid splashing or spilling acids and immediately use neutralizing pads to absorb acids.
5. Have an emergency eye wash station and first aid kit readily available in case of accidental exposure to acids or chemicals.
6. Give proper training to the personnel regarding all steps of organic acid application for carcass decontamination.
7. Store acids in corrosion-resistant, clearly labelled containers in cool and dry places.
8. Dispose/treat the used organic acid solutions/ wastewater and contaminated meat or other materials as per the standard regulations and safety protocols, in compliance with local rules.

3.7 Precautions to be Taken to Ensure Meat Safety

1. During decontamination, organic acids must be used at specific concentrations to ensure that they neither cause discoloration of carcass surfaces nor alter the texture of meat.
2. The organic acid solutions, when chosen for use as decontaminants on carcasses or meat, should neither have any adverse effects on workers' health (respiratory problems and skin irritation) nor on consumers (effect on desirable sensory properties) during their application or use.
3. The decontamination strategies must not be at the cost of replacing the general hygiene practices (GHP) and good manufacturing practices (GMP) employed during slaughtering process, but should only be considered as a secondary/complementary measure of limiting micro-organisms on carcasses in order to increase the safety of meat.
4. In the case of organic acids used at higher concentrations, secondary rinsing with potable/sterile distilled water is recommended to remove excessive acids and balance the sensory qualities of carcass/raw meat.
5. Prolonged usage of particular organic acid(s) as a strategy to decontaminate carcasses should be closely monitored and controlled, as microorganisms may become resistant to organic acids and facilitate the development of resistant strains.
6. While selecting a particular organic acid for application, several aspects, including the occurrence of residues, impact on physicochemical parameters, and shelf-life of carcass or meat must be considered.

References

1. Das AK, Nanda PK, Das A, Biswas S (2019) Hazards and safety issues of meat and meat products. In: Food safety and human health. Elsevier, Academic Press, London, UK, pp 145–168
2. Boulahlib CY, Aggoun M, Arhab R, Gagaoua M, Boulahlib CY, Aggoun M, Arhab R, Gagaoua M (2024) Potential applications of biosurfactants in animal production and meat research. AIMSAGRI 9:237–253. https://doi.org/10.3934/agrfood.2024014
3. Gadekar YP, Das AK, Thomas R, Banerjee R, Gireesh-Babu P, Naveena BM, Kandeepan G, Barbuddhe SB (2021) Importance of food safety management in meat value chain. Indian J Animal Health 60. https://doi.org/10.36062/ijah.2021.spl.01621
4. WHO (2024) Food safety. https://www.who.int/news-room/fact-sheets/detail/food-safety. Accessed 11 Nov 2024
5. Worldbank (2018) Food-borne Illnesses Cost US$ 110 Billion Per Year in Low- and Middle-Income Countries. World Bank, https://www.worldbank.org/en/news/press-release/2018/10/23/food-borne-illnesses-cost-us-110-billion-per-year-in-low-and-middle-income-countries
6. FDA (2024) People at risk of foodborne illness. FDA. https://www.fda.gov/food/consumers/people-risk-foodborne-illness
7. CDC (2023) Foodborne illness source attribution estimates for *Salmonella*, *Escherichia coli* O157, and *Listeria monocytogenes* — United States, 2021
8. Lee Y, Yoon Y (2024) Principles and applications of non-thermal technologies for meat decontamination. Food Sci Anim Resour 44:19–38. https://doi.org/10.5851/kosfa.2023.e72
9. Roobab U, Chacha JS, Abida A, Rashid S, Muhammad Madni G, Lorenzo JM, Zeng XA, Aadil RM (2022) Emerging trends for nonthermal decontamination of raw and processed meat: ozonation, high-hydrostatic pressure and cold plasma. Foods 11. https://doi.org/10.3390/foods11152173
10. Nkosi DV, Bekker JL, Hoffman LC (2021) The use of organic acids (lactic and acetic) as a microbial decontaminant during the slaughter of meat animal species: a review. Foods 10:2293. https://doi.org/10.3390/foods10102293
11. Sinhamahapatra M, Biswas S, Das AK, Bhattacharyya D (2004) Comparative study of different surface decontaminants on chicken quality. Br Poult Sci 45:624–630. https://doi.org/10.1080/00071660400006552
12. Rossi GAM, Link DT, Bertolini AB, Tobias FL, Mioni M de SR (2023) A descriptive review of the use of organic acids and peracetic acid as a decontaminating strategy for meat. eFood 4:e104. https://doi.org/10.1002/efd2.104
13. Dittoe DK, Feye KM, Ovall C, Thompson HA, Ricke SC (2024) Exploiting the microbiota of organic and inorganic acid-treated raw poultry products to improve shelf-life. Front Microbiol 15. https://doi.org/10.3389/fmicb.2024.1348159
14. Dickson JS, Hardin MD, Acuff GR (1996) Organic Acid Rinses. In: Microbiology Intervention Methods, Proceedings of the 49th Annual Reciprocal Meat Conference, American Meat Science Association, pp 129–131.https://earthwormexpress.com/wp-content/uploads/2019/01/Microbiology-Intervention-Methods-Organic-Acid-Rinses.pdf
15. Nielsen B, Colle MJ, Ünlü G (2021) Meat safety and quality: a biological approach. Int J Food Sci Technol 56:39–51. https://doi.org/10.1111/ijfs.14602
16. Gad SE, Sullivan DW (2014) Generally Recognized as Safe (GRAS). In: Encyclopedia of toxicology, 3rd edn. pp 706–709. https://doi.org/10.1016/B978-0-12-386454-3.00848-4
17. FDA (2022) Generally Recognized as Safe (GRAS) Substances (SCOGS) database. U.S. Food and Drug Administration. https://www.fda.gov/food/generally-recognized-safe-gras/gras-substances-scogs-database#:~:text=This%20database%20allows%20access%20to,shown%20in%20the%20table%20below.
18. Van Ba H, Seo HW, Pil-Nam S, Kim YS, Park BY, Moon SS, Kang SJ, Choi YM, Kim JH (2018) The effects of pre-and post-slaughter spray application with organic acids on microbial population reductions on beef carcasses. Meat Sci 137:16–23. https://doi.org/10.1016/j.meatsci.2017.11.006
19. Benli H (2014) Surface decontamination treatments for improving the safety of meat and poultry. Food Eng Ser 155–174. https://doi.org/10.1007/978-1-4939-1378-7_6
20. Nkosi DV, Bekker JL, Hoffman LC (2021) The use of organic acids (lactic and acetic) as a microbial decontaminant during the slaughter of meat animal species: a review. Foods 10:2293. https://doi.org/10.3390/foods10102293
21. Kalchayanand N, Arthur TM, Bosilevac JM, Schmidt JW, Wang R, Shackelford S, Wheeler TL (2015) Efficacy of antimicrobial compounds on surface decontamination of seven shiga toxin-producing *Escherichia coli* and *Salmonella* inoculated onto fresh beef†. J Food Prot 78:503–510. https://doi.org/10.4315/0362-028X.JFP-14-268

Check for updates

Chapter 16

Combined Methods Application in Fresh Meat: Active Packaging

Renata Tieko Nassu, Lucimara Aparecida Forato, Cecilia de Souza Cordeiro and Vanessa Cristina Francisco

Abstract

Edible coatings, a practical solution for preserving or improving meat quality, have been widely studied. Among the biopolymers, many compounds based on protein, lipids, or carbohydrates can be used. Adding plasticizers, such as vegetable oils and oleic acid, can make these coatings more flexible with better mechanical properties. Natural antioxidants and antimicrobial compounds can be added to give them additional properties. Here, we describe a practical method for producing edible coatings based on various materials, such as alginate, chitosan, or zein, combined with antioxidants and/or antimicrobial compounds for application on fresh meat.

Keywords Edible coatings, Chitosan, Alginate, Zein, Antioxidants, Antimicrobials

1 Introduction

Meat is highly perishable and, therefore, susceptible to spoilage, and one alternative to extend the shelf life of food products, including meat, is the application of edible coatings. Among the materials used to obtain edible coatings are lipids, carbohydrates and proteins. Polysaccharides are interesting biopolymers to prepare edible coatings. One example of a polysaccharide-based biopolymer is chitosan, derived from the deacetylation process of chitin, which is considered the second most abundant polysaccharide in nature [1]. This material is a low-cost, renewable, biodegradable, non-toxic, biofunctional, and biocompatible source. It is mainly obtained from waste from the fishing industry extracted from the exoskeleton of crustaceans such as lobsters, crabs, and shrimp [2]. One of the characteristics of chitosan is the presence of free amino groups, which increases its ability to react with various molecules, making the biopolymer more available in functional groups and can be used in powder, sphere or film forms [3]. One of its main properties is its strong antimicrobial

Renata Ernlund Freitas de Macedo and Carlos Adam Conte-Junior (eds.), *Meat Products Preservation*, Methods and Protocols in Food Science, https://doi.org/10.1007/978-1-0716-5068-4_16,

and antifungal action. These polymers, in their various chemical derivatives, inhibit the growth of microorganisms, such as *Listeria monocytogenes, Bacillus cereus, Staphylococcus aureus, Escherichia coli, Pseudomonas fluorescens, Salmonella typhimurium*, and *Vibrio parahaemolyticus* [4, 5]. The antimicrobial activity of chitosan is conferred by its amino groups, which, once in contact with physiological fluids, are probably protonated and bind to anionic groups of these microorganisms, resulting in the agglutination of microbial cells and causing growth inhibition [6]. The application of chitosan has been studied in meat and meat products, such as fish [7], ground beef [8], and fresh beef [8, 9], where the antimicrobial effects of chitosan have been reported, which inhibited the growth of spoilage bacteria, reduced lipid oxidation and also resulted in better sensory characteristics, having an effect on the development of the red color during storage, in the case of beef.

Another polymer widely used to obtain edible coating is sodium alginate. Alginate is a naturally occurring polysaccharide commercially extracted from species of brown algae [10]. The use of alginates as films in food applications has increased in recent years, mainly due to their gelling and thickening action, resulting from their ability to complex with calcium ions (Ca^{2+}) [11]. This polymer presents biodegradability, biocompatibility, and lack of toxicity [12]. In the food industry, alginates are widely used as additives that can increase viscosity and stabilize and emulsify gel aqueous solutions, such as ice cream and salad dressings [13]. Some studies on meat and meat products have also been reported. Comaposada et al. [14] used alginate film as a casing for stuffed meats. Molayi et al. [15] observed an antimicrobial effect in an alginate coating incorporating lactoperoxidase. Chidanandaiah et al. [16] used alginate in the coating of steaks stored for up to 21 days, and a decrease in lipid oxidation values and microbiological contamination was observed without altering the sensory characteristics.

In general, protein-based coatings have superior mechanical properties than coatings produced from polysaccharides. The formation of protein-based coatings has been studied for their use in the production of food packaging materials. Protein-based coatings act as barriers that control the transfer of moisture, oxygen, carbon dioxide, oils and fats, and volatile compounds to prevent spoilage and increase the shelf life of products. Zein, the maize storage protein, is a hydrophobic protein being an alternative material that can be used as an edible coating in meat. This protein can be obtained from the byproduct of the humid grounding of corn gluten and has been applied to produce coatings for highly perishable food substances and microcapsules for the pharmaceutical industry [17]. Generally, zein-based films show an excellent barrier to oxygen, carbon dioxide, and other volatile compounds. Zein may show fragile and low flexibility films, suggestive of poor mechanical properties, mainly with humidity and temperature

[17]. Plasticizers, organic compounds combined with polymeric materials with high fusion points and low volatility, can be added to improve film properties. Plasticizers are essential in the formation of films and coatings because, when they are added to another material, they can change their physical and mechanical properties [18]. Vegetable oils can be added to coatings, thus acting as emulsifiers and plasticizers, increasing the films' flexibility and minimizing the high water permeability [19].

In this chapter, the production of filmogenic solutions of alginate, chitosan, and zein combined or not with antioxidants or antimicrobial compounds is described for application as edible coatings in fresh meat.

2 Materials

2.1 Chitosan Coating

Chitosan (Sigma Aldrich), glycerol PA, and lactic acid.

2.2 Alginate Coating

Sodium Alginate - TICA Algin® HG 400 powder for food application (Ingredion, Westchester, IL, USA), calcium chloride P.A.

2.3 Zein Coating

Zein (Sigma Aldrich), Ethanol PA, glycerol P.A., oleic acid, olive oil, pink pepper oil, and rosemary oil.

3 Methods

3.1 Preparing Coatings[3, 4]

3.1.1 Chitosan

Chitosan is commonly dissolved in acetic or lactic acid aqueous solutions [20]. After complete dissolution under stirring, plasticizers can be added, and after, if desirable, essential oils can be added. In this case, it will be necessary to homogenize the mixture with efficient systems such as Turrax homogenizer. The following steps describe how the chitosan filmogenic solution is obtained.

a) Dissolve chitosan (0.5%–2.0% (v/v) in 1% acetic acid or lactic acid (v/v) aqueous solution under stirring for 3 h at room temperature (Solution A).

b) Add 0.5%–0.75% of glycerol (w/w) under stirring for 10 min and filter with Whatman n°3 filter paper (Solution B).

c) If this is the case, add separately essential oil (1%) and Tween 80 (1%), which is a surfactant (Solution C).

d) Add solution C to solution B at 21,600 rpm for 1 min (Chitosan filmogenic solution).

3.1.2 Alginate

For the alginate coating, it is necessary to produce two solutions, as follows:

a) Sodium alginate solution: add 1.5% sodium alginate in sterile distilled water at room temperature.
b) Homogenize the alginate solution in a high rotation homogenizer (Skimsen, model TA-02-N, Brusque, SC).
c) Calcium chloride solution: dissolve 1.5% calcium chloride in distilled water.

3.1.3 Zein Coating

As zein is insoluble in water, it is commonly solubilized in ethanol aqueous solution (70%–95%) to prepare coatings based on this protein.

a) Dissolve zein (4%) (w/w) in 70% ethanol solution—under stirring at 50°C.
b) Add oleic acid (from 0.25–1%), or vegetable oil such as pink pepper, olive, or rosemary, for example[1], from 0.25%–1% (w/w).
c) Mixtures should be homogenized in a beaker with a magnetic agitator at 50°C for 2 h [21]—filmogenic solution.

3.2 Coating Procedure for Meat

Among the methods for applying edible coatings, the dipping process is suitable for irregular surfaces, ensuring the formation of a more homogeneous surface; however, the spray application can also be suitable. Here, we describe the dipping procedure for chitosan and zein coatings and the spraying procedure for alginate coatings[2]

3.2.1 Dipping

(a) Meat samples should be cut into steaks or cubes.
(b) Dip/immerse each steak in the filmogenic solution for 5 s (Fig. 1a, b).
(c) Keep the steaks/cubes in a net for 30 min in a low oxygen demand chamber at 4–6°C to dry partially and get a homogeneous covering (Fig. 1c).

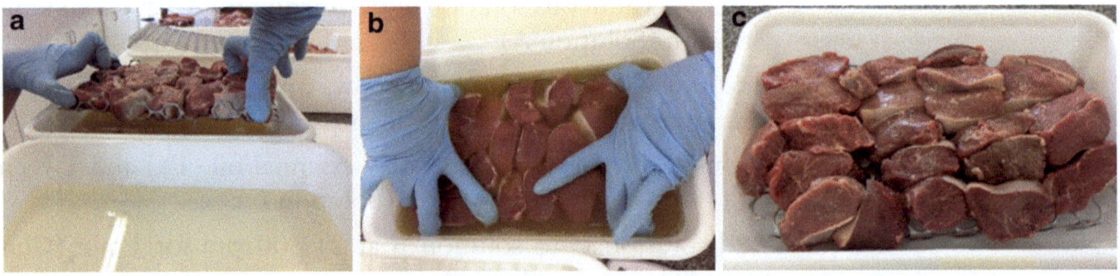

Fig. 1 Dipping procedure for edible coating in meat

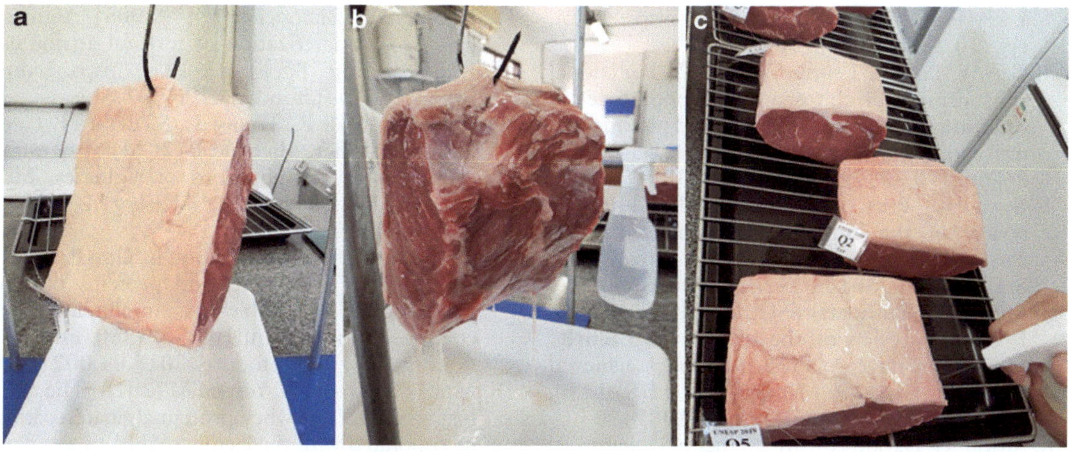

Fig. 2 Spraying procedure for edible coating in meat

3.2.2 Spraying

(a) Steaks should be hung in some apparatus so both sides of them can be sprayed (Fig. 2a, b).

(b) Spray equally the alginate solution using a manual pulverizer on both sides of the steak for a given defined time (e.g., 10 s) depending on the steak size (Fig. 2c).

(c) Follow the procedure for spraying the calcium chloride solution to create a reticulated film.

(d) Place the steaks in a net for 60 min in a cold chamber at 2°C to obtain a partial drying.

4 Notes

1. Depending on the natural antioxidant, concentrations may vary, and antimicrobial compounds should be used.
2. Sensory analysis is always desirable when using edible coatings.
3. Materials for edible coatings must be approved for human consumption.
4. Vacuum-pack the steaks in appropriate plastic bags using a vacuum packer.

References

1. Campos CA, Gerschenson LN, Flores SK (2011) Development of edible films and coatings with antimicrobial activity. Food Bioprocess Technol 4(6):849–75. https://doi.org/10.1007/s11947-010-0434-1

2. Muñoz-Tebar N, Pérez-Álvarez JA, Fernández-López J, Viuda-Martos M (2023) Chitosan edible films and coatings with added bioactive compounds: antibacterial and antioxidant properties and their application to food products:

a review. Polymers 15(2):396. https://doi.org/10.3390/polym150203963

3. Gao W, Lai JC, Leung SW (2012) Functional enhancement of chitosan and nanoparticles in cell culture, tissue engineering, and pharmaceutical applications. Front Physiol 3:321. https://doi.org/10.3389/fphys.2012.003214
4. Goy RC, Britto DD, Assis OB (2009) A review of the antimicrobial activity of chitosan. Polímeros 19:241–247. https://doi.org/10.1590/S0104-14282009000300013
5. Muthu M, Pushparaj SSC, Gopal J, Sivanesan I (2023) A review on the antimicrobial activity of chitosan microspheres: milestones achieved and miles to go. J Mar Sci Eng 11(8):1480. https://doi.org/10.3390/jmse11081480
6. Ardean C, Davidescu CM, Nemeş NS, Negrea A, Ciopec M, Duteanu N, Negrea P, Duda-Seiman D, Musta V (2021) Factors influencing the antibacterial activity of chitosan and chitosan modified by functionalization. Int J Mol Sci 22(14):7449. https://doi.org/10.3390/ijms22147449
7. da Silva Santos FM, da Silva AIM, Vieira CB, de Araújo MH, da Silva ALC, Carneiro-da-Cunha MDG, de Souza BWS, de Souza Bezerra R (2017) Use of chitosan coating in increasing the shelf life of liquid smoked Nile tilapia (*Oreochromis niloticus*) fillet. J Food Sci Technol 54(5):1304–1311. https://doi.org/10.1007/s13197-017-2570-3
8. Darmadji P, Izumimoto M (1994) Effect of chitosan in meat preservation. Meat Sci 38(2):243–254. https://doi.org/10.1016/0309-1740(94)90114-7
9. Duran A, Kahve HI (2020) The effect of chitosan coating and vacuum packaging on the microbiological and chemical properties of beef. Meat Sci 162:107961. https://doi.org/10.1016/j.meatsci.2019.107961
10. Senturk Parreidt T, Müller K, Schmid M (2018) Alginate-based edible films and coatings for food packaging applications. Foods 7(10):170. https://doi.org/10.3390/foods7100170
11. Costa MJ, Marques AM, Pastrana LM, Teixeira JA, Sillankorva SM, Cerqueira MA (2018) Physicochemical properties of alginate-based films: effect of ionic crosslinking and mannuronic and guluronic acid ratio. Food Hydrocoll 81: 442–8. https://doi.org/10.1016/j.foodhyd.2018.03.014
12. Roopa BS, Bhattacharya S (2008) Alginate gels: I. Characterization of textural attributes. J Food Eng 85(1):123–131. https://doi.org/10.1016/j.jfoodeng.2007.07.012
13. Abka-Khajouei R, Tounsi L, Shahabi N, Patel AK, Abdelkafi S, Michaud P (2022) Structures, properties and applications of alginates. Mar Drugs 20(6):364. https://doi.org/10.3390/md20060364
14. Comaposada J, Marcos B, Bou R, Gou P (2018) Influence of surfactants and proteins on the properties of wet edible calcium alginate meat coatings. Food Res Int 108:539–550. https://doi.org/10.1016/j.foodres.2018.04.002
15. Molayi R, Ehsani A, Yousefi M (2018) The antibacterial effect of whey protein-alginate coating incorporated with the lactoperoxidase system on chicken thigh meat. Food Sci Nutr 6(4):878–883. https://doi.org/10.1002/fsn3.634
16. Chidanandaiah, Keshri RC, Sanyal MK (2009) Effect of sodium alginate coating with preservatives on the quality of meat patties during refrigerated (4 ± 1C) storage. J Muscle Foods 20(3):275–292. https://doi.org/10.1111/j.1745-4573.2009.00147.x
17. Forato LA, Britto DD, Scramin JA, Colnago LA, Assis OB (2013) Mechanical and wetting properties of zein films extracted from corn gluten meal. Polímeros 23:42–48. https://doi.org/10.1590/S0104-14282012005000075
18. Banker GS (1966) Film coating – theory and practice. J Pharm Sci 55(1):81–89. https://doi.org/10.1002/jps.2600550118
19. García MA, Martino MN, Zaritzky NE (2000) Lipid addition to improve barrier properties of edible starch-based films and coatings. J Food Sci 65:941–944. https://doi.org/10.1111/j.1365-2621.2000.tb09397.x
20. Goñi MG, Tomadoni B, Roura SI, Moreira MDR (2017) Lactic acid as potential substitute of acetic acid for dissolution of chitosan: preharvest application to Butterhead lettuce. J Food Sci Technol 54(3):620–626. https://doi.org/10.1007/s13197-016-2484-5
21. Cordeiro CS, Forato LA, Bernardes Filho R, Nassu RT (2019) Utilization of zein-based coatings containing vegetable oils on the physicochemical characteristics of vacuum-packaged lamb meat. Cienc Rural 49:1. https://doi.org/10.1590/0103-8478cr20190310

Subject Index

Renata Ernlund Freitas de Macedo and Carlos Adam Conte-Junior (eds.), *Meat Products Preservation*, Methods and Protocols in Food Science, https://doi.org/10.1007/978-1-0716-5068-4,